Statistics for Business and Economics

Fifth Edition

Statistics for Business and Economics

Fifth Edition

David R. Anderson
University of Cincinnati

Dennis J. Sweeney
University of Cincinnati

Thomas A. Williams
Rochester Institute of Technology

West Publishing Company
Minneapolis/St. Paul   New York   Los Angeles   San Francisco

To Marcia, Cherri, and Robbie

Text Design: John Rokusek, Rokusek Design

Copyediting and Proofreading: Luana Richards

Composition: Carlisle Communications

Artwork: DBA Design and Illustration

Cover Illustration and Design: John Rokusek, Rokusek Design

Indexing: Schroeder Indexing

Production, Prepress, Printing, and Binding by West Publishing Company

Minitab is a registered trademark of Minitab, Inc., 3081 Enterprise Drive, State College, PA, 16801 (telephone 814/238–3280; telex 881612; fax 814/238–4383).

West's Commitment to the Environment

In 1906, West Publishing Company began recycling materials left over from the production of books. This began a tradition of efficient and responsible use of resources. Today, up to 95 percent of our legal books and 70 percent of our college texts are printed on recycled, acid-free stock. West also recycles nearly 22 million pounds of scrap paper annually—the equivalent of 181,717 trees. Since the 1960s, West has devised ways to capture and recycle waste inks, solvents, oils, and vapors created in the printing process. We also recycle plastics of all kinds, wood, glass, corrugated cardboard, and batteries, and have eliminated the use of styrofoam book packaging. We at West are proud of the longevity and the scope of our commitment to our environment.

Library of Congress Cataloging-in-Publication Data

Anderson, David Ray, 1941–
    Statistics for business and economics / David R. Anderson, Dennis
J. Sweeney, Thomas A. Williams. — 5th ed.
        p.   cm.
    Includes bibliographical references and index.
    ISBN 0-314-01244-3
    1. Commercial statistics.   2. Economics—Statistical methods.
3. Statistics.   I. Sweeney, Dennis J.   II. Williams, Thomas Arthur,
1944– .   III. Title.
HF1017.A6   1993
519.5—dc20   92-18129
                                                                  CIP

About the Authors

David R. Anderson. David R. Anderson is Professor and Head of the Department of Quantitative Analysis in the College of Business Administration at the University of Cincinnati. Born in Grand Forks, North Dakota, he earned his B.S., M.S., and Ph.D. degrees from Purdue University. Professor Anderson has served as Associate Dean of the College of Business Administration. In addition, he was the coordinator of the College's first Executive Program.

At the University of Cincinnati, Professor Anderson has taught introductory statistics for business students as well as graduate level courses in regression analysis, multivariate analysis, and management science. He has also taught statistical courses at the Department of Labor in Washington, D.C. He has been honored with nominations and awards for excellence in teaching and excellence in service to student organizations.

Professor Anderson has coauthored six textbooks in the areas of statistics, management science, linear programming, and production/operations management. He is an active consultant in the field of sampling and statistical methods.

Dennis J. Sweeney. Dennis J. Sweeney is Professor of Quantitative Analysis at the University of Cincinnati. Born in Des Moines, Iowa, he earned a B.S.B.A. degree from Drake University, graduating summa cum laude. He received his M.B.A. and D.B.A. degrees from Indiana University where he was an NDEA Fellow. Since receiving his doctorate in 1971, Professor Sweeney has spent all but 2 years at the University of Cincinnati. During 1978–79, he spent a year working in the management science group at Procter & Gamble; during 1981–82, he was a visiting professor at Duke University. Professor Sweeney served 5 years as Head of the Department of Quantitative Analysis and 4 years as Associate Dean at the University of Cincinnati.

Professor Sweeney has published over 30 articles in the general area of management science. The National Science Foundation, IBM, Procter & Gamble, and Cincinnati Gas & Electric have funded his research, which has been published in *Management Science, Operations Research, Mathematical Programming, Decision Science,* and other journals.

Professor Sweeney has coauthored six textbooks in the areas of statistics, management science, linear programming, and production and operations management.

Thomas A. Williams. Thomas A. Williams is Professor of Management Science in the College of Business at Rochester Institute of Technology. Born in Elmira, New York, he earned his B.S. degree at Clarkson University. He did his graduate work at Rensselaer Polytechnic Institute, where he received his M.S. and Ph.D. degrees.

Before joining the College of Business at RIT, Professor Williams served for 7 years as a faculty member in the College of Business Administration at the University of

Cincinnati, where he developed the undergraduate program in Information Systems and then served as its coordinator. At RIT he was the first chairman of the Decision Sciences Department. He teaches courses in management science and statistics, as well as more advanced courses in regression and decision analysis.

Professor Williams is the coauthor of seven textbooks in the area of management science, statistics, production and operations management, and mathematics. He has been a consultant for numerous Fortune 500 companies and has worked on projects ranging from the use of elementary data analysis to the development of large-scale regression models. His current research focuses on the application of total quality management in an academic setting.

Contents

CHAPTER 3   Descriptive Statistics II: Measures of Location and Dispersion, 55

CHAPTER 4   Introduction to Probability, 98

CHAPTER 5   Discrete Probability Distributions, 142

CHAPTER 6   Continuous Probability Distributions, 177

CHAPTER 9   Hypothesis Testing, 284

CHAPTER **10**   Statistical Inference about Means and Proportions With Two Populations, 336

CHAPTER **11**   Inferences about Population Variances, 370

CHAPTER 14   Simple Linear Regression and Correlation, 471

CHAPTER 15   Multiple Regression, 547

CHAPTER 16   Regression Analysis: Model Building, 600

Contents

xix

CHAPTER 19   Nonparametric Methods, 708

CHAPTER 20   Statistical Methods for Quality Control, 742

Appendixes

Preface

The purpose of this book is to provide students, primarily in the fields of business administration and economics, with a conceptual introduction to the field of statistics and its many applications. Applications-oriented and written with the needs of the nonmathematician in mind, the mathematical prerequisite for this text is knowledge of algebra.

Applications and Methodology

Applications of data analysis and statistical methodology are an integral part of the organization and presentation of the text material. The discussion and development of each technique is centered around an application setting with the statistical results providing insights to decisions and solutions to problems.

Although the book is applications oriented, we have taken care to provide a sound methodological development and to utilize notation that is generally accepted for the topic being covered. Thus, students will find that this text provides good preparation for the study of more advanced statistical material. A bibliography that provides a guide to further study has been included as an appendix.

Changes in the Fifth Edition

We appreciate the acceptance and positive response to the earlier editions of this text. Accordingly, in making modifications for this new edition, we have maintained the presentation style of previous editions. The more significant changes are summarized below.

Statistics in Practice

To emphasize the application of statistics, each chapter opens with an actual situation supplied by practitioners from business and economics. Each Statistics in Practice briefly describes an organization and a problem in which the statistical methodology introduced in the chapter was applied. Procter & Gamble, Polaroid, Monsanto, Xerox, Mead, Dow Chemical, and Colgate-Palmolive are just a few of the companies that have contributed Statistics-in-Practice features. We feel that this feature will help motivate students by placing the chapter material in a real-world context. The table at the end of this preface provides a list of the organizations and applications described in this feature.

More Examples and Problems Based on Real Data

We have continued the emphasis on helping students understand the wide range of statistical applications by expanding the number of examples and problems based on real statistical studies. Sources such as *The Wall Street Journal, Business Week, U.S.A. Today, Fortune, Forbes, Fnancial World, Barrons,* and others have been used to provide referenced applications and problems which demonstrate uses of statistics in business and economics. The use of real data means that students not only learn about the statistical methodology but also learn about the content of applications encountered in practice.

Methods Exercises and Applications Exercises

The end-of-section exercises have been split into two parts. With the Methods exercises, students are required to make sure they can handle the formulas and make the necessary computations. In the Applications exercises, students are given problems that require using the chapter material in real-world situations. With this approach, students first focus on the computational "nuts and bolts," then move on to the subtleties of statistical application and interpretation.

Self-Test Exercises

Certain exercises are identified as self-test exercises. Completely worked-out solutions for these exercises are provided in an appendix at the end of the text. Students can attempt the self-test exercises and immediately check the solution to evaluate their understanding of the concepts presented in the chapter.

Computer Cases

Many chapters have computer cases, which contain problem scenarios accompanied by modest-sized data sets. Computer solution by Minitab, The Data Analyst, or another statistical software package is required. Each case outlines a managerial report that the student prepares to summarize statistical results as well as present interpretations and recommendations. The data sets for all computer cases are available on the data disk that accompanies the text.

A New Chapter on Index Numbers

The use of index numbers to describe changes in business and economic conditions over time is presented in a new Chapter 17. Several popular indexes, including the Consumer Price Index, the Producer's Price Index, and the Dow Jones Averages, are discussed. This chapter and the times series presentation in Chapter 18 show how business and economic conditions can be forecasted through the use of a time series of index numbers.

A New Chapter on Sampling and Survey Methodology

Chapter 21 introduces sampling and survey methodology. A discussion of issues involved in designing and conducting survey research is presented. The probability sampling methods of simple random sampling, stratified simple random sampling, cluster sampling, and systematic sampling are covered. The nature and control of sampling and nonsampling error are treated.

Analysis of Variance and Experimental Design

The presentation of analysis of variance and experimental design (Chapter 13) has been revised with a new application used to introduce students to the concepts. The chapter begins with analysis of variance and its use in testing for the equality of means with three or more populations. Procedures for making multiple comparisons have been expanded and now appear immediately following the material on testing for the equality of means. Experimental design concepts are presented after the introduction to analysis of variance. Completely randomized, randomized block, and factorial experimental designs complete the coverage. Formulas for the analysis of variance computations appear in the chapter appendix.

Regression Analysis

The three-chapter sequence on regression analysis (Chapters 14, 15, and 16) has been enhanced with an expanded discussion of the use and interpretation of dummy variables and interaction terms. The best-subset selection technique and the use of normal probability plots are now included, and the discussion of residual analysis has been expanded.

Other Additions

Comments and suggestions made by our adopters, such as introducing the chi-square test for independence earlier in the text (now Chapter 12), adding the Poisson approximation of binomial probabilities, expanding the discussion on counting rules for probability, adding a tree diagram for Bayes' theorem, and expanding the binomial probability tables, have been incorporated into this edition.

Features and Pedagogy

We have continued many of the features that appeared in earlier editions. Some of the more important ones include the following.

Notes & Comments

At the end of many sections, we have provided Notes & Comments designed to give the student additional insights about the statistical methodology and its application. Various Notes & Comments include warnings or limitations about the methodology, recommendations for application, brief descriptions of additional technical considerations, and so on.

Data and Descriptive Statistics

The text begins with a discussion of how data are acquired and classified. The classification of qualitative and quantitative data and the various scales of measurement are covered in Chapter 1. Chapters 2 and 3 include the graphical and numerical procedures used to summarize data. Outliers, procedures for identifying outliers, and the exploratory data analysis concepts of stem-and-leaf displays and box plots are covered.

Hypothesis Testing

Chapter 9 emphasizes the proper formulation of hypotheses and making the proper interpretation when the hypothesis test leads to "Reject H_0 or "Do not reject H_0." The appropriate uses and limitations of hypothesis testing in research studies, testing assumptions, and decision making are discussed. The use and interpretation of *p*-values is also covered.

Statistical Quality Control

Chapter 20 provides an introduction to the important role of statistics in quality control. Acceptance sampling and control charts are presented. This material may be covered either as a stand-alone chapter or in association with earlier material in the text. For example, acceptance sampling in Section 20.1 could follow the presentation of the binomial probability distribution in Chapter 5. In addition, the use of control charts in Section 20.2 could follow the material on hypothesis testing in Chapter 9.

Computer Software

The text contains numerous examples and discussions of the important role statistical software packages play in the computation and presentation of statistical results. More examples and problems based on the Minitab statistical package have been added. Use and interpretation of the information provided by Minitab printouts are emphasized. Available to adopters is The Data Analyst version 2.0, an IBM-compatible microcomputer software package developed by the authors. The Data Analyst may be ordered shrink-wrapped with the text. The Data Analyst, Minitab, or other statistical packages can be used to solve problems and computer cases appearing in the text.

Data Disk

Data sets for text examples, problems and computer cases are available on a special Data Disk accompanying the text. The Data Disk contains the data sets in a format acceptable to Minitab and in a format acceptable to The Data Analyst software package.

Flexibility

There is a reasonable amount of instructor flexibility in selecting material to satisfy specific course needs. A possible outline for a two-quarter sequence is as follows:

Possible Two-Quarter Course Outline

First Quarter	Second Quarter
Data, Measurement, and Statistics (Chapter 1)	Hypothesis Testing (Chapter 9)
Descriptive Statistics (Chapters 2 and 3)	Two-Population Cases (Chapter 10)
Introduction to Probability (Chapter 4)	Inferences about Population Variances (Chapter 11)
Probability Distributions (Chapters 5 and 6)	Tests of Goodness of Fit and Independence (Chapter 12)
Sampling and Sampling Distributions (Chapter 7)	Analysis of Variance and Experimental Design (Chapter 13)
Interval Estimation (Chapter 8)	Regression and Correlation (Chapters 14 and 15)

Other possibilities exist for such a course, depending upon the time available and the background of the students. Topics such as Model Building in Regression Analysis (Chapter 16), Index Numbers (Chapter 17), Time Series and Forecasting (Chapter 18), Nonparametric Methods (Chapter 19), Quality Control (Chapter 20), Sample Survey (Chapter 21), and Decision Analysis (Chapter 22) can be selected at the option of the instructor to meet the special needs of a particular academic program. However, it is not possible to cover all the material in one semester or in two quarters unless some of the topics have been previously studied.

Acknowledgments

We owe a debt to many of our colleagues and friends for their helpful comments and suggestions in the development of this text. Among these are:

Mohammad Ahmadi	C. Thomas Innis	James R. Schwenke
Cheryl Asher	Ben Isselhardt	Bill Seaver
Robert Balough	Jeffrey Jarrett	William E. Stein
Harry Benham	Robert Bruce Jones	Willban Terpening
Michael Bernkopf	David Krueger	Hiroki Tsurumi
John Bryant	Martin S. Levy	Ebenge Usip
Peter Bryant	Thomas McCullough	Cindy van Es
Terri L. Byczkowski	Bette Midgarden	Andrew Welki
Robert Cochran	Glenn Milligan	Ari Wijetunga
John Cooke	David Muse	J. E. Willis
David W. Cravens	Roger Myerson	Donald Williams
George Dery	Richard O'Connell	Roy Williams
Gopal Dorai	Al Palachek	Mustafa Yilmaz
Edward Fagerlund	Diane Petersen-Salameh	Gary Yoshimoto
Nicholas Farnum	Tom Pray	Charles Zimmerman
Jerome Geaun	Ruby Ramirez	Greg Zimmerman
Jamshid C. Hasseini	Tom Ryan	

Special thanks go to Abdul Basti who offered excellent feedback on the set of Demonstration Problems/Transparency Masters that accompany this edition of the text. Also, associates from business and industry who supplied the Statistics in Practice features also made a major contribution, and we recognize them individually by a credit line on each Statistics in Practice. In addition, we would like to acknowledge the cooperation of Minitab, Inc., for permitting the use of Minitab statistical software in this text. Finally, we are also indebted to our editor, Mary C. Schiller, and others at West Publishing Company for their editorial counsel and support during the preparation of this text.

David R. Anderson
Dennis J. Sweeney
Thomas A. Williams

February 1993

An Overview of Statistics in Practice Features

Chapter Number	Chapter Title	Organization Featured	Application Topic
1	Data, Measurement, and Statistics	Kings Island	Consumer Profile Sample Survey
2	Descriptive Statistics I. Tabular and Graphical Approaches	Colgate-Palmolive	Quality Assurance for Heavy Duty Detergents
3	Descriptive Statistics II. Measures of Location and Dispersion	Barnes Hospital	Time Spent in Hospice Program
4	Introduction to Probability	Morton International	Evaluation of Customer Service Testing Program
5	Discrete Probability Distributions	Xerox	Performance Test of an On-Line Computerized Publication System
6	Continuous Probability Distributions	Procter & Gamble	Manufacturing Strategy
7	Sampling and Sampling Distributions	Mead	Estimating the Value of Mead Forest Ownership
8	Interval Estimation	Dollar General	Sampling for Estimation of LIFO Inventory Costs
9	Hypothesis Testing	Harris	Testing for Defective Plating
10	Statistical Inference about Means and Proportions with Two Populations	Pennwalt	Evaluation of New Drugs
11	Inferences about Population Variances	U.S. General Accounting Office	Water Pollution Control
12	Tests of Goodness of Fit and Independence	United Way	Determining Community Perceptions of Charities
13	Analysis of Variance and Experimental Design	Burke Marketing Services	New Product Design
14	Simple Linear Regression and Correlation	Polaroid Corporation	Aging Study of Film
15	Multiple Regression	Champion International	Control of Pulp Bleaching Process
16	Regression Analysis: Model Building	Monsanto	Feed Development for Chickens
17	Index Numbers	U.S. Department of Labor, Bureau of Labor Statistics	Consumer Price Index
18	Time Series Analysis and Forecasting	Cincinnati Gas & Electric	Forecasting Demand for Electricity
19	Nonparametric Methods	West Shell Realtors	Comparison of Real Estate Prices across Neighborhoods
20	Statistical Methods for Quality Control	Dow Chemical U.S.A.	Statistical Process Control
21	Sample Survey	Cincinnati Gas & Electric	Survey of Commercial Customers
22	Decision Analysis	Ohio Edison	Choice of Best Type of Particulate Control Equipment

Data, Measurement, and Statistics

Kings Island, Inc.*

KINGS ISLAND, OHIO

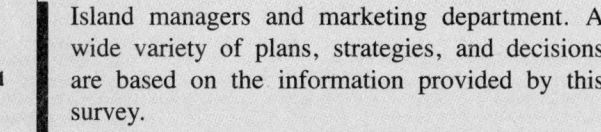

The Kings Island Family Entertainment Center is a 1600-acre year-round recreational, sports, and shopping complex located in southwestern Ohio. The Kings Island theme park, one of America's top parks, provides rides, entertainment, and other attractions that draw approximately 3 million people annually. In addition to the theme park, the Kings Island Entertainment Center includes the Jack Nicklaus Sports Center, site of both professional golf and professional tennis tournaments; the Resort Inn by Kings Island; the Kings Island Campground; and the Factory Outlet Mall. In short, Kings Island Entertainment Center is unmatched in the Midwest for entertainment, sports, shopping, and relaxation.

Behind the scenes at Kings Island, executives, managers, and staff plan the strategies and make the business decisions that continue to make Kings Island a successful business venture. The company's research group is devoted to learning about the Kings Island consumer's behavior, attitudes, perceptions, and preferences. A majority of the work done by the research group is statistical in nature, and statistical samples of park visitors are selected on a routine basis. Questionnaires are used to obtain data on consumer attitudes and characteristics. In total, the research group, through its numerous special studies and samples, interviews approximately 30,000 visitors annually.

The research group uses the Kings Island consumer profile survey to learn about park visitors. As visitors enter the park, interviewers ask sampled individuals to answer a short 2- or 3-minute questionnaire about themselves and why they came to the park. Data are collected on such categories as home zip code, distance traveled to the park, type of admission ticket, group size, number of previous visits to the park, and respondent's age. Descriptive statistics are used to summarize the sample results for the Kings Island managers and marketing department. A wide variety of plans, strategies, and decisions are based on the information provided by this survey.

Other sample surveys collect data on visitor attitudes toward park features such as food services, rides, shows, and games. To the visitor, Kings Island is a vast area of fun and excitement far removed from the busy world of business. Behind the scenes, however, an efficient business organization uses sampling and other statistical methods to provide the information necessary to make the decisions that keep Kings Island one of the country's top family entertainment centers.

In this chapter you will learn about statistical samples, the types of data that can be collected, and how descriptive statistics and statistical inference can be used to put the data in a form that is meaningful and easily interpreted.

Children enjoy a visit to the Yogi Bear Fountain in the Kings Island Storybook Kingdom of Hanna-Barbera Land.

*Bill Mefford, Manager of Marketing Communications, and Tom Russell, Market Representative for Planning and Analysis, provided the Kings Island Statistics in Practice information.

F requently, we see news articles with statements such as the following:

- Consumer prices rose 0.2% in January (*Business Week,* February 24, 1992)
- The average fuel economy of new cars sold in the United States is 27.5 miles per gallon (*The Wall Street Journal,* April 8, 1992)
- The average starting salary for business administration graduates is $24,369 per year (*College Placement Council,* March 1992)
- 45% believe taxes will be higher next year while only 9% believe taxes will be lower (*Money,* April 1992)
- Employment in the 500 largest U.S. industrial corporations dropped 3.9% last year (*Fortune,* April 22, 1992)
- The median time an employee stays in one occupation is 6.6 years (*Young Executive,* Spring 1992)
- The Dow Jones Industrial Average closed at 3338.77 (*The Wall Street Journal,* April 23, 1992)
- The Tokyo Nikkei stock market index reached a 5-year low at 19,636.99 (*Barron's,* March 23, 1992)

The numerical facts in the preceding statements (0.2%, 27.5, $24,369, and so on) are called statistics. Thus, in everyday usage, the term *statistics* refers to numerical facts. However, the field, or subject, of statistics involves much more than simply the calculation and presentation of numerical facts.

The subject of **statistics** involves the study of how data are collected, how they are analyzed, how they are presented, and how they are interpreted. Particularly in business and economics, a major reason for collecting, analyzing, presenting, and interpreting data is to provide the information necessary to make good decisions. In this text we shall present the use of statistics in a decision-making context.

Chapter 1 opens with some illustrations of the applications of statistics in business and economics. The remainder of the chapter is devoted to a discussion of how *data,* the raw material of statistics, are acquired, measured, and used. The two uses of data referred to as descriptive statistics and statistical inference are the subjects of Sections 1.5 and 1.6.

1.1   Statistical Applications in Business and Economics

The understanding and use of statistics are becoming more and more important in today's business world. Tremendous amounts of statistical information are available. The most successful managers and decision makers are the ones who can understand the information and make use of it effectively. In some cases, a statistical study will be undertaken to provide help with a current problem and pending decision. In other cases, statistics provide valuable information about the general business and economic environment within which decisions must be made. In this section we provide examples that illustrate some uses of statistics in business and economics.

Accounting

Public accounting firms use statistical sampling procedures in performing audits for their clients. For instance, suppose an accounting firm wants to determine if the amount of accounts receivable shown on the client's balance sheet fairly represents the actual amount of accounts receivable. Usually, the number of individual accounts receivable is so large that it would be too time-consuming and expensive to validate every account. In situations such as this, it is common practice for the audit staff to select and validate a sample of the accounts. After reviewing the sample results, the auditor will make a decision as to whether or not the amount shown on the client's balance sheet should be accepted. In this case, the sampling and statistical information is used directly in the decision-making process.

Finance

Financial planners and advisors use a variety of statistical information to guide investment decisions. These individuals study financial characteristics such as price-earnings ratios and dividend yields. By comparing the information for an individual stock with information on stock market averages, the financial manager will have help determining if an individual stock is over- or undervalued. For example, on April 10, 1989, *Barrons* reported that the average price-earnings ratio for the 30 stocks in the Dow Jones Industrial Average was 10.7. On the same day, Phillip Morris provided a price-earnings ratio of 12. The statistical information on price-earnings ratios showed that Phillip Morris had a higher price compared to its earnings than the average for the Dow Jones stocks. As a result, the financial manager might conclude Phillip Morris was slightly overpriced. This and other information about Phillip Morris can help the manager make buy and sell recommendations for the stock.

Marketing

Before marketing a new product nationally, companies conduct research studies designed to learn about potential market reaction and consumer acceptance of the product. Test markets and/or consumer panels may be identified. Statistical information on sales in the test markets and statistical summaries of consumer opinions will help the company decide whether to market the product nationally.

Production

Quality-control and lot-acceptance sampling are common applications of statistics in production. In recent years, U.S. manufacturers have been criticized for producing poor-quality products — especially when compared to Japanese manufacturers. Lot-acceptance sampling is a quality-control procedure used to ensure that purchased materials are of satisfactory quality. For instance, in a shipment of 10,000 components, a company might only test 100 in order to reach the decision of whether to accept the entire shipment. If a large number of components in the sample of 100 perform correctly, the entire shipment will be accepted. If this is not the case, the shipment will be rejected.

Economics

Economists are frequently asked to provide forecasts about the future of the economy or some part of it. In making such forecasts, a variety of statistical information is used. For instance, in forecasting inflation rates, economists use statistics on indicators such as the Producer Price Index, unemployment rate, and manufacturing capacity utilization. Often such statistics are used in computerized forecasting models, which predict inflation rates.

Applications of statistics, such as those just described, are an integral part of this text. Such examples provide an overview of the breadth of statistical applications. To supplement these examples, we have asked practitioners from the fields of business and economics to provide a chapter-opening Statistics-in-Practice feature that serves to introduce the material will be covered. We believe that these actual applications of statistics will provide an appreciation of the importance of statistics in several different types of decision-making situations. Table 1.1 provides an overview of these Statistics-in-Practice features.

1.2   Data

Data are the facts and figures that are collected, analyzed, and summarized. All the data collected in a particular study is referred to as the *data set* for the study. Table 1.2 shows a data set for a group of shadow stocks with dividend reinvestment plans. The term *shadow* is used to indicate that the stocks are for small- to medium-sized firms that are not followed closely by the major brokerage houses. The data set in Table 1.2 was provided by the American Association of Individual Investors, *AAII Journal,* November 1988.

Elements, Variables, and Observations

The *elements* are the entities on which data are collected. For the data set in Table 1.2, each individual stock is an element. With 25 stocks, there are 25 elements in the data set.

A *variable* is a characteristic of interest for the elements. Thus, the data set in Table 1.2 shows there are five variables, which are defined as follows:

- *Exchange:* The stock exchange listing the stock—NYSE (New York Stock Exchange), AMEX (American Stock Exchange), and OTC (Over The Counter).
- *Ticker Symbol:* The abbreviation used to identify the stock on the exchange listing.
- *Earnings Per Share:* Net income after all expenses and taxes divided by the number of common shares outstanding.
- *Dividend Yield:* Most recent indicated annual dividend per share divided by share price.
- *Price-Earnings Ratio:* Market price per share divided by most recent 12 months' earnings per share.

Data are obtained by collecting measurements on each variable for every element in the study. The set of measurements collected for a particular element is called an *observation*. Referring to Table 1.2, we see that the first element (Acme Electric) has the observation NYSE, ACE, 17, 5, and 40. With 25 elements, there are 25 observations in the data set.

Most data that statisticians collect and work with are numeric. However, as the data set in Table 1.2 shows, data for some variables may be numeric and data for other variables may be nonnumeric. Specifically, the first two variables, Exchange and Ticker Symbol, have nonnumeric data. The nonnumeric symbols used are abbreviations for the stock exchange and stock identification of each element. The other three variables—earnings per share, dividend yield, and price-earnings ratio—have numeric data.

1.3   Scales of Measurement

The type of statistical analysis appropriate for the data on a particular variable depends upon the scale of measurement used for the variable. There are four scales of measurement: nominal, ordinal, interval, and ratio. The scale of measurement determines the amount of information contained in the data and indicates the data summarization and statistical analyses that are most appropriate. We describe each of the four scales of measurement.

	Chapter/Chapter Title	Organization	Statistics in Practice
TABLE 1.1 An Overview of the Statistics in Practice Features			
	1 Data, Measurement, and Statistics	Kings Island, Inc.	Consumer Profile Sample Survey
	2 Descriptive Statistics I. Tabular and Graphical Approaches	Colgate-Palmolive Company	Quality Assurance for Heavy Duty Detergents
	3 Descriptive Statistics II. Measures of Location and Dispersion	Barnes Hospital	Time Spent in Hospice Program
	4 Introduction to Probability	Morton International	Evaluation of Customer Service Testing Program
	5 Discrete Probability Distributions	Xerox Corporation	Performance Test of an On-line Computerized Publication System
	6 Continuous Probability Distributions	Procter & Gamble	Manufacturing Strategy
	7 Sampling and Sampling Distributions	Mead Corporation	Estimating the Value of Mead Forest Ownership
	8 Interval Estimation	Dollar General Corporation	Sampling for Estimation of LIFO Inventory Costs
	9 Hypothesis Testing	Harris Corporation	Testing for Defective Plating
	10 Statistical Inference about Means and Proportions with Two Populations	Pennwalt Corporation	Evaluation of New Drugs
	11 Inferences about Population Variances	U.S. General Accounting Office	Water Pollution Control
	12 Tests of Goodness of Fit and Independence	United Way	Determining Community Perceptions of Charities
	13 Analysis of Variance and Experimental Design	Burke Marketing	New Product Design
	14 Simple Linear Regression and Correlation	Polaroid	Aging Study of Film
	15 Multiple Regression	Champion International	Control of Pulp Bleaching Process
	16 Model Building	Monsanto	Evaluating New Feed for Chickens
	17 Index Numbers	U.S. Bureau of Labor Statistics	Consumer Price Index
	18 Time Series Analysis and Forecasting	Cincinnati Gas & Electric Company	Forecasting Demand for Electricity
	19 Nonparametric Methods	West Shell Realtors	Comparison of Real Estate Prices across Neighborhoods
	20 Statistical Methods for Quality Control	Dow Chemical U.S.A.	Quality Control for a Chemical Process
	21 Sample Survey Methods	Cincinnati Gas & Electric Company	Survey of Commercial Customers
	22 Decision Analysis	Ohio Edison Company	Choice of Best Type of Particulate Control Equipment

TABLE 1.2
A Data Set for 25 Shadow Stocks with Dividend Reinvestment Plans

Stock	Exchange	Ticker Symbol	Earnings per Share ($)	Dividend Yield (%)	Price-Earnings Ratio
Acme Electric	NYSE	ACE	0.17	5	40
American Filtrona	OTC	AFIL	1.73	4	13
Amer Recreation Ctrs	OTC	AMRC	0.24	2	33
Berkshire Gas	OTC	BGAS	1.35	8	12
Colonial Gas	OTC	CGES	1.76	9	11
Champion Products	AMEX	CH	2.65	1	13
Connecticut Energy	NYSE	CNE	2.32	8	9
Connecticut Wtr Sv	OTC	CTWS	1.91	7	11
Eastern Co.	AMEX	EML	1.74	4	8
Engraph Inc.	OTC	ENGH	0.81	1	18
Energy North Inc.	OTC	ENNI	1.41	7	10
Gorman-Rupp	AMEX	GRC	2.04	4	11
Marsh Supermarkets	OTC	MARS	1.45	2	13
Mobile Gas Service	OTC	MBLE	1.35	5	12
Medalist Indus	OTC	MDIN	1.43	4	11
Middlesex Water	OTC	MSEX	2.07	7	12
New Jersey Resources	NYSE	NJR	1.39	7	13
NECO Enterprises	AMEX	NPT	1.51	8	12
Peerless Tube	AMEX	PLS	0.39	6	18
Public Sv No. Car.	OTC	PSNC	1.65	7	9
United Cities Gas	OTC	UCIT	1.23	7	10
Upper Peninsula Pwr.	OTC	UPEN	3.73	9	7
UNTIL Corp.	AMEX	UTL	3.29	7	8
Valley Resources	AMEX	VR	1.15	3	20
Vulcan Corp.	AMEX	VUL	0.16	4	137

Source: American Association of Individual Investors Journal, November, 1988.

Nominal Scale

The scale of measurement for a variable is *nominal* when the data obtained for the variable are simply labels used to identify an attribute of the element. For example, referring again to the data set in Table 1.2, we see that the first variable, Exchange, is measured on a nominal scale. This is so because NYSE, AMEX, and OTC are labels used to identify the exchange that lists the stock. The second variable, Ticker Symbol, is also measured on a nominal scale with the abbreviation providing the identification label for the stock.

In cases where the scale of measurement is nominal, a numeric code as well as a nonnumeric symbol may be used. For example, the data for the Exchange variable could use a numeric code with a value of 1 for the New York Stock Exchange, a value of 2 for the American Stock Exchange, and a value of 3 for Over The Counter. Such a code would facilitate recording and computer processing of the exchange data. However, it is important to remember that the numeric values of 1, 2, and 3 are simply labels used to identify the exchange. Thus, the scale of measurement is nominal.

Other examples of variables where data have a nominal scale are as follows:

- Sex (male, female)
- Marital status (single, married, widowed, divorced)
- Religious affiliation (many possibilities)

- Part-identification code (A13622, 12B63)
- Employment status (employed, unemployed)
- House number (5654, 2712, 624)
- Occupation (many possibilities)

As the preceding examples show, the key feature of the nominal scale is that the data obtained are labels used to identify an attribute of the element.

Finally, it is important to note that arithmetic operations such as addition, subtraction, multiplication, and division *do not* make sense for nominal data. Thus, even if the nominal data are numeric, arithmetic operations such as addition, subtraction, and averaging are inappropriate.

Ordinal Scale

The scale of measurement for a variable is *ordinal* when

1. The data have the properties of nominal data, and
2. The order or rank of the data is meaningful.

Let us use the following example to illustrate the ordinal scale of measurement. Some restaurants place questionnaires on the tables to solicit customer opinions concerning the restaurant's performance in terms of food, service, atmosphere, and so on. A questionnaire used by the Lobster Pot Restaurant in Redington Shores, Florida, is shown in Figure 1.1. Note that the customers completing the questionnaire are asked to provide ratings for

FIGURE 1.1

Customer Opinion Questionnaire Used by the Lobster Pot Restaurant, Redington Shores, Florida (*Used with Permission*)

The LOBSTER Pot
RESTAURANT

	Excellent	Good	Poor	Comments
Food				
Drinks				
Service				
Waiter				
Captain				
Hostess				

Waiter's Name _____

Captain's Name _____

Other Comments _____

six different variables: food, drinks, service, waiter, captain, and hostess. The response categories are excellent, good, and poor for each variable. The observations for each variable possess characteristics of nominal data in that each response rating is a label for excellent, good, or poor quality. In addition, the data can be ranked, or ordered, with respect to quality. For example, consider the food quality variable . After collecting the data, we can rank the data in order of food quality by beginning with excellent, followed by good, and, finally, followed by poor. With only three response categories, we can expect to see many data with tied rankings. Nonetheless, the data can be ranked in terms of food quality.

Like data obtained from a nominal scale, data obtained from an ordinal scale may be either nonnumeric or numeric. For the Lobster Pot questionnaire, the nonnumeric letters of E for excellent, G for good, and P for poor could be used to record the observations. Or, a numeric code with values of 1 for excellent, 2 for good, and 3 for poor could be used equally well. Finally, as with nominal data, it is important to remember that arithmetic operations *do not* make sense for ordinal data. Thus, even if the ordinal data are numeric, the arithmetic operations such as addition, subtraction, or averaging are inappropriate.

Interval Scale

The scale of measurement for a variable is *interval* when

1.   The data have the properties of ordinal data, and
2.   The difference or interval between data values indicates how much more or less of a variable one element possesses when compared to another element.

Interval data are always *numeric*. Subtracting the interval data value for one element from the interval data value for another element provides the difference between the two elements in terms of the variable being considered. Temperature is a good example of a variable that is measured on an interval scale. The high temperatures on February 12, 1992 (*USA Today*), for Phoenix, Arizona, Charlotte, North Carolina, and Minneapolis, Minnesota are in Table 1.3. The interval Fahrenheit-scale temperatures are numeric with the property of ordinal data in that Phoenix, Charlotte, and Minneapolis can be ranked in terms of temperature with Phoenix being the warmest (highest temperature) and Minneapolis being the coldest (lowest temperature). In addition, the difference between data values is meaningful with Phoenix being 67 − 60 = 7 degrees warmer than Charlotte and 67 − 26 = 41 degrees warmer than Minneapolis. Charlotte was 60 − 26 = 34 degrees warmer than Minneapolis.

Using the Celsius scale for temperature, the above high temperatures on February 12, 1992, would have been Phoenix (19.4), Charlotte (15.6), and Minneapolis (−3.3). These interval-scaled data can again be ordered or ranked from highest to lowest with the differences meaningful in that Phoenix was 19.4 − (−3.3) = 22.7 degrees warmer than Minneapolis on the Celsius scale.

Scholastic Aptitude Test (SAT) scores are another example of interval-scaled data. Three students with SAT scores of 1120, 1050, and 970 can be ranked or ordered in terms of best to poorest performance on the SAT. In addition, the differences are meaningful, showing student 1 scored 1120 − 1050 = 70 points more than student 2, while student 2 scored 1050 − 970 = 80 points more than student 3.

Finally, with interval-scaled data, the arithmetic operations of addition, substraction, and averaging are meaningful. As a result, data obtained using this scale lend themselves to more alternatives for statistical analysis than do data obtained from nominal or ordinal scales.

TABLE **1.3**

High Temperatures for February 12, 1992

City	High Temperature
Phoenix	67
Charlotte	60
Minneapolis	26

Ratio Scale

The scale of measurement for a variable is a *ratio* scale when

1. The data have all the properties of interval data, and
2. The ratio of two data values is meaningful.

Variables such as distance, height, weight, and time use the ratio scale of measurement. A requirement of this scale is that a zero value is inherently defined in the scale. Specifically, the zero value must indicate nothing exists for the variable at the zero point. Whenever we collect data on the cost of something, we use a ratio scale of measurement. Consider a variable indicating the cost of an automobile. The zero point is inherently defined in that a zero cost indicates the automobile is free (no cost). Then, comparing the $10,000 cost of one automobile with the $5000 cost of a second automobile, the ratio property of the data shows that the first automobile is 10,000/5000 = 2 times, or twice, the cost of the second automobile.

Since ratio data have all of the properties of interval data, ratio data are *always numeric*. Arithmetic operations such as addition, subtraction, multiplication, division, and averaging are possible with ratio data. Thus, as with interval data, ratio data lend themselves to more alternatives for statistical analysis than do data obtained from nominal or ordinal scales.

Table 1.4 provides a summary of the relationship between nonnumeric and numeric data and the four scales of measurement. We note that nominal and ordinal scales can generate both nonnumeric and numeric data, but that interval and ratio scales generate only numeric data.

The amount of information in the data varies with the scale of measurement. Nominal data contain the least amount of information, followed by ordinal, interval, and then ratio data. Since arithmetic operations are meaningful only for interval and ratio data, it is important to know the measurement scale used in order to employ the most appropriate statistical procedures. There are many statistical procedures that can be used with interval and ratio data that are not meaningful with nominal and ordinal data.

TABLE 1.4 The Relationship between Nonnumeric and Numeric Data and Scales of Measurement	Scale of Measurement	Nonnumeric Data	Numeric Data
	Nominal	Description indicates the category for the element	Numeric value indicates the category for the element
	Ordinal	Description permits ranking or ordering of the data	Numeric value permits ranking or ordering of data
	Interval		Numeric values* are defined such that the interval between data values is meaningful
	Ratio		Numeric values* have an inherently defined zero, and the ratio of data values is meaningful

*Arithmetic operations are meaningful for these kinds of data.

Qualitative and Quantitative Data

Data can also be classified as being either qualitative or quantitative. *Qualitative data* provide labels or names for categories of like items. The categories for qualitative data may be identified by either nonnumeric descriptions or by numeric codes. Qualitative data are obtained from either a nominal or an ordinal scale of measurement. On the other hand, *quantitative data* indicate either "how much" or "how many" of something. Quantitative data are always numeric and are obtained from either an interval or a ratio scale of measurement.

In terms of the statistical methods used for summarizing data, qualitative data provided by nominal and ordinal scales employ similar methods, whereas quantitative data provided by interval and ratio scales employ similar methods. Tabular and graphical methods for summarizing qualitative data are presented in Section 2.1. Tabular and graphical methods for summarizing quantitative data are presented in Sections 2.2 and 2.3.

**NOTES &
COMMENTS**

1. An element is an entity on which measurements are obtained. It could, for example, be a company, a person, or an automobile. An observation is the set of measurements obtained for each element, so the number of observations and the number of elements in a data set will always be the same. The number of measurements obtained on each element is the number of variables. Thus, the total number of data values in a data set is the number of elements times the number of variables.
2. For purposes of statistical analysis, the most important distinguishing characteristic of the types of data is that ordinary arithmetic operations are meaningful *only* with quantitative (interval- and ratio-scaled) data.

1.4   Data Acquisition

We have introduced the concepts of elements, variables, and observations and of how data collected in a particular study form a data set. In this section we describe the sources of data that are available and how the data can be acquired. The data needed for statistical analysis may be obtained through already-existing sources of data or through specially designed statistical studies that are undertaken to obtain new data.

Existing Sources of Data

In some cases, data needed for a particular application are already available within the firm or organization. For example, all companies maintain a variety of data on their employees and their business operations. Data on employee salaries, ages, and years of experience can usually be obtained relatively easily from internal personnel records. Data on sales, advertising costs, distribution costs, inventory levels, and production quantities are available from other internal information and record-keeping systems. Table 1.5 shows some of the data that are routinely available from internal information sources.

In addition, substantial amounts of business and economic data are routinely collected and published in a variety of external sources. In some instances, a company may obtain access to a large data base through a leasing arrangement with an organization that specializes in maintaining a specific information system. Mead Corporation, Dun & Bradstreet, and Dow Jones & Company are some of the firms that provide data base services.

TABLE 1.5
Some Data Available from Internal
Company Records

Company Record	Some of the Data Typically Available
Employee records	Name, address, social security number, salary, number of vacation days, number of sick days, and bonus
Production records	Part or product number, quantity produced, direct labor cost, and materials cost
Inventory records	Part or product number, number of units on hand, reorder level, economic order quantity, and discount schedule
Sales records	Product number, sales volume, sales volume by region, and sales volume by customer type
Customer credit	Customer name, address, phone number, credit limit, and accounts receivable balance

Governmental agencies provide another important source of existing data. For instance, the U.S. Department of Labor maintains considerable data on characteristics such as employment rates, wage rates, size of the labor force, and union memberships. Table 1.6 shows other selected governmental agencies and some of the data they provide. Data are also available from a variety of industry associations and special-interest organizations. The Travel Industry Association of America maintains travel-related information such as the number of tourists and travel expenditures by state. Such data would be of interest to firms and individuals participating in the travel industry. The Graduate Management Admission Council maintains data on MBA student characteristics and graduate management education programs. Most of the data from sources such as these are available to qualified users at a modest cost.

Statistical Studies

Sometimes data are not readily available from existing internal or external sources. If the data are considered to be necessary, a statistical study will have to be conducted in order to obtain the data. Such statistical studies can be classified as being either *experimental* or *observational*.

In an experimental study, variables of interest are identified. Then one or more factors in the study are controlled so that data may be obtained about how the factors influence the variables. For example, a pharmaceutical firm might be interested in conducting an experiment designed to learn about how a new drug affects blood pressure. Blood pressure is the variable of interest in the study. The new drug is the factor that influences the blood pressure. To obtain data about the effect of the new drug, a sample of individuals will be selected. The dosage level of the new drug will be controlled, with different groups of individuals being given different dosage levels. Data on blood pressure will be collected for each group. Statistical analysis of the experimental data will help determine how the new drug affects blood pressure.

In nonexperimental, or observational, statistical studies, no attempt is made to control or influence the variables of interest. A survey is perhaps the most common type of observational study. In a survey, research questions are identified. A questionnaire is then designed and administered to a sample of individuals. In this way, data are obtained about the research variables but no attempt is made to control the factors that influence the variables.

TABLE 1.6	Governmental Agency	Some of the Data Available
Selected Governmental Agencies and Some of the Data Available	Bureau of the Census	Population data and its distribution, data on number of households and their distribution, data on household income and its distribution
	Federal Reserve Board	Data on the money supply, installment credit, exchange rates, and discount rates
	Office of Management and Budget	Data on revenue, expenditures, and debt of the federal government
	Department of Commerce	Data on business activity—value of shipments by industry, level of profits by industry, and data on growing and declining industries

Managers wishing to use data and statistical analyses as an aid to decision making must always be aware of the time and cost required to obtain the data. The use of existing data sources is desirable when data must be obtained in a relatively short period of time. If the data are not available from an existing source, the additional time and costs involved in collecting the data must be taken into account. In all cases, it is desirable for the decision maker to consider the contribution of the statistical analysis to the decision-making process. The cost of data acquisition and the subsequent statistical study should not be more than the savings generated by using the statistical information to make a better decision.

Possible Data-Acquisition Errors

Managers should always be aware of the possibility of data errors in their use of statistical studies. Using erroneous data and the misleading statistical analyses generated from such data would be worse than not using the data and statistical information at all. An error in data acquisition occurs whenever the data value obtained is not equal to the true or actual value that would have been obtained with a correct procedure. Such errors can occur in a number of ways. For example, an interviewer might make a recording error such as writing the age of a 24-year-old person as 42. Or, the person answering an interview question might misinterpret the question and make an incorrect response.

Experienced data analysts take great care in collecting and recording data to ensure that errors are not made. Special procedures may be used to check for internal consistency of the data. Such procedures would indicate the analyst should review data for a respondent who is shown to be 22 years of age but who reports 20 years of work experience. Data analysts also review data for unusually large and small values, called *outliers*, which are candidates for possible data errors. In Chapter 3, we present some of the methods statisticians use to identify outliers.

The point of this discussion is to alert the users of data to the possibility of errors occurring during data acquisition. Blindly using any data that happen to be available or using data that were acquired with little care can lead to poor and misleading information. However, taking steps to acquire accurate data can help provide reliable and valuable decision-making information.

1.5   Descriptive Statistics

Most of the statistical information in newspapers, magazines, reports, and other publications comes from data that have been summarized and presented in a form that is easy for the reader to understand. These summaries of data, which may be tabular, graphical, or numerical, are referred to as *descriptive statistics*.

Refer again to the data set in Table 1.2 where 25 shadow stocks are listed. Methods of descriptive statistics can be used to provide summaries of the information in this data set. For example, a tabular summary of the data for the exchange variable is shown in Table 1.7. A graphical summary of the same data is shown in Figure 1.2. The purpose of tabular and graphical summaries such as these is to make it easier to interpret the data. Referring to Table 1.7 and Figure 1.2, we can see easily that the majority of the stocks in the data set are listed on the Over-The-Counter exchange. On a percentage basis, 56% of the stocks are listed on the Over-The-Counter exchange. Also, 32% are listed on the American Stock Exchange, and only 12% are listed on the New York Stock Exchange.

A graphical summary of the data on dividend yield of the stocks in Table 1.2 is provided by the histogram in Figure 1.3. From the histogram, it is easy to see that the dividend yields range from 1 to 9%, with the highest concentrations of dividend yields being 4 and 7%.

TABLE 1.7

Frequencies and Percentages for the Exchange Listings of 25 Shadow Stocks with Dividend Reinvestment Plans

Exchange	Frequency	Percent
New York Stock Exchange	3	12
American Stock Exchange	8	32
Over The Counter	14	56
Total	25	100

FIGURE 1.2

Bar Graph of Exchange Listings for Shadow Stocks with Dividend Reinvestment Plans

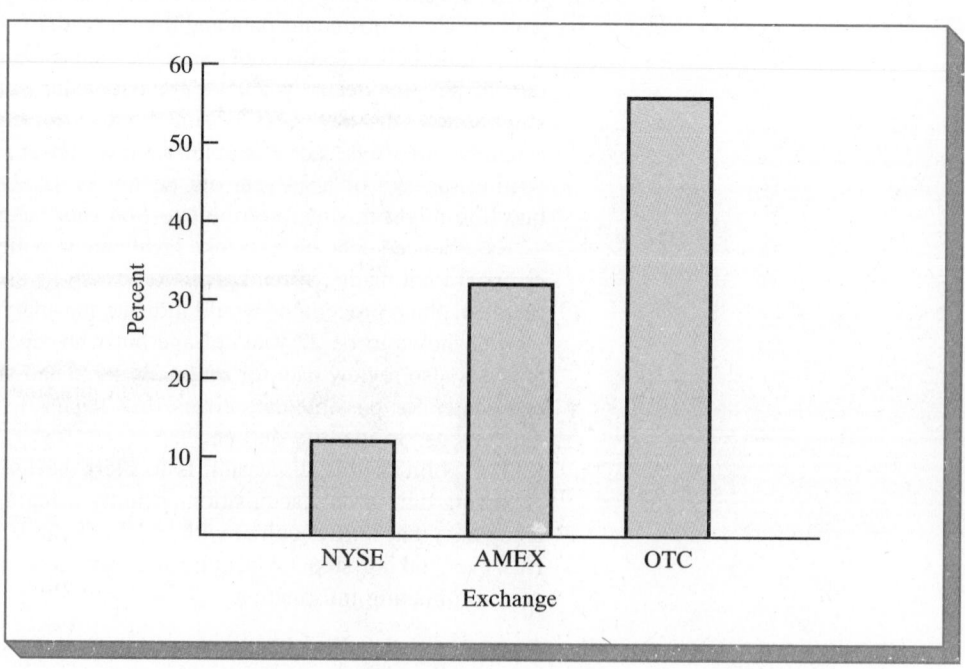

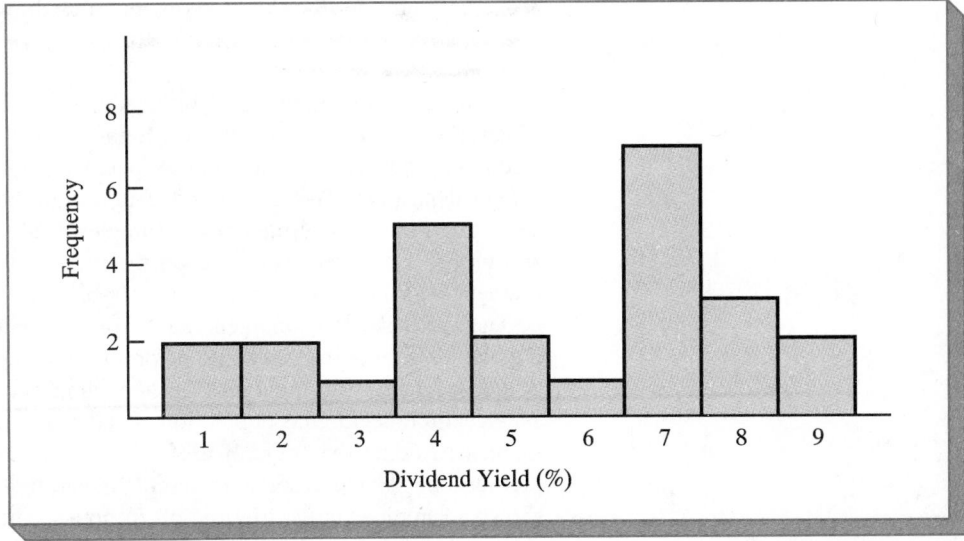

In addition to tabular and graphical displays, numerical descriptive statistics are often used to summarize data. The most common numerical descriptive statistic is the *average,* or *mean.* Using the data on dividend yield in Table 1.2, we can compute the average dividend yield by adding the dividend yields for all 25 stocks and dividing the sum by 25. Doing so tells us that the average dividend yield for the stocks is 5.44%. This average is taken as a measure of the central value, or central location, of the data.

In recent years there has been a growing interest in statistical methods that can be used for developing and presenting descriptive statistics for data sets. Chapters 2 and 3 are devoted to the tabular, graphical, and numerical methods of descriptive statistics.

1.6   Statistical Inference

In many situations there exists a large group of elements (individuals, stocks, voters, households, products, customers and so on) about which data are sought. Because of time, cost, and other considerations, data are collected from only a small portion of the group. The larger group of elements in a particular study is called the *population,* and the smaller group is called the *sample.* Formally, we will use the following definitions:

Population

A *population* is the collection of all elements of interest in a particular study.

Sample

A *sample* is a subset of the population.

A major contribution of statistics is that it enables us to use data from a sample to make estimates and test claims about the characteristics of a population. This process is referred to as *statistical inference*.

As an example of statistical inference, let us consider the study conducted by Norris Electronics. Norris manufactures a high-intensity light bulb that is used in a variety of electrical products. In an attempt to increase the useful life of the light bulb, the product design department has developed a new light-bulb filament. In order to evaluate the advantages of the new filament, a sample of 200 new-filament bulbs was manufactured and tested. Data were collected on the number of hours each bulb operated before the filament burned out and are shown in Table 1.8.

The available data was generated from the sample of 200 bulbs. The corresponding population is all bulbs that could be produced using the new filament. Suppose that Norris is interested in using the data from the sample to make an inference about the mean or average lifetime for the population of all bulbs that could be produced using the new filament. Adding the hours of useful life for the 200 bulbs and dividing the total by 200 provides the sample average, or sample mean, lifetime. Doing so for the data in Table 1.8 shows a sample average lifetime of 76 hours. Thus, we would use this sample result to estimate that the average lifetime for the bulbs in the population would be 76 hours.

Whenever statisticians make an estimate, as in the Norris Electronics example, they usually provide a statement of the precision associated with the estimate. For the Norris example, the statistician might state that the estimate of the average lifetime for the population of new light bulbs is 76 hours with a precision of ± 4 hours. Thus, 72–80 hours would be the interval estimate for the average lifetime of all bulbs that could be produced using the new filament. Using concepts from probability theory, the statistician can also state how confident he or she is that the interval of 72–80 hours contains the true

TABLE 1.8 Hours until Burnout for a Sample of 200 Bulbs for Norris Electronics									
107	73	68	97	76	79	94	59	98	57
54	65	71	70	84	88	62	61	79	98
66	62	79	86	68	74	61	82	65	98
62	116	65	88	64	79	78	79	77	86
74	85	73	80	68	78	89	72	58	69
92	78	88	77	103	88	63	68	88	81
75	90	62	89	71	71	74	70	74	70
65	81	75	62	94	71	85	84	83	63
81	62	79	83	93	61	65	62	92	65
83	70	70	81	77	72	84	67	59	58
78	66	66	94	77	63	66	75	68	76
90	78	71	101	78	43	59	67	61	71
96	75	64	76	72	77	74	65	82	86
66	86	96	89	81	71	85	99	59	92
68	72	77	60	87	84	75	77	51	45
85	67	87	80	84	93	69	76	89	75
83	68	72	67	92	89	82	96	77	102
74	91	76	83	66	68	61	73	72	76
73	77	79	94	63	59	62	71	81	65
73	63	63	89	82	64	85	92	64	73

FIGURE 1.4

The Process of Statistical Inference for the Norris Electronics Example

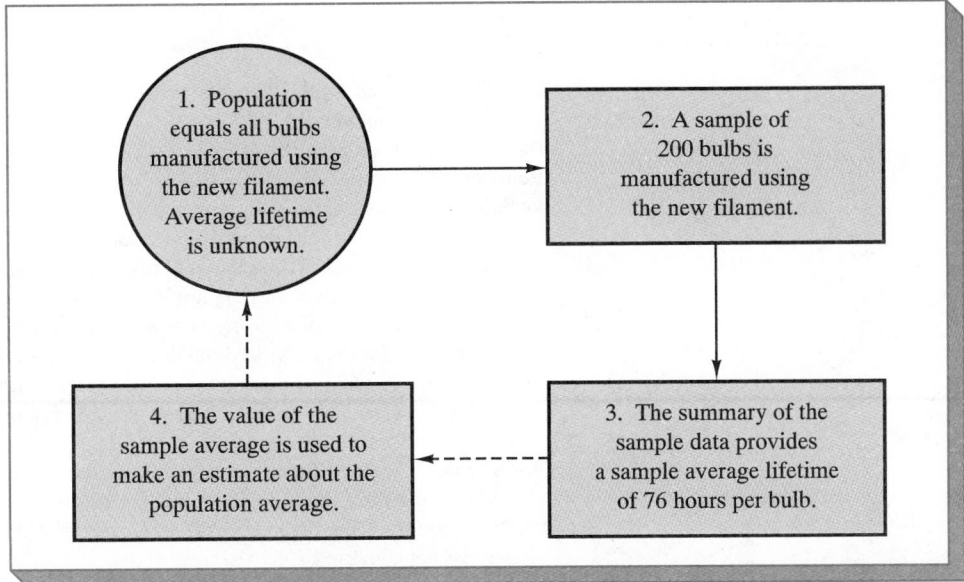

average lifetime for the population. At this point, management should be ready to decide whether the results justify going ahead with the production of the new bulbs.

Figure 1.4 provides a graphical summary of the statistical inference process used in the Norris Electronics example. Chapters 7–16, 19, 20, and 21 describe a variety of statistical inference methods available to a decision maker.

Summary

In everyday usage, the term *statistics* is often taken to mean numerical facts. However, the field of statistics requires a broader definition. Statistics is the science of collecting, analyzing, presenting, and interpreting data. Because nearly every college student majoring in business or economics is required to take a course in statistics, we began the chapter by describing typical statistical applications in business and economics.

Data were described as being the raw material of statistics. Thus, data are the facts and figures that are collected, analyzed, presented, and interpreted. A significant portion of the chapter was devoted to data, including how data are measured and how data are acquired.

Four scales of measurement are available for obtaining data on a particular variable: nominal, ordinal, interval, and ratio. The amount of information contained in the data depends upon the scale of measurement. Nominal data contain the least amount of information, followed by ordinal, interval, and ratio data. We showed that nominal and ordinal data are classified as qualitative data and may be recorded by either a nonnumeric description or a numeric code. Interval and ratio data are classified as quantitative data. Quantitative data are always numeric and indicate how much or how many for the variable of interest. Ordinary arithmetic operations are meaningful only if the data are *quantitative*. As a result, many statistical procedures that use quantitative data are not appropriate for qualitative data.

In Sections 1.5 and 1.6 we introduced the topics of descriptive statistics and statistical inference. Descriptive statistics are the tabular, graphical, and numerical methods used to summarize data. Statistical inference is the process of using data obtained from a sample to make estimates or test claims about the characteristics of a population.

Glossary

Data The facts and figures that are collected, analyzed, presented, and interpreted. Data may be numeric or nonnumeric.

Data set All the data collected in a particular study.

Elements The entities on which data are collected.

Variable A characteristic of interest for the elements.

Observation The set of measurements or data obtained for a single element.

Nominal scale A scale of measurement that uses a label or category to define an attribute of an element. Nominal data may be nonnumeric or numeric.

Ordinal scale A scale of measurement that has the properties of a nominal scale and can be used to rank or order the data. Ordinal data may be nonnumeric or numeric.

Interval scale A scale of measurement that has the properties of an ordinal scale and where the interval between data values is expressed in terms of a fixed unit of measure. Interval data are always numeric.

Ratio scale A scale of measurement that has the properties of an interval scale and that is meaningful in the ratio of data. Ratio data are always numeric.

Qualitative data Data obtained with a nominal or ordinal scale of measurement. Qualitative data may be nonnumeric or numeric.

Quantitative data Data obtained with an interval or ratio scale of measurement. Quantitative data are always numeric and indicate how much or how many for the variable of interest.

Descriptive statistics Tabular, graphical, and numerical methods used to summarize data.

Population The collection of all elements of interest in a particular study.

Sample A subset of the population.

Statistical inference The process of using data obtained from a sample to make estimates or test claims about the characteristics of a population.

☐ ☐ Exercises

1. Discuss the difference between the concept of statistics as numerical facts and the concept of statistics as a discipline or field of study.

2. Table 1.9 shows the executive compensation, sales, return on equity, and profit rating for seven firms in the aerospace industry (*Business Week,* May 1, 1989).
a. How many elements are in this data set?
b. How many variables are in this data set?
c. How many observations are in this data set?
d. Which of the variables are qualitative and which are quantitative?
e. What measurement scale is being used for each of the variables?

3. Refer to the CEO salary and profit-rating data in Table 1.9 (*Business Week,* May 1, 1989).
a. Compute the average salary for the chief executive officers.
b. What proportion of the firms received a rating of 3 on pay versus corporate profitability?
c. Why is it inappropriate to compute an average for the data on pay versus corporate profitability?

4. *Fortune* magazine provides annual data on how the 500 largest U.S. industrial corporations rank in terms of sales, profits, and investment performance. A data set based on a sample of the Fortune 500 companies is shown in Table 1.10 (*Fortune,* April 20, 1992).
a. How many elements are in the data set?
b. What is the population from which this sample is drawn?
c. Compute the average sales for the sample.
d. Use the result of (c) to make an estimate of the average sales for the population.

TABLE 1.9

Information Concerning Executive Compensation and Profitability for Seven Firms in the Aerospace Industry

Company	CEO Salary ($000s)	Sales ($000,000s)	Return on Equity (%)	Pay versus Corporation Profit Rating*
Boeing	846	16,962.0	11.4	3
General Dynamics	1,041	9,551.0	19.7	3
Lockheed	1,146	10,590.0	17.9	3
Martin Marietta	839	5,727.5	26.6	2
McDonnell Douglas	681	15,069.0	11.0	3
Parker Hannifin	765	2,397.3	12.4	3
United Technology	1,148	18,000.1	13.7	4

*CEOs are assigned to five groupings according to pay versus corporate profitability. Those in group 1 are given a rating of 1, those in group 2 a rating of 2, and so on. A rating of 1 is the best rating; a rating of 5 is the worst.

Source: Business Week, May 1, 1989.

TABLE 1.10

A Sample of 10 of the Fortune 500 Largest U.S. Industrial Corporations

Company	Sales ($000,000)	Rank	Profit ($000,000)	Assets ($000,000)	Industry Code
Raytheon	9356	51	592	6087	7
Texas Instruments	6812	77	−409	5009	7
Hershey Foods	2902	160	220	2342	8
Polaroid	2096	203	684	1889	22
Zenith Electronics	1322	282	−52	687	7
Brown-Foreman	1126	315	145	1083	3
Cooper Tire	1002	338	79	671	21
Quaker State	814	392	23	752	18
Standard Register	694	432	33	464	20
La-Z-Boy Chair	610	464	23	363	10

Source: Fortune, April 20, 1992

 SELF TEST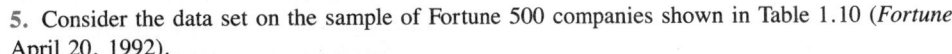

5. Consider the data set on the sample of Fortune 500 companies shown in Table 1.10 (*Fortune,* April 20, 1992).
   a. How many variables are in the data set?
   b. Which of the variables are qualitative and which are quantitative?
   c. What type of measurement scale is being used for each of the variables?

6. Columbia House provides CDs, tapes, and records to its mail-order club members. A Columbia House Music Survey conducted in 1992 asked new club members to complete an 11-question survey. Some questions asked were as follows:
   a. How many albums (CDs, tapes, or records) have you bought in the last 12 months?
   b. Are you currently a member of a national mail-order book club? (Yes or No)
   c. What is your age?
   d. Including yourself, how many people (adults and children) are in your household?
   e. What kind of music are you interested in buying? (There were 15 categories listed, including hard rock, soft rock, adult contemporary, heavy metal, rap, and country.)

Comment on each question as to whether it provides qualitative or quantitative data.

7. A California state agency classifies worker occupations as professional, white collar, or blue collar.
a. The variable is worker occupation. Is it a qualitative or quantitative variable?
b. What type of measurement scale is being used for this variable?

8. A Bruskin-Goldring research poll asked 1000 people, Where is the best place to meet a "suitable companion"? (*USA Today,* February 13, 1992) Response categories were religious gathering, friend's home, evening class, work, nightspot, or other.
a. What was the sample size in this survey?
b. What measurement scale is being used?
c. Would it make more sense to use averages or percentages as a summary of the data in this survey?
d. Nightspot was the least preferred choice with 170 people selecting it as the best place to meet a "suitable companion." What percentage was reported for the nightspot category?

9. A study of last-minute holiday shoppers conducted by Best Products, Inc. (*USA Today,* December 22, 1988) asked individuals to indicate when they completed their holiday shopping. The response alternatives were as follows:

> Before Halloween   A few days before Christmas
> Before Thanksgiving   Christmas Eve

a. The variable of interest is the date by which holiday shopping is completed. Is this a qualitative or quantitative variable?
b. What measurement scale is being used?
c. Would it make more sense to use averages or percentages as a summary of the data in this study? Explain.

10. State whether each of the following variables is qualitative or quantitative, and indicate the measurement scale that is appropriate for each.
a. Age   b. Sex   c. Class rank   d. Make of automobile
e. Number of people favoring the death penalty

11. State whether each of the following variables is qualitative or quantitative and indicate the measurement scale being used.
a. Annual sales   b. Soft-drink size (small, medium, or
c. Employee classification (GS1 through GS18)   large)
e. Method of payment (cash, check, credit card)   d. Earnings per share

12. The Hawaii Visitors Bureau collects data on visitors to Hawaii. The following questions were among the 16 questions asked in a questionnaire handed out to passengers during incoming airline flights in July 1992.

> 1. This trip to Hawaii is my: 1st, 2nd, 3rd, 4th, etc.
> 2. The primary reason for this trip is: (10 categories including vacation, convention, honeymoon)
> 3. Where I plan to stay: (11 categories including hotel, apartment, relatives, camping)
> 4. Total days in Hawaii

a. What is the population being studied?
b. Is the use of questionnaires for passengers during incoming airline flights a good way to reach this population?
c. Comment on each of the four questions above in terms of whether it will provide qualitative or quantitative data.

13. A manager of a large corporation has recommended a $10,000 raise be given in order to keep a valued subordinate from moving to another company. What internal and external sources of data might be used to decide whether such a salary increase is appropriate?

14. The marketing group at your company has come up with a new diet soft drink that it claims will capture a large share of the young adult market.

a. What data would you want to see before deciding to invest substantial funds in introducing the new product into the marketplace?

b. How would you expect the data mentioned in (a) to be obtained?

S E L F   T E S T ▶ **15.** A *1988 Newsweek/Gallup* poll investigated whether adults preferred staying home or going out as their favorite way of spending time in the evening. The poll of 1500 adults concluded that the majority of adults (70%) indicated that "staying at home with family" was the favorite evening activity.

a. What is the population of interest in this study?

b. What is the variable being studied?

c. Is the variable being studied qualitative or quantitative?

d. What was the size of the sample used?

e. Where was a descriptive statistic used in this study?

f. Describe the process of statistical inference in this study.

16. In a recent study of causes of death in males 60 years of age and older, a sample of 120 men indicated that 48 died due to some form of heart disease.

a. Develop a descriptive statistic that can be used as an estimate of the percentage of males 60 years of age or older who die from some form of heart disease.

b. Is the data on cause of death qualitative or quantitative?

c. Discuss the role of statistical inference in this type of medical research.

17. The 25th Annual Report on Shoplifting in Supermarkets (*Commercial Service Systems, Inc., 1987*) used 391 supermarkets in southern California to compile the following statistics on supermarket shoplifters caught in the act:

- ■ Most items stolen were valued between $1 and $5
- ■ Most shoplifters were male (56%)
- ■ Most frequent age category of shoplifters was "under 30" (51%)

Answer the following questions assuming that the purpose of the study was to present statistical data on the national trend and impact of shoplifting in supermarkets.

a. Cite two descriptive statistics reported.

b. What warning would you issue if the results of the study were to be used to make a statistical inference about the national trend and impact of shoplifting in supermarkets?

18. Select a recent copy of the newspaper *USA Today*.

a. Note four examples of statistical information.

b. For each example, indicate the descriptive statistics used and discuss any statistical inferences made.

19. A 7-year medical research study (*Journal of the American Medical Association,* December 1984) reported that women whose mothers took the drug DES during pregnancy were *twice* as likely to develop tissue abnormalities that might lead to cancer as women whose mothers did not take the drug.

a. This study involved the comparison of two populations. What were the populations involved?

b. Do you suppose the data obtained here were the result of a survey or an experiment?

c. For the population of women whose mothers took the drug DES during pregnancy, a sample of 3980 women showed 63 developed tissue abnormalities that might lead to cancer. Provide a descriptive statistic that could be used to estimate the number of women out of 1000 in this population who have tissue abnormalities.

d. For the population of women whose mothers did not take the drug DES during pregnancy, what is the estimate of the number of women out of 1000 who would be expected to have tissue abnormalities?

e. Medical studies of diseases and disease occurrence often use a relatively large sample (in this case, 3980). Why is this done?

20. A firm is interested in testing the advertising effectiveness of a new television commercial. As part of the test, the commercial is shown on a 6:30 P.M. local news program in Denver, Colorado. Two days later a market research firm conducts a telephone survey to obtain information on recall

rates (percentage of viewers who recall seeing the commercial) and impressions of the commercial.
a. What is the population for this study?   **b.** What is the sample for this study?
c. Why would a sample be used in this situation? Explain.

21. The Nielsen organization conducts weekly surveys of television viewing throughout the United States. The Nielsen statistical ratings indicate the size of the viewing audience for each major network television program. Rankings of the television programs and of the viewing-audience market shares for each network are published each week.
a. What is the Nielsen organization attempting to measure?
b. What is the population?
c. Why would a sample be used for this situation?
d. What kinds of decisions or actions are taken based on the Nielsen studies?

22. A sample of midterm grades for five students showed the following results: 72, 65, 82, 90, 76. Which of the following statements are correct, and which should be challenged as being too generalized?
a. The average midterm grade for the sample of five students is 77.
b. The average midterm grade for all students who took the exam is 77.
c. An estimate of the average midterm grade for all students who took the exam is 77.
d. More than half of the students who take this exam will score between 70 and 85.
e. If five other students are included in the sample, their grades will be between 65 and 90.

2

Descriptive Statistics I. Tabular and Graphical Approaches

Colgate-Palmolive Company*

NEW YORK, NEW YORK

The Colgate-Palmolive Company started as a small soap and candle shop in New York City in 1806. Today, Colgate-Palmolive products can be found around the world. International operations exist in 55 countries, and annual sales are in excess of $5 billion. While best known for its traditional product line of soaps, detergents, and toothpastes, subsidiary operations include the Kendall Company, Riviana Foods, Etonic, Bike Athletic Company, and others.

The Colgate-Palmolive Company uses statistics in its quality assurance program for home-laundry detergent products. One example is customer satisfaction with the quantity of product in a given carton. The regular-size carton of a detergent has a stated product weight of 20 ounces. Of particular concern is the density of the detergent powder that is placed in the carton. If the density of the powder is on the heavy side, it will not take as much powder to reach the 20-ounce-per-carton weight limit. In

this case, the company could be faced with the dilemma of having cartons filled with the desired 20 ounces even as they appear underfilled when opened by the consumer.

To address the problem of heavy detergent, the statistical procedure of sampling is used. Periodically, samples of powder are taken, and data summaries about powder densities are provided to operating personnel so that corrective action can be taken to keep the density of the detergent within the desired quality specifications.

Summary reports about the densities are sent to various managers throughout the company. A frequency distribution for the densities of 150 samples taken over a 1-week period and a histogram of the sample data are shown below. An unacceptably high density occurs around .40. These tabular and graphical summaries show that the operation is meeting its quality guidelines, with practically all of the densities less than .40. Managers viewing these statistical summaries would be pleased with the quality of the detergent production process.

In this chapter you will learn about tabular and graphical methods of descriptive statistics such as bar graphs, histograms, relative frequency distributions, stem-and-leaf displays, dot plots, and others. The goal of these techniques is to summarize data so that it can be easily understood and interpreted.

❑ Frequency Distribution of Density Data

Density	Frequency
.29–.30	30
.31–.32	75
.33–.34	32
.35–.36	9
.37–.38	3
.39–.40	1
Total Samples	150

❑ Histogram of Density Data

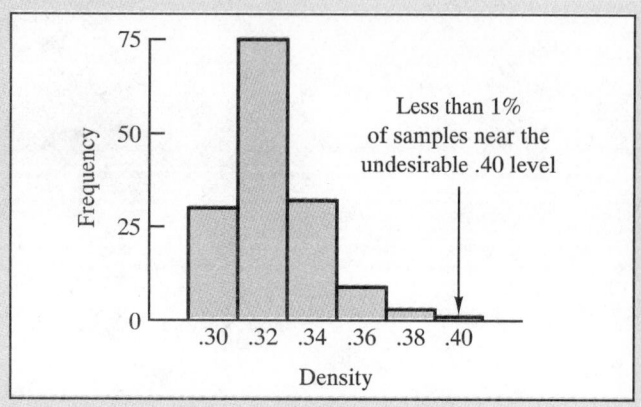

Less than 1% of samples near the undesirable .40 level

Statistical summaries of product weights aid a Colgate-Palmolive management team during a meeting on quality assurance.

*William R. Fowle, Manager of Quality Assurance, Colgate-Palmolive Company, provided this Statistics in Practice.

A s indicated in Chapter 1, data may be classified as being either *qualitative* or *quantitative*. Qualitative data provide labels or names for categories of like items. The categories for qualitative data may be identified by nonnumeric descriptions or by numeric codes. Qualitative data are measured on nominal or ordinal scales. Quantitative data indicate ''how much'' or ''how many.'' Quantitative data are always numeric and are measured on interval or ratio scales.

The purpose of this chapter is to introduce several tabular and graphical procedures commonly used to summarize both qualitative and quantitative data. Tabular and graphical summaries of data can be found in annual reports, newspaper articles, and research studies. Everyone is exposed to these types of presentations. Hence, it is important to understand how they are prepared and to know how they should be interpreted. We begin with tabular and graphical methods for summarizing qualitative data. Methods for summarizing quantitative data are presented in Section 2.2.

2.1   Summarizing Qualitative Data

Frequency Distribution

We begin the discussion of how tabular and graphical methods can be used to summarize qualitative data with the definition of a *frequency distribution*.

Frequency Distribution

A *frequency distribution* is a tabular summary of a set of data showing the frequency (or number) of items in each of several nonoverlapping classes.

The objective in developing a frequency distribution is to provide insights about the data that cannot be quickly obtained if we look only at the original data. To see how frequency distributions can be applied when dealing with qualitative data, let us consider the following illustration.

What models of automobiles are most often purchased by women? *USA Today* (December 13, 1988) reported that, based on 1988 automobile sales data, the Chevrolet Cavalier, Ford Escort, Ford Taurus, Honda Accord, and Hyundai Excel were the top choices among women buying cars. The *USA Today* article summarized the purchase choices of several thousand women. Assume that the data shown in Table 2.1 were collected from a sample of 50 women who made a recent purchase of one of the top five selling automobiles.

To develop a frequency distribution for the qualitative data in Table 2.1, we simply *count* the number of times each of the five automobiles appears in the data set. The Chevrolet Cavalier appears 9 times, the Ford Escort appears 14 times, the Ford Taurus appears 8 times, the Honda Accord appears 11 times, and the Hyundai Excel appears 8 times. These counts are summarized in the frequency distribution shown in Table 2.2.

The advantage of the frequency distribution is that it provides a better understanding of the preferences of women in the sample than does the original data shown in Table 2.1.

TABLE 2.1

Data from a Sample of 50 New-Car Purchases by Women

Honda Accord	Ford Escort	Ford Taurus
Ford Taurus	Chevrolet Cavalier	Honda Accord
Honda Accord	Ford Escort	Ford Taurus
Honda Accord	Hyundai Excel	Hyundai Excel
Ford Escort	Hyundai Excel	Chevrolet Cavalier
Ford Taurus	Ford Escort	Ford Escort
Honda Accord	Chevrolet Cavalier	Chevrolet Cavalier
Ford Escort	Ford Escort	Chevrolet Cavalier
Honda Accord	Honda Accord	Hyundai Excel
Ford Taurus	Chevrolet Cavalier	Ford Escort
Honda Accord	Hyundai Excel	Hyundai Excel
Honda Accord	Ford Escort	Ford Escort
Ford Escort	Honda Accord	Hyundai Excel
Chevrolet Cavalier	Chevrolet Cavalier	Ford Taurus
Hyundai Excel	Ford Escort	Ford Escort
Chevrolet Cavalier	Ford Escort	Honda Accord
Ford Taurus	Ford Taurus	

Using Table 2.2, we can see at a glance that the Ford Escort and the Honda Accord were the first and second choices of the women in the sample. Ford models did very well, with Escort and Taurus accounting for 14 + 8 = 22 of the 50 new-car purchases. The information about the women's new-car purchases contained in Table 2.1 was much easier to grasp after the data had been systematically summarized in the frequency distribution of Table 2.2.

Relative Frequency Distribution

A frequency distribution shows the number (frequency) of data items in each of several nonoverlapping classes. However, we are often interested in knowing the fraction, or proportion, of the data items that fall within each class. The *relative frequency* of a class is simply the fraction, or proportion, of the total number of data items belonging to the class. For a data set with a total of n observations, or items, the relative frequency of each class is given by

TABLE 2.2

Frequency Distribution of New-Car Purchases Based on a Sample of 50 Women

Automobile Purchased	Frequency
Chevrolet Cavalier	9
Ford Escort	14
Ford Taurus	8
Honda Accord	11
Hyundai Excel	8
Total	50

Relative Frequency

$$\text{Relative Frequency of a Class} = \frac{\text{Frequency of the Class}}{n} \qquad (2.1)$$

A *relative frequency distribution* is a tabular summary of a set of data showing the relative frequency in each class. Using (2.1) we can develop a *relative frequency distribution* for the data presented in Table 2.2. For example, with the sample size $n = 50$, the relative frequency for the Chevrolet Cavalier is 9/50 = .18. Computing the relative frequency for each automobile provides the relative frequency distribution shown in Table 2.3. The relative frequency of .28 for the Ford Escort shows this automobile was selected by 28% of the women in the sample. Similarly, .22, or 22%, of the women selected the Honda Accord, .16, or 16%, selected the Hyundai Excel, and so on.

TABLE 2.3
Relative Frequency Distribution of New-Car Purchases Based on a Sample of 50 Women

Automobile Purchased	Relative Frequency
Chevrolet Cavalier	.18
Ford Escort	.28
Ford Taurus	.16
Honda Accord	.22
Hyundai Excel	.16
Total	1.00

Bar Graphs and Pie Charts

A *bar graph* is a graphical device for depicting qualitative data that have been summarized in a frequency distribution or a relative frequency distribution. On the horizontal axis of the graph, we specify the labels that are used for each of the classes. Either a frequency scale or a relative frequency scale can be used for the vertical axis of the graph. Then, using a bar of fixed width drawn above each class label, we extend the height of the bar until we reach the frequency or relative frequency of the class as indicated by the vertical axis. The bars are separated to emphasize the fact that each class is a separate category. A bar graph of the frequency distribution for the 50 new-car purchases by women is shown in Figure 2.1. Note how the graphical presentation shows the Ford Escort and the Honda Accord to be the two most preferred models.

The pie chart is a commonly used graphical device for presenting relative frequency distributions for qualitative data. To draw a pie chart, first draw a circle; then use the

FIGURE 2.1
Bar Graph of New-Car Purchases Based on a Sample of 50 Women

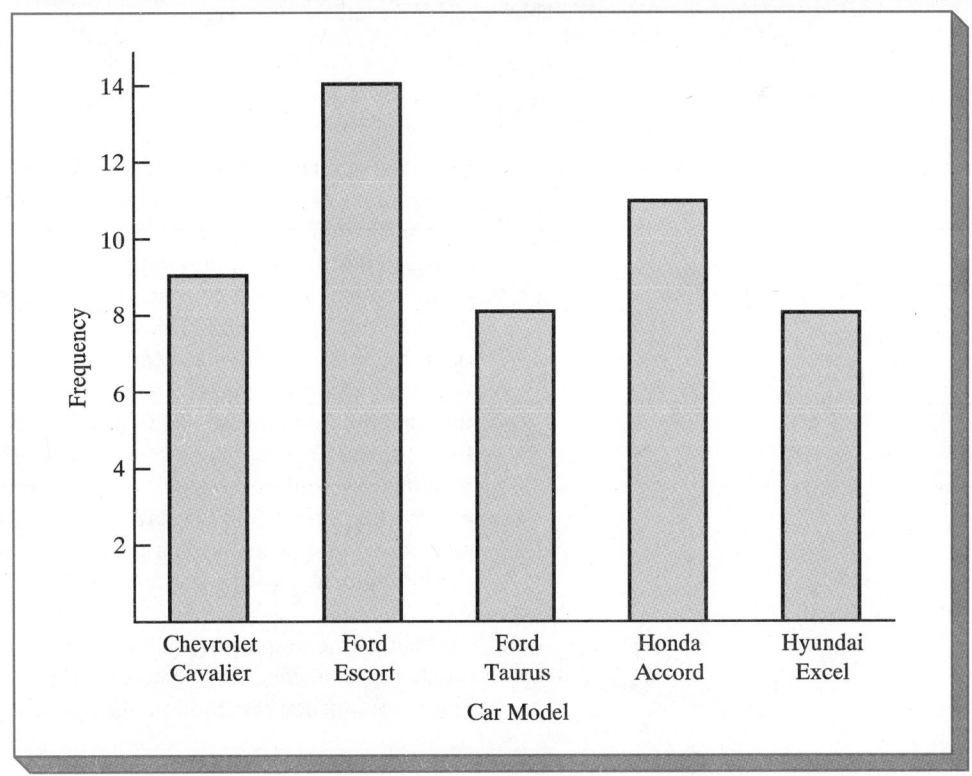

FIGURE 2.2

Pie Chart of New-Car Purchases Based on a Sample of 50 Women

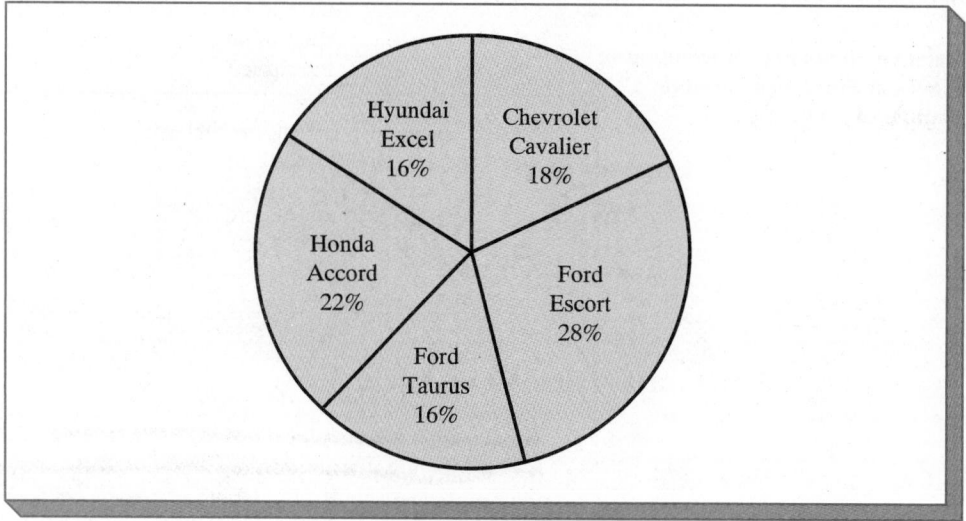

relative frequencies to subdivide the circle into sectors, or parts, that correspond to the relative frequency for each class. For example, since there are 360 degrees in a circle and since the Ford Escort has a relative frequency of .28, the sector of the pie chart labeled Ford Escort should consist of .28 × 360 = 100.8 degrees. Similar calculations for the other classes yield the pie chart shown in Figure 2.2. The numerical values shown in each sector may be frequencies, relative frequencies, or percentages, as shown in Figure 2.2.

NOTES & COMMENTS

1. Often the number of classes in a frequency distribution for qualitative data will be the same as the number of categories found in the data. This was the case for the automobile-purchase data presented in this section. Since the data considered only five automobiles, or categories, a separate frequency distribution class was defined for each automobile. If the data set had included purchases of all models of automobiles, there would have been too many models, or categories, to develop a separate class for each. In such a situation, the lower frequency categories may be grouped together to form an aggregate class. With the automobile-purchase data, models such as the Honda Civic, Chevrolet Corsica, and Toyota Camry could have been summarized in an aggregate class identified as ''other models.'' As a general guideline, most statisticians recommend that from 5 to 20 classes be used in a frequency distribution. Whenever there are too many data categories to present a separate class for each, good judgment must be exercised in deciding which categories to group together.

2. The sum of the frequencies in any frequency distribution always equals the total number of elements in the data set. The sum of the relative frequencies in any relative frequency distribution always equals 1.00.

❑ ❑ **Exercises**

Methods

1. The response to a question has three alternatives: A, B, and C. A sample of 120 provides 60 As, 24 Bs, and 36 Cs. Show the frequency and relative frequency distributions.

2. A partial relative frequency distribution is given in Table 2.4.
a. What is the relative frequency of class D?
b. The total sample size was 200. What is the frequency of class D?
c. Show the frequency distribution for the four classes.

SELF TEST ▶

3. A questionnaire provides 58 Yes answers, 42 No answers, and 20 No-Opinion answers.
a. In the construction of a pie chart, how many degrees would be in the section of the pie showing the Yes answers?
b. How many degrees would be in the section of the pie showing the No answers?
c. Construct a pie chart.
d. Construct a bar graph.

Applications

4. What is the favorite beverage of Americans? *Psychology Today* (October, 1988) provided data on the consumption of beverage products including milk (M), fruit juice (F), soft drinks (S), beer (B), and bottled water (W). The following data show the results of a sample of 30 individuals who were asked to select their most frequently consumed beverage.

M F S S B S B M W S S S F B B
S B S W S M F S S B B S F S B

a. Comment on why these are qualitative data. Is the scale of measurement nominal or ordinal?
b. Provide a frequency distribution and a relative frequency distribution summary of the data.
c. Provide a bar graph and a pie chart summary of the data.
d. Based on this sample, what is America's favorite beverage?

5. Freshmen entering the College of Business at Eastern University were asked to indicate their preferred major. The data in Table 2.5 were obtained. Summarize the data by constructing
a. a relative frequency distribution   **b.** a bar graph   **c.** a pie chart

6. What are the best-selling home videos of all time? Sales of the top-selling videos were reported by *Video Store Magazine* (*U.S. News & World Report,* February 24, 1992). Assume that sample data collected on the top six best-selling home videos are summarized with the letter codes found in Table 2.6. The following sample data are available:

F T H E H H L B E H B F B T T
B F F F H B F F H F F T F B F
L F B T F F H T E T E L E E L
T E E L E B T H B E E E T E L

a. Prepare a frequency distribution and a relative frequency distribution for the data set.
b. Rank order the top six best-selling home videos. Which video has been the most successful?

SELF TEST ▶

7. Leverock's Waterfront Steakhouse in Maderia Beach, Florida, uses a questionnaire to ask customers how they rate the server, food quality, cocktails, prices, and atmosphere at the restaurant. Each characteristic is rated on a scale from Outstanding (O), Very Good (V), Good (G), Average (A), and Poor (P). Use descriptive statistics to summarize the following data collected on food quality. What is your feeling about the food quality ratings at the restaurant?

G O V G A O V O V G O V A
V O P V O G A O O O G O V
V A G O V P V O O G O O V
O G A O V O O G V A G

TABLE 2.4

Class	Relative Frequency
A	.22
B	.18
C	.40
D	.20

TABLE 2.5

Major	Number
Management	55
Accounting	51
Finance	28
Marketing	82

TABLE 2.6

Code	Video
B	Bambi
E	ET The Extra-Terrestrial
F	Fantasia
H	Home Alone
L	The Little Mermaid
T	Batman

8. Position-by-position data for a sample of 55 members of the Baseball Hall of Fame in Cooperstown, New York, is shown below (*Sports Illustrated,* April 6, 1992). Each data item indicates the primary position played by the Hall of Famers with pitcher (P), catcher (H), 1st base (1), 2nd base (2), 3rd base (3), shortstop (S), left field (L), center field (C), and right field (R).

L	P	C	H	2	P	R	1	S	S	1	L	P	R	P
P	P	P	R	C	S	L	R	P	C	C	P	P	R	P
2	3	P	H	L	P	1	C	P	P	P	S	1	L	R
R	1	2	H	S	3	H	2	L	P					

a. Use frequency and relative frequency distributions to summarize the data for the nine positions.
b. What position provides the most Hall of Famers?
c. What position provides the fewest Hall of Famers?
d. What outfield position (L, C, or R) provides the most Hall of Famers?
e. Compare infielders (1, 2, 3, and S) to outfielders (L, C, and R).

9. Employees at Electronics Associates are on a flextime system; under this system, the employees can begin their working day at 7:00, 7:30, 8:00, 8:30, or 9:00 A.M. The following data represent a sample of the starting times selected by the employees.

7:00	8:30	9:00	8:00	7:30	7:30	8:30	8:30	7:30	7:00
8:30	8:30	8:00	8:00	7:30	8:30	7:00	9:00	8:30	8:00

Summarize the data by constructing
a. a frequency distribution   **b.** a relative frequency distribution
c. a bar graph   **d.** a pie chart
e. What do the summaries tell you about employee preferences concerning the flextime system?

10. Students in the College of Business Administration at the University of Cincinnati are asked to fill out a course-evaluation questionnaire upon completion of their courses. There are a variety of questions that use a five-category response scale. One of the questions is as follows:

> Compared to other courses that you have taken, what is the overall quality of the course you are now completing?
>
> ____   ____   ____   ____   ____
> Poor   Fair   Good   Very Good   Excellent

A sample of 60 students completing a course in business statistics during the spring quarter of 1989 provided the following responses. To aid in computer processing of the questionnaire results, a numeric scale was used with 1 = Poor, 2 = Fair, 3 = Good, 4 = Very Good, and 5 = Excellent.

3	4	4	5	1	5	3	4	5	2	4	5	3	4	4
4	5	5	4	1	4	5	4	2	5	4	2	4	4	4
5	5	3	4	5	5	2	4	3	4	5	4	3	5	4
4	3	5	4	5	4	3	5	3	4	4	3	5	3	3

a. Comment on why these are qualitative data. Is the scale of measurement nominal or ordinal?
b. Provide a frequency distribution and a relative frequency distribution summary of the data.
c. Provide a bar graph and a pie chart summary of the data.
d. Based upon your summaries, comment on the students' overall evaluation of the course.

2.2   Summarizing Quantitative Data

Frequency Distribution

As defined in Section 2.1, a frequency distribution is a tabular summary of a set of data showing the frequency (or number) of items in each of several nonoverlapping classes.

This definition holds for quantitative as well as for qualitative data. However, with quantitative data we have to be more careful in defining the nonoverlapping classes to be used in the frequency distribution.

For example, consider the quantitative data presented in Table 2.7. These data provide the time in days required to complete year-end audits for a sample of 20 clients of Sanderson and Clifford, a small public accounting firm. The three steps necessary to define the classes for a frequency distribution with quantitative data are as follows:

1. Determine the number of nonoverlapping classes.
2. Determine the width of each class.
3. Determine the class limits.

Let us demonstrate these steps by developing a frequency distribution for the audit-time data shown in Table 2.7.

Number of Classes. Classes are formed by specifying ranges of data values that will be used to group the elements in the data set. As a general guideline, we again recommend using between 5 and 20 classes. Data sets with a larger number of elements usually require a larger number of classes. Data sets with a smaller number of elements can often be summarized quite nicely with as few as 5 or 6 classes. The goal is to use enough classes to show the variation in the data, but not so many classes that there are only a few elements in many of the classes. Since the data set in Table 2.7 is relatively small ($n = 20$), we chose to develop a frequency distribution with 5 classes.

Width of the Classes. The second step in constructing a frequency distribution for quantitative data is to choose a width for the classes. As a general guideline, it is recommended that the width be the same for each class. Thus the choices of the number of classes and the width of the classes are not independent decisions. A larger number of classes means a smaller class width, and vice versa. In order to determine an approximate class width, we begin by identifying the largest and smallest data values in the data set. Then, once the desired number of classes has been specified, the following expression can be used to determine the approximate class width.

$$\text{Approximate Class Width} = \frac{\text{Largest Data Value} - \text{Smallest Data Value}}{\text{Number of Classes}} \tag{2.2}$$

The class width given by (2.2) can be adjusted to a convenient width based on the preference of the person developing the frequency distribution. For example, a computed class width of 9.28 might be adjusted to a class width of 10 simply because 10 is a more convenient class width to use in constructing a frequency distribution.

For the data set involving the year-end audit times, the largest value is 33 and the smallest value is 12. Since we have decided to summarize the data set with 5 classes, using (2.2) provides an approximate class width of $(33 - 12)/5 = 4.2$. As a result, we decided to use a class width of 5 in the frequency distribution.

In practice, the number of classes and the appropriate class width are determined by trial and error. Once a possible number of classes is chosen, (2.2) is used to find the approximate class width. The process may be repeated for a different number of classes. Ultimately, the judgment of the analyst is used to determine the combination of number of classes and class width that provides the best means for summarizing the data.

Returning to the audit-time data in Table 2.7, we have decided to use 5 classes, each with a width of 5 days, to summarize the data. The next task is to specify the class limits for each of the classes.

TABLE 2.7
Year-End Audit Times (in Days)

12	14	19	18
15	15	18	17
20	27	22	23
22	21	33	28
14	18	16	13

Class Limits. The *lower class limit* identifies the smallest possible data value assigned to the class. The *upper class limit* identifies the largest possible data value assigned to the class. Again, the judgment of the analyst is used, and a variety of acceptable class limits are possible.

For the data in Table 2.7, we defined the class limits as follows: 10–14, 15–19, 20–24, 25–29, and 30–34. The smallest data value, 12, is included in the 10–14 class. The largest data value, 33, is included in the 30–34 class. Using the 10–14 class as an example, 10 is the lower class limit and 14 is the upper class limit. The difference between the lower class limits of adjacent classes provides the class width. Using the first two lower class limits of 10 and 15, we see that the class width is $15 - 10 = 5$.

The form of each lower class limit and each upper class limit depends upon the number of places to the right of the decimal point contained in the data. Since the audit times in Table 2.7 are integer, integer class limits of 10–14, 15–19, and so on are acceptable limits. If the audit times were recorded in tenths of days, such as 12.3, 14.4, 19.3, and so on, the class limits would also be stated in tenths. In this case, class limits of 10.0–14.9, 15.0–19.9, 20.0–24.9, and so on would have been appropriate. If the data were in hundredths, which is often the case with dollar and cents data, the class limits would also 15.00–19.99, 20.00–24.99, and so on would be appropriate. Regardless of how the class limits are chosen, they should be defined in such a fashion that *each data value belongs to one and only one class*. For instance, class limits of 10–15, 15–20, and 20–25 are unacceptable, since the data values of 15 and 20 belong to two different classes.

Once the number of classes, class width, and class limits have been determined, a frequency distribution can be obtained by *counting* the number of data items belonging to each class. For example, the data in Table 2.7 shows four values—12, 14, 14, and 13—belong to the 10–14 class. Thus, the frequency for the 10–14 class is 4. Continuing this counting process for the 15–19, 20–24, 25–29, and 30–34 classes provides the frequency distribution shown in Table 2.8. Using this frequency distribution, we can observe the following:

1. The most frequently occurring audit times are in the class of 15–19 days. Eight of the 20 audit times belong to this class.
2. Only one audit required 30 or more days.

Other relevant observations are possible, depending upon the interests of the person viewing the frequency distribution. In this example, we see that the value of a frequency distribution is that it provides insights about the data that were not easily obtained by viewing the data in their original unorganized form.

Relative Frequency Distribution

We define the relative frequency distribution for quantitative data in the same manner as for qualitative data. First, recall that the relative frequency is simply the fraction or proportion of the total number of items belonging to a class. For a data set having n items,

$$\text{Relative Frequency of Class} = \frac{\text{Frequency of the Class}}{n}$$

Based on the class frequencies shown in Table 2.8 and with $n = 20$, Table 2.9 shows the relative frequency distribution for the audit-time data. Note that .40, or 40%, of the audits required from 15 to 19 days. Only .05, or 5%, of the audits required 30 or more days. Again, additional interpretations and insights are possible using Table 2.9.

T A B L E 2.8

Frequency Distribution for the Audit-Time Data

Audit Time (days)	Frequency
10–14	4
15–19	8
20–24	5
25–29	2
30–34	1
Total	20

T A B L E 2.9

Relative Frequency Distribution for the Audit-Time Data

Audit Time (days)	Relative Frequency
10–14	.20
15–19	.40
20–24	.25
25–29	.10
30–34	.05
Total	1.00

Dot Plot

One of the simplest graphical summaries of data is a *dot plot*. A horizontal axis shows the range of values for the data. Then each data value is represented by a dot placed above the axis. The dot plot for the audit-time data in Table 2.7 is shown in Figure 2.3. The three dots located at the value of 18 indicate that 18 occurs three times in the data set. Dot plots show the details of the data and are very useful for comparing two or more sets of data.

Histogram

Another common graphical presentation of quantitative data is the *histogram*. This graphical summary may be prepared for data that have been previously summarized in either a frequency distribution or a relative frequency distribution. A histogram is constructed by placing the variable of interest on the horizontal axis and the frequency or relative frequency on the vertical axis. The frequency or relative frequency of each class is shown by drawing a rectangle whose base is the class interval on the horizontal axis and whose height is the corresponding frequency or relative frequency.

A histogram for the audit-time data is shown in Figure 2.4. Note that the class with the greatest frequency is shown by the rectangle appearing above the class of 15–19 days.

FIGURE 2.3
Dot Plot for the Audit-Time Data in Table 2.7

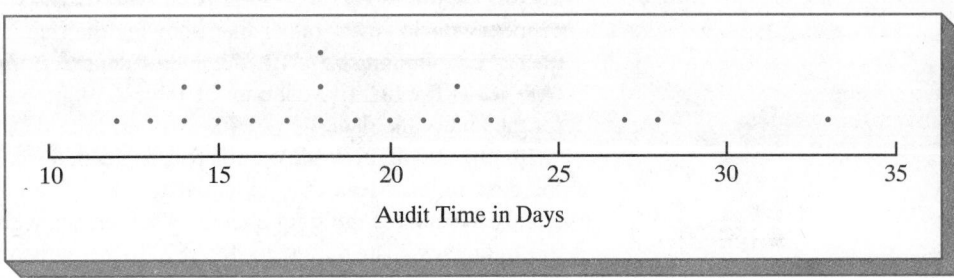

FIGURE 2.4
Histogram for the Audit-Time Data in Table 2.7

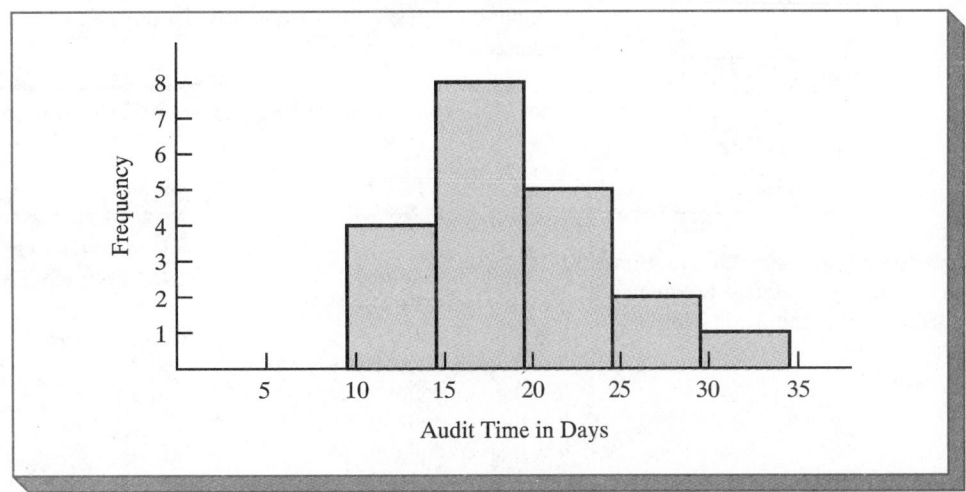

The height of the rectangle shows that the frequency of this class is 8. A histogram for the relative frequency distribution of this data would look the same as the histogram in Figure 2.4 with the exception that the vertical axis would be labeled with relative frequency values.

As Figure 2.4 shows, the adjacent rectangles of a histogram touch one another. Unlike a bar graph, there is no separation between the rectangles of adjacent classes. This is the usual convention for histograms. Since the class limits for the audit-time data were stated as 10–14, 15–19, 20–24, 25–29, and 30–34, there appear to be one-unit intervals of 14 to 15, 19 to 20, 24 to 25, and 29 to 30 between the classes. These spaces are eliminated by drawing the vertical lines of the histogram halfway between the class limits. For example, the vertical lines for the 15–19 class are drawn above the values 14.5 and 19.5. Using this procedure for all classes, the vertical lines for the histogram in Figure 2.4 are drawn above the values 9.5, 14.5, 19.5, 24.5, 29.5, and 34.5. This minor adjustment to eliminate the spaces in a histogram helps show that, even though the data are rounded, all values between the lower limit of the first class and the upper limit of the last class are possible.

Cumulative Frequency and Cumulative Relative Frequency Distributions

A variation of the frequency distribution that provides another tabular summary of quantitative data is the *cumulative frequency distribution*. The cumulative frequency distribution uses the number of classes, class widths, and class limits that were developed for the frequency distribution. However, rather than showing the frequency of each class, the *cumulative* frequency distribution shows the number of items *less than or equal to the upper class limit* of each class. The first two columns of Table 2.10 provide the cumulative frequency distribution for the audit-time data.

To understand how the cumulative frequencies are determined, consider the class with the description "less than or equal to 24." The cumulative frequency for this class is simply the sum of the frequencies for all classes with data less than or equal to 24. Using the frequency distribution in Table 2.8, the sum of the frequencies for classes 10–14, 15–19, and 20–24 indicate that there are 4 + 8 + 5 = 17 items with values less than or equal to 24. Hence, the cumulative frequency for this class is 17. Other observations based on the cumulative frequency distribution in Table 2.10 show that 12 audits were completed in less than or equal to 19 days and 19 audits were completed in less than or equal to 29 days.

As a final point, we note that a *cumulative relative frequency distribution* shows the fraction, or proportion, of items with values less than or equal to the upper limit of each class.

TABLE 2.10 Cumulative Frequency Distribution and Cumulative Relative Frequency Distribution for the Audit-Time Data	Audit Time (days)	Cumulative Frequency	Cumulative Relative Frequency
	less than or equal to 14	4	.20
	less than or equal to 19	12	.60
	less than or equal to 24	17	.85
	less than or equal to 29	19	.95
	less than or equal to 34	20	1.00

The cumulative relative frequency distribution can be computed either by summing the relative frequencies in the relative frequency distribution or by dividing the cumulative frequencies by the total number of items. Using the latter approach, the cumulative relative frequencies in column 3 of Table 2.10 are found by dividing the cumulative frequencies in column 2 by the total number of items ($n = 20$). The cumulative relative frequency distribution shows that .85, or 85%, of the audits were completed in less than or equal to 24 days; .95, or 95%, of the audits were completed in less than or equal to 29 days; and so on.

Ogive

A graph of a cumulative frequency distribution or a cumulative relative frequency distribution is called an *ogive*. The data values are shown on the horizontal axis and either the cumulative frequencies or cumulative relative frequencies are shown on the vertical axis. The ogive for the cumulative frequencies of the audit-time data shown in Table 2.10 is shown in Figure 2.5.

The ogive is constructed by plotting a point corresponding to the cumulative frequency of each class. Since the class limits for the audit-time data were 10–14, 15–19, 20–24, and so on, there appear to be one-unit intervals for 14 to 15, 19 to 20, and so on. As with the histogram, these spaces are eliminated by plotting points halfway between the class limits. Thus, 14.5 is used for the 10–14 class, 19.5 is used for the 15–19 class, and so on. Thus, the "less than or equal to 14" class with a cumulative frequency of 4 is shown on the ogive in Figure 2.5 by the point located at 14.5 on the horizontal axis and 4 on the vertical axis. The "less than or equal to 19" class with a cumulative frequency of 12 is shown by the point located at 19.5 on the horizontal axis and 12 on the vertical axis. Note that one additional point is plotted at the left end of the ogive. This point starts the ogive by showing that there are no data values below the 10–14 class. Thus, the point is plotted at 9.5 on the horizontal axis and 0 on the vertical axis. The plotted points are connected by straight lines to complete the ogive.

F IGURE 2.5
Ogive for the Audit-Time Data of Table 2.10.

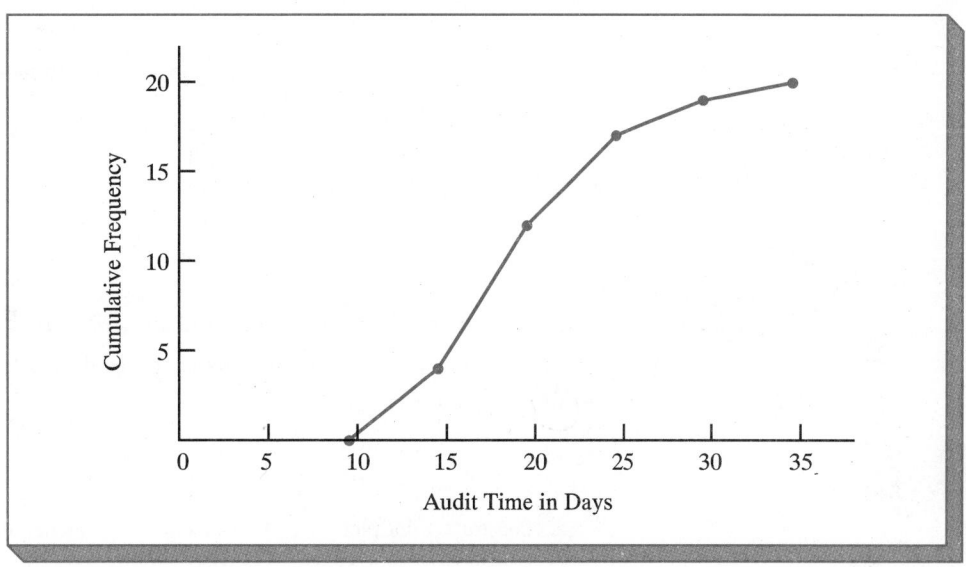

NOTES &
COMMENTS

1. In some applications, we will want to know the *midpoints* of the classes in a frequency distribution. Each class midpoint is simply halfway between the lower and upper class limits. For example, with the class limits of 10–14, 15–19, 20–24, 25–29, and 30–34 in the audit-time example, the five class midpoints would be 12, 17, 22, 27, and 32, respectively.

2. An *open-end* class is one that only has a lower class limit or an upper class limit. For example, in the audit-time data of Table 2.7, suppose two of the audits had taken 58 and 65 days. Rather than continue with the classes of width 5 with class intervals 35–39, 40–44, 45–49, and so on, it would simplify the frequency distribution to show an open-end class of "35 or more." This class would have a frequency of 2. Most often the open-end class will appear at the upper end of the distribution. Sometimes an open-end class may appear at the lower end of the distribution, and occasionally such classes appear at both ends.

3. The last entry in a cumulative frequency distribution is always the total number of elements in the data set. The last entry in a cumulative relative frequency distribution is always 1.00.

❑ ❑ Exercises

Methods

11. Consider the following data:

$$3, \quad 5, \quad 12, \quad 14, \quad 8, \quad 9, \quad 2, \quad 10, \quad 18, \quad 7.$$

Develop a frequency distribution and relative frequency distribution using class limits of 0–4, 5–9, 10–14, and 15–19.

SELF TEST ▶ **12.** Consider the following frequency distribution:

Class	Frequency
10–19	8
20–29	12
30–39	15
40–49	5

Construct a cumulative frequency distribution and a cumulative relative frequency distribution.

13. Construct a histogram and an ogive for the data in Exercise 12.

14. Consider the following data:

8.9	10.2	11.5	7.8	10.0	12.2	13.5	14.1	10.0	12.2
6.8	9.5	11.5	11.2	14.9	7.5	10.0	6.0	15.8	11.5

a. Construct a dot plot.   **b.** Construct a frequency distribution.
c. Construct a relative frequency distribution.

Applications

SELF TEST ▶

— **15.** A doctor's office staff has studied the waiting times for patients who arrive at the office with a request for emergency service. The following data were collected over a 1-month period (the waiting times are in minutes):

$$2, 5, 10, 12, 4, 4, 5, 17, 11, 8, 9, 8, 12, 21, 6, 8, 7, 13, 18, 3.$$

Use classes of 0–4, 5–9, and so on.
a. Show the frequency distribution.
b. Show the relative frequency distribution.
c. Show the cumulative frequency distribution.
d. Show the cumulative relative frequency distribution.
e. What proportion of patients needing emergency service have a waiting time of 9 minutes or less?

 AUTOCOST

16. Data on selling prices of new automobiles and used automobiles were provided in *U.S. News & World Report* (September 9, 1991). Assume the following sample data apply with values stated in thousands of dollars.

New Automobiles:

| 10.3 | 24.7 | 14.7 | 10.8 | 14.4 | 9.8 | 16.3 |
| 17.8 | 20.8 | 17.6 | 17.1 | 20.1 | 16.5 | 13.8 |

Used Automobiles:

| 5.0 | 4.4 | 5.6 | 10.4 | 9.4 | 7.0 |
| 9.3 | 4.5 | 4.2 | 7.7 | 11.7 | 11.2 |

Use frequency distributions, relative frequency distributions, and histograms to summarize the data. For comparative purposes, use classes of 0.0–4.9, 5.0–9.9, 10.0–14.9, and so on for both data sets. What comparisons can you make about the selling prices of new and used automobiles?

17. National Airlines accepts flight reservations by phone. Shown in Table 2.11 are the call durations (in minutes) for a sample of 20 phone reservations. Construct the frequency and relative frequency distributions for the data. Also provide a histogram.

TABLE 2.11

2.1	4.8	5.5	10.4
3.3	3.5	4.8	5.8
5.3	5.5	2.8	3.6
5.9	6.6	7.8	10.5
7.5	6.0	4.5	4.8

EXSALARY

18. How much do executives of some of the largest corporations get paid? *Business Week* (May 1, 1989) reported executive compensation for 1988, including salary and bonus. The data for 25 chief executive officers were reported in thousands of dollars and are as follows:

Company	Compensation	Company	Compensation
Boeing	846	Delta Airlines	457
Whirlpool	563	Chrysler	1466
Bank of Boston	1200	Coca-Cola	2164
Sherwin-Williams	746	DuPont	1611
Bristol-Myers	824	Motorola	824
General Mills	1310	Marriott	1007
Sara Lee	1367	Honeywell	575
Eastman Kodak	1252	Exxon	1354
Apple Computer	2479	Scott Paper	1238
Bausch & Lomb	927	CBS	1253
K Mart	925	AT&T	1284
Goodyear	1279	Philip Morris	1660
Teledyne	860		

Summarize the data by constructing

a. a frequency distribution   **b.** a relative frequency distribution
c. a histogram   **d.** a cumulative frequency distribution
e. Using these summaries, comment on what you learned about executive salaries from the sample data.

19. The data given in Table 2.12 are the number of units produced by a production employee for the most recent 20 days. Summarize the data by constructing

a. a frequency distribution
b. a relative frequency distribution
c. a cumulative frequency distribution
d. a cumulative relative frequency distribution
e. an ogive

20. *Car and Driver* magazine (January 1989) selected the Honda Civic as one of the 10 best cars of the year. Data provided about the Civic included fuel-economy information stated in miles per gallon. Assume that the following miles-per-gallon data were obtained from a sample of actual mileage tests with the Civic.

30.2	29.0	27.5	28.3	29.2	32.1	33.8	25.2	34.3	30.6
30.5	28.3	26.0	28.5	29.4	30.3	30.8	29.2	25.9	26.4
27.7	33.9	30.4	29.4	29.4	30.2	28.8	27.5	30.8	30.0

Summarize the data by constructing

a. a frequency distribution
b. a relative frequency distribution
c. a histogram
d. *Car and Driver* magazine reported the fuel economy of the Honda Accord as being in the range of 22–27 miles per gallon. Which of the two, the Accord or the Civic, appears to have better fuel economy?

21. The personal computer has brought computer convenience and power into the home environment. But just how many hours a week are people actually using their home computers? A study designed to determine the usage of personal computers at home (*U.S. News & World Report*, December 26, 1988) provided data in hours per week:

.5	1.2	4.8	10.3	7.0	13.1	16.0	12.7	11.6	5.1
2.2	8.2	.7	9.0	7.8	2.2	1.8	12.8	12.5	14.1
15.5	13.6	12.2	12.5	12.8	13.5	1.3	5.5	5.0	10.8
2.5	3.9	6.5	4.2	8.8	2.8	2.5	14.4	16.0	12.4
2.8	9.5	1.5	10.5	2.2	7.5	10.5	14.1	14.9	.3

Summarize the data using

a. a frequency distribution (use a class width of 3 hours)
b. a relative frequency distribution
c. a histogram
d. an ogive
e. Comment on what the data indicate about personal computer usage at home.

2.3   The Role of the Computer

Computers play an important role in providing statistical summaries of data. Prior to 1970, there were relatively few statistical packages for analyzing data using a computer. Since that time, however, the situation has changed dramatically. Today, the user of statistics has a choice of packages such as Minitab, SAS (Statistical Analysis System), SPSS (Statistical Package for the Social Sciences), and BMDP (UCLA Biomedical sta-

---

TABLE 2.12

160	170	181	156	176
148	198	179	162	150
162	156	179	178	151
157	154	179	148	156

 MPGDATA

 COMPUTER

tistical package), to name a few. In this and future chapters we provide illustrations showing how statistical computing can assist in the analysis and interpretation of data.

Throughout the text, we will use the Minitab system to illustrate the application of statistical computing packages. Minitab is a widely used, general-purpose system that can be installed on a variety of mainframe and personal computers. It has been designed for users who have had little or no previous computer experience. Although easy to use, it still has a large capacity for data summarization and analysis.

Minitab provides a worksheet of columns and rows that is used to store the data set. Each column of the worksheet corresponds to a variable. The columns are identified by labels of C1 for column 1, C2 for column 2, and so on. However, the user may select the NAME command of Minitab to rename the columns (variables) so that the variables can be easily identified. Each row of the worksheet corresponds to an element of the data set. The data in each row provide the observation for the corresponding element.

Let us illustrate the use of the Minitab system by showing how Minitab can be used to summarize the audit-time data shown in Table 2.7. Specifically, we will show how the Minitab commands of DOTPLOT and HISTOGRAM can be used to quickly, easily, and accurately develop a dot plot, frequency distribution, and histogram for the data. Refer to Figure 2.6. We will assume that the user has loaded the Minitab system on the computer. The MTB > symbol appears on the computer screen, which indicates that the Minitab system is waiting for a command from the user. The user begins with the command READ C1, which indicates that the system should take the data from the input lines that follow and store the data in column 1 of the worksheet. Note that the next line shows Minitab's response, DATA > , indicating that the user is to enter the first data value. The data input process continues with one data value being entered per line. When all the data have been input, the user responds with the command END, which terminates the data input process. At this point, the data reside in column 1 of the Minitab worksheet.

With the data in the worksheet, the user can continue with a variety of statistical summary and analysis commands. Let us assume that the user would like to view a dot plot of the data. The Minitab command DOTPLOT C1 is all that is needed. This command instructs the Minitab system to construct a dot plot for the data in column C1. As Figure 2.6 shows, the dot plot is displayed immediately following the command.

The next three commands in Figure 2.6 show that the user has instructed the computer to develop a frequency distribution and histogram for the data. The user would like a frequency distribution with a class width of 5 and class limits of 10–14, 15–19, 20–24, 25–29, and 30–34. Minitab will develop such a frequency distribution and its histogram after receiving the following user commands:

> HISTOGRAM C1;
> START 12;
> INCREMENT 5.

The command HISTOGRAM C1 requests a frequency distribution and a histogram for the data stored in column 1. The semicolon following C1 is used to indicate that user instructions about the format of the frequency distribution and histogram follow. The next two lines are *subcommands* and are requested following the Minitab prompt SUBC > . Minitab uses the *midpoint* of each class, the value halfway between the class limits, to identify the class. Accordingly, the first class of 10–14 has a midpoint of $(10 + 14)/2 = 12$; the second class of 15–19 has a midpoint of $(15 + 19)/2 = 17$; and so on. The Minitab subcommand START 12 tells the system that the first class of the frequency distribution will have a midpoint at 12. The semicolon following the 12 indi-

FIGURE 2.6
**Computer Summaries of the
Audit-Time Data Using Minitab**

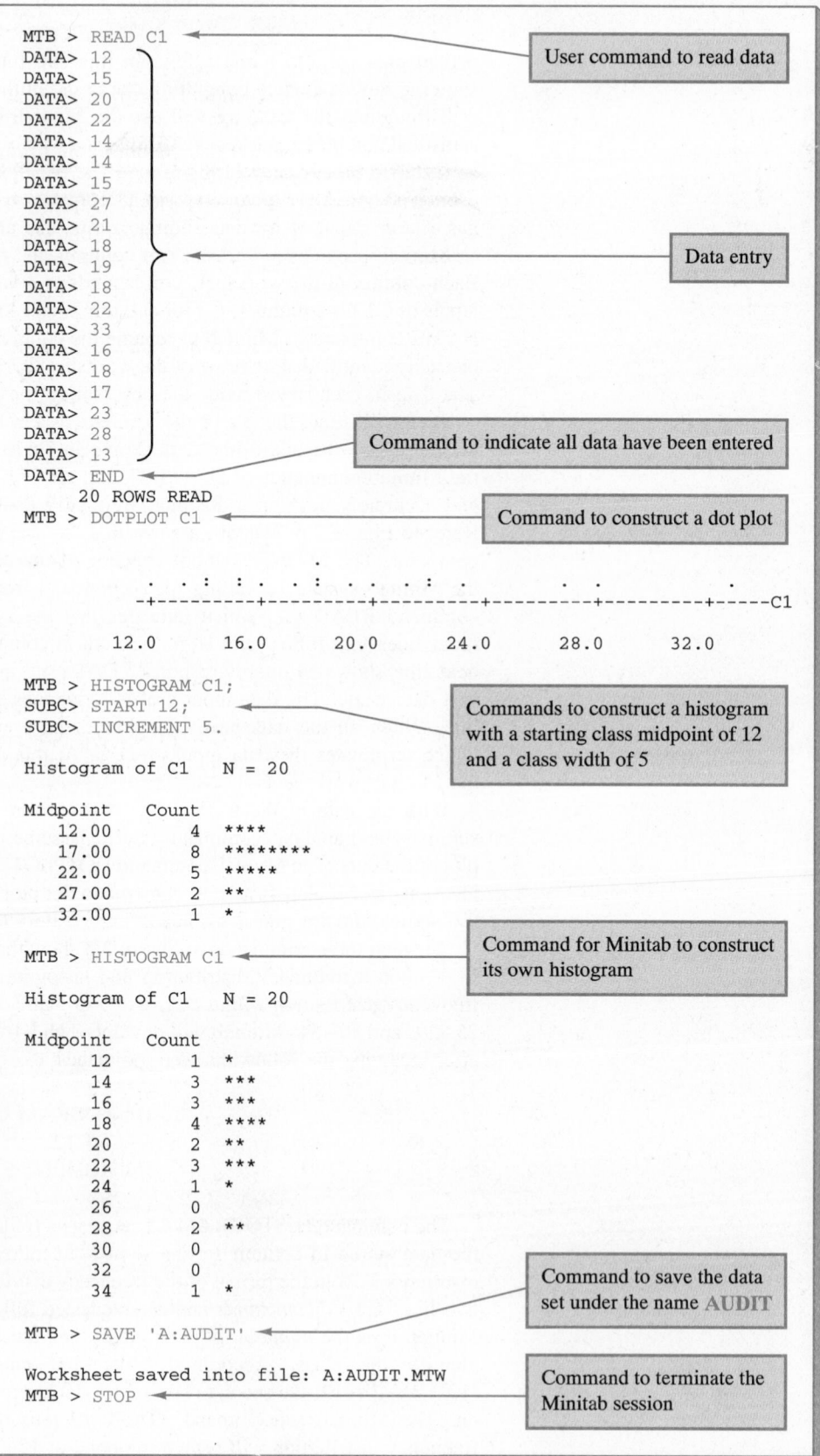

```
MTB > READ C1                          ◄───────────  User command to read data
DATA> 12  ┐
DATA> 15  │
DATA> 20  │
DATA> 22  │
DATA> 14  │
DATA> 14  │
DATA> 15  │
DATA> 27  │
DATA> 21  │
DATA> 18  ├──────────────────────────────────────  Data entry
DATA> 19  │
DATA> 18  │
DATA> 22  │
DATA> 33  │
DATA> 16  │
DATA> 18  │
DATA> 17  │
DATA> 23  │
DATA> 28  │
DATA> 13  ┘
DATA> END        ◄──────────  Command to indicate all data have been entered
     20 ROWS READ
MTB > DOTPLOT C1       ◄──────────────  Command to construct a dot plot


                               .
         . . : : : . . . : . . . : .          . .            .
      -+---------+---------+---------+---------+---------+-----C1
        12.0      16.0      20.0      24.0      28.0      32.0


MTB > HISTOGRAM C1;
SUBC>  START 12;            ◄──────    Commands to construct a histogram
SUBC>  INCREMENT 5.                    with a starting class midpoint of 12
                                       and a class width of 5
Histogram of C1    N = 20

Midpoint   Count
   12.00       4   ****
   17.00       8   ********
   22.00       5   *****
   27.00       2   **
   32.00       1   *


MTB > HISTOGRAM C1     ◄────────   Command for Minitab to construct
                                   its own histogram
Histogram of C1    N = 20

Midpoint   Count
   12          1   *
   14          3   ***
   16          3   ***
   18          4   ****
   20          2   **
   22          3   ***
   24          1   *
   26          0
   28          2   **
   30          0
   32          0                   Command to save the data
   34          1   *               set under the name AUDIT

MTB > SAVE 'A:AUDIT'    ◄──────────┘

Worksheet saved into file: A:AUDIT.MTW    Command to terminate the
MTB > STOP     ◄──────────────            Minitab session
```

cates that another subcommand will follow. The last command of INCREMENT 5 indicates the desired class width. The period after the 5 is used to indicate that the subcommands have been completed.

When the user enters the above three commands, Minitab responds with the frequency distribution and histogram shown in Figure 2.6. The label for the output reminds the user that the histogram command has been requested for the variable in column 1 and that this variable has $N = 20$ observations. Note that the class midpoints are used to identify the requested classes of 10–14, 15–19, 20–24, 25–29, and 30–34. The frequencies of each class are provided in the column labeled Count. The histogram is shown by the pattern of asterisks appearing to the right of the counts. Note that the results are identical to the frequency distribution and histogram provided in Section 2.2 except that the histogram is on its side. Relative frequencies, cumulative frequencies, and cumulative relative frequencies can be easily computed from the information in Figure 2.6.

If the user had input the command HISTOGRAM C1 and had omitted the START and INCREMENT subcommands, Minitab would have selected its own format for the frequency distribution and histogram. Such a command is shown next in Figure 2.6. Note that Minitab has selected 12 classes with a class width of 2. Since the output is produced directly after the command is given, the user can quickly look at the output, form some initial judgments about the data, enter another command to modify the format for the frequency distribution and histogram, and continue. Data analysis done interactively provides quick computer response and is an important reason why statistical packages are so valuable to the analyst.

Note that the Figure 2.6 session of Minitab continues with the user command SAVE 'A:AUDIT' which saves the data set on a data disk in computer drive A under the file name AUDIT.MTW. Later sessions of Minitab can use the data set by retrieving the file AUDIT.MTW from the data disk. The Minitab computer session concludes with the user command STOP.

NOTES & COMMENTS

This section has focused on the use of the computer to summarize quantitative data. Statistical software packages often handle qualitative data by using numeric codes to represent the data. For example, the data set for new-car purchases by women in Table 2.1 could have been represented by 1 for Chevrolet Cavalier, 2 for Ford Escort, 3 for Ford Taurus, 4 for Honda Accord, and 5 for Hyundai Excel. Using this coding, the Minitab command HISTOGRAM develops the following frequency distribution and bar graph information.

```
          Histogram of C1    N = 50


          Midpoint    Count
                 1        9   *********
                 2       14   **************
                 3        8   ********
                 4       11   ***********
                 5        8   ********
```

Interpretation of the midpoint column requires the user to be aware of the numeric codes used and to recall that 1 corresponds to the Chevrolet Cavalier and so on.

2.4   Exploratory Data Analysis

 APTEST

TABLE 2.13
Number of Questions Answered Correctly on an Aptitude Test

112	72	69	97	107
73	92	76	86	73
126	128	118	127	124
82	104	132	134	83
92	108	96	100	92
115	76	91	102	81
95	141	81	80	106
84	119	113	98	75
68	98	115	106	95
100	85	94	106	119

The techniques of *exploratory data analysis* focus on how simple arithmetic and easy-to-draw pictures can be used to summarize data quickly. In this section we study how one of these techniques—referred to as a *stem-and-leaf display*—can be used to rank order data and provide an idea of the shape of the distribution of a set of quantative data.

One simple method of displaying data involves arranging the data in ascending or descending order. This process, referred to as rank ordering data, provides some degree of organization. However, such an approach provides little insight concerning the shape of the distribution of data values. A stem-and-leaf display is a device that provides a display of both rank order and shape simultaneously.

To illustrate the use of a stem-and-leaf display in data analysis, consider the data set presented in Table 2.13. These data show the results of a 150-question aptitude test given to 50 individuals who were recently interviewed for a position at Haskens Manufacturing. The data indicate the number of questions answered correctly.

To develop a stem-and-leaf display for the data in Table 2.13, we proceed as follows. The first digits of each data item are arranged to the left of a vertical line. To the right of the vertical line, we record the last digit for each item as we pass through the scores in the order they were recorded. The last digit for each item is placed on the horizontal line corresponding to its first digit:

```
 6 | 9  8
 7 | 2  3  6  3  6  5
 8 | 6  2  3  1  1  0  4  5
 9 | 7  2  2  6  2  1  5  8  8  5  4
10 | 7  4  8  0  2  6  6  0  6
11 | 2  8  5  9  3  5  9
12 | 6  8  7  4
13 | 2  4
14 | 1
```

Given this organization of the data, it is a simple matter to rank order the digits on each horizontal line. Doing so leads to the following stem-and-leaf display of the data:

```
 6 | 8  9
 7 | 2  3  3  5  6  6
 8 | 0  1  1  2  3  4  5  6
 9 | 1  2  2  2  4  5  5  6  7  8  8
10 | 0  0  2  4  6  6  6  7  8
11 | 2  3  5  5  8  9  9
12 | 4  6  7  8
13 | 2  4
14 | 1
```

Each line in this display is referred to as a *stem,* and each piece of information on the stem is a *leaf.* For example, consider the first line

6 | 8   9

The meaning attached to this line is that there are two items in the data set whose first digit is six: 68 and 69. Similarly, the second line,

7 | 2   3   3   5   6   6

specifies that there are six items whose first digit is seven: 72, 73, 73, 75, 76, and 76. Thus we see that the data values in this stem-and-leaf display are separated into two parts. The label for each stem is the one- or two-digit first part of the number (that is, 6, 7, 8, 9, 10, 11, 12, 13, or 14) and the leaf is the single-digit second part (that is, 0, 1, 2, . . . , 8, 9). The vertical line simply serves to separate the two parts of each number listed.

To focus on the shape indicated by the stem-and-leaf display, let us use a rectangle to depict the "length" of each stem. Doing so we obtain the following:

6	8 9
7	2 3 3 5 6 6
8	0 1 1 2 3 4 5 6
9	1 2 2 2 4 5 5 6 7 8 8
10	0 0 2 4 6 6 6 7 8
11	2 3 5 5 8 9 9
12	4 6 7 8
13	2 4
14	1

Rotating this page counterclockwise onto its side provides a picture of the data very similar to that provided by a histogram of the data with classes of 60–69, 70–79, 80–89, and so on.

Although the stem-and-leaf display may appear at first glance to offer little more information about the data set than that provided by a histogram, there are two primary advantages:

1. The stem-and-leaf display is easier to construct.
2. Within an interval, the stem-and-leaf display provides more information than the histogram, since the stem-and-leaf shows the actual data values.

Just as there is no right number of classes in a frequency distribution or histogram, there is no right number of rows or stems in a stem-and-leaf display. If we believe that our original stem-and-leaf display has condensed the data too much, it is a simple matter to stretch the display by using two or more stems for each starting point. For example, to use two lines for each starting point, we would place all data values ending in 0, 1, 2, 3, or 4 on one line and all values ending in 5, 6, 7, 8, and 9 on a second line. To illustrate this approach, consider the following stretched stem-and-leaf display:

```
 6 |
 6 | 8  9
 7 | 2  3  3
 7 | 5  6  6
 8 | 0  1  1  2  3  4
 8 | 5  6
 9 | 1  2  2  2  4
 9 | 5  5  6  7  8  8
10 | 0  0  2  4
10 | 6  6  6  7  8
11 | 2  3
11 | 5  5  8  9  9
12 | 4
12 | 6  7  8
13 | 2  4
13 |
14 | 1
14 |
```

Note that data 72, 73, and 73 have leaves in the 0–4 range and are shown with the first stem value of 7. The data 75, 76 and 76 have leaves in the 5–9 range and are shown with the second stem value of 7. This stretched stem-and-leaf display is similar to a frequency distribution with intervals of 60–64, 65–69, 70–74, 75–79, and so on.

Figure 2.7 shows a portion of the stem-and-leaf output provided by the Minitab computer software package. The aptitude test scores in Table 2.13 have been read into column 1 of a Minitab worksheet. The command STEM-AND-LEAF C1 requests the stem-and-leaf display. Note that the output is identical to the stem-and-leaf display we have just discussed.

The aptitude test scores were used to generate a stem-and-leaf display for data having up to three digits. Stem-and-leaf displays for data with more digits are possible. For example, consider the following data, which show the number of hamburgers sold by a fast-food restaurant for each of 15 weeks.

1852	1644	1766	1888	1912	2044	1812	1790
1679	2008	1565	1852	1967	1954	1733	

A stem-and-leaf display of these data is as follows:

Leaf Unit = 10

```
15 | 6
16 | 4  7
17 | 3  6  9
18 | 1  5  5  8
19 | 1  5  6
20 | 0  4
```

Note that only the first three digits of each data value are used in the display. The wording "Leaf Unit = 10" is used to indicate that a leaf value of 1 represents 10–19, a

FIGURE 2.7
Stem-and-Leaf Display of the Aptitude Test Scores Using Minitab

```
MTB > STEM-AND-LEAF C1

Stem-and-leaf of C1          N = 50

      6   89
      7   223
      7   566
      8   011234
      8   56
      9   12224
      9   556788
     10   0024
     10   66678
     11   23
     11   55899
     12   4
     12   678
     13   24
     13
     14   1
```

leaf value of 2 represents 20–29, and so on. For example, the leaf value of 6 on the first line of the display represents a number from 60 to 69 indicating that the data values have four digits and that the data value corresponding to the first line is between 1560 and 1569.

Exercises

Methods

22. Construct a stem-and-leaf display for the following data:

70	72	75	64	58	83	80	82
76	75	68	65	57	78	85	72

SELF TEST ▶ 23. Construct a stem-and-leaf display for the following data:

11.3	9.6	10.4	7.5	8.3	10.5	10.0
9.3	8.1	7.7	7.5	8.4	6.3	8.8

24. Construct a stem-and-leaf display for the following data. Use the first two digits as the stem and the third digit as the leaf.

1161	1206	1478	1300	1604	1725	1361	1422
1221	1378	1623	1426	1557	1730	1706	1689

Applications

SELF TEST ▶ 25. A psychologist developed a new test of adult intelligence. The test was administered to 20 individuals, and the following data were obtained.

| 114 | 99 | 131 | 124 | 117 | 102 | 106 | 127 | 119 | 115 |
| 98 | 104 | 144 | 151 | 132 | 106 | 125 | 122 | 118 | 118 |

Construct a stem-and-leaf display for these data.

26. The earnings-per-share data for a sample of 20 companies from the *Fortune* 500 largest U.S. industrial corporations are as follows (*Fortune*, April 20, 1992):

Company	Earnings per share	Company	Earnings per share
Procter & Gamble	4.92	Sara Lee	2.15
Goodyear	1.61	General Dynamics	12.06
Ralston Purina	3.34	Eli Lilly	4.50
Chiquita Brands	2.55	Compaq Computer	1.49
Hershey Foods	2.43	Sunstrand	3.02
Data General	2.62	Briggs & Stratton	2.52
Helene Curtis	.18	Interlake	1.31
Huffy	1.52	Dell Computer	1.36
Quaker State	.84	Harley-Davidson	2.08
Anchor Glass	3.30	Zenith Electronics	1.79

Develop a stem-and-leaf display for this data. Comment on what you learned about the earnings per share for these companies.

 JOBSAT

27. In a study of job satisfaction, a series of tests were administered to 50 subjects. The following data were obtained; higher scores represent greater dissatisfaction.

87	76	67	58	92	59	41	50	90	75	80	81	70
73	69	61	88	46	85	97	50	47	81	87	75	60
65	92	77	71	70	74	53	43	61	89	84	83	70
46	84	76	78	64	69	76	78	67	74	64		

Construct a stem-and-leaf display for these data.

28. The net profit margins for oil companies were reported in the *Forbes* 41st Annual Report on American Industry (*Forbes*, January 9, 1989). Data on net profit margin are as follows:

Company	Net Profit Margin	Company	Net Profit Margin
Exxon	6.8	Phillips Petroleum	4.8
AMOCO	9.8	Quaker State	1.6
Du Pont	6.6	Ashland	2.2
Chevron	7.1	Union Pacific	8.9
Mobil	4.1	Kerr McGee	3.4
Occidental	1.8	Crown Central	6.5
Getty	2.1	Pacific Resources	1.9
Union Texas	9.2	American Petrofina	4.6
Atlantic Richfield	8.9	Coastal Corp	1.5

a. Develop a stem-and-leaf display for these data.

b. Use the results of the stem-and-leaf display to develop a frequency distribution and relative frequency distribution for the data.

FIGURE **2.8**
Tabular and Graphical Procedures for Summarizing Data

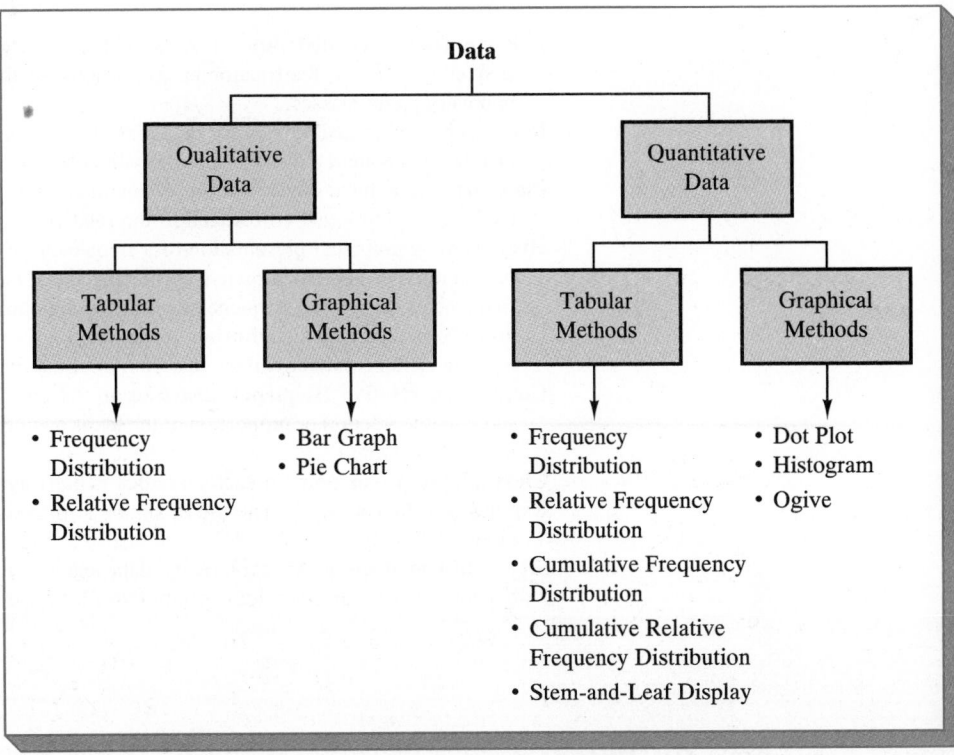

Summary

A set of data, even if modest in size, is often difficult to interpret directly in the form in which it is gathered. Tabular and graphical procedures provide means of organizing and summarizing the data so that patterns are revealed and the data are more easily interpreted. Frequency distributions, relative frequency distributions, bar graphs, and pie charts were presented as tabular and graphical procedures for summarizing qualitative data. Frequency distributions, relative frequency distributions, dot plots, histograms, cumulative frequency distributions, cumulative relative frequency distributions, and ogives were presented as ways of summarizing quantitative data. The chapter concluded with an introduction to exploratory data analysis. A stem-and-leaf display was presented as an exploratory data analysis technique that can be used to summarize quantitative data. Figure 2.8 provides a summary of the tabular and graphical methods presented in this chapter.

Glossary

Qualitative data Data that provide labels or names for categories of like items. These data are provided by either a nominal or an ordinal scale of measurement.

Quantitative data Data that indicate how much or how many. These data are provided by either an interval or a ratio scale of measurement.

Frequency distribution A tabular summary of a set of data showing the frequency (or number) of items in each of several nonoverlapping classes.

Relative frequency distribution A tabular summary of a set of data showing the relative frequency—that is, the fraction or proportion—of the total number of items in each of several nonoverlapping classes.

Bar graph A graphical device for depicting the information presented in a frequency distribution or relative frequency distribution of qualitative data.

Pie chart A graphical device for presenting qualitative data summaries based upon subdividing a circle into sectors that correspond to the relative frequency for each class.

Histogram A graphical presentation of a frequency distribution or relative frequency distribution of quantitative data constructed by placing the class intervals on the horizontal axis and the frequencies or relative frequencies on the vertical axis.

Cumulative frequency distribution A tabular summary of a set of quantitative data showing the number of items having values less than or equal to the upper class limit of each class.

Cumulative relative frequency distribution A tabular summary of a set of quantitative data showing the fraction or proportion of the items having values less than or equal to the upper class limit of each class.

Class midpoint The point in each class that is halfway between the lower and upper class limits.

Exploratory data analysis The use of simple arithmetic and easy-to-draw pictures to present data more effectively.

Stem-and-leaf display An exploratory data analysis technique that simultaneously rank orders quantitative data and provides insight into the shape of the distribution.

Key Formulas

Relative Frequency

$$\frac{\text{Frequency of the Class}}{n} \tag{2.1}$$

Approximate Class Width

$$\frac{\text{Largest Data Value} - \text{Smallest Data Value}}{\text{Number of Classes}} \tag{2.2}$$

❏ ❏ Supplementary Exercises

29. The Gallup Poll News Service selected a random sample of adults to learn what sports fans select as their favorite sport to watch, in person or on television. (*USA Today,* December 12, 1990). The sample results shown below are consistent with the findings of the Gallup poll. In the data, B is baseball, K is basketball, F is football, I is ice hockey, T is tennis, and O is other sports.

```
O  F  B  O  B  F  F  K  O  K  F  F  O  F  O
T  F  F  B  K  F  F  O  F  B  F  O  O  B  K
I  F  O  B  F  K  B  K  O  O
```

a. Show a frequency distribution.
b. Show a relative frequency distribution. What is the favorite spectator sport?
c. Show a pie chart summary of the data.

30. Each of the *Fortune* 500 companies is classified as belonging to one of several industries (*Fortune,* April 20, 1992). Shown at the top of the next page is a sample of 20 companies with their corresponding industry classification.

Company	Industry Classification	Company	Industry Classification
Coca-Cola	(Beverage)	McDonnell Douglas	(Aerospace)
Union Carbide	(Chemicals)	Morton Thiokol	(Chemicals)
General Electric	(Electronics)	Quaker Oats	(Food)
Motorola	(Electronics)	Pepsico	(Beverage)
Beatrice	(Food)	Maytag	(Electronics)
Kellogg	(Food)	Pillsbury	(Food)
Dow Chemical	(Chemicals)	Lockheed	(Aerospace)
Campbell Soup	(Food)	RJR Nabisco	(Food)
Square D	(Electronics)	Westinghouse	(Electronics)
Ralston Purina	(Food)	TRW	(Electronics)

a. Provide a frequency distribution showing the number of companies in each industry.
b. Provide a relative frequency distribution.
c. Provide a bar graph for this data.

31. Airline travelers were asked to indicate the airline they believed offered the best overall service. The four choices were American Air (A), East Coast Air (E), Suncoast (S), and Great Western (W). The following data were obtained.

```
E   A   E   S   W   W   E   S   W   E   W   E   E   A   S
S   W   E   A   W   W   S   E   E   A   E   E   S   W   A
S   E   A   W   A   A   W   E   S   W
```

Summarize the data by constructing
a. a frequency distribution   **b.** a relative frequency distribution
c. a bar graph   **d.** a pie chart
e. Which airline appears to offer the best service?

32. Voters participating in a recent election exit poll in Michigan were asked to state their political party affiliation. Coding the data 1 for Democrat, 2 for Republican, and 3 for Independent, the collected data are as follows:

```
1   2   2   1   3   1   2   2   2   1   2   3   2   3   2   1   1   2   1   2
2   1   1   1   2   1   2   3   1   1   2   1   3   1   1   2   1   2   3   2
```

a. Show a frequency distribution and a relative frequency distribution for the data.
b. Show a bar graph for the data.
c. Comment on what the data suggest about the strengths of the political parties in this voting area.

33. What are the favorite movies of the year? The biggest box-office successes for 1988 were listed in *U.S. News & World Report,* December 26, 1988. Assume that sample data collected on those movie preferences were summarized with the letter codes shown in Table 2.14. The following sample data are available:

```
V   R   O   R   B   A   D   V   V   R   R   D   A   B   V
R   R   A   B   V   A   B   R   R   V   R   A   V   B   V
D   R   A   V   B   A   O   R   R   B   R   A   D   R   R
B   A   R   O   A   A   V   A   D   A   B   R   B   R   B
```

a. Prepare a frequency distribution and a relative frequency distribution for the data set.
b. Prepare a bar graph for the data set.
c. Rank order the top five motion pictures for 1988. What motion picture appears to have been the most successful?

34. An international survey was conducted by the Union Bank of Switzerland in order to obtain data on the hourly wages of blue- and white-collar workers throughout the world (*Newsweek,*

T A B L E 2.14

Code	Movie
A	Coming to America
B	Big
D	Crocodile Dundee II
R	Who Framed Roger Rabbit
V	Good Morning, Vietnam
O	Other motion picture preferred

LAWAGES

February 17, 1992). Workers in Los Angeles ranked 7th in the world in terms of highest hourly wage. Assume that the following 25 values indicate hourly wages for workers in Los Angeles.

11.50	8.40	11.75	10.05	10.25	8.00	13.65	7.05	9.05
11.90	9.90	6.85	15.35	11.10	14.70	13.15	13.10	6.65
13.10	9.20	9.15	12.05	8.45	5.85	9.80		

a. Construct a frequency distribution using classes of 4.00–5.99, 6.00–7.99, and so on.
b. Construct a relative frequency distribution.
c. Construct cumulative frequency and cumulative relative frequency distributions.
d. Use these distributions to comment on what you have learned about the hourly wages of workers in Los Angeles.

35. The data in Table 2.15 represent sales in millions of dollars for 17 companies in the health care services industry (*The 1992 Business Week 1000*).

a. Construct a frequency distribution to summarize these data. Use a class width of 1000.
b. Develop a relative frequency distribution for the data.
c. Construct a cumulative frequency distribution for the data.
d. Construct a cumulative relative frequency distribution for the data.
e. Construct a histogram as a graphical representation of the data.

36. Given below are the closing prices of 40 common stocks. (*Investor's Daily*, April 25, 1992).

29⅝	34	43¼	8¾	37⅞	8⅝	7⅝	30⅜	35¼	19⅜
9¼	16½	38	53⅜	16⅝	1¼	48⅜	18	9⅜	9¼
10	37	18	8	28½	24¼	21⅝	18½	33⅝	31⅛
32¼	29⅝	79⅜	11⅜	38⅞	11½	52	14	9	33½

a. Construct frequency and relative frequency distributions for these data.
b. Construct cumulative frequency and cumulative relative frequency distributions for these data.
c. Construct a histogram for the data.
d. Using your summaries, make comments and observations about the price of common stock.

37. The grade point averages for 30 students majoring in economics are given below.

2.21	3.01	2.68	2.68	2.74	2.60	1.76	2.77	2.46	2.49
2.89	2.19	3.11	2.93	2.38	2.76	2.93	2.55	2.10	2.41
3.53	3.22	2.34	3.30	2.59	2.18	2.87	2.71	2.80	2.63

a. Construct a relative frequency distribution for the data.
b. Construct a cumulative relative frequency distribution for the data.
c. Construct a histogram for the data.

38. The annual cost of room and board at a sample of 50 U.S. colleges and universities is given in the table that follows (*America's Best Colleges, U.S. News & World Report*, June 5, 1992).

TABLE 2.15

6099	1709	847
3973	604	2104
166	282	170
233	868	230
225	491	2452
2301	393	

COMSTOCK

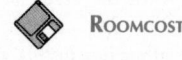
GRADEAVE

ROOMCOST

College	Annual Cost of Room and Board	College	Annual Cost of Room and Board
Auburn University	3,167	University of Missouri	3,004
University of Alaska, Fairbanks	2,860	Montana State University	3,278
Northern Arizona University	2,800	University of Nebraska	2,800
University of Arkansas	3,150	University of Nevada at Las Vegas	4,850
California State University, Fullerton	3,249	University of New Hampshire	3,600
Colorado State University	3,462	Fairleigh Dickinson	5,166
University of Connecticut	4,522	University of New Mexico	3,274
University of Delaware	3,540	Syracuse University	5,860
Georgetown University	5,732	Duke University	4,960

—Table continues at top of next page

College	Annual Cost of Room and Board	College	Annual Cost of Room and Board
—Continued from previous page			
Howard University	4,040	North Dakota State University	2,436
University of Florida	3,790	Ohio State University	3,639
University of Georgia	2,988	University of Oklahoma	3,000
University of Hawaii at Manoa	3,072	Oregon State University	2,950
DePaul University	4,333	University of Pittsburgh	3,514
Indiana University	3,730	Providence College	5,000
Iowa State University	2,850	Clemson University	3,153
University of Kansas	2,684	University of South Dakota	2,302
University of Kentucky	3,734	University of Tennessee	3,166
Tulane University	5,505	University of Texas	3,300
University of Maine	4,241	University of Vermont	4,358
University of Maryland	4,712	University of Virginia	3,312
Amherst College	4,400	University of Washington	3,684
University of Michigan	3,853	West Virginia University	3,846
University of Minnesota	3,400	University of Wisconsin	3,721
University of Mississippi	3,004	University of Wyoming	3,262

Develop the following summaries:

a. a frequency distribution   **b.** a relative frequency distribution   **c.** a histogram

d. Use your summaries to make some comments about the annual room-and-board costs associated with attending college.

39. There were 79 new shadow stocks reported by the American Association of Individual Investors, *AAII Journal*, April 1992. The term *shadow* is used to indicate that the stocks are for small- to medium-sized firms that are not followed closely by the major brokerage houses. The stock exchange listing the stock—New York Stock Exchange (NYSE), American Stock Exchange (AMEX), and Over the Counter Exchange (OTC)—the earnings per share, and the price-earnings ratio are provided for the following sample of 15 shadow stocks.

Stock	Exchange	Earnings per share	Price-Earnings Ratio
Selas Corp of America	AMEX	1.61	7.1
CE Software Holdings	OTC	4.13	7.5
Shult Homes Corp	AMEX	1.05	11.0
Basic American Medical	OTC	1.06	12.0
Titan Corporation	NYSE	.24	18.3
First Team Sports	OTC	.51	18.6
Cooker Restaurant Corp	OTC	.76	43.1
Chempower	OTC	.22	20.5
Benchmark Electronics	AMEX	.62	25.0
Mylex Corp	OTC	.11	37.5
Pharmacy Mgmt Service	OTC	.22	50.0
Arrow Automotive Indus.	AMEX	.23	38.6
U.S. Filter Corp	AMEX	.23	73.4
Sun Sportswear	OTC	.21	28.6
Village Supermarket	OTC	.61	13.5

a. Provide frequency and relative frequency distributions for the exchange data. What exchange carries the most shadow stocks?

b. Provide frequency and relative frequency distributions for the earnings-per-share and price-earnings ratio data. Use class limits of .00–.49, .50–.99, etc., for the earnings-per-share data and class limits of 0.0–9.9, 10.0–19.9, and so on for the price-earnings ratio data. What observations and comments can you make about the shadow stocks?

 STATES

40. A state-by-state listing of per capita incomes for 1991 is shown below (*The Wall Street Journal*, April 23, 1992).

State	Income	State	Income	State	Income
Ala.	$15,567	Ky.	$15,539	N.D.	$16,088
Alaska	21,932	La.	15,143	Ohio	17,916
Ariz.	16,401	Maine	17,306	Okla.	15,827
Ark.	14,753	Md.	22,080	Ore.	17,592
Calif.	20,952	Mass.	22,897	Pa.	19,128
Colo.	19,440	Mich.	18,679	R.I.	18,840
Conn.	25,881	Minn.	19,107	S.C.	15,420
Del.	20,349	Miss.	13,343	S.D.	16,392
D.C.	24,439	Mo.	17,842	Tenn.	16,325
Fla.	18,880	Mont.	16,043	Texas	17,305
Ga.	17,364	Neb.	17,852	Utah	14,529
Hawaii	21,306	Nev.	19,175	Vt.	17,747
Idaho	15,401	N.H.	20,951	Va.	19,976
Ill.	20,824	N.J.	25,372	Wash.	19,442
Ind.	17,217	N.M.	14,844	W.Va.	14,174
Iowa	17,505	N.Y.	22,456	Wis.	18,046
Kan.	18,511	N.C.	16,642	Wyo.	17,118

Develop a frequency distribution, a relative frequency distribution, and a histogram for these data.

41. The conclusion from a 40-state poll conducted by the Joint Council on Economic Education (*Time*, January 9, 1989) is that students do not learn enough economics. The findings were based on test results from 11th- and 12th-grade students who took a 46-question, multiple-choice test on basic economic concepts such as profit and the law of supply and demand. The sample data in Table 2.16 represent data on the number of questions answered correctly. Summarize these data using

a. a stem-and-leaf display   **b.** a frequency distribution
c. a relative frequency distribution   **d.** a cumulative frequency distribution
e. Based on these data, do you agree with the claim that students are not learning enough economics? Explain.

42. The daily high and low temperatures for 24 cities follow (*USA Today*, April 6, 1992).

T A B L E 2.16

12	16	22	17	18	23
31	18	24	20	24	28
24	25	19	26	18	18
22	16	13	33	19	14
8	15	16	14	22	19
10	21	15	9	16	
14	30	12	12	17	

 CITIES

City	High	Low	City	High	Low
Tampa	80	58	Birmingham	68	32
Kansas City	69	40	Minneapolis	62	39
Boise	58	35	Portland	50	41
Los Angeles	71	57	Memphis	67	41
Philadelphia	56	35	Buffalo	44	28
Milwaukee	47	29	Cincinnati	55	29

—Table continues at top of next page

City	High	Low	City	High	Low
—Continued from previous page					
Chicago	52	25	Charlotte	61	37
Albany	50	28	Boston	50	35
Houston	63	50	Tulsa	73	50
Salt Lake City	61	49	Washington, D.C.	56	35
Miami	79	56	Las Vegas	80	53
Cheyenne	66	35	Detroit	52	29

a. Prepare a stem-and-leaf display for the high temperatures.

b. Prepare a stem-and-leaf display for the low temperatures.

c. Compare the stem-and-leaf display from (a) and (b) and make some comments about the differences between daily high and low temperatures.

d. Use the stem-and-leaf display from (b) to determine the number of cities having a low temperature of freezing (32°F) or below.

e. Provide frequency distributions for both the high- and low-temperature data.

Computer Case   *Consolidated Foods, Inc.*

Consolidated Foods, Inc. operates a chain of supermarkets located in New Mexico, Arizona, and California. A recent promotional campaign has advertised the chain's offering of a new credit-card policy in which Consolidated Foods' customers have the option of paying for their purchases with credit cards such as Visa and MasterCard in addition to the usual options of cash or personal check.

TABLE 2.17

Purchase Amount and Method of Payment* for a Random Sample of 100 Consolidated Foods' Customers

Cash	Personal Check	Credit Card	Cash	Personal Check	Credit Card
7.40	27.60	50.30	5.08	52.87	69.77
5.15	30.60	33.76	20.48	78.16	48.11
4.75	41.58	25.57	16.28	25.96	
15.10	36.09	46.24	15.57	31.07	
8.81	2.67	46.13	6.93	35.38	
1.85	34.67	14.44	7.17	58.11	
7.41	58.64	43.79	11.54	49.21	
11.77	57.59	19.78	13.09	31.74	
12.07	43.14	52.35	16.69	50.58	
9.00	21.11	52.63	7.02	59.78	
5.98	52.04	57.55	18.09	72.46	
7.88	18.77	27.66	2.44	37.94	
5.91	42.83	44.53	1.09	42.69	
3.65	55.40	26.91	2.96	41.10	
14.28	48.95	55.21	11.17	40.51	
1.27	36.48	54.19	16.38	37.20	
2.87	51.66	22.59	8.85	54.84	
4.34	28.58	53.32	7.22	58.75	
3.31	35.89	26.57		17.87	
15.07	39.55	27.89		69.22	

*The above data are based on actual bills and types of payments reported for grocery purchases (*The Wall Street Journal,* April 9, 1992).

The new policy is being implemented on a trial basis with the hope that the credit-card option will encourage customers to make larger purchases.

After the first month of operation, a random sample of 100 customers was selected over a 1-week period. Data were collected on the method of payment and how much was spent by each of the 100 customers. The sample data are shown in Table 2.17. Prior to the new credit-card policy, approximately 50% of Consolidated Foods' customers paid in cash, and approximately 50% paid by personal check.

Managerial Report

Use the tabular and graphical methods of descriptive statistics presented in Chapter 2 to summarize the sample data in Table 2.17. Your report should contain summaries such as the following:

1. a frequency and relative frequency distribution for the method of payment
2. a bar graph or pie chart for the method of payment
3. frequency and relative frequency distributions for the amount spent using each method of payment
4. histograms and/or stem-and-leaf plots for the amount spent using each method of payment

What preliminary insights do you have about the amounts spent and method of payment at Consolidated Foods? The data set for this computer case is available in the data file CONSOLID (see Appendix D).

 CONSOLID

3

Descriptive Statistics II. Measures of Location and Dispersion

Barnes Hospital*

ST. LOUIS, MISSOURI

Established in 1914, Barnes Hospital, at the Washington University Medical Center, is the leading provider of health care for the people of St. Louis and neighboring areas. The 1200-bed hospital is nationally recognized as one of the best in the United States. The Hospice Program at Barnes Hospital focuses upon improving the quality of life for terminally ill patients and their families. The hospice team consists of a medical director, coordinator, RN supervisor, home and inpatient RNs, home health aids, social workers, chaplains, dietitians, trained volunteers, and professionals from other ancillary services, as needed. Through the coordinated efforts of the hospice team, patients and families are given the guidance and support necessary to cope with the strains created by serious illness, separation, and death.

In the coordination and administration of the hospice program, monthly reports and quarterly summaries help team members review the ongoing services. Statistical summaries of performance data are utilized as a basis for planning and implementing policy changes.

For example, data are collected on the length of time patients stay in the hospice program. A sample of 67 patient records showed that the time in the program varied from 1 day to 185 days. A frequency distribution was helpful in summarizing and communicating the length-of-stay data to others. In addition, the following numerical measures of descriptive statistics were used to provide valuable information about the patient time in the program.

Mean:	35.7 days
Median:	17 days
Mode:	1 day

Interpretation of these statistics shows that the mean, or average, time a patient stays in the program is 35.7 days, or slightly over a month. However, the median shows that half of the patients are in the program 17 days or less, whereas half are in the program 17 days or more. The mode of 1 day shows the most frequent data value and indicates many patients have a short stay in the program.

Other statistical summaries about the hospice program include the number of admissions, the number of days spent at home versus the number of days in the inpatient unit, the number of discharges from the inpatient unit, and the number of patient deaths at home and in the inpatient unit. These summaries are analyzed according to patient age and Medicare coverage. Overall, descriptive statistics provide valuable information about the hospice services.

In this chapter you will learn how to compute and interpret the statistical measures used by Barnes Hospital. In addition to mean, median, and mode, you will learn about other descriptive statistics such as range, variance, standard deviation, percentiles, quartiles, and others. These numerical measures will provide assistance in the understanding and interpretation of data.

*Ms. Paula H. Gianino, Hospice Coordinator at Barnes Hospital, provided this Statistics in Practice.

In Chapter 2 we discussed tabular and graphical methods used to summarize data. These procedures are effective in written reports and as visual aids when making presentations to individuals or groups. In this chapter, we present several numerical measures of location and dispersion that provide additional alternatives for summarizing data.

In Chapter 3, we consider data sets consisting of a single variable. Whenever the data for a single variable, such as age, salary, or the like, has been obtained from a sample of n elements, the data set will contain n items, or data values. The numerical measures of location and dispersion will be computed using the n data values. If there is more than one variable, the numerical measures presented here must be computed separately for each variable.

Several numerical measures of location and dispersion are introduced. If the measures are computed for data from a sample, they are called *sample statistics*. If the measures are computed for data from a population, they are called *population parameters*.

3.1   Measures of Location

Mean

Perhaps the most important numerical measure of location is the *mean,* or average value, for a variable. The mean provides a measure of central location. It is obtained by adding all the data values and dividing by the number of items. If the data are from a sample, the mean is denoted by $\bar{x}$; if the data are from a population, the mean is denoted by the Greek letter μ.

In specifying statistical formulas, it is customary to denote the value of the first data item by x_1, the value of the second data item by x_2, and so on. In general, the ith data value is denoted by x_i. Using this notation, the formula for the sample mean is as follows:

Sample Mean

$$\bar{x} = \frac{\sum x_i}{n} \qquad\qquad (3.1)$$

---

where n = number of items in the sample. In this formula, the numerator is the sum of the n data values. That is,

$$\sum x_i = x_1 + x_2 + \cdots + x_n$$

The Greek letter Σ is the summation sign.

In order to illustrate the computation of a sample mean, let us consider the following class-size data for a sample of five college classes:

$$46 \quad 54 \quad 42 \quad 46 \quad 32$$

Using the notation x_1, x_2, x_3, x_4, x_5 to represent the number of students in each of the five classes, we have:

$$x_1 = 46 \qquad x_2 = 54 \qquad x_3 = 42 \qquad x_4 = 46 \qquad x_5 = 32$$

Thus to compute the sample mean, we can write

$$\bar{x} = \frac{\sum x_i}{n} = \frac{x_1 + x_2 + x_3 + x_4 + x_5}{5} = \frac{46 + 54 + 42 + 46 + 32}{5} = 44$$

For the five classes sampled, the mean class size is 44 students.

Another illustration of the computation of a sample mean is given in the following situation. Suppose that a college placement office sent a questionnaire to a sample of business school graduates requesting information on starting salaries. Table 3.1 shows the

TABLE 3.1
Monthly Starting Salaries for a Sample of 12 Business School Graduates

Graduate	Monthly Salary	Graduate	Monthly Salary
1	2050	7	2090
2	2150	8	2330
3	2250	9	2140
4	2080	10	2525
5	1955	11	2120
6	1910	12	2080

data that have been collected. The mean monthly starting salary for the sample of 12 business college graduates is computed as shown:

$$\bar{x} = \frac{\sum x_i}{n} = \frac{x_1 + x_2 + \cdots + x_{12}}{12}$$

$$= \frac{2050 + 2150 + \cdots + 2080}{12}$$

$$= \frac{25{,}680}{12} = 2140$$

Equation (3.1) shows how the mean is computed for a sample with n items. The formula for computing the mean of a population is the same, but we use different notation to indicate that we are dealing with the entire population. The number of elements in the population is denoted by N, and, as we mentioned previously, the symbol for the population mean is μ.

Population Mean

$$\mu = \frac{\sum x_i}{N}$$

(3.2)

Trimmed Mean

Occasionally, a variable will have one or more unusually small and/or unusually large data values that significantly influence the value of the mean. With this influence, the mean may provide a poor description of the central location of the data. In order to remove the effect of the unusually small and/or large data values, we can eliminate, or trim, a percentage of small and large data values from the data set. The mean of the remaining data is called the *trimmed mean*. The intent is for the trimmed mean to be a better indicator of the central location of the data. For example, a 5% trimmed mean removes the smallest 5% of the data values *and* the largest 5% of the data values. The 5% trimmed mean is then computed as the mean of the middle 90% of the data. In general, an α percent trimmed mean is obtained by trimming α percent of the items from each end of the data and computing the mean for the remaining items.

For example, consider the 5% trimmed mean for the monthly starting salaries of 12 business school graduates. The 5% trimmed mean indicates that the smallest 5% and the largest 5% of the data values are to be trimmed from the data set. Here, 5% of 12 is

.05(12) = .60 data values. In computing the trimmed mean, we simply round the number of data values trimmed to its nearest integer value; thus, in this case .60 is rounded to 1, indicating 1 item will be removed from each end of the data set. After eliminating the smallest data value (1910) and the largest data value (2525), the 5% trimmed mean based on the 10 remaining data values can be shown to be 21,245/10 = 2124.50.

Median

The *median* is another measure of central location for data. The median is the value falling in the middle when the data items are arranged in ascending order (rank ordered from smallest to largest). If there is an odd number of items, the median is the middle item. If there is an even number of items, there is no single middle value. In this case, we follow the convention of defining the median to be the average of the middle two values. For convenience the definition of the median is restated as follows.

Median

If there is an odd number of items, the median is the value of the middle item when all items are arranged in ascending order.

If there is an even number of items, the median is the average value of the two middle items when all items are arranged in ascending order.

Let us apply this definition to compute the median class size for the sample of five college classes. Arranging the five data values in ascending order provides the following rank-ordered list.

$$32 \quad 42 \quad 46 \quad 46 \quad 54$$

Since $n = 5$ is odd, the median is the middle item in the rank-ordered list. Thus the median is 46 students. Even though there are two values of 46, each value is treated as a separate item when we arrange the data in ascending order and determine the median.

Suppose we also compute the median starting salary for the business college graduates shown in Table 3.1. Arranging the 12 items in ascending order provides the following:

1910   1955   2050   2080   2080   2090   2120   2140   2150   2250   2330   2525

Middle Two Values

Since $n = 12$ is even, we have identified the middle two items. The median is the average of these two values:

$$\text{Median} = \frac{2090 + 2120}{2} = 2105$$

Although the mean is the more commonly used measure of central location, there are some situations in which the median is preferred. As we stated previously, the mean is influenced by extremely small and large values. For instance, suppose that one of the graduates had earned a starting salary of $10,000 per month (maybe the individual's family owns the company). If we change the highest monthly starting salary in Table 3.1

from \$2525 to \$10,000 and recompute the mean, the sample mean changes from $\bar{x}, = 2140$ to $\bar{x} = 2763$. The median, however, is unchanged, since 2090 and 2120 are still the middle two items. With the extremely high starting salary included, the median provides a better measure of central location than the mean. We can generalize to say that whenever there are extreme data values, the median is often the preferred measure of central location.

Mode

A third measure of location is the *mode*. The mode is defined as follows.

Mode

The mode is the data value that occurs with greatest frequency.

To illustrate the identification of the mode, consider the sample of five class sizes. The only value that occurs more than once is 46. Since this value, occurring with a frequency of 2, has the greatest frequency, it is the mode. As another illustration, consider the sample of starting salaries for the business school graduates. The only monthly starting salary that occurs more than once is 2080. Since this value has the greatest frequency, it is the mode.

Situations can arise for which the greatest frequency occurs at two or more different values. In these instances more than one mode exists. If the data have exactly two modes, we say that the data are *bimodal*. If data have more than two modes, we say that the data are *multimodal*. In multimodal cases the mode is almost never reported, since listing three or more modes would not do a very good job of describing a location for the data.

The mode is an important measure of location for qualitative data. For example, the qualitative data set in Table 2.1 resulted in the following frequency distribution for automobile purchases by women.

Automobile Purchase	Frequency
Chevrolet Cavalier	9
Ford Escort	14
Ford Taurus	8
Honda Accord	11
Hyundai Excel	8
Total	50

The mode, or most frequently purchased automobile, is the Ford Escort. For this type of data it obviously makes no sense to speak of the mean or median. But the mode does provide what we are interested in, the most frequently purchased automobile.

Percentiles

A *percentile* is a measure that locates values in the data set that are not necessarily central locations. A percentile provides information regarding how the data items are spread over the interval from the smallest value to the largest value. With data that do not have numerous repeated values, the pth percentile divides the data into two parts. Approximately p percent of the items have values less than the pth percentile; approximately $(100 - p)$ percent of the items have values greater than the pth percentile. The pth percentile is formally defined as follows:

Percentile

The pth percentile is a value such that *at least p percent* of the items take on this value or less and *at least* $(100 - p)$ percent of the items take on this value or more.

Admission test scores for colleges and universities are frequently reported in terms of percentiles. For instance, suppose an applicant obtains a raw score of 54 on the verbal portion of an admissions test. It may not be readily apparent how this student performed relative to other students taking the same test. However, if the raw score of 54 corresponds to the 70th percentile, then we know that approximately 70% of the students had scores less than this individual and approximately 30% of the students had scores greater than this individual.

The following procedure can be used to compute the pth percentile.

Calculating the pth Percentile

Step 1. Arrange the data in ascending order (rank order from smallest value to largest value).

Step 2. Compute an index i as follows:

$$i = \left(\frac{p}{100}\right)n$$

where p is the percentile of interest and n is the number of items.

Step 3. (a) If i *is not an integer, round up.* The next integer value *greater* than i denotes the position of the pth percentile.
(b) If i *is an integer,* the pth percentile is the average of the data values in positions i and $i + 1$.

As an illustration of this procedure, let us determine the 85th percentile for the starting salary data shown in Table 3.1.

Step 1.   Arrange the 12 data values in ascending order:

1910, 1955, 2050, 2080, 2080, 2090, 2120, 2140, 2150, 2250, 2330, 2525

Step 2.

$$i = \left(\frac{p}{100}\right)n = \left(\frac{85}{100}\right)12 = 10.2$$

Step 3.   Since i is not an integer, *round up.* The position of the 85th percentile is the next integer greater than 10.2, the 11th position.

Returning to the data, we see that the 85th percentile corresponds to the 11th data position, or 2330.

As another illustration of this procedure, let us consider the calculation of the 50th percentile. Applying step 2, we obtain

$$i = \left(\frac{50}{100}\right)12 = 6$$

Since i is an integer, step 3(b) states that the 50th percentile is the average of the 6th and 7th data values; thus the 50th percentile is $(2090 + 2120)/2 = 2105$. Note that the *50th percentile is also the median*.

Quartiles and Hinges

It is often desired to divide data into four parts, with each part containing approximately one-fourth, or 25%, of the items. Figure 3.1 shows a data set divided into four parts. The division points are referred to as the *quartiles* and are defined as follows:

Q_1 = first quartile, or 25th percentile

Q_2 = second quartile, or 50th percentile (also the median)

Q_3 = third quartile, or 75th percentile

The monthly starting salary data are again arranged in ascending order, as follows. Q_2, the median, has already been identified as 2105.

1910, 1955, 2050, 2080, 2080, 2090, 2120, 2140, 2150, 2250, 2330, 2525

The computations of Q_1 and Q_3 require the use of the rule for finding the 25th and 75th percentiles. These calculations are as follows:

For Q_1,

$$i = \left(\frac{p}{100}\right)n = \left(\frac{25}{100}\right)12 = 3$$

Since i is an integer, step 3(b) indicates that the first quartile, or 25th percentile, is the average of the 3rd and 4th data values; thus, $Q_1 = (2050 + 2080)/2 = 2065$.

For Q_3,

$$i = \left(\frac{p}{100}\right)n = \left(\frac{75}{100}\right)12 = 9$$

FIGURE 3.1
Location of the Quartiles

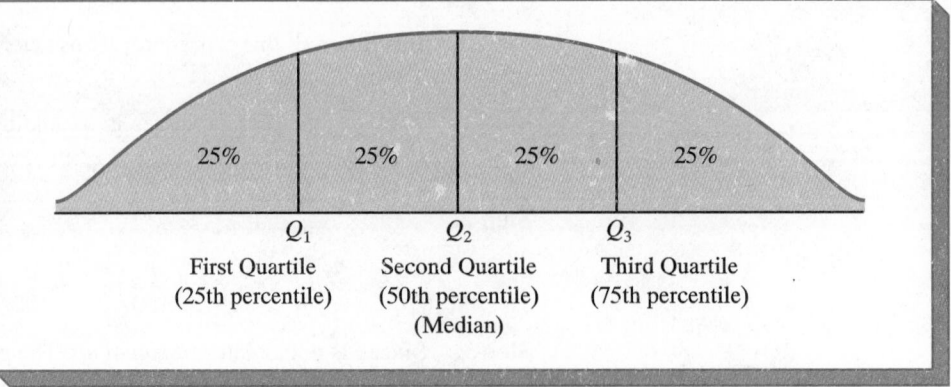

Again, since i is an integer, step 3(b) indicates that the third quartile, or 75th percentile is the average of the 9th and 10th data values; thus, $Q_3 = (2150 + 2250)/2 = 2200$.

As shown below, the quartiles have divided the 12 data values into four parts, with each part consisting of 25% of the items.

| 1910 | 1955 | 2050 | 2080 | 2080 | 2090 | 2120 | 2140 | 2150 | 2250 | 2330 | 2525 |

$$Q_1 = 2065 \qquad Q_2 = 2105 \qquad Q_3 = 2200$$
$$\text{(Median)}$$

We have defined the quartiles as the 25th, 50th, and 75th percentiles. Thus, we have computed the quartiles in the same fashion as for other percentiles. However, there are some variations in the conventions used to compute quartiles; the actual value computed may vary slightly depending on the convention used. Nevertheless, the objective of all procedures for computing quartiles is to divide data into roughly four equal parts.

Another approach used to divide data into four equal parts has been recently developed by proponents of exploratory data analysis. A *lower hinge* (lower 25%) and an *upper hinge* (upper 25%) are computed. To find these hinges, the items are first arranged in ascending order. Then the data are divided into two equal parts: data in positions less than or equal to the median position and data in positions greater than or equal to the median position. The median for the data in positions *less than or equal to the median position* is the lower hinge. The median for the data in positions *greater than or equal to the median position* is the upper hinge.

Referring to the sample of 12 monthly starting salaries for business school graduates, the median has been found to be 2105. From the listing of the data in ascending order, the position of the median is 6.5, or halfway between the 6th and 7th data values. The data in positions less than or equal to 6.5 are the data in positions 1 through 6:

$$1910 \quad 1955 \quad 2050 \quad 2080 \quad 2080 \quad 2090$$

Following the rule for determining a median, we see that the median of these six values is $(2050 + 2080)/2 = 2065$. Thus, the *lower hinge* of the data is 2065.

Again with the median position at 6.5, the data in positions greater than or equal to 6.5 are the data in positions 7 to 12:

$$2120 \quad 2140 \quad 2150 \quad 2250 \quad 2330 \quad 2525$$

The median of these six values is $(2150 + 2250)/2 = 2200$. Thus, the *upper hinge* of the data is 2200. The lower hinge, the median, and the upper hinge can be used to divide the data into four parts.

For the salary data, the lower hinge is equal to the first quartile and the upper hinge is equal to the third quartile. However, this should not be expected to hold for every data set. In some cases, the hinges and the quartiles take on slightly different values because they are based on slightly different computational procedures.

NOTES &
COMMENTS

In computing the hinges for data with an odd number of items, the median position is included in the computation of the lower hinge *and* in the computation of the upper hinge. For example, with 9 elements, the median position is 5. The median of the data in positions 1 to 5, the data value in position 3, is the lower hinge, and the median of the data in positions 5 to 9, the data value in position 7, is the upper hinge. The median position 5 is used in both computations.

❑ ❑ Exercises

Methods

1. Consider the sample of size 5 with data values as follows:

10, 20, 12, 17, 16.

Compute the mean and median.

2. Consider the sample of size 6 with data values as follows:

10, 20, 21, 17, 16, 12.

Compute the mean and median.

SELF TEST ▶ 3. Consider the sample of size 8 with data values as follows:

27, 25, 20, 15, 30, 34, 28, 25.

Compute the 20th, 25th, 65th, and 75th percentiles.

4. Given a sample of 36 items, how many items should be trimmed from each end of the data set in order to compute the 5% trimmed mean?

Applications

5. A *College Placement Council* survey of starting salaries for college graduates (March 1992) found that 1992 starting salaries were up approximately 3% compared with the preceding year. A sample of typical starting salaries for business administration graduates is shown here. Data are shown in thousands of dollars.

STARTSAL

24.8	26.2	29.8	30.1	23.5	22.5	19.7
22.5	24.0	31.2	28.1	19.3	23.5	24.5
20.8	20.0	22.6	24.2	24.0	21.0	24.0
23.5	24.3	26.6	28.5			

a. What is the mean annual starting salary of business administration graduates?
b. What is the median annual starting salary?
c. What is the mode?
d. What is the first quartile?
e. What is the third quartile?

6. Manufacturers of Japanese automobiles established export quotas for automobiles to be shipped to the United States. While the quotas pleased the Detroit automobile manufacturers, the quotas meant Japanese cars would be in short supply and more expensive for the U.S. consumer. *The Wall Street Journal* (January 11, 1989) listed the Japanese export quotas for each of the 12 months of 1988. The data shown below are in terms of thousands of automobiles.

145, 135, 100, 220, 170, 145, 190, 155, 210, 200, 205, 180.

a. What are the mean, median and mode for the automobile quota data?
b. Compute and interpret the first and third quartiles for this data set.

7. A quality control inspector found the following number of defective parts on 16 different days:

11, 14, 18, 14, 21, 17, 13, 21, 25, 19, 17, 13, 28, 13, 17, 18.

Compute the mean, median, mode, and 90th percentile.

SELF TEST ▶ 8. *American Demographics* (December 1988) reported that 25 million Americans get up each morning and go to work in their offices at home. The growing use of personal computers is suggested as one of the reasons more people can operate at-home businesses. The article presented data on the ages of individuals who work at home. Assume the following is a sample of age data for these individuals.

22	58	24	50	29	52	57	31	30	41
44	40	46	29	31	37	32	44	49	29

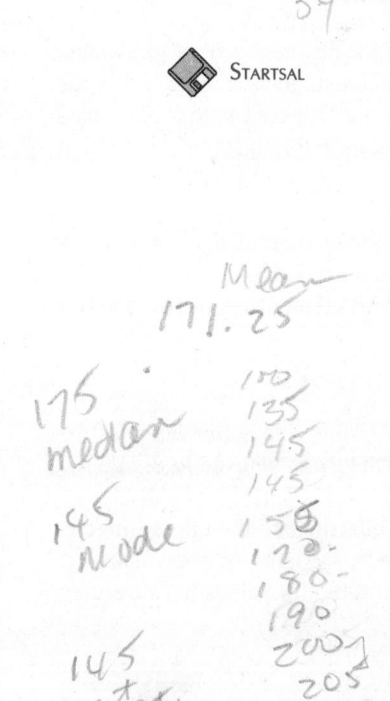

a. Compute the mean and mode.

b. Compute a 5% and a 10% trimmed mean.

c. The median age of the population of all adults is 40.5 years. Use the median age of the preceding data to comment on whether the at-home workers tend to be younger or older than the population of all adults.

d. Compute the first and third quartiles.

e. Compute and interpret the 32nd percentile.

9. The American Association of Advertising Agencies records data on nonprogramming minutes per half-hour of prime-time television programming (*U.S. News & World Report,* April 13, 1992). Representative data are shown here for a sample of prime time programs on major networks at 8:30 P.M.

$$
\begin{array}{cccccccccc}
6.0 & 6.6 & 5.8 & 7.0 & 6.3 & 6.2 & 7.2 & 5.7 & 6.4 & 7.0 \\
6.5 & 6.2 & 6.0 & 6.5 & 7.2 & 7.3 & 7.6 & 6.8 & 6.0 & 6.2
\end{array}
$$

a. Compute the mean and median.

b. Compute the first and third quartiles.

c. Using the sample mean, what percentage of viewing time is spent on prime-time advertisements, promotions, and credits? What percentage of viewing time is spent on the programs themselves?

10. A bowler has the following scores for six games:

$$182, 168, 184, 190, 170, 174.$$

Using these data as a sample, compute the following descriptive statistics.

a. mean   **b.** median   **c.** mode   **d.** 75th percentile

11. Monthly sales data for car telephone units for the RC Radio Corporation are:

$$80, 115, 82, 102, 94, 90, 88, 91, 89, 95, 105, 108.$$

Compute the mean, median, and mode for monthly sales.

12. The *Los Angeles Times* regularly reports the air quality index for various areas of southern California. Index ratings of 0–50 are considered good, 51–100 are considered moderate, 101–200 unhealthy, 201–275 very unhealthy, and over 275, hazardous. Recent air quality indexes for Pomona were 28, 42, 58, 48, 45, 55, 60, 49, and 50.

a. Compute the mean, median, and mode for the data. Could the Pomona air quality index be considered good?

b. Compute the 25th percentile and 75th percentile for the Pomona air quality data.

c. Compute the lower and upper hinges. Compare your result with the answers to (b).

13. The following data show the number of automobiles arriving at a toll booth during 20 intervals, each of 10 minutes' duration. Compute the mean, median, mode, first quartile, and third quartile for the data.

$$
\begin{array}{cccccccccc}
26 & 26 & 58 & 24 & 22 & 22 & 15 & 33 & 19 & 27 \\
21 & 18 & 16 & 20 & 34 & 24 & 27 & 30 & 31 & 33
\end{array}
$$

14. In automobile mileage and gasoline-consumption testing, 13 automobiles were road tested for 300 miles in both city and country driving conditions. The following data were recorded for miles-per-gallon performance:

City: 16.2, 16.7, 15.9, 14.4, 13.2, 15.3, 16.8, 16.0, 16.1, 15.3, 15.2, 15.3, 16.2
Country: 19.4, 20.6, 18.3, 18.6, 19.2, 17.4, 17.2, 18.6, 19.0, 21.1, 19.4, 18.5, 18.7

Use the mean, median, and mode to make a statement about the difference in performance for city and country driving.

15. A sample of 15 college seniors showed the following credit hours taken during the final term of the senior year:

$$15, 21, 18, 16, 18, 21, 19, 15, 14, 18, 17, 20, 18, 15, 16.$$

a. What are the mean, median, and mode for credit hours taken? Compute and interpret.
b. Compute the first and third quartiles.
c. Compute the lower and upper hinges. Compare your answers to (b).
d. Compute and interpret the 70th percentile.

16. The Nielsen organization provides data on television viewing in the United States. A study (*USA Today*, December 13, 1988) reported that the mean number of hours of television viewing per week was increasing. Suppose that the following data provide the hours of television viewing per week for a sample of 16 college students.

$$14, 9, 12, 4, 20, 26, 17, 15, 18, 15, 10, 6, 16, 15, 8, 5.$$

a. Compute the mean, median, and mode.
b. Compute the 10th and 80th percentiles.
c. Compute the quartiles and hinges.

3.2   Measures of Dispersion

Whenever data are collected, it is desirable to consider the dispersion, or variability, in the data values. For example, assume that you are a purchasing agent for a large manufacturing firm and that you regularly place orders with two different suppliers. Both suppliers indicate that approximately 10 working days are required to fill your orders. After several months of operation you find that the number of days required to fill orders is indeed averaging around 10 days for both of the suppliers. The histograms summarizing the number of working days required to fill orders from the suppliers are shown in Figure 3.2. Although the mean number of days required to fill orders is roughly 10 for both suppliers, do both suppliers possess the same degree of reliability in terms of making delivery on schedule? Note the dispersion, or variability, in the histograms. Which supplier would you prefer?

FIGURE 3.2   Historical Data Showing the Number of Days Required to Fill Orders

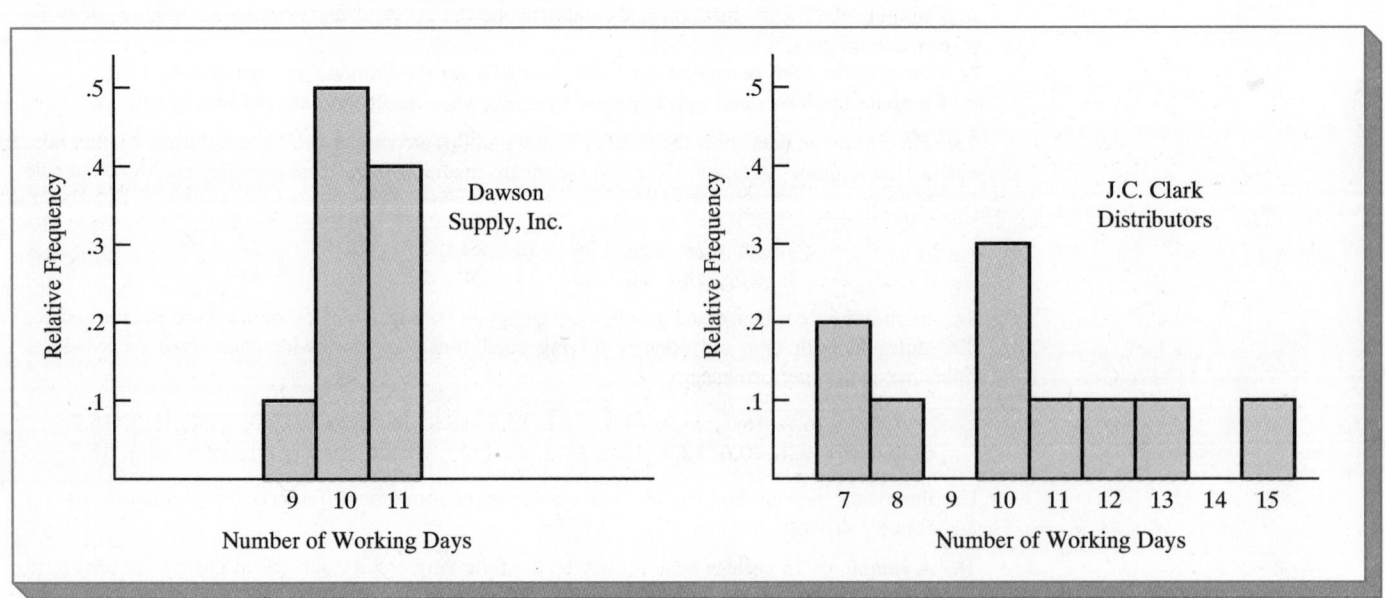

For most firms, receiving materials and supplies on schedule is an important part of the purchasing agent's responsibility. The 7- or 8-day deliveries shown for J. C. Clark Distributors might indicate that the purchasing agent is doing a good job. However, a few of the slow 13- to 15-day deliveries could be disastrous in terms of keeping a work force busy and production on schedule. This example illustrates a situation where the dispersion, or variability, in the delivery times may be an overriding consideration in selecting a supplier. For most purchasing agents, the lower dispersion shown for Dawson Supply, Inc. would make Dawson the more consistent and preferred supplier.

We turn now to a discussion of some commonly used numerical measures of the dispersion, or variability, in data.

Range

Perhaps the simplest measure of dispersion for data is the *range:*

Range

The range is the difference between the largest and smallest data values.

Let us refer to the data on monthly starting salaries for business school graduates in Table 3.1. The largest starting salary is 2525, and the smallest is 1910. The range is $2525 - 1910 = 615$.

Although the range is the easiest of the measures of dispersion to compute, it is not widely used. The reason is that the range is based on only two of the items and thus is influenced too much by extreme data values. Suppose, as we did in the previous section, that one of the graduates obtained a starting salary of $10,000. In this case the range would be $10,000 - 1910 = 8090$. This large value for the range would not be very descriptive of the variability in the data, since 11 of the 12 starting salaries are closely grouped between 1910 and 2330.

Interquartile Range

A measure of disperson that overcomes the dependency upon extreme data values is the *interquartile range* (IQR). This measure of dispersion is simply the difference between the third quartile, Q_3, and the first quartile, Q_1. In other words, the interquartile range is the range for the middle 50% of the data.

Interquartile Range

$$IQR = Q_3 - Q_1 \qquad\qquad (3.3)$$

For the data on monthly starting salaries, we found the quartiles were $Q_3 = 2200$ and $Q_1 = 2065$. Thus the interquartile range is $2200 - 2065 = 135$.

Variance

The *variance* is a measure of dispersion that utilizes all the data values. The variance is based on the difference between each data value and the mean. The difference between each data value x_i and the mean ($\bar{x}$ for a sample, μ for a population) is called a *deviation about the mean*. For a sample, a deviation is written $(x_i - \bar{x})$; for a population, it is written $(x_i - \mu)$. In the computation of the variance, the deviations about the mean are *squared*.

If the data set involved is a population, the average of the squared deviations is called the *population variance*. The population variance is denoted by the Greek symbol σ^2. Given a population of N items and using μ to denote the population mean, the definition of the population variance is as follows:

Population Variance

$$\sigma^2 = \frac{\Sigma (x_i - \mu)^2}{N} \tag{3.4}$$

In most statistical applications, the data set being analyzed is a sample. When we compute a measure of variability for a sample, we are often interested in using the sample statistic obtained to estimate the population parameter σ^2. At this point it might seem that the average of the squared deviations about the sample mean would provide a good estimate of the population variance. However, statisticians have found that the average squared deviation for the sample has the undesirable feature of providing a biased estimate of the population variance σ^2; specifically, it tends to underestimate the population variance.

 Although it is beyond the scope of this text, it can be shown that if the sum of the squared deviations about the sample mean is divided by $n - 1$, and not n, then the resulting value provides an unbiased estimate of the population variance. For this reason, the *sample variance,* denoted by s^2, is defined as follows.

Sample Variance

$$s^2 = \frac{\Sigma (x_i - \bar{x})^2}{n - 1} \tag{3.5}$$

To illustrate the computation of the variance for a sample, we will use the data on class size for the sample of five college classes. A summary of the data, including the computation of the deviations about the mean and the squared deviations about the mean is shown in Table 3.2. The sum of squared deviations about the mean is $\Sigma (x_i - \bar{x})^2 = 256$. Hence, with $n - 1 = 4$, the sample variance is

$$s^2 = \frac{\Sigma (x_i - \bar{x})^2}{n - 1} = \frac{256}{4} = 64$$

Before moving on, let us note that the units associated with the sample variance often cause confusion. Since the values being summed in the variance calculation, $(x_i - \bar{x})^2$, are squared, the units associated with the sample variance are also *squared*. For instance, the sample variance for the class-size data is $s^2 = 64$ (student)^2. The squared units associated with variance make it difficult to obtain an intuitive understanding and interpretation for the numerical value of the variance. We recommend that you think of the variance as a measure useful in comparing the amount of dispersion in 2 data sets. In comparing two samples, the one with the larger variance has more dispersion. Further interpretation of the value of the variance is not necessary.

As another illustration of computing a sample variance, consider the starting salaries for the 12 business school graduates provided in Table 3.1. In Section 3.1, we showed that the sample mean starting salary was 2140. The computation of the sample variance ($s^2 = 27,440.91$) is shown in Table 3.3.

TABLE 3.2
Computation of Deviations and Squared Deviations about the Mean for the Class-Size Data

Number of Students in Class (x_i)	Mean Class Size $\bar{x}$	Deviation About the Mean ($x_i - \bar{x}$)	Squared Deviation about the Mean ($x_i - \bar{x}$)^2
46	44	2	4
54	44	10	100
42	44	−2	4
46	44	2	4
32	44	−12	144
		0	256
		$\Sigma (x_i - \bar{x})$	$\Sigma (x_i - \bar{x})^2$

TABLE 3.3
Computation of the Sample Variance for the Starting Salary Data

Monthly Salary (x_i)	Sample Mean ($\bar{x}$)	Deviation About the Mean ($x_i - \bar{x}$)	Squared Deviation About the Mean ($x_i - \bar{x}$)^2
2050	2140	−90	8,100
2150	2140	10	100
2250	2140	110	12,100
2080	2140	−60	3,600
1955	2140	−185	34,225
1910	2140	−230	52,900
2090	2140	−50	2,500
2330	2140	190	36,100
2140	2140	0	0
2525	2140	385	148,225
2120	2140	−20	400
2080	2140	−60	3,600
		0	301,850
		$\Sigma (x_i - \bar{x})$	$\Sigma (x_i - \bar{x})^2$

Using (3.5),

$$s^2 = \frac{\Sigma (x_i - \bar{x})^2}{n - 1} = \frac{301,850}{11} = 27,440.91$$

Note that in Tables 3.2 and 3.3 we show both the sum of the deviations about the mean and the sum of the squared deviations about the mean. For any data set, the sum of the deviations about the mean will *always equal zero*. Thus, as shown in Tables 3.2 and 3.3, $\Sigma (x_i - \bar{x}) = 0$. This is true because the positive deviations and negative deviations always cancel each other, causing the sum of the deviations about the mean to equal zero.

Standard Deviation

The *standard deviation* is defined to be the positive square root of the variance. Following the notation we adopted for a sample variance and a population variance, we use s to denote the sample standard deviation and σ to denote the population standard deviation. The standard deviation is derived from the variance in the following manner:

Standard Deviation	
Sample Standard Deviation = $s = \sqrt{s^2}$	**(3.6)**
Population Standard Deviation = $\sigma = \sqrt{\sigma^2}$	**(3.7)**

Recall that the sample variance for the sample of class sizes in five college classes is $s^2 = 64$. Thus the sample standard deviation is $s = \sqrt{64} = 8$ students. For the data set consisting of starting salaries, the sample standard deviation is $s = \sqrt{27,440.91} = 165.65$.

What is gained by converting the variance to its corresponding standard deviation? Recall that the units associated with the variance are squared. For example, the sample variance for the starting salary data of business graduates was $s^2 = 27,440.91$ (dollars)^2. Since the standard deviation is simply the square root of the variance, the units of the variance, dollars squared, are converted to dollars in the standard deviation. Thus, the standard deviation of the starting salary data is \$165.65. In other words, the standard deviation is measured in the same units as the original data. For this reason the standard deviation is more easily compared to the mean and other statistics that are measured in the same units as the original data.

Coefficient of Variation

In some situations we may be interested in a descriptive statistic that indicates how large the standard deviation is relative to the mean. This statistic is called the *coefficient of variation* and is computed as follows:

Coefficient of Variation	
$\dfrac{\text{Standard Deviation}}{\text{Mean}} \times 100$	**(3.8)**

For the class-size data, we found a sample mean of 44 and a sample standard deviation of 8. The coefficient of variation is $(8/44) \times 100 = 18.2$. In words, coefficient of variation tells us the standard deviation of the sample is 18.2% of the value of the sample mean. Using the starting-salary data with a sample mean of 2140 and a sample standard deviation of 165.65, the coefficient of variation, $(165.65/2140) \times 100 = 7.7$, tells us the standard deviation for this sample is only 7.7% of the value of the sample mean. In general, the coefficient of variation is a useful statistic when comparing the dispersion in data sets having different standard deviations and different means.

Notes & Comments

1. Rounding the value of the sample mean $\bar{x}$ and the values of the squared deviations $(x_i - \bar{x})^2$ may introduce rounding errors in the values of the variance and standard deviation. To reduce rounding errors, we recommend carrying at least six significant digits during intermediate calculations. The resulting variance or standard deviation may then be rounded to fewer digits.

2. An alternative formula for the computation of the sample variance is

$$s^2 = \frac{\Sigma x_i^2 - n\bar{x}^2}{n - 1}$$

where $\Sigma x_i^2 = x_1^2 + x_2^2 + \cdots + x_n^2$. Using this formula eases the computational burden slightly and helps reduce rounding errors. Exercise 23 requires this alternative formula to compute the sample variance.

❑ ❑ Exercises

Methods

17. Consider the sample of size 5 with data values as follows:

$$10, 20, 12, 17, 16.$$

Compute the range and interquartile range.

18. Consider the sample of size 5 with data values as follows:

$$10, 20, 12, 17, 16.$$

Compute the variance and standard deviation.

Self Test ▶

19. Consider the sample of size 8 with data values as follows:

$$27, 25, 20, 15, 30, 34, 28, 25.$$

Compute the range, interquartile range, variance, and standard deviation.

Applications

20. Data on selling prices of new and used automobiles were provided in *U.S. News & World Report* (September 9, 1991). Assume the following sample data apply, with data in thousands of dollars.

New Automobiles:

10.3	24.7	14.7	10.8	14.4	9.8	16.3
17.8	20.8	17.6	17.1	20.1	16.5	13.8

Used Automobiles:

| 5.0 | 4.4 | 5.6 | 10.4 | 9.4 | 7.0 | 9.3 | 4.5 | 4.2 | 7.7 | 11.7 | 11.2 |

a. Compute the mean and median selling prices for the new and used automobiles.
b. Compute the range, interquartile range, and standard deviation in the selling prices for new and used automobiles.
c. What comparisons can you make about the selling prices of new and used cars?

21. The *Washington Post* (January 7, 1989) reported on overcrowding in the Virginia prison system. Use the following data as the population of capacities of the five Virginia state prisons.

$$233, 164, 587, 52, 175.$$

Compute the range, variance, and standard deviation for this population.

22. The *Los Angeles Times* regularly reports the air quality index for various areas of Southern California. A sample of air quality index values for Pomona provided the following data: 28, 42, 58, 48, 45, 55, 60, 49, and 50.
a. Compute the range and interquartile range.
b. Compute the sample variance and sample standard deviation.
c. A sample of air quality index readings for Anaheim provided a sample mean of 48.5, a sample variance of 136, and a sample standard deviation of 11.66. What comparisons can you make between the air quality in Pomona and Anaheim based on these descriptive statistics?

23. The Davis Manufacturing Company has just completed five weeks of operation using a new process that is supposed to increase productivity. The numbers of parts produced each week are

$$410, 420, 390, 400, 380.$$

Compute the sample variance and sample standard deviation using the definition of sample variance (3.5) as well as the alternative formula provided in the Notes & Comments.

24. Assume that the data used to construct the histograms of the number of days required to fill orders for Dawson Supply, Inc. and J. C. Clark Distributors (see Figure 3.2) are as follows:

Dawson Supply Days for Delivery: 11, 10, 9, 10, 11, 11, 10, 11, 10, 10
Clark Distributors Days for Delivery: 8, 10, 13, 7, 10, 11, 10, 7, 15, 12

Use the range and standard deviation to support the earlier observation that Dawson Supply provides the more consistent and reliable delivery times.

SELF TEST ▶

25. A bowler's scores for six games were as follows:

$$182, 168, 184, 190, 170, 174.$$

Using these data as a sample, compute the following descriptive statistics:
a. range   **b.** variance
c. standard deviation   **d.** coefficient of variation

 LAWAGES

26. The Union Bank of Switzerland conducted a survey to obtain data on the hourly wages of blue- and white-collar workers throughout the world (*Newsweek,* February 1992). A sample of 25 workers from the Los Angeles area provided the data shown here.

11.50	8.40	11.75	10.05	10.25	8.00	13.65	7.05	9.05
11.90	9.90	6.85	15.35	11.10	14.70	13.15	13.10	6.65
13.10	9.20	9.15	12.05	8.45	5.85	9.80		

Provide the following descriptive statistics:
a. mean   **b.** median
c. range   **d.** interquartile range
e. variance   **f.** standard deviation

27. A production department uses a sampling procedure to test the quality of newly produced items. The department employs the following decision rule at an inspection station: If a sample of 14 items has a variance of more than .005, the production line must be shut down for repairs. Suppose that the following data have just been collected:

$$3.43 \quad 3.45 \quad 3.43 \quad 3.48 \quad 3.52 \quad 3.50 \quad 3.39$$
$$3.48 \quad 3.41 \quad 3.38 \quad 3.49 \quad 3.45 \quad 3.51 \quad 3.50$$

Should the production line be shut down? Why or why not?

28. The following times were recorded by the quarter-mile and mile runners of a university track team (times are in minutes):

Quarter-mile Times: .92, .98, 1.04, .90, .99
Mile Times: 4.52, 4.35, 4.60, 4.70, 4.50

After viewing this sample of running times, one of the coaches commented that the quarter-milers turned in the more consistent times. Use the standard deviation and the coefficient of variation to summarize the variability in the data. Does the use of the coefficient of variation measure indicate that the coach's statement should be qualified?

3.3   Some Uses of the Mean and the Standard Deviation

We have described several measures of location and dispersion for data. The mean is the most widely used measure of location, while the standard deviation and variance are the most widely used measures of dispersion. Using only the mean and the standard deviation, we can learn much about a data set.

z-Scores

Using the mean and standard deviation, we can determine the relative location of any data value. Suppose we have a sample of n items, with the values denoted by $x_1, x_2, \ldots, x_n$. In addition, assume that the sample mean, $\bar{x}$, and the sample standard deviation, s, have been computed. Associated with each data value, x_i, is another value called its *z-score*. Equation (3.9) shows how the z-score is computed for data value x_i.

---

z-Score

$$z_i = \frac{x_i - \bar{x}}{s} \tag{3.9}$$

where z_i = the z-score for item i

$\bar{x}$ = the sample mean

s = the sample standard deviation

The z-score is often called the *standardized value* for the item. The standardized value, or z-score, z_i, can be interpreted as the *number of standard deviations x_i is from the mean $\bar{x}$.* For example, $z_1 = 1.2$ would indicate x_1 is 1.2 standard deviations above, or larger than, the sample mean. Similarly, $z_2 = -.5$ would indicate x_2 is .5, or 1/2, standard deviation below, or less than, the sample mean. As can be seen from Equation (3.9), z-scores greater than zero occur for items with values greater than the sample mean, and z-scores less than zero occur for items with values less than the mean. A z-score of zero indicates that the value of the item is equal to the mean.

The z-score for any item can be interpreted as a measure of the relative location of the item in a data set. Indeed, items in two different data sets with the same z-score can be said to have the same relative location in terms of being the same number of standard deviations from the mean.

The z-scores for the class-size data are shown in Table 3.4. Recall that the sample mean $\bar{x} = 44$ and sample standard deviation $s = 8$ have been computed previously. The z-score of -1.50 for the fifth item shows it is farthest from the mean, falling 1.50 standard deviations below the mean.

Chebyshev's Theorem

Chebyshev's theorem permits us to make statements about the percentage of items that must be within a specified number of standard deviations from the mean. The statement of Chebyshev's theorem is as follows:

Chebyshev's Theorem

At least $(1 - 1/k^2)$ of the items in any data set must be within k standard deviations of the mean, where k is any value greater than 1.

Some of the implications of this theorem, using $k = 2$, 3, and 4 standard deviations, are as follows:

- At least .75, or 75%, of the items must be within $k = 2$ standard deviations of the mean.
- At least .89, or 89%, of the items must be within $k = 3$ standard deviations of the mean.
- At least .94, or 94%, of the items must be within $k = 4$ standard deviations of the mean.

For an example using Chebyshev's theorem, assume that the midterm test scores for 100 students in a college business statistics course had a mean of 70 and a standard deviation of 5. How many students had test scores between 60 and 80? How many students had test scores between 58 and 82?

For the test scores between 60 and 80, we note that the value of 60 is two standard deviations below the mean and the value of 80 is two standard deviations above the mean. Using Chebyshev's theorem we see that at least 75% of the items must have values within

TABLE 3.4 z-Scores for the Class-Size Data	**Number of Students in Class (x_i)**	**Deviation about the Mean ($x_i - \bar{x}$)**	**z-Score** $\left(\dfrac{x_i - \bar{x}}{s}\right)$
	46	2	2/8 = .25
	54	10	10/8 = 1.25
	42	−2	−2/8 = −.25
	46	2	2/8 = .25
	32	−12	−12/8 = −1.50

two standard deviations of the mean. Thus, at least 75 of the 100 students must have scored between 60 and 80.

For the test scores between 58 and 82, we see that $(58 - 70)/5 = -2.4$ indicates 58 is 2.4 standard deviations below the mean and that $(82 - 70)/5 = +2.4$ indicates 82 is 2.4 standard deviations above the mean. Applying Chebyshev's theorem with $k = 2.4$, we have

$$\left(1 - \frac{1}{k^2}\right) = \left[1 - \frac{1}{(2.4)^2}\right] = .826$$

At least 82.6% of the 100 students must have test scores between 58 and 82.

The Empirical Rule

One of the advantages of Chebyshev's theorem is that it applies to any data set regardless of the shape of the distribution of the data. In practical applications, however, it has been found that many data sets have a mound-shaped, or bell-shaped, distribution like the one shown in Figure 3.3. When it is believed that the data approximates this distribution, the *empirical rule* can be used to determine the percentage of items that must be within a specified number of standard deviations of the mean.*

Empirical Rule

For data having a bell-shaped distribution,

■ Approximately 68% of the items will be within one standard deviation of the mean.

■ Approximately 95% of the items will be within two standard deviations of the mean.

■ Almost all of the items will be within three standard deviations of the mean.

FIGURE 3.3

A Mound-Shaped, or Bell-Shaped, Distribution

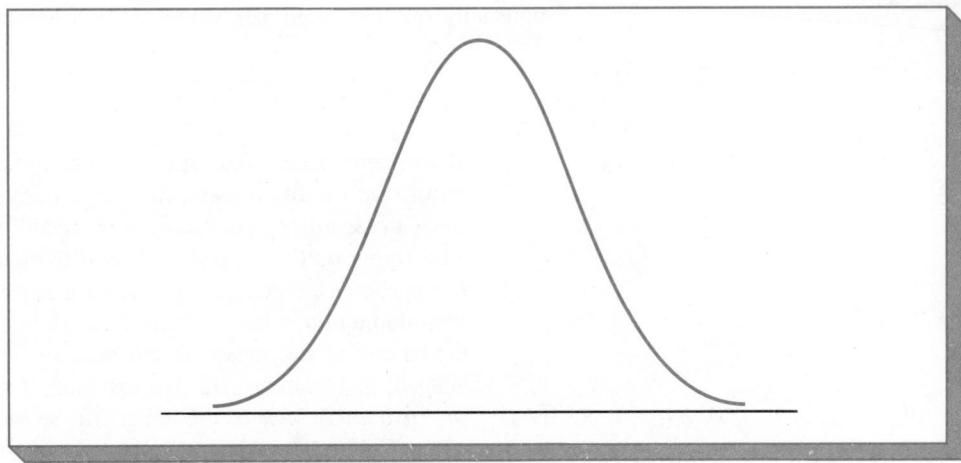

---

*The empirical rule is based on the normal probability distribution, which is presented in detail in Chapter 6.

For example, liquid-detergent cartons are filled automatically on a production line. Filling weights frequently have a bell-shaped distribution. If the mean filling weight is 16 ounces and the standard deviation is .25 ounces, we can use the empirical rule to conclude the following:

- Approximately 68% of the filled items will have weights between 15.75 and 16.25 ounces (that is, within one standard deviation of the mean).
- Approximately 95% of the filled items will have weights between 15.50 and 16.50 ounces (that is, within two standard deviations of the mean).
- Almost all filled items will have weights between 15.25 and 16.75 ounces (that is, within three standard deviations of the mean).

Detecting Outliers

Sometimes a set of data will have one or more items with unusually large or unusually small values. Extreme values such as these are called *outliers*. Experienced statisticians take steps to identify outliers and then review each one carefully. An outlier may be an item for which the value has been incorrectly recorded. If so, the value can be corrected before proceeding with further analysis. An outlier may also be an item that was incorrectly included in the data set; if so it can be removed. Finally, an outlier may just be an unusual item that has been correctly recorded and does belong in the data set. In such cases the item should remain.

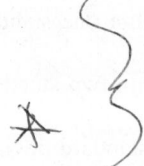

Standardized values (z-scores) can be used to help identify outliers. Recall that the empirical rule allows us to conclude that for data with a bell-shaped distribution, almost all the items will be within three standard deviations of the mean. Thus, when using z-scores to identify outliers, we recommend treating any item with a z-score less than -3 or greater than $+3$ as an outlier. Such items can then be reviewed for accuracy and to determine whether or not they belong in the data set.

Refer to the z-scores for the class size data shown in Table 3.4. The z-score of -1.50 shows the fifth item is farthest from the mean. However, this standardized value is well within the -3 to $+3$ guideline for outliers. Thus the z-scores indicate outliers with unusually small or large data values are not present in the class size data.

N O T E S &
C O M M E N T S

1. Before analyzing a data set, statisticians usually make a variety of checks to ensure the validity of data. In a large study it is not uncommon for errors to be made in recording data values or in inputting the values at a computer terminal. Identifying outliers is one tool used to check the validity of data.

2. Chebyshev's theorem is applicable for any data set and makes a statement about the minimum number of items that will be within a certain number of standard deviations of the mean. If the data set is known to be approximately bell-shaped, more can be said. For instance, the empirical rule allows us to say that *approximately* 95% of the items will be within two standard deviations of the mean; Chebyshev's theorem allows us to conclude only that at least 75% of the items will be in this interval.

☐ ☐ Exercises

Methods

29. Consider the sample of size 5 with data values as follows:

$$10, 20, 12, 17, 16.$$

Compute the z-score for each of the five data values.

30. Consider a sample with a mean of 500 and a standard deviation of 100. What is the z-score for each of the following data values: 520, 650, 500, 450, and 280?

SELF TEST ▶ 31. Consider a sample with a mean of 30 and a standard deviation of 5. Use Chebyshev's theorem to determine the proportion, or percentage, of the data within each of the following ranges:
a. 20 to 40   b. 15 to 45   c. 22 to 38   d. 18 to 42   e. 12 to 48

32. Data that have a bell-shaped distribution have a mean of 30 and a standard deviation of 5. Use the empirical rule to determine the proportion, or percentage, of data within each of the following ranges:
a. 20 to 40   b. 15 to 45   c. 25 to 35

Applications

33. The average household income in Tucson, Arizona, is $41,747 (*U.S. News & World Report*, April 6, 1992). If the standard deviation is $12,200, what is the z-score for a household with an annual income of $25,000? What is the z-score for a household with an annual income of $125,000? Interpret these z-scores and use the empirical rule to comment on whether or not either of these income values should be considered an outlier.

SELF TEST ▶ 34. A sample of 10 NCAA men's college basketball scores (*USA Today*, January 5, 1989) provided the following winning teams and the number of points scored:

Winner	Score	Winner	Score
Rutgers	87	Tulsa	70
Niagara	79	Texas-El Paso	82
Mississippi	80	Stanford	83
Western Kentucky	64	Iowa	93
Purdue	75	Montana	62

a. Compute the sample mean and sample standard deviation for this data.
b. In another game, Penn State beat Massachusetts by a score of 110 to 79. Use the z-score to determine if the Penn State score should be considered an outlier. Explain.
c. Assume that the distribution of the points scored by winning teams has a mound-shaped distribution. Estimate the percentage of all NCAA basketball games in which the winning team will score 87 or more points. Estimate the percentage of all NCAA basketball games in which the winning team will score 58 or less points.

35. A 1989 salary survey (*Working Woman*, January 1989) listed the average salary of elementary and secondary school teachers as $28,085. Assume that the standard deviation of salaries is $4500.
a. Janice Herbranson was identified as a teacher in a one-room schoolhouse in McLeod, North Dakota. Her salary was reported to be $8100 per year. What is the z-score associated with $8100? Comment on whether or not this salary figure is an outlier.
b. Compute the z-score for each of the following salaries: $33,500, $25,200, $28,985 and $39,000. Should any of these be reviewed as possible outliers?

36. Use the salary data in Exercise 35 and Chebyshev's theorem to find the percentage of elementary school teachers that must have salaries in the following ranges:
   a. $19,085 to $37,085   **b.** $14,585 to $41,585
   c. Repeat (a) and (b) if it can be assumed that the distribution of teacher salaries is approximately bell-shaped.

37. The IQ scores and birth rates were discussed in an article in the *Atlantic Monthly,* May 1989. IQ scores have a bell-shaped distribution with a mean of 100 and a standard deviation of 15.
   a. What percentage of the population should have an IQ score between 85 and 115?
   b. What percentage of the population should have an IQ score between 70 and 130?
   c. What percentage of the population should have an IQ score of more than 130?
   d. A person with an IQ score of more than 145 is considered a genius. Does the empirical rule support this statement? Explain.

38. The average fuel economy of new cars sold in the United States is 27.5 miles per gallon (*The Wall Street Journal,* April 8, 1992). Assume that the standard deviation is 3.5 miles per gallon.
   a. Use Chebyshev's theorem to calculate the percentage of new cars sold with miles-per-gallon ratings between 20.5 and 34.5 miles per gallon; between 18.75 and 36.25 miles per gallon; and between 17 and 38 miles per gallon.
   b. If it were reasonable to assume miles-per-gallon ratings for new cars followed a bell-shaped distribution, what can be said about the percentage of new cars sold with miles-per-gallon ratings between 20.5 and 34.5 miles per gallon? Between 17 and 38 miles per gallon?

39. Cruise ship inspections (*St. Petersburg Times,* December 16, 1990) are performed by the (federal) Center for Environmental Health and Injury Control. General areas of inspection include potable water, food preparation and holding, general cleanliness, and storage. Ships scoring 85 or less are reinspected more quickly than those with higher ratings. Inspection scores for 20 cruise ships are as follows:

Ship	Score	Ship	Score
Americana	98	Regent Sun	91
Costa Riviera	89	Royal Princess	87
Crown Princess	87	Seaward	94
Daphne	78	Song of America	96
Dolphin IV	95	Song of Norway	91
Fair Princess	76	Starship Atlantic	93
Jubilee	93	Starship Oceanic	97
Meridian	92	Sun Viking	88
Nordic Prince	93	Tropicana	91
Pegasus	62	Viking Princess	86

   a. Compute the mean and median.
   b. Compute the first and third quartiles.
   c. Compute the standard deviation.
   d. What are the z-scores associated with the Fair Princess and Pegasus? What is your interpretation of these values?
   e. Are there any outliers? Explain.

3.4   Exploratory Data Analysis

In Chapter 2 we introduced exploratory data analysis. Recall that the focus of exploratory data analysis is on using simple arithmetic and easy-to-draw pictures to summarize data.

In this section we continue our introduction of exploratory data analysis by considering five-number summaries and box plots.

Five-Number Summary

In a five-number summary, the following five numbers are used to summarize the data:

1. Smallest value
2. First quartile (Q_1)
3. Median
4. Third quartile (Q_3)
5. Largest value

The monthly starting salaries for a sample of 12 business school graduates were shown in Table 3.1. These data are as follows:

2050	2150	2250	2080	1955	1910
2090	2330	2140	2525	2120	2080

The median of 2105 and the quartiles $Q_1 = 2065$ and $Q_3 = 2200$ were computed in Section 3.1. Reviewing the preceding data shows a smallest value of 1910 and a largest value of 2525. Thus the five-number summary for the salary data is

$$1910, 2065, 2105, 2200, 2525.$$

Approximately one-fourth, or 25%, of the data values are between adjacent numbers in a five-number summary.

Box Plot

The *box plot* is a relatively recent development in the area of graphical summaries of data. Key to the development of a box plot is the computation of the median and the quartiles, Q_1 and Q_3. The interquartile range, IQR $= Q_3 - Q_1$, is also used. Figure 3.4 shows the box plot for the monthly starting salary data. The steps used to construct the box plot are as follows:

1. A box is drawn with the ends of the box located at the first and third quartiles. For the salary data, $Q_1 = 2065$ and $Q_3 = 2200$. This box contains the middle 50% of the data.

FIGURE 3.4
Box Plot of the Monthly Starting Salaries of Business School Graduates with Lines Showing the Inner and Outer Fences

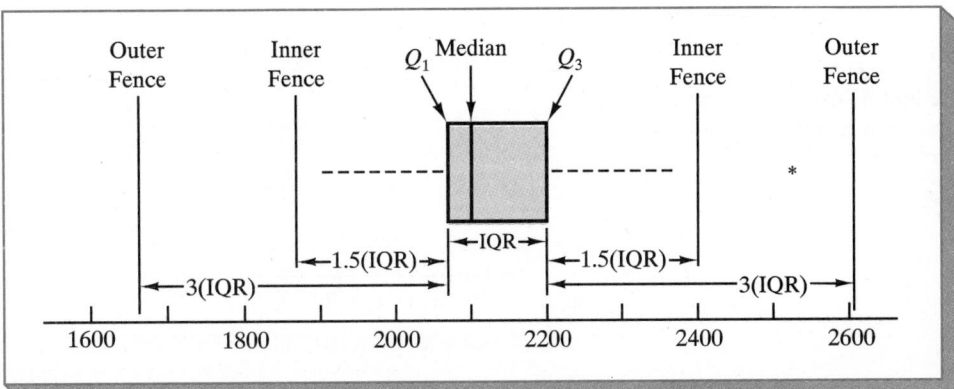

2. A vertical line is drawn in the box at the location of the median (2105 for the salary data). Thus the median line divides the data into two equal parts.

3. Using the interquartile range, IQR = $Q_3 - Q_1$, fences are located. The *inner fences* are located 1.5(IQR) below Q_1 and 1.5(IQR) above Q_3. The *outer fences* are located 3(IQR) below Q_1 and 3(IQR) above Q_3. For the salary data, IQR = $Q_3 - Q_1 = 2200 - 2065 = 135$. Thus, the inner fences are $2065 - 1.5(135) = 1862.5$ and $2200 + 1.5(135) = 2402.5$. The outer fences are $2065 - 3(135) = 1660$ and $2200 + 3(135) = 2605$. The fences are important aids in identifying outliers. Data falling between the inner and outer fences are considered *mild outliers*. Data falling outside the outer fences are considered *extreme outliers*.

4. The dashed lines in Figure 3.4 are called *whiskers*. The whiskers are drawn from the ends of the box to the smallest and largest data values *inside the inner fences*. Thus the whiskers end at salary data values of 1910 and 2330.

5. Finally, the location of mild outliers are shown with the symbol * and extreme outliers are shown with the symbol °. In Figure 3.4 we see there is one mild outlier—the data value 2525. There are no extreme outliers in the salary data.

In Figure 3.4 we have included lines showing the location of the fences. These lines were drawn to show how fences are computed and where they are located for the salary data. Although the fences are always computed, generally they are not drawn on the box plots. Figure 3.5 shows the usual appearance of a box plot for the salary data.

NOTES & COMMENTS

1. When using fences to identify outliers, we may or may not select the same items as when using z-scores less than -3 and greater than $+3$ to identify outliers. However, the objective of both approaches is simply to identify items that should be reviewed to ensure the validity of the data. Thus, outliers identified using either procedure should be reviewed.

2. An advantage of the exploratory data analysis procedures is that they are easy to use; few numerical calculations are necessary. We simply need to put the items in ascending order and identify the median and quartiles Q_1 and Q_3 in order to obtain the five-number summary. The fences and the box plot can then easily be determined. It is not necessary to compute the mean and the standard deviation for the data.

FIGURE 3.5

Box Plot of the Monthly Starting Salaries of Business School Graduates

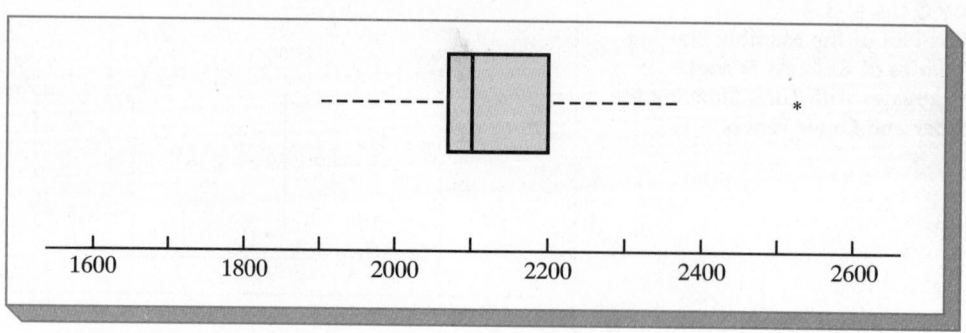

☐ ☐ **Exercises**

Methods

40. Consider the sample of size 8 with data values as follows:

27, 25, 20, 15, 30, 34, 28, 25.

Provide the five-number summary for this data.

41. Show the box plot for the data in Exercise 40.

42. Show the five-number summary and the box plot for the following data:

5, 15, 18, 10, 8, 12, 16, 10, 6.

43. A data set has a first quartile of 42 and a third quartile of 50. Compute the inner and outer fences. Should a data value of 65 be considered an outlier?

Applications

44. *Fortune* (April 20, 1992) published its annual report on the 500 largest U.S. industrial corporations. Data on the percent growth in sales for the past 12 months for a sample of 26 companies are as follows:

16.1	49.9	23.7	15.6	1.9	10.8	20.4
12.2	22.4	4.9	13.4	6.8	15.8	12.1
19.3	10.0	46.1	27.0	7.0	12.5	6.1
15.6	6.3	16.7	10.1	55.9		

a. Provide a five-number summary.
b. Compute the inner and outer fences.
c. Do there appear to be outliers? How would this information be helpful to a financial analyst?
d. Show a box plot.

45. Annual sales in millions of dollars for 17 companies in the chemical industry are as follows (*Forbes*, January 9, 1989):

484	2,731	598	2,472	3,261	1,220
4,514	32,249	15,980	8,030	3,122	8,258
1,061	2,188	2,366	1,049	636	

a. Provide a five-number summary of these data.
b. Compute the inner and outer fences.
c. Do there appear to be outliers? What does outlier information tell you in this case?
   Note: Companies associated with some of the preceding data are as follows: Valspar (484), Dow Chemical (15,980), and Du Pont (32,249).
d. Show a box plot.

46. *Consumer Reports* provides performance and quality ratings for numerous consumer products. The overall ratings were provided for a sample of 16 midpriced VCRs in the *Consumer Reports 1992 Buying Guide*. The manufacturer brands and overall scores are shown in Table 3.5.
a. Provide the mean and median overall rating.
b. Compute the first and third quartiles.
c. Provide the five-number summary.
d. Similar ratings for camcorders showed a mean of 82.56, standard deviation of 6.39, and a five-number summary of 75, 77, 82, 86, 93. Compare the *Consumer Reports* ratings data for VCRs and camcorders. Show the box plots for both.
e. Are there any outliers in the VCR data? Explain.

TABLE 3.5

Manufacturer	Score
Fisher	77
General Electric	81
Hitachi	89
J. C. Penney	78
JVC	79
Magnavox	80
Montgomery Ward	78
Mitsubishi	90
Panasonic	77
Phillips	73
Quasar	72
Radio Shack	76
RCA	79
Sanyo	75
Sony	86
Toshiba	79

INJURY

47. The Highway Loss Data Institute "Injury and Collision Loss Experience" (September 1988) rates car models based on the number of insurance claims filed after accidents. Index ratings near 100 are considered average. Lower ratings are better, and the car model is considered safer. Shown are ratings for 20 midsize cars and 20 small cars.

Midsize Cars:	81	91	93	127	68	81	60	51	58	75
	100	103	119	82	128	76	68	81	91	82
Small Cars:	73	100	127	100	124	103	119	108	109	113
	108	118	103	120	102	122	96	133	80	140

Summarize the data for the midsize and small cars separately.

a. Provide a five-number summary for midsize cars and for small cars.

b. Show the box plots.

c. Make a statement about what your summaries indicate about the safety of midsize cars compared to small cars.

WORLD

48. Morgan Stanley Capital International in Geneva, Switzerland, provided percent changes in stock markets around the world (*Barrons*, March 30, 1992). Data shown are percentage changes over the preceding 1-year period.

Country	Percent Change	Country	Percent Change
Australia	29.1	Japan	8.3
Austria	−13.4	Luxembourg	6.7
Belgium	9.3	Netherlands	13.9
Canada	8.3	New Zealand	12.6
Denmark	15.3	Norway	−15.4
Europe	10.0	Pacific	10.3
Finland	−16.7	Portugal	−7.2
France	15.8	Singapore	22.7
Germany	6.3	Spain	11.6
Hong Kong	42.8	Sweden	9.1
Italy	−4.1	United Kingdom	11.6
Ireland	9.6	United States	27.2

a. What are the mean and median percent changes among these world markets?

b. What are the first and third quartiles?

c. Are there any outliers? Show a box plot.

d. What percentile would you report for the United States?

3.5   The Role of the Computer

In this chapter we have introduced more than 10 numerical measures of descriptive statistics that can be used to summarize and interpret data. When the size of a data set increases, the computational task of computing these numerical measures can become overwhelming. Fortunately, statistical software packages are available to do the computations. After the user has entered the data, a few simple commands instruct the computer to generate the desired descriptive statistics. In Chapter 2, we described a session with the Minitab computer system, showing how frequency distributions, histograms, and dot plots could be quickly, easily, and accurately developed. In this section, we show how Minitab can be used to develop additional descriptive statistics, including a box plot.

FIGURE 3.6   **Descriptive Statistics for Starting Salary Data Using Minitab**

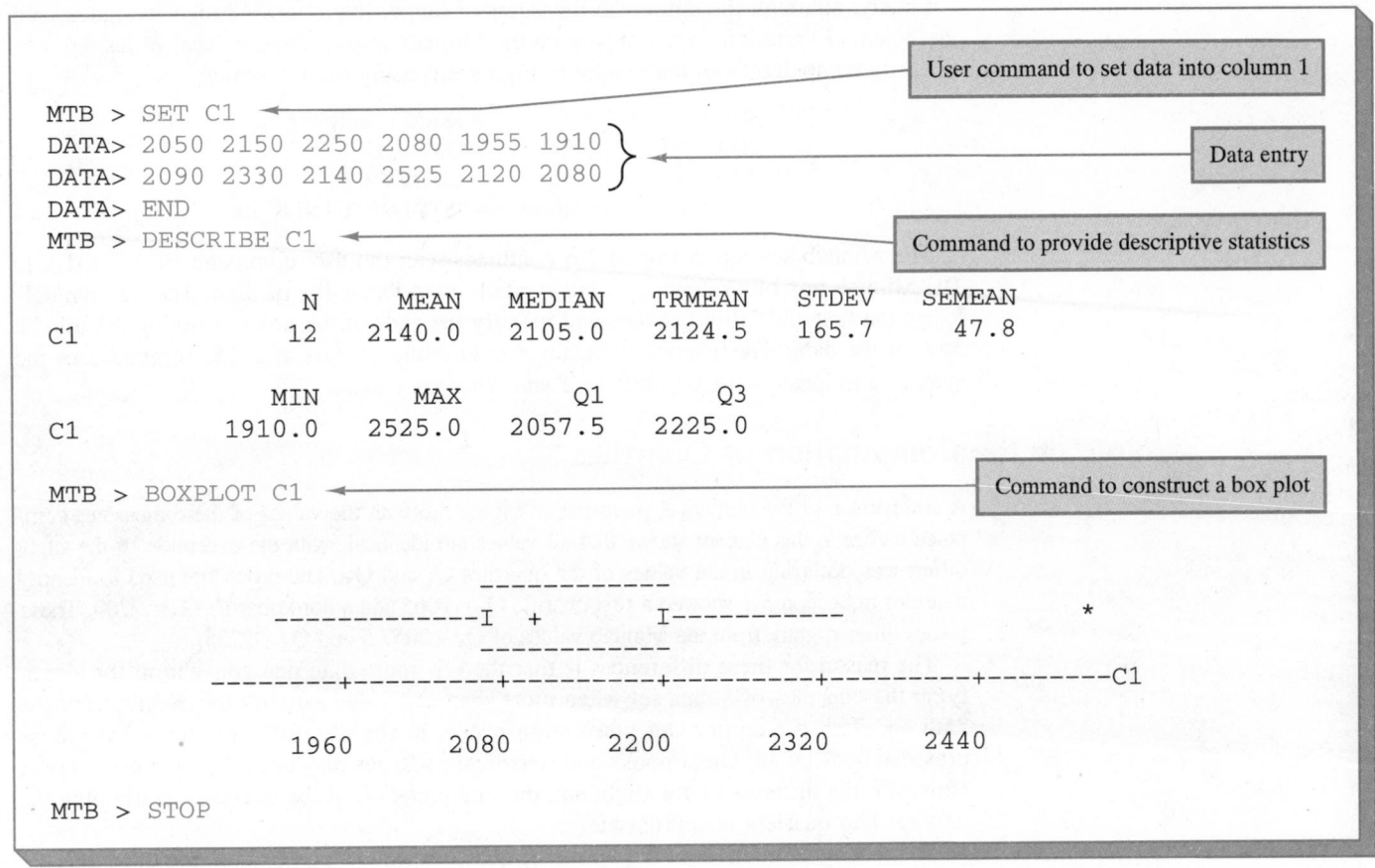

Consider the Minitab session shown by the computer printout in Figure 3.6. The starting salary data for 12 business school students is used in the example. At the Minitab prompt, MTB >, the user selects the Minitab command SET C1 to enter the data into column 1 of the Minitab worksheet. With the SET command, the user may enter any number of values on a single line, provided the individual values are separated by one or more spaces. The command END indicates that the data input process is complete.

The command DESCRIBE C1 instructs Minitab to develop the numerical measures of descriptive statistics as shown in Figure 3.6. These measures are defined as follows:

N	the number of data values
MEAN	the mean
MEDIAN	the median
TRMEAN	the 5% trimmed mean
STDEV	the standard deviation
MIN	the minimum data value
MAX	the maximum data value
Q1	the first quartile
Q3	the third quartile

We have not discussed the numerical measure label SEMEAN, which refers to the standard error of the mean. This measure is computed by dividing the stardard deviation by the square root of N. The interpretation and use of this measure are discussed in Chapter 7 when we introduce the topics of sampling and sampling distributions.

Finally, although the numerical measures of range, interquartile range, variance, and coefficient of variation do not appear on the Minitab output, these values, if desired, can be easily computed from the results in Figure 3.6 using the following:

Range	= MAX − MIN
IQR	= Q3 − Q1
Variance	= (STDEV)^2
Coefficient of Variation	= (STDEV/MEAN) × 100

The Minitab session in Figure 3.6 continues with the user command BOXPLOT C1. The Minitab box plot output uses the symbol + to locate the median. The "I" symbols locate the first and third quartiles and identify the ends of the box containing the middle 50% of the data. The asterisk indicates a mild outlier occurs at 2525. Minitab uses the symbol o to locate extreme outliers if and when they exist.

A Note on the Computation of Quartiles

A comparison of the numerical measures in Figure 3.6 with the values of these measures computed earlier in this chapter shows that all values are identical, with the exception of the slight differences occurring in the values of the quartiles Q_1 and Q_3. The procedure used to identify quartiles in Section 3.1 showed a first quartile $Q_1 = 2065$ and a third quartile $Q_3 = 2200$. These values differ slightly from the Minitab values of $Q_1 = 2057.5$ and $Q_3 = 2225$.

The reason for these differences is that there is more than one convention for identifying the quartiles of a data set when more than one value satisfies the definition of the 25th and 75th percentile. Our convention results in slightly different values than those provided by Minitab. Other books and statistical packages may use different conventions. However, the differences are slight and the interpretation of the quartiles as dividing the data set into quarters is appropriate.

To see why differences in the values of quartiles can occur, consider the third quartile for the starting-salary data set. The ordered data and the location of the third quartile are as follows:

1910   1955   2050   2080   2080   2090   2120   2140   2150   2250   2330   2525

Q_3 is located between 2150 and 2250.

The definition of the third quartile, Q_3, indicates that 75% of the data, or 9 data values, are less than or equal to Q_3 and that 25% of the data, or 3 data values, are greater than or equal to Q_3. Thus, in the preceding diagram we see that *any value from* 2150 *to* 2250 correctly identifies Q_3. Which value from 2150 to 2250 should be reported as the value of Q_3? Our computational procedure in Section 3.1 placed Q_3 halfway between 2150 and 2250, identifying Q_3 as 2200. The Minitab procedure for the third quartile, or 75th percentile, uses the following position-location formula:

$$\frac{(P)}{100}(n + 1) = \frac{75}{100}(12 + 1) = 9.75$$

To find the value associated with position location 9.75, Minitab interpolates between 2150 and 2250 using .75 to indicate the interpolated value is .75, or 3/4, of the way between 2150 and 2250. Thus, the value $Q_3 = 2225$ is reported. Note that both 2200 and 2225 provide the correct third-quartile interpretation, with 75% of the data below these values and 25% of the data above these values.

3.6   Computing Measures of Location and Dispersion for Grouped Data (Optional)

In most cases, measures of central location and dispersion are computed using the individual data values. However, sometimes we only have data in a grouped or frequency distribution form. This section describes how approximations of the mean, variance, and standard deviation can be obtained directly from a frequency distribution.

Mean

Recall that in order to compute the sample mean using the individual data values, we simply sum all the values and divide by n, the sample size. If the data are available only in frequency distribution form, we will have to approximate the sum of the data values.

To do this, we treat the midpoint of each class as if it were the mean of the items in the class. Let M_i denote the midpoint for class i and f_i denote the frequency of class i. Then an approximation to the sum of the items in class i is given by $f_i M_i$. Summing these values over all classes, we obtain $\Sigma f_i M_i$, which approximates the sum of all the data values.

Once this approximation of the sum of all the data values is obtained, an approximation of the mean is computed by dividing this sum by the total number of data items. The following formula can be used to compute the sample mean for grouped data.

Sample Mean for Grouped Data

$$\overline{x} = \frac{\Sigma f_i M_i}{n} \qquad (3.10)$$

We do not expect the calculations based on the ungrouped data and grouped data to provide exactly the same numerical result. Thus $\overline{x}$ calculated using (3.10) is an approximation of $\overline{x}$ calculated when all data values are available. The difference between the two values of $\overline{x}$ is known as *grouping error*.

In Section 2.2 we provided a frequency distribution of the time in days required to complete year-end audits for the public accounting firm of Sanderson and Clifford. The frequency distribution of audit times based on a sample of 20 clients was developed. This frequency distribution is shown again in Table 3.6.

What is the sample mean audit time based on the grouped data shown in Table 3.6? Recall that the class midpoints, M_i, are located halfway between the class limits. Thus, the first class of 10–14 has a midpoint located at $(10 + 14)/2 = 12$. Since this class has a frequency $f_i = 4$, the value $f_i M_i = 4 \times 12 = 48$. The five class midpoints and the computations necessary to determine the sample mean using (3.10) are shown in Table 3.7. As can be seen, the sample mean audit time is 19 days.

TABLE 3.6
Frequency Distribution of Audit Times

Audit Time (Days)	Frequency
10–14	4
15–19	8
20–24	5
25–29	2
30–34	1
Total	20

Variance

The approach to compute the variance for grouped data is to use a slightly altered version of the formula for the variance provided in Equation (3.5). In (3.5), the squared deviations of the data values about the sample mean $\overline{x}$ were written $(x_i - \overline{x})^2$. However, with grouped

TABLE 3.7

Computation of the Sample Mean Audit Time for Grouped Data

Audit Time (Days)	Frequency f_i	Class Midpoint M_i	f_iM_i
10–14	4	12	48
15–19	8	17	136
20–24	5	22	110
25–29	2	27	54
30–34	1	32	32
	20		380
			Σf_iM_i

$$\text{Sample mean } \bar{x} = \frac{\Sigma f_iM_i}{n} = \frac{380}{20} = 19 \text{ days}$$

data, the individual data values, x_i, are not known. In this case, we treat the class midpoint, M_i, as being a representative value for the x_i values in the corresponding class. Thus the squared deviations about the sample mean, $(x_i - \bar{x})^2$, are replaced by $(M_i - \bar{x})^2$. Then, just as we did with the sample mean calculations for grouped data, we weight each value by the frequency of the class, f_i, and sum for all classes. The sum of the squared deviations about the mean for all the data is approximated by $\Sigma f_i(M_i - \bar{x})^2$. Using this approach, the formula for the sample variance for grouped data is

Sample Variance for Grouped Data

$$s^2 = \frac{\Sigma f_i(M_i - \bar{x})^2}{n - 1} \tag{3.11}$$

The calculation of the sample variance for audit times based on the grouped data from Table 3.6 is shown in Table 3.8.

TABLE 3.8

Computation of the Sample Variance of Audit Times for Grouped Data (Sample Mean $\bar{x} = 19$)

Audit Time (Days)	Frequency f_i	Class Midpoint M_i	Deviation $(M_i - \bar{x})$	Squared Deviation $(M_i - \bar{x})^2$	$f_i(M_i - \bar{x})^2$
10–14	4	12	−7	49	196
15–19	8	17	−2	4	32
20–24	5	22	3	9	45
25–29	2	27	8	64	128
30–34	1	32	13	169	169
	20				570
					$\Sigma f_i(M_i - \bar{x})^2$

$$\text{Sample variance } s^2 = \frac{\Sigma f_i(M_i - \bar{x})^2}{n - 1} = \frac{570}{19} = 30$$

Standard Deviation

The standard deviation for grouped data is simply the square root of the variance for grouped data. For the audit-time data, the sample standard deviation is $s = \sqrt{30} = 5.48$.

Population Summaries for Grouped Data

Before closing this section on computing measures of location and dispersion for grouped data, we note that the formulas in this section were presented only for data sets constituting a sample. Population summary measures are computed in a similar manner. The grouped data formulas for a population mean and variance are as follows.

Population Mean for Grouped Data

$$\mu = \frac{\Sigma f_i M_i}{N} \tag{3.12}$$

Population Variance for Grouped Data

$$\sigma^2 = \frac{\Sigma f_i (M_i - \mu)^2}{N} \tag{3.13}$$

NOTES & COMMENTS

An alternative formula for the computation of the sample variance for grouped data is as follows:

$$s^2 = \frac{\Sigma f_i M_i^2 - n \bar{x}^2}{n - 1}$$

where $\Sigma f_i M_i^2 = f_1 M_1^2 + f_2 M_2^2 + \cdots + f_k M_k^2$ and k is the number of classes used to group the data. Using this formula may ease the computations slightly.

Exercises

Methods

SELF TEST ▶ Consider the sample data in the following frequency distribution:

Class	Midpoint	Frequency
3–7	5	4
8–12	10	7
13–17	15	9
18–22	20	5

Compute the sample mean.

SELF TEST ▶ 50. Compute the sample variance and sample standard deviation for the grouped data in Exercise 49.

Applications

51. The *Journal of Personal Selling and Sales Management* (August 1988) reported a study investigating the use of a persuasion technique in selling. A sample of 56 participants was used in the study. Assume that the frequency distribution in Table 3.9 shows the number of sales presentations made per week by the 56 participants.
a. What is the mean number of sales presentations made per week for participants in this study?
b. What are the variance and the standard deviation?

52. The following frequency distribution for the first examination in operations management was posted on the department bulletin board.

Examination Grade	Frequency
40–49	3
50–59	5
60–69	11
70–79	22
80–89	15
90–99	6
Total	62

Treating these data as a sample, compute the mean, variance, and standard deviation.

53. *Psychology Today* (November 1988) reported on the characteristics of individuals caught in the act of shoplifting in a supermarket. Data from 9832 cases showed 56% of the shoplifters were male and 44% were female. The value of most items stolen was in the $1 to $5 range. The frequency distribution in Table 3.10 shows the age of the shoplifters.
a. What is the sample mean age of supermarket shoplifters?
b. What are the variance and standard deviation of the ages?

54. A service station has recorded the following frequency distribution for the number of gallons of gasoline sold per car in a sample of 680 cars.

Gasoline (gallons)	Frequency
0–4	74
5–9	192
10–14	280
15–19	105
20–24	23
25–29	6
Total	680

Compute the mean, variance, and standard deviation for these grouped data. If the service station expects to service about 120 cars on a given day, what is an estimate of the total number of gallons of gasoline that will be sold?

55. Scores obtained by a sample of patients on a depression level test are summarized in the frequency distribution in Table 3.11. Using the given grouped data, compute the following.
a. mean   b. variance   c. standard deviation

TABLE 3.9

Number of Presentations	Frequency
10–12	5
13–15	9
16–18	22
19–21	12
22–24	8
Total	56

TABLE 3.10

Age of Shoplifter	Frequency
6–11	364
12–17	1249
18–29	3392
30–59	3962
60–80	865
Total	9832

TABLE 3.11

Depression Level Score	Frequency
25–34	3
35–44	1
45–54	2
55–64	6
65–74	4
75–84	6
85–94	2
95–104	1
Total	25

Summary

In this chapter we introduced several statistical measures that can be used to describe the location and dispersion of data. Unlike the tabular and graphical procedures for summarizing data, the measures introduced in this chapter summarize the data in terms of numerical values. When the numerical values obtained are for a sample, they are called sample statistics. When the numerical values obtained are for a population, they are called population parameters. Some of the notation used for sample statistics and population parameters are summarized as follows.

	Sample Statistic	Population Parameter
Mean	$\bar{x}$	μ
Variance	s^2	σ^2
Standard deviation	s	σ

As measures of location, we defined the mean, median, and mode for both sample and population data. Then the concept of a percentile was used to describe the location of other values in the data. Next, we presented the range, interquartile range, variance, standard deviation, and coefficient of variation as statistical measures of variability or dispersion.

A discussion of two exploratory data analysis techniques that can be used to summarize data more effectively was included in Section 3.4. Specifically, we showed how to develop a five-number summary and a box plot in order to provide simultaneous information about the location, dispersion, and shape of the distribution. A session with the software package Minitab was used to illustrate how statistical computing systems can support the analysis and interpretation of data. Finally, we described how the mean, variance, and standard deviation could be computed for grouped data. However, we recommend using the measures based on the individual data values unless the grouped format is the only manner in which the data are available.

Glossary

Population parameter A numerical value used as a summary measure for a population of data (e.g., the population mean, μ, the population variance, σ^2, and the population standard deviation, σ).

Sample statistic A numerical value used as a summary measure for a sample (e.g., the sample mean, $\bar{x}$, the sample variance, s^2, and the sample standard deviation, s).

Mean A measure of central location for a data set. It is computed by summing all the data values and dividing by the number of items.

Trimmed mean The mean of the data remaining after α percent of the smallest and α percent of the largest items have been removed. The purpose of a trimmed mean is to provide a measure of central location that has eliminated the effect of extremely large and extremely small data values.

Median A measure of central location. It is the value that splits the data into two equal groups—one with values greater than or equal to the median and one with values less than or equal to the median.

Mode A measure of location, defined as the most frequently occurring data value.

Percentile A value such that at least p percent of the items are less than or equal to this value and at least $(100 - p)$ percent of the items are greater than or equal to this value. The 50th percentile is the median.

Quartiles The 25th, 50th, and 75th percentiles referred to as the first quartile, the second quartile (median), and third quartile, respectively. The quartiles can be used to divide the data set into four parts, with each part containing approximately 25% of the data.

Hinges The value of the lower hinge is approximately the first quartile, or 25th percentile. The value of the upper hinge is approximately the third quartile, or 75th percentile. The values of the hinges and quartiles may differ slightly due to differing computational conventions.

Range A measure of dispersion, defined to be the difference between the largest and smallest data values.

Interquartile range (IQR) A measure of dispersion, defined to be the difference between the third and first quartiles.

Variance A measure of dispersion for a data set based on the squared deviations of the data values about the mean.

Standard deviation A measure of dispersion for a data set, found by taking the positive square root of the variance.

Coefficient of variation A measure of relative dispersion for a data set, found by dividing the standard deviation by the mean and multiplying by 100.

z-Score For each data item, a value found by dividing the deviation about the mean $(x_i - \bar{x})$ by the standard deviation s. A z-score is referred to as a standardized value and denotes the number of standard deviations a data value x_i is from the mean.

Chebyshev's theorem A theorem applying to any data set that can be used to make statements about the percentage of items that must be within a specified number of standard deviations of the mean.

Empirical rule A rule that states the percentages of items that are within one, two, and three standard deviations from the mean for mound-shaped, or bell-shaped, distributions.

Outlier An unusually small or unusually large data value.

Five-number summary An exploratory data analysis technique that uses the following five numbers to summarize the data set: smallest value, first quartile, median, third quartile and largest value.

Box plot A graphical summary of data. A box, drawn from the first to the third quartiles, shows the location of the middle 50% of the data. Dashed lines, called whiskers, extending from the ends of the box, show the location of data greater than the third quartile and data less than the first quartile. The locations of any outliers are also noted.

Fences Values used to identify outliers. Inner fences are located 1.5(IQR) below the first quartile and 1.5(IQR) above the third quartile. Outer fences are located 3(IQR) below the first quartile and 3(IQR) above the third quartile. Data falling between the inner and outer fences are considered mild outliers. Data falling outside the outer fences are considered extreme outliers.

Grouped data Data available in class intervals as summarized by a frequency distribution. Individual values of the original data are not recorded.

Key Formulas

Sample Mean

$$\bar{x} = \frac{\sum x_i}{n}$$

(3.1)

Population Mean

$$\mu = \frac{\sum x_i}{N}$$

(3.2)

Interquartile Range

$$IQR = Q_3 - Q_1$$

(3.3)

Population Variance

$$\sigma^2 = \frac{\sum (x_i - \mu)^2}{N}$$

(3.4)

Sample Variance

$$s^2 = \frac{\Sigma(x_i - \overline{x})^2}{n - 1} \qquad (3.5)$$

Standard Deviation

$$\text{Sample Standard Deviation} = s = \sqrt{s^2} \qquad (3.6)$$

$$\text{Population Standard Deviation} = \sigma = \sqrt{\sigma^2} \qquad (3.7)$$

Coefficient of Variation

$$\left(\frac{\text{Standard Deviation}}{\text{Mean}}\right) \times 100 \qquad (3.8)$$

z-Score

$$z_i = \frac{x_i - \overline{x}}{s} \qquad (3.9)$$

Sample Mean for Grouped Data

$$\overline{x} = \frac{\Sigma f_i M_i}{n} \qquad (3.10)$$

Sample Variance for Grouped Data

$$s^2 = \frac{\Sigma f_i (M_i - \overline{x})^2}{n - 1} \qquad (3.11)$$

Population Mean for Grouped Data

$$\mu = \frac{\Sigma f_i M_i}{N} \qquad (3.12)$$

Population Variance for Grouped Data

$$\sigma^2 = \frac{\Sigma f_i (M_i - \mu)^2}{N} \qquad (3.13)$$

❑ ❑ Supplementary Exercises

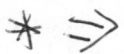 

PRICES

56. A sample of economists predicted what would happen to consumer prices during 1992 (*Business Week,* December 30, 1991). Their predictions about the percentage increase in consumer prices were as follows

4.6	3.4	2.3	2.9	2.7	3.2	2.9
3.7	3.7	3.3	3.7	2.9	3.3	2.9
3.6	3.7	3.3	3.5	3.4	2.7	3.3
2.9	3.0	3.0	1.8			

a. Compute the mean, median, and mode.
b. Compute the first and third quartiles.
c. Compute the range and interquartile range.
d. Compute the variance and standard deviation.
e. Are there any outliers?

57. In 1989, a sample of six home mortgage loans showed the following interest rates:

$$12.5, 13.2, 11.2, 13.0, 12.0, 12.5.$$

Compute the following descriptive statistics for the data set.
a. mean   **b.** median   **c.** mode
d. 25th percentile   **e.** range
f. interquartile range   **g.** variance
h. standard deviation   **i.** coefficient of variation

 58. *Time* (January 9, 1989) published an article on the academic ability of college athletes. The article noted that some of the most successful athletic programs (citing the University of Notre Dame and Duke University) have athletes with very good college board scores. Assume that the following sample data are typical of college board scores for Notre Dame football players:

$$1100, 970, 1000, 1250, 880, 790, 1300, 1050, 900, 950, 1120.$$

a. Compute the mean, median, and mode.
b. Compute the range and interquartile range.
c. Compute the variance and standard deviation.
d. Using *z*-scores, state whether or not there are any outliers in this data set.

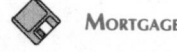 MORTGAGE

59. The following data show home mortgage loan amounts handled by one loan officer at the Westwood Savings and Loan Association. Data are in thousands of dollars.

20.0	38.5	33.0	27.5	34.0	12.5	25.9	43.2	37.5	36.2
25.2	30.9	23.8	28.4	13.0	31.0	33.5	25.4	33.5	20.2
39.0	38.1	30.5	45.5	30.5	52.0	40.5	51.6	42.5	44.8

a. Find the mean, median, and mode.
b. Find the first and third quartiles.

60. *Newsweek* (January 9, 1989) reported statistics on the number of visits a couple makes to a therapist while undergoing marriage counseling. With data based on this article, assume that the following show the number of visits for a sample of 9 couples:

$$12, 8, 3, 13, 18, 20, 10, 9, 18.$$

Compute the following descriptive statistics for this data.
a. mean   **b.** median   **c.** mode   **d.** 40th percentile
e. range   **f.** variance   **g.** standard deviation

61. A sample of 10 stocks on the New York Stock Exchange (May 22, 1992) shows the following price-earnings ratios:

$$9, 4, 6, 7, 3, 11, 4, 6, 4, 7.$$

Using these data, compute the mean, median, mode, range, variance, and standard deviation.

62. The Hawaii Visitors' Bureau reports visitors coming to Hawaii from the U.S. mainland spend an average of $102 per day (*St. Petersburg Times,* December 11, 1988). Assume the standard deviation is $27 and the distribution of expenditures is approximately bell-shaped. Answer the following:
a. What percentage of visitors will have an average daily expenditure of between $75 and $129 per day?
b. What percentage of visitors will have an average daily expenditure of between $48 and $156 per day?
c. Should an average daily expenditure of $225 per day be considered an outlier?

63. The cost of new homes in selected cities throughout the United States was reported in *U.S. News & World Report* (April 6, 1992). Assume the cost of homes in Chicago, Illinois, had a mean of $100,000 and a standard deviation of $40,000.

a. Should a home selling for $200,000 be considered an outlier? Explain.

b. Use Chebyshev's theorem to determine the percentage of homes selling between $40,000 and $160,000.

64. Public transportation and an automobile are two methods an employee has of getting to work each day. Samples of times recorded for each method are shown. Times are in minutes.

$$\textit{Public Transportation:} \quad 28, 29, 32, 37, 33, 25, 29, 32, 41, 34$$
$$\textit{Automobile:} \quad 29, 31, 33, 32, 34, 30, 31, 32, 35, 33$$

a. Compute the sample mean time to get to work for each method.

b. Compute the sample standard deviation for each method.

c. Based on your results from (a) and (b), which method of transportation should be preferred? Explain.

d. Develop a box plot for each method. Does a comparison of the box plots support your conclusion in (c)?

 EXAM

65. Final examination scores for 25 statistics students are as follows

56	77	84	82	42	61	44	95	98	84
93	62	96	78	88	58	62	79	85	89
89	97	53	76	75

a. Provide a five-number summary.

b. Provide a box plot.

66. The following data show the total yardage accumulated during the NCAA college football season for a sample of 20 receivers.

| 744 | 652 | 576 | 1112 | 971 | 451 | 1023 | 852 | 809 | 596 |
| 941 | 975 | 400 | 711 | 1174 | 1278 | 820 | 511 | 907 | 1251 |

a. Provide a five-number summary.

b. Provide a box plot.

c. Identify any outliers.

67. A frequency distribution for the duration of 20 long-distance telephone calls (rounded to the nearest minute) is shown in Table 3.12. Compute the mean, variance, and standard deviation for the data.

68. Dinner check amounts at La Maison French Restaurant have the following frequency distribution:

TABLE **3.12**

Call Duration	Frequency
4–7	4
8–11	5
12–15	7
16–19	2
20–23	1
24–27	1
Total	20

Dinner Check (Dollars)	Frequency
25–34	2
35–44	6
45–54	4
55–64	4
65–74	2
75–84	2
Total	20

Compute the mean, variance, and standard deviation for the given data.

69. Automobiles traveling on the New York State Thruway are checked for speed by a state police radar system. A frequency distribution of speeds is shown:

Speed (Miles per Hour)	Frequency
45–49	10
50–54	40
55–59	150
60–64	175
65–69	75
70–74	15
75–79	10
Total	475

a. What is the mean speed of the automobiles traveling on the New York State Thruway?

b. Compute the variance and the standard deviation.

 DUKE

70. In the 1992 NCAA Division I Basketball Championships the Duke Blue Devils became the first team since the 1973 UCLA Bruins to win back-to-back national championships. Duke's season record was 34–2, with its only losses coming at North Carolina and Wake Forest. The scores of the 36 games, with Duke's score listed first, follow (*NCAA Final Four Program*, April 1992).

Opponent	Score	Opponent	Score
East Carolina	103–75	Louisiana State	77–67
Harvard	118–65	Georgia Tech	71–62
St. John's	91–81	North Carolina State	71–63
Canisius	96–60	Maryland	91–89
Michigan	88–85	Wake Forest	68–72
William & Mary	97–61	Virginia	76–67
Virginia	68–62	UCLA	75–65
Florida State	86–70	Clemson	98–97
Maryland	83–66	North Carolina	89–77
Georgia Tech	97–84	Maryland	94–87
North Carolina State	110–75	Georgia Tech	89–76
N.C.—Charlotte	104–82	North Carolina	94–74
Boston University	95–85	Campbell	82–56
Wake Forest	84–68	Iowa	75–62
Clemson	112–73	Seton Hall	81–69
Florida State	75–62	Kentucky	104–103
Notre Dame	100–71	Indiana	81–78
North Carolina	73–75	Michigan	71–51

a. Compute the mean and median scores for Duke and its opponents.

b. Compute the range and interquartile range for Duke and its opponents.

c. Provide box plots for Duke and its opponents.

d. Comment on what you learned.

Computer Case 1   *Consolidated Foods, Inc.*

Consolidated Foods, Inc., operates a chain of supermarkets located in New Mexico, Arizona, and California. (See Computer Case, Chapter 2). Data in Table 3.13 show the dollar amounts and method of payment for a sample of 100 customers. Consolidated's management requested the

T A B L E 3.13	Cash	Personal Check	Credit Card	Cash	Personal Check	Credit Card

Purchase Amount and Method of Payment for a Random Sample of 100 Consolidated Foods' Customers

Cash	Personal Check	Credit Card	Cash	Personal Check	Credit Card
7.40	27.60	50.30	5.08	52.87	69.77
5.15	30.60	33.76	20.48	78.16	48.11
4.75	41.58	25.57	16.28	25.96	
15.10	36.09	46.24	15.57	31.07	
8.81	2.67	46.13	6.93	35.38	
1.85	34.67	14.44	7.17	58.11	
7.41	58.64	43.79	11.54	49.21	
11.77	57.59	19.78	13.09	31.74	
12.07	43.14	52.35	16.69	50.58	
9.00	21.11	52.63	7.02	59.78	
5.98	52.04	57.55	18.09	72.46	
7.88	18.77	27.66	2.44	37.94	
5.91	42.83	44.53	1.09	42.69	
3.65	55.40	26.91	2.96	41.10	
14.28	48.95	55.21	11.17	40.51	
1.27	36.48	54.19	16.38	37.20	
2.87	51.66	22.59	8.85	54.84	
4.34	28.58	53.32	7.22	58.75	
3.31	35.89	26.57		17.87	
15.07	39.55	27.89		69.22	

sample be taken in an attempt to learn about payment practices for the store's customers. In particular, management was interested in learning about how a new credit-card-payment option was related to the customers' purchase amounts.

Managerial Report

Use the methods of descriptive statistics presented in Chapter 3 to summarize the sample data. Provide summaries of the dollar purchase amounts for cash customers, personal-check customers, and credit-card customers separately. Your report should contain summaries and discussions such as the following:

a. a comparison and interpretation of means and medians
b. a comparison and interpretation of measures of dispersion such as the range and standard deviation
c. the identification and interpretation of the five-number summaries for each method of payment
d. box plots for each method of payment

CONSOLID

Use the summary section of your report to provide a discussion of what you have learned about the method of payment and the amounts of payments for Consolidated Foods' customers. The data set for this computer case is in the data file CONSOLID.

Computer Case 2   *National Health Care Association*

The National Health Care Association is concerned about the shortage of nurses the health care profession is projecting for the future. To learn the current degree of job satisfaction among nurses, the association has sponsored a study of hospital nurses throughout the country. As part of this study, a sample of 50 nurses were asked to indicate their degree of satisfaction in their work, their pay, and their opportunities for promotion. Each of the three aspects of satisfaction was measured on a scale from 0 to 100, with larger values indicating higher degrees of satisfaction. The data in Table 3.14 were collected and are available in the data set HEALTH-1.

HEALTH-1

TABLE 3.14
Work, Pay, and Promotion Job-Satisfaction Scores for a Sample of 50 Nurses

Work	Pay	Promotion	Work	Pay	Promotion
71	49	58	72	76	37
84	53	63	71	25	74
84	74	37	69	47	16
87	66	49	90	56	23
72	59	79	84	28	62
72	37	86	86	37	59
72	57	40	70	38	54
63	48	78	86	72	72
84	60	29	87	51	57
90	62	66	77	90	51
73	56	55	71	36	55
94	60	52	75	53	92
84	42	66	74	59	82
85	56	64	76	51	54
88	55	52	95	66	52
74	70	51	89	66	62
71	45	68	85	57	67
88	49	42	65	42	68
90	27	67	82	37	54
85	89	46	82	60	56
79	59	41	89	80	64
72	60	45	74	47	63
88	36	47	82	49	91
77	60	75	90	76	70
64	43	61	78	52	72

TABLE 3.15
Work, Pay, and Promotion Job-Satisfaction Scores for Nurses in Private, VA, and University Hospitals.

Private Hospitals			VA Hospitals			University Hospitals		
Work	Pay	Promotion	Work	Pay	Promotion	Work	Pay	Promotion
72	57	40	71	49	58	84	53	63
90	62	66	84	74	37	87	66	49
84	42	66	72	37	86	72	59	79
85	56	64	63	48	78	88	55	52
71	45	68	84	60	29	74	70	51
88	49	42	73	56	55	85	89	46
72	60	45	94	60	52	79	59	41
88	36	47	90	27	67	69	47	16
77	60	75	72	76	37	90	56	23
64	43	61	86	37	59	77	90	51
71	25	74	86	72	72	71	36	55
84	28	62	95	66	52	75	53	92
70	38	54	65	42	68	76	51	54
87	51	57	82	37	54	89	80	64
74	59	82	82	60	56			
89	66	62	90	76	70			
85	57	67	78	52	72			
74	47	63						
82	49	91						

HEALTH-2

In addition, the sample data in Table 3.14 were broken down by the types of hospitals employing the nurses. The types of hospitals considered were private (P), Veterans Administration (VA), and university (U). The data in Table 3.15 are available in the data set HEALTH-2.

Managerial Report

Use methods of descriptive statistics to summarize these data. Present the summaries that will be beneficial in communicating the results to others. Discuss your findings. Specifically, comment on the following:

1. Using the entire data set and the three job-satisfaction variables, what aspect of the job is most satisfying for the nurses? What appears to be the least satisfying? In what area(s), if any, do you feel improvements should be made? Discuss.

2. Using descriptive measures of dispersion, what measure of job satisfaction appears to generate the greatest difference of opinion among the nurses? Explain.

3. What can be learned about the types of hospitals? Does any particular type of hospital seem to have better levels of job satisfaction than the other types of hospitals? Do your results suggest any recommendations for learning about and/or improving job satisfaction? Discuss.

4. What additional descriptive statistics and insights can you point to in terms of learning about and possibly improving job satisfaction?

Introduction to Probability

Morton International*

CHICAGO, ILLINOIS

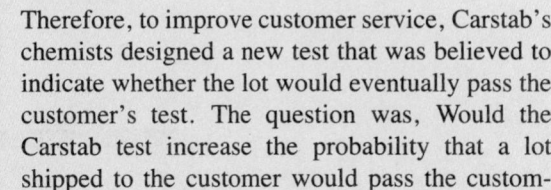

Morton International is a company with strong businesses in salt and household products, rocket motors, and specialty chemicals. Carstab Corporation, a subsidiary of Morton, produces a variety of specialty chemical products designed to meet the unique specifications of its customers. One such product, an expensive catalyst used in chemical processing, met the customer's exact specifications in only some, but not all, of the lots produced by Carstab.

Carstab's customer agreed to test each lot as it was received to determine whether the catalyst would perform the desired function. If the catalyst did not pass the test, the lot would be returned to Carstab. The problem was that only 60% of the lots sent to the customer were passing the test. This meant that, although the product was still good and usable by other customers, an unacceptably high percentage (40%) of the shipments to this particular customer were being returned.

Carstab explored the possibility of duplicating the customer's test, but equipment cost made this infeasible. Therefore, to improve customer service, Carstab's chemists designed a new test that was believed to indicate whether the lot would eventually pass the customer's test. The question was, Would the Carstab test increase the probability that a lot shipped to the customer would pass the customer's test?

A sample of lots was tested using both the customer's procedure and Carstab's proposed procedure, and 55% of the lots passed Carstab's test while 50% of the lots passed both tests. Probability analysis of these results indicated that if a lot passed Carstab's new test, there was a .909 probability of it passing the customer's test. This was good supporting evidence for the use of Carstab's new test, and the new test procedure was implemented by Carstab. Results showed an immediate improvement in customer service and a reduction in shipping costs for returned lots.

The probability computed by Carstab is called a conditional probability. In this chapter we shall see how this and other probabilities useful to decision making are computed.

*The authors are indebted to Michael Haskell of Morton International for providing this Statistics in Practice.

Throughout our lives we are faced with decision-making situations that involve uncertainty. Perhaps you will be asked for an analysis of one of the following situations:

1. What is the "chance" that sales will decrease if the price of the product is increased?
2. What is the "likelihood" that the new assembly method will increase productivity?
3. How "likely" is it that the project will be completed on time?
4. What are the "odds" that the new investment will be profitable?

The subject matter most useful in effectively dealing with such uncertainties is contained under the heading of probability. In everyday terminology, *probability* can be thought of as a numerical measure of the chance or likelihood that a particular event will occur. For example, if we consider the event "rain tomorrow," we understand that, when the television weather report indicates "a near-zero probability of rain," there is almost no chance of rain. However, if a 90% probability of rain is reported, we know that it is very likely or almost certain that rain will occur. A 50% probability indicates that rain is just as likely to occur as not.

Probability values are always assigned on a scale from 0 to 1. A probability near 0 indicates that the event is very unlikely to occur; a probability near 1 indicates that the event is almost certain to occur. Other probabilities between 0 and 1 represent varying degrees of likelihood that the event will occur. Figure 4.1 depicts this view of probability.

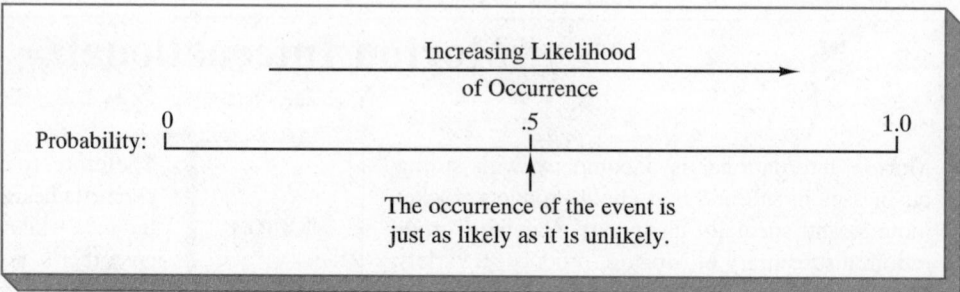

Probability is important in decision making because it provides a mechanism for measuring, expressing, and analyzing the uncertainties associated with future events. In this chapter we introduce the fundamental concepts of probability and begin to illustrate their use as decision-making tools. In subsequent chapters we will extend these basic notions of probability and demonstrate the important role that probability plays in statistical inference.

4.1   Experiments, the Sample Space, and Counting Rules

Using the terminology of probability, we define an *experiment* to be any process which generates well-defined outcomes. By this we mean that, on any single repetition of the experiment, *one and only one* of the possible experimental outcomes will occur. Several examples of experiments and their associated outcomes are as follows:

Experiment	Experimental Outcomes
Toss a coin	Head, tail
Select a part for inspection	Defective, nondefective
Conduct a sales call	Purchase, no purchase
Roll a die	1, 2, 3, 4, 5, 6
Play a football game	Win, lose, tie

The first step in analyzing a particular experiment is to carefully define the experimental outcomes. When we have defined *all* possible experimental outcomes, we have identified the *sample space* for the experiment. That is, the sample space is defined as the set of all possible experimental outcomes. Any one particular experimental outcome is referred to as a *sample point* and is an element of the sample space.

Let us consider the experiment of tossing a coin, as mentioned above. The experimental outcomes are defined by the upward face of the coin—a head or a tail. If we let *S* denote the sample space, we can use the following notation to describe the sample space and sample points for the coin-tossing experiment:

$$S = \{\text{Head, Tail}\}$$

Using this notation, the experiment of selecting a part for inspection would have a sample space with sample points as follows:

$$S = \{\text{Defective, Nondefective}\}$$

Finally, suppose that we consider the experiment of rolling a die, where the experimental outcomes are defined as the number of dots appearing on the upward face of the die. In this experiment, the numerical values 1, 2, 3, 4, 5, and 6 represent the possible experimental outcomes or sample points. Thus the sample space is denoted by

$$S = \{1, 2, 3, 4, 5, 6\}$$

Counting Rules

The ability to identify and count the sample points of an experiment is an important step in understanding what may happen when an experiment occurs. Consider an experiment of tossing two coins, with the experimental outcomes defined in terms of the pattern of heads and tails appearing on the upward faces of the two coins. How many experimental outcomes (sample points) are possible for this experiment?

We can view the experiment of tossing two coins as a two-step experiment: step 1 corresponds to tossing the first coin and step 2 corresponds to tossing the second coin. The *tree diagram* is a graphical device that is helpful in visualizing a multiple-step experiment and enumerating the experimental outcomes. Figure 4.2 provides a tree diagram for the coin-tossing experiment with branches of either a head or a tail shown for the first coin (step 1) followed by the branches of either a head or a tail shown for the second coin (step 2). The notation (H, H) is used to denote the outcome corresponding to a head on the first coin and a head on the second coin. Similarly, (H, T) is used to denote the outcome of a head on the first coin and a tail on the second coin. Thus, each of the points on the right-hand side of the tree corresponds to a sample point—an experimental outcome. Thus, we see that there are four experimental outcomes for the coin-tossing experiment, and the sample space for the experiment can be written as follows:

$$S = \{(H, H), (H, T), (T, H), (T, T)\}$$

FIGURE 4.2

Tree Diagram for the Experiment of Tossing Two Coins

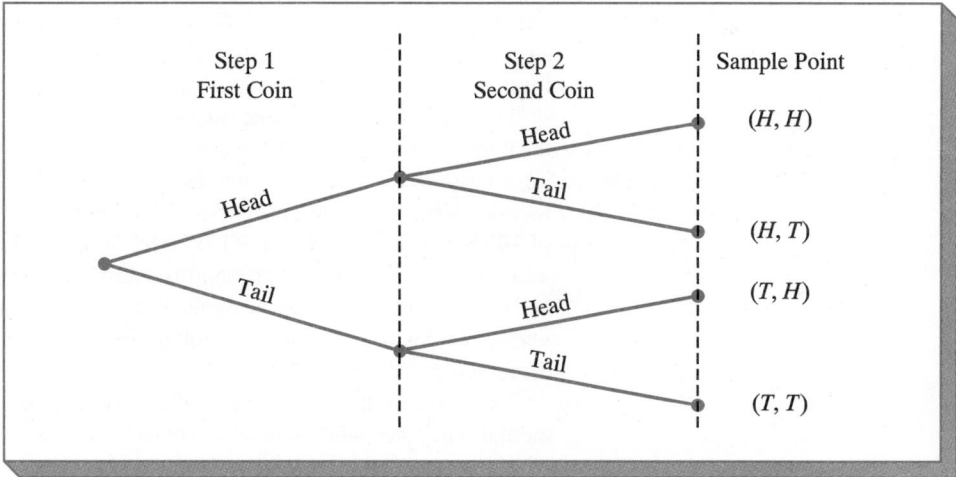

A rule that is helpful in determining the number of sample points for a multiple-step experiment is as follows:

A Counting Rule for Multiple-Step Experiments

If an experiment can be described as a sequence of k steps in which there are n_1 possible outcomes on the first step, n_2 possible outcomes on the second step, and so on, then the total number of experimental outcomes is given by $(n_1)(n_2) \ldots (n_k)$. That is, the number of outcomes for the overall experiment is found by multiplying the number of outcomes on each step.

Referring again to the experiment of tossing two coins, we have stated that it can be viewed as a two-step experiment. With two outcomes possible on the first toss of the coin ($n_1 = 2$) and two outcomes possible on the second toss of the coin ($n_2 = 2$), the counting rule states that there are $(n_1)(n_2) = (2)(2) = 4$ experimental outcomes.

Let us now see how the counting rule for multiple-step experiments can be used in the analysis of a capacity expansion problem faced by the Kentucky Power & Light Company (KP&L). The project KP&L is starting work on is designed to increase the generating capacity of one of its plants in northern Kentucky. The project is divided into two sequential stages or steps: stage 1 (design) and stage 2 (construction). While each stage will be scheduled and controlled as closely as possible, management cannot predict beforehand the exact time required to complete each stage of the project. An analysis of similar construction projects has shown completion times for the design stage of 2, 3, or 4 months and completion times for the construction stage of 6, 7, or 8 months. In addition, because of the critical need for additional electrical power, management has set a goal of 10 months for the completion of the entire project.

Since there are three possible completion times for the design stage (step 1) and three possible completion times for the construction stage (step 2), the counting rule for multiple-step experiments can be applied here to determine that there are a total of $(3)(3) = 9$ experimental outcomes. To describe the experimental outcomes, we will use a two-number notation; for instance, (2, 6) will indicate that the design stage is completed in 2 months and the construction stage is completed in 6 months. This experimental outcome results in a total of $2 + 6 = 8$ months to complete the entire project. Table 4.1 summarizes the nine experimental outcomes for the KP&L problem. The tree diagram in Figure 4.3 shows how the nine outcomes or sample points occur.

The counting rule and tree diagram have been used to help the project manager identify the experimental outcomes and determine the possible project completion times. From the information in Figure 4.3, we see that the project will be completed in from 8 to 12 months, with six of the nine experimental outcomes providing the desired completion time of 10 months or less. While it has been helpful to identify the experimental outcomes, we will need to consider how probability values can be assigned to the experimental outcomes before making a final assessment of the probability that the project will be completed within the desired 10 months. Probability assignments for experimental outcomes will be discussed in Section 4.2.

Another counting rule that is often useful allows one to count the number of experimental outcomes when n objects are to be selected from a set of N objects. It is called the counting rule for combinations.

TABLE 4.1
Listing of Experimental Outcomes or Sample Points for the KP&L Problem

Completion Time (months)			
Stage 1 (Design)	Stage 2 (Construction)	Notation for Experimental Outcome	Total Project Completion Time (months)
2	6	(2, 6)	8
2	7	(2, 7)	9
2	8	(2, 8)	10
3	6	(3, 6)	9
3	7	(3, 7)	10
3	8	(3, 8)	11
4	6	(4, 6)	10
4	7	(4, 7)	11
4	8	(4, 8)	12

FIGURE 4.3
Tree Diagram for the KP&L Project

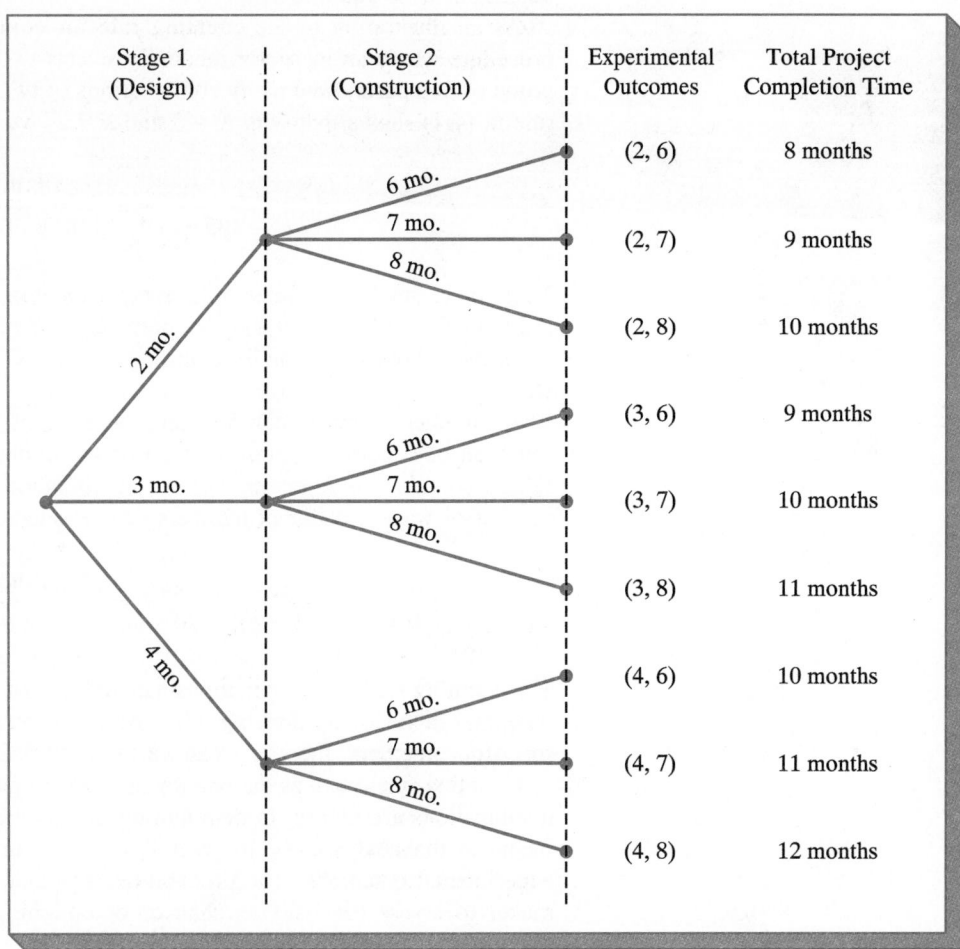

Counting Rule for Combinations

The number of combinations of N objects taken n at a time is as follows:

$$\binom{N}{n} = \frac{N!}{n!(N-n)!} \qquad (4.1)$$

where

$$N! = N(N-1)(N-2)\ldots(2)(1)$$
$$n! = n(n-1)(n-2)\ldots(2)(1)$$

and

$$0! = 1$$

The notation ! means *factorial;* for example, 5 factorial is $5! = (5)(4)(3)(2)(1) = 120$. By definition, 0! is equal to 1.

As an illustration of the counting rule for combinations, consider a quality control procedure where an inspector randomly selects two of five parts to test for defects. In a group of five parts, how many combinations of two parts may be selected? The counting rule in (4.1) shows that with $N = 5$ and $n = 2$, we have

$$\binom{5}{2} = \frac{5!}{2!(5-2)!} = \frac{(5)(4)(3)(2)(1)}{(2)(1)(3)(2)(1)} = \frac{120}{12} = 10$$

Thus, there are 10 outcomes for the experiment of randomly selecting two parts from a group of five. If we label the five parts as A, B, C, D, and E, the 10 combinations or experimental outcomes can be identified as AB, AC, AD, AE, BC, BD, BE, CD, CE, and DE.

As another example, consider that the State of Ohio lottery system uses the random selection of 6 numbers from a group of 44 numbers in order to determine the weekly lottery winner. The counting rule for combinations (4.1) can be used to determine the number of ways 6 different numbers can be selected from a group of 44 numbers.

$$\binom{44}{6} = \frac{44!}{6!(44-6)!} = \frac{44!}{6!38!} = \frac{(44)(43)(42)(41)(40)(39)}{(6)(5)(4)(3)(2)(1)} = 7,059,052$$

The counting rule for combinations has told us that there are over 7 million experimental outcomes in the lottery drawing. If an individual buys a lottery ticket and selects six of the forty-four numbers, there is 1 chance in 7,059,052 that the individual will win.

Counting rules such as the one for multiple-step experiments and the counting rule for combinations are helpful in determining the number of outcomes in a statistical experiment. In the next section we will study ways in which probabilities are assigned to experimental outcomes. The probabilities assigned will enable the manager or decision maker to assess the risks or chances of obtaining favorable, as well as unfavorable, outcomes.

NOTES & COMMENTS

In statistics, the notion of an experiment is somewhat different than the notion of an experiment in the physical sciences. In the physical sciences, an experiment is usually conducted in a laboratory or a controlled environment in order to learn about a scientific occurrence. When physical science experiments are repeated under identical conditions, the same outcome is expected to occur. In statistical experiments, the outcomes are determined by chance. Even though the experiment is repeated in exactly the same way, an entirely different outcome may occur. Because of this difference in outcomes, the experiments of statistics are sometimes called random experiments.

❑ ❑ Exercises

Methods

1. An experiment has three steps with three outcomes possible for the first step, two outcomes possible for the second step, and four outcomes possible for the third step. How many experimental outcomes exist for the entire experiment?

SELF TEST ▶ 2. How many ways can three items be selected from a group of six items? Use the letters A, B, C, D, E, and F to identify the items, and list each of the different combinations of three items.

Applications

3. Consider the experiment of conducting three sales calls. On each of the calls there will be either a purchase or no purchase.
   a. Construct a tree diagram for this three-step experiment.
   b. Identify each sample point and the sample space. How many sample points are there?
   c. How many sample points would there be if the experiment consisted of four sales calls?

4. A major resort in Florida is concerned about weather conditions in the Northeast as well as in Florida. In characterizing the temperature in both areas, the following three categories are used: below average, average, or above average. A combination of below-average temperatures in the Northeast with above-average temperatures in Florida means an increased volume of business for the resort. For the experiment of observing the weather conditions on a particular day, answer the following questions.
   a. How many experimental outcomes are possible?
   b. Develop a tree diagram for this experiment.

5. In the city of Milford, applications for zoning changes go through a two-step process: a review by the planning commission and a final decision by the city council. At step 1 the planning commission will review the zoning change request and make a positive or negative recommendation concerning the change. At step 2 the city council will review the planning commission's recommendation and then vote to approve or to disapprove the zoning change. In some instances, the city council vote has agreed with the planning commission's recommendation. However, in other instances, the council vote has been the opposite of the planning commission's recommendation. An application for a zoning change has just been submitted by the developer of an apartment complex. Consider the application process as an experiment.
   a. How many sample points are there for this experiment? List the sample points.
   b. Construct a tree diagram for the experiment.

6. An investor has two stocks: stock A and stock B. Each stock may increase in value, decrease in value, or remain unchanged. Consider the experiment of investing in the two stocks and observing the change (if any) in value.

　　a. How many experimental outcomes are possible?

　　b. Show a tree diagram for the experiment.

　　c. How many of the experimental outcomes result in an increase in value for at least one of the two stocks?

　　d. How many of the experimental outcomes result in an increase in value for both of the stocks?

7. An investor is reviewing the performance of six stocks and will select two for investment. How many alternative combinations of two stocks does the investor have to consider?

8. How many 5-card poker hands are possible with a deck of 52 cards?

S E L F　T E S T

9. Simple random sampling uses a sample of size n from a population of size N in order to obtain data that can be used to make inferences about the characteristics of a population. Suppose we have a population of 50 bank accounts and wish to take a random sample of 4 accounts in order to learn about the population. How many different random samples of 4 accounts are possible?

10. Williams, Inc. will be forming a three-member long-range planning committee that will be charged with developing a strategic 5-year plan for the company's entry into a new product market. The president has identified seven experienced managers as viable candidates for the committee. How many ways can the three-member committee be formed?

11. A company is planning to expand its operation by building two plants in the western region of the country. Eight possible locations have been identified and are being evaluated. How many combinations of two locations are possible from the eight locations?

12. Many states design their automobile license plates such that space is available for up to six letters or numbers.

　　a. If a state decides to use only numerical values for the license plates, how many different license plate numbers are possible? Assume that 000000 is an acceptable license plate number, although it will be used only for display purposes at the license bureau.

　　b. If the state decides to use two letters followed by four numbers, how many different license plate numbers are possible? Assume that the letters I and O will not be used because of their similarity to numbers 1 and 0.

　　c. Would larger states, such as New York and California, tend to use more or fewer letters in license plates? Explain.

4.2　Assigning Probabilities to Experimental Outcomes

We have an understanding of the concept of an experiment and of the sample space as the set of all experimental outcomes. Let us now see how probabilities for the experimental outcomes (sample points) can be determined. Recall the discussion at the beginning of this chapter which stated that the probability of an experimental outcome is a numerical measure of the likelihood that the experimental outcome will occur. In the assigning of probabilities to the experimental outcomes, there are various acceptable approaches; however, regardless of the approach taken, the following two basic requirements must be satisfied:

1. The probability values assigned to each experimental outcome (sample point) must be between 0 and 1. That is, if we let E_i indicate experimental outcome i and $P(E_i)$ indicate the probability of this experimental outcome, we must have

$$0 \le P(E_i) \le 1 \quad \text{for all } i \tag{4.2}$$

2. The sum of *all* of the experimental outcome probabilities must be 1. For example, if a sample space has k experimental outcomes, we must have

$$P(E_1) + P(E_2) + \cdots + P(E_k) = \Sigma P(E_i) = 1 \tag{4.3}$$

Any method of assigning probability values to the experimental outcomes which satisfies these two requirements and results in reasonable numerical measures of the likelihood of the outcomes is acceptable. In practice, one of the following three methods can be used:

1. Classical method
2. Relative frequency method
3. Subjective method

Classical Method

To illustrate the classical method of assigning probabilities, let us again consider the experiment of flipping a coin. On any one flip, we will observe one of two experimental outcomes: head or tail. It would seem reasonable to assume that the two outcomes are equally likely. Therefore, since one of the two equally likely outcomes is a head, we should logically conclude that the probability of observing a head is ½, or .50. Similarly, the probability of observing a tail is also .50. When the assumption of equally likely outcomes is used as a basis for assigning probabilities, the approach is referred to as the *classical method*. If an experiment has n possible outcomes, the classical method would assign a probability of $1/n$ to each experimental outcome.

As another illustration of the classical method, consider again the experiment of rolling a die. In Section 4.1 we described the sample space and sample points for this experiment with the following notation:

$$S = \{1, 2, 3, 4, 5, 6\}$$

It would seem reasonable to conclude that the six experimental outcomes are equally likely, and hence each outcome is assigned a probability of ⅙. Thus, if $P(1)$ denotes the probability that one dot appears on the upward face of the die, then $P(1) = ⅙$. Similarly, $P(2) = ⅙$, $P(3) = ⅙$, $P(4) = ⅙$, $P(5) = ⅙$, and $P(6) = ⅙$. Note that this probability assignment satisfies the two basic requirements for assigning probabilities. In fact, requirements (4.2) and (4.3) are always satisfied when the classical method is used, since each of the n sample points is assigned a probability of $1/n$.

The classical method was developed originally in the analysis of gambling problems, where the assumption of equally likely outcomes is often reasonable. In many business problems, however, this assumption is not valid. Hence alternative methods of assigning probabilities are required.

Relative Frequency Method

As an illustration of the relative frequency method, consider a firm that is preparing to market a new product. In order to estimate the probability that a customer will purchase the product, a test market evaluation has been set up wherein salespeople will call on potential customers. For each sales call conducted, there are two possible outcomes: the customer purchases the product or the customer does not purchase the product. Since there is no reason to assume that the two experimental outcomes are equally likely, the classical method of assigning probabilities is inappropriate.

Suppose that, in the test market evaluation of the product, 400 potential customers were contacted; 100 actually purchased the product, but 300 did not. In effect, then, we have repeated the experiment of contacting one customer 400 times and have found that the product was purchased 100 times. Thus we might decide to use the relative frequency as an estimate of the probability of a customer making a purchase. Hence we could assign a probability of 100/400 = .25 to the experimental outcome of purchasing the product.

Similarly, 300/400 = .75 could be assigned to the experimental outcome of not purchasing the product. This approach to assigning probabilities is referred to as the *relative frequency method*.

Subjective Method

The classical and relative frequency methods cannot be applied to all situations where probability assessments are desired. For example, there are situations where the experimental outcomes are not equally likely and where relative frequency data are unavailable. For example, consider the next football game that the Pittsburgh Steelers will play. What is the probability that the Steelers will win? The experimental outcomes of a win, a loss, or a tie are not necessarily equally likely. Also, since the teams involved have not played several times previously this year, there are no relative frequency data available that are relevant to the upcoming game. Thus if we want an estimate of the probability of the Steelers winning, we must use a subjective opinion of its value.

With the subjective method of assigning probabilities to the experimental outcomes, we may use any data available as well as our experience and intuition. However, after we consider all available information, a probability value that expresses the *degree of belief* that the experimental outcome will occur must be specified. This method of assigning probability is referred to as the *subjective method*. Since subjective probability expresses a person's "degree of belief," it is personal. Different people can be expected to assign different probabilities to the same event. Nonetheless, care must be taken when using the subjective method to ensure that requirements (4.2) and (4.3) are satisfied. That is, regardless of a person's "degree of belief," the probability value assigned to each experimental outcome must be between 0 and 1 and the sum of all the experimental outcome probabilities must equal 1.

Even in situations where either the classical or relative frequency approach can be applied, management may want to provide subjective probability estimates. In such cases, the best probability estimates often are obtained by combining the estimates from the classical or relative frequency approaches with the subjective probability estimates.

Probabilities for the KP&L Problem

To perform further analysis on the KP&L problem, we must develop probabilities for each of the nine experimental outcomes listed in Table 4.1. Based on experience and judgment, management concluded that the experimental outcomes were not equally likely. Hence the classical method of assigning probabilities could not be used. Management then decided to conduct a study of the completion times for similar projects undertaken by KP&L over the past three years. The results of a study of 40 similar projects are summarized in Table 4.2.

After reviewing the results of the study, management decided to employ the relative frequency method of assigning probabilities. Management could still have provided subjective probability estimates, but it was felt that the current project was quite similar to the 40 previous projects. Thus the relative frequency method was judged best.

In using the data in Table 4.2 to compute probabilities, we note that outcome (2, 6) — stage 1 completed in 2 months and stage 2 completed in 6 months — occurred 6 times in the 40 projects. Thus we can use the relative frequency method to assign a probability of 6/40 = .15 to this outcome. Similarly, outcome (2, 7) also occurred in 6 of the 40 projects, providing a 6/40 = .15 probability. Continuing in this manner, we obtain the probability assignments for the sample points of the KP&L project shown in Table 4.3. Note that $P(2, 6)$ represents the probability of the sample point (2, 6), $P(2, 7)$ represents the probability of the sample point (2, 7), and so on.

TABLE 4.2
Completion Results for 40 KP&L Projects

Completion Time (months)			Number of Past Projects Having These Completion Times
Stage 1	Stage 2	Sample Point	
2	6	(2, 6)	6
2	7	(2, 7)	6
2	8	(2, 8)	2
3	6	(3, 6)	4
3	7	(3, 7)	8
3	8	(3, 8)	2
4	6	(4, 6)	2
4	7	(4, 7)	4
4	8	(4, 8)	6
			Total 40

TABLE 4.3
Probability Assignments for the KP&L Problem Based on the Relative Frequency Method

Sample Point	Project Completion Time	Probability of Sample Point
(2, 6)	8 months	$P(2, 6) = 6/40 = .15$
(2, 7)	9 months	$P(2, 7) = 6/40 = .15$
(2, 8)	10 months	$P(2, 8) = 2/40 = .05$
(3, 6)	9 months	$P(3, 6) = 4/40 = .10$
(3, 7)	10 months	$P(3, 7) = 8/40 = .20$
(3, 8)	11 months	$P(3, 8) = 2/40 = .05$
(4, 6)	10 months	$P(4, 6) = 2/40 = .05$
(4, 7)	11 months	$P(4, 7) = 4/40 = .10$
(4, 8)	12 months	$P(4, 8) = 6/40 = .15$
		Total 1.00

☐ ☐ Exercises

Methods

13. Suppose an experiment has five equally likely outcomes: E_1, E_2, E_3, E_4, E_5. Assign probabilities to each outcome and show that conditions (4.2) and (4.3) are satisfied. What method did you use?

SELF TEST ▶ **14.** An experiment with three outcomes has been repeated 50 times and it was learned that E_1 occurred 20 times, E_2 occurred 13 times, and E_3 occurred 17 times. Assign probabilities to the outcomes. What method did you use?

15. A decision maker has subjectively assigned the following probabilities to the four outcomes of an experiment: $P(E_1) = .10$, $P(E_2) = .15$, $P(E_3) = .40$, and $P(E_4) = .20$. Are these valid probability assignments? Check to see if (4.2) and (4.3) are satisfied.

16. Consider the experiment of selecting a card from a deck of 52 cards.
a. How many sample points are possible?
b. Which method (classical, relative frequency, or subjective) would you recommend for assigning probabilities to the sample points?
c. What is the probability assignment for each card?
d. Show that your probability assignments satisfy the two basic requirements for assigning probabilities.

Applications

SELF TEST ▶

17. In a survey of new matriculants to MBA programs ("School Selection by Students," *GMAC Occasional Papers*, March 1988), the following data were obtained on the marital status of the students.

Marital status	Frequency
Never married	1106
Married	826
Other (separated, widowed, divorced)	106
Total	2038

Consider the experiment of interviewing a new MBA student and recording her or his marital status. Show your probability assignments.

18. A small-appliance store in Madeira has collected data on refrigerator sales for the last 50 weeks. The data are shown in Table 4.4.

TABLE 4.4

Number of Refrigerators Sold	Number of Weeks
0	6
1	12
2	15
3	10
4	5
5	2
Total	50

Suppose that we are interested in the experiment of observing the number of refrigerators sold in 1 week of store operations.
a. How many experimental outcomes are there?
b. Which approach would you recommend for assigning probabilities to the experimental outcomes?
c. Assign probabilities and verify that your assignments satisfy the two basic requirements.

19. Strom Construction has made a bid on two contracts. The owner has identified the possible outcomes and subjectively assigned probabilities as follows:

Experimental Outcome	Obtain Contract 1	Obtain Contract 2	Probability
1	Yes	Yes	.15
2	Yes	No	.15
3	No	Yes	.30
4	No	No	.25

a. Are these valid probability assignments? Why or why not?
b. What would have to be done to make the probability assignments valid?

20. An investor forecasts that the probabilities that a certain stock will either go down, remain the same, or go up are .20, .60, and .30, respectively. Does this seem reasonable? Explain.

21. Faced with the question of determining the probability of obtaining either 0 heads, 1 head, or 2 heads when flipping a coin twice, an individual argued that, since it seems reasonable to treat the outcomes as equally likely, the probability of each event is $\frac{1}{3}$. Do you agree? Explain.

22. A company that manufactures toothpaste is studying five different package designs. Assuming that one design is just as likely to be selected by a consumer as any other design, what probability would you assign to each of the package designs being selected? In an actual experiment, 100 consumers were asked to pick the design they preferred. The data in Table 4.5 were obtained. Do the data appear to confirm the belief that one design is just as likely to be selected as another? Explain.

TABLE 4.5

Design	Number of Times Preferred
1	5
2	15
3	30
4	40
5	10

4.3   Events and Their Probabilities

Until now we have used the term *event* much as it would be used in everyday language. We must now introduce the formal definition of an *event* as it relates to probability.

Event

An *event* is a collection of sample points.

For an example, let us return to the KP&L problem and assume that the project manager is interested in the event that the entire project can be completed in 10 months or less. Referring to Table 4.3, we see that six sample points (2, 6), (2, 7), (2, 8), (3, 6), (3, 7), and (4, 6) provide a project completion time of 10 months or less. Let C denote the event that the project is completed in 10 months or less; we write

$$C = \{(2, 6), (2, 7), (2, 8), (3, 6), (3, 7), (4, 6)\}$$

Event C is said to occur if *any one* of the six sample points shown above appears as the experimental outcome.

Other events that might be of interest to KP&L management include the following:

L = the event that the project is completed in *less* than 10 months

M = the event that the project is completed in *more* than 10 months

Using the information in Table 4.3 we see that these events consist of the following sample points:

$$L = \{(2, 6), (2, 7), (3, 6)\}$$

$$M = \{(3, 8), (4, 7), (4, 8)\}$$

A variety of additional events can be defined for the KP&L problem, but in each case the event must be identified as a collection of sample points for the experiment.

Given the probabilities of the sample points shown in Table 4.3, we can use the following definition to compute the probability of any event that KP&L management might want to consider:

Probability of an Event

The probability of any event is equal to the sum of the probabilities of the sample points in the event.

Using this definition, we calculate the probability of a particular event by adding the probabilities of the experimental outcomes that make up the event. We can now compute the probability that the project will take 10 months or less to complete. Since this event is given by $C = \{(2, 6), (2, 7), (2, 8), (3, 6), (3, 7), (4, 6)\}$, the probability of event C is shown below (note that P is used to denote the probability of the corresponding event or sample point):

$$P(C) = P(2, 6) + P(2, 7) + P(2, 8) + P(3, 6) + P(3, 7) + P(4, 6)$$

Refer to the sample point probabilities in Table 4.3; we have

$$P(C) = .15 + .15 + .05 + .10 + .20 + .05 = .70$$

Similarly, since the event that the project is completed in less than 10 months is given by $L = \{(2, 6), (2, 7), (3, 6)\}$, the probability of this event is given by

$$P(L) = P(2, 6) + P(2, 7) + P(3, 6)$$
$$= .15 \quad + \quad .15 \quad + \quad .10 \; = .40$$

Finally, for the project to be completed in more than 10 months, we have $M = \{(3, 8), (4, 7), (4, 8)\}$ and thus

$$P(M) = P(3, 8) + P(4, 7) + P(4, 8)$$
$$= .05 \quad + \quad .10 \quad + \quad .15 \quad = .30$$

Using the above probability results, we can now tell KP&L management that there is a .70 probability that the project will be completed in 10 months or less, a .40 probability that the project will be completed in less than 10 months, and a .30 probability the project will be completed in more than 10 months. This procedure of computing event probabilities can be repeated for any event of interest to the KP&L management.

Any time that we can identify all the sample points of an experiment and assign the corresponding sample point probabilities, we can use the definition to compute the probability of an event. However, in many experiments the number of sample points is large and the identification of the sample points, as well as determining their associated probabilities, becomes extremely cumbersome, if not impossible. In the remaining sections of this chapter, we present some basic probability relationships that can often be used to compute the probability of an event without requiring knowledge of all sample point probabilities.

NOTES & COMMENTS

1. The sample space, S, is an event. Since it contains all the experimental outcomes, it has a probability of 1; that is, $P(S) = 1$.
2. When the classical method is used to assign probabilities, the assumption is that the experimental outcomes are equally likely. In such cases, the probability of an event can be computed by counting the number of experimental outcomes in the event and dividing the result by the total number of experimental outcomes.

❑ ❑ Exercises

Methods

23. An experiment has three outcomes with $P(E_1) = .35$, $P(E_2) = .40$, $P(E_3) = .25$.
   a. What is the probability that E_1 or E_2 occurs?
   b. What is the probability that E_1 or E_3 occurs?
   c. What is the probability that E_1, E_2, or E_3 occurs?

24. An experiment has four equally likely outcomes.
a. What is the probability that E_2 occurs?
b. What is the probability that any two of the outcomes occurs (e.g. E_1 or E_3)?
c. What is the probability that any three of the outcomes occurs (e.g. E_1 or E_2 or E_4)?

SELF TEST ▶ 25. In Exercise 16 we considered the experiment of selecting a card from a deck of 52 cards. Each card corresponded to a sample point with a 1/52 probability.
a. List the sample points in the event an ace is selected.
b. List the sample points in the event a club is selected.
c. List the sample points in the event a face card (jack, queen, or king) is selected.
d. Find the probabilities associated with each of the events in (a), (b), and (c).

26. Consider the experiment of rolling a pair of dice. Suppose that we are interested in the sum of the face values showing on the dice.
a. How many sample points are possible? (Hint: use the counting rule for multiple-step experiments.)
b. List the sample points.
c. What is the probability of obtaining a value of 7?
d. What is the probability of obtaining a value of 9 or greater?
e. Since there are six possible even values (2, 4, 6, 8, 10, and 12) and only five possible odd values (3, 5, 7, 9, and 11), the dice should show even values more often than odd values. Do you agree with this statement? Explain.
f. What method did you use to assign the probabilities requested above?

Applications

SELF TEST ▶ 27. Use the KP&L sample point and sample point probabilities in Table 4.3 to answer the following:
a. The design stage (stage 1) will run over budget if it takes 4 months to complete. List the sample points in the event the design stage is over budget.
b. What is the probability that the design stage is over budget?
c. The construction stage (stage 2) will run over budget if it takes 8 months to complete. List the sample points in the event the construction stage is over budget.
d. What is the probability that the construction stage is over budget?
e. What is the probability that both stages are over budget?

28. Suppose that a manager of a large apartment complex provides the subjective probability estimates found in Table 4.6 about the number of vacancies that will exist next month. List the sample points in each of the following events and provide the probability of the event:
a. No vacancies.   b. At least four vacancies.   c. Two or fewer vacancies.

TABLE 4.6

Vacancies	Probability
0	.05
1	.15
2	.35
3	.25
4	.10
5	.10

29. The manager of a furniture store sells from 0 to 4 china hutches each week. Based on past experience, the following probabilities are assigned to sales of 0, 1, 2, 3, or 4 hutches:

$$
\begin{aligned}
P(0) &= .08 \\
P(1) &= .18 \\
P(2) &= .32 \\
P(3) &= .30 \\
P(4) &= \underline{.12} \\
& 1.00
\end{aligned}
$$

a. Are these valid probability assignments? Why or why not?
b. Let A be the event that 2 or fewer are sold in one week. Find $P(A)$.
c. Let B be the event that 4 or more are sold in one week. Find $P(B)$.

30. A sample of 100 customers of Montana Gas and Electric resulted in the following frequency distribution of monthly charges

Amount ($)	Number
0–49	13
50–99	22
100–149	34
150–199	26
200–249	5

TABLE 4.7

Number of activities	Frequency
0	8
1	20
2	12
3	6
4	3
5	1

a. Let A be the event that monthly charges are $150 or more. Find $P(A)$.

b. Let B be the event that monthly charges are less than $150. Find $P(B)$.

31. A survey of 50 students at Tarpon Springs College regarding the number of extracurricular activities resulted in the data in Table 4.7.

a. Let A be the event that a student participates in at least 1 activity. Find $P(A)$.

b. Let B be the event that a student participates in 3 or more activities. Find $P(B)$.

c. What is the probability a student participates in exactly 2 activities?

4.4   Some Basic Relationships of Probability

Complement of an Event

Given an event A, the *complement* of A is defined to be the event consisting of all sample points that are *not* in A. The complement of A is denoted by A^c. Figure 4.4 provides a diagram, known as a *Venn diagram*, which illustrates the concept of a complement. The rectangular area represents the sample space for the experiment and as such contains all possible sample points. The circle represents event A and contains only the sample points that belong to A. The shaded region of the diagram contains all sample points not in event A, which is by definition the complement of A.

In any probability application, either event A or its complement A^c must occur. Therefore, we have

$$P(A) + P(A^c) = 1$$

Solving for $P(A)$, we obtain the following result:

Computing Probability Using the Complement

$$P(A) = 1 - P(A^c) \tag{4.4}$$

Equation (4.4) shows that the probability of an event A can be easily computed if the probability of its complement, $P(A^c)$, is known.

As an example, consider the case of a sales manager who, after reviewing sales reports, states that 80% of new customer contacts result in no sale. By allowing A to denote the event of a sale and A^c denote the event of no sale, the manager is stating that $P(A^c) = .80$. Using (4.4), we see that

$$P(A) = 1 - P(A^c) = 1 - .80 = .20$$

FIGURE 4.4
Complement of Event A

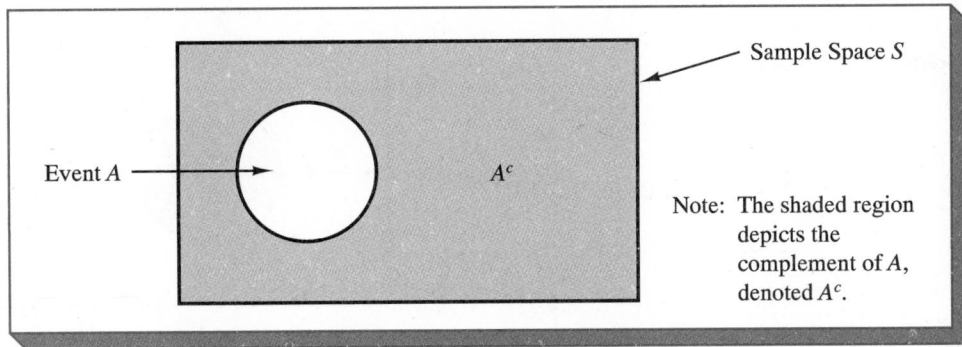

Note: The shaded region depicts the complement of A, denoted A^c.

This shows that there is a .20 probability that a sale will be made on a new customer contact.

In another example, a purchasing agent states that there is a .90 probability that a supplier will send a shipment that is free of defective parts. Using the complement, we can conclude that there is a $1 - .90 = .10$ probability that the shipment will contain defective parts.

Addition Law

The addition law is a helpful probability relationship when we have two events and are interested in knowing the probability that at least one of the events occurs. That is, with events A and B we are interested in knowing the probability that event A or event B or both occur.

Before we present the addition law, we need to discuss two concepts concerning the combination of events: the *union* of events and the *intersection* of events.

Given two events A and B, the union of A and B is defined as follows:

Union of Two Events

The *union* of A and B is the event containing *all* sample points belonging to A *or B or both*. The union is denoted by $A \cup B$.

The Venn diagram shown in Figure 4.5 depicts the union of events A and B. Note that the shaded region contains all the sample points in event A as well as all the sample points in event B. The fact that the circles overlap indicates that there are some sample points contained in both A and B.

The definition of the intersection of two events A and B is as follows:

Intersection of Two Events

Given two events A and B, the *intersection* of A and B is the event containing the sample points belonging to *both A and B*. The intersection is denoted by $A \cap B$.

FIGURE 4.5
Union of Events *A* and *B*

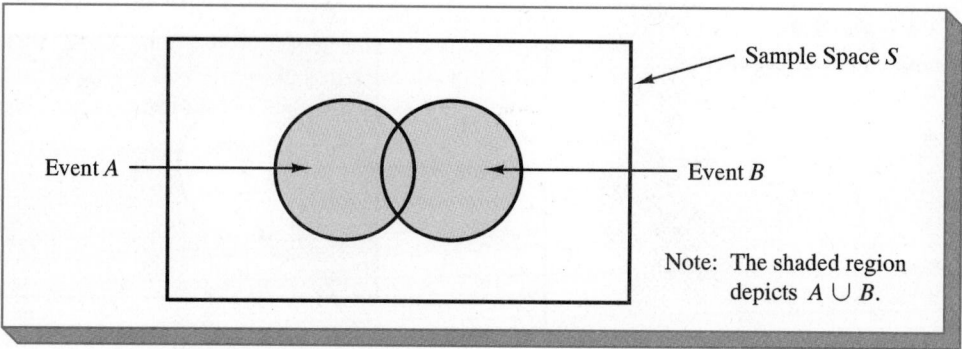

Note: The shaded region depicts $A \cup B$.

The Venn diagram depicting the intersection of the two events is shown in Figure 4.6. The area where the two circles overlap is the intersection; it contains the sample points that are in both *A* and *B*.

Let us now continue with a discussion of the addition law. The addition law provides a way to compute the probability of event *A* or *B* or both *A* and *B* occurring. In other words, the addition law is used to compute the probability of the union of two events, $A \cup B$. The addition law is written as follows:

Addition Law

$$P(A \cup B) = P(A) + P(B) - P(A \cap B) \qquad (4.5)$$

To grasp the addition law intuitively, note that the first two terms in the addition law, $P(A) + P(B)$, account for all the sample points in $A \cup B$. However, since the sample points in the intersection $A \cap B$ are in both *A* and *B*, when we compute $P(A) + P(B)$, we are in effect counting each of the sample points in $A \cap B$ twice. We correct for this by subtracting $P(A \cap B)$.

In order to present an application of the addition law, let us consider the case of a small assembly plant with 50 employees. Each worker is expected to complete work assignments on time and in such a way that the assembled product will pass a final inspection. On occasion, some of the workers fail to meet the performance standards by completing

FIGURE 4.6
Intersection of Events *A* and *B*

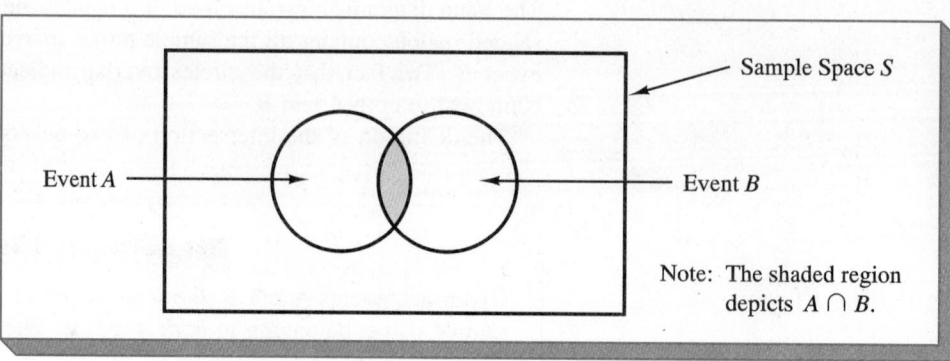

Note: The shaded region depicts $A \cap B$.

work late and/or assembling defective products. At the end of a performance evaluation period, the production manager found that 5 of the 50 workers had completed work late, 6 of the 50 workers had assembled defective products, and 2 of the 50 workers had both completed work late *and* assembled defective products. Let

$$L = \text{the event that the work is completed late}$$

$$D = \text{the event that the assembled product is defective}$$

The above relative frequency information leads to the following probabilities:

$$P(L) = \frac{5}{50} = .10$$

$$P(D) = \frac{6}{50} = .12$$

$$P(L \cap D) = \frac{2}{50} = .04$$

After reviewing the performance data, the production manager decided to assign a poor performance rating to any employee whose work was either late or defective; thus the event of interest is $L \cup D$. What is the probability that the production manager assigned an employee a poor performance rating?

Note that the probability question is about the union of two events. Specifically, we want to know $P(L \cup D)$. Here is where the addition law can be helpful. Using (4.5), we have

$$P(L \cup D) = P(L) + P(D) - P(L \cap D)$$

Knowing values for the three probabilities on the right-hand side of this expression, we can write

$$P(L \cup D) = .10 + .12 - .04 = .18$$

This tells us that there is a .18 probability that an employee will receive a poor performance rating.

As another example of the addition law, consider a recent study conducted by the personnel manager of a major computer software company. It was found that 30% of the employees that left the firm within 2 years did so primarily because they were dissatisfied with their salary, 20% left because they were dissatisfied with their work assignments, and 12% of the former employees said that *both* dissatisfaction with their salary and their work assignments were primary reasons for leaving. What is the probability that an employee that leaves within 2 years does so because of dissatisfaction with salary, dissatisfaction with the work assignment, or both?

Let

$$S = \text{the event that the employee leaves due to salary}$$

$$W = \text{the event that the employee leaves due to work assignment}$$

We have $P(S) = .30$, $P(W) = .20$, and $P(S \cap W) = .12$. Using (4.5), the addition law, we have

$$P(S \cup W) = P(S) + P(W) - P(S \cap W) = .30 + .20 - .12 = .38$$

This shows that there is a .38 probability that an employee leaves because of salary or work assignment reasons.

Before we conclude our discussion of the addition law, let us consider a special case that arises for *mutually exclusive events:*

FIGURE 4.7
Mutually Exclusive Events

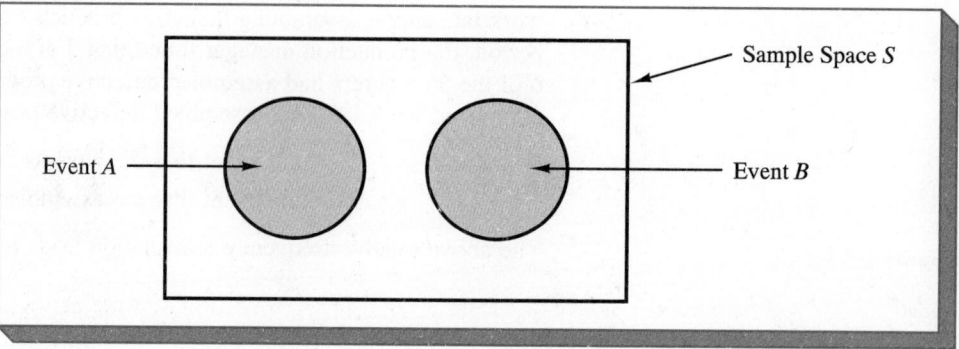

Sample Space S

Event A

Event B

Mutually Exclusive Events

Two events are said to be *mutually exclusive* if the events have no sample points in common.

That is, events A and B are mutually exclusive if, when one event occurs, the other cannot occur. Thus a requirement for A and B to be mutually exclusive is that their intersection must contain no sample points. The Venn diagram depicting two mutually exclusive events A and B is shown in Figure 4.7. In this case $P(A \cap B) = 0$; hence the addition law can be written as follows:

Addition Law for Mutually Exclusive Events

$$P(A \cup B) = P(A) + P(B)$$

❏ ❏ Exercises

Methods

32. Suppose that we have a sample space with five equally likely experimental outcomes: E_1, E_2, E_3, E_4, E_5. Let

$$A = \{E_1, E_2\}$$
$$B = \{E_3, E_4\}$$
$$C = \{E_2, E_3, E_5\}$$

a. Find $P(A)$, $P(B)$, and $P(C)$.
b. Find $P(A \cup B)$. Are A and B mutually exclusive?
c. Find A^c, C^c, $P(A^c)$, and $P(C^c)$.
d. Find $A \cup B^c$ and $P(A \cup B^c)$.
e. Find $P(B \cup C)$.

SELF TEST ▶ 33. Suppose that we have a sample space $S = \{E_1, E_2, E_3, E_4, E_5, E_6, E_7\}$, where $E_1, E_2, \ldots,$ E_7 denote the sample points. The following probability assignments apply:

$$P(E_1) = .05$$
$$P(E_2) = .20$$
$$P(E_3) = .20$$
$$\rightarrow P(E_4) = .25$$
$$P(E_5) = .15$$
$$P(E_6) = .10$$
$$P(E_7) = \underline{.05}$$
$$1.00$$

Let

$$A = \{E_1, E_4, E_6\}$$
$$B = \{E_2, E_4, E_7\}$$
$$C = \{E_2, E_3, E_5, E_7\}$$

a. Find $P(A)$, $P(B)$ and $P(C)$.
b. Find $A \cup B$ and $P(A \cup B)$. $A \cup B = \{E_1, E_2, E_4, E_6, E_7\}$
c. Find $A \cap B$ and $P(A \cap B)$. $.05 + .2 + .25 + .1 + .05$
d. Are events A and C mutually exclusive?
e. Find B^c and $P(B^c)$.

Applications

34. The public accounting firm Grant Thornton conducted a survey to see how executives felt about the 1992 recession and the potential for recovery (*Journal of Accountancy*, February, 1992). The probability that an executive indicated a recession existed was .74. If we were to choose one of the executives, what is the probability that she or he would indicate a recession does not exist?

35. A pharmaceutical company conducted a study to evaluate the effectiveness of an allergy relief medicine; 250 patients with symptoms that included itchy eyes and a skin rash were given the new drug. The results of the study were as follows: 90 of the patients treated experienced eye relief, 135 had their skin rash clear up, and 45 experienced both relief from itchy eyes and the skin rash. What is the probability that a patient who takes the drug will experience relief for at least one of the two symptoms?

36. In a study of 100 students who had been awarded university scholarships, it was found that 40 had part-time jobs, 25 had made the dean's list the previous semester, and 15 had both a part-time job and had made the dean's list. What was the probability that a student had a part-time job or was on the dean's list?

37. A survey of the subscribers to *Fortune* magazine showed that 46% have mutual funds, 63% have money market funds, and 74% have mutual funds and/or money market funds (*Fortune Subscriber Portrait*, 1988). What is the probability a subscriber will have investments in both money market and mutual funds? What is the probability a subscriber will not have investments in either type of fund?

SELF TEST ▶ 38. The survey of subscribers to *Fortune* magazine referred to in Exercise 37 (*Fortune Subscriber Portrait*, 1988) showed 54% rented a car in the past 12 months for business reasons, 51% rented a car for personal reasons, and 72% rented a car for either business or personal reasons.
a. What is the probability a subscriber rented a car in the past 12 months for business reasons and for personal reasons?

b. What is the probability a subscriber did not rent a car in the past 12 months?

c. What is the probability a subscriber rented a car for business reasons only during the past 12 months?

39. Let

$$A = \text{the event that a person runs 5 miles or more per week}$$
$$B = \text{the event that a person dies of heart disease}$$
$$C = \text{the event that a person dies of cancer}$$

Further, suppose that $P(A) = .01$, $P(B) = .25$, and $P(C) = .20$.

a. Are events A and B mutually exclusive? Can you find $P(A \cap B)$?

b. Are events B and C mutually exclusive? Find the probability that a person dies of heart disease or cancer.

c. Find the probability that a person dies from causes other than cancer.

40. During winter in Cincinnati, Mr. Krebs experiences difficulty in starting his two cars. The probability that the first car starts is .80, and the probability that the second car starts is .40. There is a probability of .30 that both cars start.

a. Define the events involved and use probability notation to show the probability information given above.

b. What is the probability that at least one car starts?

c. What is the probability that Mr. Krebs cannot start either of the two cars?

41. Let A be an event that a person's primary method of transportation to and from work is an automobile and B be an event that a person's primary method of transportation to and from work is a bus. Suppose that in a large city we find $P(A) = .45$ and $P(B) = .35$.

a. Are events A and B mutually exclusive? What is the probability that a person uses an automobile or a bus in going to and from work?

b. Find the probability that a person's primary method of transportation is something other than a bus.

4.5   Conditional Probability

Often, the probability of an event is influenced by whether or not a related event has occurred. Suppose that we have an event A with probability $P(A)$. If we obtain new information and learn that a related event, denoted by B, has occurred, we will want to take advantage of this information in calculating a new probability for event A. This new probability of event A is written $P(A \mid B)$. The notation $\mid$ is used to denote the fact that we are considering the probability of event A *given* the condition that event B has occurred. Thus the notation $P(A \mid B)$ is read "the probability of A given B."

As an illustration of the application of *conditional probability*, consider the situation of the promotional status of male and female officers of a major metropolitan police force in the eastern United States. The police force consists of 1200 officers, 960 men and 240 women. Over the past 2 years, 324 officers on the police force have been awarded promotions. The specific breakdown of promotions for male and female officers is shown in Table 4.8.

After reviewing the promotional record, a committee of female officers raised a discrimination case on the basis that 288 male officers had received promotions, but only 36 female officers had received promotions. The police administration has argued that the relatively low number of promotions for female officers is due not to discrimination, but to the fact that there are relatively few female officers on the police force. Let us show how conditional probability could be used to analyze the discrimination charge.

TABLE 4.8
Promotional Status of Police Officers over the Past 2 Years

	Men	Women	Totals
Promoted	288	36	324
Not Promoted	672	204	876
Totals	960	240	1,200

Let

$$M = \text{event an officer is a man}$$
$$W = \text{event an officer is a woman}$$
$$A = \text{event an officer is promoted}$$
$$A^c = \text{event an officer is not promoted}$$

Dividing the data values in Table 4.8 by the total of 1200 officers permits us to summarize the available information in the following probability values:

$$P(M \cap A) = 288/1200 = .24 = \text{probability that a randomly selected officer is a man } and \text{ is promoted}$$

$$P(M \cap A^c) = 672/1200 = .56 = \text{probability that a randomly selected officer is a man } and \text{ is not promoted}$$

$$P(W \cap A) = 36/1200 = .03 = \text{probability that a randomly selected officer is a woman } and \text{ is promoted}$$

$$P(W \cap A^c) = 204/1200 = .17 = \text{probability that a randomly selected officer is a woman } and \text{ is not promoted}$$

Since each of these values gives the probability of the intersection of two events, the probabilities are given the name of *joint probabilities*. Table 4.9, which provides a summary of the probability information for the police officer promotion situation, is referred to as a *joint probability table*.

The values in the margins of the joint probability table provide the probabilities of each event separately. That is, $P(M) = .80$, $P(W) = .20$, $P(A) = .27$, and $P(A^c) = .73$. These probabilities are referred to as *marginal probabilities* because of their location in the margins of the joint probability table. We note that the marginal probabilities are given by

TABLE 4.9
Joint Probability Table for Promotions

Joint probabilities appear in the body of the table

	Men (M)	Women (W)	Totals
Promoted (A)	.24	.03	.27
Not Promoted (A^c)	.56	.17	.73
Totals	.80	.20	1.00

Marginal probabilities appear in the margins of the table

summing the joint probabilities across the appropriate row or column of the joint probability table. For instance, the marginal probability of being promoted is $P(A) = P(M \cap A) + P(W \cap A) = .24 + .03 = .27$. From the marginal probabilities, we see that 80% of the force is male, 20% of the force is female, 27% of all officers received promotions, and 73% were not promoted.

Let us begin the conditional probability analysis by computing the probability that an officer is promoted given that the officer is a man. In conditional probability notation we are attempting to determine $P(A \mid M)$. In order to calculate $P(A \mid M)$, we first realize that this notation simply means that we are considering the probability of the event A (promotion) given that the condition designated as event M (the officer is a man) is known to exist. Thus $P(A \mid M)$ tells us that we are now concerned only with the promotional status of the 960 male officers. Hence since 288 of the 960 male officers received promotions, the probability of being promoted given that the officer is a man is 288/960 = .30. In other words, given that an officer is a man, there has been a 30% chance of receiving a promotion over the past 2 years.

The above procedure was easy to apply in our illustration because the data values in Table 4.8 show the number of officers in each category. We now want to demonstrate how conditional probabilities such as $P(A \mid M)$ can be computed directly from probability information rather than the frequency data of Table 4.8.

We have shown that $P(A \mid M) = 288/960 = .30$. Let us now divide both the numerator and denominator of this fraction by 1200, the total number of officers in the study. Thus,

$$P(A \mid M) = \frac{288}{960} = \frac{288/1200}{960/1200} = \frac{.24}{.80} = .30$$

We now see that the conditional probability $P(A \mid M)$ can be computed as .24/.80. Refer to the joint probability table (Table 4.9). Note in particular that .24 is the joint probability of A and M; that is, $P(A \cap M) = .24$. Also note that .80 is the marginal probability that a randomly selected officer is a man—that is, $P(M) = .80$. Thus the conditional probability $P(A \mid M)$ can be computed as the ratio of the joint probability $P(A \cap M)$ to the marginal probability $P(M)$. That is,

$$P(A \mid M) = \frac{P(A \cap M)}{P(M)} = \frac{.24}{.80} = .30$$

The fact that conditional probabilities can be computed as the ratio of a joint probability to a marginal probability provides the following general formula for conditional probability calculations for two events A and B.

Conditional Probability

$$P(A \mid B) = \frac{P(A \cap B)}{P(B)} \tag{4.6}$$

or

$$P(B \mid A) = \frac{P(A \cap B)}{P(A)} \tag{4.7}$$

Let us return to the issue of discrimination against the female officers. The probabilities in Table 4.9 show that the probability of promotion of an officer is $P(A) = .27$ (regardless of whether that officer is male or female). However, the critical issue in the discrimination case involves the two conditional probabilities $P(A \mid M)$ and $P(A \mid W)$. That is, what is the probability of a promotion *given* that the officer is a man, and what is the probability of a promotion *given* that the officer is a woman? If these two probabilities are equal, there is no basis for a discrimination argument, since the chances of a promotion are the same for male and female officers. However, if the two conditional probabilities differ, there will be support for the position that male and female officers are treated differently when it comes to promotions.

We have already determined that $P(A \mid M) = .30$. Let us now use the probability values in Table 4.9 and the basic relationship of conditional probability (4.6) to compute the probability that a randomly selected officer is promoted given that the officer is a woman; that is, $P(A \mid W)$. Using (4.6), we obtain

$$P(A \mid W) = \frac{P(A \cap W)}{P(W)} = \frac{.03}{.20} = .15$$

What conclusions do you draw? The probability of a promotion given that the officer is a man is .30, twice the .15 probability of a promotion given that the officer is a woman. While the use of conditional probability does not in itself prove that discrimination exists in this case, the conditional probability values offer support for the argument presented by the female officers.

Independent Events

In the preceding illustration we saw that $P(A) = .27$, $P(A \mid M) = .30$, and $P(A \mid W) = .15$. This shows that the probability of a promotion (event A) is affected or influenced by whether the officer is male or female. Particularly, since $P(A \mid M) \neq P(A)$, we would say that events A and M are *dependent* events. That is, the probability of event A (promotion) is altered or affected by knowing whether or not M (the officer is a man) occurs. Similarly, with $P(A \mid W) \neq P(A)$, we would say that events A and W are *dependent* events. On the other hand, if the probability of event A were not changed by the existence of event M—that is, $P(A \mid M) = P(A)$—we would say that events A and M are *independent* events. This leads us to the following definition of the independence of two events:

Independent Events

Two events A and B are independent if

$$P(A \mid B) = P(A) \tag{4.8}$$

or

$$P(B \mid A) = P(B) \tag{4.9}$$

otherwise, the events are dependent.

Multiplication Law

While the addition law of probability is used to compute the probability of a union of two events, we can now show how the multiplication law can be used to find the probability of an intersection of two events. The multiplication law is based upon the definition of conditional probability. Using (4.6) and (4.7) and solving for $P(A \cap B)$, we obtain the *multiplication law:*

Multiplication Law

$$P(A \cap B) = P(B)P(A \mid B) \tag{4.10}$$

or

$$P(A \cap B) = P(A)P(B \mid A) \tag{4.11}$$

To illustrate the use of the multiplication law, consider a newspaper circulation department where it is known that 84% of the newspaper's customers subscribe to the daily edition of the paper. If we let D denote the event that a customer subscribes to the daily edition, $P(D) = .84$. In addition, it is known that the probability that a customer who already holds a daily subscription also subscribes to the Sunday edition (event S) is .75; that is, $P(S \mid D) = .75$. What is the probability that a customer subscribes to both the Sunday and daily editions of the newspaper? Using the multiplication law, we compute the desired $P(S \cap D)$ as follows:

$$P(S \cap D) = P(D)P(S \mid D) = .84(.75) = .63$$

This tells us that 63% of the newspaper's customers take both the Sunday and daily editions.

Before concluding this section, let us consider the special case of the multiplication law when the events involved are independent. Recall that earlier in this section we defined independent events to exist whenever $P(A \mid B) = P(A)$ or $P(B \mid A) = P(B)$. Hence using (4.10) and (4.11) for the special case of independent events, the multiplication law becomes

---

Multiplication Law for Independent Events

$$P(A \cap B) = P(A)P(B) \tag{4.12}$$

Thus to compute the probability of the intersection of two independent events, we simply multiply the corresponding probabilities. Note that the multiplication law for independent events provides another way to determine if A and B are independent. That is, if $P(A \cap B) = P(A)P(B)$, then A and B are independent; if $P(A \cap B) \neq P(A)P(B)$, then A and B are dependent.

As an application of the multiplication law for independent events, consider the situation of a service station manager who knows from past experience that 80% of the customers use a credit card when they purchase gasoline. What is the probability that the next two customers purchasing gasoline will each use a credit card? If we let

A = the event that the first customer uses a credit card
B = the event that the second customer uses a credit card

then the event of interest is $A \cap B$. Given no other information, it seems reasonable to assume that A and B are independent events. Thus

$$P(A \cap B) = P(A)P(B) = (.80)(.80) = .64$$

NOTES & COMMENTS

> Do not confuse the notion of mutually exclusive events with that of independent events. Two events with nonzero probabilities cannot be both mutually exclusive and independent. If one mutually exclusive event is known to occur, the probability of the other occurring is reduced to zero. They are therefore dependent.

❏ ❏ Exercises

Methods

SELF TEST ▶ 42. Suppose that we have two events A and B with $P(A) = .5$, $P(B) = .60$ and $P(A \cap B) = .40$.
a. Find $P(A \mid B)$.   **b.** Find $P(B \mid A)$.
c. Are A and B independent? Why or why not?

43. Assume that we have two events, A and B, that are mutually exclusive. Assume further that it is known that $P(A) = .30$ and $P(B) = .40$.
a. What is $P(A \cap B)$?   **b.** What is $P(A \mid B)$?
c. A student in statistics argues that the concepts of mutually exclusive events and independent events are really the same and that, if events are mutually exclusive, they must be independent. Do you agree with this statement? Use the probability information in this problem to justify your answer.
d. What general conclusion would you make about mutually exclusive and independent events given the results of this problem?

Applications

44. A Daytona Beach nightclub has the following data on the age and marital status of 140 customers:

		Marital Status	
		Single	*Married*
	Under 30	77	14
Age			
	30 or Over	28	21

a. Develop a joint probability table using this data.
b. Use the marginal probabilities to comment on the age of customers attending the club.
c. Use the marginal probabilities to comment on the marital status of customers attending the club.
d. What is the probability of finding a customer who is single and under the age of 30?
e. If a customer is under 30, what is the probability that he or she is single?
f. Is marital status independent of age? Explain, using probabilities.

SELF TEST ▶ 45. In a survey of MBA students, the following data were obtained on "Students' first reason for application to the school in which they matriculated" ("School Selection by Students," *GMAC Occasional Papers*, Stolzenberg and Giarrusso, March, 1988).

		Reason for Application			
		School Quality	School Cost or Convenience	Other	Totals
Enrollment Status	Full time	421	393	76	890
	Part time	400	593	46	1039
Totals		821	986	122	1929

a. Develop a joint probability table using this data.
b. Use the marginal probabilities of school quality, cost/convenience, and other to comment on the most important reason for choosing a school.
c. If a student goes full time, what is the probability school quality will be the first reason for choosing a school?
d. If a student goes part time, what is the probability school quality will be the first reason for choosing a school?
e. Let A be the event that a student is full-time and let B be the event that the student lists school quality as the first reason for applying. Are events A and B independent? Justify your answer.

46. A survey of automobile ownership was conducted for 200 families in Houston. The results of the study showing ownership of automobiles of United States and foreign manufacture are summarized as follows:

		Do You Own a U.S. Car?		
		Yes	No	Totals
Do you own a foreign car?	Yes	30	10	40
	No	150	10	160
Totals		180	20	200

a. Show the joint probability table for the above data.
b. Use the marginal probabilities to compare U.S. and foreign car ownership.
c. What is the probability that a family will own both a U.S. car and a foreign car?
d. What is the probability that a family owns a car, U.S. or foreign?
e. If a family owns a U.S. car, what is the probability that it also owns a foreign car?
f. If a family owns a foreign car, what is the probability that it also owns a U.S. car?
g. Are U.S. and foreign car ownership independent events? Explain.

47. The probability that Ms. Smith will get an offer on the first job she applies for is .5, and the probability that she will get an offer on the second job she applies for is .6. She thinks that the probability that she will get an offer on both jobs is .15.
a. Define the events involved, and use probability notation to state the probability information given above.
b. What is the probability that Ms. Smith gets an offer on the second job given that she receives an offer for the first job?
c. What is the probability that Ms. Smith gets an offer on at least one of the jobs she applies for?
d. What is the probability that Ms. Smith does not get an offer on either of the two jobs she applies for?
e. Are the job offers independent? Explain.

48. Shown are data from a sample of 80 families in a midwestern city. The data shows the record of college attendance by fathers and their oldest sons.

| | | Son | |
		Attended College	Did Not Attend College
Father	Attended College	18	7
	Did Not Attend College	22	33

a. Show the joint probability table.
b. Use the marginal probabilities to comment on the comparison between fathers and sons in terms of attending college.
c. What is the probability that a son attends college given that his father attended college?
d. What is the probability that a son attends college given that his father did not attend college?
e. Is attending college by the son independent of whether or not his father attended college? Explain, using probability values.

49. The Texas Oil Company provides a limited partnership arrangement whereby small investors can pool resources in order to invest in large-scale oil exploration programs. In the exploratory drilling phase, locations for new wells are selected based on the geologic structure of the proposed drilling sites. Experience shows that there is a .40 probability of a type A structure present at the site given a productive well. It is also known that 50% of all wells are drilled in locations with type A structure. Finally, 30% of all wells drilled are productive.
a. What is the probability of a well being drilled in a type A structure *and* being productive?
b. If the drilling process begins in a location with a type A structure, what is the probability of having a productive well at the location?
c. Is finding a productive well independent of the type A geologic structure? Explain.

50. The Grant Thornton public accounting firm conducted a survey to see how executives felt about the 1992 recession and recovery potential (*Journal of Accountancy,* February, 1992). Results showed that the probability of an executive indicating the existence of a recession was .74. The probability of an executive stating that both a recession existed and a recovery would occur within 6 months was .41.
a. Given that an executive indicated the existence of a recession, what is the probability that the executive felt that a recovery would occur within 6 months?
b. Assume that if an executive denied the existence of a recession, he or she also believed that a recovery had already begun. Construct a joint probability table for "recession," "no recession," and "recovery," "no recovery." Those that feel a recovery is already underway and those who feel one will occur in 6 months should be put in the same category for this table.
c. Use the joint probability table in (b) to find the marginal probability of a recovery.

51. A purchasing agent has placed a rush order for a particular raw material with two different suppliers, A and B. If neither order arrives in 4 days, the production process must be shut down until at least one of the orders arrives. The probability that supplier A can deliver the material in 4 days is .55. The probability that supplier B can deliver the material in 4 days is .35.
a. What is the probability that both suppliers deliver the material in 4 days? Since two separate suppliers are involved, we are willing to assume independence.
b. What is the probability that at least one supplier delivers the material in 4 days?
c. What is the probability the production process is shut down in 4 days because of a shortage in raw material (that is, both orders are late)?

52. In a 1992 study of the consumer's view of the economy, the probability that a consumer would buy a house during the year was .033 and the probability that a consumer would buy a car during the year was .168 (*U.S. News & World Report,* April 13, 1992). Assume that there was only a .004 probability that a consumer would buy a house and a car during the year.

   a. What is the probability that a consumer would buy either a car or a house during the year?
   b. What is the probability that a consumer would buy a car during the year given that the consumer purchased a house during the year?
   c. Are buying a car and buying a house independent events? Explain.

4.6   Bayes' Theorem

In the discussion of conditional probability, we indicated that revising probabilities when new information is obtained is an important phase of probability analysis. Often, we begin our analysis with initial or *prior* probability estimates for specific events of interest. Then, from sources such as a sample, a special report, a product test, and so on, we obtain some additional information about the events. Given this new information, we update the prior probability values by calculating revised probabilities, referred to as *posterior probabilities. Bayes' theorem* provides a means for making these probability calculations. The steps in this probability revision process are shown in Figure 4.8.

As an application of Bayes' theorem, consider a manufacturing firm that receives shipments of parts from two different suppliers. Let A_1 denote the event that a part is from supplier 1 and A_2 denote the event that a part is from supplier 2. Currently, 65% of the parts purchased by the company are from supplier 1, while the remaining 35% are from supplier 2. Thus if a part is selected at random, we would assign the prior probabilities $P(A_1) = .65$ and $P(A_2) = .35$.

The quality of the purchased parts varies with the source of supply. Based upon historical data, the quality ratings of the two suppliers are as shown in Table 4.10. Thus if we let G denote the event that a part is good and B denote the event that a part is bad, the information in Table 4.10 provides the following conditional probability values:

$$P(G \mid A_1) = .98 \qquad P(B \mid A_1) = .02$$
$$P(G \mid A_2) = .95 \qquad P(B \mid A_2) = .05$$

FIGURE 4.8
Probability Revision Using Bayes' Theorem

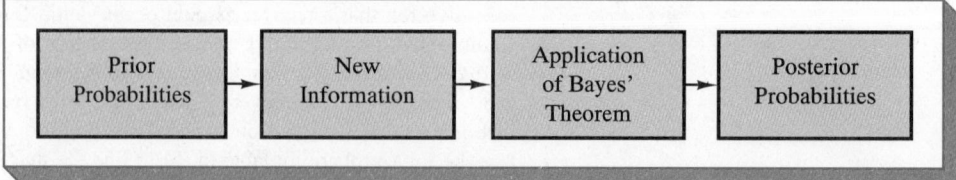

TABLE 4.10
Conditional Probabilities Based on Historical Quality Levels of Two Suppliers

	Percentage Good Parts	Percentage Bad Parts
Supplier 1	98	2
Supplier 2	95	5

The tree diagram in Figure 4.9 depicts the process of the firm receiving a part from one of the two suppliers and then discovering that the part is good or bad as a two-step experiment. We see that there are four experimental outcomes; two correspond to the part being good and two correspond to the part being bad.

Each of the experimental outcomes is the intersection of two events, so we can use the multiplication rule to compute the probabilities. For instance,

$$P(A_1, G) = P(A_1 \cap G) = P(A_1)P(G \mid A_1)$$

The process of computing these joint probabilities can be depicted in what is sometimes called a probability tree (see Figure 4.10). Moving from left to right through the tree, the

FIGURE 4.9
Two-Step Tree Diagram

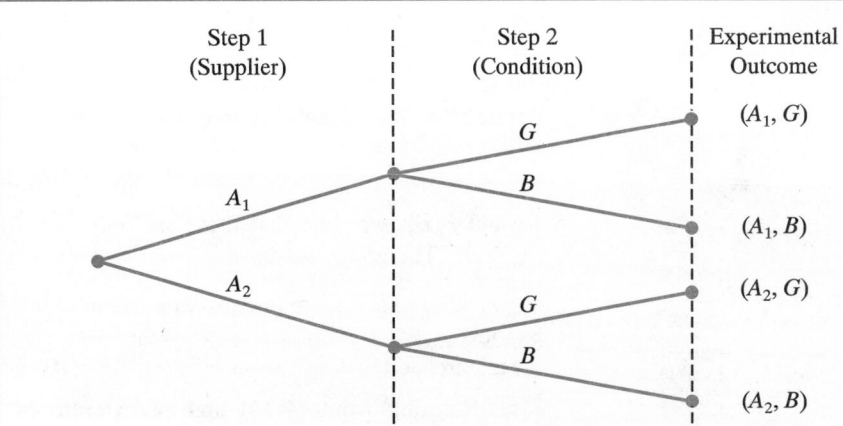

Note: Step 1 shows that the part comes from one of two suppliers, and Step 2 shows whether the part is good or bad.

FIGURE 4.10
Probability Tree for Two-Supplier Example

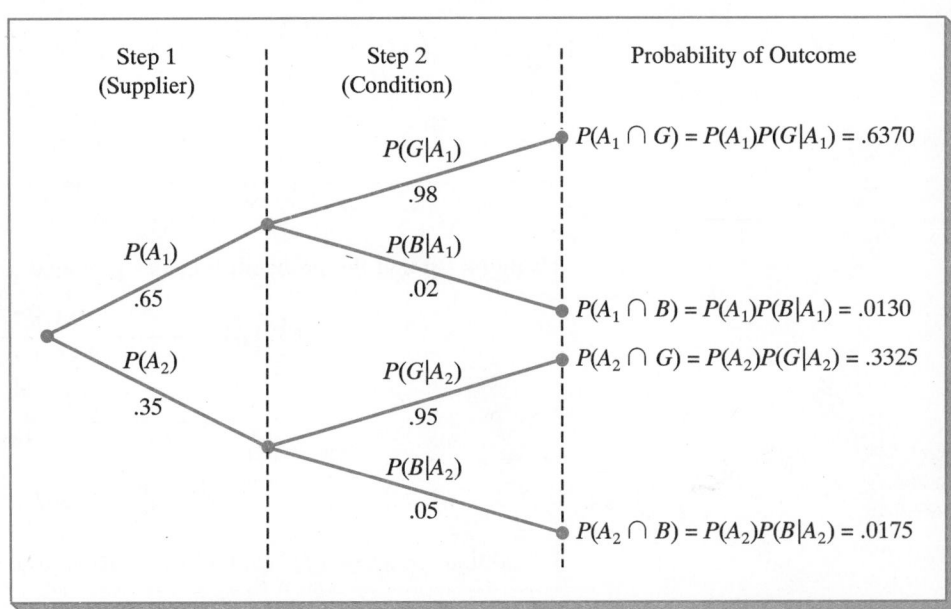

probabilities for each of the branches at step 1 are the prior probabilities, and the probabilities for each branch at step 2 are conditional probabilities. To find the probabilities of each experimental outcome, we simply multiply the probabilities on the branches leading to the outcome. Each of these joint probabilities is shown in Figure 4.10 along with the known probabilities for each branch.

Suppose now that the parts from the two suppliers are used in the firm's manufacturing process and that a machine breaks down because it attempts to process a bad part. Given the information that the part is bad, what is the probability it came from supplier 1 and what is the probability it came from supplier 2? With the information in the probability tree (Figure 4.10), Bayes' theorem can be used to answer these questions.

Letting B denote the event the part is bad, we are looking for the posterior probabilities $P(A_1 \mid B)$ and $P(A_2 \mid B)$. From the law of conditional probability, we know that

$$P(A_1 \mid B) = \frac{P(A_1 \cap B)}{P(B)} \tag{4.13}$$

Referring to the probability tree, we see that

$$P(A_1 \cap B) = P(A_1)P(B \mid A_1) \tag{4.14}$$

To find $P(B)$, we note that there are only two ways event B can occur: $(A_1 \cap B)$ and $(A_2 \cap B)$. Therefore, we have

$$P(B) = P(A_1 \cap B) + P(A_2 \cap B)$$
$$= P(A_1)P(B \mid A_1) + P(A_2)P(B \mid A_2) \tag{4.15}$$

Substituting from (4.14) and (4.15) into (4.13) and writing a similar result for $P(A_2 \mid B)$ we obtain Bayes' theorem for the case of two events.

Bayes' Theorem (Two-Event Case)

$$P(A_1 \mid B) = \frac{P(A_1)P(B \mid A_1)}{P(A_1)P(B \mid A_1) + P(A_2)P(B \mid A_2)} \tag{4.16}$$

$$P(A_2 \mid B) = \frac{P(A_2)P(B \mid A_2)}{P(A_1)P(B \mid A_1) + P(A_2)P(B \mid A_2)} \tag{4.17}$$

Using (4.16) and the probability values provided in our example, we have

$$P(A_1 \mid B) = \frac{P(A_1)P(B \mid A_1)}{P(A_1)P(B \mid A_1) + P(A_2)P(B \mid A_2)}$$

$$= \frac{(.65)(.02)}{(.65)(.02) + (.35)(.05)} = \frac{.0130}{.0130 + .0175}$$

$$= \frac{.0130}{.0305} = .4262$$

In addition, using (4.17), we find $P(A_2 \mid B)$ as follows:

$$P(A_2 \mid B) = \frac{(.35)(.05)}{(.65)(.02) + (.35)(.05)}$$

$$= \frac{.0175}{.0130 + .0175} = \frac{.0175}{.0305} = .5738$$

Note that in this application we initially started with a probability of .65 that a part selected at random was from supplier 1. However, given information that the part is bad, the probability that the part is from supplier 1 drops to .4262. In fact, if the part is bad, there is a better than 50–50 chance that the part came from supplier 2; that is, $P(A_2 \mid B) = .5738$.

Bayes' theorem is applicable when the events for which we want to compute posterior probabilities are mutually exclusive and their union is the entire sample space.* Bayes' theorem can be extended to the case where there are n mutually exclusive events $A_1, A_2, \ldots, A_n$ whose union is the entire sample space. In such a case Bayes' theorem for the computation of any posterior probability $P(A_i \mid B)$ appears as follows:

Bayes' Theorem

$$P(A_i \mid B) = \frac{P(A_i)P(B \mid A_i)}{P(A_1)P(B \mid A_1) + P(A_2)P(B \mid A_2) + \cdots + P(A_n)P(B \mid A_n)} \qquad (4.18)$$

With prior probabilities $P(A_1), P(A_2), \ldots, P(A_n)$ and the appropriate conditional probabilities $P(B \mid A_1), P(B \mid A_2), \ldots, P(B \mid A_n)$, Equation (4.18) can be used to compute the posterior probability of the events $A_1, A_2, \ldots, A_n$.

The Tabular Approach

A tabular approach is helpful in conducting the Bayes' theorem calculations. Such an approach is shown in Table 4.11 for the parts supplier problem. The computations shown there are conducted as follows:

Step 1. Prepare the following three columns:

Column 1—The mutually exclusive events for which posterior probabilities are desired.

Column 2—The prior probabilities for the events.

Column 3—The conditional probabilities of the new information *given* each event.

Step 2. In column 4, compute the joint probabilities for each event and the new information B by using the multiplication law. These joint probabilities are found by multiplying the prior probabilities in column 2 by the corresponding conditional probabilities in column 3—that is, $P(A_i \cap B) = P(A_i)P(B \mid A_i)$.

Step 3. Sum the joint probabilities in column 4. The sum is the probability of the new information, $P(B)$. Thus we see that in the above example there is a .0130

---

*If the union of events is the entire sample space, the events are often called *collectively exhaustive*.

TABLE 4.11
Summary of Bayes' Theorem
Calculations for the Two-Supplier
Problem

(1) Events A_i	(2) Prior Probabilities $P(A_i)$	(3) Conditional Probabilities $P(B \mid A_i)$	(4) Joint Probabilities $P(A_i \cap B)$	(5) Posterior Probabilities $P(A_i \mid B)$
A_1	.65	.02	.0130	.0130/.0305 = .4262
A_2	.35	.05	.0175	.0175/.0305 = .5738
	1.00		$P(B) = .0305$	1.0000

probability of a bad part from supplier 1 and there is a .0175 probability of a bad part from supplier 2. Since these are the only two ways in which a bad part can be obtained, the sum .0130 + .0175 shows that there is an overall probability of .0305 of finding a bad part from the combined shipments of both suppliers.

Step 4. In column 5, compute the posterior probabilities using the basic relationship of conditional probability

$$P(A_i \mid B) = \frac{P(A_i \cap B)}{P(B)}$$

Note that the joint probabilities $P(A_i \cap B)$ are found in column 4, whereas the probability $P(B)$ appears as the sum of column 4.

NOTES & COMMENTS

1. Bayes' theorem is used extensively in decision analysis (Chapter 22). The prior probabilities are often subjective estimates provided by a decision maker. Sample information is obtained and posterior probabilities are computed for use in developing a decision strategy.

2. An event and its complement are mutually exclusive, and their union is the entire sample space. Thus, Bayes' theorem is always applicable for computing posterior probabilities of an event and its complement.

☐ ☐ **Exercises**

Methods

53. The prior probabilities for events A_1 and A_2 are $P(A_1) = .40$ and $P(A_2) = .60$. It is also known that $P(A_1 \cap A_2) = 0$. Suppose $P(B \mid A_1) = .20$ and $P(B \mid A_2) = .05$.

SELF TEST ▶

a. Are A_1 and A_2 mutually exclusive? Why or why not?
b. Compute $P(A_1 \cap B)$ and $P(A_2 \cap B)$.
c. Compute $P(B)$.
d. Apply Bayes' theorem to compute $P(A_1 \mid B)$ and $P(A_2 \mid B)$.

54. The prior probabilities for events A_1, A_2, and A_3 are $P(A_1) = .20$, $P(A_2) = .50$, and $P(A_3) = .30$. The conditional probabilities of event B given A_1, A_2, and A_3 are $P(B \mid A_1) = .50$, $P(B \mid A_2) = .40$, and $P(B \mid A_3) = .30$.

a. Compute $P(B \cap A_1)$, $P(B \cap A_2)$, and $P(B \cap A_3)$.
b. Apply Bayes' theorem, Equation (4.18), to compute the posterior probability $P(A_2 \mid B)$.
c. Use the tabular approach to applying Bayes' theorem to compute $P(A_1 \mid B)$, $P(A_2 \mid B)$, and $P(A_3 \mid B)$.

Applications

55. A consulting firm has submitted a bid for a large research project. The firm's management initially felt there was a 50–50 chance of getting the bid. However, the agency to which the bid was submitted has subsequently requested additional information on the bid. Past experience indicates that on 75% of the successful bids and 40% of the unsuccessful bids the agency requested additional information.

a. What is your prior probability the bid will be successful (that is, prior to receiving the request for additional information)?

b. What is the conditional probability of a request for additional information given that the bid will ultimately be successful?

c. Compute a posterior probability that the bid will be successful given that a request for additional information has been received.

SELF TEST

56. A local bank is reviewing its credit-card policy with a view toward recalling some of its credit cards. In the past approximately 5% of cardholders have defaulted, and the bank has been unable to collect the outstanding balance. Thus management has established a prior probability of .05 that any particular cardholder will default. The bank has further found that the probability of missing one or more monthly payments for those customers who do not default is .20. Of course the probability of missing one or more payments for those who default is 1.

a. Given that a customer has missed a monthly payment, compute the posterior probability that the customer will default.

b. The bank would like to recall its card if the probability that a customer will default is greater than .20. Should the bank recall its card if the customer misses a monthly payment? Why or why not?

57. In a major eastern city, 60% of the automobile drivers are 30 years of age or older, and 40% of the drivers are under 30 years of age. Of all drivers 30 years of age or older, 4% will have a traffic violation in a 12-month period. Of all drivers under 30 years of age, 10% will have a traffic violation in a 12-month period. Assume that a driver has just been charged with a traffic violation; what is the probability that the driver is under 30 years of age?

58. A certain college football team plays 55% of its games at home and 45% of its games away. Given that the team has a home game, there is a .80 probability that it will win. Given that the team has an away game, there is a .65 probability that it will win. If the team wins on a particular Saturday, what is the probability that the game was played at home?

59. *M.D. Computing* (Vol. 8, No. 5, 1991) describes the use of Bayes' theorem and the use of conditional probability in medical diagnosis. Prior probabilities of diseases are based on the physician's assessment of such things as geographical location, seasonal influence, occurrence of epidemics, and so forth. Assume that a patient is believed to have one of two diseases, denoted D_1 and D_2, with $P(D_1) = .60$ and $P(D_2) = .40$ and that medical research has shown there is a probability associated with each symptom that may accompany the diseases. Suppose that, given diseases D_1 and D_2, the probabilities a patient will have symptoms S_1, S_2, or S_3 are as follows:

		Symptoms		
		S_1	S_2	S_3
Disease	D_1	.15	.10	.15 ← $P(S_3 \mid D_1)$
	D_2	.80	.15	.03

After finding a certain symptom is present, the medical diagnosis may be aided by finding the revised probabilities the patient has each particular disease. Compute the posterior probabilities of each disease given the following medical findings:

a. The patient has symptom S_1.

b. The patient has symptom S_2.

c. The patient has symptom S_3.

d. For the patient with symptom S_1 in (a), suppose we also find symptom S_2 present? What are the revised probabilities of D_1 and D_2?

Summary

In this chapter we have introduced basic probability concepts and illustrated how probability analysis can be used to provide helpful information for decision making. We described how probability can be interpreted as a numerical measure of the likelihood that an event will occur. In addition, we saw that the probability of an event could be computed either by summing the probabilities of the experimental outcomes (sample points) comprising the event or by using the relationships established by the addition, conditional probability, and multiplication laws of probability. For cases where additional information is available, we showed how Bayes' theorem could be used to obtain revised or posterior probabilities.

Glossary

Probability A numerical measure of the likelihood that an event will occur.

Experiment Any process which generates well-defined outcomes.

Sample space The set of all possible sample points (experimental outcomes).

Sample points The individual outcomes of an experiment.

Tree diagram A graphical device helpful in defining sample points of an experiment involving multiple steps.

Basic requirements of probability Two requirements which restrict the manner in which probability assignments can be made:

    a. For each experimental outcome E_i we must have $0 \le P(E_i) \le 1$.

    b. If there are k experimental outcomes, then $\Sigma P(E_i) = 1$.

Classical method A method of assigning probabilities which assumes that the experimental outcomes are equally likely.

Relative frequency method A method of assigning probabilities based upon experimentation or historical data.

Subjective method A method of assigning probabilities based upon judgment.

Event A collection of sample points.

Complement of event A The event containing all sample points that are not in A.

Venn diagram A graphical device for representing symbolically the sample space and operations involving events.

Union of events A and B The event containing all sample points that are in both A, in B, or in both. The union is denoted $A \cup B$.

Intersection of A and B The event containing all sample points that are in both A and B. The intersection is denoted $A \cap B$.

Addition law A probability law used to compute the probability of a union, $P(A \cup B)$. It is $P(A \cup B) = P(A) + P(B) - P(A \cap B)$. For mutually exclusive events, since $P(A \cap B) = 0$, it reduces to $P(A \cup B) = P(A) + P(B)$.

Mutually exclusive events Events that have no sample points in common; that is, $A \cap B$ is empty and $P(A \cap B) = 0$.

Conditional probability The probability of an event given that another event has occurred. The conditional probability of A given B is $P(A \mid B) = P(A \cap B)/P(B)$.

Independent events Two events A and B where $P(A \mid B) = P(A)$ or $P(B \mid A) = P(B)$; that is, the events have no influence on each other.

Multiplication law A probability law used to compute the probability of an intersection, $P(A \cap B)$. It is $P(A \cap B) = P(A)P(B \mid A)$ or $P(A \cap B) = P(B)P(A \mid B)$. For independent events it reduces to $P(A \cap B) = P(A)P(B)$.

Prior probabilities Initial estimates of the probabilities of events.

Posterior probabilities Revised probabilities of events based on additional information.

Bayes' theorem A method used to compute posterior probabilities.

Key Formulas

Counting Rule for Combinations

$$\binom{N}{n} = \frac{N!}{n!(N-n)!} \tag{4.1}$$

Computing Probability Using the Complement

$$P(A) = 1 - P(A^c) \tag{4.4}$$

Addition Law

$$P(A \cup B) = P(A) + P(B) - P(A \cap B) \tag{4.5}$$

Conditional Probability

$$P(A \mid B) = \frac{P(A \cap B)}{P(B)} \tag{4.6}$$

$$P(B \mid A) = \frac{P(A \cap B)}{P(A)} \tag{4.7}$$

Multiplication Law

$$P(A \cap B) = P(B)P(A \mid B) \tag{4.10}$$

$$P(A \cap B) = P(A)P(B \mid A) \tag{4.11}$$

Multiplication Law for Independent Events

$$P(A \cap B) = P(A)P(B) \tag{4.12}$$

Bayes' Theorem

$$P(A_i \mid B) = \frac{P(A_i)P(B \mid A_i)}{P(A_1)P(B \mid A_1) + P(A_2)P(B \mid A_2) + \cdots + P(A_n)P(B \mid A_n)} \tag{4.18}$$

❏   ❏   ## Supplementary Exercises

60. The Food and Drug Administration (FDA) places new drug applications in one of three categories:

> A = Potential breakthrough
>
> B = Improvement over existing product
>
> C = Me-too drug

During 1985, the FDA approved 3 As, 15 Bs, and 12 Cs (*Financial World,* January 24, 1989).

a. Consider the experiment of observing the category to which a new FDA-approved drug is assigned. How many experimental outcomes are there?

b. Using the data for 1985, assign probabilities to the experimental outcomes.

61. A financial manager has just made two new investments—one in the oil industry and one in municipal bonds. After a 1-year period, each of the investments will be classified as either successful or unsuccessful. Consider the making of the two investments as an experiment.

a. How many sample points exist for this experiment?

b. Show a tree diagram and list the sample points.

c. Let O = the event that the oil investment is successful and M = the event that the municipal bond investment is successful. List the sample points in O and in M.

d. List the sample points in the union of the events $(O \cup M)$.

e. List the sample points in the intersection of the events $(O \cap M)$.

f. Are events O and M mutually exclusive? Explain.

62. Consider an experiment where eight experimental outcomes exist. We will denote the experimental outcomes as $E_1, E_2, \ldots, E_8$. Suppose that the following events are identified:

$$A = \{E_1, E_2, E_3\}$$

$$B = \{E_2, E_4\}$$

$$C = \{E_1, E_7, E_8\}$$

$$D = \{E_5, E_6, E_7, E_8\}$$

Determine the sample points making up the following events:

a. $A \cup B$ **b.** $C \cup D$ **c.** $A \cap B$

d. $C \cap D$ **e.** $B \cap C$ **f.** A^c

g. D^c **h.** $A \cup D^c$ **i.** $A \cap D^c$

j. Are A and B mutually exclusive?

k. Are B and C mutually exclusive?

63. Referring to Exercise 62 and assuming that the classical method is an appropriate way of establishing probabilities, find the following probabilities:

a. $P(A)$, $P(B)$, $P(C)$, and $P(D)$

b. $P(A \cap B)$ **c.** $P(A \cup B)$ **d.** $P(A \mid B)$

e. $P(B \mid A)$ **f.** $P(B \cap C)$ **g.** $P(B \mid C)$

h. Are B and C independent events?

64. A survey of 2125 subscribers to *Fortune* magazine indicated the following with respect to the number of subscribers owning various credit cards (*Fortune Subscriber Portrait*, 1988).

Type of Card	Number Holding
American Express	1,360
Diners Club	234
Telephone Credit Card	1,530
Gasoline Credit Card	1,424
MasterCard	1,466
VISA	1,679
Discover	425

a. If 2083 subscribers indicated they have at least one credit card, what is the probability a *Fortune* subscriber does not have any credit cards?

b. What is the probability a subscriber holds an American Express card?

c. Suppose it were known that 55% have both a MasterCard and a VISA card. What is the probability an individual holds one or the other or both?

d. After studying the data, an analyst concluded that *at least* 956, or about 45%, of the subscribers must hold both an American Express and VISA card. Does this make sense? Why or why not?

TABLE 4.12

Number of Schools	Number of Students
1	1,230
2	304
3	184
4	118
5	78
6	51
7	25
8	13
9	20
10	8
11	9
12	6
Total	2,046

65. A survey of new matriculants to MBA programs was conducted by the Graduate Management Admissions Council during 1985 ("School Selection by Students," *GMAC Occasional Papers*, March 1988). Table 4.12 shows the number of schools to which students applied.

a. Use these data to assign probabilities to the number of schools to which an MBA student applies.

b. What is the probability a student will apply to only one school?

c. What is the probability a student will apply to three or more schools?

d. What is the probability a student will apply to more than six schools?

66. A telephone survey was used to determine viewer response to a new television show. The following data were obtained.

Rating	Frequency
Poor	4
Below average	8
Average	11
Above average	14
Excellent	13

50

a. What is the probability that a randomly selected viewer rates the new show as average or better?

b. What is the probability that a randomly selected viewer rates the new show below average or worse?

67. A bank has observed that credit-card account balances have been growing over the past year. A sample of 200 customer accounts resulted in the data found in Table 4.13.

a. Let A be the event that a customer's balance is less than $200. Find $P(A)$.

b. Let B be the event that a customer's balance is $300 or more. Find $P(B)$.

TABLE 4.13

Amount Owed ($)	Frequency
0–99	62
100–199	46
200–299	24
300–399	30
400–499	26
500 and over	12

68. Data for 2018 students from the GMAC MBA new-matriculants survey (*GMAC Occasional Papers*, March, 1988) shows the following:

		Applied to More than One School	
		Yes	No
Age Group	23 and under	207	201
	24–26	299	379
	27–30	185	268
	31–35	66	193
	36 and over	51	169

a. Prepare a joint probability table for the experiment consisting of observing the age and number of schools to which a randomly selected MBA student applies.

b. What is the probability an applicant will be 23 or under?

c. What is the probability an applicant will be older than 26?

d. What is the probability an applicant applies to more than one school?

69. Refer again to the data from the GMAC new-matriculants survey in Exercise 68.

a. Given that a person applied to more than one school, what is the probability the person is 24–26 years old?

b. Given that a person is in the 36-and-over age group, what is the probability the person applies to more than one school?

c. What is the probability a person is 24–26 years old *or* applies to more than one school?

d. Suppose a person is known to have applied to only one school. What is the probability the person is 31 or more years old?

e. Is the number of schools applied to independent of age? Explain.

70. Suppose that $P(A) = .30$, $P(B) = .25$, and $P(A \cap B) = .20$.

a. Find $P(A \cup B)$, $P(A \mid B)$, and $P(B \mid A)$.

b. Are events A and B independent? Why or why not?

71. Suppose that $P(A) = .40$, $P(A \mid B) = .60$, and $P(B \mid A) = .30$.

a. Find $P(A \cap B)$ and $P(B)$.

b. Are events A and B independent? Why or why not?

72. Suppose that $P(A) = .60$, $P(B) = .30$, and events A and B are mutually exclusive.

a. Find $P(A \cup B)$ and $P(A \cap B)$.

b. Are events A and B independent?

c. Can you make a general statement about whether mutually exclusive events can be independent?

73. A market survey of 800 people found the following facts about the ability to recall a television commercial for a particular product and the actual purchase of the product:

	Could Recall Television Commercial	Could Not Recall Television Commercial	Totals
Purchased	160	80	240
Did Not Purchase	240	320	560
Totals	400	400	800

Let T be the event of the person recalling the television commercial and B the event of buying or purchasing the product.

a. Find $P(T)$, $P(B)$, and $P(T \cap B)$.

b. Are T and B mutually exclusive events? Use probability values to explain.

c. What is the probability that a person who could recall seeing the television commercial has actually purchased the product?

d. Are T and B independent events? Use probability values to explain.

e. Comment on the value of the commercial in terms of its relationship to purchasing the product.

74. A research study investigating the relationship between smoking and heart disease in a sample of 1000 men over 50 years of age provided the following data:

	Smoker	Nonsmoker	Totals
Record of Heart Disease	100	80	180
No Record of Heart Disease	200	620	820
Totals	300	700	1000

a. Show a joint probability table that summarizes the results of this study.

b. What is the probability a man over 50 years of age is a smoker and has a record of heart disease?

c. Compute and interpret the marginal probabilities.

d. Given that a man over 50 years of age is a smoker, what is the probability that he has heart disease?

Condition

Focus

e. Given that a man over 50 years of age is a nonsmoker, what is the probability that he has heart disease?

f. Does the research show that heart disease and smoking are independent events? Use probability to justify your answer.

g. What conclusion would you draw about the relationship between smoking and heart disease?

75. A large consumer goods company has been running a television advertisement for one of its soap products. A survey was conducted. On the basis of this survey probabilities were assigned to the following events:

$$B = \text{individual purchased the product}$$

$$S = \text{individual recalls seeing the advertisement}$$

$$B \cap S = \text{individual purchased the product and recalls seeing the advertisement}$$

The probabilities assigned were $P(B) = .20$, $P(S) = .40$, and $P(B \cap S) = .12$. The following problems *relate to this situation*:

a. What is the probability of an individual's purchasing the product given that the individual recalls seeing the advertisement? Does seeing the advertisement increase the probability the individual will purchase the product? As a decision maker, would you recommend continuing the advertisement (assuming that the cost is reasonable)?

b. Assume that those individuals who do not purchase the company's soap product buy from its competitors. What would be your estimate of the company's market share? Would you expect that continuing the advertisement will increase the company's market share? Why or why not?

c. The company has also tested another advertisement and assigned it values of $P(S) = .30$ and $P(B \cap S) = .10$. What is $P(B \mid S)$ for this other advertisement? Which advertisement seems to have had the bigger effect on customer purchases?

76. A large company has done a careful analysis of a price promotion that it is currently testing. Some 20% of the people in a large sample of individuals in the test market were both aware of the promotion and made a purchase. It was further found that 80% were aware of the promotion and that prior to the promotion 25% of all people in the sample were purchasers of the product.

a. What is the probability that a person will make a purchase given that he or she is aware of the price promotion?

b. Are the events "made a purchase" and "aware of the price promotion" independent? Why or why not?

c. On the basis of these results, would you recommend that the company introduce this promotion on a national scale? Why or why not?

77. Cooper Realty is a small real estate company located in Albany, New York, specializing primarily in residential listings. They have recently become interested in the possibility of determining the likelihood of one of their listings being sold within a certain number of days. An analysis of company sales of 800 homes for the previous years produced the accompanying data:

	Days Listed Until Sold			
Initial Asking Price	*Under 30*	*31–90*	*Over 90*	**Totals**
Under $50,000	50	40	10	100
$50,000–99,999	20	150	80	250
$100,000–150,000	20	280	100	400
Over $150,000	10	30	10	50
Totals	100	500	200	800

a. If A is defined as the event that a home is listed for over 90 days before being sold, estimate the probability of A.

b. If B is defined as the event that the initial asking price is under $50,000, estimate the probability of B.

c. What is the probability of $A \cap B$?

d. Assuming that a contract has just been signed to list a home that has an initial asking price of less than $50,000, what is the probability the home will take Cooper Realty more than 90 days to sell?

e. Are events A and B independent?

78. In the evaluation of a sales training program, a firm found that of 50 salespersons making a bonus last year, 20 had attended a special sales training program. The firm has 200 salespersons. Let B = the event that a salesperson makes a bonus and S = the event a salesperson attends the sales training program.

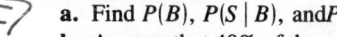

a. Find $P(B)$, $P(S \mid B)$, and $P(S \cap B)$.

b. Assume that 40% of the salespersons have attended the training program. What is the probability that a salesperson makes a bonus given that the salesperson attended the sales training program, $P(B \mid S)$?

c. If the firm evaluates the training program in terms of the effect it has on the probability of a salesperson's making a bonus, what is your evaluation of the training program? Comment on whether B and S are dependent or independent events.

79. A company has studied the number of lost-time accidents occurring at its Brownsville, Texas, plant. Historical records show that 6% of the employees had lost-time accidents last year. Management believes that a special safety program will reduce such accidents to 5% during the current year. In addition, it is estimated that 15% of those employees who had lost-time accidents last year will have a lost-time accident during the current year.

a. What percentage of the employees will have lost-time accidents in both years?

b. What percentage of the employees will have at least one lost-time accident over the 2-year period?

80. In a study of television viewing habits among married couples, a researcher found that, for a popular Saturday night program, 25% of the husbands viewed the program regularly and 30% of the wives viewed the program regularly. The study found that, for couples where the husband watches the program regularly, 80% of the wives also watch regularly.

a. What is the probability that both the husband and wife watch the program regularly?

b. What is the probability that at least one—husband or wife—watches the program regularly?

c. What percentage of married couples do not have at least one regular viewer of the program?

81. A statistics professor has noted from past experience that students who do the homework for the course have a .90 probability of passing the course. On the other hand, students who do not do the homework for the course have a .25 probability of passing the course. The professor estimates that 75% of the students in the course do the homework. Given a student who passes the course, what is the probability that she or he completed the homework?

82. A salesperson for Business Communication Systems, Inc. sells automatic envelope-addressing equipment to medium- and small-size businesses. The probability of making a sale to a new customer is .10. During the initial contact with a customer, sometimes the salesperson will be asked to call back later. Of the 30 most recent sales, 12 were made to customers who initially told the salesperson to call back later. Of 100 customers who did not make a purchase, 17 had initially asked the salesperson to call back later. If a customer asks the salesperson to call back later, should the salesperson do so? What is the probability of making a sale to a customer who has asked the salesperson to call back later?

83. Migliori Industries, Inc. manufactures a gas-saving device for use on natural gas forced-air residential furnaces. The company is currently trying to determine the probability that sales of this product will exceed 25,000 units during next year's winter sales period. The company believes that sales of the product depend to a large extent on the winter conditions. Management's best estimate is that the probability that sales will exceed 25,000 units if the winter is severe is .8. This probability drops to .5 if the winter conditions are moderate. If the weather forcast is .7 for a severe winter and .3 for moderate conditions, what is Migliori's best estimate that sales will exceed 25,000 units?

84. The Dallas IRS auditing staff is concerned with identifying potential fraudulent tax returns. From past experience they believe that the probability of finding a fraudulent return given that the return contains deductions for contributions exceeding the IRS standard is .20. Given that the deductions for contributions do not exceed the IRS standard, the probability of a fraudulent return decreases to .02. If 8% of all returns exceed the IRS standard for deductions due to contributions, what is the best estimate of the percentage of fraudulent returns?

85. An oil company has purchased an option on land in Alaska. Preliminary geologic studies have assigned the following prior probabilities:

$$P(\text{high-quality oil}) = .50$$

$$P(\text{medium-quality oil}) = .20$$

$$P(\text{no oil}) = .30$$

a. What is the probability of finding oil?

b. After 200 feet of drilling on the first well, a soil test is taken. The probabilities of finding the particular type of soil identified by the test are as follows:

$$P(\text{soil} \mid \text{high-quality oil}) = .20$$

$$P(\text{soil} \mid \text{medium-quality oil}) = .80$$

$$P(\text{soil} \mid \text{no oil}) = .20$$

How should the firm interpret the soil test? What are the revised probabilities, and what is the new probability of finding oil?

86. In the setup of a manufacturing process, a machine is either correctly or incorrectly adjusted. The probability of a correct adjustment is .90. When correctly adjusted, the machine operates with a 5% defective rate. However, if it is incorrectly adjusted, a 75% defective rate occurs.

a. After the machine starts a production run, what is the probability that a defect is observed when one part is tested?

b. Suppose that the one part selected by an inspector is found to be defective. What is the probability that the machine is incorrectly adjusted? What action would you recommend?

c. Before your recommendation in (b) was followed, a second part is tested and found to be good. Using your revised probabilities from (b) as the most recent prior probabilities, compute the revised probability of an incorrect adjustment given that the second part is good. What action would you recommend now?

87. The Wayne Manufacturing Company purchases a certain part from three suppliers A, B, and C. Supplier A supplies 60% of the parts, B 30%, and C 10%. The quality of parts is known to vary among suppliers, with A, B, and C parts having .25%, 1%, and 2% defective rates, respectively. The parts are used in one of the company's major products.

a. What is the probability that the company's major product is assembled with a defective part?

b. When a defective part is found, which supplier is the likely source?

88. A Bayesian approach can be used to revise probabilities that a prospect field will produce oil (*Oil & Gas Journal*, January 11, 1988). In one case, geological assessment indicates a 25% chance the field will produce oil. Further, there is an 80% chance that a particular well will strike oil given that oil is present on the prospect field.

a. Suppose that one well is drilled on the field and it comes up dry. What is the probability the prospect field will produce oil?

b. If two wells come up dry, what is the probability the field will produce oil?

c. The oil company would like to keep looking as long as the chances of finding oil are greater than 1%. How many dry wells must be drilled before the field will be abandoned?

5

Discrete Probability Distributions

Contents

Xerox Corporation*

STAMFORD, CONNECTICUT

Xerox Corporation is in the information products and systems business worldwide. Everyone is familiar with the Xerox copy machines, but the company is involved in many other businesses as well. Xerox's Multinational Documentation & Training Services (MD&TS) provides customers with timely, cost-effective, and high-quality communication services. The four basic services of MD&TS are documentation, training, translation, and publishing.

The professional writers and translators working for MD&TS use an on-line computerized publication system. For this system, management was interested in determining the effect of different system configurations on performance. Specifically, for a given system configuration, management was interested in determining the following:

1. The probability of a user being refused access by the system because of an excess number of users.
2. The probability of any specific number of users being on the system simultaneously.

A computer simulation model was developed for the purpose of determining these probabilities. In order to build the simulation model, it was necessary to identify the probability distribution for two key random variables:

1. The On Time per session (the length of time a user is on the system).
2. The Idle Time per session (the length of time between sessions).

Based on a survey of users, the probability distribution of on time per session was approximated as shown in the table below. Another probability distribution was developed for the random variable indicating idle time per session. These two probability distributions were key inputs to the simulation model. The results from the simulation study helped MD&TS to determine a system configuration that ensured a near-zero probability that a user would be refused access to the system.

Probability distributions, such as the ones used by Xerox Corporation in this simulation study, are studied in this chapter. We will learn about some of the discrete probability distributions that are available and about when these distributions should be used.

❏ Probability Distribution of On Time per Session

x	$f(x)$
10	.05
20	.06
30	.08
40	.20
50	.25
60	.20
70	.08
80	.06
90	.02

The actual probability distribution used in the simulation study has been modified to protect proprietary information and to simplify the discussion.

*The authors are indebted to Soterios M. Flouris for providing this Statistics in Practice.

I n this chapter we continue the study of probability by introducing the concepts of random variables and probability distributions. The focus of this chapter is on discrete probability distributions. Three special discrete probability distributions—the binomial, Poisson, and hypergeometric—are studied.

5.1   Random Variables

In Chapter 4 we defined the concept of an experiment and its associated experimental outcomes. A random variable provides a means of assigning numerical values to experimental outcomes. The definition of a random variable is as follows.

Random Variable

A *random variable* is a numerical description of the outcome of an experiment.

A random variable provides a means of associating a numerical value with each possible experimental outcome. The particular numerical value that the random variable takes on depends upon the outcome of the experiment. That is, the value of the random variable is not known until the experimental outcome is observed.

Suppose we consider the experiment of selling automobiles for one day at a particular dealership. In this case if we let x = number of cars sold, then x is a random variable whose possible values are $0, 1, 2, \ldots$. For another example, consider the fact that to receive state certification as a medical lab technician, candidates must pass a series of three examinations. If we let x = number of examinations passed, then x is a random variable that may assume the values 0, 1, 2, and 3. Some additional examples of experiments and associated random variables are given in Table 5.1.

Although many experiments such as those listed in Table 5.1 have experimental outcomes that are numerical values, others do not. For example, the outcome for the experiment of tossing a coin one time is either a head or a tail, neither of which has a natural numerical value. However, we still may want to express the outcome numerically. Thus,

---

TABLE 5.1

Examples of Random Variables

Experiment	Random Variable (x)	Possible Values for the Random Variable
Make 100 sales calls	Total number of sales	$0, 1, 2, \ldots, 100$
Inspect a shipment of 70 radios	Number of defective radios	$0, 1, 2, \ldots, 70$
Work 1 year on a project to build a new library	Percentage of project completed after 6 months	$0 \leq x \leq 100$
Operate a restaurant	Number of customers entering in one day	$0, 1, 2, \ldots$
Observe cars passing a checkpoint	Number of cars passing the checkpoint	$0, 1, 2, \ldots$

we need a rule that can be used to assign a numerical value to each of the experimental outcomes. One possibility is to let the random variable $x = 1$ if the experimental outcome is a head and $x = 0$ if the experimental outcome is a tail. While the numerical values for the random variable x are arbitrary, they are acceptable in terms of the definition of a random variable—namely, x is a random variable because it provides a numerical description of the outcome of the experiment.

A random variable can be classified as either *discrete* or *continuous* depending upon the numerical values it can assume. A random variable that may assume either a finite number of values or an infinite sequence (e.g., 1, 2, 3, . . .) of values is referred to as a *discrete random variable*. The number of units sold, the number of defects observed, and the number of customers that enter a bank during one day of operation are examples of discrete random variables. The first two and last two random variables listed in Table 5.1 are discrete random variables. Random variables such as weight, time, and temperature, which may take on all values in a certain interval or collection of intervals, are referred to as *continuous random variables*. For instance, the third random variable in Table 5.1 (percentage of project completed after 6 months) is a continuous random variable because it may take on any value in the interval from 0 to 100 (e.g., 56.33 or 64.227). Continuous random variables will be discussed further in Chapter 6.

NOTES & COMMENTS

> One way to determine whether a random variable is discrete or continuous is to think of the values of the random variable as points on a line segment. If the entire line segment between any two of these points also represents values the random variable may assume, the random variable is continuous.

❑ ❑ Exercises

Methods

✓ SELF TEST ▶   **1.** Consider the experiment of tossing a coin twice.
a. List the experimental outcomes.
b. Define a random variable that represents the number of heads occurring on the two tosses.
c. Show what value the random variable would assume for each of the experimental outcomes.
d. Is this random variable discrete or continuous?

2. Consider the experiment of a worker assembling a product, and record how long it takes.
a. Define a random variable that represents the time in minutes required to assemble the product.
b. What values may the random variable assume?
c. Is the random variable discrete or continuous?

Applications

✓ SELF TEST ▶   **3.** Three students have interviews scheduled for summer employment at the Brookwood Institute. In each case the result of the interview will either be that a position is offered or not offered. Experimental outcomes are defined in terms of the results of the three interviews.
a. List the experimental outcomes.
b. Define a random variable that represents the number of offers made. Is this a discrete or continuous random variable?
c. Show the value of the random variable for each of the experimental outcomes.

4. Subscribers of the Turner Broadcasting System may select one or more of the following services: Cable News Network, Superstation WTBS, Headline News, and Turner Network Television (*Business Week*, April 17, 1989). Suppose an experiment is designed to take a sample of five subscribers' selections. The random variable x is defined as the number in the sample who receive the Cable News Network service. What values may the random variable take on?

5. In order to perform a certain type of blood analysis, lab technicians have to perform two procedures. The first procedure requires either 1 or 2 separate steps, and the second procedure requires either 1, 2, or 3 steps.
   a. List the experimental outcomes associated with performing an analysis.
   b. If the random variable of interest is the total number of steps required to do the complete analysis, show what value the random variable will assume for each of the experimental outcomes.

6. Listed is a series of experiments and associated random variables. In each case, identify the values that the random variable can take on and state whether the random variable is discrete or continuous.

Experiment	Random Variable (x)
a. Take a 20-question examination	Number of questions answered correctly
b. Observe cars arriving at a tollbooth for 1 hour	Number of cars arriving at tollbooth
c. Audit 50 tax returns	Number of returns containing errors
d. Observe an employee's work	Number of nonproductive hours in an 8-hour work day
e. Weigh a shipment of goods	Number of pounds

5.2   Discrete Probability Distributions

TABLE 5.2

Probability Distribution for the Number of Children per Household

x	$f(x)$
0	.18
1	.39
2	.24
3	.14
4	.04
5	.01
Total	1.00

The *probability distribution* for a random variable describes how the probabilities are distributed over the values of the random variable. For a discrete random variable x, the probability distribution is defined by a *probability function,* denoted by $f(x)$. The probability function provides the probability for each value of the random variable. As an illustration of a discrete random variable and its probability distribution, we consider a study of 300 households in a village on the coast of Maine. As part of this study, data were collected showing the number of children in each household. The following results were obtained: 54 of the households had no children, 117 had 1 child, 72 had 2 children, 42 had 3 children, 12 had 4 children, and 3 had 5 children.

Suppose we consider the experiment of randomly selecting one of these households to participate in a follow-up study. If we let x denote the number of children in the household selected, possible values of x are 0, 1, 2, 3, 4, and 5. Thus, $f(0)$ provides the probability that a randomly selected household has no children, $f(1)$ provides the probability that a randomly selected household has 1 child, and so on. Since 54 of the 300 households have no children, we assign the value $54/300 = .18$ to $f(0)$. Similarly, since 117 of the 300 households have 1 child, we assign the value $117/300 = .39$ to $f(1)$. Continuing in this fashion for the other values of the random variable x, we obtain the discrete probability distribution shown in Table 5.2.

In the development of the probability function for a discrete random variable, the following two conditions must be satisfied.

FIGURE 5.1

Graphical Representation of the Probability Distribution for Number of Children per Household

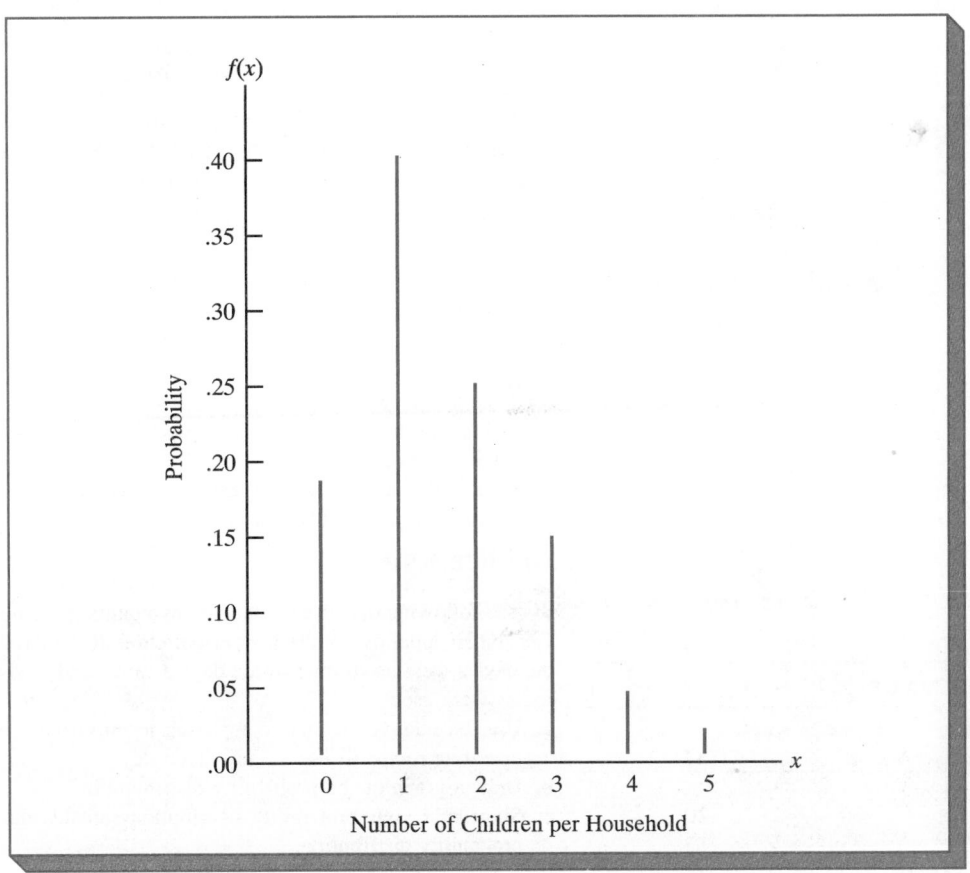

Number of Children per Household

Required Conditions for a Discrete Probability Function

$$f(x) \geq 0 \qquad\qquad (5.1)$$

$$\Sigma\, f(x) = 1 \qquad\qquad (5.2)$$

TABLE 5.3

x	$f(x)$
1	1/10
2	2/10
3	3/10
4	4/10

Table 5.2 shows that the probabilities for the random variable x satisfy condition (5.1), since $f(x)$ is greater than or equal to 0 for all values of x. In addition, since the probabilities sum to 1 and (5.2) is satisfied, the probability function is a valid discrete probability function.

We can also present probability distributions graphically. In Figure 5.1 the values of the random variable x are shown on the horizontal axis and the probability that x assumes these values is shown on the vertical axis.

For some discrete random variables, the probability distribution can be given as a formula that yields $f(x)$ for every possible value of x. Consider the random variable x and its probability distribution, as shown in Table 5.3. This probability distribution can also be given by the formula

$$f(x) = \frac{x}{10} \quad \text{for } x = 1, 2, 3, 4$$

The more widely used discrete probability distributions are usually specified by formulas. Three important cases are the binomial, Poisson and hypergeometric probability distributions, which are discussed later in the chapter.

☐ ☐ Exercises

Methods

SELF TEST ▶ 7. Table 5.4 shows the probability distribution of the random variable x.
a. Is this a proper probability distribution? Check to see that (5.1) and (5.2) are satisfied.
b. What is the probability $x = 30$?
c. What is the probability x is less than or equal to 25?
d. What is the probability x is greater than 30?

Applications

SELF TEST ▶ 8. The following data were collected by counting the number of operating rooms in use at Tampa General Hospital over a 20-day period: On 3 of the days only 1 operating room was used, on 5 of the days 2 were used, on 8 of the days 3 were used, and on 4 days all 4 of the hospital's operating rooms were used.

TABLE 5.4

x	$f(x)$
20	.20
25	.15
30	.25
35	.40
Total	1.00

a. Use the relative frequency approach to construct a probability distribution for the number of operating rooms in use on any given day.
b. Draw a graph of the probability distribution.
c. Show that your probability distribution satisfies the required conditions for a valid discrete probability distribution.

9. According to *Dataquest,* the personal computer market during 1986 was as follows:

Company	Company Identification Code	Market Share (Percentage)
IBM	1	29.8
Apple	2	9.2
COMPAQ	3	6.6
Zenith	4	3.5
Tandy	5	6.6
Commodore	6	1.9
Other	7	42.4

Let x be a random variable based on the company identification code.
a. Use these data to develop a probability distribution. Specify the values for the random variable and the corresponding values for the probability function, $f(x)$.
b. Draw a graph of the probability distribution.
c. Show that the probability distribution satisfies (5.1) and (5.2).

TABLE 5.5

x	$f(x)$
$148,000	.20
$150,000	.40
$152,000	.40

10. QA Properties is considering making an offer to purchase an apartment building. Management has subjectively assessed a probability distribution for x, the purchase price, as shown in Table 5.5.
a. Determine if this is a proper probability distribution by checking (5.1) and (5.2).
b. What is the probability that the apartment house can be purchased for $150,000 or less?

TABLE 5.6

x	f(x)
1,000	.15
1,100	.20
1,200	.30
1,300	.25
1,400	.10

11. The cleaning and changeover operation for a production system requires from 1, 2, 3, or 4 hours, depending upon the specific product that will begin production. Let x be a random variable indicating the time in hours required to make the changeover. The following probability function can be used to compute the probability associated with any changeover time x:

$$f(x) = \frac{x}{10} \quad \text{for } x = 1, 2, 3, \text{ or } 4$$

a. Show that the probability function meets the required conditions of (5.1) and (5.2).
b. What is the probability that the changeover will take 2 hours?
c. What is the probability that the changeover will take more than 2 hours?
d. Graph the probability distribution for the changeover times.

12. The director of admissions at Lakeville Community College has subjectively assessed a probability distribution for x, the number of entering students, as shown in Table 5.6.
a. Is this a valid probability distribution?
b. What is the probability there will be 1200 or fewer entering students?

13. A psychologist has determined that the number of hours required to obtain the trust of a new patient is either 1, 2, or 3. Let x be a random variable indicating the time in hours required to gain the patient's trust. The following probability function has been proposed.

$$f(x) = \frac{x}{6} \quad \text{for } x = 1, 2, \text{ or } 3$$

a. Is this a valid probability function? Explain.
b. What is the probability that it takes exactly 2 hours to gain the patient's trust?
c. What is the probability that it takes at least 2 hours to gain the patient's trust?

14. Table 5.7 shows a partial probability distribution for the MRA Company's projected profits (x = profit in $1000s) for the first year of operation (the negative value denotes a loss).
a. What is the value of f(200)? What is your interpretation of this value?
b. What is the probability that MRA will be profitable?
c. What is the probability that MRA will make at least $100,000?

TABLE 5.7

x	f(x)
−100	.10
0	.20
50	.30
100	.25
150	.10
200	

5.3   Expected Value and Variance

Expected Value

TABLE 5.8

Expected Value of the Number of Children per Household

x	f(x)	xf(x)
0	.18	0(.18) = .00
1	.39	1(.39) = .39
2	.24	2(.24) = .48
3	.14	3(.14) = .42
4	.04	4(.04) = .16
5	.01	5(.01) = .05
		1.50

$$E(x) = \mu = \Sigma\, xf(x)$$

The *expected value*, or mean, of a random variable is a measure of the central location for the random variable. The mathematical expression for the expected value of a discrete random variable x is as follows.

> **Expected Value of a Discrete Random Variable**
>
> $$E(x) = \mu = \Sigma xf(x) \qquad (5.3)$$

Both the notations $E(x)$ and μ can be used to denote the expected value of a random variable.

Equation (5.3) shows that in order to compute the expected value of a discrete random variable, we must multiply each value of the random variable by the corresponding probability $f(x)$ and then add the resulting products. Recall the problem involving a study of 300 households in a village on the coast of Maine. Table 5.8 shows the calculation of the expected value of the number of children in those households. We see that the expected value or mean is 1.50 children per household.

Variance

While the expected value provides the mean value for the random variable, we often need a measure of dispersion, or variability, for the random variable. Just as we used variance in Chapter 3 to summarize the dispersion in a data set, we now use the variance measure to summarize the variability in the values of a random variable. The mathematical expression for the variance of a discrete random variable is as follows.

Variance of a Discrete Random Variable

$$\text{Var}(x) = \sigma^2 = \Sigma(x - \mu)^2 f(x) \tag{5.4}$$

As (5.4) shows, an essential part of the variance formula is the deviation, $x - \mu$, which measures how far a particular value of the random variable is from the expected value or mean, μ. In computing the variance of a random variable, the deviations are squared and then weighted by the corresponding value of the probability function. The sum of these weighted squared deviations for all values of the random variable is referred to as the *variance*. Both the notations $\text{Var}(x)$ and σ^2 are used to denote the variance of a random variable.

The calculation of the variance for the probability distribution of the number of children per household is summarized in Table 5.9. We see that the variance for the number of children per household is 1.25. The *standard deviation,* σ, is defined as the positive square root of the variance. Thus, the standard deviation of the number of children per household is

$$\sigma = \sqrt{1.25} = 1.118$$

The standard deviation is measured in the same units as the random variable ($\sigma = 1.118$ children per household); for this reason σ is often preferred in describing the variability of a random variable. The variance σ^2 is measured in squared units and is thus more difficult to interpret.

TABLE 5.9

Calculation of Variance for Number of Children per Household

x	$x - \mu$	$(x - \mu)^2$	$f(x)$	$(x - \mu)^2 f(x)$
0	$0 - 1.50 = -1.50$	2.25	.18	$2.25(.18) = .4050$
1	$1 - 1.50 = -.50$	.25	.39	$.25(.39) = .0975$
2	$2 - 1.50 = .50$	.25	.24	$.25(.24) = .0600$
3	$3 - 1.50 = 1.50$	2.25	.14	$2.25(.14) = .3150$
4	$4 - 1.50 = 2.50$	6.25	.04	$6.25(.04) = .2500$
5	$5 - 1.50 = 3.50$	12.25	.01	$12.25(.01) = \underline{.1225}$
				1.2500

$$\sigma^2 = \Sigma(x - \mu)^2 f(x)$$

Expected Value of the Sum of Random Variables

Occasionally, an analyst wants to compute the expected value of a sum of random variables. For instance, DiCarlo Motors has dealerships in Saratoga and Albany. Let x = the number of cars sold per day at the Saratoga dealership and y = the number of cars sold per day at the Albany dealership. Suppose DiCarlo is interested in the expected total daily sales for both dealerships. To compute this expected value, we need to calculate the expected value of the sum of the random variables x and y (i.e., $E(x + y)$). Equation (5.5) shows that the expected value of the sum of two random variables is given by the sum of expected values.

Expected Value of the Sum of Two Random Variables

$$E(x + y) = E(x) + E(y) \tag{5.5}$$

Suppose that the expected number of daily sales at the Saratoga dealership is 2.50 automobiles and the expected number of daily sales at the Albany dealership is 2.85 automobiles. Using (5.5), the expected total daily sales for the dealerships is

$$E(x + y) = E(x) + E(y)$$
$$= 2.50 + 2.85 = = 5.35$$

Equation (5.5) can be extended to provide a method for computing the expected value of the sum of any number of random variables. The result is that the expected value of the sum of any number of random variables is equal to the sum of their individual expected values. For example, if DiCarlo Motors had a third dealership and if we let z represent the daily sales at the third dealership, then the expected value of the total daily sales for all three dealerships would be $E(x + y + z) = E(x) + E(y) + E(z)$.

Variance of the Sum of Independent Random Variables

In the previous chapter, we said that two events were independent if the occurrence of one of the events did not affect the probability of the other event occurring. Similarly, two random variables are said to be independent if the value that one takes on does not affect the probabilities associated with the value that the other can take on. Equation (5.6) provides the variance for the sum of two independent random variables.

Variance of the Sum of Two Independent Random Variables

$$\text{Var}(x + y) = \text{Var}(x) + \text{Var}(y) \tag{5.6}$$

As an illustration of (5.6), let us compute the variance of DiCarlo's total sales at the Saratoga and Albany dealerships. Suppose that the variance of daily sales at Saratoga is 1.55 and at Albany is 2.13. If DiCarlo feels that the sales at the dealerships are independent, the variance of total sales is given by

$$\text{Var(total sales)} = \text{Var(Saratoga sales)} + \text{Var(Albany sales)}$$
$$= 1.55 + 2.13 = 3.68$$

Note that the standard deviation of total sales is given by $\sigma = \sqrt{3.68} = 1.92$; it is not the sum of the individual standard deviations.

Equation (5.6) can be generalized to provide a method for computing the variance of the sum of any number of independent random variables. The result is that the variance of the sum is equal to the sum of the variances. For example, if DiCarlo Motors had a third dealership and we let z be the daily sales at the third dealership, then the variance of the total sales is $\text{Var}(x + y + z) = \text{Var}(x) + \text{Var}(y) + \text{Var}(z)$. Of course, in order to apply this result to DiCarlo Motors, we must be convinced that daily sales of all three dealerships are independent.

□ □ **Exercises**

Methods

15. Table 5.10 shows a probability distribution for the random variable x.
a. Compute $E(x)$, the expected value of x.
b. Compute σ^2, the variance of x.
c. Compute σ, the standard deviation of x.

SELF TEST ▶ 16. Shown below is a probability distribution for the random variable y.

TABLE 5.10

x	$f(x)$
3	.25
6	.50
9	.25
Total	1.00

y	$f(y)$
2	.20
4	.30
7	.40
8	.10
Total	1.00

a. Compute $E(y)$.
b. Compute $\text{Var}(y)$ and σ.
c. Compute $E(x + y)$ using the probability distribution in Exercise 15 for x and the one here for y.
d. Compute $\text{Var}(x + y)$ and the standard deviation of $x + y$ using the probability distribution in Exercise 15 for x and the one here for y. (Assume x and y are independent.)

Applications

17. A volunteer ambulance service handles from 0 to 5 service calls on any given day. Assume the probability distribution for the number of service calls is as shown in Table 5.11.
a. What is the expected number of service calls?
b. What is the variance in the number of service calls? What is the standard deviation?

SELF TEST ▶ 18. Individual investors allocated their funds among four different investment categories in 1988 (*Money*, January 1989). The annual return for the categories and the proportion of investments in each category are as follows:

TABLE 5.11

Number of Service Calls	Probability
0	.10
1	.15
2	.30
3	.20
4	.15
5	.10

Investment Category	Annual Return	Proportion of Total Investments
Stocks	26.0%	.31
Bonds	8.6%	.23
CDs and Money Funds	7.7%	.45
Real Estate and Gold	−2.9%	.01

Let x be a random variable indicating the annual return percentage for a $1 investment. Assume the probabilities of each investment category, $f(x)$, are provided by the proportion of total investments data.

a. What is the expected annual return percentage on a $1 investment?

b. What are the variance and standard deviation?

c. Suppose the annual return on stocks drops to 5%. Recompute the expected return and the standard deviation of the return on a $1 investment.

19. The actual shooting records of the 1992 NCAA championship final four teams (*NCAA Final Four Program,* April 1992) showed the probability of making a 2-point basket was .50 while the probability of making a 3-point basket was .39.

a. What is the expected value of a 2-point shot for these teams?

b. What is the expected value of a 3-point shot for these teams?

c. Since the probability of making a 2-point basket is greater than the probability of making a 3-point basket, why do coaches allow some players to shoot the 3-point shot if they have the opportunity? Use expected value to explain your answer.

20. The probability distribution for damage claims paid by the Newton Automobile Insurance Company on collision insurance is shown in Table 5.12.

a. Use the expected collision payment to determine the collision insurance premium that would allow the company to break even.

b. The insurance company charges an annual rate of $260 for the collision coverage. What is the expected value of the collision policy for a policyholder? (Hint: It is the expected payments from the company minus the cost of coverage.) Why does the policyholder purchase a collision policy with this expected value?

TABLE 5.12

Payment ($)	Probability
0	.90
400	.04
1000	.03
2000	.01
4000	.01
6000	.01

21. The number of dots observed on the upward face of a die has the following probability function.

$$f(x) = \frac{1}{6} \quad \text{for } x = 1, 2, 3, 4, 5, 6$$

a. Show that this probability function possesses the properties necessary for probability distributions.

b. Draw a graph of the probability distribution.

c. What is the expected value? What is the interpretation of this value?

d. What are the variance and the standard deviation for the number of dots?

TABLE 5.13

Unit demand	Probability
300	.20
400	.30
500	.35
600	.15

22. The demand for a product of Carolina Industries varies greatly from month to month. Based on the past 2 years of data, the probability distribution in Table 5.13 shows the company's monthly demand.

a. If the company places monthly orders based on the expected value of the monthly demand, what should Carolina's monthly order quantity be for this product?

b. Assume that each unit demanded generates $70 in revenue and that each unit ordered costs $50. How much will the company gain or lose in a month if it places an order based on your answer to part (a) and the actual demand for the item is 300 units?

23. What are the variance and the standard deviation for the number of units demanded in Exercise 22?

24. The J. R. Ryland Computer Company is considering a plant expansion that will enable the company to begin production of a new computer product. The company's president must determine whether to make the expansion a medium- or large-scale project. An uncertainty involves the demand for the new product, which for planning purposes may be low demand, medium demand, or high demand. The probability estimates for the demands are .20, .50, and .30, respectively. Letting x indicate the annual profit in $1000s, the firm's planners have developed the following profit forecasts for the medium- and large-scale expansion projects:

		Medium-Scale Expansion Profits		Large-Scale Expansion Profits	
		x	f(x)	y	f(y)
Demand	Low	50	.20	0	.20
	Medium	150	.50	100	.50
	High	200	.30	300	.30

a. Compute the expected value for the profit associated with the two expansion alternatives. Which decision is preferred for the objective of maximizing the expected profit?

b. Compute the variance for the profit associated with the two expansion alternatives. Which decision is preferred for the objective of minimizing the risk or uncertainty?

25. A brokerage firm has offices in Houston and Dallas. The daily commissions at the Houston offices show $E(x) = \$2000$ and a standard deviation of $300, whereas daily commissions at Dallas show $E(y) = \$1500$ and a standard deviation of $400.

a. What are the total expected daily commissions for the two offices?

b. Assuming that daily commissions at the two offices are independent, determine the standard deviation of total sales for the two offices.

26. The expected production times and variances for a three-stage assembly are as follows:

Stage	Expected Time	Variance
A	5.4	1.00
B	3.2	.64
C	8.4	1.69

a. What is the expected assembly time for a part passing through all three stages?

b. Assuming the assembly stages operate independently, what are the variance and standard deviation of the total assembly time?

5.4   The Binomial Probability Distribution

The *binomial probability distribution* is a discrete probability distribution that has many applications. It is associated with a multiple-step experiment that we call the binomial experiment.

Properties of a Binomial Experiment

For a probability experiment to be classified as a *binomial experiment*, it must have the following four properties:

1. The experiment consists of a sequence of n identical trials.
2. Two outcomes are possible on each trial. We refer to one outcome as a *success* and the other as a *failure*.
3. The probability of a success on one trial, denoted by p, does not change from trial to trial. Consequently, the probability of failure on one trial, denoted by $1 - p$, does not change from trial to trial.
4. The trials are independent.

If Properties 2, 3, and 4 are present, we say the trials are generated by a Bernoulli process. If, in addition, Property 1 is present, we say we have a *binomial experiment*. Figure 5.2 depicts one possible sequence of outcomes of a binomial experiment involving 8 trials.

In a binomial experiment, our interest is in the *number of successes occurring in the n trials*. If we let x denote the number of successes occurring in the n trials, we see that x can assume the values of 0, 1, 2, 3, . . . , n. Since the number of values is finite, x is a *discrete* random variable. The probability distribution associated with this random variable is called the *binomial probability distribution*. For an example, consider the experiment of tossing a coin 5 times and on each toss observing whether the coin lands with a head or a tail on its upward face. Suppose we are interested in counting the number of heads appearing during the 5 tosses. Does this experiment have the properties of a binomial experiment? What is the random variable of interest? Note that:

1. The experiment consists of 5 identical trials, where each trial involves the tossing of one coin.
2. There are two outcomes possible for each trial. The possible outcomes are a head and a tail. We can designate head a success and tail a failure.
3. The probability of a head and the probability of a tail are the same for each trial, with $p = .5$ and $1 - p = .5$.
4. The trials or tosses are independent, since the outcome on any one trial is not affected by what happens on other trials or tosses.

Thus, the properties of a binomial experiment are satisfied. The random variable of interest is $x =$ the number of heads appearing in the 5 trials. In this case, x can take on the values of 0, 1, 2, 3, 4, or 5.

FIGURE 5.2

Diagram of an 8-Trial Binomial Experiment

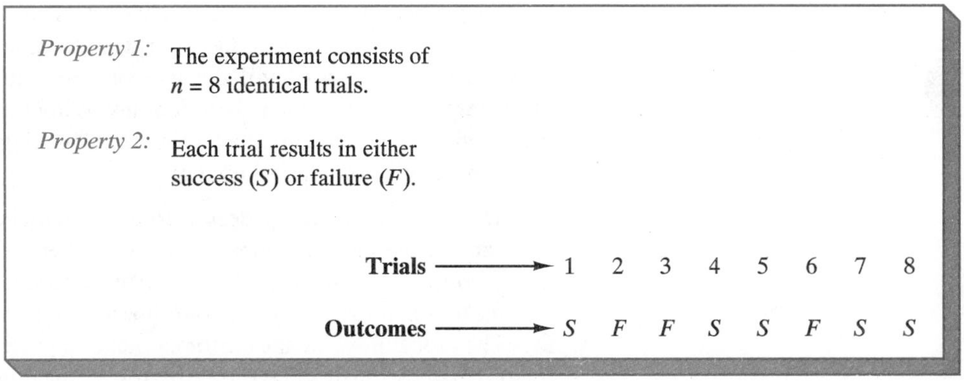

Property 1: The experiment consists of $n = 8$ identical trials.

Property 2: Each trial results in either success (S) or failure (F).

Trials ⟶	1	2	3	4	5	6	7	8
Outcomes ⟶	S	F	F	S	S	F	S	S

As another example, consider an insurance salesperson who pays a visit to 10 randomly selected families. The outcome associated with each visit is classified as a success if the family purchases an insurance policy and a failure if the family does not. From past experience, the salesperson knows the probability that a randomly selected family purchases an insurance policy is .10. Checking the properties of a binomial experiment, we observe the following:

1. The experiment consists of 10 identical trials, where each trial involves contacting one family.
2. There are two outcomes possible on each trial: The family purchases a policy (success) or the family does not purchase a policy (failure).
3. The probabilities of a purchase and a nonpurchase are assumed to be the same for each sales call, with $p = .10$ and $1 - p = .90$.
4. The trials are independent since the families are randomly selected.

Since the four assumptions are satisfied, this is a binomial experiment. The random variable of interest is the number of sales obtained in contacting the 10 families. In this case, x can assume the values of 0, 1, 2, 3, 4, 5, 6, 7, 8, 9, and 10.

Property 3 of the binomial experiment is called the *stationarity assumption* and is sometimes confused with Property 4, independence of trials. To see how they differ, consider again the case of the salesperson calling on families to sell insurance policies. If, as the day wore on, the salesperson got tired and lost enthusiasm, then the probability of success (selling a policy) might drop to .05, for example, by the tenth call. In such a case, Property 3 (stationarity) would not be satisfied, and we would not have a binomial experiment. This would be true even if Property 4 held—that is, the purchase decisions of each family were made independently.

In applications involving binomial experiments, a special mathematical formula, called the *binomial probability function,* can be used to compute the probability of x successes in the n trials. Using probability concepts introduced in Chapter 4, we will show how the formula can be developed in the context of an illustrative problem.

The Nastke Clothing Store Problem

Let us consider the purchase decisions of the next three customers who enter the Nastke Clothing Store. Based on past experience, the store manager estimates the probability that any one customer will make a purchase is .30. What is the probability that two of the next three customers will make a purchase?

Using a tree diagram (Figure 5.3), we can see that the experiment of observing the three customers each making a purchase decision has 8 possible outcomes. Using S to denote success (a purchase) and F to denote failure (no purchase), we are interested in experimental outcomes involving two successes in the three trials (purchase decisions). Next, let us verify that the experiment involving the sequence of three purchase decisions can be viewed as a binomial experiment. Checking the four requirements for a binomial experiment, we note the following:

1. The experiment can be described as a sequence of three identical trials, one trial for each of the three customers who will enter the store.
2. Two outcomes—the customer makes a purchase (success) or the customer does not make a purchase (failure)—are possible for each trial.
3. The probability that the customer makes a purchase (.30) or does not make a purchase (.70) is assumed to be the same for all customers.
4. The purchase decision of each customer is independent of the decisions of the other customers.

FIGURE 5.3
**Tree Diagram for the Nastke
Clothing Store Problem**

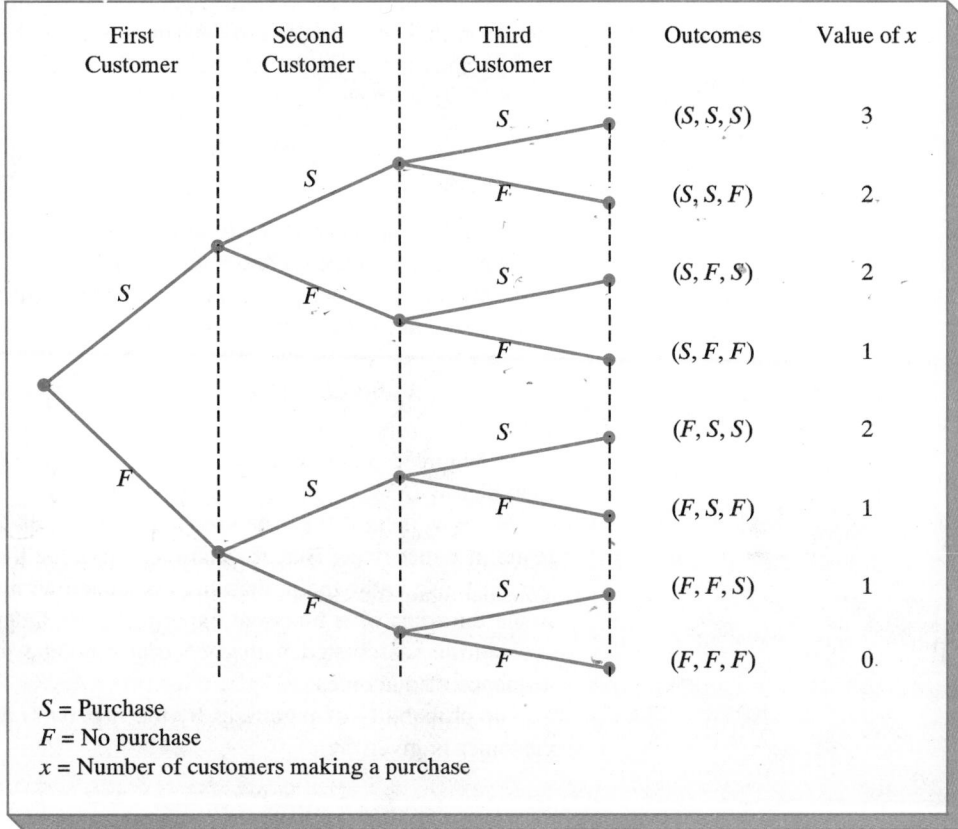

			Outcomes	Value of x
			(S, S, S)	3
			(S, S, F)	2
			(S, F, S)	2
			(S, F, F)	1
			(F, S, S)	2
			(F, S, F)	1
			(F, F, S)	1
			(F, F, F)	0

S = Purchase
F = No purchase
x = Number of customers making a purchase

Thus the properties of a binomial experiment are present.

The number of experimental outcomes resulting in exactly x successes in n trials can be computed from the following formula.*

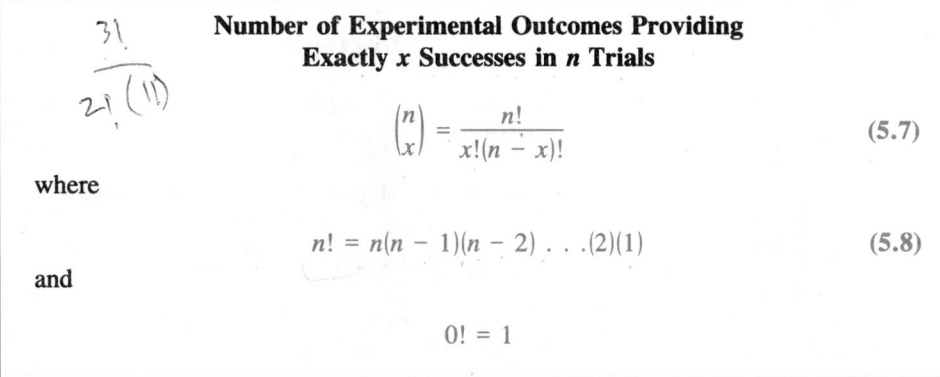

**Number of Experimental Outcomes Providing
Exactly x Successes in n Trials**

$$\binom{n}{x} = \frac{n!}{x!(n - x)!} \tag{5.7}$$

where

$$n! = n(n - 1)(n - 2) \ldots (2)(1) \tag{5.8}$$

and

$$0! = 1$$

*This is the formula introduced in Chapter 4 (4.1) to determine the number of combinations of n objects selected x at a time. For the binomial experiment, this combinatorial formula provides the number of experimental outcomes (sequences of n trials) resulting in x successes.

Now let us return to the Nastke experiment involving three customers' purchase decisions. Equation (5.7) can be used to determine the number of experimental outcomes involving two purchases; that is, the number of ways of obtaining $x = 2$ successes in the $n = 3$ trials. From (5.7) we have

$$\binom{n}{x} = \binom{3}{2} = \frac{3!}{2!(3-2)!} = \frac{(3)(2)(1)}{(2)(1)(1)} = \frac{6}{2} = 3$$

Formula (5.7) shows that three of the outcomes yield two successes. From Figure 5.3 we see these three outcomes are denoted by *SSF, SFS,* and *FSS.*

Using (5.7) to determine how many experimental outcomes have three successes (purchases) in the three trials, we obtain:

$$\binom{n}{x} = \binom{3}{3} = \frac{3!}{3!(3-3)!} = \frac{3!}{3!0!} = \frac{(3)(2)(1)}{(3)(2)(1)(1)} = \frac{6}{6} = 1$$

From Figure 5.3 we see that the one experimental outcome with three successes is identified by *SSS.*

We know that (5.7) can be used to determine the number of experimental outcomes that result in x successes. But, if we are to determine the probability of x successes in n trials, we must also know the probability associated with each of these experimental outcomes. Since the trials of a binomial experiment are independent, we can simply multiply the probabilities associated with each trial outcome to find the probability of a particular sequence of outcomes.

The probability of purchases by the first two customers and no purchase by the third customer is given by

$$pp(1 - p)$$

With a .30 probability of a purchase on any one trial, the probability of a purchase on the first two trials and no purchase on the third is given by

$$(.30)(.30)(.70) = (.30)^2(.70) = .063$$

There are two other sequences of outcomes resulting in two successes and one failure. The probabilities for all three sequences involving two successes are shown.

Trial Outcomes				
1st Customer	2nd Customer	3rd Customer	Success-Failure Notation	Probability of Experimental Outcome
Purchase	Purchase	No purchase	SSF	$pp(1 - p) = p^2(1 - p)$ $= (.30)^2(.70) = .063$
Purchase	No purchase	Purchase	SFS	$p(1 - p)p = p^2(1 - p)$ $= (.30)^2(.70) = .063$
No purchase	Purchase	Purchase	FSS	$(1 - p)pp = p^2(1 - p)$ $= (.30)^2(.70) = .063$

Observe that all three outcomes with two successes have exactly that same probability. This observation holds in general. In any binomial experiment, each sequence of trial outcomes yielding x successes in n trials has the *same probability* of occurrence. The probability of each sequence of trials yielding x successes in n trials is as follows:

$$\text{Probability of a Particular Sequence of Trial Outcomes with } x \text{ Successes in } n \text{ Trials} = p^x(1 - p)^{(n-x)} \tag{5.9}$$

For the Nastke Clothing Store, this formula shows that any outcome with two successes has a probability of $p^2(1 - p)^{(3-2)} = p^2(1 - p)^1 = (.30)^2(.70)^1 = .063$, as shown.

Since (5.7) shows the number of outcomes in a binomial experiment with x successes and since (5.9) gives the probability for each sequence involving x successes, we combine (5.7) and (5.9) to obtain the following *binomial probability function*.

Binomial Probability Function

$$f(x) = \binom{n}{x} p^x (1 - p)^{(n-x)} \tag{5.10}$$

where

$f(x)$ = the probability of x successes in n trials

n = the number of trials

$$\binom{n}{x} = \frac{n!}{x!(n-x)!}$$

p = the probability of a success on any one trial

$(1 - p)$ = the probability of a failure on any one trial

In the Nastke Clothing Store example, let us compute the probability of each of the following: no customer makes a purchase, exactly one customer makes a purchase, exactly two customers make a purchase, and all three customers make a purchase. The calculations are summarized in Table 5.14. This table provides a tabular presentation for the probability distribution of number of customer purchases. A graph of this probability distribution is shown in Figure 5.4.

The binomial probability function can be applied to *any* binomial experiment. If we are satisfied that a situation has the properties of a binomial experiment and if we know the values of n, p, and $(1 - p)$, (5.10) can be used to compute the probability of x successes in the n trials.

If we consider variations of the Nastke experiment, such as 10 customers rather than 3 entering the store, the binomial probability function given by (5.10) is still applicable. For example, the probability of making exactly 4 sales to 10 potential customers entering the store is

$$f(4) = \frac{10!}{4!6!}(.30)^4(.70)^6 = .2001$$

This is a binomial experiment with $n = 10$, $x = 4$, and $p = .30$.

TABLE 5.14

Probability Distribution for the Number of Customers Making a Purchase

x	$f(x)$
0	$\frac{3!}{0!3!}(.30)^0(.70)^3 = .343$
1	$\frac{3!}{1!2!}(.30)^1(.70)^2 = .441$
2	$\frac{3!}{2!1!}(.30)^2(.70)^1 = .189$
3	$\frac{3!}{3!0!}(.30)^3(.70)^0 = \underline{.027}$
	1.000

FIGURE 5.4
Graphical Representation of the Probability Distribution for the Nastke Clothing Store Problem

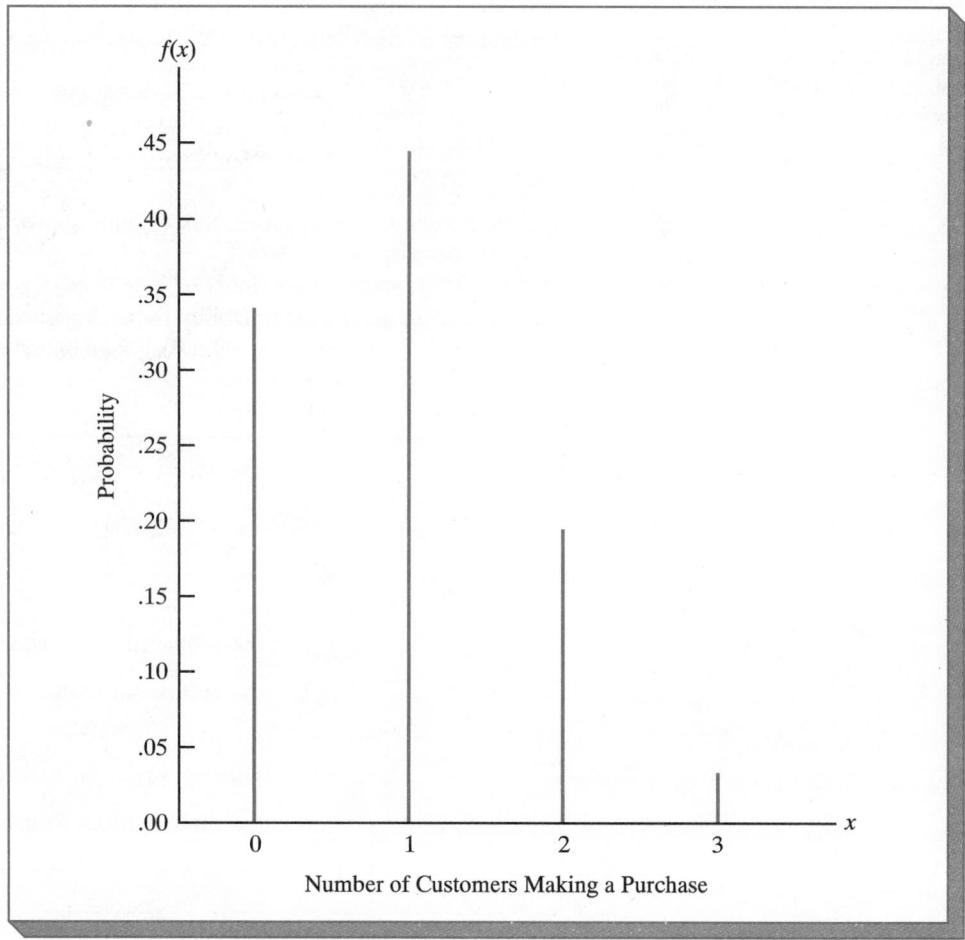

Using Tables of Binomial Probabilities

Tables have been developed that give the probability of x successes in n trials for a binomial experiment. These tables are generally easy to use and quicker than (5.10), especially when the number of trials involved is large. A table of binomial probabilities is provided as Table 5 of Appendix B. A portion of this table is given in Table 5.15. In order to use this table, it is necessary to specify the values of n, p, and x for the binomial experiment of interest. In the example at the top of Table 5.15, we see that the probability of $x = 3$ successes in a binomial experiment with $n = 10$ and $p = .40$ is .2150. You might want to use (5.10) to verify that this is the answer you would obtain using the binomial probability function directly.

Also check the use of this table by using it to verify the probability of 4 successes in 10 trials for the Nastke Clothing Store problem. Note that the value of $f(4) = .2001$ can be read directly from the table of binomial probabilities, with $n = 10$, $x = 4$, and $p = 30$, making it unnecessary to perform the calculations required by (5.10).

While the tables of binomial probabilities are relatively easy to use, it is impossible to have tables that show all possible values of n and p that might be encountered in a binomial experiment. However, with today's calculators, it is not too difficult to calculate the desired probability using (5.10), especially if the number of trials is not too large. In

TABLE 5.15
Selected Values from the Binomial Probability Table
Example: $n = 10$, $x = 3$, $p = .40$;
$f(3) = .2150$

							p				
n	x	.05	.10	.15	.20	.25	.30	.35	.40	.45	.50
9	0	.6302	.3874	.2316	.1342	.0751	.0404	.0207	.0101	.0046	.0020
	1	.2985	.3874	.3679	.3020	.2253	.1556	.1004	.0605	.0339	.0176
	2	.0629	.1722	.2597	.3020	.3003	.2668	.2162	.1612	.1110	.0703
	3	.0077	.0446	.1069	.1762	.2336	.2668	.2716	.2508	.2119	.1641
	4	.0006	.0074	.0283	.0661	.1168	.1715	.2194	.2508	.2600	.2461
	5	.0000	.0008	.0050	.0165	.0389	.0735	.1181	.1672	.2128	.2461
	6	.0000	.0001	.0006	.0028	.0087	.0210	.0424	.0743	.1160	.1641
	7	.0000	.0000	.0000	.0003	.0012	.0039	.0098	.0212	.0407	.0703
	8	.0000	.0000	.0000	.0000	.0001	.0004	.0013	.0035	.0083	.0176
	9	.0000	.0000	.0000	.0000	.0000	.0000	.0001	.0003	.0008	.0020
10	0	.5987	.3487	.1969	.1074	.0563	.0282	.0135	.0060	.0025	.0010
	1	.3151	.3874	.3474	.2684	.1877	.1211	.0725	.0403	.0207	.0098
	2	.0746	.1937	.2759	.3020	.2816	.2335	.1757	.1209	.0763	.0439
	3	.0105	.0574	.1298	.2013	.2503	.2668	.2522	.2150	.1665	.1172
	4	.0010	.0112	.0401	.0881	.1460	.2001	.2377	.2508	.2384	.2051
	5	.0001	.0015	.0085	.0264	.0584	.1029	.1536	.2007	.2340	.2461
	6	.0000	.0001	.0012	.0055	.0162	.0368	.0689	.1115	.1596	.2051
	7	.0000	.0000	.0001	.0008	.0031	.0090	.0212	.0425	.0746	.1172
	8	.0000	.0000	.0000	.0001	.0004	.0014	.0043	.0106	.0229	.0439
	9	.0000	.0000	.0000	.0000	.0000	.0001	.0005	.0016	.0042	.0098
	10	.0000	.0000	.0000	.0000	.0000	.0000	.0000	.0001	.0003	.0010
11	0	.5688	.3138	.1673	.0859	.0422	.0198	.0088	.0036	.0014	.0005
	1	.3293	.3835	.3248	.2362	.1549	.0932	.0518	.0266	.0125	.0054
	2	.0867	.2131	.2866	.2953	.2581	.1998	.1395	.0887	.0531	.0269
	3	.0137	.0710	.1517	.2215	.2581	.2568	.2254	.1774	.1259	.0806
	4	.0014	.0158	.0536	.1107	.1721	.2201	.2428	.2365	.2060	.1611
	5	.0001	.0025	.0132	.0388	.0803	.1321	.1830	.2207	.2360	.2256
	6	.0000	.0003	.0023	.0097	.0268	.0566	.0985	.1471	.1931	.2256
	7	.0000	.0000	.0003	.0017	.0064	.0173	.0379	.0701	.1128	.1611
	8	.0000	.0000	.0000	.0002	.0011	.0037	.0102	.0234	.0462	.0806
	9	.0000	.0000	.0000	.0000	.0001	.0005	.0018	.0052	.0126	.0269
	10	.0000	.0000	.0000	.0000	.0000	.0000	.0002	.0007	.0021	.0054
	11	.0000	.0000	.0000	.0000	.0000	.0000	.0000	.0000	.0002	.0005

the exercises, you should practice using (5.10) to compute the binomial probabilities unless the problem specifically requests that you use the binomial probability table.

The Expected Value and Variance for the Binomial Probability Distribution

In Section 5.3 we provided formulas for computing the expected value and variance of a discrete random variable. In the special case where the random variable has a binomial probability distribution with a known number of trials n and a known probability of success p, the general formulas for the expected value and variance can be simplified. The results are as follows.

Expected Value and Variance for the Binomial Probability Distribution

$$E(x) = \mu = np \tag{5.11}$$

$$\text{Var}(x) = \sigma^2 = np(1 - p) \tag{5.12}$$

For the Nastke Clothing Store problem with 3 customers, we may utilize (5.11) to compute the expected number of customers making a purchase.

$$E(x) = np = 3(.30) = .9$$

Suppose that for the next month Nastke's Clothing Store forecasts 1000 customers will enter the store. What is the expected number of customers who will make a purchase? The answer is $\mu = np = (1000)(.3) = 300$. Thus, to increase the expected number of sales, Nastke's must induce more customers to enter the store and/or somehow increase the probability that any individual customer will make a purchase after entering.

For the Nastke Clothing Store problem with three customers, we see that the variance and standard deviation for the number of customers making a purchase are

$$\sigma^2 = np(1 - p) = 3(.3)(.7) = .63$$

$$\sigma = \sqrt{.63} = .79$$

For the next 1000 customers entering the store, the variance and standard deviation for the number of customers making a purchase are

$$\sigma^2 = np(1 - p) = 1000(.3)(.7) = 210$$

$$\sigma = \sqrt{210} = 14.49$$

NOTES & COMMENTS

1. Some binomial tables only show values of p up to and including $p = .50$. Thus, it would appear these tables cannot be used when the probability of success exceeds $p = .50$. However, such tables can be used by noting that the probability of $n - x$ failures is also the probability of x successes. So when the probability of success is greater than $p = .50$, one can compute the probability of $n - x$ failures instead. The probability of failure, $1 - p$, will then be less than .50 when $p > .50$.

2. Some sources present binomial tables in a cumulative form. Using these, one must subtract to find the probability of x successes in n trials. For example, $f(2) = P(x \leq 2) - P(x \leq 1)$. Our tables provide these probabilities directly. To compute cumulative probabilities using our tables, one simply sums the individual probabilities. For example, to compute $P(x \leq 2)$ using our tables, we sum $f(0) + f(1) + f(2)$.

☐ ☐   **Exercises**

SELF TEST ▶
Methods

27. Consider a binomial experiment with 2 trials and $p = .4$.
   a. Draw a tree diagram showing this as a 2-trial experiment (see Figure 5.3).
   b. Compute the probability of 1 success, $f(1)$.
   c. Compute $f(0)$.
   d. Compute $f(2)$.
   e. Find the probability of at least 1 success.
   f. Find the expected value, variance, and standard deviation.

28. Consider a binomial experiment with $n = 10$ and $p = .10$. Use the binomial tables (Table 5 of Appendix B) to answer (a) through (d).
   a. Find $f(0)$.   **b.** Find $f(2)$.   **c.** Find $P(x \leq 2)$.
   d. Find $P(x \geq 1)$.   **e.** Find $E(x)$.   **f.** Find Var(x) and σ.

29. Consider a binomial experiment with $n = 20$ and $p = .70$. Use the binomial tables (Table 5 of Appendix B) to answer (a) through (d).
   a. Find $f(12)$.   **b.** Find $f(16)$.   **c.** Find $P(x \geq 16)$.
   d. Find $P(x \leq 15)$.   **e.** Find $E(x)$.   **f.** Find Var(x) and σ.

Applications

30. The greatest number of complaints by owners of two-year-old automobiles is in the area of electrical system performance (*Consumer Report 1992 Buyers Guide*). In an annual questionnaire of owners of over 300 makes and models of automobiles, *Consumer Report* found that 10% of the owners of two-year-old automobiles found trouble spots in the electrical system that included the starter, alternator, battery, switch controls, instruments, wiring, lights, and radio.
   a. What is the probability that a sample of 12 owners of two-year-old automobiles will find exactly 2 owners with electrical system problems?
   b. What is the probability that a sample of 12 owners of two-year-old automobiles will find at least 2 owners with electrical system problems?
   c. What is the probability that a sample of 20 owners of two-year-old automobiles will find at least one electrical system problem?

31. The New York State Bar Examination is the basis for admitting law school graduates into the law profession. Historically, 30% of the individuals taking the examination pass on their first attempt. Suppose that a group of 15 individuals will be taking the examination for the first time and that we are interested in the number of individuals in this group who will pass the exam. Describe the conditions necessary for this situation to be a binomial experiment.

SELF TEST ▶
32. When a new machine is functioning properly, only 3% of the items produced are defective. Assume that we will randomly select two parts produced on the machine and that we are interested in the number of defective parts found.
   a. Describe the conditions under which this situation would be a binomial experiment.
   b. Draw a tree diagram similar to Figure 5.3 showing this as a two-trial experiment.
   c. How many experimental outcomes result in exactly one defect being found?
   d. Compute the probabilities associated with finding no defects, exactly 1 defect, and 2 defects.

33. Ten percent of American truck drivers are women (*Statistical Abstract of the United States, 1987*). Suppose 10 truck drivers are randomly selected to be interviewed about quality of work conditions.
   a. Is the selection of the ten drivers a binomial experiment? Explain.
   b. What is the probability that three of the drivers will be women?
   c. What is the probability that none will be women?
   d. What is the probability that at least one will be a woman?

34. The book *100% American* by Daniel Evan Weiss reports over 1000 statistical facts about the United States and its people. One fact reported is that 64% of the people live in the state where they were born.
   a. What is the probability that a random sample of 10 people will find at least 8 people living in the state where they were born?
   b. What is the probability that a random sample of 3 people will find exactly 1 person living in the state where she or he was born?

35. National Oil Company conducts exploratory oil drilling operations in the southwestern United States. To fund the operation, investors form partnerships, which provide the financial support necessary to drill a fixed number of oil wells. Each well drilled is classified as a producer well or a dry well. Past experience shows that this type of exploratory operation provides producer wells for 15% of all wells drilled. A newly formed partnership has provided the financial support for drilling at 12 exploratory locations.
   a. What is the probability that all 12 wells will be producer wells?
   b. What is the probability that all 12 wells will be dry wells?
   c. What is the probability that exactly 1 well will be a producer well?
   d. In order to make the partnership venture profitable, at least 3 of the exploratory wells must be producer wells. What is the probability that the venture will be profitable?

36. Military radar and missile detection systems are designed to warn a country against enemy attacks. A reliability question deals with the ability of the detection system to identify an attack and issue a warning. Assume that a particular detection system has a .90 probability of detecting a missile attack. Answer the following questions using the binomial probability distribution.
   a. What is the probability that a single detection system will detect an attack?
   b. If two detection systems are installed in the same area and operate independently, what is the probability that at least one of the systems will detect the attack?
   c. If three systems are installed, what is the probability that at least one of the systems will detect the attack?
   d. Would you recommend that multiple detection systems be used? Explain.

37. Assume that the binomial experiment applies for the case of a college basketball player shooting free throws. Late in a basketball game, a team will sometimes foul intentionally in the hope that the player shooting the free throw will miss and the team committing the foul will get the ball. Assume that the best player on the opposing team has a .82 probability of making a free throw and that the worst player has a .56 probability of making a free throw.
   a. What are the probabilities that the best player makes 0, 1, and 2 points if fouled and given two free throws?
   b. What are the probabilities that the worst player makes 0, 1, and 2 points if fouled and given two free throws?
   c. Does it make sense for a coach to have a preset plan about which player to intentionally foul late in a basketball game? Explain.

38. A firm estimates the probability of having employee disciplinary problems on any day to be .10.
   a. What is the probability that the company experiences 5 days without a disciplinary problem?
   b. What is the probability of exactly 2 days with disciplinary problems in a 10-day period?
   c. What is the probability of at least 2 days with disciplinary problems in a 20-day period?

39. At a particular university it has been found that 20% of the students withdraw without completing the introductory statistics course. Assume that 20 students have registered for the course this quarter.
   a. What is the probability that 2 or fewer will withdraw?
   b. What is the probability that exactly 4 will withdraw?
   c. What is the probability that more than 3 will withdraw?
   d. What is the expected number of withdrawals?

40. For the special case of a binomial random variable, we stated that the variance measure could be computed from the formula $\sigma^2 = np(1 - p)$. For the Nastke Clothing Store problem data in Table 5.14, we found $\sigma^2 = np(1 - p) = .63$. Use the general definition of variance for a discrete random variable, Equation (5.4), and the data in Table 5.14 to verify that the variance is in fact .63.

41. Suppose a salesperson makes a sale on 20% of customer contacts. A normal work week will enable the salesperson to contact 25 customers. What is the expected number of sales for the week? What is the variance for the number of sales for the week? What is the standard deviation for the number of sales for the week?

42. Of the next-day express mailings handled by the U.S. Postal Service, 85% are actually received by the addressee 1 day after the mailing. What is the expected value and variance for the number of 1-day deliveries in a group of 250 express mailings?

43. Betting on the color red in the game of roulette has a $^{18}/_{38}$ chance of winning. What is the expected value and variance for the number of wins in a series of 100 bets on red?

5.5   The Poisson Probability Distribution

In this section we consider a discrete random variable that is often useful when dealing with the number of occurrences over a specified interval of time or space. For example, the random variable of interest might be the number of arrivals at a car wash in 1 hour, the number of repairs needed in 10 miles of highway, or the number of leaks in 100 miles of pipeline. If two assumptions are satisfied, the number of occurrences is a random variable described by the *Poisson probability function*. The two assumptions are

1. The probability of an occurrence is the same for any two intervals of equal length.
2. The occurrence or nonoccurrence in any interval is independent of the occurrence or nonoccurrence in any other interval.

The Poisson probability function is given by Equation (5.13).

Poisson Probability Function

$$f(x) = \frac{\mu^x e^{-\mu}}{x!}$$

(5.13)

where

$f(x)$ = the probability of x occurrences in an interval

μ = expected value or average number
of occurrences in an interval

e = 2.71828

Before we consider a specific example to see how the Poisson distribution can be applied, note that there is no upper limit on x, the number of occurrences. It is a discrete random variable that may assume an infinite sequence of values ($x = 0, 1, 2, \ldots$). The Poisson random variable has no upper limit.

A Poisson Example Involving Time Intervals

Suppose that we are interested in the number of arrivals at the drive-in teller window of a bank during a 15-minute period on weekday mornings. If we can assume that the probability of a car arriving is the same for any two time periods of equal length and that the arrival or nonarrival of a car in any time period is independent of the arrival or nonarrival in any other time period, the Poisson probability function is applicable. Suppose that these assumptions are satisfied and that an analysis of historical data shows that the average number of cars arriving in a 15-minute period of time is 10; then the following probability function applies:

$$f(x) = \frac{10^x e^{-10}}{x!}$$

The random variable here is x = number of cars arriving in any 15-minute period.

If management wanted to know the probability of exactly 5 arrivals in 15 minutes, we would set $x = 5$ and thus obtain*

$$\text{Probability of Exactly} \atop \text{5 Arrivals in 15 Minutes} = f(5) = \frac{10^5 e^{-10}}{5!} = .0378$$

Although the above probability was determined by evaluating the probability function with $\mu = 10$ and $x = 5$, it is often easier to refer to tables for the Poisson probability distribution. These tables provide probabilities for specific values of x and μ. We have included such a table as Table 7 of Appendix B. For convenience we have reproduced a portion of this table as Table 5.16. Note that, in order to use the table of Poisson probabilities, we need know only the values of x and μ. Thus, from Table 5.16 we see that the probability of 5 arrivals in a 15-minute period is found by locating the value in the row of the table corresponding to $x = 5$ and the column of the table corresponding to $\mu = 10$. Hence we obtain $f(5) = .0378$.

Although our illustration has involved computing the probability of 5 arrivals in a 15-minute period, other time periods may be used. Suppose we wanted to compute the probability of 1 arrival in a 3-minute period. Since 10 is the expected number of arrivals in a 15-minute period, we see that $^{10}/_{15} = ^2/_3$ is the expected number of arrivals in a 1-minute period and that $(2/3)(3\text{minutes}) = 2$ is the expected number of arrivals in a 3-minute period. Thus, the probability of x arrivals in a 3-minute time period with $\mu = 2$ is given by the following Poisson probability function:

$$f(x) = \frac{2^x e^{-2}}{x!}$$

To find the probability of 1 arrival in a 3-minute period, we can either use Table 7 in Appendix B or compute the probability directly.

$$\text{Probability of Exactly} \atop \text{1 Arrival in 3 Minutes} = f(1) = \frac{2^1 e^{-2}}{1!} = .2707$$

A Poisson Example Involving Length or Distance Intervals

Let us illustrate an application not involving time intervals where the Poisson probability distribution is useful. Suppose that we are concerned with the occurrence of major defects

---

*Values of $e^{-\mu}$ can be found in Table 6 of Appendix B. Most calculators can also provide this.

TABLE 5.16
Selected Values from the Poisson Probability Tables
Example: $\mu = 10$, $x = 5$;
$f(5) = .0378$

					μ					
x	9.1	9.2	9.3	9.4	9.5	9.6	9.7	9.8	9.9	10
0	.0001	.0001	.0001	.0001	.0001	.0001	.0001	.0001	.0001	.0000
1	.0010	.0009	.0009	.0008	.0007	.0007	.0006	.0005	.0005	.0005
2	.0046	.0043	.0040	.0037	.0034	.0031	.0029	.0027	.0025	.0023
3	.0140	.0131	.0123	.0115	.0107	.0100	.0093	.0087	.0081	.0076
4	.0319	.0302	.0285	.0269	.0254	.0240	.0226	.0213	.0201	.0189
5	.0581	.0555	.0530	.0506	.0483	.0460	.0439	.0418	.0398	.0378
6	.0881	.0851	.0822	.0793	.0764	.0736	.0709	.0682	.0656	.0631
7	.1145	.1118	.1091	.1064	.1037	.1010	.0982	.0955	.0928	.0901
8	.1302	.1286	.1269	.1251	.1232	.1212	.1191	.1170	.1148	.1126
9	.1317	.1315	.1311	.1306	.1300	.1293	.1284	.1274	.1263	.1251
10	.1198	.1210	.1219	.1228	.1235	.1241	.1245	.1249	.1250	.1251
11	.0991	.1012	.1031	.1049	.1067	.1083	.1098	.1112	.1125	.1137
12	.0752	.0776	.0799	.0822	.0844	.0866	.0888	.0908	.0928	.0948
13	.0526	.0549	.0572	.0594	.0617	.0640	.0662	.0685	.0707	.0729
14	.0342	.0361	.0380	.0399	.0419	.0439	.0459	.0479	.0500	.0521
15	.0208	.0221	.0235	.0250	.0265	.0281	.0297	.0313	.0330	.0347
16	.0118	.0127	.0137	.0147	.0157	.0168	.0180	.0192	.0204	.0217
17	.0063	.0069	.0075	.0081	.0088	.0095	.0103	.0111	.0119	.0128
18	.0032	.0035	.0039	.0042	.0046	.0051	.0055	.0060	.0065	.0071
19	.0015	.0017	.0019	.0021	.0023	.0026	.0028	.0031	.0034	.0037
20	.0007	.0008	.0009	.0010	.0011	.0012	.0014	.0015	.0017	.0019
21	.0003	.0003	.0004	.0004	.0005	.0006	.0006	.0007	.0008	.0009
22	.0001	.0001	.0002	.0002	.0002	.0002	.0003	.0003	.0004	.0004
23	.0000	.0001	.0001	.0001	.0001	.0001	.0001	.0001	.0002	.0002
24	.0000	.0000	.0000	.0000	.0000	.0000	.0000	.0001	.0001	.0001

in a section of highway 1 month after resurfacing. We will assume that the probability of a defect in this section of highway is the same for any two intervals of equal length and that the occurrence or nonoccurrence of a defect in any one interval is independent of the occurrence or nonoccurrence in any other interval. Thus the Poisson probability distribution can be applied.

Suppose we learn that major defects 1 month after resurfacing occur at the average rate of 2 per mile. Let us find the probability that there will be no major defects in a particular 3-mile section of the highway. Since we are interested in an interval with a length of 3 miles, $\mu = (2 \text{ defects/mile})(3 \text{ miles}) = 6$ represents the expected number of major defects over the 3-mile section of highway. By using (5.13) or Table 7 in Appendix B, we see that the probability of no major defects is .0025. Thus, it is very unlikely that there will be no major defects in the 3-mile section. In fact, there is a $1 - .0025 = .9975$ probability of at least one major defect in the highway section.

A Poisson Approximation to the Binomial Probability Distribution

The Poisson probability distribution can be used as an approximation to the binomial probability distribution when p, the probability of success, is small and n, the number of trials, is large. Simply set $\mu = np$ and use the Poisson tables. As a rule of thumb, the approximation will be good whenever $p \leq .05$ and $n \geq 20$. Binomial tables are often not available for large values of n, so in these cases the approximation can be useful.

As an example, suppose we want to compute the binomial probability of $x = 3$ successes in $n = 50$ trials with $p = .05$. To use the Poisson approximation, we set $\mu = np = 50(.05) = 2.5$. Referring to the Poisson probability tables (Table 7 in Appendix B), we find $f(3) = .2138$. So, we would approximate the binomial probability of 3 successes as $f(3) = .2138$.

☐ ☐ Exercises

Methods

44. Consider a Poisson probability distribution with $\mu = 3$.
a. Write the appropriate Poisson probability function.
b. Find $f(2)$.   **c.** Find $f(1)$.   **d.** Find $P(x \geq 2)$.

SELF TEST ▶ **45.** Consider a Poisson probability distribution with an average number of occurrences per time period of 2.
a. Write the appropriate Poisson probability function.
b. What is the average number of occurrences in 3 time periods?
c. Write the appropriate Poisson probability function to determine the probability of x occurrences in 3 time periods.
d. Find the probability of 2 occurrences in 1 time period.
e. Find the probability of 6 occurrences in 3 time periods.
f. Find the probability of 5 occurrences in 2 time periods.

Applications

SELF TEST ▶ **46.** Phone calls arrive at the rate of 48 per hour at the reservation desk for Regional Airways.
a. Find the probability of receiving 3 calls in a 5-minute interval of time.
b. Find the probability of receiving exactly 10 calls in 15 minutes.
c. Suppose no calls are currently on hold. If it takes the agent 5 minutes to complete processing the current call, how many callers do you expect to be waiting by that time? What is the probability none will be waiting?
d. If there are no calls currently being processed, what is the probability the agent can take 3 minutes for personal time without being interrupted?

47. During the period of time phone-in reservations are being taken at a local university, calls come in at the rate of 1 every 2 minutes.
a. What is the expected number of calls in 1 hour?
b. What is the probability of 3 calls in 5 minutes?
c. What is the probability of no calls in a 5-minute period?

48. A certain restaurant has a reputation for good food. Restaurant management boasts that on a Saturday night, groups of customers arrive at the rate of 15 groups every half-hour.
a. What is the probability that 5 minutes will pass with no groups of customers arriving?
b. What is the probability that 8 groups of customers will arrive in 10 minutes?
c. What is the probability that more than 5 groups will arrive in a 10-minute period of time?

Hmwk **49.** During rush hours, accidents occur in a particular metropolitan area at the rate of 2 per hour. The morning rush period lasts for 1 hour 30 minutes and the evening rush period lasts for 2 hours.

a. On a particular day what is the probability that there will be no accidents during the morning rush period?

b. What is the probability of 2 accidents during the evening rush period?

c. What is the probability of 4 or more accidents during the morning rush period?

d. On a particular day what is the probability there will be no accidents during both the morning and evening rush periods? $E(x)$ $V(x)$

50. Airline passengers arrive randomly and independently at the passenger-screening facility at a major international airport. The mean arrival rate is 10 passengers per minute.

a. What is the probability of no arrivals in a 1-minute period?

b. What is the probability that three or fewer passengers arrive in a 1-minute period?

c. What is the probability of no arrivals in a 15-second period?

d. What is the probability of at least one arrival in a 15-second period? $E(x)$ $V(x)$

51. Williams Company has observed that calculators fail and need to be replaced at the rate of 3 every 25 days.

a. What is the expected number of calculators that will fail in 30 days?

b. What is the probability that at least 2 will fail in 50 days?

c. What is the probability that exactly 3 will fail in 10 days?

5.6   The Hypergeometric Probability Distribution

The hypergeometric probability distribution is closely related to the binomial probability distribution. The key difference between the two probability distributions is that with the hypergeometric distribution, the trials are not independent; thus the probability of success changes from trial to trial.

To illustrate the hypergeometric distribution, suppose that a 5-member committee consists of 3 women and 2 men; 2 of the committee members are expected to represent the group at a meeting in Las Vegas. The committee has decided to choose randomly the 2 members that will attend the meeting. What is the probability that both persons chosen will be women? Since 3 of the 5 committee members are women, the probability that the first person randomly selected is a woman is $3/5 = .60$. However, if the first person chosen is a woman, then the probability of randomly selecting another woman from the 4 remaining committee members drops to $2/4 = .50$. Therefore, the probability that 2 women will be selected to attend the meeting is $(.60)(.50) = .30$.

One of the most important applications of the hypergeometric probability distribution involves sampling without replacement from a finite population. The objective is to choose a random sample of n items out of a population of N items, under the condition that once an item has been selected, it is not returned to the population. Thus, on the next selection, the probability of selecting an item of that type goes down.

The usual notation in applications of the hypergeometric probability distribution is to let r denote the number of items in the population that are labeled success and $N - r$ denote the number of items in the population that are labeled failure. The hypergeometric probability function is used to compute the probability that in a random sample of n items, selected without replacement, we will obtain x items labeled success and $n - x$ items labeled failure. Note that for this to occur, we must obtain x successes from the r successes in the population and $n - x$ failures from the $N - r$ failures. The following hypergeometric probability function provides $f(x)$, the probability of obtaining x successes in a sample of size n.

Hypergeometric Probability Function

$$f(x) = \frac{\binom{r}{x}\binom{N-r}{n-x}}{\binom{N}{n}} \qquad \text{for } x \le r \qquad (5.14)$$

where

$f(x)$ = the probability of x successes in n trials

n = the number of trials

N = number of elements in the population

r = number of elements in the population labeled success

Note that $\binom{N}{n}$ simply represents the number of ways a sample of size n can be selected from a population of size N; $\binom{r}{x}$ represents the number of ways x successes can be selected from a total of r successes in the population; and $\binom{N-r}{n-x}$ represents the number of ways $n - x$ failures can be selected from a total of $N - r$ failures in the population. To illustrate the computations involved in using (5.14), let us reconsider the problem of selecting 2 of 5 committee members to send to Las Vegas.

Recall that the objective is to select two members from the 5-member committee consisting of 3 women and 2 men. To determine the probability of obtaining a sample that consists of 2 women, we can use (5.14) with $N = 5$, $r = 3$, and $x = 2$.

$$f(2) = \frac{\binom{3}{2}\binom{2}{0}}{\binom{5}{2}} = \frac{\left(\frac{3!}{2!1!}\right)\left(\frac{2!}{2!0!}\right)}{\left(\frac{5!}{2!3!}\right)} = \frac{3}{10} = .30$$

Note that this is the same answer we obtained previously. Suppose, however, that we learn that 3 committee members will be allowed to make the trip. The probability that exactly 2 of the 3 members will be women is

$$f(2) = \frac{\binom{3}{2}\binom{2}{1}}{\binom{5}{2}} = \frac{\left(\frac{3!}{2!1!}\right)\left(\frac{2!}{1!1!}\right)}{\left(\frac{5!}{2!3!}\right)} = \frac{6}{10} = .60$$

As another illustration, suppose a population consists of 10 items, 4 of which are classified as defective and 6 of which are classified as acceptable. What is the probability that a random sample of size 3 will contain 2 defective items? For this problem we can think of obtaining a defective item as a "success" and obtaining an acceptable item as a "failure." Thus, $N = 10$, $r = 4$, $n = 3$, and $x = 2$. Using (5.14) we can compute $f(2)$ as follows:

$$f(2) = \frac{\binom{4}{2}\binom{6}{1}}{\binom{10}{3}} = \frac{\left(\frac{4!}{2!2!}\right)\left(\frac{6!}{1!5!}\right)}{\left(\frac{10!}{3!7!}\right)} = \frac{36}{120} = .30$$

The hypergeometric distribution is used to determine the probability of obtaining a certain sample outcome when sampling without replacement. When sampling with replacement, each item that is selected is returned to the population before another item is selected. The binomial probability distribution can then be used to compute the probability of x successes.

☐ ☐ Exercises

Methods

SELF TEST
52. Suppose $N = 10$ and $r = 3$. Compute the hypergeometric probabilities for the following values of n and x.
   a. $n = 4, x = 1$ b. $n = 2, x = 2$ c. $n = 2, x = 0$ d. $n = 4, x = 2$

53. Suppose $N = 15$ and $r = 4$. What is the probability of $x = 3$ for $n = 10$?

54. What is the probability of being dealt three aces in a seven-card poker hand?

Applications

55. There are 25 students (14 boys and 11 girls) in the sixth-grade class at St. Andrew School. Five students were absent Thursday.
   a. What is the probability that 2 of the absent students were girls?
   b. What is the probability that 2 of the absent students were boys?
   c. What is the probability that all were boys?
   d. What is the probability that none were boys?

SELF TEST
56. Axline Computers manufactures personal computers at two plants; one is in Las Vegas, the other in Hawaii. There are 40 employees at the Las Vegas plant and 20 in Hawaii. A random sample of 10 different employees is to be asked to fill out a benefits' questionnaire.
   a. What is the probability that none will be from the plant in Hawaii?
   b. What is the probability that 1 will be from the plant in Hawaii?
   c. What is the probability that 2 or more will be from the plant in Hawaii?
   d. What is the probability that 9 will be from the plant in Las Vegas?

Summary

The concept of a random variable was introduced in order to provide a numerical description of the outcome of an experiment. We saw that the probability distribution for a random variable describes how the probabilities are distributed over the values the random variable can take on. For any discrete random variable x, the probability distribution is defined by a probability function, denoted by $f(x)$, which provides the probability associated with each value of the random variable. Once the probability function has been defined, we can then compute the expected value and the variance for the random variable.

The binomial probability distribution can be used to determine the probability of x successes in n trials whenever the experiment has the following properties:

1. The experiment consists of a sequence of n identical trials.
2. Two outcomes are possible on each trial, one called success and the other failure.
3. The probability of a success p does not change from trial to trial. Consequently, the probability of failure, $1 - p$, does not change from trial to trial.
4. The trials are independent.

When the above conditions hold, a binomial probability function, or a table of binomial probabilities, can be used to determine the probability of x successes in n trials. Formulas were also presented for the mean and variance of the binomial probability distribution.

The Poisson probability distribution is used when it is desirable to determine the probability of obtaining x occurrences over an interval of time or space. The following assumptions are required for the Poisson distribution to be applicable:

1. The probability of an occurrence of the event is the same for any two intervals of equal length.
2. The occurrence or nonoccurrence of the event in any interval is independent of the occurrence or nonoccurrence in any other interval.

A third discrete probability distribution, the hypergeometric, was introduced in Section 5.6. Like the binomial, it is used to compute the probability of x successes in n trials. But, unlike the binomial, the probability of success changes from trial to trial.

Glossary

Random variable A numerical description of the outcome of an experiment.

Discrete random variable A random variable that can assume only a finite or infinite sequence of values.

Continuous random variable A random variable that may assume all values in an interval or collection of intervals.

Probability distribution A description of how the probabilities are distributed over the values the random variable can take on.

Probability function A function, denoted by $f(x)$, that provides the probability that x takes on a particular value for a discrete random variable.

Expected value A measure of the mean, or central location, value of a random variable.

Variance A measure of the dispersion, or variability, of a random variable.

Standard deviation The positive square root of the variance.

Binomial experiment A probability experiment possessing the four properties stated in Section 5.4.

Binomial probability distribution A probability distribution showing the probability of x successes in n trials of a binomial experiment.

Binomial probability function The function used to compute probabilities in a binomial experiment.

Poisson probability distribution A probability distribution showing the probability of x occurrences of an event over a specified interval of time or space.

Poisson probability function The function used to compute Poisson probabilities.

Hypergeometric probability function The function used to compute the probability of x successes in n trials when the trials are dependent.

Key Formulas

Expected Value of a Discrete Random Variable

$$E(x) = \mu = \Sigma x f(x) \tag{5.3}$$

Variance of a Discrete Random Variable

$$\text{Var}(x) = \sigma^2 = \Sigma(x - \mu)^2 f(x) \tag{5.4}$$

Expected Value of the Sum of Two Random Variables

$$E(x + y) = E(x) + E(y) \tag{5.5}$$

Variance of the Sum of Two Independent Random Variables

$$\text{Var}(x + y) = \text{Var}(x) + \text{Var}(y) \tag{5.6}$$

Number of Experimental Outcomes Providing Exactly x Successes in n Trials

$$\binom{n}{x} = \frac{n!}{x!(n - x)!} \tag{5.7}$$

Binomial Probability Function

$$f(x) = \binom{n}{x}p^x(1 - p)^{(n - x)} \tag{5.10}$$

Expected Value for the Binomial Probability Distribution

$$E(x) = \mu = np \tag{5.11}$$

Variance for the Binomial Probability Distribution

$$\text{Var}(x) = \sigma^2 = np(1 - p) \tag{5.12}$$

Poisson Probability Function

$$f(x) = \frac{\mu^x e^{-\mu}}{x!} \tag{5.13}$$

Hypergeometric Probability Function

$$f(x) = \frac{\binom{r}{x}\binom{N - r}{n - x}}{\binom{N}{n}} \tag{5.14}$$

❑ ❑ Supplementary Exercises

57. Which of the following are, and which are not, probability distributions? Explain.

x	$f(x)$	y	$f(y)$	z	$f(z)$
0	.20	0	.25	−1	.20
1	.30	2	.05	0	.50
2	.25	4	.10	1	−.10
3	.35	6	.60	2	.40

TABLE 5.17

x	$f(x)$
0	.15
1	.30
2	.40
3	.10
4	.05

58. An automobile agency located in Beverly Hills specializes in the rental of luxury automobiles. Assume that the probability distribution of daily demand at their agency is as shown in Table 5.17.
a. Compute the expected value of daily demand.
b. If the daily rental cost for an automobile is $75, what is the expected value of daily automobile rental?

59. At a large university, the number of student problems handled by the dean for student affairs varies from semester to semester. Assume that the number of student problems (x) handled by the dean has the following probability distribution.

x	f(x)
0	.10
1	.15
2	.30
3	.25
4	.10
5	.10

What are the mean and variance of the number of student problems handled by the dean each semester?

60. The number of weekly lost-time injuries at a particular plant (x) has the probability distribution shown in Table 5.18.
a. Compute the expected value.
b. Compute the variance.

TABLE 5.18

x	f(x)
0	.05
1	.20
2	.40
3	.20
4	.15

61. Assume that the plant in Exercise 60 initiated a safety training program and that the number of lost-time injuries during the 20 weeks following the training program was as follows:

Number of Injuries	Number of Weeks
0	2
1	8
2	6
3	3
4	1
Total	20

a. Construct a probability distribution for weekly lost-time injuries based on these data.
b. Compute the expected value and the variance and use both to evaluate the effectiveness of the safety training program.

TABLE 5.19

Price of Stock (x)	f(x)
16	.35
17	.25
18	.25
19	.10
20	.05

62. The Hub Real Estate Investment stock is currently selling for $16 per share. An investor plans to buy shares and hold the stock for 1 year. Let x be the random variable indicating the price of the stock after 1 year. The probability distribution for x is shown in Table 5.19.
a. Show that the probability distribution in Table 5.19 possesses the properties of all probability distributions.
b. What is the expected price of the stock after 1 year?
c. What is the expected gain per share of the stock over the 1-year period? What percent return on the investment is reflected by this expected value?
d. What is the variance in the price of the stock over the 1-year period?
e. Another stock with a similar expected return has a variance of 3. Which stock appears to be the better investment in terms of minimizing risk or uncertainty associated with the investment? Explain.

63. The budgeting process for a midwestern college resulted in expense forecasts for the coming year (in 1,000,000s) of $9, $10, $11, $12, and $13. Since the actual expenses are unknown, the following respective probabilities are assigned: .3, .2, .25, .05, and .2.
a. Show the probability distribution for the expense forecast.
b. What is the expected value of the expenses for the coming year?
c. What is the variance in the expenses for the coming year?
d. If income projections for the year are estimated at $12 million, comment on the financial position of the college.

64. The probability function for *x*, the hours required to change over a production system, is as follows:

$$f(x) = \frac{x}{10} \qquad \text{for } x = 1, 2, 3, \text{ or } 4$$

a. What is the expected value of the changeover time?

b. What is the variance of the changeover time?

65. Students who complete a statistics course are required to take a midterm exam and a final exam. The expected value or average on the midterm is 72 with a standard deviation of 12. The expected value or average on the final is 68 with a standard deviation of 14.

a. What is the expected value of the total number of points scored on the two exams?

b. What are the variance and standard deviation of the total number of points scored on the two exams? Assume that the number of points scored on the exams are independent of one another.

66. Customers entering a fast-food store average spending $2.28 per person. The standard deviation in the amount spent is $.75.

a. Compute the mean, variance, and standard deviation of the total food bill for 10 randomly selected customers. Assume that the customers order independently.

b. Repeat part (a) for 100 randomly selected customers.

67. A study conducted at the University of Southern California investigated the use of music in television commercial messages (*New York,* March 23, 1992). The study found 42% of commercials make use of music. Consider a sample of 12 commercials.

Hmwk

a. What is the probability that exactly 6 of the commercials use music?

b. What is the probability that exactly 3 of the commercials use music?

c. What is the probability that at least 3 of the commercials use music?

$E(x)$

$v(x)$

68. In October 1986, *Better Homes and Gardens* published the results of a reader survey to which over 30,000 readers had responded. The findings indicated that 34% of the women who responded worked full-time outside the home and 24% worked part-time outside the home.

a. For a random sample of five women who are *Better Homes and Gardens* readers, what is the probability that three work full-time outside the home?

b. For a random sample of five women readers, what is the probability that two are part-time workers?

c. Suppose a random sample of five women (not necessarily *Better Homes and Gardens* readers) was taken. Would your answers to parts (a) and (b) change? Why or why not?

69. Refer again to the *Better Homes and Gardens* survey in Exercise 68. In 65% of the two-parent households, the wife does most of the child care; in 1%, the husband does most. For a sample of four respondents from two-parent households answer the following:

a. What is the probability that none of the respondents will say the wife does most of the child care?

b. What is the probability that none of the respondents will say the husband does most of it?

c. What is the probability that three or more of the respondents will say the wife does most of it?

d. What is the probability that one of the respondents will say the husband does most of it?

70. A new clothes-washing compound is found to remove excess dirt and stains satisfactorily on 88% of the items washed. Assume that 10 items are to be washed with the new compound.

a. What is the probability of satisfactory results on all 10 items?

b. What is the probability at least 2 items are found with unsatisfactory results?

71. In an audit of a company's billings, an auditor randomly selects 5 bills. If 3% of all bills contain an error, what is the probability that the auditor will find the following?

a. exactly one bill in error

b. at least one bill in error

72. Many companies use a quality-control technique referred to as *acceptance sampling* in order to monitor incoming shipments of parts, raw materials, and so on. In the electronics industry, it is common to have component parts shipped from suppliers in large lots. Inspection of a sample of *n* components can be viewed as the *n* trials of a binomial experiment. The outcome for each component tested (trial) will be that the component is good or defective. Reynolds Electronics accepts lots from a particular supplier as long as the percent defective in the lot is not greater than 1%. Suppose a random sample of five items from a recent shipment has been tested.

a. Assume that 1% of the shipment is defective. Compute the probability that no items in the sample are defective.

b. Assume that 1% of the shipment is defective. Compute the probability that exactly one item in the sample is defective.

c. What is the probability of observing one or more defective items in the sample if 1% of the shipment is defective?

d. Would you feel comfortable accepting the shipment if one item was found defective? Why or why not?

73. Cars arrive at a carwash at the average rate of 15 cars per hour. If the number of arrivals per hour follows a Poisson distribution, what is the probability of 20 or more arrivals during any given hour of operation? Use the Poisson probability table.

74. A new automated production process has been experiencing an average of 1.5 breakdowns per day. Because of the cost associated with a breakdown, management is concerned about the possibility of having three or more breakdowns during a given day. Assume that the number of breakdowns per day follows a Poisson distribution. What is the probability of observing three or more breakdowns?

75. A regional director responsible for business development in Pennsylvania is concerned about the number of businesses that end as failures. If the average number of failures per month is 10, what is the probability that exactly 4 businesses will fail during a given month? Assume that the number of businesses failing per month follows a Poisson distribution.

76. The arrivals of customers at a bank follow the Poisson distribution. Answer the following questions assuming a mean arrival rate of three per minute.

a. What is the probability of exactly three arrivals in a 1-minute period?

b. What is the probability of at least three arrivals in a 1-minute period?

77. During the registration period at a local university, students consult advisors with questions about course selection. A particular advisor noted that, during the registration period, an average of eight students per hour ask questions, although the exact arrival times of the students were random in nature. Use the Poisson distribution to answer the following questions:

a. What is the probability that exactly eight students come in for consultation during a particular 1-hour period?

b. What is the probability that three students come in for consultation during a particular ½-hour period?

6

Continuous Probability Distributions

Procter & Gamble*

CINCINNATI, OHIO

Procter & Gamble (P&G) is in the consumer-products business worldwide. P&G produces and markets such products as detergents, disposable diapers, over-the-counter pharmaceuticals, dentifrices, bar soaps, mouthwashes, and paper towels. It has the leading brand in more categories than any other consumer products company.

As a leader in the application of quantitative methodologies to decision making, P&G employs people with diverse academic backgrounds: engineering, statistics, operations research, and business. The major quantitative technologies these people provide support for are probabilistic decision and risk analysis, advanced simulation, quality improvement, and quantitative methods (e.g., linear programming, regression analysis, probability analysis).

The Industrial Chemicals Division of P&G is a major supplier of fatty alcohols derived from natural substances such as coconut oil and from petroleum-based derivatives. The division wanted to know the economic risks and opportunities of expanding its fatty-alcohol production facilities, and P&G's experts in probabilistic decision and risk analysis were called in to help. After structuring and modeling the problem, it was determined that the key to profitability was the cost difference between the petroleum- and coconut-based raw materials. Future costs were unknown, but the analysts were able to represent them with the following continuous random variables.

x = the coconut oil price per pound of fatty alcohol

and

y = the petroleum-based raw material price per pound of fatty alcohol

Since the key to profitability was the difference between these two random variables, a third derived random variable $d = x - y$ was used in the analysis. Interviews with experts were used to determine the probability distribution for x and y. In turn, this information was used to develop a continuous probability distribution for the difference d. Using the continuous probability distribution, it was observed that there was a .90 probability that the price difference would be $.0655 or less and that there was a .50 probability that the price difference would be $.035 or less. In addition, there was only a .10 probability that the price difference would be $.0045 or less.[†]

The Industrial Chemicals Division thought that being able to quantify the impact of raw material price differences was key to reaching a consensus. The numbers obtained were used in a sensitivity analysis of the raw material price difference. The analysis yielded sufficient insight to form the basis for a recommendation to management.

The use of random variables and continuous probability distributions was helpful to P&G in analyzing the economic risks associated with its fatty-alcohol production. In this chapter, you will gain an understanding of continuous random variables and their probability distributions including one of the most important probability distributions in statistics, the normal distribution.

[†]The price differences here have been modified to protect proprietary data

*The authors are indebted to Mr. Joel Kahn of Procter & Gamble for providing this Statistics in Practice.

I n the previous chapter we discussed discrete random variables and their probability distributions. In this chapter we turn our attention to the study of continuous random variables. Specifically, we discuss three continuous probability distributions: the uniform, the normal, and the exponential probability distributions.

To understand the difference between discrete and continuous random variables, first recall that for a discrete random variable, we can compute the probability of the random variable taking on a particular value. For continuous probability distributions, the situation is much different. A continuous random variable may assume any value in an interval on the real line or in a collection of intervals. Since there are an infinite number of values in any interval, it is no longer possible to talk about the probability that the random variable will take on a specific value; instead, we must think in terms of the probability that a continuous random variable will lie within a given interval.

In our discussion of discrete probability distributions, we introduced the concept of a probability function $f(x)$. Recall that this function provided the probability that the random variable x assumed some specific value. In the continuous case, the counterpart of the probability function is the *probability density function,* also denoted by $f(x)$. For a continuous random variable, the probability density function provides the height or value of the function at any particular value of x; it does not directly provide the probability of the random variable taking on some specific value. However, the area under the graph of $f(x)$ corresponding to some interval provides the probability that the continuous random variable will take on a value in that interval. In Section 6.1 we demonstrate these concepts for a continuous random variable that has a uniform probability distribution.

6.1   The Uniform Probability Distribution

Consider the random variable x that represents the total flight time of an airplane traveling from Chicago to New York. Suppose that the flight time can be any value in the interval from 120 minutes to 140 minutes. Since the random variable x can take on any value in the interval from 120 to 140 minutes, x is a continuous rather than a discrete random variable. Let us assume that sufficient actual flight data are available to conclude that the probability of a flight time within any 1-minute interval is the same as the probability of a flight time within any other 1-minute interval up to and including 140 minutes. With every 1-minute interval being equally likely, the random variable x is said to have a *uniform probability distribution*. The *probability density function,* which defines the uniform probability distribution for the flight time random variable, is

$$f(x) = \begin{cases} 1/20 & \text{for } 120 \le x \le 140 \\ \\ 0 & \text{elsewhere} \end{cases}$$

A graph of this probability density function is shown in Figure 6.1. In general, the uniform probability density function for a random variable x is as follows (next page):

F I G U R E 6.1
Uniform Probability Density Function for Flight Time

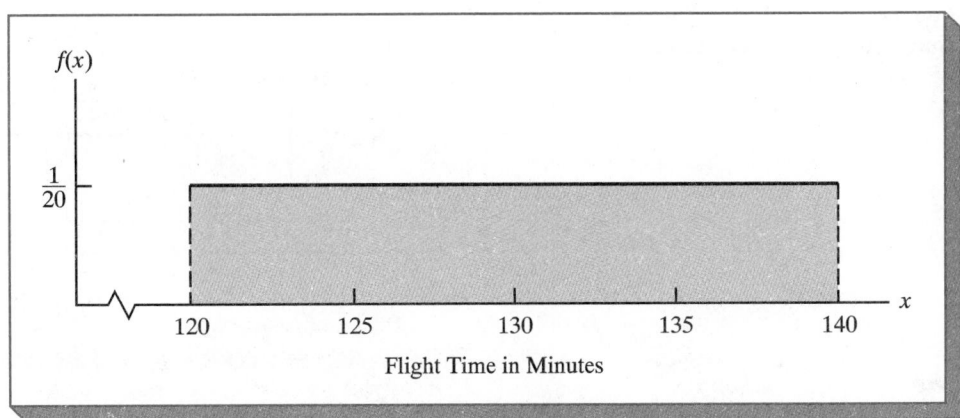

Uniform Probability Density Function

$$f(x) = \begin{cases} \dfrac{1}{b-a} & \text{for } a \leq x \leq b \\ 0 & \text{elsewhere} \end{cases} \qquad (6.1)$$

In the flight-time example, $a = 120$ and $b = 140$.

The graph of the probability density function $f(x)$ provides the height or value of the function at any particular value of x. Note that for a *uniform* probability density function, the height or value of the function is the same for each value of x. For example, in the flight-time example, $f(x) = \frac{1}{20}$ for all values of x between 120 and 140. *In general, the probability density function $f(x)$, unlike the probability function for a discrete random variable, does not represent probability. Rather it simply provides the height of the function at any particular value of x.*

For a continuous random variable, we consider probability only in terms of the likelihood that a random variable has a value within a *specified interval*. In our flight-time problem, an acceptable probability question is, What is the probability that the flight time is between 120 and 130 minutes? That is, what is $P(120 \leq x \leq 130)$? Since the flight time must be between 120 and 140 minutes and since the probability was described as being uniform over this interval, we feel comfortable saying $P(120 \leq x \leq 130) = .50$. In the following subsection we will show that this probability can be computed as the area under the graph of $f(x)$ from 120 to 130.

Area as a Measure of Probability

Let us make an observation about the graph shown in Figure 6.2. Consider the *area* under the graph of $f(x)$ in the interval from 120 to 130. The region is rectangular in shape, and the area of a rectangle is simply the width times the height. With the width of the interval equal to $130 - 120 = 10$ and the height equal to the value of probability density function $f(x) = \frac{1}{20}$, we have area = width × height = $10(\frac{1}{20}) = \frac{10}{20} = .50$.

FIGURE **6.2**

Area Provides Probability of Flight Time between 120 and 130 Minutes

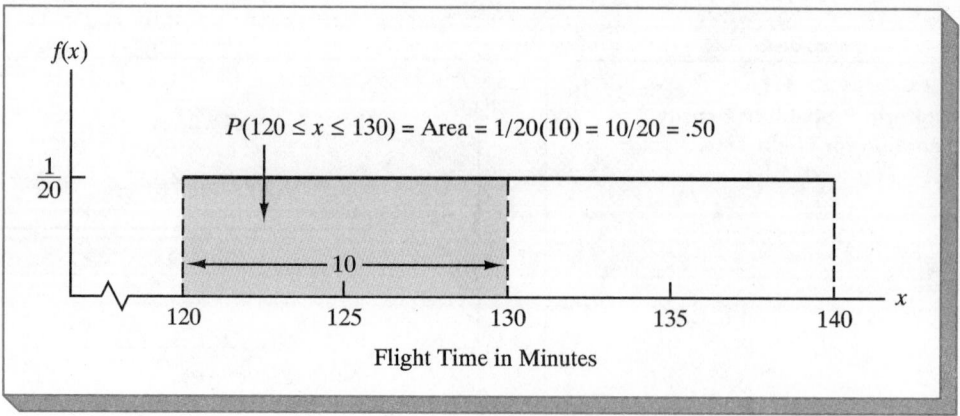

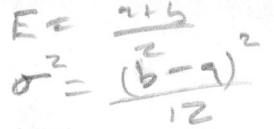

What observation can you make about the area under the graph of $f(x)$ and probability? They are identical! Indeed, this is true for all continuous random variables. Namely, once a probability density function $f(x)$ has been identified, the probability that x takes on a value between some lower value x_1 and some higher value x_2 can be found by computing the *area* under the graph of $f(x)$ over the interval x_1 to x_2.

Once we have the appropriate probability distribution and accept the interpretation of area as probability, we can answer any number of probability questions. For example, what is the probability of a flight time between 128 and 136 minutes? The width of the interval is $136 - 128 = 8$. With the uniform height of $1/20$, we see that $P(128 \leq x \leq 136) = 8/20 = .40$.

Note that $P(120 \leq x \leq 140) = 20(1/20) = 1$. That is, the total area under the graph of $f(x)$ is equal to 1. This property holds for all continuous probability distributions and is the analog of the condition that the sum of the probabilities has to equal 1 for a discrete probability function. For a continuous probability density function, we must also require that $f(x) \geq 0$ for all values of x. This is the analog of the requirement that $f(x) \geq 0$ for discrete probability functions.

When we deal with continuous random variables and probability distributions, two major differences stand out as compared to the treatment of their discrete counterparts:

1. We no longer talk about the probability of the random variable taking on a particular value. Instead, we talk about the probability of the random variable taking on a value within some given interval.

2. The probability of the random variable taking on a value within some given interval from x_1 to x_2 is defined to be the area under the graph of the probability density function between x_1 and x_2. *This implies that the probability that a continuous random variable takes on any particular value exactly is zero, since the area under the graph of $f(x)$ at a single point is zero.*

The calculation of the mean and variance for a continuous random variable is analogous to that for a discrete random variable. However, since the computational procedure involves integral calculus, we leave the derivation of the appropriate formulas to more advanced texts.

For the uniform continuous probability distribution introduced in this section, the formulas for the expected value and variance are

$$E(x) = \frac{a + b}{2}$$

$$\text{Var}(x) = \frac{(b - a)^2}{12}$$

In these formulas, a is the smallest value and b is the largest value that the random variable may take on.

Applying these formulas to the uniform probability distribution for flight times from Chicago to New York, we obtain

$$E(x) = \frac{(120 + 140)}{2} = 130$$

$$\text{Var}(x) = \frac{(140 - 120)^2}{12} = 33.33$$

The standard deviation of flight times can be found by taking the square root of the variance. Thus $\sigma = 5.77$ minutes.

1. Since for a continuous random variable the probability of any particular value is zero, we have $P(a \leq x \leq b) = P(a < x < b)$. This shows that the probability of a random variable assuming a value in any interval is the same whether or not the endpoints are included.

2. To see more clearly why the height of a probability density function is not a probability, think about a random variable with the following uniform probability distribution:

$$f(x) = \begin{cases} 2 & \text{for } 0 \leq x \leq .5 \\ 0 & \text{elsewhere} \end{cases}$$

The height of the probability density function is 2 for values of x between 0 and .5. But, we know probabilities can never be greater than 1.

☐ ☐ Exercises

Methods

SELF TEST ▶

1. The random variable x is known to be uniformly distributed between 1.0 and 1.5.
   a. Show the graph of the probability density function.
   b. Find $P(x = 1.25)$.
   c. Find $P(1.0 \leq x \leq 1.25)$.
   d. Find $P(1.20 < x < 1.5)$.

2. The random variable x is known to be uniformly distributed between 10 and 20.
   a. Show the graph of the probability density function.
   b. Find $P(x < 15)$.   **c.** Find $P(12 \leq x \leq 18)$.
   d. Find $E(x)$.   **e.** Find Var(x).

Applications

3. Delta Airlines quotes a flight time of 1 hour, 52 minutes for its flights from Cincinnati to Tampa. Suppose that we believe actual flight times are uniformly distributed between the quoted time and 2 hours, 10 minutes.
   a. Show the graph of the probability density function for flight times.
   b. What is the probability the flight will be no more than 5 minutes late?
   c. What is the probability the flight will be more than 10 minutes late?
   d. What is the expected flight time?

SELF TEST ▶

4. Most computer languages have a function that can be used to generate random numbers. In Microsoft's QuickBASIC, the RND function can be used to generate random numbers between 0 and 1. If we let x denote the random number generated, then x is a continuous random variable with the following probability density function:

$$f(x) = \begin{cases} 1 & \text{for } 0 \leq x \leq 1 \\ 0 & \text{elsewhere} \end{cases}$$

a. Graph the probability density function.
b. What is the probability of generating a random number between .25 and .75?
c. What is the probability of generating a random number with values less than or equal to .30?
d. What is the probability of generating a random number with a value greater than .60?

5. The total time to process a loan application is uniformly distributed between 3 and 7 days.
a. Give a mathematical expression for the probability density function.
b. What is the probability that the loan application will be processed in fewer than 3 days?
c. Compute the probability that a loan application will be processed in 5 days or less.
d. Find the expected processing time and the standard deviation.

6. The label on a bottle of liquid detergent shows contents to be 12 ounces per bottle. The production operation fills the bottle uniformly according to the following probability density function:

$$f(x) = \begin{cases} 8 & \text{for } 11.975 \leq x \leq 12.10 \\ 0 & \text{elsewhere} \end{cases}$$

a. What is the probability that a bottle will be filled with between 12 and 12.05 ounces?
b. What is the probability that a bottle will be filled with 12.02 or more ounces?
c. Quality control accepts production that is within .02 ounces of the number of ounces shown on the container label. What is the probability that a bottle of this liquid detergent will fail to meet the quality control standard?

7. Suppose we are interested in bidding on a piece of land and we know there is one other bidder.* The seller has announced that the highest bid in excess of $10,000 will be accepted. Assume that the competitor's bid x is a random variable that is uniformly distributed between $10,000 and $15,000.
a. Suppose you bid $12,000. What is the probability your bid will be accepted?
b. Suppose you bid $14,000. What is the probability your bid will be accepted?
c. What amount should you bid to maximize the probability you get the property?
d. Suppose you know someone who is willing to pay you $16,000 for the property. Would you consider bidding less than the amount in (c)? Why or why not?

6.2   The Normal Probability Distribution

Perhaps the most important probability distribution used to describe a continuous random variable is the *normal probability distribution*. The normal probability distribution has been applied in a wide variety of practical applications in which the random variables involved are heights and weights of people, IQ scores, scientific measurements, amounts of rainfall, and so on. In order to use this probability distribution, the random variable must be continuous. However, as we shall see, a continuous normal random variable can also be used as an approximation in situations involving discrete random variables.

The Normal Curve

The form, or shape, of the normal probability distribution is illustrated by the bell-shaped curve shown in Figure 6.3. The probability density function that defines the bell-shaped curve of the normal probability distribution is as follows.

---

*This is based on a problem suggested to us by Professor Roger Myerson of Northwestern University.

FIGURE 6.3
**Bell-Shaped Curve for the Normal
Probability Distribution**

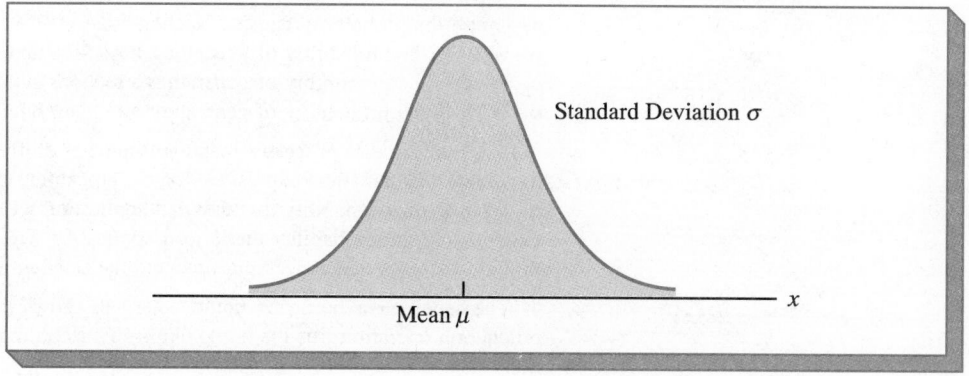

Normal Probability Density Function

$$f(x) = \frac{1}{\sqrt{2\pi}\sigma} e^{-(x-\mu)^2/2\sigma^2}$$

(6.2)

where μ = mean

σ = standarddeviation

π = 3.14159

e = 2.71828

We make some observations about the characteristics of the normal probability distribution:

1. There is an entire family of normal probability distributions with each specific normal distribution being differentiated by its mean μ and its standard deviation σ.
2. The highest point on the normal curve occurs at the mean, which is also the median and mode of the distribution.
3. The mean of the distribution can be any numerical value: negative, zero, or positive. Three normal curves with the same standard deviation but three different means (-10, 0, and 20) are shown below.

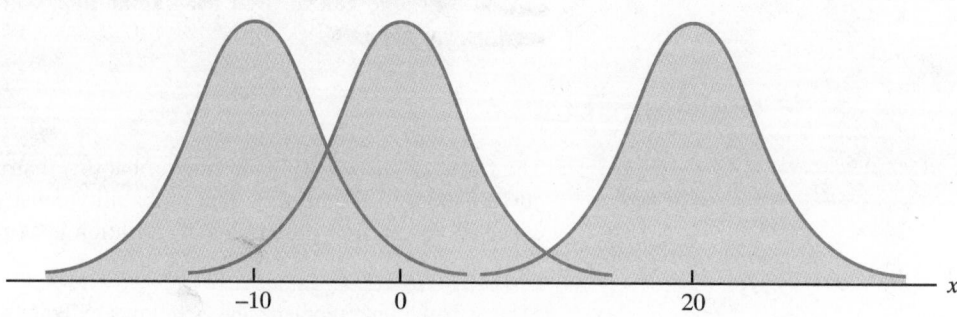

4. The normal probability distribution is symmetric, with the shape of the curve to the left of the mean a mirror image of the shape of the curve to the right of the mean. The tails of the curve extend to infinity in both directions and theoretically never touch the horizontal axis.

5. The standard deviation determines the width of the curve. Larger values of the standard deviation result in wider, flatter curves, showing more dispersion in the data. Two normal distributions with the same mean but with different standard deviations are shown below.

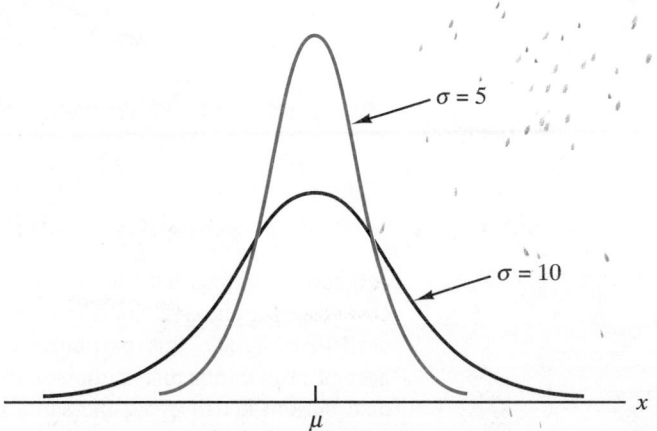

6. The total area under the curve for the normal probability distribution is 1. (This is true for all continuous probability distributions.)

7. Probabilities for the normal random variable are given by areas under the curve. Probabilities for some commonly used intervals are:

    a. 68.26% of the time, a normal random variable assumes a value within plus or minus 1 standard deviation of its mean.

    b. 95.44% of the time, a normal random variable assumes a value within plus or minus 2 standard deviations of its mean.

    c. 99.72% of the time, a normal random variable assumes a value within plus or minus 3 standard deviations of its mean. Figure 6.4 shows properties (a), (b), and (c) graphically.

FIGURE 6.4
Areas under the Curve for Any Normal Probability Distribution

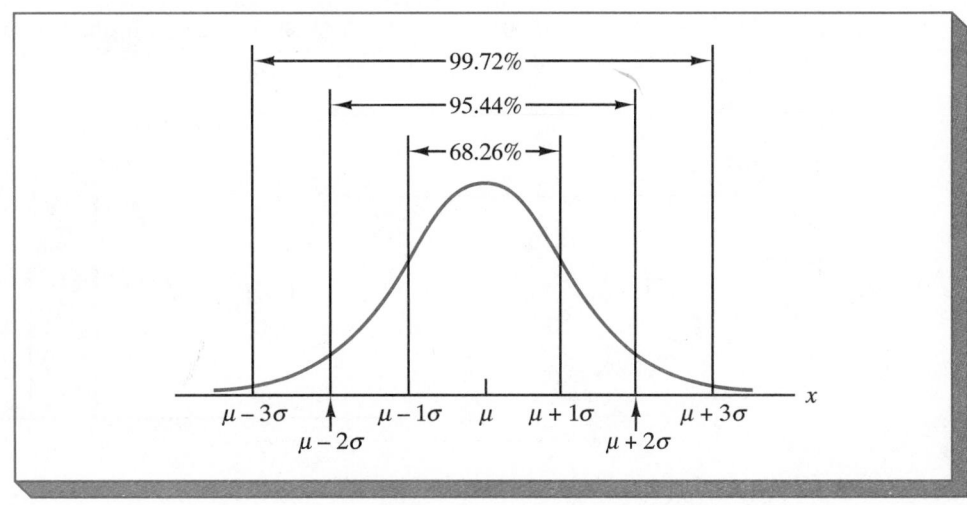

FIGURE **6.5**
**The Standard Normal Probability
Distribution**

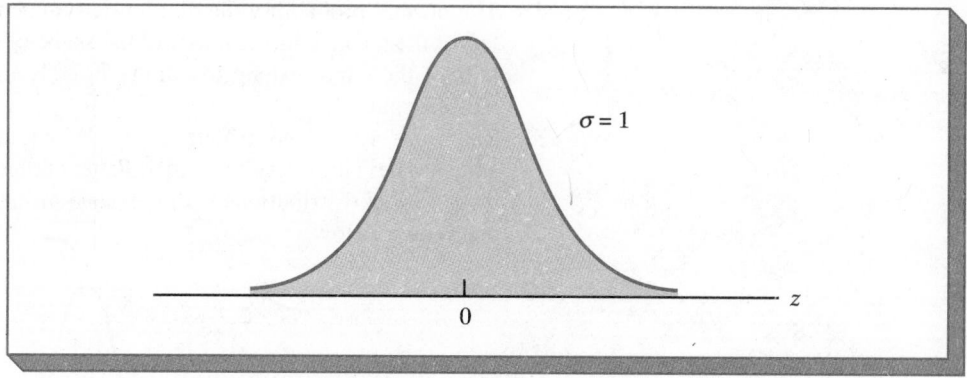

The Standard Normal Probability Distribution

A random variable that has a normal distribution with a mean of 0 and a standard deviation of 1 is said to have a *standard normal probability distribution*. The letter z is commonly used to designate this particular normal random variable. The graph of the standard normal probability distribution is shown in Figure 6.5. This too is a normal probability distribution; hence it has the same general appearance as other normal distributions but with the special properties of $\mu = 0$ and $\sigma = 1$.

As with other continuous random variables, probability calculations with any normal probability distribution are made by computing areas under the graph of the probability density function. Thus, to find the probability that a normal random variable lies within any specific interval, we must compute the area under the normal curve over that interval. For the standard normal probability distribution, areas under the normal curve have been computed and are available in tables that can be used in computing probabilities. Table 6.1 is such a table; it is also available as Table 1 of Appendix B and inside the back cover of this text.

Let us show how the table of areas under the curve for the standard normal probability distribution (Table 6.1) can be used to find probabilities by considering some examples. Later, we will see how this same table can be used to compute probabilities for any normal distribution. To begin with, let us see how we can compute the probability that the z value for the standard normal random variable will be between .00 and 1.00; that is, $P(.00 \leq z \leq 1.00)$. The shaded region in the following graph shows this area or probability.

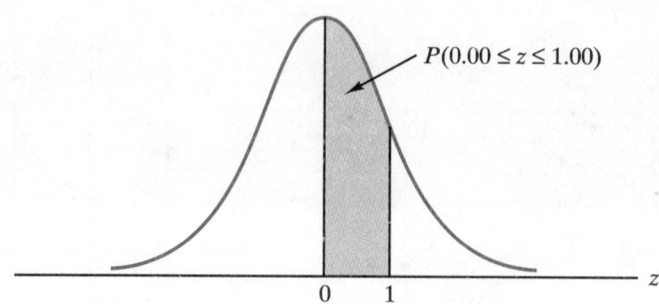

TABLE 6.1

Areas, or Probabilities, for the Standard Normal Distribution

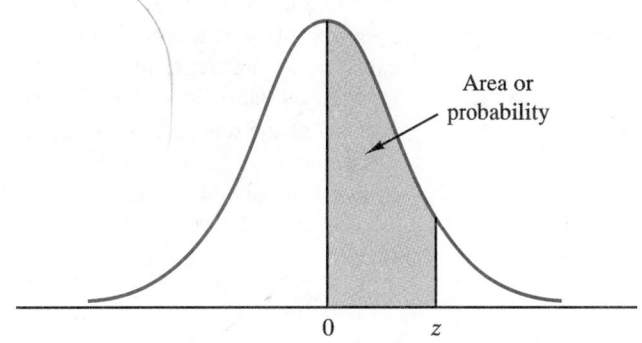

z	.00	.01	.02	.03	.04	.05	.06	.07	.08	.09
.0	.0000	.0040	.0080	.0120	.0160	.0199	.0239	.0279	.0319	.0359
.1	.0398	.0438	.0478	.0517	.0557	.0596	.0636	.0675	.0714	.0753
.2	.0793	.0832	.0871	.0910	.0948	.0987	.1026	.1064	.1103	.1141
.3	.1179	.1217	.1255	.1293	.1331	.1368	.1406	.1443	.1480	.1517
.4	.1554	.1591	.1628	.1664	.1700	.1736	.1772	.1808	.1844	.1879
.5	.1915	.1950	.1985	.2019	.2054	.2088	.2123	.2157	.2190	.2224
.6	.2257	.2291	.2324	.2357	.2389	.2422	.2454	.2486	.2518	.2549
.7	.2580	.2612	.2642	.2673	.2704	.2734	.2764	.2794	.2823	.2852
.8	.2881	.2910	.2939	.2967	.2995	.3023	.3051	.3078	.3106	.3133
.9	.3159	.3186	.3212	.3238	.3264	.3289	.3315	.3340	.3365	.3389
1.0	.3413	.3438	.3461	.3485	.3508	.3531	.3554	.3577	.3599	.3621
1.1	.3643	.3665	.3686	.3708	.3729	.3749	.3770	.3790	.3810	.3830
1.2	.3849	.3869	.3888	.3907	.3925	.3944	.3962	.3980	.3997	.4015
1.3	.4032	.4049	.4066	.4082	.4099	.4115	.4131	.4147	.4162	.4177
1.4	.4192	.4207	.4222	.4236	.4251	.4265	.4279	.4292	.4306	.4319
1.5	.4332	.4345	.4357	.4370	.4382	.4394	.4406	.4418	.4429	.4441
1.6	.4452	.4463	.4474	.4484	.4495	.4505	.4515	.4525	.4535	.4545
1.7	.4554	.4564	.4573	.4582	.4591	.4599	.4608	.4616	.4625	.4633
1.8	.4641	.4649	.4656	.4664	.4671	.4678	.4686	.4693	.4699	.4706
1.9	.4713	.4719	.4726	.4732	.4738	.4744	.4750	.4756	.4761	.4767
2.0	.4772	.4778	.4783	.4788	.4793	.4798	.4803	.4808	.4812	.4817
2.1	.4821	.4826	.4830	.4834	.4838	.4842	.4846	.4850	.4854	.4857
2.2	.4861	.4864	.4868	.4871	.4875	.4878	.4881	.4884	.4887	.4890
2.3	.4893	.4896	.4898	.4901	.4904	.4906	.4909	.4911	.4913	.4916
2.4	.4918	.4920	.4922	.4925	.4927	.4929	.4931	.4932	.4934	.4936
2.5	.4938	.4940	.4941	.4943	.4945	.4946	.4948	.4949	.4951	.4952
2.6	.4953	.4955	.4956	.4957	.4959	.4960	.4961	.4962	.4963	.4964
2.7	.4965	.4966	.4967	.4968	.4969	.4970	.4971	.4972	.4973	.4974
2.8	.4974	.4975	.4976	.4977	.4977	.4978	.4979	.4979	.4980	.4981
2.9	.4981	.4982	.4982	.4983	.4984	.4984	.4985	.4985	.4986	.4986
3.0	.4986	.4987	.4987	.4988	.4988	.4989	.4989	.4989	.4990	.4990

The entries in Table 6.1 give the area under the standard normal curve between the mean, $z = 0$, and a specified positive value of z (see the graph at the top of the table). In this case we are interested in the area between $z = 0$ and $z = 1.00$. Thus, we must find the entry in the table corresponding to $z = 1.00$. To do this, we first find 1.0 in the left-hand column of the table and then find .00 in the top row of the table. By looking in the body of the table, we find that the 1.0 row of the table and the .00 column of the table intersect at the value of .3413. We have found the desired probability; $P(.00 \leq z \leq 1.00) = .3413$. A portion of Table 6.1 showing these steps is shown below.

z	.00	.01	.02
.			
.			
.			
.9	.3159	.3186	.3212
1.0	.3413	.3438	.3461
1.1	.3643	.3665	.3686
1.2	.3849	.3869	.3888
.			
.			
.		$P(.00 \leq z \leq 1.00)$	

Using the same approach, we can find $P(.00 \leq z \leq 1.25)$. We first locate the 1.2 row and then move across to the .05 column. Doing so, we find $P(.00 \leq z \leq 1.25) = .3944$.

As a third example of the use of the table of areas for the standard normal distribution, we compute the probability of obtaining a z value between $z = -1.00$ and $z = 1.00$; that is, $P(-1.00 \leq z \leq 1.00)$.

Note that we have already used Table 6.1 to show that the probability of a z value between $z = .00$ and $z = 1.00$ is .3413, and recall that the normal probability distribution is *symmetric*. Thus, the probability of a z value between $z = .00$ and $z = -1.00$ is the *same* as the probability of a z value between $z = .00$ and $z = +1.00$. Hence, the probability of a z value between $z = -1.00$ and $z = +1.00$ is

$$P(-1.00 \leq z \leq .00) + P(.00 \leq z \leq 1.00) = .3413 + .3413 = .6826$$

This area is shown graphically as follows.

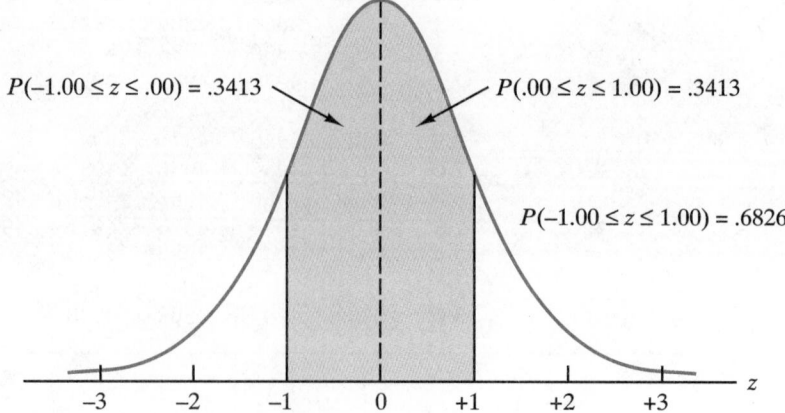

$P(-1.00 \leq z \leq .00) = .3413$ $P(.00 \leq z \leq 1.00) = .3413$

$P(-1.00 \leq z \leq 1.00) = .6826$

In a similar manner, we can use the values in Table 6.1 to show that the probability of a z value between -2.00 and $+2.00$ is $.4772 + .4772 = .9544$ and that the probability of a z value between -3.00 and $+3.00$ is $.4986 + .4986 = .9972$. Since we know that the total probability or total area under the curve for any continuous random variable must be 1.0000, the probability $.9972$ tells us that the value of z will almost always fall between -3.00 and $+3.00$.

Next, we compute the probability of obtaining a z value of at least 1.58; that is, $P(z \geq 1.58)$. First, we use the $z = 1.5$ row and the $.08$ column of Table 6.1 to find that $P(.00 \leq z \leq 1.58) = .4429$. Now, since the normal probability distribution is symmetric and the total area under the curve equals 1, we know that 50% of the area must be above the mean (i.e., $z = 0$) and 50% of the area must be below the mean. Since $.4429$ is the area between the mean and $z = 1.58$, the area or probability corresponding to $z \geq 1.58$ must be $.5000 - .4429 = .0571$. This probability is shown in the following figure.

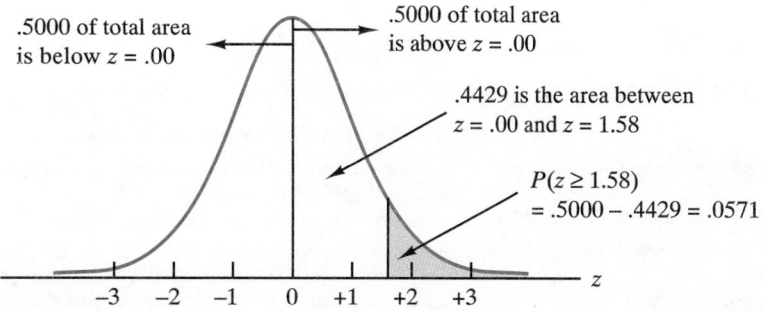

As another illustration, consider the probability the random variable z assumes a value of $-.50$ or larger; that is, $P(z \geq -.50)$. To make this computation, we note that the probability we are seeking can be written as the sum of two probabilities: $P(z \geq -.50) = P(-.50 \leq z \leq .00) + P(z \geq 0.00)$. We have previously seen that $P(z \geq .00) = .50$. Also, we know that, since the normal distribution is symmetric, $P(-.50 \leq z \leq .00) = P(.00 \leq z \leq .50)$. Referring to Table 6.1, we find that $P(.00 \leq z \leq .50) = .1915$. Therefore $P(-.50 \leq z \leq .00) = .1915$. Thus $P(z \geq -.50) = P(-.50 \leq z \leq .00) + P(z \geq .00) = .1915 + .5000 = .6915$. The graph below shows this area.

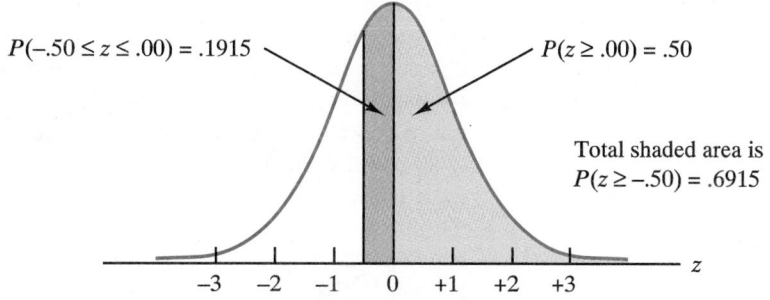

Next, we compute the probability of obtaining a z value between 1.00 and 1.58; that is, $P(1.00 \leq z \leq 1.58)$. From our previous examples, we know that there is a $.3413$

probability of a z value between $z = 0.00$ and $z = 1.00$ and that there is a .4429 probability of a z value between $z = 0.00$ and $z = 1.58$. Thus, there must be a $.4429 - .3413 = .1016$ probability of a z value between $z = 1.00$ and $z = 1.58$. Thus, $P(1.00 \leq z \leq 1.58) = .1016$. This situation is shown graphically in the following figure.

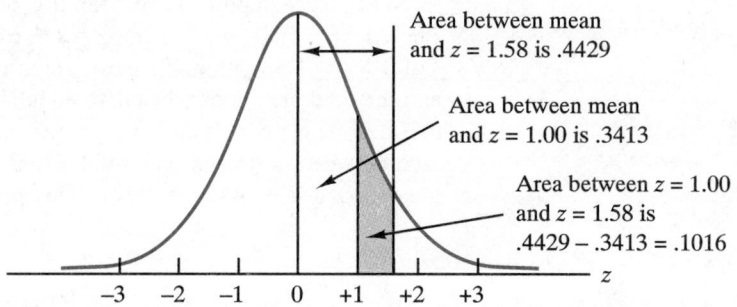

As a final illustration, let us find a z value such that the probability of obtaining a larger z value is only .10. This situation is shown graphically as follows:

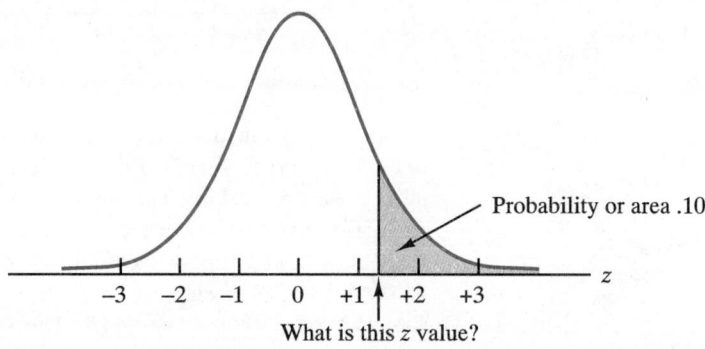

This problem is the reverse of the examples we have considered thus far. Previously we specified the z value of interest and then found the corresponding probability, or area. In this example, we are given the probability, or area, information and asked to find the corresponding z value. This can be found by using the table of areas for the standard normal probability distribution (Table 6.1) a little differently.

Recall that the body of Table 6.1 gives the area under the curve between the mean and a particular z value. In the above example, we are given the information that the area in the upper tail of the curve is .10. Thus, we must determine how much of the area is between the mean and the z value of interest. Since we know .5000 of the area is above the mean, $.5000 - .1000 = .4000$ must be the area under the curve *between* the mean and the desired z value. Scanning the body of the table, we find .3997 as the probability value closest to .4000. The section of the table providing this result is shown next:

z	.06	.07	.08	.09
.				
.				
.				
1.0	.3554	.3577	.3599	.3621
1.1	.3770	.3790	.3810	.3830
1.2	.3962	.3980	.3997	.4015
1.3	.4131	.4147	.4162	.4177
1.4	.4279	.4292	.4306	.4319
.				
.	Area value in body			
.	of table closest to .4000			

Reading the z value from the left column and the top row of the table, we find that the corresponding z value is 1.28. Thus, there will be an area of approximately .4000 (actually .3997) between the mean and z = 1.28.* In terms of the question originally asked, there is approximately a .10 probability of a z value larger than 1.28.

The examples illustrate that the table of areas for the standard normal probability distribution can be used to find probabilities associated with values of the standard normal random variable z. Two types of questions can be asked. The first type of question specifies a value, or values, for z and asks us to use the table to determine the corresponding areas, or probabilities. The second type of question provides an area, or probability, and asks us to use the table to determine the corresponding z value. Thus, we need to remain flexible in terms of using the standard normal probability table to answer the desired probability question. In most cases, sketching a graph of the standard normal probability distribution and shading the appropriate area helps to visualize the situation and aids in determining the correct answer.

Computing Probabilities for Any Normal Probability Distribution

The reason that we have been discussing the standard normal distribution so extensively is that probabilities for all normal distributions are computed using the standard normal distribution. That is, when we have a normal distribution with any mean μ and any standard deviation σ, we answer probability questions about the distribution by first converting to the standard normal distribution. Then we can use Table 6.1 and the appropriate z values to find the desired probabilities. The formula used to convert any normal random variable x with mean μ and standard deviation σ to the standard normal distribution is as follows.

---

*We could use interpolation in the body of the table to get a better approximation of the z value that cuts off an area of .4000. Doing so to provide one more decimal place of accuracy would yield a z value of 1.282. However, in most practical situations sufficient accuracy is obtained by simply using the table value closest to the desired probability.

Converting to the Standard Normal Distribution

$$z = \frac{x - \mu}{\sigma} \tag{6.3}$$

A value of x equal to its mean μ results in $z = (\mu - \mu)/\sigma = 0$. Thus, we see that a value of x equal to its mean μ corresponds to a value of z at its mean 0. Now suppose that x is one standard deviation above its mean; that is, $x = \mu + \sigma$. Applying Equation (6.3), we see that the corresponding z value is $z = [(\mu + \sigma) - \mu]/\sigma = \sigma/\sigma = 1$. Thus, a value of that is one standard deviation above its mean yields $z = 1$. In other words, we can interpret the z value as *the number of standard deviations that the normal random variable x is from its mean μ.*

To see how this conversion enables us to compute probabilities for any normal distribution, suppose we have a normal distribution with $\mu = 10$ and $\sigma = 2$. What is the probability that the random variable x is between 10 and 14? Using (6.3) we see that at $x = 10$, $z = (x - \mu)/\sigma = (10 - 10)/2 = 0$ and that at $x = 14$, $z = (14 - 10)/2 = 4/2 = 2$. Thus, the answer to our question about the probability of x being between 10 and 14 is given by the equivalent probability that z is between 0 and 2 for the standard normal distribution. In other words, the probability that we are seeking is the probability that the random variable x is between its mean and two standard deviations above the mean. Using $z = 2.00$ and Table 6.1, we see that the probability is .4772. Hence the probability that x is between 10 and 14 is .4772.

The Grear Tire Company Problem

Let us look at an application of the use of the normal probability distribution. Suppose that the Grear Tire Company has just developed a new steel-belted radial tire that will be sold through a national chain of discount stores. Since the tire is a new product, Grear's management believes that the mileage guarantee offered with the tire will be an important factor in the acceptance of the product. Before finalizing the tire mileage guarantee policy, Grear's management would like some probability information concerning the number of miles the tires will last.

From actual road tests with the tires, Grear's engineering group has estimated the mean tire mileage at $\mu = 36{,}500$ miles and the standard deviation at $\sigma = 5000$. In addition, the data collected indicate that a normal distribution is a reasonable assumption.

Using the normal distribution, what percentage of the tires can be expected to last more than 40,000 miles? In other words, what is the probability that the tire mileage will exceed 40,000? This question can be answered by finding the area of the shaded region in Figure 6.6.

At $x = 40{,}000$, we have

$$z = \frac{x - \mu}{\sigma} = \frac{40{,}000 - 36{,}500}{5000} = \frac{3500}{5000} = .70$$

Refer now to the bottom of Figure 6.6. We see that a value of $x = 40{,}000$ on the Grear Tire normal distribution corresponds to a value of $z = .70$ on the standard normal distribution and $x = 36{,}500$ corresponds to $z = 0$.

Using Table 6.1 we see that the area between the mean and $z = .70$ is .2580. Referring again to Figure 6.6, we see that this means the area between $x = 36{,}500$ and $x = 40{,}000$

FIGURE 6.6
Grear Tire Company Mileage Distribution

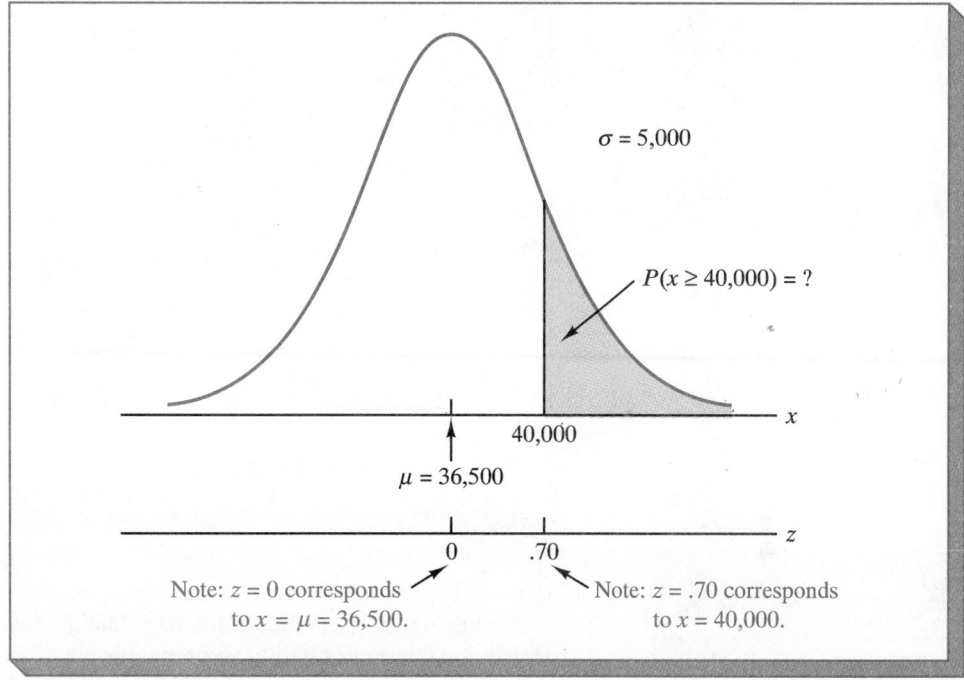

on the Grear Tire normal distribution is also .2580. Thus, .5000 − .2580 = .2420 is the probability that *x* will exceed 40,000. We can conclude that about 24.20% of the tires will exceed 40,000 in mileage.

Let us now assume that Grear is considering a guarantee that will provide a discount on a new set of tires if the original tires do not exceed the mileage stated in the guarantee. What should the guarantee mileage be if Grear would like no more than 10% of the tires to be eligible for the discount guarantee? This question is interpreted graphically in Figure 6.7.

According to Figure 6.7, 40% of the area must be between the mean and the unknown guarantee mileage. We look up .4000 in the body of Table 6.1 and see that this area occurs at approximately 1.28 standard deviations *below the mean*. That is, $z = -1.28$ is the value of the standard normal random variable corresponding to the desired mileage guarantee on the Grear Tire normal distribution. To find the mileage *x* corresponding to $z = -1.28$, we have

$$z = \frac{x - \mu}{\sigma} = -1.28$$

$$x - \mu = -1.28\sigma$$

$$x = \mu - 1.28\sigma$$

or, with $\mu = 36,500$ and $\sigma = 5000$,

$$x = 36,500 - 1.28(5000) = 30,100$$

Thus, a guarantee of 30,100 miles will meet the requirement that approximately 10% of the tires will be eligible for the guarantee. Perhaps, with this information, the firm will set its tire mileage guarantee policy at 30,000 miles.

FIGURE 6.7
Grear's Discount Guarantee

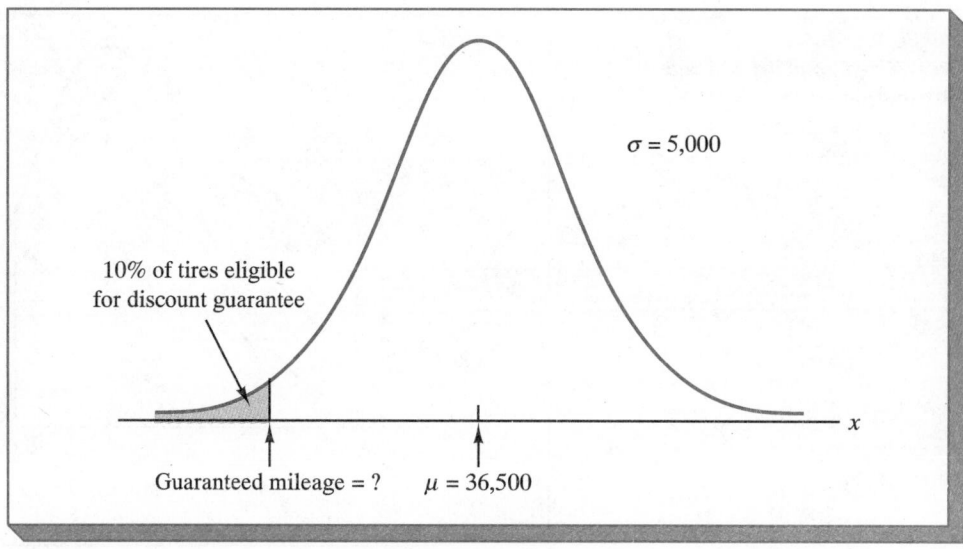

$\sigma = 5,000$

10% of tires eligible
for discount guarantee

Guaranteed mileage = ? $\mu = 36,500$

x

Again, we see the important role that probability distributions play in providing decision-making information. Namely, once a probability distribution is established for a particular problem situation, it can be used to quickly and easily provide probability data about the problem. While the data do not make a decision recommendation directly, they do provide information that helps the decision maker better understand the problem. Ultimately, this information may assist the decision maker in reaching a good decision.

Normal Approximation of Binomial Probabilities

As discussed in the previous chapter, binomial probability tables for large values of n usually are not available. A normal distribution approximation of binomial probabilities is considered acceptable when $np \geq 5$ and $n(1 - p) \geq 5$.

When using the normal approximation to the binomial, we set $\mu = np$ and $\sigma = \sqrt{np(1 - p)}$ in the definition of the normal curve. Let us illustrate the normal approximation to the binomial by supposing that a particular company has a history of making errors in 10% of its invoices. A sample of 100 invoices has been taken, and we would like to compute the probability that 12 invoices contain errors. That is, we would like to find the binomial probability of 12 successes in 100 trials. Since the binomial tables in Appendix B are not tabulated for values of n greater than 20, we will use the normal approximation to compute the desired probability.

In applying the normal approximation to the binomial, we set $\mu = np = (100)(.1) = 10$ and $\sigma = \sqrt{np(1 - p)} = \sqrt{(100)(.1)(.9)} = 3$. A normal distribution with $\mu = 10$ and $\sigma = 3$ is shown in Figure 6.8.

Recall that, with a continuous probability distribution, probabilities are computed as areas under the probability density function. As a result, the probability of any single value for the random variable is zero. Thus, to approximate the binomial probability of 12 successes, we must compute the area under the corresponding normal curve between 11.5 and 12.5. The .5 that we added and subtracted from 12 is called a *continuity correction factor*. It is introduced because a continuous distribution is being used to approximate a discrete distribution. Thus, $P(x = 12)$ for the *discrete* binomial distribution is approximated by $P(11.5 \leq x \leq 12.5)$ for the *continuous* normal distribution.

FIGURE 6.8
**Normal Approximation to a
Binomial Probability Distribution
with $n = 100$ and $p = .10$ Showing
the Probability of 12 Errors**

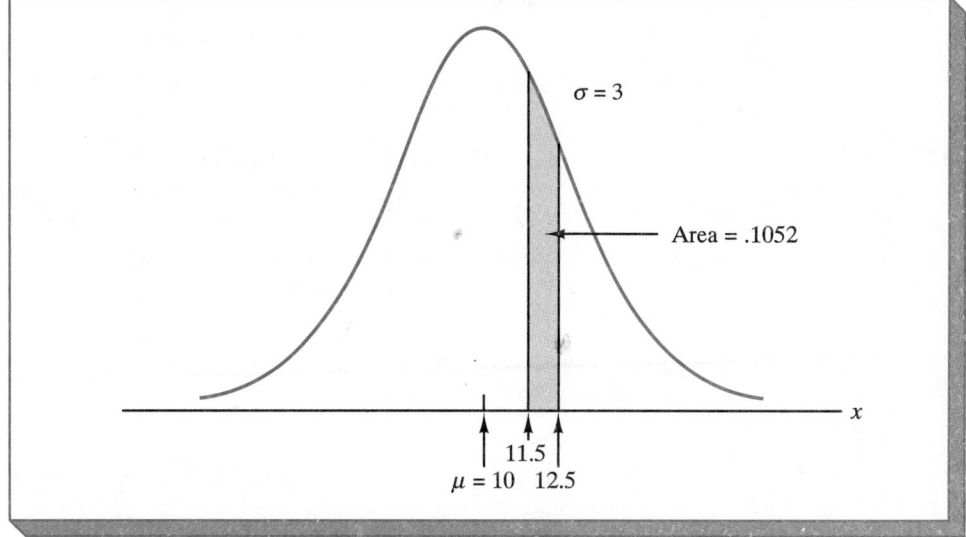

Converting to the standard normal distribution in order to compute $P(11.5 \leq x \leq 12.5)$, we have

$$z = \frac{x - \mu}{\sigma} = \frac{12.5 - 10.0}{3} = .83 \qquad \text{at } x = 12.5$$

$$z = \frac{x - \mu}{\sigma} = \frac{11.5 - 10.0}{3} = .50 \qquad \text{at } x = 11.5$$

From Table 6.1 we find the area under the curve (in Figure 6.8) between 10 and 12.5 is .2967. Similarly, the area under the curve between 10 and 11.5 is .1915. Therefore, the area between 11.5 and 12.5 is .2967 − .1915 = .1052. The normal approximation to the probability of 12 successes in 100 trials thus is .1052.

For another illustration, suppose that we want to compute the probability of 13 or fewer errors in the sample of 100 invoices. Figure 6.9 shows the area under the normal curve which approximates this probability. Note that the use of the continuity correction factor

FIGURE 6.9
**Normal Approximation to a
Binomial Probability Distribution
with $n = 100$ and $p = .10$ Showing
the Probability of 13 or Fewer
Errors**

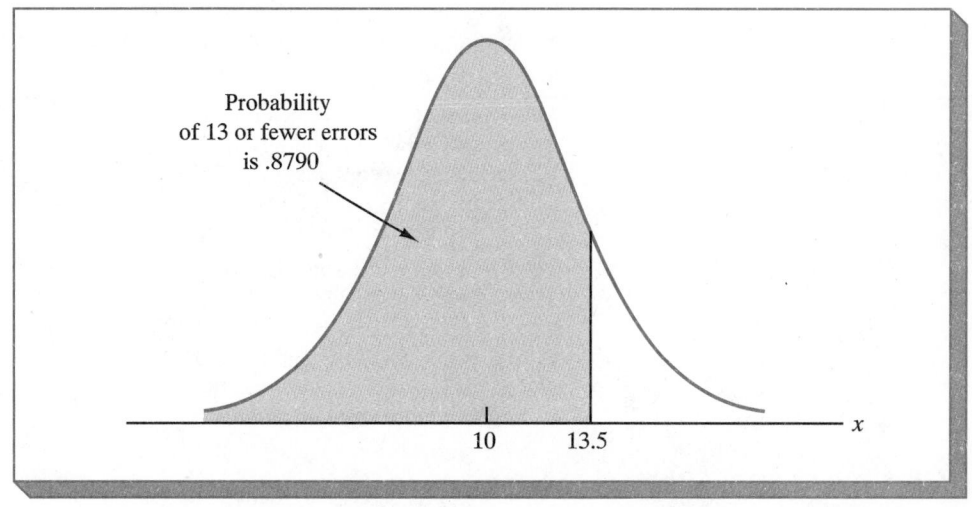

results in the value of 13.5 being used to compute the desired probability. The z value corresponding to $x = 13.5$ is

$$z = \frac{13.5 - 10.0}{3.0} = 1.17$$

Table 6.1 shows that the area under the standard normal curve between 0 and 1.17 is .3790. Thus, the area under the normal curve approximating the number of errors in 100 invoices between 10 and 13.5 is .3790. Hence, the shaded portion of the graph in Figure 6.9 represents an area, or probability, of .3790 + .5000 = .8790.

☐ ☐ Exercises

Methods

8. Using Figure 6.4 as a guide, sketch a normal curve for a random variable x that has a mean $\mu = 100$ and a standard deviation $\sigma = 10$. Label the horizontal axis with values of 70, 80, 90, 100, 110, 120, and 130.

9. The length of time required to complete a college examination is normally distributed with a mean of $\mu = 50$ minutes and a standard deviation of $\sigma = 5$ minutes.
a. Sketch a normal curve for the length of the examination. Label the horizontal axis with values of 35, 40, 45, 50, 55, 60, and 65 minutes. Figure 6.4 shows that the normal curve almost touches the horizontal line at three standard deviations below and at three standard deviations above the mean (in this case at 35 and 65).
b. What is the probability that a student will take between 45 and 55 minutes to complete the exam?
c. What is the probability that a student will take between 40 and 60 minutes to complete the exam?

10. Given that z is the standard normal random variable, sketch the standard normal curve. Label the horizontal axis at values of -3, -2, -1, 0, 1, 2, and 3. Then use the table of probabilities for the standard normal distribution to compute the following probabilities.
a. $P(0 \leq z \leq 1)$ **b.** $P(0 \leq z \leq 1.5)$ **c.** $P(0 < z < 2)$ **d.** $P(0 < z < 2.5)$

11. Given that z is the standard normal random variable, compute the following probabilities.
a. $P(-1 \leq z \leq 0)$ **b.** $P(-1.5 \leq z \leq 0)$ **c.** $P(-2 < z < 0)$
d. $P(-2.5 \leq z \leq 0)$ **e.** $P(-3 < z \leq 0)$

12. Given that z is the standard normal random variable, compute the following probabilities.
a. $P(0 \leq z \leq .83)$ **b.** $P(-1.57 \leq z \leq 0)$ **c.** $P(z > .44)$
d. $P(z \geq -.23)$ **e.** $P(z < 1.20)$ **f.** $P(z \leq -.71)$

SELF TEST ▶ **13.** Given that z is the standard normal random variable, compute the following probabilities.
a. $P(-1.98 \leq z \leq .49)$ **b.** $P(.52 \leq z \leq 1.22)$ **c.** $P(-1.75 \leq z \leq -1.04)$

14. Given that z is the standard normal random variable, find z for each situation.
a. The area between 0 and z is .4750.
b. The area between 0 and z is .2291.
c. The area to the right of z is .1314.
d. The area to the left of z is .6700.

SELF TEST ▶ **15.** Given that z is the standard normal random variable, find z for each situation.
a. The area to the left of z is .2119.
b. The area between $-z$ and z is .9030.
c. The area between $-z$ and z is .2052.
d. The area to the left of z is .9948.
e. The area to the right of z is .6915.

16. Given that z is the standard normal random variable, find z for each situation.
a. The area to the right of z is .01.
b. The area to the right of z is .025.
c. The area to the right of z is .05.
d. The area to the right of z is .10.

Applications

17. The demand for a new product is assumed to be normally distributed with $\mu = 200$ and $\sigma = 40$. Letting x be the number of units demanded, find the following:
a. $P(180 \leq x \leq 220)$ b. $P(x \geq 250)$ c. $P(x \leq 100)$ d. $P(225 \leq x \leq 250)$

SELF TEST ▶ 18. The mean cost for employee alcohol rehabilitation programs involving hospitalization is $10,000 (*USA Today*, September 12, 1991). Assume that rehabilitation program cost has a normal probability distribution with a standard deviation of $2200. Answer the following questions:
a. What is the probability that a rehabilitation program will cost at least $12,000?
b. What is the probability that a rehabilitation program will cost at least $6000?
c. What is the cost range for the 10% most expensive rehabilitation programs?

19. The Webster National Bank is reviewing its service charges and interest-paying policies on checking accounts. The bank has found that the average daily balance on personal checking accounts is $550.00, with a standard deviation of $150.00. In addition, the average daily balances have been found to be normally distributed.
a. What percentage of personal checking account customers carry average daily balances in excess of $800.00?
b. What percentage of the bank's customers carry average daily balances below $200.00?
c. What percentage of the bank's customers carry average daily balances between $300.00 and $700.00?
d. The bank is considering paying interest to customers carrying average daily balances in excess of a certain amount. If the bank does not want to pay interest to more than 5% of its customers, what is the minimum average daily balance it should be willing to pay interest on?

20. Miami University reported admission statistics for 3339 students who were admitted as freshmen for the fall semester of 1991. Of these students, 1590 had taken the Scholastic Aptitude Test (SAT). Assume the SAT verbal test scores were normally distributed with a mean of 530 and a standard deviation of 70.
a. What percentage of students were admitted with SAT verbal scores between 500 and 600?
b. What percentage of students were admitted with SAT verbal scores of 600 or more?
c. What percentage of students were admitted with SAT verbal scores of 480 or less?

21. Mensa is the international high-IQ society. To be a Mensa member, you have to have an IQ of 132 or above (*USA Today*, February 13, 1992). If IQ scores are normally distributed with a mean of 100 and a standard deviation of 15, what percentage of the population qualifies for membership in Mensa?

22. General Hospital's patient account division has compiled data on the age of accounts receivables. The data collected indicate that the age of the accounts follows a normal distribution, with $\mu = 28$ days and $\sigma = 8$ days.
a. What portion of the accounts are between 20 and 40 days old—that is, $P(20 \leq x \leq 40)$?
b. The hospital administrator is interested in sending reminder letters to the oldest 15% of accounts. How many days old should an account be before a reminder letter is sent?
c. The hospital administrator would like to give a discount to those accounts that pay their balance by the 21st day. What percentage of the accounts will receive the discount?

23. The time required to complete a final examination in a particular college course is normally distributed, with a mean of 80 minutes and a standard deviation of 10 minutes. Answer the following questions:

a. What is the probability of completing the exam in 1 hour or less?

b. What is the probability a student will complete the exam in more than 60 minutes but less than 75 minutes?

c. Assume that the class has 60 students and that the examination period is 90 minutes in length. How many students do you expect will be unable to complete the exam in the allotted time?

24. The useful life of a computer terminal at a university computer center is known to be normally distributed, with a mean of 3.25 years and a standard deviation of .5 years.

a. Historically, 22% of the terminals have had a useful life less than the manufacturer's advertised life. What is the manufacturer's advertised life for the computer terminals?

b. What is the probability that a computer terminal will have a useful life of at least 3 but less than 4 years?

25. From past experience, the management of a well-known fast-food restaurant estimates that the number of weekly customers at a particular location is normally distributed, with a mean of 5000 and a standard deviation of 800 customers.

a. What is the probability that on a given week the number of customers will be between 4760 and 5800?

b. What is the probability of a week with more than 6500 customers?

c. For 90% of the weeks, the number of customers should exceed what amount?

26. To obtain cost savings, a company is considering offering an early retirement incentive for its older management personnel. The consulting firm that designed the early retirement program has found that approximately 22% of the employees qualifying for the program will select early retirement during the first year of eligibility. Assume that the company offers the early retirement program to 50 of its management personnel.

a. What is the expected number of employees who will select early retirement in the first year?

b. What is the probability that at least 8 but not more than 12 employees will select early retirement in the first year?

c. What is the probability that 15 or more employees will select the early retirement option in the first year?

d. For the program to be judged successful, the company believes that it should entice at least 10 management employees to select early retirement in the first year. What is the probability that the program is successful?

27. Suppose that 54% of a large population of registered voters favor the Democratic candidate for state senator. A public opinion poll uses randomly selected samples of voters and asks each person in the sample his or her preference: the Democratic candidate or the Republican candidate. The weekly poll is based on the response of 100 voters.

a. What is the expected number of voters who will favor the Democratic candidate?

b. What is the variance in the number of voters who will favor the Democratic candidate?

c. What is the probability that 49 or fewer individuals in the sample express support for the Democratic candidate?

28. Homes in Chicago, Illinois, in the price range of $95,000–$130,000, are on the market an average of 70 days prior to sale (*U.S. News & World Report*, April 6, 1992). Assume that the distribution of days on the market is normal with a standard deviation of 25 days.

a. What is the probability a house will be on the market 100 days or more?

b. What is the probability a house sells during the second month it is on the market? That is, $P(31 \le x \le 60)$?

c. How many days are the fastest selling 20% on the market?

29. A Myrtle Beach resort hotel has 120 rooms. In the spring months, hotel room occupancy is approximately 75%. Use the normal approximation to the binomial distribution to answer the following questions: $u = 90$

a. What is the probability that at least half the rooms are occupied on a given day?

b. What is the probability that 100 or more rooms are occupied on a given day?

c. What is the probability that 80 or fewer rooms are occupied on a given day?

30. It is known that 30% of all customers of a major national charge card pay their bills in full before any interest charges are incurred. Use the normal approximation to the binomial distribution to answer the following questions for a group of 150 credit card holders:

a. What is the probability that between 40 and 60 customers pay their account charges before any interest charges are incurred? That is, find $P(40 \leq x \leq 60)$.

b. What is the probability that 30 or fewer customers pay their account charges before any interest charges are incurred?

6.3   The Exponential Probability Distribution

A continuous probability distribution that is often useful in describing the time it takes to complete a task is the *exponential probability distribution*. The exponential random variable can be used to describe such things as the time between arrivals at a car wash, the time required to load a truck, the distance between major defects in a highway, and so on. The exponential probability density function is as follows.

Exponential Probability Density Function

$$f(x) = \frac{1}{\mu}e^{-x/\mu} \qquad \text{for } x \geq 0, \; \mu > 0 \qquad\qquad (6.4)$$

As an example of the exponential probability distribution, assume that the time it takes to load a truck at the Schips loading dock follows an exponential probability distribution. If the mean, or average, time to load a truck is 15 minutes ($\mu = 15$), then the appropriate probability density function is

$$f(x) = \frac{1}{15}e^{-x/15}$$

The graph of this density function is shown in Figure 6.10.

Computing Probabilities for the Exponential Distribution

As with any continuous probability distribution, the area under the curve corresponding to some interval provides the probability that the random variable takes on a value in that interval. In the Schips loading dock example, the probability that it takes 6 minutes or less $(x \leq 6)$ to load a truck is defined to be the area under the curve from $x = 0$ to $x = 6$. Similarly, the probability that a truck is loaded in 18 minutes or less $(x \leq 18)$ is the area under the curve from $x = 0$ to $x = 18$. Note also that the probability that it takes between 6 minutes and 18 minutes $(6 \leq x \leq 18)$ to load a truck is given by the area under the curve from $x = 6$ to $x = 18$.

In order to compute exponential probabilities such as those described above, we make use of the following formula, which provides the probability of obtaining a value for the exponential random variable of less than or equal to some specific value of x, denoted by x_0.

FIGURE 6.10

Exponential Probability Distribution for the Schips Loading Dock Example

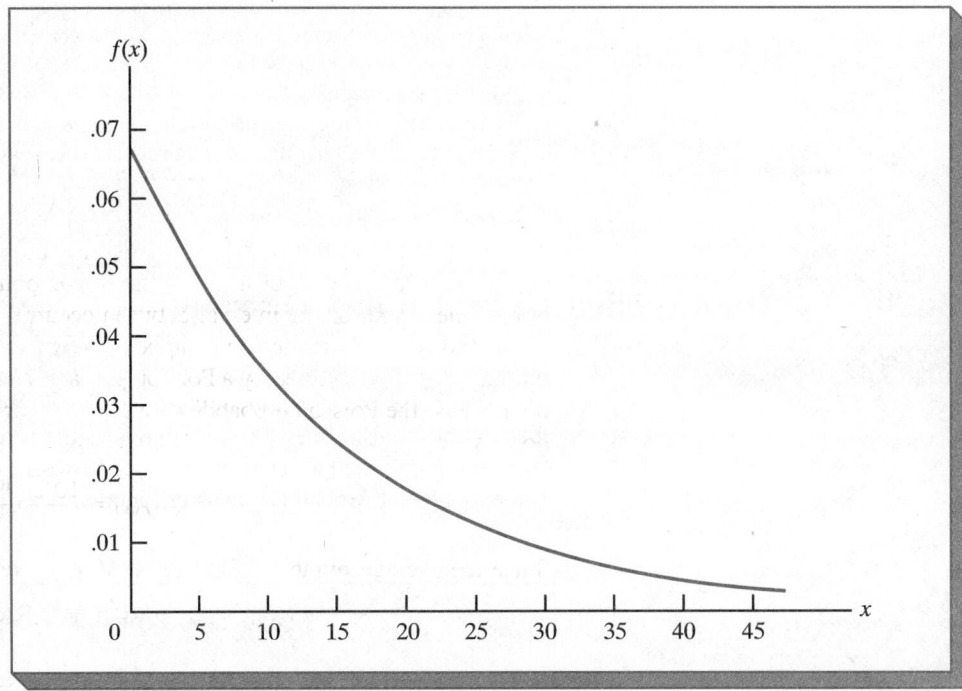

Exponential Distribution Probabilities

$$P(x \leq x_0) = 1 - e^{-x_0/\mu} \tag{6.5}$$

Thus, for the Schips loading dock example, (6.5) can be written as

$$P(\text{loading time} \leq x_0) = 1 - e-x_0/15$$

Hence, the probability that it takes 6 minutes or less ($x \leq 6$) to load a truck is

$$P(\text{loading time} \leq 6) = 1 - e^{-6/15} = .3297$$

Note also the probability that it takes 18 minutes or less ($x \leq 18$) to load a truck is

$$P(\text{loading time} \leq 18) = 1 - e^{-18/15} = .6988$$

Thus, we see that the probability that it takes between 6 minutes and 18 minutes to load a truck is equal to .6988 − .3297 = .3691. Probabilities for any other interval can be computed in a similar manner.

Relationship between the Poisson and Exponential Distributions

In Section 5.5 we introduced the Poisson distribution as a discrete probability distribution that is often useful when dealing with the number of occurrences of an event over a specified interval of time or space. Recall that the Poisson probability function is

$$f(x) = \frac{\mu^x e^{-\mu}}{x!}$$

where

$$\mu = \text{expected value or mean number of}$$
$$\text{occurrences in an interval}$$

The continuous exponential probability distribution is related to the discrete Poisson distribution in that, if the Poisson distribution provides an appropriate description of the number of occurrences per interval, then the exponential distribution provides a description of the length of the interval between occurrences.

To illustrate this relationship, suppose that the number of cars that arrive at a car wash during 1 hour is described by a Poisson probability distribution with a mean of 10 cars per hour. Thus, the Poisson probability function that provides the probability of x arrivals per hour is

$$f(x) = \frac{10^x e^{-10}}{x!}$$

Since the average number of arrivals is 10 cars per hour, the average time between cars arriving is

$$\frac{1 \text{ hour}}{10 \text{ cars}} = .1 \text{ hour/car}$$

Thus, the corresponding exponential distribution that describes the time between the arrival of cars has a mean of $\mu = .1$ hours per car; the appropriate exponential probability density function is

$$f(x) = \frac{1}{.1} e^{-x/.1} = 10 e^{-10x}$$

Exercises

Methods

31. Answer the following questions concerning the given exponential probability density function

$$f(x) = \frac{1}{8} e^{-x/8} \qquad \text{for } x \geq 0$$

a. Find $P(x \leq 6)$.   **b.** Find $P(x \leq 4)$.
c. Find $P(x \geq 6)$.   **d.** Find $P(4 \leq x \leq 6)$.

SELF TEST   ▶   **32.** Consider the following exponential probability density function:

$$f(x) = \frac{1}{3} e^{-x/3} \qquad \text{for } x \geq 0$$

a. Write the formula for $P(x \leq x_0)$.   **b.** Find $P(x \leq 2)$.
c. Find $P(x \geq 3)$.   **d.** Find $P(x \leq 5)$.
e. Find $P(2 \leq x \leq 5)$.

Applications

33. There were 34 traffic fatalities in Clermont County, Ohio, during 1987 (The *Cincinnati Enquirer*, December 8, 1988). Assume that, given an average of 34 fatalities per year, an exponential distribution accurately describes the time between fatalities.

a. What is the probability that the time between fatalities is one month or less?
b. What is the probability that the time between fatalities is one week or more?

SELF TEST ▶

34. The time between arrivals of vehicles at a particular intersection follows an exponential probability distribution with a mean of 12 seconds.
a. Sketch this exponential probability distribution.
b. What is the probability that the arrival time between vehicles is 12 seconds or less?
c. What is the probability that the arrival time between vehicles is 6 seconds or less?
d. What is the probability that there will be 30 or more seconds between arriving vehicles?

35. The lifetime (hours) of an electronic device is a random variable with the following exponential probability density function:

$$f(x) = \frac{1}{50} e^{-x/50} \qquad \text{for } x \geq 0$$

a. What is the mean lifetime of the device?
b. What is the probability the device fails in the first 25 hours of operation?
c. What is the probability the device operates 100 or more hours before failure?

36. A new automated production process has been averaging 2 breakdowns per day, where the number of breakdowns per day follows a Poisson probability distribution.
a. What is the mean time between breakdowns, assuming 8 hours of operation per day?
b. Show the exponential probability density function that can be used for the time between breakdowns.
c. What is the probability that the process will run 1 hour or more before another breakdown?
d. What is the probability that the process can run a full 8-hour shift without a breakdown?

37. The time in minutes for a student using a computer terminal at the computer center of a major university follows an exponential probability distribution with a mean of 36 minutes. Assume a second student arrives at the terminal just as another student is beginning to work on the terminal.
a. What is the probability that the wait for the second student will be 15 minutes or less?
b. What is the probability that the wait for the second student will be between 15 and 45 minutes?
c. What is the probability the second student will have to wait an hour or more?

Summary

This chapter extended the discussion of probability distributions to the case of continuous random variables. The major conceptual difference between discrete and continuous probability distributions is in the method of computing probabilities. With discrete distributions, the probability function $f(x)$ provides the probability that the random variable x assumes various values. With continuous probability distributions, we associate a probability density function, denoted by $f(x)$. The difference is that the probability density function does not provide probability values for a continuous random variable directly. Probabilities are given by areas under the curve or graph of the probability density function $f(x)$. Since the area under the curve above a single point is zero, we observe that the probability of any particular value is zero for a continuous random variable.

Three continuous probability distributions—the uniform, normal, and exponential distributions—were treated in detail. The normal probability distribution is used widely in statistical inference and will be used extensively in the remainder of the text.

Glossary

Uniform probability distribution A continuous probability distribution where the probability that the random variable will assume a value in any interval of equal length is the same for each interval.

Probability density function The function that defines the probability distribution of a continuous random variable.

Normal probability distribution A continuous probability distribution. Its probability density function is bell shaped and determined by the mean μ and standard deviation σ.

Standard normal probability distribution A normal distribution with a mean of 0 and a standard deviation of 1.

Continuity correction factor A value of .5 that is added and/or subtracted from a value of x when the continuous normal probability distribution is used to approximate the discrete binomial probability distribution.

Exponential probability distribution A continuous probability distribution that is useful in computing probabilities for the time, or space, between occurrences of an event.

Key Formulas

Uniform Probability Density Function

$$f(x) = \begin{cases} \dfrac{1}{b-a} & \text{for } a \le x \le b \\\\ 0 & \text{elsewhere} \end{cases} \tag{6.1}$$

Normal Probability Density Function

$$f(x) = \frac{1}{\sqrt{2\pi}\sigma} e^{-(x-\mu)^2/2\sigma^2} \tag{6.2}$$

Converting to the Standard Normal Distribution

$$z = \frac{x-\mu}{\sigma} \tag{6.3}$$

Exponential Probability Density Function

$$f(x) = \frac{1}{\mu} e^{-x/\mu} \qquad \text{for } x \ge 0, \ \mu > 0 \tag{6.4}$$

Exponential Distribution Probabilities

$$P(x \le x_0) = 1 - e^{-x_0/\mu} \tag{6.5}$$

Supplementary Exercises

38. In an office building the waiting time for an elevator is found to be uniformly distributed between 0 and 5 minutes.
   a. What is the probability density function $f(x)$ for this uniform distribution?
   b. What is the probability of waiting longer than 3.5 minutes?
   c. What is the probability that the elevator arrives in the first 45 seconds?
   d. What is the probability of a waiting time between 1 and 3 minutes?
   e. What is the expected waiting time?

39. The time required to complete a particular assembly operation is uniformly distributed between 30 and 40 minutes.
   a. What is the mathematical expression for the probability density function?
   b. Compute the probability that the assembly operation will require more than 38 minutes to complete.

c. If management wants to set a time standard for this operation, what time should be selected such that 70% of the time the operation will be completed within the time specified?

d. Find the expected value and standard deviation for the assembly time.

40. A particular make of automobile is listed as weighing 4000 pounds. Because of weight differences due to the options ordered with the car, the actual weight varies uniformly between 3900 and 4100 pounds.

a. What is the mathematical expression for the probability density function?

b. What is the probability that the car will weigh less than 3950 pounds?

41. Given that z is a standard normal random variable, compute the following probabilities.

a. $P(-.72 \leq z \leq 0)$ **b.** $P(-.35 \leq z \leq .35)$

c. $P(.22 \leq z \leq .87)$ **d.** $P(z \leq -1.02)$

42. Given that z is a standard normal random variable, compute the following probabilities.

a. $P(z \geq -.88)$ **b.** $P(z \geq 1.38)$

c. $P(-.54 \leq z \leq 2.33)$ **d.** $P(-1.96 \leq z \leq 1.96)$

43. Given that z is a standard normal random variable, find z if it is known that

a. the area between $-z$ and z is .90

b. the area to the right of z is .20

c. the area between -1.66 and z is .25

d. the area to the left of z is .40

e. the area between z and 1.80 is .20

44. Motorola used the normal distribution to determine the probability of defects and the number of defects expected in a production process (*APICS—The Performance Advantage*, July 1991). Assume a production process is designed to produce items with a weight of 10 ounces and that the process mean is 10. Calculate the probability of a defect and the expected number of defects for a 1000-unit production run under the following situations.

a. The process standard deviation is .15 and the process control is set at plus or minus 1 standard deviation. Units with weight less than 9.85 or greater than 10.15 ounces will be classified as defects.

b. Through process design improvements, the process standard deviation can be reduced to .05. Assume the process control remains the same with weights less than 9.85 or greater than 10.15 ounces being classified as defects.

c. What is the advantage of reducing process variation and setting process control limits at a greater number of standard deviations from the mean?

45. In 1985, the average household income for Americans was $23,618 (*Louis Rukeyser's Business Almanac,* Simon and Schuster, New York, 1988).

a. It was noted that 7.7% of the households earned less than $5000. Assuming that household income is normally distributed, what is the standard deviation of household income?

b. It was also noted that 14.8% of households earned more than $50,000. Does this seem reasonable given the standard deviation computed in part (a)? Explain.

c. In part (a) we said to assume that household income is normally distributed. Does this assumption appear to be reasonable? Explain.

46. A soup company markets eight varieties of homemade soup throughout the Eastern states. The standard-size soup can holds a maximum of 11 ounces, while the label on each can advertises contents of 10¾ ounces. The extra ¼ ounce is to allow for the possibility of the automatic filling machine placing more soup than the company actually wants in a can. Past experience shows that the number of ounces placed in a can is approximately normally distributed, with a mean of 10¾ ounces and a standard deviation of .1 ounce. What is the probability that the machine will attempt to place more than 11 ounces in a can, causing an overflow to occur?

47. The sales of High-Brite Toothpaste are believed to be approximately normally distributed, with a mean of 10,000 tubes per week and a standard deviation of 1500 tubes per week.

a. What is the probability that more than 12,000 tubes will be sold in any given week?

b. In order to have a .95 probability that the company will have sufficient stock to cover the weekly demand, how many tubes should be produced?

48. Points scored by the winning team in NCAA college football games are approximately normally distributed, with a mean of 24 and a standard deviation of 6.

a. What is the probability that a winning team in a football game scores between 20 and 30 points; that is, $P(20 \leq x \leq 30)$?

b. How many points does a winning team have to score to be in the highest 20% of scores for college football games?

49. Ward Doering Auto Sales is considering offering a special service contract that will cover the total cost of any service work required on leased vehicles. From past experience, the company manager estimates that yearly service costs are approximately normally distributed, with a mean of $150 and a standard deviation of $25.

a. If the company offers the service contract to customers for a yearly charge of $200, what is the probability that any one customer's service costs will exceed the contract price of $200?

b. What is Ward's expected profit per service contract?

50. The attendance at football games at a certain stadium is normally distributed, with a mean of 45,000 and a standard deviation of 3000.

a. What percentage of the time should attendance be between 44,000 and 48,000?

b. What is the probability of the attendance exceeding 50,000?

c. For 80% of the time the attendance should be at least how many?

51. Assume that the test scores from a college admissions test are normally distributed, with a mean of 450 and a standard deviation of 100.

a. What percentage of the people taking the test score between 400 and 500?

b. Suppose that someone receives a score of 630. What percentage of the people taking the test score better? What percentage score worse?

c. If a particular university will not admit anyone scoring below 480, what percentage of the persons taking the test would be acceptable to the university?

52. The Office Products Group of the former Burroughs Corporation manufactures plastic credit cards used in automatic bank teller machines. Any card with a length of less than 3.365 inches is considered defective. One of the dies used in making the credit cards is producing cards with a mean length of 3.367 inches. The lengths are normally distributed with a standard deviation of .001 inch.

a. What is the probability of obtaining a defective card using this die?

b. The company does not want to use any die that produces more than 1% defective cards. What should the company do in this instance?

c. Assuming the standard deviation stays at .001 inch, what is the smallest acceptable mean length for cards manufactured? (Hint: For what mean length will no more than 1% of the cards be shorter than 3.365 inches?)

53. A machine fills containers with a particular product. The standard deviation of filling weights is known from past data to be .6 ounces. If only 2% of the containers hold less than 18 ounces, what is the mean filling weight for the machine? That is, what must μ equal? Assume the filling weights have a normal distribution.

54. Consider a multiple-choice examination with 50 questions. Each question has 4 possible answers. Assume that a student who has done the homework and attended lectures has a .75 probability of answering any question correctly.

a. A student must answer 43 or more questions correctly in order to obtain a grade of A. What percentage of the students who have done their homework and attended lectures will obtain a grade of A on this multiple-choice examination?

b. A student who answers 35–39 questions correctly will receive a grade of C. What percentage of students who have done their homework and attended lectures will obtain a grade of C on this multiple-choice examination?

c. A student must answer 30 or more questions correctly in order to pass the examination. What percentage of the students who have done their homework and attended lectures will pass the examination?

d. Assume that a student has not attended class and has not done the homework for the course. Furthermore, assume that the student will simply guess at the answer to each question. What is the probability that this student answers 30 or more questions correctly and passes the examination?

55. The book *100% American* by Daniel Evan Weiss reports that 64% of Americans live in the state where they were born. What is the probability that a random sample of 100 people will find between 60 and 70 people living in the state where they were born? That is, find $P(60 \leq x \leq 70)$.

56. The time (in minutes) between telephone calls at an insurance claims office has the following exponential probability distribution:

$$f(x) = .50e^{-.50x} \qquad \text{for } x \geq 0$$

a. What is the mean time between telephone calls?
b. What is the probability that there are 30 seconds or less between telephone calls?
c. What is the probability that there is 1 minute or less between telephone calls?
d. What is the probability of going 5 or more minutes without a telephone call?

57. The time (in minutes) a checkout lane is idle between customers at a supermarket follows an exponential probability distribution with a mean of 1.2 minutes.
a. Show the probability density function for this distribution.
b. What is the probability that the next customer arrives between .5 to 1.0 minutes after a customer is served?
c. What is the probability of the checkout lane being idle for more than a minute between customers?

7

Sampling and Sampling Distributions

Mead Corporation*

DAYTON, OHIO

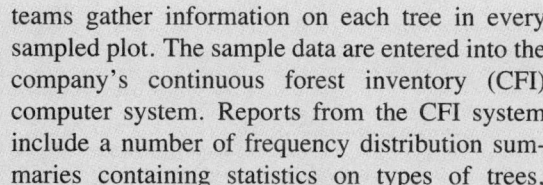

Mead Corporation, located in Dayton, Ohio, is a diversified paper and forest products company that manufactures paper, pulp, and lumber and converts paperboard into shipping containers and beverage carriers. The company's distribution capability is used to market many of its own products, including paper, school supplies, and stationery. The use of sampling by the company's internal consulting group for decision analysis provides a variety of information that enables Mead to obtain significant productivity benefits and remain competitive in its industry.

For example, Mead maintains large woodland holdings which provide the trees that are the raw material for many of the company's products. Management needs reliable and accurate information about the timberlands and forests in order to evaluate the company's ability to provide its future raw material needs. What is the present volume in the forests? What is the past growth of the forests? What is the projected future growth of the forests? All of these are important questions that must be answered. With answers to these questions, Mead's management can develop plans for the future including long-term planting and harvesting schedules for the trees.

How does Mead obtain the information it needs about its vast forest holdings? Data collected from sample plots located throughout the forests provide the basis for learning about the population of trees owned by the company. In order to identify the sample plots, the timberland holdings are first divided into three sections based on location and types of trees. Using maps and tables of random numbers, Mead analysts identify random samples of 1/5 to 1/7 acre plots in each section of the forest. The sample plots provide the locations where Mead foresters collect data and learn about the forest population.

Foresters throughout the organization participate in the field data collection process. Periodically, two-person teams gather information on each tree in every sampled plot. The sample data are entered into the company's continuous forest inventory (CFI) computer system. Reports from the CFI system include a number of frequency distribution summaries containing statistics on types of trees, present forest volume, past forest growth rates, and a projection of the future forest growth and volume. Sampling and the associated statistical summaries of the sample data provide the reports that are essential for the effective management of Mead's forests and timberland assets.

In this chapter you will learn about simple random sampling and the sample selection process. In addition, you will learn how statistics such as the sample mean and sample proportion are used to provide estimates of the population mean and population proportion. The important concept of a sampling distribution is also introduced.

Mead's paper machine known as the "Spirit of Escanaba" is one of the largest and most modern paper machines in the world.

*Dr. Edward P. Winkofsky, Mead Corporation, provided this Statistics in Practice.

n Chapter 1, we defined a *population* and a *sample* as two important aspects of a statistical study. The definitions are restated here:

1. A *population* is the collection of all the elements of interest.
2. A *sample* is a subset of the population.

As also stated in Chapter 1, the purpose of *statistical inference* is to provide information about a population based upon information contained in a sample. Let us begin by citing two situations where sampling is conducted to provide a manager or decision maker with information about a population.

1. A tire manufacturer is considering producing a new tire, which is designed to provide an increase in mileage over the firm's current line of tires. In order to evaluate the new tire design, management needs an estimate of the mean number of miles provided by the new tires. The manufacturer selects a sample of 120 new tires for testing. The test results in a sample mean of 36,500 miles. Thus 36,500 miles is used as an estimate of the mean tire life for the population of new tires.
2. Members of a political party are considering supporting a particular candidate for election to the United States Senate. To decide whether or not to enter the candidate in the upcoming primary election, party leaders need an estimate of the proportion of registered voters favoring the candidate. The time and cost associated with contacting every individual in the population of registered voters are prohibitive. Thus, a sample of 400 registered voters is selected. If 160 of the 400 voters indicate a preference for the candidate, an estimate of the proportion of the population of registered voters favoring the candidate is 160/400 = .40.

The preceding examples show how sampling and the subsequent sample results can be used to develop estimates of population characteristics. Note that in the tire mileage example, collecting the data on tire life results in wearing out each tire tested. As a result, it is not feasible to test every tire in the population; a sample is the only realistic way to obtain the desired tire mileage data. In the example involving the primary election, it is theoretically possible to contact every registered voter in the population; however, the time and cost involved in doing so are prohibitive. Thus, a sample of registered voters is preferred.

The above examples point out some of the reasons for using samples. However, it is important to realize that sample results only provide *estimates* of the values of the population characteristics. That is, we do not expect the sample mean of 36,500 miles to *exactly equal* the mean mileage for all tires in the population; neither do we expect *exactly* 40% of the population of registered voters to favor the candidate. We cannot expect this simply because the sample only contains a portion of the population. However, we believe that with proper sampling methods the sample results will provide "good" estimates of the population characteristics. But how good can we expect the sample results to be? Fortunately, statistical procedures are available for answering this question.

In this chapter we show how simple random sampling can be used to select a sample from a population. We then show how data obtained from a simple random sample can be used to compute estimates of a population mean, a population standard deviation, and a population proportion. In addition, we introduce the important concept of a sampling distribution. As we shall show, knowledge of the appropriate sampling distribution is what enables us to make statements about the goodness of the sample results.

7.1   The Electronics Associates Sampling Problem

Electronics Associates, Inc. (EAI) is an international company that manufactures a diverse line of products in plants located throughout the United States, Canada, and Europe. The firm's director of personnel has been assigned the task of developing a profile of the company's 2500 managers. The characteristics that are to be identified include the mean annual salary for the managers and the proportion of managers having completed the company's management training program.

Using the 2500 managers as the population for this study, we can find the annual salary and the training program status for each individual in the population by referring to the firm's personnel records. Let us assume that this has been done and that we have obtained the corresponding annual salary and management training program participation information for all 2500 managers in the population.

Using the formulas for a population mean and a population standard deviation that were presented in Chapter 3, we can compute the mean and standard deviation of annual salary for the population. Assume that these calculations have been performed with the following results for annual salary:

$$\text{Population mean:} \quad \mu = \frac{\Sigma x_i}{2500} = \$51{,}800$$

$$\text{Population standard deviation:} \quad \sigma = \sqrt{\frac{\Sigma (x_i - \mu)^2}{2500}} = \$4000$$

Furthermore, assume that a review of the 2500 records shows that 1500 managers have completed the training program. Letting p denote the proportion of the population having completed the training program, we see that $p = 1500/2500 = .60$.

Whenever a numerical characteristic of a population is calculated using observations from all elements of the population, the numerical characteristic is called a *parameter* of the population. Thus, the population mean annual salary ($\mu = \$51{,}800$), the population standard deviation of annual salary ($\sigma = \$4000$), and the population proportion having completed the training program ($p = .60$) are parameters of the population of EAI managers.

The question that we would now like to consider is how the firm's director of personnel could have obtained estimates of these population parameters by using a sample of managers rather than all 2500 managers in the population. Assume that a sample of 30 managers will be used. Clearly, the time and the cost required to develop a profile for 30 managers would be substantially less than those for the entire population. If the personnel director could be assured that a sample of 30 managers would provide adequate information about the population of 2500 managers, working with a sample would be preferred to working with the entire population. Let us explore the possibility of using a sample for the EAI study by first considering how we could identify a sample of 30 managers.

7.2   Simple Random Sampling

There are several methods that can be used to select a sample from a population; one of the most common of these is *simple random sampling*. The definition of a simple random sample and the process of selecting a simple random sample depends upon whether the size of the population is *finite* or *infinite*. Since the EAI sampling problem introduced in the previous section involves a finite population of 2500 managers, we first consider sampling from finite populations.

Sampling from Finite Populations

A simple random sample of size n from a finite population of size N is defined as follows.

Simple Random Sample (Finite Population)

A simple random sample of size n from a finite population of size N is a sample selected such that each possible sample of size n has the same probability of being selected.

Fortunately, there is a relatively easy and straightforward procedure for identifying a simple random sample from a finite population. This method of selecting a simple random sample enables us to choose, or select, the elements for the sample *one at a time*. At each selection we make sure that each of the elements remaining in the population has the *same probability* of being selected for the sample. Sampling n elements in this fashion will satisfy the definition of a simple random sample from a finite population.

Let us demonstrate the selection of a simple random sample from a finite population by referring to the EAI problem. First, we will assume that the 2500 EAI managers have been numbered sequentially (i.e., 1, 2, 3, . . . , 2499, 2500) in the order that their names appear in the EAI personnel file. We could then write the numbers from 1 to 2500 on equal-sized pieces of paper. The 2500 pieces of paper could then be placed in a container and mixed thoroughly. We would begin the process of identifying managers for the sample by reaching into the container and selecting one piece of paper *randomly*. The number on the chosen piece of paper would correspond to one of the numbered managers in the file of 2500 managers; thus, that manager would be selected for the sample. The remaining 2499 pieces of paper would be thoroughly mixed again, after which another piece of paper would be selected. This second number corresponds to another EAI manager to be included in the sample. The process continues until 30 managers have been selected from the population. The 30 managers identified in this manner form a simple random sample from the population.

In this procedure we did not place a selected (sampled) piece of paper back into the container after it was drawn. Thus, we are selecting a simple random sample *without replacement*. Certainly we could have followed the sampling procedure of *replacing* each sampled element before selecting subsequent elements. This form of sampling, referred to as sampling *with replacement*, would have made it possible for some elements to appear in the sample more than once. While sampling with replacement is a valid way of identifying a simple random sample, sampling *without replacement* is the sampling procedure used most often. Whenever we refer to simple random sampling, we will assume that the sampling is done without replacement.

This procedure for selecting a simple random sample of 30 EAI managers requires the labeling of 2500 pieces of paper. In practice, tables of random numbers can be used to provide the same results much more easily. Tables of random numbers are available from a variety of handbooks* that contain page after page of random numbers. We have included one such page of random numbers in Table 8 of Appendix B. A portion of this

---

*For example, The Rand Corporation, *A Million Random Digits with 100,000 Normal Deviates*. New York: The Free Press, 1983.

TABLE 7.1
Random Numbers

63271	59986	71744	51102	15141	80714	58683	93108	13554	79945
88547	09896	95436	79115	08303	01041	20030	63754	08459	28364
55957	57243	83865	09911	19761	66535	40102	26646	60147	15702
46276	87453	44790	67122	45573	84358	21625	16999	13385	22782
55363	07449	34835	15290	76616	67191	12777	21861	68689	03263
69393	92785	49902	58447	42048	30378	87618	26933	40640	16281
13186	29431	88190	04588	38733	81290	89541	70290	40113	08243
17726	28652	56836	78351	47327	18518	92222	55201	27340	10493
36520	64465	05550	30157	82242	29520	69753	72602	23756	54935
81628	36100	39254	56835	37636	02421	98063	89641	64953	99337
84649	48968	75215	75498	49539	74240	03466	49292	36401	45525
63291	11618	12613	75055	43915	26488	41116	64531	56827	30825
70502	53225	03655	05915	37140	57051	48393	91322	25653	06543
06426	24771	59935	49801	11082	66762	94477	02494	88215	27191
20711	55609	29430	70165	45406	78484	31639	52009	18873	96927
41990	70538	77191	25860	55204	73417	83920	69468	74972	38712
72452	36618	76298	26678	89334	33938	95567	29380	75906	91807
37042	40318	57099	10528	09925	89773	41335	96244	29002	46453
53766	52875	15987	46962	67342	77592	57651	95508	80033	69828
90585	58955	53122	16025	84299	53310	67380	84249	25348	04332
32001	96293	37203	64516	51530	37069	40261	61374	05815	06714
62606	64324	46354	72157	67248	20135	49804	09226	64419	29457
10078	28073	85389	50324	14500	15562	64165	06125	71353	77669
91561	46145	24177	15294	10061	98124	75732	00815	83452	97355
13091	98112	53959	79607	52244	63303	10413	63839	74762	50289

page of random numbers is shown in Table 7.1. The first line of this table begins as follows:

63271   59986   71744   51102   15141   80714

Each digit shown 6, 3, 2, . . . , is a random selection of the digits 0, 1, . . . , 9 with each digit having an equal chance of occurring. The grouping of the numbers into sets of five is simply for the convenience of making the table easy to read.

Let us see how the numbers in this random number table can be used to select a simple random sample of 30 EAI managers. As we did using the pieces of paper, we want to select numbers from 1 to 2500 such that every number has an equal chance of being selected. Since the largest number in the EAI population, 2500, has 4 digits, we will select random numbers from the table in sets or groups of 4 digits. While we could select 4-digit numbers from any portion of the random number table, suppose we start by using the first row of random numbers appearing in Table 7.1. The 4-digit grouping of the first 28 random numbers in the first row provides

6327   1599   8671   7445   1102   1514   1807

Since the numbers in the table are random, the preceding 4-digit numbers are all equally probable, or equally likely.

We can now use the equally likely 4-digit random numbers to give each element in the population an equal chance of being included in the sample. The first number, 6327, is greater than 2500. It does not correspond to an element in the population, and thus it is discarded. The second number, 1599, is between 1 and 2500. Thus the first individual selected for the sample is manager 1599 on the list of EAI managers. Continuing the process, we ignore 8671 and 7445 before identifying individuals 1102, 1514, and 1807 as the next managers to be included in the sample. This process of selecting managers continues until the desired simple random sample of size 30 has been obtained. We note that with this random number procedure for simple random sampling, a random number previously used to identify an element for the sample may reappear in the random number table. Since we want to select the simple random sample *without replacement,* previously used random numbers are ignored because the corresponding element is already included in the sample.

As a final comment, note that random numbers can be selected from anywhere in the random number table. We used the first row of the table in the above example. However, we could have started at any other point in the table and continued in any direction. Once the arbitrary starting point is selected, it is recommended that a predetermined systematic procedure, such as reading across rows or down columns, be used to determine the subsequent random numbers.

Sampling from Infinite Populations

To this point we have restricted our attention to selecting a simple random sample from a finite population. Although most sampling situations in business and economics involve finite populations, there are situations in which the population is either infinite or so large that for practical purposes it must be treated as infinite. In sampling from an infinite population, we must provide a new definition of a simple random sample. In addition, since the elements in an infinite population cannot be numbered, we must use a different process for selecting elements for the sample.

Let us consider an example requiring a simple random sample from an infinite population. Suppose we want to estimate the average time between placing an order and receiving food for customers at a fast-food restaurant during the 11:30 A.M. to 1:30 P.M. lunch period. If we consider the population as being all possible customer visits, we see that it would not be feasible to specify a finite limit on the number of possible visits. In fact, if we define the population as being all customer visits that could *conceivably* occur during the lunch period, we can consider the population as being infinite. Our task is now to select a simple random sample of *n* customers from this population. With this situation in mind, we now state the definition of a simple random sample from an infinite population.

Simple Random Sample (Infinite Population)

A simple random sample from an infinite population is a sample selected such that the following conditions are satisfied:

1. Each element selected comes from the same population.
2. Each element is selected independently.

For the problem of selecting a simple random sample of customer visits at a fast-food restaurant, we find that the first condition defined above is satisfied by any customer visit occurring during the 11:30 A.M. to 1:30 P.M. lunch period while the restaurant is operating with its regular staff under ''normal'' operating conditions. The second condition is satisfied by ensuring that the selection of a particular customer does not influence the selection of any other customer. That is, the customers are selected independently.

A well-known fast-food restaurant has implemented a simple random sampling procedure for just such a situation. The sampling procedure is based on the fact that some customers will present discount coupons which provide special prices on sandwiches, drinks, french fries, and so on. Whenever a customer presents a discount coupon, the *next* customer served is selected for the sample. Since the customers present discount coupons in a random and independent fashion, the firm is satisfied that the sampling plan satisfies the two conditions for a simple random sample from an infinite population.

NOTES & COMMENTS

1. Finite populations are often defined by lists such as organization membership rosters, enrolled students, credit-card accounts, inventory product numbers, and so on. Infinite populations are often defined by an ongoing process where the elements of the population consist of items generated if the process were to operate indefinitely under the same conditions; in such cases, it is impossible to obtain a list of all items in the population. For example, populations consisting of all possible parts to be manufactured, all possible customer visits, all possible bank transactions, and so on can be classified as infinite populations.

2. The number of different simple random samples of size n that can be selected from a finite population of size N is:

$$\frac{N!}{n!(N - n)!}$$

In this formula, $N!$ and $n!$ refer to the factorial computations discussed in Chapters 4 and 5. For the EAI problem with $N = 2500$ and $n = 30$, this expression can be used to show that there are approximately 2.75×10^{69} different simple random samples of 30 EAI managers.

3. Simple random sampling from a finite population is generally done without replacement. However, if a simple random sample is selected from a finite population *with replacement,* the elements are selected from the same population and the elements are selected independently. Since these conditions satisfy the requirements of a simple random sample from an infinite population, simple random sampling with replacement from a finite population is equivalent to simple random sampling from an infinite population.

Exercises

Methods

SELF TEST ▶

1. Consider a small finite population with five items labeled A, B, C, D, and E. There are ten possible simple random samples of size 2 that can be selected.
   a. List the 10 samples beginning with AB, AC, and so on.
   b. Using simple random sampling, what is the probability that each sample of size 2 is selected?
   c. Assume random number 1 corresponds to A, random number 2 corresponds to B, and so on. List the elements in a simple random sample of two items that will be selected using the random digits 8 0 5 7 5 3 2?

2. Assume a finite population has 350 items. Using the last 3 digits of the 5-digit random numbers shown below, read across the row, and determine the first four units that will be selected for the simple random sample.

98601   73022   83448   02147   34229   27553   84147   93289   14209

Applications

SELF TEST ▶

3. *Business Week* publishes data on sales, profits, assets, dividends, shares, and earnings per share for America's 1000 most valuable companies (*The 1992 Business Week 1000*). The companies are ranked and then listed in numerical order based on stock-market value. Assume that you wanted to select a simple random sample of 12 companies from the list of 1000 companies. Use column 9 of Table 7.1 beginning with 554. Read down the column and identify the numbers of the 12 companies that would be selected from the *Business Week 1000*.

4. Based on sales (*The Wall Street Journal*, October 13, 1988), the top 10 athletic footwear manufacturers are

1. Reebok	6. L. A. Gear
2. Nike	7. Etonic/Tretorn
3. Converse	8. New Balance
4. Avia	9. ASICS Tiger
5. Adidas	10. British Knights

a. Beginning with the first random digit in Table 7.1 (6) and reading down (8, 5, 4, and so on), use single-digit random numbers from the first column to select a simple random sample of five footwear manufacturers from the population of the top 10 footwear manufacturers.
b. If the random number 1 corresponds to Reebok, 2 corresponds to Nike, and so on, what single-digit random digit would have to appear in the first column in order to select British Knights for the sample?
c. How many different simple random samples of size 5 can be selected from this population of 10 manufacturers?

5. A student government organization is interested in estimating the proportion of students who favor a mandatory "pass-fail" grading policy for elective courses. A list of names and addresses of the 645 students enrolled during the current quarter is available from the registrar's office. Using row 10 of Table 7.1 and moving across the row from left to right, identify the first 10 students who would be selected by simple random sampling. When every digit in row 10 is used the three-digit random numbers begin with 816, 283, and 610.

6. The *County and City Data Book*, published by the Bureau of Census, lists information on 3139 counties throughout the United States. Assume that a national study will collect data from 30 randomly selected counties. Use 4-digit random numbers from the last column of Table 7.1 to identify the numbers corresponding to the first 5 counties selected for the sample. Ignore the first digit in the last column and begin with the 4-digit random numbers 9945, 8364, 5702, and so on.

7. Assume that we wish to identify a simple random sample of 12 of the 372 doctors located in a particular city. The doctors' names are available from a local medical organization. Use the eighth column of 5-digit random numbers in Table 7.1 to identify the 12 doctors for the sample. Ignore the first 2 random digits in each 5-digit grouping of the random numbers. This process begins with random number 108 and proceeds down the column of random numbers.

8. *Business Week*, February 20, 1989, provided detailed information for 640 mutual funds available to investors. Data about the funds included assets, fees, return on investment, price/earnings ratios, and more. Assume that you would like to do a statistical study of the financial characteristics for the population of 640 mutual funds by using a simple random sample of 12 mutual funds. Use the third column of 5-digit random numbers in Table 7.1, beginning with 71744. Ignore the 44 and only use the first 3 digits, 717. Reading down the column of 3-digit random numbers, identify the numbers corresponding to the 12 mutual funds to be included in the sample.

9. Schuster's Interior Design, Inc., specializes in a variety of home decorating services for its clients. During the previous year the firm provided major decorating consultation for 875 homes. Schuster's management was interested in obtaining information about customer satisfaction 6–12 months after the project was complete. To obtain this information, the firm decided to sample 30 of the 875 clients and interview the group to learn about client satisfaction and ways that Schuster might improve its service. Using the last 3 digits in column 10 of Table 7.1, and moving down the column, the random number sequence would be 945, 364, 702, and so on. Use this procedure to identify the first 10 clients that would be included in the sample. Assume that the 875 clients are numbered sequentially in the order in which the decorating projects were conducted.

10. Haskell Public Opinion Poll, Inc., conducts telephone surveys concerning a variety of political and general public interest issues. The households included in the survey are identified by taking a simple random sample from telephone directories in selected metropolitan areas. The telephone directory for a major Midwest area contains 853 pages with 400 lines per page.
    a. Describe a two-stage random selection procedure that could be used to identify a simple random sample of 200 households. The selection process should involve first selecting a page at random (Stage 1) and then selecting a line on the sampled page (Stage 2). Use the random numbers in Table 7.1 to illustrate this process. Select your own arbitrary starting point in the table.
    b. What would you do if the line selected in part (a) was clearly inappropriate for the study (that is, the line provided the phone number of a business, restaurant, etc.)?

11. Read the Kings Island Statistics in Practice at the beginning of Chapter 1.
    a. Assume that the Kings Island research group treats the population of consumer visits as an infinite population. Is this acceptable? Explain.
    b. Assume that immediately after completing an interview with a consumer, the interviewer returns to the entrance gate and begins counting individuals as they enter the park. The 25th individual counted is selected as the next person to be sampled for the survey. After completing this interview, the interviewer returns to the entrance and again selects the 25th individual entering the park. Does this sampling process appear to provide a simple random sample? Explain.

12. Indicate whether the populations listed below should be considered finite or infinite:
    a. All the registered voters in the state of California.
    b. All the television sets that could be produced by the Allentown, Pennsylvania, plant of the TV-M Company.
    c. All orders that could be processed by a mail-order firm.
    d. All emergency telephone calls that could come into a local police station.
    e. All items that were produced on the second shift on May 17.

7.3   Point Estimation

Now that we have described how to select a simple random sample, let us return to the EAI problem. We will assume that a simple random sample of 30 managers has been selected and that the corresponding data on annual salary and management training program participation are as shown in Table 7.2. The notation x_1, x_2, and so on, is used to denote the annual salary of the first manager in the sample, the second manager in the sample, and so on. Participation in the management training program is indicated by a Yes in the management training program column.

In order to estimate the value of a population parameter, we compute a corresponding characteristic of the sample, referred to as a *sample statistic*. For example, in order to estimate the population mean μ and the population standard deviation σ for the annual salary of EAI managers, we simply use the data in column 1 of Table 7.2 to calculate the corresponding sample statistics: the sample mean $\bar{x}$ and the sample standard deviation s.

TABLE 7.2
Annual Salary and Training Program Status for a Simple Random Sample of 30 Managers

Annual Salary ($)	Management Training Program?	Annual Salary ($)	Management Training Program?
x_1 = 49,094.30	Yes	x_{16} = 51,766.00	Yes
x_2 = 53,263.90	Yes	x_{17} = 52,541.30	No
x_3 = 49,643.50	Yes	x_{18} = 44,980.00	Yes
x_4 = 49,894.90	Yes	x_{19} = 51,932.60	Yes
x_5 = 47,621.60	No	x_{20} = 52,973.00	Yes
x_6 = 55,924.00	Yes	x_{21} = 45,120.90	Yes
x_7 = 49,092.30	Yes	x_{22} = 51,753.00	Yes
x_8 = 51,404.40	Yes	x_{23} = 54,391.80	No
x_9 = 50,957.70	Yes	x_{24} = 50,164.20	No
x_{10} = 55,109.70	Yes	x_{25} = 52,973.60	No
x_{11} = 45,922.60	Yes	x_{26} = 50,241.30	No
x_{12} = 57,268.40	No	x_{27} = 52,793.90	No
x_{13} = 55,688.80	Yes	x_{28} = 50,979.40	Yes
x_{14} = 51,564.70	No	x_{29} = 55,860.90	Yes
x_{15} = 56,188.20	No	x_{30} = 57,309.10	No

Using the formulas for a sample mean and a sample standard deviation as presented in Chapter 3, the sample mean is

$$\bar{x} = \frac{\Sigma x_i}{n} = \frac{1,554,420}{30} = \$51,814.00$$

and the sample standard deviation is

$$s = \sqrt{\frac{\Sigma(x_i - \bar{x})^2}{n-1}} = \sqrt{\frac{325,009,260}{29}} = \$3347.72$$

In addition, by computing the proportion of managers in the sample who have responded Yes, we can estimate the proportion of managers in the population who have completed the management training program. Column 2 of Table 7.2 shows that 19 of the 30 managers in the sample have completed the training program. Thus, the sample proportion, denoted by $\bar{p}$, is given by

$$\bar{p} = \frac{19}{30} = .63$$

This value is used as the estimate of the population proportion p.

By making the preceding computations, we have completed the statistical procedure called *point estimation*. In point estimation we use the data from the sample to compute a value of a sample statistic that serves as an estimate of a population parameter. Using the terminology of point estimation, we would refer to $\bar{x}$ as the *point estimator* of the population mean μ, s as the *point estimator* of the population standard deviation σ, and $\bar{p}$ as the *point estimator* of the population proportion p. The actual numerical value obtained for $\bar{x}$, s, or $\bar{p}$ in a particular sample is called the *point estimate* of the parameter. Thus, based upon the sample of 30 EAI managers, $51,814.00 is the point estimate of μ, $3,347.72 is the point estimate of σ, and .63 is the point estimate of p. Table 7.3 provides a summary of the sample results and compares the point estimates to the actual values of the population parameters.

TABLE 7.3
Summary of Point Estimates Obtained from a Simple Random Sample of 30 EAI Managers

Population Parameter	Parameter Value	Point Estimator	Point Estimate
μ = Population mean annual salary	$51,800.00	$\bar{x}$ = Sample mean annual salary	$51,814.00
σ = Population standard deviation for annual salary	$ 4,000.00	s = Sample standard deviation for annual salary	$ 3,347.72
p = Population proportion having completed the management training program	.60	$\bar{p}$ = Sample proportion having completed the management training program	.63

NOTES & COMMENTS

In our discussion of point estimators, we use $\bar{x}$ to denote a sample mean and $\bar{p}$ to denote a sample proportion. Our use of $\bar{p}$ is based on the fact that the sample proportion is also a *sample mean*. For instance, suppose that in a sample of n items with data values $x_1, x_2, \ldots, x_n$, we let $x_i = 1$ when a characteristic of interest is present for the ith item and $x_i = 0$ when the characteristic is not present. Then the sample proportion is computed by $\Sigma x_i/n$, which is the formula for a sample mean. We also like the consistency of using the bar over the letter to remind the reader that the sample proportion $\bar{p}$ estimates the population proportion similar to the way the sample mean $\bar{x}$ estimates the population mean. Some texts in statistics use $\hat{p}$ instead of $\bar{p}$ to denote the sample proportion.

Exercises

Methods

SELF TEST ▶ **13.** The following data have been collected from a simple random sample.

$$5 \quad 8 \quad 10 \quad 7 \quad 10 \quad 14$$

a. What is the point estimate of the population mean?
b. What is the point estimate of the population standard deviation?

14. A survey question with 150 responses showed 75 Yes responses, 55 No responses, and 20 No Opinions.
a. What is the point estimate of the proportion in the population who will respond Yes?
b. What is the point estimate of the proportion in the population who will respond No?

Applications

SELF TEST ▶ **15.** A simple random sample of 5 months of sales data provided the following:

Month:	1	2	3	4	5
Units Sold:	94	100	85	94	92

a. What is the point estimate of the population mean number of units sold per month?
b. What is the point estimate of the population standard deviation?

16. Forty-nine television commercials were shown during the 1989 Super Bowl Game. A sample of 60 viewers was used to rate each of the commercials (*USA Today,* January 23, 1989). Each viewer in the sample provided a rating on a scale of 1 to 10, with a higher rating indicating a more preferred commercial. Based on the sample mean ratings, American Express and Diet Pepsi provided the two best-liked commercials shown during the Super Bowl game. Suppose that the following data represent a portion of the ratings obtained for the American Express commercial.

10   9   10   9   7   8   10   10   9   6   10   8   7   10   9   10

Assume these ratings are from a simple random sample of 16 viewers selected from the population of all viewers of the Super Bowl game.

a. What is the point estimate of the population mean rating of the American Express commercial?

b. What is the point estimate of the population standard deviation of the American Express commercial?

17. The California Highway Patrol maintains records showing the time between an accident report being received and an officer arriving at the accident scene. A simple random sample of 10 records shows the following times in minutes:

12.6   3.4   4.8   5.0   6.8   2.3   3.6   8.1   2.5   10.3

a. What is a point estimate of the population mean time between receipt of accident report and officer arrival?

b. What is a point estimate of the population standard deviation of the time between receipt of accident report and officer arrival?

18. An official for United Airlines reported that out of 104 United Airlines flights arriving at Chicago's O'Hare Airport, 3 flights arrived more than 15 minutes late (*The Wall Street Journal,* November 7, 1988). Assuming that the 104 flights are a random sample of all United Airline flights into O'Hare, what is the point estimate of the proportion of all United Airline flights into O'Hare that are more than 15 minutes late?

19. A survey of 400 women college students was conducted to determine future plans concerning career, marriage, and family. The following results were recorded:

- 310 women answered Yes to the question, Do you plan to begin a full-time career immediately following graduation?
- 225 women answered Yes to the question, Do you plan to marry before the age of 30?
- 175 women answered Yes to the question, Do you plan to have children?

Use the survey results to provide point estimates for each of the following population parameters.

a. The proportion of college women planning to begin full-time careers immediately following graduation.

b. The proportion of college women planning to marry before the age of 30.

c. The proportion of college women planning to have children.

7.4   Introduction to Sampling Distributions

In the previous section a simple random sample of 30 EAI managers was used to develop point estimates of the mean and standard deviation of annual salary for the population of all EAI managers as well as the proportion of the managers in the population that have completed the company's management training program. In this section we consider the point estimates that would be observed if different simple random samples, each of size 30, were selected. The resulting analysis will introduce the important concept of a *sampling distribution*.

Suppose we were to select another simple random sample of 30 EAI managers. Let us assume that this has been done and that an analysis of the data from the second simple random sample provides the following:

Sample Mean $\bar{x}$ = \$52,669.70
Sample Standard Deviation s = \$4,239.07
Sample Proportion $\bar{p}$ = .70

These results show that different values of $\bar{x}$, s, and $\bar{p}$ have been obtained with the second sample. In general, this is to be expected because this second simple random sample will most likely not contain the same data values that were in the first sample. Let us imagine carrying out the same process of selecting a new simple random sample of 30 managers over and over again, each time computing values of $\bar{x}$, s, and $\bar{p}$. In this way, we could begin to identify the variety of values that these point estimators can take on. To illustrate this, we repeated the simple random sampling process for the EAI problem until we obtained 500 samples of 30 managers each and their corresponding $\bar{x}$, s, and $\bar{p}$ values. A portion of the results are shown in Table 7.4. Table 7.5 shows the frequency and relative frequency distributions for the 500 $\bar{x}$ values. Figure 7.1 shows the relative frequency histogram for the $\bar{x}$ results.

Recall that in Chapter 5 we defined a random variable as a numerical description of the outcome of an experiment. If we consider the process of selecting a simple random as an experiment, the sample mean $\bar{x}$ is the numerical description of the outcome of the experiment. Thus, the sample mean $\bar{x}$ is a random variable. As a result $\bar{x}$, just like other random variables, has a mean or expected value, a variance, and a probability distribu-

TABLE 7.4
Values of $\bar{x}$, s, and $\bar{p}$ from 500 Simple Random Samples of 30 EAI Managers

Sample Number	Sample Mean $\bar{x}$	Sample Standard Deviation s	Sample Proportion $\bar{p}$
1	\$51,814.00	\$3,347.72	.63
2	\$52,669.70	\$4,239.07	.70
3	\$51,780.30	\$4,433.43	.67
4	\$51,587.90	\$3,985.32	.53
.	.	.	.
.	.	.	.
.	.	.	.
500	\$51,752.00	\$3,857.82	.50

TABLE 7.5
Frequency Distribution of $\bar{x}$ from 500 Simple Random Samples of 30 EAI Managers

Mean Annual Salary ($)	Frequency	Relative Frequency
49,500.00–49,999.99	2	.004
50,000.00–50,499.99	16	.032
50,500.00–50,999.99	52	.104
51,000.00–51,499.99	101	.202
51,500.00–51,999.99	133	.266
52,000.00–52,499.99	110	.220
52,500.00–52,999.99	54	.108
53,000.00–53,499.99	26	.052
53,500.00–53,999.99	6	.012
Totals	500	1.000

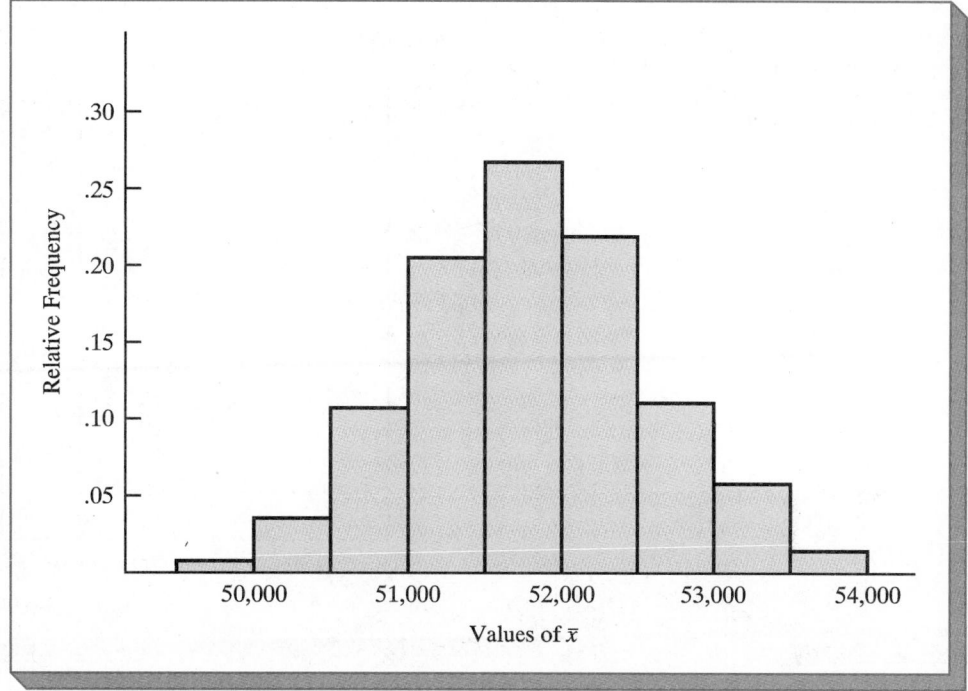

tion. Since the various possible values of $\bar{x}$ are the result of different simple random *samples,* the probability distribution of $\bar{x}$ is called the *sampling distribution* of $\bar{x}$. Knowledge of this sampling distribution and its properties will enable us to make probability statements about how close the sample mean $\bar{x}$ is to the population mean μ.

Let us return to Figure 7.1. We would need to enumerate every possible sample of 30 managers and compute each sample mean to completely determine the sampling distribution of $\bar{x}$. However, the histogram of 500 $\bar{x}$ values gives an approximation of this sampling distribution. From this approximation we observe the bell-shaped appearance of the distribution. We also note that the mean of the 500 $\bar{x}$ values is near the population mean $\mu = \$51,800$. We will describe the properties of the sampling distribution of $\bar{x}$ more fully in the next section.

The 500 values of the sample standard deviation s and the 500 values of the sample proportion $\bar{p}$ are summarized by the relative frequency histograms in Figures 7.2 and 7.3. As in the case of $\bar{x}$, both s and $\bar{p}$ are random variables that provide numerical descriptions of the outcome of a simple random sample. If every possible sample of size 30 were selected from the population and if a value of s and a value of $\bar{p}$ were computed for each sample, the resulting probability distributions would be called the sampling distribution of s and the sampling distribution of $\bar{p}$, respectively. The relative frequency histograms of the 500 sample values shown in Figures 7.2 and 7.3 provide a general idea of the appearance of these two sampling distributions.

Before closing this section, let us note that in practice we select *only one simple random sample* from the population. The reason that we repeated the sampling process 500 times in this section was to illustrate that many different samples are possible and that the different samples generate a variety of values for the sample statistics $\bar{x}$, s, and $\bar{p}$. The

FIGURE 7.2
Relative Frequency Histogram of *s* Values from 500 Simple Random Samples of Size 30 Each

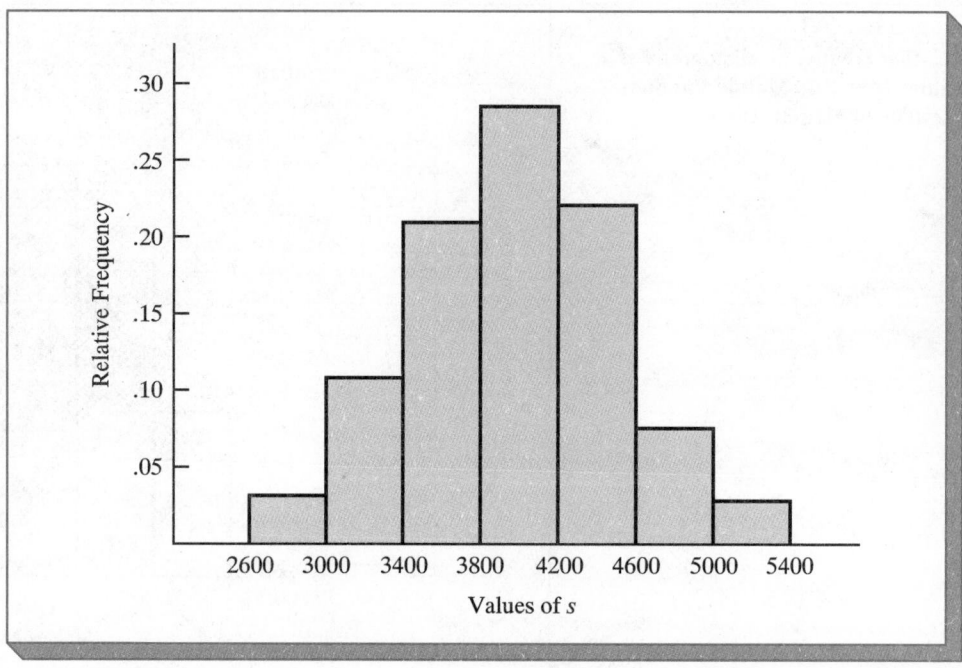

FIGURE 7.3
Relative Frequency Histogram of $\bar{p}$ Values from 500 Simple Random Samples of Size 30 Each

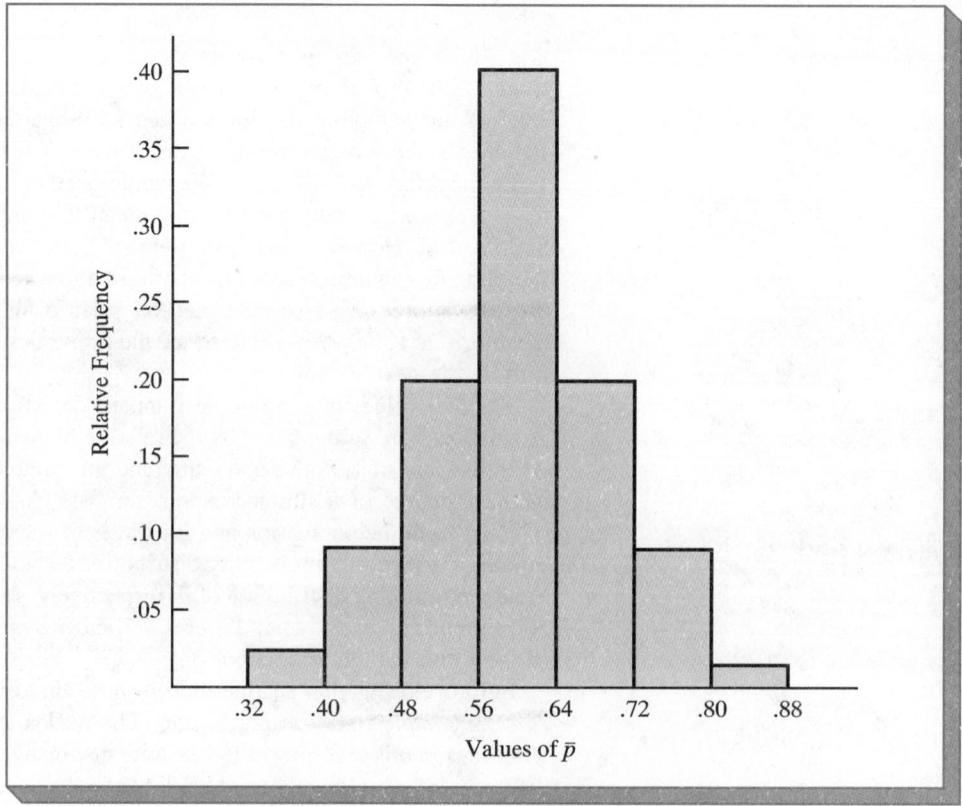

probability distribution of any particular sample statistic is called the sampling distribution of the statistic. In Section 7.5 we show that the characteristics of the sampling distribution of $\bar{x}$ are known even before we sample from the population. In Section 7.6 we show the characteristics of the sampling distribution of $\bar{p}$. We defer further discussion of the sampling distribution of s until we consider sampling distributions pertaining to sample variances, which are covered in Chapter 11.

7.5   Sampling Distribution of $\bar{x}$

In the previous section we stated that the sample mean $\bar{x}$ is a random variable and that the probability distribution of $\bar{x}$ is referred to as the sampling distribution of $\bar{x}$. The purpose of this section is to describe the properties of the sampling distribution of $\bar{x}$, including the expected value or mean of $\bar{x}$, the standard deviation of $\bar{x}$, and the shape or form of the sampling distribution itself. As we shall see, knowledge of the sampling distribution of $\bar{x}$ will enable us to make probability statements about the error involved when $\bar{x}$ is used to estimate μ. Let us begin by considering the mean of all possible $\bar{x}$ values or, simply, the expected value of $\bar{x}$.

Expected Value of $\bar{x}$

As we saw in the EAI sampling problem, different simple random samples result in a variety of values for the sample mean (for example, \$51,814.00, \$52,669.70, \$51,780.30, \$51,587.90, and so on). Since many different values of the random variable $\bar{x}$ are possible, we are often interested in the mean of all possible values of $\bar{x}$ that can be generated by the various simple random samples. First note that the mean of the $\bar{x}$ random variable is simply the expected value of $\bar{x}$. As a result, we are able to use concepts from sampling theory to calculate the expected value of $\bar{x}$ without having to go through the process of computing all possible values of $\bar{x}$ and then computing their mean. Let $E(x)$ represent the expected value of $\bar{x}$, or simply the mean of all possible $\bar{x}$ values, and μ equal the population mean. It can be shown that when using simple random sampling, these two values are the same.

Expected Value of $\bar{x}$

$$E(\bar{x}) = \mu \qquad (7.1)$$

where

$E(\bar{x})$ = the expected value of the random variable $\bar{x}$

μ = the population mean

This result, which is derived in the appendix to this chapter, shows that with simple random sampling, the expected value or mean for $\bar{x}$ is equal to the mean of the population. Refer to the EAI study and recall that in Section 7.1 we saw that the mean annual salary for the population of EAI managers was μ = \$51,800. Thus, according to (7.1) the mean of all possible sample means for the EAI study also is \$51,800.

Standard Deviation of $\bar{x}$

As we have stated, various simple random samples can be expected to generate a variety of $\bar{x}$ values. Let us now explore what sampling theory tells us about the standard deviation of all possible $\bar{x}$ values. We will use the following notation:

$\sigma_{\bar{x}}$ = the standard deviation of all possible $\bar{x}$ values
σ = the population standard deviation
n = the sample size
N = the population size

It can be shown that with simple random sampling, the standard deviation of $\bar{x}$ depends upon whether the population is finite or infinite. The two expressions for the standard deviation of $\bar{x}$ are as follows.

Standard Deviation of $\bar{x}$

Finite Population	*Infinite Population*	
$\sigma_{\bar{x}} = \sqrt{\dfrac{N-n}{N-1}}\left(\dfrac{\sigma}{\sqrt{n}}\right)$	$\sigma_{\bar{x}} = \dfrac{\sigma}{\sqrt{n}}$	(7.2)

A derivation of the formulas for $\sigma_{\bar{x}}$ is discussed in the appendix to this chapter. In comparing the two expressions in (7.2), we see that the factor $\sqrt{(N-n)/(N-1)}$ is required for the finite population but not for the infinite population case. This factor is commonly referred to as the *finite population correction factor*. In many practical sampling situations, we find that the population involved, although finite, is "large," whereas the sample size is relatively "small." In such cases the finite population correction factor $\sqrt{(N-n)/(N-1)}$ is close to 1. As a result the difference between the values of the standard deviation of $\bar{x}$ for the finite and infinite population cases becomes negligible. When this occurs, $\sigma_{\bar{x}} = \sigma/\sqrt{n}$ becomes a very good approximation to the standard deviation of $\bar{x}$ even though the population is finite. As a general guideline or rule of thumb for computing the standard deviation of $\bar{x}$, we state the following.

Use the following expression to calculate the standard deviation of $\bar{x}$

$$\sigma_{\bar{x}} = \frac{\sigma}{\sqrt{n}} \qquad\qquad (7.3)$$

whenever

1. The population is infinite; or
2. The population is finite, *and* the sample size is less than or equal to 5% of the population size—that is, $n/N \le .05$.

In cases where $n/N > .05$, the finite population version of (7.2) should be used in the computation of $\sigma_{\bar{x}}$. Note that unless specifically noted, throughout the text we will be assuming that the population size is "large," the finite population correction factor is unnecessary, and (7.3) can be used to compute $\sigma_{\bar{x}}$.

Now let us return to the EAI study and determine the standard deviation of all possible sample means that can be generated with samples of 30 EAI managers. Recall that in Section 7.1, we identified the population standard deviation for the annual salary data to be $\sigma = 4000$. In this case the population is finite, with $N = 2500$. However, with a sample size of 30, we have $n/N = 30/2500 = .012$. Following the rule of thumb given in (7.3), we can ignore the finite population correction factor and use (7.3) to compute the standard deviation of $\bar{x}$:

$$\sigma_{\bar{x}} = \frac{\sigma}{\sqrt{n}} = \frac{4000}{\sqrt{30}} = 730.30$$

Later we will see that the value of $\sigma_{\bar{x}}$ is helpful in determining how far the sample mean may be from the population mean. Because of the role that $\sigma_{\bar{x}}$ plays in computing possible estimation errors, $\sigma_{\bar{x}}$ is referred to as the *standard error of the mean*.

Central Limit Theorem

The final step in identifying the characteristics of the sampling distribution of $\bar{x}$ is to determine the form of the probability distribution of $\bar{x}$. We consider two cases: one where the population distribution is unknown and one where the population distribution is known to be normal.

For the situation where the population distribution is unknown, we rely on one of the most important theorems in statistics — the *central limit theorem*. A statement of the central limit theorem as it applies to the sampling distribution of $\bar{x}$ is as follows.

Central Limit Theorem

In selecting simple random samples of size n from a population, the sampling distribution of the sample mean $\bar{x}$ can be approximated by a *normal probability distribution* as the sample size becomes large.

Figure 7.4 shows how the central limit theorem works for three different populations; in each case the population clearly is not normal. However, note what begins to happen to the sampling distribution of $\bar{x}$ as the sample size is increased. When the samples are of size 2, we see that the sampling distribution of $\bar{x}$ begins to take on an appearance different than the population distribution. For samples of size 5, we see all three sampling distributions beginning to take on a bell-shaped appearance. Finally, the samples of size 30 show all three sampling distributions to be approximately normal. Thus, for sufficiently large samples, the sampling distribution of $\bar{x}$ can be approximated by a normal probability distribution. However, how large must the sample size be to assume that the central limit theorem applies? Statistical researchers have investigated this question by studying the sampling distribution of $\bar{x}$ for a variety of populations and a variety of sample sizes. Whenever the population distribution is mound-shaped and symmetrical, sample sizes as small as 5–10 can be enough for the central limit theorem to apply. However, if the population distribution is highly skewed and clearly nonnormal, larger sample sizes are needed. General statistical practice is to assume that for most applications, the sampling distribution of $\bar{x}$ can be approximated by a normal probability distribution whenever the *sample size is 30 or more*. In effect, a sample size of 30 or more is assumed to satisfy the

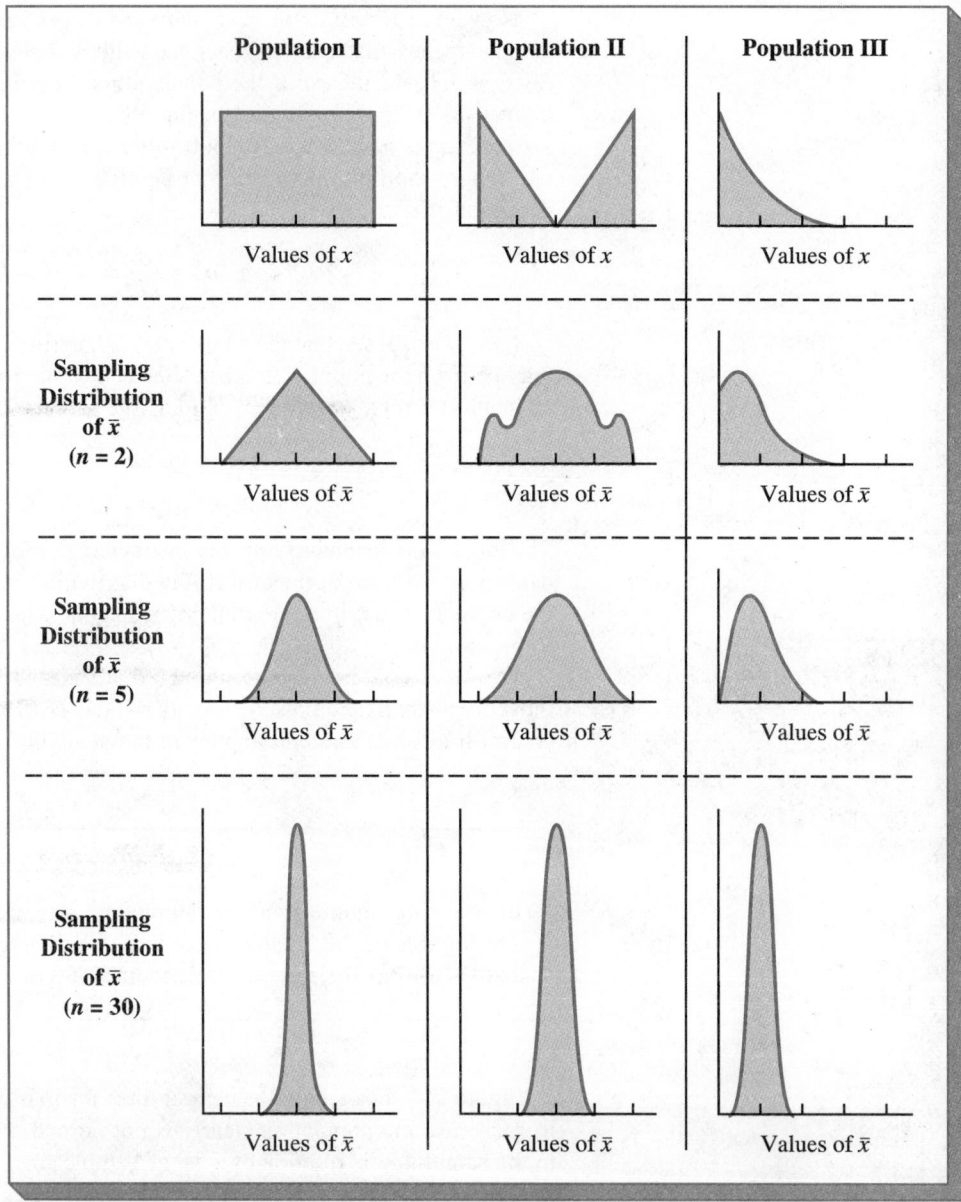

large sample condition of the central limit theorem. This observation is so important that we restate it as follows.

> The sampling distribution of $\bar{x}$ can be approximated by a normal probability distribution whenever the sample size is large. The large-sample-size condition can be assumed for simple random samples of size 30 or more.

The central limit theorem is the key to identifying the form of the sampling distribution of $\bar{x}$ whenever the population distribution is unknown. However, we may encounter some sampling situations where the population is assumed or believed to have a normal distribution. When this condition occurs, the following result identifies the form of the sampling distribution of $\bar{x}$.

> Whenever the population has a normal probability distribution, the sampling distribution of $\bar{x}$ is a normal probability distribution for any sample size.

In summary, if we use a large (n $\geq$ 30) simple random sample, the central limit theorem enables us to conclude that the sampling distribution of $\bar{x}$ can be approximated by a normal probability distribution. In cases where the simple random sample is small ($n < 30$), the sampling distribution of $\bar{x}$ can be considered normal only if we assume that the population has a normal probability distribution.

Sampling Distribution of $\bar{x}$ for the EAI Problem

Let us draw upon our knowledge of the sampling distribution of $\bar{x}$ to determine the properties of the sampling distribution of $\bar{x}$ for the EAI study. Previously, we have shown that $E(\bar{x}) = 51,800$ and $\sigma_{\bar{x}} = 730.30$. Since we are using a simple random sample of 30 managers, the central limit theorem enables us to conclude that the sampling distribution of $\bar{x}$ is approximately normal. Thus, the sampling distribution of $\bar{x}$ for the EAI problem is as shown in Figure 7.5. Although we do not have actual data available for all possible $\bar{x}$ values, we do have the 500 values of $\bar{x}$ that were obtained from the 500 simple random samples referred to in Section 7.4 (see Figure 7.1). In Figure 7.6 we compare the theoretical sampling distribution of $\bar{x}$ with the relative frequency histogram of the 500 $\bar{x}$ values that we have actually observed. Note how close the distribution of the 500 $\bar{x}$ values is to the theoretical normal probability distribution, as indicated by the central limit theorem.

FIGURE 7.5

Sampling Distribution of $\bar{x}$ for the Mean Annual Salary of a Simple Random Sample of 30 Managers

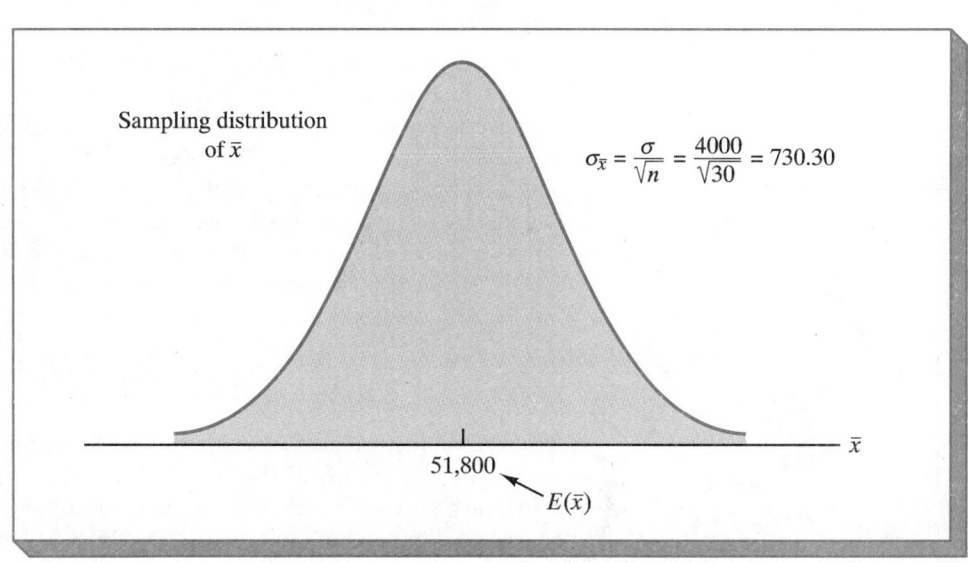

FIGURE 7.6
**Comparison of the Theoretical
Sampling Distribution of $\bar{x}$ and the
Relative Frequency Histogram of $\bar{x}$
from 500 Simple Random Samples**

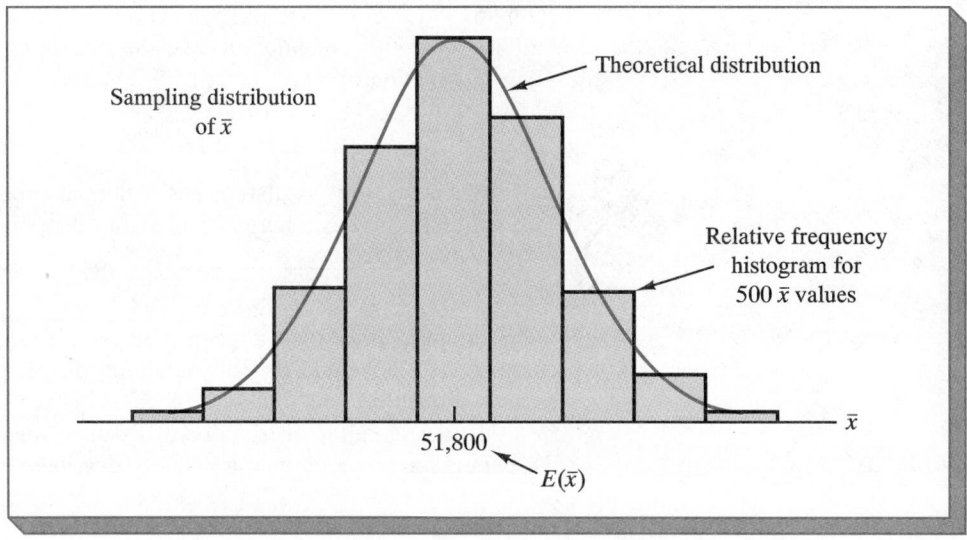

Practical Value of the Sampling Distribution of $\bar{x}$

Whenever a simple random sample is selected and the value of the sample mean $\bar{x}$ is used to estimate the value of the population mean μ, we cannot expect the sample mean to *equal exactly* the population mean. The absolute value of the difference between the value of the sample mean $\bar{x}$ and the value of the population mean μ, $|\bar{x} - \mu|$, is called the *sampling error.* Before putting faith or confidence in the sample mean, we would like to know something about the potential sampling error when $\bar{x}$ is used as an estimator of μ.

The practical reason that we are interested in the sampling distribution of $\bar{x}$ is that this distribution can be used to provide probability information about the sampling error. To demonstrate this, let us return to the EAI problem. Suppose that the personnel director believes the sample mean will be an acceptable estimate of the population mean if the sample mean is within $500 of the population mean. In probability terms, the personnel director is really concerned with the following question: What is the probability that the sample mean we obtain from a simple random sample of 30 EAI managers will be within $500 of the population mean?

Since we have identified the sampling distribution for $\bar{x}$ (see Figure 7.5), we will use this distribution to answer the probability question. Refer to the sampling distribution of $\bar{x}$ shown again in Figure 7.7. The personnel director is asking about the probability that the sample mean is between $51,300 and $52,300. If the value of the sample mean $\bar{x}$ is in this interval, the value of $\bar{x}$ will be within $500 of the population mean. The probability is given by the shaded area of the sampling distribution shown in Figure 7.7. Since the sampling distribution is normal, with mean 51,800 and standard deviation 730.30, we can use the standard normal probability distribution table to find the area or probability. At $\bar{x} = 51,300$, we have

$$z = \frac{51,300 - 51,800}{730.30} = -.68$$

Referring to the standard normal probability distribution table, we find an area between $z = 0$ and $z = -.68$ of .2518. Similar calculations for $\bar{x} = 52,300$ show an area between $z = 0$ and $z = +.68$ of .2518. Thus, the probability of the value of the sample mean being between 51,300 and 52,300 is .2518 + .2518 = .5036.

FIGURE 7.7
Shaded Area Is the Probability of a Sample Mean Being within $500 of the Population Mean

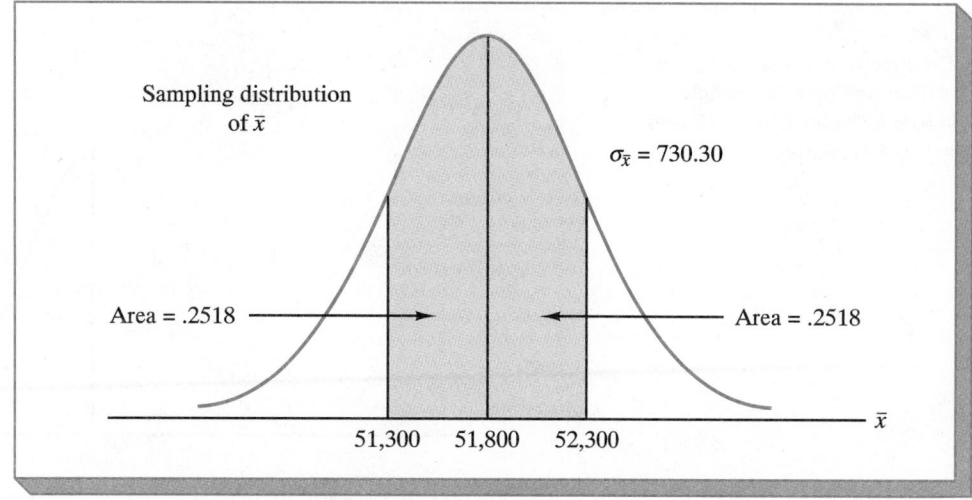

The preceding computations show that a simple random sample of 30 EAI managers has a .5036 probability of providing a sample mean $\bar{x}$ that is within $500 of the population mean. Thus, there is a $1 - .5036 = .4964$ probability that the sample mean will miss the population mean by more than $500. In other words, a simple random sample of 30 EAI managers has roughly a 50–50 chance of providing a sample mean within the allowable $500 margin of error. Perhaps a larger sample size should be considered. Let us explore this possibility by considering the relationship between the sample size and the sampling distribution of $\bar{x}$.

The Relationship between the Sample Size and the Sampling Distribution of $\bar{x}$

Suppose that in the EAI sampling problem, we selected a simple random sample of 100 EAI managers instead of the 30 originally considered. Intuitively, it would seem that with more data provided by the larger sample size, the sample mean based on $n = 100$ should tend to provide a better estimate of the population mean than the sample mean based on $n = 30$. To see that this is true, let us consider the relationship between the sample size and the sampling distribution of $\bar{x}$.

First note that $E(\bar{x}) = \mu$ regardless of the sample size. Thus, the mean of all possible values of $\bar{x}$ is equal to the population mean μ regardless of the sample size n. However, note that the standard error of the mean, $\sigma_{\bar{x}} = \sigma/\sqrt{n}$, is inversely proportional to the sample size. That is, whenever the sample size is increased, the standard error of the mean $\sigma_{\bar{x}}$ is decreased. With $n = 30$, the standard error of the mean for the EAI problem was 730.30. However, with the increase in the sample size to $n = 100$, the standard error of the mean is decreased to

$$\sigma_{\bar{x}} = \frac{\sigma}{\sqrt{n}} = \frac{4000}{\sqrt{100}} = 400$$

The sampling distributions of $\bar{x}$ with $n = 30$ and $n = 100$ are shown in Figure 7.8. Since the sampling distribution with $n = 100$ has a smaller standard error, the various values of $\bar{x}$ have less variation and tend to fall closer to the population mean than do the values of $\bar{x}$ with $n = 30$.

FIGURE 7.8

A Comparison of the Sampling Distributions of $\bar{x}$ for Simple Random Samples of $n = 30$ and $n = 100$ EAI Managers

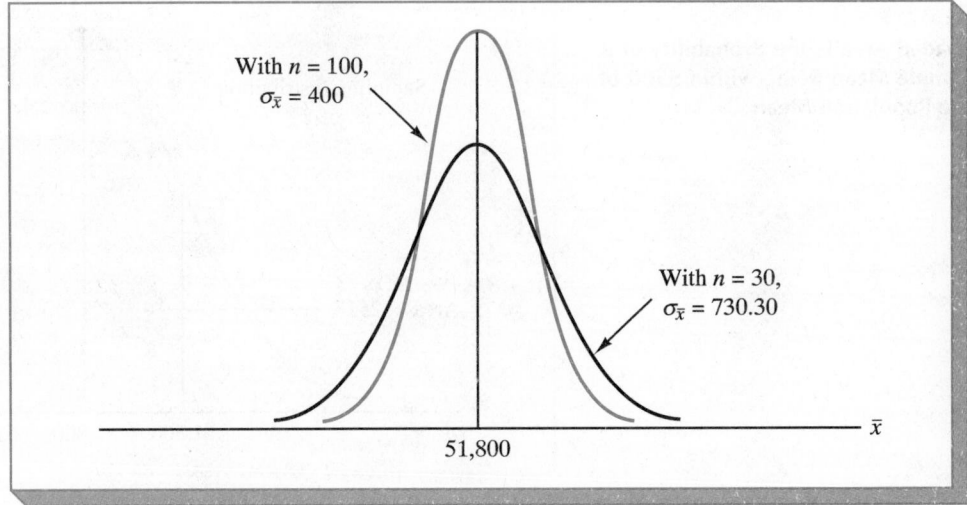

We can use the sampling distribution of $\bar{x}$ for the case with $n = 100$ to compute the probability that a simple random sample of 100 EAI managers will provide a sample mean that is within \$500 of the population mean. Since the sampling distribution is normal, with mean 51,800 and standard deviation 400, we can use the standard normal probability distribution table to find the area or probability. At $\bar{x} = 51,300$ (Figure 7.9), we have

$$z = \frac{51,300 - 51,800}{400} = -1.25$$

Referring to the standard normal probability distribution table, we find an area between $z = 0$ and $z = -1.25$ of .3944. With a similar calculation for $\bar{x} = 52,300$, we see that the probability of the value of the sample mean being between 51,300 and 52,300 is .3944 + .3944 = .7888. Thus, by increasing the sample size from 30 to 100 EAI managers, the probability of obtaining a sample mean within \$500 of the population mean has increased from .5036 to .7888.

FIGURE 7.9

Shaded Area Is the Probability of a Sample Mean Being within \$500 of the Population Mean when a Simple Random Sample of 100 EAI Managers Is Used

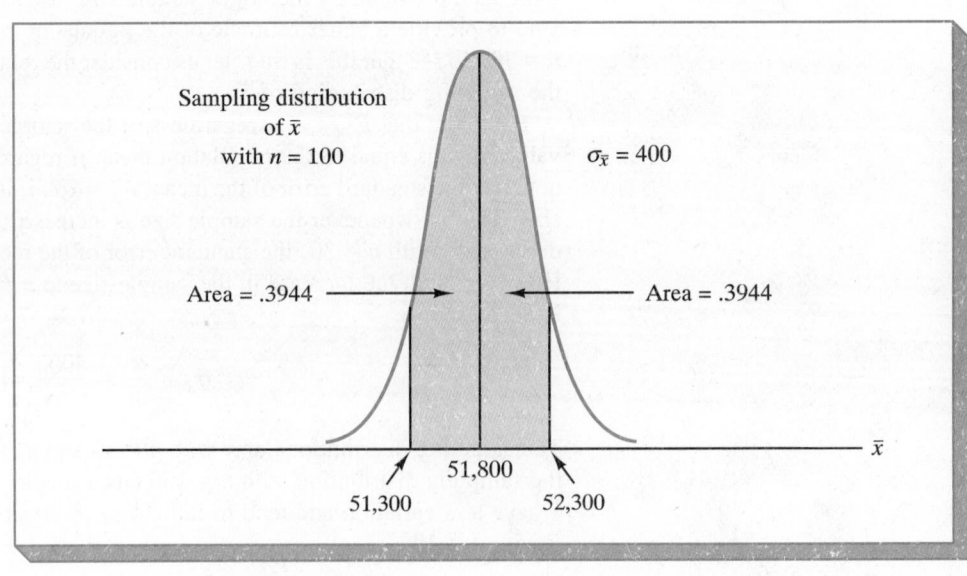

Perhaps an even larger sample size should be considered. However, the important point in this discussion is that as the sample size is increased, the standard error of the mean is decreased. As a result, the sampling distribution of $\bar{x}$ will have less variation. In effect, the larger sample size will provide a higher probability that the value of the sample mean is within a specified distance of the population mean.

NOTES & COMMENTS

1. In presenting the sampling distribution of $\bar{x}$ for the EAI problem we took advantage of the fact that the population mean $\mu = 51{,}800$ and the population standard deviation $\sigma = 4000$ were provided in the discussion of Section 7.1. However, in general, the values of the population mean μ and the population standard deviation σ that are needed to determine the sampling distribution of $\bar{x}$ will be unknown. In Chapter 8 we will show how the sample mean $\bar{x}$ and the sample standard deviation s from a simple random sample are used when μ and σ are unknown.

2. The theoretical proof of the central limit theorem requires independent observations or items in the sample. This condition exists for infinite populations and for finite populations where sampling is done with replacement. Although the central limit theorem does not directly address sampling without replacement from finite populations, general statistical practice has been to apply the findings of the central limit theorem in this situation provided that the population size is large.

❑ ❑ **Exercises**

Methods

20. A population has a mean of 200 and a standard deviation of 50. A simple random sample of size 100 will be taken with the sample mean $\bar{x}$ used to estimate the population mean.
a. What is the expected value of $\bar{x}$?
b. What is the standard deviation of $\bar{x}$?
c. Show the sampling distribution of $\bar{x}$.
d. What does the sampling distribution of $\bar{x}$ show?

21. What important role does the central limit theorem serve whenever $\bar{x}$ is used to estimate μ?

SELF TEST   ▶ **22.** A population has a mean of 200 and a standard deviation of 50. Suppose a simple random sample of size 100 is selected and $\bar{x}$ is used to estimate μ.
a. What is the probability that the sample mean will be within ± 5 of the population mean?
b. What is the probability that the sample mean will be within ± 10 of the population mean?

23. Assume that $\mu = 32$ and standard deviation $\sigma = 5$. Furthermore, assume that the population has 1000 items and that a simple random sample of 30 items is used to obtain information about this population.
a. What is the expected value of $\bar{x}$?
b. What is the standard deviation of $\bar{x}$?

24. Assume the population standard deviation is $\sigma = 25$. Compute the standard error of the mean, $\sigma_{\bar{x}}$, for sample sizes of 50, 100, 150, and 200. What can you say about the size of the standard error of the mean as the sample size is increased?

25. Suppose a simple random sample of size 50 is selected from a population with $\sigma = 10$. Find the value of the standard error of the mean in each of the following cases (use the finite population correction factor if appropriate).

a. The population size is infinite.
b. The population size is $N = 50,000$.
c. The population size is $N = 5000$.
d. The population size is $N = 500$.

26. A population has a mean of 400 and a standard deviation of 50. The probability distribution of the population is unknown.
a. A research study will use simple random samples of either 10, 20, 30, or 40 items to collect data about the population. In which of these sample-size alternatives will we be able to use a normal probability distribution to describe the sampling distribution of $\bar{x}$? Explain.
b. Show the sampling distribution of $\bar{x}$ for the instances where the normal probability distribution is appropriate.

27. A population has a mean of 100 and a standard deviation of 16. What is the probability that a sample mean will be within ± 2 of the population mean for each of the following sample sizes?
a. $n = 50$ b. $n = 100$ c. $n = 200$ d. $n = 400$
e. What is the advantage of a larger sample size?

Applications

28. Refer to the EAI sampling problem. Suppose that the simple random sample had contained 60 managers.
a. Sketch the sampling distribution of $\bar{x}$ when simple random samples of size 60 are used.
b. What happens to the sampling distribution of $\bar{x}$ if simple random samples of size 120 are used?
c. What general statement can you make about what happens to the sampling distribution of $\bar{x}$ as the sample size is increased? Does this seem logical? Explain.

SELF TEST ▶ ✓ 29. In the EAI sampling problem, we showed that for $n = 30$, there was .5036 probability of obtaining a sample mean within $\pm \$500$ of the population mean.
a. What is the probability that $\bar{x}$ is within \$500 of the population mean if a sample of size 60 is used?
b. Answer part (a) for a sample of size 120.

30. Statistics help computer scientists and data-processing specialists understand the operating characteristics of computer systems. Statistical information includes waiting time, running time, central processor time, number of disk accesses, and so on. For a particular class of jobs run on the Amdahl 5880 mainframe computer (*Technical Update*, University of Cincinnati, Spring 1989), the population mean running time is 12.55 minutes per job. The population standard deviation is 4.0 minutes. Assume that a simple random sample of 40 jobs will be used to monitor the running time for the jobs.
a. Show the sampling distribution of $\bar{x}$ where $\bar{x}$ is the sample mean running time.
b. What is the probability that a simple random sample of 40 jobs will provide a sample mean within 1 minute of the population mean?
c. What is the probability that a simple random sample of 40 jobs will provide a sample mean within 30 seconds of the population mean?

31. A statistics class has 80 students. The mean score on the midterm exam was $\mu = 72$ and the standard deviation was $\sigma = 12$. Assume that a simple random sample of 20 students will be selected and the sample mean exam score $\bar{x}$ will be computed. What is the expected value and standard deviation of $\bar{x}$?

✓ 32. Weights for males between the ages of 20 and 30 have a mean $\mu = 170$ pounds with a standard deviation of $\sigma = 28$ pounds. If a simple random sample of 40 males in this age group is to be selected and the sample mean weight $\bar{x}$ computed, what are the values of $E(\bar{x})$ and $\sigma_{\bar{x}}$?

33. Annual surveys of starting salaries for college graduates are conducted by the College Placement Council. The mean annual starting salary for accounting majors is \$26,542 (*USA Today*, September 9, 1991). Assume the population of graduates with accounting majors has a mean of \$26,542 and a standard deviation of \$2000.
a. What is the probability that a simple random sample of accounting graduates will have a sample mean within $\pm \$250$ of the population mean for each of the following sample sizes: 30, 50, 100, 200, and 400?

b. What is the advantage of taking a larger sample size when attempting to estimate a population mean?

34. *Money* (February 1989) listed the national mean interest rate for new auto loans as 11.28. This was up from the 10.81 mean interest rate in 1988. Assume that the mean interest rate for the population of all new auto loans is 11.28 and that the population standard deviation is 1.5. Suppose that a simple random sample of 50 loans will be used to monitor interest rates of auto loans.

a. What is the probability that a simple random sample of 50 loans will provide a mean interest rate within .2 of the population mean?

b. What is the probability that a simple random sample of 50 loans will provide a mean interest rate within .1 of the population mean?

35. In a study of the growth rate of a certain plant, a botanist is planning to use a simple random sample of 25 plants for data-collection purposes. After analyzing the data on plant growth rate, the botanist believes that the standard error of the mean is too large. What size simple random sample should the botanist use in order to reduce the standard error to one-half its current value?

36. The population mean price for a new automobile is $16,012 (*U.S. News & World Report,* September 9, 1991). Assume that the population standard deviation is $4200 and that a sample of 100 new automobile purchases will be selected.

a. Show the sampling distribution for the sample mean price for new automobiles based on the sample of 100.

b. What is the probability that the sample mean for the 100 purchases will be within $1000 of the population mean?

c. Repeat part (b) for sampling errors of $500, $250, and $100.

d. If it were desired to estimate the population mean price to within ±$250 or ±$100, what would you recommend?

37. An automatic machine used to fill cans of soup has the following characteristics: $\mu = 15.9$ ounces and $\sigma = .5$ ounces.

a. Show the sampling distribution of $\bar{x}$, where $\bar{x}$ is the sample mean for 40 cans selected randomly by a quality control inspector.

b. What is the probability of finding a sample of 40 cans with a mean $\bar{x}$ greater than 16 ounces?

38. The mean wage rate for workers at General Motors is $14.25 per hour (*Detroit Daily News,* April 14, 1989). Assume that the population standard deviation is $2.00. Answer the following questions if a simple random sample of 50 workers is selected from the population of workers at General Motors.

a. Show the sampling distribution of $\bar{x}$, where $\bar{x}$ is the sample mean hourly wage rate.

b. What is the probability that the sample mean is at least $13.80 per hour?

c. What is the probability that the sample mean is within $.25 of the population mean of $14.25 per hour?

d. Answer parts (b) and (c) with the sample size increased to 100 workers.

39. In a population of 4000 employees, a simple random sample of 40 employees is selected in order to estimate the mean age for the population.

a. Would you use the finite population correction factor in calculating the standard error of the mean? Explain.

b. If the population standard deviation is $\sigma = 8.2$ years, compute the standard error both with and without using the finite population correction factor. What is the rationale behind ignoring the finite population correction factor whenever $n/N \leq .05$?

c. What is the probability that the sample mean age of the employees will be within ±2 years of the population mean age?

40. A library checks out an average of $\mu = 320$ books per day, with a standard deviation of $\sigma = 75$ books. Consider a sample of 30 days of operation, with $\bar{x}$ being the sample mean number of books checked out per day.

a. Show the sampling distribution of $\bar{x}$.

b. What is the standard deviation of $\bar{x}$?

c. What is the probability that the sample mean for the 30 days will be between 300 and 340 books?

d. What is the probability that the sample mean will show 325 or more books checked out?

7.6   Sampling Distribution of $\bar{p}$

As we observed in Section 7.4, the sample proportion $\bar{p}$ is a random variable. To determine how close the sample proportion $\bar{p}$ is to the population proportion p, we need to understand the properties of the sampling distribution of $\bar{p}$: the expected value of $\bar{p}$, the standard deviation of $\bar{p}$, and the shape of the sampling distribution of $\bar{p}$.

Expected Value of $\bar{p}$

It can be shown that the expected value of $\bar{p}$ (i.e., the mean of all possible values of $\bar{p}$) is as follows.

Expected Value of $\bar{p}$

$$E(\bar{p}) = p \tag{7.4}$$

where

$E(\bar{p})$ = the expected value of the random variable $\bar{p}$

p = the population proportion

Equation (7.4) shows that the mean of all possible $\bar{p}$ values is equal to the population proportion p. Recall that in Section 7.1 we showed that $p = .60$ for the EAI population, where p was the proportion of the population of managers who had participated in the company's management training program. Thus, the expected value of $\bar{p}$ for the EAI sampling problem is .60.

Standard Deviation of $\bar{p}$

Different simple random samples generate a variety of values for $\bar{p}$. We now are interested in determining the standard deviation of $\bar{p}$, which is referred to as the *standard error of the proportion*. Just as we found for the sample mean $\bar{x}$, the standard deviation of $\bar{p}$ depends upon whether the population is finite or infinite. The two expressions for the standard deviation of $\bar{p}$ are as follows.

Standard Deviation of $\bar{p}$

Finite Population   *Infinite Population*

$$\sigma_{\bar{p}} = \sqrt{\frac{N-n}{N-1}}\sqrt{\frac{p(1-p)}{n}} \qquad \sigma_{\bar{p}} = \sqrt{\frac{p(1-p)}{n}} \tag{7.5}$$

Comparing the two expressions in (7.5), we see that the only difference is the use of the finite population correction factor $\sqrt{(N-n)/(N-1)}$.

As was the case with the sample mean $\bar{x}$, we find that the difference between the above expressions for the finite population and the infinite population becomes negligible if the size of the finite population is large compared to the sample size. We follow the same rule of thumb that we recommended for the sample mean. That is, if the population is finite with $n/N \leq .05$, we will use $\sigma_{\bar{p}} = \sqrt{p(1-p)/n}$. However, if the population is finite and if $n/N > .05$, the finite population correction factor should be used, as shown in (7.5). Again, unless specifically noted, throughout the text we will be assuming that the population size is large relative to the sample size and that the finite population correction factor is unnecessary.

For the EAI study we know that the population proportion of managers that have participated in the management training program is $p = .60$. With $n/N = 30/2500 = .012$, we can ignore the finite population correction factor when we compute the standard deviation of $\bar{p}$. For the simple random sample of 30 managers, $\sigma_{\bar{p}}$ is

$$\sigma_{\bar{p}} = \sqrt{\frac{p(1-p)}{n}} = \sqrt{\frac{.60(1-.60)}{30}} = \sqrt{.008} = .0894$$

Form of the Sampling Distribution of $\bar{p}$

Now that we know the mean and standard deviation of $\bar{p}$, we want to consider the form of the sampling distribution of $\bar{p}$. Applying the central limit theorem as it relates to the $\bar{p}$ produces the following result.

> The sampling distribution of $\bar{p}$ can be approximated by a normal probability distribution whenever the sample size is large.

With $\bar{p}$, the sample size can be considered large whenever the following two conditions are satisfied:

$$np \geq 5$$
$$n(1-p) \geq 5$$

Recall that for the EAI sampling problem, we know that the population proportion of managers that have participated in the training program is $p = .60$. Let us use this value to determine the characteristics of the sampling distribution of $\bar{p}$ for the EAI study. With a simple random sample of size 30, we have $np = 30(.60) = 18$ and $n(1-p) = 30(.40) = 12$. Thus, according to the above rule of thumb, the sampling distribution of $\bar{p}$ can be approximated by a normal probability distribution. Figure 7.10 shows this sampling distribution for the EAI study. Figure 7.11 shows the close agreement between this theoretical sampling distribution of $\bar{p}$ and the histogram of the 500 $\bar{p}$ values we obtained from the 500 repeated samples of size 30 (see Figure 7.3).

Practical Value of the Sampling Distribution of $\bar{p}$

Whenever a simple random sample is selected and the value of the sample proportion $\bar{p}$ is used to estimate the value of the population proportion p, we anticipate some sampling error. In this case, the sampling error is the absolute value of the difference between the value of the sample proportion $\bar{p}$ and the value of the population proportion p. The practical value of the sampling distribution of $\bar{p}$ is that it can be used to provide probability information about the sampling error.

FIGURE 7.10

Sampling Distribution of $\bar{p}$ for the Proportion of EAI Managers Having Participated in the Management Training Program

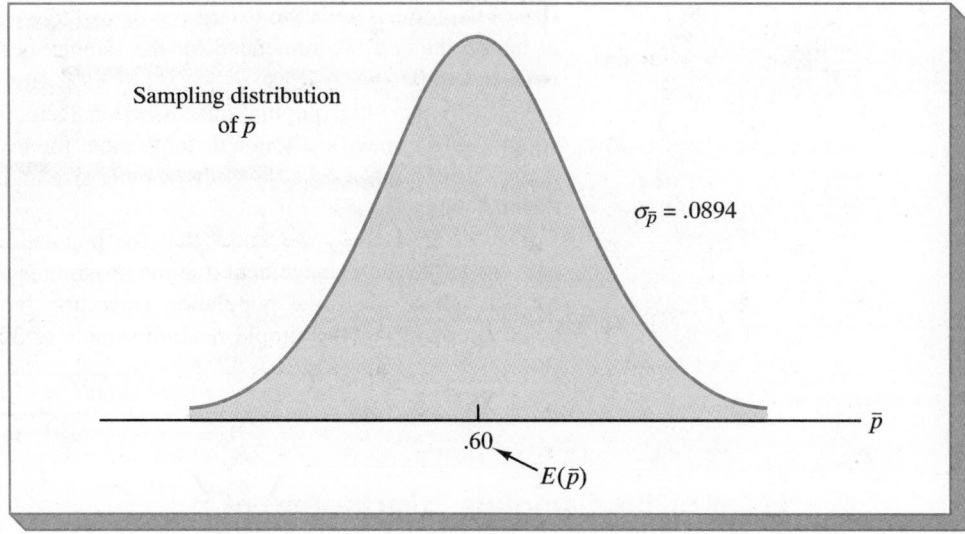

FIGURE 7.11

Comparison of the Theoretical Sampling Distribution of $\bar{p}$ and the Relative Frequency Histogram of $\bar{p}$ from 500 Simple Random Samples

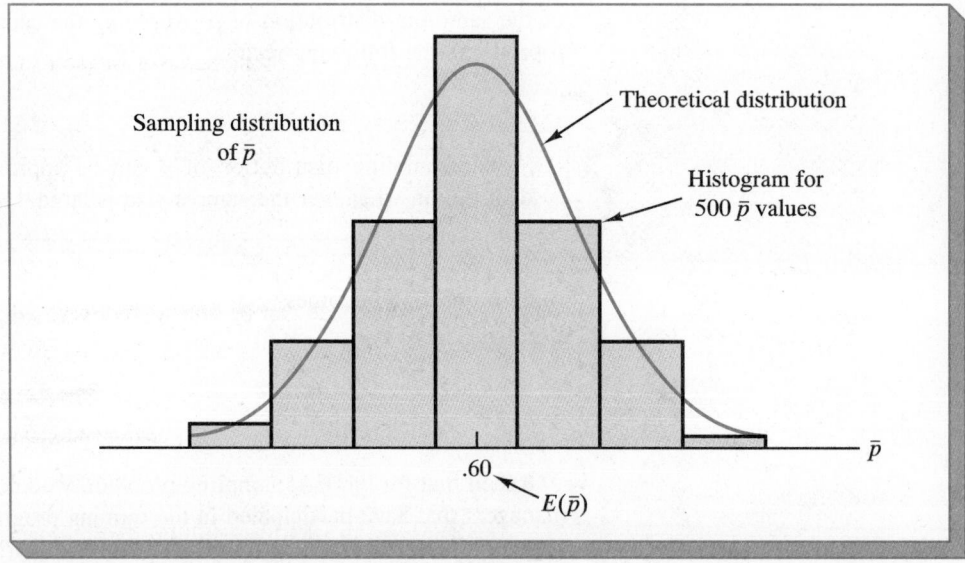

Suppose, in the EAI problem, the personnel director wanted to know the probability of obtaining a value of $\bar{p}$ that is within .05 of the population proportion of EAI managers who have participated in the training program. That is, what is the probability of obtaining a sample with a sample proportion $\bar{p}$ between .55 and .65? The shaded area in Figure 7.12 shows this probability. Using the fact that the sampling distribution of $\bar{p}$ is approximately normal with mean .60 and standard deviation $\sigma_{\bar{p}} = .0894$, the standard normal random variable corresponding to $\bar{p} = .55$ has a value of $z = (.55 - .60)/.0894 = -.56$. Referring to the standard normal probability distribution table, the area between $z = -.56$ and $z = 0$ is .2123. Similarly, at $\bar{p} = .65$ we find an area between $z = 0$ and $z = .56$ of .2123. Thus, the probability of selecting a sample providing a sample proportion $\bar{p}$ within .05 of the population proportion p is .2123 + .2123 = .4246.

As we stated in Section 7.5, there is a relationship between the sample size and the sampling distribution of $\bar{x}$ — namely, as the sample size is increased, the standard error of the mean is decreased. As a result, for larger sample sizes there is a higher probability that

Figure 7.12
Sampling Distribution of $\bar{p}$ for the EAI Sampling Problem

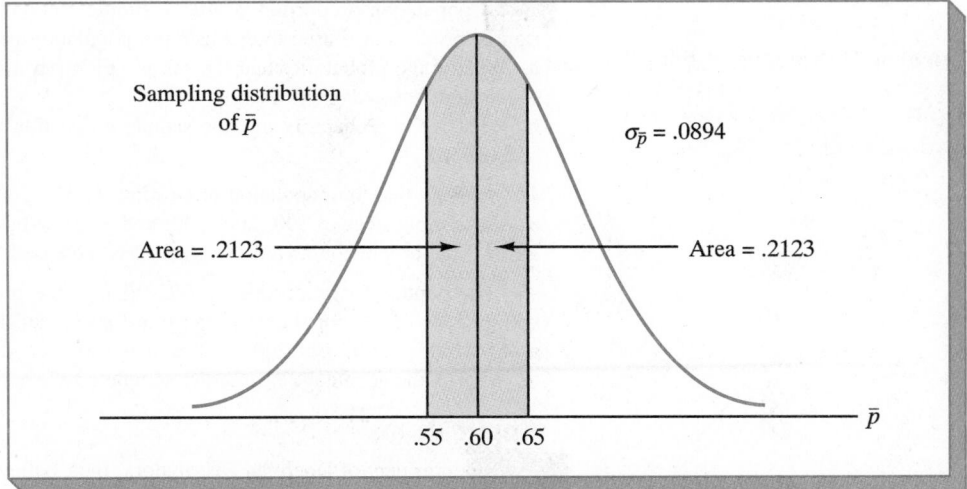

the value of the sample mean will be within a stated distance from the population mean. A similar relationship exists between the sample size and the sampling distribution of $\bar{p}$. The formula for the standard error of the proportion is given by

$$\sigma_{\bar{p}} = \sqrt{\frac{p(1-p)}{n}}$$

Thus, we see that the standard error is inversely proportional to the sample size.

With a sample size of $n = 30$, the standard error of the proportion for the EAI problem is $\sigma_{\bar{p}} = .0894$. If we consider increasing the sample size to $n = 100$, the standard error of the proportion becomes

$$\sigma_{\bar{p}} = \sqrt{\frac{.60(1-.60)}{100}} = \sqrt{.0024} = .0490$$

With a sample size of 100 EAI managers, the probability of the sample proportion having a value within .05 of the population proportion can now be computed. Since the sampling distribution is approximately normal, with mean .60 and standard deviation .0490, we can use the standard normal probability distribution table to find the area or probability. At $\bar{p} = .55$, we have $z = (.55 - .60)/.0490 = -1.02$. Referring to the standard normal probability distribution table, the area between $z = -1.02$ and 0 is .3461. Similarly, at .65 the area between $z = 0$ and $z = 1.02$ is .3461. Thus, if the sample size is increased from 30 to 100, the probability that the sample proportion $\bar{p}$ is within .05 of the population proportion p will be increased from .4246 to .3461 + .3461 = .6922.

□ □ Exercises

Methods

41. A simple random sample of size 100 is selected from a population with $p = .40$.
a. What is the expected value of $\bar{p}$?
b. What is the standard deviation of $\bar{p}$?
c. Show the sampling distribution of $\bar{p}$.
d. What does the sampling distribution of $\bar{p}$ show?

42. A population proportion is .40. A simple random sample of size 200 will be taken with the sample proportion $\bar{p}$ used to estimate the population proportion.
    a. What is the probability that the sample proportion will be within ±.03 of the population proportion?
    b. What is the probability that the sample proportion will be within ±.05 of the population proportion?

43. Assume that the population proportion is .55. Compute the standard error of the proportion, $\sigma_{\bar{p}}$, for sample sizes of 100, 200, 500, and 1000. What can you say about the size of the standard error of the proportion as the sample size is increased?

44. The population proportion is .30. What is the probability that a sample proportion will be within ±.04 of the population proportion for each of the following sample sizes?
    a. $n = 100$ b. $n = 200$ c. $n = 500$ d. $n = 1000$
    e. What is the advantage of a larger sample size?

Applications

45. The president of Doerman Distributors, Inc., believes that 30% of the firm's orders come from new or first-time customers. A simple random sample of 100 orders will be used to estimate the proportion of new or first-time customers. The results of the sample will be used to verify the president's claim of $p = .30$.
    a. Assume that the president is correct and $p = .30$. What is the sampling distribution of $\bar{p}$ for this study?
    b. What is the probability that the sample proportion $\bar{p}$ will be between .20 and .40?
    c. What is the probability that the sample proportion will be within ±.05 of the population proportion $p = .30$?

46. A March 1989 *Newsweek* poll conducted by The Gallup Organization used a national sample of 756 adults to obtain information on the American public views of the job being done by President George Bush (*Newsweek*, March 20, 1989). The poll reported a margin of error of ±4 percentage points.
    a. For the question, Do you approve or disapprove of the way George Bush is handling his job? assume that 60% of the population would approve; that is, $p = .60$. What is the probability the sample proportion $\bar{p}$ from the sample of 756 adults would be within 4 percentage points of this population proportion?
    b. What is the probability the sample proportion from the sample of 756 adults would be within 2 percentage points of the population proportion?
    c. Why did the Newsweek poll quote a 4% margin of error rather than a 2% margin of error?

47. Louis Harris & Associates, Inc. conducted a survey of 1253 adults to learn how individuals feel about the United States' position in the global economy (*Business Week*, April 6, 1992). One question asked how concerned the individual was about U.S. industry becoming less competitive in the global economy. Assume that for the entire population, 55% of the adults are very concerned about U.S. industry becoming less competitive. Let $\bar{p}$ be the sample proportion of the polled adults who are very concerned on this issue.
    a. Show the sampling distribution of $\bar{p}$ if the population proportion is $p = .55$.
    b. What is the probability that the Harris poll sample proportion will have a sampling error of ±.02 or less?
    c. What is the probability that the Harris poll sample proportion will have a sampling error of ±.03 or less?
    d. Comment on why the Harris poll stated, "Results should be accurate to within 3 percentage points."

48. *American Association of Individual Investors* (April 1992) provided results from a survey on a range of investment-related topics. One question was, Do you favor relaxed SEC rules for financial reporting by foreign corporations to allow more foreign stocks to be traded in the United States? Assume that the population proportion is .42 and that the sample size is 300.

a. What is the probability that the sample proportion will be within ±.03 of the population proportion?

b. What is the probability that the sample proportion will be .45 or greater?

c. What is the probability that the sample proportion will show 50% or more of the population favoring relaxed SEC rules for foreign corporations?

49. A particular county in West Virginia has a 9% unemployment rate. A monthly survey of 800 individuals is conducted by a state agency in order to monitor the unemployment rate of the county.

a. Assume that $p = .09$. What is the sampling distribution of $\bar{p}$ when a sample of size 800 is used?

b. What is the probability that a sample proportion $\bar{p}$ of at least .08 will be observed?

50. Surveys of subscribers of *The Wall Street Journal* and *Investor's Daily* in 1988 provided reader profile information for these two business publications (*Investor's Daily,* March 23, 1989). The profile information included median reader age, percentage of college graduates, percentage in top management, mean income, and mean net worth. Based on data provided by these surveys, assume that 80% of the population of all subscribers of these business publications are college graduates. Answer the following questions about the results of the simple random sample of subscribers. The statistic of interest is the sample proportion of subscribers found to be college graduates.

a. Show the sampling distribution of $\bar{p}$ if a simple random sample of 400 subscribers is used. Assume $p = .80$.

b. With a sample size of 400, what is the probability that the margin of error will be 3% or less? That is, what is the probability the sample proportion of college graduates among the 400 subscribers will be within .03 of the population proportion $p = .80$?

c. Answer part (b) if the size of the simple random sample is increased to 750 subscribers.

51. What is the probability that the EAI estimate of the proportion of managers having completed the firm's training program, $\bar{p}$, is within ±.05 of the population proportion $p = .60$? Use samples of size 60 and 120.

52. Assume that 15% of the items produced in an assembly line operation are defective, but that the firm's production manager is not aware of this situation. Assume further that 50 parts are tested by the quality assurance department in order to determine the quality of the assembly operation. Let $\bar{p}$ be the sample proportion defective found by the quality assurance test.

a. Show the sampling distribution for $\bar{p}$.

b. What is the probability that the sample proportion will be within ±.03 of the population proportion defective?

c. If the test shows $\bar{p} = .10$ or more, the assembly line operation will be shut down to check for the cause of the defects. What is the probability that the sample of 50 parts will lead to the conclusion that the assembly line should be shut down?

53. Baskin Robbins offers frozen yogurt at 900 of its 2500 outlets (*Advertising Age,* April 10, 1989).

a. What is the population proportion of Baskin Robbins outlets offering frozen yogurt?

b. Assume that a simple random sample of 40 Baskin Robbins outlets will be used and that the same proportion $\bar{p}$ will be used to estimate the population proportion of Baskin Robbins outlets offering frozen yogurt. What is the probability that the sample proportion will be within ±.05 of the population proportion?

c. What is the probability that the sample proportion will be within ±.05 of the population proportion if the sample size is increased to 120 outlets?

7.7   Properties of Point Estimators

In this chapter we have shown how sample statistics such as a sample mean $\bar{x}$, a sample standard deviation s, and a sample proportion $\bar{p}$, can be used as point estimators of their corresponding population parameters μ, σ, and p. It is intuitively appealing that each sample statistic should be the point estimator of its corresponding population parameter. However, before using a sample statistic as a point estimator, statisticians check to see whether the sample statistic has some of the properties associated with good point estimators. In this section we discuss the properties of good point estimators referred to as unbiasedness, efficiency, and consistency.

Since there are several different sample statistics that can be used as point estimators of different population parameters, we will use the following general notation in this section:

$$\theta = \text{the population parameter of interest}$$

$$\hat{\theta} = \text{the sample statistic or point estimator of } \theta$$

The notation θ is the Greek letter theta, and the notation $\hat{\theta}$ is pronounced theta-hat. In general, θ represents any population parameter such as a population mean, population standard deviation, population proportion, and so on; $\hat{\theta}$ represents the corresponding sample statistic such as the sample mean, sample standard deviation, and sample proportion.

Unbiasedness

If the expected value of the sample statistic is equal to the population parameter being estimated, the sample statistic is said to be an *unbiased* estimator of the population parameter. The property of unbiasedness is defined as follows.

Unbiasedness

The sample statistic $\hat{\theta}$ is an unbiased estimator of the population parameter θ if

$$E(\hat{\theta}) = \theta \tag{7.6}$$

where

$$E(\hat{\theta}) = \text{expected value of the sample statistic } \hat{\theta}$$

The unbiasedness property states that the expected value, or mean, of all possible values of the sample statistic is equal to the population parameter being estimated.

Figure 7.13 shows the cases of unbiased and biased point estimators. In the illustration showing the unbiased estimator, the mean of the sampling distribution is equal to the value of the population parameter. The sampling errors balance out in this case, since sometimes the value of the point estimator $\hat{\theta}$ may be less than θ and other times it may be greater than θ. In the case of the biased estimator, the mean of the sampling distribution is less than or greater than the value of the population parameter. In the illustration shown in Figure 7.13(b), $E(\hat{\theta}) > \theta$; thus, the sample statistic has a high probability of overestimating the value of the population parameter. This tendency to overestimate the parameter is the biased property of the estimator. The amount of the bias is shown in the figure.

In discussing the sampling distributions of the sample mean and the sample proportion, we stated that $E(\bar{x}) = \mu$ and $E(\bar{p}) = p$. Thus, both $\bar{x}$ and $\bar{p}$ are unbiased estimators of their corresponding population parameters μ and p.

FIGURE 7.13

Examples of Unbiased and Biased Point Estimators

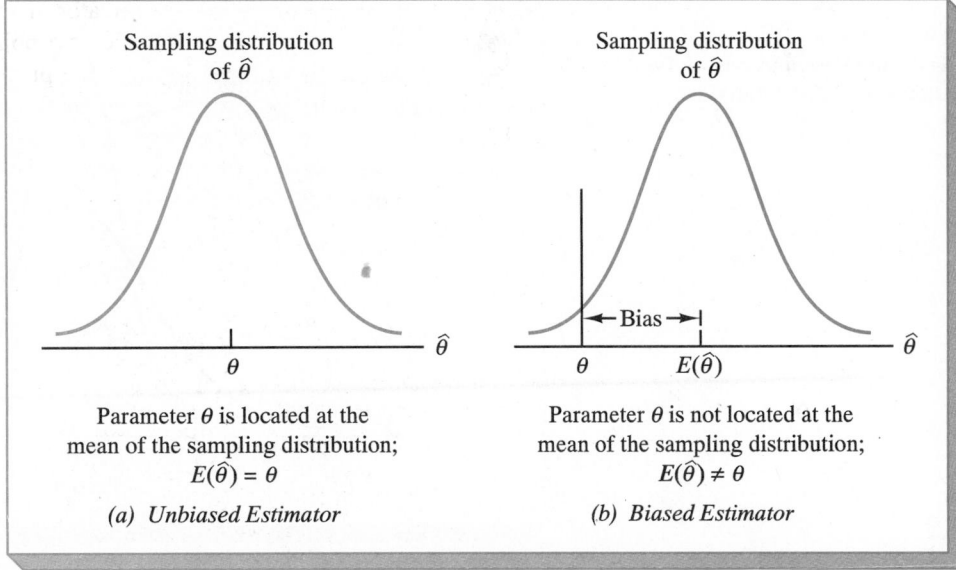

Sampling distribution of $\hat{\theta}$

Sampling distribution of $\hat{\theta}$

←— Bias —→|

θ $E(\hat{\theta})$

Parameter θ is located at the mean of the sampling distribution; $E(\hat{\theta}) = \theta$

Parameter θ is not located at the mean of the sampling distribution; $E(\hat{\theta}) \neq \theta$

(a) Unbiased Estimator

(b) Biased Estimator

We delay a more complete discussion of the sampling distribution of the sample standard deviation s and the sample variance s^2 until Chapter 11. However, it can be shown that $E(s^2) = \sigma^2$. Thus, we conclude that the sample variance s^2 is an unbiased estimator of the population variance σ^2. In fact, when we first presented the formulas for the sample variance and the sample standard deviation in Chapter 3, $n - 1$ rather than n was used in the denominators. The reason for using $n - 1$ rather than n is to make the sample variance an unbiased estimator of the population variance. If we had used n in the denominator, the sample variance would have been a biased estimator, tending to slightly underestimate the population variance.

Efficiency

Assume that a simple random sample of n elements can be used to provide two unbiased point estimators of the same population parameter. In this situation, we would prefer to use the point estimator with the smaller variance, since it tends to provide estimates closer to the population parameter. The point estimator with the smaller variance is said to have greater *relative efficiency* than the other.

Figure 7.14 shows the sampling distributions of two unbiased point estimators, $\hat{\theta}_1$ and $\hat{\theta}_2$. Note that the variance of $\hat{\theta}_1$ is less than the variance of $\hat{\theta}_2$; thus, values of $\hat{\theta}_1$ have a greater chance of being close to the parameter θ than do values of $\hat{\theta}_2$. Since the variance of point estimator $\hat{\theta}_1$ is less than the variance of point estimator $\hat{\theta}_2$, $\hat{\theta}_1$ is relatively more efficient than $\hat{\theta}_2$; thus $\hat{\theta}_1$ is the preferred point estimator.

Consistency

A third property associated with good point estimators is *consistency*. Loosely speaking, a point estimator is consistent if the values of the point estimator tend to become closer to the population parameter as the sample size becomes larger. In other words, a larger sample size tends to provide a better point estimate than does a smaller sample size. Note

FIGURE 7.14
**Sampling Distributions of Two
Unbiased Point Estimators**

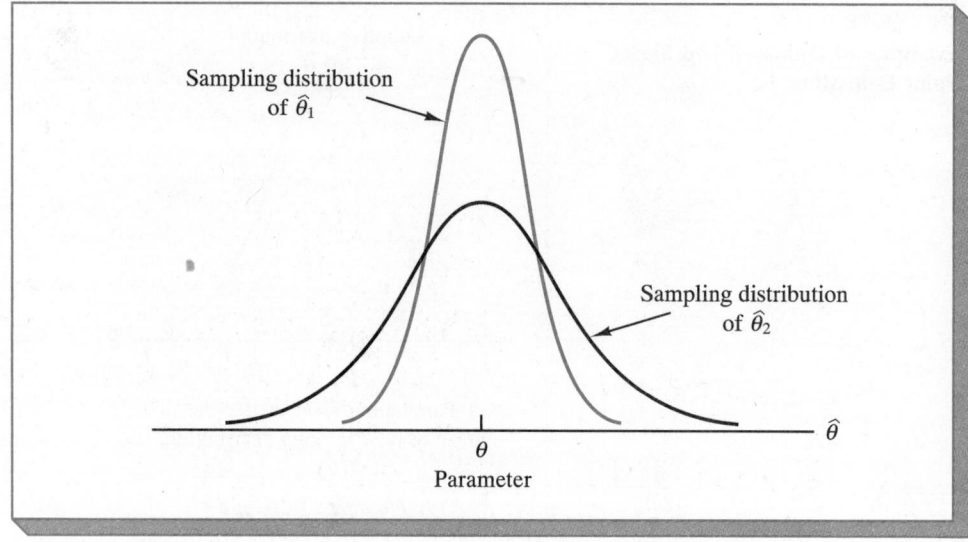

that for the sample mean $\overline{x}$, we showed that the standard deviation of $\overline{x}$ is given by $\sigma_{\overline{x}} = \sigma/\sqrt{n}$. Thus, we observed that $\sigma_{\overline{x}}$ is inversely proportional to the sample size. As a result, larger sample sizes provide smaller values for $\sigma_{\overline{x}}$. Thus, we conclude that a larger sample size tends to provide point estimates closer to the population mean μ. In this sense, we can say that the sample mean $\overline{x}$ is a consistent estimator of the population mean μ. Using a similar rationale, we can also conclude that the sample proportion $\overline{p}$ is a consistent estimator of the population proportion p.

**NOTES &
COMMENTS**

In Chapter 3 we stated that the mean and the median are two measures of central location. In this chapter we discussed only the mean. The reason for this is that in sampling from a normal population, where the population mean and population median are identical, the standard error of the median is approximately 25% larger than the standard error of the mean. Recall that in the EAI problem where $n = 30$, the standard error of the mean $\sigma_{\overline{x}} = 730.30$. The standard error of the median for this problem would have been approximately $1.25 \times (730.30) = 913$. As a result, the more efficient sample mean will have a higher probability of being within a specified distance of the population mean.

Summary

In this chapter we presented the concepts of simple random sampling and sampling distributions. We demonstrated how a simple random sample can be selected and how the data collected for the sample can be used to develop point estimates of population parameters. Since different simple random samples provided a variety of different values for the point estimators, we saw that point estimators, such as $\overline{x}$ and $\overline{p}$, are random variables. The probability distribution of such a random variable is called a sampling distribution. In particular, we described the sampling distributions of the sample mean $\overline{x}$ and the sample proportion $\overline{p}$.

In considering the characteristics of the sampling distributions of $\overline{x}$ and $\overline{p}$, we stated that $E(\overline{x}) = \mu$ and $E(\overline{p}) = p$. After developing the standard deviation or standard error formulas for these estimators, we showed how the central limit theorem provided the basis for using a normal probability distribution to approximate these sampling distributions in the large-sample case. Rules of thumb were provided for determining when large-sample-size conditions were satisfied. We then discussed some properties of point estimators, including unbiasedness, efficiency, and consistency.

Glossary

Parameter A population characteristic, such as a population mean μ, a population standard deviation σ, a population proportion p, and so on.

Simple random sampling Finite population: a sample selected such that each possible sample of size n has the same probability of being selected. Infinite population: a sample selected such that each element comes from the same population and the successive elements are selected independently.

Sampling without replacement Once an item from the population has been included in the sample, it is removed from the population and thus cannot be selected a second time.

Sampling with replacement As each item is selected for the sample, it is returned to the population. It is possible that a previously selected item may be selected again and therefore appear in the sample more than once.

Sample statistic A sample characteristic, such as a sample mean $\bar{x}$, a sample standard deviation s, a sample proportion $\bar{p}$, and so on. The value of the sample statistic is used to estimate the value of the population parameter.

Sampling distribution A probability distribution consisting of all possible values of a sample statistic.

Point estimate A single numerical value used as an estimate of a population parameter.

Point estimator The sample statistic, such as $\bar{x}$, s, and $\bar{p}$, that provides the point estimate of the population parameter.

Finite population correction factor The term $\sqrt{(N - n)/(N - 1)}$ that is used in the formulas for $\sigma_{\bar{x}}$ and $\sigma_{\bar{p}}$ whenever a finite population, rather than an infinite population, is being sampled. The generally accepted rule of thumb is to ignore the finite population correction factor whenever $n/N \leq .05$.

Standard error The standard deviation of a point estimator.

Central limit theorem A theorem that allows us to use the normal probability distribution to approximate the sampling distribution of $\bar{x}$ and $\bar{p}$ whenever the sample size is large.

Unbiasedness A property of a point estimator that occurs whenever the expected value of the point estimator is equal to the population parameter it estimates.

Relative efficiency Given two unbiased point estimators of the same population parameter, the point estimator with the smaller variance is said to have greater efficiency than the other.

Consistency A property of a point estimator that occurs whenever larger sample sizes tend to provide point estimates closer to the population parameter.

Key Formulas

Expected Value of $\bar{x}$

$$E(\bar{x}) = \mu \tag{7.1}$$

Standard Deviation of $\bar{x}$

Finite Population *Infinite Population*

$$\sigma_{\bar{x}} = \sqrt{\frac{N - n}{N - 1}} \left(\frac{\sigma}{\sqrt{n}}\right) \qquad \sigma_{\bar{x}} = \frac{\sigma}{\sqrt{n}} \tag{7.2}$$

Expected Value of $\bar{p}$

$$E(\bar{p}) = p \tag{7.4}$$

Standard Deviation of $\bar{p}$

Finite Population   Infinite Population

$$\sigma_{\bar{p}} = \sqrt{\frac{N - n}{N - 1}} \sqrt{\frac{p(1 - p)}{n}} \qquad \sigma_{\bar{p}} = \sqrt{\frac{p(1 - p)}{n}} \tag{7.5}$$

☐ ☐ Supplementary Exercises

54. Nationwide Supermarkets has 4800 retail stores located in 32 states. At the end of each year, a sample of 35 stores is selected for physical inventories. Results from the inventory samples are used in annual tax reports. Assume that the retail stores are listed sequentially on a computer printout. Begin at the bottom of the second column of the random numbers shown in Table 7.1. Using 4-digit random numbers beginning with 8112, read *up* the column to identify the first five stores to be included in the simple random sample.

55. A study of the time from computer program submission until program return (i.e., "turn-around" time) was conducted at a university computer center. Assume that under standard operating conditions the population mean is 120 minutes, with a population standard deviation of 40 minutes.
a. Future studies of turnaround time are to be based on simple random samples of 30 programs. Show the sampling distribution of the sample mean turnaround time.
b. What is the probability that the sample mean for 30 programs will be less than 100 minutes? Over 125 minutes?

56. *Survey and Analysis of Salary Trends,* 1988, from the research department of the American Federation of Teachers, lists the average annual salary for elementary/secondary school teachers as $28,085. Answer the following questions using this value as the population mean and 3200 as the population standard deviation.
a. What is the probability that a simple random sample of 100 teachers will provide a sample mean annual salary within $200 of the population mean?
b. What is the probability that a simple random sample of 300 teachers will provide a sample mean annual salary within $200 of the population mean?
c. Assuming that we would like a 90% chance of having a sample mean annual salary within $200 of the population mean, how many teachers should be included in the simple random sample?

57. An electrical component is designed to provide a mean service life of 3000 hours, with a standard deviation of 800 hours. A customer purchases a batch of 50 components; assume that this batch can be considered a simple random sample of the population of components. What is the probability that the mean life for the group of 50 components will be at least 2750 hours? At least 3200 hours?

58. The population mean household income in Pittsburgh, Pennsylvania, is $49,000 (*U.S. News & World Report,* April 6, 1992). Assume that the population standard deviation is $12,000. Furthermore, assume that a simple random sample of households in the Pittsburgh area will be selected and the sample mean will be used to estimate the population mean.
a. Show the sampling distribution of the sample mean if a sample of 100 households is used.
b. What is the probability that a sample of 100 will provide a sampling error of $1000 or less?
c. What is the probability that a sample of 200 will provide a sampling error of $1000 or less?
d. What is the probability that a sample of 400 will provide a sampling error of $1000 or less?
e. How large of a sample would be required if we wanted a .95 probability of having a sampling error of $1000 or less?

59. The time it takes a fire department to respond to a request for emergency aid has a mean of $\mu = 14$ minutes with a standard deviation of $\sigma = 4$ minutes. Suppose we randomly sample 50

emergency requests over a 2-month period. Records of aid-request times and arrival times will be used to compute a sample mean response time for the 50 requests.
a. Show the sampling distribution of $\bar{x}$.
b. What role does the central limit theorem play in identifying this sampling distribution?
c. What is the probability that the sample mean will be 15 minutes or less?
d. What is the probability that the sample mean will be within $\pm.5$ minutes of the mean time for the population?

60. The speed of automobiles on a section of I-75 in northern Florida has a mean of $\mu = 67$ miles per hour with a standard deviation of $\sigma = 6$ miles per hour. Answer the following questions if the population can be assumed to have a *normal distribution* and if a sample of 16 automobiles will be selected to compute a sample mean automobile speed.
a. What is the expected value of $\bar{x}$?
b. What is the value of the standard error of the mean?
c. Show the sampling distribution of $\bar{x}$.
d. What is the probability that the value of the sample mean will be 65 miles per hour or more?
e. What is the probability that the value of the sample mean will be between 66 and 68 miles per hour?

61. Consider a population of size $N = 500$ with a mean $\mu = 200$ and a standard deviation $\sigma = 40$. Assume that a simple random sample of size $n = 100$ will be selected from this population.
a. Should the finite population correction factor be used in computing the standard error of the mean?
b. What is the value of the standard error of the mean for this problem?
c. What is the probability of selecting a simple random sample that provides a value of $\bar{x}$ that is within ± 5 of the population mean μ?

62. In a population of 5000 students, a simple random sample of 50 students is selected to estimate the mean grade point average for the population.
a. Would you use the finite population correction factor in calculating the standard error of the mean? Explain.
b. If the population standard deviation is $\sigma = .4$ years, compute the standard error of the mean, first with and then without the finite population correction factor. What is the rationale for ignoring the finite population correction factor whenever $n/N \leq .05$?
c. What is the probability that the sample grade point average for 50 students will be within $\pm.10$ of the population mean grade point average?

63. During a complete review of 2 months of billings, an accountant found the following values for the mean and standard deviation of the dollar amounts per billing: $\mu = \$22.00$ and $\sigma = \$7.00$. The company's controller believes that the accountant could have obtained very good estimates of the mean billing amount by taking a simple random sample of 50 billings. Assume that a simple random sampling procedure was conducted.
a. Explain how the sample mean billing $\bar{x}$ would have a sampling distribution.
b. Show the sampling distribution of $\bar{x}$.
c. What is the standard error of the mean?
d. What would happen to the sampling distribution of $\bar{x}$ if a sample size of 100 was considered?

64. In the EAI study the population of managers had annual salaries with $\mu = \$51,800$ and $\sigma = \$4,000$. Samples of size 30 provided a .5036 probability of selecting a sample with $\bar{x}$ within $\pm\$500$ of the population mean. How large a sample should be selected if the personnel director wishes the probability of a sample mean $\bar{x}$ being within $\pm\$500$ of μ to be .95?

65. Three firms have inventories that vary in size. Firm A has a population of 2000 items, firm B has a population of 5000 items, and firm C has a population of 10,000 items. The population standard deviation for the cost of the items is $\sigma = 144$. A statistical consultant recommends that each firm take a sample of 50 items from their respective populations to provide statistically valid estimates of the average cost per item. Management of the small firm states that since it has the smallest population, it should be able to obtain the data from a much smaller sample size than required by the larger firms. However, the consultant states that to obtain the same standard error

and thus the same precision in the sample results, all firms should take the same sample size regardless of population size.

a. Using the finite population correction factor, compute the standard error for each of the three firms given a sample of size 50.

b. What is the probability that for each firm, the sample mean $\bar{x}$ will be within ± 25 of the population mean μ?

66. A survey reports its results by stating that the standard error of the mean is 20. The population standard deviation is 500.

a. How large is the sample used in this survey?

b. What is the probability that the estimate would be within ± 25 of the population mean?

67. A production process is checked periodically by a quality control inspector. The inspector selects simple random samples of 30 finished products and computes the sample mean product weights $\bar{x}$. If test results over a long period of time show that 5% of the $\bar{x}$ values are over 2.1 pounds and 5% are under 1.9 pounds, what are the mean and the standard deviation for the population produced with this process?

68. The grade point average for all juniors at Strausser College has a standard deviation of .50.

a. A random sample of 20 students is to be used to estimate the population mean grade point average. What assumption is necessary to compute the probability of obtaining a sample mean within $\pm .2$ of the population mean?

b. Provided that this assumption can be made, what is the probability of $\bar{x}$ being within $\pm .2$ of the population mean?

c. If this assumption cannot be made, what would you recommend doing?

69. Assume that the proportion of persons having a college degree is $p = .35$.

a. Explain how the sampling distribution of $\bar{p}$ results from random samples of size 80 being used to estimate the proportion of individuals having a college degree.

b. Show the sampling distribution for $\bar{p}$ in this case.

c. If the sample size is increased to 200, what happens to the sampling distribution of $\bar{p}$? Compare the standard error for the $n = 80$ and $n = 200$ alternatives.

70. CinemaScore is a movie research firm that uses a sample of film audiences to learn why members of the audience chose to attend a particular motion picture (*USA Today,* April 11, 1989). Possible responses include the subject matter, the cast, and other reasons. Assume that for a particular new motion picture, 65% of the audience population chose the movie due to subject matter. In addition, assume that a simple random sample will be used to estimate the proportion of the population that chose the movie due to subject matter.

a. What is the probability that a simple random sample of 100 people will provide a sample proportion within .04 of the population proportion?

b. What is the probability that a simple random sample of 200 people will provide a sample proportion within .04 of the population proportion?

c. Assume that we would like to select a simple random sample that will provide a 90% chance of obtaining a sample proportion within .04 of the population proportion. How many people should be included in the simple random sample?

71. A market research firm conducts telephone surveys with a 40% historical response rate. What is the probability that in a new sample of 400 telephone numbers, at least 150 individuals will cooperate and respond to the questions? In other words, what is the probability the sample proportion will be at least $150/400 = .375$?

72. A production run is not acceptable for shipment to customers if a sample of 100 items contains 5% or more defective items. If a production run has a population proportion defective of $p = .10$, what is the probability that $\bar{p}$ will be at least .05?

73. The proportion of individuals insured by the All-Driver Automobile Insurance Company that have received at least one traffic ticket during a 5-year period is .15.

a. Show the sampling distribution of $\bar{p}$ if a random sample of 150 insured individuals is used to estimate the proportion having received at least one ticket.

b. What is the probability that the sample proportion will be within $\pm.03$ of the population proportion?

74. Historical records show that .50 of all orders placed at Big Burger fast-food restaurants include a soft drink. With a simple random sample of 40 orders, what is the probability that between .45 and .55 of the sampled orders will include a soft drink?

75. Lori Jeffrey is a successful sales representative for a major publisher of college textbooks. Historically, Lori obtains a book adoption on 25% of her sales calls. Viewing her sales calls for 1 month as a sample of all possible sales calls, a statistical analysis of the data yields a standard error of the proportion of .0625.

a. How large was the sample used in this analysis? That is, how many sales calls did Lori make during the month?

b. Let $\bar{p}$ indicate the sample proportion of book adoptions obtained during the month. Show the sampling distribution $\bar{p}$.

c. Using the sampling distribution of $\bar{p}$, compute the probability that Lori will obtain book adoptions on 30% or more of her sales calls during the 1-month period.

APPENDIX

The Expected Value and Standard Deviation of $\bar{x}$

In this appendix we present the mathematical basis for the expressions for the expected value of $\bar{x}$, $E(\bar{x})$, as given by (7.1) and the standard deviation of $\bar{x}$, $\sigma_{\bar{x}}$, as given by (7.2).

Expected Value of $\bar{x}$

Assume a population with mean μ and variance σ^2. A simple random sample of size n is selected with individual observations denoted $x_1, x_2, \ldots, x_n$. A sample mean $\bar{x}$ is computed as follows:

$$\bar{x} = \frac{\Sigma x_i}{n}$$

With repeated simple random samples of size n, $\bar{x}$ is a random variable that takes on different numerical values depending upon the specific n items selected. The expected value of the random variable $\bar{x}$, or the mean of all possible $\bar{x}$ values, is as follows:

$$\text{Mean of } \bar{x} = E(\bar{x}) = E\left(\frac{\Sigma x_i}{n}\right)$$

$$= \frac{1}{n}\left[E(x_1 + x_2 + \cdots + x_n)\right]$$

$$= \frac{1}{n}\left[E(x_1) + E(x_2) + \cdots + E(x_n)\right]$$

Since for any x_i, we have $E(x_i) = \mu$, we can write

$$E(\overline{x}) = \frac{1}{n}(\mu + \mu + \cdots + \mu)$$

$$= \frac{1}{n}(n\mu) = \mu$$

The above expression shows that the mean of all possible $\overline{x}$ values is the same as the population mean μ. That is, $E(\overline{x}) = \mu$.

Standard Deviation of $\overline{x}$

Again assume a population with mean μ, variance σ^2, and a sample mean given by

$$\overline{x} = \frac{\Sigma x_i}{n}$$

With repeated simple random samples of size n, we know that $\overline{x}$ is a random variable that takes on different numerical values depending upon the specific n items selected. Shown below is the derivation of the expression for the standard deviation of the $\overline{x}$ values, $\sigma_{\overline{x}}$, for the case where the population is infinite. The derivation of the expression for $\sigma_{\overline{x}}$ for a finite population when sampling is done without replacement is more difficult and is beyond the scope of this text.

Returning to the infinite population case, recall that a simple random sample from an infinite population consists of observations $x_1, x_2, \ldots, x_n$ that are independent. The following two expressions are general formulas concerning the variance of random variables:

$$\mathrm{Var}(ax) = a^2\,\mathrm{Var}(x) \tag{A.1}$$

where a is a constant and x is a random variable, and

$$\mathrm{Var}(x + y) = \mathrm{Var}(x) + \mathrm{Var}(y) \tag{A.2}$$

where x and y are *independent* random variables. Using (A.1) and (A.2), we can develop the expression for the variance of the random variable $\overline{x}$ as follows:

$$\mathrm{Var}(\overline{x}) = \mathrm{Var}\!\left(\frac{\Sigma x_i}{n}\right) = \mathrm{Var}\!\left(\frac{1}{n}\Sigma x_i\right)$$

Using (A.1) with $1/n$ viewed as the constant, we have

$$\mathrm{Var}(\overline{x}) = \left(\frac{1}{n}\right)^2 \mathrm{Var}(\Sigma x_i)$$

$$= \left(\frac{1}{n}\right)^2 \mathrm{Var}(x_1 + x_2 + \cdots + x_n)$$

With the infinite population case, the random variables $x_1, x_2, \ldots, x_n$ are independent. Thus, (A.2) enables us to write

$$\mathrm{Var}(\overline{x}) = \left(\frac{1}{n}\right)^2\!\left[\mathrm{Var}(x_1) + \mathrm{Var}(x_2) + \cdots + \mathrm{Var}(x_n)\right]$$

Since for any x_i, we have $\mathrm{Var}(x) = \sigma^2$, we have

$$\text{Var}(\bar{x}) = \left(\frac{1}{n}\right)^2 \underbrace{(\sigma^2 + \sigma^2 + \cdots + \sigma^2)}_{n \text{ items}}$$

With n values of σ^2 in this expression, we have

$$\text{Var}(\bar{x}) = \left(\frac{1}{n}\right)^2 (n\sigma^2) = \frac{\sigma^2}{n}$$

Taking the square root provides the formula for the standard deviation of $\bar{x}$ for the infinite population case:

$$\sigma_{\bar{x}} = \sqrt{\text{Var}(\bar{x})} = \frac{\sigma}{\sqrt{n}}$$

This expression, which appeared as (7.2), shows that as the sample size n is increased, the standard deviation of the various sample means $\bar{x}$ will decrease. In effect, the larger samples tend to provide better estimates of the population mean.

Interval Estimation

Dollar General Corporation*

NASHVILLE, TENNESSEE

Dollar General Corporation was founded in 1939 as a dry goods wholesale company. After World War II, the company began opening retail locations in rural southcentral Kentucky. Today Dollar General Corporation operates more than 1300 neighborhood stores in 23 states. Serving predominately low- and middle-income customers, Dollar General markets soft goods, health, beauty, and cleaning supplies at low, everyday prices.

Being in an inventory-intense business with approximately 17,000 different products, Dollar General made the decision to adopt the LIFO (last-in-first-out) method of inventory valuation. This method provides a better match of current costs against current revenues, which minimizes the effect of radical price changes on profit and loss results. In addition, the LIFO method reduces net income and thereby income taxes during periods of inflation. This in turn brings disposable cash generated from sales in line with income and allows for the replacement of inventory at current costs.

Accounting practices require that a LIFO index be established for inventory under the LIFO method of evaluation. For example, a LIFO index of 1.048 indicates that the company's inventory at current costs contains a 4.8% increase in value due to the inflation occurring over the most recent one-year period.

The establishment of a LIFO index requires that the year-end inventory count for each product be evaluated at the current year-end cost and at the preceding year-end cost. To avoid counting the inventory of every product in over 1300 retail locations, a random sample of 800 products is selected from 75 retail locations and 3 warehouses. Physical inventories for the 800 sampled products are taken at the end of the year. Accounting personnel then provide the current-year and preceding-year costs needed to construct the LIFO index.

For a recent year, the LIFO index was 1.070. However, since this index is only a sample estimate of the population's LIFO index, a statement about the precision of the estimate was required. Using the sample results and a 95% confidence level, the sampling error was computed to be .006. Thus, the interval from 1.064 to 1.076 provides the 95% confidence interval estimate of the population LIFO index. This precision was judged to be very good.

In this chapter you will learn how to make a probability statement about the sampling error associated with the sample mean and sample proportion. Then, you will learn how to use the sampling error to construct and interpret confidence interval estimates of a population mean and a population proportion. You will also learn how to determine the sample size needed to ensure that the sampling errors will be within acceptable limits.

*Mr. Robert S. Knaul, Controller, Dollar General Corporation, provided this Statistics in Practice.

In Chapter 7 we showed that the value of the sample mean $\bar{x}$ provides a point estimate of the population mean μ and that the value of the sample proportion $\bar{p}$ provides a point estimate of the population proportion p. Since some degree of error due to sampling is anticipated, we cannot expect the value of a point estimate to be *exactly* equal to the corresponding population parameter.

Point estimates of population parameters do not provide information about the *precision,* or magnitude of the sampling error, present in the estimation process. Interval estimates of population parameters have an advantage over point estimates in that interval estimates provide the desired precision information. Often the precision information is essential in evaluating and interpreting the sample results. For example, assume that a sample is used to estimate the mean annual starting salary for recent college graduates with degrees in business administration. Suppose the sample mean for this study is

$\bar{x} = \$25,500$. The sample mean of \$25,500 provides the point estimate of the population mean annual starting salary. If the margin of error associated with the estimate is $\pm\$10,000$, the interval estimate of \$15,500 to \$35,500 shows that the point estimate has limited use due to the wide margin of error. On the other hand, if the margin of error is $\pm\$100$, the information is extremely useful in that the population mean annual salary for graduates is provided by the interval estimate of \$25,400 to \$25,600.

In this chapter we will make use of the sampling distributions of $\bar{x}$ and $\bar{p}$ presented in Chapter 7 to develop interval estimates of the population mean μ and the population proportion p. Let us introduce the procedure for interval estimation of a population mean by considering a sampling study conducted by Statewide Insurance Company.

8.1   Interval Estimation of a Population Mean: Large-Sample Case

The central limit theorem, introduced in Chapter 7, allows us to conclude that the sampling distribution of $\bar{x}$ can be approximated by a normal probability distribution whenever the sample size is large. Recall that the large-sample condition is assumed to be satisfied whenever a simple random sample of size 30 or more is used. As an illustration of interval estimation of a population mean for the large-sample case, let us consider a sample of 36 Statewide Insurance policyholders.

The Statewide Insurance Company provides a variety of life, health, disability, and business insurance policies for customers located throughout the United States. As part of an annual review of life insurance policies, a simple random sample of 36 Statewide policyholders is selected. The corresponding 36 life insurance policies are reviewed in terms of the amount of the coverage, the cash value of the policy, the disability options, and so on. For the current policy review study, the project manager has requested information on the ages of the life insurance policyholders. Table 8.1 shows the age data collected from the simple random sample of 36 policyholders. Let us use these data to develop an interval estimate of the mean age of the population of life insurance policyholders covered by Statewide.

The interval estimation procedure that we will develop is based on the assumption that the value of the population standard deviation is *known*. For the Statewide study, previous studies on policyholder ages permit us to use a known population standard deviation of $\sigma = 7.2$ years. Later in this section we will show how to develop an interval estimate when the value of the population standard deviation is unknown.

Let x_1 indicate the age of the first policyholder in the sample, x_2 the age of the second policyholder, and so on. The sample mean $\bar{x}$ provides a point estimate of the population mean μ. Using the data in Table 8.1, we obtain

$$\bar{x} = \frac{\Sigma x_i}{n} = \frac{1422}{36} = 39.5$$

Thus, the point estimate of the mean age of the population of Statewide life insurance policyholders is 39.5 years.

In discussing the practical value of the sampling distribution of $\bar{x}$ in Chapter 7, we indicated that we cannot expect the value of a sample mean $\bar{x}$ to *exactly* equal the value of the population mean μ. Thus, anytime a sample mean is used to provide a point estimate of a population mean, someone may ask, How good is the estimate? The "how good" question is a way of asking about the error involved when the value of $\bar{x}$ is used as a point estimate of the population mean μ. In general, we refer to the absolute value of the difference between an unbiased point estimator and the population parameter as the

TABLE 8.1

Ages of Life Insurance Policyholders from a Simple Random Sample of 36 Statewide Policyholders

Policyholder	Age	Policyholder	Age	Policyholder	Age
1	32	13	39	25	23
2	50	14	46	26	36
3	40	15	45	27	42
4	24	16	39	28	34
5	33	17	38	29	39
6	44	18	45	30	34
7	45	19	27	31	35
8	48	20	43	32	42
9	44	21	54	33	53
10	47	22	36	34	28
11	31	23	34	35	49
12	36	24	48	36	39

sampling error. For the case of a sample mean estimating a population mean, the sampling error is:

$$\text{Sampling Error} = |\bar{x} - \mu| \tag{8.1}$$

Note that even after a sample is selected and the sample mean is computed ($\bar{x} = 39.5$ in the Statewide example), we will not be able to use (8.1) to find the value of the sampling error because the population mean μ is unknown. However, we shall see that the sampling distribution of $\bar{x}$ developed in Chapter 7 can be used to make probability statements about the sampling error.

From the central limit theorem we know that whenever the sample size is large ($n \geq 30$), the sampling distribution of $\bar{x}$ can be approximated by a normal probability distribution. For the Statewide Insurance study, where $\sigma = 7.2$ years and $n = 36$, this theorem enables us to conclude that the sampling distribution of $\bar{x}$ is approximately normal with mean μ and standard deviation $\sigma_{\bar{x}} = \sigma/\sqrt{n} = 7.2/\sqrt{36} = 1.2$ years.* This sampling distribution is shown in Figure 8.1.

Although the population mean μ is unknown, the sampling distribution in Figure 8.1 shows how the $\bar{x}$ values are distributed around μ. In effect, this distribution is providing us with information about the possible differences between $\bar{x}$ and μ and, as a result, about the possible sampling error.

Probability Statements about the Sampling Error

Since the sampling distribution of $\bar{x}$ can be approximated by a normal probability distribution, we can use the table of normal probabilities to make probability statements about the sampling error. For example, using the table of areas for a standard normal probability distribution, we find that 95% of the values of a normally distributed random variable lie within ± 1.96 standard deviations of the mean. Hence, for the sampling distribution of $\bar{x}$ shown in Figure 8.1, 95% of all $\bar{x}$ values are within $\pm 1.96\sigma_{\bar{x}}$ of the mean μ. Since $1.96\sigma_{\bar{x}} = 1.96(1.2) = 2.35$, we can state that 95% of all sample means lie within ± 2.35 years of the population mean μ. The location of all the sample means that provide

---

*In this chapter we will be assuming that $n/N \leq .05$. Thus, the finite population correction factor is not needed in the computation of $\sigma_{\bar{x}}$, and we use $\sigma_{\bar{x}} = \sigma/\sqrt{n}$.

FIGURE 8.1
Sampling Distribution of the Sample Mean Age from Simple Random Samples of 36 Statewide Policyholders

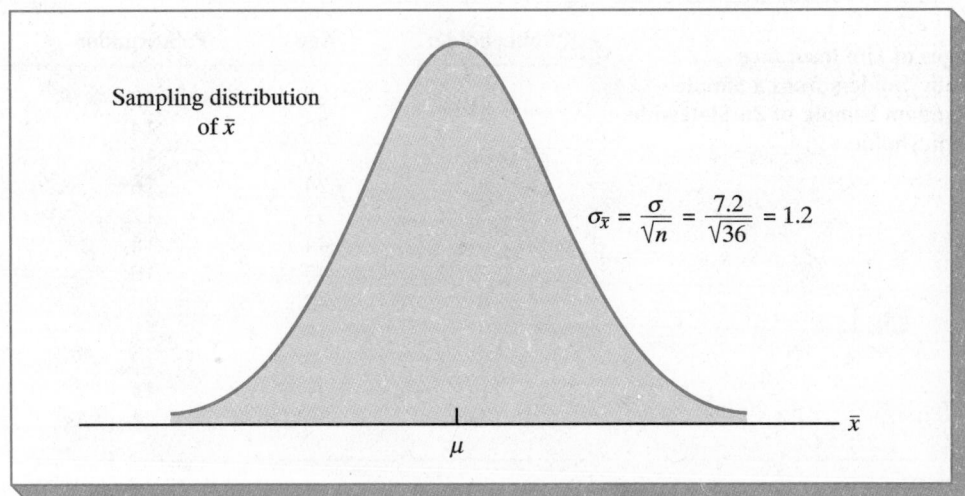

$$\sigma_{\bar{x}} = \frac{\sigma}{\sqrt{n}} = \frac{7.2}{\sqrt{36}} = 1.2$$

a sampling error of 2.35 years or less is shown in Figure 8.2. It is possible for a sample mean to fall in one of the two tails of the sampling distribution, which would result in a sampling error greater than 2.35 years. However, we see from Figure 8.2 that the probability of this occurring is only $1 - .95 = .05$. Thus, knowledge of the sampling distribution of $\bar{x}$ enables us to make the following probability statement about the sampling error whenever a simple random sample of 36 Statewide policyholders is used to provide a point estimate of the mean age of the population:

> There is a .95 probability that the sample mean will provide a sampling error of 2.35 years or less.

The above probability statement about the sampling error is a statement of the *precision* of the estimate. If the project manager is not satisfied with this degree of precision, a

FIGURE 8.2
Sampling Distribution of $\bar{x}$ Showing the Location of Sample Means that Provide a Sampling Error of 2.35 Years or Less

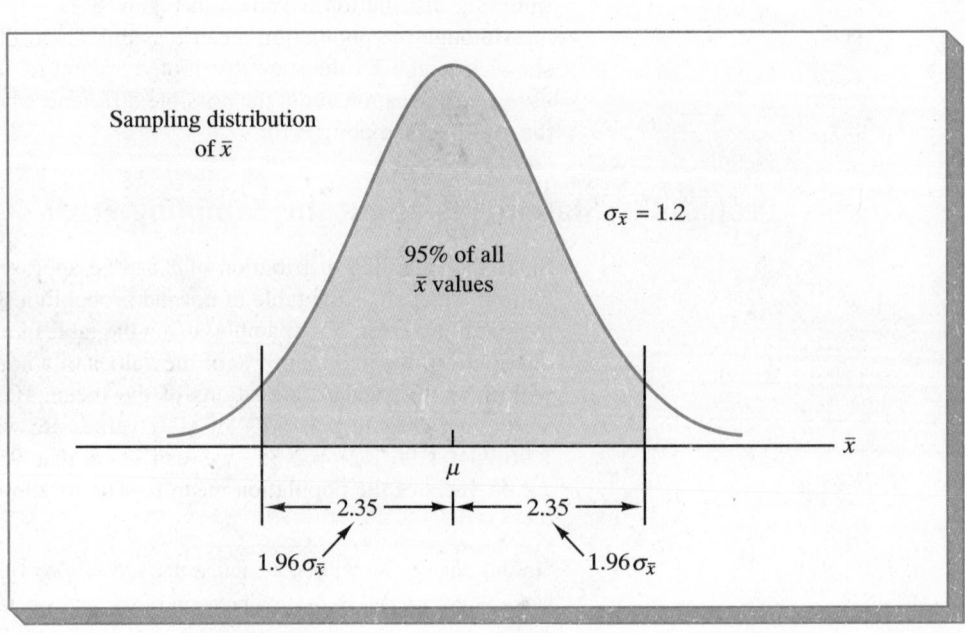

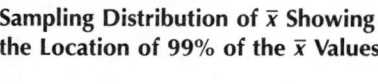

FIGURE 8.3

Sampling Distribution of $\bar{x}$ Showing the Location of 99% of the $\bar{x}$ Values

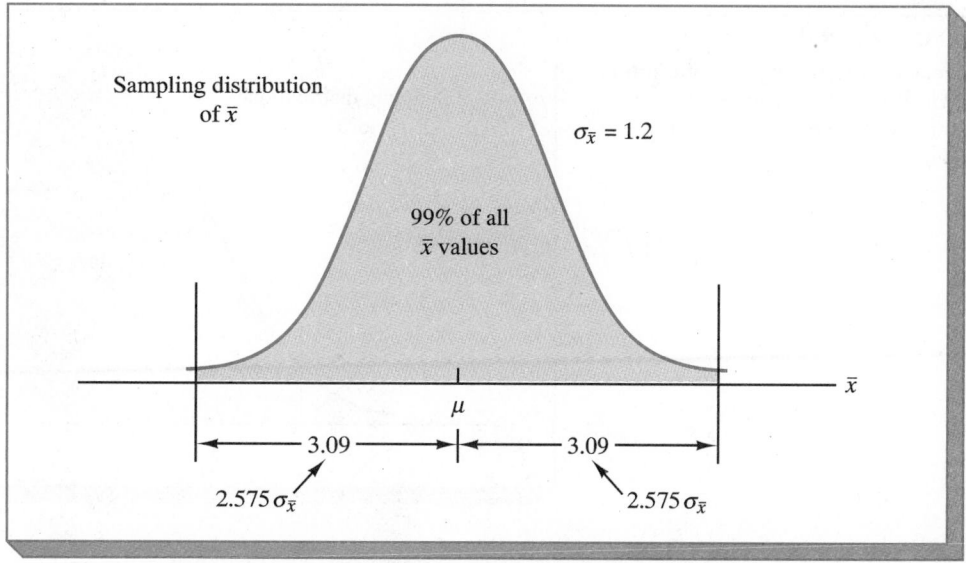

larger sample size will be necessary. We shall discuss a procedure for determining the sample size necessary to obtain a desired precision in Section 8.3.

Note that in the above analysis, the .95 probability used in the statement about the sampling error was arbitrary. Although a .95 probability is frequently used in making such statements, other probability values can be selected. Probabilities of .90 and .99 are suggested alternatives. Let us consider what would have happened if a probability of .99 had been selected. Figure 8.3 shows the location of 99% of the sample means for the Statewide Insurance sampling problem. From the standard normal probability distribution table, we find that 99% of the $\bar{x}$ values lie within ±2.575 standard deviations of the mean μ. Since $2.575\sigma_{\bar{x}} = 2.575(1.2) = 3.09$, we can make the following statement about the sampling error whenever a simple random sample of 36 Statewide policyholders is used to provide a point estimate of the mean age of the population:

> There is a .99 probability that the sample mean will provide a sampling error of 3.09 years or less.

A similar calculation with a .90 probability shows that there is a .90 probability that the sample mean will provide a sampling error of $1.645\sigma_{\bar{x}} = 1.645(1.2) = 1.97$ years or less.

These results show that there are various probability statements that can be made about the sampling error. They also show that there is a trade-off between the probability specified and the stated limit on the sampling error. In particular, note that the higher probability statements possess larger values for the sampling error.

Let us now generalize the procedure we are using to make probability statements about the sampling error whenever a sample mean is used to provide a point estimate of a population mean. We will use the Greek letter α (alpha) to indicate the probability that a sampling error is *larger* than the sampling error mentioned in the precision statement. Refer to Figure 8.4. We see that $\alpha/2$ will be the area or probability in each tail of the distribution, and $1 - \alpha$ will be the area or probability that a sample mean will provide a sampling error *less than or equal to* the sampling error used in the precision statement.

Refer to the Statewide Insurance example. The statement that there is a .95 probability that the value of a sample mean will provide a sampling error of 2.35 years or less is based

FIGURE 8.4

Areas of a Sampling Distribution of $\bar{x}$ Used to Make Probability Statements about the Sampling Error

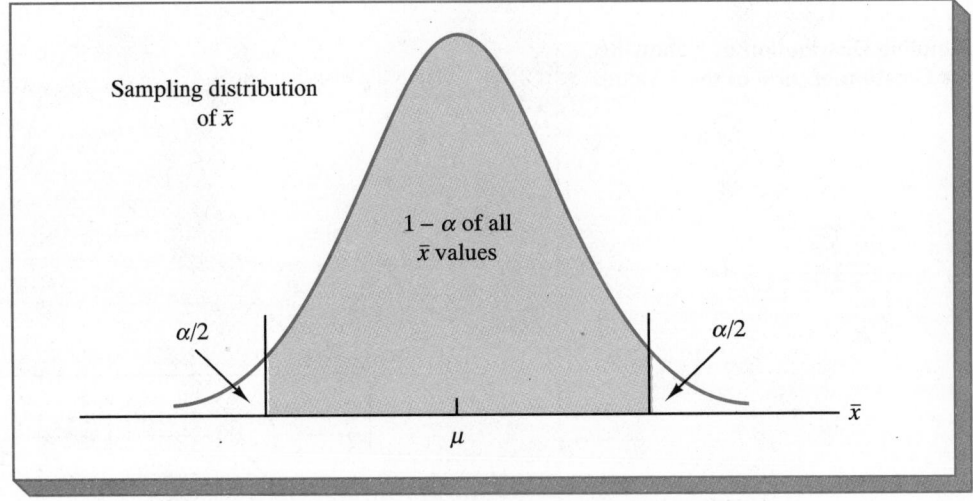

on $\alpha = .05$ and $1 - \alpha = .95$. The area in each tail of the sampling distribution is $\alpha/2 = .025$ (see Figure 8.2).

Using the z notation for the standard normal random variable, we will place a subscript on the z value to denote the *area in the upper tail* of the probability distribution. Thus, $z_{.025}$ will correspond to the z value with .025 of the area in the upper tail of the probability distribution. As can be found in the standard normal probability distribution table, $z_{.025} = 1.96$. If we wanted a .99 probability statement $\alpha = .01$, in this case we would be interested in an area of $\alpha/2 = .005$ in the upper tail of the distribution and hence, $z_{.005} = 2.575$.

In general, $z_{\alpha/2}$ denotes the value of the standard normal random variable corresponding to an area of $\alpha/2$ in the upper tail of the distribution. Also let $\sigma_{\bar{x}}$ denote the standard deviation of the sampling distribution of $\bar{x}$ (also called the standard error of the mean). We now have the following general procedure for making a probability statement about the sampling error whenever $\bar{x}$ is used to estimate μ.

Probability Statement about the Sampling Error

There is a $1 - \alpha$ probability that the value of a sample mean will provide a sampling error of $z_{\alpha/2}\sigma_{\bar{x}}$ or less.

Calculating an Interval Estimate

We have the ability to make probability statements about the sampling error. We can now combine the point estimate with the probability information about the sampling error to obtain an *interval estimate* of the population mean. The rationale for the interval-estimation procedure is as follows: We have already stated that there is a $1 - \alpha$ probability that the value of a sample mean will provide a sampling error of $z_{\alpha/2}\sigma_{\bar{x}}$ or less. This means that there is a $1 - \alpha$ probability that the sample mean *will not miss* the population mean *by more than* $z_{\alpha/2}\sigma_{\bar{x}}$. Thus, if we form an interval by subtracting $z_{\alpha/2}\sigma_{\bar{x}}$ from the sample mean $\bar{x}$ and then adding $z_{\alpha/2}\sigma_{\bar{x}}$ to the sample mean $\bar{x}$, we would have a $1 - \alpha$ probability of obtaining an interval that *includes* the population mean μ. This condition is stated as follows:

> There is a $1 - \alpha$ probability that the interval formed by $\bar{x} \pm z_{\alpha/2}\sigma_{\bar{x}}$ will contain the population mean μ.

Let us return to the Statewide Insurance study. We previously stated that there is a .95 probability that the value of a sample mean will provide a sampling error of 2.35 years or less. Look at the sampling distribution of $\bar{x}$ as shown in Figure 8.5. Let us consider possible values of the sample mean $\bar{x}$ that could be obtained from three different simple random samples, each containing 36 policyholders. Remember, in each case we will form an interval estimate of the population mean by subtracting 2.35 from $\bar{x}$ and adding 2.35 to $\bar{x}$.

Consider what happens if the first sample mean turns out to have the value shown in Figure 8.5 as $\bar{x}_1$. Note that in this case, the interval formed by subtracting 2.35 from $\bar{x}_1$ and adding 2.35 to $\bar{x}_1$ includes the population mean μ. Now consider what happens if the sample mean turns out to have the value shown in Figure 8.5 as $\bar{x}_2$. While this next sample mean is different from the first sample mean, we see that the interval based on $\bar{x}_2$ also includes the population mean μ. However, the interval based on the third sample mean, denoted by $\bar{x}_3$, does not include the population mean. The reason for this is that the sample mean $\bar{x}_3$ lies in a tail of the probability distribution at a distance further than 2.35 from μ. Thus, subtracting and adding 2.35 to $\bar{x}_3$ forms an interval that does not include μ.

Now think of repeating the sampling process many times, each time computing the value of the sample mean and then forming an interval from $\bar{x} - 2.35$ to $\bar{x} + 2.35$. Any sample mean $\bar{x}$ that falls between the vertical lines in Figure 8.5 will provide an interval

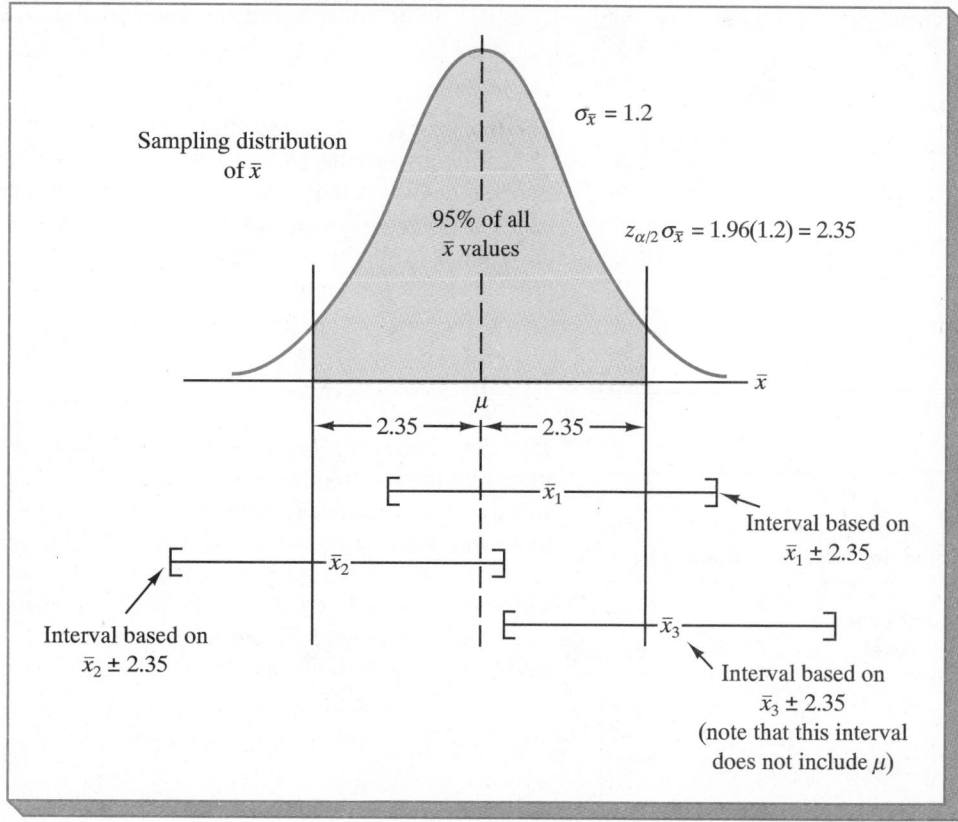

FIGURE 8.5

Intervals Formed from Selected Sample Means at Locations $\bar{x}_1$, $\bar{x}_2$, and $\bar{x}_3$

that includes the population mean μ. Since 95% of the sample means are in the shaded region, 95% of all intervals that could be formed will include μ. As a result, we say that we are 95% confident that an interval constructed from $\bar{x} - 2.35$ to $\bar{x} + 2.35$ will include the population mean. In common statistical terminology, the interval is referred to as a *confidence interval*. With 95% of the sample means leading to a confidence interval including μ, we say the interval is established at the 95% *confidence level*. The value .95 is referred to as the *confidence coefficient*.

Recall that we previously found the sample mean age for 36 Statewide life insurance policyholders to be $\bar{x} = 39.5$. A 95% confidence interval estimate of the mean age for the population of statewide life insurance policyholders is 39.5 ± 2.35, or 37.15 years to 41.85 years. Thus, at a 95% confidence level, Statewide can conclude that the mean age for the population of life insurance policyholders is between 37.15 and 41.85 years.

Let us now state the general procedure for computing an interval estimate of a population mean. As previously noted, there is a $1 - \alpha$ probability that the interval formed by $\bar{x} \pm z_{\alpha/2}\sigma_{\bar{x}}$ includes the population mean μ. Using the fact that $\sigma_{\bar{x}} = \sigma/\sqrt{n}$, the general procedure for calculating the *interval estimate* of population mean using a large sample can be written as follows.

**Interval Estimate of a Population Mean
Large-Sample Case (n ≥ 30)**

$$\bar{x} \pm z_{\alpha/2}\frac{\sigma}{\sqrt{n}} \qquad (8.2)$$

where $1 - \alpha$ is the confidence coefficient and $z_{\alpha/2}$ is the z value providing an area of $\alpha/2$ in the upper tail of the standard normal probability distribution.

The values of $z_{\alpha/2}$ for the most commonly used confidence levels are shown in Table 8.2.

A difficulty in using (8.2) is that in most sampling situations the value of the population standard deviation σ is unknown. In the large-sample case $(n \geq 30)$, we simply use the value of the sample standard deviations as the point estimate of the population standard deviation σ to obtain the confidence interval, $\bar{x} \pm z_{\alpha/2} \, s/\sqrt{n}$. In the following example, we provide an illustration to how a confidence interval is developed when σ is unknown and $n \geq 30$.

An Example: College Room-and-Board Costs

The *1992 America's Best Colleges* (*U.S. News & World Report,* June 5, 1992) lists a variety of information on colleges and universities in the United States. A sample of 50 colleges and universities from this population was used to estimate the mean annual room-and-board cost associated with attending college. The sample of 50 provided a sample mean of $\bar{x} = \$3685$ per year and a sample standard deviation of $s = \$849$. Develop a 90% confidence interval estimate of the population mean annual room-and-board cost.

At 90% confidence, Table 8.2 shows that $z_{.05} = 1.645$. Using the sample standard deviation $s = 849$ to estimate σ, (8.2) can be used to provide the confidence interval

$$3685 \pm 1.645\left(\frac{849}{\sqrt{50}}\right)$$

$$3685 \pm 198$$

TABLE 8.2

Values of $z_{\alpha/2}$ for the Most Commonly Used Confidence Level

Confidence Level	α	$\alpha/2$	$z_{\alpha/2}$
90%	.10	.05	1.645
95%	.05	.025	1.96
99%	.01	.005	2.575

Thus, we can be 90% confident that the interval $3487 to $3883 contains the mean annual room-and-board cost for the population of colleges and universities.

NOTES &
COMMENTS

1. In using (8.2) to develop an interval estimate of the population mean, we specify the desired confidence coefficient $(1 - \alpha)$ before selecting the sample. Thus, prior to selecting the sample, we conclude that there is a $1 - \alpha$ probability that the confidence interval we eventually compute will contain the population mean μ. However, once the sample is taken, the sample mean $\bar{x}$ is computed, and the particular interval estimate is determined, the resulting interval *may or may not* contain μ. If $1 - \alpha$ is reasonably large, we can be confident that the resulting interval contains μ because we know that if we use this procedure long term, $100(1 - \alpha)$ percent of all possible intervals developed in this manner will contain μ.

2. Note that the sample size n appears in the denominator of the interval estimation expression (8.2). Thus, if a particular sample size provides too wide an interval to be of any practical use, we may want to consider increasing the sample size. With n in the denominator, a larger sample size will reduce the margin of error, resulting in a narrower interval and a greater precision. The procedure for determining the size of a simple random sample required to obtain a desired precision is discussed in Section 8.3.

3. The justification for (8.2) comes from the fact that the central limit theorem enables us to *approximate* the sampling distribution of $\bar{x}$ by a normal probability distribution whenever the sample size is large. In addition, in most applications the sample standard deviation s is used to *approximate* the population standard deviation σ. Thus, statisticians often refer to (8.2) as an *approximate* confidence interval for a population mean.

□ □ **Exercises**

Methods

1. A simple random sample of 40 items resulted in a sample mean of 25. The population standard deviation is $\sigma = 5$.
   a. What is the standard error of the mean?
   b. At a 95% probability, what can be said about the size of the sampling error?

SELF TEST ▶ 2. A simple random sample of 50 items resulted in a sample mean of 32 and a sample standard deviation of 6.
   a. Provide a 90% confidence interval for the population mean.
   b. Provide a 95% confidence interval for the population mean.
   c. Provide a 99% confidence interval for the population mean.

3. A sample of 60 items resulted in a sample mean of 80 and a sample standard deviation of 15.
   a. Compute the 95% confidence interval for the population mean.
   b. Assume that the same sample mean and sample standard deviation were obtained from a sample of 120 items. Provide a 95% confidence interval for the population mean.
   c. What is the effect of a larger sample size on the interval estimate of a population mean?

4. A 95% confidence interval for a population mean is reported to be 122 to 130. If the sample mean was 126 and the sample standard deviation was 16.07, what sample size was used in this study?

Applications

SELF TEST ▸

5. In an effort to estimate the mean amount spent per customer for dinner meals at a major Atlanta restaurant, data were collected for a sample of 49 customers over a 3-week period.
a. Assume a population standard deviation of $2.50. What is the standard error of the mean?
b. With a .95 probability, what statement can be made about the sampling error?
c. If the sample mean is $12.60, what is the 95% confidence interval for the population mean?

6. Data on the automotive industry and automobile purchasing characteristics was presented in *Financial World*, April 14, 1992. The mean car payment per month was reported as $310. Assume that this result was based on a sample of 250 car payment records and that the sample standard deviation was $100. Compute the 95% confidence interval for the mean monthly car payments.

7. The results of an annual survey of 1404 mutual funds were presented in *Forbes*, September 2, 1991. A sample of 75 funds showed a mean return over the previous 12 months of 10.2%. The sample standard deviation was 3%. Provide a 95% confidence interval for the mean return for the population of mutual funds.

8. The mean annual income of U.S. factory workers is $24,000 (*Barron's*, April 10, 1989). Assume that this estimate was based on a sample of 250 U.S. factory workers and that the population standard deviation was $\sigma = 5000.
a. Compute the 90% confidence interval for the population mean.
b. Compute the 95% confidence interval for the population mean.
c. Compute the 99% confidence interval for the population mean.
d. Discuss what happens to the width of the interval estimate as the confidence level is increased. Why does this seem reasonable?

9. A production filling operation has a historical standard deviation of 5.5 ounces. A quality control inspector periodically selects 36 containers at random and uses the sample mean filling weight to estimate the population mean filling weight for the production process.
a. What is the standard error of the mean, $\sigma_{\bar{x}}$?
b. With .75, .90, and .99 probabilities, what statements can be made about the sampling error? What happens to the statement about the sampling error when the probability is increased? Why does this happen?
c. What is the 99% confidence interval for the population mean filling weight for the process if a sample mean is 48.6 ounces?

10. A survey of readers of *Money* magazine found that the sample mean age of men was 47 years and the sample mean age of women was 44 years (*Money Extra*, Fall 1988). All together, 454 people were included in the reader poll— 340 men and 114 women. Assume that the population standard deviation of age for both men and women is 8 years.
a. Develop a 95% confidence interval for the mean age of the population of men who read *Money* magazine.
b. Develop a 95% confidence interval for the mean age of the population of women who read *Money* magazine.
c. Compare the widths of the two interval estimates from parts (a) and (b). Did the estimate of the mean age of men or the mean age of women have the better precision? Why?

11. E. Lynn and Associates is an energy research firm that provides estimates of monthly heating costs for new homes based on style of house, square footage, insulation, and so on. The firm's service is used by both builders and potential buyers of new homes who wish advance information on heating costs. For winter months the standard deviation in the home heating bills for residential homes in a certain area is $100. Assume that a sample of 36 homes in a particular subdivision will be used to estimate the mean monthly heating bill for the population all homes in this type of subdivision.
a. What is the standard error of the mean, $\sigma_{\bar{x}}$?
b. Show the sampling distribution for the sample mean heating bill.
c. At an 80% probability, what can be said about the sampling error? Show this probability on the graph of the sampling distribution in part (b).
d. What is the 98% confidence interval for the population mean monthly heating bill if the sample mean is $196.50?

12. *Consumer Research*, April 1989, reports information on the time required for caffeine from products such as coffee and soft drinks to leave the body after consumption. Assume that the 95% confidence interval of the population mean time for adults is 5.6 hours to 6.4 hours.

a. What is the point estimate of the mean time for caffeine to leave the body of adults after consumption?

b. If the population standard deviation is 2 hours, how large a sample was used to provide the interval estimate?

13. Data were collected on the golf-ball driving distances by professional golfers in a recent tournament. Using data on the first drives of 30 randomly selected golfers, it was found that the sample mean distance was 250 yards and the sample standard deviation was 10 yards. Develop a 95% confidence interval for the mean driving distance for the population.

14. Researchers at the University of Illinois used a sample of 70 students from five high schools in a large metropolitan area to learn about performance of students on the American College Test (ACT) (*Journal of College Student Development*, February 1989). Of the students in the sample, 47 went on to college and 23 did not. The ACT scores for the sample of students who went on to college had a mean $\bar{x} = 25.23$ and a standard deviation $s = 3.21$.

a. Provide a 95% confidence interval for the population mean ACT score for students going on to college.

b. Provide a 98% confidence interval for the population mean ACT score for students going on to college.

8.2   Interval Estimation of a Population Mean: Small-Sample Case

The central limit theorem plays an important role in the development of the interval estimate shown in (8.2). Specifically, with a large sample ($n \geq 30$), the central limit theorem enables us to approximate the sampling distribution of $\bar{x}$ by a normal probability distribution. This normal distribution approximation is what makes the use of $z_{\alpha/2}$ appropriate in (8.2) when $n \geq 30$. However, what happens to the interval-estimation procedure when the sample size is small ($n < 30$) and we cannot use the central limit theorem to justify the use of a normal probability distribution to approximate the sampling distribution of $\bar{x}$?

In the small-sample case, the sampling distribution of $\bar{x}$ depends upon the distribution of the population. *If the population has a normal probability distribution*, the methodology presented in this section can be used to develop a confidence interval for a population mean. However, if the assumption of a normal probability distribution for the population is not appropriate, the only alternative is to increase the sample size to $n \geq 30$ and rely on the large-sample interval-estimation procedure given by (8.2).

If the population has a normal probability distribution, the sampling distribution of $\bar{x}$ will be normal regardless of the sample size. In this case, if the population standard deviation σ is *known*, (8.2) can still be used to compute an interval estimate of a population mean even with a small sample. However, if the population standard deviation σ is *unknown*, which is the usual situation, and the sample standard deviation s is used to estimate σ, the resulting confidence interval $\bar{x} \pm z_{\alpha/2} s/\sqrt{n}$ is only appropriate provided $n \geq 30$; in the small sample case, the appropriate confidence interval is based upon a probability distribution known as the *t distribution*.

The *t* distribution is actually a family of similar probability distributions, with a specific *t* distribution depending upon a parameter known as the *degrees of freedom*. That is, there is a unique *t* distribution with 1 degree of freedom, with 2 degrees of freedom, with 3 degrees of freedom, and so on. As the number of degrees of freedom increases, the difference between the *t* distribution and the standard normal probability distribution becomes smaller and smaller. Figure 8.6 shows *t* distributions with 10 and 20 degrees of

FIGURE **8.6**

Comparison of the Standard Normal Distribution with *t* Distributions Having 10 and 20 Degrees of Freedom

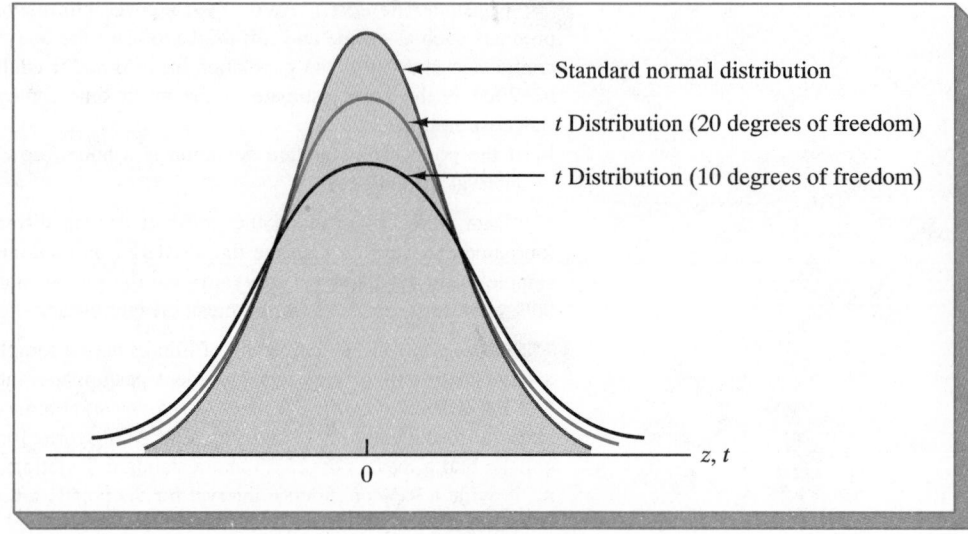

Standard normal distribution

t Distribution (20 degrees of freedom)

t Distribution (10 degrees of freedom)

freedom and their relationship to the standard normal probability distribution. Note that a *t* distribution with more degrees of freedom has less dispersion and more closely resembles the standard normal distribution. Note also that the mean of the distribution is zero.

We will use a subscript for *t* to indicate the area in the upper tail of the *t* distribution. For example, just as we used $z_{.025}$ to indicate the z value providing a .025 area in the upper tail of a standard normal probability distribution, we will use $t_{.025}$ to indicate a .025 area in the upper tail of the *t* distribution. In general, we will use the notation $t_{\alpha/2}$ to represent a *t* value with an area of $\alpha/2$ in the upper tail of the *t* distribution. See Figure 8.7.

A table for the *t* distribution is provided in Table 8.3. This table is also shown inside the back cover of the text. Note, for example, that for a *t* distribution with 10 degrees of freedom, $t_{.025} = 2.228$. Similarly, for a *t* distribution with 20 degrees of freedom, $t_{.025} = 2.086$. As the degrees of freedom continue to increase, $t_{.025}$ approaches $z_{.025}$.

FIGURE **8.7**

***t* Distribution with $\alpha/2$ Area or Probability in the Upper Tail**

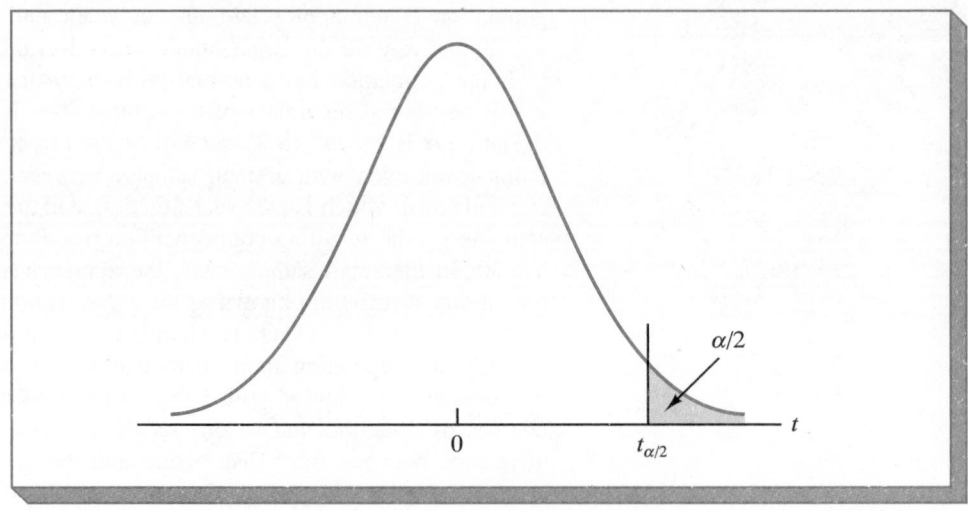

$\alpha/2$

0 $t_{\alpha/2}$ t

TABLE 8.3

t Distribution Table for Areas in the Upper Tail. Example: with 10 Degrees of Freedom $t_{.025} = 2.228$

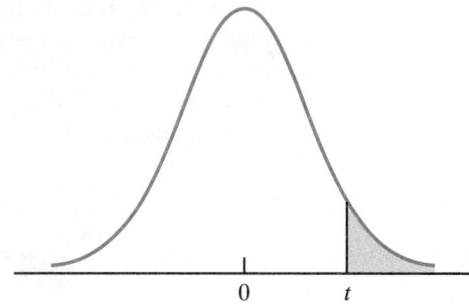

Degrees of Freedom	Upper-Tail Area (Shaded)				
	.10	.05	.025	.01	.005
1	3.078	6.314	12.706	31.821	63.657
2	1.886	2.920	4.303	6.965	9.925
3	1.638	2.353	3.182	4.541	5.841
4	1.533	2.132	2.776	3.747	4.604
5	1.476	2.015	2.571	3.365	4.032
6	1.440	1.943	2.447	3.143	3.707
7	1.415	1.895	2.365	2.998	3.499
8	1.397	1.860	2.306	2.896	3.355
9	1.383	1.833	2.262	2.821	3.250
10	1.372	1.812	2.228	2.764	3.169
11	1.363	1.796	2.201	2.718	3.106
12	1.356	1.782	2.179	2.681	3.055
13	1.350	1.771	2.160	2.650	3.012
14	1.345	1.761	2.145	2.624	2.977
15	1.341	1.753	2.131	2.602	2.947
16	1.337	1.746	2.120	2.583	2.921
17	1.333	1.740	2.110	2.567	2.898
18	1.330	1.734	2.101	2.552	2.878
19	1.328	1.729	2.093	2.539	2.861
20	1.325	1.725	2.086	2.528	2.845
21	1.323	1.721	2.080	2.518	2.831
22	1.321	1.717	2.074	2.508	2.819
23	1.319	1.714	2.069	2.500	2.807
24	1.318	1.711	2.064	2.492	2.797
25	1.316	1.708	2.060	2.485	2.787
26	1.315	1.706	2.056	2.479	2.779
27	1.314	1.703	2.052	2.473	2.771
28	1.313	1.701	2.048	2.467	2.763
29	1.311	1.699	2.045	2.462	2.756
30	1.310	1.697	2.042	2.457	2.750
40	1.303	1.684	2.021	2.423	2.704
60	1.296	1.671	2.000	2.390	2.660
120	1.289	1.658	1.980	2.358	2.617
∞	1.282	1.645	1.960	2.326	2.576

Now that we have an idea of what the t distribution is, let us show how it is used to develop an interval estimate of a population mean. Assume that the population has a normal probability distribution and that the sample standard deviation s is used as a point estimate of the population standard deviation σ. The following interval-estimation procedure is applicable.

Interval Estimate of a Population Mean
Small-Sample Case (n < 30)

$$\overline{x} \pm t_{\alpha/2} \frac{s}{\sqrt{n}} \qquad (8.3)$$

where $1 - \alpha$ is the confidence coefficient, $t_{\alpha/2}$ is the t value providing an area of $\alpha/2$ in the upper tail of a t distribution with $n - 1$ *degrees of freedom,* and s is the sample standard deviation. It is assumed that the population has a normal probability distribution.*

The reason the number of degrees of freedom associated with the t value in (8.3) is $n - 1$ has to do with the use of s as an estimate of the population standard deviation σ. The expression for the sample standard deviation is

$$s = \sqrt{\frac{\Sigma(x_i - \overline{x})^2}{n - 1}}$$

Degrees of freedom here refers to the number of independent pieces of information that go into the computation of $\Sigma(x_i - \overline{x})^2$. The pieces of information involved in computing $\Sigma(x_i - \overline{x})^2$ are $x_1 - \overline{x}, x_2 - \overline{x}, \ldots, x_n - \overline{x}$. In Section 3.2 we indicated that $\Sigma(x_i - \overline{x}) = 0$ for any data set. Thus, only $n - 1$ of the $x_i - \overline{x}$ values are independent; if we know $n - 1$ of the values, the remaining value can be determined exactly by using the condition that the $x_i - \overline{x}$ values must sum to 0. Thus, $n - 1$ is the number of degrees of freedom associated with $\Sigma(x_i - \overline{x})^2$ and hence the t distribution used in (8.3).

Let us demonstrate the use of the above interval estimate by considering the training program evaluation conducted by Scheer Industries. Scheer's director of manufacturing is interested in a computer-assisted training program that can be used to train the firm's maintenance employees for machine-repair operations. It is anticipated that the computer-assisted training will reduce training time and costs. To evaluate the training method, the director of manufacturing has requested an estimate of the mean training time required with the computer-assisted training program.

Suppose that management has agreed to train 15 employees with the new approach. The data on training days required for each employee in the sample are shown in Table 8.4. The sample mean and sample standard deviation for these data are as follows:

---

*The population standard deviation σ is usually unknown and is estimated by the sample standard deviation s. However, if σ is known and if the population has a normal probability distribution, the small-sample case would use z rather than t; $\overline{x} \pm z_{\alpha/2}(\sigma/\sqrt{n})$ could then be used to develop the interval estimate of the population mean.

$$\overline{x} = \frac{\Sigma x_i}{n} = \frac{808}{15} = 53.87 \text{ days}$$

$$s = \sqrt{\frac{\Sigma(x_i - \overline{x})^2}{n-1}} = \sqrt{\frac{651.73}{14}} = 6.82 \text{ days}$$

The point estimate of the mean training time for the population of employees is 53.87 days. We can obtain information about the precision of this estimate by developing an interval estimate of the population mean. Since the population standard deviation is unknown, we will use the sample standard deviation $s = 6.82$ days as the point estimate of σ. With the small-sample size, $n = 15$, we will use (8.3) to develop interval estimate of the population mean at 95% confidence. Assuming that the population of training times has a normal probability distribution, the t distribution with $n - 1 = 14$ degrees of freedom is the appropriate probability distribution for the interval-estimation procedure. We see from Table 8.3 that with 14 degrees of freedom, $t_{\alpha/2} = t_{.025} = 2.145$. Using (8.3), we have

$$\overline{x} \pm t_{.025}\frac{s}{\sqrt{n}}$$

$$53.87 \pm 2.145\left(\frac{6.82}{\sqrt{15}}\right)$$

$$53.87 \pm 3.78$$

Thus, the 95% confidence interval estimate of the population mean training time is 50.09 days to 57.65 days.

The above approach, which uses the t distribution to develop an interval estimate of μ, is applicable whenever the population standard deviation is unknown and the population being sampled has a normal probability distribution. However, statistical research has shown that (8.3) is applicable even if the population being sampled is not quite normal. That is, confidence intervals based on the t distribution give good results when the population distribution does not differ extensively from a normal probability distribution. The fact that the t distribution can give satisfactory results even when the population distribution is not normal is referred to as the *robustness* property of the t distribution.

Computer-Generated Confidence Intervals

Computer software packages are available for computing confidence intervals for population means. The Minitab output shown in Figure 8.8 illustrates how Minitab can be used to develop the confidence intervals for the two examples discussed previously in this chapter: the mean annual college room-and-board cost and the mean training time for Scheer Industries.

TABLE 8.4

Training Time in Days for the Computer-Assisted Training Program at Scheer Industries

Employee	Time	Employee	Time	Employee	Time
1	52	6	59	11	54
2	44	7	50	12	58
3	55	8	54	13	60
4	44	9	62	14	62
5	45	10	46	15	63

F I G U R E   **8.8**   **Minitab Confidence Intervals for the Mean Annual College Room-and-Board Cost and the Mean Training Time for Scheer Industries**

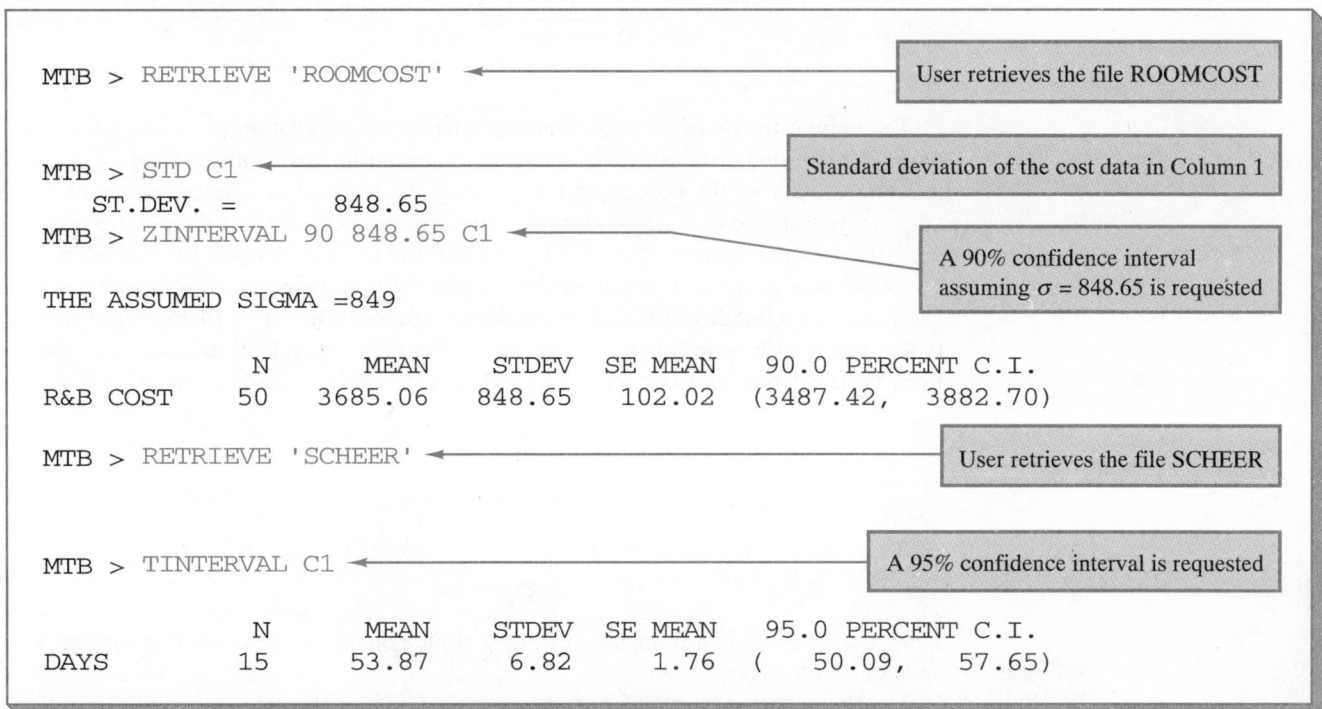

Minitab has two interval-estimation commands, ZINTERVAL and TINTERVAL. In Figure 8.8, we see that the user selects the Minitab RETRIEVE command to obtain the previously stored data on annual room-and-board costs for a sample of 50 colleges and universities. The STD C1 command is used to compute the sample standard deviation for the annual room-and-board cost which is shown to be $848.65. Since the sample size of 50 is considered large, the user continues the Minitab session by selecting the ZINTERVAL procedure to develop the confidence interval. The specific command ZINTERVAL 90 848.65 C1 indicates that a 90% confidence interval is desired with the population standard deviation estimated by the sample standard deviation of 848.65. The C1 notes that the cost data are stored in Column 1 of the Minitab worksheet. The printout shown in Figure 8.8 is self-explanatory. After providing the sample size, the sample mean of 3685, the sample standard deviation of 848.65, and the standard error of the mean of 120.02, the printout shows the 90% confidence interval for the population mean annual room-and-board cost to the nearest dollar is $3487 to $3883.

Referring to Figure 8.8, we see that the user continues by retrieving the previously stored Scheer Industries training time data which was presented in Table 8.4. With the small sample size of $n = 15$, the user selects the TINTERVAL command to indicate that the t distribution is to be used to develop the confidence interval. Whenever the ZINTERVAL or TINTERVAL command does not specify a confidence level, Minitab automatically provides a 95% confidence interval. Thus, since the Scheer Industries example specified a 95% confidence interval, the user does not have to include 95 in the TINTERVAL command. Finally, since the TINTERVAL command does not specify a value for the standard deviation, the sample standard deviation is automatically computed from the data. Thus, the printout as shown in Figure 8.8 indicates that the 95% confidence interval estimate of the population mean training time is 50.09 days to 57.65 days.

NOTES &
COMMENTS

We would like to point out that the t distribution is not restricted to the small-sample situation. Actually, the t distribution is applicable whenever the population is normal or near normal and whenever the sample standard deviation is used to estimate the population standard deviation. If these conditions exist, the t distribution can be used for any sample size. However, (8.2) shows that with a large sample ($n \geq 30$), interval estimation of a population mean can be based on the standard normal probability distribution and the value $z_{\alpha/2}$. Thus, with (8.2) available for the large-sample-size case, we generally do not consider the use of the t distribution until we encounter a small-sample-size case.

☐ ☐ **Exercises**

Methods

15. For a t distribution with 12 degrees of freedom, find the area, or probability, that lies in each region.
a. to the left of 1.782
b. to the right of -1.356
c. to the right of 2.681
d. to the left of -1.782
e. between -2.179 and $+2.179$
f. between -1.356 and $+1.782$

16. Find the t value for each of the following:
a. upper tail area of .05 with 18 degrees of freedom
b. lower tail area of .10 with 22 degrees of freedom
c. upper tail area of .01 with 5 degrees of freedom
d. 90% of the area is between these two t values with 14 degrees of freedom
e. 95% of the area is between these two t values with 28 degrees of freedom

SELF TEST ▷

17. The following data have been collected from a sample of eight items:

$$10, 8, 12, 15, 13, 11, 6, 5.$$

a. What is the point estimate of the population mean?
b. What is the point estimate of the population standard deviation?
c. What is the 95% confidence interval for the population mean?

18. A simple random sample of 20 items resulted in a sample mean of 17.25 and a sample standard deviation of 3.3.
a. Develop a 90% confidence interval for the population mean.
b. Develop a 95% confidence interval for the population mean.
c. Develop a 99% confidence interval for the population mean.

Applications

SELF TEST ▷

19. In the testing of a new production method, 18 employees were randomly selected and asked to try the new method. The sample mean production rate for the 18 employees was 80 parts per hour, and the sample standard deviation was 10 parts per hour. Provide 90% and 95% confidence intervals for the mean production rate for the new method assuming the population has a normal distribution.

20. The Money & Investing section of the *The Wall Street Journal* contains a summary of the daily investment performances for the New York Stock Exchange, the American Stock Exchange, overseas markets, options, commodities, futures, and so on. In the New York Stock Exchange section, information is provided on each stock's 52-week high price per share, 52-week low price per share, dividend rate, yield, P/E ratio, daily volume, daily high price per share, daily low price per share, closing price per share, and daily net change. The P/E ratio for each stock is determined by dividing the price of a share of stock by the earnings per share reported by the company for the most recent four quarters. A sample of 10 stocks taken from *The Wall Street Journal,* May 19, 1992, provided the following data on P/E ratios:

5, 7, 9, 10, 14, 23, 20, 15, 3, 26.

a. What is the point estimate of the mean P/E ratio for the population of all stocks listed on the New York Stock Exchange?
b. What is the point estimate of the standard deviation of the P/E ratios for the population of all stocks listed on the New York Stock Exchange?
c. Using a .95 confidence coefficient, what is the interval estimate of the mean P/E ratio for the population of all stocks listed on the New York Stock Exchange?
d. Comment on the precision of the results.

21. The following data are family sizes from a simple random sample of households in a new test market area:

Household	Family Size	Household	Family Size
1	4	7	3
2	3	8	2
3	2	9	3
4	2	10	6
5	4	11	3
6	5	12	2

Provide a 95% confidence interval for the mean family size for the population.

22. The American Association of Advertising Agencies records data on nonprogramming minutes per half hour of prime-time television programming (*U.S. News & World Report,* April 13, 1992). Representative data for a sample of prime-time programs on major networks at 8:30 P.M. are shown in Table 8.5. Provide a point estimate and a 95% confidence interval for the mean number of minutes of nonprogramming on half-hour prime-time television shows at 8:30 P.M.

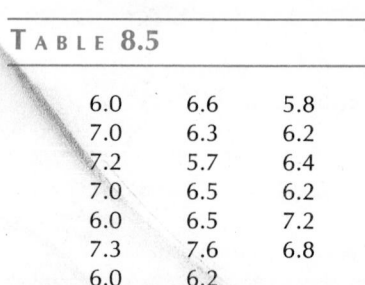

TABLE 8.5

6.0	6.6	5.8
7.0	6.3	6.2
7.2	5.7	6.4
7.0	6.5	6.2
6.0	6.5	7.2
7.3	7.6	6.8
6.0	6.2	

23. Hertz announced that it would impose a surcharge of $15 to $56 per day on residents of Queens, Brooklyn, and the Bronx who rent cars in the New York metropolitan area (*The New York Times,* January 3, 1992). Hertz, the nation's largest auto-rental company, was imposing the charge to recoup losses it had suffered under a New York state law which makes a car rental agency liable for damage caused by a renter. Typical daily rental costs without the surcharge are shown in the following sample of 12 rentals:

65, 58, 40, 48, 52, 60, 75, 38, 50, 51, 59, 40.

a. What is the point estimate of the mean daily car rental cost in the New York area?
b. What is a 95% confidence interval for the mean daily car rental cost in the New York area?

24. Sales personnel for Skillings Distributors are required to submit weekly reports listing the customer contacts made during the week. A sample of 61 weekly contact reports showed a mean of 22.4 customer contacts per week for the sales personnel. The sample standard deviation was 5 contacts.

a. Develop a 95% confidence interval for the mean number of weekly customer contacts for the population of sales personnel.
b. Assume that the population of weekly contact data has a normal distribution. Use the *t* distribution with 60 degrees of freedom to develop a 95% confidence interval for the mean number of weekly customer contacts.
c. Compare your answers for parts (a) and (b). Comment on why in the large-sample case it is permissible to base interval estimates on the procedure used in part (a) even though the *t* distribution may also be applicable.

25. Researchers at the University of Georgia used a sample of college students to study psychological issues associated with attending college (*Journal of College Student Development*, January 1989). One variable of interest was how the students progressed in terms of working on autonomy. The autonomy score was lowest in the fall quarter, better in the winter quarter, and highest in the spring quarter. A sample of 16 students for the fall quarter resulted in a sample mean of $\bar{x} = 51$ and sample standard deviation of $s = 10.18$.

  a. Compute the 95% confidence interval for the population mean autonomy score for the fall quarter.

  b. What assumption about the population was necessary in order to obtain an answer to part (a)?

  c. Assume that it was desirable to estimate the population mean autonomy score with a sampling error of ±3 points. Does the statistical data provide this desired level of precision? What action, if any, would you recommend be taken?

26. The duration (in minutes) for a sample of 20 flight-reservation telephone calls is shown in Table 8.6.

  a. What is the point estimate of the population mean time for flight-reservation phone calls?

  b. Assuming that the population has a normal distribution, develop a 95% confidence interval for the population mean time.

TABLE 8.6

2.1	4.8	5.5
10.4	3.3	3.5
4.8	5.8	5.3
5.5	2.8	3.6
5.9	6.6	7.8
10.5	7.5	6.0
4.5	4.8	

8.3   Determining the Sample Size

In Section 8.1 we were able to make the following probability statement about the sampling error whenever a sample mean was used to provide a point estimate of a population mean:

> There is a $1 - \alpha$ probability that the value of the sample mean will provide a error of $z_{\alpha/2}\sigma_{\bar{x}}$ or less.

Since $\sigma_{\bar{x}} = \sigma/\sqrt{n}$, we can rewrite this statement as follows:

> There is a $1 - \alpha$ probability that the value of the sample mean will provide a sampling error of $z_{\alpha/2}(\sigma/\sqrt{n})$ or less.

From this statement we see that the values of $z_{\alpha/2}$, σ, and the sample size n combine to determine the sampling error mentioned in the precision statement. Once we select a confidence coefficient or probability of $1 - \alpha$, $z_{\alpha/2}$ can be determined. Given values for $z_{\alpha/2}$ and σ, we can determine the sample size n needed to provide any sampling error. The formula used to compute the required sample size n is developed as follows.

Let E = the sampling error mentioned in the statement about the desired precision. We have

$$E = z_{\alpha/2}\frac{\sigma}{\sqrt{n}} \tag{8.4}$$

Using (8.4) to solve for $\sqrt{n}$, we have

$$\sqrt{n} = \frac{z_{\alpha/2}\sigma}{E}$$

Squaring both sides of this equation, we obtain the following equation for the sample size.

Sample Size for an Interval Estimate of a Population Mean

$$n = \frac{(z_{\alpha/2})^2\sigma^2}{E^2} \tag{8.5}$$

This sample size will provide a precision statement with a $1 - \alpha$ probability that the sampling error will be *E or less*.

In (8.5) the value *E* is the maximum sampling error that the user is willing to accept, and the value of $z_{\alpha/2}$ follows directly from the confidence level to be used in developing the interval estimate. Although user preference must be considered, 95% confidence is the most frequently chosen value ($z_{.025} = 1.96$).

Finally, use of (8.5) requires a value for the population standard deviation σ. In most cases, σ will be unknown. However, to be able to use (8.5), we must have a preliminary or *planning value* for σ. In practice, one of the following procedures can be used.

1. Use the sample standard deviation from a previous sample of the same or similar units.
2. Use a pilot study to select a preliminary sample of units. The sample standard deviation from the preliminary sample can be used as the planning value for σ.
3. Use judgment or a ''best guess'' for the value of σ. For example, we might begin by estimating the largest and smallest data values in the population. The difference between the largest and smallest values provides an estimate of the range for the data. Finally, the range divided by 4 is often suggested as a rough approximation of the standard deviation and thus, an acceptable planning value for σ.

Let us return to the Scheer Industries example in Section 8.2 to see how (8.5) can be used to determine the sample size for the study. Previously we showed that with a 95% level of confidence, a sample of 15 Scheer employees generated a population mean training time estimate of 53.87 ± 3.78 days. Assume that after viewing these results, Scheer's director of manufacturing is not satisfied with this degree of precision, feeling that a sampling error of ± 3.78 days is too large. Furthermore, suppose that the director makes the following statement about the desired precision: ''I would like a .95 probability that the value of the sample mean will provide a sampling error of 2 days or less.'' We can see that the director is specifying a maximum sampling error of $E = 2$ days. In addition, the .95 probability indicates that a 95% confidence level is to be used; thus, $z_{\alpha/2} = z_{.025} = 1.96$. Lastly, we need a planning value for σ to use (8.5) to determine the sample size. Do we have a planning value for σ in the Scheer Industries example? Although σ is unknown, let us take advantage of the data provided for the 15 employees in Section 8.2. We can view these employees as being in a pilot study with the sample standard deviation $s = 6.82$ days providing the planning value for σ. Thus, using (8.5), we have

$$n = \frac{(z_{\alpha/2})^2 \sigma^2}{E^2} = \frac{(1.96)^2 (6.82)^2}{2^2} = 44.67$$

In cases where the computed *n* is a fraction, we round up to the next integer value; thus, the recommended sample size for the Scheer Industries example is 45 employees. Since Scheer already has test data for 15 employees, an additional $45 - 15 = 30$ employees should be tested if the director wishes to obtain the desired precision of ± 2 days at a 95% confidence level.

Finally, note that in the Scheer Industries example, $z_{.025}$ was used to determine the sample size even though the original computations for 15 employees had employed the *t* distribution. The reason for the use of $z_{.025}$ is that since the sample size is yet to be determined, we are anticipating that *n* will be larger than 30, making $z_{.025}$ the appropriate value. In addition, if *n* is yet to be determined, we do not know the $(n - 1)$ degrees of freedom necessary to use the *t* distribution. Thus, the use of (8.5) to determine the sample size will always be based on a *z* value rather than a *t* value.

☐ ☐ **Exercises**

Methods

27. How large of a sample should be selected to be 95% confident that the sampling error is 5 or less? Assume that the population standard deviation is 25.

SELF TEST ▶ **28.** The range for a set of data is estimated to be 36.
a. What is the planning value for the population standard deviation?
b. How large a sample should be taken to be 95% confident that the sampling error is 3 or less?
c. How large a sample should be taken to be 95% confident that the sampling error is 2 or less?

Applications

SELF TEST ▶ **29.** What sample size would have been recommended for the Scheer Industries example if the director of manufacturing had specified a .95 probability for a sampling error of 1.5 days or less? How large a sample would have been necessary if the precision statement had specified a .90 probability for a sampling error of 2 days or less?

30. In Section 8.1 the Statewide Insurance Company used a simple random sample of 36 policyholders to estimate the mean age of the population of policyholders. The resulting precision statement was reported to have a .95 probability that the value of the sample mean provided a sampling error of 2.35 years or less. This statement was based on a known population standard deviation of 7.2 years.
a. How large a simple random sample would have been necessary to reduce the sampling error to 2 years or less? To 1.5 years or less? To 1 year or less?
b. Would you recommend that Statewide attempt to estimate the mean age of the policyholders with $E = 1$ year? Explain.

31. Starting annual salaries for college graduates with business administration degrees are believed to have a standard deviation of approximately $2000. Assume that a 95% confidence interval estimate of the mean annual starting salary is desired. How large a sample size should be taken if the size of the sampling error in the precision statement is
a. $500   **b.** $200   **c.** $100

32. The mean number of days a house is on the market prior to selling was reported for 100 different cities (*U.S. News & World Report,* April 6, 1992). In a particular city the standard deviation of the number of days a house is on the market prior to selling is 20. How many house sales records would have to be collected to estimate the population mean to within ±2 days? Use a 95% level of confidence. $\sigma = 20$ $\alpha = .05$ $z_{.025} = 1.96$

33. Refer to Exercise 8, which provided the mean annual income for U.S. factory workers (*Barron's,* April 10, 1989). The population standard deviation of annual income is $\sigma = \$5000$. If we wanted to estimate the mean annual income for the population of U.S. factory workers with a $500 margin of error, what sample size should be used? Assume 95% confidence.

34. From Exercise 20, the sample standard deviation of P/E ratios for stocks listed on the New York Stock Exchange is $s = 7.8$. (*The Wall Street Journal,* May 19, 1992). Assume that we are interested in estimating the population mean P/E ratio for all stocks listed on the New York Stock Exchange. How many stocks should be included in the sample if we would like a .95 probability that the sampling error is 2 or less?

35. A gasoline service station shows a standard deviation of $6.25 for the charges made by the credit-card customers. Assume that the station's management would like to estimate the population mean gasoline bill for its credit-card customers to within ±$1.00. For a 95% confidence level, how large a sample would be necessary?

36. A national survey research firm has past data that indicate that the interview time for a consumer opinion study has a standard deviation of 6 minutes.
a. How large a sample should be taken if the firm desires a .98 probability of estimating the mean interview time to within 2 minutes or less?

b. Assume that the simple random sample you recommended in part (a) is taken and that the mean interview time for the sample is 32 minutes. What is the 98% confidence interval estimate for the mean interview time for the population of interviews?

8.4   Interval Estimation of a Population Proportion

In Section 8.2 we presented the Scheer Industries example, which involved estimating the mean employee training time for a new machine-repair training program. To evaluate the program from a different perspective, management has requested that some measure of program quality be developed. The degree of success of the training program has previously been measured by the score the employee obtains on a standard examination given at the end of the training program. From previous experience, the company has found that an individual passing the examination has an excellent chance of high performance on the job. After some discussion, management has agreed to evaluate the program quality for the new training method based on the proportion of employees that pass the examination. Let us assume that Scheer implemented the sample-size recommendation of the preceding section. Thus, we now have a sample of 45 employees which can be used to develop an interval estimate for the proportion of the population that pass the examination.

In Chapter 7 we learned that a sample proportion $\bar{p}$ is an unbiased estimator of a population proportion p and that the large-sample approximation of the sampling distribution of $\bar{p}$ is normal as shown in Figure 8.9. Recall that the use of the normal distribution as an approximation of the sampling distribution of $\bar{p}$ is based on the condition that both np and $n(1 - p)$ are 5 or more. We will be using our knowledge of the sampling distribution of $\bar{p}$ to make probability statements about the sampling error whenever a sample proportion $\bar{p}$ is used to estimate a population proportion p. In this case, the sampling error is defined as the absolute value of the difference between $\bar{p}$ and p, written $|\bar{p} - p|$.

The probability statements that we can make about the sampling error for the proportion take the following form:

> There is a $1 - \alpha$ probability that the value of the sample proportion will provide a sampling error of $z_{\alpha/2}\sigma_{\bar{p}}$ or less.

FIGURE 8.9

Normal Approximation of the Sampling Distribution of $\bar{p}$ when Both $np \geq 5$ and $n(1 - p) \geq 5$

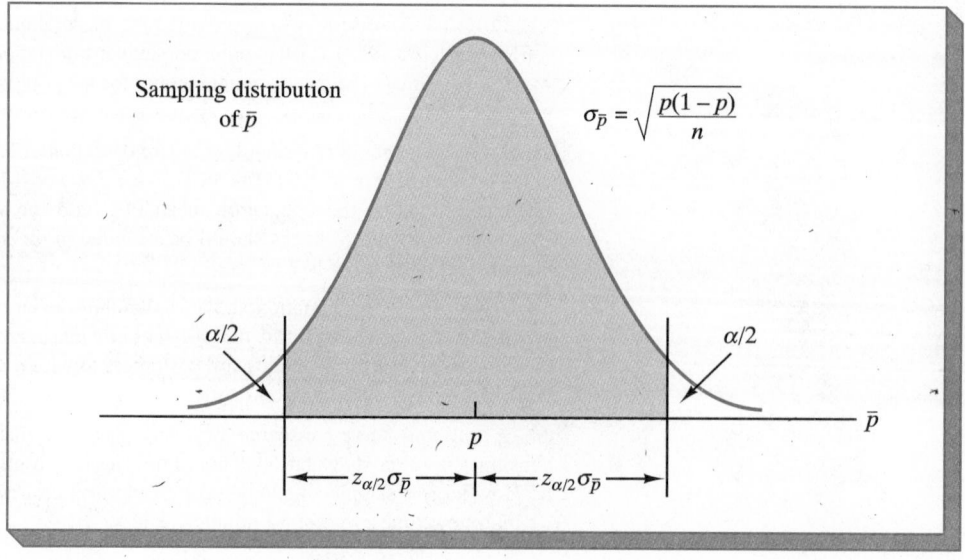

Sampling distribution of $\bar{p}$

$$\sigma_{\bar{p}} = \sqrt{\frac{p(1-p)}{n}}$$

$\alpha/2$ $\alpha/2$

p $\bar{p}$

$z_{\alpha/2}\sigma_{\bar{p}}$ $z_{\alpha/2}\sigma_{\bar{p}}$

The rationale for the preceding statement is the same as we used when the value of a sample mean was used as an estimate of a population mean. Namely, since we know that the sampling distribution of $\bar{p}$ can be approximated by a normal probability distribution, we can use the value of $z_{\alpha/2}$ and the value of the standard error of the proportion $\sigma_{\bar{p}}$ to make the probability statement about the sampling error.

Once we see that the probability statement concerning the sampling error is based on $z_{\alpha/2}\sigma_{\bar{p}}$, we can subtract and add this value to $\bar{p}$ to obtain an interval estimate of the population proportion. Such an interval estimate is given by

$$\bar{p} \pm z_{\alpha/2}\sigma_{\bar{p}} \tag{8.6}$$

where $1 - \alpha$ is the confidence coefficient. Since $\sigma_{\bar{p}} = \sqrt{p(1 - p)/n}$, we can rewrite (8.6) as

$$\bar{p} \pm z_{\alpha/2}\sqrt{\frac{p(1 - p)}{n}} \tag{8.7}$$

However, in order to use (8.7) to develop an interval estimate of a population proportion p, the value of p would have to be *known*. Since the value of p is *unknown*, we simply substitute the sample proportion $\bar{p}$ for p. As a result, the general expression for a confidence interval estimate of a population proportion is as follows.*

Interval Estimate of a Population Proportion

$$\bar{p} \pm z_{\alpha/2}\sqrt{\frac{\bar{p}(1 - \bar{p})}{n}} \tag{8.8}$$

where $1 - \alpha$ is the confidence coefficient and $z_{\alpha/2}$ is the z value providing an area of $\alpha/2$ in the upper tail of the standard normal probability distribution.

Let us return to the Scheer Industries example. Assume that in the sample of 45 employees who completed the new training program, .36 scored passed the examination. Thus, the point estimate of the proportion in the population that pass the examination is $\bar{p} = 36/45 = .80$. Using (8.8) and a .95 confidence coefficient, the interval estimate for the population proportion is given by

$$\bar{p} \pm z_{.025}\sqrt{\frac{\bar{p}(1 - \bar{p})}{n}}$$

$$.80 \pm 1.96\sqrt{\frac{.80(1 - .80)}{45}}$$

$$.80 \pm .12$$

Thus, at the 95% confidence level, the interval estimate of the population proportion is .68 to .92.

---

*An unbiased estimate of $\sigma_{\bar{p}}^2$ is $\bar{p}(1 - \bar{p})/(n - 1)$ which suggests $\sqrt{\bar{p}(1 - \bar{p})/(n - 1)}$ should be used in place of $\sqrt{\bar{p}(1 - \bar{p})/n}$ in (8.8). However, the bias introduced by using n in the denominator does not cause any difficulty because large samples are generally used in making estimates concerning population proportions and in such cases the numerical difference between the results using n and $n - 1$ is negligible.

Determining the Sample Size

Let us consider the question of how large the sample size should be to obtain an estimate of a population proportion at a specified level of precision. The rationale for the sample-size determination in developing interval estimates of p is very similar to the rationale used in Section 8.3 to determine the sample size for estimating a population mean.

Earlier in this section we provided the following probability statement about the sampling error:

> There is a $1 - \alpha$ probability that the value of the sample proportion will provide a sampling error of $z_{\alpha/2}\sigma_{\bar{p}}$ or less.

With $\sigma_{\bar{p}} = \sqrt{p(1 - p)/n}$, the sampling error in this statement is based on the values of $z_{\alpha/2}$, the population proportion p, and the sample size n. For a given confidence coefficient $1 - \alpha$, $z_{\alpha/2}$ can be determined. Then, since the value of the population proportion is fixed, the sampling error mentioned in the precision statement is determined by the sample size n. Larger sample sizes again provide better precision.

Let E = the maximum sampling error that the user is willing to accept; thus

$$E = z_{\alpha/2} \sqrt{\frac{p(1 - p)}{n}}$$

Solving the preceding equation for n provides the following formula for the required sample size.

Sample Size for an Interval Estimate of a Population Proportion

$$n = \frac{(z_{\alpha/2})^2 p(1 - p)}{E^2} \qquad (8.9)$$

In (8.9), the value of the sampling error E must be specified by the user; however, in most cases, E is .10 or less. User preference also specifies the confidence level and thus the corresponding value of $z_{\alpha/2}$. Finally, use of (8.9) requires a planning value for the population proportion p. In practice, one of the following procedures can be used to select this planning value.

1. Use the sample proportion from a previous sample of the same or similar units.
2. Use a pilot study to select a preliminary sample of units. The sample proportion from this sample can be used as the planning value for p.
3. Use judgment or a "best guess" for the value of p.
4. If the none of the above alternatives apply, use $p = .50$.

Let us return to the Scheer Industries example where we were interested in estimating the proportion of employees that pass the training program examination. How large a sample of employees should be used if Scheer's director of manufacturing would like to estimate the population proportion with a sampling error of .10 or less at a 95% confidence level? With $E = .10$ and $z_{.025} = 1.96$, we need a planning value for p to answer the sample-size question. Earlier in this section we reported that 36 of the 45 employees who took the examination passed. Thus, $\bar{p} = 36/45 = .80$ can be used as the planning value for p. Using (8.9), we obtain

$$n = \frac{(1.96)^2 .80(1 - .80)}{(.10)^2} = 61.47$$

Thus, a sample size of 62 employees is recommended. Since 45 employees are in the current sample, Scheer would need an additional $62 - 45 = 17$ employees to meet the precision requirement of $\pm .10$ at a 95% confidence level.

Before leaving this section, let us comment on the fourth alternative suggested for selecting a planning value for p—namely, using $p = .50$. This value of p is frequently used when there is no other information available. To understand why, note that the numerator of (8.9) shows the sample size is proportional to the quantity $p(1 - p)$. A larger value for the quantity $p(1 - p)$ will result in a larger sample size. Table 8.7 shows some possible values of $p(1 - p)$. Note that the largest value of $p(1 - p)$ occurs when $p = .50$. Thus, if there is uncertainty regarding an appropriate planning value for p, we know that $p = .50$ will provide the largest sample-size recommendation. In effect, we are being on the safe or conservative side in recommending the largest possible sample size. If the proportion turns out to be different than the .50 planning value, the precision statement will be better than anticipated. In any case, in using $p = .50$, we are guaranteeing that the sample size will be sufficient to obtain the desired level of precision.

In the Scheer Industries example, a planning value of $p = .50$ would have provided the following recommended sample size:

$$n = \frac{(1.96)^2 .50(1 - .50)}{(.10)^2} = 96$$

The larger recommended sample size reflects the caution inherent in using the conservative planning value for the population proportion.

NOTES & COMMENTS

> The desired sampling error or margin of error for estimating a population proportion is almost always .10 or less. In national public opinion polls conducted by organizations such as Gallup and Harris, a .03 or .04 margin of error is generally reported. The use of these margins of error (E in Equation 8.9) will generally provide a sample size that is large enough to satisfy the central limit theorem requirements of $np \geq 5$ and $n(1 - p) \geq 5$.

TABLE 8.7
Some Possible Values for $p(1 - p)$

p	$p(1 - p)$	
.10	$(.10)(.90) = .09$	
.30	$(.30)(.70) = .21$	
.40	$(.40)(.60) = .24$	
.50	$(.50)(.50) = .25$	← Largest value for $p(1 - p)$
.60	$(.60)(.40) = .24$	
.70	$(.70)(.30) = .21$	
.90	$(.90)(.10) = .09$	

☐ ☐ Exercises

Methods

SELF TEST ▶

37. A simple random sample of 400 items provides 100 Yes responses.

a. What is the point estimate of the proportion of the population who would provide Yes responses?
b. What is the standard error of the proportion?
c. Compute the 95% confidence interval for the population proportion.

38. A simple random sample of 800 units generates a sample proportion $\bar{p} = .70$.

a. Provide a 90% confidence interval for the population proportion.
b. Provide a 95% confidence interval for the population proportion.

39. In a survey, the planning value for the population proportion p is given as .35. How large a sample should be taken to be 95% confident that the sample proportion is within $\pm.05$ of the population proportion?

40. How large a sample should be taken to be 95% confident that the sampling error for the estimation of a population proportion is .03 or less? Assume past data is not available for developing a planning value for p.

Applications

SELF TEST ▶

41. A Louis Harris survey of 400 senior executives found that 248 of the executives stated that the U.S. legal system significantly hampers the ability of U.S. companies to compete with Japanese and European companies (*Business Week,* April 13, 1992).

a. What is the point estimate of the population proportion of executives who believe the legal system hampers the ability to compete?
b. What is the 90% confidence interval for the population proportion?

42. What is the public opinion toward the U.S. Supreme Court ruling on abortion that stated that women may end a pregnancy during the first 3 months? A survey of 1227 adults in 12 southern states found that only 429 adults favored the Supreme Court ruling (*Atlanta Constitution,* April 13, 1989). What is the 95% confidence interval for the proportion of adults in the 12 southern states favoring the Supreme Court ruling? What is your interpretation of this interval estimate?

43. A *USA Today* poll (January 11, 1990) reported that 79% of Americans say that, if they had evidence, they would turn in a relative who killed someone. Experts were not surprised by the poll results, saying the poll reflects the ''socially desirable response'' rather than real-life action. The telephone poll of 305 adults was conducted by the Gordon S. Black Corporation. What is the 95% confidence interval for the population proportion?

44. Medical researchers at Cornell University studied the effect of tight neckties on the flow of blood to the head and the possible decrease in the brain's ability to respond to visual information (*Medical Self Care,* July–August 1988). Results of a sample of businessmen found that 67% wear their ties too tight. Assuming a sample size of 250 businessmen, what is the 98% confidence interval estimate of the proportion of the population of businessmen who wear their ties too tight?

45. In an election campaign, a campaign manager requests that a sample of voters be polled to determine public support for the candidate. In a sample of 120 voters, 64 expressed plans to support the candidate.

a. What is the point estimate of the proportion of voters in the population who will support the candidate?
b. Develop and interpret the 95% confidence interval for the proportion of voters in the population who will support the candidate.
c. Given the result from part (b), is the campaign manager justified in feeling confident that the candidate has the support of at least 50% of the voters? Explain.

d. How many voters should be sampled if we want to estimate the population proportion with a sampling error of 5% or less? Continue to use the 95% confidence level.

46. It is estimated that 29% of all minimum-wage workers are teenagers (*Boston Globe*, April 12, 1989). Using a 95% confidence level, provide an interval estimate of the proportion of the population of minimum-wage workers who are teenagers if the estimate is based on

a. a simple random sample of 200
b. a simple random sample of 600
c. a simple random sample of 1000
d. In general, what happens to the interval estimate of a population proportion as the sample size is increased?

47. The Tourism Institute for the State of Florida plans to sample visitors at major beaches throughout the state in order to estimate the proportion of beach visitors who are not residents of Florida. Preliminary estimates are that 55% of the beach visitors are not residents of Florida.

a. How large a sample should be taken to estimate the proportion of out-of-state visitors to within ±3% of the actual value? Use a 95% confidence level.
b. How large a sample should be taken if the error is increased to ±6%?

48. Where do people place their investment dollars? A sample conducted by *The Wall Street Journal* showed that 47% of all investors own some real estate (*The Wall Street Journal*, December 2, 1988).

a. Compute a 95% confidence interval for the proportion of the population of investors who own some real estate. Assume that a sample size of 250 investors was used in the preceding study.
b. How large a sample of investors should be used if a 95% confidence level that the sampling error is 5% or less is desired?

49. A firm provides national survey and interview services designed to estimate the proportion of the population who have certain beliefs or preferences. Typical questions seek to find the proportion favoring gun control, abortion, a particular political candidate, and so on. Assume that all interval estimates of population proportions are conducted at the 95% confidence level. How large a sample size would you recommend if the firm desired the sampling error to be

a. 3% or less?   **b.** 2% or less?   **c.** 1% or less?

50. A Gallup poll asked the following question: "Would you favor or oppose federal legislation banning the manufacture, sale and possession of semiautomatic assault guns, such as the AK-47?" (*The Wall Street Journal*, April 7, 1989). If it is believed approximately 75% of the population would favor such legislation, how many individuals should be sampled for each of the following margins of error? Use a 95% confidence level.

a. 10% margin of error
b. 7.5% margin of error
c. 5% margin of error
d. 3% margin of error
e. In general, what happens to the sample size as the margin of error decreases?

51. The National Automobile Dealers Association collects data on sales numbers, prices, and usage of both new and used automobiles. One statistic of interest is the percentage of automobiles that are still on the road after 10 years (*U.S. News & World Report*, September 9, 1991).

a. How large a sample should be taken if we want to be 95% confident that the sample percentage is within ±2.5% of the actual percentage of automobiles that are still on the road after 10 years? Use $p = .25$ as a planning value for the population proportion.
b. Using your sample size in part (a), assume that 357 of the automobiles sampled in 1991 were still on the road after 10 years. Provide the point estimate and a 95% confidence interval for the population proportion.
c. In 1980, only 21% of automobiles were still on the road after 10 years. What conclusion can you make after viewing the confidence interval results for 1991 in part (b)?

Summary

In this chapter we presented methods for developing a confidence interval for a population mean μ and a population proportion p. The purpose of developing a confidence interval is to provide the user with a better understanding of the sampling error that may be present. A wide confidence interval indicates poor precision in that a large sampling error may be present; in such cases, the sample size can be increased to reduce the width of the interval and improve the precision of the estimate.

Figure 8.10 summarizes the interval-estimation procedures for a population mean and provides a practical guide for computing the interval estimate. The figure shows that the expression used to compute an interval estimate depends upon whether the sample size is large ($n \geq 30$) or small ($n < 30$), whether the population standard deviation is known, and in some cases, whether or not

FIGURE 8.10   Summary of Interval Estimation Procedures for a Population Mean

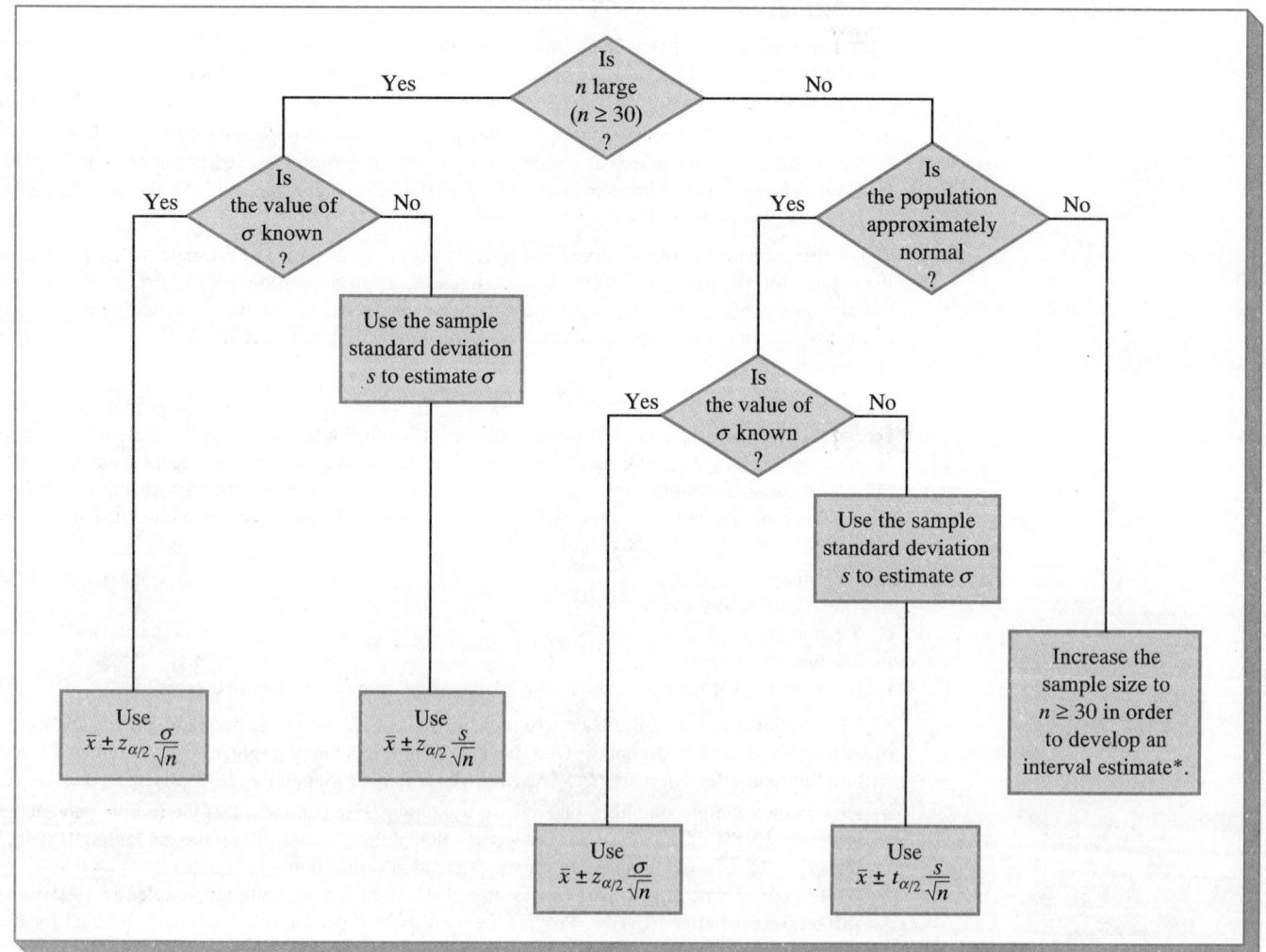

*In some cases methods from nonparametric statistics may be used to develop confidence intervals for location parameters of a population. However, these procedures are beyond the scope of this text, so increasing the sample size to $n \geq 30$ is recommended here.

the population has a normal or approximately normal probability distribution. If the sample size is large, no assumption is required about the distribution of the population and $z_{\alpha/2}$ is used in the computation of the interval estimate. If the sample size is small, the population must have a normal or approximately normal probability distribution to compute an interval estimate of μ. If this is the case, $z_{\alpha/2}$ is used in the computation of the interval estimate when σ is known, while $t_{\alpha/2}$ is used when σ is estimated by the sample standard deviation s. Finally, if the sample size is small and the assumption of a normally distributed population is inappropriate, we recommend increasing the sample size to $n \geq 30$ to develop an interval estimate of the population mean.

In addition, we showed how to determine the sample size so that interval estimates of μ and p would possess a specified level of precision. In practice, the sample sizes required for interval estimates of a population proportion are generally large. Thus, we provided the large-sample interval-estimation formulas for a population proportion where both $np \geq 5$ and $n(1 - p) \geq 5$.

Glossary

Interval estimate An estimate of a population parameter that provides an interval believed to contain the value of the parameter.

Sampling error The absolute value of the difference between the value of an unbiased point estimator, such as the sample mean $\bar{x}$, and the value of the population parameter it estimates, such as the population mean μ; in this case the sampling error is $|\bar{x} - \mu|$. In the case of the population proportion, the sampling error is $|\bar{p} - p|$.

Precision A probability statement about the sampling error.

Confidence level The confidence associated with an interval estimate. For example, if an interval-estimation procedure provides intervals such that 95% of the intervals developed will include the population parameter, an interval estimate is said to be constructed at the 95% confidence level; note that .95 is referred to as the *confidence coefficient*.

_t_ Distribution A family of probability distributions which can be used to develop interval estimates of a population mean whenever the population standard deviation is unknown and the population has a normal or near-normal probability distribution.

Degrees of freedom A parameter of the t distribution. When the t distribution is used in the computation of an interval estimate of a population mean, the appropriate t distribution has $n - 1$ degrees of freedom, where n is the size of the simple random sample.

Key Formulas

Sampling Error when Estimating μ

$$|\bar{x} - \mu| \tag{8.1}$$

Interval Estimate of a Population Mean (Large-Sample Case)

$$\bar{x} \pm z_{\alpha/2} \frac{\sigma}{\sqrt{n}} \tag{8.2}$$

Interval Estimate of a Population Mean (Small-Sample Case)

$$\bar{x} \pm t_{\alpha/2} \frac{s}{\sqrt{n}} \tag{8.3}$$

Sample Size for an Interval Estimate of a Population Mean

$$n = \frac{(z_{\alpha/2})^2 \sigma^2}{E^2} \tag{8.5}$$

Interval Estimate of a Population Proportion

$$\bar{p} \pm z_{\alpha/2} \sqrt{\frac{\bar{p}(1 - \bar{p})}{n}} \tag{8.8}$$

Sample Size for an Interval Estimate of a Population Proportion

$$n = \frac{(z_{\alpha/2})^2 p(1 - p)}{E^2} \tag{8.9}$$

❑ ❑ Supplementary Exercises

52. In the United States, 99% of the households have at least one television set, and 98% of Americans watch some television every day (*In Health,* January 1992). Part of the study indicated a mean of 2.25 television sets per household. Assume that the mean number of sets per households was based on a sample of 300 households and that the sample standard deviation was 1.2 television sets. Provide a 95% confidence interval estimate of the population mean number of television sets per household.

53. The North Carolina Savings and Loan Association would like to develop an estimate of the mean size of home improvement loans granted by its member institutions. A sample of 100 loans granted by member institutions resulted in a sample mean of $3400 and a sample standard deviation of $650. With these data develop a 98% confidence interval for the population mean dollar amount of home improvement loans.

54. A sample of 1033 recreational fishermen was used in a study reported in the *Journal of Marketing Research,* November 1987. Data on temperature were collected to study the relationship between temperature and a fisherman's decision to go fishing. The sample mean temperature was 55.6° and the sample standard deviation was 7.37°. Construct a 95% confidence interval for the population mean temperature for the recreational fishermen considered in this study.

55. Dailey Paints, Inc., implemented a long-term test study designed to check the wear resistance of its major brand of paint. The test consisted of painting eight houses in various parts of the United States and observing the number of months until signs of peeling were observed. The following data were obtained from a normal population:

House	1	2	3	4	5	6	7	8
Months until signs of peeling	60	51	64	45	48	62	54	56

a. What is a point estimate of the mean number of months until signs of peeling are observed?
b. Develop a 95% confidence interval to estimate the population mean number of months until signs of peeling are observed.
c. Develop a 99% confidence interval for the population mean.

56. What is the mean annual salary/bonus paid to the chief executive officers of the largest firms in the United States? A random sample of seven firms provided the annual salary/bonus data (*Business Week,* May 1, 1989) shown in Table 8.8.

a. What is the point estimate of the population mean annual salary/bonus for chief executives?
b. What is the point estimate of the population standard deviation for annual salaries?

TABLE 8.8

Firm	Annual Salary/ Bonus ($000s)
Bank of Boston	1200
Citicorp	1798
DuPont	1611
Abbott Laboratories	1920
Teledyne	860
Emerson Electric	1681
Conagra	1312

c. What is the 95% confidence interval estimate of the population mean annual salary/bonus for chief executives?

57. The Atlantic Fishing and Tackle Company has developed a new synthetic fishing line. To estimate the breaking strength of this line (pounds), testers subjected six lengths of line to breakage testing. The following data were obtained from a normal population.

Line	1	2	3	4	5	6
Breaking Strength (pounds)	18	24	19	21	20	18

Develop a 95% confidence interval for the mean breaking strength of the new line.

58. Sample assembly times for a particular manufactured part were 8, 10, 10, 12, 15, and 17 minutes. If the mean of the sample is used to estimate the mean of the population of assembly times, provide a point estimate and a 90% confidence interval for the population mean. Assume that the population has a normal distribution.

59. A utility company finds that a sample of 100 delinquent accounts yields an average amount owed of $131.44, with a sample standard deviation of $16.19. Develop a 90% confidence interval for the population mean amount owed.

60. In Exercise 59 the utility company sampled 100 delinquent accounts to estimate the mean amount owed by these accounts. The sample standard deviation was $16.19. How large a sample should be taken if the company wants to be 90% confident that the estimate of the population mean will have a sampling error of $1 or less?

61. Consider the Atlantic Fishing and Tackle Company problem presented in Exercise 57. How large a sample would be necessary to estimate the mean breaking strength of the new line with a .99 probability of a sampling error of 1 pound or less?

62. Mileage tests are conducted for a particular model of automobile. If the desired precision is stated such that there is a .98 probability of a sampling error of 1 mile per gallon or less, how many automobiles should be used in the test? Assume that preliminary mileage tests indicate the standard deviation for the automobiles to be 2.6 miles per gallon.

63. In developing patient appointment schedules, a medical center desires an estimate of the mean time that a staff member spends with each patient. How large a sample should be taken if the precision of the estimate is to be ± 2 minutes at a 95% level of confidence? How large a sample should be taken for a 99% level of confidence? Use a planning value for the population standard deviation of 8 minutes.

64. Exercise 56 provided annual salary/bonus data for chief executives of firms in the United States (*Business Week,* May 1, 1989). The sample standard deviation was $375, with data provided in thousands of dollars. How many chief executives should be in the sample if we would like to estimate the population mean annual salary/bonus with a margin of error of $100,000? Note that the margin of error should be stated as $100 since the data were provided in thousands of dollars. Use a 95% confidence level.

65. The New Orleans Beverage Company has been experiencing problems with the automatic machine that places labels on bottles. The company desires an estimate of the percentage of bottles that have improperly applied labels. A simple random sample of 400 bottles resulted in 18 bottles with improperly applied labels. Using these data, develop a 90% confidence interval for the population proportion of bottles with improperly applied labels.

66. H. G. Forester and Company is a distributor of lumber supplies throughout the southwest United States. Management at H. G. Forester would like to check a shipment of over 1 million pine boards in order to determine if excessive warpage exists for the boards. A sample of 50 boards resulted in the identification of 7 boards with excessive warpage. With these data develop a 95% confidence interval for the proportion of boards defective in the whole shipment.

67. A University of Michigan study (January 1992) reported that drug use among college students had continued its decade-long decline. However, the survey showed alcohol consumption remained steady. Among the 1400 college students surveyed, 602 indicated they had five or more drinks within the two weeks prior to the survey. Compute a 95% confidence interval for the proportion of all college students who have had five or more drinks within the two-week period.

68. Towers Perrin, a compensation consultant, asked 500 U.S. companies if they encouraged quality by giving top performers recognition and/or rewards such as cash and stock (*Business Week,* December 1991). Results showed that 56% of the companies used recognition to encourage quality, while 26% of the companies used cash and stock rewards. Develop 95% confidence intervals for the population in terms of both the proportion that uses recognition and the proportion that uses cash and stock rewards to encourage quality.

69. A Time/CNN telephone poll of 1400 American adults asked ''Where would you rather go in your spare time?'' (*Time,* April 6, 1992.) The top response by 504 adults was a shopping mall.
a. What is the point estimate of the proportion of adults who would prefer going to a shopping mall in their spare time?
b. At 95% confidence, what is the sampling error associated with this estimate?

70. A well-known bank credit-card firm is interested in estimating the proportion of credit-card holders that carry a nonzero balance at the end of the month and incur an interest charge. Assume that the desired precision for the proportion estimate is ±3% at a 98% confidence level.
a. How large a sample should be recommended if it is anticipated that roughly 70% of the firm's cardholders carry a nonzero balance at the end of the month?
b. How large a sample would be recommended if no planning value for the population proportion can be specified?

71. A sample of 200 people were asked to identify their major source of news information; 110 stated that their major source was television news coverage.
a. Construct a 95% confidence interval for the proportion of people in the population that consider television their major source of news information.
b. How large a sample would be necessary to estimate the population proportion with a sampling error of .05 or less at a 95% confidence level?

72. A Gallup poll of labor-management negotiations found that 55% of the cases studied showed that labor is taking a more aggressive bargaining stance (*Barron's,* April 10, 1989). If a sample of 500 firms was used in the poll, what is the 90% confidence interval for the proportion on negotiations where labor is taking a more aggressive bargaining stance?

73. *Newsweek* (April 6, 1992) reported data on the percentage of adults who smoke in the United States. Assume that the study designed to collect the data for this report had a preliminary estimate that 30% of the population smoke.
a. How large a sample should be taken to estimate the current proportion of smokers in the population to within ±2% at 95% confidence?
b. Assume that the study used your sample-size recommendation in part (a) and found 555 smokers. What is the point estimate of the proportion of smokers in the population, and what is the 95% confidence interval?

74. A survey of executives will be used to learn about how people in business view the quality of education provided by public school systems (*Fortune,* March, 1989). Using a 90% confidence level, the study is to be designed to have a 4% margin of error for questions where the population proportion is approximately .60.
a. How many executives should be included in the sample?
b. Using your sample size, what is the 90% confidence interval for the population proportion rating public schools ''fair/poor'' if 313 executives in the sample respond ''fair/poor''?
c. Using your sample size, what is the 90% confidence interval for the population proportion responding ''public schools have deteriorated in the last 10 years'' if 260 executives respond Yes to this statement?
d. Do the interval estimates in parts (b) and (c) provide the desired 4% margin of error? Explain.

Computer Case   *Metropolitan Research, Inc.*

Metropolitan Research, Inc., is a consumer research organization that takes surveys designed to evaluate a wide variety of products and services available to consumers. In one particular study, Metropolitan was interested in learning about consumer satisfaction with the performance of automobiles produced by a major Detroit manufacturer. A questionnaire sent to owners of one of the manufacturer's full-sized car revealed several complaints about early transmission problems. To learn more about the transmission failures, Metropolitan used a sample of actual transmission repairs provided by a transmission repair firm located in the Detroit area. The following data show the actual number of miles that 50 vehicles had been driven at the time of transmission failure. The data are available in the data set AUTO.

 Auto

85,092	32,609	59,465	77,437	32,534	64,090	32,464	59,902
39,323	89,641	94,219	116,803	92,857	63,436	65,605	85,861
64,342	61,978	67,998	59,817	101,769	95,774	121,352	69,568
74,276	66,998	40,001	72,069	25,066	77,098	69,922	35,662
74,425	67,202	118,444	53,500	79,294	64,544	86,813	116,269
37,831	89,341	73,341	85,288	138,114	53,402	85,586	82,256
77,539	88,798						

Managerial Report

1. Use appropriate descriptive statistics to summarize the transmission failure data.
2. Develop a 95% confidence interval for the mean number of miles driven until transmission failure for the population of automobiles that have experienced transmission failure. Provide a managerial interpretation of the interval estimate.
3. Discuss the implication of your statistical finding in terms of the feeling that some owners of the automobiles have experienced early transmission failures.
4. How many repair records should be sampled if the research firm would like the population mean number of miles driven until transmission failure to be estimated to within ±5000 miles at 95% confidence?
5. What other information would you like to gather to more fully evaluate the transmission failure problem?

9

Hypothesis Testing

Harris Corporation*

MELBOURNE, FLORIDA

Harris Corporation's RF Communications Division, located in Melbourne, Florida, is a major manufacturer of point-to-point radio communications equipment. It is a horizontally integrated manufacturing company with a multiplant facility. Most of the Harris products require medium- to high-volume production operations including printed circuit assembly, final product assembly, and testing.

One of the company's high-volume products uses an assembly called an RF deck. Each RF deck consists of 16 electronic components soldered to a machined casting which forms the plated surface on the deck. During a manufacturing run, a problem developed in the soldering process; the flow of solder onto the deck did not meet the quality criteria established for the product. After considering a variety of factors that might affect the soldering process, an engineer made the preliminary determination that the soldering problem was most likely due to defective platings.

The question raised by the engineer was, Did the proportion of defective platings in the Harris inventory exceed the supplier's design specification? Letting p indicate the proportion of defective platings in the Harris inventory and p_0 indicate the proportion of defective platings under the supplier's design specifications, the following hypotheses were formulated:

$$H_0: p \leq p_0$$
$$H_a: p > p_0$$

Hypothesis H_0 indicates that the Harris inventory has a defective plating proportion less than or equal to the design specification. In this case, the defective proportion would be judged acceptable, and the engineer would need to look for other causes of the soldering problem. However, the alternative hypothesis H_a indicates that the Harris inventory has a defective plating proportion greater than the design specification. In this case, excessive defective platings may well be the cause of the soldering problem; action should be taken to determine why the defective proportion in inventory is so high.

Tests made on a sample of platings from the Harris inventory showed a defective sample proportion of .15. This proportion resulted in the rejection of H_0. As a result, it was concluded that the H_a was true and that the platings in inventory had a higher defective rate than should exist under the supplier's design specification. Further investigation of the inventory area led to the conclusion that the underlying problem was shelf contamination during storage. By altering the storage environment, the engineer was able to solve the problem.

In this chapter you will learn how to formulate hypotheses about a population mean and a population proportion. Through the analysis of sample data, you will be able to determine whether the hypothesis should be rejected. Appropriate conclusions and actions will be demonstrated for testing research hypotheses, testing the validity of assumptions, and decision making.

*The authors are indebted to Richard A. Marshall of the Harris Corporation for providing this Statistics in Practice.

In Chapters 7 and 8 we showed how a sample could be used to develop point and interval estimates of population parameters. In this chapter we continue the discussion of statistical inference by showing how *hypothesis testing* can be used to determine whether a statement about the value of a population parameters should be rejected.

In hypothesis testing we begin by making a tentative assumption about a population parameter. This tentative assumption is called the *null hypothesis* and is denoted by H_0. We then define another hypothesis, called the *alternative hypothesis,* which is the opposite of what is stated in the null hypothesis. This alternative hypothesis is denoted by H_a. The hypothesis-testing procedure involves using data from a sample to test the two competing statements indicated by H_0 and H_a.

The situation encountered in hypothesis testing is similar to the one encountered in a criminal trial. In a criminal trial the assumption is that the defendant is innocent. Thus, the null hypothesis is one of innocence. The opposite of the null hypothesis is the alternative hypothesis—that the defendant is guilty. Thus, the hypotheses for a criminal trial would be written

H_0: The defendant is innocent

H_a: The defendant is guilty

To test these competing statements, or hypotheses, a trial is held. The testimony and evidence obtained during the trial provide the sample information. If the sample information is not inconsistent with the assumption of innocence, the null hypothesis that the defendant is innocent cannot be rejected. However, if the sample information is inconsistent with the assumption of innocence, the null hypothesis will be rejected. In this case, action will be taken based upon the alternative hypothesis that the defendant is guilty.

The purpose of this chapter is to show how hypothesis tests can be conducted about a population mean and a population proportion. We begin by providing examples that illustrate approaches to developing null and alternative hypotheses.

9.1   Developing Null and Alternative Hypotheses

In the introduction we stated that in hypothesis testing we develop two competing statements, or hypotheses, about a population parameter. One statement is called the null hypothesis H_0, and the other statement is called the alternative hypothesis H_a. In some applications it may not be obvious how the null and alternative hypotheses should be formulated. Care must be taken to be sure that the hypotheses are structured appropriately and that the hypothesis-testing conclusion provides the information that the researcher or decision maker desires.

Guidelines for establishing the null and alternative hypotheses will be given for three types of situations that frequently employ hypothesis-testing procedures. A discussion of each of these situations follows.

Testing Research Hypotheses

Consider a particular model automobile that currently obtains an average of 24 miles per gallon. A product-research group has developed a new carburetor specifically designed to increase the miles-per-gallon rating. To evaluate the new carburetors, several carburetors will be manufactured, installed in automobiles, and subjected to research-controlled driving tests. Note that the product-research group is looking for evidence to conclude that the new design *increases* the mean miles-per-gallon rating. In this case, the research hypothesis is that the new carburetor will provide a mean miles-per-gallon rating exceeding 24; that is, $\mu > 24$. As a general guideline, a research hypothesis such as this should be formulated as the *alternative hypothesis*. Thus, the appropriate null and alternative hypotheses for the study are as follows:

$$H_0: \mu \leq 24$$
$$H_a: \mu > 24$$

If the sample results indicate that H_0 cannot be rejected, we will not be able to conclude that the new carburetor is better. Perhaps more research and subsequent testing should be conducted. However, if the sample results indicate H_0 can be rejected, the inference can be made that H_a: $\mu > 24$ is true. With this conclusion, the researcher has the statistical support necessary to conclude that the new carburetor increases the mean number of miles per gallon. Thus, we see that action is taken if the null hypothesis is rejected.

In research studies such as these, the null and alternative hypotheses should be formulated so that the rejection of H_0 will provide the researcher with the conclusion and action being sought. Thus, the research hypothesis should be expressed as the alternative hypothesis.

Testing the Validity of a Claim

As an illustration of testing the validity of claims that companies make about their products, consider the situation of a manufacturer of soft drinks who states that 2-liter containers of its products have an average of at least 67.6 fluid ounces. A sample of 2-liter containers will be selected, and the contents will be measured to test the manufacturer's claim. In this type of hypothesis-testing situation, we generally follow the rationale suggested by the criminal trial analogy. That is, the manufacturer's claim should be assumed true (innocent) unless the sample evidence proves otherwise (guilty). Using this approach for the soft-drink example, the null and alternative hypotheses would be stated as follows:

$$H_0: \mu \geq 67.6$$

$$H_a: \mu < 67.6$$

If the sample results indicate H_0 cannot be rejected, the manufacturer's claim cannot be challenged. However, if the sample results indicate H_0 can be rejected, the inference will be made that H_a: $\mu < 67.6$ is true. With this conclusion, statistical evidence indicates that the manufacturer's claim is incorrect and that the soft-drink containers are being filled with a mean less than the claimed 67.6 ounces. Appropriate action against the manufacturer may be considered. Note that, as was true for research hypotheses, action is only taken if H_0 is rejected.

In any situation which involves testing the validity of a product claim, the null hypothesis is generally formulated based on the assumption that the claim is true. The alternative hypothesis is then formulated so that rejection of H_0 will provide the statistical evidence that the stated assumption is incorrect. Action to correct the claim should be considered whenever H_0 is rejected.

Testing in Decision-Making Situations

In testing research hypotheses or testing the validity of a claim, action is only taken if H_0 is rejected. In many instances, however, action must be taken if H_0 cannot be rejected or H_0 can be rejected. In general, this type of situation occurs when a decision maker must choose between two courses of action, one associated with the null hypothesis and another associated with the alternative hypothesis. For example, on the basis of a sample of parts from a shipment that has just been received, a quality-control inspector must decide whether to accept the entire shipment or to return the shipment to the supplier because it does not meet specifications. Assume that specifications for a particular part indicate a mean length of 2 inches per part is required. If the average length of the parts is greater or less than the 2-inch standard, the parts will cause quality problems in the assembly operation. In this case, the null and alternative hypotheses would be formulated as follows:

$$H_0: \mu = 2$$

$$H_a: \mu \neq 2$$

If the sample results indicate H_0 cannot be rejected, the quality-control inspector will have no reason to doubt that the shipment meets specifications, and the shipment will be accepted. However, if the sample results indicate that H_0 should be rejected, the conclusion can be made that the parts do not meet specifications. In this case, the quality-control inspector has sufficient evidence to return the shipment to the supplier. Thus, we see that for these types of situations, action is taken if H_0 cannot be rejected or H_0 can be rejected.

A Summary of Forms for Null and Alternative Hypotheses

Let μ_0 denote the specific numerical value being considered in the null and alternative hypotheses. In general, a hypothesis test concerning the values of a population mean μ must take one of the following three forms:

$$H_0: \mu \geq \mu_0 \qquad H_0: \mu \leq \mu_0 \qquad H_0: \mu = \mu_0$$
$$H_a: \mu < \mu_0 \qquad H_a: \mu > \mu_0 \qquad H_a: \mu \neq \mu_0$$

In many situations, the choice of H_0 and H_a is not obvious; in such cases, judgment on the part of the user is needed to select the proper form of H_0 and H_a. However, as the above forms show, the equality part of the expression (either $\geq$, $\leq$, or $=$) *always* appears in the null hypothesis. In selecting the proper form of H_0 and H_a, keep in mind that the alternative hypothesis is what the sampling study is attempting to establish. Thus, asking whether the user is looking for evidence to support $\mu < \mu_0$, $\mu > \mu_0$, or $\mu \neq \mu_0$ will help determine H_a. The following exercises are designed to provide practice in choosing the proper form for a hypothesis test.

☐ ☐ Exercises

1. The manager of the Danvers-Hilton Resort Hotel has stated that the mean guest bill for a weekend is $400 or less. A member of the hotel's accounting staff has noticed that the total charges for guest bills have been increasing in recent months. The accountant will use a sample of weekend guest bills to test the manager's claim.
a. Which form of the hypotheses should be used to test the manager's claim? Explain.

$$H_0: \mu \geq 400 \qquad H_0: \mu \leq 400 \qquad H_0: \mu = 400$$
$$H_a: \mu < 400 \qquad H_a: \mu > 400 \qquad H_a: \mu \neq 400$$

b. What conclusion is appropriate when H_0 cannot be rejected?
c. What conclusion is appropriate when H_0 can be rejected?

SELF TEST ▶ 2. The manager of an automobile dealership is considering a new bonus plan that is designed to increase sales volume. Currently, the mean sales volume is 14 automobiles per month. The manager would like to conduct a research study to see if there is evidence that the new bonus plan increases sales volume. To collect data on the plan, a sample of sales personnel will be allowed to sell under the new bonus plan for a 1-month period.
a. Develop the null and alternative hypotheses that are most appropriate for this research situation.
b. Comment on the conclusion when H_0 cannot be rejected.
c. Comment on the conclusion when H_0 can be rejected.

3. A production-line operation is designed to fill cartons of laundry detergent with a mean weight of 32 ounces. A sample of cartons is periodically selected and weighed to determine if underfilling or overfilling exists. If the sample data generate a conclusion of underfilling or overfilling, the production line will be shut down and adjusted to obtain proper filling.
a. Formulate the null and alternative hypotheses that will help in deciding whether or not to shut down and adjust the production line.

b. Comment on the conclusion and the decision when H_0 cannot be rejected.

c. Comment on the conclusion and the decision when H_0 can be rejected.

4. Because of high production-changeover time and costs, a director of manufacturing must convince management that a proposed manufacturing method reduces costs before the new method can be implemented. The current production method operates with a mean cost of $220 per hour. A research study will be conducted with the cost of the new method measured over a sample production period.

a. Develop the null and alternative hypotheses that are most appropriate for this study.

b. Comment on the conclusion when H_0 cannot be rejected.

c. Comment on the conclusion when H_0 can be rejected.

9.2   Type I and Type II Errors

The null and alternative hypotheses are competing statements about the true state of nature. Either the null hypothesis H_0 is true or the alternative hypothesis H_a is true, but not both. Ideally the hypothesis testing procedure should lead to the acceptance of H_0 when H_0 is the true state of nature and the rejection of H_0 when H_a is the true state of nature. Unfortunately, these results are not always possible. Since hypothesis tests are based upon sample information, we must allow for the possibility of errors. Table 9.1 illustrates the two kinds of errors that can be made in hypothesis testing.

The first row of Table 9.1 shows what can happen if we make the conclusion to accept H_0. If H_0 is the true state of nature, this conclusion is correct. However, if H_a is the true state of nature, we have made a *Type II error;* that is, we have accepted H_0 when it is false.

The second row of Table 9.1 shows what can happen if we make the conclusion to reject H_0. If the true state of nature is H_0, we have made a *Type I error;* that is, we rejected H_0 when it is true. However, if H_a is the true state of nature, then rejecting H_0 is correct.

Although we cannot eliminate the possibility of errors in hypothesis testing, we can consider the probability of their occurrence. Using common statistical notation, we denote the probabilities of making the two errors as follows:

$$\alpha = \text{the probability of making a Type I error}$$

$$\beta = \text{the probability of making a Type II error}$$

For example, recall the hypothesis testing illustration discussed in Section 9.1 in which an automobile product-research group had developed a new carburetor designed to increase the miles-per-gallon rating of a particular automobile. With the current model obtaining an average of 24 miles per gallon, the hypothesis test was formulated as follows:

TABLE 9.1

Errors and Correct Conclusions in Hypothesis Testing

		State of Nature	
		H_0 *True*	H_a *True*
Conclusion	*Accept H_0*	Correct Conclusion	Type II Error
	Reject H_0	Type I Error	Correct Conclusion

$$H_0: \mu \leq 24$$

$$H_a: \mu > 24$$

The alternative hypothesis, H_a: $\mu > 24$, indicates that the researchers are looking for sample evidence that will permit the rejection of H_0, thus supporting H_a and the conclusion that the mean miles per gallon is greater than 24.

In this application, the Type I error of rejecting H_0 when it is true corresponds to the researchers claiming that the new carburetor improves the miles per gallon rating ($\mu > 24$) when in fact the new carburetor is not any better than the current carburetor. On the other hand, the Type II error of accepting H_0 when it is false corresponds to the researchers accepting that the new carburetor is not any better than the current carburetor ($\mu \leq 24$) when in fact the new carburetor provides an improved miles-per-gallon performance.

In practice, the person conducting the hypothesis test specifies the maximum allowable probability of making a Type I error, called the *level of significance* for the test. Common choices for the level of significance are .05 and .01. Referring to the second row of Table 9.1, note that the conclusion to *reject H_0* indicates that either a Type I error or a correct conclusion has been made. Thus, if the probability of making a Type I error is controlled for by selecting a low value for the level of significance, we have a high degree of confidence that the conclusion to reject H_0 is correct. In such cases, we have statistical support to conclude that H_0 is false and H_a is true. Any action suggested by the alternative hypothesis H_a is appropriate.

Although most applications of hypothesis testing control the probability of making a Type I error, they do not always control for the probability of making a Type II error. Thus, whenever we decide to accept H_0 we cannot determine how confident we can be with the decision to accept H_0. Because of the uncertainty associated with making a Type II error, statisticians often recommend that we use the statement *do not reject H_0* instead of *accept H_0*. Using the statement *do not reject H_0* carries the recommendation to withhold both judgment and action. In effect, by never directly accepting H_0, the statistician avoids the risk of making a Type II error. Whenever the probability of making a Type II error has not been determined and controlled, we will not make the conclusion to accept H_0. In such cases, only two conclusions are possible: *do not reject H_0* or *reject H_0*.

Just because it is not common to control for the Type II error in hypothesis testing, this does not mean that it is not possible to do so. In fact, in Sections 9.7 and 9.8, we will illustrate procedures for determining and controlling for the probability of making a Type II error. If proper controls have been established for this error, it can be appropriate to take action based on the decision to accept H_0.

NOTES & COMMENTS

1. The statement *do not reject H_0* is used to avoid accepting that H_0 is true; thus, this eliminates the possibility of making a Type II error. In other words, the statement *do not reject H_0* indicates that the test results are inconclusive. In this case, statisticians often recommend that further study be conducted to clarify the situation.

2. Many applications of hypothesis testing have a decision-making goal. The *reject H_0* provides the statistical support to conclude that H_a is true and take whatever action is appropriate. The statement *do not reject H_0*, although inconclusive, often forces management to behave as if H_0 is true. In this case, management needs to be aware of the fact that such behavior may result in a Type II error.

☐ ☐ Exercises

SELF TEST ▶ ✓5. The average American buys 6.08 books per year (*Louis Rukeyser's Business Almanac*, 1988). A University of Iowa researcher believes that adults in Des Moines purchase books at an annual rate higher than the national average. The following null and alternative hypotheses have been formulated by the researcher.

$$H_0: \mu \leq 6.08$$

$$H_a: \mu > 6.08$$

[handwritten: $H_a > 6.08$]

a. What is the Type I error in this situation? What are the consequences of making this error?
b. What is the Type II error in this situation? What are the consequences of making this error?

✓6. The label on a 3-quart container of orange juice claims that the orange juice contains, an average of 1 gram of fat or less. Answer the following questions for a hypothesis test that could be used to test the claim on the label.
a. Develop the appropriate null and alternative hypotheses.
b. What is the Type I error in this situation? What are the consequences of making this error?
c. What is the Type II error in this situation? What are the consequences of making this error?

7. Carpetland salespersons have been selling an average of $8000 of carpeting per week. Steve Contois, the firm's vice president, has proposed a compensation plan with new selling incentives. Steve hopes that the results of a trial selling period will enable him to conclude that the compensation plan increases the average sales per salesperson.
a. Develop the appropriate null and alternative hypotheses.
b. What is the Type I error in this situation? What are the consequences of making this error?
c. What is the Type II error in this situation? What are the consequences of making this error?

8. Suppose that the new production method will be implemented if a hypothesis test supports the conclusion that the new method reduces the mean operating cost per hour.
a. State the appropriate null and alternative hypotheses if the mean cost for the current production method is $220 per hour.
b. What is the Type I error in this situation? What are the consequences of making this error?
c. What is the Type II error in this situation? What are the consequences of making this error?

9.3   One-Tailed Hypothesis Tests About a Population Mean: Large-Sample Case

The Federal Trade Commission (FTC) periodically conducts studies designed to test the claims manufacturers make about their products. For example, the label on a large can of Hilltop Coffee states that the can contains at least 3 pounds of coffee. Suppose that we wish to test this claim using hypothesis testing.

The first step is to develop the null and the alternative hypotheses. We begin by tentatively assuming that the manufacturer's claim is correct. Note that if the population of coffee cans has a mean weight of 3 or more pounds per can, Hilltop's claim about its product is correct. However, if the population of coffee cans has a mean weight less than 3 pounds per can, Hilltop's claim is invalid.

With μ denoting the mean weight for the population, the null and the alternative hypotheses are formulated as follows:

$$H_0: \mu \geq 3$$

$$H_a: \mu < 3$$

If the sample data indicate that H_0 cannot be rejected, the statistical evidence does not support the conclusion that a label violation has occurred. Thus, no action would be taken against Hilltop. However, if sample data indicate that H_0 can be rejected, we will conclude that the alternative hypothesis, H_a: $\mu < 3$, is true. In this case, an FTC claim of under-filling and a charge of a label violation would be appropriate.

Suppose that a random sample of 36 cans of coffee is selected. Note that if the mean filling weight for the sample of 36 cans is less than 3 pounds, the sample results will begin to cast doubt on the null hypothesis H_0: $\mu \geq 3$. But how much less than 3 must $\bar{x}$ be before we would be willing to risk making a Type I error and falsely accuse the company of a label violation?

To answer this question, let us tentatively assume that the null hypothesis is true with $\mu = 3$. From our study of sampling distributions in Chapter 7, we know that whenever the sample size is large ($n \geq 30$), the sampling distribution of $\bar{x}$ can be approximated by a normal probability distribution. Figure 9.1 shows the sampling distribution of $\bar{x}$ when the null hypothesis is true at $\mu = 3$.

The value of $z = (\bar{x} - 3)/\sigma_{\bar{x}}$ gives the number of standard deviations $\bar{x}$ is from $\mu = 3$. For hypothesis tests about a population mean, we will use z as a *test statistic* to determine whether $\bar{x}$ deviates enough from $\mu = 3$ to justify rejecting the null hypothesis. Note that a value of $z = -1$ means that $\bar{x}$ is 1 standard deviation below $\mu = 3$, a value of $z = -2$ means that $\bar{x}$ is 2 standard deviations below $\mu = 3$, and so on. Obtaining a value of $z < -3$ is very unlikely if the null hypothesis is true. The key question is: How small must the test statistic z be before we have enough evidence to reject the null hypothesis?

Figure 9.2 shows that the probability of observing a value of $\bar{x}$ more than 1.645 standard deviations below the mean of $\mu = 3$ is .05. Thus, if we were to reject the null hypothesis whenever the value of the test statistic $z = (\bar{x} - 3)/\sigma_{\bar{x}}$ is less than -1.645, the probability of making a Type I error would be .05. If the FTC considered .05 to be an acceptable level for the probability of making a Type I error, then we would reject the null hypothesis whenever the test statistic indicated that the sample mean was more than 1.645 standard deviations below $\mu = 3$. Thus, we would reject H_0 if $z < -1.645$.

The methodology of hypothesis testing requires that we specify the maximum allowable probability of a Type I error. As noted in the previous section, this maximum probability is called the level of significance for the test; it is denoted by α, and it represents the probability of making a Type I error when the null hypothesis is true as an equality. Specifying the level of significance is a job for the manager. If the cost of making

FIGURE 9.1

Sampling Distribution of $\bar{x}$ for the Hilltop Coffee Study when the Null Hypothesis is True ($\mu = 3$)

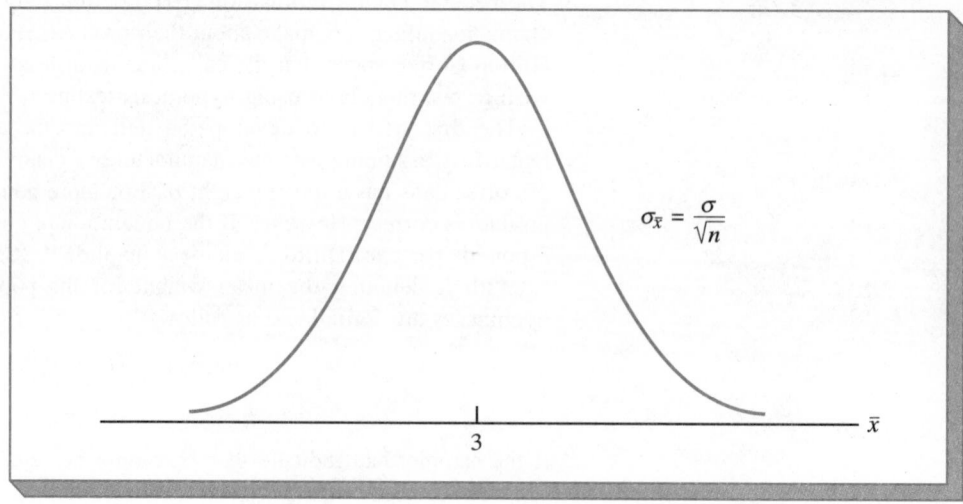

$$\sigma_{\bar{x}} = \frac{\sigma}{\sqrt{n}}$$

$\bar{x}$

3

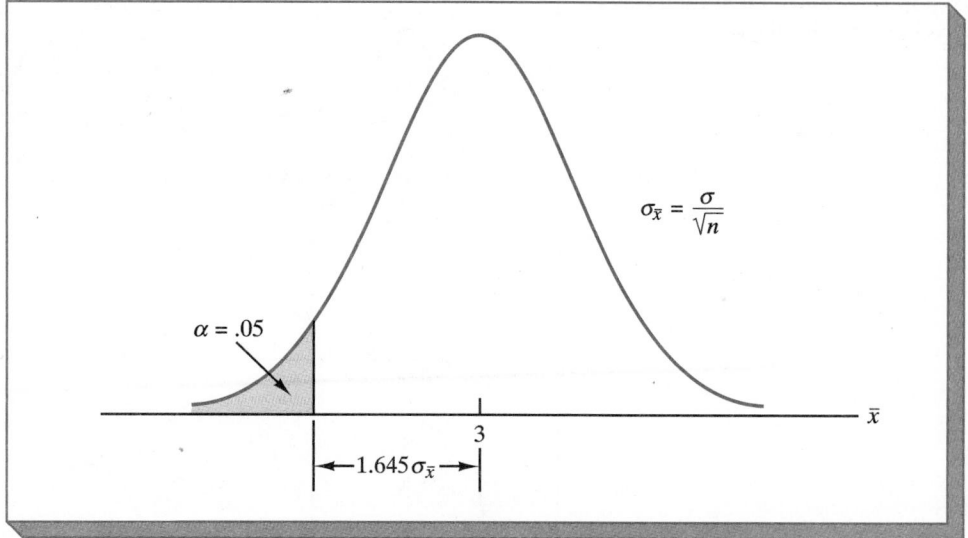

a Type I error is high, then a small value should be chosen for the level of significance. If the cost is not too great, a larger value may be appropriate.

In the Hilltop Coffee study, the director of the weight-testing program has made the following statement: "If the company is meeting its weight specifications exactly ($\mu = 3$), I would like a 99% chance of not taking any action against the company. While I do not want to accuse the company wrongly of underfilling its product, I am willing to live with a 1% chance of making this error."

From the director's statement, the maximum probability of a Type I error is .01. Thus, the level of significance for the hypothesis test is $\alpha = .01$. Figure 9.3 shows both the sampling distributions of $\bar{x}$ and $z = (\bar{x} - \mu)/\sigma_{\bar{x}}$ for the Hilltop Coffee example. Note that when the null hypothesis is true at $\mu = 3$, the probability is .01 that $\bar{x}$ is more than 2.33 standard deviations below the mean of 3. Therefore, we establish the following rejection rule:

$$\text{Reject } H_0 \text{ if } z = \frac{\bar{x} - \mu}{\sigma_{\bar{x}}} < -2.33$$

If the value of $\bar{x}$ is such that the test statistic z is in the rejection region, then we reject H_0 and conclude that H_a is true. On the other hand, if the value of $\bar{x}$ is such that the test statistic z is not in the rejection region, then we cannot reject H_0. Note that the rejection region shown in Figure 9.3 is in only one tail of the sampling distribution. Whenever this occurs, we say the test is a *one-tailed* hypothesis test.

Suppose that a sample of 36 cans provides a mean of $\bar{x} = 2.92$ pounds and that it is known from previous studies that the population standard deviation is $\sigma = .18$. With $\sigma_{\bar{x}} = \sigma/\sqrt{n}$, the value of the test statistic is given by

$$z = \frac{\bar{x} - 3}{\sigma/\sqrt{n}} = \frac{2.92 - 3}{.18/\sqrt{36}} = -2.67$$

Figure 9.4 shows that the value of the test statistic is in the rejection region. We are now justified in concluding that $\mu < 3$ at a .01 level of significance. The director now has the statistical justification to take action against Hilltop Coffee for underfilling its product.

FIGURE 9.3
**Hilltop Coffee Rejection Rule Has a
Level of Significance of $\alpha = .01$**

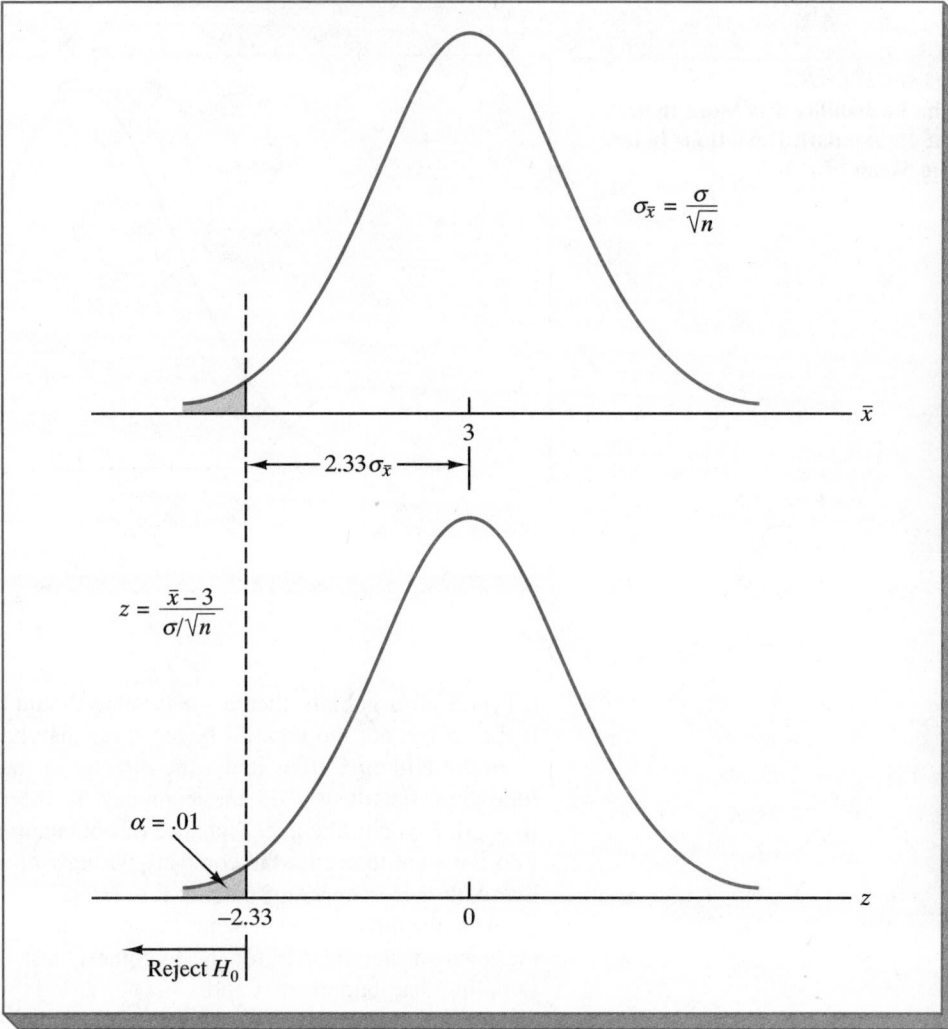

FIGURE 9.4
**Value of the Test Statistic for
$\bar{x} = 2.92$ Is in the Rejection Region**

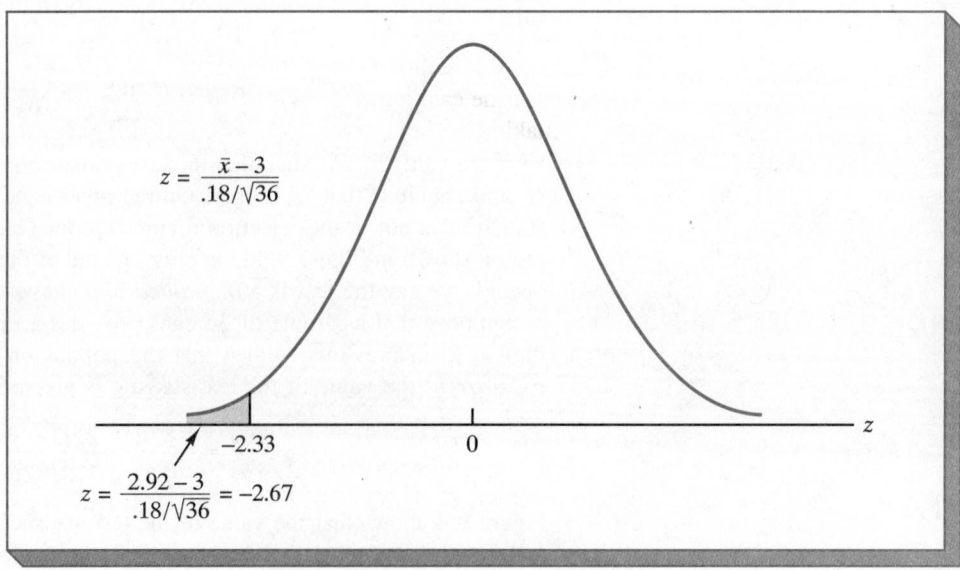

FIGURE 9.5
Value of the Test Statistic for $\bar{x} = 2.97$ Is Not in the Rejection Region

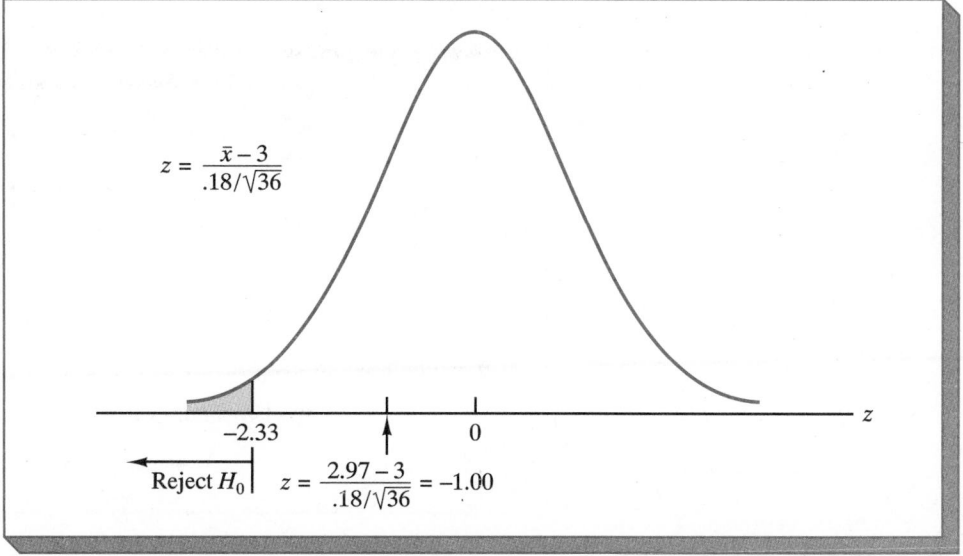

Suppose, instead, that the sample of 36 cans had provided a sample mean of $\bar{x} = 2.97$. In this case, the value of the test statistic would be:

$$z = \frac{\bar{x} - 3}{\sigma/\sqrt{n}} = \frac{2.97 - 3}{.18/\sqrt{36}} = -1.00$$

Since $z = -1.00$ is greater than -2.33, the value of the test statistic is not in the rejection region (see Figure 9.5). Hence we cannot reject the null hypothesis. No further inference can be made, and no statistical justification has been provided to take action against Hilltop Coffee.

The value of z that establishes the boundary of the rejection region is called the *critical value*. In establishing the critical value, we tentatively assume the null hypothesis is true. But, for Hilltop Coffee, the null hypothesis is true whenever $\mu \geq 3$, and we considered only the case when $\mu = 3$. What about the case when $\mu > 3$? If $\mu > 3$, the probability of making a Type I error will be less than it is when $\mu = 3$; that is, in this case, it is even less likely that we will find a value of the test statistic that is in the rejection region. Since the objective of the hypothesis-testing procedure is to limit the maximum probability of making a Type I error, the critical value for the test is established assuming $\mu = 3$.

Summary: One-Tailed Tests About a Population Mean

Let us generalize the hypothesis-testing procedure for one-tailed tests about a population mean. We restrict ourselves here to the large-sample case ($n \geq 30$) where the central limit theorem permits us to assume a normal sampling distribution for $\bar{x}$. In this large-sample case when σ is unknown, we simply substitute the sample standard deviation s for σ in computing the test statistic. The general form of a lower-tailed test is as follows where μ_0 is a stated value for the population mean.

Large-Sample ($n \geq 30$) Hypothesis Test about a Population Mean for a One-Tailed Test of the Form

$$H_0: \mu \geq \mu_0$$

$$H_a: \mu < \mu_0$$

Test Statistic:

$$z = \frac{\overline{x} - \mu_0}{\sigma/\sqrt{n}}$$

If σ is unknown, substitute s for σ in computing z.

Rejection Rule at a Level of Significance of α

Reject H_0 if $z < -z_\alpha$ (9.1)

A second form of the one-tailed test rejects the null hypothesis when the test statistic is in the upper tail of the sampling distribution. This one-tailed test and rejection rule are summarized next (see Figure 9.6). Again, we are considering the large-sample case; when σ is unknown, s may be substituted for σ in the computation of the test statistic z.

Large-Sample ($n \geq 30$) Hypothesis Test about a Population Mean for a One-Tailed Test of the Form

$$H_0: \mu \leq \mu_0$$

$$H_a: \mu > \mu_0$$

Test Statistic:

$$z = \frac{\overline{x} - \mu_0}{\sigma/\sqrt{n}}$$

If σ is unknown, substitute s for σ in computing z.

Rejection Rule at a Level of Significance of α

Reject H_0 if $z > z_\alpha$ (9.2)

The Use of *p*-Values

We now consider another approach that is sometimes used in hypothesis testing. This approach is based upon what is called a *p-value*. We will show how the *p*-value for a sample can be computed from the value of the test statistic z. The *p*-value can also be used to make the decision whether or not to reject H_0.

Assuming that the null hypothesis is true, the *p*-value is the probability of obtaining a sample result that is at least as unlikely as what is observed. In the Hilltop Coffee example, the rejection region is in the lower tail; therefore the *p*-value is the probability of observing a sample mean less than or equal to what is observed.

FIGURE 9.6

**Rejection Region for an
Upper-Tailed Hypothesis Test about
a Population Mean**

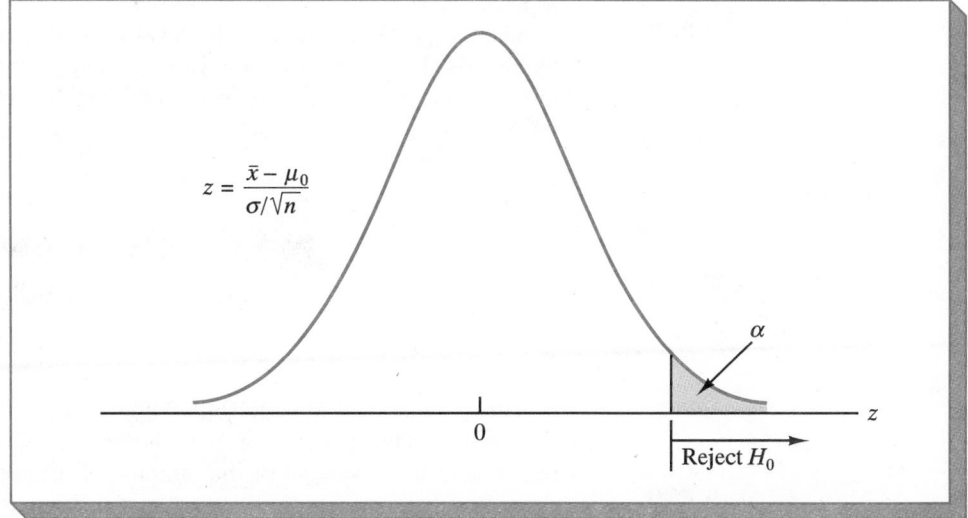

Let us compute the p-value associated with the sample mean $\bar{x} = 2.92$ in the Hilltop Coffee example. The p-value in this case is the probability of obtaining a value for the sample mean that is less than or equal to the observed value of $\bar{x} = 2.92$, given the hypothesized value for the population mean of $\mu = 3$. Previously we showed that the test statistic $z = -2.67$ corresponded to $\bar{x} = 2.92$. Thus, as shown in Figure 9.7, the p-value is the area in the tail of the standard normal probability distribution for $z = -2.67$. Using the standard normal probability distribution table, we find that the area between the mean and $z = -2.67$ is .4962. Thus, there is a $.5000 - .4962 = .0038$ probability of obtaining a sample mean that is less than or equal to the observed $\bar{x} = 2.92$. The p-value is therefore .0038. This p-value shows us that there is a very small probability of obtaining a sample mean $\bar{x} = 2.92$ when sampling from a population with $\mu = 3$.

The p-value can be used to make the decision for a hypothesis test by noting that if the *p-value is less than the level of significance α*, the value of the test statistic must be in the *rejection region*. Similarly, if the p-value is *greater than or equal to α*, the value of the

FIGURE 9.7

**p-Value for the Hilltop Coffee Study
when $\bar{x} = 2.92$ and $z = -2.67$**

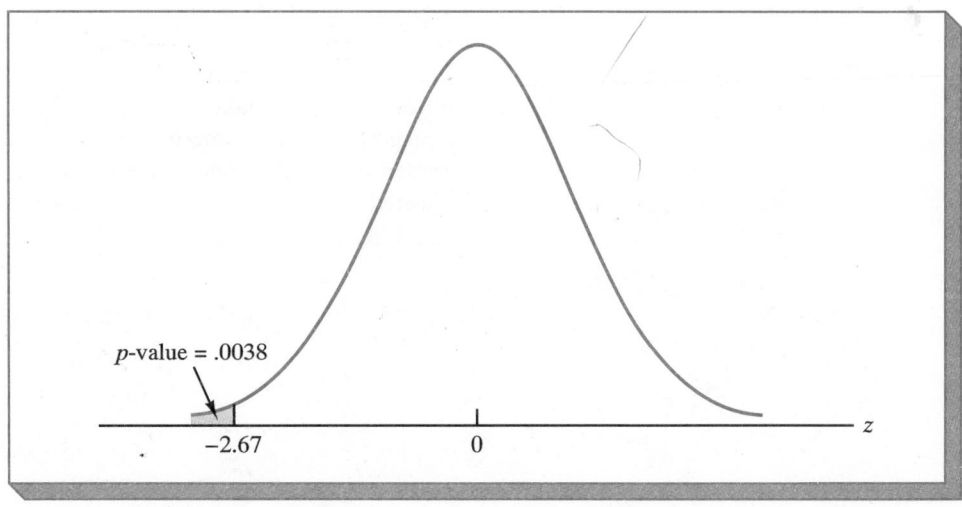

test statistic is not in the rejection region. For the Hilltop Coffee example, the fact that the p-value of .0038 is less than the level of significance, $\alpha = .01$, indicates that the null hypothesis should be rejected. Given the stated level of significance α for any hypothesis test, the decision of whether or not to reject H_0 can be made in terms of the p-value as follows.

---

p-Value Criterion for Hypothesis Testing

Reject H_0 if the p-value $< \alpha$

The p-value and the corresponding test statistic will always provide the same hypothesis-testing conclusion at a chosen level of significance α. When the rejection region is in the lower tail of the sampling distribution, the p-value is the area under the curve less than or equal to the test statistic. When the rejection region is in the upper tail of the sampling distribution, the p-value is the area under the curve greater than or equal to the test statistic. A small p-value thus indicates a sample result that is unusual given the assumption that H_0 is true. Thus, small p-values lead to rejection of H_0. On the other hand, a relatively large p-value indicates that the null hypothesis cannot be rejected.

The Steps of Hypothesis Testing

In conducting the Hilltop Coffee hypothesis test, we carried out the steps that are required for any hypothesis-testing procedure. A summary of the steps that can be applied to any hypothesis test are as follows.

Steps of Hypothesis Testing

1. Determine the null and alternative hypotheses that are appropriate for the application.
2. Select the test statistic that will be used to decide whether or not to reject the null hypothesis.
3. Specify the level of significance α for the test.
4. Use the level of significance to develop the rejection rule that indicates the values of the test statistic that will lead to the rejection of H_0.
5. Collect the sample data, and compute the value of the test statistic.
6. Compare the value of the test statistic to the critical value(s) specified in the rejection rule to determine whether or not H_0 should be rejected.
7. If desired, compute the p-value for the test.

For Hilltop Coffee, the null and alternative hypotheses (step 1) were as follows:

$$H_0: \mu \geq 3$$

$$H_a: \mu < 3$$

With the large sample ($n \geq 30$) and the population standard deviation given as $\sigma = .18$, the test statistic (step 2) was

$$z = \frac{\bar{x} - \mu}{\sigma/\sqrt{n}}$$

The level of significance (step 3) was given as $\alpha = .01$. Using $\alpha = .01$, the corresponding rejection rule for the test statistic (step 4) was reject H_0 if $z < -2.33$.

With a simple random sample of $n = 36$ cans of coffee and a sample mean of $\bar{x} = 2.92$ pounds, the value of the test statistic (step 5) was computed to be

$$z = \frac{2.92 - 3.00}{.18/\sqrt{36}} = -2.67$$

A comparison of $z = -2.67$ to the critical value specified in the rejection rule (step 6), showed that H_0 should be rejected. Finally, the p-value (step 7) associated with $z = -2.67$ was shown to be .0038. Since $.0038 < \alpha$, H_0 should be rejected.

**NOTES &
COMMENTS**

> The p-value is often called the *observed level of significance* for the test. It is a measure of how unlikely the sample results are assuming the null hypothesis is true. The smaller the p-value, the less likely the sample results. Most statistical software packages print the p-value associated with a hypothesis test.

❑ ❑ Exercises

Methods

9. Consider the following hypothesis test:

$$H_0: \mu \geq 10$$

$$H_a: \mu < 10$$

A sample of 50 provides a sample mean of 9.46 and sample standard deviation of 2.
   a. Using $\alpha = .05$, what is the critical value for z? What is the rejection rule?
   b. Compute the value of the test statistic z. What is your conclusion?

SELF TEST ✓ 10. Consider the following hypothesis test:

$$H_0: \mu \leq 15$$

$$H_a: \mu > 15$$

A sample of 40 provides a sample mean of 16.5 and sample standard deviation of 7.
   a. Using $\alpha = .02$, what is the critical value for z, and what is the rejection rule?
   b. Compute the value of the test statistic z.
   c. What is the p-value?
   d. What is your conclusion?

11. Consider the following hypothesis test:

$$H_0: \mu \geq 25$$

$$H_a: \mu < 25$$

A sample of 100 is used and the population standard deviation is 12. Use $\alpha = .05$. Provide the value of the test statistic z and your conclusion for each of the following sample results:

a. $\bar{x} = 22.0$ **b.** $\bar{x} = 24.0$ **c.** $\bar{x} = 23.5$ **d.** $\bar{x} = 22.8$

12. Consider the following hypothesis test:

$$H_0: \mu \le 5$$

$$H_a: \mu > 5$$

Assume the test statistics are as shown below. Compute the corresponding p-values and make the appropriate conclusions based on $\alpha = .05$.

a. $z = 1.82$ **b.** $z = .45$ **c.** $z = 1.50$
d. $z = 3.30$ **e.** $z = -1.00$

Applications

SELF TEST ▶

13. According to *Business Week* (February 6, 1989), the mean cost of a heart-bypass operation is $26,100, and approximately 230,000 operations are performed annually. A sample of 36 bypass operations in a particular city showed a mean cost of $\bar{x} = \$25,000$ and $s = \$2400$.
 a. Develop appropriate hypotheses to test whether or not the mean bypass operation cost is less than $26,100 in this city.
 b. Use $\alpha = .05$. What is your conclusion?
 c. What is the p-value for this test?

14. In 1990, *The Motor Vehicle Manufacturers Association*, Detroit, Michigan, reported statistics on the average number of years passenger cars were being used. In 1980, the population mean was reported to be 6.5 years. Assume that the 1990 data were obtained from a sample of 100 passenger cars and showed a sample mean of 7.8 years and a sample standard deviation of 2.2 years.
 a. Formulate the null and alternative hypotheses if the researcher is looking for evidence to show that individuals are driving cars longer in 1990.
 b. Does the data support the conclusion that individuals are driving cars longer? Use a .01 level of significance.
 c. What implications does this have for vehicle manufacturers?

15. The average annual income in the United States is $19,780 (*St. Petersburg Times*, December 1989). A sample of 150 individuals in Japan resulted in a sample mean income of $21,040 in U.S. dollars. Assuming a sample standard deviation of $6000, do these data support the conclusion that the mean annual income in Japan is greater than the mean annual income in the United States? Use a .05 level of significance. What is the p-value? What is your conclusion?

16. Fightmaster and Associates Real Estate, Inc., advertises that the mean selling time of a residential home is 40 days or less after it is listed with the company. A sample of 50 recently sold residential homes shows a sample mean selling time of 45 days and a sample standard deviation of 20 days. Using a .02 level of significance, test the validity of the company's claim.

17. Fowle Marketing Research, Inc., bases charges to a client on the assumption that telephone surveys can be completed with a mean time of 15 minutes or less. If a greater mean survey time is required, a premium rate is charged the client. Suppose that a sample of 35 surveys shows a sample mean of 17 minutes and a sample standard deviation of 4 minutes. Is the premium rate justified? Test at the $\alpha = .01$ level of significance.

18. New tires manufactured by a company in Findlay, Ohio, are designed to provide a mean of at least 28,000 miles. Tests with 30 tires show a sample mean of 27,500 miles with a sample standard deviation of 1000 miles. Using a .05 level of significance, test whether or not there is sufficient evidence to reject the claim of a mean of at least 28,000 miles. What is the p-value?

19. A company currently pays its production employees a mean wage of $15.00 per hour. The company is planning to build a new factory, and several locations are being considered. The availability of labor at a rate less than $15.00 per hour is a major factor in the location decision. For

one location, a sample of 40 workers showed a current mean hourly wage of $\bar{x}$ = \$14.00 and a sample standard deviation of s = \$2.40.

a. Using a .10 level of significance, does the sample data indicate that the location has a mean wage rate significantly below the \$15.00 per hour rate?

b. What is the p-value?

20. A new diet program claims that participants will lose on average at least 8 pounds during the first week of the program. A random sample of 40 people participating in the program showed a sample mean weight loss of 7 pounds. The sample standard deviation was 3.2 pounds.

a. What is the rejection rule with α = .05?

b. What is your conclusion about the claim made by the diet program?

c. What is the p-value?

9.4 Two-Tailed Hypothesis Tests About a Population Mean: Large-Sample Case

Two-tailed hypothesis tests differ from one-tailed tests in that the rejection region is placed in both the lower and the upper tails of the sampling distribution. Let us introduce an example to show how and why two-tailed tests are conducted.

The United States Golf Association (USGA) has established rules which manufacturers of golf equipment must meet to have their products acceptable for use in USGA events. One of the rules regarding the manufacture of golf balls states that "A brand of golf ball, when tested on apparatus approved by the USGA on the outdoor range at the USGA Headquarters . . . shall not cover an average distance in carry and roll exceeding 280 yards. . . ." Suppose that Superflight, Inc., has recently developed a high-technology manufacturing method which can produce golf balls that have an average distance in carry and roll of 280 yards.

Superflight realizes, however, that if the new manufacturing process goes out of adjustment, the process may produce balls with an average distance of less than 280 yards or with an average distance of greater than 280 yards. In the former case, Superflight may experience a downturn in sales as a result of marketing an inferior product, and in the latter case, Superflight may have their golf balls rejected by the USGA. As a result, management at Superflight has instituted a quality-control program to carefully monitor the new manufacturing process.

As part of the quality-control program, an inspector periodically selects a sample of balls from the production line and subjects them to tests which are equivalent to those performed by the USGA. With no good reason to doubt that the manufacturing process is functioning correctly, we establish the following null and alternative hypotheses:

$$H_0: \mu = 280$$

$$H_a: \mu \neq 280$$

As usual, we make the tentative assumption that the null hypothesis is true. A rejection region must be established for the test statistic z. We want to reject the claim that μ = 280 when the z value indicates that the sample mean $\bar{x}$ is significantly less than 280 yards or when the z value indicates that the sample mean is significantly greater than 280 yards. Thus, H_0 should be rejected for values of the test statistic in either the lower tail or the upper tail of the sampling distribution. As a result the test will be referred to as a *two-tailed* hypothesis test.

Following the hypothesis-testing procedure developed in the previous sections, we first specify a level of significance by determining a maximum allowable probability of making a Type I error. Suppose we choose $\alpha = .05$ as the level of significance. This means that there will be a .05 probability of concluding that the mean distance is not 280 yards when in fact it is. The test statistic is

$$z = \frac{\bar{x} - \mu}{\sigma/\sqrt{n}}$$

Figure 9.8 shows the sampling distribution of z with the two-tailed rejection region for $\alpha = .05$. With two-tailed hypothesis tests, we will always determine the rejection region by placing an area or probability of $\alpha/2$ in each tail of the distribution. The values of z that provide an area of .025 in each tail can be found from the standard normal probability distribution table. We see in Figure 9.8 that $-z_{.025} = -1.96$ identifies an area of .025 in the lower tail and $z_{.025} = +1.96$ identifies an area of .025 in the upper tail. Referring to Figure 9.8, we can establish the following rejection rule:

Reject H_0 if $z < -1.96$ or if $z > 1.96$

Suppose that a sample of 36 golf balls provides a sample mean distance of $\bar{x} = 278.5$ yards and a sample standard deviation of $s = 12$ yards. Using the value of μ from the null hypothesis and the sample standard deviation of $s = 12$ as an estimate of the population standard deviation σ, the value of the test statistic is

$$z = \frac{\bar{x} - \mu}{\sigma/\sqrt{n}} = \frac{278.5 - 280}{12/\sqrt{36}} = -.75$$

According to the rejection rule, H_0 cannot be rejected. The sample results indicate that the quality-control manager has no reason to doubt the assumption that the manufacturing process is producing golf balls with a mean distance of 280 yards.

FIGURE 9.8

Rejection Region (Shaded Area) for the Two-Tailed Hypothesis Test for Superflight, Inc.

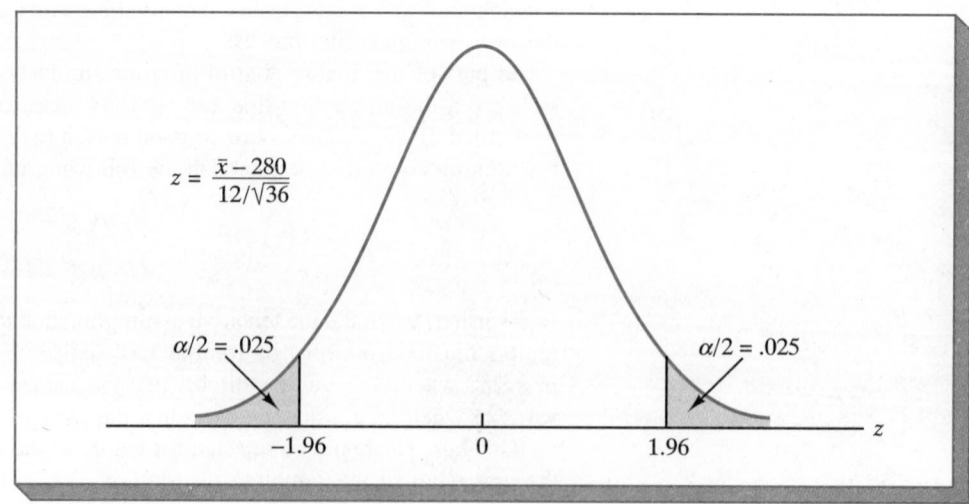

Summary: Two-Tailed Tests About a Population Mean

Let μ_0 represent the value of the mean as claimed in the null hypothesis. The general form of the two-tailed hypothesis test about a population mean is as follows.

Large-Sample ($n \geq 30$) Hypothesis Test about a Population Mean for a Two-Tailed Test of the Form

$$H_0: \mu = \mu_0$$

$$H_a: \mu \neq \mu_0$$

Test Statistic:

$$z = \frac{\bar{x} - \mu_0}{\sigma/\sqrt{n}}$$

If σ is unknown, substitute s for σ in computing z.

Rejection Rule at a Level of Significance of α

Reject H_0 if $z < -z_{\alpha/2}$ or if $z > z_{\alpha/2}$ (9.3)

---

p-Values for Two-Tailed Tests

Assuming that the null hypothesis is true, the *p*-value is the probability of obtaining a sample result that is at least as unlikely as what is observed. A small *p*-value indicates the sample result is unusual given the assumption that H_0 is true. Thus, as with the one-tailed hypothesis tests, a small *p*-value leads to the rejection of H_0.

Let us compute the *p*-value for the Superflight golf ball example. The sample mean of $\bar{x} = 278.5$ has a corresponding z value of $-.75$. The table for the standard normal probability distribution shows that the area between the mean and $z = -.75$ is .2734. Thus, the area in the lower tail less than or equal to $\bar{x} = 278.5$ and $z = -.75$ is $.5000 - .2734 = .2266$. Looking at Figure 9.8, we see that the lower-tailed portion of the rejection region has an area or probability of $\alpha/2 = .05/2 = .025$. Thus, with $.2266 > .025$, the test statistic does not fall in the rejection region, and the null hypothesis cannot be rejected.

One question remains: What value should we report as the *p*-value for the two-tailed test? At first glance, you may be inclined to say the *p*-value is .2266. If this is your choice, you will have to remember two different rules: one for the one-tailed test, which is to reject H_0 if the *p*-value $< \alpha$, and another for the two-tailed test, which is to reject H_0 if the *p*-value $< \alpha/2$. Alternatively, suppose we define the *p*-value for a two-tailed test as *double* the area found in the tail of the distribution? Thus, for the Superflight example, we would define the *p*-value to be $2(.2266) = .4532$. The advantage of this definition of the *p*-value for a two-tailed test is that the *p*-value can be compared directly to the level of significance α. Thus, with $.4532 > .05$, we see that the null hypothesis cannot be rejected. By remembering that the *p*-value for a two-tailed test is simply double the area found in the tail of the distribution, our previous rule to reject H_0 if the *p*-value $< \alpha$ will be true for all hypothesis tests.

FIGURE 9.9   **Minitab Output for the Superflight Golf Ball Hypothesis Test**

```
MTB > ZTEST 280 12 C1


TEST OF MU = 280.000 VS MU N.E. 280.000
THE ASSUMED SIGMA = 12.0


              N      MEAN    STDEV    SE MEAN       Z    P VALUE
C1           36   278.500   12.000     2.000    -0.75      0.45
```

Computer Software and Hypothesis Testing

Computer software packages are helpful in performing the computations required for a hypothesis test. The Minitab printout for the Superflight golf ball hypothesis test is shown in Figure 9.9. The data for the yards driven by each of the 36 golf balls has previously been entered into a Minitab worksheet and stored in the computer. Since the sample size is large, the user selects the command ZTEST to indicate a z test statistic is to be used in the hypothesis test computation. The full command ZTEST 280 12 C1 indicates the test is to be conducted for the null hypothesis H_0: $\mu = 280$ with the population standard deviation estimated by the sample standard deviation $s = 12$. The C1 indicates that the data for the 36 golf balls is stored in Column 1 of the Minitab worksheet.

The output in Figure 9.9 is easily interpreted by the user. The number of items in the data set, the sample mean, the sample standard sample deviation, and the standard error of the mean are provided. The z value of -0.75 and the associated p-value of 0.45 show that at the .05 level of significance, the null hypothesis of $\mu = 280$ cannot be rejected.

The Relationship Between Interval Estimation and Hypothesis Testing

In Chapters 8 and 9 we have discussed statistical procedures that can be used to make inferences about the value of a population mean. In Chapter 8 we discussed interval estimation, and in Chapter 9 we have focused on hypothesis testing. In the case of interval estimation, the population mean μ was unknown. Once the sample was selected and the sample mean $\bar{x}$ computed, we developed an interval around the value of $\bar{x}$ that had a good chance of including the value of the parameter μ. The interval estimate computed was referred to as a confidence interval with $1 - \alpha$ defined as the confidence coefficient. In the large-sample case, the interval estimate of a population mean was given by

$$\bar{x} \pm z_{\alpha/2} \frac{\sigma}{\sqrt{n}} \qquad (9.4)$$

Conducting a hypothesis test requires us first to make an assumption about the value of a population parameter. In the case of the population mean, the two-tailed hypothesis test has the form

$$H_0: \mu = \mu_0$$

$$H_a: \mu \neq \mu_0$$

where μ_0 is the hypothesized value for the population mean. Using the rejection rule provided by (9.3), we see that the region over which we do not reject H_0 includes all values of the sample mean $\bar{x}$ that are within $-z_{\alpha/2}$ and $+z_{\alpha/2}$ standard errors of μ_0. Thus, the following expression provides the do-not-reject region for the sample mean $\bar{x}$ in a two-tailed hypothesis test with a level of significance of α:

$$\mu_0 \pm z_{\alpha/2} \frac{\sigma}{\sqrt{n}} \tag{9.5}$$

A close look at (9.4) and (9.5) will provide insight into the relationship between the estimation and hypothesis-testing approaches to statistical inference. Note in particular that both procedures require the computation of the values $z_{\alpha/2}$ and $\sigma/\sqrt{n}$. Focusing on α, we see that a confidence coefficient of $(1 - \alpha)$ for interval estimation corresponds to a level of significance of α in hypothesis testing. For example, a 95% confidence interval for estimation corresponds to a .05 level of significance for hypothesis testing. Furthermore, (9.4) and (9.5) show that since $z_{\alpha/2}\,(\sigma/\sqrt{n})$ is the plus or minus value for both expressions, if $\bar{x}$ falls in the do-not-reject region defined by (9.5), the hypothesized value μ_0 will be in the confidence interval defined by (9.4). Conversely, if the hypothesized value μ_0 falls in the confidence interval defined by (9.4), the sample mean $\bar{x}$ will be in the do-not-reject region for the hypothesis H_0: $\mu = \mu_0$. These observations lead to the following procedure for using confidence interval results to draw hypothesis-testing conclusions:

A Confidence Interval Approach to Testing a Hypothesis of the Form

$$H_0\colon \mu = \mu_0$$

$$H_a\colon \mu \neq \mu_0$$

1. Select a simple random sample from the population, and use the value of the sample mean $\bar{x}$ to develop the confidence interval for the population mean μ.

$$\bar{x} \pm z_{\alpha/2} \frac{\sigma}{\sqrt{n}}$$

2. If the confidence interval contains the hypothesized value μ_0, do not reject H_0. Otherwise, reject H_0.

Let us return to the Superflight golf ball study discussed earlier to demonstrate the use of a confidence interval for hypothesis testing. The Superflight golf ball study resulted in the following two-tailed test:

$$H_0\colon \mu = 280$$

$$H_a\colon \mu \neq 280$$

To test this hypothesis with a level of significance of $\alpha = .05$, we sampled 36 golf balls and found a sample mean distance of $\bar{x} = 278.5$ yards and a sample standard deviation of $s = 12$ yards. Using these results with $z_{.025} = 1.96$, the 95% confidence interval estimate of the population mean becomes

$$\bar{x} \pm z_{.025} \frac{\sigma}{\sqrt{n}}$$

$$278.5 \pm 1.96 \frac{12}{\sqrt{36}}$$

$$278.5 \pm 3.92$$

or

$$274.58 \text{ to } 282.42$$

This finding enables the quality-control manager to conclude with 95% confidence that the mean distance for the population of golf balls is between 274.58 and 282.42 yards. Since the hypothesized value for the population mean, $\mu_0 = 280$, is in this interval, the hypothesis-testing conclusion is that the null hypothesis, H_0: $\mu = 280$, cannot be rejected.

Note that this discussion and example have been devoted to two-tailed hypothesis tests about a population mean. However, the same confidence interval and hypothesis-testing relationship exists for other population parameters as well. In addition, the relationship can be extended to make one-tailed tests about population parameters. However, this requires the development of one-sided confidence intervals.

**NOTES &
COMMENTS**

1. The p-value is called the observed level of significance; it depends only on the sample outcome. However, it is necessary to know whether the hypothesis test being investigated is one-tailed or two-tailed. Given the value of $\bar{x}$ in a sample, the p-value for a two-tailed test will always be *twice* the area in the tail of the sampling distribution at the value of $\bar{x}$.
2. The interval-estimation approach to hypothesis testing helps to highlight the role of the sample size. From (9.4) it can be seen that larger sample sizes n lead to more narrow confidence intervals. Thus, for a given level of significance α, a larger sample is less likely to lead to an interval containing μ_0 when the null hypothesis is false. That is, the larger sample size will provide a higher probability of rejecting H_0 when H_0 is false.

☐ ☐ **Exercises**

Methods

21. Consider the following hypothesis test:

$$H_0: \mu = 10$$

$$H_a: \mu \neq 10$$

A sample of 36 provides a sample mean of 11 and sample standard deviation of 2.5.
a. Using $\alpha = .05$, what is the rejection rule?
b. Compute the value of the test statistic z. What is your conclusion?

SELF TEST ▶ √**22.** Consider the following hypothesis test:

$$H_0: \mu = 15$$

$$H_a: \mu \neq 15$$

A sample of 50 gives a sample mean of 14.2 and sample standard deviation of 5.
a. Using $\alpha = .02$, what is the rejection rule?
b. Compute the value of the test statistic z.
c. What is the p-value?
d. What is your conclusion?

23. Consider the following hypothesis test:

$$H_0: \mu = 25$$

$$H_a: \mu \neq 25$$

A sample of 80 is used, and the population standard deviation is 10. Use $\alpha = .05$. Compute the value of the test statistic z and specify your conclusion for each of the following sample results:
a. $\bar{x} = 22.0$ b. $\bar{x} = 27.0$ c. $\bar{x} = 23.5$ d. $\bar{x} = 28.0$

24. Consider the following hypothesis test:

$$H_0: \mu = 5$$

$$H_a: \mu \neq 5$$

Assume the test statistics are as shown below. Compute the corresponding p-values and specify your conclusions based on $\alpha = .05$.
a. $z = 1.80$ b. $z = -.45$ c. $z = 2.05$
d. $z = -3.50$ e. $z = -1.00$

Applications

25. The U.S. Bureau of the Census reported mean hourly earnings in the wholesale trade industry of $9.70 per hour in 1987. A sample of 49 wholesale trade workers in a particular city showed a sample mean hourly wage of $\bar{x} = \$9.30$ with a standard deviation of $s = \$1.05$. Use $H_0: \mu = 9.70$ and $H_a: \mu \neq 9.70$.
a. Test to see if wage rates in the city differ significantly from the reported $9.70. Use $\alpha = .05$.
b. What is the p-value for this test?

26. A study of the operation of a city-owned parking garage shows a historical mean parking time of 220 minutes per car. The garage area has recently been remodeled, and the parking charges have been increased. The city manager would like to know if these changes have had any effect on the mean parking time. Test the hypotheses $H_0: \mu = 220$ and $H_a: \mu \neq 220$ at a .05 level of significance.
a. What is your conclusion if a sample of 50 cars showed $\bar{x} = 208$ and $s = 80$?
b. What is the p-value?

27. A production line operates with a filling weight standard of 16 ounces per container. Overfilling or underfilling is a serious problem, and the production line should be shut down if either occurs. From past data, σ is known to be .8 ounces. A quality-control inspector samples 30 items every 2 hours and at that time makes the decision of whether or not to shut the line down for adjustment.
a. With a .05 level of significance, what is the rejection rule for the hypothesis-testing procedure?
b. If a sample mean of $\bar{x} = 16.32$ ounces occurs, what action would you recommend?
c. If $\bar{x} = 15.82$ ounces, what action would you recommend?
d. What is the p-value for parts (b) and (c)?

28. An automobile assembly-line operation has a scheduled mean completion time of 2.2 minutes. Because of the effect of completion time on both earlier and later assembly operations, it is important to maintain the 2.2-minute standard. A random sample of 45 times shows a sample mean completion time of 2.39 minutes, with a sample standard deviation of .20 minutes. Use a .02 level of significance and test whether or not the operation is meeting its 2.2-minute standard.

29. Historically, evening long-distance phone calls from a particular city have averaged 15.20 minutes per call. In a random sample of 35 calls, the sample mean time was 14.30 minutes per call, with a sample standard deviation of 5 minutes. Use this sample information to test whether or not

there has been a change in the mean duration of long distance phone calls. Use a .05 level of significance. What is the *p*-value?

30. The mean salary for full professors at public universities is $45,300 (from the *Statistical Abstract of the United States*, 1988). A sample of 36 full professors at business colleges showed $\bar{x} = \$52,000$ and $s = \$5,000$. Choose $H_0: \mu = 45,300$ and $H_a: \mu \neq 45,300$.
   a. Develop a 95% confidence interval for the mean salary of business college professors using the sample data.
   b. Use the confidence interval to conduct the hypothesis test. What is your conclusion?

31. At Western University the historical mean scholarship examination score of entering students has been 900, with a standard deviation of 180. Each year a sample of applications is taken to see if the examination scores are at the same level as in previous years. The null hypothesis tested is $H_0: \mu = 900$. A sample of 200 students in this year's class shows a sample mean score of 935. Use a .05 level of significance.
   a. Use a confidence interval approach to conduct a hypothesis test.
   b. Use a test statistic to test this hypothesis.
   c. What is the *p*-value for this test?

32. An industry pays an average wage rate of $9.00 per hour. A sample of 36 workers from one company showed a mean wage of $\bar{x} = \$8.50$ and a sample standard deviation of $s = \$.60$.
   a. A one-sided confidence interval uses the sample results to establish either an upper limit or a lower limit for the value of the population parameter. For this exercise establish an upper 95% confidence limit for the hourly wage rate paid by the company. The form of this one-sided confidence interval requires that we be 95% confident that the population mean is this value or less. What is the 95% confidence statement for this one-sided confidence interval?
   b. Use the one-sided confidence interval result to test the hypothesis $H_0: \mu \geq 9$. What is your conclusion? Explain.

9.5   Hypothesis Tests about A Population Mean: Small-Sample Case

The methods of hypothesis testing that we have discussed thus far have required sample sizes of at least 30. The reason for this is that in the large-sample situation ($n \geq 30$), the central limit theorem enables us to approximate the sampling distribution of $\bar{x}$ with a normal probability distribution. Thus

$$z = \frac{\bar{x} - \mu_0}{\sigma/\sqrt{n}} \tag{9.6}$$

is a standard normal random variable and is used as the test statistic. The sample standard deviation *s* can be used in (9.6) when the population standard deviation σ is unknown.

Assume that the sample size is small ($n < 30$) and that the sample standard deviation *s* is used to estimate the population standard deviation σ.* If it is also reasonable to assume that the population has a normal distribution, the *t* distribution can be used to make inferences about the value of a population mean. In using the *t* distribution for hypothesis tests about a population mean, the test statistic is

$$t = \frac{\bar{x} - \mu_0}{s/\sqrt{n}} \tag{9.7}$$

This test statistic has a *t* distribution with $n - 1$ degrees of freedom. Noting the similarities of the test statistics defined by (9.6) and (9.7), it should not be surprising that the

---

*If $n < 30$, if the population has a normal distribution and if the population standard deviation σ is *known*, (9.6) can be used as the test statistic for the small-sample hypothesis test.

small-sample procedure for hypothesis tests about a population mean is very similar to the large-sample procedures presented in Sections 9.3 and 9.4.

Consider the following example: A company that specializes in products for home gardening has developed a new plant food that has been designed to increase the growing height of plants. The new plant food was tested on a sample of 12 plants of a type known to have a mean growing height of 18 inches. Results showed a sample mean height of 19.4 inches and a sample standard deviation of 3 inches. Assume that the heights of the plants are normally distributed.

Using a .10 level of significance, is there reason to believe that the new plant food increases plant height? The null and alternative hypotheses are as follows:

$$H_0: \mu \le 18$$

$$H_a: \mu > 18$$

The new plant food will be judged to increase plant height if H_0 can be rejected.

The rejection region is located in the upper tail of the sampling distribution. With $n - 1 = 12 - 1 = 11$ degrees of freedom, Table 2 of Appendix B shows that $t_{.10} = 1.363$. Thus, the rejection rule is

$$\text{Reject } H_0 \text{ if } t > 1.363$$

Using (9.7) with $\bar{x} = 19.4$ and $s = 3$, we have the following value for the test statistic:

$$t = \frac{\bar{x} - \mu_0}{s/\sqrt{n}} = \frac{19.4 - 18}{3/\sqrt{12}} = 1.62$$

Since 1.62 is greater than 1.363, the null hypothesis is rejected. At the .10 level of significance, it can be concluded that the mean plant height exceeds 18 inches when the new plant food is used. Figure 9.10 shows that the value of the test statistic is in the rejection region.

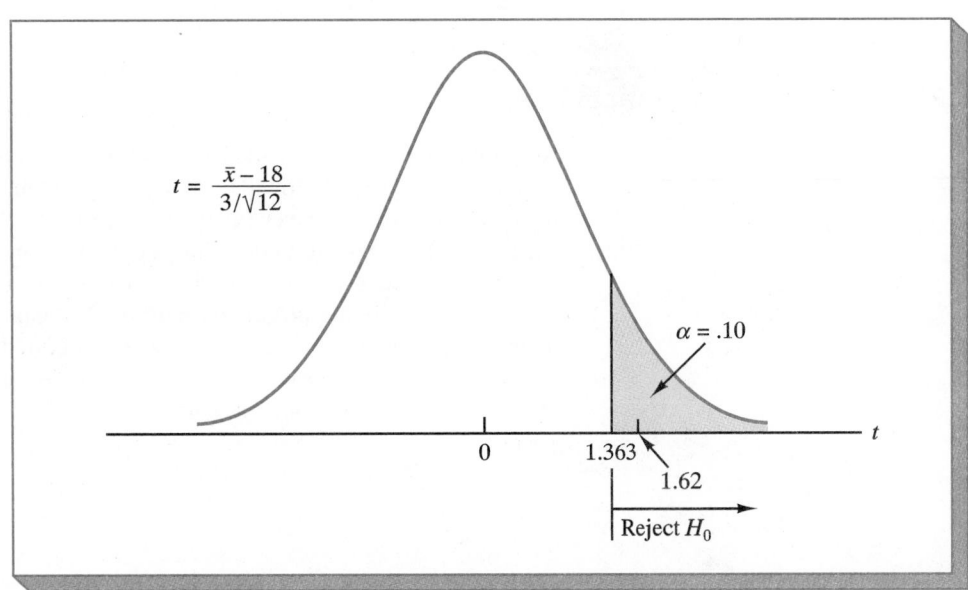

FIGURE 9.10

Value of the Test Statistic ($t = 1.62$) for the Plant Food Hypothesis Test

p-Values and the t Distribution

Let us consider the p-value for the plant food hypothesis test. As discussed in Sections 9.3 and 9.4, the p-value can be interpreted as the observed level of significance for the test. The usual rule applies: If the p-value is less than the level of significance α, the null hypothesis can be rejected. Unfortunately, the format of the t distribution table provided in most statistics textbooks does not have sufficient detail to determine the exact p-value for the test. However, we can still use the t distribution table to identify a range for the p-value. For example, the t distribution used in the plant food hypothesis test has 11 degrees of freedom. Referring to Table 2 of Appendix B, we see that row 11 provides the following information about a t distribution with 11 degrees of freedom.

Area in Upper Tail	.10	.05	.025	.01	.005
t Value	1.363	1.796	2.201	2.718	3.106

The computed t value for the hypothesis test was $t = 1.62$. The p-value is the area in the tail corresponding to $t = 1.62$. From the information above, we see 1.62 is between 1.363 and 1.796. Thus, although we cannot determine the exact p-value associated with $t = 1.62$, we do know that the p-value must be between .10 and .05. With a level of significance of $\alpha = .10$, we know that the p-value must be less than .10; thus, the null hypothesis is rejected.

An advantage of computer software packages is that computer routines can be used to compute the exact p-value for the t distribution test. The Minitab printout for the plant food hypothesis test is shown in Figure 9.11. The TTEST command is selected so that the t distribution will be used in the computations. The null hypothesis value of 18 is specified in the TTEST command. The Minitab subcommand of ALTERNATIVE = 1 indicates that the one-tailed test has the alternative hypothesis H_a: $\mu > 18$. The subcommand ALTERNATIVE = -1 would have been used for the alternative hypothesis H_a: $\mu < 18$. The printout shown in Figure 9.11 provides summary statistics, the value of the test statistic $t = 1.62$, and the exact p-value = 0.067. Thus, since the p-value (0.067) is less than the level of significance (.10), the null hypothesis can be rejected.

A Two-Tailed Test

As an example of a two-tailed hypothesis test about a population mean using a small sample, consider the following production problem. A production process is designed to fill containers with a mean filling weight of $\mu = 16$ ounces. An undesirable condition exists if the process is underfilling containers and the consumer is not receiving the amount of product indicated on the container label. In addition, an equally undesirable condition exists if the process is overfilling containers; in this case, the firm is losing money since the process is placing more product in the container than is required. Quality-assurance personnel periodically select a simple random sample of eight containers and test the following two-tailed hypotheses:

$$H_0: \mu = 16$$

$$H_a: \mu \neq 16$$

FIGURE 9.11 **Minitab Output for the Plant Food Hypothesis Test**

```
MTB > TTEST 18 C1;
SUBC> ALTERNATIVE = 1.

TEST OF MU = 18.000 VS MU G.T. 18.000

              N      MEAN     STDEV    SE MEAN       T     P VALUE
C1           12    19.400     3.000      0.866    1.62       0.067
```

If H_0 is rejected, the production manager will request that the production process be stopped and that the mechanism for regulating filling weights be readjusted to provide a mean filling weight of 16 ounces. If the sample provides data values of 16.02, 16.22, 15.82, 15.92, 16.22, 16.32, 16.12, and 15.92 ounces, what action should be taken at a .05 level of significance? Assume that the population of filling weights is normally distributed.

Since the data have not been summarized, we must first compute the sample mean and sample standard deviation. Doing so provides the following results:

$$\bar{x} = \frac{\Sigma x_i}{n} = \frac{128.56}{8} = 16.07 \text{ ounces}$$

and

$$s = \sqrt{\frac{\Sigma (x_i - \bar{x})^2}{n - 1}} = \sqrt{\frac{.22}{7}} = .18 \text{ ounces}$$

With a two-tailed test and a level of significance of $\alpha = .05$, $-t_{.025}$ and $t_{.025}$ determine the rejection region for the test. Using the table for the t distribution, we find that with $n - 1 = 8 - 1 = 7$ degrees of freedom, $-t_{.025} = -2.365$ and $t_{.025} = +2.365$. Thus, the rejection rule is written

$$\text{Reject } H_0 \text{ if } t < -2.365 \text{ or if } t > 2.365$$

Using $\bar{x} = 16.07$ and $s = .18$, we have

$$t = \frac{\bar{x} - \mu_0}{s/\sqrt{n}} = \frac{16.07 - 16.00}{.18/\sqrt{8}} = 1.10$$

Since $t = 1.10$ is not in the rejection region, the null hypothesis $\mu = 16$ ounces cannot be rejected. There is not enough evidence to stop the production process.

Using Table 2 of Appendix B and the row for 7 degrees of freedom, we see that the computed t value of 1.10 has an upper tail area of *more than* .10. Although the format of the t distribution table prevents us from being more specific, we can at least conclude that the two-tailed p-value is greater than $2(.10) = .20$. Since this is greater than the .05 level of significance, we see that the p-value leads to the same conclusion; that is, do not reject H_0. The computer solution for this hypothesis test shows $t = 1.10$ and the exact p-value = .31.

❑ ❑ Exercises

Methods

33. Consider the following hypothesis test:

$$H_0: \mu \leq 10$$

$$H_a: \mu > 10$$

A sample of 16 provides a sample mean of 11 and sample standard deviation of 3.
a. Using $\alpha = .05$, what is the rejection rule?
b. Compute the value of the test statistic t. What is your conclusion?

SELF TEST ▶ ✓34. Consider the following hypothesis test:

$$H_0: \mu = 20$$

$$H_a: \mu \neq 20$$

Data from a sample of six items are as follows:

18, 20, 16, 19, 17, 18.

a. Compute the sample mean.
b. Compute the sample standard deviation.
c. Using $\alpha = .05$, what is the rejection rule?
d. Compute the value of the test statistic t.
e. What is your conclusion?

35. Consider the following hypothesis test:

$$H_0: \mu \geq 15$$

$$H_a: \mu < 15$$

A sample of 22 is used and the sample standard deviation is 8. Use $\alpha = .05$. Provide the value of the test statistic t and your conclusion for each of the following sample results:
a. $\bar{x} = 13.0$ b. $\bar{x} = 11.5$ c. $\bar{x} = 15.0$ d. $\bar{x} = 19.0$

36. Consider the following hypothesis test:

$$H_0: \mu \leq 50$$

$$H_a: \mu > 50$$

Assume a sample of 16 items provides the test statistics as shown below. What can you say about the p-values in each case? What are your conclusions based on $\alpha = .05$.
a. $t = 2.602$ b. $t = 1.341$ c. $t = 1.960$
d. $t = 1.055$ e. $t = 3.261$

Applications

SELF TEST ▶ ✓37. City Homes bought 101 badly deteriorated Baltimore rowhouses and turned them into housing for the poor. The mean monthly rental for these houses was $200 (*Financial World*, November 29, 1988). Nine hundred homes in another city were built and turned over to a foundation to rent. A sample of 10 of these homes showed the following monthly rental rates:

$220, $190, $250, $230, $185, $210, $240, $260, $200, $195.

a. Using $\alpha = .05$, test to see if the mean rental rate for the population of 900 homes exceeds the $200 per month mean rental rate in Baltimore. What is your conclusion?
b. What is the p-value for this test.

✓ **38.** The average hourly wage in the United States is \$10.05 (*The Tampa Tribune*, December 15, 1991). Assume that a sample of 25 individuals in Phoenix, Arizona, showed a sample mean wage of \$10.83 per hour with a sample standard deviation of \$3.25 per hour. Test the hypotheses H_0: $\mu = 10.05$ and H_a: $\mu \neq 10.05$ to see if the population mean in Phoenix differs from the mean throughout the United States. Using a .05 level of significance, what is your conclusion?

39. It is estimated that, on the average, a housewife with a husband and two children works 55 hours or less per week on household-related activities. Shown below are the hours worked during a week for a sample of eight housewives.

$$58, 52, 64, 63, 59, 62, 62, 55.$$

a. Use $\alpha = .05$ to test the hypotheses H_0: $\mu \leq 55$, H_a: $\mu > 55$. What is your conclusion about the mean number of hours worked per week?

b. What can you say about the p-value?

40. A study of a drug designed to reduce blood pressure used a sample of 25 men between the ages of 45 and 55. With μ indicating the mean change in blood pressure for the population of men receiving the drug, the hypotheses in the study were written: H_0: $\mu \geq 0$ and H_a: $\mu < 0$. Rejection H_0 shows that the mean change is negative, indicating that the drug is effective in lowering blood pressure.

a. At a .05 level of significance, what conclusion should be drawn if $\bar{x} = -10$ and $s = 15$?

b. What can you say about the p-value?

41. Last year the number of lunches served at an elementary-school cafeteria was normally distributed with a mean of 300 lunches per day. At the beginning of the current year, the price of a lunch was raised by 25¢. A sample of 6 days during the months of September, October, and November provided the following number of children being served lunches: 290, 275, 310, 260, 270, and 275. Do these data indicate that the mean number of lunches per day has dropped compared to last year? Test the hypothesis H_0: $\mu \geq 300$ against the alternative hypothesis H_a: $\mu < 300$ at a .05 level of significance.

42. Joan's Nursery specializes in custom-designed landscaping for residential areas. The labor cost associated with a particular landscaping proposal is estimated based on the number of plantings of trees, shrubs, and so on associated with the project. For cost-estimating purposes, management figures 2 hours of labor time for the planting of a medium-size tree. Actual times from a sample of 10 plantings during the past month are as follows (times in hours):

$$1.9, 1.7, 2.8, 2.4, 2.6, 2.5, 2.8, 3.2, 1.6, 2.5.$$

Using a .05 level of significance, test to see if the mean tree-planting time exceeds 2 hours. What is your conclusion, and what recommendations would you consider making to management?

9.6   Hypothesis Tests about a Population Proportion

Using the symbol p to denote the population proportion and p_0 to denote a particular hypothesized value for the population proportion, the three forms for a hypothesis test about a population proportion are as follows:

$$H_0: p \geq p_0 \qquad H_0: p \leq p_0 \qquad H_0: p = p_0$$

$$H_a: p < p_0 \qquad H_a: p > p_0 \qquad H_a: p \neq p_0$$

The first two forms are one-tailed tests, whereas the third form is a two-tailed test. The specific form used depends upon the application of interest.

Hypothesis tests about a population proportion are based on the difference between the sample proportion $\bar{p}$ and the hypothesized value p_0. The methods used to conduct the tests are very similar to the procedures used for hypothesis tests about a population mean. The only difference is that we use the sample proportion $\bar{p}$ and its standard deviation $\sigma_{\bar{p}}$ in developing the test statistic. We begin by formulating null and alternative hypotheses about the value of the population proportion. Then, using the value of the sample proportion $\bar{p}$ and its standard deviation $\sigma_{\bar{p}}$, we compute a value for the test statistic z. Comparing the value of the test statistic to the critical value enables us to determine whether or not the null hypothesis should be rejected.

Let us illustrate hypothesis testing for a population proportion by considering the situation faced by Pine Creek golf course. Over the past few months, 20% of the players at Pine Creek have been female. In an effort to increase the proportion of female players, course management used a special promotion for female players. After one week, a random sample of 400 players showed 300 male and 100 female players. Course management would like to determine if the data support the conclusion that there has been an increase in the proportion of female players at Pine Creek.

Since we want to determine if the effect of the promotion has been to increase the proportion of women golfers, the null alternative hypotheses are as follows:

$$H_0: p \le .20$$

$$H_a: p > .20$$

As usual, we tentatively begin the hypothesis-testing procedure by assuming that H_0 is true with $p = .20$. Using the sample proportion $\bar{p}$ to estimate p, we next consider the sampling distribution of $\bar{p}$. Since $\bar{p}$ is an unbiased estimator of p, we know that if $p = .20$, the mean of the sampling distribution of $\bar{p}$ is .20. In addition, we know from Chapter 7 that the standard deviation of $\bar{p}$ is given by

$$\sigma_{\bar{p}} = \sqrt{\frac{p(1-p)}{n}}$$

With the assumed value of $p = .20$ and a sample size of $n = 400$, the standard deviation of $\bar{p}$ for the Pine Creek golf course problem is

$$\sigma_{\bar{p}} = \sqrt{\frac{.20(1-.20)}{400}} = .02$$

In Chapter 7 we saw that the sampling distribution of $\bar{p}$ can be approximated by a normal probability distribution if both np and $n(1-p)$ are greater than or equal to 5. In the Pine Creek case, $np = 400(.20) = 80$ and $n(1-p) = 400(.80) = 320$; thus, the normal distribution approximation is appropriate. The sampling distribution of $\bar{p}$ for the Pine Creek problem is shown in Figure 9.12.

Since the sampling distribution of $\bar{p}$ is approximately normal, the following test statistic can be used.

Test Statistic for Hypothesis Tests about Proportions

$$z = \frac{\bar{p} - p_0}{\sigma_{\bar{p}}} \tag{9.8}$$

where

$$\sigma_{\bar{p}} = \sqrt{\frac{p_0(1 - p_0)}{n}} \tag{9.9}$$

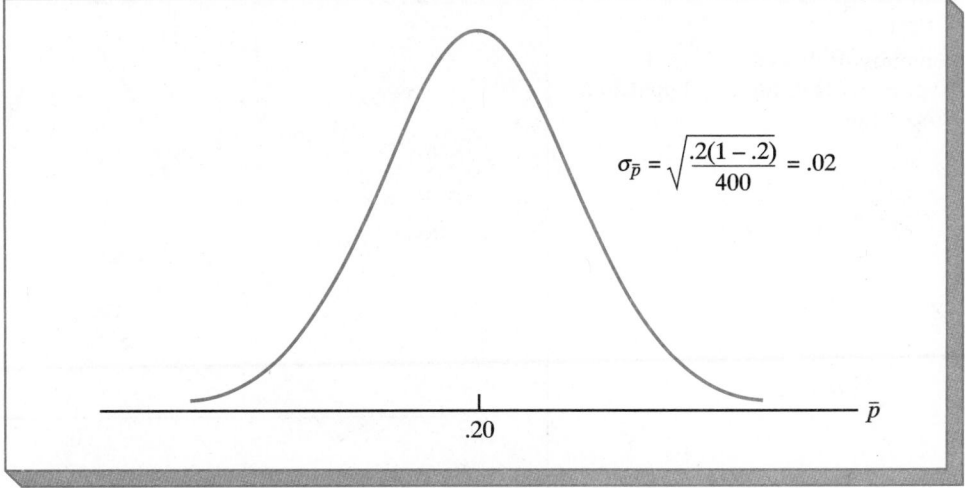

$$\sigma_{\bar{p}} = \sqrt{\frac{.2(1 - .2)}{400}} = .02$$

Let us assume that $\alpha = .05$ has been selected as the level of significance for the test. With $z_{.05} = 1.645$, the upper-tail rejection region for the hypothesis test (see Figure 9.13) provides the following rejection rule:

$$\text{Reject } H_0 \text{ if } z > 1.645$$

Once the rejection rule has been determined, it is necessary to collect the data, compute the value of the point estimate $\bar{p}$, and compute the corresponding value of the test statistic z. By comparing the value of z to the critical value ($z_{.05} = 1.645$), a decision on whether or not to reject the null hypothesis can be made.

Since 100 of the 400 players during the promotion were female, we obtain $\bar{p} = 100/400 = .25$. With $\sigma_{\bar{p}} = .02$, the value of the test statistic is

$$z = \frac{\bar{p} - p_0}{\sigma_{\bar{p}}} = \frac{.25 - .20}{.02} = 2.5$$

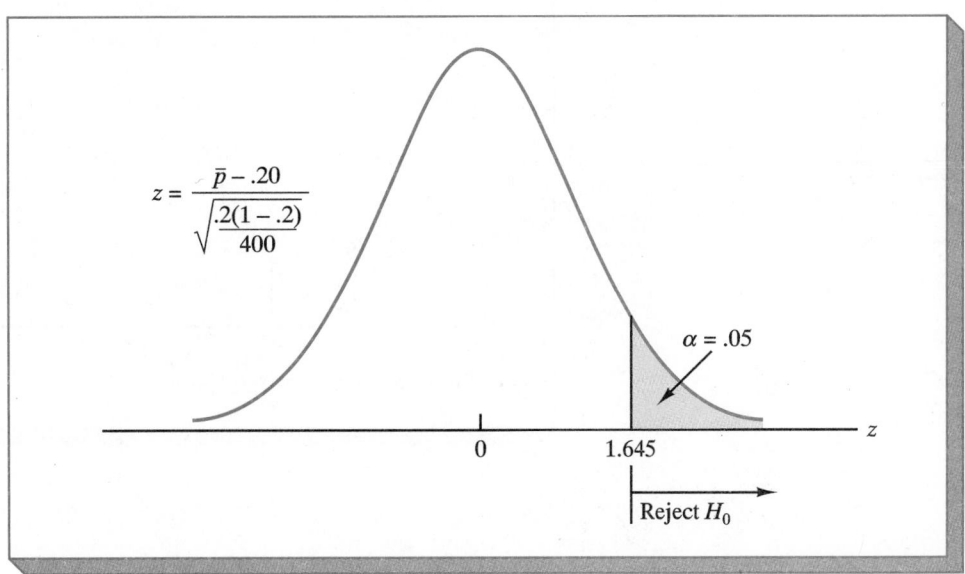

$$z = \frac{\bar{p} - .20}{\sqrt{\dfrac{.2(1 - .2)}{400}}}$$

$\alpha = .05$

Reject H_0

F I G U R E   9.14
**Summary of Rejection Rules for
Hypothesis Tests about a Population
Proportion**

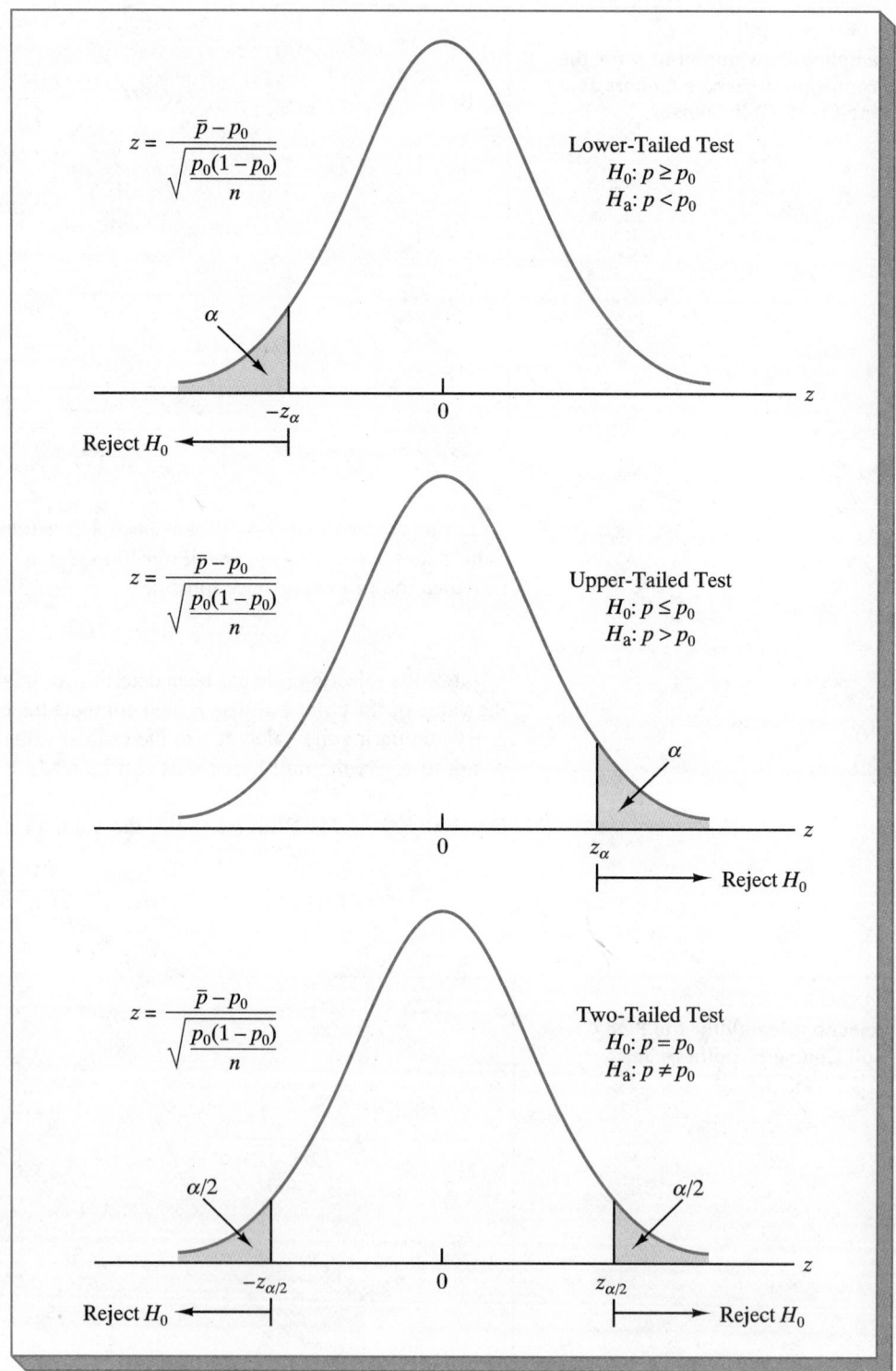

Thus, the rejection rule tells us to reject H_0. As a result, Pine Creek can conclude that there has been an increase in the proportion of female players.

Using the table of areas for the standard normal probability distribution, we find that the p-value for the test can also be computed. For example, with $z = 2.50$, the table of areas shows a .4938 area or probability between the mean and $z = 2.50$. Thus, the p-value for the test is $.5000 - .4938 = .0062$. With a p-value less than α, the rule for hypothesis tests using p-values shows that the null hypothesis can be rejected.

We see that hypothesis tests about a population proportion and a population mean are similar; the primary difference being that the test statistic is based on the sampling distribution of $\bar{x}$ when the hypothesis test involves a population mean and on the sampling distribution of $\bar{p}$ when the hypothesis test involves a population proportion. The tentative assumption that the null hypothesis is true, the use of the level of significance to establish the critical value, and the comparison of the test statistic to the critical value are identical for both testing procedures. A summary of the decision rules for hypothesis tests about a population proportion is presented in Figure 9.14. We assume the large-sample case ($np \geq 5$ and $n(1 - p) \geq 5$) where the normal probability distribution is a good approximation to the sampling distribution of $\bar{p}$.

NOTES & COMMENTS

We have not shown the procedure for small-sample hypothesis tests involving population proportions. In the small-sample case, the sampling distribution of $\bar{p}$ follows the binomial distribution and thus the normal approximation is not applicable. More advanced texts show how hypothesis tests are conducted for this situation. However, in practice small-sample tests are rarely conducted for a population proportion.

☐ ☐ Exercises

Methods

43. Consider the following hypothesis test:

$$H_0: p \leq .50$$

$$H_a: p > .50$$

A sample of 200 provided a sample proportion $\bar{p} = .57$.
a. Using $\alpha = .05$, what is the rejection rule?
b. Compute the value of the test statistic z. What is your conclusion?

SELF TEST ▷ 44. Consider the following hypothesis test:

$$H_0: p = .20$$

$$H_a: p \neq .20$$

A sample of 400 provided a sample proportion of $\bar{p} = .175$.
a. Using $\alpha = .05$, what is the rejection rule?
b. Compute the value of the test statistic z.
c. What is the p-value?
d. What is your conclusion?

45. Consider the following hypothesis test:

$$H_0: p \geq .75$$

$$H_a: p < .75$$

A sample of 300 is selected. Use $\alpha = .05$. Provide the value of the test statistic z, the p-value, and your conclusion for each of the following sample results:

a. $\bar{p} = .68$ **b.** $\bar{p} = .72$ **c.** $\bar{p} = .70$ **d.** $\bar{p} = .77$

Applications

46. The Honolulu Board of Water Supply suggested the water-by-request rule be adopted at restaurants on the island of Oahu to conserve water. A restaurateur stated that 30% of the patrons never touch their water (*The Honolulu Advertiser,* December 28, 1991). Thus, the conservation of water would come not only from the unused water in each glass, but also the water saved in washing the glasses. Test $H_0: p = .30$ versus $H_a: p \neq .30$. Assume a sample of 480 patrons at restaurants showed 128 patrons never touched their water. Test the restauranteur's claim at a .05 level of significance. What is the p-value and what is your conclusion?

SELF TEST   ▶  **47.** The American Association of Individual Investors conducted a survey to learn about member responses to the October 1987 stock market crash. (*AAII Journal,* May 1988). Results indicated that 11.3% of the AAII members owned stock on margin. A sample of 200 individual clients with a large brokerage firm showed that 29 owned stock on margin. Let p be the proportion of all the brokerage firm's clients owning stock on margin. Use $H_0: p \leq .113$ and $H_a: p > .113$ in a hypothesis test. Does the sample information indicate that the proportion of the brokerage firm's clients owning stock on margin is greater than a .113? Use $\alpha = .05$.

48. The director of a college placement office claims that at least 80% of graduating seniors have made employment commitments 1 month prior to graduation. At a .05 level of significance, what is your conclusion if a sample of 100 seniors shows that 75 actually made employment commitments 1 month prior to graduation? Should the director's claim be rejected? What is the p-value?

49. A magazine claims that 25% of its readers are college students. A random sample of 200 readers is taken. It is found that 42 of these readers are college students. Use a .10 level of significance to test $H_0: p = .25$ and $H_a: p \neq .25$. What is the p-value?

50. A new television series must prove that it has more than 25% of the viewing audience after its initial 13-week run in order to be judged successful. Assume that in a sample of 400 households, 112 were watching the series.

a. At a .10 level of significance, can the series be judged successful based on the sample information?

b. What is the p-value for the sample results? What is your hypothesis-testing conclusion?

51. An accountant believes that the company's cash-flow problems are a direct result of the slow collection of accounts receivable. The accountant claims that at least 70% of the current accounts receivable are over 2 months old. A sample of 120 accounts receivable shows 78 over 2 months old. Test the accountant's claim at the .05 level of significance.

52. In 1987, it was reported that 26.8% of adults in California smoked (*Newsweek,* April 6, 1992). In 1988, California passed Proposition 99, a popular initiative designed to help discourage smoking. Researchers conducted a study in 1991 to learn about California's smoking rate and to identify the impact of Proposition 99 on the smoking population. Assume that a sample of 1000 adults showed 222 smokers. At a .01 level of significance, can it be concluded that the proportion of Californians smoking in 1991 is less than the proportion of Californians smoking in 1987? What is your conclusion?

53. A television station predicts election winners based on the following hypothesis test, where p is the proportion of voters selecting the leading candidate.

$$H_0: p \leq .50$$

$$H_a: p > .50$$

If H_0 can be rejected, the station will predict that the leading candidate is the winner.
a. What is the Type I error? What are the consequences of making this error?
b. What is the Type II error? What are the consequences of making this error?
c. Which value, $\alpha = .05$ or $\alpha = .01$, makes more sense as the level of significance?

54. A fast-food restaurant plans to initiate a special offer that will enable customers to purchase specially designed drink glasses featuring well-known cartoon characters. If more than 15% of the customers will purchase the glasses, the special offer will be initiated. A preliminary test has been set up at several locations, and 88 of 500 customers have purchased the glasses. Should the special glass offer be made? Conduct a hypothesis test that will support your decision. Use a .01 level of significance. What is your recommendation?

55. It has been claimed that 5% of the students at one college make blood donations during a given year. If, in a random sample, 10 of 250 students have given blood during the past year, test $H_0: p = .05$ versus $H_a: p \neq .05$. Use a .05 level of significance in reaching your conclusion. What is the p-value?

9.7   Hypothesis Testing and Decision Making

In Section 9.1 we noted three types of situations where hypothesis testing is used:

1. Testing research hypotheses.
2. Testing the validity of a claim.
3. Testing in decision making situations.

In the first two situations, action is taken only when the null hypothesis H_0 is rejected and hence the alternative hypothesis H_a concluded to be true. In the third situation—decision making—it is necessary to take action when the null hypothesis is not rejected as well as when it is rejected.

The hypothesis-testing procedures presented thus far have limited applicability in a decision-making situation because it is not considered appropriate to accept H_0 and take action based on the conclusion that H_0 is true. The reason given for not taking action when the test results indicate *do not reject* H_0 is that the decision to accept H_0 exposes the decision maker to the risk of making a Type II error; that is, accepting H_0 when it is false. With the hypothesis-testing procedures described in the previous sections, the probability of a Type I error was controlled by establishing a level of significance for the test. However, the probability of making the Type II error was not controlled.

Clearly, in certain decision-making situations the decision maker may want—and in some cases may be forced—to take action under both the conclusion *do not reject* H_0 and the conclusion *reject* H_0. A good illustration of this situation is lot-acceptance sampling, a topic we will discuss in more depth in Chapter 20. For example, a quality-control manager must decide to accept a shipment of batteries from a supplier or to return the shipment due to poor quality. Assume that design specifications require batteries from the supplier to have a mean useful life of at least 120 hours. To evaluate the quality of an incoming shipment, a sample of 36 batteries will be selected and tested. On the basis of the sample, a decision must be made to accept the shipment of batteries or to return it to the supplier because of poor quality. Let μ denote the mean hours of useful life for batteries in the shipment. The null and alternative hypotheses about the population mean are as follows:

$$H_0: \mu \geq 120$$

$$H_a: \mu < 120$$

If H_0 is rejected, the alternative hypothesis is concluded to be true. This conclusion indicates that the appropriate decision is to return the shipment to the supplier. However, if H_0 is not rejected, the decision maker must still make a decision regarding what action should be taken. Thus, without directly concluding that H_0 is true but merely by not rejecting it, the decision maker will have made the decision to accept the shipment as being of satisfactory quality.

In decision-making situations such as these, it is recommended that the hypothesis-testing procedure be extended to include a consideration of the probability of making a Type II error. Since a decision will be made and action taken when we do not reject H_0, knowledge of the probability of making a Type II error will determine the confidence we can have in the conclusion to accept that H_0 is true. Sections 9.8 and 9.9 are devoted to computing the probability of making a Type II error and to understanding how the sample size can be adjusted to help control the probability of making a Type II error.

9.8   Calculating the Probability of Type II Errors

In this section we show how to calculate the probability of making a Type II error for a hypothesis test concerning a population mean. We will illustrate the procedure using the lot-acceptance example described in Section 9.7. The null and alternative hypotheses about the mean hours of useful life for a shipment of batteries were written as follows:

$$H_0: \mu \geq 120$$

$$H_a: \mu < 120$$

If H_0 is rejected, the decision will be made to return the shipment to the supplier because the mean hours of useful life is less than the specified 120 hours. If H_0 is not rejected, the decision will be made to accept the shipment.

Suppose that a level of significance of $\alpha = .05$ is used to conduct the hypothesis test. The test statistic is

$$z = \frac{\bar{x} - \mu}{\sigma/\sqrt{n}} = \frac{\bar{x} - 120}{\sigma/\sqrt{n}}$$

With $z_{.05} = 1.645$, the rejection rule for the lower-tailed test becomes

Reject H_0 if $z < -1.645$

Suppose that a sample of 36 batteries will be selected and that it is known from previous testing that the standard deviation for the population is $\sigma = 12$ hours. The preceding rejection rule indicates we will reject H_0 when

$$z = \frac{\bar{x} - 120}{12/\sqrt{36}} < -1.645$$

Solving for $\bar{x}$ in the preceding expression indicates that we will reject H_0 whenever

$$\bar{x} < 120 - 1.645\left(\frac{12}{\sqrt{36}}\right) = 116.71$$

Rejecting H_0 when $\bar{x} < 116.71$ means that we will make the decision to accept the shipment whenever

$$\bar{x} \geq 116.71$$

Based on this information, we are now ready to compute probabilities associated with making a Type II error. First, recall that we make a Type II error whenever the true shipment mean is less than 120 hours and we make the decision to accept H_0: $\mu \geq 120$. Thus, to compute the probability of making a Type II error, we must select a value of μ less than 120 hours. For example, suppose that the shipment is considered to be of poor quality if the batteries have a mean life of $\mu = 112$ hours. If $\mu = 112$ is really true, what is the probability of accepting H_0: $\mu \geq 120$, and hence committing a Type II error? To find this probability, note that it is the probability that the sample mean $\bar{x}$ is greater than or equal to 116.71 when $\mu = 112$.

Figure 9.15 shows the sampling distribution of $\bar{x}$ when the mean is $\mu = 112$. The shaded area in the upper tail shows the probability of obtaining $\bar{x} \geq 116.71$. Using the standard normal probability distribution calculation based on Figure 9.15, we see that

$$z = \frac{\bar{x} - \mu}{\sigma/\sqrt{n}} = \frac{116.71 - 112}{12/\sqrt{36}} = 2.36$$

The standard normal probability distribution table shows that with $z = 2.36$, the area in the upper tail is $.5000 - .4909 = .0091$. This is the probability of accepting H_0 when $\mu = 112$. In other words, this is the probability of making a Type II error when $\mu = 112$. Denoting the probability of making a Type II error as β, we see that when $\mu = 112$, $\beta = .0091$. Therefore, we can conclude that if the mean of the population is 112 hours, the probability of making a Type II error is only .0091.

We can repeat these calculations for other values of μ less than 120. Doing this will show that there is a different probability of making a Type II error for each value of μ less than 120. For example, suppose that we consider the case where the shipment of batteries has a mean useful life of $\mu = 115$ hours. Since we will accept H_0 whenever $\bar{x} \geq 116.71$, the z value for $\mu = 115$ is given by

$$z = \frac{\bar{x} - \mu}{\sigma/\sqrt{n}} = \frac{116.71 - 115}{12/\sqrt{36}} = .86$$

FIGURE 9.15
Probability of a Type II Error when
$\mu = 112$

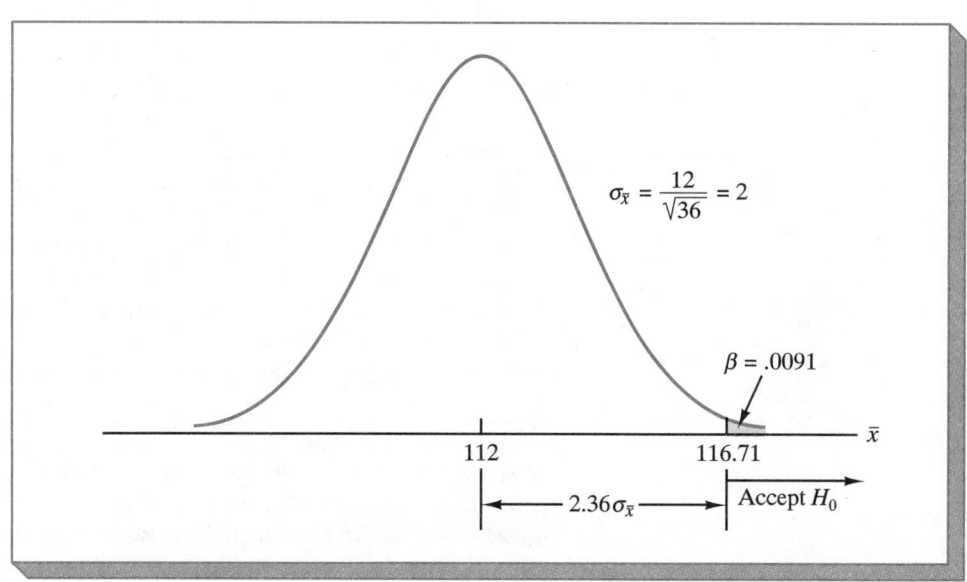

TABLE 9.2
Probabilities of Making a Type II Error for the Lot-Acceptance Hypothesis Test

Value of μ	$z = \dfrac{116.71 - \mu}{12/\sqrt{36}}$	Probability of Making a Type II Error (β)	Power ($1 - \beta$)
112	2.36	.0091	.9909
114	1.36	.0869	.9131
115	.86	.1949	.8051
116.71	.00	.5000	.5000
117	−.15	.5596	.4404
118	−.65	.7422	.2578
119.999	−1.645	.9500	.0500

From the standard normal probability distribution table, we find that the area in the upper tail of the standard normal probability distribution for $z = .86$ is $.5000 - .3051 = .1949$. Thus, the probability of making a Type II error is $\beta = .1949$ when the true mean is $\mu = 115$.

In Table 9.2 we show the probability of making a Type II error for a variety of values of μ less than 120. Note that as μ increases toward 120, the probability of making a Type II error increases toward an upper bound of .95. However, as μ decreases to values further below 120, the probability of making a Type II error becomes smaller and smaller. This is what we should expect. When the true population mean μ is close to the null hypothesis value of $\mu = 120$, there is a high probability that we will accept H_0 and make a Type II error. However, when the true population mean μ is far below the null hypothesis value of $\mu = 120$, there is a low probability that we will accept H_0 and make a Type II error.

The probability of correctly rejecting H_0 when it is false is called the *power* of the test. For any particular value of μ, the power is $1 - \beta$. That is, the probability of correctly rejecting the null hypothesis is 1 minus the probability of making a Type II error. Values of power are also shown in Table 9.2. Using these values, the power associated with each value of μ is shown graphically in Figure 9.16. Such a graph is called a *power curve*. Note that the power curve extends over the values of μ for which the null hypothesis is false. The height of the power curve at any value of μ provides the probability of correctly rejecting H_0 when H_0 is false.*

In summary, the following step-by-step procedure can be used to compute the probability of making a Type II error in hypothesis tests about a population mean.

1. Formulate the null and alternative hypotheses.
2. Use the level of significance α to establish a rejection rule based on the test statistic.
3. Using the rejection rule, solve for the value of the sample mean that identifies the rejection region for the test.
4. Use the results from step 3 to state the values of the sample mean that lead to the acceptance of H_0; this defines the acceptance region for the test.
5. Using the sampling distribution of $\bar{x}$ for any value of μ from the alternative hypothesis, and the acceptance region from step 4, compute the probability that the sample mean will be in the acceptance region. This probability is the probability of making a Type II error at the chosen value of μ.
6. Repeat step 5 for other values of μ from the alternative hypothesis.

---

*Another graph, called the *operating characteristic curve*, is sometimes used to provide information about the probability of making a Type II error. The operating characteristic curve shows the probability of accepting H_0 and thus provides β for the values of μ where the null hypothesis is false. The probabilities of making Type II errors can be read directly from this graph. The power curve and operating characteristic curve are discussed further in Chapter 20.

FIGURE 9.16

**Power Curve for the
Lot-Acceptance Hypothesis Test**

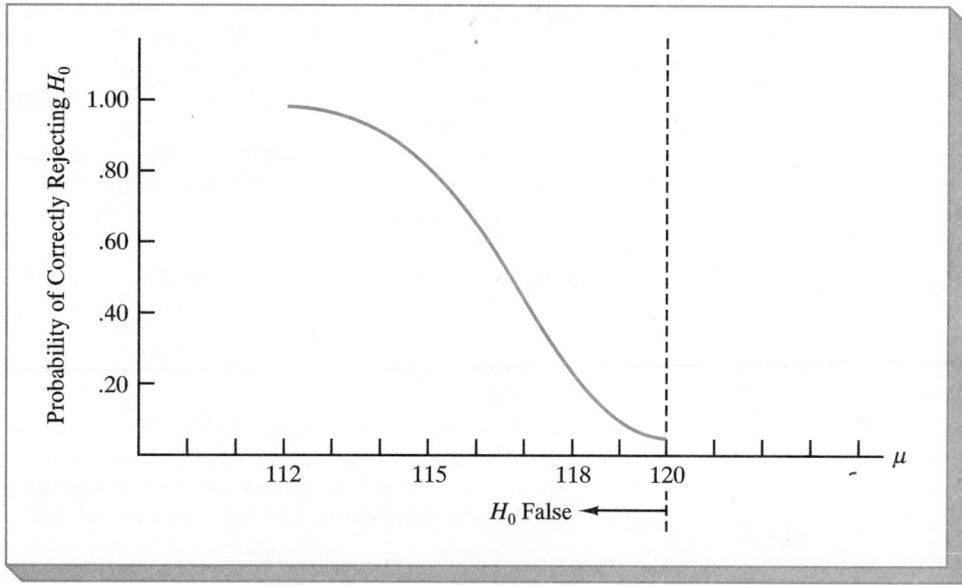

Exercises

SELF TEST ▶

Methods

56. Consider the following hypothesis test:

$$H_0: \mu \geq 10$$

$$H_a: \mu < 10$$

The sample size is 120, and the population standard deviation is 5. Use $\alpha = .05$.
a. If the actual population mean is 9, what is the probability that the sample mean leads to the conclusion *do not reject H_0*?
b. What type of error would be made if the actual population mean is 9 and we conclude that $H_0: \mu \geq 10$ is true?
c. What is the probability of making a Type II error if the actual population mean is 8?

57. Consider the following hypothesis test:

$$H_0: \mu = 20$$

$$H_a: \mu = 20$$

A sample of 200 items will be taken, and the population standard deviation is 10. Use $\alpha = .05$. Compute the probability of making a Type II error if the population mean is as shown below:
a. $\mu = 18.0$ b. $\mu = 22.5$ c. $\mu = 21.0$

Applications

58. Fowle Marketing Research, Inc., bases charges to a client on the assumption that telephone surveys can be completed within 15 minutes or less. If more time is required, a premium rate is charged to the client. Using a sample of 35 surveys, a standard deviation of 4 minutes, and a level of significance of .01, the sample mean will be used to test the null hypothesis $H_0: \mu \leq 15$.
a. What is your interpretation of the Type II error for this problem? What is its impact on the firm?
b. What is the probability of making a Type II error when the actual mean time is $\mu = 17$ minutes?
c. What is the probability of making a Type II error when the actual mean time is $\mu = 18$ minutes?
d. Sketch the general shape of the power curve for this test.

SELF TEST

59. A consumer-research group is interested in testing an automobile manufacturer's claim that a new economy model will provide at least 25 miles per gallon of gasoline. That is, H_0: $\mu \geq 25$.
   a. Using a .02 level of significance and a sample of 30 cars, what is the rejection rule based on the value of $\bar{x}$ for the test to determine if the manufacturer's claim should be rejected? Assume that σ is 3 miles per gallon.
   b. What is the probability of committing a Type II error if the actual mileage is 23 miles per gallon?
   c. What is the probability of committing a Type II error if the actual mileage is 24 miles per gallon?
   d. What is the probability of committing a Type II error if the actual mileage is 25.5 miles per gallon?

60. *Young Adult* magazine states the following hypotheses about the mean age of its subscribers:

$$H_0: \mu = 28$$

$$H_a: \mu \neq 28$$

   a. What would it mean to make a Type II error in this situation?
   b. The population standard deviation is known at $\sigma = 6$ years, and the sample size is 100. With $\alpha = .05$, what is the probability of accepting H_0 for μ equal to 26, 27, 29, and 30?
   c. What is the power at $\mu = 26$? What does this tell you?

61. A production line operation is tested for filling-weight accuracy with the following hypotheses:

Hypothesis	Conclusion and Action
H_0: $\mu = 16$	Filling okay; keep running
H_a: $\mu \neq 16$	Filling off standard; stop and adjust machine

The sample size is 30 and the population standard deviation is considered known at $\sigma = .8$. Use $\alpha = .05$.
   a. What would a Type II error mean in this situation?
   b. What is the probability of making a Type II error when the machine is overfilling by .5 ounces?
   c. What is the power of the statistical test when the machine is overfilling by .5 ounces?
   d. Show the power curve for this hypothesis test. What information does it contain for the production manager?

62. Refer to Exercise 58. Repeat parts (b) and (c) if the firm selects a sample of 50 surveys. What observation can you make about how increasing the sample size affects the probability of making a Type II error?

63. Sparr Investments, Inc., specializes in tax-deferred investment opportunities for its clients. Recently Sparr has offered a payroll deduction investment program for the employees of a particular company. Sparr estimates that the employees are currently averaging $100 or less per month in tax deferred investments. A sample of 40 employees will be used to test Sparr's hypothesis about the current level of investment activity among the population of employees. Assume that the employee monthly tax-deferred investment amounts have a standard deviation of $75 and that a .05 level of significance will be used in the hypothesis test.
   a. What is the Type II error in this situation?
   b. What is the probability of the Type II error if the actual mean employee monthly investment is $120?
   c. What is the probability of the Type II error if the actual mean employee monthly investment is $130?
   d. Repeat parts (b) and (c) if a sample size of 80 employees is used.

9.9   Determining the Sample Size for a Hypothesis Test about a Population Mean

Assume that a hypothesis test is to be conducted about the value of a population mean. The user's specified level of significance determines the probability of making a Type I error for the test. By controlling the sample size, the user can also control the probability of making a Type II error. Let us use the following example to show how a sample size can be determined for the following one-tailed test about a population mean.

$$H_0: \mu \geq \mu_0$$

$$H_a: \mu < \mu_0$$

where μ_0 is the hypothesized value for the population mean.

The upper part of Figure 9.17 shows the sampling distribution of $\bar{x}$ when H_0 is true and $\mu = \mu_0$. Note that the user's specified level of significance α determines the rejection region for the test. Let c denote the critical value such that $\bar{x} < c$ determines the rejection region for the test. Using the upper part of Figure 9.17 with z_α indicating the z value corresponding to an area of α in the tail of the standard normal probability distribution, c is computed as follows:

$$c = \mu_0 - z_\alpha \frac{\sigma}{\sqrt{n}} \tag{9.10}$$

FIGURE 9.17

Determining the Sample Size for Specified Levels of the Type I (α) and Type II (β) Errors.

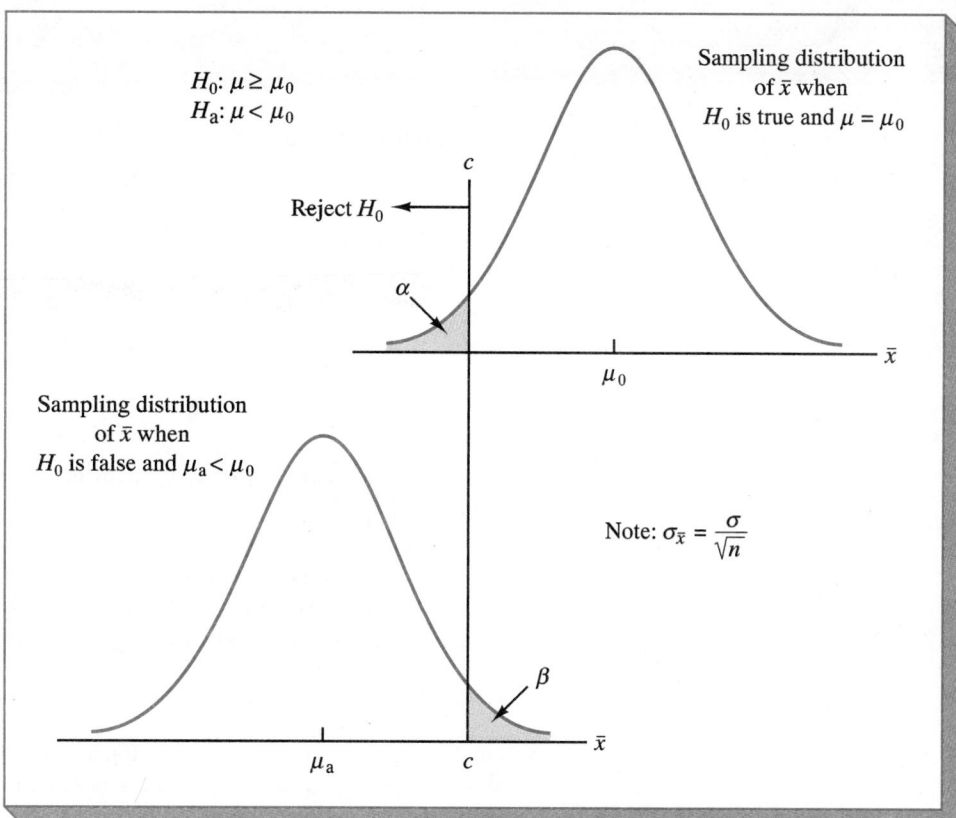

Now consider the sampling distribution shown in the lower part of Figure 9.17. Specifically, we have selected a value of the population mean, denoted by μ_a, which corresponds to the case when H_0 is false and H_a is true with $\mu_a < \mu_0$. Let us assume that the user specifies the probability of a Type II error that can be tolerated if the true population mean is μ_a. This probability is shown as β in Figure 9.17. Using the lower part of Figure 9.17 with z_β indicating the z value corresponding to an area of β in the tail of the standard normal probability distribution, the critical value c can be computed as follows:

$$c = \mu_a + z_\beta \frac{\sigma}{\sqrt{n}} \tag{9.11}$$

Since (9.10) and (9.11) provide two expressions for c, we know they must be equal, and thus the following must be true.

$$\mu_0 - z_\alpha \frac{\sigma}{\sqrt{n}} = \mu_a + z_\beta \frac{\sigma}{\sqrt{n}}$$

To determine the expression that will provide the desired sample size, we first solve for the $\sqrt{n}$ as follows:

$$\mu_0 - \mu_a = z_\alpha \frac{\sigma}{\sqrt{n}} + z_\beta \frac{\sigma}{\sqrt{n}}$$

$$\mu_0 - \mu_a = \frac{(z_\alpha + z_\beta)\sigma}{\sqrt{n}}$$

and

$$\sqrt{n} = \frac{(z_\alpha + z_\beta)\sigma}{(\mu_0 - \mu_a)}$$

Squaring both sides of the above expression provides the following sample-size formula for a one-tailed hypothesis test about a population mean.

Recommended Sample Size for a One-Tailed Hypothesis Test

$$n = \frac{(z_\alpha + z_\beta)^2 \sigma^2}{(\mu_0 - \mu_a)^2} \tag{9.12}$$

where

z_α = z value providing an area of α in the tail of a standard normal distribution
z_β = z value providing an area of β in the tail of a standard normal distribution
σ = the population standard deviation
μ_0 = the value of the population mean in the null hypothesis
μ_a = the value of the population mean used for the Type II error
Note: In a two-tailed hypothesis test, use (9.12) with $z_{\alpha/2}$ replacing z_α.

Although the logic of (9.12) was developed for the hypothesis test shown in Figure 9.17, it holds for any one-tailed test about a population mean. Note that in a two-tailed hypothesis test about a population mean, $z_{\alpha/2}$ is used instead of z_α in (9.12).

Let us return to the lot-acceptance example from Sections 9.7 and 9.8. The design specification for the shipment of batteries indicated a mean useful life of at least 120 hours for the batteries. Shipments were rejected if the hypothesis H_0: $\mu \geq 120$ was rejected. Let us assume that the quality-control manager makes the following statements about the allowable probabilities for the Type I and Type II errors:

> Type I error statement: If the mean life of the batteries in the shipment is $\mu = 120$, we are willing to risk an $\alpha = .05$ probability of rejecting the shipment.

> Type II error statement: If the mean life of the batteries in the shipment is 5 hours under the specification (i.e., $\mu = 115$), we are willing to risk a $\beta = .10$ probability of accepting the shipment.

These statements are based on the judgment of the manager. Someone else might specify different restrictions on the probabilities. However, statements about the allowable probabilities of both errors must be made before the sample size can be determined.

Thus, in our example, $\alpha = .05$ and $\beta = .10$. Using the standard normal probability distribution, we have $z_{.05} = 1.645$ and $z_{.10} = 1.28$. From the statements about the error probabilities, we note that $\mu_0 = 120$ and $\mu_a = 115$. Finally, the population standard deviation was assumed known at a value of $\sigma = 12$. Using (9.12), the recommended sample size for the lot-acceptance example is

$$n = \frac{(1.645 + 1.28)^2 (12)^2}{(120 - 115)^2} = 49.3$$

Rounding up, the recommended sample size is 50.

Since both the Type I and Type II error probabilities have been controlled at allowable levels with $n = 50$, the quality-control manager is now justified in using the accept H_0 and reject H_0 statements for the hypothesis test. The accompanying inferences of H_0 true or H_a true are made with known and allowable probabilities of error.

Before closing this section, let us make some comments about the relationship among α, β, and the sample size n.

1. Once two of the three values are known, the other can be computed.
2. For a given level of significance α, increasing the sample size will reduce β.
3. For a given sample size, decreasing α will increase β, whereas increasing α will decrease β.

The third observation is something to keep in mind when the probability of a Type II error is not being controlled. It suggests that one should not choose unnecessarily small values for the level of significance. For a given sample size, choosing a smaller level of significance α means more exposure to a Type II error. Inexperienced users of hypothesis testing often think that smaller values of α are always better. This is true if we are concerned only about the Type I error. However, smaller values of α have the disadvantage of increasing the risk of making a Type II error.

 Exercises

Methods

SELF TEST ▶ 64. Consider the following hypothesis test:

$$H_0: \mu \geq 10$$

$$H_a: \mu < 10$$

The sample size is 120, and the population standard deviation is 5. Use $\alpha = .05$. If the actual population mean is 9, the probability of a Type II is .2912. Suppose that the researcher would like to reduce the probability of a Type II to .10 when the actual population mean is 9. What sample size is recommended?

65. Consider the following hypothesis test:

$$H_0: \mu = 20$$

$$H_a: \mu \neq 20$$

The population standard deviation is 10. Use $\alpha = .05$. How large a sample should be taken if the researcher is willing to accept a .05 probability of making a Type II error when the actual population mean is 22?

Applications

66. Suppose that the project director for the Hilltop Coffee study (see Section 9.3) had asked for a .10 probability of claiming that Hilltop was in not violation when it really was underfilling by 1 ounce ($\mu = 2.9375$ pounds). What sample size would have been recommended?

SELF TEST ▶  67. A special industrial battery must have a life of at least 400 hours. A hypothesis test is to be conducted with a .02 level of significance. If the batteries from a particular production run have an actual mean use life of 385 hours, the production manager would like a sampling procedure that erroneously concludes the batch is acceptable only 10% of the time. What sample size is recommended for the hypothesis test? Use 30 hours as an estimate of the population standard deviation.

68. For the *Young Adult* magazine study $H_0: \mu = 28$ years was tested at a .05 level of significance. If the manager conducting the test will permit only a .15 probability of making a Type II error when the true mean age is 29, what sample size should be selected? Assume $\sigma = 6$.

69. An automobile mileage study tested the following hypotheses:

Hypothesis	Conclusion
$H_0: \mu \geq 25$ mpg	Manufacturer's claim supported
$H_a: \mu < 25$ mpg	Manufacturer's claim rejected; average mileage per gallon less than stated

For $\sigma = 3$ and a .02 level of significance, what sample size would be recommended if the tester desires an 80% chance of detecting that μ is less than 25 miles per gallon when actually $\mu = 24$?

Summary

Hypothesis testing is a statistical procedure that uses sample data to determine whether or not a statement about the value of a population parameter should be rejected. The hypotheses, which come from a variety of sources, must be two competing statements: a null hypothesis H_0 and an alternative hypothesis H_a. In some applications it is not obvious how the null and alternative hypotheses should be formulated. We suggested guidelines for developing hypotheses based on three types of situations most frequently encountered.

1. Testing research hypotheses: The research hypothesis should be formulated as the alternative hypothesis. The null hypothesis is based on an established theory or the statement that the research treatment will have no effect. Whenever the sample data contradict the null hypothesis, the null hypothesis is rejected. In this case, the alternative, or research, hypothesis is supported and can be claimed true.

2. Testing the validity of a claim: Generally, this situation corresponds to the ''innocent until proven guilty'' analogy. The claim made is chosen as the null hypothesis; the challenge to the claim is chosen as the alternative hypothesis. Action against the claim will be taken whenever the sample data contradict the null hypothesis. When this occurs, the challenge implied by the alternative hypothesis is concluded to be true.

3. Testing in decision making situations: This occurs when a decision maker must choose between two courses of action, one associated with the null hypothesis and one associated with the alternative hypothesis. In these situations it is often suggested that the hypotheses be formulated such that the Type I error is the more serious error. However, whenever an action must be taken based on the decision to accept H_0, the hypothesis-testing procedure should be extended to control the probability of making a Type II error. These procedures were discussed in Sections 9.8 and 9.9.

Figure 9.18 summarizes the test statistics used in hypothesis tests about a population mean and provides a practical guide for selecting the hypothesis-testing procedure. The figure shows that the

FIGURE 9.18 Summary of the Test Statistics to Be Used in a Hypothesis Test about a Population Mean

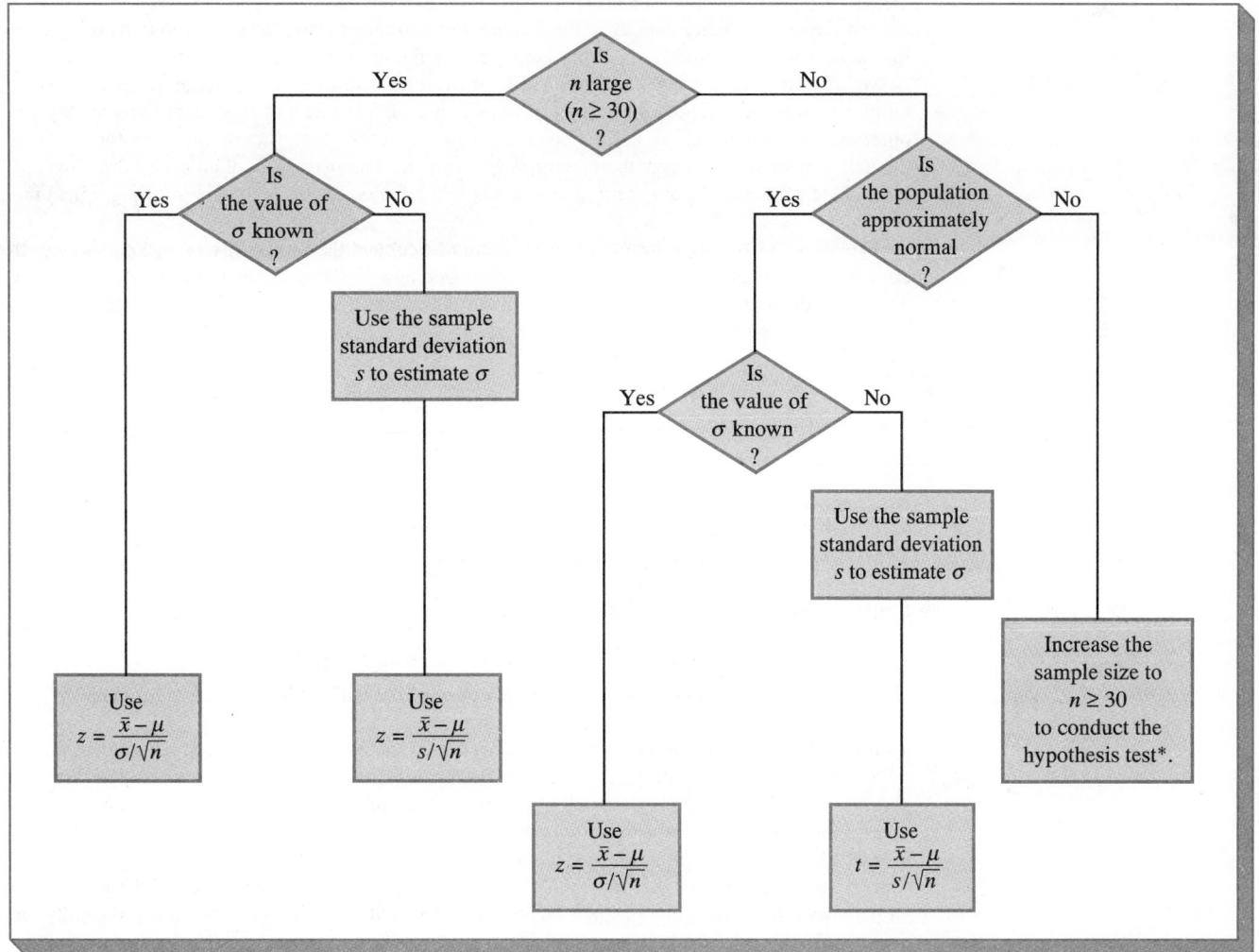

*In some cases, methods of nonparametric statistics may be used to conduct a hypothesis test. Nonparametric methods are discussed in Chapter 19.

test statistic depends on whether the sample size is large ($n \geq 30$) or small ($n < 30$), whether the population standard deviation is known, and in some cases, whether or not the population has a normal or approximately normal probability distribution. If the sample size is large, the z test statistic is used to conduct the hypothesis test. If the sample size is small, the population must have a normal or approximately normal distribution to conduct a hypothesis test about the value of μ. If this is the case, the z test statistic is used if σ is known, while the t test statistic is used if σ is estimated by the sample standard deviation s. Finally, note that if the sample size is small and the assumption of a normally distributed population is inappropriate, we recommend increasing the sample size of $n \geq 30$ to develop an interval estimate of the population mean.

Hypothesis tests concerning a population proportion were developed for the large-sample case ($np \geq 5$ and $n(1 - p) \geq 5$). The test statistic used is

$$z = \frac{\bar{p} - p_0}{\sqrt{\dfrac{p_0(1 - p_0)}{n}}}$$

The rejection rule for all the hypothesis-testing procedures involves comparing the value of the test statistic with a critical value. For lower-tail tests, the null hypothesis is rejected if the value of the test statistic is less than the critical value. For upper-tail tests, the null hypothesis is rejected if the value of the test statistic is greater than the critical value. For two-tailed tests, the null hypothesis is rejected for values of the test statistic in either tail of the sampling distribution.

We also saw that p-values could be used for hypothesis testing. The p-value yields the probability, when the null hypothesis is true, of obtaining a sample result at least as unlikely as what is observed. When p-values are used to conduct a hypothesis test, the rejection rule calls for rejecting the null hypothesis whenever the p-value is less than α. The p-value is often called the observed level of significance because the null hypothesis will be rejected for any value of α larger than the p-value.

Extensions of the hypothesis-testing procedure to control the probability of making a Type II error were also presented. In Section 9.8, we showed how to compute the probability of making a Type II error. In Section 9.9 we showed how to determine a sample size that would enable us to control the probability of making a Type II error.

Glossary

Null hypothesis The hypothesis tentatively assumed true in the hypothesis-testing procedure.

Alternative hypothesis The hypothesis concluded to be true if the null hypothesis is rejected.

Type I error The error of rejecting H_0 when it is true.

Type II error The error of accepting H_0 when it is false.

Critical value A value that is compared with the test statistic to determine whether or not H_0 should be rejected.

Level of significance The maximum probability of a Type I error.

One-tailed test A hypothesis test in which rejection of the null hypothesis occurs for values of the test statistic in one tail of the sampling distribution.

Two-tailed test A hypothesis test in which rejection of the null hypothesis occurs for values of the test statistic in either tail of the sampling distribution.

p-Value The probability, when the null hypothesis is true, of obtaining a sample result that is at least as unlikely as what is observed. It is often called the observed level of significance.

Power The probability of correctly rejecting H_0 when it is false.

Power curve A graph of the probability of rejecting H_0 for all possible values of the population parameter not satisfying the null hypothesis. The power curve provides the probability of correctly rejecting the null hypothesis.

Key Formulas

Test Statistic about a Population Mean (Large-Sample Case)

$$z = \frac{\bar{x} - \mu_0}{\sigma/\sqrt{n}} \tag{9.6}$$

When σ is unknown, substitute s.

Test Statistic about a Population Mean (Small-Sample Case with s Used to Estimate σ)

$$t = \frac{\bar{x} - \mu_0}{s/\sqrt{n}} \tag{9.7}$$

Test Statistic about a Population Proportion

$$z = \frac{\bar{p} - p_0}{\sigma_{\bar{p}}} \tag{9.8}$$

where

$$\sigma_{\bar{p}} = \sqrt{\frac{p_0(1 - p_0)}{n}} \tag{9.9}$$

Sample Size for a One-Tailed Hypothesis Test about a Population Mean

$$n = \frac{(z_\alpha + z_\beta)^2 \sigma^2}{(\mu_0 - \mu_a)^2} \tag{9.12}$$

In a two-tailed test, replace z_α with $z_{\alpha/2}$.

❑ ❑ **Supplementary Exercises**

70. The Graduate Management Admission Council conducted an extensive survey of first-year students in MBA programs ("An Overview of Demographic and Family Characteristics of First-Year Students in U.S. MBA Programs," *GMAC Occasional Papers,* April 1988). This study reported that the mean age of first-year MBA students was 27 years. A sample of 45 first-year MBA students at a particular Eastern school had a mean age of 27.6 years and a standard deviation of $s = 1.5$ years.
   a. Conduct a hypothesis test to see if the mean age for the population of students at the school was greater than the GMAC reported mean age. Use $\alpha = .01$. What is your conclusion?
   b. What is the *p*-value for the hypothesis test?
   c. Construct a 95% confidence interval estimate of the mean age for MBA students admitted to this school.

71. The manager of the Keeton Department Store has assumed that the mean annual income of the store's credit-card customers is at least $28,000 per year. A sample of 58 credit-card customers shows a sample mean of $27,200 and a sample standard deviation of $3000. At the .05 level of significance should this assumption be rejected? What is the *p*-value?

72. The chamber of commerce of a Florida Gulf Coast community advertises area residential property available at a mean cost of $25,000 or less per lot. Using a .05 level of significance, test the validity of this claim. Suppose a sample of 32 properties provided a sample mean of $26,000 per lot and a sample standard deviation of $2500. What is the *p*-value?

73. A bath soap manufacturing process is designed to produce a mean of 120 bars of soap per batch. Quantities over or under this standard are undesirable. A sample of ten batches shows the following numbers of bars of soap:

$$108, \; 118, \; 120, \; 122, \; 119, \; 113, \; 124, \; 122, \; 120, \; 123.$$

Using a .05 level of significance, test to see if the sample results indicate that the manufacturing process is functioning properly.

74. The monthly rent for a two-bedroom apartment in a particular city is reported to average \$350. Suppose we would like to test the hypothesis H_0: $\mu = 350$ versus the hypothesis H_a: $\mu \neq 350$. A sample of 36 two-bedroom apartments is selected. The sample mean turns out to be $\bar{x} = \$362$, with a sample standard deviation of $s = \$40$.

a. Conduct this hypothesis test with a .05 level of significance.

b. Compute the p-value.

c. Use the sample results to construct a 95% confidence interval for the population mean. What hypothesis-testing conclusion would you draw based on the confidence interval result?

75. Stout Electric Company operates a fleet of trucks which provide electrical service to the construction industry. Monthly mean maintenance costs have been \$75 per truck. A random sample of 40 trucks shows a sample mean maintenance cost of \$82.50 per month, with a sample standard deviation of \$30. Management would like a test to determine whether or not the mean monthly maintenance cost has increased.

a. With a .05 level of significance, what is the rejection rule for this test?

b. What is your conclusion based on the sample mean of \$82.50?

c. What is the p-value associated with this sample result? What is your conclusion based on the p-value?

76. In making bids on building projects, Sonneborn Builders, Inc., assumes construction workers are idle no more than 15% of the time. For a normal 8-hour shift, this means that the mean idle time per worker should be 72 minutes or less per day. A sample of 30 construction workers provided a mean idle time of 80 minutes per day. The sample standard deviation was 20 minutes. Suppose a hypothesis test is to be designed to test the validity of the company's assumption.

a. What is the p-value associated with the sample result?

b. Using a .05 level of significance and the p-value, test the hypothesis H_0: $\mu \leq 72$. What is your conclusion?

77. Sixty percent of Americans believe that business profits are distributed unfairly (*General Social Surveys,* National Opinion Research Center, University of Chicago). Suppose a sample of 40 Midwesterners showed that 27 believe business profits are distributed unfairly.

a. Do these results justify making the inference that a larger proportion of Midwesterners believe that business profits are distributed unfairly? Use $\alpha = .05$.

b. What is the p-value?

78. In the past, the Dumont Clothing Store has recorded 72% charge purchases and 28% cash purchases. A sample of 200 recent purchases shows that 160 were charge purchases. Does this suggest a change in the paying practices of the Dumont customers? Test with $\alpha = .05$. What is the p-value?

79. The manager of K-Mark Supermarkets assumes that at least 30% of the Saturday customers purchase the price-reduced special advertised in the Friday newspaper. Use $\alpha = .05$ and test the validity of the manager's assumption if a sample of 250 Saturday customers shows that 60 purchased the advertised special.

80. The Gallup Organization conducted a survey of 1350 people for the National Occupational Information Coordinating Committee, a panel that Congress created to make better use of job information (*The Arizona Republic,* January 12, 1990). A research question related to the study was, Do individuals hold jobs that they planned to hold or do they hold jobs for such reasons as chance or lack of choice? Let p indicate the population proportion of individuals who hold jobs that they planned to hold.

a. If the hypotheses are stated H_0: $p \geq .50$ and H_a: $p < .50$, discuss the research hypothesis H_a terms of what the researcher is investigating.

b. The Gallup poll reported 41% of the respondents hold jobs that they planned to hold. What is your conclusion at a .01 level of significance? Discuss.

81. A well-known doctor hypothesized that 75% of women wear shoes that are too small. A 1991 study of 356 women by the American Orthopedic Foot and Ankle Society found 313 women wore shoes that were at least a size too small (*New York Times,* March 10, 1991). Test H_0: $p = .70$ and H_a: $p \neq .70$ at $\alpha = .01$. What is your conclusion?

82. The filling machine for a production operation must be adjusted if more than 8% of the items being produced are underfilled. A random sample of 80 items from the day's production contained 9 underfilled items. Does the sample evidence indicate that the filling machine should be adjusted? Use $\alpha = .02$. What is the p-value?

83. A radio station in a major resort area announced that at least 90% of the hotels and motels would be full for the Memorial Day weekend. The station went on to advise listeners to make reservations in advance if they planned to be in the resort over the weekend. On Saturday night a sample of 58 hotels and motels showed 49 with a no-vacancy sign and 9 with vacancies. What is your reaction to the radio station's claim based on the sample evidence? Use $\alpha = .05$ in making this statistical test. What is the p-value for the sample results?

84. It is assumed that at least 90% of juvenile first-time criminals are given probation upon the admission of guilt. Test this hypothesis at a .02 level of significance, if a sample of 92 juvenile criminal convictions shows 78 juveniles receiving probation. What is the p-value?

85. Refer again to Exercise 76.
a. What is the probability of making a Type II error when the population mean idle time is 80 minutes?
b. What is the probability of making a Type II error when the population mean idle time is 75 minutes?
c. What is the probability of making a Type II error when the population mean idle time is 70?
d. Sketch the power curve for this problem.

86. A federal funding program is available to low-income neighborhoods. To qualify for the funding a neighborhood must have a mean household income of less than $7000 per year. Neighborhoods with mean annual household incomes of $7000 or more do not qualify. Funding decisions are based on a sample of residents in the neighborhood. A hypothesis test with a .02 level of significance is conducted. If the funding guidelines call for a maximum probability of .05 of not funding a neighborhood with a mean annual household income of $6500, what sample size should be used in the funding decision study? Use $\sigma = \$2000$ as a planning value.

87. The bath soap production process uses the hypothesis H_0: $\mu = 120$ and H_a: $\mu \neq 120$ to test whether or not the production process is meeting the standard output of 120 bars per batch. Use a .05 level of significance for the test and a planning value of 5 for the standard deviation.
a. If the mean output drops to 117 bars per batch, the firm would like a 98% chance of concluding that the standard production output is not being met. How large a sample should be selected?
b. Using your sample size from part (a), what is the probability of concluding that the process is operating satisfactorily for each of the following actual mean outputs: 117, 118, 119, 121, 122, and 123 bars per batch? That is, what is the probability of a Type II error in each case?

Computer Case: *Quality Associates, Inc.*

Quality Associates, Inc., is a consulting firm that advises its clients about sampling and statistical procedures that can be used to control their clients' manufacturing processes. In one particular application, a client supplied Quality Associates with a sample of 800 observations taken during a time in which this client's process was operating satisfactorily. The sample standard deviation for these data was .21; thus, the population standard deviation was assumed to be .21. Quality Associates then suggested that random samples of size 30 be taken periodically to monitor the process in an ongoing manner. By analyzing the new samples, the client could quickly learn if the process was operating satisfactorily. When the process was not operating satisfactorily, corrective action could be taken to eliminate the problem. The design specification indicated the process's mean should be 12. The hypotheses test suggested by Quality Associates was as follows:

$$H_0: \mu = 12$$

$$H_a: \mu \neq 12$$

Corrective action will be taken any time H_0 is rejected.

The following samples were collected at hourly intervals during the first day of operation of the new statistical process-control procedure. These data are available in the data set named QUALITY.

 QUALITY

Sample 1	Sample 2	Sample 3	Sample 4
11.55	11.62	11.91	12.02
11.62	11.69	11.36	12.02
11.52	11.59	11.75	12.05
11.75	11.82	11.95	12.18
11.90	11.97	12.14	12.11
11.64	11.71	11.72	12.07
11.80	11.87	11.61	12.05
12.03	12.10	11.85	11.64
11.94	12.01	12.16	12.39
11.92	11.99	11.91	11.65
12.13	12.20	12.12	12.11
12.09	12.16	11.61	11.90
11.93	12.00	12.21	12.22
12.21	12.28	11.56	11.88
12.32	12.39	11.95	12.03
11.93	12.00	12.01	12.35
11.85	11.92	12.06	12.09
11.76	11.83	11.76	11.77
12.16	12.23	11.82	12.20
11.77	11.84	12.12	11.79
12.00	12.07	11.60	12.30
12.04	12.11	11.95	12.27
11.98	12.05	11.96	12.29
12.30	12.37	12.22	12.47
12.18	12.25	11.75	12.03
11.97	12.04	11.96	12.17
12.17	12.24	11.95	11.94
11.85	11.92	11.89	11.97
12.30	12.37	11.88	12.23
12.15	12.22	11.93	12.25

Managerial Report

1. Conduct the hypothesis test for each sample at the .01 level of significance and determine what action if any should be taken. Provide the test statistic and p-value for each test.
2. Consider the standard deviation for each of the four samples. Does the assumption of .21 for the population standard deviation appear reasonable?

3. Compute limits for the sample mean $\bar{x}$ around $\mu = 12$ such that, as long as a new sample mean falls within these limits, the process will be considered to be operating satisfactorily. If $\bar{x}$ exceeds the upper limit or if $\bar{x}$ is less than the lower limit, corrective action will be taken. These limits are referred to as upper and lower control limits for quality-control purposes.

4. Discuss the implications of changing the level of significance to a larger value. What mistake or error could increase if this were done?

Statistical Inference about Means and Proportions with Two Populations

---

Pennwalt Corporation*

ROCHESTER, NEW YORK

Pennwalt Corporation of Rochester, New York, was founded in 1850 as a chemical company manufacturing caustic soda. The company steadily expanded its products to cover a variety of markets, such as fruit sizing, dental products, agricultural needs, food processing, and pharmaceuticals. Today, Pennwalt Corporation is an international organization that employs over 9000 people.

Pennwalt's Pharmaceutical Division uses statistical procedures to test new drugs. The testing process consists of three stages: (1) preclinical testing, (2) testing for long-term usage and safety, and (3) final clinical testing. At each successive stage the chances that the drug survives the rigorous testing decreases; however, the cost of further testing increases dramatically. As a result, it is important to reject or weed out unsuccessful new drugs in the early stages of the testing process.

For example, preclinical testing uses a two-population study to determine if the drug should continue to be studied in the long-term usage and safety program. The preclinical testing process begins when a new drug is sent to the pharmacology department for testing of efficacy—the capacity of the drug to produce the desired effects. As part of the process, a statistician is asked to design an experiment that can be used to test the new drug. The design must specify the sample size and the statistical methods of analysis that will be used. One sample is used to obtain data on the efficacy of the new drug (population 1) while a second sample is used to obtain data on the efficacy of a standard drug (population 2). Based on the intended use, the new and standard drugs are tested in disciplines such as neurology, cardiology, and immunology. In most studies, the statistical method employed involves hypothesis testing for the difference between the means of the new drug population and the standard drug population. If a new drug lacks efficacy or produces any undesirable effects compared to the standard drug, the new drug is rejected and withdrawn from further testing. Only new drugs that show promising comparisons to standard drugs are forwarded to the long-term usage and safety testing program.

Further data collection and two-population studies are conducted in the long-term usage and safety testing program as well as in the final clinical testing program. If the new drug meets all requirements relative to the standard drug, a new drug application is filed with the FDA.

In this chapter you will learn how to construct interval estimates and make hypothesis tests about the means and proportions with two populations. Techniques will be presented for analyzing independent random samples as well as matched samples.

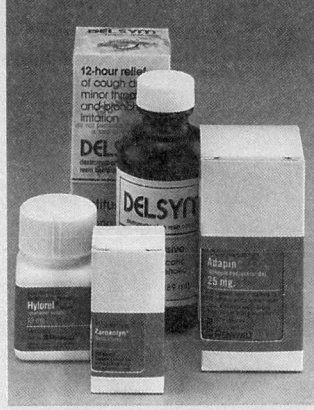

Some of the drug products manufactured by the pharmaceutical division of Pennwalt Corporation.

*The authors are indebted to Dr. M. C. Trivedi, Pennwalt Corporation, Pharmaceutical Division for providing this Statistics in Practice.

I n Chapters 8 and 9 we presented statistical methods for developing interval estimates and conducting hypothesis tests for population means and population proportions. However, the statistical procedures we have discussed thus far have considered only single-population situations. In this chapter we consider cases where *two populations* are involved. For example, we may want to compare a population of men versus a population of women, a population of parts from supplier A versus a population of parts from supplier B, a population of oil-industry firms versus a population of steel-industry firms, and so on. In cases such as these, we will be using simple random sampling to develop both interval estimates and test hypotheses about the differences between the means and between the proportions of two populations.

10.1   Estimation of the Difference between the Means of Two Populations: Independent Samples

In many practical situations we are faced with two separate populations where the difference between the means of the two populations is of prime importance. We know from Chapters 7 and 8 that we can take a simple random sample from a single population and use the sample mean $\bar{x}$ as a point estimator of the population mean. In the two-population case, we will select two separate and independent simple random samples, one from population 1 and another from population 2. Let

μ_1 = mean of population 1

μ_2 = mean of population 2

$\bar{x}_1$ = sample mean for the simple random sample from population 1

$\bar{x}_2$ = sample mean for the simple random sample from population 2

The difference between the two population means is $\mu_1 - \mu_2$. The point estimator of $\mu_1 - \mu_2$ is as follows.

Point Estimator of the Difference between the Means of Two Populations

$$\bar{x}_1 - \bar{x}_2 \qquad (10.1)$$

Thus, we see that the point estimator of the difference between two population means is the difference between the sample means of the two independent simple random samples.

Sampling Distribution of $\bar{x}_1 - \bar{x}_2$

In the study of the difference between the means of two populations, $\bar{x}_1 - \bar{x}_2$ is the point estimator of interest. This point estimator, just like the point estimators discussed previously, has its own sampling distribution. If we can identify the sampling distribution of

$\bar{x}_1 - \bar{x}_2$, we can use it to develop an interval estimate of the difference between the two population means in much the same way that we used the sampling distribution of $\bar{x}$ for interval estimation about a single population mean. The properties of the sampling distribution of $\bar{x}_1 - \bar{x}_2$ are as follows.

Sampling Distribution of $\bar{x}_1 - \bar{x}_2$

$$\text{Expected Value: } E(\bar{x}_1 - \bar{x}_2) = \mu_1 - \mu_2 \qquad (10.2)$$

$$\text{Standard Deviation: } \sigma_{\bar{x}_1 - \bar{x}_2} = \sqrt{\frac{\sigma_1^2}{n_1} + \frac{\sigma_2^2}{n_2}} \qquad (10.3)$$

where

σ_1 = standard deviation of population 1

σ_2 = standard deviation of population 2

n_1 = sample size for the simple random sample from population 1

n_2 = sample size for the simple random sample from population 2

Distribution form: If the sample sizes are both *large* ($n_1 \geq 30$ and $n_2 \geq 30$), the sampling distribution of $\bar{x}_1 - \bar{x}_2$ can be approximated by a normal probability distribution.

Figure 10.1 shows the sampling distribution of $\bar{x}_1 - \bar{x}_2$ and its relationship to the individual sampling distributions of $\bar{x}_1$ and $\bar{x}_2$.

Let us use the sampling distribution of $\bar{x}_1 - \bar{x}_2$ to develop an interval estimate of the difference between the means of two populations. We shall consider two cases, one where the sample sizes are large ($n_1 \geq 30$ and $n_2 \geq 30$) and the other where one or both sample sizes are small ($n_1 < 30$ and/or $n_2 < 30$). Let us consider the large-sample case first.

Large-Sample Case

Greystone Department Stores, Inc., operates two stores in Buffalo, New York, one located in the inner city and one located in a suburban shopping center. Over a period of time the regional manager has noticed that products that sell extremely well in one store do not always sell well in the other. One plausible explanation for the differences in sales at the two stores is that there are differences between the customers who shop at the two locations. Customer differences may be noticeable in age, education, income, and so on. While we could elect to study any of these characteristics, let us assume that the manager has asked about the difference between the mean ages of the customers who shop at the two stores. Figure 10.2 provides an illustration of this two-population situation.

FIGURE 10.1

Sampling Distribution of $\bar{x}_1 - \bar{x}_2$ and Its Relationship to the Individual Sampling Distributions of $\bar{x}_1$ and $\bar{x}_2$

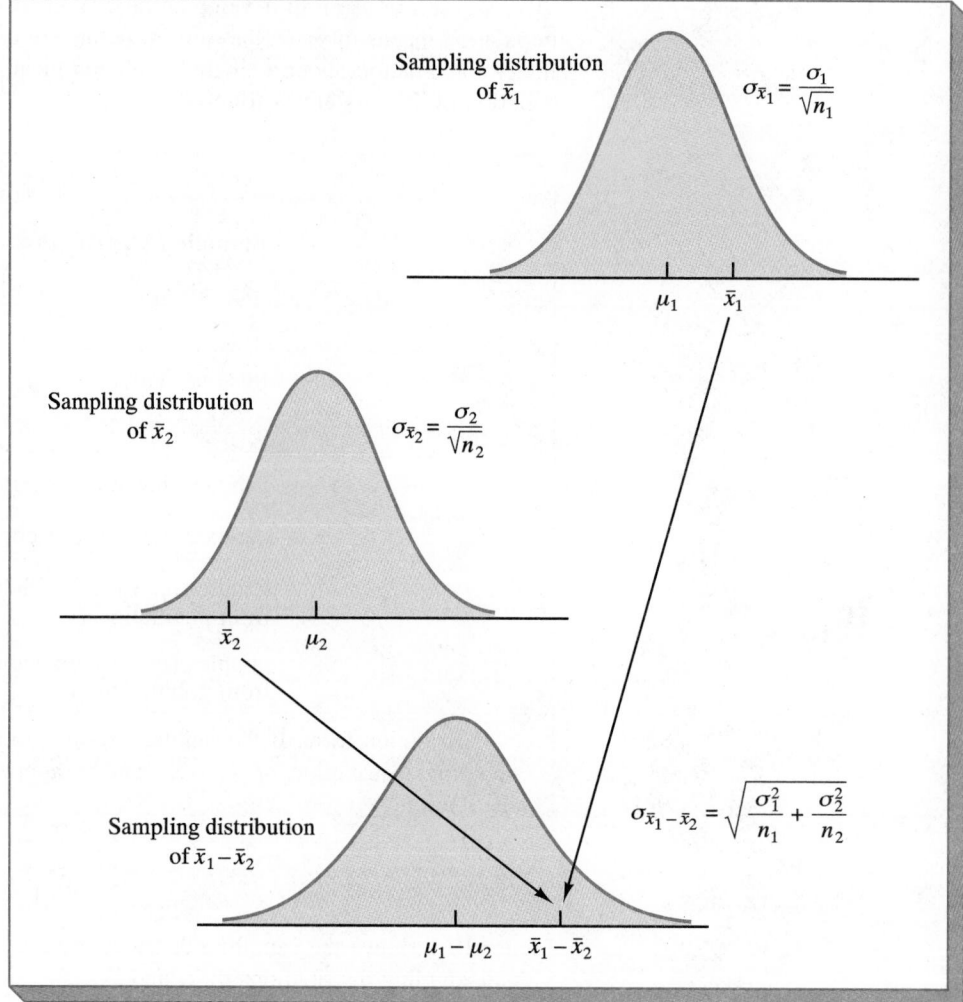

FIGURE 10.2

Two Populations for the Greystone Department Store

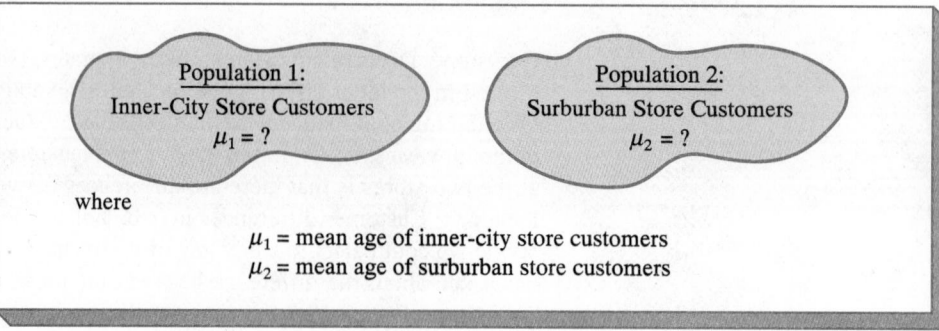

Let us suppose that Greystone conducts a survey of customers at each store. The customer age data collected from two independent random samples provide the following results:

Store	Number of Customers Sampled	Sample Mean Age	Sample Standard Deviation
Inner city	36	$\bar{x}_1 = 40$ years	$s_1 = 9$ years
Suburban	49	$\bar{x}_2 = 35$ years	$s_2 = 10$ years

Using (10.1), the point estimate of the difference between mean ages of the two populations is $\bar{x}_1 - \bar{x}_2 = 40 - 35 = 5$ years. Thus, we are led to believe that the customers at the inner-city store have a mean age approximately 5 years greater than the mean age of the suburban store customers. However, as with all point estimates, we know that 5 years is only an approximate value of the difference between the mean ages of the two populations. Thus, we will compute a confidence interval estimate of the difference between the means of the two populations.

In the large-sample-size case, we know that the sampling distribution of $\bar{x}_1 - \bar{x}_2$ can be approximated by a normal probability distribution. With this approximation we can use the following expression to develop an interval estimate of the difference between the means of the two populations.

**Interval Estimate of the Difference between
the Means of Two Populations
(Large-Sample Case with $n_1 \geq 30$ and $n_2 \geq 30$)**

$$\bar{x}_1 - \bar{x}_2 \pm z_{\alpha/2}\sigma_{\bar{x}_1 - \bar{x}_2} \qquad (10.4)$$

where $1 - \alpha$ is the confidence coefficient.

Let us use (10.4) to develop a confidence interval for the difference between the mean ages for the two populations in the Greystone Department Store study. Since the population standard deviations σ_1 and σ_2 are unknown, we cannot use (10.3) to calculate $\sigma_{\bar{x}_1 - \bar{x}_2}$. However, we can use the sample standard deviations as estimates of the population standard deviations and estimate $\sigma_{\bar{x}_1 - \bar{x}_2}$ as follows.

Point Estimator of $\sigma_{\bar{x}_1 - \bar{x}_2}$

$$s_{\bar{x}_1 - \bar{x}_2} = \sqrt{\frac{s_1^2}{n_1} + \frac{s_2^2}{n_2}} \qquad (10.5)$$

With large sample sizes, $s_{\bar{x}_1 - \bar{x}_2}$ can be accepted as a good estimate of $\sigma_{\bar{x}_1 - \bar{x}_2}$.

Using (10.5) to estimate $\sigma_{\bar{x}_1 - \bar{x}_2}$ for the Greystone Department Store problem, we have

$$s_{\bar{x}_1 - \bar{x}_2} = \sqrt{\frac{(9)^2}{36} + \frac{(10)^2}{49}} = \sqrt{4.29} = 2.07$$

With this value as the estimate of $\sigma_{\bar{x}_1 - \bar{x}_2}$ and with $z_{\alpha/2} = z_{.025} = 1.96$, (10.4) provides the following 95% confidence interval:

$$5 \pm (1.96)(2.07)$$

or

$$5 \pm 4.06$$

Thus, at a 95% level of confidence the interval estimate for the difference between the mean ages of the two Greystone populations is .94 years to 9.06 years.

Small-Sample Case

Let us now consider the interval-estimation procedure for the difference between the means of two populations whenever one or both sample sizes are less than 30 — that is, $n_1 < 30$ and/or $n_2 < 30$. This will be referred to as the small-sample case.

In Chapter 8 we presented a procedure for interval estimation of the mean for a single population whenever a small sample was used. Recall that the procedure required the assumption that the population had a normal probability distribution. With the sample standard deviation s used as an estimate of the population standard deviation σ, the t distribution was used to develop an interval estimate of the population mean.

To develop interval estimates for the two-population small-sample case, we will make two assumptions about the two populations and the samples selected from the two populations:

1. Both populations have normal probability distributions.
2. The variances of the populations are equal ($\sigma_1^2 = \sigma_2^2 = \sigma^2$).

Given these assumptions, the sampling distribution of $\bar{x}_1 - \bar{x}_2$ is normal regardless of the sample sizes involved. The expected value of $\bar{x}_1 - \bar{x}_2$ is $\mu_1 - \mu_2$. Because of the equal variances assumption, (10.3) can be written

$$\sigma_{\bar{x}_1 - \bar{x}_2} = \sqrt{\frac{\sigma^2}{n_1} + \frac{\sigma^2}{n_2}} = \sqrt{\sigma^2 \left(\frac{1}{n_1} + \frac{1}{n_2} \right)} \tag{10.6}$$

The sampling distribution of $\bar{x}_1 - \bar{x}_2$ is shown in Figure 10.3.

If the variance σ^2 of the populations is known, (10.4) can be used to develop the interval estimate of the difference between the two population means. However, in most cases, σ^2 is unknown; thus, the two sample variances s_1^2 and s_2^2 must be used to develop the estimate of σ^2 in (10.6). Since (10.6) is based on the assumption that $\sigma_1^2 = \sigma_2^2 = \sigma^2$, we do not need separate estimates of σ_1^2 and σ_2^2. In fact, we can combine the data from both samples to provide the best single estimate of σ^2. The process of combining the results of two independent random samples to provide one estimate of

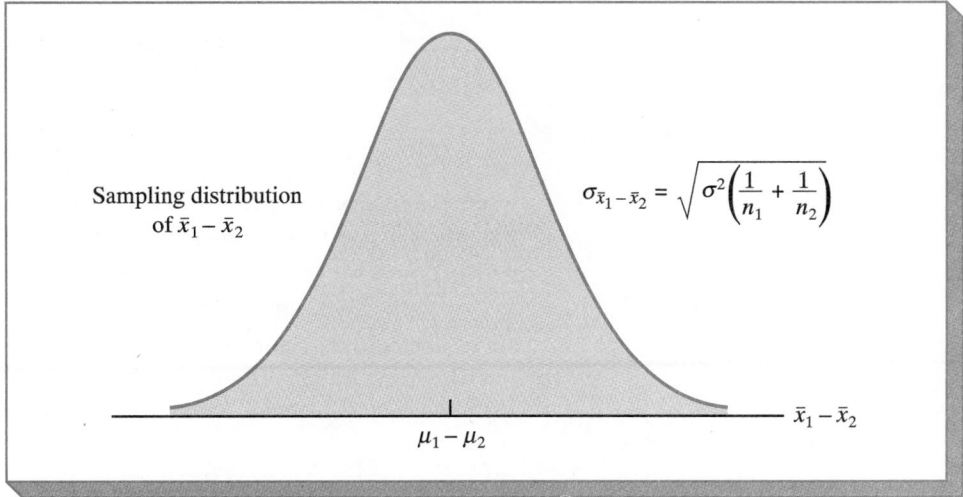

σ^2 is referred to as *pooling*. The *pooled estimator* of σ^2, denoted by s^2, is a weighted average of the two sample variances s_1^2 and s_2^2. The formula for the pooled estimator of σ^2 is as follows.

Pooled Estimator of σ^2

$$s^2 = \frac{(n_1 - 1)s_1^2 + (n_2 - 1)s_2^2}{n_1 + n_2 - 2}$$

(10.7)

With s^2 as the pooled estimator of σ^2 and using (10.6), the following estimator of the standard deviation of $\bar{x}_1 - \bar{x}_2$ can be obtained:

Point Estimator of $\sigma_{\bar{x}_1 - \bar{x}_2}$ when $\sigma_1^2 = \sigma_2^2$

$$s_{\bar{x}_1 - \bar{x}_2} = \sqrt{s^2\left(\frac{1}{n_1} + \frac{1}{n_2}\right)}$$

(10.8)

The t distribution can now be used to compute an interval estimate of the difference between the means of the two populations. Since there are $n_1 - 1$ degrees of freedom associated with the random sample from population 1 and $n_2 - 1$ degrees of freedom associated with the random sample from population 2, the t distribution will have $n_1 + n_2 - 2$ degrees of freedom. The interval-estimation procedure is as follows.

> ### Interval Estimate of the Difference between the Means of Two Populations
> ### (Small-Sample Case with $n_1 < 30$ and/or $n_2 < 30$)
>
> $$\bar{x}_1 - \bar{x}_2 \pm t_{\alpha/2} s_{\bar{x}_1 - \bar{x}_2} \qquad (10.9)$$
>
> where the t value is based on a t distribution with $n_1 + n_2 - 2$ degrees of freedom and where $1 - \alpha$ is the confidence coefficient.

Let us demonstrate the interval-estimation procedure for the sampling study conducted by the Clearview National Bank. Independent random samples of checking account balances for customers at two Clearview branch banks show the following results:

Branch Bank	Number of Checking Accounts	Sample Mean Balance	Sample Standard Deviation
Cherry Grove	12	$\bar{x}_1 = \$1000$	$s_1 = \$150$
Beechmont	10	$\bar{x}_2 = \$\ 920$	$s_2 = \$120$

Let us use these data to develop a 90% confidence interval for the difference between the mean checking account balances for the two branch banks. Using (10.7), the pooled estimate of the population variance becomes

$$s^2 = \frac{(n_1 - 1)s_1^2 + (n_2 - 1)s_2^2}{n_1 + n_2 - 2} = \frac{(11)(150)^2 + (9)(120)^2}{12 + 10 - 2} = 18,855$$

The corresponding estimate of the standard deviation of $\bar{x}_1 - \bar{x}_2$ is

$$s_{\bar{x}_1 - \bar{x}_2} = \sqrt{s^2\left(\frac{1}{n_1} + \frac{1}{n_2}\right)} = \sqrt{18,855\left(\frac{1}{12} + \frac{1}{10}\right)} = 58.79$$

The appropriate t distribution for the interval-estimation procedure has $n_1 + n_2 - 2 = 12 + 10 - 2 = 20$ degrees of freedom. With $\alpha = .10$, $t_{\alpha/2} = t_{.05} = 1.725$. Thus, using (10.9), the interval estimate becomes

$$\bar{x}_1 - \bar{x}_2 \pm t_{.05} s_{\bar{x}_1 - \bar{x}_2}$$

$$1000 - 920 \pm (1.725)(58.79)$$

$$80 \pm 101.41$$

At a 90% level of confidence, the interval estimate of the difference between the mean account balances at the two branch banks is $-\$21.41$ to $\$181.41$. The fact that the interval includes a negative range of values indicates that the actual difference between the two means, $\mu_1 - \mu_2$, may be negative. Thus, μ_2 could actually be larger than μ_1, indicating that the population mean balance could be greater for the Beechmont branch even though the results show a greater sample mean balance at the Cherry Grove branch. The fact that the confidence interval contains the value 0 can be interpreted as indicating that we do not have sufficient evidence to conclude that the population mean account balances differ at the two branches.

NOTES &
COMMENTS

1. The use of the t distribution in the small-sample procedure presented in this section is based upon the assumptions that both populations have a normal probability distribution and that $\sigma_1^2 = \sigma_2^2$. Fortunately, this procedure is a *robust* statistical procedure, meaning that it is relatively insensitive to these assumptions. For instance, if $\sigma_1^2 \neq \sigma_2^2$, the procedure provides acceptable results if n_1 and n_2 are approximately equal.

2. The t distribution is not restricted to the small-sample situation. However, (10.4) and (10.5) show how to determine an interval estimate of the difference between the means of two populations whenever the sample sizes are large. In this case, the use of the t distribution and its corresponding assumptions are not required. As a result, we do not need to refer to the t distribution until we have a small-sample case.

☐ ☐ Exercises

Methods

SELF TEST ▸ 1. Consider the results in Table 10.1 for two independent random samples taken from two populations.
  a. What is the point estimate of the difference between the two population means?
  b. Provide a 90% confidence interval for the difference between the two population means.
  c. Provide a 95% confidence interval for the difference between the two population means.

2. Consider the following results for two independent samples random taken from two populations.

TABLE 10.1

Sample 1	Sample 2
$n_1 = 50$	$n_2 = 35$
$\bar{x}_1 = 13.6$	$\bar{x}_2 = 11.6$
$s_1 = 2.2$	$s_2 = 3.0$

Sample 1	Sample 2
$n_1 = 10$	$n_2 = 8$
$\bar{x}_1 = 22.5$	$\bar{x}_2 = 20.1$
$s_1 = 2.5$	$s_2 = 2.0$

  a. What is the point estimate of the difference between the two population means?
  b. What is the pooled estimate of the population variance?
  c. Develop a 95% confidence interval for the difference between the two population means.

3. Consider the data in Table 10.2 for two independent random samples taken from two populations.
  a. Compute the two sample means.
  b. Compute the two sample standard deviations.
  c. What is the point estimate of the difference between the two population means?
  d. What is the pooled estimate of the population variance?
  e. Develop a 95% confidence interval for the difference between the two population means.

TABLE 10.2

Sample 1	Sample 2
10	8
12	8
9	6
7	7
7	4
9	9

Applications

4. A survey of salaries for professors of statistics was conducted at the University of Iowa (*Amstat News*, December 1989). The survey compared annual salaries for faculty at the ranks of associate professor and assistant professor. A sample of 41 professors that have held the rank of associate professor for 1–2 years showed a sample mean of $40,800. A sample of 71 professors that have held the rank of assistant professor for 1–2 years showed a sample mean of $36,000. Assume that the standard deviations for associate professors and assistant professors were $3600 and $1400, respectively.

a. What is the point estimate of the difference between the mean annual salaries for the population of associate professors and the population of assistant professors?

b. What is the 95% confidence interval for the difference between the two population means?

5. A college admissions board is interested in estimating the difference between the mean grade point averages of students from two high schools. Independent simple random samples of students at the two high schools provided the results shown in Table 10.3.

a. What is the point estimate of the difference between the means of the two populations?

b. Develop a 90% confidence interval for the difference between the two population means.

c. Develop a 95% confidence interval for the difference between the two population means.

6. *Working Woman* (January 1989) reported on a survey of salary information for women in a variety of occupations. The mean entry level salary for women accountants in large public accounting firms was $23,750. The mean entry level salary for women accountants in large corporations was $21,000. Assume that the following sample sizes and sample standard deviations were available:

Public Accounting	Corporations
$n_1 = 220$	$n_2 = 250$
$s_1 = \$1,200$	$s_2 = \$1,000$

Provide a 95% confidence interval for the difference between the mean entry level salaries for women in large public accounting firms and in large corporations.

7. The Butler County Bank and Trust Company is interested in estimating the difference between the mean credit-card balances at two of its branch banks. Independent random samples of credit-card customers generated the results in Table 10.4.

a. Develop a point estimate of the difference between the mean balances at the two branches.

b. Develop a 99% confidence interval for the difference between the mean balances.

8. An urban-planning group is interested in estimating the difference between the mean household income for two neighborhoods in a large metropolitan area. Independent random samples of households in the neighborhoods provided the following results:

Neighborhood 1	Neighborhood 2
$n_1 = 8$	$n_2 = 12$
$\bar{x}_1 = \$15,700$	$\bar{x}_2 = \$14,500$
$s_1 = \$700$	$s_2 = \$850$

a. Develop a point estimate of the difference between the mean incomes in the two neighborhoods.

b. Develop a 95% confidence interval for the difference between the mean incomes in the two neighborhoods.

c. What assumptions were made to compute the interval estimates in part (b)?

9. Production quantities for two assembly-line workers are shown in Table 10.5. Each data value indicates the amount produced during a randomly selected 1-hour period.

a. Develop a point estimate of the difference between the mean hourly production rates of the two workers. Which worker appears to have the higher mean production rate?

b. Develop a 90% confidence interval for the difference between the mean production rates of the two workers. Does the confidence interval provide support for the conclusion that the worker having the higher sample mean production rate is actually the worker with the overall higher production rate? Explain.

TABLE 10.3

Mt. Washington	Country Day
$n_1 = 46$	$n_2 = 33$
$\bar{x}_1 = 3.02$	$\bar{x}_2 = 2.72$
$s_1 = .38$	$s_2 = .45$

TABLE 10.4

Branch 1	Branch 2
$n_1 = 32$	$n_2 = 36$
$\bar{x}_1 = \$500$	$\bar{x}_2 = \$375$
$s_1 = \$150$	$s_2 = \$130$

SELF TEST ▶

TABLE 10.5

Worker 1	Worker 2
20	22
18	18
21	20
22	23
20	24

10. A sample of 15 graduates from Eastern University showed that the mean time until they received their first job promotion was 5.2 years, with a sample standard deviation of 1.4 years. A sample of 12 graduates from Midwestern University showed a sample mean of 2.7 years, with a standard deviation of 1.1 years.

a. What is the pooled estimate of the population variance?

b. Develop a 95% confidence interval for the difference between the mean time until the first job promotion for the populations of Eastern and Midwestern University graduates.

10.2   Hypothesis Tests about the Difference between the Means of Two Populations: Independent Samples

In this section we present procedures that can be used to test hypotheses about the difference between the means of two populations. The methodology is again divided into large-sample ($n_1 \geq 30$, $n_2 \geq 30$) and small-sample ($n_1 < 30$ and/or $n_2 < 30$) cases.

Large-Sample Case

As part of a study to evaluate differences between the educational quality of two training centers, a standardized examination is given to individuals who were trained at the two centers. The examination scores are a major factor in assessing any quality differences between the centers.

Let

μ_1 = the mean examination score for the population of individuals trained at center A

μ_2 = the mean examination score for the population of individuals trained at center B

We begin with the tentative assumption that there is no difference between the training quality provided at the two centers. Thus, in terms of the mean examination scores, the null hypothesis is that $\mu_1 - \mu_2 = 0$. If sample evidence leads to the rejection of this hypothesis, we will conclude that the mean examination scores differ for the two populations. This conclusion indicates that a quality differential exists between the two centers and that a follow-up study investigating the reasons for the differential may be warranted. The null and alternative hypotheses are written as follows:

$$H_0: \mu_1 - \mu_2 = 0$$

$$H_a: \mu_1 - \mu_2 \neq 0$$

Following the hypothesis-testing procedure from Chapter 9, we will make the tentative assumption that H_0 is true. Using the difference between the sample means as the point estimator of the difference between the population means, we consider the sampling distribution of $\overline{x}_1 - \overline{x}_2$ when H_0 is true. Assuming the large-sample case, this distribution is as shown in Figure 10.4. Since the sampling distribution is approximately normal, the following test statistic can be used:

$$z = \frac{(\overline{x}_1 - \overline{x}_2) - (\mu_1 - \mu_2)}{\sqrt{\sigma_1^2/n_1 + \sigma_2^2/n_2}} \tag{10.10}$$

Whenever $n \geq 30$ and $n_2 \geq 30$, we will use s_1^2 and s_2^2 as estimates of σ_1^2 and σ_2^2 to compute the above test statistic.

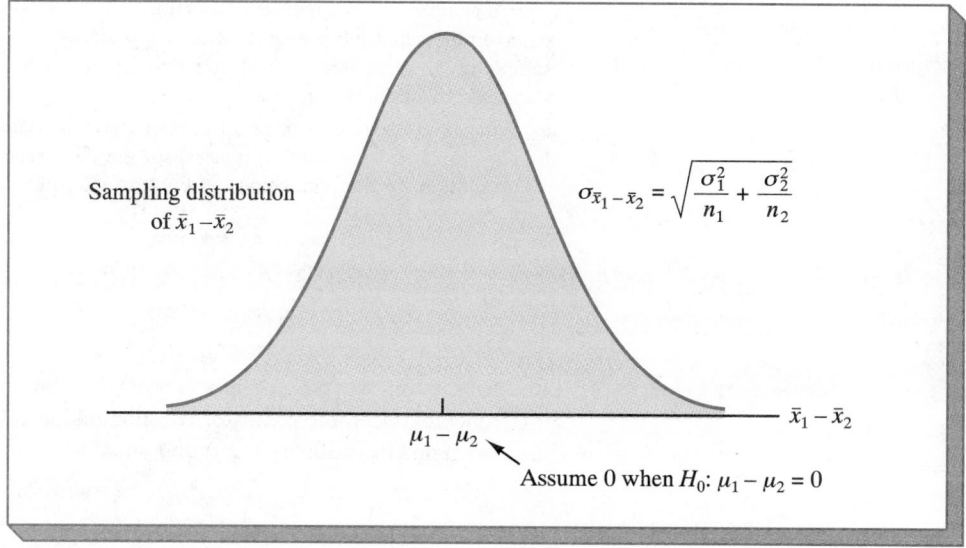

Sampling distribution
of $\bar{x}_1 - \bar{x}_2$

$$\sigma_{\bar{x}_1 - \bar{x}_2} = \sqrt{\frac{\sigma_1^2}{n_1} + \frac{\sigma_2^2}{n_2}}$$

$\bar{x}_1 - \bar{x}_2$

$\mu_1 - \mu_2$

Assume 0 when H_0: $\mu_1 - \mu_2 = 0$

The value of z given by (10.10) can be interpreted as the number of standard deviations $\bar{x}_1 - \bar{x}_2$ is from the value of $\mu_1 - \mu_2$ specified in H_0. For $\alpha = .05$ and thus $z_{\alpha/2} = z_{.025} = 1.96$, the rejection region for the two-tailed hypothesis test is shown in Figure 10.5. The rejection rule is as follows:

$$\text{Reject } H_0 \text{ if } z < -1.96 \text{ or if } z > +1.96$$

TABLE **10.6**
Examination Score Results

Training Center A	Training Center B
$n_1 = 30$	$n_2 = 40$
$\bar{x}_1 = 82.5$	$\bar{x}_2 = 78$
$s_1 = 8$	$s_2 = 10$

Let us assume that independent random samples of individuals trained at the two centers provide the examination score results shown in Table 10.6. Using s_1^2 and s_2^2 to estimate σ_1^2 and σ_2^2, the test statistic z given by (10.10) for the null hypothesis H_0: $\mu_1 - \mu_2 = 0$ becomes

$$z = \frac{(82.5 - 78) - 0}{\sqrt{(8)^2/30 + (10)^2/40}} = 2.09$$

With this value of z, the conclusion is to reject H_0. Thus, the sample scores lead the firm to conclude that the educational quality differs at the two centers.

With $z = 2.09$, the standard normal probability distribution table can be used to compute the p-value for this two-tailed test. With an area of .4817 between the mean and $z = 2.09$, the p-value is $2(.5000 - .4817) = .0366$.

In this hypothesis test, we were interested in determining if a difference exists between the means of the two populations. Since we did not have a prior belief that one mean might be greater than or less than the other, the hypotheses H_0: $\mu_1 - \mu_2 = 0$ and H_a: $\mu_1 - \mu_2 \neq 0$ were appropriate. In other hypothesis tests about the difference between the means of two populations, we may want to test whether one of the means is greater than or perhaps less than the other mean. In these cases, a one-tailed hypothesis test would be appropriate. The two forms of a one-tailed test about the difference between two population means are as follows:

$$H_0: \mu_1 - \mu_2 \leq 0 \qquad H_0: \mu_1 - \mu_2 \geq 0$$

$$H_a: \mu_1 - \mu_2 > 0 \qquad H_a: \mu_1 - \mu_2 < 0$$

These hypotheses are tested using the test statistic z given by (10.10). The rejection region is determined in a manner identical to the one-tailed approach presented in Chapter 9.

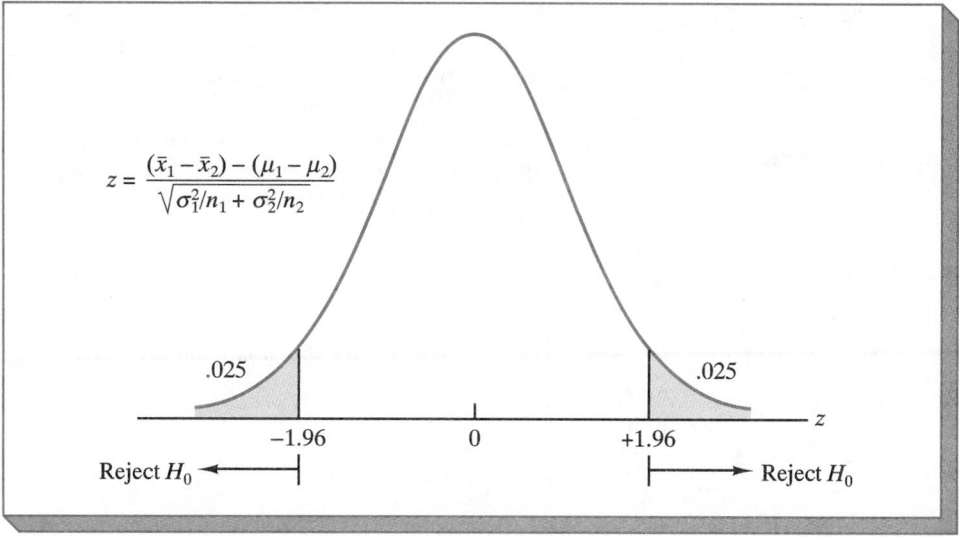

Small-Sample Case

Let us now consider hypothesis tests about the difference between the means of two populations for the small-sample case; that is, where $n_1 < 30$ and/or $n_2 < 30$. The procedure we will use is based on the t distribution with $n_1 + n_2 - 2$ degrees of freedom. As discussed in Section 10.1, assumptions are made that both populations have normal probability distributions and that the variances of the populations are equal.

A new computer software package has been developed to help systems analysts reduce the time required to design, develop, and implement an information system. To evaluate the benefits of the new software package, a random sample of 24 systems analysts was selected. Each analyst was given specifications for a hypothetical information system, and 12 of the analysts were instructed to produce the information system using current technology. The other 12 analysts were first trained in the use of the new software package and then instructed to use it to produce the information system.

In this study, there are two populations: a population of systems analysts using the current technology and a population of systems analysts using the new software package. In terms of the time required to complete the information systems design project, the population means are as follows:

μ_1 = the mean project-completion time for analysts using
      the current technology
μ_2 = the mean project-completion time for analysts using
      the new software package

The researcher in charge of the new software-evaluation project hopes to show that the new software package will provide a smaller mean project-completion time. Thus, the researcher is looking for evidence to conclude that μ_2 is less than μ_1; in this case, the difference between the two population means $\mu_1 - \mu_2$ will be greater than zero. In formulating the hypotheses, the research hypothesis $\mu_1 - \mu_2 > 0$ is stated as the alternative hypothesis. Thus we have

$$H_0: \mu_1 - \mu_2 \leq 0$$
$$H_a: \mu_1 - \mu_2 > 0$$

In tentatively assuming H_0 is true, we are taking the position that using the new software package takes the same time or perhaps even longer than the current technology. The researcher is looking for evidence to reject H_0 and conclude that the new software package possesses a smaller mean completion time.

Let us assume that the 24 analysts completed the study with the results shown in Table 10.7. Under the assumption that the variances of the populations are equal, (10.7) is used to compute the pooled estimate of σ^2.

TABLE 10.7

Results of Study

Current Technology	New Software Package
$n_1 = 12$	$n_2 = 12$
$\bar{x}_1 = 325$ hours	$\bar{x}_2 = 288$ hours
$s_1 = 40$ hours	$s_2 = 44$ hours

$$s^2 = \frac{(n_1 - 1)s_1^2 + (n_2 - 1)s_2^2}{(n_1 + n_2 - 2)} = \frac{11(40)^2 + 11(44)^2}{(12 + 12 - 2)} = 1768$$

The test statistic for the small-sample case is

$$t = \frac{(\bar{x}_1 - \bar{x}_2) - (\mu_1 - \mu_2)}{\sqrt{s^2\left(\frac{1}{n_1} + \frac{1}{n_2}\right)}} \qquad (10.11)$$

In the case of two independent random samples of sizes n_1 and n_2, the t distribution will have $n_1 + n_2 - 2$ degrees of freedom. For $\alpha = .05$, the t distribution table shows that with $12 + 12 - 2 = 22$ degrees of freedom, $t_{.05} = 1.717$. Thus, the rejection region for the one-tailed test is as follows:

$$\text{Reject } H_0 \text{ if } t > 1.717$$

The sample data and (10.11) provide the following value for the test statistic:

$$t = \frac{(325 - 288) - 0}{\sqrt{1768\left(\frac{1}{12} + \frac{1}{12}\right)}} = 2.16$$

Checking the rejection region, we see that $t = 2.16$ allows the rejection of H_0 at the .05 level of significance. Thus, the sample results permit the researcher to conclude that the the new software package provides a lower mean completion time.

Computer Solution

Minitab and other computer software packages provide the capability for testing hypotheses concerning the difference between the means of two populations. The Minitab output in Figure 10.6 shows the hypothesis-testing results for the illustration involving the new software package used to produce an information system. The completion times for the 12 systems analysts who used current technology to produce the information system have been entered into column C1 of the Minitab worksheet, while the completion times for the 12 systems analysts who used the new software package have been entered into column C2. The command TWOSAMPLE C1 C2 requests a two-independent-sample test for a difference between the means of the data in columns C1 and C2. The Minitab subcommand ALTERNATIVE = 1 indicates a one-tailed test with H_a: $\mu_1 - \mu_2 > 0$, and the subcommand POOLED indicates that the statistical analysis is to be based on the pooled estimate of σ^2. The first part of the output shows the mean completion time and the standard deviation of completion time for the two samples. The row TTEST provides the hypothesis-testing results. With $t = 2.16$ and a p-value = 0.021, the null hypothesis can be rejected at the .05 level of significance; thus, the conclusion is that the new software package provides a lower mean completion time.

FIGURE 10.6.   **Minitab Output for the Hypothesis Test About the Current and New Software Technology**

```
MTB > TWOSAMPLE C1 C2;
SUBC> ALTERNATIVE = 1;
SUBC> POOLED.

TWOSAMPLE T FOR CURRENT VS NEW
               N        MEAN       STDEV      SE MEAN
CURRENT    12        325.0        40.0          12
NEW        12        288.0        44.0          13

95 PCT CI FOR MU CURRENT - MU NEW: (1, 73)

TTEST MU CURRENT = MU NEW (VS GT): T= 2.16   P=0.021   DF=  22

POOLED STDEV =       42.1
```

Note that the computer solution also provides the 95% confidence interval for the difference between the two population means. Although the hypothesis test enabled us to conclude that the new software is better, the 95% confidence interval shows that on average the improvement that can be expected with the new software may be as little as 1 hour or as high as 73 hours. The wide confidence interval suggests that further study may be desirable to obtain a more precise estimate of how much improvement can be anticipated with the new software package.

NOTES & COMMENTS

1. In the previous section we stated that using the t distribution for inferences about the means of two populations is fairly insensitive to the assumptions of normal populations and equal variances. However, if a user feels strongly that these assumptions are not appropriate for a particular application, one of the following actions should be taken:

   a. Consider the nonparametric Wilcoxon rank sum test presented in Chapter 19.

   b. If the populations are approximately normal but the variances may not be equal ($\sigma_1^2 \neq \sigma_2^2$), use (10.5) to estimate $\sigma_{\bar{x}_1 - \bar{x}_2}$. The t distribution can still be used with the degrees of freedom given by

   $$df = \frac{(1/n_1 + s_2^2/s_1^2 n_2)^2}{[1/n_1^2(n_1 - 1)] + [s_2^2/s_1^2 n_2^2(n_2 - 1)]}$$

   c. Increase both sample sizes to the large-sample case with $n_1 \geq 30$ and $n_2 \geq 30$.

2. In hypothesis tests about the difference between the means of two populations, the null hypothesis almost always contains the condition that there is no difference between the means. Thus, null hypotheses are as follows:

   $$H_0: \mu_1 - \mu_2 = 0 \qquad H_0: \mu_1 - \mu_2 \leq 0 \qquad H_0: \mu_1 - \mu_2 \geq 0$$

Continued on next page

Continued from previous page

> are possible choices. In some instances, we may wish to determine if a nonzero difference D_0 exists between the population means. The specific value chosen for D_0 depends on the application under study. However, in this case, the null hypothesis may be of the following forms:
>
> $$H_0: \mu_1 - \mu_2 = D_0 \qquad H_0: \mu_1 - \mu_2 \leq D_0 \qquad H_0: \mu_1 - \mu_2 \geq D_0$$
>
> The hypothesis-testing computations remain the same with the exception that D_0 is used for the value of $\mu_1 - \mu_2$ in (10.10) and (10.11).

Exercises

SELF TEST ▶

Methods

11. Consider the following hypothesis test:

$$H_0: \mu_1 - \mu_2 \leq 0$$

$$H_a: \mu_1 - \mu_2 > 0$$

The results in Table 10.8 are for two independent samples taken from the two populations.
a. Using $\alpha = .05$, what is your hypothesis-testing conclusion?
b. What is the p-value?

TABLE 10.8

Sample 1	Sample 2
$n_1 = 40$	$n_2 = 50$
$\bar{x}_1 = 25.2$	$\bar{x}_2 = 22.8$
$s_1 = 5.2$	$s_2 = 6.0$

12. Consider the following hypothesis test:

$$H_0: \mu_1 - \mu_2 = 0$$

$$H_a: \mu_1 - \mu_2 \neq 0$$

The following results are for two independent samples taken from the two populations.

Sample 1	Sample 2
$n_1 = 80$	$n_2 = 70$
$\bar{x}_1 = 104$	$\bar{x}_2 = 106$
$s_1 = 8.4$	$s_2 = 7.6$

a. Using $\alpha = .05$, what is your hypothesis-testing conclusion?
b. What is the p-value?

13. Consider the following hypothesis test:

$$H_0: \mu_1 - \mu_2 = 0$$

$$H_a: \mu_1 - \mu_2 \neq 0$$

The results in Table 10.9 are for two independent samples taken from the two populations. Using $\alpha = .05$, what is your hypothesis-testing conclusion?

TABLE 10.9

Sample 1	Sample 2
$n_1 = 8$	$n_2 = 7$
$\bar{x}_1 = 1.4$	$\bar{x}_2 = 1.0$
$s_1 = 0.4$	$s_2 = 0.6$

Applications

14. Are starting salary differentials present for male and female graduates of business schools? Miami University in Oxford, Ohio, reported that a sample of 30 males entering public accounting received an average starting salary of $29,300 while a sample of 36 females entering public

accounting received an average starting salary of \$28,800 (*Miami University Class Profile,* 1990). If the sample standard deviations were \$2000 and \$1800, respectively, is there any statistical support for the conclusion that a differential exists between the mean starting salaries for the population of males and the population of females? Test $H_0: \mu_1 - \mu_2 \leq 0$ and $H_a: \mu_1 - \mu_2 > 0$ at a .05 level of significance. What is the *p*-value, and what is your conclusion?

SELF TEST ▶ 15. The Greystone Department Store study in Section 10.1 supplied the following data on customer ages from independent random samples taken at two store locations:

Inner-City Store	Suburban Store
$n_1 = 36$	$n_2 = 49$
$\bar{x}_1 = 40$ years	$\bar{x}_2 = 35$ years
$s_1 = 9$ years	$s_2 = 10$ years

For $\alpha = .05$, test the hypothesis $H_0: \mu_1 - \mu_2 = 0$ against the alternative hypothesis $H_a: \mu_1 - \mu_2 \neq 0$. What is your conclusion about the mean ages of the populations of customers at the two stores?

16. The Educational Testing Service conducted a study to investigate differences between the scores of males and females on the Scholastic Aptitude Test (*Journal of Educational Measurement,* Spring 1987). The study identified a random sample of 562 females and 852 males who had achieved the same high score on the mathematics portion of the test. That is, both the females and males were viewed as having similar high abilities in mathematics. The SAT verbal scores for the two samples are summarized in Table 10.10. Do the data support the conclusion that given a population of females and a population of males with similar high mathematical abilities, the females will have a significantly higher verbal ability? Test at a .02 level of significance. What is your conclusion?

TABLE 10.10

Females	Males
$\bar{x}_1 = 547$	$\bar{x}_2 = 525$
$s_1 = 83$	$s_2 = 78$

17. A firm is studying the delivery times for two raw material suppliers. The firm is basically satisfied with supplier A and is prepared to stay with this supplier provided that the mean delivery time is the same as or less than that of supplier B. However, if the firm finds that the mean delivery time from supplier B is less than that of supplier A, it will begin making raw material purchases from supplier B.

a. What are the null and alternative hypotheses for this situation?

b. Assume that independent samples show the following delivery time characteristics for the two suppliers:

Supplier A	Supplier B
$n_1 = 50$	$n_2 = 30$
$\bar{x}_1 = 14$ days	$\bar{x}_2 = 12.5$ days
$s_1 = 3$ days	$s_2 = 2$ days

TABLE 10.11

Male Employees	Female Employees
$n_1 = 44$	$n_2 = 32$
$\bar{x}_1 = \$9.25$	$\bar{x}_2 = \$8.70$
$s_1 = \$1.00$	$s_2 = \$.80$

Using $\alpha = .05$, what is your conclusion for the hypotheses from part (a)? What action do you recommend in terms of supplier selection?

18. In a wage discrimination case involving male and female employees, independent samples of male and female employees with 5 years or more experience provided the hourly wage results shown in Table 10.11. The null hypothesis is stated such that male employees have a mean hourly wage less than or equal to that of the female employees. Rejection of H_0 leads to the conclusion that male employees have a mean hourly wage exceeding the female employees' mean hourly wage. Test the hypothesis with $\alpha = .01$. Does wage discrimination appear to exist in this case?

19. A production line is designed on the assumption that the difference between mean assembly times for two operations is 5 minutes. Independent tests for the two assembly operations show the following results:

Operation A	Operation B
$n_1 = 100$	$n_2 = 50$
$\bar{x}_1 = 14.8$ minutes	$\bar{x}_2 = 10.4$ minutes
$s_1 = .8$ minutes	$s_2 = .6$ minutes

TABLE 10.12

Accounting	Finance
28.8	26.3
25.3	23.6
26.2	25.0
27.9	23.0
27.0	27.9
26.2	24.5
28.1	29.0
24.7	27.4
25.2	23.5
29.2	26.9
29.7	26.2
29.3	24.0

Using $\alpha = .02$, test the hypothesis that the difference between the mean assembly times is $\mu_1 - \mu_2 = 5$ minutes.

20. Starting salary data for college graduates is reported by the College Placement Council (*USA Today,* April 6, 1992). Annual salaries in thousands of dollars for a sample of accounting majors and a sample of finance majors are shown in Table 10.12.

a. Use a .05 level of significance to test the hypothesis that there is no difference between the mean annual starting salary for accounting majors and the mean annual starting salary for finance majors. What is your conclusion?

b. Provide the point estimate and the 95% confidence interval for the difference between the mean starting salaries for the two majors.

10.3 Inferences about the Difference between the Means of Two Populations: Matched Samples

Suppose that a manufacturing company has two methods available for employees to perform a production task. To maximize production output, the company would like to identify the method with the smallest mean completion time per unit. Let μ_1 denote the mean completion time for production method 1 and μ_2 denote the mean completion time for production method 2. With no preliminary indication of the preferred production method, we begin by tentatively assuming that the two production methods have the same mean completion time. Thus, the null hypothesis becomes H_0: $\mu_1 - \mu_2 = 0$. If this hypothesis is rejected, we can conclude that a difference between the mean completion times exists. In this case, the method providing the smaller mean completion time would be recommended. The null and alternative hypotheses are written as follows:

$$H_0: \mu_1 - \mu_2 = 0$$

$$H_a: \mu_1 - \mu_2 \neq 0$$

In designing the sampling procedure that will be used to collect production time data and test the above hypotheses, we consider two alternative designs. One design is based on *independent samples,* and the other design is based on *matched samples.* The designs are described as follows:

1. *Independent-sample design:* A simple random sample of workers is selected, and each worker uses method 1. A second independent simple random sample of workers is selected and each worker uses method 2. The test of the difference between means is based on the procedures of Section 10.2.

2. *Matched-sample design:* One simple random sample of workers is selected with each worker first using one method and then using the other method. The order of the two methods is assigned randomly to the workers, with some workers performing method 1 first and others performing method 2 first. Each worker provides a pair of data values, one value for method 1 and another value for method 2.

Our interest in the matched-sample design is that since both production methods are tested under similar conditions (i.e., same workers), this design often leads to a smaller sampling error than the independent sample design. The primary reason for this is that in a matched-sample design variation between workers is eliminated as a source of the sampling error.

Let us demonstrate the analysis of a matched-sample design by assuming that this method is used to test the difference between the two production methods. A random sample of 6 workers is used. The data on completion times for the 6 workers are shown in Table 10.13. Note that each worker provides a pair of data values, one for each production method. Also note that the last column contains the difference in completion times d_i for each worker in the sample.

The key to the analysis of the matched-sample design is to realize that we consider only the column of differences in the two methods. As a result we have six data values (.6, −.2, .5, .3, .0, and .6) that will be used in the analysis of the difference between means of the two production methods.

Let μ_d = the mean of the *difference* values for the population of workers. With this notation the null and alternative hypotheses are rewritten as follows:

$$H_0: \mu_d = 0$$

$$H_a: \mu_d \neq 0$$

If H_0 can be rejected, we can conclude that a difference between the mean completion times exists.

The d notation is a reminder that the matched sample provides *difference* data. The sample mean and sample standard deviation for the six difference values in Table 10.13 are as follows:

$$\bar{d} = \frac{\Sigma d_i}{n} = \frac{1.8}{6} = .30$$

$$s_d = \sqrt{\frac{\Sigma (d_i - \bar{d})^2}{n - 1}} = \sqrt{\frac{.56}{5}} = .335$$

In Chapter 9 we stated that if the population can be assumed to be normally distributed, the t distribution with $n - 1$ degrees of freedom can be used to test the null hypothesis about a population mean. With difference data, the test statistic becomes

TABLE 10.13	Worker	Completion Time for Method 1 (minutes)	Completion Time for Method 2 (minutes)	Difference in Completion Times (d_i)
Task-Completion Times for a Matched-Sample Design	1	6.0	5.4	.6
	2	5.0	5.2	−.2
	3	7.0	6.5	.5
	4	6.2	5.9	.3
	5	6.0	6.0	.0
	6	6.4	5.8	.6

$$t = \frac{\bar{d} - \mu_d}{s_d/\sqrt{n}} \qquad (10.12)$$

With $\alpha = .05$ and $n - 1 = 5$ degrees of freedom ($t_{.025} = 2.571$), the rejection rule for the two-tailed test becomes

Reject H_0 if $t < -2.571$ or if $t > 2.571$

With $\bar{d} = .30$, $s_d = .335$, and $n = 6$, the value of the test statistic for the null hypothesis $H_0: \mu_d = 0$ is

$$t = \frac{\bar{d} - \mu_d}{s_d/\sqrt{n}} = \frac{.30 - 0}{.335/\sqrt{6}} = 2.19$$

Since $t = 2.19$ does not fall in the rejection region, the sample data do not provide sufficient evidence to reject H_0.

With the above sample results, an interval estimate of the difference between the two population means can be based on the single-population methodology of Chapter 8. Doing so provides the following:

$$0.3 \pm t_{\alpha/2}\frac{s_d}{\sqrt{n}}$$

$$0.3 \pm 2.571\frac{.335}{\sqrt{6}}$$

$$0.3 \pm .35$$

Thus, the 95% confidence interval for the difference between the means of the two production methods is $-.05$ minutes to $.65$ minutes.

NOTES &
COMMENTS

1. In the example presented in this section, workers performed the production task using first one method and then the other method. This is an example of a matched-sample design, where each sampled item (worker) provides a pair of data values. Although this is often the procedure used in the matched-sample analysis, it is possible to use different but "similar" items to provide the pair of data values. In this sense, a worker at one location could be matched with a similar worker at another location (similarity based on age, education, sex, experience, etc.). The pairs of workers would provide the difference data that could be used in the matched-sample analysis.

2. Since a matched-sample procedure for inferences about two population means generally provides better precision than the independent-sample approach, it is the recommended design. However, in some applications the matching cannot be achieved, or perhaps the time and cost associated with matching is excessive. In these cases, the independent-sample design should be used.

3. The example presented in this section used a sample size of 6 workers. As such, the small-sample case existed, and the t distribution was used in both the test of hypothesis and interval-estimation computations. If the sample size is large ($n \geq 30$), the use of the t distribution is unnecessary; in such cases, statistical inferences can be based on the z values of the standard normal probability distribution.

☐   ☐   **Exercises**

Methods

SELF TEST ▶ 21. Consider the following hypothesis test:

$$H_0: \mu_d \leq 0$$

$$H_a: \mu_d > 0$$

The following data are from matched samples taken from two populations.

	Population	
Element	1	2
1	21	20
2	28	26
3	18	18
4	20	20
5	26	24

a. Compute the difference value for each element.
b. Compute $\bar{d}$.
c. Compute the standard deviation s_d.
d. Test the hypothesis using $\alpha = .05$. What is your conclusion?

22. The data in Table 10.14 are from matched samples taken from two populations.
a. Compute the difference value for each element.
b. Compute $\bar{d}$.
c. Compute the standard deviation s_d.
d. What is the point estimate of the difference between the two population means?
e. Provide a 95% confidence interval for the difference between the two population means.

TABLE 10.14

	Population	
Element	1	2
1	11	8
2	7	8
3	9	6
4	12	7
5	13	10
6	15	15
7	15	14

Applications

SELF TEST ▶ 23. A market research firm used a sample of individuals to rate the purchase potential for a particular product before and after the individuals saw a new television commercial about the product. The purchase-potential ratings were based on a 0 to 10 scale, with higher values indicating a higher purchase potential. The null hypothesis stated that the mean rating "after" would be less than or equal to the mean rating "before." Rejection of this hypothesis would provide the conclusion that the commercial improved the mean purchase-potential rating. Use $\alpha = .05$ and the following data to test the hypothesis and comment on the value of the commercial:

Individual	**Purchase Rating** After	Before	**Individual**	**Purchase Rating** After	Before
1	6	5	5	3	5
2	6	4	6	9	8
3	7	7	7	7	5
4	4	3	8	6	6

TABLE 10.15

Corporation	1987	1988
IBM	8.72	9.27
Sears Roebuck	4.35	2.72
Chevron	3.65	5.17
Walt Disney	2.85	3.80
American Express	1.20	2.31
McDonalds	2.89	3.43
Anheuser-Busch	2.04	2.45
Kellogg	3.20	3.90
J. C. Penney	4.11	6.02
Motorola	2.39	3.43

TABLE 10.16

Respondent	Television	Reading
1	10	6
2	14	16
3	16	8
4	18	10
5	15	10
6	14	8
7	10	14
8	12	14
9	4	7
10	8	8
11	16	5
12	5	10
13	8	3
14	19	10
15	11	6

TABLE 10.17

	Weekly Sales	
Salesperson	Before	After
1	15	18
2	12	14
3	18	19
4	15	18
5	16	18

24. Earnings per share for a sample of 10 corporations (*Business Week*, 1989 Bonus Issue) are presented in Table 10.15 for the years 1987 and 1988.

a. Based on the preceding data, is it appropriate to conclude that the mean earnings per share was significantly higher in 1988? Test at a .05 level of significance.

b. Provide a 95% confidence interval for the increase in mean earnings per share for the 1-year period.

25. Transportation costs from the airport to the downtown area depend on the method of transportation. One-way costs for taxi and shuttle bus transportation for a sample of 10 major cities are shown below (*USA Today*, February 13, 1992). Provide a 95% confidence interval for the mean cost increase associated with taxi transportation.

City	Taxi	Shuttle Bus	City	Taxi	Shuttle Bus
Atlanta	15	7	Minneapolis	16.5	7.5
Chicago	22	12.5	New Orleans	18	7
Denver	11	5	New York (LaGuardia)	16	8.5
Houston	15	4.5	Philadelphia	20	8
Los Angeles	26	11	Washington, D.C.	10	5

26. A survey was made of Book-of-the-Month-Club members to see if members spend more time watching television than they do reading (*The Cincinnati Enquirer*, November 21, 1991). Assume a small sample of respondents in this survey provided the weekly hours of watching television and weekly hours of reading as shown in Table 10.16. Using a .05 level of significance, can it be concluded that even Book-of-the-Month-Club members spend more time per week, on average, watching television than reading?

27. A manufacturer produces both a deluxe and a standard model automatic sander designed for home use. Selling prices obtained from a sample of retail outlets are as follows:

Retail Outlet	Model Price Deluxe	Model Price Standard	Retail Outlet	Model Price Deluxe	Model Price Standard
1	39	27	5	40	30
2	39	28	6	39	34
3	45	35	7	35	29
4	38	30			

a. The manufacturer's suggested retail prices for the two models show a $10 differential in prices. Using a .05 level of significance, test that the mean difference between prices of the two models is $10.

b. What is the 95% confidence interval for the difference between the mean prices for the two models?

28. A company attempts to evaluate the potential for a new bonus plan by selecting a random sample of 5 salespersons to use the bonus plan for a trial period. The weekly sales volumes before and after implementing the bonus plan are shown in Table 10.17.

a. Use $\alpha = .05$ and test to see if the bonus plan will result in an increase in the mean weekly sales.

b. Provide a 90% confidence interval for the mean increase in weekly sales that can be expected if a new bonus plan is implemented.

29. Word-processing systems are often justified on the basis of improved efficiencies for a secretarial staff. Shown below are typing rates in words per minute for 7 secretaries who previously used electronic typewriters and who are now using computer-based word processors. Test at the .05 level of significance to see if there has been an increase in the mean typing rate due to the word-processing system.

Secretary	Electronic Typewriter	Word Processor	Secretary	Electronic Typewriter	Word Processor
1	72	75	5	52	55
2	68	66	6	55	57
3	55	60	7	64	64
4	58	64			

10.4   Inferences about the Difference between the Proportions of Two Populations

A tax preparation firm is interested in comparing the quality of work at two of its regional offices. By randomly selecting samples of tax returns prepared at each office and having the sample returns verified for accuracy, the firm will be able to estimate the proportion of erroneous returns prepared at each office. Of particular interest is the difference between these proportions.

Let

p_1 = proportion of erroneous returns for population 1 (office 1)
p_2 = proportion of erroneous returns for population 2 (office 2)
$\bar{p}_1$ = sample proportion for a simple random sample from population 1
$\bar{p}_2$ = sample proportion for a simple random sample from population 2

The difference between the two population proportions is given by $p_1 - p_2$. The point estimator of $p_1 - p_2$ is as follows.

**Point Estimator of the Difference
between the Proportions of Two Populations**

$$\bar{p}_1 - \bar{p}_2$$

Thus, the point estimator of the difference between two population proportions is the difference between the sample proportions of two independent simple random samples.

Sampling Distribution of $\bar{p}_1 - \bar{p}_2$

In the study of the difference between two population proportions, $\bar{p}_1 - \bar{p}_2$ is the point estimator of interest. As we have seen in several previous cases, the sampling distribution of the point estimator is a key factor in developing interval estimates and in testing hypotheses about the parameters of interest. The properties of the sampling distribution of $\bar{p}_1 - \bar{p}_2$ are as follows.

Sampling Distribution of $\bar{p}_1 - \bar{p}_2$

Expected Value: $E(\bar{p}_1 - \bar{p}_2) = p_1 - p_2$ (10.13)

Standard Deviation: $\sigma_{\bar{p}_1 - \bar{p}_2} = \sqrt{\dfrac{p_1(1 - p_1)}{n_1} + \dfrac{p_2(1 - p_2)}{n_2}}$ (10.14)

where

n_1 = sample size for the simple random sample from population 1
n_2 = sample size for the simple random sample from population 2

Distribution form: If the sample sizes are large (i.e., $n_1 p_1$, $n_1(1 - p_1)$, $n_2 p_2$, and $n_2(1 - p_2)$ are all greater than or equal to 5), the sampling distribution of $\bar{p}_1 - \bar{p}_2$ can be approximated by a normal probability distribution.

Figure 10.7 shows the sampling distribution of $\bar{p}_1 - \bar{p}_2$.

Interval Estimation of $p_1 - p_2$

Let us assume that independent simple random samples of tax returns from the two offices show the following:

Office 1	Office 2
$n_1 = 250$	$n_2 = 300$
Number of returns with errors = 35	Number of returns with errors = 27

FIGURE 10.7
Sampling Distribution of $\bar{p}_1 - \bar{p}_2$

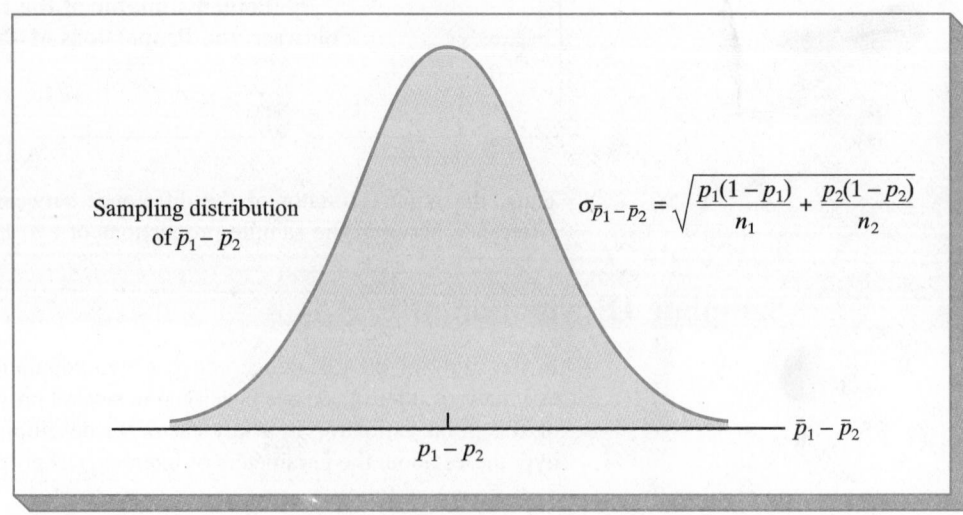

Sampling distribution of $\bar{p}_1 - \bar{p}_2$

$\sigma_{\bar{p}_1 - \bar{p}_2} = \sqrt{\dfrac{p_1(1 - p_1)}{n_1} + \dfrac{p_2(1 - p_2)}{n_2}}$

$p_1 - p_2$

$\bar{p}_1 - \bar{p}_2$

The sample proportions for the two offices are as follows:

$$\bar{p}_1 = \frac{35}{250} = .14$$

$$\bar{p}_2 = \frac{27}{300} = .09$$

The point estimate of the difference between the proportion of erroneous tax returns for the two populations is $\bar{p}_1 - \bar{p}_2 = .14 - .09 = .05$. Specifically, we are led to believe that office 1 possesses a 5% greater error rate than office 2. However, as with all point estimates, we know that the .05 difference is only one of many possible sample values for the difference between the two population proportions. The following expression can be used to develop an interval estimate of the difference between the proportions of the two populations.

Interval Estimate of the Difference between the Proportions of Two Populations (Large-Sample Case with $n_1 p_1$, $n_1(1 - p_1)$, $n_2 p_2$, and $n_2(1 - p_2) \geq 5$)

$$\bar{p}_1 - \bar{p}_2 \pm z_{\alpha/2}\sigma_{\bar{p}_1 - \bar{p}_2} \tag{10.15}$$

where $1 - \alpha$ is the confidence coefficient.

Let us use this procedure to develop an interval estimate of the difference between the population proportions of erroneous tax returns existing at the two offices. Since p_1 and p_2 are unknown, we cannot use (10.14) to calculate $\sigma_{\bar{p}_1 - \bar{p}_2}$. However, using $\bar{p}_1$, the point estimator of p_1, and $\bar{p}_2$, the point estimator of p_2, we can estimate $\sigma_{\bar{p}_1 - \bar{p}_2}$ as follows.

Point Estimator of $\sigma_{\bar{p}_1 - \bar{p}_2}$

$$s_{\bar{p}_1 - \bar{p}_2} = \sqrt{\frac{\bar{p}_1(1 - \bar{p}_1)}{n_1} + \frac{\bar{p}_2(1 - \bar{p}_2)}{n_2}} \tag{10.16}$$

The above expression provides an estimate of $\sigma_{\bar{p}_1 - \bar{p}_2}$ and can be used in (10.15) to obtain an interval estimate of $p_1 - p_2$.

Let us make these calculations. Using (10.16) we have

$$s_{\bar{p}_1 - \bar{p}_2} = \sqrt{\frac{.14(.86)}{250} + \frac{.09(.91)}{300}} = .0275$$

With a 90% confidence interval $z_{\alpha/2} = z_{.05} = 1.645$, (10.15) provides the following interval estimate:

$$(.14 - .09) \pm 1.645(.0275)$$
$$.05 \pm .045$$

Thus, the 90% confidence interval for the difference in error rates at the two offices is .005 to .095.

Hypothesis Tests about $p_1 - p_2$

As an example of hypothesis tests concerning the difference between the proportions of two populations, let us consider the data collected in the preceding example and assume that the firm is attempting to determine whether a difference exists in the error proportions at the two offices. Let us illustrate the statistical analysis we could use to test the following hypotheses:

$$H_0: p_1 - p_2 = 0$$
$$H_a: p_1 - p_2 \neq 0$$

Figure 10.8 shows the sampling distribution of $\bar{p}_1 - \bar{p}_2$ based on the assumption that there is no difference between the two population proportions. That is, $p_1 - p_2 = 0$. With the sampling distribution approximately normal, the test statistic for the difference between two population proportions can be written

$$z = \frac{(\bar{p}_1 - \bar{p}_2) - (p_1 - p_2)}{\sigma_{\bar{p}_1 - \bar{p}_2}} \qquad (10.17)$$

Using $\alpha = .10$ and $z_{\alpha/2} = z_{.05} = 1.645$, the rejection rule is

Reject H_0 if $z < -1.645$ or if $z > 1.645$

The computation of z in (10.17) requires a value for the standard error of the difference between proportions $\sigma_{\bar{p}_1 - \bar{p}_2}$. Since this standard error is unknown, it will have to be estimated from the sample data. While we may be tempted to use $\bar{p}_1$ and $\bar{p}_2$ in (10.16) as we did with the interval-estimation procedure, in hypothesis testing we often adjust (10.16) to a slightly different form. For the *special case* where the hypotheses involve no difference between the population proportions (i.e., either $H_0: p_1 - p_2 = 0$, $H_0: p_1 - p_2 \leq 0$, or $H_0: p_1 - p_2 \geq 0$), (10.16) is modified to reflect the fact that when we assume H_0 to be true at the equality, we are assuming $p_1 = p_2$. When this occurs, we

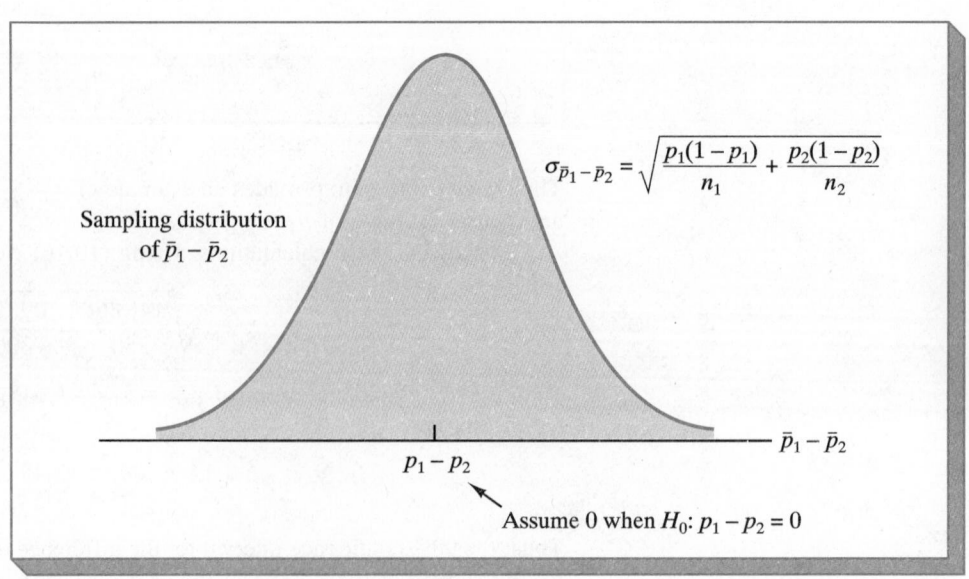

FIGURE **10.8**
Sampling Distribution of $\bar{p}_1 - \bar{p}_2$ with H_0: $p_1 - p_2 = 0$

combine or *pool* the two sample proportions to provide one estimate. This pooled estimator, denoted by $\bar{p}$, is as follows:

$$\bar{p} = \frac{n_1\bar{p}_1 + n_2\bar{p}_2}{n_1 + n_2} \tag{10.18}$$

With $\bar{p}$ used in place of both $\bar{p}_1$ and $\bar{p}_2$, (10.16) is revised to

$$s_{\bar{p}_1-\bar{p}_2} = \sqrt{\bar{p}(1-\bar{p})\left(\frac{1}{n_1} + \frac{1}{n_2}\right)} \tag{10.19}$$

Using (10.18) and (10.19), we can now proceed with the calculations as follows:

$$\bar{p} = \frac{250(.14) + 300(.09)}{550} = \frac{62}{550} = .113$$

$$s_{\bar{p}_1-\bar{p}_2} = \sqrt{(.113)(.887)\left(\frac{1}{250} + \frac{1}{300}\right)} = .0271$$

Using (10.17), the value of the test statistic becomes

$$z = \frac{(\bar{p}_1 - \bar{p}_2) - (p_1 - p_2)}{s_{\bar{p}_1-\bar{p}_2}} = \frac{(.14 - .09) - 0}{.0271} = 1.85$$

Since $1.85 > 1.645$, at the .10 level of significance the null hypothesis is rejected. The sample evidence indicates that there is a difference between the error proportions at the two offices.

As we saw with the hypothesis tests about differences between two population means, one-tailed tests can also be developed for the difference between two population proportions. The one-tailed rejection regions are established in a manner similar to the one-tailed hypothesis-testing procedures for a single-population proportion.

☐ ☐ Exercises

Methods

TABLE 10.18

Sample 1	Sample 2
$n_1 = 400$	$n_2 = 300$
$\bar{p}_1 = .48$	$\bar{p}_2 = .36$

30. Consider the results in Table 10.18 for two independent samples taken from two populations.
a. What is the point estimate of the difference between the two population proportions?
b. Develop a 90% confidence interval for the difference between the two population proportions.
c. Develop a 95% confidence interval for the difference between the two population proportions.

SELF TEST ▶

31. Consider the following hypothesis test: $H_0: p_1 - p_2 \geq 0$
$H_a: p_1 - p_2 < 0$

TABLE 10.19

Sample 1	Sample 2
$n_1 = 200$	$n_2 = 300$
$\bar{p}_1 = .22$	$\bar{p}_2 = .16$

The results in Table 10.19 are for two independent samples taken from the two populations.
a. Using $\alpha = .05$, what is your hypothesis-testing conclusion?
b. What is the *p*-value?

Applications

32. *Business Week*/Harris polls (*Business Week,* April 6, 1992) compared the views adults had about their children's future in 1989 with the views adults held in 1992. In a 1989 poll, 59% of the adults sampled felt their children would have a better life than they had. In a 1992 poll, 34% of the adults sampled felt their children would have a better life. Assume that 1250 adults were used in both polls. Provide a 95% confidence interval estimate of the difference between the proportions in 1989 and 1992. What is your interpretation of the interval estimate and the difference shown?

SELF TEST ▶

33. During the primary elections of 1988, a particular presidential candidate had the preelection voter support in Wisconsin and Illinois shown in Table 10.20. Compute a 95% confidence interval for the difference between the proportion of voters favoring the candidate in the two states.

TABLE 10.20

State	Voters Surveyed	Voters Favoring the Candidate
Wisconsin	500	270
Illinois	360	162

34. Leo J. Shapiro & Associates, a Chicago market-research firm, surveys consumers on a variety of issues (*The Wall Street Journal*, October 17, 1988). Every year in late summer and early fall, the firm surveys consumers to learn about spending plans for the forthcoming holiday season. In 1988, 36% of the consumers surveyed indicated they planned to ''spend less'' during the 1988 season. In 1987, 28% of the consumers surveyed indicated they planned to ''spend less'' during the 1987 season. Assume that 400 consumers were surveyed each year.

a. Compare the results for the two years. Use $\alpha = .05$ to see if there has been a significant increase in the proportion of consumers planning to spend less during the 1988 season. What is your conclusion? What is the p-value?

b. Assume you are a retailer in October 1988. Comment on the value of the 2-year study and what the results might mean to your business.

c. Develop an interval estimate of the 1-year increase in the proportion of consumers who indicate they will spend less during the 1988 season. Use a 95% confidence level.

35. Two loan officers at the North Ridge National Bank show the following data for defaults on loans that they have approved (the data are based on samples of loans granted over the past 5 years):

Loan Officer	Loans Reviewed in the Sample	Defaulted Loans
A	60	9
B	80	6

Using $\alpha = .05$, test the hypothesis that the default rates are the same for the two loan officers.

36. A Media General/Associated Press Poll (*USA Today*, April 27, 1989) reported that 16% of men and 5% of women would want to be president of the United States. Assume 500 men and 500 women participated in the poll. Test H_0: $p_1 - p_2 = 0$ versus H_a: $p_1 - p_2 \neq 0$. Use $\alpha = .05$. What is your conclusion about the proportions of men and women wanting to be president of the United States?

37. A survey firm conducts door-to-door surveys on a variety of issues. Some individuals cooperate with the interviewer and complete the interview questionnaire, while others do not. The sample data in Table 10.21 are available.

a. Using $\alpha = .05$, test the hypothesis that the response rate is the same for both men and women.

b. Compute the 95% confidence interval for the difference between the proportions of men and women that cooperate with the survey.

TABLE 10.21

Respondents	Sample Size	Number Cooperating
Men	200	110
Women	300	210

38. In a test of the quality of two television commercials, each commercial was shown in a separate test area six times over a 1-week period. The following week a telephone survey was conducted to identify individuals who had seen the commercials. The individuals who had seen the commercials were asked to state the primary message in the commercial. The following results were recorded:

Commercial	Number Who Saw Commercial	Number Who Recalled Primary Message
A	150	63
B	200	60

a. Using $\alpha = .05$, test the hypothesis that there is no difference in the recall proportions for the two commercials.

b. Compute a 95% confidence interval for the difference between the recall proportions for the two populations.

39. *The Los Angeles Times* polled Californians to learn whether they felt the state government was on the right track in terms of economic and social service programs (*Business Week*, December 30, 1991). In May of 1991, 520 of 1679 respondents felt the state was on the right track. In December of 1991, 293 of 1629 respondents felt the state was on the right track.

a. Considering the populations in May and December as two populations, was there a shift in the proportion who felt the state was on the right track with its economic and social service programs? Test H_0: $p_1 - p_2 = 0$ and H_a: $p_1 - p_2 \neq 0$. Use a .01 level of significance.

b. Provide a 95% confidence interval for any difference between the proportion who felt the state was on the right track. What is your interpretation of this interval estimate?

Summary

In this chapter we discussed procedures for developing interval estimates and conducting hypothesis tests involving two populations. First, we showed how to make inferences about the differences between the means of two populations when independent simple random samples are selected. We considered both the large- and small-sample cases. The z values from the standard normal probability distribution are used for inferences about the difference between two population means when the sample sizes are large. The t distribution is used for the inferences when the populations are assumed normal with equal variances. This permits inferences in the small-sample case.

Inferences about the difference between the means of two populations were then discussed for the matched-sample design. In the matched-sample design each element provides a pair of data values, one from each population. The difference between the paired data values is then used in the statistical analysis. The matched-sample design is generally preferred over the independent-sample design because the matched-sample procedure often reduces variability, thus tending to reduce the sampling error and to improve the precision of the estimate.

Finally, interval estimation and hypothesis testing about the difference between two population proportions were discussed. Statistical procedures for analyzing the difference between proportions for two populations are similar to the procedures for analyzing the difference between means for two populations.

Glossary

Pooled variance An estimate of the variance of a population based on the combination of two (or more) sample results. The pooled variance estimate is appropriate whenever the variances of two (or more) populations are assumed equal.

Independent samples Samples selected from two (or more) populations where the elements making up one sample are chosen independently of the elements making up the other sample(s).

Matched samples Samples where each data value in one sample is matched with a corresponding data value in the other sample.

Key Formulas

Expected Value of $\bar{x}_1 - \bar{x}_2$

$$E(\bar{x}_1 - \bar{x}_2) = \mu_1 - \mu_2 \qquad (10.2)$$

Standard Deviation of $\bar{x}_1 - \bar{x}_2$

$$\sigma_{\bar{x}_1 - \bar{x}_2} = \sqrt{\frac{\sigma_1^2}{n_1} + \frac{\sigma_2^2}{n_2}} \tag{10.3}$$

Interval Estimate of the Difference between the Means of Two Populations (Large-Sample Case with $n_1 \geq 30$ and $n_2 \geq 30$)

$$\bar{x}_1 - \bar{x}_2 \pm z_{\alpha/2}\sigma_{\bar{x}_1 - \bar{x}_2} \tag{10.4}$$

Point Estimator of $\sigma_{\bar{x}_1 - \bar{x}_2}$

$$s_{\bar{x}_1 - \bar{x}_2} = \sqrt{\frac{s_1^2}{n_1} + \frac{s_2^2}{n_2}} \tag{10.5}$$

Standard Deviation of $\bar{x}_1 - \bar{x}_2$ when $\sigma_1^2 = \sigma_2^2 = \sigma^2$

$$\sigma_{\bar{x}_1 - \bar{x}_2} = \sqrt{\frac{\sigma^2}{n_1} + \frac{\sigma^2}{n_2}} = \sqrt{\sigma^2\left(\frac{1}{n_1} + \frac{1}{n_2}\right)} \tag{10.6}$$

Pooled Estimator of σ^2

$$s^2 = \frac{(n_1 - 1)s_1^2 + (n_2 - 1)s_2^2}{n_1 + n_2 - 2} \tag{10.7}$$

Point Estimator of $\sigma_{\bar{x}_1 - \bar{x}_2}$ when $\sigma_1^2 = \sigma_2^2$

$$s_{\bar{x}_1 - \bar{x}_2} = \sqrt{s^2\left(\frac{1}{n_1} + \frac{1}{n_2}\right)} \tag{10.8}$$

Interval Estimate of the Difference between the Means of Two Populations (Small-Sample Case with $n_1 < 30$ and/or $n_2 < 30$)

$$\bar{x}_1 - \bar{x}_2 \pm t_{\alpha/2}s_{\bar{x}_1 - \bar{x}_2} \tag{10.9}$$

Test Statistic for Hypothesis Tests about the Difference between the Means of Two Populations

$$z = \frac{(\bar{x}_1 - \bar{x}_2) - (\mu_1 - \mu_2)}{\sqrt{\sigma_1^2/n_1 + \sigma_2^2/n_2}} \tag{10.10}$$

Sample Mean for Matched Samples

$$\bar{d} = \frac{\Sigma d_i}{n}$$

Sample Standard Deviation for Matched Samples

$$s_d = \sqrt{\frac{\Sigma(d_i - \bar{d})^2}{n - 1}}$$

Test Statistic for Matched Samples

$$t = \frac{\bar{d} - \mu_d}{s_d/\sqrt{n}} \tag{10.12}$$

Expected Value of $\bar{p}_1 - \bar{p}_2$

$$E(\bar{p}_1 - \bar{p}_2) = p_1 - p_2 \tag{10.13}$$

Standard Deviation of $\bar{p}_1 - \bar{p}_2$

$$\sigma_{\bar{p}_1 - \bar{p}_2} = \sqrt{\frac{p_1(1 - p_1)}{n_1} + \frac{p_2(1 - p_2)}{n_2}} \tag{10.14}$$

Interval Estimate of the Difference between the Proportions of Two Populations (Large-Sample Case with $n_1 p_1$, $n_1(1 - p_1)$, $n_2 p_2$, and $n_2(1 - p_2) \geq 5$)

$$\bar{p}_1 - \bar{p}_2 \pm z_{\alpha/2}\sigma_{\bar{p}_1 - \bar{p}_2} \tag{10.15}$$

Point Estimator of $\sigma_{\bar{p}_1 - \bar{p}_2}$

$$s_{\bar{p}_1 - \bar{p}_2} = \sqrt{\frac{\bar{p}_1(1 - \bar{p}_1)}{n_1} + \frac{\bar{p}_2(1 - \bar{p}_2)}{n_2}} \tag{10.16}$$

Test Statistic for Hypothesis Tests about the Difference between Proportions of Two Populations

$$z = \frac{(\bar{p}_1 - \bar{p}_2) - (p_1 - p_2)}{\sigma_{\bar{p}_1 - \bar{p}_2}} \tag{10.17}$$

Pooled Estimator of the Population Proportion

$$\bar{p} = \frac{n_1\bar{p}_1 + n_2\bar{p}_2}{n_1 + n_2} \tag{10.18}$$

Point Estimator of $\sigma_{\bar{p}_1 - \bar{p}_2}$ when $p_1 = p_2$

$$s_{\bar{p}_1 - \bar{p}_2} = \sqrt{\bar{p}(1 - \bar{p})\left(\frac{1}{n_1} + \frac{1}{n_2}\right)} \tag{10.19}$$

☐ ☐ Supplementary Exercises

TABLE 10.22

Master's Degree	Bachelor's Degree
$n_1 = 60$	$n_2 = 80$
$\bar{x}_1 = \$23{,}000$	$\bar{x}_2 = \$21{,}000$
$s_1 = \$2{,}500$	$s_2 = \$2{,}000$

TABLE 10.23

Instructor A	Instructor B
$n_1 = 12$	$n_2 = 15$
$\bar{x}_1 = 72$	$\bar{x}_2 = 78$
$s_1 = 8$	$s_2 = 10$

40. Starting annual salaries for individuals with master's and bachelor's degrees in business were collected in two independent random samples. Use the data shown in Table 10.22 to develop a 90% confidence interval estimate of the increase in starting salary that can be expected upon completion of the master's degree.

41. Safegate Foods, Inc., is redesigning the checkout lanes in its supermarkets throughout the country. Two designs have been suggested. Tests on customer checkout times have been collected at two stores where the two new systems have been installed. The sample data are as follows:

System A	System B
$n_1 = 120$	$n_2 = 100$
$\bar{x}_1 = 4.1$ minutes	$\bar{x}_2 = 3.3$ minutes
$s_1 = 2.2$ minutes	$s_2 = 1.5$ minutes

Test at the .05 level of significance to determine if there is a difference between the mean checkout times for the two systems. Which system is preferred?

42. Samples of final examination scores for two statistics classes with different instructors provided the results shown in Table 10.23. With $\alpha = .05$, test whether these data are sufficient to conclude that the mean grades differ for the two classes.

43. In a study of job attitudes and job satisfaction, a sample of 50 men and 50 women were asked to rate their overall job satisfaction on a 1 to 10 scale. A high rating indicates a higher degree of job satisfaction. Using the sample results shown below, does there appear to be a significant difference between the levels of job satisfaction for men and women? Use $\alpha = .05$.

Men	Women
$\bar{x}_1 = 7.2$	$\bar{x}_2 = 6.4$
$s_1 = 1.7$	$s_2 = 1.4$

44. Figure Perfect, Inc., is a women's figure salon that specializes in weight-reduction programs. Weights for a sample of clients before and after a 6-week introductory program are shown in Table 10.24. Using $\alpha = .05$, test to determine if the introductory program provides a statistically significant weight loss.

TABLE 10.24

Client	Weight	
	Before	After
1	140	132
2	160	158
3	210	195
4	148	152
5	190	180
6	170	164

45. A cable television firm is considering submitting bids for rights to operate in two regions of the state of Florida. Surveys of the two regions provided the following data on customer acceptance of the cable television service:

Region I	Region II
$n_1 = 500$	$n_2 = 800$
Number indicating an intent to purchase = 175	Number indicating an intent to purchase = 360

Develop a 99% confidence interval for the difference between population proportions of customers acceptance in the two regions.

46. A group of physicians from Denmark conducted a yearlong study on the effectiveness of nicotine chewing gum in helping people stop smoking (*New England Journal of Medicine,* 1988). The 113 people who participated in the study were all smokers. Of these, 60 were given chewing gum with 2 milligrams of nicotine, and 53 were given a placebo chewing gum with no nicotine content. No one in the study knew which type of gum he or she had been given. All were told to use the gum and refrain from smoking.

a. Define the null and alternative hypotheses that would be appropriate if the researchers hoped to show that the group given nicotine chewing gum had a higher proportion of nonsmokers 1 year after the study began.

b. Results showed that 23 of the smokers given nicotine chewing gum had remained nonsmokers for the 1-year period while 12 of the smokers given the placebo had remained nonsmokers during the same period. Do these results support the conclusion that nicotine gum can help stop smoking? Test using $\alpha = .05$. What is the p-value?

47. A large automobile-insurance company selected samples of single and married male policyholders and recorded the number who had made an insurance claim over the previous 3-year period:

Single Policyholders	Married Policyholders
$n_1 = 400$	$n_2 = 900$
Number making claims = 76	Number making claims = 90

a. Using $\alpha = .05$, test to determine if the claim rates differ between single and married male policyholders.

b. Provide a 95% confidence interval for the difference between the proportions for the two populations.

48. Medical tests were conducted to learn about drug-resistant cases of tuberculosis (*The New York Times*, January 24, 1992). Of 142 cases tested in New Jersey, 9 were found to be drug-resistant. Of 268 cases tested in Texas, 5 were found to be drug-resistant. Do these data suggest a statistically significant difference between the proportion of drug-resistant cases in the two states? Test $H_0: p_1 - p_2 = 0$ at the .02 level of significance. What is the *p*-value, and what is your conclusion?

Computer Case: *Par, Inc.*

Par, Inc., is a major manufacturer of golf equipment. Management believes that Par's market share could be increased with the introduction of a cut-resistant, longer-lasting golf ball. As a result, the research group at Par has been investigating a new golf ball coating that is designed to resist cuts and provide a more durable ball. The tests with the coating have been very promising.

A concern raised by one of the researchers has to do with the effect that the new coating will have on driving distances. Par would like the new cut-resistant ball to still offer good driving distances when compared to their current-model golf ball. To compare the driving distances for the two balls, 40 balls of both the new and current models were subjected to distance tests. The testing was performed with a mechanical hitting machine so that if a difference exists between the mean distances for the two models, if can be attributed to a difference in the design. The results of the tests, with distances measured to the nearest yard, are shown below. These data are available on the data disk in the file named GOLF.

GOLF

Model		**Model**		**Model**		**Model**	
Current	New	Current	New	Current	New	Current	New
264	277	270	272	263	274	281	283
261	269	287	259	264	266	274	250
267	263	289	264	284	262	273	253
272	266	280	280	263	271	263	260
258	262	272	274	260	260	275	270
283	251	275	281	283	281	267	263
258	262	265	276	255	250	279	261
266	289	260	269	272	263	274	255
259	286	278	268	266	278	276	263
270	264	275	262	268	264	262	279

Managerial Report

1. Formulate and present the rationale for a hypothesis test that Par could use to compare the driving distances of the current and new golf ball products.
2. Analyze the data to provide the hypothesis-testing conclusion. What is the *p*-value for your test? What is your recommendation for Par, Inc.?
3. Provide descriptive statistical summaries of the data for each model.
4. What are the 95% confidence intervals for the population mean of each model and what is the 95% confidence interval for the difference between the means of the two populations?
5. Do you feel there is a need for larger sample sizes and more testing with the golf balls? Discuss.

Inferences about Population Variances

U.S. General Accounting Office*

WASHINGTON, D.C.

The U.S. General Accounting Office (GAO) is an independent, nonpolitical audit organization in the legislative branch of the federal government. The GAO evaluators determine the effectiveness of existing or proposed federal programs. To carry out their duties, evaluators must be proficient in records review, legislative research, and statistical analysis techniques.

In one case, the GAO evaluators studied a Department of Interior program established to help clean up the nation's rivers and lakes. As part of this program, federal grants were made to small cities scattered throughout the United States. Congress asked the GAO to determine how effectively these programs were operating. To do this, the GAO examined records and visited the sites of several waste treatment plants.

One objective of the GAO audit was to ensure that the effluent (treated sewage) at the plants met certain standards. Among other things the audits reviewed sample data on the oxygen content, the pH level, and the amount of suspended solids in the effluent. A requirement of the program was that a variety of tests be taken daily at each plant with the collected data periodically sent to the state engineering department. The GAO's investigation of the data was concerned with determining whether various characteristics of the effluent were within acceptable limits.

For example, the mean or average pH level of the effluent was looked at carefully. In addition, the variance in the reported pH levels was also reviewed. The follow-

ing hypothesis test was conducted concerning the variance in pH level for the population of effluent:

$$H_0: \sigma^2 = \sigma_0^2$$
$$H_a: \sigma^2 \neq \sigma_0^2$$

In this test, σ_0^2 was the population variance in pH level expected at a properly functioning plant. In one particular plant, the null hypothesis was rejected. Further analysis showed that this plant had a variance in pH level that was significantly less than normal.

The auditors visited the plant to examine the measuring equipment and to discuss their statistical findings with the plant manager. The auditors found that the measuring equipment was not being used because the operator did not know how to operate it. Instead, the operator had been told by an engineer what an acceptable pH level was and had simply recorded similar values without actually conducting the test. The unusually low variance in this plant's data had resulted in rejecting H_0. The GAO suspected that similar problems might exist at other plants and thus they recommended an operator training program to improve the data-collection aspect of the pollution-control program.

In this chapter you will learn how to conduct statistical inferences about the variances of one and two populations. Two new distributions, the chi-square distribution and the F distribution, will be introduced to make interval estimates and hypothesis tests about population variances.

*Thanks to Mr. Art Foreman and Mr. Dale Ledman of the U.S. General Accounting Office for providing this Statistics in Practice.

I n the previous four chapters we focused on methods of statistical inference involving population means and population proportions. In this chapter we expand the discussion to situations involving inferences about population variances. As an example of a case where a variance can provide important decision-making information, consider the production process of filling containers with a liquid detergent product. The filling mechanism for the process is adjusted so that the mean filling quantity is 16 ounces per container. While a mean of 16 ounces is desired, the variance of the fillings is also critical. That is, even with the filling mechanism properly adjusted for the mean of 16 ounces, we cannot expect every container to have exactly 16 ounces. By selecting a sample of containers, we can compute a sample variance for the filling quantities. This value will serve as an estimate of the variance for the population of containers being filled by the production process. If the sample variance is modest, the production process will be continued. However, if the sample variance is excessive, overfilling and underfilling may exist even though the mean may be correct at 16 ounces. In this case, the filling mechanism will be readjusted in an attempt to reduce the filling variance for the containers.

In the following section we consider inferences about the variance of a single population. Later we will discuss procedures that can be used to make inferences about the variances of two populations.

11.1   Inferences about a Population Variance

In earlier chapters we used the sample variance

$$s^2 = \frac{\Sigma(x_i - \overline{x})^2}{n - 1} \qquad (11.1)$$

as the point estimator of the population variance σ^2. In using the sample variance as a basis for making inferences about a population variance, the sampling distribution of the quantity $(n - 1)s^2/\sigma^2$ is very helpful. This sampling distribution is described as follows.

Sampling Distribution of $(n - 1)s^2/\sigma^2$

Whenever a simple random sample of size n is selected from a *normal population*, the quantity

$$\frac{(n - 1)s^2}{\sigma^2} \qquad (11.2)$$

has a *chi-square distribution* with $n - 1$ degrees of freedom.

A graph of the sampling distribution of $(n - 1)s^2/\sigma^2$ is shown in Figure 11.1.

In previous chapters we have shown that knowledge of a sampling distribution is essential for computing interval estimates and conducting hypothesis tests about a population parameter. For inferences about a population variance we find it convenient to work with the sampling distribution of $(n - 1)s^2/\sigma^2$. The reason for this is that with normal populations, the sampling distribution of $(n - 1)s^2/\sigma^2$ is known to have a chi-square probability distribution with $n - 1$ degrees of freedom. Since tables of areas or probabilities are readily available for the chi-square distribution (see Table 3 of Appendix B), it becomes relatively easy to use the chi-square distribution to develop interval estimates and conduct hypothesis tests about a population variance.

From the three chi-square distributions shown in Figure 11.1, we see that the chi-square distribution is not symmetric and that the shape of a particular chi-square distribution depends upon the number of degrees of freedom. Also note that chi-square values can never be negative.

Interval Estimation of σ^2

Let us show how the chi-square distribution can be used to develop a confidence interval estimate of a population variance σ^2. As an illustration of this process, let us assume that we are interested in estimating the population variance for the production filling process

Figure 11.1
Examples of the Sampling
Distribution of $(n-1)s^2/\sigma^2$
(a Chi-Square Distribution)

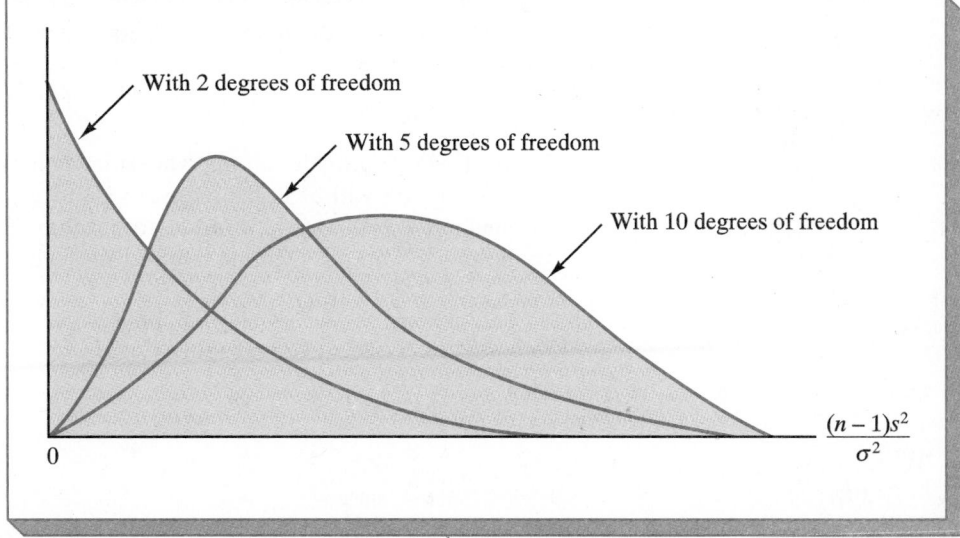

mentioned at the beginning of this chapter. A sample of 20 containers is taken, and the sample variance for the filling quantities is found to be $s^2 = .0025$. However, we know we cannot expect the variance of a sample of 20 containers to provide the exact value of the variance for the population of containers filled by the production process. Thus, our interest will be in developing an interval estimate for the population variance.

We will use the notation χ^2_α to denote the value for the chi-square distribution that provides an area or probability of α to the *right* of the stated χ^2_α value. For example, in Figure 11.2 the chi-square distribution with 19 degrees of freedom is shown with $\chi^2_{.025} = 32.85$ indicating that 2.5% of the chi-square values are to the right of 32.85, and $\chi^2_{.975} = 8.91$ indicating that 97.5% of the chi-square values are to the right of 8.91. Refer to Table 3 of Appendix B and verify that these chi-square values with 19 degrees of freedom (19th row of the table) are correct.

From the graph in Figure 11.2 we see that .95, or 95%, of the chi-square values are between $\chi^2_{.975}$ and $\chi^2_{.025}$. That is, there is a .95 probability of obtaining a χ^2 value such that

$$\chi^2_{.975} \leq \chi^2 \leq \chi^2_{.025}$$

Figure 11.2
A Chi-Square Distribution with 19
Degrees of Freedom

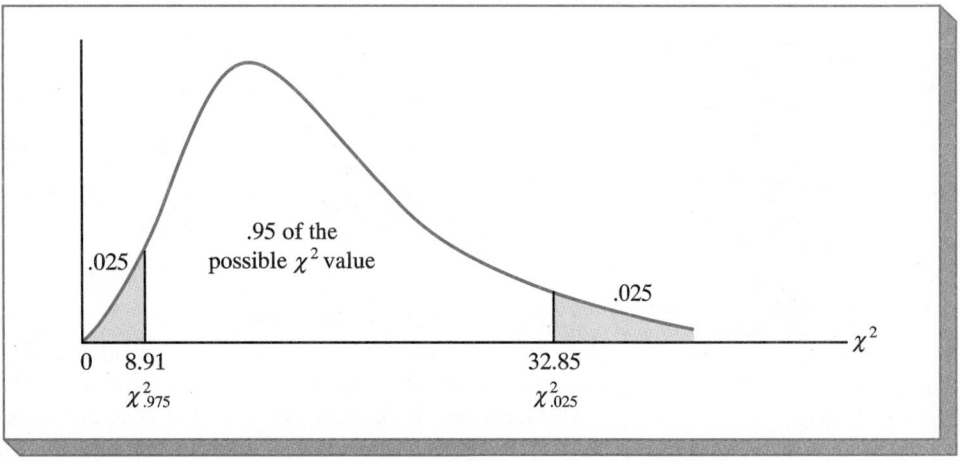

However, since we stated in (11.2) that $(n - 1)s^2/\sigma^2$ follows a chi-square distribution, we can substitute $(n - 1)s^2/\sigma^2$ for the above χ^2 and write

$$\chi^2_{.975} \le \frac{(n - 1)s^2}{\sigma^2} \le \chi^2_{.025} \tag{11.3}$$

In effect, (11.3) provides an interval estimate in that .95, or 95%, of all possible values for $(n - 1)s^2/\sigma^2$ will be in the interval $\chi^2_{.975}$ to $\chi^2_{.025}$. We now need to do some algebraic manipulations with (11.3) in order to develop an interval estimate for the population variance σ^2. Working with the leftmost inequality in (11.3), we have

$$\chi^2_{.975} \le \frac{(n - 1)s^2}{\sigma^2}$$

Thus

$$\sigma^2 \chi^2_{.975} \le (n - 1)s^2$$

or

$$\sigma^2 \le \frac{(n - 1)s^2}{\chi^2_{.975}} \tag{11.4}$$

Performing similar algebraic manipulations with the rightmost inequality in (11.3) gives

$$\frac{(n - 1)s^2}{\chi^2_{.025}} \le \sigma^2 \tag{11.5}$$

Finally, the results of (11.4) and (11.5) can be combined to provide

$$\frac{(n - 1)s^2}{\chi^2_{.025}} \le \sigma^2 \le \frac{(n - 1)s^2}{\chi^2_{.975}} \tag{11.6}$$

Since (11.3) will be true for 95% of the $(n - 1)s^2/\sigma^2$ values, (11.6) provides a 95% confidence interval for the population variance σ^2.

Let us return to the problem of providing an interval estimate of the population variance of filling quantities. Recall that the sample of 20 containers provided a sample variance of $s^2 = .0025$. With a sample of 20, we have 19 degrees of freedom. As shown in Figure 11.2, we have already determined that $\chi^2_{.975} = 8.91$ and $\chi^2_{.025} = 32.85$. Using these values in (11.6) provides the following interval estimate for the population variance:

$$\frac{(19)(.0025)}{32.85} \le \sigma^2 \le \frac{(19)(.0025)}{8.91}$$

or

$$.0014 \le \sigma^2 \le .0053$$

Taking the square root of the above values gives the following 95% confidence interval for the population standard deviation:

$$.0374 \le \sigma \le .0728$$

Thus, we have illustrated the process of using the chi-square distribution to establish interval estimates of a population variance and a population standard deviation. Note specifically that since $\chi^2_{.975}$ and $\chi^2_{.025}$ were used, the interval estimate has a .95 confidence coefficient. Extending (11.6) to the general case of any confidence coefficient, we have the following interval estimate of a population variance.

> **Interval Estimate of a Population Variance**
>
> $$\frac{(n-1)s^2}{\chi^2_{\alpha/2}} \le \sigma^2 \le \frac{(n-1)s^2}{\chi^2_{(1-\alpha/2)}} \tag{11.7}$$
>
> where the χ^2 values are based on a chi-square distribution with $n-1$ degrees of freedom and where $1-\alpha$ is the confidence coefficient.

Hypothesis Testing

Let us now consider an example and the statistical methodology necessary to test hypotheses concerning the value of a population variance. The St. Louis Metro Bus Company has recently made a concerted effort to promote an image of reliability by encouraging its drivers to maintain consistent schedules. As a standard policy the company expects arrival times at various bus stops to have low variability. In terms of the variance of arrival times, the company standard specifies an arrival time variance of 4 or less with arrival times measured in minutes. Periodically, the company collects arrival-time data at various bus stops to determine if the variability guideline is being maintained. The sample results are used to test the following hypotheses:

$$H_0: \sigma^2 \le 4$$
$$H_a: \sigma^2 > 4$$

In tentatively assuming H_0 true, we are assuming that the variance for arrival times is within the company guidelines. We will reject H_0 only if the sample evidence indicates that the guidelines are not being maintained. In this sense, rejection of H_0 suggests that follow-up steps are necessary to reduce the arrival time variability.

Assume for the moment that a random sample of 10 bus arrivals will be taken at a particular downtown intersection. If the population of arrival times has a normal probability distribution, we know from (11.2) that the quantity $(n-1)s^2/\sigma^2$ has a chi-square distribution with $n-1$ degrees of freedom. With the null hypothesis assumed true at $\sigma^2 = 4$ and with a sample size of $n = 10$, we can conclude that

$$\frac{(n-1)s^2}{\sigma^2} = \frac{9s^2}{4}$$

has a chi-square distribution with $n-1 = 9$ degrees of freedom. Thus, once the sample data are obtained and the sample variance s^2 computed, the following equation will provide an observed chi-square (χ^2) value

$$\chi^2 = \frac{9s^2}{4} \tag{11.8}$$

The chi-square distribution showing the rejection region for this one-tailed test is shown in Figure 11.3. Note that we will reject H_0 only if the sample variance s^2 leads to a large χ^2 value. With $\alpha = .05$, Table 3 of Appendix B shows that with 9 degrees of freedom $\chi^2_{.05} = 16.92$. With this as the critical value for the test, the rejection rule is as follows:

$$\text{Reject } H_0 \text{ if } \chi^2 > 16.92$$

Let us assume that the sample of arrival times for 10 buses shows a sample variance of $s^2 = 4.8$. Is this sample evidence sufficient to reject H_0 and conclude that the buses are

FIGURE 11.3

Rejection Region for the St. Louis Metro Bus Test with $\alpha = .05$

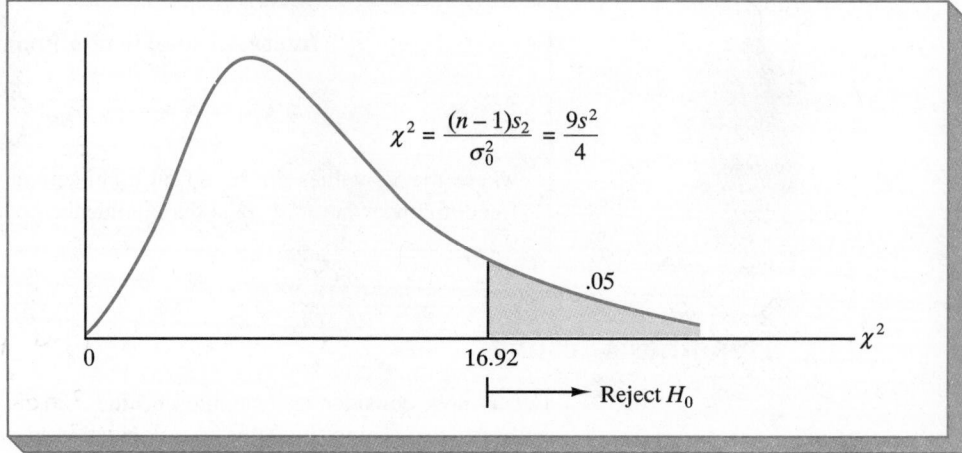

not meeting the company's arrival-time variance guideline? With $s^2 = 4.8$ and (11.8) we obtain the following χ^2 value:

$$\chi^2 = \frac{9(4.8)}{4} = 10.80$$

Since $\chi^2 = 10.80$ is less than 16.92, we cannot reject H_0. Thus, the sample variance of $s^2 = 4.8$ is insufficient evidence to conclude that the arrival-time variance is not meeting the company standard.

One-tailed tests concerning the value of a population variance as just demonstrated are in practice perhaps the most frequently encountered tests about population variances. That is, in situations involving arrival times, production times, filling weights, part dimensions, and so on, low variances are generally desired, whereas large variances tend to be unacceptable. With a statement about the maximum allowable variance, we frequently test the null hypothesis that the variance is less than or equal to this value against the alternative hypothesis that the variance is greater than this value. We now outline the decision rule for making this one-tailed test about a population variance.

One-Tailed Test about a Population Variance

$$H_0: \sigma^2 \leq \sigma_0^2$$
$$H_a: \sigma^2 > \sigma_0^2$$

Test Statistic

$$\chi^2 = \frac{(n-1)s^2}{\sigma_0^2}$$

Rejection Rule

Reject H_0 if $\chi^2 > \chi_\alpha^2$

where σ_0^2 is the hypothesized value for the population variance, α is the level of significance for the test, and the value of χ_α^2 is based on a chi-square distribution with $n-1$ degrees of freedom.

However, just as we saw with population means and population proportions, other forms of the hypotheses can be developed. The one-tailed test for H_0: $\sigma^2 \geq \sigma_0^2$ is similar to the test shown above with the exception that the one-tailed rejection region occurs in the lower tail at a critical value of $\chi^2_{(1-\alpha)}$. The two-tailed test with H_0: $\sigma^2 = \sigma_0^2$, as with other two-tailed tests, places an area of $\alpha/2$ in each tail to establish the two critical values for the test. The decision rule for conducting a two-tailed test about the value of a population variance is summarized as follows.

Two-Tailed Test about a Population Variance

$$H_0\text{: } \sigma^2 = \sigma_0^2$$
$$H_a\text{: } \sigma^2 \neq \sigma_0^2$$

Test Statistic

$$\chi^2 = \frac{(n-1)s^2}{\sigma_0^2}$$

Rejection Rule

Reject H_0 if $\chi^2 < \chi^2_{(1-\alpha/2)}$ or if $\chi^2 > \chi^2_{\alpha/2}$

where σ_0^2 is the hypothesized value for the population variance, α is the level of significance for the test, and the values of $\chi^2_{(1-\alpha/2)}$ and $\chi^2_{\alpha/2}$ are based on a chi-square distribution with $n-1$ degrees of freedom.

Let us demonstrate the use of the chi-square distribution in conducting a two-tailed test about a population variance by considering a situation faced by a bureau of motor vehicles. Historically, the variance in test scores for individuals applying for driver's licenses has been $\sigma^2 = 100$. A new examination with new test questions has been developed. Administrators of the bureau of motor vehicles believe that it is desirable for the variance in the test scores for the new examination to remain at the historical level. To evaluate the variance in the new-examination test scores, the following two-tailed hypothesis test has been proposed:

$$H_0\text{: } \sigma^2 = 100$$
$$H_a\text{: } \sigma^2 \neq 100$$

Rejection of H_0 will indicate that some questions in the new examination may need revision to make the variance of the new test scores similar to the variance of the old test scores. A sample of 30 applicants for driver's licenses will be given the new version of the examination.

The chi-square distribution can be used to conduct this two-tailed test. With a .05 level of significance, the critical regions will be based on $\chi^2_{.975}$ and $\chi^2_{.025}$. With $n-1 = 29$ degrees of freedom, Table 3 of Appendix B shows that $\chi^2_{.975} = 16.05$ and $\chi^2_{.025} = 45.72$. The rejection rule for the two-tailed test becomes

Reject H_0 if $\chi^2 < 16.05$ or if $\chi^2 > 45.72$

What is the appropriate conclusion if the sample of 30 new-examination scores provides a sample variance of $s^2 = 64$? With H_0: $\sigma^2 = 100$, the value of the χ^2 statistic is computed to be

$$\chi^2 = \frac{(n-1)s^2}{\sigma_0^2} = \frac{29(64)}{100} = 18.56$$

Since 18.56 does not fall in the rejection region, we are not able to reject H_0. There is no statistical evidence that the variance in the new-examination scores differs from the historical variance in examination scores.

☐ ☐ Exercises

Methods

1. Find the following chi-square distribution values from Table 3 of Appendix B.
   a. $\chi^2_{.05}$ with df = 12   b. $\chi^2_{.025}$ with df = 15
   c. $\chi^2_{.975}$ with df = 20   d. $\chi^2_{.01}$ with df = 10
   e. $\chi^2_{.95}$ with df = 18

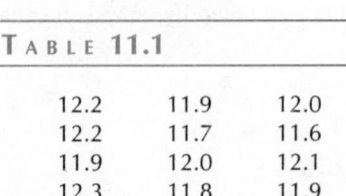

2. A sample of 20 items provides a sample standard deviation of 5.
   a. Compute the 90% confidence interval for the population variance.
   b. Compute the 95% confidence interval for the population variance.
   c. Compute the 95% confidence interval for the population standard deviation.

3. A sample of 16 items provides a sample standard deviation of 8. Test the following hypotheses using $\alpha = .05$. What is your conclusion?

$$H_0: \sigma^2 \le 50$$
$$H_a: \sigma^2 > 50$$

Applications

4. The variance in drug weights is critical in the pharmaceutical industry. For a specific drug, with weights measured in grams, a sample of 18 units provided a sample variance of $s^2 = .36$.
   a. Construct a 90% confidence interval for the population variance for the weights of this drug.
   b. Construct a 90% confidence interval for the population standard deviation.

5. A sample of 12 cans of soup produced by Carle Foods provides the weights, measured in ounces, as shown in Table 11.1. Provide 95% confidence intervals for the variance and the standard deviation of the population.

6. A study of worker attitudes about their jobs was conducted for the airline industry (*Industrial Relations*, Winter 1988). A sample of airline employees provided a sample mean age of 40 years and a sample standard deviation of 9.5 years. For purposes of illustration, compute the 95% confidence interval estimate of the population standard deviation under the assumption the following sample sizes were used:
   a. $n = 30$ b. $n = 51$ c. $n = 101$
   d. What happens to the interval estimate of the population standard deviation as the sample size increases?

7. In the St. Louis Metro Bus Company example, the sample of 10 bus arrivals showed a sample variance of $s^2 = 4.8$.
   a. Provide a 95% confidence interval for the population variance of arrival times.
   b. Assume that the sample variance of $s^2 = 4.8$ had been obtained from a sample of 25 bus arrivals. Provide a 95% confidence interval for the population variance of arrival times.

TABLE 11.1

12.2	11.9	12.0
12.2	11.7	11.6
11.9	12.0	12.1
12.3	11.8	11.9

c. What effect does a larger sample size have on the interval estimate of a population variance? Does this seem reasonable?

8. *Barron's* (March 30, 1992) reported the percent changes in major investment markets around the world. Based on market performances during 1991, which ranged from $+42.8\%$ in Hong Kong to -16.7% in Finland, the variance in the percent returns over a 3-month period was 48. A sample of 24 markets during the first 3 months of 1992 showed a sample variance in the percent returns to be 67.6.

a. Test at a .05 level of significance to determine if the variance in percent returns for the investment markets appears to have increased in the first 3 months of 1992. What is your conclusion?

b. Compute the 95% confidence interval for the variance in percent returns over a 3-month period based on the 1992 sample.

c. Compute the 95% confidence interval for the standard deviation in the percent returns.

SELF TEST ▶ 9. A certain part must be machined to very close tolerances, or it is not acceptable to customers. Production specifications call for a maximum variance in the lengths of the parts of .0004. Suppose that the sample variance for 30 parts turns out to be $s^2 = .0005$. Using $\alpha = .05$, test to see if the population variance specification is being violated.

10. City Trucking, Inc., claims consistent delivery times for its routine customer deliveries. A sample of 22 truck deliveries shows a sample variance of 1.5. Test to determine if H_0: $\sigma^2 \leq 1$ can be rejected. Use $\alpha = .10$.

11. The variance in the filling amounts of cups of soft drink from an automatic drink machine is an important consideration to the owner of the soft-drink service. If the variance is too large, overfilling and underfilling of cups will cause customer dissatisfaction with the service. An acceptable variance in filling amounts is $\sigma^2 \leq .25$ where filling amounts are measured in ounces. In a test of filling amounts for a particular machine, a sample of 18 cups showed a sample variance of .40.

a. Do the sample results indicate that the filling mechanism on the machine should be adjusted due to a large variance in filling amounts? Use a .05 level of significance.

b. Provide a 90% confidence interval for the variance in the filling amounts for this machine.

12. From a sample of 9 days over the past 6 months, a dentist has seen the following number of patients: 22, 25, 20, 18, 15, 22, 24, 19, and 26. Assuming that the number of patients seen per day is normally distributed, would an analysis of this sample data reject the hypothesis that the variance in the number of patients seen per day is equal to 10? Use a .10 level of significance. What is your conclusion?

11.2   Inferences about the Variances of Two Populations

In some statistical applications it is desirable to compare the variances of two populations. For instance, we might want to compare the variances in product quality resulting from two different production processes, the variances in assembly times for two assembly methods, or the variances in temperatures for two heating devices. In making comparisons about the variances of two populations, we will be using data collected from two independent random samples, one from population 1 and another from population 2. In using the two sample variances s_1^2 and s_2^2 as a basis for making inferences about the two population variances σ_1^2 and σ_2^2, the sampling distribution of the ratio of the two sample variances s_1^2/s_2^2 is very helpful. The sampling distribution of this ratio is described as follows.

Sampling Distribution of s_1^2/s_2^2

Whenever independent simple random samples of sizes n_1 and n_2 are selected from normal populations with equal variances, the ratio

$$\frac{s_1^2}{s_2^2} \tag{11.9}$$

has an F distribution with $n_1 - 1$ degrees of freedom for the numerator and $n_2 - 1$ degrees of freedom for the denominator; s_1^2 is the sample variance for the random sample of n_1 items from population 1 and s_2^2 is the sample variance for the random sample of n_2 items from population 2.

A graph of the F distribution with 20 degrees of freedom for both the numerator and denominator is shown in Figure 11.4. As can be seen from this graph, the F distribution is not symmetric, and the F values can never be negative. The actual shape of any particular F distribution depends upon its numerator and denominator degrees of freedom.

We will use F_α to denote the value for the F distribution that provides an area or probability of α to the *right* of the stated F_α value. For example, as noted in Figure 11.4, $F_{.05}$ denotes the upper 5% of the F values for an F distribution with 20 degrees of freedom for the numerator and 20 degrees of freedom for the denominator. Table 4 of Appendix B shows that for this particular F distribution, $F_{.05} = 2.12$. Let us show how the F distribution can be used for a hypothesis test concerning the variances of two populations.

Dullus County Schools is renewing its school bus service contract for the coming year and must select one of two bus companies, the Milbank Company or the Gulf Park Company. We will be interested in using the variance of the arrival or pickup/delivery times as a primary measure of the quality of the bus service. Low variance values indicate the more consistent and higher quality service. If the variances of arrival times associated with the two services are equal, Dullus School administrators will select the company offering the better financial terms. However, if the sample data on bus arrival times for the two companies indicate that a significant difference exists between the variances, the

FIGURE 11.4

F **Distribution with 20 Degrees of Freedom for the Numerator and 20 Degrees of Freedom for the Denominator**

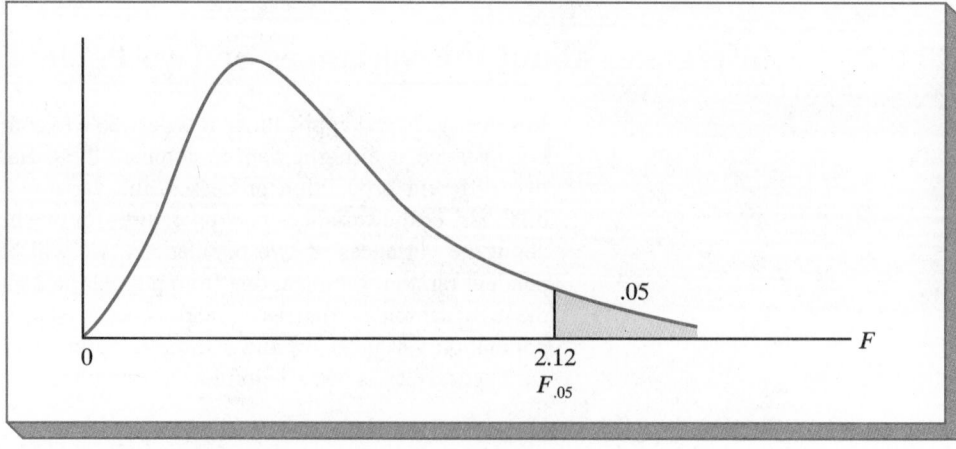

administrators may want to give special consideration to the company with the better or lower variance service. The appropriate hypotheses are as follows:

$$H_0: \sigma_1^2 = \sigma_2^2$$
$$H_a: \sigma_1^2 \neq \sigma_2^2$$

If H_0 can be rejected, the conclusion of unequal service qualities is appropriate. In this case, the company with the lower sample variance would be preferred.

Assume that the above hypothesis test will be conducted with $\alpha = .10$. Furthermore, assume that we obtain samples of arrival times from school systems currently using the two school bus services. A sample of 25 arrival times is available for the Milbank service (population 1) and a sample of 16 arrival times is available for the Gulf Park service (population 2). The graph of the F distribution with $n_1 - 1 = 24$ degrees of freedom for the numerator and $n_2 - 1 = 15$ degrees of freedom for the denominator is shown in Figure 11.5. Note that the two-tailed rejection region is indicated by the critical values at $F_{.95}$ and $F_{.05}$.

Let us assume that the two samples of bus arrival times resulted in sample variances of $s_1^2 = 48$ for the Milbank service and $s_2^2 = 20$ for the Gulf Park service. What conclusion is now appropriate concerning the quality of the two bus services? Assume that the two populations of arrival times have normal distributions, and assume that H_0 is true with $\sigma_1^2 = \sigma_2^2$. The F distribution can now be used to reach a decision. Specifically, we compute $F = s_1^2/s_2^2$ and use the rejection region shown in Figure 11.5. Thus, we find

$$F = \frac{s_1^2}{s_2^2} = \frac{48}{20} = 2.40$$

Using Table 4 of Appendix B, we find that the upper-tail critical value with 24 numerator degrees of freedom and 15 denominator degrees of freedom is $F_{.05} = 2.29$. While Table 4 does not provide $F_{.95}$ values, note that the determination of this lower-tail critical value is not necessary. We can already observe that $F = 2.40$ exceeds $F_{.05} = 2.29$. Thus, at the .10 level of significance, H_0 is rejected. This leads us to the conclusion that the two bus services differ in terms of pickup/delivery time variances. Specifically, it is recommended that the Dullus School administrators give special consideration to the better or lower variance service offered by the Gulf Park Company.

FIGURE 11.5
Rejection Region for the Dullus County School Bus Example with $\alpha = .10$

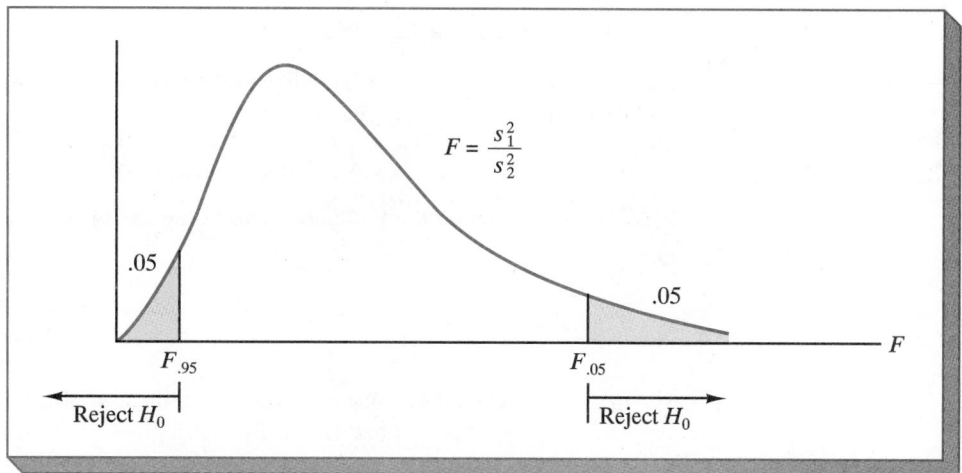

In the Dullus School example, you might feel that we were lucky in carrying out the test because the lower-tail critical value $F_{.95}$, which was not provided in the F distribution table, was not even necessary. Thus, we were able to draw the appropriate hypothesis-testing conclusion without knowing the value of $F_{.95}$. If you ever need to know a lower-tail F value, $F_{(1-\alpha)}$, it can be determined by using an upper-tail F_α value and the following relationship:

$$F_{(1-\alpha), \mathrm{df}_1, \mathrm{df}_2} = \frac{1}{F_{\alpha, \mathrm{df}_2, \mathrm{df}_1}} \tag{11.10}$$

Thus, $F_{.95}$ with 24 degrees of freedom in the numerator ($\mathrm{df}_1 = 24$) and 15 degrees of freedom in the denominator ($\mathrm{df}_2 = 15$) can be computed by reversing the degrees of freedom and using the $F_{.05}$ value with 15 degrees of freedom in the numerator and 24 degrees of freedom in the denominator. Using the F distribution table, we find that $F_{.05}$ with 15 numerator and 24 denominator degrees of freedom is 2.11. Thus, the value of $F_{.95}$ with 24 degrees of freedom in the numerator and 15 degrees of freedom in the denominator is

$$F_{.95} = \frac{1}{2.11} = .47$$

While (11.10) can be used to compute a lower-tail F value, common practice is to conduct the hypothesis test computations so that only upper-tail F values are needed. In hypothesis tests with $H_0: \sigma_1^2 = \sigma_2^2$, we simply denote the population with the *larger sample variance* as population 1. That is, which population we denote as either population 1 or 2 is arbitrary. By labeling the population with the larger sample variance as population 1, we guarantee that $F = s_1^2/s_2^2$ will be greater than or equal to 1. Thus, a rejection of H_0 can only occur in the *upper tail*. While the lower-tail critical value still exists, we do not need to know its value simply because the convention of using the population with the largest sample variance as population 1 always places the ratio s_1^2/s_2^2 in the upper-tail direction. In the Dullus School example, population 1, the Milbank bus service, possessed the largest sample variance. Thus, we proceeded directly with the test. If the Gulf Park bus service had provided the largest sample variance, we simply would have denoted Gulf Park as population 1 and followed the same statistical testing procedure. A summary of the two-tailed test for the equality of two population variances with this procedure is as follows.

Two-Tailed Test about the Variances of Two Populations

$$H_0: \sigma_1^2 = \sigma_2^2$$
$$H_a: \sigma_1^2 \neq \sigma_2^2$$

Denote the population providing the *largest sample variance* as population 1.

Test Statistic	Rejection Rule
$F = \dfrac{s_1^2}{s_2^2}$	Reject H_0 if $F > F_{\alpha/2}$

where the value of $F_{\alpha/2}$ is based on an F distribution with $n_1 - 1$ degrees of freedom for the numerator and $n_2 - 1$ degrees of freedom for the denominator.

One-tailed tests involving two population variances are also possible. Again the F distribution is used, with the one-tailed rejection region enabling us to conclude whether one population variance is significantly greater or significantly less than the other. Only upper-tail F values are needed. For any one-tailed test we set up the null hypothesis so that the rejection region is in the upper tail. This can be accomplished by labeling the population with the larger variance in H_a as population 1. The general procedure is as follows.

One-Tailed Test about the Variances of Two Populations

$$H_0: \sigma_1^2 \leq \sigma_2^2$$
$$H_a: \sigma_1^2 > \sigma_2^2$$

Test Statistic   *Rejection Rule*

$$F = \frac{s_1^2}{s_2^2}$$ Reject H_0 if $F > F_\alpha$

where the value of F_α is based on an F distribution with $n_1 - 1$ degrees of freedom for the numerator and $n_2 - 1$ degrees of freedom for the denominator.

Let us demonstrate the use of the F distribution to conduct a one-tailed test about the variances of two populations by considering the following public opinion survey. Samples of 31 men and 41 women will be used to study attitudes about current political issues. The researcher conducting the study would like to test to see if the sample data indicate that women demonstrate a greater variation in attitude on political issues than men. Using the form of the one-tailed hypothesis test shown earlier, women will be denoted as population 1 and men will be denoted as population 2. The hypothesis test will be stated

$$H_0: \sigma_{\text{women}}^2 \leq \sigma_{\text{men}}^2$$
$$H_a: \sigma_{\text{women}}^2 > \sigma_{\text{men}}^2$$

If H_0 is rejected, the researcher will have the statistical support necessary to conclude that women demonstrate a greater variation in attitude on political issues.

With the sample variance for women in the numerator and the sample variance for men in the denominator, the F distribution with $41 - 1 = 40$ degrees of freedom in the numerator and $31 - 1 = 30$ degrees of freedom in the denominator will be used to conduct the one-tailed test. Using a .05 level of significance, the rejection region is based on $F_{.05}$. Using Table 4 of Appendix B, we find that $F_{.05} = 1.79$. Thus, the rejection rule becomes

$$\text{Reject } H_0 \text{ if } F > 1.79$$

where F is computed from the ratio of the two sample variances s_1^2/s_2^2.

Assume that the survey shows a sample variance of $s_1^2 = 120$ for the 41 women and a sample variance of $s_2^2 = 80$ for the 31 men. What is the appropriate statistical conclusion? The F statistic becomes

$$F = \frac{s_1^2}{s_2^2} = \frac{120}{80} = 1.50$$

Since 1.50 is less than 1.79, H_0 cannot be rejected. Thus, the sample results do not support the conclusion that women have a greater variance than men in their attitudes on political issues.

❑ ❑ Exercises

Methods

13. Find the following F distribution values from Table 4 of Appendix B.
a. $F_{.05}$ with degrees of freedom 12 and 10   **b.** $F_{.025}$ with degrees of freedom 20 and 15
c. $F_{.01}$ with degrees of freedom 8 and 12   **d.** $F_{.975}$ with degrees of freedom 10 and 20

14. A sample of 16 from population 1 has a sample variance $s_1^2 = 5.8$ and a sample of 20 from population 2 has a sample variance $s_2^2 = 2.4$. Test the following hypotheses at the .05 level of significance.

$$H_0: \sigma_1^2 \leq \sigma_2^2$$
$$H_a: \sigma_1^2 > \sigma_2^2$$

What is your conclusion?

SELF TEST ▶ **15.** Consider the following hypothesis test:

$$H_0: \sigma_1^2 = \sigma_2^2$$
$$H_a: \sigma_1^2 \neq \sigma_2^2$$

What is your conclusion if $n_1 = 25$, $s_1^2 = 4.0$, $n_2 = 21$, and $s_2^2 = 8.2$? Use $\alpha = .05$.

Applications

16. The average price of a new automobile in 1991 was $16,700, an increase of 4.3% over the 1990 average price of $16,012 (*U.S. News & World Report,* September 9, 1991). Assume that samples of 121 new 1991 automobiles showed a sample standard deviation in price of $4200 and 121 new 1990 automobiles showed a sample standard deviation in price of $3850. Using a .05 level of significance, can it be concluded that the variance in prices of new automobiles also increased in 1991?

SELF TEST ▶ **17.** Most individuals are aware of the fact that the average annual repair cost for an automobile depends on the age of the automobile. For example, the average annual repair cost for automobiles 4 years old ($400) is almost twice as large as the average annual repair cost for automobiles 2 years old ($220), (*Consumer Reports 1992 Buyers Guide*). A researcher is interested in studying the variance of the annual repair costs to see if the variance in the repair costs also increases with the age of the automobile. A sample of 25 automobiles 4 years old showed a sample standard deviation for annual repair costs of $170 while a sample of 25 automobiles 2 years old showed a sample standard deviation for annual repair costs of $100.
a. State the null and alternative hypotheses for the research hypothesis that the variance in annual repair costs is larger for the older automobiles.
b. Using a .01 level of significance, what is your conclusion? Discuss the reasonableness of your findings.

18. The Educational Testing Service has conducted studies designed to identify differences between the scores of males and females on the Scholastic Aptitude Test (*Journal of Educational Measurement,* Spring 1987). For a sample of females, the standard deviation of test scores was 83 on the verbal portion of the SAT. For a sample of males, the standard deviation was 78 on the same test. Assume that standard deviations were based on random samples of 121 females and 121 males. Do the data indicate there are differences between the variances of females and males on the verbal portion of the SAT? Use $\alpha = .05$.

19. Independent random samples of parts manufactured by two suppliers provided the following results:

Supplier	Sample Size	Sample Variance of Part Sizes
Durham Electric	41	$s_1^2 = 3.8$
Raleigh Electronics	31	$s_2^2 = 2.0$

The firm making the supplier-selection decision is prepared to use the Durham supplier unless the test results show that the Raleigh supplier provides a significantly lower variance in part sizes. Use $\alpha = .05$ and conduct the statistical test that will help the firm select a supplier. Which supplier do you recommend?

20. The following sample data have been collected from two independent random samples:

Population	Sample Size	Sample Mean	Sample Variance
A	$n_A = 25$	$\bar{x}_A = 40$	$s_A^2 = 5$
B	$n_B = 21$	$\bar{x}_B = 50$	$s_B^2 = 11$

Test for the equality of the variances of population A and population B. Use $\alpha = 10$. What is your conclusion?

21. Two secretaries are each given eight typing assignments of equal difficulty. The sample standard deviations of the completion times were 3.8 minutes and 5.2 minutes, respectively. Do the data suggest that there is a difference in the variability of completion times for the two secretaries? Test the hypothesis at a .10 level of significance.

22. A research hypothesis is that the variance of stopping distances of automobiles on wet pavement is substantially greater than the variance of stopping distances of automobiles on dry pavement. In the research study, 16 automobiles traveling at the same speeds are tested with respect to stopping distances on wet pavement and then tested with respect to stopping distances on dry pavement. On wet pavement, the standard deviation of stopping distances was 32 feet. On dry pavement, the standard deviation was 16 feet.

a. At a .05 level of significance, do the sample data justify the conclusion that the variance in stopping distances on wet pavement is greater than the variance in stopping distances on dry pavement?

b. What are the implications of your statistical conclusions in terms of driving safety recommendations?

Summary

In this chapter we have presented statistical procedures that can be used to make inferences about population variances. In the process we have introduced two new probability distributions: the chi-square distribution and the F distribution. The chi-square distribution can be used as the basis for interval estimation and hypothesis tests concerning the variance of a normal population. In particular, we showed that for simple random samples of size n selected from a normal population, the quantity $(n - 1)s^2/\sigma^2$ has a chi-square distribution with $n - 1$ degrees of freedom.

We illustrated the use of the F distribution in making hypothesis tests concerning the variances of two normal populations. In particular, we showed that with independent simple random samples of sizes n_1 and n_2 selected from two normal populations with equal variances $\sigma_1^2 = \sigma_2^2$, the sampling distribution of the ratio of the two sample variances s_1^2/s_2^2 has an F distribution with $n_1 - 1$ degrees of freedom for the numerator and $n_2 - 1$ degrees of freedom for the denominator.

Key Formulas

Interval Estimate of a Population Variance

$$\frac{(n-1)s^2}{\chi_{\alpha/2}^2} \le \sigma^2 \le \frac{(n-1)s^2}{\chi_{(1-\alpha/2)}^2} \tag{11.7}$$

Sampling Distribution of s_1^2/s_2^2 when $\sigma_1^2 = \sigma_2^2$

$$F = \frac{s_1^2}{s_2^2} \tag{11.9}$$

☐ ☐ Supplementary Exercises

23. Because of staffing decisions, management of the Gibson-Marimont Hotel is interested in the variability for the number of rooms occupied per day during a particular season of the year. A sample of 20 days of operation shows a sample mean of 290 rooms occupied per day and a sample standard deviation of 30 rooms.
  a. What is the point estimate of the population variance?
  b. Provide a 90% confidence interval for the population variance.
  c. Provide a 90% confidence interval for the population standard deviation.

24. Initial public offerings (IPOs) of stocks are on average underpriced. However, in some cases, the IPOs are actually overpriced (*Financial Analysts Journal,* December 1987). The standard deviation measures the dispersion or variation in the underpricing-overpricing indicators. A sample of 13 Canadian IPOs that were subsequently traded on the Toronto Stock Exchange had a standard deviation of 14.95. Develop a 95% confidence interval for the population standard deviation for the underpricing-overpricing indicator.

25. Historical delivery times for Buffalo Trucking, Inc., have had a mean of 3 hours and a standard deviation of .5 hours. A sample of 22 deliveries over the past month provides a sample mean of 3.1 hours and a sample standard deviation of .75 hours.
  a. Use a hypothesis test to determine if the sample results lead to rejection of the historical delivery variance of H_0: $\sigma^2 = (.5)^2 = .25$. Use $\alpha = .05$.
  b. Compute the 95% confidence intervals for the population variance and the population standard deviation.

26. Part variability is critical in the manufacturing of ball bearings. Large variances in the size of the ball bearings cause bearing failure and rapid wearout. Production standards call for a maximum variance of .0001 when the bearing sizes are measured in inches. A sample of 15 bearings shows a sample standard deviation of .014 inches.
  a. Using $\alpha = .10$, determine if the sample indicates that the maximum variance is being exceeded.
  b. Compute the 90% confidence interval estimate for the variance of the ball bearings in the population.

27. The filling variance for boxes of cereal is designed to be .02 or less. A sample of 41 boxes of cereal shows a sample standard deviation of .16 ounces. Using $\alpha = .05$, determine if the variance in the cereal box fillings is exceeding the standard.

28. A sample standard deviation for the number of passengers taking a particular airline flight is 8. A 95% confidence interval estimate for the standard deviation is 5.86 passengers to 12.62 passengers.
   a. Was a sample size of 10 or 15 used in the above statistical analysis?
   b. If the sample standard deviation of $s = 8$ had been based on a sample of 25 flights, what change would you expect in the confidence interval for the population standard deviation? Compute a 95% confidence interval for σ if a sample of size 25 had been used.

29. A firm gives a mechanical aptitude test to all job applicants. A sample of 20 male applicants shows a sample variance of 80 for the test scores. A sample of 16 female applicants shows a sample variance of 220. Using $\alpha = .05$, determine if the test score variances differ for male and female job applicants. If a difference in variances exists, which group has the higher variance in mechanical aptitude?

30. The grade point averages of 352 students who completed a college course in financial accounting has a standard deviation of .940 (*The Accounting Review*, January 1988). The grade point averages of 73 students who dropped out of the same course has a standard deviation of .797. Do the data indicate that there is a difference between the variances of grade point averages for students who complete financial accounting courses and students who drop out? Use a .05 level of significance. Note: $F_{.025}$ with 351 and 72 degrees of freedom is approximately 1.45.

31. The accounting department analyzes the variance of the weekly unit costs reported by two production departments. A sample of 16 cost reports for each of the two departments shows cost variances of 2.3 and 5.4, respectively. Is this sample sufficient to conclude that the two production departments differ in terms of unit cost variances? Use $\alpha = .10$.

32. Two new assembly methods are tested with the variances in assembly times shown in Table 11.2. Using $\alpha = .10$, test for equality of the two population variances.

TABLE 11.2

Method	Sample Size	Sample Variance
A	31	$s_1^2 = 25$
B	25	$s_2^2 = 12$

Computer Case: *Air Force Training Program*

An Air Force introductory course in electronics uses a personalized system of instruction whereby each student views a video-taped lecture and then is given a programmed instruction text. The students work independently with the text until they have completed the training and passed a test. Of concern is the varying pace at which the students complete this portion of their training program. Some students are able to cover the programmed instruction text relatively quickly, while other students work much longer with the text and require additional time to complete the course. The faster students wait until the slower students complete the introductory course before the entire group proceeds together with other aspects of their training.

A proposed method of instruction involves use of computer-assisted instruction. Under this method, each student will view the same video-taped lecture and then be assigned to a computer terminal for further instruction. Each student will work independently with the computer guiding the student through the self-training portion of the course.

To compare the proposed and current methods of instruction, an entering class of 122 students was randomly assigned to one of the two methods. One group of 61 students used the current programmed-text method, and the other group of 61 students used the proposed computer-assisted method. The time in hours was recorded for each student in the study. The following data are provided on the data disk in the file named TRAINING.

Managerial Report

1. Use appropriate descriptive statistics to summarize the training-time data for each method. What similarities and/or differences do you observe from the sample data?
2. Use the methods of Chapter 10 to comment on any difference between the population means for the two methods. Discuss your findings.

3. Compute the standard deviation and variance for each training method. Conduct a hypothesis test about the equality of population variances for the two training methods. Discuss your findings.

4. What conclusion can you reach about any differences between the two methods? What is your recommendation? Explain.

5. Can you suggest other data or testing that might be desirable before making a final decision on the training program to be used in the future?

Course-Completion Times (hours) for Current Training Method

76	76	77	74	76	74	74	77	72	78	73
78	75	80	79	72	69	79	72	70	70	81
76	78	72	82	72	73	71	70	77	78	73
79	82	65	77	79	73	76	81	69	75	75
77	79	76	78	76	76	73	77	84	74	74
69	79	66	70	74	72					

TRAINING

Course-Completion Times (hours) for Proposed Computer-Assisted Method

74	75	77	78	74	80	73	73	78	76	76
74	77	69	76	75	72	75	72	76	72	77
73	77	69	77	75	76	74	77	75	78	72
77	78	78	76	75	76	76	75	76	80	77
76	75	73	77	77	77	79	75	75	72	82
76	76	74	72	78	71					

12

Tests of Goodness of Fit and Independence

United Way*

ROCHESTER, NEW YORK

The United Way of Greater Rochester is a non-profit fund-raising and social planning organization dedicated to improving the quality of life of residents in the six counties it serves. The annual United Way/Red Cross campaign, conducted each spring, helps support more than 140 human service agencies. These agencies meet a wide variety of human needs—physical, mental, and social—and serve people of all ages, backgrounds, and economic means. Because of widespread volunteer involvement, United Way is able to hold its operating costs to less than nine cents of every dollar raised.

The United Way of Greater Rochester decided to conduct a survey to learn more about community perceptions of charities. Focus-group interviews were conducted with professional, service, and general worker groups to get preliminary information on perceptions. The information obtained was then used to help develop the questionnaire for the survey. The questionnaire was pretested, modified, and distributed to 440 individuals; 323 completed questionnaires were obtained.

A variety of descriptive statistics, including frequency distributions and two-way tabular summaries, were provided from the data collected. An important part of the analysis involved the use of contingency tables and chi-square tests of independence. One such use of this statistical test was to determine whether perceptions concerning administrative expenses were independent of occupation.

The hypotheses for the test of independence were as follows:

H_0: Perception of United Way administrative expenses is independent of the occupation of the respondent

H_a: Perception of United Way administrative expenses is not independent of the occupation of the respondent

Two questions in the survey provided the data for the statistical test. One question obtained data on perceptions concerning the percentage of funds going to administrative expenses (up to 10%, 11–20%, and 21% or more). The other question asked for the occupation of the respondent.

The chi-square test at a 5% level of significance led to rejection of the null hypothesis of independence and the conclusion that perceptions concerning United Way's administrative expenses did vary by occupation. Actual administrative expenses were less than 9%, but 35% of the respondents perceived that administrative expenses were 21% or more. Thus, many had very inaccurate perceptions concerning administrative costs. In this group, production-line employees, clerical, sales, and professional-technical employees have more inaccurate perceptions than other groups.

In this chapter, you will learn how a statistical test of independence such as that described above is conducted. The study described was helpful to United Way of Rochester in developing adjustments to its program and fund-raising activities.

A poster child for the annual United Way campaign.

*The authors are indebted to Dr. Philip R. Tyler, Marketing Consultant to the United Way, for providing this Statistics in Practice.

I n Chapter 11 we introduced the chi-square distribution and illustrated how it could be used in estimation and hypothesis tests about a population variance. In this chapter we introduce two additional hypothesis-testing procedures, both based on the use of the chi-square distribution. As with other hypothesis-testing procedures, these tests compare sample results with those that are expected when the null hypothesis is true. The conclusion of the hypothesis test is based upon how ''close'' the sample results are to the expected results.

In the following section we introduce a goodness of fit test involving a multinomial population. Later we discuss the test for independence using contingency tables and then show goodness of fit tests for Poisson and normal probability distributions.

12.1   Goodness of Fit Test: A Multinomial Population

In this section we consider the case where each element of a population is assigned to one and only one of several classes or categories. Such a population is a *multinomial population*. For example, consider the market analysis being conducted by the J. Scott and Associates market research firm. The study involves a market-share evaluation. Over the past year market shares have stabilized, with 30% for company A, 50% for company B, and 20% for company C. Recently company C has developed a ''new and improved'' product that will replace its current entry in the market. Management of company C has asked J. Scott and Associates to determine if the new product will cause a shift in the market shares of the three competitors.

In this case, the population of interest is a multinomial population, since each customer is classified as buying from company A, company B, or company C. Thus, we have a multinomial population with three classifications or categories. Let us define the following notation:

$$p_A = \text{market share for company A}$$
$$p_B = \text{market share for company B}$$
$$p_C = \text{market share for company C}$$

Based on the assumption that company C's new product will not alter the market shares, the null and alternative hypotheses would be stated as follows:

$$H_0: p_A = .30, \; p_B = .50, \text{ and } p_C = .20$$

$$H_a: \text{The population proportions are not}$$
$$p_A = .30, \; p_B = .50, \text{ and } p_C = .20$$

If the sample results lead to the rejection of H_0, the firm will have evidence that the introduction of the new product has had an impact on the market shares.

Let us assume that the market research firm will use a consumer panel of 200 customers for the study. Each individual will be asked to specify a purchase preference among the three alternatives: company A's product, company B's product, and company C's new product. The 200 purchase preference responses are summarized below:

Company A's Product	Company B's Product	Company C's New Product
48	98	54

We now want to demonstrate a *goodness of fit test* that will determine if the sample of 200 customer purchase preferences is consistent with the null hypothesis. The goodness of fit test is based on a comparison of the sample of *observed* results such as those shown above with the *expected* results under the assumption that the null hypothesis is true. Thus, the next step is to compute expected purchase preferences for the 200 customers under the assumption that $p_A = .30$, $p_B = .50$, and $p_C = .20$. Doing this provides the expected results:

Company A's Product	Company B's Product	Company C's New Product
$200(.30) = 60$	$200(.50) = 100$	$200(.20) = 40$

Thus, we see that the expected frequency for each category is found by multiplying the sample size of 200 by the hypothesized proportion for the category.

The goodness of fit test now focuses on the differences between the observed frequencies and the expected frequencies. Large differences between observed and expected frequencies cast doubt on the assumption that the hypothesized proportions or market shares are correct. Whether the differences between the observed and expected frequencies are "large" or "small" is a question answered with the aid of the following test statistic.

Test Statistic for Goodness of Fit

$$\chi^2 = \sum_{i=1}^{k} \frac{(f_i - e_i)^2}{e_i} \tag{12.1}$$

where

f_i = observed frequency for category i

e_i = expected frequency for category i based on the assumption that H_0 is true

k = the number of categories

Note: The test statistic has a chi-square distribution with $k - 1$ degrees of freedom provided the expected frequencies are 5 *or more* for all categories.

Let us return to the market-share data for the three companies. Since the expected frequencies are all 5 or more, we can proceed with the computation of the chi-square test statistic as follows:

$$\chi^2 = \frac{(48 - 60)^2}{60} + \frac{(98 - 100)^2}{100} + \frac{(54 - 40)^2}{40} = 2.40 + .04 + 4.90 = 7.34$$

Suppose that we test the null hypothesis that the multinomial population has the proportions of $p_A = .30$, $p_B = .50$, and $p_C = .20$ at the $\alpha = .05$ level of significance. Since we will reject the null hypothesis if the differences between the observed and expected frequencies are *large,* we will place a rejection area of .05 in the upper tail of chi-square distribution. Checking the chi-square distribution table (Table 3 of Appendix B), we find that with $k - 1 = 3 - 1 = 2$ degrees of freedom $\chi^2_{.05} = 5.99$. Since $7.34 > 5.99$, we reject H_0. In rejecting H_0 we are concluding that the introduction of the new product by company C will alter the current market-share structure. While the goodness of fit test itself permits no further conclusions, we can informally compare the observed and expected frequencies to obtain an idea of how the market-share structure has changed.

Considering company C, we find that the observed frequency of 54 is larger than the expected frequency of 40. Since the expected frequency was based on current market shares, the larger observed frequency suggests that the new product will have a positive effect on company C's market share. Comparisons of the observed and expected frequencies for the other two companies indicate that company C's gain in market share will hurt company A more than company B.

As illustrated in the example, the goodness of fit test uses the chi-square distribution to determine if a hypothesized multinomial probability distribution for a population provides a good fit. The hypothesis test is based upon differences between observed frequencies in a sample and the expected frequencies based on the assumed population distribution. Let us outline the general steps that can be used to conduct a goodness of fit test for any hypothesized multinomial population distribution:

1. Formulate a null hypothesis indicating a hypothesized multinomial distribution for the population.
2. Use a simple random sample of n items and record the observed frequencies for each of k classes or categories.
3. Based upon the assumption that the null hypothesis is true, determine the probability or proportion associated with each of the classes.
4. Multiply the category proportions in step 3 by the sample size to determine the expected class frequencies.
5. Use the observed and expected frequencies in (12.1) to compute a χ^2 value for the test.
6. Complete the test by using the following rejection rule:

$$\text{Reject } H_0 \text{ if } \chi^2 > \chi^2_\alpha$$

where α is the level of significance for the test.

□ □ **Exercises**

SELF TEST ▶

Methods

1. Conduct a test of the following hypotheses using the χ^2 goodness of fit test.

$$H_0: p_A = .40, p_B = .40, \text{ and } p_C = .20$$

$$H_a: \text{The population proportions are not} \\ p_A = .40, p_B = .40, p_C = .20$$

A sample of size 200 yielded 60 in category A, 120 in category B, and 20 in category C. Use $\alpha = .01$ and test to see if the proportions are as stated in H_0.

2. Suppose we have a multinomial population with four categories: A, B, C, and D. The null hypothesis is that the proportion of items in each category is the same. The null hypothesis is

$$H_0: \quad p_A = p_B = p_C = p_D = .25$$

A sample of size 300 yielded the following numbers in each category:

A: 85, B: 95, C: 50, D: 70.

Use $\alpha = .05$ and the χ^2 test to determine if H_0 should be rejected.

Applications

SELF TEST ▶

3. During the first 13 weeks of the television season, the Saturday evening 8:00 P.M. to 9:00 P.M. audience proportions were recorded as ABC, 29%; CBS, 28%; NBC, 25%; and independents, 18%. A sample of 300 homes 2 weeks after a Saturday night schedule revision showed the following viewing audience data: ABC, 95 homes; CBS, 70 homes; NBC, 89 homes; and independents, 46 homes. Test with $\alpha = .05$ to determine if the viewing audience proportions have changed.

4. Where do America's millionaires live? Assume a sample of 300 millionaires showed 63 living in the Northeast, 62 living in the Midwest, 100 living in the South, and 75 living in the West. The sample results are based on data from *Louis Rukeyser's Business Almanac*, 1988. Conduct a hypothesis test for H_0: $p_1 = p_2 = p_3 = p_4 = .25$ with $\alpha = .05$. What is your conclusion about where America's millionaires live?

5. The four major competitors in the computer-workstation market are Sun Microsystems (29%), Hewlett-Packard (18.8%), IBM (16%), and Digital Equipment (11.6%) with other manufacturers holding 24.6% of the market (*USA Today*, February 13, 1992). Assume 1 year later a survey of 400 computer workstations found 106 Sun, 72 Hewlett-Packard, 80 IBM, 48 Digital and 94 other systems in use. Do the data suggest any changes have occurred during the 1-year period? Test at a .05 level of significance.

6. A new container design has been adopted by a manufacturer. Color preferences indicated in a sample of 150 individuals are as follows:

Red	Blue	Green
40	64	46

Test using $\alpha = .10$ to see if the color preferences are different. Hint: Formulate the null hypothesis as H_0: $p_1 = p_2 = p_3 = \frac{1}{3}$.

7. Consumer panel preferences for three proposed store displays are as follows:

Display A	Display B	Display C
43	53	39

Use $\alpha = .05$ and test to see if there is a difference in preference among the three display designs.

8. Grade-distribution guidelines for a statistics course at a major university are as follows: 10% A, 30% B, 40% C, 15% D, and 5% F. A sample of 120 statistics grades at the end of a semester showed 18 A's, 30 B's, 40 C's, 22 D's, and 10 F's. Use $\alpha = .05$ and test to see if the actual grades deviate significantly from the grade-distribution guidelines.

12.2   Test of Independence: Contingency Tables

Another important application of the chi-square distribution involves using sample data to test for the independence of two variables. Let us illustrate the test of independence by considering the study conducted by the Alber's Brewery of Tucson, Arizona. Alber's manufactures and distributes three types of beers: a light beer, a regular beer, and a dark beer. In an analysis of the market segments for the three beers, the firm's market research group has raised the question of whether or not preferences for the three beers differ among male and female beer drinkers. If beer preference is independent of the sex of the beer drinker, one advertising campaign will be initiated for all of Alber's beers. However, if beer preference depends upon the sex of the beer drinker, the firm will tailor its promotions toward different target markets.

A test of independence addresses the question of whether or not the beer preference (light, regular, or dark) is independent of the sex of the beer drinker (male, female). The hypotheses for this test of independence are as follows:

H_0: Beer preference is independent of the sex of the beer drinker

H_a: Beer preference is not independent of the sex of the beer drinker

TABLE 12.1
Contingency Table for Beer Preference and Sex of Beer Drinker

	Beer Preference		
Sex	Light	Regular	Dark
Male	(cell 1)	(cell 2)	(cell 3)
Female	(cell 4)	(cell 5)	(cell 6)

Table 12.1 can be used to describe the situation being studied. By identifying the population to be all male and female beer drinkers, a sample can be selected and each individual asked to state his or her preference for the three Alber's beers. Every individual in the sample will be classified in one of the 6 cells in the table. For example, an individual may be a male preferring regular beer (cell 2), a female preferring light beer (cell 4), a female preferring dark beer (cell 6), and so on. Since we have listed all possible combinations of beer preference and sex or, in other words, listed all possible contingencies, Table 12.1 is called a *contingency table*. The test of independence makes use of the contingency table format and for this reason is sometimes referred to as a *contingency table test*.

Let us assume that a simple random sample of 150 beer drinkers has been selected. After they have taste-tested each beer, the individuals in the sample are asked to state their preference or first choice. The contingency table in Table 12.2 summarizes the responses for the study. As we see in Table 12.2, the data for the test of independence are collected in terms of counts or frequencies for each cell or category. Thus, of the 150 individuals in the sample, 20 were men who favored light beer, 40 were men who favored regular beer, 20 were men who favored dark beer, and so on.

Note that the data in Table 12.2 contain the sample or observed frequencies for each of 6 classes or categories. If we can determine the expected frequencies under the assumption of independence between beer preference and sex of the beer drinker, we can use the chi-square distribution, just as we did in the previous section, to determine whether or not there is a significant difference between observed and expected frequencies.

Expected frequencies for the cells of the contingency table are based on the following rationale: First, we assume that the null hypothesis of independence between beer preference and sex of the beer drinker is true. Then, we note that the sample of 150 beer

TABLE 12.2
Sample Results of Beer Preferences for Male and Female Beer Drinkers (Observed Frequencies)

		Beer Preference			
		Light	Regular	Dark	**Total**
Sex	Male	20	40	20	80
	Female	30	30	10	70
	Total	50	70	30	150

drinkers showed a total of 50 preferring light beer, 70 preferring regular beer, and 30 preferring dark beer. In terms of fractions we conclude that $50/150 = 1/3$ of the beer drinkers prefer light beer, $70/150 = 7/15$ prefer regular beer, and $30/150 = 1/5$ prefer dark beer. If the *independence* assumption is valid, we argue that these same fractions must be applicable to both male and female beer drinkers. Thus, under the assumption of independence, we would expect the sample of 80 male beer drinkers to show that $(1/3)80 = 26.67$ prefer light beer, $(7/15)80 = 37.33$ prefer regular beer, and $(1/5)80 = 16$ prefer dark beer. Application of these same fractions to the 70 female beer drinkers provides the expected frequencies as shown in Table 12.3.

Let e_{ij} denote the expected frequency for the contingency table category in row i and column j. With this notation let us reconsider the expected frequency calculation for males (row $i = 1$) who prefer regular beer (column $j = 2$) — that is, expected frequency e_{12}. Following the previous argument for the computation of expected frequencies, we showed that

$$e_{12} = (7/15)80 = 37.33$$

Writing this slightly differently, we find

$$e_{12} = (7/15)80 = (70/150)80 = \frac{(80)(70)}{150} = 37.33$$

Note that 80 in the above expression is the total number of males (row 1 total), 70 is the total number of males and females preferring regular beer (column 2 total), and 150 is the total sample size. Thus, we see that

$$e_{12} = \frac{(\text{Row 1 Total})(\text{Column 2 Total})}{\text{Sample Size}}$$

Generalization of the above expression shows that the following formula provides the expected frequencies for a contingency table in the test of independence.

Expected Frequencies for Contingency Tables under the Assumption of Independence

$$e_{ij} = \frac{(\text{Row } i \text{ Total})(\text{Column } j \text{ Total})}{\text{Sample Size}} \tag{12.2}$$

Using the above formula for male beer drinkers who prefer dark beer, we find an expected frequency of $e_{13} = (80)(30)/150 = 16.00$, as shown previously in Table 12.3.

TABLE 12.3

Expected Frequencies if Beer Preference Is Independent of the Sex of the Beer Drinker

		Beer Preference			
		Light	Regular	Dark	Total
Sex	Male	26.67	37.33	16.00	80
	Female	23.33	32.67	14.00	70
	Total	50	70	30	150

Use (12.2) to verify that it provides the other expected frequencies shown in Table 12.3.

The test procedure for comparing the observed frequencies of Table 12.2 with the expected frequencies of Table 12.3 is similar to the goodness of fit calculations made in the previous section. Specifically, the χ^2 value based on the observed and expected frequencies is computed as follows.

Test Statistic for Independence

$$\chi^2 = \sum_i \sum_j \frac{(f_{ij} - e_{ij})^2}{e_{ij}} \tag{12.3}$$

where

　　f_{ij} = observed frequency for contingency table category in row i and column j

　　e_{ij} = expected frequency for contingency table category in row i and column j based on the assumption of independence

Note: With n rows and m columns in the contingency table, the test statistic has a chi-square distribution with $(n - 1)(m - 1)$ degrees of freedom provided the expected frequencies are 5 *or more* for all categories.

The double summation in (12.3) is used to indicate that the calculation must be made for all the cells in the contingency table.

By reviewing the expected frequencies in Table 12.3, we see that the expected frequencies are 5 or more for each category. Thus, we proceed with the computation of the chi-square test statistic. The resulting value is as follows:

$$\chi^2 = \frac{(20 - 26.67)^2}{26.67} + \frac{(40 - 37.33)^2}{37.33} + \cdots + \frac{(10 - 14.00)^2}{14.00}$$

$$= 1.67 + .19 + \cdots + 1.14 = 6.13$$

The number of degrees of freedom for the appropriate chi-square distribution is computed by multiplying the *number of rows minus 1* times the *number of columns minus 1*. With 2 rows and 3 columns, we have $(2 - 1)(3 - 1) = (1)(2) = 2$ degrees of freedom for the test of independence of beer preference and the sex of the beer drinker. With $\alpha = .05$ for the level of significance of the test, Table 3 of Appendix B shows an upper-tail χ^2 value of $\chi^2_{.05} = 5.99$. Note here that we are again using the upper-tail value because we will reject the null hypothesis only if the differences between observed and expected frequencies provide a large χ^2 value. In our example, $\chi^2 = 6.13$ is greater than the critical value of $\chi^2_{.05} = 5.99$. Thus, we reject the null hypothesis of independence and conclude that the preference for the beers is not independent of the sex of the beer drinkers.

Although the test for independence allows only the above conclusion, again we can informally compare the observed and expected frequencies to obtain an idea of how the dependence between the beer preference and sex of the beer drinker comes about. Refer to Tables 12.2 and 12.3. We see that male beer drinkers have higher observed than expected frequencies for both regular and dark beers, while female beer drinkers have a higher observed than expected frequency for only the light beer. These observations give us an insight into the beer preference differences between the male and female beer drinkers.

□ □ Exercises

Methods

SELF TEST ▶

9. Shown in Table 12.4 is a 2 × 3 contingency table with observed frequencies for a sample of 200. Test for independence of the row and column factors using the χ^2 test with $\alpha = .025$.

10. Shown below is a 3 × 3 contingency table with observed frequencies from a sample of 240. Test for independence of the row and column factors using the χ^2 test with $\alpha = .05$.

TABLE 12.4

	Column Factor		
Row Factor	A	B	C
P	20	44	50
Q	30	26	30

	Column Factor		
Row Factor	A	B	C
P	20	30	20
Q	30	60	25
R	10	15	30

Applications

SELF TEST ▶

11. The 1992 NCAA basketball championship final four teams were Duke, Michigan, Indiana, and Cincinnati. Data below show the season 3-point shooting records (*NCAA Final Four Program,* April 1992) for the four teams. At the .05 level of significance, is there a difference in the 3-point shooting abilities among the four teams? What is your conclusion?

3-Point Shooting	Duke	Michigan	Indiana	Cincinnati
Made	160	113	154	202
Missed	214	228	215	331

12. The number of units sold by three salespersons over a 3-month period are shown in Table 12.5. Use $\alpha = .05$ and test for the independence of salesperson and type of product. What is your conclusion?

13. Starting positions for business and engineering graduates are classified by industry as shown below:

TABLE 12.5

	Product		
Salesperson	A	B	C
Troutman	14	12	4
Kempton	21	16	8
McChristian	15	5	10

	Industry			
Degree Major	Oil	Chemical	Electrical	Computer
Business	30	15	15	40
Engineering	30	30	20	20

Use $\alpha = .01$ and test for independence of degree major and industry type.

14. A *CBS News/New York Times* poll (February 24, 1991) asked a sample of individuals a series of questions about the involvement of the United States in the Persian Gulf war. One question asked men and women the following: Do you think the United States did the right thing in starting the ground war against Iraq, or should the United States have waited longer to see if bombing from the air worked? Assume the responses for men and women were summarized in the contingency table

TABLE 12.6

Response	Men	Women
Right thing	243	207
Waited longer	48	66
Not sure	9	27

shown in Table 12.6. Use the chi-square test of independence to analyze this data. What is your conclusion at a .05 level of significance?

15. Medical researchers at Harvard and Boston University randomly assigned 227 General Electric employees with alcohol problems to one of three alcohol-treatment groups: a 28-day hospitalization followed by Alcoholics Anonymous (AA) meetings, AA meetings only with no hospitalization, or a choice of the two programs (*USA Today,* September 12, 1991). Two years later, the researchers identified the patients who had remained sober after completing the program. Assume that the data are as shown in the following contingency table:

Status/Program	Hospitalization	AA only	Choice
Remained sober	28	13	12
Did not remain sober	48	63	63

Use the chi-square test of independence with a .01 level of significance to analyze the data. What is your conclusion and recommendation?

16. The *GMAC Occasional Papers* (March 1988) provided data on the primary reason for application to an MBA program by full-time and part-time students. Do the data suggest full-time and part-time students differ in their reason for applying to MBA programs? Explain. Use $\alpha = .01$.

Student Status	Primary Reason for Application		
	Program Quality	Convenience/Cost	Other
Full-time	421	393	76
Part-time	400	593	46

17. A sport preference poll shows the following data for men and women:

Sex	Favorite Sport		
	Baseball	Basketball	Football
Men	19	15	24
Women	16	18	16

Use $\alpha = .05$ and test for similar sport preferences by men and women. What is your conclusion?

18. Three suppliers provide the following data on defective parts.

Supplier	Part Quality		
	Good	Minor Defect	Major Defect
A	90	3	7
B	170	18	7
C	135	6	9

Use α = .05 and test for independence between supplier and part quality. What does the result of your analysis tell the purchasing department?

19. A study of educational levels of voters and their political party affiliations showed the following results:

| | Party Affiliation | | |
Educational Level	Democratic	Republican	Independent
Did not complete high school	40	20	10
High school degree	30	35	15
College degree	30	45	25

Use α = .01 and test if party affiliation is independent of the educational level of the voters.

20. *Personnel Administrator* (January 1984) provided the following data as an example of selection among 40 male and 40 female applicants for 12 open positions:

Applicant	Selected	Not Selected	Total
Male	7	33	40
Female	5	35	40

a. The chi-square test of independence was suggested as a way of determining if the decision to hire 7 males and 5 females should be interpreted as having a selection bias in favor of males. Conduct the test of independence using α = .10. What is your conclusion?
b. Using the same test, would the decision to hire 8 males and 4 females suggest concern for a selection bias?
c. How many males could be hired for the 12 open positions before the procedure would suggest concern for a selection bias?

12.3   Goodness of Fit Test: Poisson and Normal Distributions

In Section 12.1 we introduced the goodness of fit test involving a multinomial population. In general, the goodness of fit test can be used with any hypothesized probability distribution. In this section we illustrate the goodness of fit test procedure for cases where the population is hypothesized to have a Poisson or a normal distribution. As we shall see, the goodness of fit test and the use of the chi-square distribution for these tests follow the same general procedure used for the goodness of fit test in Section 12.1.

A Poisson Distribution

Let us illustrate the goodness of fit test for the case where the hypothesized population distribution is a Poisson distribution. As an example, consider the arrivals of customers at Dubek's Food Market in Tallahassee, Florida. Dubek's management makes staffing decisions involving the number of clerks, the number of checkout lanes, and so on based on the anticipated arrivals of customers at the store. Because of some recent staffing problems, Dubek's management has asked a local consulting firm to assist with the scheduling of clerks for the checkout lanes. The general objective of the consulting firm's work is to provide enough clerks to achieve a good level of service while maintaining a reasonable total payroll cost.

After reviewing the checkout lane operation, the consulting firm makes a recommendation for a clerk-scheduling procedure. The procedure, based on a mathematical analysis of waiting lines, is applicable only in situations where the number of customers arriving during a specified time period follows the Poisson probability distribution. Thus, before the scheduling process is implemented, data on customer arrivals must be collected and a statistical test conducted to see if an assumption of a Poisson distribution for arrivals appears reasonable.

We define the arrivals at the store in terms of the *number of customers* entering the store during 5-minute intervals. Thus, the following null and alternative hypotheses are appropriate for the Dubek's Food Market study:

H_0: The number of customers entering the store during 5-minute intervals has a Poisson probability distribution
H_a: The number of customers entering the store during 5-minute intervals does not have a Poisson distribution

If a sample of customer arrivals indicates H_0 cannot be rejected, Dubek's will proceed with the implementation of the consulting firm's scheduling procedure. However, if the sample leads to the rejection of H_0, the assumption of the Poisson distribution for the arrivals cannot be made, and other scheduling procedures will have to be considered.

To test the assumption of a Poisson distribution for the number of arrivals during weekday morning hours, a store employee randomly selects a sample of 128 5-minute intervals during weekday mornings over a 3-week period. For each 5-minute interval in the sample, the store employee records the number of customer arrivals. In summarizing the data, the employee determines the number of 5-minute intervals having no arrivals, the number of 5-minute intervals having one arrival, the number of 5-minute intervals having two arrivals, and so on. These data are summarized in Table 12.7. The number of customers arriving is the category description, with the observed frequencies recorded in the column showing the number of 5-minute intervals having the corresponding number of customer arrivals.

Table 12.7 provides the observed frequencies for the ten categories. We now want to use a goodness of fit test to determine whether or not the sample of 128 time periods supports the hypothesized Poisson probability distribution. In order to conduct the goodness of fit test, we need to consider the expected frequency for each of the 10 categories under the assumption that the Poisson distribution of arrivals is true. That is, we need to compute the expected number of time periods that no customers, 1 customer, 2 customers, and so on would arrive if, in fact, the customer arrivals have a Poisson distribution.

The Poisson probability function, which was first introduced in Chapter 5, is as follows:

$$f(x) = \frac{\mu^x e^{-\mu}}{x!} \tag{12.4}$$

In this function, μ represents the mean or expected number of customers arriving per 5-minute period, and x is the random variable indicating the number of customers arriving during a 5-minute period. In this case, x may be equal to 0, 1, 2, and so on. Finally, $f(x)$ is the probability that x customers will arrive in a 5-minute interval.

Before we use (12.4) to compute Poisson probabilities, we must obtain an estimate of μ, the mean number of customer arrivals during a 5-minute time period. The sample mean for the data in Table 12.7 provides this estimate. With no customers arriving in two 5-minute time periods, one customer arriving in eight 5-minute time periods, and so on,

TABLE 12.7

Observed Frequencies of Dubek's Customer Arrivals for a Sample of 128 5-Minute Time Periods

Category Description: Number of Customers Arriving	Observed Frequency
0	2
1	8
2	10
3	12
4	18
5	22
6	22
7	16
8	12
9	6
Total	128

the total number of customers that arrived during the sample of 128 5-minute time periods is given by $0(2) + 1(8) + 2(10) + \cdots + 9(6) = 640$. The 640 customer arrivals over the sample of 128 periods provides a mean arrival rate of $\mu = 640/128 = 5$ customers per 5-minute period. With this value for the mean of the Poisson probability distribution, an estimate of the Poisson probability function for Dubek's Food Market is

$$f(x) = \frac{5^x e^{-5}}{x!} \tag{12.5}$$

Assume that the Poisson probability distribution is appropriate for Dubek's customer arrivals. The above probability function then can be evaluated for different values of x to determine the probability associated with each category of arrivals. These probabilities, which can also be found in Table 7 of Appendix B, are shown in Table 12.8. For example, the probability of 0 customers arriving during a 5-minute interval is $f(0) = .0067$, the probability of 1 customer arriving during a 5-minute interval is $f(1) = .0337$, and so on. As we saw in Section 12.1, the expected frequencies for the categories are found by multiplying the probabilities by the sample size. For example, the expected number of periods with 0 arrivals is given by $(.0067)(128) = .8576$, the expected number of periods with 1 arrival is given by $(.0337)(128) = 4.3136$, and so on.

Before we make the usual chi-square calculations to compare the observed and expected frequencies, note that in Table 12.8, four of the categories have an expected frequency less than 5. This condition violates the requirements necessary for use of the chi-square distribution. However, expected category frequencies less than 5 cause no difficulty, since adjacent categories can be combined to satisfy the "at least 5" expected frequency requirement. In particular, we will combine 0 and 1 into a single category and then combine 9 with "10 or more" into another single category. Thus, the rule of a minimum expected frequency of 5 in each category is satisfied.

Let us combine the categories for the observed results in Table 12.7 accordingly. Now, we can compare the observed frequencies with the expected frequencies. This comparison is summarized in Table 12.9.

As in Section 12.1 the goodness of fit test focuses on the differences between observed and expected frequencies, $f_i - e_i$. Obviously, large differences cast doubt on the assumption that the customer arrivals have a Poisson distribution.

TABLE 12.8	Category Description: Number of Customers Arriving (x)	Poisson Probability $f(x)$	Expected Number of 5-Minute Time Periods with x Arrivals, 128 $f(x)$
Expected Frequencies of Dubek's Customer Arrivals, Assuming a Poisson Probability Distribution with $\mu = 5$	0	.0067	.8576
	1	.0337	4.3136
	2	.0842	10.7776
	3	.1404	17.9712
	4	.1755	22.4640
	5	.1755	22.4640
	6	.1462	18.7136
	7	.1044	13.3632
	8	.0653	8.3584
	9	.0363	4.6464
	10 or more	.0318	4.0704
		Total	128.0000

	TABLE 12.9	**Category Description:** **Number of** **Customers Arriving**	**Observed** **Frequency** (f_i)	**Expected** **Frequency** (e_i)	**Difference** $(f_i - e_i)$

TABLE 12.9

Comparison of the Observed and Expected Frequencies for Dubek's Customer Arrivals

Category Description: Number of Customers Arriving	Observed Frequency (f_i)	Expected Frequency (e_i)	Difference $(f_i - e_i)$
0 or 1	10	5.1712	4.8288
2	10	10.7776	−.7776
3	12	17.9712	−5.9712
4	18	22.4640	−4.4640
5	22	22.4640	−.4640
6	22	18.7136	3.2864
7	16	13.3632	2.6368
8	12	8.3584	3.6416
9 or more	6	8.7168	−2.7168
Total	128	128.0000	

Using the observed and expected frequencies shown in Table 12.9, we again compute the chi-square test statistic:

$$\chi^2 = \sum_{i=1}^{k} \frac{(f_i - e_i)^2}{e_i}$$

Doing so, we have

$$\chi^2 = \frac{(4.8288)^2}{5.1712} + \frac{(-.7776)^2}{10.7776} + \cdots + \frac{(-2.7168)^2}{8.7168}$$

$$= 4.5091 + .0561 + \cdots + .8468 = 10.98$$

We need to determine the appropriate degrees of freedom associated with this goodness of fit test. In general, the chi-square distribution for a goodness of fit test has $k - p - 1$ degrees of freedom, where k is the number of categories and p is the number of population parameters estimated from the sample data. For the Poisson distribution goodness of fit test we are considering, Table 12.9 shows $k = 9$ categories. Since the sample data were used to estimate the mean of the Poisson distribution, $p = 1$. Thus, $k - p - 1 = 9 - 1 - 1 = 7$ degrees of freedom exist for the chi-square distribution in the Dubek's Food Market study.

Suppose that we test the hypothesis that the probability distribution for the customer arrivals is Poisson with a .05 level of significance ($\alpha = .05$). This is an upper-tail test, since we will reject the null hypothesis only if the difference between the observed and expected frequencies—and thus the χ^2 value—becomes large. We check the χ^2 values in Table 3 of Appendix B and find that with 7 degrees of freedom $\chi^2_{.05} = 14.07$. As with similar one-tailed tests, we will reject the null hypothesis only if the computed value of χ^2 exceeds the value of χ^2_{α}.

Checking the previous calculations for the Dubek's Food Market study, we find a computed $\chi^2 = 10.98$. Since this value is less than the critical value of 14.07, we cannot reject the null hypothesis. Thus, for this analysis, the assumption of a Poisson probability distribution for weekday morning customer arrivals cannot be rejected. With this statistical finding, Dubek's will proceed with the consulting firm's scheduling procedure for weekday mornings.

A Normal Distribution

TABLE 12.10

Chemline Employee Aptitude Test Scores for 50 Randomly Chosen Job Applicants

71	66	61	65	54	93
60	86	70	70	73	73
55	63	56	62	76	54
82	79	76	68	53	58
85	80	56	61	61	64
65	62	90	69	76	79
77	54	64	74	65	65
61	56	63	80	56	71
79	84				

The goodness of fit test for a normal probability distribution is also based on the use of the chi-square distribution. It is similar to the procedure we have just discussed for the Poisson distribution. In particular, observed frequencies for several categories of sample data will be compared to expected frequencies under the assumption that the population has a normal distribution. Since the normal probability distribution is continuous, we must modify the way that the categories are defined and how the expected frequencies are computed. Let us demonstrate the goodness of fit test for a normal probability distribution by considering the job applicant test data for Chemline, Inc., shown in Table 12.10.

Chemline hires approximately 400 new employees annually for its four plants located throughout the United States. Standardized tests are given by the personnel department, with performance on the test being a major factor in the employee-hiring decision. With numerous tests being given annually, the personnel director has asked if a normal distribution could be applied to the population of test scores. If such a distribution can be applied, use of the distribution would be most helpful in evaluating specific test scores. That is, scores in the upper 20%, lower 40%, and so on, could quickly be identified. Thus, we would like to test the null hypothesis that the population of aptitude test scores follows a normal probability distribution.

Let us first use the data in Table 12.10 to develop estimates of the mean and standard deviation of the normal distribution that will be considered in the null hypothesis. We use the sample mean $\bar{x}$ and the sample standard deviation s as point estimators of the mean and standard deviation of the normal distribution. Calculations are as follows:

$$\bar{x} = \frac{\Sigma x_i}{n} = \frac{3421}{50} = 68.42$$

$$s = \sqrt{\frac{\Sigma(x_i - \bar{x})^2}{n - 1}} = \sqrt{\frac{5310.0369}{49}} = 10.41$$

Using these values, we state the following hypotheses about the distribution of the job applicant test scores:

H_0: The population of test scores has a normal distribution with mean 68.42 and standard deviation 10.41

H_a: The population of test scores does not have a normal distribution with mean 68.42 and standard deviation 10.41

The hypothesized normal distribution is shown in Figure 12.1.

Now let us consider a way of defining the categories for a goodness of fit test involving a normal distribution. For the discrete probability distribution in the Poisson distribution test, the categories were readily defined in terms of the number of customers arriving such as 0, 1, 2, and so on. However, with a continuous probability distribution, such as the normal, we will have to come up with a different procedure for defining the categories. We will need to define the categories in terms of *intervals* of test scores.

Recall the rule of thumb for an expected frequency of at least 5 in each interval or category. We will have to define the categories of test scores such that the expected frequencies will be at least 5 for each category. With a sample size of 50, one way of doing this is to divide the normal distribution into 10 equal-probability intervals (see Figure 12.2). With a sample size of 50, we would expect 5 outcomes in each interval or category, and the rule of thumb for expected frequencies would be satisfied. This is the procedure we will follow for determining the number of categories for the goodness of fit test whenever a continuous probability distribution is being considered. Namely, we will

FIGURE 12.1

Hypothesized Normal Distribution of Test Scores for the Chemline Job Applicants

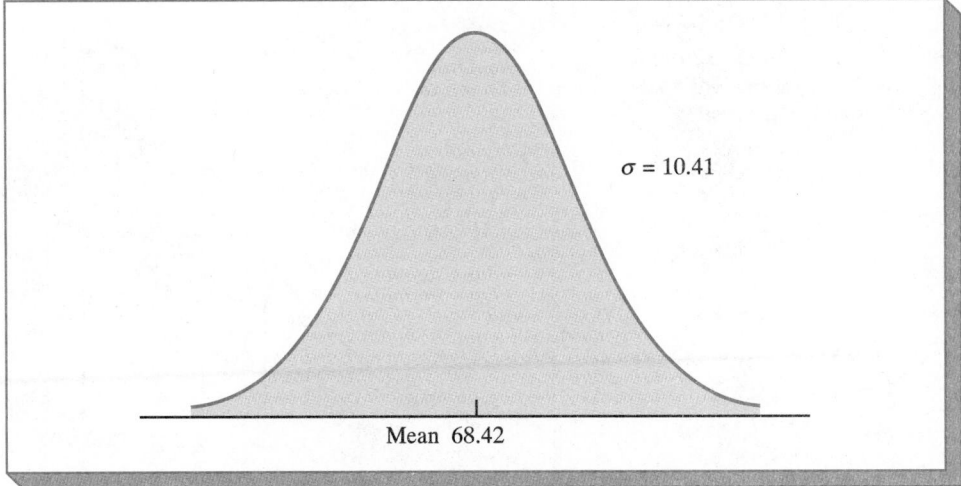

$\sigma = 10.41$

Mean 68.42

break the assumed population distribution into equal-probability intervals such that at least 5 observations are expected in each.

Let us look more closely at the procedure used to calculate the category boundaries. Since the normal probability distribution is being assumed, the standard normal probability tables can be used to determine these boundaries. First consider the test score cutting off the lowest 10% of the test scores. From Table 1 of Appendix B we find that the z value for this test score is approximately -1.28. Therefore, the test score of $x = 68.42 - 1.28(10.41) = 55.10$ provides this cutoff value for the lowest 10% of the scores. For the lowest 20%, we find $z = -.84$, and thus $x = 68.42 - .84(10.41) = 59.68$. Working through the normal distribution in a similar manner provides the following test score values:

Lower 10%:	$68.42 - 1.28(10.41) =$	55.10
Lower 20%:	$68.42 - .84(10.41) =$	59.68
Lower 30%:	$68.42 - .52(10.41) =$	63.01
Lower 40%:	$68.42 - .25(10.41) =$	65.82
Mid-score:	$68.42 + 0(10.41) =$	68.42
Upper 40%:	$68.42 + .25(10.41) =$	71.02
Upper 30%:	$68.42 + .52(10.41) =$	73.83
Upper 20%:	$68.42 + .84(10.41) =$	77.16
Upper 10%:	$68.42 + 1.28(10.41) =$	81.74

These cutoff or interval boundary points have been identified on the graph in Figure 12.2.

With the categories or intervals of test scores now defined and with the known expected frequencies of 5 per category, we can return to the sample data of Table 12.10 and determine the observed frequencies for the categories. Doing so provides the results in Table 12.11. Also note that Table 12.11 contains a column of differences between the observed and expected frequencies.

With the results in Table 12.11, the goodness of fit calculations proceed exactly as before. Namely, we compare the observed and expected results by computing a χ^2 value as follows:

$$\chi^2 = \sum_{i=1}^{k} \frac{(f_i - e_i)^2}{e_i} = \frac{0^2}{5} + \frac{0^2}{5} + \frac{4^2}{5} + \cdots + \frac{1^2}{5} = 7.2$$

FIGURE 12.2

Normal Probability Distribution for the Chemline Example with 10 Equal-Probability Intervals

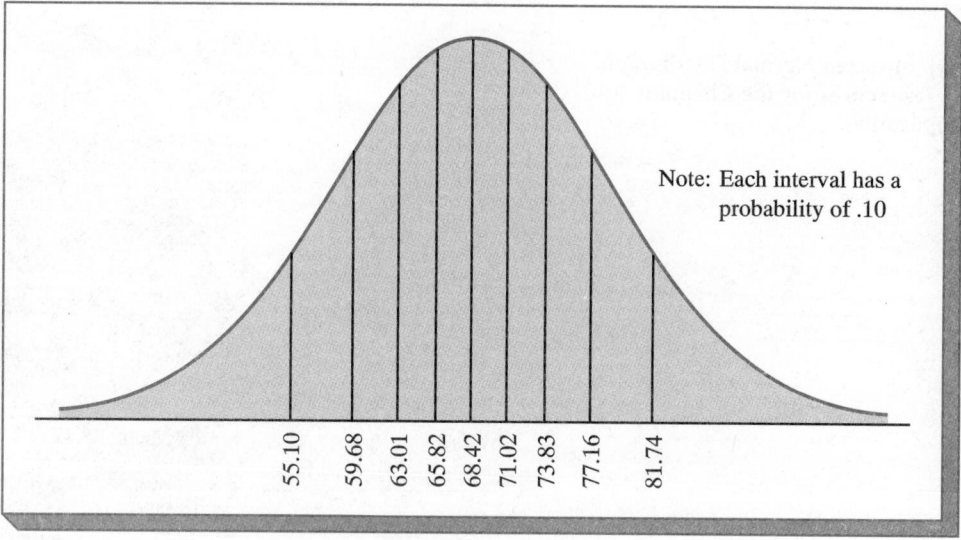

Note: Each interval has a probability of .10

55.10 59.68 63.01 65.82 68.42 71.02 73.83 77.16 81.74

TABLE 12.11

Observed and Expected Frequencies for Chemline Job-Applicant Test Scores

Test Score Interval	Observed Frequency (f_i)	Expected Frequency (e_i)	Difference $(f_i - e_i)$
Less than 55.10	5	5	0
55.10 to 59.68	5	5	0
59.68 to 63.01	9	5	4
63.01 to 65.82	6	5	1
65.82 to 68.42	2	5	−3
68.42 to 71.02	5	5	0
71.02 to 73.83	2	5	−3
73.83 to 77.16	5	5	0
77.16 to 81.74	5	5	0
81.74 and Over	6	5	1
Total	50	50	

To determine whether the computed χ^2 value of 7.20 is large enough to reject H_0, we need to refer to the appropriate chi-square probability distribution tables. Using the rule for computing the number of degrees of freedom for the goodness of fit test, we have $k - p - 1 = 10 - 2 - 1 = 7$ degrees of freedom, where there are $k = 10$ categories and $p = 2$ parameters (mean and standard deviation) estimated from the sample data. Using a .10 level of significance for this hypothesis test, we have $\chi^2_{.10} = 12.017$ for the upper-tail rejection region. With $7.20 < 12.017$ we conclude that the null hypothesis cannot be rejected. Thus, the hypothesis that the probability distribution for the Chemline job applicant test scores is a normal probability distribution cannot be rejected.

☐ ☐ **Exercises**

Methods

SELF TEST

21. Shown in Table 12.12 are data on the number of occurrences per time period and observed frequencies. Use $\alpha = .05$ and use the goodness of fit test to see if the data fit a Poisson distribution.

22. The following data are believed to have come from a normal probability distribution. Use the goodness of fit test and $\alpha = .025$ to test this claim.

17	23	22	24	19	23	18	22	20	13	11	21	18	20	21
21	18	15	24	23	23	43	29	27	26	30	28	33	23	29

TABLE 12.12

Number of Occurrences	Observed Frequency
0	39
1	30
2	30
3	18
4	3

Applications

23. The number of automobile accidents occurring per day in a particular city is believed to have a Poisson distribution. A sample of 80 days during the past year gives the data shown in Table 12.13. Do these data support the belief that the number of accidents per day has a Poisson distribution? Use $\alpha = .05$.

24. The number of incoming phone calls occurring at a company switchboard during 1-minute intervals is believed to have a Poisson distribution. Use $\alpha = .10$ and the following data to test the assumption that the incoming phone calls have a Poisson distribution:

TABLE 12.13

Number of Accidents	Observed Frequency (days)
0	34
1	25
2	11
3	7
4	3

Number of Incoming Phone Calls During a 1-Minute Interval	Observed Frequency
0	15
1	31
2	20
3	15
4	13
5	4
6	2
Total	100

SELF TEST

25. The weekly demand for a product is believed to be normally distributed. Use a goodness of fit test and the data in the sample shown in Table 12.14 to test this assumption. Use $\alpha = .10$. The sample mean is 24.5, and the sample standard deviation is 3.

TABLE 12.14

18	20	22	27	22
25	22	27	25	24
26	23	20	24	26
27	25	19	21	25
26	25	31	29	25
25	28	26	28	24

26. Use $\alpha = .01$ and conduct a goodness of fit test to see if the following sample appears to have been selected from a normal distribution:

55	86	94	58	55	95	55	52	69	95	90	65	87	50	56
55	57	98	58	79	92	62	59	88	65					

After you complete the goodness of fit calculations, construct a histogram of the data. Does the histogram representation support the conclusion reached with the goodness of fit test? (Note: $\bar{x} = 71$ and $s = 17$.)

Summary

In this chapter we introduced the goodness of fit test and the test of independence, both of which are based on the use of the chi-square distribution. The purpose of the goodness of fit test is to determine whether a hypothesized probability distribution can be rejected as a distribution for a particular population of interest. The computations for conducting the goodness of fit test involve comparing observed frequencies from a sample with expected frequencies when the hypothesized probability distribution is assumed true. A chi-square distribution is used to determine if the differences in observed and expected frequencies are sufficient to reject the hypothesized probability distribution. We illustrated the goodness of fit test for assumed multinomial, Poisson, and normal probability distributions.

A test of independence for two variables is a straightforward extension of the methodology employed in the goodness of fit test for a multinomial population. A contingency table is used to determine the observed and expected frequencies. Then a chi-square value is computed. Large chi-square values, caused by large differences between observed and expected frequencies, lead to the rejection of the null hypothesis of independence.

Glossary

Goodness of fit test A statistical test conducted to determine whether or not to reject a hypothesized probability distribution for a population.

Contingency table A table used to summarize observed and expected frequencies for a test of independence.

Key Formulas

Test Statistic for Goodness of Fit

$$\chi^2 = \sum_{i=1}^{k} \frac{(f_i - e_i)^2}{e_i} \tag{12.1}$$

Expected Frequencies for Contingency Tables under the Assumption of Independence

$$e_{ij} = \frac{(\text{Row } i \text{ Total})(\text{Column } j \text{ Total})}{\text{Sample Size}} \tag{12.2}$$

Test Statistic for Independence

$$\chi^2 = \sum_i \sum_j \frac{(f_{ij} - e_{ij})^2}{e_{ij}} \tag{12.3}$$

❑ ❑ Supplementary Exercises

27. In setting sales quotas, the marketing manager makes the assumption that order potentials are the same in each of four sales territories. A sample of 200 sales shows the following number of orders from each region:

Sales Territories			
I	II	III	IV
60	45	59	36

Should the manager's assumption be rejected? Use $\alpha = .05$.

28. An October 1988 poll sponsored by the *Cincinnati Post* used a random sample of 606 registered voters throughout the state of Ohio to determine how Ohioans rate their local schools. The rating categories and results shown in Table 12.15 were reported. Assume Ohio school administrators had hypothesized rating percentages of 20% excellent, 40% good, 25% fair, and 15% poor for the population of Ohio registered voters. Use the goodness of fit test and the survey data to determine if the administrator's hypothesis should be rejected. Use $\alpha = .05$.

29. At Ontario University entering freshmen have historically selected the following colleges:

College	Historical Percentage
Business	15
Education	20
Engineering	30
Liberal Arts	25
Science	10

Data obtained for the most recent class show that 68 students selected business, 110 selected education, 140 selected engineering, 122 selected liberal arts, and 60 selected science. Use $\alpha = .10$ to see if the historical percentages have changed.

30. A regional transit authority was concerned about the number of riders on one of its bus routes. In setting up the route, the assumption was that the number of riders was uniformly distributed from Monday through Friday. Using the data shown in Table 12.16, test with $\alpha = .05$ to determine if the transit authority's assumption appears to be incorrect.

31. A sample of parts provided the following contingency table data concerning part quality and production shift:

Shift	Number Good	Number Defective
First	368	32
Second	285	15
Third	176	24

Use $\alpha = .05$ and test the hypothesis that part quality is independent of the production shift. What is your conclusion?

32. The Graduate Management Admission Council (GMAC) sponsored a survey of MBA students to learn about characteristics of the population of students interested in graduate education in business administration. The following table was published in the *GMAC Occasional Papers* (March 1988). Do the data suggest males and females differ in the reason for application to MBA programs? Explain. Use $\alpha = .05$.

TABLE 12.15

School Rating Category	Frequency
Excellent	135
Good	234
Fair	139
Poor	98

TABLE 12.16

Day	Number of Riders
Monday	13
Tuesday	16
Wednesday	28
Thursday	17
Friday	16

MBA Students	Primary Reason for Application		
	Program Quality	Convenience/Cost	Other
Male	519	599	86
Female	298	390	36

TABLE 12.17

Loan Officer	Loan Approval Decision	
	Approved	Rejected
Miller	24	16
McMahon	17	13
Games	35	15
Runk	11	9

33. A lending institution supplied the data shown in Table 12.17 regarding loan approvals by four different loan officers. Use $\alpha = .05$ and test to determine if the loan approval decision is independent of the loan officer reviewing the loan application.

34. An analysis of attendance records and performance on the final examination was made for a first-year mathematics course. The following results were obtained:

Number of Classes Missed	Grade on Final		
	80 or Above	70s	Below 70
None	18	11	6
1–5	14	12	6
More than 5	3	9	20

Use $\alpha = .05$ and test for independence between number of classes missed and the grade on the final examination. What is your conclusion?

35. As part of the standard course evaluation, students are asked to rate the course as either poor, good, or excellent. The course evaluation form also asks students to indicate whether the course taken was a required part of their academic program or was taken as an elective. The dean of the college is interested in determining if the rating of the course is independent of the reason for taking the course. The following results were obtained:

Reason for Taking the Course	Rating		
	Poor	Good	Excellent
Required	16	38	16
Elective	4	10	16

How would you respond to the dean? Use $\alpha = .01$.

36. The following data were collected on the number of emergency ambulance calls for an urban county and a rural county in Virginia (*Journal of The Operational Research Society*, November 1986):

		Day of Week							Total
		Sun	Mon	Tue	Wed	Thur	Fri	Sat	
County	Urban	61	48	50	55	63	73	43	393
	Rural	7	9	16	13	9	14	10	78
Total		68	57	66	68	72	87	53	471

Conduct a test for independence using $\alpha = .05$. What is your conclusion?

37. A random sample of final examination grades for a college course is shown below:

55	85	72	99	48	71	88	70	59	98	80	74	93	85	74
82	90	71	83	60	95	77	84	73	63	72	95	79	51	85
76	81	78	65	75	87	86	70	80	64					

Use $\alpha = .05$ and test to determine if a normal distribution should be rejected as being representative of the population's distribution of grades.

38. Office occupancy rates were reported for 1991 in four California metropolitan areas (*Business Week*, 1991). Do the following data suggest that the office vacancies are independent of metropolitan area? Use a .05 level of significance. What is your conclusion?

Occupancy Status	Los Angeles	San Diego	San Francisco	San Jose
Occupied	160	116	192	174
Vacant	40	34	33	26

T A B L E 12.18

Number of Sales	Observed Frequency (days)
0	30
1	32
2	25
3	10
4	3
Total	100

39. A salesperson makes four calls per day. A sample of 100 days gives the frequencies of sales volumes shown in Table 12.18. Assume the population is a binomial distribution with a probability of purchase equal to $p = .30$. Recall that in Chapter 5 the binomial probabilities were given by

$$f(x) = \frac{n!}{x!(n-x)!} p^x(1-p)^{n-x}$$

For this exercise $n = 4$, $p = .30$, and $x = 0, 1, 2, 3,$ and 4.

a. Compute the expected frequencies for $x = 0, 1, 2, 3,$ and 4 using the binomial probability function. Combine categories if necessary to satisfy the requirement that the expected frequencies are 5 or more for all categories.

b. Should the assumption of a binomial distribution be rejected? Use $\alpha = .05$.

13

Analysis of Variance and Experimental Design

Burke Marketing Services, Inc.*

CINCINNATI, OHIO

Burke Marketing Services, Inc., is one of the most experienced market research firms in the industry. Burke writes more proposals, on more projects, every day than any other market research company in the world. Supported by state-of-the-art technology, Burke offers a wide variety of research capabilities, providing answers to nearly any marketing question.

In one study, Burke was retained by a firm to evaluate potential new versions of a children's dry cereal. To maintain confidentiality, we will refer to the cereal manufacturer as the Anon Company. The four key ingredients that Anon's product developers thought would enhance the taste of the cereal were:

1. Ratio of wheat to corn in the cereal flake.
2. Type of sweetness: sugar, honey, or artificial.
3. Presence or absence of flavor bits with a fruit taste.
4. Short or long cooking time.

An experiment was designed to determine what effects these four factors had on cereal taste. For example, one test cereal was made using a specified ratio of wheat to corn, sugar as the sweetener, flavor bits, and a short cooking time; another test cereal was made using a different ratio of wheat to corn with the other three factors held the same, and so on. Groups of children then taste-tested these cereals and stated what they thought about the taste of each.

The data obtained from the taste tests were analyzed using analysis of variance as the statistical method. The results of the analysis showed that:

- The flake composition and sweetener type were very influential in taste evaluation.
- The flavor bits actually detracted from the taste of the cereal.
- The cooking time had no impact on the taste.

This information helped Anon identify the factors that would lead to the best-tasting cereal.

The experimental design employed by Burke and the subsequent analysis of variance were helpful in making a recommendation concerning product design. In this chapter, we will see how such procedures are carried out.

Burke's in-store research provides valuable information for clients.

*The authors are indebted to Dr. Ronald Tatham of Burke Marketing Services for providing this Statistics in Practice.

I n Chapter 8 we showed how to develop an interval estimate of a population mean; then, in Chapter 9 we showed how to conduct hypothesis tests about a population mean. The extension of these concepts to cases involving two populations was presented in Chapter 10. In this chapter we present a statistical technique called *analysis of variance* (ANOVA), which can be used to test the hypothesis that the means of three or more populations are equal. In addition, we will discuss the process of designing experimental studies that result in the collection of data; this process is referred to as *experimental design*. Specifically, we cover the completely randomized, randomized block, and factorial experimental designs.

13.1   An Introduction to Analysis of Variance

National Computer Products, Inc. (NCP), manufactures printers and fax machines at plants located in Charlotte, Houston, and San Diego. To measure how much employees at these plants know about total quality management, a random sample of 6 employees was selected from each plant and given a quality-awareness examination. The examination scores obtained for these 18 employees are shown in Table 13.1. The corresponding sample means, sample variances, and sample standard deviations for each group are provided. Management would like to use these data to test the hypothesis that the mean examination score is the same at each plant.

We will define population 1 as all employees at the Charlotte plant, population 2 as all employees at the Houston plant, and population 3 as all employees at the San Diego plant. Let

μ_1 = mean examination score for population 1

μ_2 = mean examination score for population 2

μ_3 = mean examination score for population 3

Although we will never know the actual values of μ_1, μ_2, and μ_3, we would like to use the sample results to test the following hypotheses:

H_0: $\mu_1 = \mu_2 = \mu_3$

H_a: Not all population means are equal

As we will demonstrate shortly, analysis of variance is a statistical procedure for testing these hypotheses.

Assumptions for Analysis of Variance

To test the above hypotheses using analysis of variance, the following three assumptions must be made:

1. **For each population, the response variable is normally distributed.** Implication: In the NCP example the examination scores (response variable) must be normally distributed at each plant.

TABLE 13.1 Examination Scores for 18 Employees	Observation	Plant 1 Charlotte	Plant 2 Houston	Plant 3 San Diego
	1	85	71	59
	2	75	75	64
	3	82	73	62
	4	76	74	69
	5	71	69	75
	6	85	82	67
	Sample mean	79	74	66
	Sample variance	34	20	32
	Sample standard deviation	5.83	4.47	5.66

2. **The variance of the response variable, denoted σ^2, is the same for each population.** Implication: In the NCP example the variance of examination scores must be the same at each plant.
3. **The observations must be independent.** Implication: In the NCP example the examination score for each employee must be independent of the examination score for any other employee.

A Conceptual Overview

Suppose that the assumptions for analysis of variance are satisfied in the NCP example. If the null hypothesis is true ($\mu_1 = \mu_2 = \mu_3 = \mu$), each sample observation would have been drawn from the same normal probability distribution with mean μ and variance σ^2. To provide a visual perspective of the situation, consider the Minitab dotplots for the NCP data shown in Figure 13.1. Does it appear that the observations in each sample have been drawn from populations with the same mean? Although this is purely a subjective observation, you might agree that the employees at the Charlotte plant appear to have higher examination scores, while the employees at the San Diego plant have lower examination scores.

If the means for the three populations are equal, we would expect the three sample means to be close together. In fact, the closer the three sample means are to one another, the more evidence we have supporting the conclusion that the population means are equal. Alternatively, the more the sample means differ, the more evidence we have supporting the conclusion that the population means are not equal. In other words, if the variability among the sample means is "small," this is support for H_0; if the variablity among the sample means is "large," this is support for H_a.

If the null hypothesis, H_0: $\mu_1 = \mu_2 = \mu_3$, is true, we can use the variability among the sample means to develop an estimate of σ^2. And, if the assumptions for analysis of variance are satisfied, each sample will have come from the same normal probability distribution with mean μ and variance σ^2. Recall from Chapter 7 that the sampling distribution of the sample mean $\bar{x}$ for a simple random sample of size n from a normal population will be normally distributed with mean μ and variance σ^2/n. Figure 13.2 illustrates such a sampling distribution.

If the null hypothesis is true, we can think of each of the three sample means, $\bar{x}_1 = 79$, $\bar{x}_2 = 74$, and $\bar{x}_3 = 66$ from Table 13.1 as values drawn at random from the sampling distribution shown in Figure 13.2. In this case, the mean and variance of the three $\bar{x}$ values can be used to estimate the mean and variance of the sampling distribution.

FIGURE **13.1**   **Minitab Dotplot for Examination Scores**

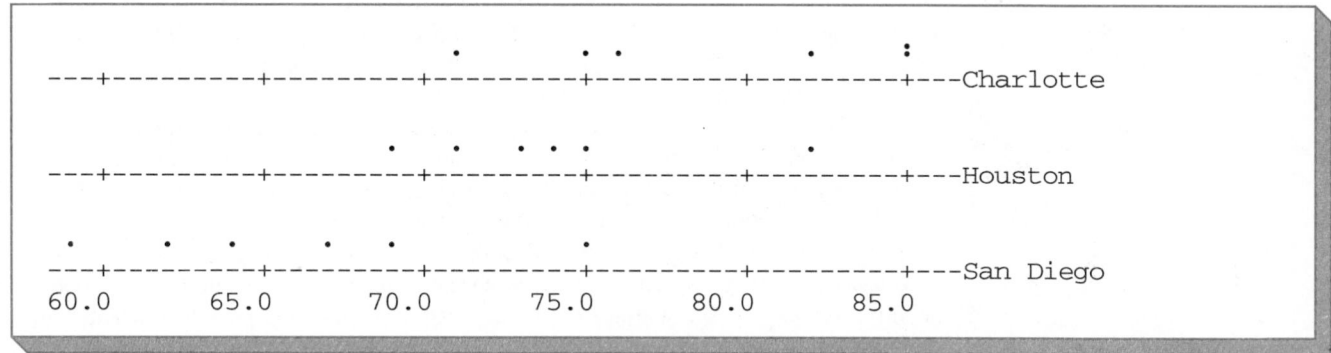

FIGURE 13.2
Sampling Distribution of $\bar{x}$ Given
H_0 Is True

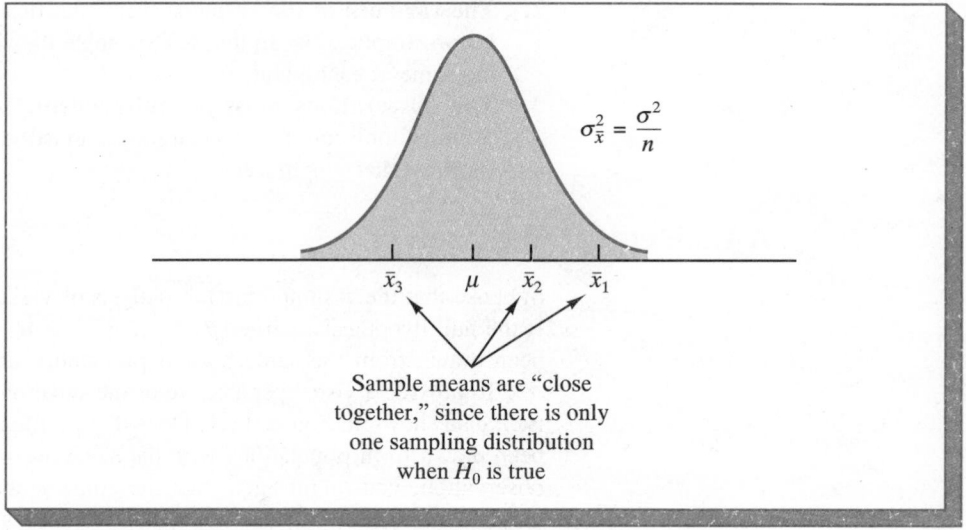

$$\sigma_{\bar{x}}^2 = \frac{\sigma^2}{n}$$

$\bar{x}_3$ μ $\bar{x}_2$ $\bar{x}_1$

Sample means are "close
together," since there is only
one sampling distribution
when H_0 is true

In the NCP example the best estimate of the mean of the sampling distribution of $\bar{x}$ is the mean or average of the three sample means. That is, $(79 + 74 + 66)/3 = 73$. We refer to this estimate as the *overall sample mean*. To estimate the variance of the sampling distribution of $\bar{x}$, we compute the variance using the three sample means:

$$s_{\bar{x}}^2 = \frac{(79 - 73)^2 + (74 - 73)^2 + (66 - 73)^2}{3 - 1} = \frac{86}{2} = 43$$

Since $\sigma_{\bar{x}}^2 = \sigma^2/n$, solving for σ^2 gives

$$\sigma^2 = n\sigma_{\bar{x}}^2$$

Hence,

$$\text{Estimate of } \sigma^2 = n \ (\text{Estimate of } \sigma_{\bar{x}}^2) = ns_{\bar{x}}^2 = 6(43) = 258$$

The result, $ns_{\bar{x}}^2 = 258$, is referred to as the *between-samples* estimate of σ^2.

The between-samples estimate of σ^2 is based on the assumption that the null hypothesis is true. In this case, each sample comes from the same population, and there is only one sampling distribution of $\bar{x}$. To illustrate what happens when H_0 is false, suppose that the population means *all differ*. Note that since the three samples are from normal populations with different means, there will be different sampling distributions. Figure 13.3 shows that in this case, the sample means are not as close together as they were when H_0 was true. Thus, $s_{\bar{x}}^2$ will be larger, causing the between-samples estimate of σ^2 to be larger. In general, when the population means are not equal, the between-samples estimate will overestimate the population variance σ^2.

The variation within each of the samples can also have an effect on the conclusion we reach in analysis of variance. When a simple random sample is selected from each population, each of the sample variances provides an unbiased estimate of σ^2. Thus, we can combine or pool the individual estimates of σ^2 into one overall estimate. The estimate of σ^2 obtained in this fashion is called the *pooled* or *within-samples* estimate of σ^2. Because each sample variance provides an estimate of σ^2 based only on the variation within each sample, the within-samples estimate of σ^2 is not affected by whether or not the population means are equal. When the sample sizes are equal, the within-samples

FIGURE 13.3
Sampling Distribution for $\bar{x}$ Given H_0 Is False

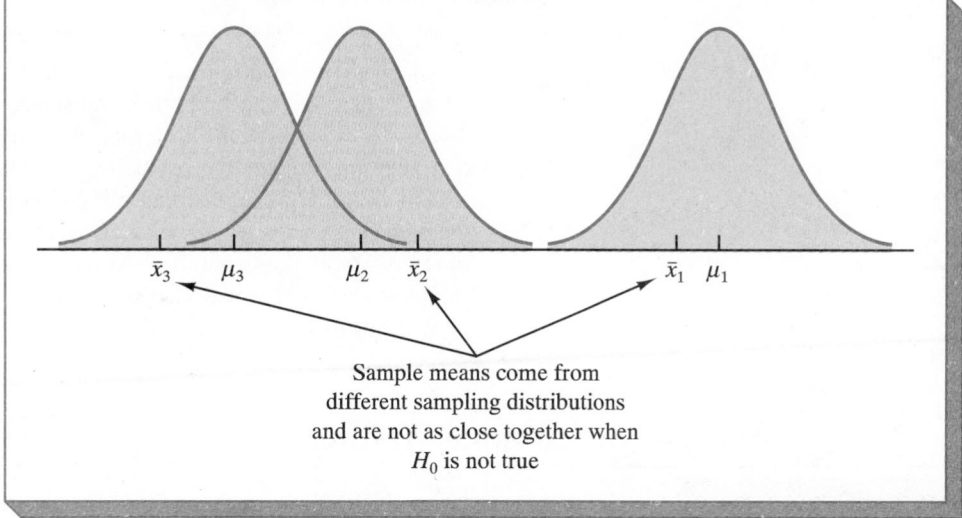

estimate of σ^2 can be obtained by computing the average of the individual sample variances. For the NCP example we obtain

$$\text{Within-Samples Estimate of } \sigma^2 = \frac{34 + 20 + 32}{3} = \frac{86}{3} = 28.67$$

In the NCP example the between-samples estimate of σ^2 (258) is much larger than the within-samples estimate of σ^2 (28.67). In fact, the ratio of these two estimates is $258/28.67 = 9.00$. Recall, however, that the between-samples approach provides a good estimate of σ^2 only if the null hypothesis is true; if the null hypothesis is false, the between-samples approach *overestimates* σ^2. The within-samples approach provides a good estimate of σ^2 in either case. Thus, if the null hypothesis is true, both estimates will be similar and their ratio will be close to 1. If the null hypothesis is false, the between-samples estimate will be larger than the within-samples estimate, and their ratio will be large. In the next section we will show how large this ratio must be to reject H_0.

In summary then, the logic behind ANOVA is based on the development of two independent estimates of the common population variance σ^2. One estimate of σ^2 is based on the variability among the sample means themselves, and the other estimate of σ^2 is based on the variability of the data within each sample. By comparing these two estimates of σ^2, we will be able to determine if the population means are equal. Since the methodology uses a comparison of variances, it is referred to as analysis of variance.

NOTES &
COMMENTS

1. In Chapter 10 we presented statistical methods for testing the hypothesis that the means of two populations are equal. Although in the introduction to this chapter we stated that the analysis of variance is a statistical technique for testing the hypothesis that the means of three or more populations are equal, ANOVA can also be used to test the hypothesis that the means of two populations are equal. In practice, however, analysis of variance is usually thought of as a technique for testing for the equality of three or more population means.

Continued on next page

Continued from previous page

2. In Chapter 10 we discussed how to test for the equality of two population means whenever one or both sample sizes are less than 30. As part of that discussion we illustrated the process of combining the results of two independent random samples to provide one estimate of σ^2; this process was referred to as pooling, and the resulting sample variance we obtained was referred to as the pooled estimator of σ^2. In analysis of variance, the within-samples estimate of σ^2 is simply the generalization of this concept to the case of more than two samples; this is why we also referred to the within-samples estimator as the pooled estimator of σ^2.

13.2   Analysis of Variance: Testing for the Equality of k Population Means

In general, analysis of variance can be used to test for the equality of k population means. The general form of the hypotheses tested is

$$H_0: \mu_1 = \mu_2 = \cdots = \mu_k$$

$$H_a: \text{Not all the means are equal}$$

where

$$\mu_j = \text{mean of the } j\text{th population}$$

We assume that a simple random sample of size n_j has been selected from each of the k populations. Let

$$x_{ij} = \text{the } i\text{th observation in the } j\text{th sample}$$

$$n_j = \text{the number of observations in the } j\text{th sample}$$

$$\overline{x}_j = \text{the mean of the } j\text{th sample}$$

$$s_j^2 = \text{the variance of the } j\text{th sample}$$

$$s_j = \text{the standard deviation of the } j\text{th sample}$$

The formulas for the jth sample mean and variance are as follows:

$$\overline{x}_j = \frac{\displaystyle\sum_{i=1}^{n_j} x_{ij}}{n_j} \tag{13.1}$$

$$s_j^2 = \frac{\displaystyle\sum_{i=1}^{n_j} (x_{ij} - \overline{x}_j)^2}{n_j - 1} \tag{13.2}$$

The overall sample mean, denoted $\overline{\overline{x}}$, is the sum of all the observations divided by the total number of observations. That is,

$$\overline{\overline{x}} = \frac{\displaystyle\sum_{j=1}^{k}\sum_{i=1}^{n_j} x_{ij}}{n_T} \tag{13.3}$$

where

$$n_T = n_1 + n_2 + \cdots + n_k \tag{13.4}$$

If the size of each sample is n, $n_T = kn$; in this case (13.3) reduces to

$$\overline{\overline{x}} = \frac{\sum\limits_{j=1}^{k} \sum\limits_{i=1}^{n_j} x_{ij}}{nk} = \frac{\sum\limits_{j=1}^{k} \sum\limits_{i=1}^{n_j} x_{ij}/n}{k} = \frac{\sum\limits_{j=1}^{k} \overline{x}_j}{k} \tag{13.5}$$

In other words, whenever the sample sizes are the same, the overall sample mean is just the average of the k sample means.

Since each sample in the NCP example consists of $n = 6$ observations, the overall sample mean can be computed using (13.5). For the data shown in Table 13.1 we obtained the following result:

$$\overline{\overline{x}} = \frac{79 + 74 + 66}{3} = 73$$

Thus, if the null hypothesis is true, the overall sample mean of 73 is the best estimate of the population mean μ.

Between-Samples Estimate of Population Variance

In the previous section we introduced the concept of a between-samples estimate of σ^2. This estimate of σ^2 is called the *mean square between* and is denoted MSB. The formula for computing MSB is as follows:

$$\text{MSB} = \frac{\sum\limits_{j=1}^{k} n_j(\overline{x}_j - \overline{\overline{x}})^2}{k - 1} \tag{13.6}$$

The numerator in (13.6) is called the *sum of squares between* and is denoted SSB. The denominator, $k - 1$, represents the degrees of freedom associated with SSB. Thus, the mean square between can be computed as follows.

Mean Square Between

$$\text{MSB} = \frac{\text{SSB}}{k - 1} \tag{13.7}$$

where

$$\text{SSB} = \sum\limits_{j=1}^{k} n_j(\overline{x}_j - \overline{\overline{x}})^2 \tag{13.8}$$

If H_0 is true, MSB provides an unbiased estimate of σ^2. However, if the means of the k populations are not equal, MSB is not an unbiased estimate of σ^2; in fact, in this case, MSB should overestimate σ^2.

For the NCP data shown in Table 13.1, we obtain the following results:

$$\text{SSB} = \sum_{j=1}^{k} n_j(\bar{x}_j - \bar{\bar{x}})^2 = 6(79 - 73)^2 + 6(74 - 73)^2 + 6(66 - 73)^2 = 516$$

$$\text{MSB} = \frac{\text{SSB}}{k - 1} = \frac{516}{2} = 258$$

Within-Samples Estimate of Population Variance

The second estimate of σ^2 is based on the variation of the sample observations within each sample. This estimate of σ^2 is called the *mean square within* and is denoted MSW. The formula for computing MSW is as follows:

$$\text{MSW} = \frac{\sum_{j=1}^{k} (n_j - 1)s_j^2}{n_T - k} \tag{13.9}$$

The numerator in (13.9) is called the *sum of squares within* and is denoted SSW. The denominator of MSW is referred to as the degrees of freedom associated with SSW. Thus, the formula for MSW can also be stated as follows.

Mean Square Within

$$\text{MSW} = \frac{\text{SSW}}{n_T - k} \tag{13.10}$$

where

$$\text{SSW} = \sum_{j=1}^{k} (n_j - 1)s_j^2 \tag{13.11}$$

Note that MSW is based on the variation within each of the samples; it is not influenced by whether or not the null hypothesis is true. Thus, MSW always provides an unbiased estimate of σ^2.

For the NCP data in Table 13.1 we obtain the following results:

$$\text{SSW} = \sum_{j=1}^{k} (n_j - 1)s_j^2 = (6 - 1)34 + (6 - 1)20 + (6 - 1)32 = 430$$

$$\text{MSW} = \frac{\text{SSW}}{n_T - k} = \frac{430}{18 - 3} = \frac{430}{15} = 28.67$$

Comparing the Variance Estimates: The *F* Test

Let us assume for the moment that the null hypothesis is true. In this case, MSB and MSW provide two independent, unbiased estimates of σ^2. Recall from Chapter 11 that for normal populations, the sampling distribution of the ratio of two independent estimates of σ^2 follows an F distribution. Thus, if the null hypothesis is true and the ANOVA assumptions are valid, the sampling distribution of MSB/MSW is an F distribution with numerator degrees of freedom equal to $k - 1$ and denominator degrees of freedom equal to $n_T - k$.

If the means of the k populations are not equal, the value of MSB/MSW will be inflated because MSB overestimates σ^2. Hence, we will reject H_0 if the resulting value of MSB/MSW appears to be too large to have been selected at random from an F distribution with degrees of freedom $k - 1$ in the numerator and $n_T - k$ in the denominator. The value of MSB/MSW that will cause us to reject H_0 depends upon α, the level of significance. Once α is selected, a critical value can be determined. Figure 13.4 shows the sampling distribution of MSB/MSW and the rejection region associated with a level of significance equal to α where F_α denotes the critical value.

Suppose the manager responsible for making the decision in the National Computer Products example was willing to accept a probability of a Type I error of $\alpha = .05$. From Table 4 of Appendix B we can determine the critical F value by locating the value corresponding to numerator degrees of freedom equal to $k - 1 = 3 - 1 = 2$ and denominator degrees of freedom equal to $n_T - k = 18 - 3 = 15$. Thus, we obtain the value $F_{.05} = 3.29$. Note that this tells us that if we were to select a value at random from an F distribution with 2 numerator degrees of freedom and 15 denominator degrees of freedom, only 5% of the time would we observe a value greater than 3.29. Moreover, the theory behind the analysis of variance tells us that if the null hypothesis is true, the ratio of MSB/MSW would be a value from this F distribution. Hence, the appropriate rejection rule for the NCP example is written

$$\text{Reject } H_0 \text{ if MSB/MSW} > 3.29$$

Recall that MSB = 258 and MSW = 28.67. Since MSB/MSW = 258/28.67 = 9.00 is greater than the critical value, $F_{.05} = 3.29$, there is sufficient evidence to reject the null hypothesis that the means of the three groups are equal. In other words, analysis of variance supports the conclusion that the population mean examination scores at the three NCP plants are not identical.

The ANOVA Table

The results of the preceding calculations can be conveniently displayed in a table referred to as the *analysis of variance table*. Table 13.2 shows the analysis of variance table for the

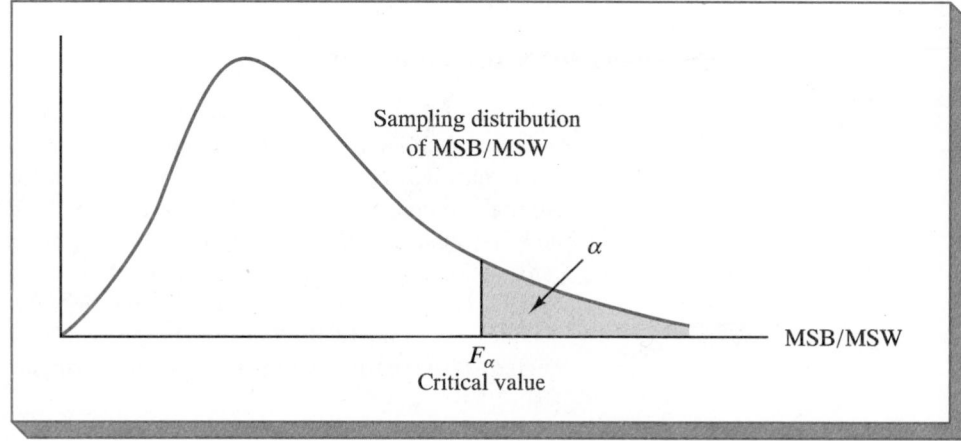

FIGURE 13.4

Sampling Distribution of MSB/MSW; the Critical Value for Rejecting the Null Hypothesis of Equality of Means Is F_α

TABLE 13.2	Source of Variation	Sum of Squares	Degrees of Freedom	Mean Square	F
Analysis of Variance Table for the NCP Example	Between	516	2	258.00	9.00
	Within	430	15	28.67	
	Total	946	17		

National Computer Products example. The sum of squares associated with the source of variation referred to as "Total" is called the total sum of squares (SST). Note that the results for the NCP example suggest that SST = SSB + SSW, and that the degrees of freedom associated with this total sum of squares is the sum of the degrees of freedom associated with the between-samples estimate of σ^2 and the within-samples estimate of σ^2.

We should point out that SST divided by its degrees of freedom $n_T - 1$, is nothing more than the overall sample variance which would be obtained if we treated the entire set of 18 observations as one data set. Using the entire data set as one sample, the formula for computing the total sum of squares, SST, is as follows:

$$SST = \sum_{j=1}^{k} \sum_{i=1}^{n_j} (x_{ij} - \overline{\overline{x}})^2 \qquad (13.12)$$

It can be shown that the results we observed for the analysis of variance table for the NCP example also apply to other problems. That is,

$$SST = SSB + SSW \qquad (13.13)$$

In other words, SST can be partitioned into two sums of squares: the sum of squares between and the sum of squares within. Note also that the degrees of freedom corresponding to SST, $n_T - 1$, can be partitioned into the degrees of freedom corresponding to SSB, $k - 1$, and the degrees of freedom corresponding to SSW, $n_T - k$. The analysis of variance can be viewed as the process of partitioning the total sum of squares and degrees of freedom into their corresponding sources: between and within. Dividing the sum of squares by the appropriate degrees of freedom provides the variance estimates and the F value used to test the hypothesis of equal population means.

Computer Results for Analysis of Variance

Because of the widespread availability of statistical computer packages, analysis of variance computations with large sample sizes and/or a large number of populations can be easily handled. In Figure 13.5 we show output for the NCP example obtained from the Minitab computer package. The first part of the computer output contains the familiar ANOVA table format. Comparing Figure 13.5 with Table 13.2, we see that the same information is available, although some of the headings are a little different. The heading SOURCE is used for the source of variation column, FACTOR identifies the between-samples row, and ERROR identifies the within-samples row. The sum of squares and degrees of freedom columns are also interchanged, and a p-value is provided for the F test.

FIGURE 13.5   **Minitab Output for the NCP Analysis of Variance**

```
ANALYSIS OF VARIANCE
SOURCE      DF         SS        MS        F         p
FACTOR       2      516.0     258.0      9.00     0.003
ERROR       15      430.0      28.7
TOTAL       17      946.0
                                  INDIVIDUAL 95 PCT CI'S FOR MEAN
                                  BASED ON POOLED STDEV
  LEVEL      N       MEAN      STDEV    ---+---------+---------+---------+---
PLANT 1      6     79.000      5.831                   (------*------)
PLANT 2      6     74.000      4.472             (------*-----)
PLANT 3      6     66.000      5.657    (-----*------)
                                        ---+---------+---------+---------+---
POOLED STDEV =     5.354              63.0      70.0      77.0      84.0
```

Note that below the ANOVA table, the computer output contains the respective sample sizes, the sample means and the standard deviations. In addition, Minitab provides a figure which shows individual 95% confidence interval estimates of each population mean. In developing these confidence interval estimates, Minitab uses MSW as the estimate of σ^2. Thus, the square root of MSW provides the best estimate of the population standard deviation σ. This estimate of σ is referred to on the computer output as the POOLED STDEV; it is equal to 5.354. To provide an illustration of how these interval estimates are developed, we will compute a 95% confidence interval estimate of the population mean for the Charlotte plant, identified as PLANT 1 in the computer output.

From our study of interval estimation in Chapter 8, we know that the general form of an interval estimate of a population mean is

$$\bar{x} \pm t_{\alpha/2} \frac{s}{\sqrt{n}}$$

(13.14)

where s is the estimate of the population standard deviation σ. Since in the analysis of variance the best estimate of σ is provided by the square root of MSW or the POOLED STDEV, we use a value of 5.354 for s in (13.14). The degrees of freedom for the t value is 15, the degrees of freedom associated with the within-samples estimate of σ^2. Thus, with $t_{.025} = 2.131$ we obtain

$$79 \pm 2.131 \frac{5.354}{\sqrt{6}} = 79 \pm 4.66$$

From this calculation, we see that the figure shown on the Minitab output for Plant 1 depicts an interval that goes from 74.34 to 83.66. Since the sample sizes are all equal for the NCP example, the confidence intervals for Plants 2 and 3 are also constructed by adding and subtracting 4.66 from each sample mean. Thus, in the figure provided by Minitab we see that the width of each confidence interval is the same.

NOTES &
COMMENTS

1. The overall sample mean can also be computed as a weighted average of the k sample means.

$$\overline{\overline{x}} = \frac{n_1\overline{x}_1 + n_2\overline{x}_2 + \cdots + n_k\overline{x}_k}{n_T}$$

In problems where the sample means have been provided, this formula is simpler than (13.3) for computing the overall mean.

2. If each sample consists of n observations, (13.6) can be written as

$$\text{MSB} = \frac{n\sum_{j=1}^{k}(\overline{x}_j - \overline{\overline{x}})^2}{k - 1} = n\left[\frac{\sum_{j=1}^{k}(\overline{x}_j - \overline{\overline{x}})^2}{k - 1}\right] = ns_{\overline{x}}^2$$

Note that this is the same result that we presented in Section 13.1 when we introduced the concept of the between-samples estimate of σ^2. Equation (13.6) is simply a generalization of this result to the unequal sample-size case.

3. If each sample has n observations, $n_T = kn$; thus, $n_T - k = k(n - 1)$, and (13.9) can be rewritten as

$$\text{MSW} = \frac{\sum_{j=1}^{k}(n - 1)s_j^2}{k(n - 1)} = \frac{(n - 1)\sum_{j=1}^{k}s_j^2}{k(n - 1)} = \frac{\sum_{j=1}^{k}s_j^2}{k}$$

In other words, if the sample sizes are the same, the within-samples estimate of σ^2 is just the average of the k sample variances. Note that this is the result we used in Section 13.1 when we introduced the concept of the within-samples estimate of σ^2.

❑ ❑ Exercises

Methods

SELF TEST

1. Samples of 5 observations were selected from each of three populations. The data obtained are shown below.

Observation	Sample 1	Sample 2	Sample 3
1	32	44	33
2	30	43	36
3	30	44	35
4	26	46	36
5	32	48	40
Sample mean	30	45	36
Sample variance	6.00	4.00	6.50

a. Develop the dotplots for these data. Based on your subjective evaluation of the dotplots, does it appear that the observations in each sample have been drawn from the same population?

b. Compute the between-samples estimate of σ^2.

c. Compute the within-samples estimate of σ^2.

d. At the $\alpha = .05$ level of significance, can we reject the null hypothesis that the means of the three populations are equal.

e. Set up the ANOVA table for this problem.

2. Four observations were selected from each of three populations. The data obtained is shown below:

Observation	Sample 1	Sample 2	Sample 3
1	165	174	169
2	149	164	154
3	156	180	161
4	142	158	148
Sample mean	153	169	158
Sample variance	96.67	97.33	82.00

a. Compute the between-samples estimate of σ^2.

b. Compute the within-samples estimate of σ^2.

c. At the $\alpha = .05$ level of significance, can we reject the null hypothesis that the three population means are equal? Explain.

d. Set up the ANOVA table for this problem.

3. Samples were selected from three populations. The data obtained are shown in Table 13.3.

a. Compute the between-samples estimate of σ^2.

b. Compute the within-samples estimate of σ^2.

c. At the $\alpha = .05$ level of significance, can we reject the null hypothesis that the three population means are equal? Explain.

d. Set up the ANOVA table for this problem.

TABLE 13.3

	Sample 1	Sample 2	Sample 3
	93	77	88
	98	87	75
	107	84	73
	102	95	84
		85	75
		82	
$\bar{x}_j$	100	85	79
s_j^2	35.33	35.60	43.50

4. A random sample of 16 observations was selected from each of four populations. A portion of the ANOVA table is shown below:

Source of Variation	Sum of Squares	Degrees of Freedom	Mean Square	F
Between			400	
Within				
Total	1500			

a. Complete the missing entries in the ANOVA table.

b. At the $\alpha = .05$ level of significance, can we reject the null hypothesis that the means of the four populations are equal.

5. Random samples of 25 observations were selected from each of three populations. For these data, SSB = 120 and SSW = 216.

a. Set up the ANOVA table for this problem.

b. At the $\alpha = .05$ level of significance, what is the critical F value?

c. At the $\alpha = .05$ level of significance, can we reject the null hypothesis that the three population means are equal?

Applications

6. To test if the mean time needed to mix a batch of material is the same for machines produced by three manufacturers, the Jacobs Chemical Company obtained the data in Table 13.4 on the time (in minutes) needed to mix the material. Use these data to test if the population mean times needed to mix a batch of material differ for the three manufacturers. Use $\alpha = .05$.

7. A 100-question science test was given to 13-year-old children from several different countries (*Newsweek*, February 17, 1992); a portion of the data showing the number of questions that were answered correctly for three samples of six students is shown below.

SELF TEST ▶

TABLE 13.4

	Manufacturer		
	1	**2**	**3**
	20	28	20
	26	26	19
	24	31	23
	22	27	22
$\bar{x}_j$	23	28	21
s_j^2	6.67	4.67	3.33

	South Korea	**Soviet Union**	**United States**
	81	71	63
	71	78	61
	85	62	69
	70	71	70
	74	68	75
	87	76	64
$\bar{x}_j$	78	71	67
s_j^2	53.6	32.8	27.6

At the $\alpha = .05$ level of significance, test if the population mean test scores differ.

8. Managers at all levels of an organization need to have the information necessary to perform their respective tasks. A recent study investigated the effect the source has on the dissemination of the information (*Journal of Management Information Systems*, Fall 1988). In this particular study the sources of information were a superior, a peer, and a subordinate. In each case, a measure of dissemination was obtained with higher values indicating greater dissemination of information. Using $\alpha = .05$, and the data shown in Table 13.5, test whether or not the source of information significantly affects dissemination. What is your conclusion, and what does this suggest about the use and dissemination of information?

9. A study investigated the perception of corporate ethical values among individuals specializing in marketing (*Journal of Marketing Research*, July 1989). Suppose that the data shown below were obtained in a similar study (higher scores indicate higher ethical values). Using $\alpha = .05$, test to see if there are significant differences in perception for the three groups of specialists.

INFO

TABLE 13.5

	Superior	**Peer**	**Subordinate**
	8	6	6
	5	6	5
	4	7	7
	6	5	4
	6	3	3
	7	4	5
	5	7	7
	5	6	5
$\bar{x}_j$	5.75	5.5	5.25
s_j^2	1.64	2.00	1.93

	Marketing Managers	**Marketing Research**	**Advertising**
	6	5	6
	5	5	7
	4	4	6
	5	4	5
	6	5	6
	4	4	6
$\bar{x}_j$	5	4.5	6
s_j^2	.8	.3	.4

MACHINES

10. To test for any significant difference in the number of hours between breakdowns for four machines, the following data were obtained:

	Machine			
	1	2	3	4
	6.4	8.7	11.1	9.9
	7.8	7.4	10.3	12.8
	5.3	9.4	9.7	12.1
	7.4	10.1	10.3	10.8
	8.4	9.2	9.2	11.3
	7.3	9.8	8.8	11.5
$\bar{x}_j$	7.1	9.1	9.9	11.4
s_j^2	1.21	.93	.70	1.02

At the $\alpha = .05$ level of significance, is there any difference in the population mean times among the four machines?

13.3   Multiple Comparison Procedures

When we use analysis of variance to test if the means of k populations are equal, we must keep in mind that rejection of the null hypothesis only allows us to conclude that the population means are *not all equal*. Sometimes we may be satisfied with this conclusion, but in other cases, we will want to go a step further and determine where the differences among means occur. The purpose of this section is to introduce several methods that can be used to conduct statistical comparisons between pairs of population means.

Fisher's LSD

Fisher's least significant difference (LSD) procedure is one of the oldest and perhaps most widely used methods for making pairwise comparisons of population means. In the NCP example there were three populations. The hypotheses were written as follows:

$$H_0: \mu_1 = \mu_2 = \mu_3$$

$$H_a: \text{Not all population means are equal}$$

Using analysis of variance, we rejected H_0 and concluded that the population mean examination scores are not the same at each plant. In this case, the follow-up question is, We believe that the plants differ, but where do the differences occur? That is, do the means of populations 1 and 2 differ? Or do those of populations 1 and 3? Or those of populations 2 and 3?

For example, let us test to see if there is a significant difference between the means of population 1 (Charlotte) and population 2 (Houston). Although Table 13.1 shows that the sample mean is 79 for the Charlotte plant and 74 for the Houston plant, is this sample information sufficient to justify the conclusion that there is a difference between the population mean examination scores for these two plants? In statistical terms, we state the following hypotheses:

$$H_0: \mu_1 = \mu_2$$

$$H_a: \mu_1 \neq \mu_2$$

In Chapter 10 we presented a statistical procedure for testing the hypothesis that the means of two populations are equal. With a slight modification in how we estimate the population variance, Fisher's LSD procedure is based on the t test statistic presented for the two-population case. The test statistic for Fisher's LSD procedure is

$$t = \frac{\bar{x}_1 - \bar{x}_2}{\sqrt{\text{MSW}\left(\frac{1}{n_1} + \frac{1}{n_2}\right)}} \qquad (13.15)$$

We reject H_0 if $t < -t_{\alpha/2}$ or $t > t_{\alpha/2}$. In the NCP example, Table 13.2 shows that the value of MSW is 28.67. This is the estimate of σ^2 and is based on 15 degrees of freedom. At the .05 level of significance, the t distribution table shows that with 15 degrees of freedom, $t_{.025} = 2.131$. Thus, if $t < -2.131$ or $t > 2.131$, we reject H_0. For the NCP data we obtain the following t value:

$$t = \frac{79 - 74}{\sqrt{28.67\left(\frac{1}{6} + \frac{1}{6}\right)}} = 1.62$$

Since $t = 1.62$, we do not have sufficient statistical evidence to reject the null hypothesis; thus, we cannot conclude that the population mean score at the Charlotte plant is different from the population mean score at the Houston plant.

Many practitioners find it easier to determine how large a difference between the sample means must exist to reject H_0. Thus, if

$$\bar{x}_1 - \bar{x}_2 > t_{\alpha/2}\sqrt{\text{MSW}\left(\frac{1}{n_1} + \frac{1}{n_2}\right)}$$

or

$$\bar{x}_1 - \bar{x}_2 < -t_{\alpha/2}\sqrt{\text{MSW}\left(\frac{1}{n_1} + \frac{1}{n_2}\right)}$$

we reject H_0. If we define the least significant difference (LSD) as

$$\text{LSD} = t_{\alpha/2}\sqrt{\text{MSW}\left(\frac{1}{n_1} + \frac{1}{n_2}\right)} \qquad (13.16)$$

we will reject H_0 if

$$\bar{x}_1 - \bar{x}_2 > \text{LSD}$$

or

$$\bar{x}_1 - \bar{x}_2 < -\text{LSD}$$

If we consider only the magnitude or absolute value of the difference, we can write the rejection rule for Fisher's LSD test as

$$\text{Reject } H_0 \text{ if } |\bar{x}_1 - \bar{x}_2| > LSD$$

To illustrate this approach we will compute the value of LSD for the NCP example.

$$\text{LSD} = 2.131\sqrt{28.67\left(\frac{1}{6} + \frac{1}{6}\right)} = 6.59$$

Note that when the sample sizes are equal, only one value for LSD is computed. In such cases, we can simply compare the magnitude of the difference between any two means with the value of LSD. For the NCP example, the difference between the sample means for population 1 and population 3 is $79 - 66 = 13$. Since this difference is greater than 6.59, we can reject the null hypothesis that the population mean examination score for the Charlotte plant is equal to the population mean examination score for the San Diego plant. Similarly, since the difference between the sample means for populations 2 and 3 is $74 - 66 = 8 > 6.59$, we also can reject the hypothesis that the population mean examination score for the Houston plant is equal to the population mean examination score at the San Diego plant. In effect, our conclusion is that the Charlotte and Houston plants both differ from the San Diego plant.

Fisher's LSD procedure can also be used to develop a confidence interval estimate of the difference between two population means. For example, a confidence interval estimate of the difference between the means of populations 1 and 2 is given by the following expression:

$$\bar{x}_1 - \bar{x}_2 \pm t_{\alpha/2} \sqrt{\text{MSW}\left(\frac{1}{n_1} + \frac{1}{n_2}\right)} \tag{13.17}$$

Using (13.16) we can write this interval as

$$\bar{x}_1 - \bar{x}_2 \pm \text{LSD} \tag{13.18}$$

If the confidence interval in (13.18) includes the value of 0, we cannot reject the hypothesis that the two population means are equal. However, if the confidence interval does not include the value 0, we conclude that there is a difference between the population means. For the NCP example, recall that LSD = 6.59 (corresponding to $t_{.025} = 2.131$). Thus, a 95% confidence interval estimate of the difference between the means of populations 1 and 2 is $79 - 74 \pm 6.59 = 5 \pm 6.59 = -1.59$ to 11.59; since this interval includes 0, we can not reject the hypothesis that the two population means are equal.

Type I Error Rates

We began the discussion of Fisher's LSD procedure with the premise that in analysis of variance we found statistical evidence to reject the null hypothesis of equal population means. In such cases, we showed how Fisher's LSD procedure can be used to determine where the differences occur. Technically, this is referred to as a *protected* or *restricted* LSD test since it is only employed if we find a significant F value using analysis of variance. To see why this is important in multiple comparison tests, we need to explain the difference between a *comparisonwise* Type I error rate and an *experimentwise* Type I error rate.

Consider the following three hypotheses tests:

Test 1	Test 2	Test 3
$H_0: \mu_1 = \mu_2$	$H_0: \mu_1 = \mu_3$	$H_0: \mu_2 = \mu_3$
$H_a: \mu_1 \neq \mu_2$	$H_a: \mu_1 \neq \mu_3$	$H_a: \mu_2 \neq \mu_3$

Assume that these three hypotheses show all possible pairwise comparisons for the problem. Suppose that we use Fisher's LSD procedure to test each of these three pairwise comparisons. If we carry out each test using a level of significance of $\alpha = .05$ and are not able to reject the null hypothesis for any test, it would seem reasonable to conclude that the three population means must be equal. Before jumping to any conclusions, however, let us consider what can happen if we follow this approach.

To begin with, let us consider using Fisher's LSD procedure for test 1. If the null hypothesis is true $(\mu_1 = \mu_2)$, the probability that we will make a Type I error is $\alpha = .05$; hence, the probability that we will not make a Type I error is $1 - .05 = .95$. Now, suppose that we follow this same procedure for test 2; the probability that we will make a Type I error for this test is also $\alpha = .05$, and hence, the probability that we will not make a Type I error for test 2 is also .95. Clearly, when performing a single statistical test, the probability of making a Type I error is $\alpha = .05$. In discussing multiple comparison procedures we refer to $\alpha = .05$ as the *comparisonwise Type I error rate*. In essence, comparisonwise Type I error rates indicate the level of significance associated with a single statistical test.

Let us now consider a slightly different question. What is the probability that in using this sequential approach to hypothesis testing, we will commit a Type I error on at least one of the first two tests? To answer this question, note that the probability that we will not make a Type I error for tests 1 and 2 is $(.95)(.95) = .9025$.* Since the sum of the probabilities of making zero, one, or two Type I errors is 1, the probability of making at least one Type I error is $1 - .9025 = .0975$. Thus, when we use Fisher's LSD procedure to sequentially test two sets of hypotheses, the Type I error rate associated with this approach is not .05, but actually .0975; we refer to this error rate as the *experimentwise Type I error rate*.

Continuing with the analysis, suppose we also use Fisher's LSD procedure for test 3. The comparisonwise error rate still remains at $\alpha = .05$; however, the probability that we will commit a Type I error on at least one of the three tests has now increased to $1 - (.95)(.95)(.95) = 1 - .8574 = .1426$. Thus, if we sequentially apply Fisher's LSD to all pairwise comparisons, the *overall* or *experimentwise Type I error rate* associated with this sequential approach is .1426. To avoid confusion, we will denote the experimentwise Type I error rate as α_{EW}.

To write a general expression for the experimentwise Type I error rate, let C denote the number of possible pairwise comparisons. For a problem with k populations, the value of C is the number of combinations of k populations taken 2 at a time; that is

$$C = \text{Number of Pairwise Comparisons} = \binom{k}{2} = \frac{k!}{(k-2)!2!} = \frac{k(k-1)}{2} \qquad \textbf{(13.19)}$$

For example, if $k = 5$, there are $[5(5-1)/2 = 10]$ possible pairwise comparisons. In general then, for a problem involving C pairwise comparisons, the probability of making at least one Type I error is $1 - (1 - \alpha)^C$. That is,

$$\alpha_{EW} = \text{Experimentwise Type I Error Rate} = 1 - (1 - \alpha)^C \qquad \textbf{(13.20)}$$

For example, when there are five populations and we want to test all possible pairwise comparisons using Fisher's LSD with a comparisonwise error rate of $\alpha = .05$, the experimentwise Type I error rate is $1 - (1 - .05)^{10} = .40$. With this large of an experimentwise Type I error rate, many practitioners look to alternatives that provide better control over the experimentwise error rate.

---

*This assumes that the two tests are independent, and hence the joint probability of the two events can be obtained by simply multiplying the individual probabilities. In fact, the two tests are not independent since MSW is used in each test; hence, the error involved is even greater than that shown.

Bonferroni Adjustment

A problem with Fisher's LSD procedure is that the experimentwise Type I error rate is really dependent upon the comparisonwise error rate α and the number of pairwise comparisons. Instead of specifying a comparisonwise error rate then, suppose we specify the value of α_{EW} that is desired and then find the value of α that will provide this value.

In the preceding discussion we stated that the probability of making at least one Type I error for problems involving k possible pairwise comparisons is

$$\alpha_{EW} = \text{Experimentwise Type I Error Rate} = 1 - (1 - \alpha)^C$$

The Italian mathematician Bonferroni proved mathematically that

$$1 - (1 - \alpha)^C \le C\alpha \tag{13.21}$$

for any value of C whenever α is between 0 and 1. Thus, since $\alpha_{EW} = 1 - (1 - \alpha)^C$, the probability of making at least one Type I error whenever C pairwise comparisons are tested is less than or equal to $C\alpha$. If we want the maximum probability of making a Type I error for the overall experiment to be α_{EW}, we simply use a comparisonwise Type I error rate equal to α_{EW}/C.

For example, suppose that we want to use Fisher's LSD procedure to test all three pairwise comparisons for the NCP example. Furthermore, suppose that we want the maximum experimentwise Type I error rate to be .05. If we set the comparisonwise error rate to be $\alpha = .05/3 = .017$, we will ensure that the overall or experimentwise error rate will be less than or equal to .05. Thus, when using Fisher's LSD procedure with the Bonferroni adjustment, the t value is $t_{.017/2} = t_{.0085}$. Since we do not have tables to determine the exact t value corresponding to this level of significance, we note that this value is approximately $t_{.01} = 2.602$. Recall that the value of LSD for the NCP example is

$$LSD = 2.131 \sqrt{28.67\left(\frac{1}{6} + \frac{1}{6}\right)} = 6.59$$

If we use the t value of 2.602 instead of 2.131, we obtain a value of LSD that we will refer to as the Bonferroni significant difference, denoted BSD:

$$BSD = 2.602 \sqrt{28.67\left(\frac{1}{6} + \frac{1}{6}\right)} = 8.04$$

Thus, to reject the hypothesis that the two means are equal using Fisher's LSD with the Bonferroni adjustment, we must observe a difference between the sample means of more than 8.04 points. Note that this is a larger difference between the two sample means than would be required when using Fisher's LSD procedure.

For the NCP example, the difference between the sample means at the Charlotte and Houston plants is $79 - 74 = 5$; since this difference does not exceed BSD = 8.04, we cannot reject the null hypothesis that the two population means are equal. However, since the difference between the Charlotte and San Diego plants is $79 - 66 = 13 > 8.04$, we can reject the hypothesis that the population mean examination score for the Charlotte plant is the same as the population mean examination score for the San Diego plant. Finally, since the difference between the mean at Houston and the mean at San Diego is $74 - 66 = 8$, we cannot conclude that the population means for these two plants are significantly different.

If a problem consisted of five populations and hence 10 possible pairwise comparisons, the Bonferroni adjustment would suggest a comparisonwise Type I error rate of .05/10 = .005. Recall from our discussion of hypothesis testing in Chapter 9 that for a fixed sample size, any decrease in the probability of making a Type I error will result in an increase in the probability of making a Type II error, which corresponds to accepting that the hypothesis that the two means of the populations are equal when in fact they are not equal. As a result, many practitioners feel uncomfortable in performing individual tests with a very low comparisonwise Type I error rate since it carries an increased risk of making a Type II error. In the following discussion we introduce a procedure that was developed to help in this regard.

Tukey's Procedure

Tukey's procedure allows an experimenter to perform tests of all possible pairwise comparisons and still maintain an overall experimentwise Type I error rate such as α_{EW} = .05. The basis for the test is a probability distribution referred to as the "studentized range" distribution. Let $\bar{x}_{max}$ denote the largest sample mean and $\bar{x}_{min}$ denote the smallest sample mean. In situations where the size of each sample is the same size (n) and the population variances are equal, the sampling distribution of

$$q = \frac{\bar{x}_{max} - \bar{x}_{min}}{\sqrt{\dfrac{MSW}{n}}} \tag{13.22}$$

follows a studentized range distribution. Table 11 of Appendix B presents critical values of the studentized range distribution for both α = .05 and α = .01. A portion of this table for α_{EW} = .05 is shown in Table 13.6. To show how the above result can be used to perform tests of all pairwise comparisons, we will illustrate the procedure for the NCP example.

Recall that in the NCP example we had three samples, each of size n = 6. Suppose we want to perform tests on all possible pairwise comparisons and ensure an overall experimentwise error rate of α_{EW} = .05. The columns in Table 13.6 correspond to the number of populations we are sampling from and the rows correspond to the degrees of freedom associated with the best estimate of σ^2, the population variance. Thus, for the NCP example we select column 3, since there are k = 3 populations. We must also select row 15, since the best estimate of σ^2 is MSW and this estimate has 15 degrees of freedom. Note that the critical value is q = 3.67.

Using q = 3.67, n = 6, and MSW = 28.67, we can use (13.22) to determine how large the difference between any two sample means has to be to reject the null hypothesis that the corresponding population means are equal. We refer to this value as *Tukey's significant difference,* denoted TSD.

$$TSD = \text{Tukey's Significant Difference} = q\sqrt{\frac{MSW}{n}} \tag{13.23}$$

For the NCP example

$$TSD = 3.67\sqrt{\frac{28.67}{6}} = 8.02$$

Thus, if the absolute value of the difference between any two sample means exceeds 8.02, there is sufficient evidence to conclude that the corresponding population means are not equal. Note that TSD is greater than LSD and slightly less than the value of BSD; however, Tukey's procedure is similar to using the Bonferroni adjustment in that a larger

TABLE 13.6
Critical Values of the Studentized Range Distribution for $\alpha_{EW} = .05$

Degrees of Freedom	Number of Populations								
	2	3	4	5	6	7	8	9	10
1	18.0	27.0	32.8	37.1	40.4	43.1	45.4	47.4	49.1
2	6.08	8.33	9.80	10.9	11.7	12.4	13.0	13.5	14.0
3	4.50	5.91	6.82	7.50	8.04	8.48	8.85	9.18	9.46
4	3.93	5.04	5.76	6.29	6.71	7.05	7.35	7.60	7.83
5	3.64	4.60	5.22	5.67	6.03	6.33	6.58	6.80	6.99
6	3.46	4.34	4.90	5.30	5.63	5.90	6.12	6.32	6.49
7	3.34	4.16	4.68	5.06	5.36	5.61	5.82	6.00	6.16
8	3.26	4.04	4.53	4.89	5.17	5.40	5.60	5.77	5.92
9	3.20	3.95	4.41	4.76	5.02	5.24	5.43	5.59	5.74
10	3.15	3.88	4.33	4.65	4.91	5.12	5.30	5.46	5.60
11	3.11	3.82	4.26	4.57	4.82	5.03	5.20	5.35	5.49
12	3.08	3.77	4.20	4.51	4.75	4.95	5.12	5.27	5.39
13	3.06	3.73	4.15	4.45	4.69	4.88	5.05	5.19	5.32
14	3.03	3.70	4.11	4.41	4.64	4.83	4.99	5.13	5.25
15	3.01	**3.67**	4.08	4.37	4.59	4.78	4.94	5.08	5.20
16	3.00	3.65	4.05	4.33	4.56	4.74	4.90	5.03	5.15
17	2.98	3.63	4.02	4.30	4.52	4.70	4.86	4.99	5.11
18	2.97	3.61	4.00	4.28	4.49	4.67	4.82	4.96	5.07
19	2.96	3.59	3.98	4.25	4.47	4.65	4.79	4.92	5.04
20	2.95	3.58	3.96	4.23	4.45	4.62	4.77	4.90	5.01
24	2.92	3.53	3.90	4.17	4.37	4.54	4.68	4.81	4.92
30	2.89	3.49	3.85	4.10	4.30	4.46	4.60	4.72	4.82
40	2.86	3.44	3.79	4.04	4.23	4.39	4.52	4.63	4.73
60	2.83	3.40	3.74	3.98	4.16	4.31	4.44	4.55	4.65
120	2.80	3.36	3.68	3.92	4.10	4.24	4.36	4.47	4.56
∞	2.77	3.31	3.63	3.86	4.03	4.17	4.29	4.39	4.47

difference between two sample means must be observed to conclude that the population means are not equal.

The difference between the sample means at the Charlotte and Houston plants is $79 - 74 = 5$; thus, we cannot reject the hypothesis that the two populations means are equal since 5 does not exceed TSD = 8.02. However, since the difference between the Charlotte and San Diego plants is $79 - 66 = 13 > 8.02$, we can reject the hypothesis that the population mean examination score for the Charlotte plant is the same as the population mean examination score for the San Diego plant. Finally, since the difference between the mean at Houston and the mean at San Diego is $74 - 66 = 8$, we cannot conclude that the population means for these two plants are significantly different.

We can also use the value of TSD to compute confidence intervals for each pair of differences. We refer to the resulting set of interval estimates as the *simultaneous confidence interval estimates*. The general expression for computing these interval estimates is as follows:

$$\bar{x}_i - \bar{x}_j \pm q \sqrt{\frac{MSW}{n}} \qquad (13.24)$$

Note that for the NCP example the value of TSD is 8.02, regardless of which two populations we are considering. Thus, the set of simultaneous confidence intervals is

Charlotte–Houston

$$(79 - 74) \pm 8.02 = -3.02 \text{ to } 13.02$$

Charlotte–San Diego

$$(79 - 66) \pm 8.02 = 4.98 \text{ to } 21.02$$

Houston–San Diego

$$(74 - 66) \pm 8.02 = -.02 \text{ to } 16.02$$

Since 0 is contained in the Charlotte–Houston interval and the Houston–San Diego interval, there is not sufficient statistical evidence to conclude that the population means of these plants differ. However, since the Charlotte–San Diego interval does not include the value of 0, we have statistical evidence to conclude that the population means of these two plants are not equal. The key point about using Tukey's procedure to form these confidence intervals is that we are 95% confident that these intervals hold true for all the intervals formed.

Recommendations for Performing Multiple Comparisons

We have introduced three procedures for performing multiple comparisons: Fisher's LSD, Bonferroni adjustment, and Tukey's procedure. In using Fisher's LSD procedure for the NCP example, we concluded that there is a significant difference between the population means for the Charlotte and San Diego plants as well as for the Houston and San Diego plants. When using Fisher's LSD with the Bonferroni adjustment or Tukey's procedure, we concluded that the only difference is between the Charlotte and San Diego plants. Thus, different conclusions can be drawn depending on which procedure is used.

There is considerable controversy in the statistical community as to which procedure is "best." The truth of the matter is that no one procedure is best for all types of problems. However, if the number of pairwise comparisons that you want to make is small, we recommend using the Bonferroni adjustment. If the number of pairwise comparisons is very large, then Tukey's procedure provides a better alternative. For example, suppose you only want to test two or three pairwise comparisons, each of which has been identified prior to looking at the data; in this case, we recommend use of the Bonferroni adjustment. Note however, that this virtually rules out the use of the Bonferroni adjustment as a method for testing all possible pairwise comparisons. If the latter is your objective, we recommend that Tukey's procedure be used.

Recent research has also suggested that in situations where you view the outcome of your analysis as nothing more than a way to suggest what really needs to be investigated in future studies, Fisher's LSD procedure is preferred. That is, in such cases, Fisher's LSD procedure allows us to base our rejection of the hypothesis that the two population means are equal on smaller observed differences than are required using the other two approaches. Thus, there is less likelihood we might discard something that may at a later time show a much more significant difference. Situations like this are often referred to as hypothesis generators.

NOTES &
COMMENTS

> Tukey's procedure is an *unprotected* or *unrestrictive* testing approach. That is, it is not necessary to find a significant difference when using analysis of variance before applying Tukey's procedure. Thus, Tukey's procedure provides an alternative to analysis of variance for testing if the means of k populations are equal. However, to use Tukey's procedure we still have to estimate the population variance using MSW.

☐ ☐ Exercises

Methods

SELF TEST ▶

11. In Exercise 1, 5 observations were selected from each of three populations. For these data, $\bar{x}_1 = 30$, $\bar{x}_2 = 45$, $\bar{x}_3 = 36$, and MSW = 5.5. At the $\alpha = .05$ level of significance, the null hypothesis of equal population means was rejected. In answering the following questions, use $\alpha = .05$.

a. Using Fisher's LSD procedure, test to see if there is a significant difference between the means of populations 1 and 2, populations 1 and 3, and populations 2 and 3.

b. Use Fisher's LSD procedure to develop a 95% confidence interval estimate of the difference between the means of populations 1 and 2.

c. Using the Bonferroni adjustment, which means appear to be different?

d. Using Tukey's procedure, which means appear to be different?

e. Use Tukey's procedure to compute 95% confidence intervals for each pair of differences.

T A B L E 13.7

	Sample 1	Sample 2	Sample 3
	63	82	69
	47	72	54
	54	88	61
	40	66	48
$\bar{x}_j$	51	77	58
s_j^2	96.67	97.34	81.99

12. Four observations were selected from each of three populations. The data obtained is shown in Table 13.7. In answering the following questions, use $\alpha = .05$.

a. Using analysis of variance, test if there is a significant difference among the means of the three populations.

b. Using Fisher's LSD procedure, which means appear to be different?

c. Using the Bonferroni adjustment, which means appear to be different?

d. Using Tukey's procedure, which means appear to be different?

Applications

SELF TEST ▶

13. Refer to Exercise 6. For these data $\bar{x}_1 = 23$, $\bar{x}_2 = 28$, $\bar{x}_3 = 21$, and MSW = 4.89. At the $\alpha = .05$ level of significance, use Fisher's LSD procedure to test for the equality of the means for manufacturers 1 and 3. What conclusion can you make after carrying out this test?

SELF TEST ▶

14. Refer to Exercise 13. Use Tukey's procedure to develop a 95% confidence interval estimate of the difference between the means of population 1 and population 2.

15. In Exercise 7 scores on a 100-question science test were provided for 13-year-old children from South Korea, the Soviet Union, and the United States; the sample means were 78, 71, and 67, respectively, and MSW = 38.0. Use Tukey's procedure to test for the equality of the means for South Korea and the United States. What conclusion can you make after carrying out the test? Use $\alpha_{EW} = .05$

16. Refer to Exercise 15. Use the Bonferroni adjustment to test for the equality of the means for South Korea and the United States.

17. Refer to Exercise 15. Use Tukey's procedure to test for the equality of the means for South Korea and the United States. Use Tukey's procedure to develop a 95% confidence interval for the mean difference between these two countries.

18. Refer to Exercise 9. At the $\alpha = .05$ level of significance, we can conclude that there are differences in the perceptions for marketing managers ($\bar{x}_1 = 5$), marketing research ($\bar{x}_2 = 4.5$), and advertising ($\bar{x}_3 = 6$); for these data MSW = .5. Use the procedures in this section to determine where the differences occur. Use $\alpha = .05$.

13.4   An Introduction to Experimental Design

Statistical studies can be classified as being either experimental or observational. In an *experimental study,* variables of interest are identified. Then, one or more factors in the study are controlled so that data may be obtained about how the factors influence the variables. In *observational* or *nonexperimental* studies, no attempt is made to control the influences of factors on the variable or variables of interest. A survey (see Chapter 21) is perhaps the most common type of observational study.

The NCP example that we used to introduce analysis of variance is an illustration of an observational statistical study. To measure how much NCP employees knew about total quality management, a random sample of 6 employees was selected from each of NCP's three plants and given a quality-awareness examination. The examination scores for these employees were then analyzed using analysis of variance to test the hypothesis that the population mean examination scores were equal for the three plants.

As an example of an experimental statistical study, let us consider the problem facing Chemitech, Inc. Chemitech has developed a new filtration system for municipal water supplies. The components for the new filtration system will be purchased from several different suppliers, and Chemitech will assemble the components at their plant in Columbia, South Carolina. The industrial engineering group has been given the responsibility of determining the best assembly method for the new filtration system. After considering a variety of possible approaches, the group has narrowed the alternatives to three possibilities: method A, method B, and method C. Each of these methods differs in terms of the sequence of steps used to assemble the product. Management at Chemitech would like to determine which assembly method results in the greatest number of filtration systems produced per week.

In the Chemitech experiment the assembly method is referred to as a *factor.* Since there are three assembly methods, or levels, corresponding to this factor, we say that there are three *treatments* associated with this experiment: one treatment corresponds to method A, another to method B, and the third to method C. In general, a factor is just a variable that the experimenter has selected for investigation, and a treatment corresponds to a level of a factor. The Chemitech problem is an example of a single-factor experiment involving a qualitative factor (method of assembly). Other experiments may consist of multiple factors; some may be qualitative and some may be quantitative.

The three assembly methods or treatments define the three populations of interest for the Chemitech experiment. One population corresponds to all Chemitech employees who use assembly method A, another corresponds to those who use method B, and the third to those who use method C. Note that for each population the random variable of interest (the response variable) is the number of filtration systems assembled per week, and the primary statistical objective for the experiment is to determine whether the mean number of units produced per week is the same for all three populations. In experimental design terminology, the random variable of interest is referred to as the *dependent variable,* the *response variable,* or simply the *response*.

Suppose that a random sample of 3 employees is selected from all assembly workers at the Chemitech production facility. In experimental design terminology, the 3 randomly selected workers are referred to as the *experimental units.* The experimental design that we will use for the Chemitech problem is referred to as a *completely randomized design.* This type of design requires that each of the three assembly methods or treatments be randomly assigned to one of the experimental units or workers. For example, method A might be assigned to the second worker, method B to the first worker, and method C to the third worker. The concept of *randomization,* as illustrated in this example, is an important principle of all experimental designs.

Note that the experiment as described above would only result in one measurement or number of units assembled for each treatment. In other words, we have a sample size of 1 corresponding to each treatment. Thus, to obtain additional data for each assembly method, we must repeat or replicate the basic experimental process. For example, suppose that instead of selecting just 3 workers at random we had selected 15 workers and then randomly assigned each of the three treatments to 5 of the workers. Since each method of assembly is assigned 5 workers, we say that 5 replicates have been obtained. The process of *replication* is another important principle of experimental design. Figure 13.6 shows the completely randomized design for the Chemitech experiment.

Data Collection

Once we are satisfied with the experimental design, we proceed by collecting and analyzing the data. In this case, the employees would be instructed in how to perform the assembly method that they have been assigned to and then would begin assembling the new filtration systems using that method. Suppose that this has been done and the number of units assembled by each employee during 1 week are as shown in Table 13.8. The sample mean number of units produced for the three assembly methods are shown below:

Assembly Method	Mean Number Produced
A	62
B	66
C	52

FIGURE 13.6
Completely Randomized Design for Evaluating the Chemitech Assembly Method Experiment

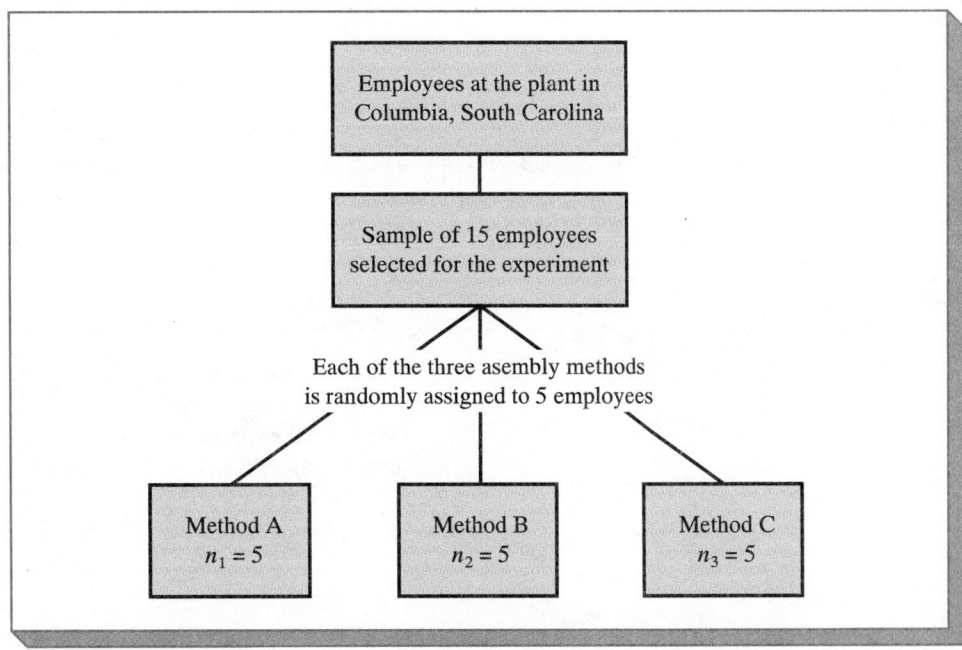

TABLE 13.8
Number of Units Produced for 15 Workers

Observation	Method		
	A	B	C
1	58	58	48
2	64	69	57
3	55	71	59
4	66	64	47
5	67	68	49
Sample mean	62	66	52
Sample variance	27.5	26.5	31.0
Sample standard deviation	5.24	5.15	5.57

From these data it appears that method B may result in higher production rates than either of the other methods.

The real issue is whether the three sample means observed are different enough for us to conclude that the means of the populations corresponding to the three methods of assembly are different. To write this question in statistical terms, we introduce the following notation:

μ_1 = mean number of units produced per week for method A

μ_2 = mean number of units produced per week for method B

μ_3 = mean number of units produced per week for method C

Although we will never know the actual values of μ_1, μ_2, and μ_3, what we want to do is use the sample means to test the following hypotheses:

$$H_0: \mu_1 = \mu_2 = \mu_3$$

$$H_a: \text{Not all population means are equal}$$

The problem that we face in analyzing data from a completely randomized experimental design is the same problem we faced when we first introduced analysis of variance as a method for testing whether the means of more than two populations are equal. In the next section we will show how analysis of variance is applied in problem situations such as this.

**NOTES &
COMMENTS**

1. Randomization in experimental design is the analog of probability sampling in an observational experiment.
2. In many medical experiments potential bias is eliminated by using a double-blind study. In these studies neither the physician applying the treatment nor the subject know which treatment is being applied. Many other types of experiments could benefit from this type of study.

13.5   Completely Randomized Designs

The hypotheses that we want to test when analyzing the data from a completely randomized design are exactly the same as the general form of the hypotheses we presented in Section 13.2.

$$H_0: \mu_1 = \mu_2 = \cdots = \mu_k$$

$$H_a: \text{Not all population means are equal}$$

where

$$\mu_j = \text{mean of the } j\text{th population}$$

We will proceed based on the assumptions stated in Section 13.1 for analysis of variance. Thus, to test for the equality of means in situations where the data have been collected using a completely randomized experimental design, we can use analysis of variance as introduced in Sections 13.1 and 13.2. Recall that analysis of variance requires the calculation of two independent estimates of the population variance σ^2.

Between-Treatments Estimate of Population Variance

In the context of experimental design the between-samples estimate of σ^2 is referred to as the *mean square due to treatments* and is denoted MSTR. This is the same as what we called mean square between, MSB, in Section 13.2. It is also referred to as the *mean square between treatments*. The formula for computing MSTR is as follows:

$$\text{MSTR} = \frac{\sum_{j=1}^{k} n_j(\overline{x}_j - \overline{\overline{x}})^2}{k - 1} \tag{13.25}$$

The numerator in (13.25) is called the *sum of squares between* or *sum of squares due to treatments* and is denoted by SSTR. The denominator $k - 1$ represents the degrees of freedom associated with SSTR.

For the Chemitech data shown in Table 13.8, we obtain the following results (note: $\overline{\overline{x}} = 60$):

$$\text{SSTR} = \sum_{j=1}^{k} n_j(\overline{x}_j - \overline{\overline{x}})^2 = 5(62 - 60)^2 + 5(66 - 60)^2 + 5(52 - 60)^2 = 520$$

$$\text{MSTR} = \frac{\text{SSTR}}{k - 1} = \frac{520}{3 - 1} = 260$$

Within-Treatments Estimate of Population Variance

The second estimate of σ^2 is based on the variation of the sample observations within each sample or treatment. In our discussion of analysis of variance, we referred to this estimate of σ^2 as the within-samples estimate of population variance. This estimate is referred to as the *mean square due to error* and is denoted MSE. This is the same as what we called mean square within, MSW, in Section 13.2. It is also referred to as the *mean square within treatments*. The formula for computing MSE is as follows:

$$\text{MSE} = \frac{\sum\limits_{j=1}^{k} (n_j - 1)s_j^2}{n_T - k} \tag{13.26}$$

The numerator in (13.26) is given the name *sum of squares within* or *sum of squares due to error* and is denoted SSE. The denominator of MSE is referred to as the degrees of freedom associated with the within-treatment variance estimate.

For the Chemitech data shown in Table 13.8, we obtain the following results:

$$\text{SSE} = \sum_{j=1}^{k} (n_j - 1)s_j^2 = 4(27.5) + 4(26.5) + 4(31) = 340$$

$$\text{MSE} = \frac{\text{SSE}}{n_T - k} = \frac{340}{15 - 3} = 28.33$$

Comparing the Variance Estimates: The F Test

If the null hypothesis is true and the ANOVA assumptions are valid, the sampling distribution of MSTR/MSE is an F distribution with numerator degrees of freedom equal to $k - 1$ and denominator degrees of freedom equal to $n_T - k$. Recall also that if the means of the k populations are not equal, the value of MSTR/MSE will be inflated because MSTR overestimates σ^2. Hence we will reject H_0 if the resulting value of MSTR/MSE appears to be too large to have been selected at random from an F distribution with degrees of freedom $k - 1$ in the numerator and $n_T - k$ in the denominator.

For the Chemitech problem the value of $F = \text{MSTR/MSE} = 260/28.33 = 9.18$. The critical F value is based upon 2 numerator degrees of freedom and 12 denominator degrees of freedom. For a .05 level of significance, Table 4 of Appendix B shows a value of $F_{.05} = 3.89$. Since the observed value of F is greater than the critical value, we reject the null hypothesis and conclude that not all the population means are equal.

The ANOVA Table for Completely Randomized Designs

Using the terminology we have introduced for the completely randomized experimental design, we can now write the result which shows how the total sum of squares, SST, is partitioned:

$$\text{SST} = \text{SSTR} + \text{SSE} \tag{13.27}$$

Note that this result also holds true for the degrees of freedom associated with each of these sums of squares; that is, the total degrees of freedom is the sum of the degrees of freedom associated with SSTR and SSE. The general form of the ANOVA table for a completely randomized design is shown in Table 13.9; Table 13.10 shows the corresponding ANOVA table for the Chemitech problem.

Pairwise Comparisons

We can use Tukey's procedure to test all possible pairwise comparisons for the Chemitech problem. At the 5% level of significance, Table 13.6 can be used to provide the critical value of the studentized range distribution with $k = 3$ populations and 12 degrees of freedom; the value obtained is $q = 3.77$. Using MSE = 28.33 in place of MSW in (13.23), we obtain Tukey's significant difference.

TABLE 13.9
ANOVA Table for a Completely Randomized Design

Source of Variation	Sum of Squares	Degrees of Freedom	Mean Square	F
Treatments	SSTR	$k - 1$	$MSTR = \dfrac{SSTR}{k - 1}$	$\dfrac{MSTR}{MSE}$
Error	SSE	$n_T - k$	$MSE = \dfrac{SSE}{n_T - k}$	
Total	SST	$n_T - 1$		

TABLE 13.10
ANOVA Table for the Chemitech Problem

Source of Variation	Sum of Squares	Degrees of Freedom	Mean Square	F
Treatments	520	2	260.00	9.18
Error	340	12	28.33	
Total	860	14		

$$TSD = q\sqrt{\frac{MSE}{n}} = 3.77\sqrt{\frac{28.33}{5}} = 8.97$$

Thus, if the magnitude of the difference between any two sample means exceeds 8.97, we can reject the hypothesis that the corresponding population means are equal. For the Chemitech data in Table 13.8, we obtain the following results:

Sample Differences	Significant?
Method A − Method B = 62 − 66 = − 4	No
Method A − Method C = 62 − 52 = 10	Yes
Method B − Method C = 66 − 52 = 14	Yes

Thus, the difference in the population means is attributable to the difference between the means for method A and method C and the difference between the means for Method B and Method C. Thus, methods A and B are preferred to method C. However, more testing should be done to compare method A with method B. Based on the current study there is not sufficient evidence to conclude that these two methods differ.

NOTES & COMMENTS

The computational aspect of the analysis of variance procedure is devoted primarily to computing the appropriate sums of squares. When a hand calculator is used to compute the sum of squares, some computational help can be obtained by using alternate forms of the sums-of-squares formulas. In Appendix 13.1 we provide a step-by-step procedure that uses these revised formulas to compute the sums of squares for a completely randomized design.

☐ ☐ **Exercises**

SELF TEST ▶

TABLE 13.11

Observation	Treatment		
	A	B	C
1	162	142	126
2	142	156	122
3	165	124	138
4	145	142	140
5	148	136	150
6	174	152	128
$\bar{x}_j$	156	142	134
s_j^2	164.4	131.2	110.4

Methods

19. The data in Table 13.11 are from a completely randomized design.
a. Compute the sum of squares between treatments.
b. Compute the mean square between treatments.
c. Compute the sum of squares due to error.
d. Compute the mean square due to error.
e. At the $\alpha = .05$ level of significance, test if the means for the three treatments are equal.

20. Refer to Exercise 19.
a. Set up the ANOVA table.
b. At the $\alpha = .05$ level of significance, use Tukey's procedure to test all possible comparisons. What conclusion can you make after carrying out this procedure?

21. In a completely randomized experimental design, 7 experimental units were used for each of the 5 levels of the factor. Complete the ANOVA table shown.

Source of Variation	Sum of Squares	Degrees of Freedom	Mean Square	F
Treatments	300			
Error				
Total	480			

22. Refer to Exercise 21.
a. What hypotheses are implied in this problem?
b. At the $\alpha = .05$ level of significance, can we reject the null hypothesis in (a)? Explain.

23. In an experiment designed to test the output levels of three different treatments, the following results were obtained: SST = 400, SSTR = 150, $n_T = 18$. Set up the ANOVA table and test for any significant difference between the mean output levels of the three treatments. Use $\alpha = .05$.

24. In completely randomized experimental design, 12 experimental units were used for the first treatment, 15 experimental units for the second treatment, 20 experimental units for the third treatment, and 10 experimental units for the fourth treatment. Complete the analysis of variance table shown. Using a .05 level of significance, is there a significant difference between the treatments?

Source of Variation	Sum of Squares	Degrees of Freedom	Mean Square	F
Treatments	1200			
Error				
Total	1800			

TABLE 13.12

	Treatment		
	A	B	C
	136	107	92
	120	114	82
	113	125	85
	107	104	101
	131	107	89
	114	109	117
	129	97	110
	102	114	120
		104	98
		89	106
$\bar{x}_j$	119	107	100
s_j^2	146.86	96.44	173.78

25. Develop the analysis of variance computations for the experimental design shown in Table 13.12. Using $\alpha = .05$, is there a significant difference between the treatment means?

Applications

26. Three different methods for assembling a product were proposed by an industrial engineer. To investigate the number of units assembled correctly using each method, 30 employees were randomly selected and randomly assigned to the three proposed methods such that 10 workers were associated with each method. The number of units assembled correctly was recorded, and the analysis of variance procedure was applied to the resulting data set. The following results were obtained: SST = 10,800, SSTR = 4560.

a. Set up the ANOVA table for this problem.

b. Using $\alpha = .05$, test for any significant difference in the means for the three assembly methods.

27. In an experiment designed to test the breaking strength of four types of cables, the following results were obtained: SST = 85.05, SSTR = 61.64, $n_T = 24$. Set up the ANOVA table and test for any significant difference in the mean breaking strength of the four cables. Use $\alpha = .05$.

28. To study the effect of temperature upon yield in a chemical process, five batches were produced under each of three temperature levels. The results are given. Construct an analysis of variance table. Using a .05 level of significance, test to see if the temperature level appears to have an effect upon the mean yield of the process.

	Temperature		
	50°C	60°C	70°C
	34	30	23
	24	31	28
	36	34	28
	39	23	30
	32	27	31
$\bar{x}_j$	33	29	28
s_j^2	32	17.5	9.5

JUDGMENT

29. Auditors must make judgments concerning various aspects of an audit based on their own direct experience, indirect experience, or a combination of the two. In a study, auditors were asked to make judgments about the frequency of errors to be found in an audit (*Journal of Accounting Research,* Autumn 1988). The judgments by the auditors were then compared to the actual results. Suppose the data in Table 13.13 were obtained from a similar study; lower scores indicate better judgments. Using $\alpha = .05$, test to see if the basis for the judgment affects the quality of the judgment. What is your conclusion?

PAINT

30. Four different paints are advertised as having the same drying time. To check the manufacturer's claims, five paint samples were tested for each make of paint. The time in minutes until the paint was dry enough for a second coat to be applied was recorded. The following data were obtained:

TABLE 13.13

	Direct	Indirect	Combination
	17.0	16.6	25.2
	18.5	22.2	24.0
	15.8	20.5	21.5
	18.2	18.3	26.8
	20.2	24.2	27.5
	16.0	19.8	25.8
	13.3	21.2	24.2
$\bar{x}_j$	17.0	20.4	25.0
s_j^2	5.01	6.26	4.01

	Paint 1	Paint 2	Paint 3	Paint 4
	128	144	133	150
	137	133	143	142
	135	142	137	135
	124	146	136	140
	141	130	131	153
$\bar{x}_j$	133	139	136	144
s_j^2	47.5	50	21	54.5

At the $\alpha = .05$ level of significance test to see if the mean drying time is the same for each type of paint.

31. Three top-of-the-line intermediate-sized automobiles manufactured in the United States have been test-driven and compared on a variety of criteria by a well-known automotive magazine. In the area of gasoline mileage performance, five automobiles of each brand were each test-driven 500 miles; the miles per gallon data obtained are shown in the table below.

	Automobile		
	A	B	C
	19	19	24
	21	20	26
	20	22	23
	19	21	25
	21	23	27
$\bar{x}_j$	20	21	25
s_j^2	1	2.5	2.5

Use the analysis of variance procedure with $\alpha = .05$ to determine if there is a significant difference in the mean miles per gallon for the three types of automobiles.

32. Refer to Exercise 29. Use Tukey's procedure to test all possible comparisons. What conclusion can you make after carrying out this procedure? Use $\alpha = .05$.

33. Refer to Exercise 31. Use Tukey's procedure to test all possible comparisons. What conclusion can you make after carrying out this procedure? Use $\alpha = .05$.

13.6   Randomized Block Design

Thus far we have considered the completely randomized experimental design. Recall that to test for a difference among treatment means, we computed an F value using the ratio

$$F = \frac{\text{MSTR}}{\text{MSE}} \qquad (13.28)$$

A problem can arise whenever differences due to extraneous factors (ones not considered in the experiment) cause the MSE term in this ratio to become large. In such cases, the F value in (13.28) can become small, signaling no difference among treatment means when in fact such a difference exists.

In this section we present an experimental design referred to as a *randomized block design*. The purpose of this design is to control some of the extraneous sources of variation by removing such variation from the MSE term. This design tends to provide a better estimate of the true error variance and leads to a more powerful hypothesis test in terms of the ability to detect differences among treatment means. To illustrate, let us consider a stress study for air traffic controllers.

Air Traffic Controller Stress Test

A study directed at measuring the fatigue and stress on air traffic controllers has resulted in proposals for modification and redesign of the controller's work station. After consideration of several designs for the work station, three specific alternatives have been selected as having the best potential for reducing controller stress. The key question is: To what extent do the three alternatives differ in terms of their effect on controller stress? To answer this question we need to design an experiment that will provide measurements of air traffic controller stress under each alternative.

In a completely randomized design a random sample of controllers would be assigned to each work station alternative. However, it is believed that controllers differ substantially in terms of their ability to handle stressful situations. What is high stress to one controller might be only moderate or even low stress to another. Thus, when considering the within-group source of variation (MSE), we must realize that this variation includes both random error and error due to individual controller differences. In fact, for this study, management expected controller variability to be a major contributor to the MSE term.

One way to separate the effect of the individual differences is to use a randomized block design. This design will identify the variability stemming from individual controller differences and remove it from the MSE term. The randomized block design calls for a single sample of controllers. Each controller in the sample is tested using each of the three work station alternatives. In experimental design terminology, the work station is the *factor of interest,* and the controllers are referred to as the *blocks*. The three treatments or populations associated with the work station factor correspond to the three work station alternatives. For simplicity, we will refer to the work station alternatives as system A, system B, and system C.

The *randomized* aspect of the randomized block design refers to the fact that the order in which the treatments (systems) are assigned to the controllers is chosen randomly. If every controller were to test the three systems in the same order, any observed difference in systems might be due to the order of the test rather than to true differences in the systems.

To provide the necessary data, the three types of work stations were installed at the Cleveland Control Center in Oberlin, Ohio. Six controllers were selected at random and assigned to operate each of the systems. A follow-up interview and a medical examination of each controller participating in the study provided a measure of the stress for each controller on each system. The data are shown in Table 13.14.

A summary of the stress data collected is shown in Table 13.15. In this table we have included column totals (treatments) and row totals (blocks) as well as some sample means that will be helpful in making the sum of squares computations for the ANOVA procedure. Since lower stress values are viewed as better, the sample data available would seem to favor system B with its mean stress rating of 13. However, the usual question remains: Do the sample results justify the conclusion that the mean stress levels for the three systems differ? That is, are the differences statistically significant? An analysis of variance computation similar to the one performed for the completely randomized design can be used to answer this statistical question.

TABLE 13.14

A Randomized Block Design for the Air Traffic Controller Stress Test

| | | Treatments | | |
		System A	*System B*	*System C*
	(Controller 1)	15	15	18
	(Controller 2)	14	14	14
Blocks	*(Controller 3)*	10	11	15
	(Controller 4)	13	12	17
	(Controller 5)	16	13	16
	(Controller 6)	13	13	13

TABLE 13.15

Summary of Stress Data for the Air Traffic Controller Stress Test

		Treatments			Row or Block Totals	Block Means
		System A	System B	System C		
Blocks	Controller 1	15	15	18	48	$\bar{x}_1. = 48/3 = 16.0$
	Controller 2	14	14	14	42	$\bar{x}_2. = 42/3 = 14.0$
	Controller 3	10	11	15	36	$\bar{x}_3. = 36/3 = 12.0$
	Controller 4	13	12	17	42	$\bar{x}_4. = 42/3 = 14.0$
	Controller 5	16	13	16	45	$\bar{x}_5. = 45/3 = 15.0$
	Controller 6	13	13	13	39	$\bar{x}_6. = 39/3 = 13.0$
Column or Treatment Totals		81	78	93	252	$\bar{\bar{x}} = \dfrac{252}{18} = 14.0$
Treatment Means		$\bar{x}._1 = \dfrac{81}{6}$ $= 13.5$	$\bar{x}._2 = \dfrac{78}{6}$ $= 13.0$	$\bar{x}._3 = \dfrac{93}{6}$ $= 15.5$		

The ANOVA Procedure for a Randomized Block Design

The ANOVA procedure for the randomized block design requires us to partition the sum of squares total (SST) into three groups: sum of squares due to treatments, sum of squares due to blocks, and sum of squares due to error. The formula for this partitioning is as follows:

$$\text{SST} = \text{SSTR} + \text{SSBL} + \text{SSE} \tag{13.29}$$

This sum of squares partition is summarized in the ANOVA table for the randomized block design as shown in Table 13.16. The notation used in this table is as follows.

$$k = \text{the number of treatments}$$

$$b = \text{the number of blocks}$$

$$n_T = \text{the total sample size } (n_T = kb)$$

Note that the ANOVA table in Table 13.16 also shows how the $n_T - 1$ total degrees of freedom are partitioned such that $k - 1$ go to treatments, $b - 1$ go to blocks, and $(k - 1)(b - 1)$ go to the error term. The mean square column shows the sum of squares divided by the degrees of freedom, and $F = \text{MSTR/MSE}$ is the F ratio used to test for a significant difference among the treatment means. The primary contribution of the randomized block design is that, by including blocks, we have removed the individual controller differences from the MSE term and obtained a more powerful test for the stress differences in the three work station alternatives.

Computations and Conclusions

To compute the F statistic needed to test for a difference among treatment means using a randomized block design, we need to compute MSTR and MSE. To calculate these two mean squares, we must first compute SSTR and SSE; in doing so, we will also compute

TABLE 13.16

ANOVA Table for the Randomized Block Design with k Treatments and b Blocks

Source of Variation	Sum of Squares	Degrees of Freedom	Mean Square	F
Treatments	SSTR	$k-1$	$MSTR = \dfrac{SSTR}{k-1}$	$\dfrac{MSTR}{MSE}$
Blocks	SSBL	$b-1$	$MSBL = \dfrac{SSBL}{b-1}$	
Error	SSE	$(k-1)(b-1)$	$MSE = \dfrac{SSE}{(k-1)(b-1)}$	
Total	SST	$n_T - 1$		

SSBL and SST. To simplify the presentation, we will perform the calculations using four steps. In addition to k, b, and n_T as previously defined, the following notation is used:

x_{ij} = value of the observation under treatment j in block i

$\bar{x}_{.j}$ = sample mean of the jth treatment

$\bar{x}_{i.}$ = sample mean for the ith block

$\bar{\bar{x}}$ = overall sample mean

Step 1.   Compute the total sum of squares (SST):

$$SST = \sum_{i=1}^{b} \sum_{j=1}^{k} (x_{ij} - \bar{\bar{x}})^2 \qquad (13.30)$$

Step 2.   Compute the sum of squares due to treatments (SSTR):

$$SSTR = b \sum_{j=1}^{k} (\bar{x}_{.j} - \bar{\bar{x}})^2 \qquad (13.31)$$

Step 3.   Compute the sum of squares due to blocks (SSBL):

$$SSBL = k \sum_{i=1}^{b} (\bar{x}_{i.} - \bar{\bar{x}})^2 \qquad (13.32)$$

Step 4.   Compute the sum of squares due to error (SSE):

$$SSE = SST - SSTR - SSBL \qquad (13.33)$$

For the air traffic controller data in Table 13.15, these steps lead to the following sum of squares:

Step 1. $SST = (15 - 14)^2 + (15 - 14)^2 + (18 - 14)^2 + \cdots + (13 - 14)^2 = 70$

Step 2. $SSTR = 6[(13.5 - 14)^2 + (13.0 - 14)^2 + (15.5 - 14)^2] = 21$

Step 3. $SSBL = 3[(16 - 14)^2 + (14 - 14)^2 + (12 - 14)^2 + (14 - 14)^2 + (15 - 14)^2 + (13 - 14)^2] = 30$

Step 4. $SSE = 70 - 21 - 30 = 19$

These sums of squares divided by their degrees of freedom provide the corresponding mean square values shown in Table 13.17. The F ratio used to test for differences between

TABLE 13.17
ANOVA Table for the Air Traffic Controller Stress Test

Source of Variation	Sum of Squares	Degrees of Freedom	Mean Square	F
Treatments	21	2	10.5	10.5/1.9 = 5.53
Blocks	30	5	6.0	
Error	19	10	1.9	
Total	70	17		

treatment means is MSTR/MSE = 10.5/1.9 = 5.53. Checking the F values in Table 4 of Appendix B, we find that the critical F value at α = .05 (2 numerator degrees of freedom and 10 denominator degrees of freedom) is 4.10. With F = 5.53, we reject the null hypothesis H_0: $\mu_1 = \mu_2 = \mu_3$ and conclude that the work station designs differ in terms of the mean stress effects on air traffic controllers.

Before leaving this section let us make some general comments about the randomized block design. The blocking as described in this section is referred to as a *complete* block design; the word "complete" indicates that each block is subjected to all k treatments. That is, all controllers (blocks) were tested using all three systems (treatments). Experimental designs employing blocking where some but not all treatments are applied to each block are referred to as *incomplete* block designs. A discussion of incomplete block designs is beyond the scope of this text.

In addition, note that in the air traffic controller stress test, each controller in the study was required to use all 3 systems. While this guarantees a complete block design, in some cases blocking is carried out with "similar" experimental units in each block. For example, assume that in a pretest of air traffic controllers, the population of controllers was divided into groups ranging from extremely high stress individuals to extremely low stress individuals. The blocking could still have been accomplished by having three controllers from each of the stress classifications participate in the study. Each block would then be formed from 3 controllers in the same stress class. The randomized aspect of the block design would be conducted by randomly assigning the 3 controllers in each block to the 3 systems.

Finally, note that the ANOVA table shown in Table 13.17 provides an F value to test for treatment effects but *not* for blocks. The reason is that the experiment was designed to test a single factor—work station design. The blocking based on individual stress differences was conducted to remove this variation from the MSE term. However, the study was not designed to test specifically for individual differences in stress.

Some analysts compute F = MSB/MSE and use this statistic to test for significance of the blocks. Then they use the result as a guide to whether this type of blocking would be desired in future experiments. However, if individual stress difference is to be a factor in the study, a different experimental design should be used. A test of significance on blocks should not be performed to attempt to draw such a conclusion about a second factor.

NOTES &
COMMENTS

1. The matched-samples t test introduced in Chapter 10 is an example of a randomized block design with 2 blocks.
2. Alternate formulas for SST, SSTR, and SSBL can be developed that can ease the computational burden when using hand calculation. In Appendix 13.2 we have included a step-by-step procedure that illustrates the use of these alternate formulas.

Exercises

Methods

34. Consider the experimental results of a randomized block design shown in Table 13.18. Make the calculations necessary to set up the analysis of variance table, and using $\alpha = .05$, test for any significant differences.

35. The following data were obtained for a randomized block design involving treatments and 3 blocks: SST = 430, SSTR = 310, SSBL = 85. Set up the ANOVA table and test for any significant differences. Use $\alpha = .05$.

36. An experiment has been conducted for 4 treatments using 8 blocks. Complete the following analysis of variance table:

TABLE 13.18

		Treatments		
---	---	A	B	C
	1	10	9	8
	2	12	6	5
Blocks	3	18	15	14
	4	20	18	18
	5	8	7	8

Source of Variation	Sum of Squares	Degrees of Freedom	Mean Square	F
Treatments	900			
Blocks	400			
Error				
Total	1800			

Using $\alpha = .05$, test for any significant differences.

Applications

37. An automobile dealer conducted a test to determine if the time needed to complete a minor engine tuneup depends on whether a computerized engine analyzer or an electronic analyzer is used. Because tuneup time varies among compact, intermediate, and full-sized cars, the three types of cars were used as blocks in the experiment. The data obtained are shown below:

		Analyzer	
		Computerized	Electronic
	Compact	50	42
Car	Intermediate	55	44
	Full-size	63	46

Using $\alpha = .05$, test for any significant differences.

38. Five different auditing procedures were compared with respect to total audit time. To control for possible variation due to the person conducting the audit, 4 accountants were selected randomly and treated as blocks in the experiment. The following values were obtained using the ANOVA procedure: SST = 100, SSTR = 45, SSBL = 36. Using $\alpha = .05$, test to see if there is any significant difference in total audit time stemming from the auditing procedure used.

39. An important factor in selecting software for word-processing and data base management systems is the time required to learn how to use a particular system. To evaluate 3 file management systems, a firm designed a test involving 5 different word-processing operators. Since operator variability was believed to be a significant factor, each of the 5 operators was trained on each of the 3 file management systems. The data obtained are shown at the top of the next page.

		System		
		A	B	C
	1	16	16	24
	2	19	17	22
Operator	3	14	13	19
	4	13	12	18
	5	18	17	22

Using $\alpha = .05$, test to see if there is any difference in training time for the three systems.

13.7   Factorial Experiments

The experimental designs we have considered thus far enable statistical conclusions to be drawn about one factor. However, in some experiments we want to draw conclusions about more than one variable or factor. *Factorial experiments* and their corresponding ANOVA computations are valuable designs when simultaneous conclusions are required about two or more factors. The term "factorial" is used because the experimental conditions include all possible combinations of the factors involved. For example, if there are *a* levels of factor A and *b* levels of factor B, the experiment will involve collecting data on *ab* treatment combinations. In this section we will show the analysis of a two-factor factorial experiment. This basic approach can be extended to experiments involving more than two factors.

As an illustration of a two-factor factorial experiment, we will consider a study involving the Graduate Management Admissions Test (GMAT), a standardized test used by graduate schools of business to evaluate an applicant's ability to pursue a graduate program in that field. Scores on the GMAT range from 200 to 800, with higher scores implying higher aptitude.

In an attempt to improve the performance of students on the GMAT exam, a major Texas university is considering offering the following 3 GMAT preparation programs:

1. A 3-hour review session covering the types of questions generally asked on the GMAT.
2. A 1-day program covering relevant exam material, along with the taking and grading of a sample exam.
3. An intensive 10-week course involving the identification of each student's weaknesses and the setting up of individualized programs for improvement.

Thus, one factor in this study is the GMAT preparation program, which has 3 treatments: 3-hour review, 1-day program, and 10-week course. Before selecting the preparation program to adopt, further study will be conducted to determine how the proposed programs affect GMAT scores.

In addition, it was noted that the GMAT is usually taken by students from three colleges: the College of Business, the College of Engineering, and the College of Arts and Sciences. Thus, also of interest in the experiment is whether or not a student's undergraduate college affects the GMAT score. This second factor, undergraduate college, also has 3 treatments: business, engineering, and arts and sciences. The factorial design for this experiment with 3 treatments corresponding to factor A, the preparation program, and 3 treatments corresponding to factor B, the undergraduate college, will have total of

TABLE 13.19
Nine Treatment Combinations for the Two-Factor GMAT Experiment

		Factor B: College		
		Business	*Engineering*	*Arts and Sciences*
Factor A:	*3-hour review*	1	2	3
Preparation	*1-day program*	4	5	6
Program	*10-week course*	7	8	9

$3 \times 3 = 9$ treatment combinations. These treatment combinations or experimental conditions are summarized in Table 13.19.

Assume that a sample of 2 students will be selected corresponding to each of the 9 treatment combinations shown in Table 13.19: 2 business students will take the 3-hour review, 2 will take the 1-day program, and 2 will take the 10-week course. In addition, 2 engineering students and 2 arts and sciences students will take each of the 3 preparation programs. In experimental design terminology, the sample size of 2 for each treatment combination indicates that we have 2 replications. Additional replications and an increased sample size could easily be made, but we elected not to do so to minimize the computational aspects for this illustration.

This experimental design requires that 6 students who plan to attend graduate school be randomly selected from *each* of the 3 undergraduate colleges. Then 2 students from each college should be assigned randomly to each preparation program, resulting in a total of 18 students being used in the study.

Let us assume that the students have been randomly selected, have participated in the preparation program, and have taken the GMAT. The scores obtained are shown in Table 13.20.

The analysis of variance computations using the data in Table 13.20 will provide answers to the following questions:

- **Main effect (factor A):** Do the preparation programs differ in terms of effect on GMAT scores?
- **Main effect (factor B):** Do the undergraduate colleges differ in terms of student ability to perform on the GMAT?
- **Interaction effect (factors A and B):** Do students in some colleges do better on one type of preparation program while others do better on a different type of preparation program?

TABLE 13.20
GMAT Scores for the Two-Factor Experiment

		Factor B: College		
		Business	*Engineering*	*Arts and Sciences*
	3-hour review	500 580	540 460	480 400
Factor A: **Preparation** **Program**	*1-day program*	460 540	560 620	420 480
	10-week course	560 600	600 580	480 410

The term *interaction* refers to a new effect that we can now study because we have used a factorial experiment. If the interaction effect has a significant impact on the GMAT scores, it will mean that the effect of the type of preparation program depends on the undergraduate college.

The ANOVA Procedure

The ANOVA procedure for the two-factor factorial experiment is similar to the completely randomized experiment and the randomized block experiment in that we once again partition the sum of squares and the degrees of freedom into their respective sources. The formula for partitioning the sum of squares for the two-factor factorial experiments is as follows:

$$SST = SSA + SSB + SSAB + SSE \qquad (13.34)$$

The partitioning of the sum of squares and degrees of freedom is summarized in Table 13.21. The following notation is used:

a = number of levels of factor A

b = number of levels of factor B

r = number of replications

n_T = total number of observations taken in the experiment; $n_T = abr$

Computations and Conclusions

To compute the F statistics needed to test for the significance of factor A, factor B, and interaction, we need to compute MSA, MSB, MSAB, and MSE. To calculate these four mean squares, we must first compute SSA, SSB, SSAB, and SSE; in doing so we will also compute SST. To simplify the presentation, we will perform the calculations using five steps. In addition to a, b, r, and n_T as previously defined, the following notation is used:

x_{ijk} = observation corresponding to the kth replicate taken from treatment i of factor A and treatment j of factor B

$\bar{x}_{i.}$ = sample mean for the observations in treatment i (factor A)

TABLE 13.21

ANOVA Table for the Two-Factor Factorial Experiment with r Replications

Source of Variation	Sum of Squares	Degrees of Freedom	Mean Square	F
Factor A	SSA	$a - 1$	$MSA = \dfrac{SSA}{a - 1}$	$\dfrac{MSA}{MSE}$
Factor B	SSB	$b - 1$	$MSB = \dfrac{SSB}{b - 1}$	$\dfrac{MSB}{MSE}$
Interaction	SSAB	$(a - 1)(b - 1)$	$MSAB = \dfrac{SSAB}{(a - 1)(b - 1)}$	$\dfrac{MSAB}{MSE}$
Error	SSE	$ab(r - 1)$	$MSE = \dfrac{SSE}{ab(r - 1)}$	
Total	SST	$n_T - 1$		

$\overline{x}_{.j}$ = sample mean for the observations in treatment j (factor B)

$\overline{x}_{ij}$ = sample mean for the observations corresponding to the combination of treatment i (factor A) and treatment j (factor B)

$\overline{\overline{x}}$ = overall sample mean of all n_T observations

Step 1. Compute the total sum of squares:

$$SST = \sum_{i=1}^{a} \sum_{j=1}^{b} \sum_{k=1}^{r} (x_{ijk} - \overline{\overline{x}})^2 \tag{13.35}$$

Step 2. Compute the sum of squares for factor A:

$$SSA = br \sum_{i=1}^{a} (\overline{x}_{i.} - \overline{\overline{x}})^2 \tag{13.36}$$

Step 3. Compute the sum of squares for factor B:

$$SSB = ar \sum_{j=1}^{b} (\overline{x}_{.j} - \overline{\overline{x}})^2 \tag{13.37}$$

Step 4. Compute the sum of squares for interaction:

$$SSAB = r \sum_{i=1}^{a} \sum_{j=1}^{b} (\overline{x}_{ij} - \overline{x}_{i.} - \overline{x}_{.j} + \overline{\overline{x}})^2 \tag{13.38}$$

Step 5. Compute the sum of squares due to error:

$$SSE = SST - SSA - SSB - SSAB \tag{13.39}$$

Table 13.22 shows the data collected in the experiment, along with the various sums that will help us with the sum of squares computations. Using (13.35) to (13.39) we have the following sum of squares for the GMAT two-factor factorial experiment:

Step 1. $SST = (500 - 515)^2 + (580 - 515)^2 + (540 - 515)^2 + \cdots + (410 - 515)^2 = 82,450$

Step 2. $SSA = (3)(2)[(493.33 - 515)^2 + (513.33 - 515)^2 + (538.33 - 515)^2] = 6100$

Step 3. $SSB = (3)(2)[(540 - 515)^2 + (560 - 515)^2 + (445 - 515)^2] = 45,300$

Step 4. $SSAB = 2[(540 - 493.33 - 540 + 515)^2 + (500 - 493.33 - 560 + 515)^2 + \cdots + (445 - 538.33 - 445 + 515)^2] = 11,200$

Step 5. $SSE = 82,450 - 6100 - 45,300 - 11,200 - 19,850$

These sums of squares divided by their corresponding degrees of freedom, as shown in Table 13.23, provide the appropriate mean square values for testing the two main effects (preparation program and undergraduate college) and the interaction effect. The F ratio used to test for differences among preparation programs is 1.38. The critical F value at $\alpha = .05$ (with 2 numerator degrees of freedom and 9 denominator degrees of freedom) is 4.26. With $F = 1.38$, we cannot reject the null hypothesis and must conclude that there is not a significant difference in the 3 preparation programs. However, for the undergraduate college effect, $F = 10.27$ exceeds the critical F value of 4.26. Thus, the analysis of variance results allow us to conclude that there is a difference in GMAT test scores among the 3 undergraduate colleges; that is, the 3 undergraduate colleges do not provide the same

TABLE 13.22 GMAT Summary Data for the Two-Factor Experiment

		Factor B: College			Row Totals	Factor A Means
		Business	Engineering	Arts and Sciences		
Treatment combination totals →						
Factor A: Preparation Program	3-hour review	500 580 1080 $\bar{x}_{11} = \frac{1080}{2} = 540$	540 460 1000 $\bar{x}_{12} = \frac{1000}{2} = 500$	480 400 880 $\bar{x}_{13} = \frac{880}{2} = 440$	2960	$\bar{x}_{1.} = \frac{2960}{6} = 493.33$
	1-day program	460 540 1000 $\bar{x}_{21} = \frac{1000}{2} = 500$	560 620 1180 $\bar{x}_{22} = \frac{1180}{2} = 590$	420 480 900 $\bar{x}_{23} = \frac{900}{2} = 450$	3080	$\bar{x}_{2.} = \frac{3080}{6} = 513.33$
	10-week course	560 600 1160 $\bar{x}_{31} = \frac{1160}{2} = 580$	600 580 1180 $\bar{x}_{32} = \frac{1180}{2} = 590$	480 410 890 $\bar{x}_{33} = \frac{890}{2} = 445$	3230	$\bar{x}_{3.} = \frac{3230}{6} = 538.33$
Column Totals		3240	3360	2670	9270 ——— Overall total	
Factor B Means		$\bar{x}_{.1} = \frac{3240}{6} = 540$	$\bar{x}_{.2} = \frac{3360}{6} = 560$	$\bar{x}_{.3} = \frac{2670}{6} = 445$		$\bar{\bar{x}} = \frac{9270}{18} = 515$

TABLE 13.23
ANOVA Table for the Two-Factor GMAT Study

Source of Variation	Sum of Squares	Degrees of Freedom	Mean Square	F
Factor A	6,100	2	3,050	3050/2206 = 1.38
Factor B	45,300	2	22,650	22,650/2206 = 10.27
Interaction	11,200	4	2,800	2800/2206 = 1.27
Error	19,850	9	2,206	
Total	82,450	17		

preparation for performance on the GMAT. Finally, the interaction F value of $F = 1.27$ (critical F value = 3.63 at $\alpha = .05$) means that we cannot identify a significant interaction effect. Thus, there is no reason to believe that the 3 preparation programs differ in their ability to prepare students from the different colleges for the GMAT.

Undergraduate college was found to be a significant factor. Checking the calculations in Table 13.22 we see that the sample means are as follows: business students $\bar{x}_{.1} = 540$, engineering students $\bar{x}_{.2} = 560$, and arts and sciences students $\bar{x}_{.3} = 445$. Tests on individual treatment means can be conducted; yet after reviewing the 3 sample means we would anticipate no difference in preparation for business and engineering graduates. However, the arts and sciences students appear to be significantly less prepared for the GMAT than students in the other colleges. Perhaps this observation will lead the university to consider other options for assisting these students in preparing for graduate management admission tests.

Because of the computational effort involved in any modest to large-size factorial experiment, the computer usually plays an important role in making and summarizing the analysis of variance computations. The computer printout for the analysis of variance of the GMAT two-factor factorial experiment is shown in Figure 13.7.

NOTES & COMMENTS

Alternate formulas for SST, SSA, SSB, SSAB, and SSE can be developed that can ease the computational burden when using hand calculation. In Appendix 13.3 we have included a step-by-step procedure that illustrates the use of these alternate formulas.

FIGURE 13.7
Computer Output for the GMAT Two-Factor Design

```
ANALYSIS OF VARIANCE   GMAT

SOURCE         DF       SS        MS
FACTOR A        2       6100      3050
FACTOR B        2       45300     22650
INTERACTION     4       11200     2800
ERROR           9       19850     2206
TOTAL          17       82450
```

❑ ❑ Exercises

Methods

SELF TEST ▶

40. A factorial experiment involving 3 levels of factor A and 2 levels of factor B resulted in the following data:

		Factor B		
		Level 1	*Level 2*	*Level 3*
Factor A	*Level 1*	135	90	75
		165	66	93
	Level 2	125	127	120
		95	105	136

Test for any significant main effect and any interaction. Use $\alpha = .05$.

41. The calculations for a factorial experiment involving 4 levels of factor A, 3 levels of factor B, and 3 replications resulted in the following data: SST = 280, SSA = 26, SSB = 23, SSAB = 175. Set up the ANOVA table and test for any significant main effects and any interaction effect. Use $\alpha = .05$.

Applications

42. A mail-order catalog firm designed a factorial experiment to test the effect of the size of a magazine advertisement and the advertisement design on the number of catalog requests received (1000s). Three advertising designs and two different-size advertisements were considered. The data obtained are shown in Table 13.24. Use the ANOVA procedure for factorial designs to test for any significant effects due to type of design, size of advertisement, or interaction. Use $\alpha = .05$.

43. An amusement park has been studying methods for decreasing the waiting time (minutes) on rides by loading and unloading riders more efficiently. Two alternative loading/unloading methods have been proposed. To account for potential differences due to the type of ride and the possible interaction between the method of loading and unloading and the type of ride, a factorial experiment was designed. Using the data shown below, test for any significant effect due to the loading and unloading method, the type of ride, and interaction. Use $\alpha = .05$.

44. Jack's Restaurant is considering a new specialty sandwich. To determine the effect of sandwich price and sandwich size on sales, the new sandwich was test-marketed in selected company restaurants. The data in terms of the number of sandwiches sold per day, are shown in Table 13.25. Test for any significant differences due to price, size, and interaction. Use $\alpha = .05$.

TABLE 13.24

		Size of Advertisement	
		Small	*Large*
Design	A	8	12
		12	8
	B	22	26
		14	30
	C	10	18
		18	14

TABLE 13.25

		Price		
		$1.49	*$1.79*	*$1.99*
Size	¹⁄₄ *pound*	955	845	820
		985	860	845
	¹⁄₃ *pound*	945	910	860
		875	905	935

		Type of Ride		
		Roller Coaster	*Screaming Demon*	*Log Flume*
Method 1		41	52	50
		43	44	46
Method 2		49	50	48
		51	46	44

Summary

In this chapter we have shown how analysis of variance can be used to test for differences among means of several populations or treatments. In addition, we introduced the single-factor completely randomized, the randomized block, and the two-factor factorial experimental designs. The completely randomized design and the randomized block designs are used to draw conclusions about differences in the means of a single factor. The primary purpose of blocking in the randomized block design is to remove extraneous sources of variation from the error term. This blocking provides a better estimate of the error variance and a better test to determine whether the population or treatment means of the factor differed significantly.

We showed that the basis for the statistical tests used in analysis of variance and experimental design is the development of independent estimates of the population variance σ^2. In the single-factor case, one estimator is based on the variation between the treatments; this estimator provides an unbiased estimate of σ^2 only if the means $\mu_1, \mu_2, \ldots, \mu_k$ are all equal. A second estimator of σ^2 is based upon the variation of the observations within each sample; this estimator will always provide an unbiased estimate of σ^2. By computing the ratio of these two estimators using the F distribution, we developed a rejection rule for determining whether or not to reject the null hypothesis that the population or treatment means are equal. In all the experimental designs considered, the partitioning of the sum of squares and degrees of freedom into their various sources enabled us to compute the appropriate values for making the analysis of variance calculations and tests. We also showed how Fisher's LSD procedure, the Bonferroni adjustment, and Tukey's procedure can be used to perform pairwise comparisons used to determine which means are different.

Glossary

Analysis of variance (ANOVA) A statistical procedure for determining whether the means of several different populations are equal.

ANOVA table A table used to summarize the analysis of variance computations and results. It contains columns showing the source of variation, the degrees of freedom, the sum of squares, the mean squares, and the F values.

Partitioning The process of allocating the total sum of squares and degrees of freedom into the various components.

Multiple comparison procedures Statistical procedures used to conduct statistical comparisons between pairs of the population means or treatments.

Factor Another word for the variable of interest in an experiment.

Treatment Different levels of a factor.

Single-factor experiment An experiment involving only one factor with k populations or treatments.

Experimental units The objects of interest in the experiment.

Completely randomized design An experimental design where the treatments are randomly assigned to the experimental units.

Mean square The sum of squares divided by its corresponding degrees of freedom. This quantity is used in the F ratio to determine if significant differences among means exist.

Blocking The process of using the same or similar experimental units for all treatments. The purpose of blocking is to remove a source of variation from the error term and hence provide a more powerful test for a difference in population or treatment means.

Randomized block design An experimental design employing blocking. The experimental unit(s) within a block are randomly ordered for the treatments.

Factorial experiments An experimental design that permits statistical conclusions about two or more factors. All levels of each factor are considered with all levels of the other factors to specify the experimental conditions for the experiment.

Replication The number of times each experimental condition is repeated in an experiment. It is the sample size associated with each treatment combination.

Interaction The effect produced when the levels of one factor interact with the levels of another factor in influencing the response variable.

Key Formulas

Analysis of Variance

jth Sample Mean

$$\bar{x}_j = \frac{\displaystyle\sum_{i=1}^{n_j} x_{ij}}{n_j} \tag{13.1}$$

jth Sample Variance

$$s_j^2 = \frac{\displaystyle\sum_{i=1}^{n_j} (x_{ij} - \bar{x}_j)^2}{n_j - 1} \tag{13.2}$$

Overall Sample Mean

$$\bar{\bar{x}} = \frac{\displaystyle\sum_{j=1}^{k} \sum_{i=1}^{n_j} x_{ij}}{n_T} \tag{13.3}$$

$$n_T = n_1 + n_2 + \cdots + n_k \tag{13.4}$$

Mean Square Between

$$MSB = \frac{SSB}{k - 1} \tag{13.7}$$

Sum of Squares Between

$$SSB = \sum_{j=1}^{k} n_j(\bar{x}_j - \bar{\bar{x}})^2 \tag{13.8}$$

Mean Square Within

$$MSW = \frac{SSW}{n_T - k} \tag{13.10}$$

Sum of Squares Within

$$SSW = \sum_{j=1}^{k} (n_j - 1)s_j^2 \tag{13.11}$$

Total Sum of Squares

$$\text{SST} = \sum_{j=1}^{k} \sum_{i=1}^{n_j} (x_{ij} - \overline{\overline{x}})^2 \qquad \textbf{(13.12)}$$

Multiple Comparison Procedures

Test Statistic for Fisher's LSD

$$t = \frac{\overline{x}_1 - \overline{x}_2}{\sqrt{\text{MSW}\left(\dfrac{1}{n_1} + \dfrac{1}{n_2}\right)}} \qquad \textbf{(13.15)}$$

Fisher's LSD

$$\text{LSD} = t_{\alpha/2} \sqrt{\text{MSW}\left(\frac{1}{n_1} + \frac{1}{n_2}\right)} \qquad \textbf{(13.16)}$$

Test Statistic for Tukey's Test

$$q = \frac{\overline{x}_{\max} - \overline{x}_{\min}}{\sqrt{\dfrac{\text{MSW}}{n}}} \qquad \textbf{(13.22)}$$

Tukey's Significant Difference

$$\text{TSD} = \, = q \sqrt{\frac{\text{MSW}}{n}} \qquad \textbf{(13.23)}$$

Completely Randomized Designs

Mean Square Due to Treatments

$$\text{MSTR} = \frac{\displaystyle\sum_{j=1}^{k} n_j (\overline{x}_j - \overline{\overline{x}})^2}{k - 1} \qquad \textbf{(13.25)}$$

Mean Square Due to Error

$$\text{MSE} = \frac{\displaystyle\sum_{j=1}^{k} (n_j - 1)s_j^2}{n_T - k} \qquad \textbf{(13.26)}$$

The F Value

$$F = \frac{\text{MSTR}}{\text{MSE}} \qquad \textbf{(13.28)}$$

Randomized Block Designs

Total Sum of Squares

$$SST = \sum_{i=1}^{b} \sum_{j=1}^{k} (x_{ij} - \overline{\overline{x}})^2 \qquad (13.30)$$

Sum of Squares Due to Treatments

$$SSTR = b \sum_{j=1}^{k} (\overline{x}_{.j} - \overline{\overline{x}})^2 \qquad (13.31)$$

Sum of Squares Due to Blocks

$$SSBL = k \sum_{i=1}^{b} (\overline{x}_{i.} - \overline{\overline{x}})^2 \qquad (13.32)$$

Sum of Squares Due to Error

$$SSE = SST - SSTR - SSBL \qquad (13.33)$$

Factorial Experiments

Total Sum of Squares

$$SST = \sum_{i=1}^{a} \sum_{j=1}^{b} \sum_{k=1}^{r} (x_{ijk} - \overline{\overline{x}})^2 \qquad (13.35)$$

Sum of Squares Due to Factor A

$$SSA = br \sum_{i=1} (\overline{x}_{i.} - \overline{\overline{x}})^2 \qquad (13.36)$$

Sum of Squares Due to Factor B

$$SSB = ar \sum_{j=1}^{b} (\overline{x}_{.j} - \overline{\overline{x}})^2 \qquad (13.37)$$

Sum of Squares Due to Interaction

$$SSAB = r \sum_{i=1}^{a} \sum_{j=1}^{b} (\overline{x}_{ij} - \overline{x}_{i.} - \overline{x}_{.j} + \overline{\overline{x}})^2 \qquad (13.38)$$

Sum of Squares Due to Error

$$SSE = SST - SSA - SSB - SSAB \qquad (13.39)$$

□ □ **Supplementary Exercises**

45. In your own words explain what the ANOVA procedure is used for.

46. What has to be true for the mean square between, MSB, to provide a good estimate of σ^2? Explain.

47. Why do we assume that the populations sampled all have the same variance when we apply the ANOVA procedure?

48. Explain why the mean square between, MSB, and the mean square within, MSW, provide two independent estimates of σ^2.

49. Explain why the mean square between, MSB, provides an inflated estimate of σ^2 when the population means are not the same.

50. A simple random sample of the asking price ($1000s) of four houses currently for sale in each of two residential areas resulted in the data shown in Table 13.26.

a. Use the procedure developed in Chapter 10 to test if the mean asking price is the same in both areas. Use $\alpha = .05$.

b. Use the ANOVA procedure to test if the mean asking price is the same. Compare your analysis with part (a). Use $\alpha = .05$.

51. Suppose that in Exercise 50 data were collected for another residential area. The asking prices for the simple random sample from the third area were as follows: $81,000, $86,000, $75,000, and $90,000; the samle mean and sample variance for these data are 83 and 42 respectively. Is the mean asking price for all three areas the same? Use $\alpha = .05$.

52. An analysis of the number of units sold by 10 salespersons in each of 4 sales territories resulted in the following data:

TABLE 13.26

	Area 1	Area 2
	92	90
	89	102
	98	96
	105	88
$\bar{x}_j$	96	94
s_j^2	50	40

		Sales Territory		
Category	1	2	3	4
Number of salespersons	10	10	10	10
Mean number sold ($\bar{x}$)	130	120	132	114
Sample variance (s^2)	72	64	69	67

Test at the $\alpha = .05$ level to see if there is any significant difference in the mean number of units sold in the 4 sales territories.

53. Suppose that in Exercise 52 the number of salespersons in each territory was as follows: $n_1 = 10$, $n_2 = 12$, $n_3 = 10$, and $n_4 = 15$. Using the same data for $\bar{x}$ and s^2 as given in Exercise 52, test at the $\alpha = .05$ level if there is any significant difference in the population mean number of units sold in the 4 sales territories.

 SERVICE

54. Executives rated service quality in each of several American industries (*Journal of Accountancy*, February 1992). Assume a sample of ratings are as shown below for the airline, retail, hotel, and automotive industries; higher scores indicate a higher service quality rating. At the $\alpha = .05$ level of significance, test for a significant difference in the population mean quality ratings for the four industries. What is your conclusion?

	Airlines	Retail	Hotel	Automotive
	59	63	70	49
	56	49	68	55
	47	60	62	48
	46	54	69	49
	55	56	59	50
	54	55		
	48			
$\bar{x}_j$	52.14	56.17	65.6	50.20
s_j^2	25.81	23.77	23.20	7.70

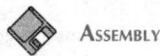

ASSEMBLY

55. Three different assembly methods have been proposed for a new product. A completely randomized experimental design was used to determine which assembly method results in the greatest number of parts produced per hour, 30 workers were randomly selected and assigned to use one of the proposed methods. The number of units produced by each worker is given below:

	Method		
	A	B	C
	97	93	99
	73	100	94
	93	93	87
	100	55	66
	73	77	59
	91	91	75
	100	85	84
	86	73	72
	92	90	88
	95	83	86
$\bar{x}_j$	90	84	81
s_j^2	98.00	168.44	159.78

Use these data and test to see if the mean number of parts produced with each method is the same. Use $\alpha = .05$.

56. In a completely randomized experimental design, three brands of paper towels were tested for their ability to absorb water. Equal-sized towels were used, with four sections of towels tested per brand. The absorbency rating data are given. Using a .05 level of significance, does there appear to be a difference in the ability of the brands to absorb water?

	Brand		
	X	Y	Z
	91	99	83
	100	96	88
	88	94	89
	89	99	76
$\bar{x}_j$	92	97	84
s_j^2	30	6	35.33

T ABLE 13.27

	Method 1	Method 2	Method 3
	58	52	48
	64	63	57
	55	65	59
	66	58	47
	67	62	49
$\bar{x}_j$	62	60	52
s_j^2	27.5	26.5	31

DowJones

57. To test to see if there is any significant difference in the mean number of units produced per week by each of three production methods, a completely randomized experimental design was used to obtain the data in Table 13.27. At the $\alpha = .05$ level of significance, is there any difference in the means for the three methods?

58. Shown below are the percent changes in the Dow Jones Industrial Averages for stock market performance in each of the four years of six presidential terms. (*The Beacon Street Financial,* February 1992). Does there appear to be any significant effect due to the year of the presidential term on stock market performance? Use $\alpha = .05$.

	First Year	Second Year	Third Year	Fourth Year
	10.9	−18.9	15.2	4.3
	−15.2	4.8	6.1	14.6
	−16.7	−27.6	38.3	17.9
	−17.3	−3.1	4.2	14.9
	−9.2	19.6	20.3	−3.7
	27.7	22.6	2.3	11.9
	27.0	−4.3	20.3	8.8
$\bar{x}_j$	1.03	−0.99	15.24	9.81
s_j^2	416.93	343.04	159.31	55.43

59. Hargreaves Automotive Parts, Inc., would like to compare the mileage for four different types of brake linings. Thirty linings of each type were produced and placed on a fleet of rental cars. The number of miles that each brake lining lasted until it no longer met the required federal safety standard was recorded, and an average value was computed for each type of lining. The following data were obtained:

Type	Sample Size	Sample Mean	Standard Deviation
A	30	32,000	1450
B	30	27,500	1525
C	30	34,200	1650
D	30	30,300	1400

Test if the corresponding population means are equal. Use $\alpha = .05$.

60. A manufacturer of batteries for electronic toys and calculators is considering three new battery designs. An attempt was made to determine if the mean lifetime in hours is the same for each of the three designs.

	Design A	Design B	Design C
	78	112	115
	98	99	101
	88	101	100
	96	116	120
$\bar{x}_j$	90	107	109
s_j^2	82.67	68.67	100.67

Test to see if the population means are equal. Use $\alpha = .05$.

61. A study was conducted to investigate the browsing activity by shoppers (*Journal of the Academy of Marketing Science,* Winter 1989). Each shopper was initially classified as nonbrowser, light browser, or heavy browser. For each shopper in the study a measure was obtained to determine how comfortable the shopper was in a store. Higher scores indicated greater comfort. Suppose the data in Table 13.28 is from a related study.

a. Using $\alpha = .05$, test for differences among comfort levels for the three types of browsers.

b. Use Tukey's procedure to compare the comfort levels of nonbrowsers and light browsers. Use $\alpha = .05$. What is your conclusion?

62. A research firm tests the miles per gallon characteristics of three brands of gasoline. Because of different gasoline performance characteristics in different brands of automobiles, five brands of automobiles are selected and treated as blocks in the experiment. That is, each brand of automobile is tested with each type of gasoline. The results of the experiment (in miles per gallon) are shown

BROWSE

TABLE 13.28

	Browser		
Nonbrowser	Light	Heavy
4	5	5
5	6	7
6	5	5
3	4	7
3	7	4
4	4	6
5	6	5
4	5	7
$\bar{x}_j$ 4.25	5.25	5.75
s_j^2 1.07	1.07	1.36

		Gasoline Brands		
		I	II	III
	A	18	21	20
	B	24	26	27
Blocks: Automobiles	C	30	29	34
	D	22	25	24
	E	20	23	24

Using $\alpha = .05$, is there a significant difference in the mean miles per gallon characteristics of the three brands of gasoline?

63. Analyze the experimental data provided in Exercise 62 using the ANOVA procedure for completely randomized designs. Compare your findings with those obtained in Exercise 62. What is the advantage of attempting to remove the block effect?

T A B L E 13.29

		Location			
		1	2	3	4
	A	99	73	85	103
Compound	B	82	72	85	97
	C	81	79	82	86

64. Three different road-repair compounds were tested at four different highway locations. At each location, three sections of the road were repaired, with each section using one of the three compounds. Data were then collected on the number of days of traffic usage before additional repair was required. These data are shown in Table 13.29. Using $\alpha = .01$, test if there are significant differences in the compounds.

65. A factorial experiment was designed to test for any significant differences in the time needed to perform English-to-foreign language translations using two computerized language translators. Since the type of language translated was also considered a significant factor, translations were made using both systems for three different languages: Spanish, French, and German. Use the following data for translation time shown in hours:

	Language		
	Spanish	French	German
System 1	8	10	12
	12	14	16
System 2	6	14	16
	10	16	22

T A B L E 13.30

	Loading System	
	Manual	Automatic
Machine 1	30	30
	34	26
Machine 2	20	24
	22	28

Test for any significant differences due to language translator, type of language, and interaction. Use $\alpha = .05$.

66. A manufacturing company designed a factorial experiment to determine if the number of defective parts produced by two machines differed and if the number of defective parts produced also depended upon whether raw material needed by each machine was loaded manually or using an automatic feed system. The data in Table 13.30 shows the number of defective parts produced. Using $\alpha = .05$, test for any significant effect due to machine, loading system, and interaction.

Computer Case: *Wentworth Medical Center*

 MEDICAL1

 MEDICAL2

As part of a long-term study of individuals 65 years of age or older, sociologists and physicians at the Wentworth Medical Center in upstate New York conducted a study to investigate the relationship between geographic location and depression. A sample of 60 individuals, all in reasonably good health, was selected; 20 individuals were residents of Florida, 20 were residents of New York, and 20 were residents of North Carolina. Each of the individuals sampled was given a standardized test to measure depression. The data collected are shown below; higher test scores indicate higher levels of depression. These data are available on the data disk in the file MEDICAL1.

A second part of the study considered the relationship between geographic location and depression for individuals 65 years of age or older who had a chronic health condition such as arthritis, hypertension, and/or heart ailment. A sample of 60 individuals with such conditions was identified. Again 20 were residents of Florida, 20 were residents of New York, and 20 were residents of North Carolina. The levels of depression recorded for this study are also shown below. These data are available on the data disk in the file MEDICAL2.

Data from MEDICAL 1			Data from MEDICAL 2		
Florida	New York	North Carolina	Florida	New York	North Carolina
3	8	10	13	14	10
7	11	7	12	9	12
7	9	3	17	15	15
3	7	5	17	12	18
8	8	11	20	16	12
8	7	8	21	4	14
8	8	4	16	18	17
5	4	3	14	14	8
5	13	7	13	15	14
2	10	8	17	17	16
6	6	8	12	20	18
2	8	7	9	11	17
6	12	3	12	23	19
6	8	9	15	19	15
9	6	8	16	17	13
7	8	12	15	14	14
5	5	6	13	9	11
4	7	3	10	14	12
7	7	8	11	13	13
3	8	11	17	11	11

Managerial Report

1. Use descriptive statistics to summarize the data in the two studies. What are your preliminary observations about the depression scores?
2. Use analysis of variance on both data sets. State the hypothesis being tested in each case. What are your conclusions?
3. Use inferences about individual treatment means where appropriate. What are your conclusions?
4. Discuss extensions of this study or other analyses that you feel might be helpful.

APPENDIX 13.1

Computational Procedure for a Completely Randomized Design

The following step-by-step procedure is designed to ease the burden in computing the appropriate sums of squares for completely randomized designs. The formulas shown below can be applied to both balanced and unbalanced designs.

Notation

$$x_{ij} = \text{value of the } i\text{th observation under treatment } j$$
$$T_j = \text{sum of all observations for treatment } j$$
$$T = \text{sum of all observations}$$
$$n_j = \text{sample size for the } j\text{th treatment}$$
$$n_T = \text{total sample size for the experiment}$$
$$k = \text{number of treatments}$$

Procedure

Step 1. Compute the total sum of squares:

$$\text{SST} = \sum_{j=1}^{k} \sum_{i=1}^{n_j} x_{ij}^2 - \frac{T^2}{n_T}$$

Step 2. Compute the sum of squares due to treatments:

$$\text{SSTR} = \sum_{j=1}^{k} \frac{T_j^2}{n_j} - \frac{T^2}{n_T}$$

Step 3. Compute the sum of squares due to error:

$$\text{SSE} = \text{SST} - \text{SSTR}$$

Example

Using this computational procedure with the Chemitech data in Table 13.8, we obtain the following results:

Step 1. SST = 54,860 − 810,000/15 = 860

Step 2. SSTR = $(310)^2/5 + (330)^2/5 + (260)^2/5 - 810,000/15 = 520$

Step 3. SSE = SST − SSTR = 860 − 520 = 340

APPENDIX 13.2

Computational Procedure for a Randomized Block Design

The following step-by-step procedure is designed to help in computing the appropriate sums of squares for randomized block designs.

Notation

$$x_{ij} = \text{value of the observation under treatment } j \text{ in block } i$$

$$T_{i\cdot} = \text{the total of all observations in block } i$$

$$T_{\cdot j} = \text{the total of all observations in treatment } j$$

$$T = \text{the total of all observations}$$

$$k = \text{the number of treatments}$$

$$b = \text{the number of blocks}$$

$$n_T = \text{the total sample size } (n_T = kb)$$

Procedure

Step 1. Compute the total sum of squares:

$$\text{SST} = \sum_{i=1}^{b} \sum_{j=1}^{k} x_{ij}^2 - \frac{T^2}{n_T}$$

Step 2. Compute the sum of squares due to treatments:

$$\text{SSTR} = \frac{\displaystyle\sum_{j=1}^{k} T_{\cdot j}^2}{b} - \frac{T^2}{n_T}$$

Step 3. Compute the sum of squares due to blocks (SSBL):

$$\text{SSBL} = \frac{\displaystyle\sum_{i=1}^{b} T_{i\cdot}^2}{k} - \frac{T^2}{n_T}$$

Step 4. Compute the sum of squares due to error (SSE):

$$\text{SSE} = \text{SST} - \text{SSTR} - \text{SSBL}$$

Example

For the air traffic controller data in Table 13.15, these steps lead to the sum of squares:

Step 1. $\text{SST} = 3598 - \dfrac{(252)^2}{18} = 70$

Step 2. $\text{SSTR} = \dfrac{(81)^2 + (78)^2 + (93)^2}{6} - \dfrac{(252)^2}{18} = 21$

Step 3. $\text{SSBL} = \dfrac{(48)^2 + (42)^2 + \cdots + (39)^2}{3} - \dfrac{(252)^2}{18} = 30$

Step 4. $\text{SSE} = 70 - 21 - 30 = 19$

APPENDIX 13.3

Computational Procedure for a Two-Factor Factorial Design

Notation

x_{ijk} = observation corresponding to the kth replicate taken from treatment i of factor A and treatment j of factor B

$T_{i.}$ = total of all observations in treatment i (factor A)

$T_{.j}$ = total of all observations in treatment j (factor B)

T_{ij} = total of all observations in the combination of treatment i (factor A) and treatment j (factor B)

T = total of all observations

a = number of levels of factor A

b = number of levels of factor B

r = number of replications

n_T = total number of observations; $n_T = abr$

Procedure

Step 1. Compute the total sum of squares:

$$\text{SST} = \sum_{i=1}^{a} \sum_{j=1}^{b} \sum_{k=1}^{r} x_{ijk}^2 - \frac{T^2}{n_T}$$

Step 2. Compute the sum of squares due to factor A:

$$\text{SSA} = \frac{\sum_{i=1}^{a} T_{i.}^2}{br} - \frac{T^2}{n_T}$$

Step 3. Compute the sum of squares due to factor B:

$$\text{SSB} = \frac{\sum_{j=1}^{b} T_{.j}^2}{ar} - \frac{T^2}{n_T}$$

Step 4. Compute the sum of squares due to interaction:

$$\text{SSAB} = \frac{\sum_{i=1}^{a} \sum_{j=1}^{b} T_{ij}^2}{r} - \frac{T^2}{n_T} - \text{SSA} - \text{SSB}$$

Step 5. Compute the sum of squares due to error:

$$\text{SSE} = \text{SST} - \text{SSA} - \text{SSB} - \text{SSAB}$$

Example

For the GMAT data in Table 13.22, these steps lead to the following sum of squares:

Step 1. $\quad$ SST $\quad = 4{,}856{,}500 - \dfrac{(9270)^2}{18} = 82{,}450$

Step 2. $\quad$ SSA $\quad = \dfrac{(2960)^2 + (3080)^2 + (3230)^2}{6} - \dfrac{(9270)^2}{18} = 6100$

Step 3. $\quad$ SSB $\quad = \dfrac{(3240)^2 + (3360)^2 + (2670)^2}{6} - \dfrac{(9270)^2}{18} = 45{,}300$

Step 4. $\quad$ SSAB $= \dfrac{(1080)^2 + (1000)^2 + \cdots + (890)^2}{2} - \dfrac{(9270)^2}{18}$

$\qquad\qquad\qquad - 6100 - 45{,}300 = 11{,}200$

Step 5. $\quad$ SSE $\quad = 82{,}450 - 6100 - 45{,}300 - 11{,}200 = 19{,}850$

CHAPTER

14

Simple Linear Regression and Correlation

Polaroid Corporation*

CAMBRIDGE, MASSACHUSETTS

Polaroid's consumer photography business began in 1947 when the company's founder, Dr. Edwin Land, announced a one-step dry process for producing a finished photograph within 1 minute after taking the picture. The first Polaroid Land camera and Polaroid Land film went on sale in 1949. Since then, Polaroid's continuous experimentation and development in chemistry, optics, and electronics have produced photographic systems of ever higher quality, reliability, and convenience.

Polaroid's other major business segment, technical and industrial photography, focuses on making Polaroid's instant photography a key component of the growing number of imaging systems used in today's visual communications environment. To this end, Polaroid markets a wide variety of instant photographic systems, cameras, components, and films for professional, industrial, scientific, and medical uses. Other businesses include magnetics, sunglasses, industrial polarizers, chemicals, custom coating, and holography.

Sensitometry, the measurement of the sensitivity of photographic materials, provides information on many characteristics of film, such as its useful exposure range. Within Polaroid's central sensitometry laboratory, scientists systematically sample and analyze instant films that have been stored at temperature and humidity levels approximating those the films will be subjected to once in the hands of the consumers. To investigate the relationship between film speed and the age of a Polaroid extended range, color professional print film, Polaroid's central sensitometry lab selected film samples ranging in age from 1 to 13 months after manufacture. The data showed that film speed decreases with age, and that a straight-line or linear relationship could be used to approximate the relationship between change in film speed and age of the film.

Using regression analysis, Polaroid was able to develop the following equation which relates the change in the film speed to the film's age:

$$\hat{y} = -19.8 - 7.6x$$

where

$$\hat{y} = \text{change in film speed}$$

$$x = \text{film age in months}$$

This equation shows that the average decrease in film speed is 7.6 units per month. The information provided by this analysis, when coupled with consumer purchase and use patterns, enables Polaroid to make manufacturing adjustments that help them produce films with the performance levels their customers require.

In this chapter you will learn how regression analysis can be used to develop an equation relating two variables, such as the change in film speed to the age of the film in the Polaroid example. In subsequent chapters we will extend this concept to cases involving more than two variables.

*The authors are indebted to Mr. Lawrence Friedman, Manager, Photographic Quality for providing this Statistics in Practice.

Managerial decisions are often based on the relationship between two or more variables. For example, after considering the relationship between advertising expenditures and subsequent sales volume, a marketing manager might attempt to predict sales volume from a known level of advertising expenditures. In another case, public utilities often use the relationship between temperature and electricity use to predict demand for electricity. Sometimes, a manager will rely on intuition to judge how two variables are related, but a more objective approach is to collect data on the two variables and then use statistical procedures to determine how the variables are related.

Regression analysis is a statistical procedure that can be used to develop a mathematical equation showing how variables are related. In regression terminology the variable that is being predicted by the mathematical equation is called the *dependent* variable. The variable or variables being used to predict the value of the dependent variable are called the *independent* variables. For example, in analyzing the effect of advertising on sales volume, a marketing manager's desire to predict sales volume would suggest making sales volume the dependent variable for the analysis. Advertising expenditure would be the independent variable used to predict the sales volume. Common statistical notation is to use y to denote the dependent variable and x to denote the independent variable.

In this chapter we consider the simplest type of regression: situations involving one independent and one dependent variable for which the relationship between the variables is approximated by a straight line. This is called *simple linear regression*. Regression analysis involving two or more independent variables is called multiple regression analysis; multiple regression is covered in Chapters 15 and 16.

Another topic discussed in this chapter is correlation. In correlation analysis we are not concerned with identifying a mathematical equation relating an independent and dependent variable; we are concerned only with determining the extent to which the variables are linearly related. Correlation analysis is a procedure for making this determination and, if such a relationship exists, for providing a measure of the relative strength of the relationship.

We caution the reader before beginning this chapter that neither regression nor correlation analysis can be interpreted as establishing cause-and-effect relationships. Regression and correlation analyses can indicate only how or to what extent variables are associated with each other. Any conclusions about a cause-and-effect relationship must be based on the *judgment of the analyst*.

14.1   The Least Squares Method

In this section we show how the least squares method can be used to develop a linear equation relating two variables. As an illustration, let us consider the problem currently being faced by the management of Armand's Pizza Parlors, a chain of Italian-food restaurants located in a five-state area. One of the most successful types of locations for Armand's has been near a college campus.

Prior to opening a new restaurant, Armand's management requires an estimate of annual sales revenues. Such an estimate is used in planning the appropriate restaurant capacity, making initial staffing decisions, and deciding whether the potential revenues justify the cost of the operation.

Suppose management believes that the size of the student population on the nearby campus is related to the annual sales revenues. On an intuitive basis, management believes restaurants located near larger campuses generate more revenue than those near small campuses. To evaluate the relationship between student population x and annual sales y, Armand's collected data from a sample of 10 of its restaurants located near college campuses. The data for these 10 observations are summarized in Table 14.1. For the ith observation, x_i represents the value of the independent variable and y_i represents the value of the dependent variable. We see that restaurant 1, with $x_1 = 2$ and $y_1 = 58$, is located near a campus with 2000 students and has annual sales of $58,000; restaurant 2 is located near a campus with 6000 students and has annual sales of $105,000; and so on.

Figure 14.1 shows graphically the data presented in Table 14.1. The size of the student population is shown on the horizontal axis, with annual sales on the vertical axis. A graph such as this is known as a *scatter diagram*. Values for the independent variable are shown on the horizontal axis and the corresponding values for the dependent variable are shown on the vertical axis. The scatter diagram provides an overview of the data and enables us to draw preliminary conclusions about a possible relationship between the variables.

TABLE 14.1
Student Population and Annual
Sales for 10 Armand's Pizza Parlors

Restaurant i	Student Population (1000s) x_i	Annual Sales ($1000s) y_i
1	2	58
2	6	105
3	8	88
4	8	118
5	12	117
6	16	137
7	20	157
8	20	169
9	22	149
10	26	202

FIGURE 14.1
Scatter Diagram of Annual Sales
versus Student Population for
Armand's Pizza Parlors

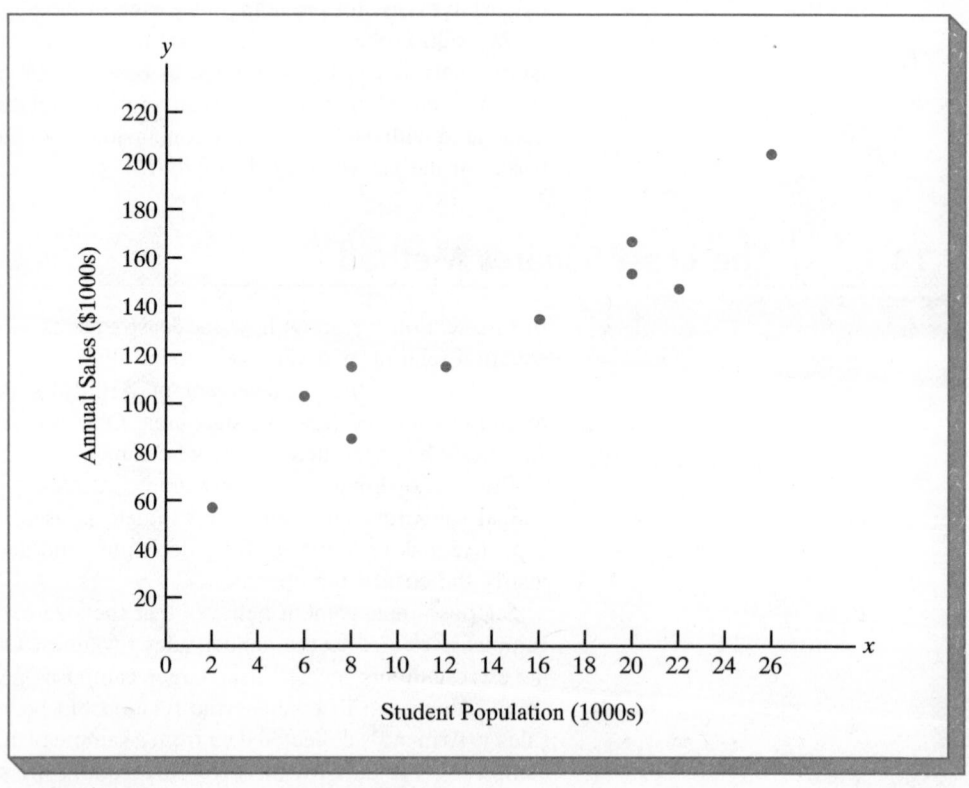

What preliminary conclusions can we draw from Figure 14.1? It appears that low sales volumes are associated with small student populations and higher sales volumes are associated with larger student populations. It also appears that the relationship between the two variables can be approximated by a straight line. In Figure 14.2 we have drawn a straight line through the data that appears to provide a good linear approximation of the relationship between the variables. The equation for this line is $y = 50 + 5.5x$. The y-intercept, the point at which the line intersects the y-axis, is 50, and the slope, the amount of change in y per unit change in x, is 5.5.

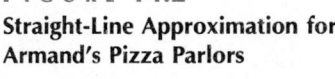

FIGURE 14.2
Straight-Line Approximation for Armand's Pizza Parlors

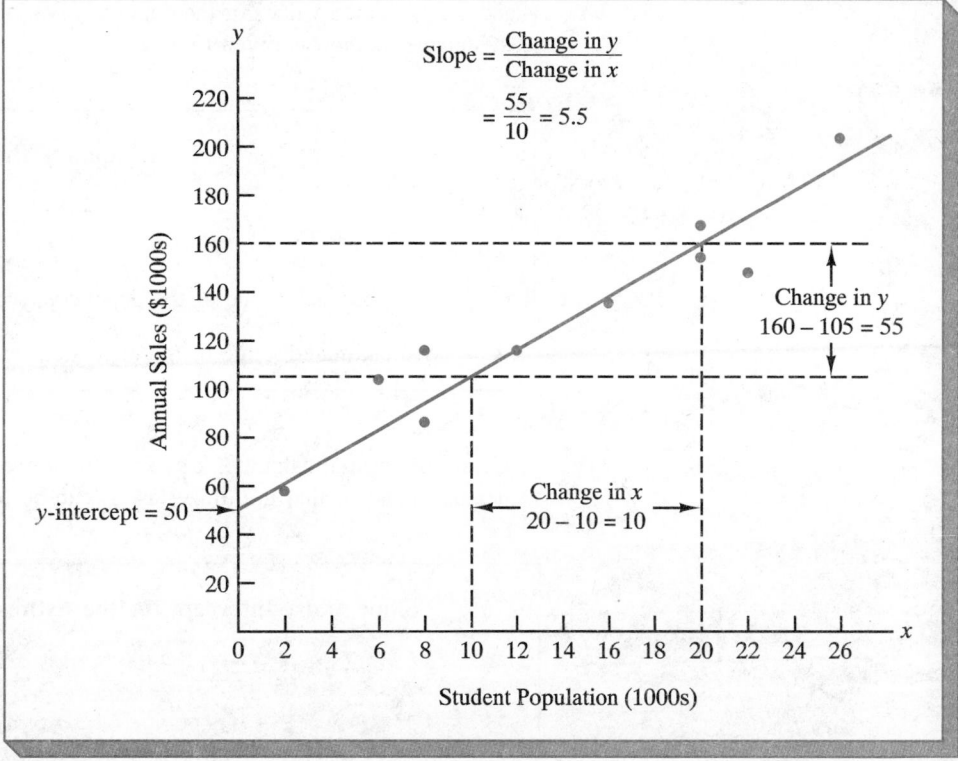

Clearly, there are many different straight lines that we could have drawn in Figure 14.2 to represent the relationship between x and y. The question is, Which of the straight lines that could be drawn best represents the relationship?

The *least squares method* is a procedure that is used to find the straight line that provides the best approximation for the relationship between the independent and dependent variables. We refer to the equation of the line developed using the least squares method as the *estimated regression line,* or the *estimated regression equation.*

Estimated Regression Equation

$$\hat{y} = b_0 + b_1 x \qquad (14.1)$$

where

b_0 = y-intercept of the line

b_1 = slope of the line

$\hat{y}$ = estimated value of the dependent variable

For any particular value of the independent variable x_i, the corresponding value on the estimated regression line is denoted by $\hat{y}_i = b_0 + b_1 x_i$.

Application of the least squares method provides the values of b_0 and b_1 that make the sum of the squared deviations between the observed values of the dependent variable

y_i and the estimated values of the dependent variable $\hat{y}_i$ a minimum. The criterion for the least squares method is given by (14.2).

Least Squares Criterion

$$\min \Sigma(y_i - \hat{y}_i)^2 \tag{14.2}$$

where

y_i = observed value of the dependent variable for the ith observation

$\hat{y}_i$ = estimated value of the dependent variable for the ith observation

---

Using differential calculus, it can be shown (see the appendix to this chapter) that the values of b_0 and b_1 that minimize (14.2) can be found using (14.3) and (14.4).

Slope and *y*-Intercept for the Estimated Regression Equation

$$b_1 = \frac{\Sigma(x_i - \overline{x})(y_i - \overline{y})}{\Sigma(x_i - \overline{x})^2} = \frac{\Sigma x_i y_i - (\Sigma x_i \, \Sigma y_i)/n}{\Sigma x_i^2 - (\Sigma x_i)^2/n} \tag{14.3}$$

$$b_0 = \overline{y} - b_1\overline{x} \tag{14.4}$$

where

x_i = value of the independent variable for the ith observation

y_i = value of the dependent variable for the ith observation

$\overline{x}$ = mean value for the independent variable

$\overline{y}$ = mean value for the dependent variable

n = total number of observations

The second form of (14.3) is normally used for computing b_1 with a calculator because it avoids the tedious calculations involving the computation of each $(x_i - \overline{x})$ and $(y_i - \overline{y})$. However, to avoid rounding errors, it is best to carry as many significant digits as possible in the calculation; we recommend carrying at least four significant digits.

Some of the calculations necessary to develop the least squares estimated regression equation for Armand's Pizza Parlors are shown in Table 14.2. In this example there are 10 restaurants, or observations; hence $n = 10$. Using (14.3) and (14.4), we can now compute the slope and intercept of the estimated regression equation for Armand's Pizza Parlors. The calculation of the slope (b_1) proceeds as follows:

$$b_1 = \frac{\Sigma x_i y_i - (\Sigma x_i \Sigma y_i)/n}{\Sigma x_i^2 - (\Sigma x_i)^2/n}$$

$$= \frac{21,040 - (140)(1300)/10}{2528 - (140)^2/10}$$

TABLE 14.2
Calculations for the Least Squares Estimated Regression Equation for Armand's Pizza Parlors

Restaurant i	x_i	y_i	$x_i y_i$	x_i^2
1	2	58	116	4
2	6	105	630	36
3	8	88	704	64
4	8	118	944	64
5	12	117	1,404	144
6	16	137	2,192	256
7	20	157	3,140	400
8	20	169	3,380	400
9	22	149	3,278	484
10	26	202	5,252	676
Totals	140	1300	21,040	2528
	Σx_i	Σy_i	$\Sigma x_i y_i$	Σx_i^2

$$= \frac{2840}{568}$$

$$= 5$$

The calculation of the y intercept (b_0) is as follows:

$$\overline{x} = \frac{\Sigma x_i}{n} = \frac{140}{10} = 14$$

$$\overline{y} = \frac{\Sigma y_i}{n} = \frac{1300}{10} = 130$$

$$b_0 = \overline{y} - b_1 \overline{x}$$

$$= 130 - 5(14)$$

$$= 60$$

Thus, the estimated regression equation found by using the method of least squares is

$$\hat{y} = 60 + 5x$$

In Figure 14.3 we show the graph of this equation.

The slope of the estimated regression equation ($b_1 = 5$) is positive, implying that as student population increases, annual sales increase. In fact, we can conclude (since sales are measured in $1000s and student population in 1000s) that an increase in the student population of 1000 is associated with an increase of $5000 in expected annual sales; that is, sales are expected to increase by $5.00 per student.

If we believe that the least squares estimated regression equation adequately describes the relationship between x and y, then it would seem reasonable to use the estimated regression equation to predict the value of y for a given value of x. For example, if we wanted to predict annual sales for a restaurant to be located near a campus with 16,000 students, we would compute

$$\hat{y} = 60 + 5(16)$$

$$= 140$$

FIGURE 14.3
Graph of the Estimated Regression Equation for Armand's Pizza Parlors: $\hat{y} = 60 + 5x$

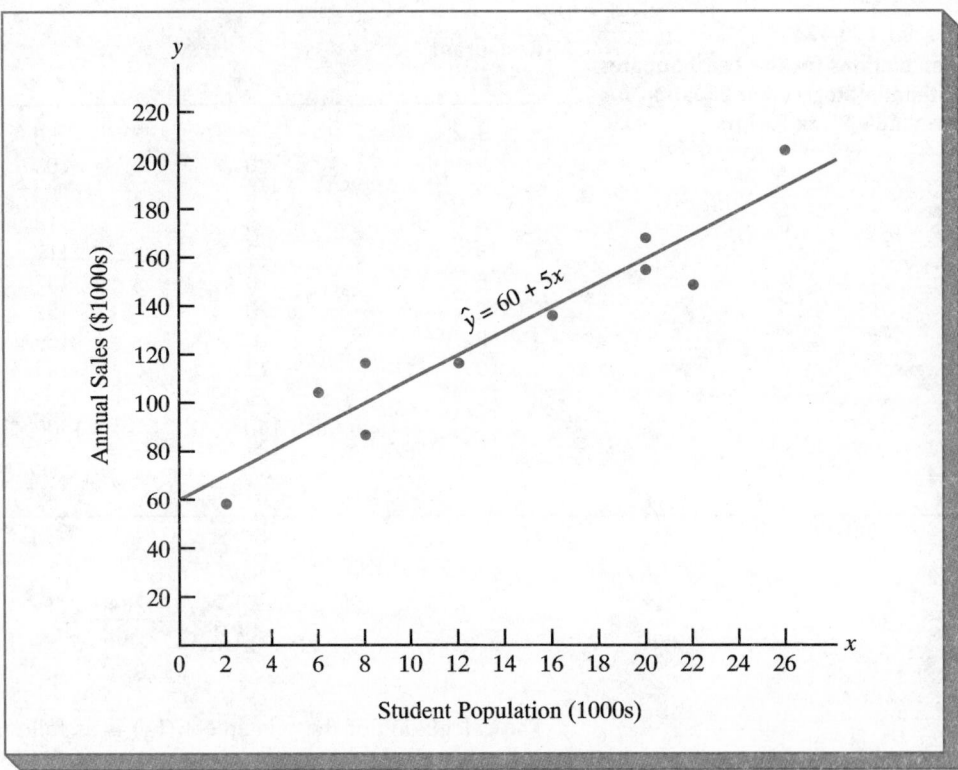

Hence, we would predict sales of $140,000 per year. In the following sections we will discuss methods for assessing the appropriateness of using the estimated regression equation for estimation and prediction.

NOTES & COMMENTS

The least squares method provides an estimated regression equation that minimizes the sum of squared deviations between the observed values of the dependent variable y_i and the estimated values of the dependent variable $\hat{y}_i$. This is the least squares criterion for choosing the equation that provides the best fit. If some other criterion were used, such as minimizing the sum of the absolute deviations between y_i and $\hat{y}_i$, a different equation would be obtained. In practice, the least squares method is the most widely used.

❏ ❏ Exercises

Methods

SELF TEST ▶

1. Given are 5 observations for two variables, x and y.

x_i	1	2	3	4	5
y_i	3	7	5	11	14

a. Develop a scatter diagram for these data.
b. What does the scatter diagram developed in (a) indicate about the relationship between the two variables?
c. Try to approximate the relationship between x and y by drawing a straight line through the data.
d. Develop the estimated regression equation by computing the values of b_0 and b_1 using (14.3) and (14.4).
e. Use the estimated regression equation to predict the value of y when $x = 4$.

2. Given are 5 observations for two variables, x and y.

x_i	2	3	5	1	8
y_i	25	25	20	30	16

a. Develop a scatter diagram for these data.
b. What does the scatter diagram developed in (a) indicate about the relationship between the two variables?
c. Try to approximate the relationship between x and y by drawing a straight line through the data.
d. Develop the estimated regression equation by computing the values of b_0 and b_1 using (14.3) and (14.4).
e. Use the estimated regression equation to predict the value of y when $x = 6$.

3. Given below are five observations collected in a regression study on two variables.

x_i	2	4	5	7	8
y_i	2	3	2	6	4

a. Develop a scatter diagram for these data.
b. Develop the estimated regression equation for these data.
c. Use the estimated regression equation to predict the value of y when $x = 4$.

Applications

SELF TEST ▶

4. The following data were collected on the height (inches) and weight (pounds) of women swimmers.

Height	68	64	62	65	66
Weight	132	108	102	115	128

a. Develop a scatter diagram for these data with height as the independent variable.
b. What does the scatter diagram developed in (a) indicate about the relationship between the two variables?
c. Try to approximate the relationship between height and weight by drawing a straight line through the data.
d. Develop the estimated regression equation by computing the values of b_0 and b_1 using (14.3) and (14.4).
e. If a swimmer's height is 63 inches, what would you estimate her weight to be?

5. The data in Table 14.3 were collected regarding the monthly starting salaries and the grade point averages (GPA) for undergraduate students who had obtained a degree in political science.
a. Develop a scatter diagram for these data with GPA as the independent variable.
b. What does the scatter diagram developed in (a) indicate about the relationship between the two variables?
c. Draw a straight line through the data to approximate a linear relationship between GPA and salary.
d. Use the least squares method to develop the estimated regression equation.
e. Predict the monthly starting salary for a student with a 3.0 GPA and for a student with a 3.5 GPA.

TABLE 14.3

GPA	Monthly Salary ($)
2.6	1400
3.4	1700
3.6	2100
3.2	1600
3.5	2000
2.9	1700

6. Performance data for a Century Coronado 21 with a 310-hp MerCruiser V-8 gasoline inboard engine was reported in *Boating*, September, 1991. Data on how the boat speed in miles per hour (mph) affected fuel consumption in gallons per hour (gph) are shown below.

Speed (mph)	Fuel Consumption (gph)
6.1	2.3
10.7	4.8
20.9	7.5
27.5	9.2
31.5	12.4

a. Develop a scatter diagram for these data.
b. What does the scatter diagram developed in (a) indicate about the relationship between these two variables?
c. Develop the estimated regression equation showing how fuel consumption is related to the boat speed.
d. What is the estimated fuel consumption if the boat speed is 25 mph?

7. Eddie's Restaurants collected the data shown in Table 14.4 on the relationship between advertising and sales using a sample of seven restaurants. Develop a scatter diagram for these data with advertising expenditures as the independent variable and sales as the dependent variable. Does there appear to be a linear relationship?

8. *Consumer Reports* uses a survey to collect data on the annual cost of repairs for over 300 makes and models of automobiles (*Consumer Reports* 1992 Buying Guide). The following data show the average annual repair cost ($) and the age of the automobile (years).

Age	1	2	3	4	5
Repair	135	175	320	300	450

a. Develop the estimated regression equation showing how annual repair cost is related to the age of an automobile.
b. What is the estimated annual repair cost for a 3-year-old automobile?

9. Table 14.5 shows some data that a sales manager has collected concerning annual sales and years of experience.
a. Develop a scatter diagram for these data with years of experience as the independent variable.
b. Develop the estimated regression equation that can be used to predict annual sales given the years of experience.
c. Use the estimated regression equation to predict annual sales for a salesperson with 9 years of experience.

10. Tyler Realty collected the following data regarding the selling price of new homes and the size of the homes measured in terms of square footage of living space.

Square Footage	Selling Price
2500	$124,000
2400	$108,000
1800	$ 92,000
3000	$146,000
2300	$110,000

TABLE 14.4

Advertising Expenditures ($1000s)	Sales ($1000s)
1.0	19.0
2.0	32.0
4.0	44.0
6.0	40.0
10.0	52.0
14.0	53.0
20.0	54.0

TABLE 14.5

Salesperson	Years of Experience	Annual Sales ($1000s)
1	1	80
2	3	97
3	4	92
4	4	102
5	6	103
6	8	111
7	10	119
8	10	123
9	11	117
10	13	136

a. Develop a scatter diagram for these data with square footage on the horizontal axis.
b. Try to approximate the relationship between square footage and selling price by drawing a straight line through the data.
c. Does there appear to be a linear relationship?
d. Develop an estimated regression equation using the least squares method.
e. Predict the selling price for a home with 2700 square feet.

11. Shown in Table 14.6 are the number of golf courses and the number of paid rounds of golf (in millions) for the Myrtle Beach, South Carolina, area over an 8-year period (*Myrtle Beach Magazine,* October 1991).
a. Develop a scatter diagram for these data.
b. Does there appear to be a linear relationship?
c. Develop the estimated regression equation relating the number of paid rounds of golf to the number of golf courses.
d. Suppose that next year 57 golf courses will be available. What is an estimate of the number of paid rounds of golf? Discuss any potential problem associated with making this estimate.

12. The following data show the percentage of women working in each company (x) and the percentage of management jobs held by women in that company (y); the data shown represent companies in retailing and trade (*Louis Rukeyser's Business Almanac*).

Company	x_i	y_i
Federated Department Stores	72	61
Kroger	47	16
Marriott	51	32
McDonald's	57	46
Sears, Roebuck	55	36

a. Develop a scatter diagram for these data.
b. What does the scatter diagram developed in (a) indicate about the relationship between x and y?
c. Develop the estimated regression equation for these data.
d. Predict the percentage of management jobs held by women in a company that has 60% women employees.
e. Use the estimated regression equation to predict the percentage of management jobs held by women in a company where 55% of the jobs are held by women. How does this predicted value compare to the 36% value observed for Sears, Roebuck, a company in which 55% of the employees are women?

13. A university medical center has developed a test designed to measure a patient's stress level. The test is designed so that higher scores on the test correspond to higher levels of stress. As part of a research study, the blood pressure (low reading) of patients who took the test was recorded. Table 14.7 gives the results.
a. Develop a scatter diagram for these data with stress test score on the horizontal axis. Does a linear relationship between the two variables appear to be appropriate?

TABLE 14.6

Number of Golf Courses	Number of Paid Rounds of Golf
26	1.0
30	1.1
31	1.2
32	1.3
33	1.4
35	1.6
38	1.8
43	2.0

TABLE 14.7

Stress Test Score	Blood Pressure
53	70
94	91
64	78
73	78
82	85
90	84

b. Develop the estimated regression equation for these data.

c. Estimate an individual's blood pressure if he or she scored 85 on the stress test.

14. The annual growth rate and cash flow for nine pharmaceutical companies are shown below (*Forbes*, May 1989).

Firm	Annual Growth Rate (%)	Cash Flow ($1,000,000s)
Pfizer	12	986
Bristol Meyers	14	957
Merck	16	1412
American Home Products	11	1074
Abbott Laboratories	18	1023
Eli Lilly	11	965
SmithKline Beckman	3	450
Upjohn	10	456
Warner-Lambert	7	437

a. Develop a scatter diagram for these data, plotting cash flow on the vertical axis. Does it appear that the two variables are related?

b. Use the least squares method to fit a straight line to the data.

c. Estimate the cash flow for a company that has a 10% growth rate. How does the predicted value compare with the observed results for Upjohn?

14.2   The Coefficient of Determination

In Section 14.1 we showed that the least squares method provides a linear approximation for the relationship between two variables. For the Armand's Pizza Parlors problem, we developed the estimated regression equation $\hat{y} = 60 + 5x$ to approximate the linear relationship between student population x and annual sales y. A question that might occur is, How good is the fit? In this section we show that the *coefficient of determination* provides a measure of goodness of fit of the estimated regression equation to the data.

Recall that the least squares method is a technique for finding the values of b_0 and b_1 that minimize the sum of squared deviations between the observed values of the dependent variable y_i and the predicted values of the dependent variable $\hat{y}_i$. The difference between y_i and $\hat{y}_i$ represents the error in using $\hat{y}_i$ to estimate y_i; the difference for the ith observation is $y_i - \hat{y}_i$. This difference is referred to as the ith *residual*. Thus, the sum of squares minimized by the least squares method is referred to as the sum of squares due to error, or the residual sum of squares. We use SSE to represent this quantity.

Sum of Squares Due to Error

$$SSE = \Sigma(y_i - \hat{y}_i)^2 \tag{14.5}$$

Table 14.8 shows the calculations required to compute SSE for Armand's Pizza Parlors. SSE = 1530 is a measure of the error involved in using the estimated regression equation $\hat{y} = 60 + 5x$ to predict the y_i values.

TABLE 14.8
Calculation of SSE for Armand's Pizza Parlors

Restaurant i	x_i = Student Population (1000s)	y_i = Annual Sales ($1000s)	$\hat{y}_i = 60 + 5x_i$	$y_i - \hat{y}_i$	$(y_i - \hat{y}_i)^2$
1	2	58	70	−12	144
2	6	105	90	15	225
3	8	88	100	−12	144
4	8	118	100	18	324
5	12	117	120	−3	9
6	16	137	140	−3	9
7	20	157	160	−3	9
8	20	169	160	9	81
9	22	149	170	−21	441
10	26	202	190	12	144
					SSE = 1530

Now suppose that we were asked to develop an estimate of annual sales without using the size of the student population. We could not use the estimated regression equation and would have to use the value of the sample mean, $\bar{y} = 130$, as the best estimate of annual sales. In Table 14.9 we show the errors that would result from using $\bar{y}$ to estimate annual sales at the 10 Armand's Pizza Parlors. The corresponding sum of squares about the mean, denoted by SST, is as follows:

$$
\boxed{
\begin{array}{c}
\textbf{Total Sum of Squares} \\[4pt]
\text{SST} = \Sigma(y_i - \bar{y})^2 \hspace{2cm} \textbf{(14.6)}
\end{array}
}
$$

For the Armand's Pizza Parlors example, SST = 15,730.

TABLE 14.9
Computation of the Total Sum of Squares for Armand's Pizza Parlors

Restaurant i	x_i = Student Population (1000s)	y_i = Annual Sales ($1000s)	$y_i - \bar{y}$	$(y_i - \bar{y})^2$
1	2	58	−72	5,184
2	6	105	−25	625
3	8	88	−42	1,764
4	8	118	−12	144
5	12	117	−13	169
6	16	137	7	49
7	20	157	27	729
8	20	169	39	1,521
9	22	149	19	361
10	26	202	72	5,184
				SST = 15,730

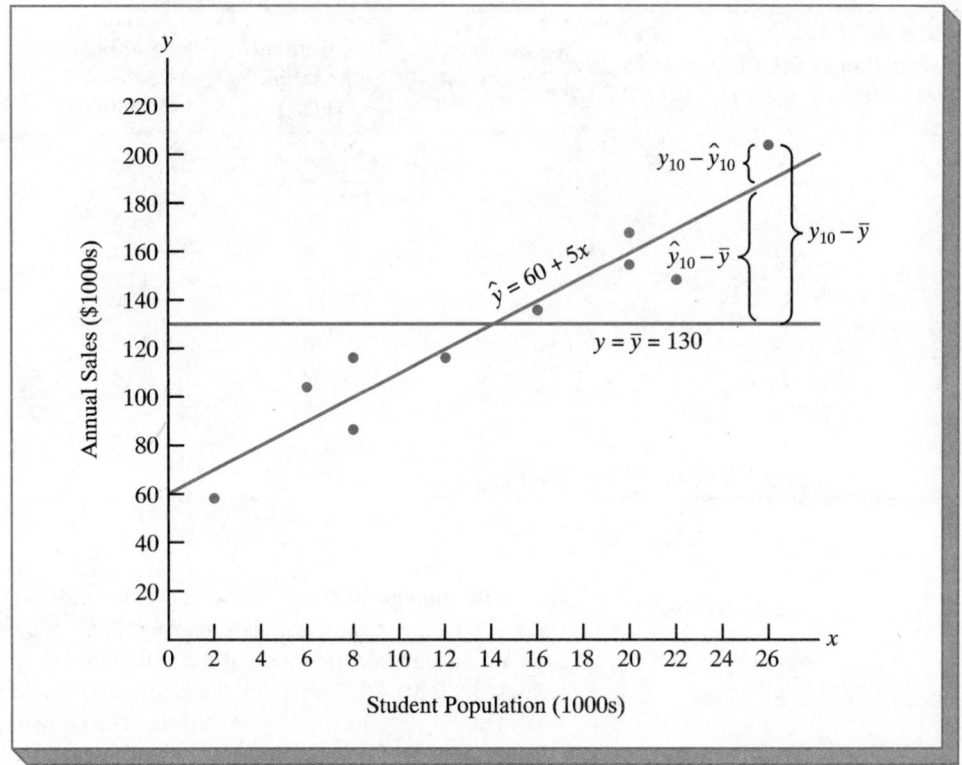

In Figure 14.4 we show the least squares regression line $\hat{y} = 60 + 5x$ and the line corresponding to $\bar{y} = 130$. Note that, in general, the points cluster more closely around the estimated regression line than they do about the line $\bar{y} = 130$. For example, for the 10th restaurant we see that the error is much larger when $\bar{y} = 130$ is used as an estimate of y_{10} than when $\hat{y}_{10} = 190$ is used. We can think of SST as a measure of how well the observations cluster about the $\bar{y}$ line and SSE as a measure of how well the observations cluster about the $\hat{y}$ line.

To measure how much the predicted values $\hat{y}$ on the estimated regression line deviate from $\bar{y}$, another sum of squares is computed. This sum of squares is called the *sum of squares due to regression* and is denoted by SSR. The sum of squares due to regression can be written as follows.

Sum of Squares Due to Regression

$$\text{SSR} = \Sigma(\hat{y}_i - \bar{y})^2 \qquad\qquad (14.7)$$

From the preceding discussion, we should expect that SSE, SST, and SSR are related. Indeed, they are. The relationship among SSE, SST, and SSR is as shown.

Relationship among SST, SSR, and SSE

$$SST = SSR + SSE \qquad (14.8)$$

where

SST = total sum of squares

SSR = sum of squares due to regression

SSE = sum of squares due to error

Using (14.8), we can conclude that the sum of squares due to regression relationship for Armand's Pizza Parlors is

$$SSR = SST - SSE = 15,730 - 1530 = 14,200$$

Now let us see how these sums of squares can be used to provide a measure of the goodness of fit for the regression relationship. We would have the best possible fit if every observation happened to lie on the least squares line; the line would pass through each point, and we would have SSE = 0. Hence, for a perfect fit, SSR must equal SST, and thus, the ratio SSR/SST = 1. On the other hand, a poorer fit to the observed data results in a larger SSE. Since SST = SSR + SSE, however, the largest SSE (and hence worst fit) occurs when SSR = 0. In this case, the estimated regression equation does not help predict y. Thus, the worst possible fit yields the ratio SSR/SST = 0.

If we were to use the ratio SSR/SST to evaluate the goodness of fit for the regression relationship, we would have a measure that could take on values between 0 and 1, with values closer to 1 implying a better fit. The fraction SSR/SST is called the *coefficient of determination* and is denoted r^2.

Coefficient of Determination

$$r^2 = \frac{SSR}{SST} \qquad (14.9)$$

The value of the coefficient of determination for Armand's Pizza Parlors is

$$r^2 = \frac{SSR}{SST} = \frac{14,200}{15,730} = .903$$

To better interpret r^2, we can think of SST as the measure of how good $\overline{y}$ is as a predictor of annual sales volume. After developing the estimated regression equation, we compute SSE as the measure of the goodness of $\hat{y}$ as a predictor of annual sales volume. Thus, SSR (the difference between SST and SSE) really measures the portion of SST that is explained by the estimated regression equation. We can think of r^2 as

$$r^2 = \frac{\text{Sum of Squares Explained by Regression}}{\text{Total Sum of Squares}}$$

When it is expressed as a percentage, r^2 can be interpreted as the percentage of the total sum of squares (SST) that can be explained using the estimated regression equation. Statisticians often use r^2 as a measure of the goodness of fit of a regression line to the data. For Armand's Pizza Parlors, we conclude that the estimated regression equation has accounted for 90.3% of the total sum of squares. We should be very pleased with such a good fit.

Computational Efficiencies

When using a calculator to compute the value of the coefficient of determination, computational efficiencies can be realized by computing SSR directly using the following formula

$$\text{SSR} = \frac{[\Sigma x_i y_i - (\Sigma x_i \ \Sigma y_i)/n]^2}{\Sigma x_i^2 - (\Sigma x_i)^2/n} \qquad (14.10)$$

In addition, we need not compute SST using the expression $\Sigma \ (y_i - \bar{y})^2$; this expression can be algebraically expanded to provide

$$\text{SST} = \Sigma y_i^2 - \frac{(\Sigma y_i)^2}{n} \qquad (14.11)$$

For Armand's Pizza Parlors, part of the calculations needed to compute SSR and SST using the above formulas are shown in Table 14.10. Using the values in this table along with (14.10) and (14.11), we can compute SSR and SST as follows:

$$\text{SSR} = \frac{[21,040 - (140)(1300)/10]^2}{2528 - (140)^2/10}$$

$$= \frac{8,065,600}{568}$$

TABLE 14.10
Calculations Used in Computing SSR and SST for Armand's Pizza Parlors

Restaurant i	x_i	y_i	$x_i y_i$	x_i^2	y_i^2
1	2	58	116	4	3,364
2	6	105	630	36	11,025
3	8	88	704	64	7,744
4	8	118	944	64	13,924
5	12	117	1,404	144	13,689
6	16	137	2,192	256	18,769
7	20	157	3,140	400	24,649
8	20	169	3,380	400	28,561
9	22	149	3,278	484	22,201
10	26	202	5,252	676	40,804
Totals	140	1300	21,040	2528	184,730
	Σx_i	Σy_i	$\Sigma x_i y_i$	Σx_i^2	Σy_i^2

$$= 14,200$$

$$\text{SST} = 184,730 - \frac{(1300)^2}{10}$$

$$= 15,730$$

Note that since these are equivalent formulas, we get the same values for SSR and SST that we obtained previously:

$$r^2 = \frac{\text{SSR}}{\text{SST}} = \frac{14,200}{15,730} = .903$$

NOTES & COMMENTS

1. In developing the least squares estimated regression equation and computing the coefficient of determination, no probabilistic assumptions and no statistical inferences have been made. Larger values of r^2 simply imply that the least squares line provides a better fit to the data; that is, the observations are more closely grouped about the least squares line. But, using only r^2, no conclusion can be made regarding whether the relationship between x and y is statistically significant. Such a conclusion must be based on considerations that involve the sample size and the properties of the appropriate sampling distributions of the least squares estimators.

2. As a practical matter, for typical data found in the social sciences, values of r^2 as low as .25 are often considered useful. For data in the physical and medical sciences, r^2 values of .60 or greater are often found; in fact, in some cases, r^2 values greater than .90 can be found.

□ □ Exercises

Methods

SELF TEST   **15.** The data from Exercise 1 are shown below:

x_i	1	2	3	4	5
y_i	3	7	5	11	14

The estimated regression equation for these data is $\hat{y} = .20 + 2.60x$.
a. Compute SSE, SST, and SSR using (14.5), (14.6), and (14.8).
b. Compute the coefficient of determination r^2. Comment on the goodness of fit.
c. Recompute SSR and SST using (14.10) and (14.11). Do you get the same results as in (a)?

16. The data from Exercise 2 are shown below:

x_i	2	3	5	1	8
y_i	25	25	20	30	16

The estimated regression equation for these data is $\hat{y} = 30.33 - 1.88x$.
a. Compute SSE, SST, and SSR.
b. Compute the coefficient of determination r^2. Comment on the goodness of fit.

17. The data from Exercise 3 are shown below:

x_i	2	4	5	7	8
y_i	2	3	2	6	4

The estimated regression equation for these data is $\hat{y} = .75 + .51x$. What percentage of the total sum of squares can be accounted for by the estimated regression equation?

Applications

SELF TEST ▶

18. In Exercise 5 data were collected regarding the monthly salaries y and the grade point averages x for undergraduate students who had obtained a degree in political science. The data from Exercise 5 are shown in Table 14.11. The estimated regression equation for these data is $\hat{y} = -109.46 + 581.08x$.

a. Compute SSE, SST, and SSR.

b. Compute the coefficient of determination r^2. Comment on the goodness of fit.

19. The data from Exercise 7 are shown below:

TABLE 14.11

GPA	Monthly Salary ($)
2.6	1400
3.4	1700
3.6	2100
3.2	1600
3.5	2000
2.9	1700

Advertising Expenditures (x_i) ($1000s)	Sales (y_i) ($1000s)
1.0	19.0
2.0	32.0
4.0	44.0
6.0	40.0
10.0	52.0
14.0	53.0
20.0	54.0

The estimated regression equation for these data is $\hat{y} = 29.4 + 1.55x$. What percentage of the total sum of squares can be accounted for by the estimated regression equation? Comment on the goodness of fit.

20. A medical laboratory at Duke University estimates the amount of protein in liver samples through the use of a regression model. A spectrometer emitting light shines through a substance containing the sample, and the amount of light absorbed is used to estimate the amount of protein in the sample. A new estimated regression equation is developed daily because of differing amounts of dye in the solution. On one day six samples with known protein concentrations gave the absorbence readings shown in Table 14.12.

a. Use these data to develop an estimated regression equation relating the light absorbance reading to milligrams of protein present in the sample.

b. Compute r^2. Would you feel comfortable using this regression model to estimate the amount of protein in a sample?

c. In a sample just received the light absorbance reading was .941. Estimate the amount of protein in the sample.

TABLE 14.12

Absorbence Reading (x_i)	Milligrams of Protein (y_i)
.509	0
.756	20
1.020	40
1.400	80
1.570	100
1.790	127

21. A list of the best-selling cars for 1987 whose sales in units varied between 175,000 and 300,000 (rounded to the nearest thousand) is shown in the following table (*The World Almanac*, 1989). The 1988 suggested retail price (in thousands of dollars, rounded to the nearest hundred dollars) is also shown.

Model	1988 Price ($1000s)	Number Sold (1000s)
Hyundai	5.4	264
Oldsmobile Ciera	11.4	245
Nissan Sentra	6.4	236
Ford Tempo	9.1	219
Chev. Corsica/Beretta	10.0	214
Pontiac Grand Am	10.3	211
Toyota Camry	11.2	187
Chevrolet Caprice	12.5	177

a. Use these data to develop an estimated regression equation that could be used to predict the number sold given the price.

b. Compute r^2. Would you feel comfortable using the estimated regression equation to estimate the number sold given the price? Explain.

14.3 The Regression Model and Its Assumptions

An important concept that must be understood before we consider testing for significance in regression analysis involves the distinction between a *deterministic model* and a *probabilistic model*. In a deterministic model the relationship between the dependent variable y and the independent variable x is such that if we specify the value of the independent variable, the value of the dependent variable can be determined *exactly*. For example, if a major oil company leases a service station under a contractual agreement of $500 per month plus 10% of the gross sales, the relationship between the dealer's monthly payment y and the gross sales value x can be expressed as

$$y = 500 + .10x$$

With this relationship, a June gross sales of $6000 would provide a monthly payment of $y = 500 + .10(6000) = \$1100$, and a July gross sales of $7200 would provide a monthly payment of $y = 500 + .10(7200) = \$1220$. This type of relationship is deterministic: Once the gross sales value x is specified, the monthly payment y is determined exactly. Figure 14.5 shows graphically the relationship between gross sales and monthly payment.

To illustrate a relationship between two variables that is probabilistic rather than deterministic, recall the Armand's Pizza Parlors example introduced in Section 14.1; the data for this problem are presented in Table 14.1. Note that restaurants 3 and 4 are both located near college campuses having 8000 students. Restaurant 3 shows annual sales of $88,000; however, restaurant 4 shows annual sales of $118,000. Thus, we see that the relationship between y and x cannot be deterministic, since different values of y are observed for the same value of x. Note that this is also the case for restaurants 7 and 8, where a given campus size generates annual sales of $157,000 for restaurant 7 and $169,000 for restaurant 8. Since the value of y cannot be determined exactly from the value of x, we say that the model relating x and y is *probabilistic*.

Next, let us reconsider Figure 14.1, the scatter diagram for the Armand's Pizza Parlors data. We concluded that the relationship between student population x and annual sales y

FIGURE **14.5**

Illustration of a Deterministic Relationship

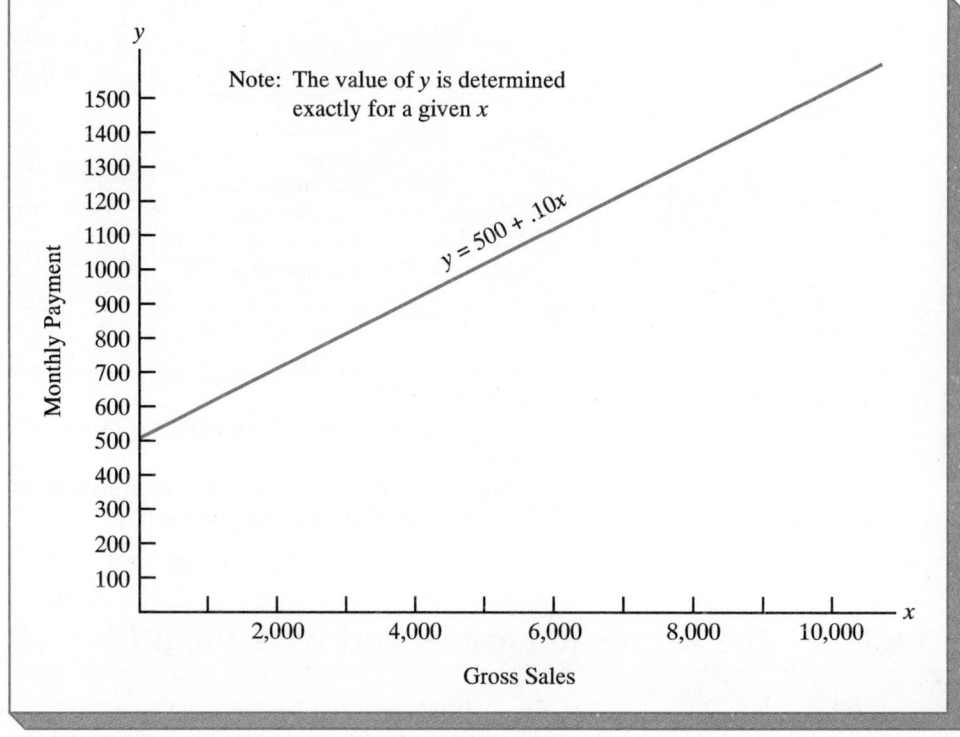

could be approximated by a straight line. As a result, we used the least squares method to develop the following estimated regression equation:

$$\hat{y} = 60 + 5x$$

However, when we graphed this estimated regression equation in Figure 14.3, we saw that the relationship described by this equation was not perfect; that is, none of the observations fell exactly on the regression line.

Since we are unable to guarantee a single value of y for each value of x, the underlying relationship for Armand's Pizza Parlors can be explained only with a probabilistic model. Based on the observation that the relationship between student population and annual sales can be approximated by a straight line, we now make the assumption that the following probabilistic model—referred to as the *regression model*—is a realistic representation of the true relationship between the two variables.

Regression Model

$$y = \beta_0 + \beta_1 x + \epsilon \tag{14.12}$$

where

$\beta_0 =$ y-intercept of the line given by $\beta_0 + \beta_1 x$

$\beta_1 =$ the slope of the line given by $\beta_0 + \beta_1 x$

$\epsilon =$ the error or deviation of the actual y value from the line given by $\beta_0 + \beta_1 x$

Using (14.12) as a model of the relationship between x and y, we are saying that we believe the two variables are related in such a fashion that the line given by $\beta_0 + \beta_1 x$ provides a good approximation of the y value at each x. However, to identify the exact value of y we must also consider the error term ϵ (the Greek letter epsilon), which is the measure of how far the actual y value is above or below the line $\beta_0 + \beta_1 x$. In the regression model, the independent variable x is treated as being known; the model is used to predict y given knowledge of x. We refer to β_0 (the y-intercept) and β_1 (the slope) as the *parameters* of the model.

The following assumptions are made about the error term ϵ in the regression model $y = \beta_0 + \beta_1 x + \epsilon$.

Assumptions about the Error Term ϵ in the Regression Model

$$y = \beta_0 + \beta_1 x + \epsilon$$

1. The error term ϵ is a random variable with mean or expected value of 0; that is, $E(\epsilon) = 0$.

   Implication: Since β_0 and β_1 are constants, $E(\beta_0) = \beta_0$ and $E(\beta_1) = \beta_1$; thus for a given value of x, the expected value of y is

   $$E(y) = \beta_0 + \beta_1 x \qquad (14.13)$$

   Equation (14.13) is referred to as the *regression equation*.

2. The variance of ϵ, denoted by σ^2, is the same for all values of x.

   Implication: The variance of y equals σ^2 and is the same for all values of x.

3. The values of ϵ are independent.

   Implication: The value of ϵ for a particular value of x is not related to the value of ϵ for any other value of x; thus, the value of y for a particular value of x is not related to the value of y for any other value of x.

4. The error term ϵ is a normally distributed random variable.

   Implication: Since y is a linear function of ϵ, y is also a normally distributed random variable.

Figure 14.6 provides an illustration of the model assumptions and their implications; note that in this graphical interpretation, the value of $E(y)$ changes according to the specific value of x considered. However, regardless of the x value, the probability distribution of ϵ and hence the probability distribution of y are normally distributed, each with the same variance. The specific value of the error ϵ at any particular point depends upon whether the actual value of y is greater than or less than $E(y)$.

At this point, we must keep in mind that we are also making an assumption or hypothesis about the relationship between x and y. That is, we have assumed that a straight line represented by $\beta_0 + \beta_1 x$ is the basis for the relationship between the variables. We must not lose sight of the fact that some other model, for instance $\beta_0 + \beta_1 x^2$, may turn out to be a better model for the underlying relationship. After using the sample data to estimate the parameters of the regression model (β_0 and β_1), we will want to conduct further analysis to determine whether the model assumed or hypothesized appears to be valid.

FIGURE 14.6
Assumptions for the Regression Model

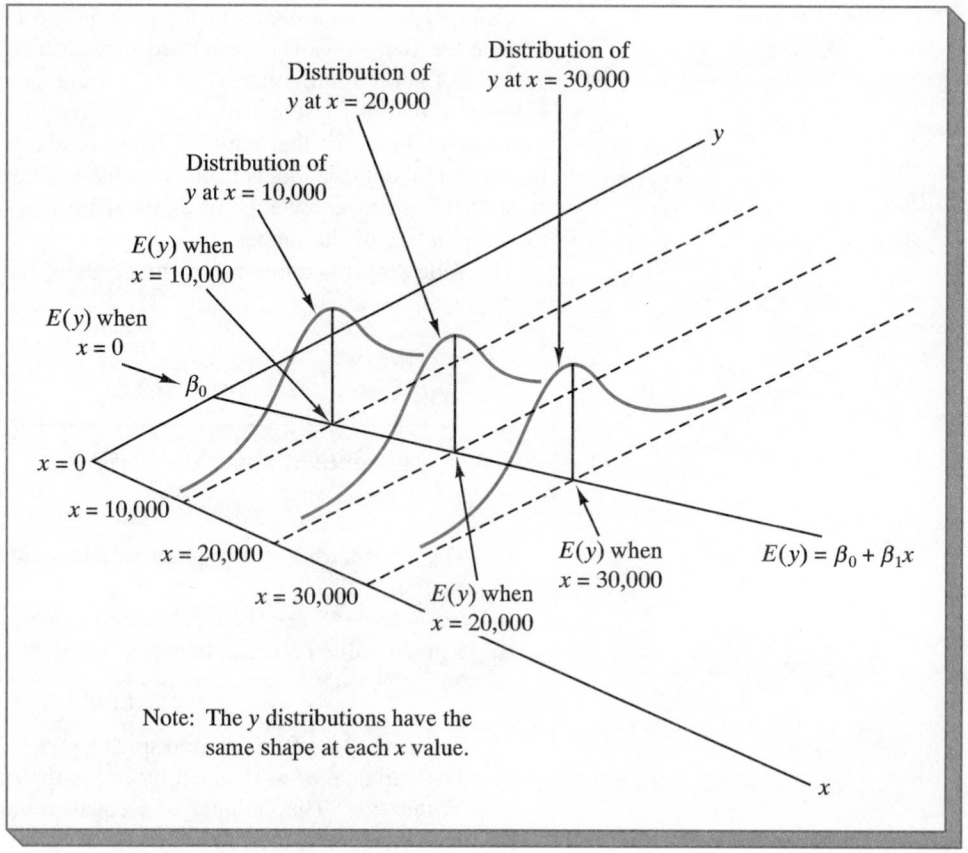

Note: The y distributions have the same shape at each x value.

The Relationship between the Regression Equation and the Estimated Regression Equation

Recall from Chapter 8 that when data were available for just one variable, the objective was to use a sample statistic (e.g., the sample mean) to make inferences about the corresponding population parameter (e.g., population mean). When we discussed the least squares method in Section 14.1, we presented formulas for computing the y-intercept (b_0) and the slope (b_1) of the estimated regression equation. The value of b_0 is a sample statistic that provides an estimate of the β_0 parameter in the regression model, and the value of b_1 is a sample statistic that provides an estimate of the β_1 parameter. Thus, since the regression equation is $E(y) = \beta_0 + \beta_1 x$, the best estimate of the regression equation is provided by the estimated regression equation $\hat{y} = b_0 + b_1 x$, and $\hat{y}$ therefore provides the estimate of $E(y)$. Figure 14.7 summarizes these concepts.

14.4   Testing for Significance

In Section 14.2 we saw how the coefficient of determination (r^2) could be used as a measure of the goodness of fit of the estimated regression line. Larger values of r^2 indicated a better fit. However, the value of r^2 does not allow us to conclude whether a regression relationship is statistically significant. To draw conclusions concerning statis-

FIGURE 14.7
Estimating the Population Regression Equation Using Sample Data

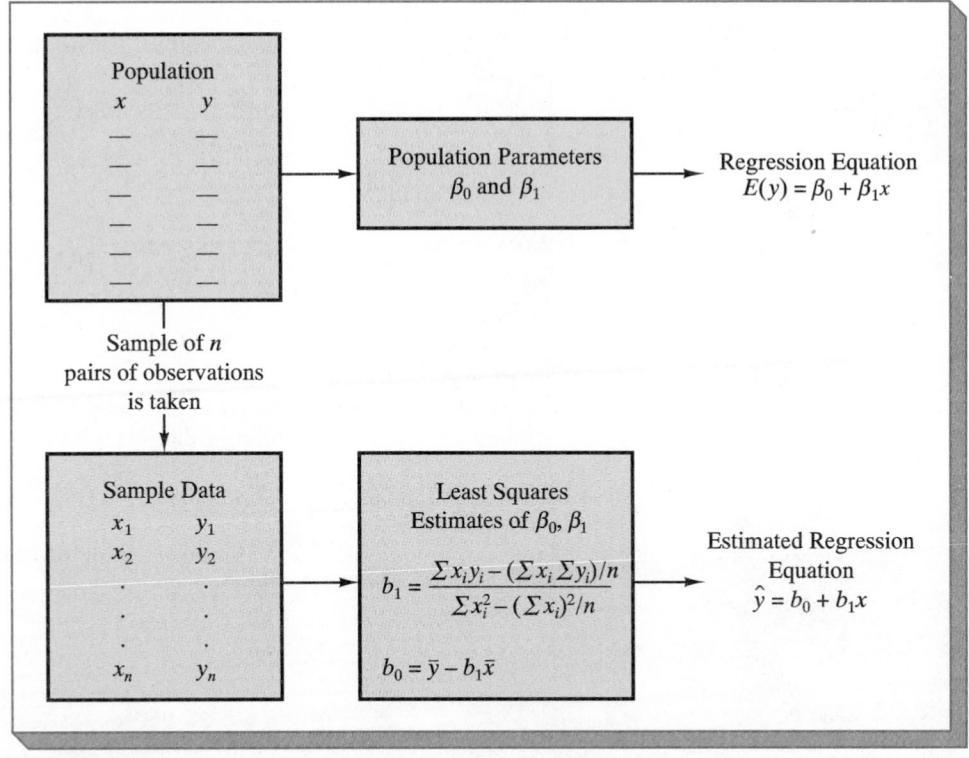

tical significance, we must, among other things, take the sample size into consideration. In this section we show how to conduct significance tests that will allow us to draw conclusions about the existence of a regression relationship.

An Estimate of σ^2

As stated in Section 14.3, σ^2 is the variance of the error term ϵ in the regression model $y = \beta_0 + \beta_1 x + \epsilon$. In the following discussion we show how to obtain an estimate of σ^2 using the sum of squares due to error, SSE. First, recall that SSE is a measure of the variability of the actual observations about the estimated regression equation. With $\hat{y}_i = b_0 + b_1 x_i$, SSE can be written:

$$\text{SSE} = \Sigma(y_i - \hat{y}_i)^2 = \Sigma(y_i - b_0 - b_1 x_i)^2$$

Every sum of squares has associated with it a number called its degrees of freedom. The degrees of freedom indicates how many independent pieces of information involving the n independent values $y_1, y_2, \ldots, y_n$ are used to compute the sum of squares. Statisticians have shown that SSE has $n - 2$ degrees of freedom since 2 parameters (β_0 and β_1) have to be estimated in order to compute SSE.

Mean square is a number computed by dividing a sum of squares by its degrees of freedom. Thus, the mean square due to error, also referred to as mean square error, is computed by dividing SSE by its degrees of freedom, $n - 2$. Statisticians have shown that the mean square error, denoted MSE, provides an unbiased estimate of σ^2. Since MSE is an estimate of σ^2, the notation s^2 is also used.

Mean Square Error (Estimate of σ^2)

$$s^2 = \text{MSE} = \frac{\text{SSE}}{n - 2} \tag{14.14}$$

From the data in Table 14.8, we see that SSE = 1530; thus, for Armand's Pizza Parlors we have

$$s^2 = \text{MSE} = \frac{1530}{8} = 191.25$$

Thus, an unbiased estimate of σ^2 is equal to $s^2 = 191.25$. To estimate σ we take the square root of s^2. The resulting value s is referred to as the *standard error of the estimate*.

Standard Error of the Estimate

$$s = \sqrt{\text{MSE}} = \sqrt{\frac{\Sigma(y_i - \hat{y}_i)^2}{n - 2}} \tag{14.15}$$

For Armand's Pizza Parlors, $s = \sqrt{\text{MSE}} = \sqrt{191.25} = 13.829$. In the discussion which follows, we will use this estimate of σ in tests for the significance of the regression equation.

t Test

Recall that the regression equation is assumed to be $E(y) = \beta_0 + \beta_1 x$. If there really exists a relationship of this form between x and y in which the value of x influences the value of y, β_1 could not equal 0. Thus, to test for a significant relationship between the two variables we use the following hypotheses:

$$H_0: \beta_1 = 0$$

$$H_a: \beta_1 \neq 0$$

Before using the t test for these hypotheses, we need to consider the properties of b_1, the least squares estimator of β_1.

First, let us consider what would have happened if we had used a different random sample for the same regression study. For example, suppose that in the Armand's Pizza Parlors example we had used the sales records of a different sample of 10 restaurants. A regression analysis of this new sample might result in an estimated regression equation similar to our previous estimated regression equation $\hat{y} = 60 + 5x$. However, it is doubtful that we would obtain exactly the same equation (with an intercept of exactly 60 and a slope of exactly 5). Indeed, b_0 and b_1, the least squares estimators, are sample statistics that have their own sampling distributions. The properties of the sampling distribution of b_1 are shown below.

Sampling Distribution of b_1

Expected Value:

$$E(b_1) = \beta_1$$

Standard Deviation:

$$\sigma_{b_1} = \frac{\sigma}{\sqrt{\Sigma x_i^2 - (\Sigma x_i)^2/n}} \qquad (14.16)$$

Distribution Form:

Normal

Note that the expected value of b_1 is equal to β_1, so b_1 is an unbiased estimator of β_1.

Since we do not know the value of σ, we develop an estimate of σ_{b_1}, denoted s_{b_1}, by estimating σ with s in (14.16). Thus, we obtain the following estimate of σ_{b_1}.

Estimated Standard Deviation of b_1

$$s_{b_1} = \frac{s}{\sqrt{\Sigma x_i^2 - (\Sigma x_i)^2/n}} \qquad (14.17)$$

For Armand's Pizza Parlors, $s = 13.829$. Thus,

$$s_{b_1} = \frac{13.829}{\sqrt{2528 - (140)^2/10}} = .5803$$

The t test regarding β_1 is based on the fact that the test statistic

$$\frac{b_1 - \beta_1}{s_{b_1}}$$

follows a t distribution with $n - 2$ degrees of freedom. If the null hypothesis is true, then $\beta_1 = 0$ and $t = b_1/s_{b_1}$. Using b_1/s_{b_1} as the test statistic, the rejection rule to test $H_0: \beta_1 = 0$ versus $H_a: \beta_1 \neq 0$ is as follows:

$$\text{Reject } H_0 \text{ if } \quad \frac{b_1}{s_{b_1}} < -t_{\alpha/2} \quad \text{ or if } \quad \frac{b_1}{s_{b_1}} > t_{\alpha/2}$$

For Armand's Pizza Parlors, we have $b_1 = 5$ and $s_{b_1} = .5803$. Thus, we have $b_1/s_{b_1} = 5/.5803 = 8.62$. From Table 2 of Appendix B we find that the t value corresponding to $\alpha = .01$ and 8 degrees of freedom is $t_{.005} = 3.355$. Since $b_1/s_{b_1} = 8.62 > 3.355$, we reject H_0 and conclude at the .01 level of significance that β_1 is not equal to zero.

F Test

The *t* test has been used to test the null hypothesis $H_0: \beta_1 = 0$. An *F* test also exists for testing this null hypothesis. In regression models with only one independent variable, the *t* test and the *F* test yield the same conclusion; that is, if the *t* test results in the rejection of H_0, the *F* test will also lead to the rejection of H_0. But with more than one independent variable, only the *F* test can be used to test for a significant relationship between a dependent variable and a set of independent variables. Here we introduce the *F* test and show that it leads to the same conclusion as the *t* test.

The hypotheses we will be testing are the same as before:

$$H_0: \beta_1 = 0$$

$$H_a: \beta_1 \neq 0$$

The logic behind the use of the *F* test for determining whether the relationship between *x* and *y* is statistically significant is based on our being able to develop two independent estimates of σ^2. We have just seen that MSE provides an estimate of σ^2. If the null hypothesis $H_0: \beta_1 = 0$ is true, the *mean square due to regression* or *mean square regression,* denoted MSR, provides another *independent* estimate of σ^2.

To compute MSR, recall that for any sum of squares that the mean square is the sum of squares divided by its degrees of freedom. Thus,

$$\text{MSR} = \frac{\text{SSR}}{\text{Regression degrees of freedom}}$$

Since the number of degrees of freedom for SSR is equal to the number of independent variables, we can write

$$\text{MSR} = \frac{\text{SSR}}{\text{Number of independent variables}} \tag{14.18}$$

In this chapter we only consider models involving one independent variable; in this case MSR = SSR/1 = SSR. Thus, for Armand's Pizza Parlors, MSR = SSR = 14,200.

If the null hypothesis ($H_0: \beta_1 = 0$) is true, MSR and MSE are two independent estimates of σ^2. In this case, the sampling distribution of MSR/MSE follows an *F* distribution with numerator degrees of freedom equal to 1 and denominator degrees of freedom equal to $n - 2$. The test concerning the significance of the regression relationship is based on the following *F* statistic:

$$F = \frac{\text{MSR}}{\text{MSE}} \tag{14.19}$$

Given any sample size, the numerator of the *F* statistic will increase as more of the variability in *y* is explained by the regression model and decrease as less is explained. Similarly, the denominator will increase if there is more variability about the estimated regression line and decrease if there is less variability. Thus one would intuitively expect large values of *F* = MSR/MSE to cast doubt on the null hypothesis and lead us to the conclusion that $\beta_1 \neq 0$. Indeed, this is correct; large values of *F* lead to rejection of H_0 and the conclusion that the relationship between *x* and *y* is statistically significant.

Let us now conduct the *F* test for the Armand's Pizza Parlors. Assume that the level of significance is $\alpha = .01$. From Table 4 of Appendix B we can determine the critical *F* value by locating the value corresponding to numerator degrees of freedom equal to 1 (the number of independent variables) and denominator degrees of freedom equal to $n - 2 = 10 - 2 = 8$. Thus, we obtain *F* = 11.26. Hence, the appropriate rejection rule for Armand's Pizza Parlors is written

$$\text{Reject } H_0 \text{ if MSR/MSE} > 11.26$$

Since MSR/MSE = 14,200/191.25 = 74.25 is greater than the critical value $F_{.01}$ = 11.26, we can reject H_0 and conclude that there is a statistically significant relationship between annual sales and the size of the student population.

We see that for Armand's Pizza Parlors, the F test leads to the same conclusion as the t test. In fact, as we stated earlier, in simple linear regression these two tests are equivalent. It is interesting to note, however, that the t test required both a lower- and upper-tailed rejection region, each corresponding to a level of significance of $\alpha/2$. However, when using the F test, the rejection region only appears in the upper tail of the F distribution. In the case of the F test, MSR/MSE will always be positive so the null hypothesis will only be rejected if we find a large value for MSR/MSE.

A Caution regarding Statistical Significance

It is important to note here that rejection of H_0 does not permit us to conclude that the relationship between x and y is *linear*. However, it is valid to conclude that x and y are related and that a linear relationship explains a significant amount of the variability in y over the range of x values observed in the sample. To illustrate this qualification, we call your attention to Figure 14.8, where an F test (on $\beta_1 = 0$) yielded the conclusion that x and y were related. The figure shows that the actual relationship is nonlinear. In the graph we see that the linear approximation is very good for the values of x used in developing the least squares line, but it is very bad for larger values of x.

Given a significant relationship, we should feel confident in using the regression equation for predictions corresponding to x values within the range of the x values for the sample. For Armand's Pizza Parlors, this corresponds to values of x between 2 and 26. But unless there are reasons to believe the model is valid beyond this range, predictions outside the range of the independent variable should be made with caution. For Armand's

FIGURE 14.8

Example of a Linear Approximation of a Nonlinear Relationship

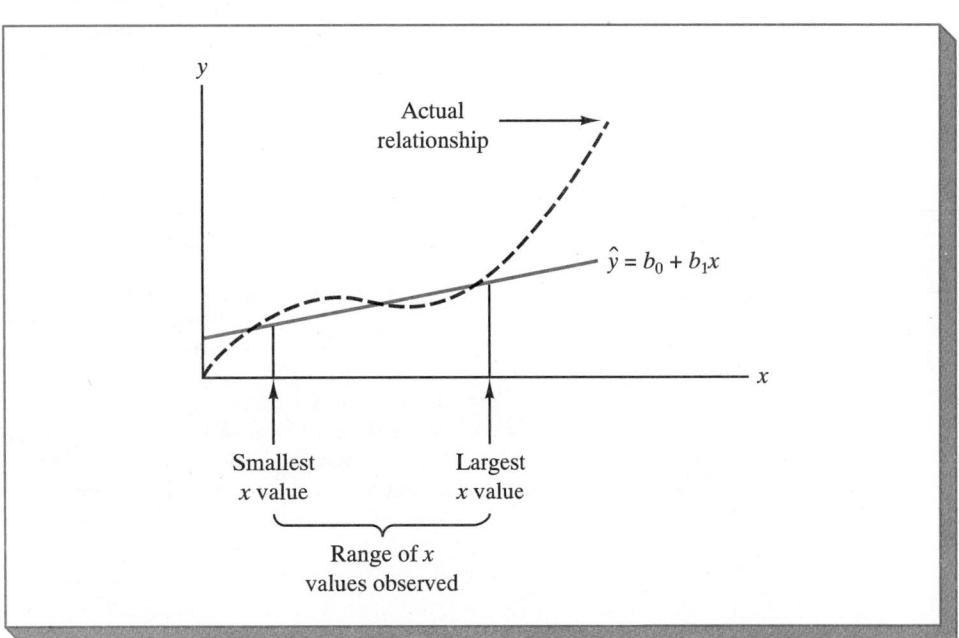

Pizza Parlors, since the regression relationship has been found significant at the .01 level, we should feel confident using it to predict sales whenever the student population is between 2000 and 26,000.

NOTES &
COMMENTS

1. The assumptions made about the error term (Section 14.3) are what permit the tests of statistical significance in this section. The properties of the sampling distribution of b_1 and the subsequent F and t tests directly follow from these assumptions.

2. Do not confuse statistical significance with practical significance. With very large sample sizes, it is possible to obtain statistically significant results for small values of b_1; in such cases, one must exercise care in concluding that the relationship has practical significance.

3. The reason that the F test and the t test yield the same result *for simple linear regression* is that $F = t^2$. The critical value for the F test is the square of the critical value for the t test, and the test statistic for the F test is the square of the test statistic for the t test.

□ □ **Exercises**

Methods

SELF TEST ▶

22. The data from Exercise 1 are shown below:

x_i	1	2	3	4	5
y_i	3	7	5	11	14

a. Compute the mean square error using (14.14).
b. Compute the standard error of the estimate using (14.15).
c. Compute the standard deviation of b_1 using (14.17).
d. Use the t test to test the following hypotheses ($\alpha = .05$):

$$H_0: \beta_1 = 0$$

$$H_a: \beta_1 \neq 0$$

e. Use the F test to test the hypotheses in (d) at the $\alpha = .05$ level of significance.

23. The data from Exercise 2 are shown below:

x_i	2	3	5	1	8
y_i	25	25	20	30	16

a. Compute the mean square error using (14.14).
b. Compute the standard error of the estimate using (14.15).
c. Compute the standard deviation of b_1 using (14.17).
d. Use the t test to test the following hypotheses ($\alpha = .05$):

$$H_0: \beta_1 = 0$$

$$H_a: \beta_1 \neq 0$$

e. Use the F test to test the hypotheses in (d) at the $\alpha = .05$ level of significance.

24. The data from Exercise 3 are shown:

x_i	2	4	5	7	8
y_i	2	3	2	6	4

Test whether x and y are related at the $\alpha = .05$ level of significance.

Applications

SELF TEST ▷

25. In Exercise 5 the data in Table 14.13 were collected.

a. Use the t test to test the following hypotheses ($\alpha = .05$):

$$H_0: \beta_1 = 0$$

$$H_a: \beta_1 \neq 0$$

b. Use the F test to test the hypotheses in (a) at the $\alpha = .05$ level of significance.

26. Refer to Exercise 10, where an estimated regression line relating square footage to selling prices of new homes was developed. Test whether selling price and square footage are related at the $\alpha = .01$ level of significance.

27. Refer to Exercise 12, where an estimated regression line relating the percentage of management jobs held by women and the percentage of women employed was developed. Test whether these two variables are related at the $\alpha = .05$ level of significance.

28. Refer to Exercise 9, where an estimated regression line relating years of experience and annual sales was developed. At the $\alpha = .05$ level of significance, determine whether annual sales and years of experience are related.

29. Refer to Exercise 20, where an estimated regression line relating light absorbence readings and milligrams of protein present in a liver sample was developed. Test whether the absorbence readings and amount of protein present are related at the $\alpha = .01$ level of significance.

TABLE 14.13

GPA	Monthly Salary ($)
2.6	1400
3.4	1700
3.6	2100
3.2	1600
3.5	2000
2.9	1700

14.5   Estimation and Prediction

For Armand's Pizza Parlors we concluded that annual sales y and the size of the student population x are related. Moreover, the estimated regression equation $\hat{y} = 60 + 5x$ describes the relationship between x and y. Now we can begin to use the estimated regression line to develop interval estimates of annual sales for a given student population.

There are two types of interval estimates to consider. The first is an interval estimate of the mean value of y for a particular value of x. We refer to this type of interval estimate as a *confidence interval estimate*. For instance, we might want a confidence interval estimate of the *expected* annual sales for all restaurants located near a campus with a student population of 10,000. In this case, the expected annual sales represents the average of the annual sales for all restaurants located near a college campus with 10,000 students.

The second type of interval estimate that we will consider is appropriate in situations where we want to predict an individual value of y corresponding to a given value of x. We refer to this type of interval estimate as a *prediction interval estimate*. For instance, we might be interested in developing a prediction interval estimate for the annual sales for one specific restaurant located near Talbot College, a school with 10,000 students. In this case, our interest is in predicting the annual sales for one specific restaurant, as opposed to predicting the average sales for all restaurants located near campuses with 10,000 students.

Confidence Interval Estimate of the Mean Value of y

Suppose we wanted to develop a confidence interval estimate of the mean or expected value of annual sales for all restaurants located near a campus with 10,000 students. First, recall that for Armand's Pizza Parlors that the expected value of annual sales is given by

$$E(y) = \beta_0 + \beta_1 x$$

So, when the student population size is 10,000, $x = 10$; hence $E(y) = \beta_0 + \beta_1(10)$.

The least squares method was used to develop estimates of β_0 and β_1 and hence an estimate of the regression equation; for Armand's Pizza Parlors the estimated regression equation was found to be $\hat{y} = 60 + 5x$. Thus, for restaurants located near a campus with 10,000 students, we would obtain $\hat{y} = 60 + 5(10) = 110$. The corresponding estimate of expected annual sales would be $110,000.

In general, the point estimate of $E(y)$ for a particular value of x is the corresponding value of $\hat{y}$ given by the estimated regression equation. We denote the particular value of x using x_p, the mean value of y at x_p using $E(y_p)$, and the estimate of $E(y_p)$ using $\hat{y}_p = b_0 + b_1 x_p$.

Since b_0 and b_1 are only estimates of β_0 and β_1, we cannot expect that the estimated value $\hat{y}_p$ will exactly equal $E(y_p)$. For instance, in our example above, we do not expect that the mean annual sales for all restaurants located near a campus with 10,000 students to exactly equal $110,000, our estimated value. If we want to make an inference about how close $\hat{y}_p$ is to the true mean value $E(y_p)$, however, we will have to consider the variability that exists when we develop estimates based on the estimated regression equation. Statisticians have developed the following estimate of the variance of $\hat{y}_p$:

$$\text{Estimated variance of } \hat{y}_p = s_{\hat{y}_p}^2 = s^2 \left[\frac{1}{n} + \frac{(x_p - \bar{x})^2}{\Sigma x_i^2 - (\Sigma x_i)^2/n} \right]$$

Hence, an estimate of the standard deviation of $\hat{y}_p$ is given by the square root of the variance.

$$s_{\hat{y}_p} = s \sqrt{\frac{1}{n} + \frac{(x_p - \bar{x})^2}{\Sigma x_i^2 - (\Sigma x_i)^2/n}} \tag{14.20}$$

For Armand's Pizza Parlors, the estimated standard deviation of $\hat{y}_p$ for a restaurant located near a campus with 10,000 students is

$$s_{\hat{y}_p} = 13.829 \sqrt{\frac{1}{10} + \frac{(10 - 14)^2}{2528 - (140)^2/10}}$$

$$= 13.829\sqrt{.1282}$$

$$= 4.95$$

The confidence interval estimate of $E(y_p)$ is as follows.

Confidence Interval Estimate of $E(y_p)$

$$\hat{y}_p \pm t_{\alpha/2} s_{\hat{y}_p} \tag{14.21}$$

where the confidence coefficient is $1 - \alpha$ and the t value has $n - 2$ degrees of freedom.

Thus, to develop a 95% confidence interval estimate of the expected annual sales for all restaurants located near a campus with 10,000 students, we need to find the t value from Table 2 of Appendix B corresponding to $n - 2 = 10 - 2 = 8$ degrees of freedom and $\alpha = .05$. Doing so, we find $t_{.025} = 2.306$. Hence, the resulting confidence interval is

$$[b_0 + b_1(10)] \pm 2.306 s_{\hat{y}_p}$$

$$[60 + 5(10)] \pm 2.306(4.95)$$

$$110 \pm 11.415$$

In dollars, the 95% confidence interval estimate is $\$110,000 \pm \$11,415$. Thus, we obtain $\$98,585$ to $\$121,415$ as a confidence interval estimate of the expected or average sales volume for all restaurants located near a campus with 10,000 students.

Note that the estimated standard deviation of $\hat{y}_p$ (see (14.20)) is smallest when the given value of $x_p = \bar{x}$. In this case, (14.20) becomes

$$s_{\hat{y}_p} = s \sqrt{\frac{1}{n} + \frac{(\bar{x} - \bar{x})^2}{\Sigma x_i^2 - (\Sigma x_i)^2/n}} = s \sqrt{\frac{1}{n}}$$

which implies that we can expect to make our best estimates of $E(y_p)$ at the mean of the independent variable.

Prediction Interval Estimate for an Individual Value of y

In the preceding discussion we showed how to develop a confidence interval estimate of the expected annual sales for all restaurants located near a campus with 10,000 students. Now we turn to the problem of developing point and interval estimates for an individual value of y corresponding to a particular value of x. Suppose we want to predict annual sales for one specific restaurant located near Talbot College, a school with 10,000 students.

The point estimate for an individual value of y is given by $\hat{y}_p = b_0 + b_1 x_p$. Hence, the point estimate of annual sales for one specific restaurant located near Talbot College is $\hat{y}_p = 60 + 5(10) = 110$. Note that this is the same as the point estimate of the mean annual sales for a restaurant located near a campus with 10,000 students.

To develop a prediction interval estimate, we must first determine the variance associated with using $\hat{y}_p$ as an estimate of a particular value of y when $x = x_p$. This variance is made up of the sum of the following two components:

1. The variance of individual y values about the mean $E(y_p)$, an estimate of which is given by s^2.

2. The variance associated with using $\hat{y}_p$ to estimate $E(y_p)$, an estimate of which is given by $s_{\hat{y}_p}^2$.

Statisticians have shown that an estimate of the variance of an individual value of y_p, which we denote s_{ind}^2, is given by

$$s_{ind}^2 = s^2 + s_{\hat{y}_p}^2$$

$$= s^2 + s^2 \left[\frac{1}{n} + \frac{(x_p - \bar{x})^2}{\Sigma x_i^2 - (\Sigma x_i)^2/n} \right]$$

$$= s^2 \left[1 + \frac{1}{n} + \frac{(x_p - \bar{x})^2}{\Sigma x_i^2 - (\Sigma x_i)^2/n} \right]$$

Hence, an estimate of the standard deviation of an individual value of y_p is given by

$$s_{ind} = s \sqrt{1 + \frac{1}{n} + \frac{(x_p - \bar{x})^2}{\sum x_i^2 - (\sum x_i)^2/n}} \qquad (14.22)$$

For Armand's Pizza Parlors, the estimated standard deviation corresponding to the prediction of annual sales for one specific restaurant located near a campus with 10,000 students is computed as follows:

$$s_{ind} = 13.829 \sqrt{1 + \frac{1}{10} + \frac{(10 - 14)^2}{2528 - (140)^2/10}}$$

$$= 13.829\sqrt{1.1282}$$

$$= 14.69$$

The prediction interval estimate of y_p is given by (14.23).

Prediction Interval Estimate of y_p

$$\hat{y}_p \pm t_{\alpha/2}s_{ind} \qquad (14.23)$$

where the confidence coefficient is $1 - \alpha$ and the t value has $n - 2$ degrees of freedom.

Thus, a 95% prediction interval for sales for one specific restaurant located near a campus with 10,000 students is

$$[b_0 + b_1(10)] \pm t_{\alpha/2}(14.69)$$

$$[60 + 5(10)] \pm 2.306(14.69)$$

$$110 \pm 33.875$$

Therefore, the 95% prediction interval for annual sales for one specific restaurant located near a campus with 10,000 students is $76,125 to $143,875. We note that this prediction interval is greater in width than the confidence interval for mean sales of all restaurants located near campuses with 10,000 students ($98,585 to $121,415). This difference simply reflects the fact that we are able to estimate the mean annual sales volume with more precision than we can the annual sales for any individual restaurant.

❑ ❑ Exercises

Methods

SELF TEST ▶ **30.** The data from Exercise 1 are shown below:

x_i	1	2	3	4	5
y_i	3	7	5	11	14

a. Use (14.20) to estimate the standard deviation of y_p when $x = 4$.

b. Use (14.21) to develop a 95% confidence interval estimate of the expected value of y when $x = 4$.

c. Use (14.22) to estimate the standard deviation of an individual value when $x = 4$.

d. Use (14.23) to develop a 95% prediction interval for $x = 4$.

31. The data from Exercise 2 are shown below:

x_i	2	3	5	1	8
y_i	25	25	20	30	16

a. Estimate the standard deviation of y_p when $x = 3$.

b. Develop a 95% confidence interval estimate of the expected value of y when $x = 3$.

c. Estimate the standard deviation of an individual value when $x = 3$.

d. Develop a 95% prediction interval when $x = 3$.

32. The data from Exercise 3 are shown below:

x_i	2	4	5	7	8
y_i	2	3	2	6	4

Develop the 95% confidence and prediction intervals when $x = 3$. Explain why these two intervals are different.

Applications

33. As an extension of Exercise 12, develop a 95% confidence interval for estimating the mean percentage of management jobs held by women for companies in which 60% of the employees are women.

SELF TEST ▶ **34.** As an extension of Exercise 5, develop a 95% confidence interval for estimating the mean starting salary for students with a 3.0 GPA.

35. As an extension of Exercise 10, develop a 95% confidence interval for predicting the mean selling price for homes with 2200 square feet of living space.

SELF.TEST ▶ **36.** As an extension of Exercise 5, develop a 95% prediction interval for estimating the starting salary of Joe Heller, who has a GPA of 3.0.

37. As an extension of Exercise 10, develop a 95% prediction interval for the selling price of a home on Highland Terrace with 2800 square feet.

38. A study conducted by a department of transportation regarding driving speed and mileage for midsize automobiles resulted in the data shown in Table 14.14.

a. Determine the estimated regression equation that relates mileage to the driving speed.

b. At the $\alpha = .05$ level of significance, determine whether mileage and driving speed are related.

c. Did the estimated regression line provide a good fit to the data?

d. Develop a 95% confidence interval for estimating the mean mileage for cars that are driven at 50 miles per hour.

e. If we were interested in one specific car that was driven at 50 miles per hour, how would our estimate of mileage change as compared to the estimate developed in (d)?

TABLE 14.14

Driving Speed (mph)	Mileage (mpg)
30	28
50	25
40	25
55	23
30	30
25	32
60	21
25	35
50	26
55	25

14.6   Computer Solution of Regression Problems

Performing all the computations associated with regression analysis can be quite time-consuming. In this section we discuss how the computational burden can be simplified by using a computer software package. The general procedure followed in using computer packages is for the user to input the data (x and y values for the sample) together with some instructions concerning the types of analyses that are required. The software package performs the analysis and prints the results in an output report. Before discussing the details of this approach, we discuss the use of the analysis of variance (ANOVA) table as a device for summarizing the calculations performed in regression analysis. The ANOVA table is an important component of the output report produced by most software packages.

The ANOVA Table

In Chapter 13 we saw how the ANOVA table could provide a convenient summary of the computational aspects of analysis of variance. In regression analysis a similar table can be developed. Table 14.15 shows the general form of the ANOVA table for two-variable regression studies and Table 14.16 shows the ANOVA table for the Armand's Pizza Parlors problem. It can be seen that the relationship that holds among the sum of squares (i.e., SST = SSR + SSE) also holds for the degrees of freedom. That is, the degrees of freedom for the total sum of squares is equal to the degrees of freedom for the regression sum of squares plus the degrees of freedom for the error sum of squares.

Computer Output

In Figure 14.9 we show the input and the Minitab computer output for the Armand's Pizza problem. Values of the independent variable, student population, are entered into column 1 and values of the dependent variable, annual sales, are entered into column 2. After the data has been input, the variables are given names. The independent variable x is labeled POP and the dependent variable is labeled SALES. The command REGRESS C2 1 C1 is given by the user to perform the regression analysis. The C2 indicates that the dependent variable is in column 2 and the 1 C1 indicates that one independent variable located in column 1 is to be used in the regression analysis. The subcommand PREDICT 10 is given

---

TABLE 14.15
General Form of the ANOVA Table for Two-Variable Regression Analysis

Source of Variation	Sum of Squares	Degrees of Freedom	Mean Square	F
Regression	SSR	1	$\text{MSR} = \dfrac{\text{SSR}}{1}$	$F = \dfrac{\text{MSR}}{\text{MSE}}$
Error	SSE	$n - 2$	$\text{MSE} = \dfrac{\text{SSE}}{n - 2}$	
Total	SST	$n - 1$		

TABLE 14.16
ANOVA Table for the Armand's Pizza Parlors Problem

Source of Variation	Sum of Squares	Degrees of Freedom	Mean Square	F
Regression	14,200	1	$\dfrac{14,200}{1} = 14,200$	$\dfrac{14,200}{191.25} = 74.25$
Error	1,530	8	$\dfrac{1530}{8} = 191.25$	
Total	15,730	9		

FIGURE 14.9   **Minitab Output for the Armand's Pizza Parlors Problem**

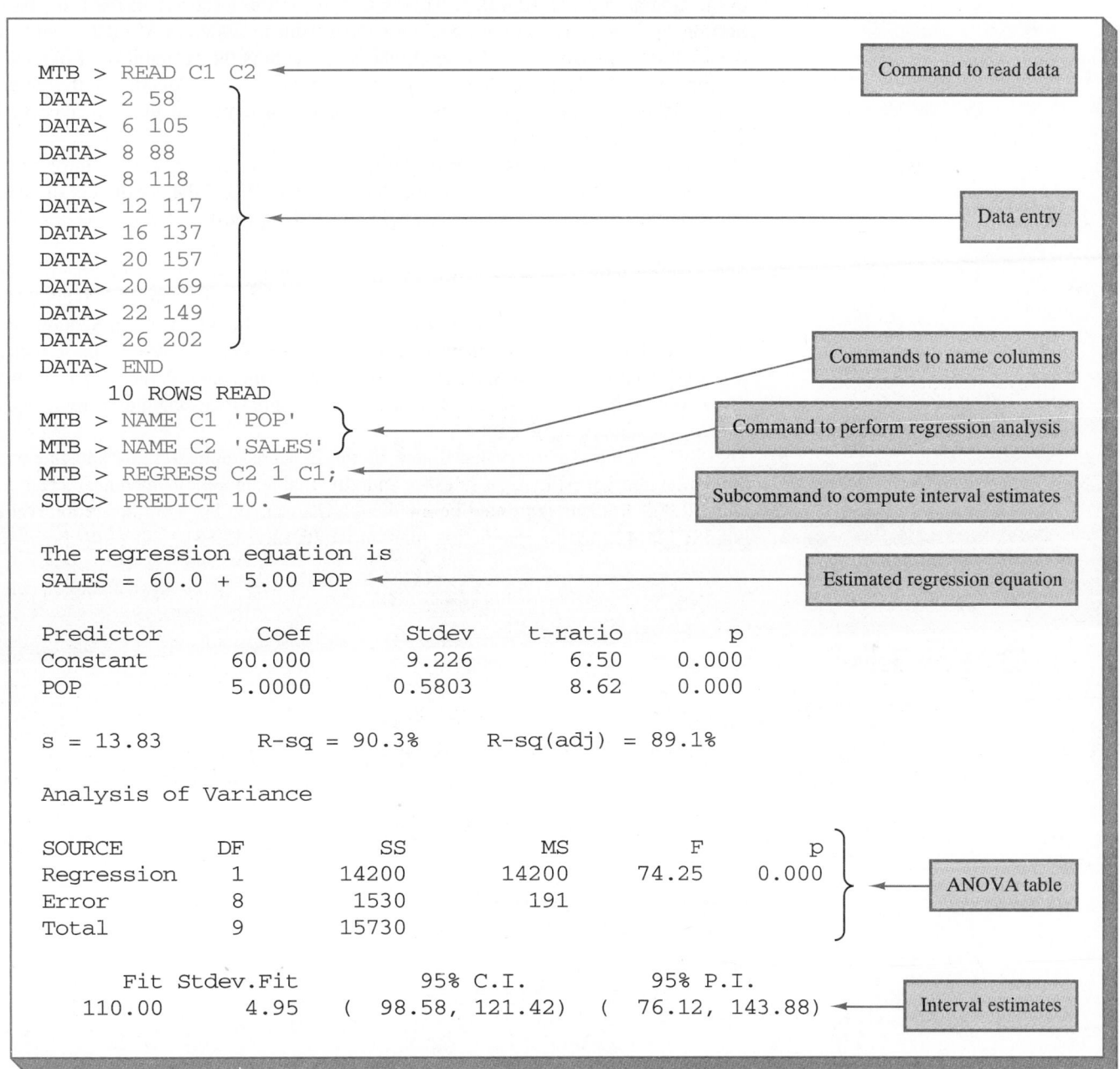

```
MTB > READ C1 C2                                          Command to read data
DATA> 2  58
DATA> 6  105
DATA> 8  88
DATA> 8  118
DATA> 12 117                                              Data entry
DATA> 16 137
DATA> 20 157
DATA> 20 169
DATA> 22 149
DATA> 26 202
DATA> END
     10 ROWS READ                                         Commands to name columns
MTB > NAME C1 'POP'
MTB > NAME C2 'SALES'                                     Command to perform regression analysis
MTB > REGRESS C2 1 C1;
SUB C> PREDICT 10.                                        Subcommand to compute interval estimates


The regression equation is
SALES = 60.0 + 5.00 POP                                   Estimated regression equation


Predictor       Coef        Stdev      t-ratio        p
Constant      60.000        9.226         6.50    0.000
POP           5.0000       0.5803         8.62    0.000


s = 13.83        R-sq = 90.3%      R-sq(adj) = 89.1%


Analysis of Variance


SOURCE         DF            SS           MS          F        p
Regression      1         14200        14200      74.25    0.000
Error           8          1530          191                          ANOVA table
Total           9         15730


      Fit Stdev.Fit          95% C.I.          95% P.I.
   110.00      4.95    ( 98.58, 121.42)  ( 76.12, 143.88)             Interval estimates
```

to direct the software package to provide interval estimates of sales when the population is 10,000. The interpretation of the output is as follows:

1. Minitab prints the estimated regression equation as SALES = 60.0 + 5.00 POP.
2. A table is printed that shows the values of the coefficients b_0 and b_1, the standard deviation of each coefficient, the t value obtained by dividing each coefficient value by its standard deviation, and the p-value associated with the t test. Thus, to test

$H_0: \beta_1 = 0$ versus $H_a: \beta_1 \neq 0$, we could compare 8.62 (located in the t-ratio column) to the appropriate critical value. This is the procedure described in the last part of Section 14.4. Alternatively, we could use the p-value provided by Minitab to perform the same test. Recall from Chapter 9 that the p-value is the probability of obtaining a sample result more unlikely than what is observed. Since the p-value in this case is zero (to three decimal places), the sample results indicate that the null hypothesis ($H_0: \beta_1 = 0$) should be rejected.

3. Minitab prints the standard error of the estimate $s = 13.83$, as well as information regarding the goodness of fit. Note that "R-sq = 90.3%" is the coefficient of determination expressed as a percentage. The output "R-Sq (adj) = 89.1%" is discussed in Chapter 15.

4. The ANOVA table is printed below the heading, Analysis of Variance. Note that DF is an abbreviation for degrees of freedom and that MSR is given as 14,200 and MSE as 191. The ratio of these two values provides the F value of 74.25; in Section 14.4 we showed how the F value can be used to determine if there is a significant statistical relationship between SALES and POP. Minitab also prints the p-value associated with this F test. Since the p-value is zero (to three decimal places), the relationship is judged statistically significant.

5. The 95% confidence interval estimate of the expected annual sales and the 95% prediction interval estimate of sales for an individual restaurant located near a campus with 10,000 students is printed below the ANOVA table. The confidence interval is (98.58, 121.42) and the prediction interval is (76.12, 143.88).

❑ ❑ **Exercises**

Methods

SELF TEST ▶

39. The data from Exercise 1 are shown below:

x_i	1	2	3	4	5
y_i	3	7	5	11	14

Develop the ANOVA table for these data.

40. The data from Exercise 2 are shown below:

x_i	2	3	5	1	8
y_i	25	25	20	30	16

Develop the ANOVA table for these data.

41. The data from Exercise 3 are shown below:

x_i	2	4	5	7	8
y_i	2	3	2	6	4

Develop the ANOVA table for these data.

Applications

SELF TEST ▶

42. The commercial division of a real estate firm is conducting a regression analysis of the relationship between x, annual gross rents ($1000s), and y, selling price ($1000s) for apartment buildings. Data have been collected on a number of properties recently sold, and the output has been obtained in a computer run:

```
The regression equation is
Y = 20.0 + 7.21X

Predictor        Coef        Stdev      t-ratio
Constant       20.000       3.2213        6.21
X               7.210       1.3626        5.29

Analysis of Variance

SOURCE          DF            SS
Regression       1        41587.3
Error            7
Total            8        51984.1
```

a. How many apartment buildings were in the sample?
b. Write the estimated regression equation.
c. What is the value of s_{b_1}?
d. Use the F statistic to test the significance of the relationship at an $\alpha = .05$ level of significance.
e. Estimate the selling price of an apartment building with gross annual rents of $50,000.

43. Shown below is a portion of the computer output for a regression analysis relating maintenance expense (dollars per month) to usage (hours per week) of a particular brand of computer terminal:

```
The regression equation is
Y = 6.1092 + .8951X

Predictor        Coef        Stdev
Constant       6.1092       0.9361
X              0.8951       0.1490

Analysis of Variance

SOURCE          DF           SS           MS

Regression       1        1575.76       1575.76
Error            8         349.14         43.64
Total            9        1924.90
```

a. Write the estimated regression equation.
b. Test to see (use a t test) if monthly maintenance expense is related to usage at the .05 level of significance.
c. Use the estimated regression equation to predict monthly maintenance expense for any terminal that is used 25 hours per week.

TABLE 14.17

Population	Value ($1000s)
1410	61
1523	92
1354	93
822	45
746	50
1281	29
1016	56
1070	45
1694	183
1910	156
1745	120
1353	75
1016	122

PRESCRIP

HOME1

44. A regression model relating x, number of salespersons at a branch office, to y, annual sales at the office ($1000s), has been developed. Shown below is the computer output from a regression analysis of the data:

```
The regression equation is
Y = 80.0 + 50.00X

Predictor        Coef        Stdev       t-ratio
Constant         80.0        11.333        7.06
X                50.0         5.482        9.12

Analysis of Variance

SOURCE        DF          SS          MS
Regression     1        6828.6      6828.6
Error         28        2298.8        82.1
Total         29        9127.4
```

a. Write the estimated regression equation.

b. How many branch offices were involved in the study?

c. Compute the F statistic and test the significance of the relationship at an $\alpha = .05$ level of significance.

d. Predict the annual sales at the Memphis branch office. This branch has 12 salespersons.

45. The data in Table 14.17 show the dollar value of prescriptions for 13 pharmacies in Iowa and the population of the city served by the given pharmacy ("The Use of Categorical Variables in Data Envelopment Analysis," R. Banker and R. Morey, *Management Science*, December 1986).

a. Use a computer package to develop a scatter diagram for these data; plot population on the horizontal axis.

b. Does there appear to be any relationship between these two variables?

c. Use the computer package to develop the estimated regression line that could be used to predict the dollar value of prescriptions given the population of the city.

d. Test for the significance of the relationship at an $\alpha = .05$ level of significance.

e. Predict the dollar value for a particular city with a population of 1500 people. Use $\alpha = .05$.

46. The National Association of Home Builders compared the median home prices with the median household incomes in cities throughout the United States (*USA Today*, September 10, 1991) with 23 of the most affordable cities listed below. Both home prices and household incomes are shown in thousands of dollars.

City	Median Income	Median Home Price	City	Median Income	Median Home Price
Amarillo, Texas	36.7	69.0	Lorain, Ohio	38.8	72.5
Brazoria, Texas	42.4	80.0	Mansfield, Ohio	35.6	58.5
Canton, Ohio	34.1	66.0	Milwaukee, Wisconsin	41.8	72.0
Davenport, Iowa	38.4	59.0	Oklahoma City, Oklahoma	34.5	63.0
Daytona Beach, Florida	31.0	63.0	Omaha, Nebraska	38.8	65.0
Detroit, Michigan	44.6	77.0	Rockford, Illinois	41.6	73.5
Fort Walton Beach, Florida	34.2	65.0	Saginaw, Michigan	39.7	61.0
Grand Rapids, Michigan	40.3	73.0	Shreveport, Louisiana	34.4	66.0
Jackson, Michigan	36.8	60.0	Toledo, Ohio	39.4	65.0
Kansas City, Missouri	41.1	77.0	Tulsa, Oklahoma	36.2	68.0
Lansing, Michigan	40.0	70.0	Winter Haven, Florida	30.2	56.0
			Youngstown, Ohio	34.9	59.0

a. Use a computer package to develop a scatter diagram for these data; plot median income on the horizontal axis.

b. Does there appear to be any relationship between these two variables?

c. Use the computer package to develop the estimated regression equation that could be used to predict the median home price given the median income.

d. Test the significance of the relationship at the $\alpha = .05$ level of significance.

e. Did the estimated regression equation provide a good fit? Explain.

f. Predict the expected median home price for cities with a median income of $35,000.

g. Predict the median home price for Elmira, New York, a city with a median income of $35,000.

14.7   Residual Analysis: Testing Model Assumptions

For each observation in a regression analysis, there is a residual; it is the difference between the observed value of the dependent variable y_i and the value predicted by the regression equation $\hat{y}_i$. The residual for observation i, $y_i - \hat{y}_i$, is an estimate of the error resulting from using the estimated regression equation to predict the value of y_i.

The analysis of residuals plays an important role in validating the assumptions made in regression analysis. In Section 14.4 we showed how hypothesis testing can be used to determine whether a regression relationship is statistically significant. Hypothesis tests concerning regression relationships are based on the assumptions made about the regression model. If the assumptions made are not satisfied, the hypothesis tests are not valid, and the estimated regression equation should not be used. However, keep in mind that the regression model is being used only as an approximation of reality, so good judgment must be used to determine whether an assumption violation is severe enough to invalidate the model.

There are two key issues in verifying that the assumptions are satisfied in a regression model. Are the four assumptions concerning the error term ϵ satisfied, and is the form we have assumed for the model appropriate?

Regression analysis begins with an assumption concerning the appropriate form of the regression model. The simple linear regression model assumes the form

$$y = \beta_0 + \beta_1 x + \epsilon$$

With this form, y is a linear function of x. The assumptions regarding the error term (presented in Section 14.3) are as follows:

1. $E(\epsilon) = 0$.

2. The variance of ϵ, denoted by σ^2, is the same for all values of x.

3. The values of ϵ are independent.

4. The error term ϵ is a normally distributed random variable.

Validating the assumptions concerning the error term ϵ means using the residuals to check to see if these assumptions seem reasonable.

Recall that the first assumption concerning ϵ implies that the regression equation is

$$E(y) = \beta_0 + \beta_1 x$$

This regression equation shows a linear relationship between x and the expected value of y, $E(y)$. Validating the assumption concerning model form means satisfying ourselves that the relationship between independent and dependent variable is adequately represented by the regression equation. It is possible that the true relationship between x and y is curvilinear and/or that more independent variables should have been included (multiple regression). We will see how the statistician uses residual analysis to recognize when this assumption concerning model form might be violated.

The residuals $y_i - \hat{y}_i$ are estimates of ϵ; with n observations in a regression analysis, we have n residuals. Residual plots are graphical presentations of the residuals that help reveal patterns and thus help determine whether the assumptions concerning ϵ are satisfied. Three of the most common residual plots are

1. A plot of the residuals against the independent variable x.
2. A plot of the residuals against the predicted value of the dependent variable $\hat{y}$.
3. A standardized plot in which each residual is replaced by its z-score (i.e., the mean is subtracted and the result is divided by the standard error).

Residual Plot against x

A residual plot against the independent variable x is constructed by placing x on the horizontal axis and the residuals on the vertical axis. A residual is plotted for each observation; the first coordinate is x_i, the second is the residual $y_i - \hat{y}_i$. Figure 14.10 shows some of the patterns statisticians look for when analyzing residuals. Panel A shows the type of plot to expect when the assumptions are satisfied. The patterns shown in Panels B and C indicate violation of one or more assumptions.

If the assumption that the variance of ϵ is the same for all values of x is valid, the residual plot should give an overall impression of a horizontal band of points. Panel A shows the type of pattern to be expected in this case. On the other hand, if the variance of ϵ is not constant—for example, the variability about the regression line is greater for larger values of x—we would observe a pattern such as that of Panel B. Finally, if we observe a residual pattern such as that of Panel C, we would conclude that the error terms are not independent. The assumption of a linear relationship between x and y is not appropriate, or perhaps a multiple regression model is needed.

A plot of the residuals against the independent variable x for Armand's Pizza Parlors is shown in Figure 14.11 (the residuals were computed in Table 14.3), and we see that the residuals appear to follow the pattern of Panel A in Figure 14.10. We thus conclude that the assumptions are satisfied and that the simple linear regression model for Armand's Pizza Parlors is valid.

Residual Plot against $\hat{y}$

A residual plot against the predicted value of the dependent variable is constructed by placing the predicted value on the horizontal axis and plotting each residual directly above the corresponding value of $\hat{y}_i$. A plot of the residuals against the predicted values for Armand's Pizza Parlors (the predicted values were also computed in Table 14.8) is shown in Figure 14.12. Note that the pattern of this residual plot is the same as the pattern of the residual plot against the independent variable x. For simple linear regression, both the residual plot against x and the plot against $\hat{y}$ provide the same information. With multiple regression models (more than one independent variable), the residual plot against $\hat{y}$ is more widely used.

FIGURE 14.10

Residual Plots From Three Regression Studies

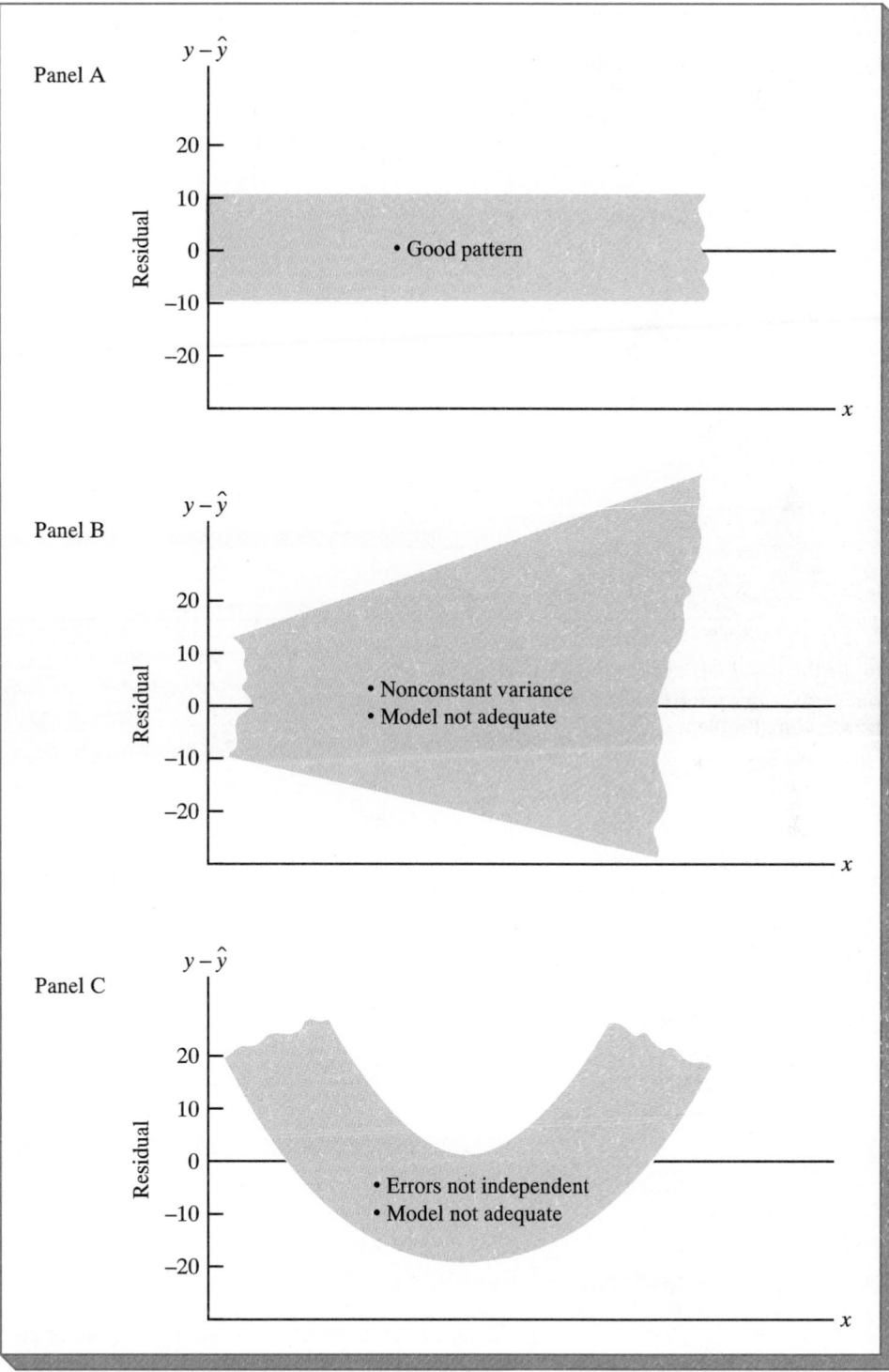

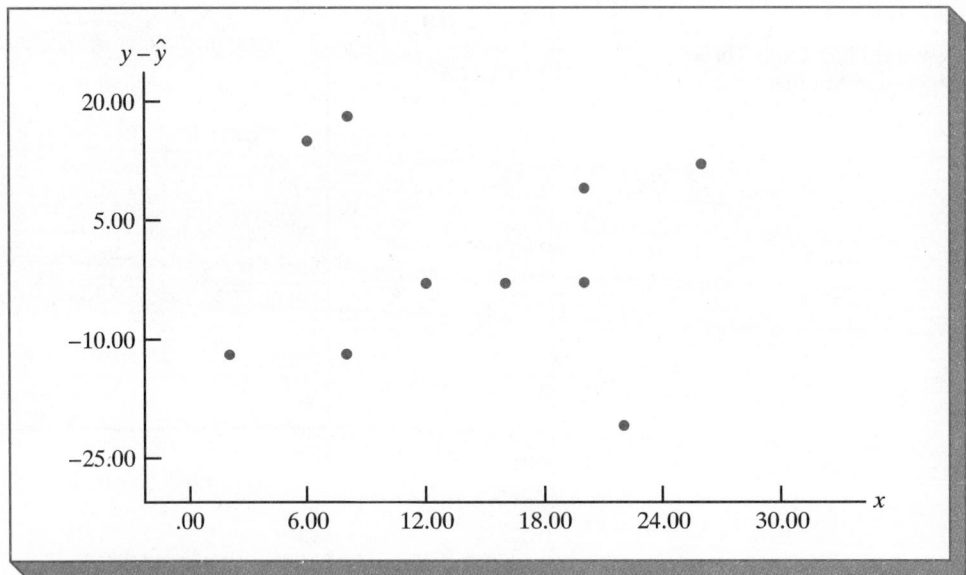

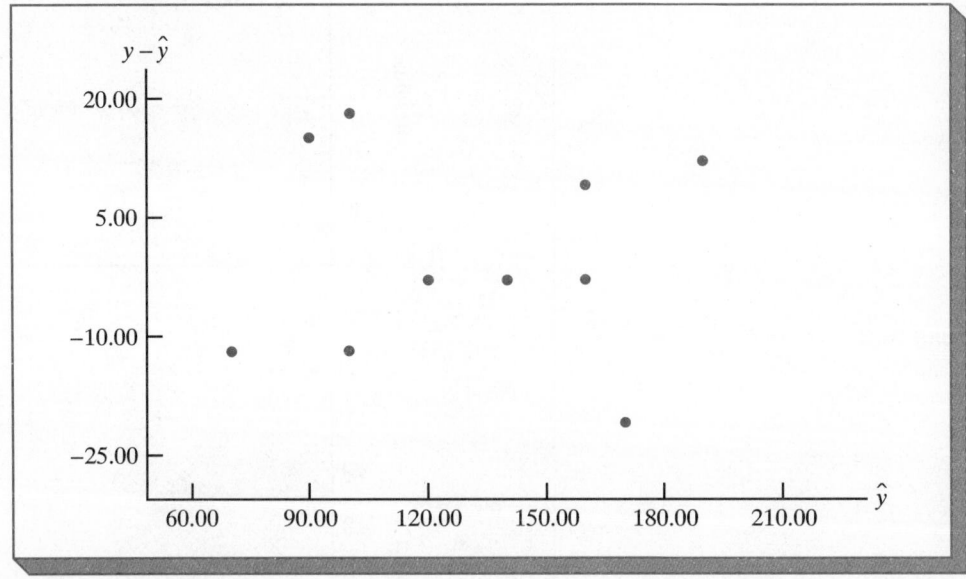

Standardized Residuals

Many of the residual plots provided by computer software packages use a standardized version of the residuals. As we have seen in earlier chapters, a random variable is standardized by subtracting its mean and dividing the result by its standard deviation. With the least squares method, the mean of the residuals is zero. Thus, simply dividing each residual by its standard deviation provides the standardized residual.

It can be shown that the standard deviation of the *i*th residual depends on $s = \sqrt{\text{MSE}}$ and the corresponding value of the independent variable.

$$
\begin{array}{c}
\textbf{Standard Deviation of } i\textbf{th Residual*} \\[6pt]
s_{y_i - \hat{y}_i} = \sqrt{s^2(1 - h_i)} \qquad\qquad \textbf{(14.24)}
\end{array}
$$

where

$$
s_{y_i} - \hat{y}_i = \text{standard deviation of residual } i
$$

$$
h_i = \frac{1}{n} + \frac{(x_i - \bar{x})^2}{\Sigma x_i^2 - (\Sigma x_i)^2/n} = \frac{1}{n} + \frac{(x_i - \bar{x})^2}{\Sigma(x_i - \bar{x})^2}
$$

Note that (14.24) shows that residuals corresponding to different values of x have different standard deviations. Once the standard deviation of each residual is calculated, we compute the standardized residual by dividing each residual by its corresponding standard deviation. Table 14.18 shows the calculations involved for Armand's Pizza Parlors. (Recall that $s^2 = 191.25$.) Figure 14.13 is a plot of the standardized residuals against x. Note that, in this case, the standardized residual plot has the same general pattern as the original residual plot shown in Figure 14.11.

Because of the effort required to compute the estimated values of $\hat{y}$, the residuals, and the standardized residuals, most statistical packages provide these values as optional regression output. As a result, the development of residual plots such as a standardized residual plot can be easily obtained. For large problems these packages provide the only practical means for developing the residual plots we have discussed in this section.

The standardized residual plot can provide insight concerning the normality assumption for ϵ (assumption 4). If the normality assumption is satisfied, the standardized

T A B L E 14.18

Computation of Standardized Residuals for the Armand's Pizza Parlors Problem

Restaurant i	$x_i - \bar{x}$	$\dfrac{(x_i - \bar{x})^2}{\Sigma x_i^2 - (\Sigma x_i)^2/10}$	h_i	$s_{y_i - \hat{y}_i}$	$y_i - \hat{y}_i$	Standardized Residual
1	-12	.2535	.3535	11.1193	-12	-1.0792
2	-8	.1127	.2127	12.2709	15	1.2224
3	-6	.0634	.1634	12.6493	-12	$-.9487$
4	-6	.0634	.1634	12.6493	18	1.4230
5	-2	.0070	.1070	13.0682	-3	$-.2296$
6	2	.0070	.1070	13.0682	-3	$-.2296$
7	6	.0634	.1634	12.6493	-3	$-.2372$
8	6	.0634	.1634	12.6493	9	.7115
9	8	.1127	.2127	12.2709	-21	-1.7114
10	12	.2535	.3535	11.1193	12	1.0792

Note: From Table 14.2, we can compute $\bar{x} = 14$ and $\Sigma x_i^2 - (\Sigma x_i^2)/10 = 568$. The values of $y_i - \hat{y}_i$ are given in Table 14.8.

*This is actually an estimate of the standard deviation of the ith residual, since s^2 is used instead of σ^2. The value of σ^2 is never known when working with real data and is always estimated by s^2.

FIGURE 14.13
Plot of the Standardized Residuals against *x* for Armand's Pizza Parlors

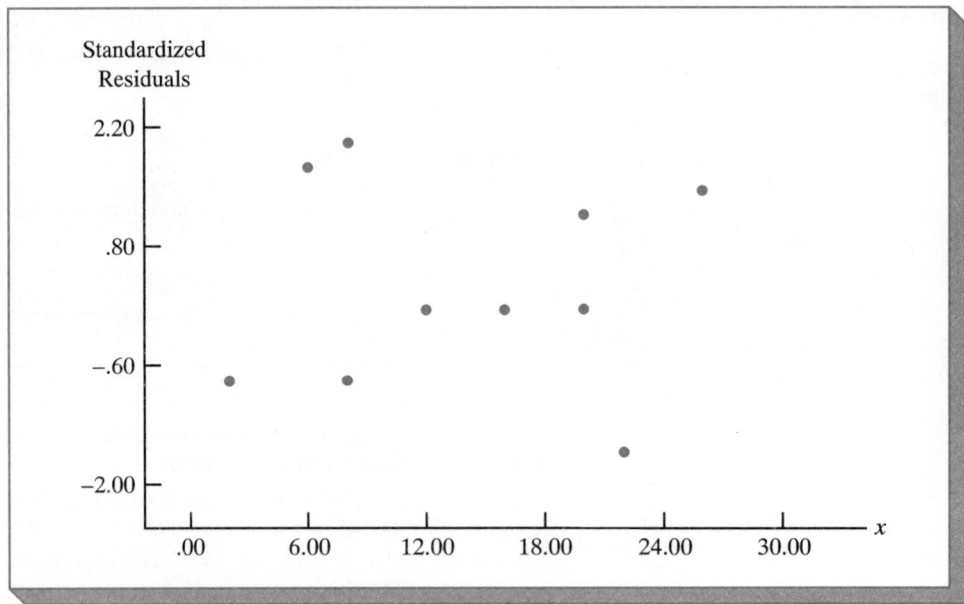

residuals should appear to come from a standard normal probability distribution.* Thus, when looking at a standardized residual plot, we should expect to see approximately 95% of the standardized residuals between -2 and $+2$. Referring to the standardized residuals for Armand's Pizza Parlors (Table 14.18), we see that they are all between -2 and $+2$. Thus, we conclude that the normality assumption is valid.

Normal Probability Plot

Another approach that can be used to test whether the error terms are normally distributed involves the development of a normal probability plot. To show how the normal probability plot is developed, we need to introduce the concept of *normal scores*.

TABLE 14.19
Normal Scores for *n* = 10

Order Statistic	Normal Score
1	-1.55
2	-1.00
3	$-.65$
4	$-.37$
5	$-.12$
6	$.12$
7	$.37$
8	$.65$
9	1.00
10	1.55

Consider an experiment in which 10 values are randomly selected from a normal probability distribution with a mean of 0 and a standard deviation of 1. Suppose this experiment is repeated over and over, and that for each sample the 10 values selected are ordered from the smallest to the largest. To begin with, let us consider only the smallest value in each sample. The *first-order statistic* is a random variable that represents the values of the smallest observation that would be obtained in repeated sampling. Since the value of the first-order statistic varies from sample to sample, the resulting probability distribution is called the sampling distribution of the first-order statistic.

Statisticians have shown that for samples of size 10, the expected value of the first-order statistic, referred to as a *normal score*, is approximately -1.55. In general, if we have a data set consisting of *n* observations, there are *n*-order statistics and hence *n* normal scores. For the case in which we have a sample of size $n = 10$, the normal scores corresponding to the order statistics from 1 to 10 are shown in Table 14.19.

To show how the normal scores can be used to determine whether the standardized residuals for Armand's Pizza Parlors are normally distributed, we begin by ordering the

*Since s^2 is substituted for σ^2 in (14.24), the probability distribution of the standardized residuals is not technically normal. However, in most regression studies, the sample size is large enough that a normal approximation is very good.

TABLE 14.20
Standardized Residuals and Normal Scores

Standardized Residual	Normal Score
−1.71	−1.55
−1.08	−1.00
−.95	−.65
−.24	−.37
−.23	−.12
−.23	.12
.71	.37
1.08	.65
1.22	1.00
1.42	1.55

10 standardized residuals shown in Table 14.18 from smallest to largest. After rounding we obtain:

$$-1.71, \ -1.08, \ -.95, \ -.24, \ -.23, \ -.23, \ .71, \ 1.08, \ 1.22, \ 1.42.$$

We next form a table in which the smallest value for each standardized residual is associated with the smallest normal score, the next smallest value of each standardized residual is associated with the next smallest normal score, and so on; the results are shown in Table 14.20. If the standardized residuals came from a normal distribution, the smallest standardized residual should be close to the smallest normal score, the next smallest standardized residual should be close to the next smallest normal score, and so on. Thus, if we were to plot the standardized residuals (vertical axis) against the normal scores (horizontal axis), the points should appear to fall on a 45° line that passes through the origin. Such a plot is referred to as a *normal probability plot*.

In Figure 14.14 we show the normal probability plot for the standardized residuals for Armand's Pizza Parlors. Clearly, the points do not all fall exactly on a 45° line that passes through the origin. Given the small sample size, this is not surprising. Thus, judgment

FIGURE 14.14
Normal Probability Plot for Armand's Pizza Parlors

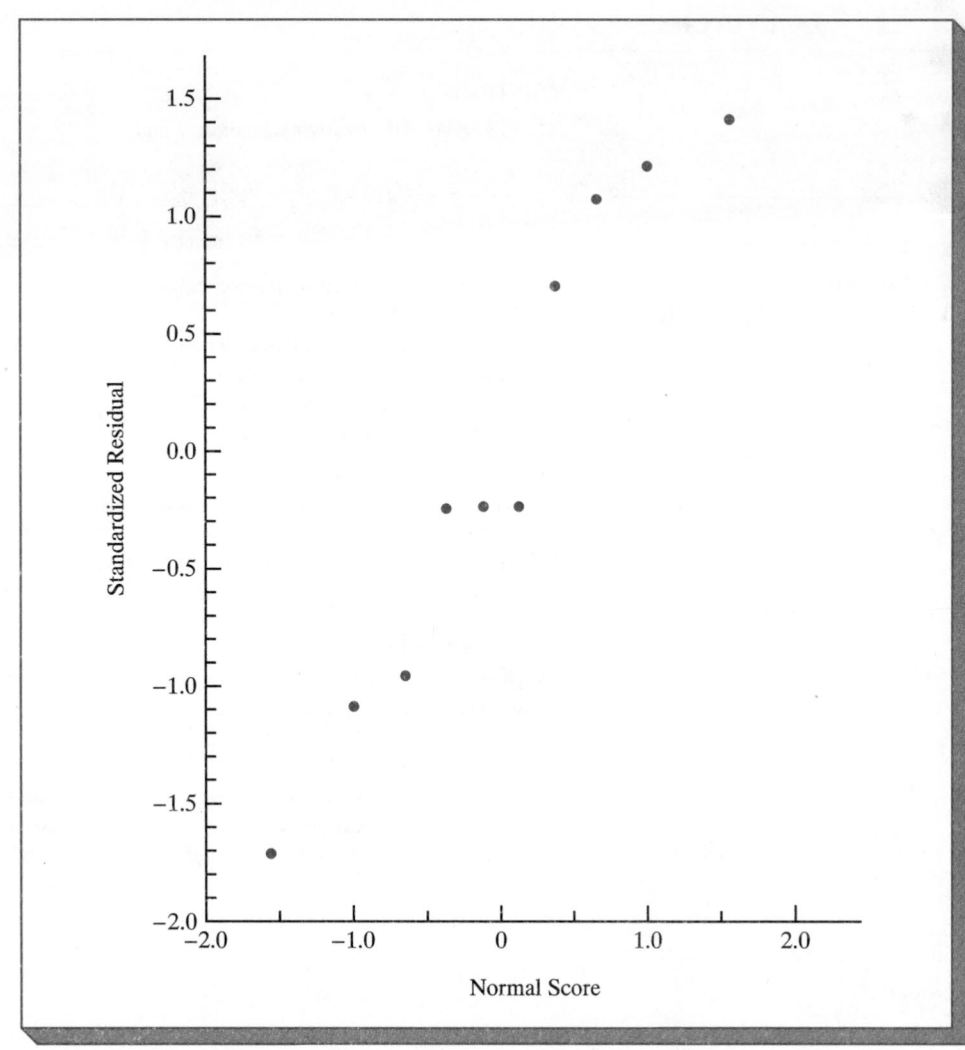

must be used to determine if the pattern observed is different enough from what we would expect if the data were normally distributed. In this case, we feel that the normal probability plot does support the conclusion that the residuals are normally distributed. In general, the straighter the plot, the stronger the evidence supporting normality. Any curvature in the normal probability plot, however, is evidence that the residuals are not normally distributed.

The computation of the normal scores for any sample size is a complex problem that involves mathematical concepts beyond the scope of this text. Fortunately, many statistical packages (e.g. Minitab) provide commands that will generate the normal scores corresponding to any set of data.

The analysis of residuals is the primary method by which statisticians verify that the assumptions are satisfied in a regression model. To validate a model, both the assumption concerning model form (simple linear, in this chapter) and the assumptions on the error term are checked for possible violations. Even if no violations are found, it does not necessarily follow that the model will yield good predictions. However, if in addition r^2 is large and the model is statistically significant, one should obtain good results.

❏ ❏ Exercises

Methods

SELF TEST ▶

47. Given are data for two variables, x and y.

x_i	6	11	15	18	20
y_i	6	8	12	20	30

a. Develop an estimated regression equation for these data.
b. Compute the residuals.
c. Develop a plot of the residuals against the independent variable x. Do the assumptions concerning the error terms seem to be satisfied?
d. Compute the standardized residuals.
e. Develop a plot of the standardized residuals against $\hat{y}$. What conclusions can you draw from this plot?

48. The data in Table 14.21 were used in a regression study.
a. Develop an estimated regression equation for this data.
b. Construct a plot of the residuals. Do the assumptions concerning the error terms seem to be satisfied?

Applications

SELF TEST ▶

49. In Exercise 7 data concerning advertising expenditures and sales at Eddie's Restaurants were given. These data are repeated here:

TABLE 14.21

Observation	x_i	y_i
1	2	4
2	3	5
3	4	4
4	5	6
5	7	4
6	7	6
7	7	9
8	8	5
9	9	11

Advertising Expenditures ($1000s)	Sales ($1000s)	Advertising Expenditures ($1000s)	Sales ($1000s)
1.0	19.0	10.0	52.0
2.0	32.0	14.0	53.0
4.0	44.0	20.0	54.0
6.0	40.0		

a. Let x equal advertising expenditures ($1000s) and y equal sales ($1000s). Use the method of least squares to develop a straight line approximation to the relationship between the two variables.

b. Test whether sales and advertising expenditures are related at the $\alpha = .05$ level of significance.

c. Prepare a residual plot of $y - \hat{y}$ versus $\hat{y}$. Use the result of (a) to obtain the values of $\hat{y}$.

d. What conclusions can you draw from residual analysis? Should this model be used, or should we look for a better one?

50. Refer to Exercise 9, where an estimated regression equation relating years of experience and annual sales was developed.

a. Compute the residuals and construct a residual plot for this problem.

b. Do the assumptions concerning the error terms seem reasonable in light of the residual plot?

51. The following data show the number of employees and the yearly revenues for the 10 largest wholesale bakers (*Louis Rukeyser's Business Almanac*).

Company	Employees	Revenues ($1,000,000s)
Nabisco Brands USA	9,500	1,734
Continental Baking Co.	22,400	1,600
Campbell Taggart, Inc.	19,000	1,044
Keebler Company	8,943	988
Interstate Bakeries Corp.	11,200	704
Flowers Industries, Inc.	10,200	557
Sunshine Biscuits, Inc.	5,000	490
American Bakeries Co.	6,600	461
Entenmann's Inc.	3,734	450
Kitchens of Sara Lee	1,550	405

a. Use a computer package to develop an estimated regression equation relating revenues y to the number of employees x.

b. Construct a residual plot of the standardized residuals against the independent variable.

c. Do the assumptions concerning the error terms and model form seem reasonable in light of the residual plot?

14.8   Residual Analysis: Outliers and Influential Observations

In Section 14.7 we showed how residual analysis could be used to determine when violations of assumptions concerning the regression model had occurred. In this section, we discuss how residual analysis can be used to identify observations that can be classified as outliers or as being especially influential in determining the estimated regression equation. Some steps that should be taken when such observations have been found are noted.

Detecting Outliers

Figure 14.15 shows a scatter diagram for a data set which has an outlier, a data point (observation) that does not fit the trend exhibited by the remaining data. Outliers represent

FIGURE 14.15

A Data Set with an Outlier

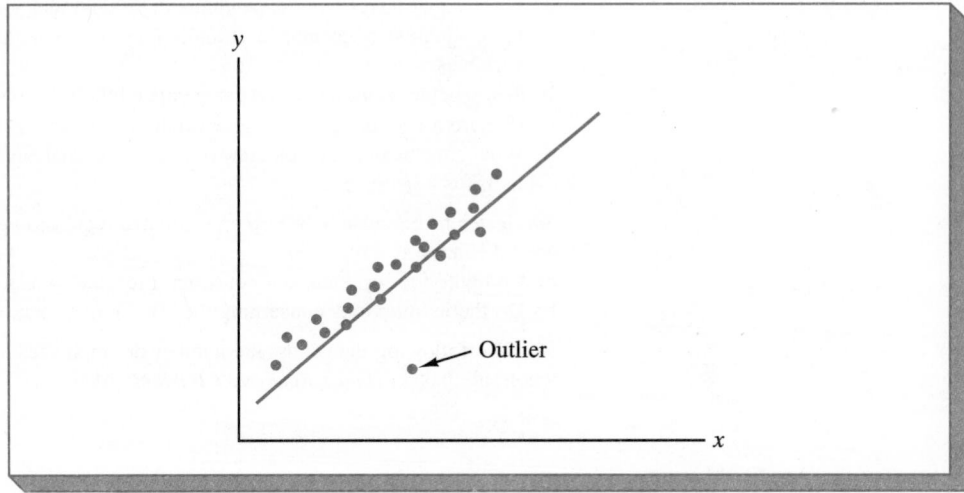

TABLE 14.22

Data Set Illustrating the Effect of an Outlier

x_i	y_i
1	45
1	55
2	50
3	75
3	40
3	45
4	30
4	35
5	25
6	15

observations that are suspect and warrant careful examination. They may represent erroneous data; if so, the data should be corrected. They may signal a violation of model assumptions; if so, other models should be considered. And, finally, they may simply be unusual values that have occurred by chance. In this case, they should be retained.

To illustrate the process of detecting outliers, consider the data set shown in Table 14.22; a scatter diagram is shown in Figure 14.16. Except for observation 4 ($x = 3$, $y = 75$), a pattern suggesting a negative linear relationship is apparent. Indeed, given the pattern of the rest of the data, we would have expected the y value for observation 4 to be much smaller and thus would identify the observation as an outlier. For the case of simple linear regression, one can usually detect outliers by simply examining the scatter diagram.

The standardized residuals can also be used to identify outliers. If the value of y for a particular x is unusually large or small (does not seem to follow the trend of the rest of the data), the corresponding standardized residual will be large in absolute value. Many computer packages automatically identify observations with standardized residuals that are large in absolute value. In Figure 14.17 we show the Minitab output from a regression analysis of the data in Table 14.22. The next to last line of the output shows that the standardized residual for observation 4 is 2.67. Minitab considers a standardized residual of less than -2 or greater than $+2$ to be an outlier; in such cases, the observation is printed on a separate line with an R next to the standardized residual, as shown in Figure 14.17. Assuming normally distributed errors, standardized residuals should fall outside these limits only approximately 5% of the time.

In deciding how to handle an outlier, we should first check to see if it is a valid observation. Perhaps an error has been made in initially recording the data or in entering the data into the computer system. For example, suppose that in checking the data for the outlier in Table 14.22, we find that an error has been made and that the correct value for observation 4 is $x = 3$, $y = 30$. Figure 14.18 shows the Minitab output obtained after correcting the value of y_4. We see that the effect of using an incorrect value for the dependent variable had a substantial effect on the goodness of fit. With the correct data, the value of r^2 has increased from 49.7 to 83.8% and the value of b_0 has decreased from 64.958 to 59.237. The slope of the line, however, has changed only from -7.331 to -6.949.

FIGURE 14.16
Scatter Diagram for Data Set of Table 14.22

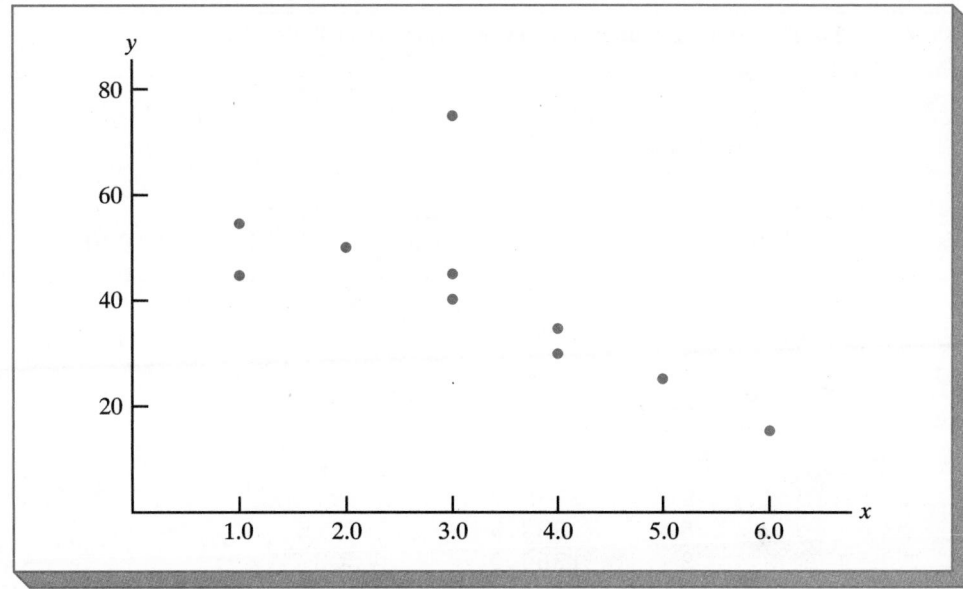

FIGURE 14.17   **Minitab Output for Regression Analysis of Data Set with Outlier (Table 14.22)**

```
The regression equation is
Y = 65.0 - 7.33 X

Predictor        Coef        Stdev       t-ratio         p
Constant       64.958        9.258          7.02     0.000
X              -7.331        2.608         -2.81     0.023

s = 12.67       R-sq = 49.7%      R-sq(adj) = 43.4%

Analysis of Variance

SOURCE          DF            SS            MS          F         p
Regression       1        1268.2        1268.2       7.90     0.023
Error            8        1284.3         160.5
Total            9        2552.5

Unusual Observations
Obs.       X            Y        Fit  Stdev.Fit   Residual    St.Resid
  4      3.00        75.00      42.97       4.04      32.03       2.67R

R denotes an obs. with a large st. resid.
```

FIGURE 14.18   **Minitab Output for Revised Data Set in Table 14.22**

```
The regression equation is
Y = 59.2 - 6.95 X

Predictor        Coef        Stdev      t-ratio           p
Constant       59.237       3.835        15.45       0.000
X              -6.949       1.080        -6.43       0.000

s = 5.248        R-sq = 83.8%      R-sq(adj) = 81.8%

Analysis of Variance

SOURCE         DF          SS           MS          F           p
Regression      1       1139.7       1139.7      41.38       0.000
Error           8        220.3        27.5
Total           9       1360.0
```

Detection of Influential Observations

In regression analysis, it sometimes happens that one or more observations have a strong influence on the results obtained. Figure 14.19 shows an example of an influential observation in simple linear regression. The estimated regression line has a negative slope. But, if the influential observation is dropped from the data set, the slope of the estimated regression line would change from negative to positive, and the y-intercept would be smaller. Clearly, this one observation is much more influential in determining the estimated regression line than any of the others; dropping one of the other observations from the data set would have very little effect on the estimated regression equation.

Influential observations can be identified from a scatter diagram when only one independent variable is present. An influential observation may be an outlier (an observation with a y value that deviates substantially from the trend); it may correspond to an x value

FIGURE 14.19
A Data Set with an Influential Observation

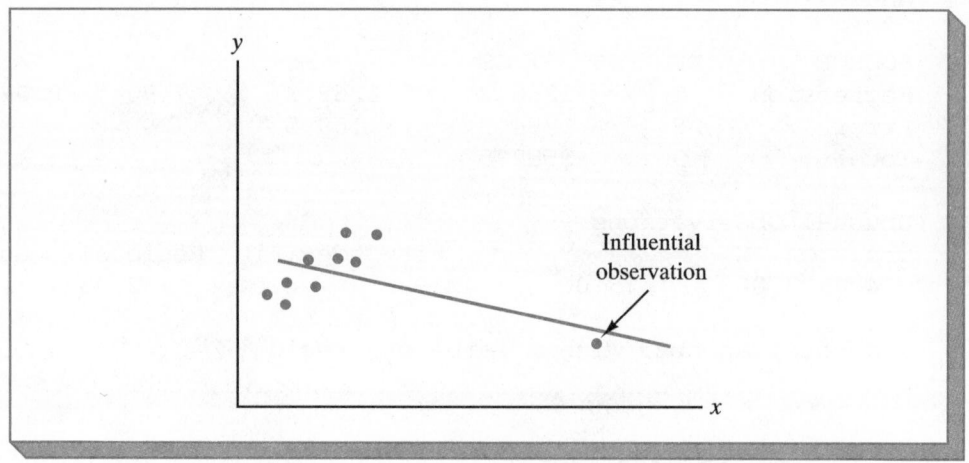

far away from its mean (e.g., see Figure 14.19); or it may be caused by a combination of the two (a somewhat off-trend y value and a somewhat extreme x value).

Since influential observations can have such a dramatic effect on the estimated regression equation, it is important that they be examined carefully. We should first check to make sure that no error has been made in collecting or recording the data. If an error has occurred, it can be corrected and a new estimated regression equation can be developed. On the other hand, if the observation is valid, we might consider ourselves fortunate to have it. Such a point, if valid, can contribute to a better understanding of the appropriate model and can lead to a better estimated regression equation. The presence of the influential observation in Figure 14.19, if valid, would suggest trying to obtain data on intermediate values of x to understand better the relationship between x and y.

Observations with extreme values for the independent variables are called high leverage points. The influential observation in Figure 14.19, caused by an extreme value of x, is a point with high leverage. The leverage of an observation is determined by how far the values of the independent variables are from their mean values. For the single-independent-variable case, the leverage of the ith observation, denoted h_i, can be computed using (14.25).

Leverage of Observation i

$$h_i = \frac{1}{n} + \frac{(x_i - \bar{x})^2}{\Sigma(x_i - \bar{x})^2} \qquad (14.25)$$

From the formula, it is clear that the farther x_i is from its mean $\bar{x}$, the higher the leverage of observation i.

Many computer packages automatically identify observations with high leverage as part of the standard regression output. To provide an illustration of how the Minitab statistical package identifies points with high leverage, let us consider the data set presented in Table 14.23.

A scatter diagram for the data set in Table 14.23 is shown in Figure 14.20. From the scatter diagram, it is clear that observation 7 ($x = 70$, $y = 100$) is an observation with an extreme value of x. Thus, we would expect it to be identified as a point with high leverage. For this observation, the leverage is computed using (14.25) as follows:

$$h_7 = \frac{1}{n} + \frac{(x_7 - \bar{x})^2}{\Sigma(x_i - \bar{x})^2} = \frac{1}{7} + \frac{(70 - 24.286)^2}{2621.43} = .94$$

For the case of simple linear regression, Minitab identifies observations as having high leverage if $h_i > 6/n$; for the data set in Table 14.23, $6/n = 6/7 = .86$. Since $h_7 = .94 > .86$, Minitab will identify observation 7 as a high leverage point. Figure 14.21 shows the Minitab output for a regression analysis of this data set. Observation 7 ($x = 70$, $y = 100$) is identified as having high leverage; it is printed on a separate line at the bottom, with an X in the right-hand margin.

Influential observations are caused by an interaction of large residuals and high leverage. Diagnostic procedures are available that take both into account in determining when an observation is influential. One such measure, called Cook's D statistic, will be discussed in Chapter 15.

TABLE 14.23

Data Set with a High Leverage Observation

x_i	y_i
10	125
10	130
15	120
20	115
20	120
25	110
70	100

FIGURE 14.20
Scatter Diagram for the Data Set in Table 14.23

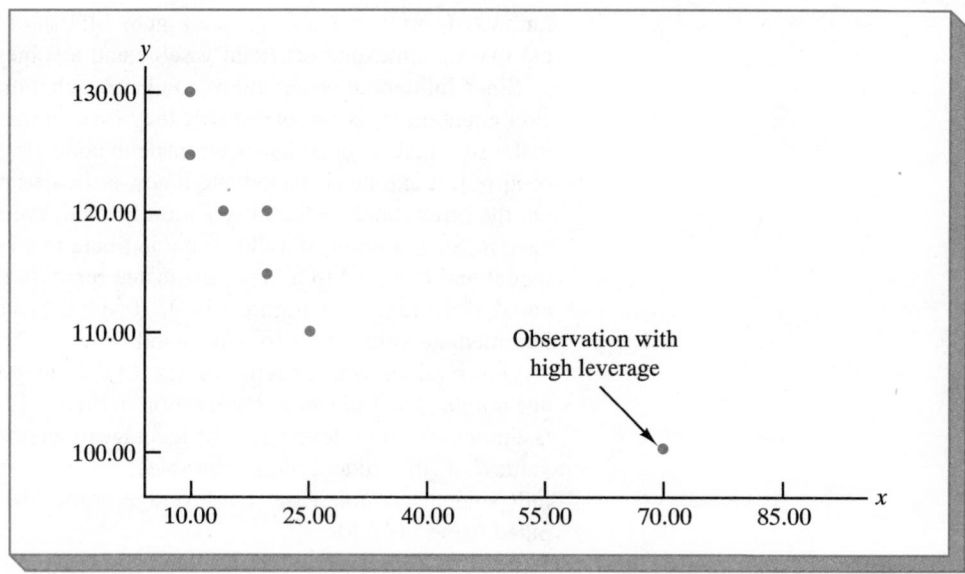

FIGURE 14.21 **Minitab Output for the Data Set in Table 14.23**

```
The regression equation is
Y = 127 -0.425 X

Predictor        Coef        Stdev      t-ratio          p
Constant       127.466       2.961        43.04      0.000
X              -0.42507      0.09537      -4.46      0.007

s = 4.883       R-sq = 79.9%     R-sq(adj) = 75.9%

Analysis of Variance

SOURCE         DF          SS           MS          F          p
Regression      1        473.65       473.65      19.87      0.007
Error           5        119.21        23.84
Total           6        592.86

Unusual Observations
Obs.        X            Y       Fit Stdev.Fit   Residual    St.Resid
  7       70.0       100.00     97.71      4.73       2.29       1.91 X

X denotes an obs. whose X value gives it large influence
```

NOTES &
COMMENTS

> Once an observation has been identified as potentially influential because of a large residual or high leverage, its impact on the estimated regression equation should be evaluated. More advanced texts discuss diagnostics for doing so. However, if one is not familiar with the more advanced material, a simple procedure is to run the regression analysis with and without the observation. Although more time-consuming, this approach will reveal the influence of the observation on the results.

Exercises

Methods

SELF TEST

52. Consider the following data for two variables, x and y:

x_i	135	110	130	145	175	160	120
y_i	145	100	120	120	130	130	110

a. Compute the standardized residuals for these data. Do there appear to be any outliers in the data? Explain.
b. Plot the standardized residuals against $\hat{y}$. Does this plot reveal any outliers?
c. Develop a scatter diagram for these data. Does the scatter diagram indicate any outliers in the data? In general, what implications does this have for simple linear regression?

53. Consider the following data for two variables, x and y.

x_i	4	5	7	8	10	12	12	22
y_i	12	14	16	15	18	20	24	19

a. Compute the standardized residuals for these data. Do there appear to be any outliers in the data? Explain.
b. Compute the leverage values for these data. Do there appear to be any influential observations in these data? Explain.
c. Develop a scatter diagram for these data. Does the scatter diagram indicate any influential observations? Explain.

TABLE 14.24

Number of Golf Courses	Number of Paid Rounds of Golf
26	1.0
30	1.1
31	1.2
32	1.3
33	1.4
35	1.6
38	1.8
43	2.0
57	2.5
67	3.0

Applications

SELF TEST

54. Table 14.24 gives the number of golf courses and the number of paid rounds of golf (in millions) for the Myrtle Beach, South Carolina, area over a 10-year period (*Myrtle Beach Magazine,* October 1991).
a. Develop the estimated regression equation for these data.
b. Use residual analysis to determine if any outliers and/or influential observations are present. Briefly summarize your findings and conclusions.

 HOME2

55. The National Association of Home Builders compared the median home prices with the median household incomes in cities throughout the United States (*USA Today,* September 10, 1991). The 25 most affordable cities are listed in the table at the top of page 524. Both home prices and household incomes are shown in thousands of dollars.

City	Median Income	Median Home Price	City	Median Income	Median Home Price
Amarillo, Texas	36.7	69.0	Milwaukee, Wisconsin	41.8	72.0
Brazoria, Texas	42.4	80.0	Minneapolis, Minnesota	48.0	91.0
Canton, Ohio	34.1	66.0	Nashua, New Hampshire	52.9	111.0
Davenport, Iowa	38.4	59.0	Oklahoma City, Oklahoma	34.5	63.0
Daytona Beach, Florida	31.0	63.0	Omaha, Nebraska	38.8	65.0
Detroit, Michigan	44.6	77.0	Rockford, Illinois	41.6	73.5
Fort Walton Beach, Florida	34.2	65.0	Saginaw, Michigan	39.7	61.0
Grand Rapids, Michigan	40.3	73.0	Shreveport, Louisiana	34.4	66.0
Jackson, Michigan	36.8	60.0	Toledo, Ohio	39.4	65.0
Kansas City, Missouri	41.1	77.0	Tulsa, Oklahoma	36.2	68.0
Lansing, Michigan	40.0	70.0	Winter Haven, Florida	30.2	56.0
Lorain, Ohio	38.8	72.5	Youngstown, Ohio	34.9	59.0
Mansfield, Ohio	35.6	58.5			

a. Develop the estimated regression equation for these data that can be used to predict the median home price given the median income.

b. Use residual analysis to determine if any outliers and/or influential observations are present. Briefly summarize your findings and conclusions.

 TEMPSC

56. The data in Table 14.25 show the temperatures for air and water in the Myrtle Beach, South Carolina area (*Myrtle Beach and South Carolina Grand Strand,* 1992).

a. Develop the estimated regression equation for these data that can be used to predict the water temperature given the air temperature.

b. Use residual analysis to determine if any outliers and/or influential observations are present. Briefly summarize your findings and conclusions.

14.9   Correlation Analysis

TABLE 14.25

Month	Air	Water
January	57	49
February	59	51
March	65	56
April	75	66
May	81	71
June	86	78
July	88	83
August	88	80
September	84	77
October	75	72
November	68	60
December	59	50

As we indicated in the introduction to this chapter, there are some situations in which the decision maker is not as concerned with the equation that relates two variables as in measuring the extent to which the two variables are related. In such cases, a statistical technique referred to as correlation analysis can be used to determine the strength of the relationship between the two variables.* The output of a correlation study is a number referred to as the correlation coefficient. Because of the way in which it is defined, values of the correlation coefficient are always between -1 and $+1$. A value of $+1$ indicates that x and y are perfectly related in a positive linear sense. That is, all the points lie on a straight line that has a positive slope. A value of -1 indicates that x and y are perfectly related in a negative linear sense. That is, all the points lie on a straight line that has a negative slope. Values of the correlation coefficient close to zero indicate that x and y are not linearly related.

To provide an illustration of correlation analysis, we consider the situation of a stereo and sound-equipment store located in San Francisco. Management would like to investigate whether there is any relationship between the number of commercials x shown on Friday evening television and the resulting sales volume on Saturday y, measured in hundreds of dollars. The sample data that were obtained are shown in Table 14.26.

*In correlation analysis it is assumed that x and y are both random variables.

	Number of	Sales Volume
Store	Commercials	($100s)
1	2	24
2	5	28
3	1	22
4	3	26
5	4	25
6	1	24
7	5	26

TABLE 14.26
Sample Data for the Stereo and Sound-Equipment Problem

In Figure 14.22 we show a scatter diagram of this data. The scatter diagram appears to indicate that there is a positive linear relationship between x and y. To measure the degree of linear association between these two variables, we first define a measure of linear association known as the *covariance*.

Covariance

Sample covariance is defined as follows.

Sample Covariance

$$s_{xy} = \frac{\sum (x_i - \overline{x})(y_i - \overline{y})}{n - 1}$$

(14.26)

FIGURE 14.22
Scatter Diagram for the Stereo and Sound-Equipment Problem

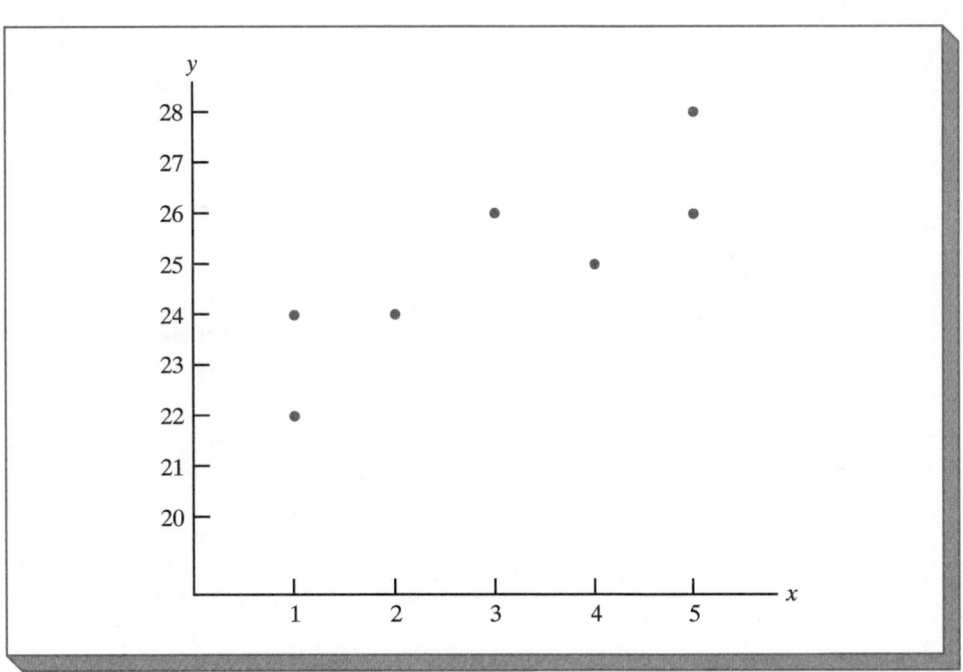

TABLE 14.27
Calculations for the Sample Covariance

x_i	y_i	$x_i - \bar{x}$	$y_i - \bar{y}$	$(x_i - \bar{x})(y_i - \bar{y})$
2	24	−1	−1	1
5	28	2	3	6
1	22	−2	−3	6
3	26	0	1	0
4	25	1	0	0
1	24	−2	−1	2
5	26	2	1	2
Totals 21	175	0	0	17

In this formula each x_i value is paired with a y_i value. We then sum the products obtained by multiplying the deviation of each x_i from its sample mean $\bar{x}$ times the deviation of the corresponding y_i from its sample mean $\bar{y}$; this sum is then divided by $n - 1$.

To measure the strength of the linear relationship between the number of commercials x and the sales volume y in the stereo and sound-equipment problem, we can use (14.26) to compute the sample covariance. The calculations shown in Table 14.27 illustrate the computations of $\Sigma(x_i - \bar{x})(y_i - \bar{y})$. Note that $\bar{x} = 21/7 = 3$ and $\bar{y} = 175/7 = 25$. Using (14.26) we obtain

$$s_{xy} = \frac{\Sigma(x_i - \bar{x})(y_i - \bar{y})}{n - 1} = \frac{17}{6} = 2.8333$$

The formula for computing the covariance of a population of size N is similar to (14.26), but we use different notation to indicate that we are dealing with the entire population.

Population Covariance

$$\sigma_{xy} = \frac{\Sigma(x_i - \mu_x)(y_i - \mu_y)}{N} \qquad (14.27)$$

In (14.27) we use the notation μ_x for the population mean of the variable x and μ_y for the population mean of the variable y. The sample covariance s_{xy} is an estimate of the population covariance σ_{xy} based on a sample of size n.

Interpretation of the Covariance

To aid in the interpretation of the *sample covariance*, consider Figure 14.23. It is the same as the scatter diagram of Figure 14.22 with a vertical line at $x = 3$ (the value of $\bar{x}$) and a horizontal line at $y = 25$ (the value of $\bar{y}$). Four quadrants have been identified on the graph. Points that fall in quadrant I correspond to x_i values greater than $\bar{x}$ and y_i values

FIGURE 14.23
Quadrants I, II, III, and IV for the Stereo and Sound-Equipment Problem

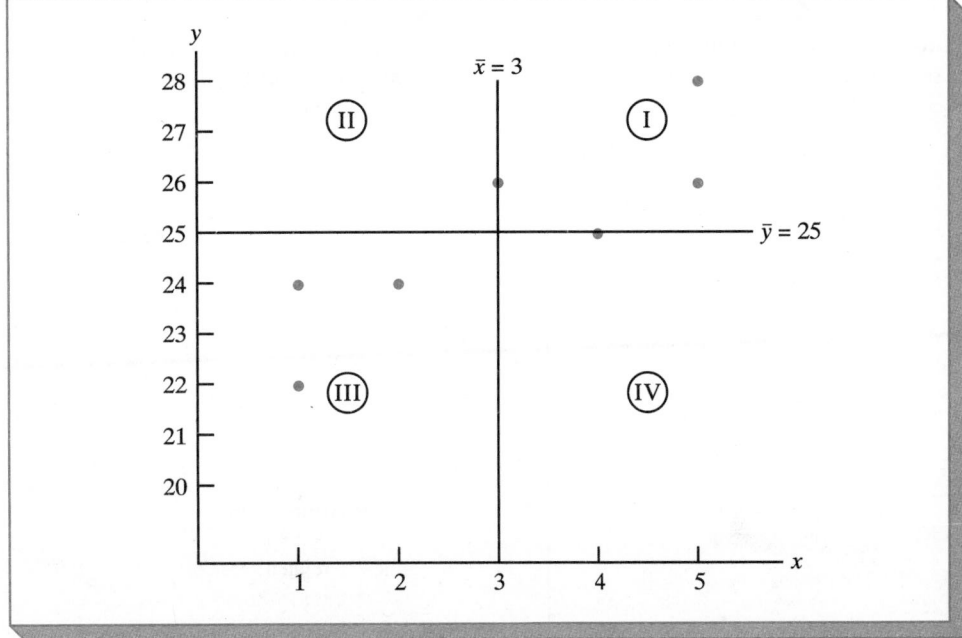

greater than $\bar{y}$; points that fall in quadrant II correspond to x_i values less than $\bar{x}$ and y_i values greater than $\bar{y}$, and so on. Thus, the value of $(x_i - \bar{x})(y_i - \bar{y})$ must be positive for points located in quadrant I, negative for points located in quadrant II, positive for points located in quadrant III, and negative for points located in quadrant IV.

If the value of s_{xy} is positive, the points that have had the greatest effect on s_{xy} must lie in quadrants I and/or III. Hence, a positive value for s_{xy} is indicative of a positive linear association between x and y; that is, as the value of x increases, the value of y increases. If the value of s_{xy} is negative, however, the points that have had the greatest effect on s_{xy} lie in quadrants II and/or IV. Hence, a negative value for s_{xy} is indicative of a negative linear association between x and y; that is, as the value of x increases, the value of y decreases. Finally, if the points are evenly distributed across all four quadrants, the value of s_{xy} will be close to zero, indicating no linear association between x and y. Figure 14.24 shows the values of s_{xy} that can be expected with these three different types of scatter diagrams.

From the previous discussion, it might appear that a large positive value for the covariance is indicative of a strong positive linear relationship and that a large negative value is indicative of a strong negative linear relationship. However, one problem with using covariance as a measure of the strength of the linear relationship is that the value we obtain for the covariance depends on the units of measurement for x and y. For example, suppose we were interested in the relationship between height x and weight y for individuals. If height is measured in inches, we will get much larger numerical values for $(x_i - \bar{x})$ than if it is measured in feet. Thus, with height measured in inches, we would obtain larger values for $\Sigma(x_i - \bar{x})(y_i - \bar{y})$—and hence a larger covariance—when, in fact, there is no difference in the relationship. A measure of relationship that avoids this difficulty is the *correlation coefficient*.

FIGURE 14.24
Interpretation of Sample
Covariance

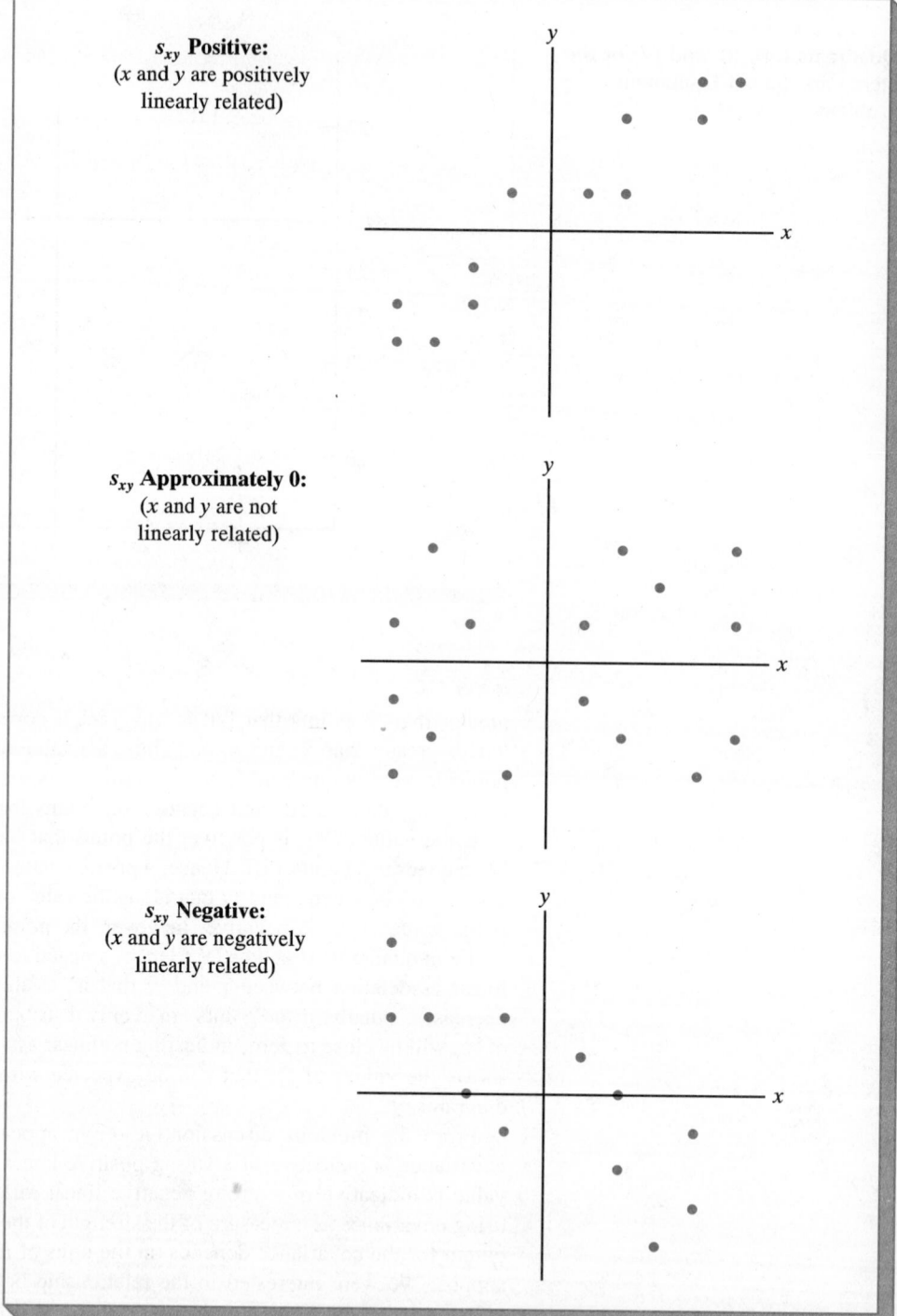

Correlation Coefficient

For sample data, the Pearson Product Moment correlation coefficient is defined as follows.

Pearson Product Moment Correlation Coefficient: Sample Data

$$r_{xy} = \frac{s_{xy}}{s_x s_y} \qquad (14.28)$$

where

r_{xy} = sample correlation coefficient

s_{xy} = sample covariance

s_x = sample standard deviation of x

s_y = sample standard deviation of y

Equation (14.28) shows that the Pearson Product Moment correlation coefficient for sample data (commonly referred to more simply as the *sample correlation coefficient*) is computed by dividing the sample covariance by the product of the standard deviation of x and the standard deviation of y. Before we consider further interpretation of the sample correlation coefficient, let us consider the use of (14.28) for the stereo and sound-equipment problem.

Using the data presented in Table 14.27, we can compute the sample correlation coefficient.

$$s_x = \sqrt{\frac{\Sigma(x_i - \overline{x})^2}{n-1}} = \sqrt{\frac{18}{6}} = 1.7321$$

$$s_y = \sqrt{\frac{\Sigma(y_i - \overline{y})^2}{n-1}} = \sqrt{\frac{22}{6}} = 1.9149$$

and, since $s_{xy} = 2.8333$, we have

$$r_{xy} = \frac{s_{xy}}{s_x s_y} = \frac{2.8333}{(1.7321)(1.9149)} = .854$$

When using a calculator to compute the sample correlation coefficient, the formula given by (14.29) is preferred because the computation of each deviation $x_i - \overline{x}$ and $y_i - \overline{y}$ is not necessary, and thus less round-off error is introduced.

Pearson Product Moment Correlation Coefficient:
Sample Data, Alternate Formula

$$r_{xy} = \frac{\Sigma x_i y_i - (\Sigma x_i \Sigma y_i)/n}{\sqrt{\Sigma x_i^2 - (\Sigma x_i)^2/n} \ \sqrt{\Sigma y_i^2 - (\Sigma y_i)^2/n}} \qquad (14.29)$$

TABLE 14.28
Computations for Using the Alternate Formula for Computing r_{xy}

x_i	y_i	$x_i y_i$	x_i^2	y_i^2
2	24	48	4	576
5	28	140	25	784
1	22	22	1	484
3	26	78	9	676
4	25	100	16	625
1	24	24	1	576
5	26	130	25	676
Totals 21	175	542	81	4397

Algebraically, Equations (14.28) and (14.29) are equivalent. In Table 14.28 we provide the calculations needed to use (14.29). Using these computations and (14.29), we obtain:

$$r_{xy} = \frac{542 - (21)(175)/7}{\sqrt{81 - (21)^2/7}\ \sqrt{4397 - (175)^2/7}} = \frac{17}{19.8997} = .854$$

Thus, we see that the value obtained for r_{xy} using (14.29) is the same as the value obtained using (14.28).

The formula for computing the correlation coefficient of a population, denoted by the Greek letter ρ_{xy} (rho, pronounced "row"), is as follows.

**Pearson Product Moment Correlation Coefficient:
Population Data**

$$\rho_{xy} = \frac{\sigma_{xy}}{\sigma_x \sigma_y} \qquad (14.30)$$

where

ρ_{xy} = population correlation coefficient

σ_{xy} = population covariance

σ_x = population standard deviation for x

σ_y = population standard deviation for y

The sample correlation coefficient r_{xy} is an estimate of the population correlation coefficient ρ_{xy}.

Interpretation of the Correlation Coefficient

First let us consider a simple example that illustrates the concept of perfect positive linear association. The scatter diagram shown in Figure 14.25 depicts the relationship between the following $n = 3$ pairs of points:

x_i	1	2	3
y_i	10	30	50

FIGURE 14.25
Scatter Diagram Depicting a Perfect Positive Linear Association

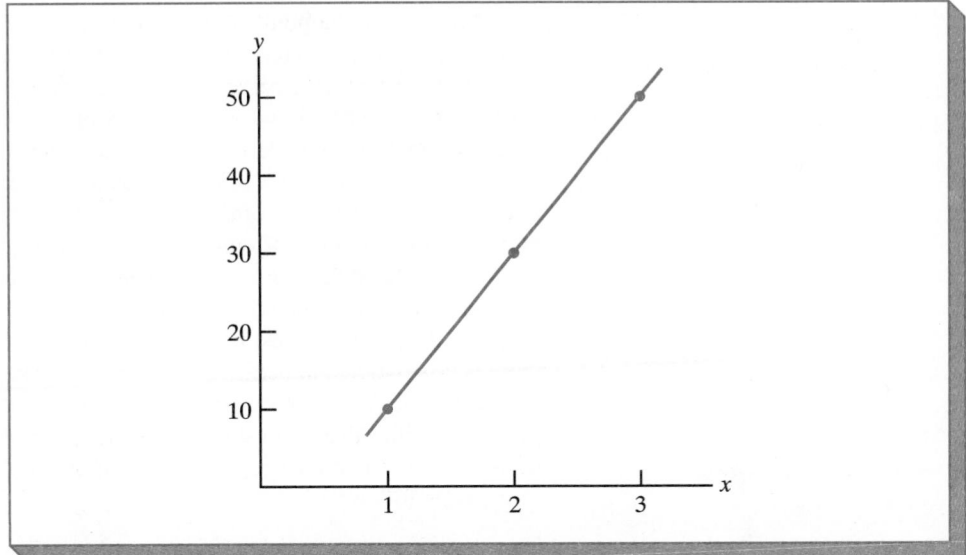

FIGURE 14.25
Scatter Diagram Depicting a Perfect Positive Linear Association

The straight line drawn through each of the three points shows that there is a perfect linear relationship between the two variables x and y. The calculations needed to compute r_{xy} are shown in Table 14.29. Using the values in this table we obtain

$$r_{xy} = \frac{\Sigma x_i y_i - (\Sigma x_i \Sigma y_i)/n}{\sqrt{\Sigma x_i^2 - (\Sigma x_i)^2/n} \sqrt{\Sigma y_i^2 - (\Sigma y_i)^2/n}}$$

$$= \frac{220 - (6)(90)/3}{\sqrt{14 - (6)^2/3} \sqrt{3500 - (90)^2/3}} = \frac{40}{40} = 1$$

Thus, we see that the value of the sample correlation coefficient for this data set is 1.

In general, it can be shown that if all the points in a data set fall on a straight line having positive slope, then the value of the sample correlation coefficient is $+1$; that is, a sample correlation coefficient of $+1$ corresponds to a perfect positive linear association between x and y. Moreover, if the points in the data set fall on a straight line having negative slope, the value of the sample correlation coefficient is -1; that is, a sample correlation coefficient of -1 corresponds to a perfect negative linear association between x and y.

Let us now suppose that for a certain data set there is a positive linear association between x and y but that the relationship is not perfect. The value of r_{xy} will be less than

TABLE 14.29
Calculations for Computing r for the Example Used to Illustrate Perfect Positive Linear Association

	x_i	y_i	$x_i y_i$	x_1^2	y_i^2
	1	10	10	1	100
	2	30	60	4	900
	3	50	150	9	2500
Totals	6	90	220	14	3500

1, indicating that the points in the scatter diagram do not all fall on a straight line. As the points in a data set deviate more and more from a perfect positive linear association, the value of r_{xy} becomes smaller and smaller. A value of r_{xy} equal to 0 indicates no linear relationship between x and y and values of r_{xy} near zero indicate a weak relationship.

Recall that for the data set involving the stereo and sound-equipment store, r_{xy} = +.854. Since r_{xy} = +.854 we conclude that there is a positive linear association between the number of commercials and Saturday sales volume. More specifically, an increase in the number of commercials is associated with an increase in sales volume.

We have stated that values of r_{xy} near +1 indicate a strong linear association between two variables and values of r_{xy} near zero indicate little or no linear association between the variables. But we must be careful not to conclude that a value of r_{xy} near zero means there is no relationship between the variables. The scatter diagram in Figure 14.26 shows a case where r_{xy} = 0 and there is no linear relationship; however, in this case, there is a perfect curvilinear relationship between the variables. Table 14.30 provides the calculations needed to compute r_{xy} for this example. Using these calculations, the computation of r_{xy} is as follows:

$$r_{xy} = \frac{210 - (42)(35)/7}{\sqrt{310 - (42)^2/7}\ \sqrt{231 - (35)^2/7}} = \frac{0}{56.9912} = 0$$

To reiterate, our last example illustrates an important concept regarding the proper interpretation of the sample correlation coefficient. The sample correlation coefficient

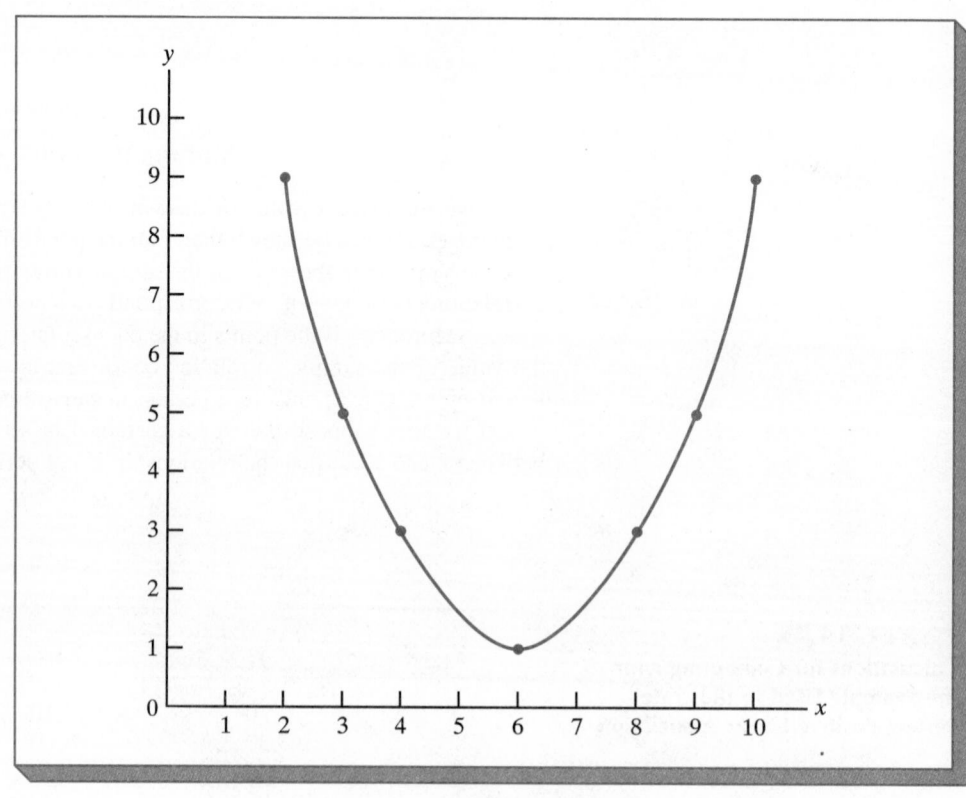

FIGURE 14.26

Even though r_{xy} = 0, There Is a Perfect Curvilinear Relationship for the Data

	x_i	y_i	x_iy_i	x_i^2	y_i^2
	2	9	18	4	81
	3	5	15	9	25
	4	3	12	16	9
	6	1	6	36	1
	8	3	24	64	9
	9	5	45	81	25
	10	9	90	100	81
Totals	42	35	210	310	231

TABLE 14.30
Calculations for the Example Illustrating $r_{xy} = 0$

measures only the degree of *linear association* between the two variables. A value of r_{xy} equal to zero cannot be interpreted as implying that there is no relationship between the two variables. One should always look at the associated scatter diagram as well as the value of the sample correlation coefficient when attempting to determine if, and how, two variables are related.

In closing this part of our discussion on correlation, we caution that while a correlation coefficient near ± 1 does imply a strong linear association between two variables, it does not imply a cause-and-effect relationship. Conclusions concerning cause-and-effect must be based on the judgment of the analyst.

Determining the Sample Correlation Coefficient from the Regression Analysis Output

In this discussion we will assume that the least squares estimated regression equation is $\hat{y} = b_0 + b_1x$. In such cases, the sample correlation coefficient can be computed using one of the following formulas.

Sample Correlation Coefficient

$$r_{xy} = (\text{sign of } b_1)\sqrt{\text{Coefficient of Determination}} = \pm \sqrt{r^2} \qquad (14.31)$$

$$r_{xy} = b_1\left(\frac{s_x}{s_y}\right) \qquad (14.32)$$

where

b_1 = slope of the estimated regression equation

s_x = sample standard deviation of x

s_y = sample standard deviation of y

Note that the sign of the sample correlation coefficient is the same as the sign of b_1, the slope of the estimated regression equation. For the Armand's Pizza Parlors problem presented earlier in this chapter, $b_1 = 5$, indicating a positive relationship. Thus, with $r^2 = .903$ we obtain

$$r_{xy} = \pm \sqrt{.903}$$

$$= +.95$$

Testing for Significance

The sample correlation coefficient is a point estimator of the population correlation coefficient. With ρ_{xy} denoting the population correlation coefficient, a statistical test for the significance of a linear association between x and y can be performed by testing the following hypotheses:

$$H_0: \rho_{xy} = 0$$

$$H_a: \rho_{xy} \neq 0$$

It can be shown that testing these hypotheses is equivalent to testing the hypotheses regarding the significance of β_1, the slope of the regression equation. Recall that the appropriate hypotheses in this case are

$$H_0: \beta_1 = 0$$

$$H_a: \beta_1 \neq 0$$

Since for the Armand's Pizza Parlors problem we earlier rejected the null hypothesis $H_0: \beta_1 = 0$ (see Section 14.4), we can also reject the null hypothesis $H_0: \rho_{xy} = 0$ and conclude that x and y are correlated. Alternatively, statisticians have developed a procedure for testing the following hypothesis without performing a regression study:

$$H_0: \rho_{xy} = 0$$

$$H_a: \rho_{xy} \neq 0$$

It can be shown that if H_0 is true, then the value of

$$r_{xy} \sqrt{\frac{n-2}{1 - r_{xy}^2}} \tag{14.33}$$

has a t distribution with $n - 2$ degrees of freedom.

For Armand's Pizza Parlors with $\alpha = .05$ and $n - 2 = 10 - 2 = 8$ degrees of freedom, we see that the appropriate t value from Table 2 of Appendix B is 2.306. Thus, if the value of (14.33) exceeds 2.306 or is less than -2.306, we must reject the null hypothesis $H_0: \rho_{xy} = 0$.

With a sample correlation coefficient of $r_{xy} = .95$, the value of (14.33) is

$$.95 \sqrt{\frac{8}{1 - .903}} = 8.63$$

Since 8.63 exceeds the t value of 2.306, we reject H_0 and hence conclude that x and y have a significant correlation. We note that this test yields the same result as the previous test on β_1.

☐ ☐ Exercises

Methods

SELF TEST ▶

57. Given are five observations taken for two variables.

x_i	4	6	11	3	16
y_i	50	50	40	60	30

a. Develop a scatter diagram with x on the horizontal axis.
b. What does the scatter diagram developed in (a) indicate about the relationship between the two variables?
c. Compute and interpret the sample covariance for these data.
d. Compute and interpret the sample correlation coefficient for these data.

58. Given are five observations taken for two variables.

x_i	6	11	15	21	27
y_i	6	9	6	17	12

a. Develop a scatter diagram for these data.
b. What does the scatter diagram indicate about a possible relationship between x and y?
c. Compute and interpret the sample covariance for these data.
d. Compute and interpret the sample correlation coefficient for these data.

59. Eight observations on two random variables are given.

x_i	2	9	6	8	4	7	5	6
y_i	11	4	6	5	9	4	9	7

a. Compute r_{xy}.
b. Test the hypotheses

$$H_0: \rho_{xy} = 0$$

$$H_a: \rho_{xy} \neq 0$$

at the $\alpha = .01$ level of significance.

Applications

60. A high-school guidance counselor collected the following data regarding the grade point average (GPA) and the SAT mathematics test score for six seniors.

GPA	2.7	3.5	3.7	3.3	3.6	3.0
SAT	440	560	720	520	640	480

a. Develop a scatter diagram for these data with GPA as the independent variable.
b. Does there appear to be any relationship between the GPA and the SAT mathematics test score? Explain.
c. Compute and interpret the sample covariance for these data.
d. Compute the sample correlation coefficient for these data. What does this value tell us about the relationship between the two variables?

TABLE 14.31

Driving Speed	Mileage
30	28
50	25
40	25
55	23
30	30
25	32
60	21
25	35
50	26
55	25

61. A study conducted by a department of transportation regarding driving speed and mileage for midsize automobiles resulted in the data shown in Table 14.31.

a. Compute and interpret the sample correlation coefficient for these data.

b. Test the hypotheses

$$H_0: \rho_{xy} = 0$$

$$H_a: \rho_{xy} \neq 0$$

at the $\alpha = .01$ level of significance.

62. A sociologist collected data regarding the ages of wives and husbands when they married.

Wife's Age	19	42	28	25	36
Husband's Age	20	32	31	24	33

a. Develop a scatter diagram for these data with the wife's age on the horizontal axis.

b. Does there appear to be a linear association? Explain.

c. Compute and interpret the sample correlation coefficient for these data.

63. The following estimated regression equation has been developed to estimate the relationship between x, the number of units produced per week, and y, the total weekly cost of production ($):

$$\hat{y} = 60 + 3.2x$$

The standard deviation of weekly production is 10 units, and the standard deviation of weekly cost is $35.00. Compute the sample correlation coefficient r_{xy}.

64. As more U.S. households receive cable television, the advertising revenue has continued to increase. The following data show the 1988 and 1987 expenditures for the top 10 cable television advertisers.

Advertiser	1988 Expenditure ($1,000,000s)	1987 Expenditure ($1,000,000s)
Procter & Gamble	30.2	23.7
Philip Morris	23.1	20.6
Anheuser-Busch	21.4	22.9
Time	21.2	16.4
General Mills	20.0	18.6
RJR Nabisco	14.3	14.7
Eastman Kodak	11.0	2.6
Clorox	10.1	6.9
Mars	10.0	14.9
Chrysler	9.5	6.1

a. Develop a scatter diagram for the data. Does it appear that the two variables are linearly related?

b. Compute the sample correlation coefficient for these data.

c. Test the hypotheses

$$H_0: \rho_{xy} = 0$$

$$H_a: \rho_{xy} \neq 0$$

at the $\alpha = .01$ level at significance.

 HIGHLOW

65. The daily high and low temperatures for 24 cities are given below (*USA Today*, April 6, 1992):

City	High	Low	City	High	Low
Tampa	80	58	Birmingham	68	32
Kansas City	69	40	Minneapolis	62	39
Boise	58	35	Portland	50	41
Los Angeles	71	57	Memphis	67	41
Philadelphia	56	35	Buffalo	44	28
Milwaukee	47	29	Cincinnati	55	29
Chicago	52	25	Charlotte	61	37
Albany	50	28	Boston	50	35
Houston	63	50	Tulsa	73	50
Salt Lake City	61	49	Washington, D.C.	56	35
Miami	79	56	Las Vegas	80	53
Cheyenne	66	35	Detroit	52	29

a. What is the correlation between the high and low temperatures?
b. Using a .05 level of significance, test for a significant correlation. What is your conclusion?

Summary

In this chapter we introduced the topics of regression and correlation analysis. We discussed how regression analysis can be used to develop an equation showing how variables are related and how correlation analysis can be used to determine the strength of the relationship between two variables. Before concluding our discussion, however, we would like to reemphasize a potential misinterpretation of these studies. Regression and correlation analyses can indicate only how or to what extent the variables are associated with each other. These techniques cannot be interpreted directly as showing cause-and-effect relationships.

Glossary

Note: The definitions here are all stated with the understanding that simple linear regression and correlation are being considered.

Dependent variable The variable that is being predicted or explained. It is denoted by y in the regression equation.

Independent variable The variable that is doing the predicting or explaining. It is denoted by x in the regression equation.

Simple linear regression The simplest kind of regression, involving only two variables that are related approximately by a straight line.

Regression equation The mathematical equation relating the independent variable to the expected value of the dependent variable; that is, $E(y) = \beta_0 + \beta_1 x$.

Estimated regression equation The estimate of the regression equation obtained by the least squares method; that is, $\hat{y} = b_0 + b_1 x$.

Scatter diagram A graph of the available data in which the independent variable appears on the horizontal axis and the dependent variable appears on the vertical axis.

Least squares method The approach used to develop the estimated regression equation which minimizes the sum of squared residuals.

Coefficient of determination (r^2) A measure of the variation explained by the estimated regression equation. It is a measure of how well the estimated regression equation fits the data.

Deterministic model A relationship between an independent variable and a dependent variable whereby specifying the value of the independent variable allows one to compute exactly the value of the dependent variable.

Probabilistic model A relationship between an independent variable and a dependent variable in which specifying the value of the independent variable is not sufficient to allow determination of the value of the dependent variable.

Residual The difference between the observed value of the dependent variable and the value predicted using the estimated regression equation; that is, $y_i - \hat{y}_i$.

Standardized residual The value obtained by dividing the residual by its standard deviation.

Sample correlation coefficient (r_{xy}) A statistical measure of the linear association between two variables.

Key Formulas

Estimated Regression Equation

$$\hat{y} = b_0 + b_1 x \tag{14.1}$$

Slope and y-Intercept for the Estimated Regression Equation

$$b_1 = \frac{\Sigma(x_i - \bar{x})(y_i - \bar{y})}{\Sigma(x_i - \bar{x})^2} = \frac{\Sigma x_i y_i - (\Sigma x_i \, \Sigma y_i)/n}{\Sigma x_i^2 - (\Sigma x_i)^2/n} \tag{14.3}$$

$$b_0 = \bar{y} - b_1 \bar{x} \tag{14.4}$$

Sum of Squares Due to Error

$$SSE = \Sigma(y_i - \hat{y}_i)^2 \tag{14.5}$$

Total Sum of Squares

$$SST = \Sigma(y_i - \bar{y})^2 \tag{14.6}$$

Sum of Squares Due to Regression

$$SSR = \Sigma(\hat{y}_i - \bar{y})^2 \tag{14.7}$$

Relationship among SST, SSR, and SSE

$$SST = SSR + SSE \tag{14.8}$$

Coefficient of Determination

$$r^2 = \frac{SSR}{SST} \tag{14.9}$$

Computational Formula for SSR

$$SSR = \frac{[\Sigma x_i y_i - (\Sigma x_i \, \Sigma y_i)/n]^2}{\Sigma x_i^2 - (\Sigma x_i)^2/n} \tag{14.10}$$

Computational Formula for SST

$$SST = \Sigma y_i^2 - (\Sigma y_i)^2/n \qquad (14.11)$$

Regression Model

$$y = \beta_0 + \beta_1 x + \epsilon \qquad (14.12)$$

Regression Equation

$$E(y) = \beta_0 + \beta_1 x \qquad (14.13)$$

Mean Square Error (Estimate of σ^2)

$$s^2 = MSE = \frac{SSE}{n-2} \qquad (14.14)$$

Standard Error of the Estimate

$$s = \sqrt{MSE} = \sqrt{\frac{\Sigma(y_i - \hat{y}_i)^2}{n-2}} \qquad (14.15)$$

Standard Deviation of b_1

$$\sigma_{b_1} = \frac{\sigma}{\sqrt{\Sigma x_i^2 - (\Sigma x_i)^2/n}} \qquad (14.16)$$

Estimated Standard Deviation of b_1

$$s_{b_1} = \frac{s}{\sqrt{\Sigma x_i^2 - (\Sigma x_i)^2/n}} \qquad (14.17)$$

Mean Square Due to Regression

$$MSR = \frac{SSR}{\text{Regression df}} = \frac{SSR}{\text{Number of independent variables}} \qquad (14.18)$$

The F Statistic

$$F = \frac{MSR}{MSE} \qquad (14.19)$$

Estimated Standard Deviation of $\hat{y}_p$

$$s_{\hat{y}_p} = s\sqrt{\frac{1}{n} + \frac{(x_p - \bar{x})^2}{\Sigma x_i^2 - (\Sigma x_i)^2/n}} \qquad (14.20)$$

Confidence Interval Estimate of $E(y_p)$

$$\hat{y}_p \pm t_{\alpha/2}s_{\hat{y}_p} \qquad (14.21)$$

Estimated Standard Deviation when Predicting an Individual Value

$$s_{\text{ind}} = s\sqrt{1 + \frac{1}{n} + \frac{(x_p - \bar{x})^2}{\Sigma x_i^2 - (\Sigma x_i)^2/n}} \qquad (14.22)$$

Prediction Interval Estimate of y_p

$$\hat{y}_p \pm t_{\alpha/2} s_{\text{ind}} \tag{14.23}$$

Standard Deviation of ith Residual

$$s_{y_i} - \hat{y}_i = \sqrt{s^2(1 - h_i)} \tag{14.24}$$

Leverage of Observation i

$$h_i = \frac{1}{n} + \frac{(x_i - \overline{x})^2}{\Sigma(x_i - \overline{x})^2} \tag{14.25}$$

Sample Covariance

$$s_{xy} = \frac{\Sigma(x_i - \overline{x})(y_i - \overline{y})}{n - 1} \tag{14.26}$$

Pearson Product Moment Correlation Coefficient:
Sample Data

$$r_{xy} = \frac{s_{xy}}{s_x s_y} \tag{14.28}$$

Pearson Product Moment Correlation Coefficient:
Sample Data, Alternate Formula

$$r_{xy} = \frac{\Sigma x_i y_i - (\Sigma x_i \Sigma y_i)/n}{\sqrt{\Sigma x_i^2 - (\Sigma x_i)^2/n} \ \sqrt{\Sigma y_i^2 - (\Sigma y_i)^2/n}} \tag{14.29}$$

Determining the Sample Correlation Coefficient from the Regression Analysis Output

$$r_{xy} = (\text{sign of } b_1)\sqrt{\text{Coefficient of Determination}} = \pm\sqrt{r^2} \tag{14.31}$$

❑ ❑ Supplementary Exercises

66. What is the difference between regression analysis and correlation analysis?

67. Does a high value of r^2 imply that two variables are causally related? Explain.

68. In your own words, explain the difference between an interval estimate of the mean value of y for a given x and an interval estimate for an individual value of y for a given x.

69. What is the purpose of testing whether or not $\beta_1 = 0$? If we reject $\beta_1 = 0$, does this imply a good fit?

70. In a manufacturing process the assembly line speed (feet per minute) was thought to affect the number of defective parts found during the inspection process. To test this theory, management devised a situation where the same batch of parts was inspected visually at a variety of line speeds. Table 14.32 gives the collected data.

    a. Develop the estimated regression equation that relates line speed to the number of defective parts found.

    b. At the $\alpha = .05$ level of significance determine whether line speed and number of defective parts found are related.

    c. Did the estimated regression equation provide a good fit to the data?

    d. Develop a 95% confidence interval to predict the mean number of defective parts for a line speed of 50 feet per minute.

TABLE **14.32**

Line Speed	Number of Defective Parts Found
20	21
20	19
40	15
30	16
60	14
40	17

71. A study was conducted by Monsanto Company to determine the relationship between the percentage of supplemental methionine used in feed and the body weight of the poultry. Using the data collected in this study, regression analysis was used to develop the following estimated regression line.

$$\hat{y} = .21 + .42x$$

where

$\hat{y}$ = estimated body weight in kilograms

x = percentage of supplemental methionine used in the feed

The coefficient of determination r^2 was .78, indicating a reasonably good fit for the data. Suppose it is known that a sample size of 30 was used for the study and that SST = 45.
a. Compute SSR and SSE.
b. Test for a significant regression relationship using $\alpha = .01$.
c. What is the value of the sample correlation coefficient?

72. The PJH&D Company is in the process of deciding whether to purchase a maintenance contract for its new word-processing system. They feel that maintenance expense should be related to usage and have collected the information shown in Table 14.33 on weekly usage (hours) and annual maintenance expense.
a. Develop the estimated regression equation that relates annual maintenance expense, in hundreds of dollars, to weekly usage.
b. Test the significance of the relationship in (a) at the $\alpha = .05$ level of significance.
c. PJH&D expects to operate the word processor 30 hours per week. Develop a 95% prediction interval for the company's annual maintenance expense.
d. If the maintenance contract costs $3000 per year, would you recommend purchasing it? Why or why not?

73. A sociologist was hired by a large city hospital to investigate the relationship between the number of unauthorized days that an employee is absent per year and the distance (miles) between home and work for the employees. A sample of 10 employees was chosen, and the following data were collected.

TABLE **14.33**

Weekly Usage (hours)	Annual Maintenance Expense ($100s)
13	17.0
10	22.0
20	30.0
28	37.0
32	47.0
17	30.5
24	32.5
31	39.0
40	51.5
38	40.0

Distance to Work (miles)	Number of Days Absent
1	8
3	5
4	8
6	7
8	6
10	3
12	5
14	2
14	4
18	2

a. Develop a scatter diagram for these data. Does a linear relationship appear reasonable? Explain.
b. Develop the least squares estimated regression equation.
c. Is there a significant relationship between the two variables? Use $\alpha = .05$.
d. Did the estimated regression equation provide a good fit? Explain.
e. Use the estimated regression equation developed in (b) to develop a 95% confidence interval estimate of the expected number of days absent for employees living 5 miles from the company.

TABLE 14.34

Number of Competitors	Sales ($)
1	3600
1	3300
2	3100
3	2900
3	2700
4	2500
5	2300
5	2000

74. The owner of a chain of fast-food restaurants would like to investigate the relationship between the daily sales volume of a company restaurant and the number of competitor restaurants within a 1-mile radius of the firm's restaurant. Table 14.34 gives the data.
  a. Develop the least squares estimated regression equation that relates daily sales volume to the number of competitor restaurants within a 1-mile radius.
  b. Is there a significant relationship between the two variables? Use $\alpha = .05$.
  c. Did the estimated regression equation provide a good fit? Explain.
  d. Use the estimated regression line developed in (a) to develop a 95% interval estimate of the daily sales volume for a particular company restaurant that has four competitors within a 1-mile radius.

75. Performance data for a Century Coronado 21 with a 310-hp MerCruiser V-8 gasoline inboard engine was reported in *Boating*, September 1991. Data on how the engine revolutions per minute (rpm) affected boat speed in miles per hour (mph) are shown below.

rpm	mph
1000	6.1
1500	10.7
2000	20.9
2500	27.5
3000	31.5
3500	33.6
4000	37.9
4500	40.2
4800	40.7

  a. Develop the estimated regression equation showing how boat speed is related to the engine revolutions per minute.
  b. Test the significance of the relationship at the $\alpha = .05$ level of significance.
  c. Develop a plot of the standardized residuals against $\hat{y}$. What conclusions can you draw from this plot?

76. The 1992 U.S. Men's Olympic Marathon Trials (Columbus, Ohio, April 11, 1992) provided marathon qualifying times and ages for 109 runners. Data below show the number of minutes that the qualifying times exceeded two hours and the age for a sample of eight runners.

Age	Qualifying Time
33	12.1
33	12.6
31	13.1
26	14.0
26	14.1
25	14.6
30	15.0
29	15.5

a. Develop the estimated regression equation showing how qualifying time is related to age.

b. Test the significance of the relationship at the $\alpha = .05$ level of significance.

c. Develop a plot of the standardized residuals against $\hat{y}$. What conclusions can you draw from this plot?

77. The regional transit authority for a major metropolitan area would like to determine if there is any relationship between the age of a bus and the annual maintenance cost. A sample of 10 buses resulted in the data shown in Table 14.35. Compute the sample correlation coefficient for the above data in the table.

78. Reconsider the regional transit authority problem presented in Exercise 77.

a. Develop the least squares estimated regression equation.

b. Test to see if the two variables are significantly related at $\alpha = .05$.

c. Did the least squares line provide a good fit to the observed data? Explain.

d. Develop a 95% prediction interval for the maintenance cost for a specific bus that is 4 years old.

79. A psychology professor at Givens College is interested in the relation between hours spent studying and total points earned in the course. Data collected on 10 students who took the course last quarter are given below:

TABLE **14.35**

Age of Bus (years)	Maintenance Cost ($)
1	350
2	370
2	480
2	520
2	590
3	550
4	750
4	800
5	790
5	950

Hours Spent Studying	Total Points Earned
45	40
30	35
90	75
60	65
105	90
65	50
90	90
80	80
55	45
75	65

Compute the sample correlation coefficient for these data.

80. Reconsider the Givens College data in Exercise 79.

a. Develop an estimated regression equation relating total points earned to hours spent studying.

b. Test the significance of the model at the $\alpha = .05$ level.

c. Predict the total points earned by Mark Sweeney. He spent 95 hours studying.

d. Develop a 95% prediction interval for the total points earned by Mark Sweeney.

81. *USA Today* publishes college basketball computer rankings that are based upon a strength rating computed for each team. The strength ratings can also be used to predict the victory margin for games. For the visiting team, the predicted score is its strength rating. For the home team, the predicted score is the strength rating plus $4\frac{1}{2}$. For each game, the actual victory margin is the winning team's score minus the losing team's score. The predicted victory margin is the predicted score for the winning team minus the predicted score for the losing team.

A sample was taken of 10 college basketball games selected from *USA Today*, December 13, 1988, to investigate the accuracy of the predicted victory margin. The following table gives the data obtained. The strength ratings are in parentheses.

Visiting Team	Score	Home Team	Score
Eastern Michigan (77.06)	57	Michigan (101.07)	80
Jackson State (65.75)	71	Iowa (96.63)	86
Georgia Southern (80.01)	80	Eastern Kentucky (61.24)	69
Seton Hall (92.66)	96	Rutgers (72.22)	70
Niagara (70.32)	78	St. Bonaventure (70.57)	81
Fairfield (63.01)	48	Connecticut (85.15)	71
S. Carolina St. (73.29)	70	Clemson (80.13)	93
Monmouth (57.43)	70	Maryland (80.53)	74
Illinois-Chicago (73.95)	74	Michigan State (83.45)	96
Oral Roberts (71.37)	75	Georgetown (87.48)	91

a. Let x be the predicted victory margin and y be the actual victory margin. Use the strength ratings and actual game scores to compute x and y for the 10 games.

b. Compute the correlation coefficient between the predicted victory margin and the actual victory margin.

c. Test for a significant relationship. Use $\alpha = .01$.

d. Develop an estimated regression equation using the predicted victory margin as the independent variable and the actual victory margin as the dependent variable.

e. What are the values of b_0 and b_1? Comment on what the values of β_0 and β_1 should be for an ideal system of predicting victory margins.

Computer Exercise: *U.S. Department of Transportation*

SAFETY

As part of a study on transportation safety, the U.S. Department of Transportation collected data on the number of fatal accidents per 1000 licenses and the percentage of licensed drivers under the age of 21 in a sample of 42 cities. Data collected over a 1-year period are shown below. These data are available on the data disk in the file named SAFETY.

Percent Under 21	Fatal Accidents per 1000 Licenses	Percent Under 21	Fatal Accidents per 1000 Licenses
13	2.962	17	4.100
12	0.708	8	2.190
8	0.885	16	3.623
12	1.652	15	2.623
11	2.091	9	0.835
17	2.627	8	0.820
18	3.830	14	2.890
8	0.368	8	1.267
13	1.142	15	3.224
8	0.645	10	1.014
9	1.028	10	0.493
16	2.801	14	1.443
12	1.405	18	3.614
9	1.433	10	1.926

—Table continues on the next page

—Table continued from previous page

Percent Under 21	Fatal Accidents per 1000 Licenses	Percent Under 21	Fatal Accidents per 1000 Licenses
10	0.039	14	1.643
9	0.338	16	2.943
11	1.849	12	1.913
12	2.246	15	2.814
14	2.855	13	2.634
14	2.352	9	0.926
11	1.294	17	3.256

Managerial Report

1. Develop numerical and graphical summaries of the data.
2. Use regression analysis to investigate the relationship between the number of fatal accidents and the percentage of drivers under the age of 21. Discuss your findings.
3. What conclusions and/or recommendations can you derive from your analysis?

APPENDIX

Calculus-Based Derivation of Least Squares Formulas

As mentioned in the chapter, the least squares method is a procedure for determining the values of b_0 and b_1 that minimize the sum of squared residuals. The sum of squared residuals is given by

$$\Sigma(y_i - \hat{y}_i)^2$$

Substituting $\hat{y}_i = b_0 + b_1x_i$, we get

$$\Sigma(y_i - b_0 - b_1x_i)^2 \tag{14A.1}$$

as the expression that must be minimized.

To minimize (14A.1) we must take the partial derivatives with respect to b_0 and b_1, set them equal to zero, and solve. Doing so we get

$$\frac{\partial \Sigma(y_i - b_0 - b_1x_i)^2}{\partial b_0} = -2\Sigma(y_i - b_0 - b_1x_i) = 0 \tag{14A.2}$$

$$\frac{\partial \Sigma(y_i - b_0 - b_1x_i)^2}{\partial b_1} = -2\Sigma x_i(y_i - b_0 - b_1x_i) = 0 \tag{14A.3}$$

Dividing (14A.2) by 2 and summing each term individually yields

$$-\Sigma y_i + \Sigma b_0 + \Sigma b_1 x_i = 0$$

Bringing Σy_i to the other side of the equal sign and noting that $\Sigma b_0 = nb_0$, we obtain

$$nb_0 + (\Sigma x_i)b_1 = \Sigma y_i \qquad \text{(14A.4)}$$

Similar algebraic simplification applied to (14A.3) yields

$$(\Sigma x_i)b_0 + (\Sigma x_i^2)b_1 = \Sigma x_i y_i \qquad \text{(14A.5)}$$

(14A.4) and (14A.5) are known as the *normal equations*. Solving (14A.4) for b_0 yields

$$b_0 = \frac{\Sigma y_i}{n} - b_1 \frac{\Sigma x_i}{n} \qquad \text{(14A.6)}$$

Using (14A.6) to substitute for b_0 in (14A.5) provides

$$\frac{\Sigma x_i \, \Sigma y_i}{n} - \frac{(\Sigma x_i)^2}{n} b_1 + (\Sigma x_i^2)b_1 = \Sigma x_i y_i \qquad \text{(14A.7)}$$

Rearranging (14A.7), we obtain

$$b_1 = \frac{\Sigma x_i y_i - (\Sigma x_i \, \Sigma y_i)/n}{\Sigma x_i^2 - (\Sigma x_i)^2/n} \qquad \text{(14A.8)}$$

Since $\bar{y} = \Sigma y_i/n$ and $\bar{x} = \Sigma x_i/n$, we can rewrite (14A.6):

$$b_0 = \bar{y} - b_1 \bar{x} \qquad \text{(14A.9)}$$

Equations (14A.8) and (14A.9) are the formulas we used in the chapter to compute the coefficients in the estimated regression equation.

15

Multiple Regression

Contents

Champion International Corporation*

STAMFORD, CONNECTICUT

Champion International Corporation is one of the largest forest product companies in the world, with over 3 million acres of timberlands in the United States. They produce building materials, such as lumber and plywood, white paper products, such as printing and writing grades of white paper, and brown paper products, such as linerboard and corrugated containers. To make these paper products, Champion's pulp mills process wood chips and chemicals to produce wood pulp. The wood pulp is then used at a paper mill to produce paper products.

When producing white paper products, the pulp must be bleached to remove any discoloration. A key bleaching agent used in the process is chlorine dioxide, which, because of its combustible nature, is usually produced at Champion pulp mill facilities and then piped in solution form into the bleaching tower of the pulp mill. To improve one of the processes which Champion uses to produce chlorine dioxide, a study was undertaken to look at process control and efficiency. One of the aspects studied was the chemical-feed rate for chlorine dioxide production.

To produce the chlorine dioxide, four chemicals flow at metered rates into the chlorine dioxide generator. The chlorine dioxide that is produced in the generator then flows to an absorber where chilled water absorbs the chlorine dioxide gas to form a chlorine dioxide solution. The solution is then piped into the paper mill. A key part of controlling the process involves the chemical-feed rates. Historically, the chemical-feed rates were simply set by experienced opera

tors; this approach, however, led to a situation in which the process was being overcontrolled by the operators. Consequently, chemical engineers at the mill requested that a set of control equations, one for each chemical feed, be developed to aid the operators in setting the rates.

Using multiple regression analysis, statistical analysts at Champion were able to develop a multiple regression equation for each of the four chemicals used in the process. Each equation related the production of carbon dioxide to the amount of chemical used and the concentration level of the carbon dioxide solution. The resulting set of four equations was programmed into a microcomputer at each mill. In the new system, operators enter the concentration of the chlorine dioxide solution and the desired production rate; the computer software then calculates the chemical feed needed to achieve the desired production rate. Since the operators have begun using the control equations, the chlorine dioxide generator efficiency has increased, and the number of times that the concentrations fall within acceptable ranges has increased significantly.

Champion used multiple regression analysis to develop its control equations. In this chapter we will discuss how statistical computer packages such as Minitab are used for such purposes. Most of the concepts introduced in Chapter 14 for simple linear regression can be directly extended to the multiple regression case.

*The authors are indebted to Marian Williams and Bill Griggs of Champion International Corporation for providing this Statistics in Practice.

I n Chapter 14 we discussed how regression analysis can be used to develop a mathematical equation representing the relationship between two variables. Recall that the variable being predicted or explained by the mathematical equation is called the dependent variable; the variable being used to predict or explain the value of the dependent variable is called the independent variable. In this chapter we continue our study of regression analysis by considering situations that involve two or more independent variables. The study of regression models involving more than one independent variable is called multiple regression analysis.

15.1   The Multiple Regression Model and Its Assumptions

The probabilistic model for multiple regression analysis is a direct extension of the one introduced in the previous chapter for simple linear regression. To show this, consider a situation involving the sale of a new product y in a certain region. Suppose that we believe sales are related to the population size x_1 and the average disposable income x_2 of people in the region by the following regression model:

$$y = \beta_0 + \beta_1 x_1 + \beta_2 x_2 + \epsilon \tag{15.1}$$

The relationship shown in (15.1) is a *multiple regression model* involving two independent variables. Note that if $\beta_2 = 0$, then x_2 is not related to y, and hence, the multiple regression model reduces to the single-independent-variable model discussed in Chapter 14; that is, $y = \beta_0 + \beta_1 x_1 + \epsilon$.

The multiple regression model of (15.1) can be extended to the case of p independent variables by simply adding more terms. Equation (15.2) shows the general case.

Multiple Regression Model

$$y = \beta_0 + \beta_1 x_1 + \beta_2 x_2 + \cdots + \beta_p x_p + \epsilon \tag{15.2}$$

Note that if $\beta_3, \beta_4, \ldots, \beta_p$ all equal zero, (15.2) reduces to the two-independent-variable multiple regression model of (15.1).

The assumptions made about the error term ϵ in Chapter 14 also apply in multiple regression analysis.

Assumptions about the Error Term ϵ
in the Regression Model $y = \beta_0 + \beta_1 x_1 + \cdots + \beta_p x_p + \epsilon$

1. The error ϵ is a random variable with mean or expected value of zero; that is, $E(\epsilon) = 0$.
   Implication: For given values of $x_1, x_2, \ldots, x_p$, the expected, or average, value of y is given by

$$E(y) = \beta_0 + \beta_1 x_1 + \beta_2 x_2 + \cdots + \beta_p x_p \tag{15.3}$$

Equation (15.3) is referred to as the *multiple regression equation*. In this equation, $E(y)$ represents the average of all possible values of y that might occur for the given values of $x_1, x_2, \ldots, x_p$.

2. The variance of ϵ is denoted by σ^2 and is the same for all values of the independent variables $x_1, x_2, \ldots, x_p$.
   Implication: The variance of y equals σ^2 and is the same for all values of $x_1, x_2, \ldots, x_p$.

Continued on next page

Continued from previous page

3. The values of ϵ are independent.
   Implication: The size of the error for a particular set of values for the independent variables is not related to the size of the error for any other set of values.

4. The error ϵ is a normally distributed random variable reflecting the deviation between the y value and the expected value of y given by $\beta_0 + \beta_1 x_1 + \beta_2 x_2 + \cdots + \beta_p x_p$.
   Implication: Since $\beta_0, \beta_1, \ldots, \beta_p$ are constants, for the given values of $x_1, x_2, \ldots, x_p$, the dependent variable y is also a normally distributed random variable.

To obtain more insight into the form of the relationship given by (15.3), consider for the moment the following two-independent-variable multiple regression equation:

$$E(y) = \beta_0 + \beta_1 x_1 + \beta_2 x_2 \tag{15.4}$$

The graph of this equation is a plane in three-dimensional space. Figure 15.1 shows such a graph with x_1 and x_2 on the horizontal axis and y on the vertical axis. Note that ϵ is shown as the difference between the actual y value and the expected value of y, $E(y)$, when $x_1 = x_1^*$ and $x_2 = x_2^*$.

In regression analysis, the term *response variable* is often used in place of the term *dependent variable*. Furthermore, since the multiple regression equation generates a plane or surface, its graph is referred to as a response surface.

In the previous chapter the least squares method was used to develop estimates of β_0 and β_1 for the simple linear regression model. In multiple regression analysis, the least

FIGURE 15.1

Graph of the Regression Equation for Multiple Regression Analysis with Two Independent Variables

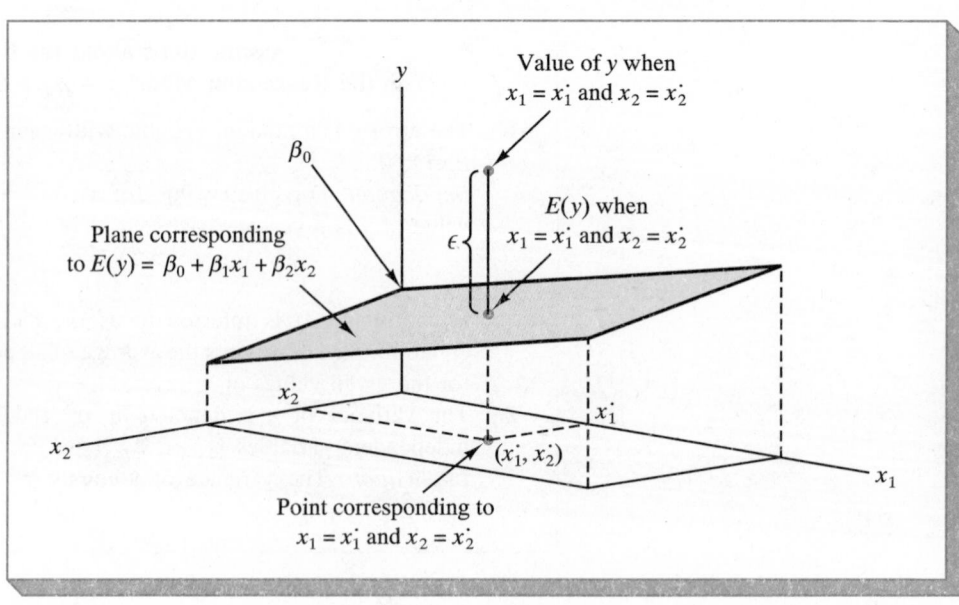

squares method is used in an analogous manner to develop estimates of the parameters $\beta_0, \beta_1, \beta_2, \ldots, \beta_p$. These estimates are denoted $b_0, b_1, b_2, \ldots, b_p$; the corresponding estimated regression equation is written as follows.

Estimated Regression Equation

$$\hat{y} = b_0 + b_1 x_1 + \cdots + b_p x_p \tag{15.5}$$

At this point, we can begin to see the similarity between the concepts of multiple regression analysis and those of the previous chapter. The concepts of simple linear regression have simply been extended to the case involving more than one independent variable.

15.2   Developing the Estimated Regression Equation

To show how an estimated regression equation can be developed in situations which involve two or more independent variables, let us consider the problem facing Butler Trucking, an independent trucking company located in southern California. A major portion of Butler's business involves deliveries throughout its local area. To develop better work schedules, management would like to use an estimated regression equation to help predict total daily travel time for its drivers.

Initially, management felt that travel time should be closely related to miles traveled. A random sample of 10 days of operation was taken; the data obtained are presented in Table 15.1 with the corresponding scatter diagram shown in Figure 15.2. The scatter diagram indicates that the number of miles traveled x_1 and the travel time y are positively related; that is, as x_1 increases, y increases. After observing this scatter diagram, management hypothesized the following regression model:

$$y = \beta_0 + \beta_1 x_1 + \epsilon \tag{15.6}$$

Note that this is nothing more than the simple linear regression model with x_1 replacing x.

In Figure 15.3 we show the Minitab computer output corresponding to the data in Table 15.1. The resulting estimated regression equation is shown on the computer output as

$$\text{TIME} = 1.27 + 0.0678 \text{ MILES}$$

where

$$\text{TIME} = \text{travel time in hours}$$

$$\text{MILES} = \text{miles traveled}$$

In other words,

$$\hat{y} = 1.27 + 0.0678 x_1$$

TABLE 15.1

Preliminary Data for Butler Trucking

Day	x_1 = Miles Traveled	y = Travel Time (hours)
1	100	9.3
2	50	4.8
3	100	8.9
4	100	6.5
5	50	4.2
6	80	6.2
7	75	7.4
8	65	6.0
9	90	7.6
10	90	6.1

FIGURE 15.2
Scatter Diagram of Preliminary Data for Butler Trucking

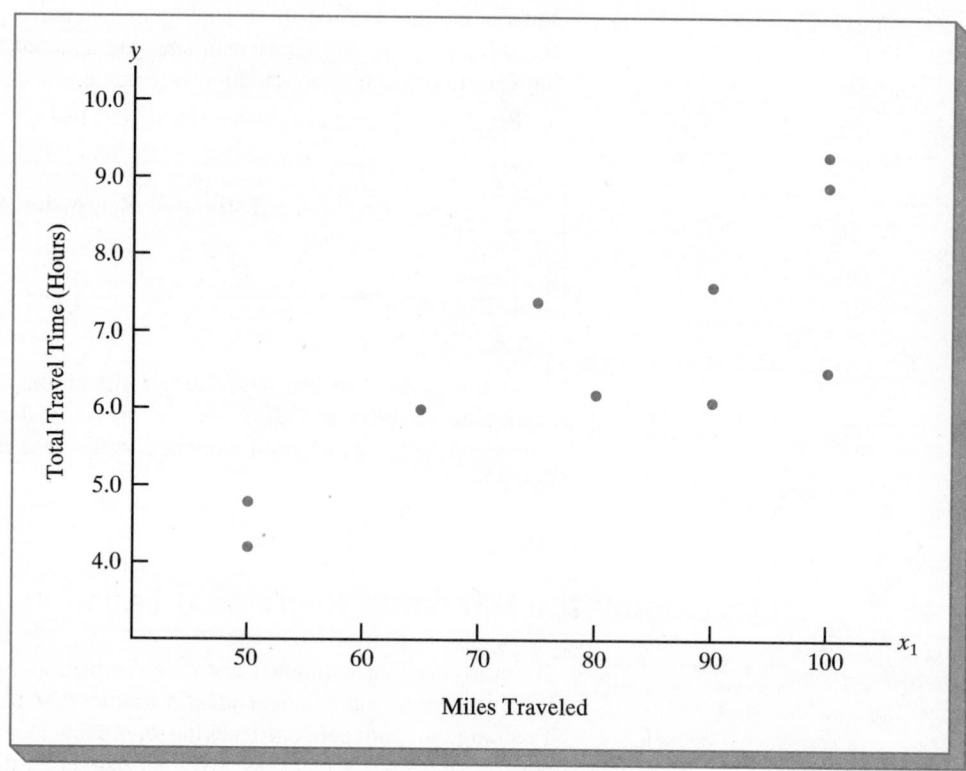

FIGURE 15.3   **Minitab Output for Butler Trucking with One Independent Variable**

```
The regression equation is
TIME = 1.27 + 0.0678 MILES

Predictor        Coef        Stdev      t-ratio          p
Constant        1.274        1.401         0.91      0.390
MILES         0.06783      0.01706         3.98      0.004

s = 1.002        R-sq = 66.4%      R-sq(adj) = 62.2%

Analysis of Variance

SOURCE          DF          SS            MS          F          p
Regression       1       15.871        15.871      15.81      0.004
Error            8        8.029         1.004
Total            9       23.900
```

The relationship between x_1 and y does appear to be statistically significant. The p-value for the t test and the F test is .004. Moreover, with a coefficient of determination of $r^2 = 0.664$, we see that 66.4% of the variability in travel time has been explained by the relationship with miles traveled. However, before reaching any final conclusions regarding the usefulness of the estimated regression equation, let us consider the standardized residual plot shown in Figure 15.4.

In Figure 15.4 the standardized residuals are shown on the vertical axis and the predicted values $\hat{y}$ are shown on the horizontal axis. The standardized residual plot indicates that the variability of the standardized residuals is increasing as $\hat{y}$ increases and that the assumption of constant variance for the error term (assumption 2) does not appear to be satisfied. Since the tests for statistical significance in regression analysis are based on the assumptions regarding the error term, our previous conclusion regarding the statistical significance of the relationship must be questioned. In addition, even though 66.4% of the variability has been explained using just miles traveled, 33.6% of the variability is still unexplained; the goodness of fit can perhaps be improved. Given the problem we observed with the standardized residual plot, we should doubt the adequacy of the existing model.

Looking for an alternative, management suggested that perhaps the number of deliveries made could also be used to help predict travel time and hence improve the regression model. The data, with the addition of the number of deliveries, are shown in Table 15.2, where x_2 denotes the number of deliveries.

FIGURE 15.4   Standardized Residual Plot for Butler Trucking with One Independent Variable

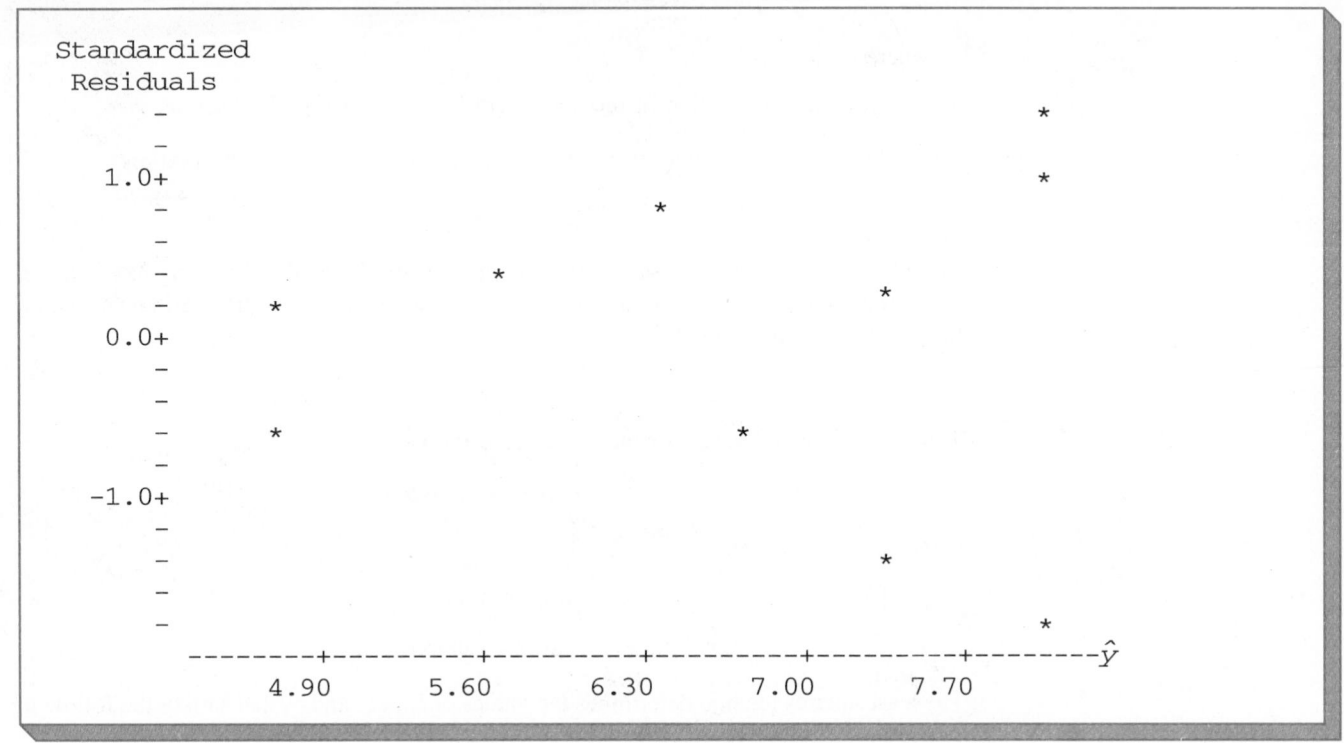

TABLE 15.2

Data for Butler Trucking with Miles Traveled (x_1) and Number of Deliveries (x_2) as the Independent Variables

BUTLER

Day	x_1 = Miles Traveled	x_2 = Number of Deliveries	y = Travel Time (hours)
1	100	4	9.3
2	50	3	4.8
3	100	4	8.9
4	100	2	6.5
5	50	2	4.2
6	80	2	6.2
7	75	3	7.4
8	65	4	6.0
9	90	3	7.6
10	90	2	6.1

Multiple Regression and the Least Squares Criterion

In Chapter 14 we showed how the least squares method can be used to find the straight line that provides the best approximation for the relationship between the independent and dependent variables. The criterion for the least squares method is given by (15.7).

Least Squares Criterion

$$\min \Sigma (y_i - \hat{y}_i)^2 \qquad (15.7)$$

where

y_i = observed value of the dependent variable for the ith observation

$\hat{y}_i$ = estimated value of the dependent variable for the ith observation

---

In multiple regression, this same criterion applies. In the Butler Trucking example, the estimated regression equation that includes the effect of the miles traveled (x_1) as well as the number of deliveries (x_2) is

$$\hat{y} = b_0 + b_1 x_1 + b_2 x_2$$

Thus, the predicted value for the ith observation is

$$\hat{y}_i = b_0 + b_1 x_{1i} + b_2 x_{2i}$$

where

$$x_{1i} = i\text{th observation for } x_1$$

$$x_{2i} = i\text{th observation for } x_2$$

The least squares method determines the values of b_0, b_1, and b_2 that satisfy the following criterion:

$$\min \Sigma (y_i - \hat{y}_i)^2 = \min \Sigma (y_i - b_0 - b_1 x_{1i} - b_2 x_{2i})^2$$

FIGURE 15.5 **Minitab Output for Butler Trucking with Two Independent Variables**

```
The regression equation is
TIME = - 0.869 + 0.0611 MILES + 0.923 DELIV

Predictor        Coef        Stdev      t-ratio          p
Constant      -0.8687       0.9515        -0.91      0.392
MILES        0.061135     0.009888         6.18      0.000
DELIV          0.9234       0.2211         4.18      0.004

s = 0.5731       R-sq = 90.4%      R-sq(adj) = 87.6%

Analysis of Variance

SOURCE         DF          SS           MS          F          p
Regression      2      21.601       10.800      32.88      0.000
Error           7       2.299        0.328
Total           9      23.900
```

In Chapter 14 we presented formulas for computing the values of b_0 and b_1 for problems involving one independent variable. In the multiple regression case, the usual presentation of the formulas for the coefficients of the estimated regression equation involves the use of matrix algebra and is beyond the scope of this text.* Thus, we will focus our attention on how computer software packages can be used to obtain the estimated regression equation. In subsequent sections we will show how computer output can be used to determine the goodness of fit, test for significance, and develop interval estimates.

Computer Solution

In Figure 15.5 we show the output from the Minitab computer package for the version of the Butler Trucking example with two independent variables, miles traveled and the number of deliveries. The estimated regression equation is

$$TIME = -0.869 + 0.0611 \text{ MILES} + 0.923 \text{ DELIV}$$

where

$$TIME = \text{travel time (hours)}$$

$$MILES = \text{miles traveled}$$

$$DELIV = \text{number of deliveries}$$

In other words,

$$\hat{y} = -0.869 + 0.0611x_1 + 0.923x_2$$

---

*In the appendix to the chapter we show how the least squares estimates can be obtained algebraically for the Butler Trucking example with two independent variables.

The least squares estimates $b_0 = -0.869$, $b_1 = 0.0611$, and $b_2 = 0.923$ are also shown in the output in the column labeled Coef; however, in this column the estimates are shown with more significant digits.

Note that the sum of squares due to regression (SSR = 21.601) plus the sum of squares due to error (SSE = 2.299) is equal to the total sum of squares (SST = 23.9); that is, SST = SSR + SSE. In addition, we see that the degrees of freedom associated with the regression sum of squares (2) plus the degrees of freedom associated with the error sum of squares (7) are equal to the degrees of freedom associated with the total sum of squares (9). In general the degrees of freedom for the regression sum of squares is the number of independent variables p. The degrees of freedom for the total sum of squares is the number of observations minus one, $n - 1$. So, the degrees of freedom for the error sum of squares is $n - p - 1$. Note also that the mean square due to regression (MSR = 10.8) is the regression sum of squares divided by the regression degrees of freedom, and that the mean square due to error (MSE = 0.328) is the error sum of squares divided by the error degrees of freedom. The F value ($F = 32.88$) is MSR/MSE.

Let us compare the computer output in Figure 15.3 for the model involving just one independent variable, miles traveled, with the output in Figure 15.5. Note that the total sum of squares (SST = 23.9) is the same in both cases. To better understand the significance of this result, recall that the total sum of squares is

$$SST = \Sigma(y_i - \overline{y})^2 \tag{15.8}$$

Since SST depends only on the observed values of the dependent variable, its value does not depend on how many independent variables we have in the regression model. Recognizing that the total sum of squares remains the same for every model is a key point in understanding what happens when we add additional independent variables to the model; that is, since SST = SSR + SSE, whenever we add additional independent variables to the model, we will see a decrease in the error sum of squares and a corresponding increase in the regression sum of squares.

In the following sections we will discuss these relationships among the sources of variation in more detail. Before doing so, however, let us consider more carefully the interpretation of the coefficients in the estimated multiple regression equation.

A Note on Interpretation of Coefficients

One observation can be made at this time concerning the relationship between the estimated regression equation with only the miles traveled as an independent variable and the equation which includes the number of deliveries as a second independent variable. The value of b_1 is not the same in both cases. In simple linear regression, we interpret b_1 as the amount of change in y for a 1-unit change in the independent variable. In multiple regression analysis, this interpretation must be modified somewhat. That is, in multiple regression analysis, we interpret each regression coefficient as follows: b_i represents an estimate of the change in y corresponding to a 1-unit change in x_i when all other independent variables are held constant. For example, in the Butler Trucking example involving two independent variables, $b_1 = .0611$. Thus, .0611 hours is an estimate of the expected increase in travel time corresponding to an increase of 1 mile in the distance traveled when the number of deliveries is held constant. Similarly, since $b_2 = .923$, an estimate of the expected increase in travel time corresponding to an increase of 1 delivery when the number of miles traveled is held constant is .923 hours.

□   □   **Exercises**

Note to student: The exercises involving data in this and subsequent sections were designed to be solved using a computer software package.

Methods

1. Shown below is the estimated regression equation for a model involving two independent variables and 10 observations.

$$\hat{y} = 29.1270 + .5906x_1 + .4980x_2$$

a. Interpret b_1 and b_2 in this estimated regression equation.

b. Estimate y when $x_1 = 180$ and $x_2 = 310$.

SELF TEST ▶   **2.** Consider the data in Table 15.3 involving a dependent variable y and two independent variables, x_1 and x_2.

a. Using these data, develop an estimated regression equation relating y to x_1. Estimate y if $x_1 = 45$.

b. Using these data, develop an estimated regression equation relating y to x_2. Estimate y if $x_2 = 15$.

c. Using these data, develop an estimated regression equation relating y to x_1 and x_2. Estimate y if $x_1 = 45$ and $x_2 = 15$.

3. In a regression analysis involving 30 observations, the following estimated regression equation was obtained.

$$\hat{y} = 17.6 + 3.8x_1 - 2.3x_2 + 7.6x_3 + 2.7x_4$$

a. Interpret b_1, b_2, and b_3 in this estimated regression equation.

b. Estimate y when $x_1 = 10$, $x_2 = 5$, $x_3 = 1$, and $x_4 = 2$.

Applications

4. A shoe store has developed the following estimated regression equation relating sales to inventory investment and advertising expenditures:

$$\hat{y} = 25 + 10x_1 + 8x_2$$

where

$$x_1 = \text{inventory investment (\$1000s)}$$

$$x_2 = \text{advertising expenditures (\$1000s)}$$

$$y = \text{sales (\$1000s)}$$

a. Estimate sales if there is a $15,000 investment in inventory and an advertising budget of $10,000.

b. Interpret b_1 and b_2 in this estimated regression equation.

5. The owner of TAI Movie Theaters, Inc., would like to investigate the effect of television advertising on weekly gross revenue. Table 15.4 gives the historical data.

a. Using these data, develop an estimated regression equation relating weekly gross revenue to television advertising expenditure.

b. Estimate the weekly gross revenue in a week in which $3500 is spent on television advertising.

SELF TEST ▶   **6.** As an extension of Exercise 5, consider the possibility of incorporating the effects of both newspaper and television advertising on weekly gross revenue. The following data were developed from historical records:

TABLE 15.3

y	x_1	x_2
94	30	12
108	47	10
112	25	17
178	51	16
94	40	5
175	51	19
170	74	7
117	36	12
142	59	13
211	76	16

TABLE 15.4

Weekly Gross Revenue ($1000s)	Television Advertising ($1000s)
96	5.0
90	2.0
95	4.0
92	2.5
95	3.0
94	3.5
94	2.5
94	3.0

MOVIE

Weekly Gross Revenue ($1000s)	Television Advertising ($1000s)	Newspaper Advertising ($1000s)
96	5.0	1.5
90	2.0	2.0
95	4.0	1.5
92	2.5	2.5
95	3.0	3.3
94	3.5	2.3
94	2.5	4.2
94	3.0	2.5

a. Find an estimated regression equation relating weekly gross revenue to television and newspaper advertising.

b. Is the coefficient for television advertising expenditures the same in Exercise 5(a) and 6(a)? Interpret this coefficient in each case.

MOWER

7. Heller Company manufactures lawn mowers and related lawn equipment. They believe that the quantity of lawnmowers sold depends on the price of their mower and the price of a competitor's mower. Let

$$y = \text{quantity sold (1000s)}$$

$$x_1 = \text{price of competitor's mower (\$)}$$

$$x_2 = \text{price of Heller's mower (\$)}$$

Management would like an estimated regression equation that relates quantity sold to the price of the Heller mower and the competitor's mower. The following data are available concerning prices in 10 different cities.

TABLE 15.5

Value ($1000s)	Population	Average Inventory Value ($)
61	1,410	8,000
92	1,523	9,000
93	1,354	13,694
45	822	4,250
50	746	6,500
29	1,281	7,000
56	1,016	4,500
45	1,070	5,000
183	1,694	27,000
156	1,910	21,560
120	1,745	15,000
75	1,353	8,500
122	1,016	18,000

Competitor's Price (x_1)	Heller's Price (x_2)	Quantity Sold (y)
120	100	102
140	110	100
190	90	120
130	150	77
155	210	46
175	150	93
125	250	26
145	270	69
180	300	65
150	250	85

a. Determine the estimated regression equation that can be used to predict the quantity sold given the competitor's price and Heller's price.

b. Interpret b_1 and b_2.

c. Predict the quantity sold in a city where Heller prices its mower at $160 and the competitor prices its mower at $170.

8. The data in Table 15.5 show the value of prescriptions sold (in $1000s) for 13 pharmacies in Iowa, the population of the city served by the given pharmacy, and the average prescription inventory value. ("The use of categorical variables in Data Envelopment Analysis," R. Banker and R. Morey, *Management Science*, December 1986).

PHARMACY

a. Determine the estimated regression equation that can be used to predict the dollar value of prescriptions y given the population size x_1 and the average inventory value x_2.

b. What other variables do you think might be useful in predicting y?

 SCHOOLS1

9. Two experts provided subjective lists of school districts that they think are among the best in the country. For each school district the average class size, the combined SAT score, and the percentage of students who attended a 4-year college were provided (*The Wall Street Journal*, March 31, 1989).

District	Average Class Size	Combined SAT Score	% Attend 4-Year College
Blue Springs, Mo.	25	1083	74
Garden City, N.Y.	18	997	77
Indianapolis, Ind.	30	716	40
Newport Beach, Calif.	26	977	51
Novi, Mich.	20	980	53
Piedmont, Calif.	28	1042	75
Pittsburg, Pa.	21	983	66
Scarsdale, N.Y.	20	1110	87
Wayne, Pa.	22	1040	85
Weston, Mass.	21	1031	89
Farmingdale, N.Y.	22	947	81
Mamaroneck, N.Y.	20	1000	69
Mayfield, Ohio	24	1003	48
Morristown, N.J.	22	972	64
New Rochelle, N.Y.	23	1039	55
Newtown Square, Pa.	17	963	79
Omaha, Neb.	23	1059	81
Shaker Heights, Ohio	23	940	82

a. Using these data, develop an estimated regression equation relating the percentage of students that attend a 4-year college to the average class size and the combined SAT score.

b. Estimate the percentage of students that attend a 4-year college if the average class size is 20 and the combined SAT score is 1000.

 HOUSING

10. Data on housing markets were provided for 100 different cities in the United States (*U.S. News & World Report*, April 6, 1992). A sample of 16 cities with data on the median cost of a new home, the number of new housing starts during 1991–1992, and the average household income are shown below. All data are in 1000s.

City	Cost	Housing Starts	Household Income	City	Cost	Housing Starts	Household Income
Chicago	181.8	12.9	61.0	West Palm Beach	130.4	8.9	58.1
Dayton, Ohio	107.8	3.8	48.4	San Antonio	72.5	1.5	57.0
Atlanta	100.6	24.2	54.7	Pittsburgh	79.5	4.9	49.2
Oklahoma City	68.9	3.3	53.2	Jacksonville	82.1	8.0	47.5
Columbia, S.C.	90.3	3.1	57.4	Cleveland	122.9	5.4	54.0
Tacoma, Wash.	96.1	4.1	51.1	Gary, Ind.	98.2	3.2	45.7
Mobile, Ala.	68.5	1.0	41.0	Scranton, Pa.	81.6	2.5	44.8
Baltimore	121.8	11.1	62.8	Richmond, Va.	102.8	6.0	64.5

a. Using these data, develop the estimated regression equation relating cost to the number of housing starts and the household income.

b. Estimate the median cost for a city with 8000 housing starts and an average household income of $50,000.

15.3   Determining the Goodness of Fit

In Chapter 14 we used the coefficient of determination (x^2) to evaluate the goodness of fit for the regression relationship. Recall that r^2 was computed as

$$r^2 = \frac{SSR}{SST}$$

In multiple regression analysis, we compute a similar quantity, called the multiple coefficient of determination.

Multiple Coefficient of Determination

$$R^2 = \frac{SSR}{SST} \tag{15.9}$$

When multiplied by 100, the multiple coefficient of determination represents the percentage of variability in y that is explained by the estimated regression equation.

In the case of Butler Trucking (refer to Figure 15.5 for SSR and SST), we find

$$R^2 = \frac{21.601}{23.900} = .9038$$

Therefore 90.38% of the variability in y is explained by the relationship with miles traveled and number of deliveries. In Figure 15.5, this is rounded to show R-sq = 90.4%

Refer now to Section 15.2. Note that the regression model with only miles traveled as the independent variable had $r^2 = .664$. Therefore, the percentage of variability explained has increased from 66.4% to 90.38%. In general, it is always true that R^2 will increase as more independent variables are added to the regression model because adding variables to the model causes the prediction errors to be smaller, hence reducing SSE. Since SST = SSR + SSE, when SSE gets smaller, SSR must get larger, causing R^2 = SSR/SST to increase.

Many analysts recommend adjusting R^2 for the number of independent variables to avoid overestimating the impact of adding an independent variable on the amount of explained variability. This *adjusted multiple coefficient of determination* is computed as follows.

Adjusted Multiple Coefficient of Determination

$$R_a^2 = 1 - (1 - R^2)\frac{n - 1}{n - p - 1} \tag{15.10}$$

For the Butler Trucking example, we obtain

$$R_a^2 = 1 - (1 - 0.9038)\frac{10 - 1}{10 - 2 - 1}$$

$$= 1 - (0.0962)(1.2857)$$

$$= 0.8763$$

Thus, after adjusting for the number of independent variables in the model, 87.63% of the variability in y has been accounted for. Note that both the value of R^2 and the value of R_a^2 are provided by the Minitab output shown in Figure 15.5 where $R_a^2 = 0.8763$ is rounded to show R-sq(adj) = 87.6%.

**NOTES &
COMMENTS**

> If the value of R^2 is small and the model contains a large number of independent variables, the adjusted coefficient of determination can take on a negative value; in such cases, Minitab sets the adjusted coefficient of determination to zero.

☐ ☐ Exercises

Methods

11. In Exercise 1 the following estimated regression equation based on 10 observations was presented:

$$\hat{y} = 29.1270 + .5906x_1 + .4980x_2$$

The values of SST and SSR are 6,724.125 and 6,216.375, respectively.
a. Find SSE.   **b.** Compute R^2.   **c.** Compute R_a^2.
d. Comment on the goodness of fit.

SELF TEST ▶

12. In Exercise 2, 10 observations were provided for a dependent variable y and two independent variables x_1 and x_2; for these data $SST = $ 15,182.9, and SSR $=$ 14,052.2.
a. Compute R^2.   **b.** Compute R_a^2.
c. Does the estimated regression equation explain a large amount of the variability in the data? Explain.

13. In Exercise 3 the following estimated regression equation based on 30 observations was developed.

$$\hat{y} = 17.6 + 3.8x_1 - 2.3x_2 + 7.6x_3 + 2.7x_4$$

The values of SST and SSR are 1805 and 1760, respectively.
a. Compute R^2.   **b.** Compute R_a^2.   **c.** Comment on the goodness of fit.

Applications

SELF TEST ▶

14. In Exercise 4 the following estimated regression equation relating sales to inventory investment and advertising expenditures was given:

$$\hat{y} = 25 + 10x_1 + 8x_2$$

The data used to develop the model came from a survey of 10 stores; for these data, SST = 16,000 and SSR = 12,000.
a. For the estimated regression equation given, compute R^2.
b. Compute R_a^2.
c. Does the model appear to explain a large amount of variability in the data? Explain.

15. Refer to Exercise 9.

a. What are R^2 and R_a^2 for this problem?

b. Does the estimated regression equation provide a good fit to the data? Explain.

16. Refer to Exercise 10.

a. What are R^2 and R_a^2 for this problem?

b. Does the estimated regression equation provide a good fit to the data? Explain.

15.4   Test for a Significant Relationship

The regression equation that we have assumed for the Butler Trucking example with two independent variables is

$$E(y) = \beta_0 + \beta_1 x_1 + \beta_2 x_2$$

Therefore, the appropriate test for determining whether there is a significant relationship among x_1, x_2, and y is as follows:

$$H_0: \beta_1 = \beta_2 = 0$$

$$H_a: \text{One or more of the parameters is not equal to zero}$$

If we reject H_0, we conclude that there is a significant relationship among x_1, x_2, and y.

The test used to determine if there is a significant relationship in the multiple regression case is an F test very similar to the one introduced in Chapter 14 for models involving one independent variable. First, we will show how it is used by applying it in the Butler Trucking example. Then we will generalize the application of the test to models involving p independent variables.

The F Test for the Butler Trucking Example

It can be shown that if the null hypothesis ($H_0: \beta_1 = \beta_2 = 0$) is true and the four underlying regression model assumptions are valid, then the sampling distribution of MSR/MSE follows an F distribution. The number of numerator degrees of freedom is equal to the degrees of freedom associated with the sum of squares due to regression, and the denominator degrees of freedom is equal to the degrees of freedom associated with the sum of squares due to error. Recall that the sum of squares due to regression (SSR) measures the amount of the variability in y explained by the regression model. Thus, we would expect large values of F = MSR/MSE to cast doubt on the null hypothesis (no relationship between the dependent and independent variables). Indeed, this is true; small values of F do not permit us to reject H_0, and large values lead to the rejection of H_0.

Let us return to the Butler Trucking example and test the significance of the multiple regression model. Refer to Figure 15.5. We see that

$$F = \text{MSR/MSE}$$

$$= 10.8/.328$$

$$= 32.88$$

To test the hypothesis that $\beta_1 = \beta_2 = 0$, we must determine whether 32.88 is a value that appears likely when random sampling from an F distribution with 2 numerator degrees of freedom and 7 denominator degrees of freedom.

Suppose that the level of significance is $\alpha = .05$. The critical value from the F distribution table (Table 4 of Appendix B) is 4.74. That is, if we are sampling randomly

Figure 15.6
Determination of the Critical Value and Rejection Region Using $\alpha = .05$

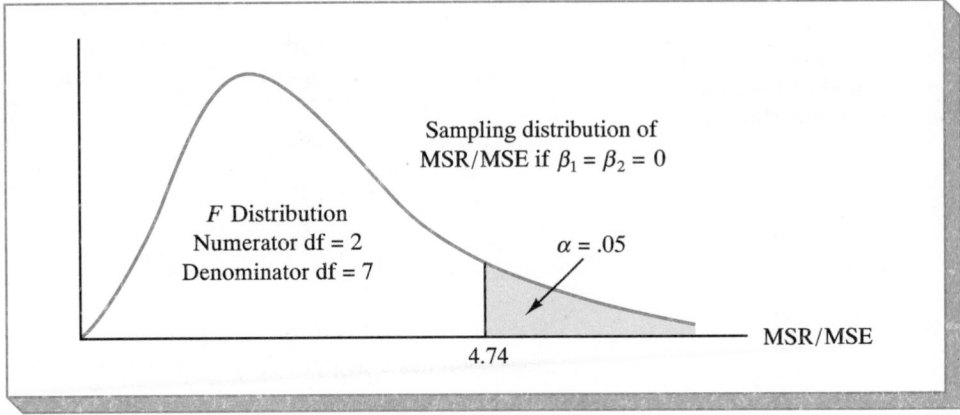

Figure 15.6
Determination of the Critical Value and Rejection Region Using $\alpha = .05$

from an F distribution with numerator and denominator degrees of freedom equal to 2 and 7, respectively (as we would be if H_0 were true), then only 5% of the time would we get a value larger than 4.74. Figure 15.6 illustrates the determination of the critical region.

We can use the above analysis to formulate the following rejection rule at the .05 significance level.

$$\text{Reject } H_0 \text{ if } \frac{\text{MSR}}{\text{MSE}} > 4.74$$

Since the value of MSR/MSE was 32.88, we can reject H_0. Hence, we conclude that there is a significant relationship between total travel time and the two independent variables. Thus, the estimated regression equation should be useful in predicting y for values of the independent variables within the range of those included in the sample.

The p-value provided by Minitab (see Figure 15.5) provides a convenient way to perform this same test. Since the p-value provides the probability of obtaining a sample result more unlikely than what is observed, the p-value of 0.000 indicates that for a level of significance of $\alpha = .05$, the null hypothesis that $\beta_1 = \beta_2 = 0$ should be rejected. In practice, the ease of use of the p-value approach to testing has made it the preferred method for performing hypothesis tests in regression analysis.

The General ANOVA Table and the F Test

Now that we know how the F test can be applied for a multiple regression model with two independent variables, let us generalize the test to the case involving a model with p independent variables. The appropriate hypothesis test to determine if there is a significant relationship is as follows:

$$H_0: \beta_1 = \beta_2 = \cdots = \beta_p = 0$$

H_a: One or more of the parameters is not equal to zero

Again, if we reject H_0, we can conclude that there is a significant relationship and that the estimated regression equation is useful for predicting or explaining the dependent variable y.

The general form of the ANOVA table for the multiple regression case involving p independent variables is shown in Table 15.6. The only change from the two-variable case

TABLE 15.6
ANOVA Table for a Multiple Regression Model with p Independent Variables

Source	Sum of Squares	Degrees of Freedom	Mean Square	F
Regression	SSR	p	$MSR = \dfrac{SSR}{p}$	$F = \dfrac{MSR}{MSE}$
Error	SSE	$n - p - 1$	$MSE = \dfrac{SSE}{n - p - 1}$	
Total	SST	$n - 1$		

is the degrees of freedom corresponding to SSR and SSE. Here the sum of squares due to regression has p degrees of freedom corresponding to the p independent variables; hence

$$MSR = \frac{SSR}{p} \qquad (15.11)$$

In addition, the sum of squares due to error has $n - p - 1$ degrees of freedom; thus

$$MSE = \frac{SSE}{n - p - 1} \qquad (15.12)$$

Hence, the F statistic for the case of p independent variables is computed as follows:

$$F = \frac{MSR}{MSE} = \frac{SSR/p}{SSE/(n - p - 1)} \qquad (15.13)$$

When looking up the critical value from the F distribution table, the numerator degrees of freedom are p and the denominator degrees of freedom are $n - p - 1$. As we mentioned earlier, the decision to reject or not to reject H_0 can be made by comparing the computed F-statistic with the critical value.

t Test for Significance of Individual Parameters

If after using the F test we conclude that the multiple regression relationship is significant (i.e., we conclude that at least one of the $\beta_i \neq 0$), it is often of interest to conduct further tests to see which individual parameters β_i are significant. The t test is a statistical method for testing the significance of the individual parameters.

The hypothesis test we wish to conduct is the same for the coefficient of each independent variable. It is stated as follows:

$$H_0: \beta_i = 0$$

$$H_a: \beta_i \neq 0$$

Recall that in Chapter 14 we learned how to conduct such a test for the case where there is only one independent variable. The hypotheses were:

$$H_0: \beta_1 = 0$$

$$H_a: \beta_1 \neq 0$$

To test these hypotheses, we computed the sample statistic b_1/s_{b_1}, where b_1 was the least squares estimate of β_1 and s_{b_1} was an estimate of the standard deviation of the sampling distribution of b_1. We learned that the sampling distribution of b_1/s_{b_1} follows a t distribution with $n - 2$ degrees of freedom. Thus, to conduct the hypothesis test, we used the following rejection rule:

$$\text{Reject } H_0 \text{ if } \frac{b_1}{s_{b_1}} > t_{\alpha/2}$$

$$\text{or if } \frac{b_1}{s_{b_1}} < -t_{\alpha/2}$$

The procedure for testing individual parameters in the multiple regression case is essentially the same. The only differences are in the number of degrees of freedom for the appropriate t distribution and in the formula for computing s_{b_i}. The number of degrees of freedom is the same as for the sum of squares due to error. Thus, we use $n - p - 1$ degrees of freedom, where p is the number of independent variables. (Note that for the case of one independent variable, this reduces to the $n - 2$ degrees of freedom used in Chapter 14.) The formula for s_{b_i} is more involved, and we do not present it here; however, s_{b_i} is calculated and printed by most computer software packages for multiple regression analysis.

Let us return now to the Butler Trucking example to test the significance of the parameters β_1 and β_2. Note that in the Minitab printout (Figure 15.5) the values of b_1, b_2, s_{b_1}, and s_{b_2} were given as

$$b_1 = 0.061135 \qquad s_{b_1} = 0.009888$$

$$b_2 = 0.9234 \qquad s_{b_2} = 0.2211$$

Therefore, for the parameters β_1 and β_2 we obtain

$$\frac{b_1}{s_{b_1}} = \frac{0.061135}{0.009888} = 6.18$$

$$\frac{b_2}{s_{b_2}} = \frac{0.9234}{0.2211} = 4.18$$

Note that both of these values were provided by the Minitab output of Figure 15.5 under the column labeled t-ratio. Using $\alpha = .05$ and $10 - 2 - 1 = 7$ degrees of freedom, we can find the appropriate t value for our hypothesis tests in Table 2 of Appendix B. We obtain

$$t_{.025} = 2.365$$

Now, since $b_1/s_{b_1} = 6.18 > 2.365$, we reject the hypothesis that $\beta_1 = 0$. Furthermore, since $b_2/s_{b_2} = 4.18 > 2.365$, we reject the hypothesis that $\beta_2 = 0$. Note also that the p-values provided by the Minitab outputs permit us to perform these tests very easily. For instance, the p-value of 0.004 for DELIV indicates that the null hypothesis H_0: $\beta_1 = 0$ should be rejected for all values of α larger than .004.

Multicollinearity

We have used the term *independent variable* in regression analysis to refer to any variable being used to predict or explain the value of the dependent variable. The term does not mean, however, that the independent variables themselves are independent in any statistical sense. Quite the contrary, most independent variables in a multiple regression problem are correlated to some degree with one another. For example, in the Butler Trucking example involving the two independent variables x_1 (miles traveled) and x_2 (number of deliveries), we could treat the miles traveled as the dependent variable and the number of deliveries as the independent variable to determine if these two variables are themselves related. We could then compute the sample correlation coefficient $r_{x_1 x_2}$ to determine the extent to which these variables are related. Doing so yields $r_{x_1 x_2} = .28$. Thus, there is some degree of linear association between the two independent variables. In multiple regression analysis we use the term *multicollinearity* to refer to the correlation among the independent variables.

To provide a better perspective of the potential problems of multicollinearity, let us consider a modification of the Butler Trucking example. Instead of x_2 being the number of deliveries, let x_2 denote the number of gallons of gasoline consumed. Clearly, x_1 (the miles traveled) and x_2 are related; that is, we know that the number of gallons of gasoline used depends on the number of miles traveled. Thus, we would conclude logically that x_1 and x_2 are highly correlated independent variables.

Assume that we obtain the equation $\hat{y} = b_0 + b_1 x_1 + b_2 x_2$ and find that the F test shows that the regression is significant. Then suppose that we were to conduct a t test on β_1 to determine if $\beta_1 \neq 0$, and we cannot reject H_0: $\beta_1 = 0$. Does this mean that travel time is not related to miles traveled? Not necessarily. What it probably means is that with x_2 already in the model, x_1 does not make a significant contribution toward determining the value of y. This would seem to make sense in our example, since if we know the amount of gasoline consumed, we do not gain much additional information useful in predicting y by knowing the miles traveled. Similarly, a t test might lead us to conclude $\beta_2 = 0$ on the grounds that, with x_1 in the model, knowledge of the amount of gasoline consumed does not add much.

To summarize, the difficulty caused by multicollinearity in conducting t tests for the significance of individual parameters is that it is possible to conclude that none of the individual parameters are significantly different from zero when an F test on the overall multiple regression equation indicates a significant relationship. This problem is avoided when there is very little correlation among the independent variables.

Ordinarily, multicollinearity does not affect the way in which we perform our regression analysis or interpret the output from a study. However, when multicollinearity is severe—that is, when two or more of the independent variables are highly correlated with one another—we can run into difficulties interpreting the results of t tests on the individual parameters. In addition to the type of problem illustrated above, severe cases of multicollinearity have been shown to result in least squares estimates that even have the wrong sign. That is, in simulated studies where researchers created the underlying regression model and then applied the least squares technique to develop estimates of β_0, β_1, β_2, and so on, it has been shown that under conditions of high multicollinearity the least squares estimates can even have a sign opposite to that of the parameter being estimated. For example, β_2 might actually be $+10$ and b_2, its estimate, might turn out to be -2. Thus, little faith can be placed in the individual coefficients themselves if multicollinearity is present to a high degree.

Statisticians have developed several tests for determining whether multicollinearity is high enough to cause these types of problems. One simple test, referred to as the rule of thumb test, says that multicollinearity is a potential problem if the absolute value of the sample correlation coefficient exceeds .7 for any two of the independent variables. The other types of tests are more advanced and beyond the scope of this text.

If possible, every attempt should be made to avoid including independent variables that are highly correlated. In practice, however, it is rarely possible to adhere to this policy strictly. Thus, the decision maker should be warned that when there is reason to believe that substantial multicollinearity is present, it is difficult to separate out the effect of the individual independent variables on the dependent variable.

☐ ☐ Exercises

Methods

SELF TEST ▶

17. In Exercise 1 the following estimated regression equation based on 10 observations was presented:

$$\hat{y} = 29.1270 + .5906x_1 + .4980x_2$$

Here SST = 6,724.125, SSR = 6,216.375, $s_{b_1} = .0813$, and $s_{b_2} = .0567$.
a. Compute MSR and MSE.
b. Compute F and perform the appropriate F test. Using $\alpha = .05$.
c. Perform a t test for the significance of β_1. Use $\alpha = .05$.
d. Perform a t test for the significance of β_2. Use $\alpha = .05$.

18. Refer to the data presented in Exercise 2. The estimated regression equation for these data is

$$\hat{y} = -18.4 + 2.01x_1 + 4.74x_2$$

a. Test for a significant relationship among x_1, x_2, and y. Use $\alpha = .05$.
b. Is β_1 significant? Use $\alpha = .05$.
c. Is β_2 significant? Use $\alpha = .05$.

19. The following estimated regression equation was developed for a model involving two independent variables:

$$\hat{y} = 40.7 + 8.63x_1 + 2.71x_2$$

After dropping x_2 from the model, the least squares method was used again to obtain an estimated regression equation involving only x_1 as an independent variable:

$$\hat{y} = 42.0 + 9.01x_1$$

a. Give an interpretation of the coefficient of x_1 in both models.
b. Could multicollinearity explain why the coefficient of x_1 differs in the two models? If so, how?

Applications

SELF TEST ▶

20. In Exercise 4 the following estimated regression equation for relating sales to inventory investment and advertising expenditures was given:

$$\hat{y} = 25 + 10x_1 + 8x_2$$

The data used to develop the model came from a survey of 10 stores; for these data SST = 16,000 and SSR = 12,000.
a. Compute SSE, MSE, and MSR.
b. Use an F test and an $\alpha = .05$ level of significance to determine if there is a relationship among the variables.

21. Refer to Exercise 6.
a. Use a level of significance of $\alpha = .01$ to test the hypotheses

$$H_0: \beta_1 = \beta_2 = 0$$

$$H_a: \beta_1 \text{ or } \beta_2 \text{ is not equal to zero}$$

for the model $y = \beta_0 + \beta_1 x_1 + \beta_2 x_2 + \epsilon$, where

$$x_1 = \text{television advertising (\$1000s)}$$

$$x_2 = \text{newspaper advertising (\$1000s)}$$

b. Use a level of significance of $\alpha = .05$ to test the significance of β_1. Should x_1 be dropped from the model?

c. Use a level of significance of $\alpha = .05$ to test the significance of β_2. Should x_2 be dropped from the model?

22. Refer to Exercise 7 involving the Heller Company. Test the significance of the overall model at $\alpha = .05$.

FORBES1

23. The following data show the price-earnings (P/E) ratio, the net profit margin, and the growth rate for 19 companies listed in "The *Forbes* 500s on Wall Street" (*Forbes*, May 1, 1989).

Firm	P/E Ratio	Profit Margin (%)	Growth Rate (%)
Exxon	11.3	6.5	10
Chevron	10.0	7.0	5
Texaco	9.9	3.9	5
Mobil	9.7	4.3	7
Amoco	10.0	9.8	8
Pfizer	11.9	14.7	12
Bristol Meyers	16.2	13.9	14
Merck	21.0	20.3	16
American Home Products	13.3	16.9	11
Abbott Laboratories	15.5	15.2	18
Eli Lilly	18.9	18.7	11
Upjohn	14.6	12.8	10
Warner-Lambert	16.0	8.7	7
Amdahl	8.4	11.9	4
Digital	10.4	9.8	19
Hewlett-Packard	14.8	8.1	18
NCR	10.1	7.3	6
Unisys	7.0	6.9	6
IBM	11.8	9.2	6

a. Determine the estimated regression equation that can be used to predict the price-earnings ratio given the net profit margin and the growth rate.

b. At the $\alpha = .05$ level of significance, determine if there is a relationship among the variables.

c. Does there appear to be any multicollinearity present in the data? Explain.

24. Refer to the data in Exercise 10. The estimated regression equation for these data is

$$\hat{y} = -5.7 + 1.54x_1 + 1.81x_2$$

where

$$\hat{y} = \text{estimated cost}$$

$$x_1 = \text{number of housing starts}$$

$$x_2 = \text{household income}$$

a. Test for a significant relationship among x_1, x_2, and y. Use $\alpha = .05$.

b. Is β_1 significant? Use $\alpha = .05$.

c. Is β_2 significant? Use $\alpha = .05$.

d. Briefly discuss the results obtained.

15.5   Estimation and Prediction

Estimating the mean value of y and predicting an individual value of y in multiple regression is similar to that for the case of regression analysis involving one independent variable. First, recall that in Chapter 14 we showed that the point estimate of the expected value of y for a given value of x was the same as the point estimate of an individual value of y. In both cases, we used $\hat{y} = b_0 + b_1 x$ as the point estimate.

In multiple regression we use the same procedure. That is, we substitute the given values of $x_1, x_2, \ldots, x_p$ into the estimated regression equation and use the corresponding value of $\hat{y}$ as the point estimate. For example, suppose that for the Butler Trucking example, we wanted to use the estimated regression equation involving x_1 (miles traveled) and x_2 (number of deliveries) to do the following:

1. Develop a *confidence interval estimate* of the mean travel time for all trucks that travel 100 miles and make two deliveries;
2. Develop a *prediction interval estimate* of the travel time for *one specific* truck that travels 100 miles and makes two deliveries.

Using the estimated regression equation $\hat{y} = -.869 + .0611 x_1 + .923 x_2$ with $x_1 = 100$ and $x_2 = 2$, we obtain the following value of $\hat{y}$:

$$\hat{y} = -.869 + .0611(100) + .923(2) = 7.09$$

Hence, the point estimate of travel time in both cases is approximately 7 hours.

To develop interval estimates for the mean value of y and for an individual value of y, we use a procedure similar to that for the case of regression analysis involving one independent variable. The formulas required, however, are beyond the scope of the text. But, computer packages for multiple regression analysis will often provide confidence intervals once the values of $x_1, x_2 \ldots, x_p$ are specified by the user. In Table 15.7 we show 95% confidence and prediction interval estimates for the Butler Trucking example for selected values of x_1 and x_2; these values were obtained using Minitab. Note that the interval estimate for an individual value of y is wider than the interval estimate for the expected value of y. This simply reflects the fact that for given values of x_1 and x_2 we can predict the mean travel time for all trucks with more precision than we can the travel time for one specific truck.

TABLE 15.7 The 95% Confidence and Prediction Interval Estimates for Butler Trucking	Value of x_1	Value of x_2	Confidence Interval		Prediction Interval	
			Lower Limit	Upper Limit	Lower Limit	Upper Limit
	50	2	3.140		2.414	5.656
	50	3	4.127	5.789	3.500	
	50	4	4.815	6.948	4.157	7.607
	100	2	6.258	7.926	5.500	8.683
	100	3	7.385	8.645	6.520	9.510
	100	4	8.135	9.742	7.362	10.515

Exercises

Methods

25. In Exercise 1 the following estimated regression equation based on 10 observations was presented:

$$\hat{y} = 29.1270 + .5906x_1 + .4980x_2$$

a. Develop a point estimate of the mean value of y when $x_1 = 180$ and $x_2 = 310$.
b. Develop a point estimate for an individual value of y when $x_1 = 180$ and $x_2 = 310$.

SELF TEST ▶   **26.** Refer to the data in Exercise 2. The estimated regression equation for these data is

$$\hat{y} = -18.4 + 2.01x_1 + 4.74x_2$$

a. Develop a 95% confidence interval estimate of the mean value of y when $x_1 = 45$ and $x_2 = 15$.
b. Develop a 95% prediction interval estimate of y when $x_1 = 45$ and $x_2 = 15$.

Applications

27. The following estimated regression equation has been developed to predict annual sales for account executives:

$$\hat{y} = 160 + 8x_1 + 15x_2$$

where

$$\hat{y} = \text{sales (\$1000s)}$$

$$x_1 = \text{years of experience}$$

$$x_2 = \text{number of sales training programs attended}$$

a. Estimate expected annual sales for an employee with 3 years of experience who did not attend any sales training program.
b. Estimate annual sales for a given employee with 2 years of experience who attended one sales training program.
c. What is the expected increase in sales as a result of attending one sales training program?

SELF TEST ▶   **28.** Refer to Exercise 7.
a. Develop a 95% confidence interval estimate of the mean quantity sold if both the competitor's price and Heller's price are $175.
b. Would a prediction interval estimate have any meaning in this situation? Explain.

29. Refer to Exercise 9.
a. Develop a 95% confidence interval estimate of the mean percentage of students that attend a 4-year college for a school district that has an average class size of 25 and whose students have a combined SAT score of 1000.
b. Suppose that a school district in Conway, South Carolina, has an average class size of 25 and a combined SAT score of 950. Develop a 95% prediction interval estimate of the percentage of students that attend a 4-year college.

15.6   The Use of Qualitative Variables

So far, the variables that we have used to build regression models have been quantitative variables; that is, variables that are measured in terms of how much or how many. Frequently, however, we will need to use independent variables that are not measured in these terms. We refer to such variables as *qualitative variables*. For instance, suppose that we were interested in predicting sales for a product which was available in either bottles or cans. Clearly the independent variable "container type" could influence the dependent variable "sales"; but, container type is a qualitative, not a quantitative variable. The distinguishing feature of qualitative variables is that there is no natural measure of how much or how many; these variables are used to represent attributes that are either present or not present.

To illustrate how qualitative variables are used in regression analysis, let us consider the problem currently being faced by Johnson Filtration, Inc., a firm which provides sales and service for industrial water-filtration systems throughout southern Florida. To investigate the relationship between the service time to repair a particular type of system and the number of months since the machine was serviced, Johnson Filtration collected the data shown in Table 15.8; a scatter diagram for these data is shown in Figure 15.7. The scatter diagram indicates that there may be a linear relationship between the number of months since the last service call and the repair time in hours. As a result, the following regression model was proposed:

$$y = \beta_0 + \beta_1 x_1 + \epsilon$$

where

$$y = \text{the repair time in hours}$$

$$x_1 = \text{the number of months since the last service call}$$

The corresponding estimated regression equation is

$$\hat{y} = b_0 + b_1 x_1$$

In Figure 15.8 we show the computer output obtained using the Minitab statistical system. The estimated regression equation is

$$\text{HOURS} = 2.15 + 0.304 \text{ MONTHS}$$

where the estimated repair time is denoted as HOURS and the number of months since the last service call is denoted as MONTHS. To test for the significance of the number of months since the last service call, the appropriate hypotheses are

$$H_0: \beta_1 = 0$$

$$H_a: \beta_1 \neq 0$$

Suppose we perform the test at the .05 level of significance. Looking at the *p*-value for either the *F* test or the *t* test (since they are equivalent procedures for simple linear regression), we see that the *p*-value of .016 is less than the level of significance; thus, we must reject $H_0: \beta_1 = 0$ and conclude that the number of months since the last service call is a significant factor in predicting the repair time. However, we also note that this variable only explains 53.4% of the variability in repair time.

After further consideration, management felt that the type of failure that occurred, mechanical or electrical, should also be considered in attempting to predict the repair time. Note that "type of repair" is an example of a qualitative variable. To incorporate the effect of type of failure into a regression model to predict repair time, we define the following variable:

$$x_2 = \begin{cases} 0 \text{ if the type of failure is mechanical} \\ 1 \text{ if the type of failure is electrical} \end{cases}$$

TABLE 15.8
Data for Johnson Filtration

Repair Time (Hours)	Months Since Last Service Call
2.9	2
3.0	6
4.8	8
1.8	3
2.9	2
4.9	7
4.2	9
4.8	8
4.4	4
4.5	6

FIGURE 15.7
Scatter Diagram for Johnson Filtration

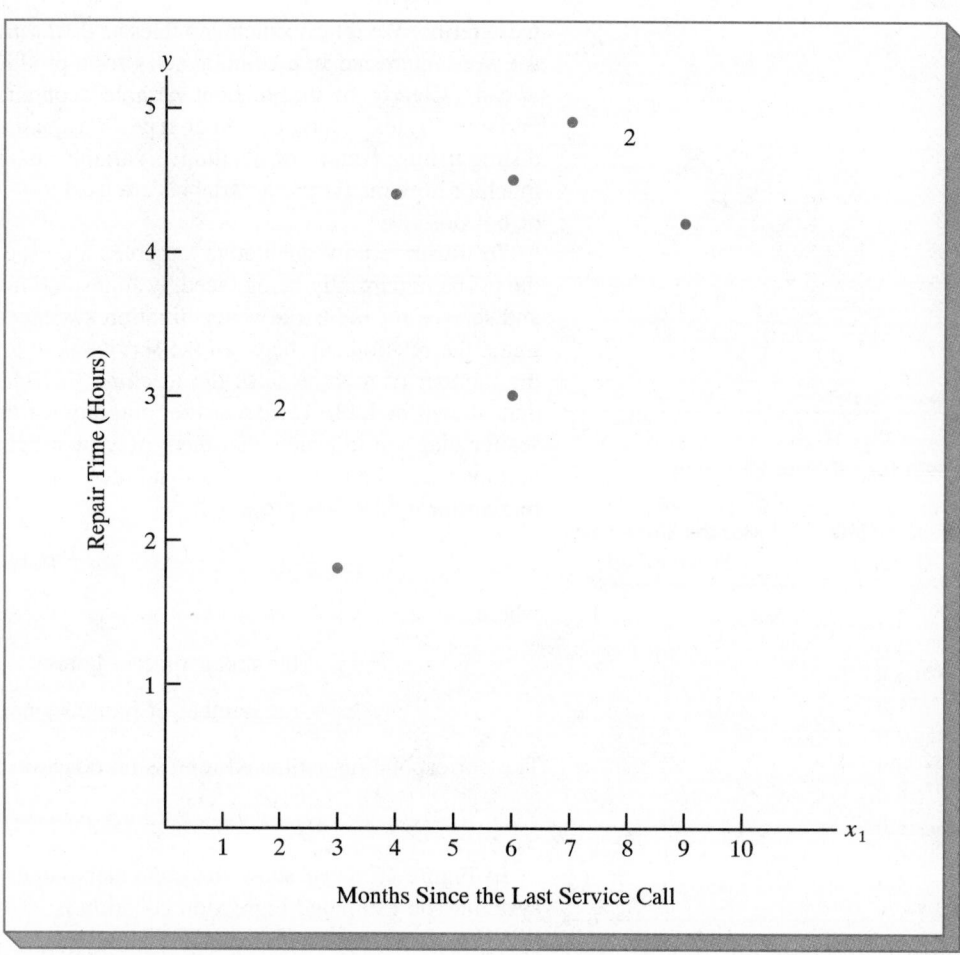

FIGURE 15.8   **Minitab Output for Johnson Filtration**

```
The regression equation is
HOURS = 2.15 + 0.304 MONTHS

Predictor      Coef       Stdev     t-ratio        p
Constant      2.1473      0.6050      3.55       0.008
MONTHS        0.3041      0.1004      3.03       0.016

s = 0.7810        R-sq = 53.4%      R-sq(adj) = 47.6%

Analysis of Variance

SOURCE        DF          SS          MS          F          p
Regression    1         5.5960      5.5960      9.17       0.016
Error         8         4.8800      0.6100
Total         9        10.4760
```

In regression analysis this way of representing a variable provides what is commonly referred to as a *dummy,* or *indicator* variable. Using this dummy variable, we can now write the following multiple regression equation to account for the effect of both the number of months since the last service call and the type of failure that occurred:

$$E(y) = \beta_0 + \beta_1 x_1 + \beta_2 x_2$$

To interpret the parameters in this equation we begin by considering what the expected value of y is when the type of failure is mechanical.

Interpreting the Parameters

Since a mechanical failure corresponds to $x_2 = 0$, we obtain

$$E(y \mid \text{mechanical failure}) = \beta_0 + \beta_1 x_1 + \beta_2(0)$$

or

$$E(y \mid \text{mechanical failure}) = \beta_0 + \beta_1 x_1 \tag{15.14}$$

The vertical line separating "y" and "mechanical failure" stands for the word "given"; thus, the expression $E(y \mid \text{mechanical failure})$ denotes the "expected value of the repair time given a mechanical failure." Equation (15.14) shows that the expected repair time for a mechanical failure is a linear function of x_1, the time since the previous service call; note that the y-intercept of this equation is β_0 and the slope is β_1.

Similarly, since an electrical failure corresponds to $x_2 = 1$, the expression for the expected value of y corresponding to an electrical failure is

$$E(y \mid \text{electrical failure}) = \beta_0 + \beta_1 x_1 + \beta_2(1)$$

or

$$E(y \mid \text{electrical failure}) = = \beta_0 + \beta_1 x_1 + \beta_2 \tag{15.15}$$

Note that (15.15) can be rewritten as

$$E(y \mid \text{electrical failure}) = = (\beta_0 + \beta_2) + \beta_1 x_1 \tag{15.16}$$

Equation (15.16) shows that the expected repair time for an electrical failure is also a linear function of x_1; although the slope of this equation is still β_1, the y-intercept has changed from β_0 to $\beta_0 + \beta_2$. Thus, for any value of x_1, β_2 is the difference between the expected repair time for a mechanical failure and the expected repair time for an electrical failure.

If β_2 is positive, the expected repair time for an electrical failure will be greater than the expected repair time for a mechanical failure; however, if β_2 is negative, the expected repair time for a mechanical failure will be greater. Finally, if β_2 is zero, there will be no difference between the expected repair time between a mechanical and an electrical failure.

We can also interpret β_2 by subtracting (15.14), the expected repair time for a mechanical failure, from (15.15), the expected repair time for an electrical failure; the result is

$$(\beta_0 + \beta_1 x_1 + \beta_2) - (\beta_0 + \beta_1 x_1) = \beta_2$$

This also shows that β_2 can be interpreted as the difference in the expected repair time between an electrical failure and a mechanical failure.

Computer Solution

Table 15.9

Johnson Filtration Data Classified for Type of Failure

Repair Time (Hours)	Months Since Last Service Call	Type of Failure*
2.9	2	1
3.0	6	0
4.8	8	1
1.8	3	0
2.9	2	1
4.9	7	1
4.2	9	0
4.8	8	0
4.4	4	1
4.5	6	1

*Type of failure is 0 if the failure is mechanical and 1 if the failure is electrical.

Table 15.9 shows the data for the Johnson Filtration problem after classifying each observation as to the type of failure. Table 15.9 shows that $x_2 = 0$ if an observation corresponded to a mechanical failure, and $x_2 = 1$ if an observation corresponded to an electrical failure. Using these data, we can now fit the following estimated regression equation:

$$\hat{y} = b_0 + b_1 x_1 + b_2 x_2$$

In Figure 15.9 we show the computer output obtained using the Minitab statistical software package; the estimated regression equation is

$$\text{HOURS} = 0.930 + 0.388 \text{ MONTHS} + 1.26 \text{ TYPE}$$

where TYPE is the name for the dummy variable x_2. Hence, the best estimate of β_2, the difference between the expected repair time for an electrical failure (TYPE = 1) and the expected repair time for a mechanical failure (TYPE = 0), is 1.26 hours.

To test for the significance of x_2 given the effect of x_1, the appropriate hypotheses are

$$H_0: \beta_2 = 0$$

$$H_a: \beta_2 \neq 0$$

Using a .05 level of significance, the p-value of .005 (corresponding to the t test for TYPE) shows that for $\alpha = .05$, we can reject H_0 and conclude that type of failure is a significant factor in predicting the repair time once the effect of the time since the last service call has been accounted for. Note also that R_a^2 of 81.9% shows that a good fit (much better than before) has been obtained using these two predictors.

To provide additional insight into the role that qualitative variables play in regression analysis, in Figure 15.10 we have shown a scatter diagram for these data in which we have

Figure 15.9   **Minitab Output for Data in Table 15.9**

```
The regression equation is
HOURS = 0.930 + 0.388 MONTHS + 1.26 TYPE


Predictor       Coef        Stdev      t-ratio         p
Constant       0.9305      0.4670         1.99     0.087
MONTHS         0.38762     0.06257        6.20     0.000
TYPE           1.2627      0.3141         4.02     0.005


s = 0.4590      R-sq = 85.9%     R-sq(adj) = 81.9%


Analysis of Variance


SOURCE         DF          SS           MS           F         p
Regression     2        9.0009       4.5005      21.36     0.001
Error          7        1.4751       0.2107
Total          9       10.4760
```

FIGURE 15.10
Scatter Diagram for Data in Table 15.9

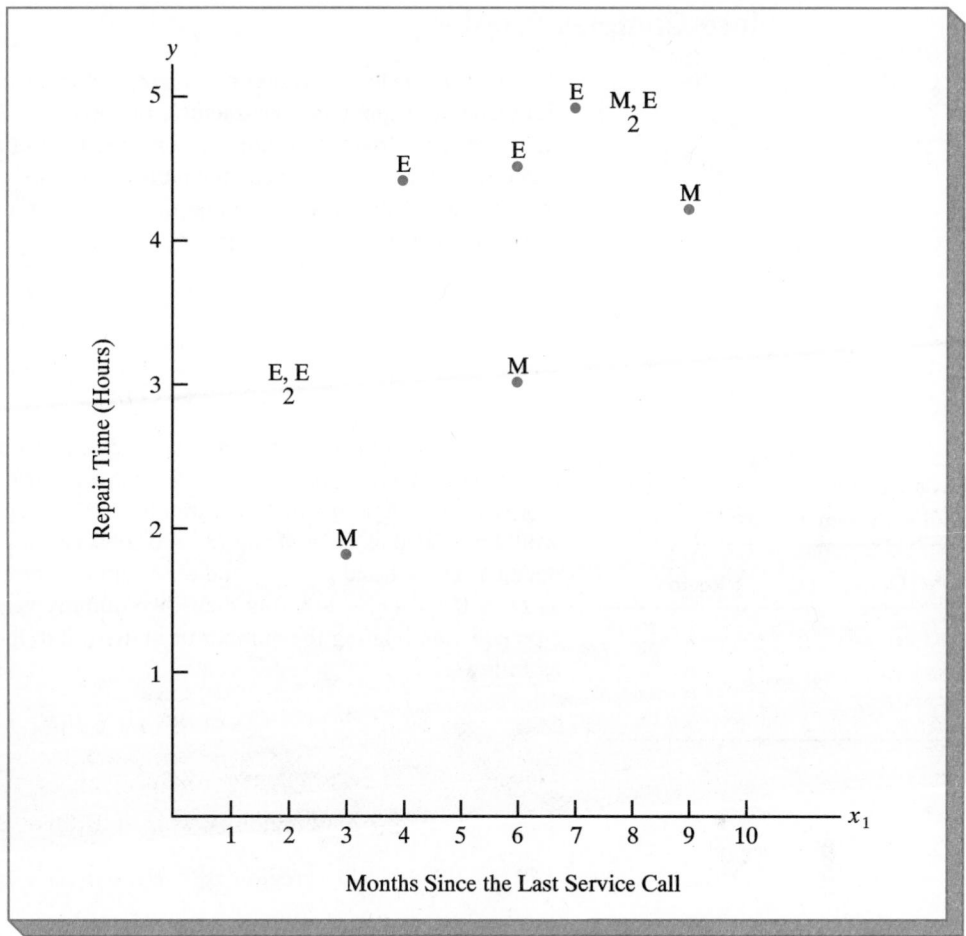

identified each point with the letter M for a mechanical failure, and the letter E for an electrical failure. Figure 15.10 suggests that there may be one underlying function relating y and x_1 for all the M points and another function relating y and x_1 for all the E points. Note that the estimated regression equation corresponding to a mechanical failure (TYPE = 0) is

$$\text{HOURS} = 0.930 + 0.388 \text{ MONTHS}$$

and the estimated regression equation corresponding to a mechanical failure (TYPE = 1) is

$$\text{HOURS} = 0.930 + 0.388 \text{ MONTHS} + 1.26$$

or

$$\text{HOURS} = 2.19 + 0.388 \text{ MONTHS}$$

Thus, by using a dummy variable, we have been able to develop two equations for predicting the repair time in hours; one corresponding to mechanical failures and one corresponding to electrical failures.

More Complex Problems

Let us now consider a problem involving a qualitative variable with more than two levels. For example, suppose a manufacturer of copy machines has organized the sales territories for a particular state into three regions: A, B, and C. Management would like to investigate the relationship between the number of units sold each week y and the region. Since region is a qualitative variable that takes on three values, we need two dummy variables to represent it. One possible specification is

$$D_1 = \begin{cases} 1 & \text{if region B} \\ 0 & \text{otherwise} \end{cases}$$

$$D_2 = \begin{cases} 1 & \text{if region C} \\ 0 & \text{otherwise} \end{cases}$$

TABLE 15.10

Coding System 1

D_1	D_2	Region
0	0	A
1	0	B
0	1	C

With this definition of dummy variables, the values of D_1 and D_2 for each of the three regions are given in Table 15.10. We have defined two dummy variables D_1 and D_2 to represent the single qualitative variable: region. Observations corresponding to region A would be coded as $D_1 = 0$ and $D_2 = 0$, observations corresponding to region B would be coded as $D_1 = 1$ and $D_2 = 0$, and observations corresponding to region C would be coded as $D_1 = 0$ and $D_2 = 1$. Using these two dummy variables, then, we can write the regression equation relating the number of units sold to the region in which the sales take place as follows:

$$E(y) = \beta_0 + \beta_1 D_1 + \beta_2 D_2$$

Thus,

$$E(y \mid \text{region A}) = \beta_0 + \beta_1(0) + \beta_2(0) = = \beta_0 \qquad (15.17)$$

$$E(y \mid \text{region B}) = \beta_0 + \beta_1(1) + \beta_2(0) = \beta_0 + \beta_1 \qquad (15.18)$$

$$E(y \mid \text{region C}) = \beta_0 + \beta_1(0) + \beta_2(1) = \beta_0 + \beta_2 \qquad (15.19)$$

To interpret β_1 we subtract (15.17) from (15.18); hence, β_1 is the difference between the expected number of units sold in region B and the expected number of units sold in region A. Similarly, by subtracting (15.17) from (15.19), we see that β_2 is the difference between the expected number of units sold in region C and the expected number of units sold in region A.

In general, whenever we want to incorporate the effect of a qualitative variable that takes on k values, we use $k - 1$ dummy variables. Note that there are several possible ways in which we can code the $k - 1$ dummy variables. For example, we could have coded the data in our sales problem as shown in Table 15.11.

TABLE 15.11

Coding System 2

D_1	D_2	Region
0	0	B
1	0	A
0	1	C

Using this coding system, observations corresponding to region B would be coded as $D_1 = 0$ and $D_2 = 0$, observations corresponding to region A would be coded as $D_1 = 1$ and $D_2 = 0$, and observations corresponding to region C would be coded as $D_1 = 0$ and $D_2 = 1$. Hence,

$$E(y \mid \text{region B}) = \beta_0 + \beta_1(0) + \beta_2(0) = \beta_0$$

$$E(y \mid \text{region A}) = \beta_0 + \beta_1(1) + \beta_2(0) = \beta_0 + \beta_1$$

$$E(y \mid \text{region C}) = \beta_0 + \beta_1(0) + \beta_2(1) = \beta_0 + \beta_2$$

Thus, β_1 denotes the difference between the expected number of units sold in region A and the expected number of units sold in region B, and β_2 denotes the difference between the expected number of units sold in region C and the expected number of units sold in region B. Note that the interpretations of β_1 and β_2 are different from the interpretation obtained using the previous coding system.

A third possibility would have been to code the data as shown in Table 15.12. It does not matter which coding system we use as long as we are consistent in defining both the data that will be analyzed and interpreting the results.

A Cautionary Note when Using Dummy Variables

TABLE 15.12
Coding System 3

D_1	D_2	Region
0	0	C
1	0	A
0	1	B

Suppose that we had a fourth region in the sales territory illustration, identified as region D. Since we have four levels or regions, we would need to use three dummy variables. A common mistake is attempting to use only two dummy variables by setting both D_1 and D_2 equal to 1 if the observation corresponds to region D. Note that if this approach is followed, the expected sales for region D is

$$E(y \mid \text{region D}) = \beta_0 + \beta_1(1) + \beta_2(1) = \beta_0 + \beta_1 + \beta_2 \qquad (15.20)$$

If we subtract (15.17) from (15.20) we see that $\beta_1 + \beta_2$ is the difference between the expected sales in region D and region A. However, since β_1 is the difference between the expected sales in regions B and A, and β_2 is the difference between the expected sales in regions C and A, this approach forces the difference between D and A to be a linear combination of the other two differences. In general, using a coding system that forces any conclusions on the data is not warranted. Instead, the proper approach is to let the data tell us what the relationship is. When using dummy variables, we can ensure a correct interpretation by always following the rule of using $k - 1$ dummy variables whenever the qualitative variable takes on k values.

☐ ☐ Exercises

Methods

S E L F T E S T ▶

30. Consider a regression study involving a dependent variable y, a quantitative independent variable x_1, and a qualitative variable with two levels (level 1 and level 2).
a. Write a multiple regression equation relating x_1 and the qualitative variable to y.
b. What is the expected value of y corresponding to level 1 of the qualitative variable?
c. What is the expected value of y corresponding to level 2 of the qualitative variable?
d. Interpret the parameters in your regression equation.

31. Consider a regression study involving a dependent variable y, a quantitative independent variable x_1, and a qualitative independent variable with three possible levels (level 1, level 2, and level 3).
a. How many dummy variables are required to represent the qualitative variable?
b. Write a multiple regression equation relating x_1 and the qualitative variable to y.
c. Interpret the parameters in your regression equation.

Applications

S E L F T E S T ▶

32. The following regression model has been proposed to predict sales at a fast-food outlet:

$$y = \beta_0 + \beta_1 x_1 + \beta_2 x_2 + \beta_3 x_3 + \epsilon$$

where

$$x_1 = \text{number of competitors within 1 mile}$$

$$x_2 = \text{population within 1 mile (1000s)}$$

$$x_3 = \begin{cases} 1 & \text{if drive-up window present} \\ 0 & \text{otherwise} \end{cases}$$

$$y = \text{sales (\$1000s)}$$

The following estimated regression equation was developed after 20 outlets were surveyed:

$$\hat{y} = 10.1 - 4.2x_1 + 6.8x_2 + 15.3x_3$$

a. What is the expected amount of sales attributable to the drive-up window?
b. Predict sales for a store with two competitors, a population of 8000 within 1 mile, and no drive-up window.
c. Predict sales for a store with one competitor, a population of 3000 within 1 mile, and a drive-up window.

 REPAIR

33. Refer to the Johnson Filtration problem introduced in this section. Suppose that in addition to the number of months since the machine was serviced and whether a mechanical or an electrical failure had occurred, management also obtained data showing which repairperson had performed the service. The revised data are shown below:

Repair Time (hours)	Months Since Previous Service Call (months)	Type of Failure	Repairperson
2.9	2	Electrical	Dave Newton
3.0	6	Mechanical	Dave Newton
4.8	8	Electrical	Bob Jones
1.8	3	Mechanical	Dave Newton
2.9	2	Electrical	Dave Newton
4.9	7	Electrical	Bob Jones
4.2	9	Mechanical	Bob Jones
4.8	8	Mechanical	Bob Jones
4.4	4	Electrical	Bob Jones
4.5	6	Electrical	Dave Newton

a. Ignore for now the months since the previous service call (x_1) and the repairperson that performed the service. Develop the estimated simple linear regression equation to predict the repair time (y) given the type of failure (x_2). Recall that $x_2 = 0$ if the failure is mechanical and 1 if the failure is electrical.
b. Does the equation that you developed in (a) provide a good fit for the observed data? Explain.
c. Ignore for now the months since the previous service call and the type of failure associated with the machine. Develop the estimated simple linear regression equation to predict the repair time given the repairperson that performed the service. Let $x_3 = 0$ if Bob Jones performed the service and $x_3 = 1$ if Dave Newton performed the service.
d. Does the equation that you developed in (c) provide a good fit for the observed data? Explain.

34. This problem is an extension of the situation described in Exercise 33.
a. Develop the estimated regression equation to predict the repair time given the number of months since the previous service call, the type of failure, and the repairperson that performed the service.
b. At the $\alpha = .05$ level of significance, test whether the estimated regression equation developed in (a) represents a significant relationship between the independent variables and the dependent variable.
c. Is the addition of the independent variable x_3, the repairperson that performed the service, statistically significant? Use $\alpha = .05$. What explanation can you give for the results observed?

Forbes2

35. The following data show the price-earnings ratio, the net profit margin, and the growth rate for 19 companies listed in "The *Forbes* 500s on Wall Street" (*Forbes,* May 1, 1989). The data in the column labeled "Industry" are simply codes used to define the industry for each company: 1 = energy-international oil, 2 = health-drugs, and 3 = electronics-computers.

Firm	P/E Ratio	Profit Margin	Growth Rate	Industry
Exxon	11.3	6.5	10	1
Chevron	10.0	7.0	5	1
Texaco	9.9	3.9	5	1
Mobil	9.7	4.3	7	1
Amoco	10.0	9.8	8	1
Pfizer	11.9	14.7	12	2
Bristol Meyers	16.2	13.9	14	2
Merck	21.0	20.3	16	2
American Home Products	13.3	16.9	11	2
Abbott Laboratories	15.5	15.2	18	2
Eli Lilly	18.9	18.7	11	2
Upjohn	14.6	12.8	10	2
Warner-Lambert	16.0	8.7	7	2
Amdahl	8.4	11.9	4	3
Digital	10.4	9.8	19	3
Hewlett-Packard	14.8	8.1	18	3
NCR	10.1	7.3	6	3
Unisys	7.0	6.9	6	3
IBM	11.8	9.2	6	3

TABLE 15.13

Risk	Age	Pressure	Smoker
12	57	152	0
24	67	163	0
13	58	155	0
56	86	177	1
28	59	196	0
51	76	189	1
18	56	155	1
31	78	120	0
37	80	135	1
15	78	98	0
22	71	152	0
36	70	173	1
15	67	135	1
48	77	209	1
15	60	199	0
36	82	119	1
8	66	166	0
34	80	125	1
3	62	117	0
37	59	207	1

a. Develop the estimated regression equation that can be used to predict the price-earnings ratio given the profit margin, growth rate, and type of industry.

b. Consider modifying the model developed in (a) to account for possible interaction involving profit margin and the type of industry. Is there significant interaction involving these variables? Explain.

Stroke

36. The American Heart Association collects data on the risk of strokes. A 10-year study provided data on how age, blood pressure, and smoking relate to the risk of strokes (*U.S. News & World Report,* April 13, 1992). Assume that data from a portion of this study are shown in Table 15.13. Risk is interpreted as the probability (times 100) that the patient will have a stroke over the next 10-year period. For the smoking variable, a 1 indicates a smoker and a 0 indicates a nonsmoker.

a. Using these data, develop an estimated regression equation that relates risk of a stroke to the person's age, blood pressure, and whether the person is a smoker.

b. Is smoking a significant factor in the risk of a stroke? Explain. Use $\alpha = .05$.

c. What is the probability of a stroke over the next 10 years for Art Speen, a 68-year-old smoker who has a blood pressure of 175? What action might the physician recommend for this patient?

15.7   Residual Analysis

In Chapter 14 we showed how residual plots could be used to determine whether the assumptions made regarding the error term and model form are valid. In the multiple regression case, these same methods can be used to validate the assumptions made for the proposed model and provide additional information concerning the adequacy of the fitted least squares equation.

The statistical tests that we discussed in Section 15.4 are based on the assumptions presented in Section 15.1 for the error term ϵ and the assumption that the population regression equation has the linear form given by $E(y) = \beta_0 + \beta_1 x_1 + \cdots + \beta_p x_p$. Residual analysis will allow us to make a judgment about whether the model assumptions appear to be satisfied. In addition, residual analysis can often provide insight as to whether a different type of model—for example, one involving more variables or a different functional form—might better describe the observed relationship.

In simple linear regression we showed how a residual plot against the independent variable x and a residual plot against the predicted value $\hat{y}$ could be used to determine if the assumptions regarding the error term are appropriate. In multiple regression analysis, these plots can be developed and interpreted similarly; the only difference in multiple regression is that plots against more than one independent variable can be developed. Statisticians usually look first at the residual plot against $\hat{y}$ and then review the plots against the independent variables if necessary.

In Chapter 14 we indicated that many of the residual plots provided by statistical software packages use a standardized version of the residuals. For simple linear regression the ith standardized residual is defined as follows.

Standardized Residual

$$\frac{y_i - \hat{y}_i}{s\sqrt{1 - h_i}} \tag{15.21}$$

where

$$h_i = \frac{1}{n} + \frac{(x_i - \overline{x})^2}{\Sigma(x_i - \overline{x})^2} \tag{15.22}$$

The comments that we made in the previous paragraph apply equally well for residual plots developed using the standardized residuals.

The computation of the predicted values, the residuals, and the standardized residuals is too complex to do by hand. Using a statistical software package such as Minitab, however, these values can be simply obtained as part of the standard regression analysis output. In Table 15.14, we show the predicted values, residuals, and the standardized residuals for the Butler Trucking example; these values were obtained using the Minitab statistical software package. Using the values in Table 15.14, the appropriate residual plots can be simply constructed. Our preference is to initially plot the standardized residuals versus the predicted values. In reviewing the subsequent standardized residual plot, we will look for the same patterns that we looked for when analyzing residuals in simple linear regression.

Figure 15.11 shows three forms of standardized residual plots versus $\hat{y}$, the value of y predicted by the regression equation. The plot in Panel A shows the type of pattern to expect when the model assumptions are satisfied.

The plot in Panel B shows a pattern that can be expected when the constant variance assumption is not satisfied. The error term gets larger as the value of $\hat{y}$ increases. Adding another independent variable will sometimes correct this problem. It was this type of

TABLE 15.14	Miles Traveled (x_1)	Deliveries (x_2)	Travel Time (y)	Predicted Value $(\hat{y})$	Residual $(y - \hat{y})$	Standardized Residual
Residual Analysis for Butler Trucking	100	4	9.3	8.93846	0.361541	0.78344
	50	3	4.8	4.95830	−0.158304	−0.34962
	100	4	8.9	8.93846	−0.038460	−0.08334
	100	2	6.5	7.09161	−0.591609	−1.30929
	50	2	4.2	4.03488	0.165121	0.38167
	80	2	6.2	5.86892	0.331083	0.65431
	75	3	7.4	6.48667	0.913331	1.68917
	65	4	6.0	6.79875	−0.798749	−1.77372
	90	3	7.6	7.40369	0.196311	0.36703
	90	2	6.1	6.48026	−0.380263	−0.77639

pattern that suggested that adding a variable might be helpful in the Butler Trucking example.

Finally, the plot in Panel C shows a case where the errors are not independent. When $\hat{y}$ is small, the error term is positive; when $\hat{y}$ assumes intermediate values, the error term is negative; and when $\hat{y}$ is large, the error term is again positive. Often a curvilinear model is needed in this case.

In Figure 15.12 we show a standardized residual plot against the predicted values $\hat{y}$ for the Butler Trucking example involving two independent variables. Note that this residual plot does not indicate any unusual abnormality; thus, we conclude that the model assumptions are reasonable.

Outliers

An outlier is a data point (observation) that does not follow the trend exhibited by the remaining data. In Chapter 14 we showed how standardized residuals can be used to identify outliers in a data set. If outliers are present, the fit provided by the estimated regression equation will not be as good as it would be if there were no outliers, and the value of s, the standard error of the estimate, will be larger. Since s appears in the denominator of (15.21), the size of the standardized residuals will be decreased as s increases. Thus, even though a residual is unusually large, it can be difficult to identify it as an outlier using the standardized residuals. An approach that can circumvent this type of problem uses a form of the residuals that are referred to as *studentized deleted residuals*.

Use of Studentized Deleted Residuals for Identifying Outliers

Suppose that the ith observation was deleted from the data set and a new regression equation was developed. Let $s_{(i)}$ denote the standard error of the estimate based upon the data set with the ith observation deleted. If we compute the standardized residuals for the original data set using $s_{(i)}$ instead of s, the resulting values are referred to as the *studentized deleted residuals*.

FIGURE 15.11
**Possible Residual Patterns and Their
Causes**

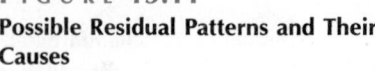

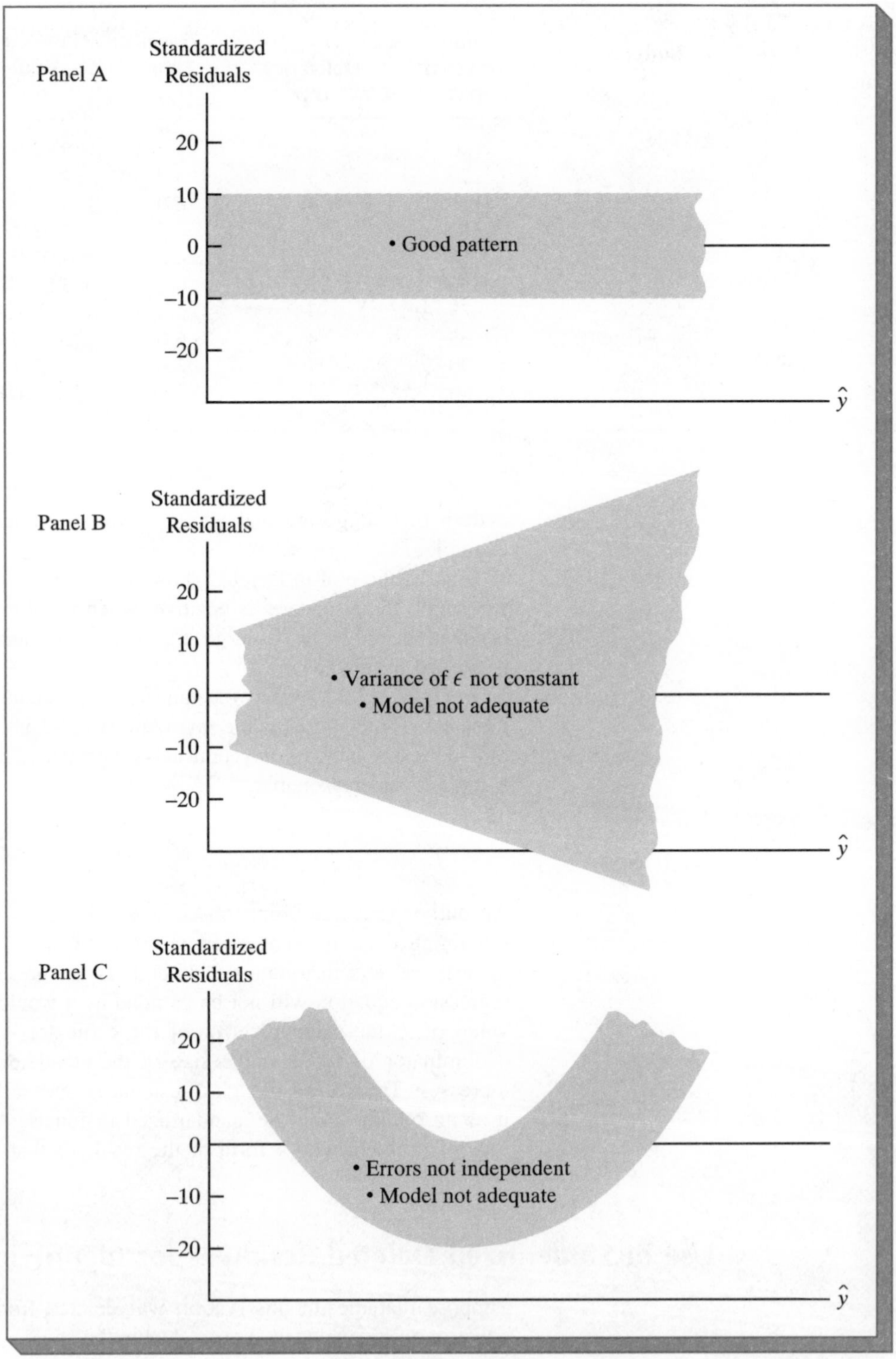

Panel A

Standardized
Residuals

20

10

0   • Good pattern

−10

−20

$\hat{y}$

Panel B

Standardized
Residuals

20

10

0   • Variance of ϵ not constant
         • Model not adequate

−10

−20

$\hat{y}$

Panel C

Standardized
Residuals

20

10

0

−10   • Errors not independent
         • Model not adequate

−20

$\hat{y}$

FIGURE 15.12 **Standardized Residual Plot for Butler Trucking with Two Independent Variables**

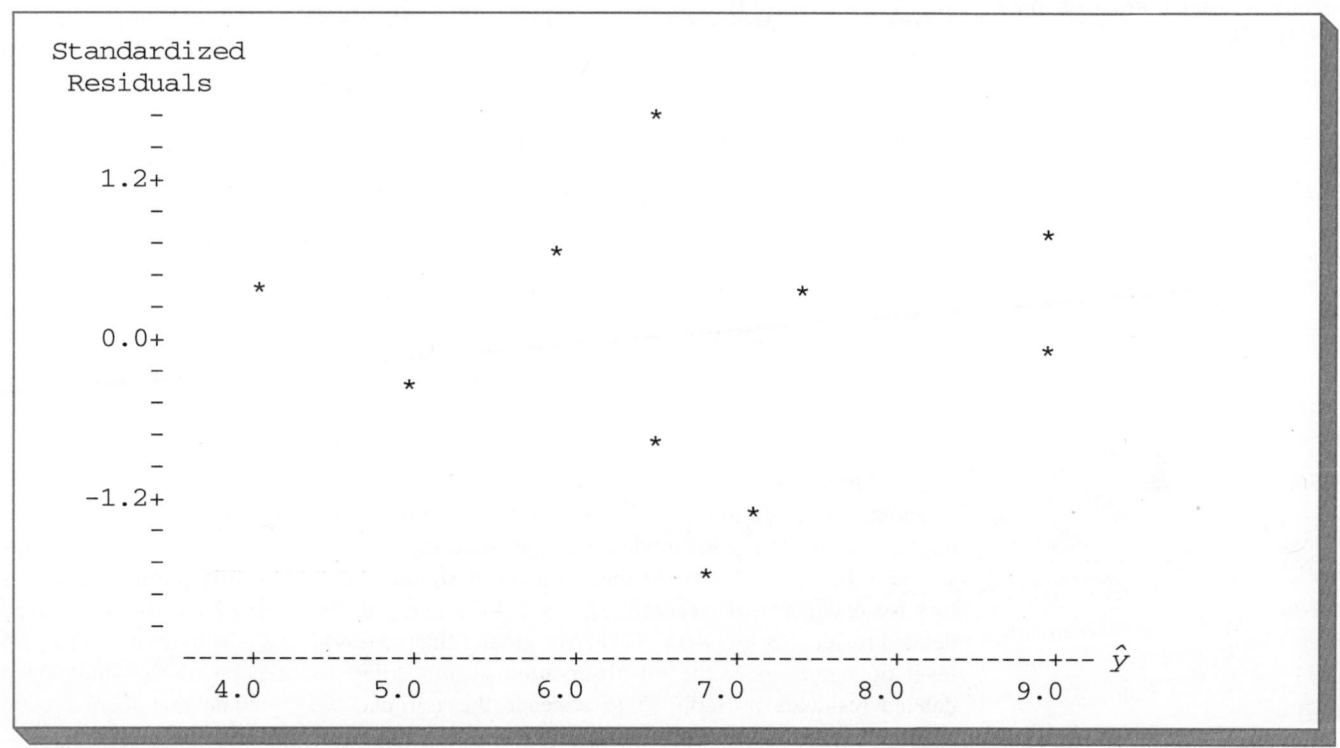

Studentized Deleted Residual

$$\frac{y_i - \hat{y}_i}{s_{(i)}\sqrt{1 - h_i}} \qquad (15.23)$$

If the ith observation is an outlier, deleting this observation will result in a smaller value for the standard error of the estimate; that is, $s_{(i)} < s$. Since the ith studentized deleted residual is computed using $s_{(i)}$, the absolute value of the ith studentized residual will be larger than the absolute value for the standardized residual.

Many statistical software packages provide options for obtaining the studentized deleted residuals as part of the regression analysis output. For example, in the Minitab statistical system, the studentized deleted residuals can be obtained using the REGRESS command with the subcommand TRESID; in Table 15.15 we show the values of the studentized deleted residuals for the Butler Trucking example involving 10 observations and two independent variables. To identify outliers in the data set, we look for large absolute values for the studentized deleted residuals.

The t distribution can be used to determine how large the ith studentized deleted residual must be to conclude that the ith observation is an outlier. Recall that p denotes the number of independent variables in the model and n denotes the number of observations in the original data set; thus, if we delete the ith observation, the total number of obser-

TABLE 15.15
Studentized Deleted Residuals for Butler Trucking

Miles Traveled (x_1)	Deliveries (x_2)	Travel Time (y)	Standardized Residual	Studentized Deleted Residual
100	4	9.3	0.78344	0.75939
50	3	4.8	−0.34962	−0.32654
100	4	8.9	−0.08334	−0.07720
100	2	6.5	−1.30929	−1.39494
50	2	4.2	0.38167	0.35709
80	2	6.2	0.65431	0.62519
75	3	7.4	1.68917	2.03187
65	4	6.0	−1.77372	−2.21314
90	3	7.6	0.36703	0.34312
90	2	6.1	−0.77639	−0.75190

vations drops to $n - 1$ and thus the error sum of squares has $n - p - 2$ degrees of freedom. For the Butler Trucking example involving two independent variables, the error degrees of freedom corresponding to the fitted model in which we delete the ith observation is $10 - 2 - 2 = 6$. At the .05 level of significance, the t distribution table shows that for 6 degrees of freedom, $t_{.025} = 2.447$. Thus, if the value of the ith studentized deleted residual is less than -2.447 or greater than $+2.447$, we conclude that at the .05 level of significance, the ith observation is an outlier. Since none of the studentized deleted residuals in Table 15.15 exceeds these limits, we conclude that there are no outliers in the data.

Influential Observations

In Chapter 14 we discussed how leverage can be used to identify observations for which the value of the independent variable has a strong influence on the results obtained. Recall that the leverage of the ith observation was defined for the case of simple linear regression as follows.

Leverage of Observation i

$$h_i = \frac{1}{n} + \frac{(x_i - \bar{x})^2}{\Sigma(x_i - \bar{x})^2}$$

(15.24)

In multiple regression analysis we also use leverage to identify observations that have a strong influence on the coefficients. In the Minitab statistical computing system, an observation is considered influential if $h_i > 3(p + 1)/n$, where p denotes the number of independent variables and n is the number of observations. As an illustration, let us consider the Butler Trucking example involving $n = 10$ observations and $p = 2$ independent variables. The critical value for leverage is

$$\frac{3(p + 1)}{n} = \frac{3(2 + 1)}{10} = \frac{9}{10} = .9$$

TABLE 15.16
Leverage Values for Butler Trucking

Miles Traveled (x_1)	Deliveries (x_2)	Travel Time (y)	Leverage (h_i)
100	4	9.3	0.351704
50	3	4.8	0.375863
100	4	8.9	0.351704
100	2	6.5	0.378451
50	2	4.2	0.430220
80	2	6.2	0.220557
75	3	7.4	0.110009
65	4	6.0	0.382657
90	3	7.6	0.129098
90	2	6.1	0.269737

The values of h_i, obtained using the Minitab statistical computing system, are shown in Table 15.16. Since none of the values exceed .9, we conclude that there are no influential observations in the data.

Use of Cook's Distance Measure to Identify Influential Observations

A problem that can arise when using leverage to identify influential observations is that an observation can be identified as having high leverage and not necessarily be influential in terms of the resulting estimated regression equation. For example, Table 15.17 shows a data set consisting of 8 observations and their corresponding leverage values (obtained using Minitab). Since the leverage for the 8th observation is .91 > .86 (the critical leverage value for simple linear regression), this observation is identified as an influential observation. Before reaching any final conclusions, however, let us consider the situation from a different perspective.

In Figure 15.13 we show the scatter diagram and the estimated regression equation corresponding to the data set in Table 15.17. Using Minitab, the following estimated regression equation was developed for these data:

$$\hat{y} = 18.2 + 1.39x$$

The straight line shown in Figure 15.13 is the graph of this equation. Now, let's delete the observation $x = 15$, $y = 39$ from the data set and fit a new estimated regression equation to the remaining 7 observations; the new estimated regression equation is

$$\hat{y} = 18.1 + 1.42x$$

TABLE 15.17

Data Set Illustrating Potential Problem Using the Leverage Criterion

x_i	y_i	Leverage h_i
1	18	0.204170
1	21	0.204170
2	22	0.164205
3	21	0.138141
4	23	0.125977
4	24	0.125977
5	26	0.127715
15	39	0.909644

We note that the y-intercept and slope of the new estimated regression equation have changed very little from the values obtained using all the data. Thus, although the 8th observation was identified as an influential observation using the leverage criterion, this observation clearly had little influence on the results obtained. Thus, in some situations using only leverage to identify influential observations can lead to wrong conclusions.

Cook's distance measure uses both the values of the residuals and the leverage values to determine if an observation is influential.

FIGURE 15.13

Scatter Diagram for the Data Set in Table 15.17

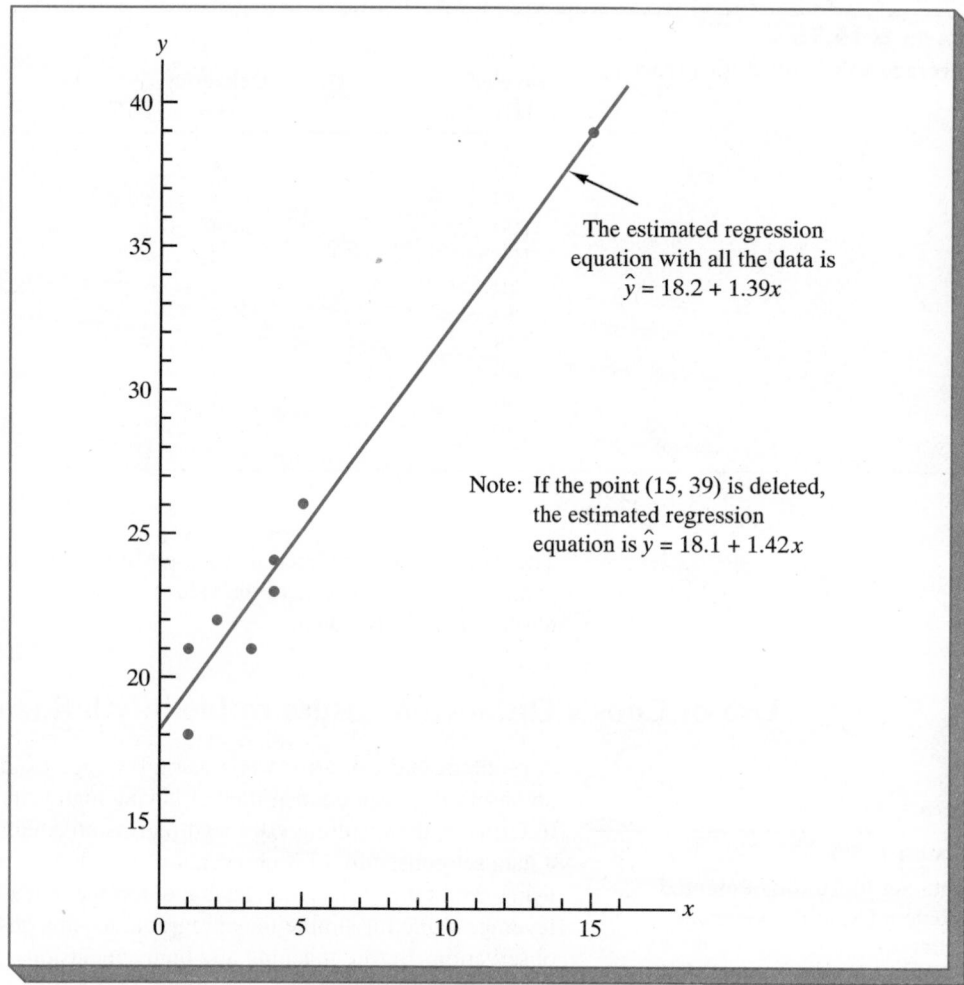

The estimated regression equation with all the data is $\hat{y} = 18.2 + 1.39x$

Note: If the point (15, 39) is deleted, the estimated regression equation is $\hat{y} = 18.1 + 1.42x$

Cook's Distance Measure

$$D_i = \frac{(y_i - \hat{y}_i)^2}{(p + 1)s^2} \left[\frac{h_i}{(1 - h_i)^2}\right] \qquad (15.25)$$

where

p = number of independent variables

s^2 = variance of the estimate (developed using all n observations)

h_i = leverage for the ith observation (developed using all n observations)

Note that the value of D_i depends upon the value of the ith residual $y_i - \hat{y}_i$ and the leverage of the ith observation h_i. Thus, the ith observation can be influential if it has (1) a large residual and a small leverage value, (2) a small residual and a large leverage value, or (3) a large residual and a large leverage value. As a rule of thumb, values of $D_i > 1$ are

TABLE 15.18

Cook's Distance Measure

x_i	y_i	Cook's Distance Measure D_i
1	18	0.265344
1	21	0.230628
2	22	0.089443
3	21	0.113389
4	23	0.029850
4	24	0.004164
5	26	0.044397
15	39	0.055598

considered large enough to conclude that the ith observation is influential and thus should be studied further. In Table 15.18 we show the values of Cook's distance measure for the data set shown in Table 15.17. Since none of the values exceeds 1, we conclude that the data set does not contain any influential observations. Recall that when we just used leverage to identify influential observations, we erroneously concluded that observation 8 was influential.

NOTES & COMMENTS

To determine if the value of Cooks's distance measure D_i is large enough to conclude that the ith observation is influential, we can also compare the value of D_i to the 50th percentile of an F distribution (denoted $F_{.50}$) with $p + 1$ numerator degrees of freedom and $n - p - 1$ denominator degrees of freedom. The F tables corresponding to a 50% level of significance must be available to carry out the test. The rule of thumb we provided $(D_i > 1)$ is based on the fact that the table value is very close to 1 for a wide variety of cases.

☐ ☐ Exercises

Methods

SELF TEST ▶

37. Shown below are data for two variables, x and y.

x_i	1	2	3	4	5
y_i	3	7	5	11	14

a. Develop the estimated regression equation for these data.
b. Plot the standardized residuals versus $\hat{y}$. Do there appear to be any outliers in these data? Explain.
c. Compute the studentized deleted residuals for these data. At the $\alpha = .05$ level of significance, can any of these observations be classified as an outlier? Explain.

38. Shown below are data for two variables, x and y.

x_i	22	24	26	28	40
y_i	12	21	31	35	70

a. Develop the estimated regression equation for these data.

b. Compute the studentized deleted residuals for these data. At the $\alpha = .05$ level of significance, can any of these observations be classified as an outlier? Explain.

c. Compute the leverage values for these data. Do there appear to be any influential observations in these data? Explain.

d. Compute Cook's distance measure for these data. Are any observations influential? Explain.

Applications

39. In Exercise 6 data were presented showing the weekly gross revenue ($1000s), the television advertising ($1000s), and the newspaper advertising ($1000s) for TAI Movie Theaters; these data are repeated below:

Weekly Gross Revenue ($1000s)	Television Advertising ($1000s)	Newspaper Advertising ($1000s)
96	5.0	1.5
90	2.0	2.0
95	4.0	1.5
92	2.5	2.5
95	3.0	3.3
94	3.5	2.3
94	2.5	4.2
94	3.0	2.5

a. Find an estimated regression equation relating weekly gross revenue to television and newspaper advertising.

b. Plot the standardized residuals against $\hat{y}$. Does the residual plot support the assumptions regarding ϵ? Explain.

c. Check for any outliers in these data. What are your conclusions?

d. Are there any influential observations? Explain.

40. Refer to the data in Exercise 23.

a. Plot the standardized residuals against $\hat{y}$. Does the residual plot support the assumptions regarding ϵ? Explain.

b. Check for any outliers in these data. What are your conclusions?

c. Are there any influential observations? Explain.

41. Refer to the data in Exercise 9. Let SIZE denote the average class size, SAT the combined SAT score, and %COLLEGE the percentage of students who attended a 4-year college.

a. Develop an estimated regression equation that can be used to predict %COLLEGE given SAT.

b. Based on the estimated regression equation developed in (a), do there appear to be any outliers and/or influential observations in these data? Explain.

c. Develop an estimated regression equation that can be used to predict %COLLEGE given SIZE and SAT.

d. Based upon the estimated regression equation developed in (c), do there appear to be any outliers and/or influential observations? Explain.

Summary

In this chapter we showed how extensions of the concepts of simple linear regression can be used to develop an estimated regression equation for predicting y that involves several independent variables. We noted that the interpretation of the coefficients had to be modified somewhat for this case. That is, we interpreted b_i as an estimate of the change in the dependent variable y that would result from a 1-unit change in independent variable x_i when the other independent variables do not change.

A key part of any multiple regression study is the use of a computer software package for carrying out the computational work. Many excellent packages exist and, after a short learning period, can be used to develop the estimated regression equation, conduct the appropriate significance tests, and prepare residual plots.

In Section 15.6, we discussed how dummy variables can be used to account for the effect of qualitative variables. Then, in Section 15.7 we extended the application of residual analysis by showing additional procedures for identifying outliers and influential observations.

Glossary

Multiple regression model A regression model in which more than one independent variable is used to predict the dependent variable.

Multiple coefficient of determination (R^2) A measure of the goodness of fit for the estimated regression equation.

Adjusted multiple coefficient of determination (R_a^2) A measure of the goodness of fit for the estimated regression equation which accounts for the number of independent variables in the model.

Multicollinearity A term used to describe the case when the independent variables in a multiple regression model are correlated.

Qualitative variable A variable that is not measured in terms of how much or how many, but instead is assigned values to represent categories.

Dummy variable A variable that takes on the values 0 or 1 and is used to incorporate the effects of qualitative variables in a regression model.

Outlier An observation with a residual that is far greater in magnitude than the rest of the residual values.

Influential observation An observation that has a great deal of influence in determining the estimated regression equation.

Key Formulas

Multiple Regression Model

$$y = \beta_0 + \beta_1 x_1 + \beta_2 x_2 + \cdots + \beta_p x_p + \epsilon \tag{15.2}$$

Multiple Regression Equation

$$E(y) = \beta_0 + \beta_1 x_1 + \cdots + \beta_p x_p \tag{15.3}$$

Estimated Regression Equation

$$\hat{y} = b_0 + b_1 x_1 + \cdots + b_p x_p \tag{15.5}$$

Least Squares Criterion

$$\min \Sigma \ (y_i - \hat{y}_i)^2 \tag{15.7}$$

Total Sum of Squares

$$\text{SST} = \Sigma \ (y_i - \overline{y})^2 \tag{15.8}$$

Multiple Coefficient of Determination

$$R^2 = \frac{\text{SSR}}{\text{SST}} \tag{15.9}$$

Adjusted Multiple Coefficient of Determination

$$R_a^2 = 1 - (1 - R^2)\left(\frac{n-1}{n-p-1}\right) \tag{15.10}$$

Mean Square Due to Regression

$$\text{MSR} = \frac{\text{SSR}}{p} \tag{15.11}$$

Mean Square Due to Error

$$\text{MSE} = \frac{\text{SSE}}{n-p-1} \tag{15.12}$$

The F Statistic

$$F = \frac{\text{MSR}}{\text{MSE}} = \frac{\text{SSR}/p}{\text{SSE}/(n-p-1)} \tag{15.13}$$

Standardized Residual

$$\frac{y_i - \hat{y}_i}{s\sqrt{1 - h_i}} \tag{15.21}$$

Studentized Deleted Residual

$$\frac{y_i - \hat{y}_i}{s_{(i)}\sqrt{1 - h_i}} \tag{15.23}$$

Leverage of Observation i

$$h_i = \frac{1}{n} + \frac{(x_i - \bar{x})^2}{\Sigma(x_i - \bar{x})^2} \tag{15.24}$$

Cook's Distance Measure

$$D_i = \frac{(y_i - \hat{y}_i)^2}{(p+1)s^2}\left[\frac{h_i}{(1 - h_i)^2}\right] \tag{15.25}$$

❏ ❏ **Supplementary Exercises**

42. The admission's officer for Clearwater College developed the following estimated regression equation relating the final college GPA to the student's SAT mathematics scores and their high-school GPA.

$$\hat{y} = -1.41 + .0235x_1 + .00486x_2$$

where

$$x_1 = \text{high-school grade point average}$$

$$x_2 = \text{SAT mathematics score}$$

$$y = \text{final college grade point average}$$

a. Interpret the coefficients in this estimated regression equation.
b. Estimate the final college GPA for a student who has a high-school average of 84 and a score of 540 on the SAT mathematics test.

43. The personnel director for Electronics Associates developed the following estimated regression equation relating an employee's score on a job satisfaction test to his or her length of service and wage rate:

$$\hat{y} = 14.4 - 8.69x_1 + 13.5x_2$$

where

$$x_1 = \text{length of service (years)}$$

$$x_2 = \text{wage rate (dollars)}$$

$$y = \text{job satisfaction test score (higher scores}$$
$$\text{indicate more job satisfaction)}$$

a. Interpret the coefficients in this estimated regression equation.
b. Develop an estimate of the job satisfaction test score for an employee that has 4 years of service and makes \$6.50 per hour.

44. In a regression analysis involving 18 observations and four independent variables, it was determined that SSR = 18,051.63 and SSE = 1014.3.
a. Determine R^2 and R_a^2.
b. Test for the significance of the relationship at the $\alpha = .01$ level of significance.

45. The following estimated regression equation involving three independent variables has been developed:

$$\hat{y} = 18.31 + 8.12x_1 + 17.9x_2 - 3.6x_3$$

Computer output indicates that $s_{b_1} = 2.1$, $s_{b_2} = 9.72$, and $s_{b_3} = .71$. There were 15 observations in the study.
a. Test $H_0: \beta_1 = 0$ at $\alpha = .05$.
b. Test $H_0: \beta_2 = 0$ at $\alpha = .05$.
c. Test $H_0: \beta_3 = 0$ at $\alpha = .05$.
d. Would you recommend dropping any of the independent variables from the model?

46. Shown is a partial computer output from a regression analysis:

```
The regression equation is
Y = 8.103 + 7.602 X1 + 3.111 X2

Predictor                 Coef        Stdev      t-ratio
Constant               _____       2.667     _____
X1                     _____       2.105     _____
X2                     _____       0.613     _____

s = 3.35        R-sq = 92.3%      R-sq(adj) = _____
```

Analysis of Variance

SOURCE	DF	SS	MS	F
Regression	_____	1612	_____	_____
Error	12	_____	_____	
Total	_____	_____		

a. Compute the appropriate t-ratios.
b. Test for the significance of β_1 and β_2 at $\alpha = .05$.
c. Compute the entries in the DF, SS, and MS = SS/DF columns.
d. Compute R_a^2.

47. Recall that in Exercise 42, the admissions' officer for Clearwater College developed the following estimated regression equation relating final college GPA to student's SAT mathematics scores and their high-school GPA.

$$\hat{y} = -1.41 + .0235x_1 + .00486x_2$$

where

$$x_1 = \text{high-school grade point average}$$

$$x_2 = \text{SAT mathematics score}$$

$$y = \text{final college grade point average}$$

A portion of the Minitab computer output is shown.

```
The regression equation is
Y = -1.41 + .0235 X1 + .00486 X2
```

Predictor	Coef	Stdev	t-ratio
Constant	-1.4053	0.4848	_____
X1	0.023467	0.008666	_____
X2	_____	0.001077	_____

s = 0.1298   R-sq = _____   R-sq(adj) = _____

Analysis of Variance

SOURCE	DF	SS	MS	F
Regression	_____	1.76209	_____	_____
Error	_____	_____	_____	
Total	9	1.88000		

a. Complete the missing entries in this output.
b. Compute F and test at the $\alpha = .05$ level to see whether a significant relationship exists.
c. Did the estimated regression equation provide a good fit to the data? Explain.
d. Use the t test and $\alpha = .05$ to test $H_0: \beta_1 = 0$ and $H_0: \beta_2 = 0$.

48. Recall that in Exercise 43 the personnel director for Electronics Associates developed the following estimated regression equation relating an employee's score on a job satisfaction test to their length of service and wage rate:

$$\hat{y} = 14.4 - 8.69x_1 + 13.5x_2$$

where

$$x_1 = \text{length of service (years)}$$

$$x_2 = \text{wage rate (dollars)}$$

$$y = \text{job satisfaction test score (higher scores indicate more job satisfaction)}$$

A portion of the Minitab computer output is shown:

```
The regression equation is
Y = 14.4 - 8.69 X1 + 13.52 X2

Predictor              Coef        Stdev        t-ratio
Constant              14.448       8.191          1.76
X1                    _____      1.555        _____
X2                    13.517       2.085        _____

s = 3.773    R-sq = _____    R-sq(adj) = _____

Analysis of Variance

SOURCE                  DF          SS          MS          F
Regression              2         _____     _____     _____
Error                 _____      71.17      _____
Total                   7         720.0
```

a. Complete the missing entries in this output.
b. Compute F and test at the $\alpha = .05$ level to see whether a significant relationship exists or not.
c. Did the estimated regression equation provide a good fit to the data? Explain.
d. Use the t test and $\alpha = .05$ to $H_0: \beta_1 = 0$ and $H_0: \beta_2 = 0$.

49. Bauman Construction Company makes bids on a variety of projects. In an effort to estimate the bid to be made by one of its competitors, Bauman has obtained data on 15 previous bids and developed the following estimated regression equation:

$$\hat{y} = 80 + 45x_1 - 3x_2$$

where

$$\hat{y} = \text{competitor's bid (\$1000s)}$$

$$x_1 = \text{square feet (1000s)}$$

$$x_2 = \text{local index of construction activity}$$

a. Estimate the competitor's bid on a project involving 50,000 square feet and an index of construction activity of 70.
b. If SSR = 19,780 and SST = 21,533, test at $\alpha = .01$ the significance of the relationship.

JOBS

50. The following data set, reported in *Louis Rukeyser's Business Almanac* (1988 Simon and Schuster, p. 47), shows the percentage of management jobs held by women in various companies and the percentage of women in each company.

Industry/Company	Management Jobs Held by Women (%)	Women Employees (%)
Industrial		
DuPont	7	22
Exxon	8	27
General Motors	6	19
Goodyear Tire and Rubber	25	39
Technology		
AT&T	32	48
General Electric	6	26
IBM	16	28
Xerox	23	38
Consumer products		
Johnson & Johnson	18	47
PepsiCo	28	46
Phillip Morris (excluding General Foods)	14	31
Proctor & Gamble	17	28
Retailing and trade		
Federated Department Stores	61	72
Kroger	16	47
Marriott	32	51
McDonald's	46	57
Sears, Roebuck	36	55
Media		
ABC (excluding Capital Cities)	36	43
Time	46	54
Times Mirror	27	37
Financial Services		
American Express	37	57
BankAmerica	64	72
Chemical Bank	34	57
Prudential Life Insurance	32	53
Wells Fargo Bank	58	71

a. Fit a simple linear regression model which can be used to predict the percentage of management jobs held by women given the percentage of women employed by the company.

b. Did the model developed in (a) provide a good fit to the data? Explain.

c. Use dummy variables to develop a model which relates the percentage of management jobs held by women to the type of industry (industrial, technology, and so on).

d. What conclusions can you reach based upon the model developed in (c)?

e. Develop a model which can be used to predict the percentage of management jobs held by women using the percentage of women employed by the company and the type of industry.

f. What final conclusions can be made regarding the percentage of management jobs held by women based upon your analyses?

51. Refer to Exercise 50.

a. Develop a standardized residual plot for the estimated regression equation developed in (a) of Exercise 50. Does the pattern of the residual plot appear acceptable? Explain.

FOOTBALL

b. Are there any outliers? Explain.

c. Are there any influential observations? If so, what effect do they have on the model?

52. The following table shows some of the data available for 14 teams in the National Football League at the end of week 15 for the 1988 season.

Team	Won–Lost	Points Scored	Rushing Yards	Passing Yards	Interceptions Made by Team	Interceptions Made by Opponent
Atlanta	5–10	305	1907	2473	19	23
Chicago	12–3	187	2134	2718	14	24
Dallas	3–12	358	1858	3386	24	10
Detroit	4–11	292	1184	1971	15	12
Green Bay	3–12	298	1274	3046	22	20
L.A. Rams	9–6	277	1882	3604	17	22
Minnesota	10–5	206	1744	3633	16	35
N. Orleans	9–6	274	1843	2963	15	17
N.Y. Giants	10–5	277	1492	3096	14	15
Philadelphia	9–6	312	1812	3247	17	29
Phoenix	7–8	372	1909	3633	19	14
San Francisco	10–5	256	2453	3131	14	21
Tampa Bay	4–11	340	1650	3169	33	18
Washington	7–8	367	1377	3930	24	14

a. Develop an estimated regression equation that can be used to predict the number of points scored given the number of interceptions made by the team.

b. Develop a standardized residual plot for the model developed in (a). Does the pattern of the residual plot appear acceptable? Explain.

c. Are there any outliers? Explain.

d. Are there any influential observations? If so, what effect do they have on the model?

53. Refer to the data set in Exercise 52.

a. Develop an estimated regression equation that can be used to predict the number of points scored given the number of interceptions made by the opponents.

b. Develop a standardized residual plot for the model developed in (a). Does the pattern of the residual plot appear acceptable? Explain.

c. Are there any outliers? Explain.

d. Are there any influential observations? If so, what effect do they have on the model?

Computer Case: *Consumer Research, Inc.*

CONSUMER

Consumer Research, Inc., is an independent agency that conducts research on consumer attitudes and behaviors for a variety of firms. In one study, a client asked for an investigation of the consumer characteristics that can be used to predict the amount charged by credit-card users. Data were collected on the annual income, household size, and annual credit-card charges for a sample of 50 consumers. The data are shown below and are on the data disk in the file named CONSUMER.

Income ($1000s)	Household Size	Amount Charged ($)	Income ($1000s)	Household Size	Amount Charged ($)
54	3	4016	54	6	5573
30	2	3159	30	1	2583
32	4	5100	48	2	3866
50	5	4742	34	5	3586
31	2	1864	67	4	5037
55	2	4070	50	2	3605
37	1	2731	67	5	5345
40	2	3348	55	6	5370
66	4	4764	52	2	3890
51	3	4110	62	3	4705
25	3	4208	64	2	4157
48	4	4219	22	3	3579
27	1	2477	29	4	3890
33	2	2514	39	2	2972
65	3	4214	35	1	3121
63	4	4965	39	4	4183
42	6	4412	54	3	3730
21	2	2448	23	6	4127
44	1	2995	27	2	2921
37	5	4171	26	7	4603
62	6	5678	61	2	4273
21	3	3623	30	2	3067
55	7	5301	22	4	3074
42	2	3020	46	5	4820
41	7	4828	66	4	5149

Managerial Report

1. Use methods of descriptive statistics to summarize the data. Comment on the findings.
2. Develop estimated regression equations first using annual income as the independent variable and then using household size as the independent variable. Which variable is the better predictor of annual credit-card charges? Discuss your findings.
3. Develop an estimated regression equation with annual income and household size as the independent variables. Discuss your findings.
4. What is the predicted annual credit-card charge for a 3-person household with an annual income of $40,000?
5. Discuss the need for additional independent variables that could be added to the model. What additional variables might be helpful?

APPENDIX

Calculus-Based Derivation and Solution of Multiple Regression Problems with Two Independent Variables

For the multiple regression analysis case involving two independent variables, the least squares criterion calls for the minimization of

$$\text{SSE} = \Sigma(y_i - b_0 - b_1 x_{1i} - b_2 x_{2i})^2 \qquad \textbf{(15A.1)}$$

To minimize (15A.1) we must take the partial derivatives of SSE with respect to b_0, b_1, and b_2. We can then set the partial derivatives equal to zero and solve for the estimated regression coefficients b_0, b_1, and b_2. Taking the partial derivatives and setting them equal to zero provides

$$\frac{\partial \text{SSE}}{\partial b_0} = -2\Sigma(y_i - b_0 - b_1 x_{1i} - b_2 x_{2i}) = 0 \qquad \textbf{(15A.2)}$$

$$\frac{\partial \text{SSE}}{\partial b_1} = -2\Sigma x_{1i}(y_i - b_0 - b_1 x_{1i} - b_2 x_{2i}) = 0 \qquad \textbf{(15A.3)}$$

$$\frac{\partial \text{SSE}}{\partial b_2} = -2\Sigma x_{2i}(y_i - b_0 - b_1 x_{1i} - b_2 x_{2i}) = 0 \qquad \textbf{(15A.4)}$$

Dividing (15A.2) by 2 and summing the terms individually yields

$$-\Sigma y_i + \Sigma b_0 + \Sigma b_1 x_{1i} + \Sigma b_2 x_{2i} = 0$$

Bringing Σy_i to the right-hand side of the equation and noting that $\Sigma b_0 = nb_0$, we obtain

$$nb_0 + (\Sigma x_{1i})b_1 + (\Sigma x_{2i})b_2 = \Sigma y_i \qquad \textbf{(15A.5)}$$

Similar algebraic simplification applied to (15A.3) and (15A.4) leads to the equations:

$$(\Sigma x_{1i})b_0 + (\Sigma x_{1i}^2)b_1 + (\Sigma x_{1i}x_{2i})b_2 = \Sigma x_{1i}y_i \qquad \textbf{(15A.6)}$$

$$(\Sigma x_{2i})b_0 + (\Sigma x_{1i}x_{2i})b_1 + (\Sigma x_{2i}^2)b_2 = \Sigma x_{2i}y_i \qquad \textbf{(15A.7)}$$

Equations (15A.5) through (15A.7) are referred to as the *normal equations*. Application of these procedures to a regression model involving p independent variables would lead to $p + 1$ normal equations of this type. However, matrix algebra is usually used for that type of derivation.

Solving the Normal Equations for Butler Trucking

Refer to Table 15A.1. Substituting the values in Table 15A.1 results in the following normal equations:

$$10b_0 + 800b_1 + 29b_2 = 67.0 \qquad \text{(15A.8)}$$

$$800b_0 + 67{,}450b_1 + 2345b_2 = 5594.0 \qquad \text{(15A.9)}$$

$$29b_0 + 2345b_1 + 91b_2 = 202.2 \qquad \text{(15A.10)}$$

By multiplying (15A.8) by 80 and subtracting the result from (15A.9), we can eliminate b_0 and obtain an equation involving only b_1 and b_2.

$$
\begin{array}{r}
800b_0 + 67{,}450.0b_1 + 2345b_2 = 5594.0 \\
-800b_0 - 64{,}000.5b_1 - 2320b_2 = -5360 \\
\hline
3450b_1 + 25b_2 = 234.0
\end{array}
\qquad \text{(15A.11)}
$$

Now multiply (15A.8) by 2.9 and subtract the result from (15A.10). This manipulation yields a second equation involving only b_1 and b_2.

$$
\begin{array}{r}
29b_0 + 2345b_1 + 91.0b_2 = 202.2 \\
-29b_0 - 2320b_1 - 84.1b_2 = -194.3 \\
\hline
25b_1 + 6.9b_2 = 7.9
\end{array}
\qquad \text{(15A.12)}
$$

With Equations (15A.11) and (15A.12), we can solve simultaneously for b_1 and b_2. Multiplying (15A.12) by 25/6.9 and subtracting the result from (15A.11) gives us an equation involving only b_1.

$$
\begin{array}{r}
3{,}450.0000b_1 + 25b_2 = 234.0000 \\
- \quad 90.5797b_1 - 25b_2 = -28.6232 \\
\hline
3{,}359.4203b_1 = 205.3768
\end{array}
\qquad \text{(15A.13)}
$$

Using (15A.13) to solve for b_1 we get

$$b_1 = \frac{205.3768}{3{,}359.4203} = .061135$$

Table 15A.1

Calculation of Coefficients for Normal Equations

y_i	x_{1i}	x_{2i}	x_{1i}^2	x_{2i}^2	$x_{1i}x_{2i}$	$x_{1i}y_i$	$x_{2i}y_i$
9.3	100	4	10,000	16	400	930	37.2
4.8	50	3	2,500	9	150	240	14.4
8.9	100	4	10,000	16	400	890	35.6
6.5	100	2	10,000	4	200	650	13.0
4.2	50	2	2,500	4	100	210	8.4
6.2	80	2	6,400	4	160	496	12.4
7.4	75	3	5,625	9	225	555	22.2
6.0	65	4	4,225	16	260	390	24.0
7.6	90	3	8,100	9	270	684	22.8
6.1	90	2	8,100	4	180	549	12.2
67.0	800	29	67,450	91	2,345	5,594	202.2

Using this value for b_1, we can substitute into (15A.12) to solve for b_2:

$$25(.061135) + 6.9b_2 = 7.9$$

$$1.528375 + 6.9b_2 = 7.9$$

$$6.9b_2 = 6.371625$$

$$b_2 = .923424$$

Now we can substitute the values obtained for b_1 and b_2 into (15A.8) thus obtaining b_0:

$$10b_0 + 800(.061135) + 29(.923424) = 67.0$$

$$10b_0 + 48.90800 \quad + 26.779296 \quad = 67.0$$

$$10b_0 \qquad\qquad\qquad\qquad = -8.687296$$

$$b_0 \qquad\qquad\qquad\qquad = -.8687296$$

Rounding, we obtain the following estimated regression equation for Butler Trucking:

$$\hat{y} = -.8687 + .0611x_1 + .9234x_2 \qquad\qquad \textbf{(15A.11)}$$

Regression Analysis: Model Building

Monsanto Company*

ST. LOUIS, MISSOURI

Monsanto Company traces its roots to one entrepreneur's investment of $5000 and a dusty warehouse on the Mississippi riverfront, where in 1901 John F. Queeney began manufacturing saccharin. Today, Monsanto is one of the nation's largest chemical companies, producing more than a thousand products ranging from industrial chemical products to synthetic playing surfaces used in sports stadiums. Monsanto is a worldwide corporation with manufacturing facilities, laboratories, technical centers, and marketing operations in 65 countries.

The Nutrition Chemicals Division of the Monsanto Company manufactures and markets a methionine supplement for use in poultry, swine, and cattle feeds. Since poultry growers in particular work with very high volumes and low profit margins, they have specific nutritional requirements for poultry. Optimal feed compositions result in more rapid growth and in a higher final body weight for a given feed intake. The chemical industry has worked closely with poultry growers to optimize poultry feed products, and their successful partnership in this endeavor is reflected in the low cost of poultry relative to other meat products.

In studies conducted by Monsanto, regression analysis has been used to relate body weight in kilograms after a fixed period of growth y to the amount of methionine added to the feed, expressed as the percentage of sulfur-containing amino acids in the feed x. Initially, the following estimated regression equation was developed:

$$\hat{y} = .21 + .42x$$

The simple linear model proved statistically significant and explained 78% of the variability in the dependent variable.

However, residual analysis indicated that a curvilinear relationship might provide a better model.

The straight-line relationship provided by the estimated regression equation suggests that increases in the level of supplemental methionine will result in increases in the weight. However, further research conducted by Monsanto has shown that at some point further increases in the level of supplemental methionine led to a leveling off in body weight; in fact, when the amount of supplemental methionine increases beyond nutritional requirements, the body weight may even decline. As a result, Monsanto concluded that the level of methionine which provides peak growth performance cannot be found using a straight-line relationship.

Further experimentation was conducted and additional data were collected. In follow-up studies Monsanto used a nonlinear regression model to account for the curvilinear pattern in the data. It was found that the optimal body weight occurs at a sulfur-containing amino acid level of 1.25%. Thus, finding the right model was essential in enabling Monsanto to determine the optimal level of methionine supplement.

Monsanto Company used model-building techniques to handle the curvilinear relationship between a feed additive and weight gain in poultry. In this chapter we will extend our discussion of regression analysis by showing how curvilinear models such as that used by Monsanto can be developed and how a variety of tools can be applied to determine which set of independent variables leads to the best estimated regression equation.

The authors are indebted to James R. Ryland and Robert M. Schisla, Senior Research Specialists, Monsanto Nutrition Chemical Division, St. Louis, Missouri, for providing this Statistics in Practice.

Model building in regression analysis is the process of developing a regression model that best describes the relationship among the independent and dependent variables. The major issues are finding the proper form (linear or curvilinear) of the relationship and selecting the variables. Variable selection involves determining which of a candidate set of independent variables should be included in the regression model.

In Chapters 14 and 15, we introduced and worked with two regression models:

Simple Linear Regression: $y = \beta_0 + \beta_1 x + \epsilon$

Multiple Regression: $y = \beta_0 + \beta_1 x_1 + \beta_2 x_2 + \cdots + \beta_p x_p + \epsilon$

Both of these models specify a linear relationship between the independent variable(s) and the dependent variable.

The primary method introduced in Chapters 14 and 15 to determine the adequacy of a regression model was residual analysis. Essentially, if the residual plot looked like a horizontal band, we concluded that the assumptions concerning model form and the error term were satisfied. Otherwise, we concluded that the model was inadequate. When the regression model is judged inadequate, it may be because we have chosen the wrong functional form for the model (e.g., the actual relationship is curvilinear) and/or the proper independent variables have not been included. In Chapter 15, we saw that when the residual plot for Butler Trucking showed a nonconstant variance, a model that corrected this deficiency was created by adding a second independent variable.

In this chapter, we focus on the model-building issues of proper model form identification and variable selection. Section 16.1, which establishes the framework for model building, introduces the concept of a *general linear model*. Surprisingly, the general linear model makes it possible for us to accommodate curvilinear relationships between the independent variables and the dependent variable with no more computational difficulty than that involved in multiple regression with linear relationships.

Section 16.2 provides the foundation for the more sophisticated computer-based procedures for variable selection. The issue of when one variable or a group of variables should be added to a regression model is examined. We show that the general approach to determining when to add or delete variables is based on the value of an F statistic. In Section 16.3 a large problem of a type often encountered by statistical analysts is introduced. It involves 25 observations on 8 independent variables. This larger problem provides an illustration for the computer-based variable-selection procedures of Section 16.4. The stepwise-regression, forward-selection, backward-elimination, and best-subsets regression procedures are explained.

In Section 16.5 we show how the Durbin-Watson test is used to detect serial or autocorrelation. The chapter concludes with a discussion of how regression analysis can be used to solve analysis of variance and experimental design problems.

16.1   The General Linear Model

Suppose that we have collected data for one dependent variable y and k independent variables $x_1, x_2, \ldots, x_k$. The objective is to use these data to develop an estimated regression equation that provides the best relationship between the dependent and independent variables. To provide a general framework for developing more complex relationships among the independent variables, we introduce the concept of a general linear model involving p independent variables.

General Linear Model

$$y = \beta_0 + \beta_1 z_1 + \beta_2 z_2 + \cdots + \beta_p z_p + \epsilon \qquad \text{(16.1)}$$

In (16.1), each of the independent variables z_j (where $j = 1, 2, \ldots, p$) is a function of $x_1, x_2, \ldots, x_k$ (the variables for which data have been collected). In some cases, each z_j may be a function of only one x variable. The simplest case occurs in which we have collected data for just one variable x_1 and we want to estimate y using a straight-line relationship. In this case $z_1 = x_1$ and (16.1) becomes

$$y = \beta_0 + \beta_1 x_1 + \epsilon \qquad (16.2)$$

Note that this is just the simple linear regression model introduced in Chapter 14 with the exception that the independent variable is labeled x_1 instead of x. In the statistical literature, this model is referred to as a *simple first-order model with one predictor variable*.

Modeling Curvilinear Relationships

 REYNOLDS

T ABLE 16.1
Data for the Reynolds Example

Scales Sold	Months Employed
275	41
296	106
317	76
376	104
162	22
150	12
367	85
308	111
189	40
235	51
83	9
112	12
67	6
325	56
189	19

More complex types of relationships can be modeled with (16.1). To illustrate how this is done, let us consider the problem facing Reynolds, Inc., a manufacturer of industrial scales and laboratory equipment. Management at Reynolds would like to investigate the relationship between length of employment of their salespeople and the number of their electronic laboratory scales sold. Table 16.1 gives the number of scales sold by 15 randomly selected salespeople for the most recent sales period and the number of months each salesperson has been employed by the firm. Figure 16.1 shows the scatter diagram for these data. The scatter diagram indicates a possible curvilinear relationship between the length of time employed and the number of units sold.

Before considering how to develop a curvilinear relationship for Reynolds, let us consider the Minitab output shown in Figure 16.2 corresponding to a simple first-order model; the estimated regression equation is

SALES = 111 + 2.38 MONTHS

where

SALES = number of electronic laboratory scales sold

MONTHS = the number of months the salesperson has been employed

Figure 16.3 shows the corresponding standardized residual plot. Although the computer output shows that a linear relationship explains a high percentage of the variability in sales (R-sq = 78.1%), the residual plot also suggests that some type of curvilinear relationship is needed. The residuals are negative for small values of MONTHS, positive for intermediate values, and negative for large values of MONTHS.

To account for the curvilinear relationship we see in the data, we set $z_1 = x_1$ and $z_2 = x_1^2$ in (16.1) to obtain the model

$$y = \beta_0 + \beta_1 x_1 + \beta_2 x_1^2 + \epsilon \qquad (16.3)$$

This model is referred to as a *second-order model with one predictor variable*. To develop an estimated regression equation corresponding to this second-order model, we need to provide the statistical software package we are using with the original data in Table 16.1, as well as that data corresponding to adding a second independent variable that is the square of the number of months the employee has been with the firm. In Figure 16.4 we show the Minitab output corresponding to the second-order model; the estimated regression equation is

SALES = 45.3 + 6.34 MONTHS − .0345 MONTHSQ

where

FIGURE 16.1
Scatter Diagram for the Reynolds Example

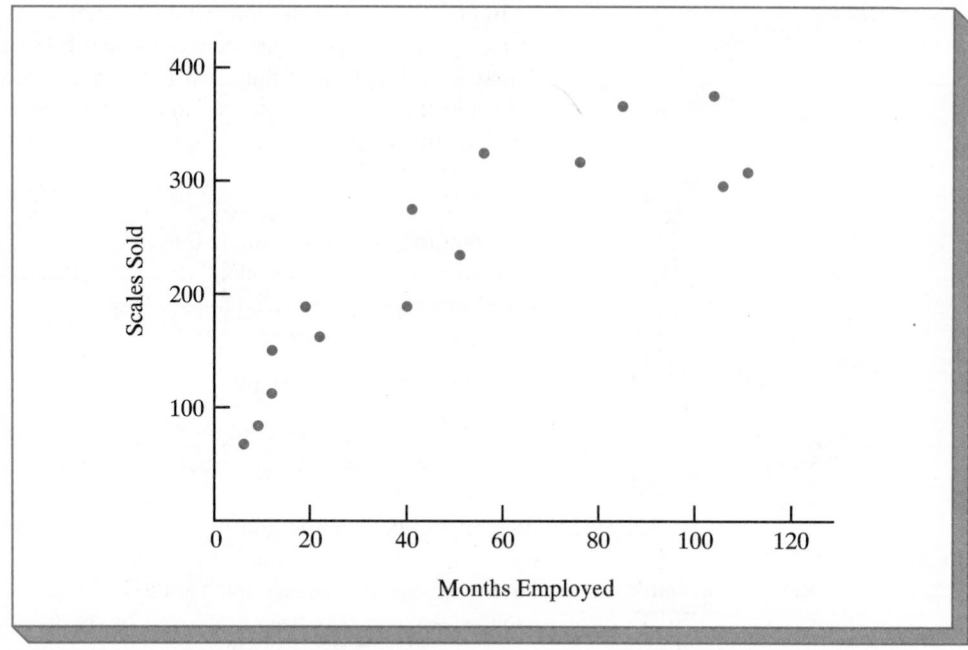

FIGURE 16.2   **Minitab Output for the Reynolds Example: First-Order Model**

```
The regression equation is
SALES = 111 + 2.38 MONTHS


Predictor        Coef        Stdev      t-ratio          p
Constant       111.23        21.63         5.14      0.000
MONTHS         2.3768       0.3489         6.81      0.000


s = 49.52         R-sq = 78.1%      R-sq(adj) = 76.4%


Analysis of Variance


SOURCE        DF           SS           MS          F         p
Regression     1       113783       113783      46.41     0.000
Error         13        31874         2452
Total         14       145657
```

MONTHSQ = the square of the number of months the
salesperson has been employed

The corresponding standardized residual plot is shown in Figure 16.5. The residual plot shows that the previous curvilinear pattern has been removed. At the .05 level of significance, the computer output shows that the overall model is significant (p-value for the *F* test is 0.000); note also that the *p*-values corresponding to the *t*-ratios for both MONTHS and MONTHSQ are less than .05, and hence, we can conclude that adding MONTHSQ

FIGURE 16.3   **Standardized Residual Plot for the Reynolds Example: First-Order Model**

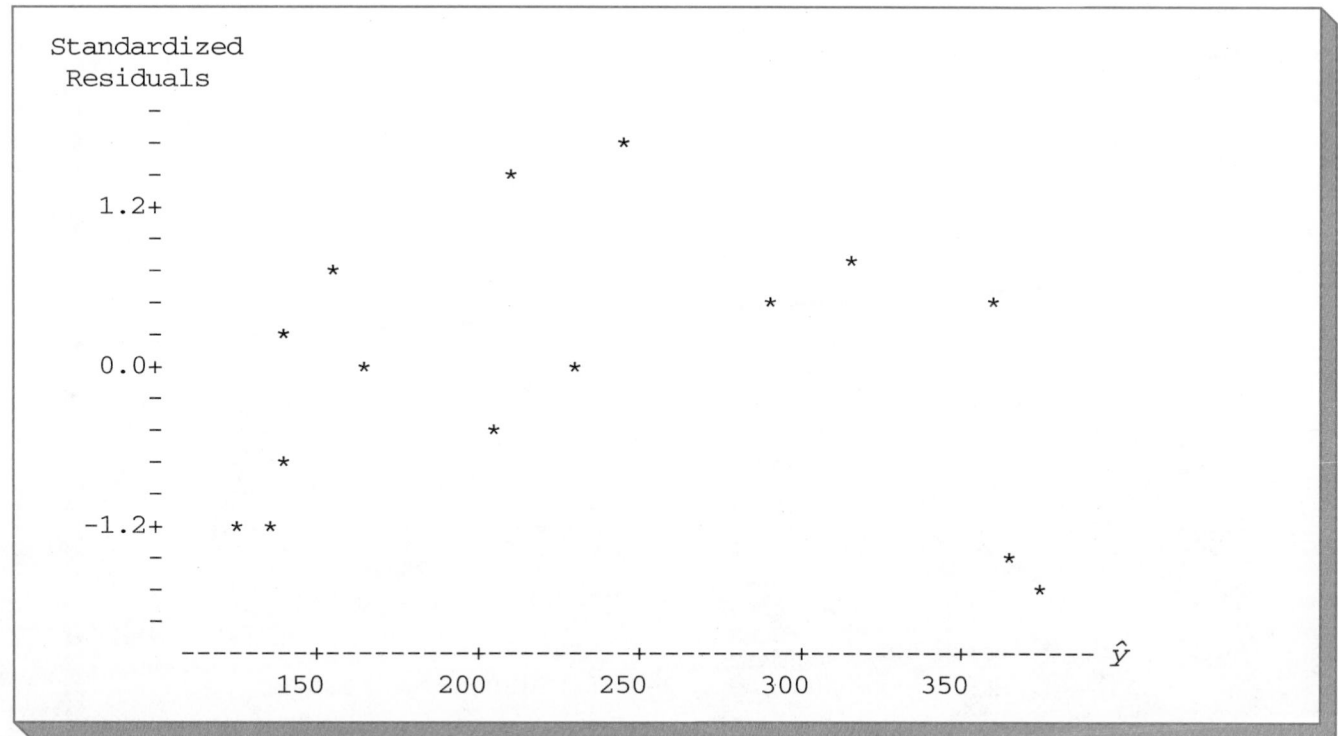

FIGURE 16.4   **Minitab Output for the Reynolds Example: Second-Order Model**

```
The regression equation is
SALES = 45.3 + 6.34 MONTHS - 0.0345 MONTHSQ

Predictor        Coef        Stdev      t-ratio          p
Constant        45.35        22.77         1.99      0.070
MONTHS          6.345        1.058         6.00      0.000
MONTHSQ     -0.034486     0.008948        -3.85      0.002

s = 34.45        R-sq = 90.2%      R-sq(adj) = 88.6%

Analysis of Variance

SOURCE       DF           SS           MS          F          p
Regression    2       131413        65707      55.36      0.000
Error        12        14244         1187
Total        14       145657
```

FIGURE **16.5**   **Standardized Residual Plot for the Reynolds Example: Second-Order Model**

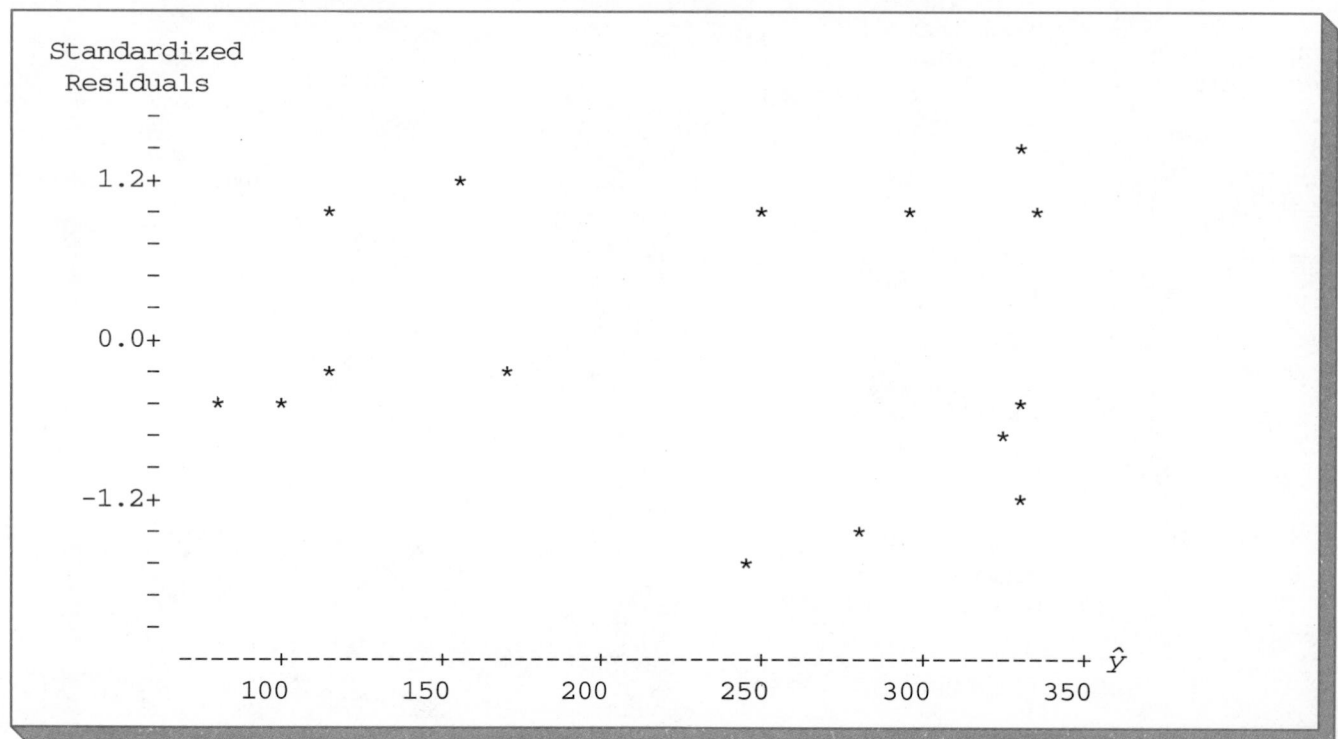

to the model involving MONTHS is significant. With an R-sq(adj) value of 88.6%, we should be pleased with the fit provided by this second-order model. More importantly, however, we now see how easy it is to handle curvilinear relationships in regression analysis.

It should now be apparent that many types of relationships can be modeled using (16.1). Thus, the regression techniques with which we have been working are definitely not limited to linear, or straight-line, relationships. In multiple regression analysis the word *linear* in the term ''general linear model'' refers only to the fact that β_0, β_1, . . . , β_p all have exponents of 1; it does not imply that the relationship between y and the x_i's is linear. Indeed, in this section we have seen one example where (16.1) can be used to model a curvilinear relationship.

Interaction

If the original data set consists of observations for y and two independent variables x_1 and x_2, we can develop a complete second-order model with two predictor variables by setting $z_1 = x_1$, $z_2 = x_2$, $z_3 = x_1^2$, $z_4 = x_2^2$, and $z_5 = x_1 x_2$ in the general linear model of (16.1). The model obtained is

$$y = \beta_0 + \beta_1 x_1 + \beta_2 x_2 + \beta_3 x_1^2 + \beta_4 x_2^2 + \beta_5 x_1 x_2 + \epsilon \qquad (16.4)$$

In this second-order model, the variable $z_5 = x_1 x_2$ is added to account for the potential effects of the two variables acting together. This type of effect is called *interaction*.

To provide an illustration of interaction and what it means, let us review the regression study conducted by Tyler Personal Care for one of its new shampoo products. Two factors

TABLE 16.2
Data for the Tyler Personal Care Example

TYLER

Price	Advertising Expenditure ($1000s)	Unit Sales (1000s)	Price	Advertising Expenditure ($1000s)	Unit Sales (1000s)
$2.00	50	478	$2.00	100	810
$2.50	50	373	$2.50	100	653
$3.00	50	335	$3.00	100	345
$2.00	50	473	$2.00	100	832
$2.50	50	358	$2.50	100	641
$3.00	50	329	$3.00	100	372
$2.00	50	456	$2.00	100	800
$2.50	50	360	$2.50	100	620
$3.00	50	322	$3.00	100	390
$2.00	50	437	$2.00	100	790
$2.50	50	365	$2.50	100	670
$3.00	50	342	$3.00	100	393

believed to have the most influence on sales were unit selling price and advertising expenditure. To investigate the effects of these two variables on sales, prices of $2.00, $2.50, and $3.00 were paired with advertising expenditures of $50,000 and $100,000 in 24 test markets. The unit sales that were observed (in 1000s) are shown in Table 16.2.

Table 16.3 provides a summary of these data. Note that the mean unit sales corresponding to a price of $2.00 and an advertising expenditure of $50,000 is 461,000, and the mean unit sales corresponding to a price of $2.00 and an advertising expenditure of $100,000 is 808,000; thus, holding price constant at $2.00, the difference in mean unit sales between an advertising expenditure of $50,000 and $100,000 is 808,000 − 461,000 = 347,000. When the price of the product is $2.50, the difference in mean unit sales is 646,000 − 364,000 = 282,000. Finally, when the price is $3.00, the difference in mean unit sales is 375,000 − 332,000 = 43,000. Clearly, the difference in mean unit sales between an advertising expenditure of $50,000 and $100,000 depends upon the price of the product. In other words, at higher selling prices, we see that the effect of increased advertising expenditure diminishes. These observations provide evidence of interaction between the price and advertising expenditure variables.

To provide another perspective of interaction, Figure 16.6 shows the mean unit sales for the six price-advertising expenditure combinations. This graph also shows that the

TABLE 16.3
Mean Unit Sales for the Tyler Personal Care Example

		Price		
		$2.00	*$2.50*	*$3.00*
Advertising	*$ 50,000*	461	364	332
Expenditure	*$100,000*	808	646	375

Mean unit sales of 808,000 when price = $2.00 and advertising expenditure = $100,000

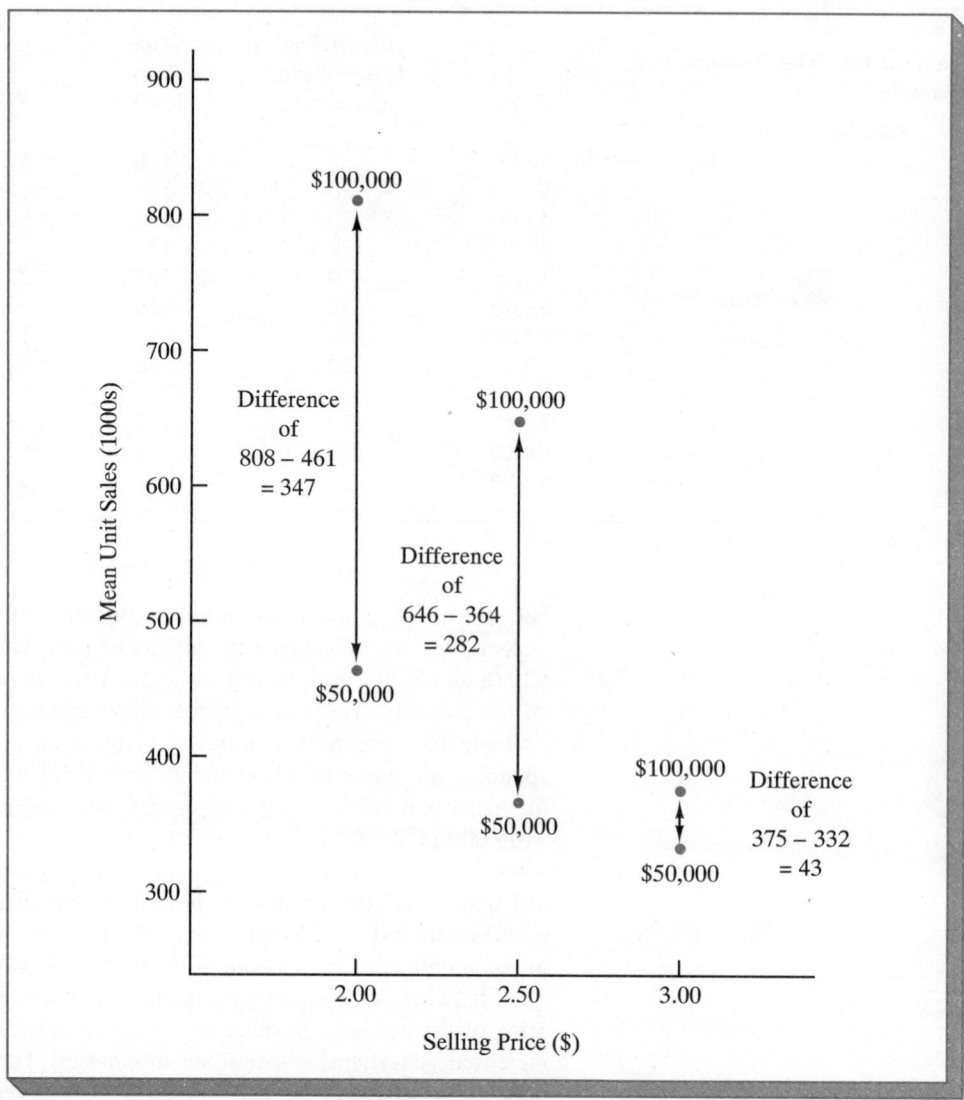

FIGURE 16.6

Mean Unit Sales as a Function of Selling Price and Advertising Expenditure

effect of advertising expenditure on mean unit sales depends on the level of the price of the product; we again see the effect of interaction. When interaction between two variables is present, we cannot study the effect of one variable on the response y independently of the other variable. In other words, meaningful conclusions can only be developed if we consider the joint effect that both variables have on the response.

To account for the effect of interaction, we will use the following regression model:

$$y = \beta_0 + \beta_1 x_1 + \beta_2 x_2 + \beta_3 x_1 x_2 + \epsilon \qquad (16.5)$$

where

$$y = \text{unit sales (1000s)}$$

$$x_1 = \text{selling price (\$)}$$

$$x_2 = \text{advertising expenditure (\$1000s)}$$

Note that this model is simply (16.4) with the curvilinear effects of x_1^2 and x_2^2 dropped from the model; it reflects Tyler's belief that the number of units sold depends linearly on selling price and advertising expenditures (accounted for by the $\beta_1 x_1$ and $\beta_2 x_2$ terms), and that there is interaction between the two variables (accounted for by the $\beta_3 x_1 x_2$ term).

To develop an estimated regression equation, a general linear model involving three independent variables (z_1, z_2, and z_3) was used.

$$y = \beta_0 + \beta_1 z_1 + \beta_2 z_2 + \beta_3 z_3 + \epsilon \tag{16.6}$$

where

$$z_1 = x_1$$

$$z_2 = x_2$$

$$z_3 = x_1 x_2$$

In Figure 16.7 we show the Minitab output corresponding to the interaction model for the Tyler Personal Care example. The resulting estimated regression equation is:

$$\text{SALES} = -276 + 175 \text{ PRICE} + 19.7 \text{ ADVER} - 6.08 \text{ PRICEADV}$$

where

$$\text{SALES} = \text{unit sales (1000s)}$$

$$\text{PRICE} = \text{price of the product (\$)}$$

$$\text{ADVER} = \text{advertising expenditure (\$1000s)}$$

$$\text{PRICEADV} = \text{interaction term (PRICE times ADVER)}$$

Since the p-value corresponding to the t test for PRICEADV is 0.000, we conclude that interaction is significant given the linear effect of the price of the product and the adver-

---

FIGURE 16.7   Minitab Output for the Tyler Personal Care Example

```
The regression equation is
SALES = - 276 + 175 PRICE + 19.7 ADVER - 6.08 PRICEADV

Predictor         Coef         Stdev      t-ratio         p
Constant         -275.8        112.8        -2.44      0.024
PRICE            175.00        44.55         3.93.     0.001
ADVER            19.680        1.427        13.79      0.000
PRICEADV         -6.0800       0.5635      -10.79      0.000

s = 28.17        R-sq = 97.8%        R-sq(adj) = 97.5%

Analysis of Variance

SOURCE        DF          SS          MS          F          p
Regression     3       709316      236439      297.87     0.000
Error         20        15875         794
Total         23       725191
```

tising expenditure. Thus, the regression results do show that the effect of advertising expenditure on unit sales depends on the level of the selling price.

Transformations Involving the Dependent Variable

MPG

TABLE 16.4
Miles per Gallon Ratings and Weights for 12 Automobiles

Miles per Gallon	Weight
28.7	2289
29.2	2113
34.2	2180
27.9	2448
33.3	2026
26.4	2702
23.9	2657
30.5	2106
18.1	3226
19.5	3213
14.3	3607
20.9	2888

In showing how the general linear model can be used to model a variety of possible relationships between the independent variables and the dependent variable, we have focused our attention on transformations involving one or more of the independent variables. Oftentimes it is worthwhile to consider transformations involving the dependent variable y. To provide an illustration of when we might want to transform the dependent variable, consider the data shown in Table 16.4. The data show the miles per gallon ratings and weights for 12 automobiles. The scatter diagram in Figure 16.8 shows a negative linear relationship between these two variables. As a result, we will use a simple first-order model to relate these two variables. The Minitab output that we obtained is shown in Figure 16.9; the resulting estimated regression equation is

$$MPG = 56.1 - 0.0116 \text{ WEIGHT}$$

where

$$MPG = \text{miles per gallon rating}$$

$$WEIGHT = \text{weight of the car in pounds}$$

The model is significant (p-value for the F test is 0.000) and the fit is very good (R-sq = 93.5%). However, we note that observation 3 has been identified as having a large standardized residual.

In Figure 16.10 we show the standardized residual plot corresponding to the first-order model. The pattern that we observe does not look like the horizontal band we should expect to observe if the assumptions regarding the error term are valid. Instead, it appears that the variability in the residuals increases as the predicted value of $\hat{y}$ increases. In other

FIGURE 16.8
Scatter Diagram for the Miles per Gallon Problem

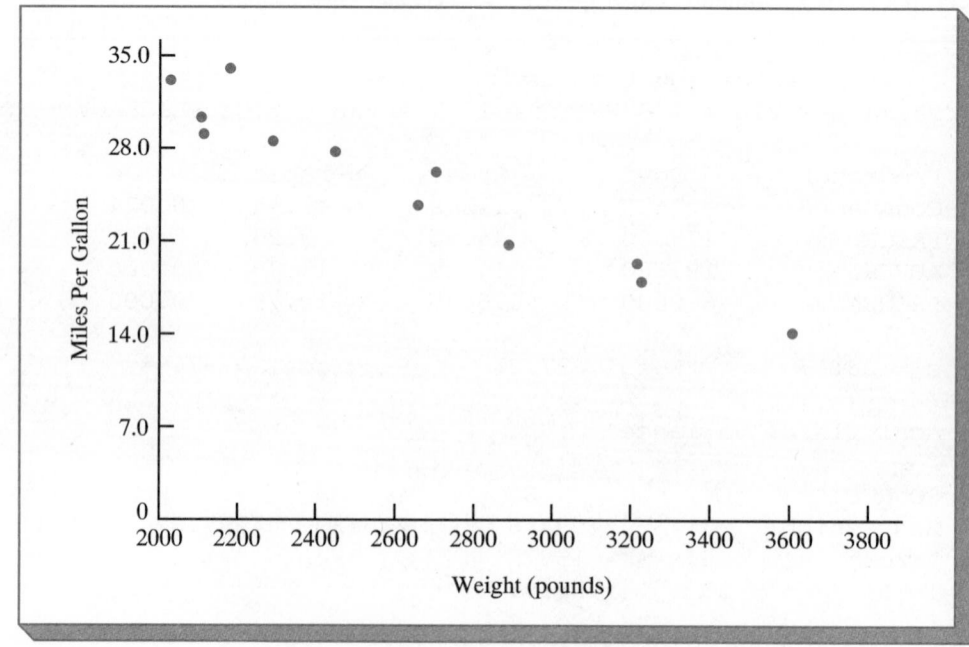

```
The regression equation is
MPG = 56.1 - 0.0116 WEIGHT


Predictor         Coef        Stdev      t-ratio         p
Constant        56.096        2.582        21.72     0.000
WEIGHT      -0.0116436    0.0009677       -12.03     0.000


s = 1.671        R-sq = 93.5%       R-sq(adj) = 92.9%

Analysis of Variance


SOURCE          DF          SS          MS          F         p
Regression       1       403.98      403.98     144.76    0.000
Error           10        27.91        2.79
Total           11       431.88

Unusual Observations
Obs.   WEIGHT       MPG       Fit  Stdev.Fit  Residual   St.Resid
  3     2180     34.200    30.713      0.644     3.487      2.26R

R denotes an obs. with a large st. resid.
```

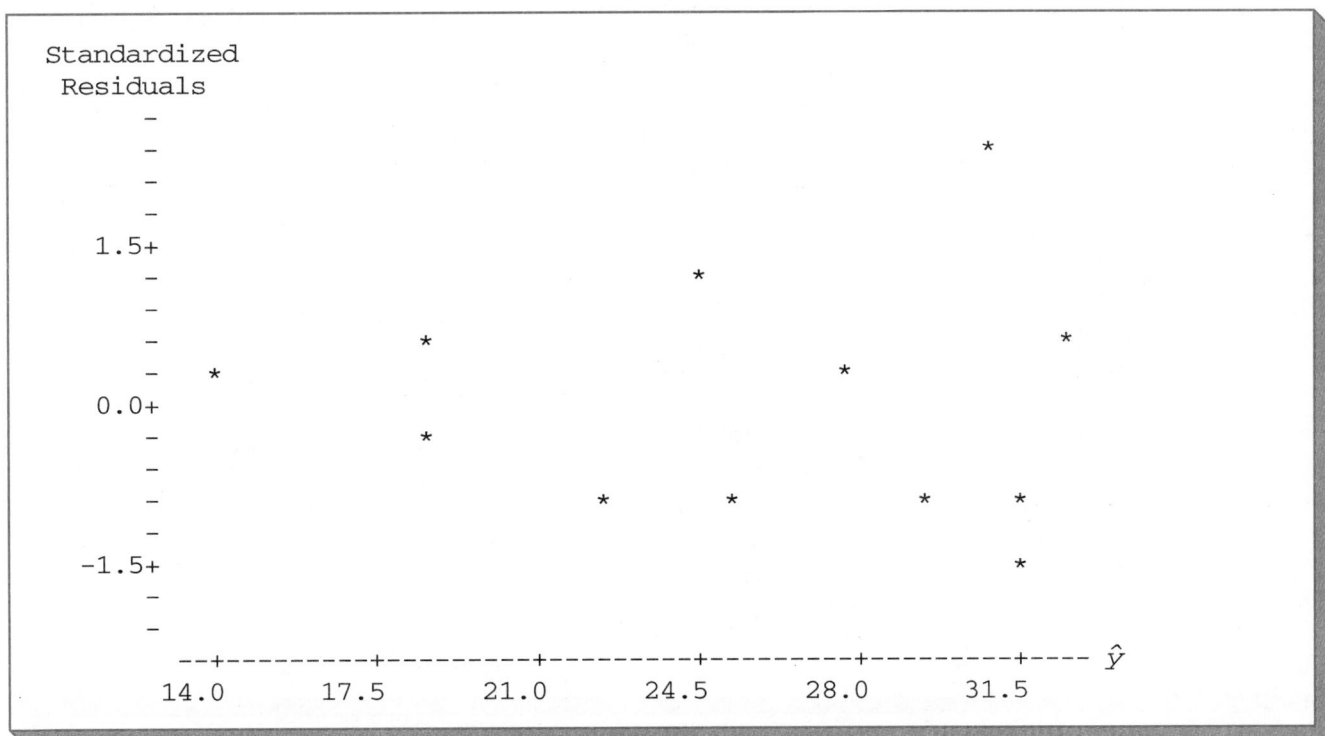

words, we have the wedged-shaped pattern referred to in Chapters 14 and 15 as being indicative of nonconstant variance. Thus, we are not justified in reaching any conclusions regarding the statistical significance of our resulting model since the underlying assumptions for the tests of significance do not appear to be satisfied.

Oftentimes the problem of nonconstant variance can be corrected by transforming the dependent variable to a different scale. For instance, if we work with the logarithm of the dependent variable instead of the original dependent variable, the effect will be to compress the values of the dependent variable and thus diminish the effects of nonconstant variance in the data. Most statistical packages provide the ability to apply logarithmic transformations using either the base 10 (common logarithm) or the base $e = 2.71828 \ldots$ (natural logarithm). We applied a natural logarithmic transformation to the miles per gallon data and developed the estimated regression equation relating weight to the natural logarithm of miles per gallon. The regression results we obtained using the natural logarithm of miles per gallon as the dependent variable, labeled LOGEMPG in the output, are shown in Figure 16.11; the corresponding standardized residual plot is shown in Figure 16.12.

Looking at the residual plot in Figure 16.12, we see that the wedged-shaped pattern has now disappeared. Moreover, none of the observations have been identified as having a large standardized residual. The model using the logarithm of miles per gallon as the dependent variable is statistically significant and provides an excellent fit to the observed data. Thus, we would recommend using the estimated regression equation

$$\text{LOGEMPG} = 4.52 - 0.000501 \text{ WEIGHT}$$

To estimate the miles per gallon rating for a car that weighs 2500 pounds, we first develop an estimate of the logarithm of the miles per gallon rating.

$$\text{LOGEMPG} = 4.52 - 0.000501(2500) = 3.2675$$

The miles per gallon estimate is obtained by finding the number whose natural logarithm is 3.2675. Using a calculator with an exponential function, or raising e to the power 3.2675, we obtain 26.2 miles per gallon.

FIGURE 16.11   **Minitab Output for the Miles per Gallon Problem: Logarithmic Transformation**

```
The regression equation is
LOGEMPG = 4.52 -0.000501 WEIGHT

Predictor        Coef       Stdev      t-ratio        p
Constant      4.52423     0.09932       45.55      0.000
WEIGHT    -0.00050110  0.00003722      -13.46      0.000

s = 0.06425     R-sq = 94.8%     R-sq(adj) = 94.2%

Analysis of Variance

SOURCE        DF          SS          MS         F        p
Regression     1     0.74822     0.74822    181.22    0.000
Error         10     0.04129     0.00413
Total         11     0.78950
```

FIGURE 16.12   **Standardized Residual Plot for the Miles per Gallon Problem: Logarithmic Transformation**

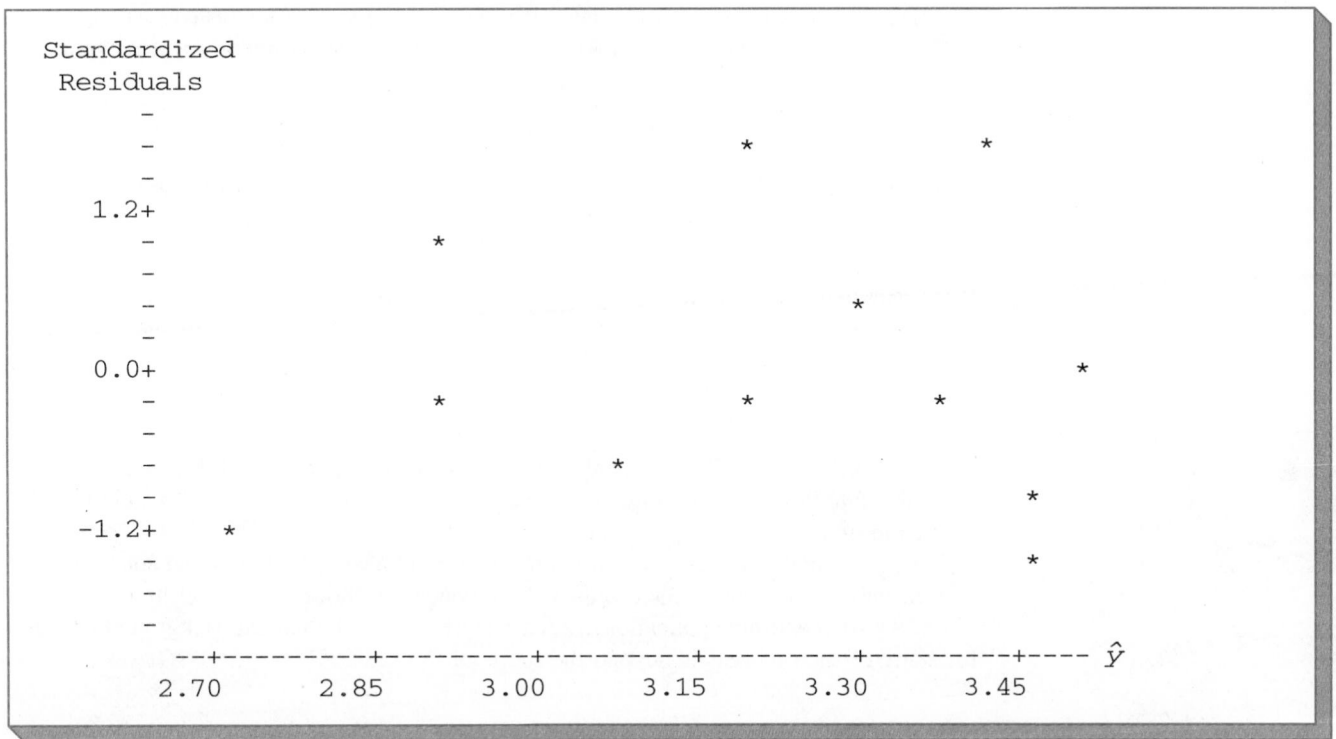

Another approach to dealing with problems of nonconstant variance involves using $1/y$ as the dependent variable instead of y. This type of transformation is called a *reciprocal transformation*. For instance, if the dependent variable is measured in miles per gallon, the reciprocal transformation would result in a new dependent variable whose units would be 1/(miles per gallon) or gallons per mile. In general, there is no way to determine whether a logarithmic transformation or a reciprocal transformation will perform best without actually trying each transformation.

Nonlinear Models that Are Intrinsically Linear

Models in which the parameters $(\beta_0, \beta_1, \ldots, \beta_p)$ have exponents other than 1 are referred to as nonlinear models. However, for the case of the exponential model, it is possible to perform a transformation of variables that will permit us to perform regression analysis using (16.1), the general linear model. The exponential model involves the following regression equation:

$$E(y) = \beta_0 \beta_1^x \tag{16.7}$$

This model is appropriate in cases where the dependent variable y increases or decreases by a constant percentage, instead of by a fixed amount, as x increases.

As an example, suppose that sales for a product y were related to advertising expenditure x (in \$1000s) according to the following exponential model:

$$E(y) = 500(1.2)^x$$

Thus, for $x = 1$, $E(y) = 500(1.2)^1 = 600$; for $x = 2$, $E(y) = 500(1.2)^2 = 720$; and for $x = 3$, $E(y) = 500(1.2)^3 = 864$. Note that $E(y)$ is not increasing by a constant amount in this case, but by a constant percentage; the percentage increase is 20%.

We can transform this nonlinear model to a linear model by taking the logarithm of both sides of (16.7):

$$\log E(y) = \log \beta_0 + x \log \beta_1 \qquad (16.8)$$

Now if we let $y' = \log E(y)$, $\beta_0' = \log \beta_0$, and $\beta_1' = \log \beta_1$, we can rewrite (16.8) as

$$y' = \beta_0' + \beta_1' x$$

It is clear that the formulas for simple linear regression can now be used to develop estimates of β_0' and β_1'. Denoting the estimates as b_0' and b_1' leads to the following estimated regression equation:

$$\hat{y}' = b_0' + b_1' x \qquad (16.9)$$

To obtain predictions of the original dependent variable y given a value of x, we would first substitute the value of x into (16.9) and compute $\hat{y}'$. The antilog of $\hat{y}'$ would be our prediction of y, or the expected value of y.

We should make it clear that there are many nonlinear models that cannot be transformed into an equivalent linear model. However, such models have had limited use in business and economic applications. Furthermore, the mathematical background needed for study of such models is beyond the scope of this text.

❏ ❏ Exercises

Methods

SELF TEST ▶ 1. Consider the following data for two variables, x and y.

x	22	24	26	30	35	40
y	12	21	33	35	40	36

a. Develop an estimated regression equation for these data of the form $\hat{y} = b_0 + b_1 x$.
b. Using the results from (a), test for a significant relationship between x and y; use $\alpha = .05$.
c. Develop a scatter diagram for these data. Does the scatter diagram suggest an estimated regression equation of the form $\hat{y} = b_0 + b_1 x + b_2 x^2$? Explain.
d. Develop an estimated regression equation for these data of the form $\hat{y} = b_0 + b_1 x + b_2 x^2$.
e. Refer to (d). Is the relationship between x, x^2, and y significant? Use $\alpha = .05$.
f. Predict the value of y when $x = 25$.

2. Consider the following data for two variables, x and y.

x	9	32	18	15	26
y	10	20	21	16	22

a. Develop an estimated regression equation for these data of the form $\hat{y} = b_0 + b_1 x$. Comment on the adequacy of this equation for predicting y.

b. Develop an estimated regression equation for these data of the form $\hat{y} = b_0 + b_1x + b_2x^2$. Comment on the adequacy of this equation for predicting y.

c. Predict the value of y when $x = 20$.

3. Consider the following data for two variables, x and y.

x	2	3	4	5	7	7	7	8	9
y	4	5	4	6	4	6	9	5	11

a. Does there appear to be a linear relationship between x and y? Explain.

b. Develop the estimated regression equation relating x and y.

c. Plot the standardized residuals versus $\hat{y}$ for the estimated regression equation developed in (b). Do the model assumptions appear to be satisfied? Explain.

d. Perform a logarithmic transformation on the dependent variable y. Develop an estimated regression equation using the transformed dependent variable. Do the model assumptions appear to be satisfied using the transformed dependent variable? Does a reciprocal transformation work better in this case? Explain.

Applications

4. The highway department is doing a study on the relationship between traffic flow and speed. The following model has been hypothesized:

$$y = \beta_0 + \beta_1x + \epsilon$$

where

$$y = \text{traffic flow in vehicles per hour}$$

$$x = \text{vehicle speed in miles per hour}$$

The data in Table 16.5 have been collected during rush hour for six highways leading out of the city.

a. Develop an estimated regression equation for these data.

b. Using $\alpha = .01$, test for a significant relationship.

5. In working further with the problem of Exercise 4, statisticians suggested the use of the following curvilinear estimated regression equation:

$$\hat{y} = b_0 + b_1x + b_2x^2$$

a. Use the data of Exercise 4 to estimate the parameters of this estimated regression equation.

b. Using $\alpha = .01$, test for a significant relationship.

c. Estimate the traffic flow in vehicles per hour at speeds of 38 miles per hour.

6. A study of emergency service facilities investigated the relationship between the number of facilities and the average distance traveled to provide the emergency service (*Management Science*, July 1988). Table 16.6 gives the data collected.

a. Develop a scatter diagram for these data treating average travel distance as the dependent variable.

b. Does a simple linear model appear to be appropriate? Explain.

c. Fit a model to the data that you believe will best explain the relationship between these two variables.

7. Performance data for a Century Coronado 21 with a 310-hp MerCruiser V-8 gasoline inboard engine was reported in *Boating*, September 1991. Data on how the engine speed in revolutions per minute (rpm) affected boat speed in miles per hour (mph) are shown below:

TABLE 16.5

Traffic Flow (y)	Vehicle Speed (x)
1256	35
1329	40
1226	30
1335	45
1349	50
1124	25

TABLE 16.6

Number of Facilities	Average Travel Distance (miles)
9	1.66
11	1.12
16	.83
21	.62
27	.51
30	.47

Engine Speed (rpm)	Boat Speed (mph)
1000	6.1
1500	10.7
2000	20.9
2500	27.5
3000	31.5
3500	33.6
4000	37.9
4500	40.2
4800	40.7

a. Develop an estimated regression equation that shows how boat speed is related to engine speed.

b. Estimate the boat speed for an engine being run at 2800 rpm.

8. The Statistics in Practice for Chapter 14 involved a study by Polaroid of the relationship between film speed and the age of Polaroid extended range, color, professional quality print film. The data for this study are shown in Table 16.7.

a. Develop a scatter diagram for these data treating change in film speed as the dependent variable.

b. Does a simple linear model appear to be appropriate?

c. Fit a model to these data that you believe will best explain the relationship between these two variables.

9. Data on housing markets was provided for 100 different cities in the United States (*U.S. News & World Report*, April 6, 1992). A sample of 16 cities with data on the median cost of a new home, the number of new housing starts during 1991–1992 and the average household income are shown below. All data are in thousands.

POLAROID

TABLE 16.7

Film Age (months)	Change in Film Speed
0	0.0
1	−20.5
2	−36.8
3	−50.8
4	−56.1
5	−67.4
6	−72.3
7	−82.2
8	−85.1
9	−85.6
10	−99.3
11	−101.4
12	−100.4
13	−111.5

HOUSING

City	Cost	Housing Starts	Household Income
Chicago	181.8	12.9	61.0
Dayton, Ohio	107.8	3.8	48.4
Atlanta	100.6	24.2	54.7
Oklahoma City	68.9	3.3	53.2
Columbia, S.C.	90.3	3.1	57.4
Tacoma, Wash.	96.1	4.1	51.1
Mobile, Ala.	68.5	1.0	41.0
Baltimore	121.8	11.1	62.8
West Palm Beach	130.4	8.9	58.1
San Antonio	72.5	1.5	57.0
Pittsburgh	79.5	4.9	49.2
Jacksonville	82.1	8.0	47.5
Cleveland	122.9	5.4	54.0
Gary, Ind.	98.2	3.2	45.7
Scranton, Pa.	81.6	2.5	44.8
Richmond, Va.	102.8	6.0	64.5

a. Using these data, develop an estimated regression equation which can be used to predict the median cost of a new home.

b. Use the estimated regression equation developed in (a) to predict the median cost of a new home in a city with 8000 housing starts and a household income of $45,000.

16.2   Determining When to Add or Delete Variables

In this section we will show how an F test can be used to determine whether it is advantageous to add one variable—or a group of variables—to a multiple regression model. This test is based on a determination of the amount of reduction in the error sum of squares resulting from adding one or more independent variables to the model. We will first illustrate how the test might be used in the context of the Butler Trucking example.

In Chapter 15 the Butler Trucking example was introduced to illustrate the use of multiple regression analysis. Recall that in this problem management wanted to develop an estimated regression equation to predict total daily travel time for their trucks using two independent variables: miles traveled and number of deliveries. With miles traveled x_1 as the only independent variable, the least squares procedure provided the following estimated regression equation:

$$\hat{y} = 1.27 + .0678x_1$$

In Chapter 15 we showed that the error sum of squares for this model was SSE = 8.029. When x_2, the number of deliveries, was added as a second independent variable, we obtained the following estimated regression equation:

$$\hat{y} = -.869 + .0611x_1 + .923x_2$$

The error sum of squares for this model was 2.299. Clearly, adding x_2 resulted in a reduction of SSE. The question we want to answer is: Does adding the variable x_2 lead to a *significant* reduction in SSE?

We will use the notation $SSE(x_1)$ to denote the error sum of squares when x_1 is the only independent variable in the model, $SSE(x_1, x_2)$ the error sum of squares when x_1 and x_2 are both in the model, and so on. Hence, the reduction in SSE resulting from adding x_2 to the model involving just x_1 is

$$SSE(x_1) - SSE(x_1, x_2) = 8.029 - 2.299 = 5.730$$

An F test is conducted to determine whether this reduction is significant.

The numerator of the F statistic is the reduction in SSE divided by the number of variables added to the original model. Here only one variable, x_2, has been added; thus, the numerator of the F statistic is

$$\frac{SSE(x_1) - SSE(x_1, x_2)}{1} = 5.730$$

The result is a measure of the reduction in SSE per variable added to the model. The denominator of the F statistic is the mean square error for the model that includes all of the independent variables. For Butler Trucking this corresponds to the model containing both x_1 and x_2; thus, $p = 2$ and hence

$$MSE = \frac{SSE(x_1, x_2)}{n - p - 1} = \frac{2.299}{7} = .3284$$

The following F statistic provides the basis for testing whether the addition of x_2 is statistically significant:

$$F = \frac{\dfrac{SSE(x_1) - SSE(x_1, x_2)}{1}}{\dfrac{SSE(x_1, x_2)}{n - p - 1}} \tag{16.10}$$

The numerator degrees of freedom for this F test is equal to the number of variables added to the model, and the denominator degrees of freedom is equal to $n - p - 1$.

For the Butler Trucking problem, we obtain

$$F = \frac{\dfrac{5.730}{1}}{\dfrac{2.299}{7}} = \frac{5.730}{.3284} = 17.45$$

Refer to Table 4 of Appendix B. We find that for a level of significance of $\alpha = .05$, $F_{.05} = 5.59$. Since $F = 17.45 > F_{.05} = 5.59$ we reject the null hypothesis that x_2 is not statistically significant; in other words, adding x_2 to the model involving only x_1 results in a significant reduction in the error sum of squares.

When we want to test for the significance of adding only one additional independent variable to an existing model, the result found with the F test just described could also be obtained by using the t test for the significance of an individual parameter (described in Section 15.4). Indeed, the F statistic we just computed is the square of the t statistic used to test the significance of an individual parameter.

Since the t test is equivalent to the F test when only one variable is being added to the model, we can now further clarify the proper use of the t test for testing the significance of an individual parameter. If an individual parameter is not significant, the corresponding variable can be dropped from the model. However, no more than one variable can ever be dropped from a model on the basis of a t test; if one variable is dropped, a second variable that was not significant initially might become significant.

We now turn to a consideration of whether the addition of more than one variable — as a set — results in a significant reduction in the error sum of squares.

The General Case

Consider the following multiple regression model involving q independent variables, where $q < p$:

$$y = \beta_0 + \beta_1 x_1 + \beta_2 x_2 + \cdots + \beta_q x_q + \epsilon \tag{16.11}$$

If we add variables $x_{q+1}, x_{q+2}, \ldots, x_p$ to this model, we obtain a model involving p independent variables:

$$y = \beta_0 + \beta_1 x_1 + \beta_2 x_2 + \cdots + \beta_q x_q \tag{16.12}$$
$$+ \beta_{q+1} x_{q+1} + \beta_{q+2} x_{q+2} + \cdots + \beta_p x_p + \epsilon$$

To test whether the addition of $x_{q+1}, x_{q+2}, \ldots, x_p$ is statistically significant, the null and alternative hypotheses can be stated as follows:

$$H_0: \beta_{q+1} = \beta_{q+2} = \cdots = \beta_p = 0$$

H_a: One or more of the parameters is not equal to zero

The following F statistic provides the basis for testing whether the additional variables are statistically significant:

$$F = \cfrac{\cfrac{SSE(x_1, x_2, \ldots, x_q) - SSE(x_1, x_2, \ldots, x_q, x_{q+1}, \ldots, x_p)}{p - q}}{\cfrac{SSE(x_1, x_2, \ldots, x_q, x_{q+1}, \ldots, x_p)}{n - p - 1}} \qquad (16.13)$$

This computed F value is then compared with F_α, the table value with $p - q$ numerator degrees of freedom and $n - p - 1$ denominator degrees of freedom. If $F > F_\alpha$, we reject H_0 and conclude that the set of additional variables is statistically significant. Note that for the special case where $q = 1$ and $p = 2$, (16.13) reduces to (16.10).

Many students find (16.13) somewhat complex. To provide a simpler description of this F ratio, we can refer to the model with the fewer number of independent variables as the reduced model and the model with the greater number of independent variables as the full model. If we let SSE(reduced) denote the error sum of squares for the reduced model and SSE(full) denote the error sum of squares for the full model, we can write the numerator of (16.13) as

$$\frac{SSE(\text{reduced}) - SSE(\text{full})}{\text{number of extra terms}} \qquad (16.14)$$

Note that "number of extra terms" denotes the difference between the number of independent variables in the full model and the number of independent variables in the reduced model. The denominator of (16.13) is the error sum of squares for the full model divided by the corresponding degrees of freedom; in other words, the denominator is the mean square error for the full model. Denoting the mean square error for the full model as MSE (full) enables us to write (16.13) as

$$F = \frac{\cfrac{SSE(\text{reduced}) - SSE(\text{full})}{\text{number of extra terms}}}{MSE(\text{full})} \qquad (16.15)$$

To illustrate the use of this F statistic, suppose that we had a regression problem involving 30 observations. One model involving the independent variables x_1, x_2, and x_3 had an error sum of squares of 150, and a second model involving the independent variables x_1, x_2, x_3, x_4, and x_5 had an error sum of squares of 100. Did the addition of the two independent variables x_4 and x_5 result in a significant reduction in the error sum of squares?

First, note that the degrees of freedom for SST is $30 - 1 = 29$ and that the degrees of freedom for the regression sum of squares for the full model is 5 (the number of independent variables in the full model). Thus, the degrees of freedom for the error sum of squares for the full model is $29 - 5 = 24$, and hence MSE(full) $= 50/29 = 1.72$. Thus, the F statistic is

$$F = \frac{\cfrac{150 - 100}{2}}{1.72} = 14.53$$

This computed F value is compared with the table F value with 2 numerator and 24 denominator degrees of freedom. At the .05 level of significance, Table 4 of Appendix B shows $F_{.05} = 3.40$. Since $F = 14.53$ is greater than 3.40, we conclude that the addition of variables x_4 and x_5 is statistically significant.

The F statistic can also be computed based upon the difference in the regression sums of squares. To show this form of the F statistic, we first note that

$$\text{SSE(reduced)} = \text{SST} - \text{SSR(reduced)}$$

$$\text{SSE(full)} = \text{SST} - \text{SSR(full)}$$

Hence

$$\text{SSE(reduced)} - \text{SSE(full)} = [\text{SST} - \text{SSR(reduced)}] - [\text{SST} - \text{SSR(full)}]$$

$$= \text{SSR(full)} - \text{SSR(reduced)}$$

Thus,

$$F = \frac{\dfrac{\text{SSR(full)} - \text{SSR(reduced)}}{\text{number of extra terms}}}{\text{MSE(full)}}$$

□ □ Exercises

Methods

10. In a regression analysis involving 27 observations, the following estimated regression equation was developed:

$$\hat{y} = 16.3 + 2.3x_1 + 12.1x_2 - 5.8x_3$$

Also, the following standard errors were obtained:

$$s_{b_1} = .53 \qquad s_{b_2} = 8.15 \qquad s_{b_3} = 1.30$$

At an $\alpha = .05$ level of significance, conduct the following hypothesis tests:
a. $H_0: \beta_1 = 0$ versus $H_a: \beta_1 \neq 0$.
b. $H_0: \beta_2 = 0$ versus $H_a: \beta_2 \neq 0$.
c. $H_0: \beta_3 = 0$ versus $H_a: \beta_3 \neq 0$.
d. Can any of the variables be dropped from the model? Why or why not?

SELF TEST ▶

11. In a regression analysis involving 30 observations, the following estimated regression equation was obtained:

$$\hat{y} = 17.6 + 3.8x_1 - 2.3x_2 + 7.6x_3 + 2.7x_4$$

For this model SST = 1805 and SSR = 1760.
a. At $\alpha = .05$ test the significance of the relationship among the variables. Suppose that variables x_1 and x_4 were dropped from the model, and the following estimated regression equation was obtained:

$$\hat{y} = 11.1 - 3.6x_2 + 8.1x_3$$

For this model SST = 1805 and SSR = 1705.
b. Compute $\text{SSE}(x_1, x_2, x_3, x_4)$.
c. Compute $\text{SSE}(x_2, x_3)$.
d. Use an F test and an $\alpha = .05$ level of significance to determine if x_1 and x_4 contribute significantly to the model.

Applications

12. The following table shows some of the data available for 14 teams in the National Football League at the end of week 15 for the 1988 season.

Team	Won–Lost	Total Points	Rushing Yards	Passing Yards	Interceptions Made by Team	Interceptions Made by Opponent
Atlanta	5–10	305	1907	2473	19	23
Chicago	12– 3	187	2134	2718	14	24
Dallas	3–12	358	1858	3386	24	10
Detroit	4–11	292	1184	1971	15	12
Green Bay	3–12	298	1274	3046	22	20
L.A. Rams	9– 6	277	1882	3604	17	22
Minnesota	10– 5	206	1744	3633	16	35
New Orleans	9– 6	274	1843	2963	15	17
N.Y. Giants	10– 5	277	1492	3096	14	15
Philadelphia	9– 6	312	1812	3247	17	29
Phoenix	7– 8	372	1909	3633	19	14
San Francisco	10– 5	256	2453	3131	14	21
Tampa Bay	4–11	340	1650	3169	33	18
Washington	7– 8	367	1377	3930	24	14

a. Develop an estimated regression equation which can be used to predict the total points scored given the number of interceptions made by the team.

b. Develop an estimated regression equation which can be used to predict the total points scored given the number of interceptions made by the team, the number of rushing yards, and the number of interceptions made by the opponents.

c. At the $\alpha = .05$ level of significance, test to see if the addition of the number of rushing yards and the number of interceptions made by the opponents contribute significantly to the estimated regression equation developed in (a)? Explain.

13. Refer to Exercise 12.

a. Develop an estimated regression equation which relates the total points to the number of passing yards, interceptions made by the team, and interceptions made by the opponents.

b. Develop an estimated regression equation using the independent variables in (a) and the number of rushing yards.

c. At the $\alpha = .05$ level of significance, did the number of rushing yards contribute significantly to the estimated regression equation developed in (a)? Explain.

14. The American Heart Association collects data on the risk of strokes. A 10-year study provided data on how age, blood pressure, and smoking relate to the risk of strokes (*U.S. News & World Report*, April 13, 1992). Assume that data from a portion of this study are shown in Table 16.8. Risk is interpreted as the probability (times 100) that the patient will have a stroke over the next 10-year period. For the smoking variable, a 1 indicates a smoker and a 0 indicates a nonsmoker.

a. Develop an estimated regression equation which can be used to predict the risk of stroke given the age and blood-pressure level.

b. Consider adding two independent variables to the model developed in (a), one involving the interaction between age and blood-pressure level and the other involving whether the person is a smoker. Develop an estimated regression equation using these four independent variables.

c. At the $\alpha = .05$ level of significance, test to see if the addition of the interaction term and whether a person is a smoker contribute significantly to the estimated regression equation developed in (a).

TABLE 16.8

Risk	Age	Blood Pressure	Smoker
12	57	152	0
24	67	163	0
13	58	155	0
56	86	177	1
28	59	196	0
51	76	189	1
18	56	155	1
31	78	120	0
37	80	135	1
15	78	98	0
22	71	152	0
36	70	173	1
15	67	135	1
48	77	209	1
15	60	199	0
36	82	119	1
8	66	166	0
34	80	125	1
3	62	117	0
37	59	207	1

16.3   First Steps in the Analysis of a Larger Problem: The Cravens Data

In introducing multiple regression analysis, we utilized the Butler Trucking example extensively. Although the small size of this problem was an advantage when exploring introductory concepts, its limited size makes it difficult to illustrate some of the variable-selection issues involved in model building. To provide an illustration of the variable-selection procedures discussed in the next section, we now introduce a data set consisting of 25 observations on 8 independent variables. Permission to use these data was provided by Dr. David W. Cravens of the Department of Marketing at Texas Christian University. Consequently, we refer to the data set as the Cravens data.*

The Cravens data involve a company that sells products in a number of sales territories, each of which is assigned to a single sales representative. A regression analysis was conducted to determine if sales in each territory could be explained using a variety of predictor (independent) variables. A random sample of 25 sales territories resulted in the data shown in Table 16.9; the variable definitions are shown in Table 16.10.

As a preliminary step, let us consider the sample correlation coefficients between each pair of variables. Figure 16.13 shows the correlation matrix obtained using the Minitab

TABLE 16.9
The Cravens Data

CRAVENS

SALES	TIME	POTEN	ADV	SHARE	CHANGE	ACCTS	WORK	RATING
3,669.88	43.10	74,065.1	4,582.9	2.51	0.34	74.86	15.05	4.9
3,473.95	108.13	58,117.3	5,539.8	5.51	0.15	107.32	19.97	5.1
2,295.10	13.82	21,118.5	2,950.4	10.91	−0.72	96.75	17.34	2.9
4,675.56	186.18	68,521.3	2,243.1	8.27	0.17	195.12	13.40	3.4
6,125.96	161.79	57,805.1	7,747.1	9.15	0.50	180.44	17.64	4.6
2,134.94	8.94	37,806.9	402.4	5.51	0.15	104.88	16.22	4.5
5,031.66	365.04	50,935.3	3,140.6	8.54	0.55	256.10	18.80	4.6
3,367.45	220.32	35,602.1	2,086.2	7.07	−0.49	126.83	19.86	2.3
6,519.45	127.64	46,176.8	8,846.2	12.54	1.24	203.25	17.42	4.9
4,876.37	105.69	42,053.2	5,673.1	8.85	0.31	119.51	21.41	2.8
2,468.27	57.72	36,829.7	2,761.8	5.38	0.37	116.26	16.32	3.1
2,533.31	23.58	33,612.7	1,991.8	5.43	−0.65	142.28	14.51	4.2
2,408.11	13.82	21,412.8	1,971.5	8.48	0.64	89.43	19.35	4.3
2,337.38	13.82	20,416.9	1,737.4	7.80	1.01	84.55	20.02	4.2
4,586.95	86.99	36,272.0	10,694.2	10.34	0.11	119.51	15.26	5.5
2,729.24	165.85	23,093.3	8,618.6	5.15	0.04	80.49	15.87	3.6
3,289.40	116.26	26,878.6	7,747.9	6.64	0.68	136.58	7.81	3.4
2,800.78	42.28	39,572.0	4,565.8	5.45	0.66	78.86	16.00	4.2
3,264.20	52.84	51,866.1	6,022.7	6.31	−0.10	136.58	17.44	3.6
3,453.62	165.04	58,749.8	3,721.1	6.35	−0.03	138.21	17.98	3.1
1,741.45	10.57	23,990.8	861.0	7.37	−1.63	75.61	20.99	1.6
2,035.75	13.82	25,694.9	3,571.5	8.39	−0.43	102.44	21.66	3.4
1,578.00	8.13	23,736.3	2,845.5	5.15	0.04	76.42	21.46	2.7
4,167.44	58.44	34,314.3	5,060.1	12.88	0.22	136.58	24.78	2.8
2,799.97	21.14	22,809.5	3,552.0	9.14	−0.74	88.62	24.96	3.9

*For details see David W. Cravens, Robert B. Woodruff, and Joe C. Stamper, "An Analytical Approach for Evaluating Sales Territory Performance," *Journal of Marketing* 36 (January 1972): 31–37.

TABLE 16.10
Minitab Variable Definitions for the Cravens Data

Variable	Definition
SALES	Total sales in units credited to the sales representative.
TIME	Length in time employed in months.
POTEN	Market potential; total industry sales in units for the sales territory.*
ADV	Advertising expenditure in the sales territory.
SHARE	Market share; weighted average for the past 4 years.
CHANGE	Change in the market share over the previous 4 years.
ACCTS	Number of accounts assigned to the sales representative.*
WORK	Work load; a weighted index based on annual purchases and concentrations of accounts.
RATING	Sales representative overall rating on eight performance dimensions; an aggregate rating on a 1–7 scale.

*These data were coded to preserve confidentiality.

correlation command. Note that the sample correlation coefficient between SALES and TIME is .623, between SALES and POTEN is .598, and so on.

Looking at the sample correlation coefficients between the independent variables, we see that the correlation between TIME and ACCTS is .758; thus, if ACCTS is used as an independent variable, TIME would not provide much more explanatory power to the model. Recall from the discussion of multicollinearity in Section 15.4 that the rule-of-thumb test says that multicollinearity can cause problems if the absolute value of the sample correlation coefficient exceeds .7 for any two of the independent variables. If possible, then, we should avoid including both TIME and ACCTS in the same regression model. The sample correlation coefficient of .549 between CHANGE and RATING is also quite high and may warrant further consideration.

Looking at the sample correlation coefficients between SALES and each of the independent variables can provide us with a quick indication of which independent variables are, by themselves, good predictors. We see that the single best predictor of SALES is ACCTS, since it has the highest sample correlation coefficient (.754). Recall that for the case of one independent variable, the square of the sample correlation coefficient is the coefficient of determination. Thus, ACCTS can explain $(.754)^2(100)$, or 56.85%, of the variability in SALES. The next most important independent variables are TIME, POTEN, and ADV, each with a sample correlation coefficient of approximately .6.

FIGURE 16.13 **Sample Correlation Coefficients for the Cravens Data (as Printed by Minitab)**

	SALES	TIME	POTEN	ADV	SHARE	CHANGE	ACCTS	WORK
TIME	0.623							
POTEN	0.598	0.454						
ADV	0.596	0.249	0.174					
SHARE	0.484	0.106	-0.211	0.264				
CHANGE	0.489	0.251	0.268	0.377	0.085			
ACCTS	0.754	0.758	0.479	0.200	0.403	0.327		
WORK	-0.117	-0.179	-0.259	-0.272	0.349	-0.288	-0.199	
RATING	0.402	0.101	0.359	0.411	-0.024	0.549	0.229	-0.277

FIGURE **16.14**   **Minitab Output for the Model Involving All Eight Independent Variables**

```
The regression equation is
SALES = - 1508 + 2.01 TIME + 0.0372 POTEN + 0.151 ADV + 199 SHARE + 291 CHANGE
          + 5.55 ACCTS + 19.8 WORK + 8 RATING

Predictor       Coef       Stdev     t-ratio        p
Constant       1507.8       778.6      -1.94      0.071
TIME            2.010       1.931       1.04      0.313
POTEN        0.037205    0.008202       4.54      0.000
ADV           0.15099     0.04711       3.21      0.006
SHARE         199.02        67.03       2.97      0.009
CHANGE        290.9        186.8        1.56      0.139
ACCTS           5.551       4.776       1.16      0.262
WORK           19.79       33.68        0.59      0.565
RATING          8.2        128.5        0.06      0.950

s = 449.0       R-sq = 92.2%      R-sq(adj) = 88.3%

Analysis of Variance

SOURCE       DF          SS          MS         F          p
Regression    8     38153568     4769196     23.65     0.000
Error        16      3225984      201624
Total        24     41379552
```

Although there are potential multicollinearity problems, let us for the moment consider developing an estimated regression equation using all eight independent variables. Using the Minitab computer package provided the results shown in Figure 16.14. The eight-variable multiple regression model has an adjusted coefficient of determination of 88.3%, which is very high for real data. Note, however, that the p column (the p-values for the t tests of individual parameters) shows that only POTEN, ADV, and SHARE are significant at the $\alpha = .05$ level, given the effect of all the others. Thus, we might be inclined to investigate the results that would be obtained if we used just these three variables. Figure 16.15 shows the Minitab results obtained for the model that uses just these three variables. We see that the model using just three independent variables has an adjusted coefficient of determination of 82.7%, which, although not quite as good as for the eight-independent-variable model, is still very high.

How can we find a model that will do the best job given the data available? One approach sometimes advocated for determining the best model is to compute all possible regressions. That is, one could develop 8 one-variable models (each of which corresponds to one of the independent variables), 28 two-variable models (the number of combinations of 8 variables taken 2 at a time), and so on. In all, for the Cravens data, there are 255 different models involving one or more independent variables that would have to be fitted to the data.

With the excellent computer packages available today, it is possible to compute all possible regressions. But, doing so involves a great deal of computation and requires the model builder to review a great deal of computer output, much of which is associated with

FIGURE 16.15   **Minitab Output for the Model Involving POTEN, ADV, and SHARE**

```
The regression equation is
SALES = - 1604 + 0.0543 POTEN + 0.167 ADV + 283 SHARE

Predictor        Coef        Stdev      t-ratio          p
Constant       -1603.6        505.6       -3.17      0.005
POTEN         0.054286     0.007474        7.26      0.000
ADV            0.16748      0.04427        3.78      0.001
SHARE          282.75         48.76        5.80      0.000

s = 545.5       R-sq = 84.9%      R-sq(adj) = 82.7%

Analysis of Variance

SOURCE        DF           SS           MS          F          p
Regression     3     35130240     11710080      39.35      0.000
Error         21      6249310       297586
Total         24     41379552
```

obviously poor models. Statisticians often prefer a more systematic approach to selecting the subset of independent variables providing the best model. In the next section, we introduce some of the more popular approaches.

16.4   Variable-Selection Procedures

In this section, we discuss four computer-based methods for selecting the independent variables in a regression model: stepwise regression, forward selection, backward elimination, and best subsets regression. Given a data set involving several possible independent variables, these methods can be used to identify which independent variables provide the best model. The first three methods are iterative; at each step a single variable is added or deleted, and the new model is evaluated. The process continues until a stopping criterion indicates that the procedure cannot find a better model. The last method (best subsets) is not a one-variable-at-a-time method; it evaluates regression models involving different subsets of the independent variables.

The criterion for selecting an independent variable to add or delete from the model at each step is based on the F statistic introduced in Section 16.2. Suppose, for instance, that we were considering adding x_3 to a model involving x_1 or deleting x_3 from a model involving x_1 and x_3. In Section 16.2 we showed that

$$F = \frac{\dfrac{SSE(x_1) - SSE(x_1, x_3)}{1}}{\dfrac{SSE(x_1, x_3)}{n - p - 1}}$$

can be used as a criterion for determining whether the presence of x_3 in the model causes a significant reduction in the error sum of squares. The value of this F statistic is the criterion used by the first three methods to determine whether a variable should be added to or deleted from the regression model at each step. It is also used to indicate when the iterative procedure should stop. The first three procedures stop when no more significant reduction in the error sum of squares can be obtained. As also noted in Section 16.2, when only one variable at a time is to be added or deleted, the t statistic (recall that $t^2 = F$) provides the same criterion.

With the stepwise-regression procedure, a variable may be added or deleted at each step. The procedure stops when no more improvement can be obtained by adding or deleting a variable. With the forward-selection procedure, a variable is added at each step, but variables are never deleted. The procedure stops when no more improvement can be obtained by adding a variable. With the backward-elimination procedure, the procedure starts with a model involving all the possible independent variables. At each step a variable is eliminated. The procedure stops when no more improvement can be obtained by deleting a variable.

Stepwise Regression

We will illustrate the stepwise-regression procedure using the Cravens data. To see how a step of the procedure is performed, suppose that after three steps the following three independent variables have been selected: ACCTS, ADV, and POTEN. At the next step, the procedure first determines if any of the variables *already in the model* should be deleted. It does so by first determining which of the three variables is the least significant addition in moving from a two- to three-independent-variable model. To determine this, an F statistic is computed for each of the three variables. The F statistic for ACCTS enables us to test whether adding ACCTS to a model that already includes ADV and POTEN leads to a significant reduction in SSE. If not, the stepwise procedure will consider dropping ACCTS from the model. Before doing so, however, the same F statistic will be computed for ADV and POTEN. The variable with the smallest F statistic makes the least significant addition in moving from a two- to three-independent-variable regression model and becomes a candidate for deletion. If any variable is to be deleted, it will be the one.

We will denote by FMIN the smallest of the F statistics for all variables in the regression model at the beginning of a new step. The variable with the smallest F statistic is the least significant addition to the model. If the value of FMIN is too small to be significant, the corresponding variable is deleted from the model. On the other hand, if FMIN is large enough to be significant, none of the variables are deleted from the model (none of the other variables can have smaller F statistics).

The user of a computer-based stepwise-regression procedure must specify a cutoff value for the F statistic so that the method can determine when FMIN is large enough to be significant. With the Minitab package, the smallest significant F value is denoted by FREMOVE. If the user does not specify a value for FREMOVE, it is automatically set equal to 4 by Minitab. Anytime FMIN < FREMOVE, the stepwise procedure of Minitab will delete the corresponding variable from the model. If FMIN ≥ FREMOVE, no variable is deleted at that step of the procedure.

If no variable can be removed from the model, the stepwise procedure next checks to see if adding a variable can improve the model. For each variable *not in the model*, an F statistic is computed. The largest of these F statistics corresponds to the variable that will cause the largest reduction in SSE. This variable then becomes a candidate for inclusion in the model. We will denote the largest F statistic for variables not currently in the model by FMAX. Again, a cutoff value for the F statistic must be used to determine if FMAX is large enough for the corresponding variable to make a significant improvement in the model.

The cutoff value for determining when to add a variable is denoted by FENTER in the Minitab computer package. The user of the package may specify a cutoff value for FENTER; if the user does not, Minitab will automatically set FENTER = 4. If FMAX > FENTER, the corresponding variable is added to the model, and the stepwise-regression procedure goes on to the next step. The procedure stops when no variables can be deleted and no variables can be added.

In summary, at each step of the stepwise-regression procedure, the first consideration is to see if any variable can be removed. If none of the variables can be removed, the procedure then checks to see if any variables can be added. Because of the nature of the stepwise procedure, it is possible for a variable to enter the model at one step, be deleted at a subsequent step, and then reenter the model at a later step. The procedure stops when FMIN ≥ FREMOVE (no variables can be deleted) and FMAX ≤ FENTER (no variables can be added).

Figure 16.16 shows the result of using the Minitab stepwise-regression procedure for the Cravens data. As we noted in Section 16.2, when only one variable is being added, the t statistic provides the same criterion as the F statistic. (One can show that $F = t^2$.) The entries in the T-RATIO row are the t statistics. The values of FREMOVE and FENTER were both automatically set equal to 4. At step 1, there are no variables to consider for deletion. The variable providing the largest value for the F statistic is ACCTS, with $F = t^2 = (5.5)^2 = 30.25$. Since $30.25 > 4$, ACCTS is added to the model. On the next three steps, ADV, POTEN, and SHARE are added to the model. After step 4, an F statistic was computed for each of the four variables in the model. The values of the F statistics were $t^2 = (3.22)^2 = 10.37$, $t^2 = 22.47$, $t^2 = 22.94$, and $t^2 = 14.59$ for ACCTS, ADV, POTEN, and SHARE, respectively. Thus, FMIN = 10.37, and the corresponding variable is ACCTS. Since, $10.37 > 4$, no variable is dropped from the model.

FIGURE 16.16
Minitab Output Using Stepwise Regression for the Cravens Data

```
STEPWISE REGRESSION OF SALES ON 8 PREDICTORS, WITH N = 25

        STEP         1         2         3         4
    CONSTANT    709.32     50.30   -327.23  -1441.93

    ACCTS         21.7      19.0      15.6       9.2
    T-RATIO       5.50      6.41      5.19      3.22

    ADV                    0.227     0.216     0.195
    T-RATIO                 4.50      4.77      4.74

    POTEN                            0.0219    0.0382
    T-RATIO                           2.53      4.79

    SHARE                                       190
    T-RATIO                                     3.82

    S              881       650       583       454
    R-SQ         56.85     77.51     82.77     90.04
```

An F statistic was then computed for each of the other four variables not in the model. Since all of these F statistics were less than 4, no variables were added to the model. The stepwise procedure stopped at this point; no variables could be deleted, and none could be added to improve the model. The results shown in Figure 16.16 were printed at this point. The estimated regression equation identified by the Minitab stepwise-regression procedure is

$$\hat{y} = -1441.93 + 9.2 \text{ ACCTS} + .195 \text{ ADV} + .0382 \text{ POTEN} + 190 \text{ SHARE}$$

Note also in Figure 16.16 that, with the error sum of squares being reduced at each step, $s = \sqrt{\text{MSE}}$ has been reduced from 881 with the best one-variable model to 454 after four steps. The value of R-sq has been increased from 56.85% to 90.04%.

Forward Selection

Forward selection is another computer-based procedure for variable selection. It is similar to the stepwise-regression procedure except that it does not permit a variable to be deleted from the model once it has been added. The forward-selection procedure starts out with no independent variables. Then it adds variables, one at a time, as long as a significant reduction in the error sum of squares (SSE) can be achieved. When no variable can be added that will cause a further significant reduction in SSE, the procedure stops and prints out the results. For the Cravens data, the stepwise-regression procedure added one variable at each step and did not delete any variables. Thus, for the Cravens data, the forward-selection procedure leads to the same model as that provided by the stepwise procedure.

Backward Elimination

The backward-elimination procedure begins with a model including all the independent variables the model builder wants considered. (Figure 16.14 shows a regression model involving all eight independent variables for the Cravens data.) It then deletes one variable at a time using the same criterion as that for removing variables using the stepwise-regression procedure. The variable with the smallest F statistic is deleted, provided F is less than the preestablished cutoff criterion (FREMOVE for Minitab). The major difference between the backward-elimination procedure and the stepwise procedure is that once a variable has been removed from the model, it cannot reenter at a later step.

Best-Subsets Regression

Stepwise regression, forward selection, and backward elimination are approaches to choosing the regression model by adding or deleting independent variables one at a time. As such, there is no guarantee that the best model for a given number of variables will be found. Thus, these one-variable-at-a-time methods are properly viewed as heuristics for selecting a good regression model.

Some software packages have a procedure called best-subsets regression that permits the user to find, given a specified number of independent variables, the best regression model. Minitab has such a procedure. In Figure 16.17, we show a portion of the computer output obtained using the best-subsets procedure for the Cravens data set.

This output identifies the two best one-variable estimated regression equations, the two best two-variable equations, the two best three-variable equations, and so on. The crite-

FIGURE 16.17
**Portion of Minitab Output Using
Best-Subsets Regression**

FIGURE 16.17
**Portion of Minitab Output Using
Best-Subsets Regression**

Vars	R-sq	Adj. R-sq	s	P T I M E	S O T A D V	H H A A N G E	A C C T S	A W C O R K	T I N G
1	56.8	55.0	881.09				X		
1	38.8	36.1	1049.3	X					
2	77.5	75.5	650.40		X		X		
2	74.6	72.3	691.11		X	X			
3	84.9	82.7	545.53		X	X	X		
3	82.8	80.3	582.66		X	X		X	
4	90.0	88.1	453.86		X	X	X	X	
4	89.6	87.5	463.95	X	X	X	X		
5	91.5	89.3	430.21	X	X	X	X	X	
5	91.2	88.9	436.75		X	X	X	X	X
6	92.0	89.4	428.00	X	X	X	X	X	X
6	91.6	88.9	438.20		X	X	X	X	X	X
7	92.2	89.0	435.67	X	X	X	X	X	X	X
7	92.0	88.8	440.29	X	X	X	X	X		X
8	92.2	88.3	449.02	X	X	X	X	X	X	X

rion used in determining which estimated regression equations are best for any number of predictors is the value of the coefficient of determination (R-sq). For instance, ACCTS, with an R-sq = 56.8%, provides the best estimated regression equation using only one independent variable; ADV and ACCTS, with an R-sq = 77.5%, provides the best estimated regression equation using two independent variables; and POTEN, ADV, and SHARE, with an R-sq = 84.9%, provides the best estimated regression equation with three independent variables. For the Cravens data the adjusted coefficient of determination (Adj. R-sq) is largest for the model with six independent variables: TIME, POTEN, ADV, SHARE, HANGE, and ACCTS. However, the best model with four independent variables (POTEN, ADV, SHARE, ACCTS) has an adjusted coefficient of determination almost as high (88.1%). All other things being equal, a simpler model with fewer variables is usually preferred.

Making the Final Choice

The analysis performed on the Cravens data to this point is good preparation for choosing a final model, but more analysis should be conducted before making the final choice. As we have noted in Chapters 14 and 15, a careful analysis of the residuals should be made. We want the residual plot for the model chosen to resemble approximately a horizontal band. For now, let us assume that there is no difficulty with the residuals and that we want to utilize the results of the best-subsets procedure to help choose the model.

The best-subsets procedure has shown us that the best four-variable model utilizes the independent variables ACCTS, ADV, POTEN, and SHARE. This also happens to be the four-variable model identified with the stepwise-regression procedure. Table 16.11 is helpful in making the final choice. It shows several possible models consisting of some or all of these four independent variables.

Table 16.11
Selected Models Involving ACCTS, ADV, POTEN, and SHARE

Model	Independent Variables	Adj. R-sq
1	ACCTS	55.0
2	ACCTS, ADV	75.5
3	POTEN, SHARE	72.3
4	ACCTS, ADV, POTEN	80.3
5	ADV, POTEN, SHARE	82.7
6	ACCTS, ADV, POTEN, SHARE	88.1

From Table 16.11, we see that the model involving just ACCTS and ADV is pretty good. The adjusted coefficient of determination is 75.5%, whereas the model using all four variables only provides a 12.6% increase. The simpler two-variable model might be preferred, for instance, if it is difficult to measure market potential (POTEN). On the other hand, if the data are readily available, the model builder would clearly prefer the model involving all four variables if highly accurate predictions of sales are needed.

Notes &
Comments

1. In the stepwise procedure, FENTER cannot be set smaller than FREMOVE. To see why, suppose this condition was not satisfied. For instance, suppose a model builder set FENTER = 2 and FREMOVE = 4 and at some step of the procedure, a variable with an F statistic of 3 was entered into the model. At the next step, any variable with an F statistic of 3 would be a candidate for removal from the model (since 3 < FREMOVE = 4). If the stepwise procedure deleted it, then at the very next step the variable would enter again, it would be deleted again, and so on. Thus, the procedure would cycle forever. To avoid this, the stepwise procedure requires that FENTER be greater than or equal to FREMOVE.

2. Functions of the independent variables may be used to create new independent variables for use with any of the procedures of this section. For instance, if we desired $x_1 x_2$ in the model to account for interaction, the data for x_1 and x_2 would be used to create the data for $z_i = x_{1i} x_{2i}$.

3. None of the procedures that add or delete variables one at a time can be guaranteed to identify the best regression model. But they are excellent approaches to finding good models—especially when there is not much multicollinearity present.

❏ ❏ **Exercises**

S E L F   T E S T

Applications

15. Two experts provided subjective lists of school districts that they think are among the best in the country. For each school district, the following data were obtained: average class size, instructional spending per student, average teacher salary, combined SAT score, percent of students taking the SAT, and the percentage of students that attended a 4-year college (*The Wall Street Journal*, March 31, 1989).

City	Average Class Size	Instructional Spending Per Student ($)	Average Teacher Salary ($)	Combined SAT Score/ (% Taking Test)	Attend 4-Year College (%)
Blue Springs, Mo.	25	3,060	29,359	1083/(8)	74
Garden City, N.Y.	18	9,700	51,000	997/(99)	77
Indianapolis, Ind.	30	3,222	30,482	716/(42)	40
Newport Beach, Calif. (Newport-Mesa)	26	4,028	37,043	977/(46)	51
Novi, Mich.	20	3,067	39,797	980/(15)	53
Piedmont, Calif. (Piedmont City)	28	4,208	37,274	1,042/(91)	75
Pittsburgh, Pa. (Fox Chapel area)	21	4,884	37,156	983/(80)	66
Scarsdale, N.Y. (Edgemont)	20	9,853	31,555	1,110/(98)	87
Wayne, Pa. (Radnor Township)	22	5,022	40,406	1,040/(95)	85
Weston, Mass.	21	4,680	39,800	1,031/(99)	89
Farmingdale, N.Y.	22	6,729	45,846	947/(75)	81
Mamaroneck, N.Y.	20	10,405	49,625	1,000/(90)	69
Mayfield, Ohio	24	5,881	36,228	1,003/(25)	48
Morristown, N.J.	22	6,300	37,000	972/(80)	64
New Rochelle, N.Y.	23	8,875	41,650	1,039/(80)	55
Newton Square, Pa. (Marple-Newton)	17	5,313	38,000	963/(75)	79
Omaha, Neb. (Westside)	23	4,815	32,500	1,059/(31)	81
Shaker Heights, Ohio	23	4,370	38,639	940/(56)	82

 SCHOOLS2

Let the dependent variable be the percentage of students that attend a 4-year college.
a. Develop the best one-variable model.
b. Use the stepwise procedure to develop the best model.
c. Use the backward-elimination procedure to develop the best model.
d. Use the best-subsets regression procedure to develop the best model.

16. Refer to Exercise 15. Let the dependent variable be the combined SAT score.
a. Develop the best one-variable model.
b. Use the stepwise procedure to develop the best model.
c. Use the backward-elimination procedure to develop the best model.
d. Use the best-subsets regression procedure to develop the best model.

17. Refer to the data in Exercise 12. Let the dependent variable be the number of wins.
a. Develop the best one-variable model.
b. Use the stepwise procedure to develop the best model.
c. Use the backward-elimination procedure to develop the best model.
d. Use the best-subsets regression procedure to develop the best model.

18. Refer to the data in Exercise 12. Let the dependent variable be the total points scored.
a. Develop the best one-variable model.
b. Use the stepwise procedure to develop the best model.
c. Use the backward-elimination procedure to develop the best model.
d. Use the best-subsets regression procedure to develop the best model.

19. The following data show the price-earnings ratio, the net profit margin, and the growth rate for 19 companies listed in "The *Forbes* 500s on Wall Street." (*Forbes*, May 1, 1989) The data in the column labeled "Industry" are simply codes used to define the industry for each company: 1 = energy-international oil; 2 = health-drugs, and 3 = electronics-computers.

Firm	P/E Ratio	Profit Margin	Growth Rate	Industry
Exxon	11.3	6.5	10	1
Chevron	10.0	7.0	5	1
Texaco	9.9	3.9	5	1
Mobil	9.7	4.3	7	1
Amoco	10.0	9.8	8	1
Pfizer	11.9	14.7	12	2
Bristol Meyers	16.2	13.9	14	2
Merck	21.0	20.3	16	2
American Home Products	13.3	16.9	11	2
Abbott Laboratories	15.5	15.2	18	2
Eli Lilly	18.9	18.7	11	2
Upjohn	14.6	12.8	10	2
Warner-Lambert	16.0	8.7	7	2
Amdahl	8.4	11.9	4	3
Digital	10.4	9.8	19	3
Hewlett-Packard	14.8	8.1	18	3
NCR	10.1	7.3	6	3
Unisys	7.0	6.9	6	3
IBM	11.8	9.2	6	3

 FORBES2

Develop the best model that can be used to predict price-earnings ratio. Briefly describe the process you used to develop a recommended estimated regression equation for these data.

20. Refer to Exercise 14. Develop a model that can be used to predict risk using age, blood pressure, whether a person is a smoker, and any interaction involving these variables. Briefly describe the process you used to develop an estimated regression equation for these data.

16.5   Residual Analysis

In Chapters 14 and 15 we showed how residual plots can be used to determine when violations of assumptions concerning the regression model occur. We looked for violations of assumptions concerning the error term ϵ and the assumed functional form of the model. Some of the actions that can be taken when such violations are detected have been discussed in this chapter. When a different functional form is needed, curvilinear and interaction terms can be included through the use of the general linear model. When other (or more) variables need to be considered, some of the variable-selection procedures of the preceding section may be appropriate.

In Chapters 14 and 15 we also discussed how residual analysis could be used to identify observations that can be classified as outliers or as being especially influential in determining the estimated regression equation. Some steps that should be taken when such observations are found were noted. In many regression studies involving economic data, a special type of correlation involving the error terms can cause problems; it is called *serial correlation,* or autocorrelation. In this section we show how residual analysis using the Durbin-Watson test can be used to determine when autocorrelation is a problem.

Autocorrelation and the Durbin-Watson Test

The data used for regression studies in business and economics have often been collected over time. In such cases, it is not uncommon for the value of y at time t, denoted by y_t, to be related to the value of y at previous time periods. When this occurs, we say autocorrelation (also called serial correlation) is present in the data. If the value of y in time period t is related to its value in time period $t - 1$, we say first-order autocorrelation is present. If the value of y in time period t is related to the value of y in time period $t - 2$, we say second-order autocorrelation is present, and so on.

When autocorrelation is present, one of the assumptions of the regression model is violated: The error terms are not independent. In the case of first-order autocorrelation, the error at time t, denoted by ϵ_t, will be related to the error at time period $t - 1$, denoted by ϵ_{t-1}. Two cases of first-order autocorrelation are shown in Figure 16.18. Panel A illustrates the case of positive autocorrelation; panel B illustrates the case of negative autocorrelation. With positive autocorrelation we expect a positive residual in one period to be followed by a positive residual in the next period, a negative residual in one period to be followed by a negative residual in the next period, and so on. With negative autocorrelation, we expect a positive residual in one period to be followed by a negative residual in the next period, then a positive residual, and so on.

When autocorrelation is present, serious errors can be made in statistical inferences about the regression model. Thus, it is important to be able to detect autocorrelation and take corrective action. We will show how the Durbin-Watson statistic can be used to detect first-order autocorrelation.

Suppose that the values of ϵ are not independent but are related in the following manner:

$$\epsilon_t = \rho \epsilon_{t-1} + z_t \qquad (16.16)$$

where ρ is a parameter with an absolute value less than 1 and z_t is a normally and independently distributed random variable with mean 0 and variance σ^2. From (16.16) we see that if $\rho = 0$, then the error terms are not related, and each has a mean of 0 and a

FIGURE 16.18
Two Data Sets with First-Order Autocorrelation

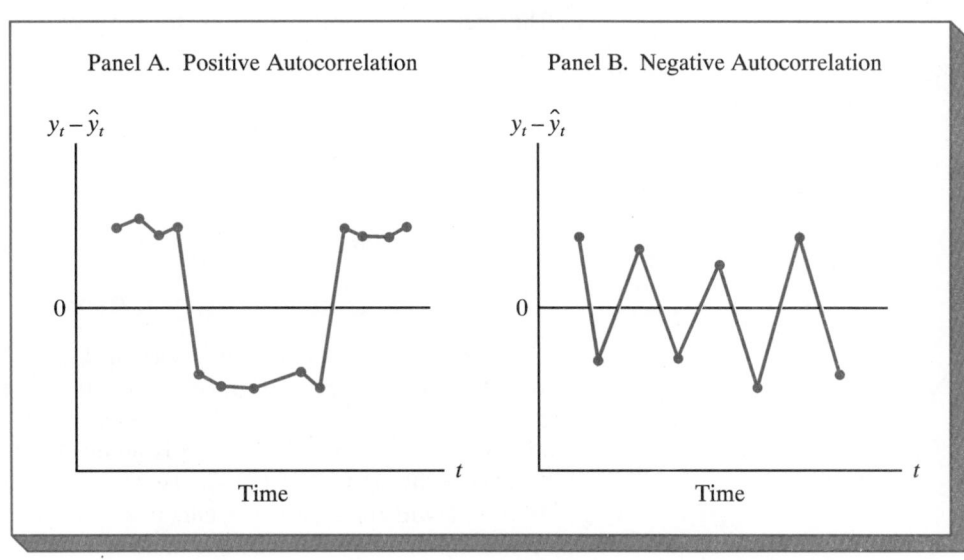

variance of σ^2. In this case, there is no autocorrelation and the regression assumptions are satisfied. If $\rho > 0$, we have positive autocorrelation; if $\rho < 0$, we have negative autocorrelation. In either of these cases, the regression assumptions concerning the error term are violated.

The Durbin-Watson test for autocorrelation uses the residuals to determine whether $\rho = 0$. To simplify the notation for the Durbin-Watson statistic, we shall denote the ith residual by $e_i = y_i - \hat{y}_i$. The Durbin-Watson statistic denoted d is given by

---

Durbin-Watson Statistic

$$d = \frac{\sum_{t=2}^{n} (e_t - e_{t-1})^2}{\sum_{t=1}^{n} e_t^2} \tag{16.17}$$

If successive values of the residuals are close together (positive autocorrelation), the Durbin-Watson statistic will be small. If successive values of the residuals are far apart (negative autocorrelation), the Durbin-Watson statistic will tend to be large.

The Durbin-Watson statistic ranges in value between 0 and 4, with a value of 2 indicating no autocorrelation is present. Durbin and Watson have developed tables that can be used to determine when their test statistic indicates the presence of autocorrelation. Table 16.12 shows lower and upper bounds (d_L and d_U) for hypothesis tests using $\alpha = .05$, $\alpha = .025$, and $\alpha = .01$; n denotes the number of observations, and k is the number of independent variables in the model. The null hypothesis to be tested is always taken to be one of no autocorrelation:

$$H_0: \rho = 0$$

The alternative hypothesis to test for positive autocorrelation is

$$H_a: \rho > 0$$

The alternative hypothesis to test for negative autocorrelation is

$$H_a: \rho < 0$$

A two-sided test is also possible. In this case the alternative hypothesis is

$$H_a: \rho \neq 0$$

Figure 16.19 shows how the values of d_L and d_U in Table 16.12 are to be used to test for autocorrelation. Panel A illustrates the test for positive autocorrelation. If $d < d_L$, we conclude positive autocorrelation is present. If $d_L \leq d \leq d_U$, we say the test is inconclusive. If $d > d_U$, we conclude there is no evidence of positive autocorrelation.

Panel B illustrates the test for negative autocorrelation. If $d > 4 - d_L$, we conclude negative autocorrelation is present. If $4 - d_U \leq d \leq 4 - d_L$, we say the test is inconclusive. If $d < 4 - d_U$, we conclude there is no evidence of negative autocorrelation.

TABLE 16.12
Critical Values for the Durbin-Watson Test for Autocorrelation

NOTE: Entries in the table give the critical values for a one-tailed Durbin-Watson test for autocorrelation. For a two-tailed test, the level of significance is doubled.

Significance Points of d_L and d_U: $\alpha = .05$
Number of Independent Variables

k	1		2		3		4		5	
n	d_L	d_U	d_L	d_U	d_L	d_U	d_L	d_U	d_L	d_U
15	1.08	1.36	0.95	1.54	0.82	1.75	0.69	1.97	0.56	2.21
16	1.10	1.37	0.98	1.54	0.86	1.73	0.74	1.93	0.62	2.15
17	1.13	1.38	1.02	1.54	0.90	1.71	0.78	1.90	0.67	2.10
18	1.16	1.39	1.05	1.53	0.93	1.69	0.82	1.87	0.71	2.06
19	1.18	1.40	1.08	1.53	0.97	1.68	0.86	1.85	0.75	2.02
20	1.20	1.41	1.10	1.54	1.00	1.68	0.90	1.83	0.79	1.99
21	1.22	1.42	1.13	1.54	1.03	1.67	0.93	1.81	0.83	1.96
22	1.24	1.43	1.15	1.54	1.05	1.66	0.96	1.80	0.86	1.94
23	1.26	1.44	1.17	1.54	1.08	1.66	0.99	1.79	0.90	1.92
24	1.27	1.45	1.19	1.55	1.10	1.66	1.01	1.78	0.93	1.90
25	1.29	1.45	1.21	1.55	1.12	1.66	1.04	1.77	0.95	1.89
26	1.30	1.46	1.22	1.55	1.14	1.65	1.06	1.76	0.98	1.88
27	1.32	1.47	1.24	1.56	1.16	1.65	1.08	1.76	1.01	1.86
28	1.33	1.48	1.26	1.56	1.18	1.65	1.10	1.75	1.03	1.85
29	1.34	1.48	1.27	1.56	1.20	1.65	1.12	1.74	1.05	1.84
30	1.35	1.49	1.28	1.57	1.21	1.65	1.14	1.74	1.07	1.83
31	1.36	1.50	1.30	1.57	1.23	1.65	1.16	1.74	1.09	1.83
32	1.37	1.50	1.31	1.57	1.24	1.65	1.18	1.73	1.11	1.82
33	1.38	1.51	1.32	1.58	1.26	1.65	1.19	1.73	1.13	1.81
34	1.39	1.51	1.33	1.58	1.27	1.65	1.21	1.73	1.15	1.81
35	1.40	1.52	1.34	1.58	1.28	1.65	1.22	1.73	1.16	1.80
36	1.41	1.52	1.35	1.59	1.29	1.65	1.24	1.73	1.18	1.80
37	1.42	1.53	1.36	1.59	1.31	1.66	1.25	1.72	1.19	1.80
38	1.43	1.54	1.37	1.59	1.32	1.66	1.26	1.72	1.21	1.79
39	1.43	1.54	1.38	1.60	1.33	1.66	1.27	1.72	1.22	1.79
40	1.44	1.54	1.39	1.60	1.34	1.66	1.29	1.72	1.23	1.79
45	1.48	1.57	1.43	1.62	1.38	1.67	1.34	1.72	1.29	1.78
50	1.50	1.59	1.46	1.63	1.42	1.67	1.38	1.72	1.34	1.77
55	1.53	1.60	1.49	1.64	1.45	1.68	1.41	1.72	1.38	1.77
60	1.55	1.62	1.51	1.65	1.48	1.69	1.44	1.73	1.41	1.77
65	1.57	1.63	1.54	1.66	1.50	1.70	1.47	1.73	1.44	1.77
70	1.58	1.64	1.55	1.67	1.52	1.70	1.49	1.74	1.46	1.77
75	1.60	1.65	1.57	1.68	1.54	1.71	1.51	1.74	1.49	1.77
80	1.61	1.66	1.59	1.69	1.56	1.72	1.53	1.74	1.51	1.77
85	1.62	1.67	1.60	1.70	1.57	1.72	1.55	1.75	1.52	1.77
90	1.63	1.68	1.61	1.70	1.59	1.73	1.57	1.75	1.54	1.78
95	1.64	1.69	1.62	1.71	1.60	1.73	1.58	1.75	1.56	1.78
100	1.65	1.69	1.63	1.72	1.61	1.74	1.59	1.76	1.57	1.78

—Table continues

Source: J. Durbin and G. S. Watson, "Testing for serial correlation in least square regression II," *Biometrika*, 38, 1951, 159–178.

TABLE 16.12
(Continued)

Significance Points of d_L and d_U: $\alpha = .025$
Number of Independent Variables

k	1		2		3		4		5	
n	d_L	d_U	d_L	d_U	d_L	d_U	d_L	d_U	d_L	d_U
15	0.95	1.23	0.83	1.40	0.71	1.61	0.59	1.84	0.48	2.09
16	0.98	1.24	0.86	1.40	0.75	1.59	0.64	1.80	0.53	2.03
17	1.01	1.25	0.90	1.40	0.79	1.58	0.68	1.77	0.57	1.98
18	1.03	1.26	0.93	1.40	0.82	1.56	0.72	1.74	0.62	1.93
19	1.06	1.28	0.96	1.41	0.86	1.55	0.76	1.72	0.66	1.90
20	1.08	1.28	0.99	1.41	0.89	1.55	0.79	1.70	0.70	1.87
21	1.10	1.30	1.01	1.41	0.92	1.54	0.83	1.69	0.73	1.84
22	1.12	1.31	1.04	1.42	0.95	1.54	0.86	1.68	0.77	1.82
23	1.14	1.32	1.06	1.42	0.97	1.54	0.89	1.67	0.80	1.80
24	1.16	1.33	1.08	1.43	1.00	1.54	0.91	1.66	0.83	1.79
25	1.18	1.34	1.10	1.43	1.02	1.54	0.94	1.65	0.86	1.77
26	1.19	1.35	1.12	1.44	1.04	1.54	0.96	1.65	0.88	1.76
27	1.21	1.36	1.13	1.44	1.06	1.54	0.99	1.64	0.91	1.75
28	1.22	1.37	1.15	1.45	1.08	1.54	1.01	1.64	0.93	1.74
29	1.24	1.38	1.17	1.45	1.10	1.54	1.03	1.63	0.96	1.73
30	1.25	1.38	1.18	1.46	1.12	1.54	1.05	1.63	0.98	1.73
31	1.26	1.39	1.20	1.47	1.13	1.55	1.07	1.63	1.00	1.72
32	1.27	1.40	1.21	1.47	1.15	1.55	1.08	1.63	1.02	1.71
33	1.28	1.41	1.22	1.48	1.16	1.55	1.10	1.63	1.04	1.71
34	1.29	1.41	1.24	1.48	1.17	1.55	1.12	1.63	1.06	1.70
35	1.30	1.42	1.25	1.48	1.19	1.55	1.13	1.63	1.07	1.70
36	1.31	1.43	1.26	1.49	1.20	1.56	1.15	1.63	1.09	1.70
37	1.32	1.43	1.27	1.49	1.21	1.56	1.16	1.62	1.10	1.70
38	1.33	1.44	1.28	1.50	1.23	1.56	1.17	1.62	1.12	1.70
39	1.34	1.44	1.29	1.50	1.24	1.56	1.19	1.63	1.13	1.69
40	1.35	1.45	1.30	1.51	1.25	1.57	1.20	1.63	1.15	1.69
45	1.39	1.48	1.34	1.53	1.30	1.58	1.25	1.63	1.21	1.69
50	1.42	1.50	1.38	1.54	1.34	1.59	1.30	1.64	1.26	1.69
55	1.45	1.52	1.41	1.56	1.37	1.60	1.33	1.64	1.30	1.69
60	1.47	1.54	1.44	1.57	1.40	1.61	1.37	1.65	1.33	1.69
65	1.49	1.55	1.46	1.59	1.43	1.62	1.40	1.66	1.36	1.69
70	1.51	1.57	1.48	1.60	1.45	1.63	1.42	1.66	1.39	1.70
75	1.53	1.58	1.50	1.61	1.47	1.64	1.45	1.67	1.42	1.70
80	1.54	1.59	1.52	1.62	1.49	1.65	1.47	1.67	1.44	1.70
85	1.56	1.60	1.53	1.63	1.51	1.65	1.49	1.68	1.46	1.71
90	1.57	1.61	1.55	1.64	1.53	1.66	1.50	1.69	1.48	1.71
95	1.58	1.62	1.56	1.65	1.54	1.67	1.52	1.69	1.50	1.71
100	1.59	1.63	1.57	1.65	1.55	1.67	1.53	1.70	1.51	1.72

—Table continues

Source: J. Durbin and G. S. Watson, "Testing for serial correlation in least squares regression II," *Biometrika*, 38, 1951, 159–178.

TABLE 16.12
(Continued)

Significance Points of d_L and d_U: $\alpha = .01$
Number of Independent Variables

	k	1		2		3		4		5	
n		d_L	d_U	d_L	d_U	d_L	d_U	d_L	d_U	d_L	d_U
15		0.81	1.07	0.70	1.25	0.59	1.46	0.49	1.70	0.39	1.96
16		0.84	1.09	0.74	1.25	0.63	1.44	0.53	1.66	0.44	1.90
17		0.87	1.10	0.77	1.25	0.67	1.43	0.57	1.63	0.48	1.85
18		0.90	1.12	0.80	1.26	0.71	1.42	0.61	1.60	0.52	1.80
19		0.93	1.13	0.83	1.26	0.74	1.41	0.65	1.58	0.56	1.77
20		0.95	1.15	0.86	1.27	0.77	1.41	0.68	1.57	0.60	1.74
21		0.97	1.16	0.89	1.27	0.80	1.41	0.72	1.55	0.63	1.71
22		1.00	1.17	0.91	1.28	0.83	1.40	0.75	1.54	0.66	1.69
23		1.02	1.19	0.94	1.29	0.86	1.40	0.77	1.53	0.70	1.67
24		1.04	1.20	0.96	1.30	0.88	1.41	0.80	1.53	0.72	1.66
25		1.05	1.21	0.98	1.30	0.90	1.41	0.83	1.52	0.75	1.65
26		1.07	1.22	1.00	1.31	0.93	1.41	0.85	1.52	0.78	1.64
27		1.09	1.23	1.02	1.32	0.95	1.41	0.88	1.51	0.81	1.63
28		1.10	1.24	1.04	1.32	0.97	1.41	0.90	1.51	0.83	1.62
29		1.12	1.25	1.05	1.33	0.99	1.42	0.92	1.51	0.85	1.61
30		1.13	1.26	1.07	1.34	1.01	1.42	0.94	1.51	0.88	1.61
31		1.15	1.27	1.08	1.34	1.02	1.42	0.96	1.51	0.90	1.60
32		1.16	1.28	1.10	1.35	1.04	1.43	0.98	1.51	0.92	1.60
33		1.17	1.29	1.11	1.36	1.05	1.43	1.00	1.51	0.94	1.59
34		1.18	1.30	1.13	1.36	1.07	1.43	1.01	1.51	0.95	1.59
35		1.19	1.31	1.14	1.37	1.08	1.44	1.03	1.51	0.97	1.59
36		1.21	1.32	1.15	1.38	1.10	1.44	1.04	1.51	0.99	1.59
37		1.22	1.32	1.16	1.38	1.11	1.45	1.06	1.51	1.00	1.59
38		1.23	1.33	1.18	1.39	1.12	1.45	1.07	1.52	1.02	1.58
39		1.24	1.34	1.19	1.39	1.14	1.45	1.09	1.52	1.03	1.58
40		1.25	1.34	1.20	1.40	1.15	1.46	1.10	1.52	1.05	1.58
45		1.29	1.38	1.24	1.42	1.20	1.48	1.16	1.53	1.11	1.58
50		1.32	1.40	1.28	1.45	1.24	1.49	1.20	1.54	1.16	1.59
55		1.36	1.43	1.32	1.47	1.28	1.51	1.25	1.55	1.21	1.59
60		1.38	1.45	1.35	1.48	1.32	1.52	1.28	1.56	1.25	1.60
65		1.41	1.47	1.38	1.50	1.35	1.53	1.31	1.57	1.28	1.61
70		1.43	1.49	1.40	1.52	1.37	1.55	1.34	1.58	1.31	1.61
75		1.45	1.50	1.42	1.53	1.39	1.56	1.37	1.59	1.34	1.62
80		1.47	1.52	1.44	1.54	1.42	1.57	1.39	1.60	1.36	1.62
85		1.48	1.53	1.46	1.55	1.43	1.58	1.41	1.60	1.39	1.63
90		1.50	1.54	1.47	1.56	1.45	1.59	1.43	1.61	1.41	1.64
95		1.51	1.55	1.49	1.57	1.47	1.60	1.45	1.62	1.42	1.64
100		1.52	1.56	1.50	1.58	1.48	1.60	1.46	1.63	1.44	1.65

Source: J. Durbin and G. S. Watson, "Testing for serial correlation in least squares regression II," *Biometrika*, 38, 1951, 159–178.

FIGURE **16.19**
Hypothesis Test for Autocorrelation Using the Durbin-Watson Test

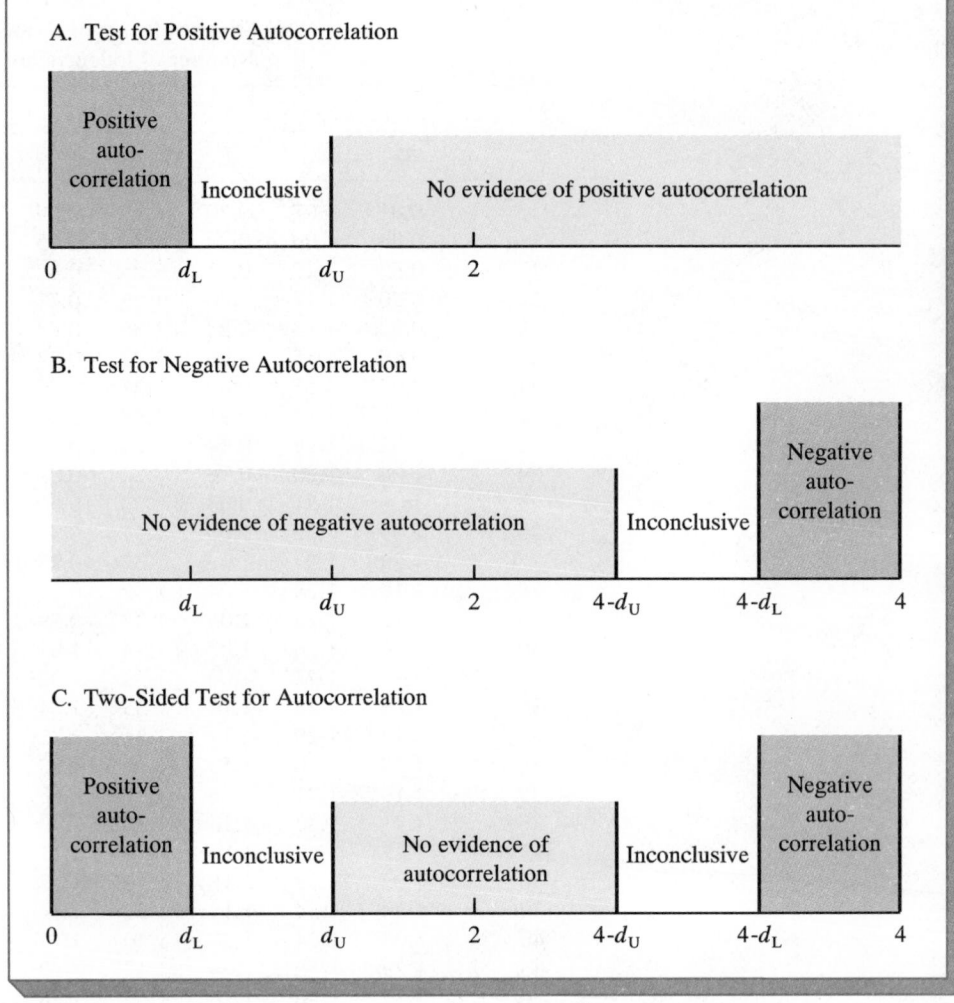

Panel C illustrates the two-sided test. If $d < d_L$ or $d > 4 - d_L$, we reject H_0 and conclude autocorrelation is present. If $d_L \leq d \leq d_U$ or $4 - d_U \leq d \leq 4 - d_L$, we say the test is inconclusive. If $d_U < d < 4 - d_U$, we conclude there is no evidence of autocorrelation.

If significant autocorrelation is identified, we should investigate whether we have omitted one or more key independent variables that have time-ordered effects on the dependent variable. If no such variables can be identified, including an independent variable that measures the time of the observation (for instance, the value of this variable could be 1 for the first observation, 2 for the second observation, and so on) will sometimes eliminate or reduce the autocorrelation. When these attempts to reduce or remove autocorrelation do not work, transformations on the dependent or independent variables can prove helpful; a discussion of such transformations can be found in more advanced texts on regression analysis.

In closing this section we note that the Durbin-Watson tables list the smallest sample size as 15. The reason for this is that the test is generally inconclusive for smaller sample sizes; in fact, many statisticians believe that the sample size should be at least 50 for the test to produce worthwhile results.

❏ ❏ Exercises

Applications

SELF TEST ▶

21. Consider the data set presented in Exercise 19.

a. Develop the estimated regression equation which can be used to predict the price-earnings ratio given the profit margin.

b. Plot the residuals obtained from the model developed in (a) as a function of the order in which the data are presented. Does there appear to be any autocorrelation present in the data? Explain.

c. At the .05 level of significance, test for any positive autocorrelation in the data.

22. Refer to the Cravens data set presented in Table 16.9. In Section 16.3 we showed that the 0model involving ACCTS, ADV, POTEN, and SHARE had an adjusted coefficient of determination of 88.1%. At the .05 level of significance, use the Durbin-Watson test to determine if positive autocorrelation is present.

16.6   Multiple Regression Approach to Analysis of Variance and Experimental Design

In Section 15.6 we discussed the use of dummy variables in multiple regression analysis. In this section we show how the use of dummy variables in a multiple regression equation can provide another approach to solving analysis of variance and experimental design problems. We will demonstrate the multiple regression approach to analysis of variance by applying it to the National Computer Products, Inc. (NCP), problem introduced in Chapter 13.

Recall that NCP manufactures printers and fax machines at plants located in Charlotte, Houston, and San Diego. To measure how much their employees know about total quality management, a random sample of six employees was selected from each plant and given a quality-awareness exam. Management would like to use the exam scores obtained for these 18 employees to determine if the mean examination score is the same at each plant.

We begin the regression approach to this problem by defining two dummy variables that will be used to indicate the plant from which each sample observation was selected. Since there are three plants or populations in the NCP problem, we need two dummy variables. In general, if the factor being investigated involves k distinct levels or populations, we need to define $k - 1$ dummy variables. For the NCP problem we define x_1 and x_2 as shown in Table 16.13.

We can use the dummy variables x_1 and x_2 to relate the score on the quality-awareness examination y to the plant at which the employee works:

$E(y)$ = Expected value of the score on the quality-awareness examination

$$= \beta_0 + \beta_1 x_1 + \beta_2 x_2$$

Thus, if we are interested in the expected value of the examination score for an employee who works at the Charlotte plant, our procedure for assigning numerical values to the dummy variables x_1 and x_2 would result in setting $x_1 = x_2 = 0$. The multiple regression equation then reduces to

$$E(y) = \beta_0 + \beta_1(0) + \beta_2(0) = \beta_0$$

Thus, we can interpret β_0 as the expected value of the examination score for employees who work at the Charlotte plant.

Next let us consider the forms of the multiple regression equation for each of the other plants. For the Houston plant, $x_1 = 1$ and $x_2 = 0$, and

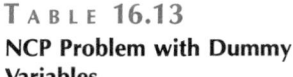

TABLE 16.13

NCP Problem with Dummy Variables

x_1	x_2	
0	0	Observation is associated with the Charlotte plant
1	0	Observation is associated with the Houston plant
0	1	Observation is associated with the San Diego plant

$$E(y) = \beta_0 + \beta_1(1) + \beta_2(0) = \beta_0 + \beta_1$$

For the San Diego plant, $x_1 = 0$ and $x_2 = 1$, and

$$E(y) = \beta_0 + \beta_1(0) + \beta_2(1) = \beta_0 + \beta_2$$

We see that $\beta_0 + \beta_1$ represents the expected value of the examination score for employees who work at the Houston plant, and $\beta_0 + \beta_2$ represents the expected value of the examination score for employees at the San Diego plant.

We now want to estimate the coefficients β_0, β_1, and β_2 and hence develop estimates of the expected value of the examination score for each plant. The sample data consisting of 18 observations of x_1, x_2, and y were entered into Minitab. The actual input data and the output from Minitab are shown in Table 16.14 and Figure 16.20, respectively.

Refer to Figure 16.20. We see that the estimates of β_0, β_1, and β_2 are $b_0 = 79$, $b_1 = -5$, and $b_2 = -13$. Thus, our best estimate of the expected value of the examination score for each plant is as follows:

Plant	Estimate of $E(y)$
Charlotte	$b_0 = 79$
Houston	$b_0 + b_1 = 79 - 5 = 74$
San Diego	$b_0 + b_2 = 79 - 13 = 66$

Note that the best estimate of the expected value of the examination score for each plant obtained from the regression analysis is the same as the sample means found earlier when applying the ANOVA procedure. That is, $\bar{x}_1 = 79$, $\bar{x}_2 = 74$, and $\bar{x}_3 = 66$.

Now let us see how we can use the output from the multiple regression package to perform the ANOVA test on the difference in the means for the three plants. First, we observe that if there is no difference in the means, then

$$E(y) \text{ for the Houston plant} - E(y) \text{ for the Charlotte plant} = 0$$

$$E(y) \text{ for the San Diego plant} - E(y) \text{ for the Charlotte plant} = 0$$

TABLE 16.14
Input Data for the NCP Problem

Observations Correspond to	x_1	x_2	y	Observations Correspond to	x_1	x_2	y
Charlotte Plant	0	0	85	San Diego Plant	0	1	59
	0	0	75		0	1	64
	0	0	82		0	1	62
	0	0	76		0	1	69
	0	0	71		0	1	75
	0	0	85		0	1	67
Houston Plant	1	0	71				
	1	0	75				
	1	0	73				
	1	0	74				
	1	0	69				
	1	0	82				

FIGURE 16.20   **Multiple Regression Output for the NCP Problem**

```
The regression equation is
Y = 79.0 - 5.00 X1 - 13.0 X2

Predictor        Coef        Stdev      t-ratio          p
Constant       79.000        2.186        36.14      0.000
X1             -5.000        3.091        -1.62      0.127
X2            -13.000        3.091        -4.21      0.001

s = 5.354        R-sq = 54.5%      R-sq(adj) = 48.5%

Analysis of Variance

SOURCE         DF           SS          MS          F          p
Regression      2       516.00      258.00       9.00      0.003
Error          15       430.00       28.67
Total          17       946.00
```

Since β_0 equals $E(y)$ for the Charlotte plant and $\beta_0 + \beta_1$ equals $E(y)$ for the Houston plant, the first difference above is equal to $(\beta_0 + \beta_1) - \beta_0 = \beta_1$. Moreover, since $\beta_0 + \beta_2$ equals $E(y)$ for the San Diego plant, the second difference above is equal to $(\beta_0 + \beta_2) - \beta_0 = \beta_2$. Hence, we would conclude that there is no difference in the three means if $\beta_1 = 0$ and $\beta_2 = 0$. Thus, the null hypothesis for a test for difference of means can be stated as

$$H_0: \beta_1 = \beta_2 = 0$$

Recall that to test this type of null hypothesis about the significance of the regression relationship, we must compare the value of MSR/MSE to the critical value from an F distribution with numerator and denominator degrees of freedom equal to the degrees of freedom for the regression sum of squares and the error sum of squares, respectively. In the current problem, the regression sum of squares has 2 degrees of freedom and the error sum of squares has 15 degrees of freedom. Thus, the values for MSR and MSE are

$$MSR = \frac{SSR}{2} = \frac{516}{2} = 258$$

$$MSE = \frac{SSE}{12} = \frac{430}{15} = 28.67$$

Hence, the computed F value is

$$F = \frac{MSR}{MSE} = \frac{258}{28.67} = 9.00$$

At the $\alpha = .05$ level of significance, the critical value of F with 2 numerator degrees of freedom and 15 denominator degrees of freedom is 3.29. Since the observed value of F is greater than the critical value of 3.29, we reject the null hypothesis $H_0: \beta_1 = \beta_2 = 0$. Hence, we conclude that the means for the three plants are different.

❑ ❑ Exercises

Methods

SELF TEST ▶ 23. Consider a completely randomized design involving four treatments: A, B, C, and D. Write a multiple regression equation which can be used to analyze these data. Define all variables.

24. Write a multiple regression equation which can be used to analyze the data for a randomized block design involving three treatments and two blocks. Define all variables.

25. Write a multiple regression equation which can be used to analyze the data for a two-factorial factorial design with two levels for factor A and three levels for factor B. Define all variables.

Applications

SELF TEST ▶ 26. The Jacobs Chemical Company wants to estimate the mean time (minutes) required to mix a batch of material on machines produced by three different manufacturers. To limit the cost of testing, four batches of material were mixed on machines produced by each of the three manufacturers. The times needed to mix the material were recorded and are as follows:

Manufacturer 1	Manufacturer 2	Manufacturer 3
20	28	20
26	26	19
24	31	23
22	27	22

a. Write a multiple regression equation that can be used to analyze the data.
b. What are the best estimates of the coefficients in your regression equation?
c. In terms of the regression equation coefficients, what hypotheses do we have to test to see if the mean time to mix a batch of material is the same for each manufacturer?
d. For the $\alpha = .05$ level of significance what conclusion should be drawn?

27. Four different paints are advertised as having the same drying time. To check the manufacturers' claims, five paint samples were tested for each brand of paint. The time in minutes until the paint was dry enough for a second coat to be applied was recorded for each sample. Table 16.15 gives the data.
a. Using $\alpha = .05$, test for any significant differences in mean drying times among the paints.
b. What is your estimate of mean drying time for paint 2? How is it obtained from the computer output?

TABLE 16.15

Paint 1	Paint 2	Paint 3	Paint 4
128	144	133	150
137	133	143	142
135	142	137	135
124	146	136	140
141	130	131	153

28. An automobile dealer conducted a test to determine if the time needed to complete a minor engine tuneup depends on whether a computerized engine analyzer or an electronic analyzer is used. Because tuneup time varies among compact, intermediate, and full-sized cars, the three types of cars were used as blocks in the experiment. The data (time in minutes) obtained are shown below:

		Car		
		Compact	Intermediate	Full-size
Analyzer	Computerized	50	55	63
	Electronic	42	44	46

Using $\alpha = .05$, test for any significant differences.

T A B L E **16.16**

		Size of Advertisement	
		Small	*Large*
	A	8 12	12 8
Design	B	22 14	26 30
	C	10 18	18 14

29. A mail-order catalog firm designed a factorial experiment to test the effect of the size of·a magazine advertisement and the advertisement design on the number of catalog requests received (1000s). Three advertising designs and two sizes of advertisements were considered. The data are shown in Table 16.16. Test for any significant effects due to type of design, size of advertisement, or interaction. Use $\alpha = .05$.

Summary

In this chapter we discussed several concepts used by model builders in identifying the best estimated regression equation. First, we introduced the concept of a general linear model to show how the methods discussed in Chapters 14 and 15 could be extended to handle curvilinear relationships and interaction effects. Then, we discussed how transformations involving the dependent variable could be used in model building to account for problems such as nonconstant variance in the error terms.

In applications of regression analysis to real problems, there are usually a large number of potential independent variables to consider. We presented a general approach, based on an F statistic for adding or deleting variables from a regression model. We then introduced a larger problem involving 25 observations and 8 independent variables. We saw that one issue encountered when solving larger problems is finding the best subset of the possible independent variables. To help in this regard, we discussed several variable-selection procedures, including stepwise regression, forward selection, backward elimination, and best-subsets regression.

In Section 16.5, we extended the applications of residual analysis to show the Durbin-Watson test for autocorrelation. The chapter concluded with a discussion of how multiple regression models could be developed to provide another approach for solving analysis of variance and experimental design problems.

Glossary

General linear model. A model of the form $y = \beta_0 + \beta_1 z_1 + \beta_2 z_2 + \cdots + \beta_p z_p + \epsilon$, where each of the independent variables z_j, $j = 1, 2, \ldots, p$, is a function of $x_1, x_2, \ldots, x_k$, the variables for which data have been collected.

Interaction The joint effect of two variables acting together.

Variable-selection procedures Computer-based methods for selecting a subset of the potential independent variables for a regression model.

Autocorrelation Correlation in the errors that arises when the error terms at successive points in time are related.

Serial correlation Same as autocorrelation.

Durbin-Watson test A test to determine whether first-order autocorrelation is present.

Key Formulas

General Linear Model

$$y = \beta_0 + \beta_1 z_1 + \beta_2 z_2 + \cdots + \beta_p z_p + \epsilon \tag{16.1}$$

General F Test for Adding or Deleting $p - q$ Variables

$$F = \frac{\dfrac{\text{SSE}(x_1, x_2, \ldots, x_q) - \text{SSE}(x_1, x_2, \ldots, x_q, x_{q+1}, \ldots, x_p)}{p - q}}{\dfrac{\text{SSE}(x_1, x_2, \ldots, x_q, x_{q+1}, \ldots, x_p)}{n - p - 1}} \tag{16.13}$$

Autocorrelated Error Terms

$$\epsilon_t = \rho \epsilon_{t-1} + z_t \tag{16.16}$$

Durbin-Watson Statistic

$$d = \frac{\sum\limits_{t=2}^{n} (e_t - e_{t-1})^2}{\sum\limits_{t=1}^{n} e_t^2} \tag{16.17}$$

❑ ❑ Supplementary Exercises

30. Refer to the Cravens data set presented in Table 16.9.

a. Develop a scatter diagram showing SALES as a function of TIME.

b. Does a linear relationship between SALES and TIME appear to be appropriate? Explain.

c. Develop a model that can be used to predict SALES using just TIME or some appropriate function of TIME.

31. A study reported in the *Journal of Accounting Research* (Vol. 2, No. 2 Autumn 1987) investigated the relationship between audit delay (AUDELAY), the length of time from a company's fiscal year-end to the date of the auditor's report, and variables that describe the client and the auditor. Some of the independent variables that were included in this study were

INDUS   A dummy variable which was coded as 1 if the firm was an industrial company or 0 if the firm was a bank, savings and loan, or insurance company.

PUBLIC   A dummy variable coded as 1 if the company was traded on an organized exchange or over the counter; otherwise coded 0.

ICQUAL   A measure of overall quality of internal controls, as judged by the auditor, using a five-point scale ranging from ''virtually none'' (1) to ''excellent'' (5).

INTFIN   A measure ranging from 1 to 4, as judged by the auditor, where 1 indicates ''all work performed subsequent to year-end'' and 4 indicates ''most work performed prior to year-end.''

Suppose that in a similar study a sample of 40 companies provided the following data:

AUDELAY	INDUS	PUBLIC	ICQUAL	INTFIN
62	0	0	3	1
45	0	1	3	3
54	0	0	2	2
71	0	1	1	2
91	0	0	1	1
62	0	0	4	4
61	0	0	3	2
69	0	1	5	2
80	0	0	1	1
52	0	0	5	3
47	0	0	3	2
65	0	1	2	3
60	0	0	1	3
81	1	0	1	2
73	1	0	2	2
89	1	0	2	1
71	1	0	5	4
76	1	0	2	2
68	1	0	1	2
68	1	0	5	2
86	1	0	2	2
76	1	1	3	1
67	1	0	2	3
57	1	0	4	2
55	1	1	3	2
54	1	0	5	2
69	1	0	3	3
82	1	0	5	1
94	1	0	1	1
74	1	1	5	2
75	1	1	4	3
69	1	0	2	2
71	1	0	4	4
79	1	0	5	2
80	1	0	1	4
91	1	0	4	1
92	1	0	1	4
46	1	1	4	3
72	1	0	5	2
85	1	0	5	1

 AUDIT

a. Develop the estimated regression equation using all of the independent variables.
b. Did the model developed in (a) provide a good fit? Explain.
c. Develop a scatter diagram which shows AUDELAY as a function of INTFIN. What does this scatter diagram indicate about the relationship between AUDELAY and INTFIN?
d. Based upon your observations regarding the relationship between AUDELAY and INTFIN, develop an alternative model to the one developed in (a) to explain as much of the variability in AUDELAY as possible.

32. The following data set, reported in *Louis Rukeyser's Business Almanac* (1988 Simon & Schuster, p. 47), shows the percentage of management jobs held by women in various companies and the percentage of women in each company.

Industry/Company	Management Jobs Held by Women (%)	Women Employees (%)
Industrial		
DuPont	7	22
Exxon	8	27
General Motors	6	19
Goodyear Tire and Rubber	25	39
Technology		
AT&T	32	48
General Electric	6	26
IBM	16	28
Xerox	23	38
Consumer products		
Johnson & Johnson	18	47
PepsiCo	28	46
Phillip Morris (excluding General Foods)	14	31
Proctor & Gamble	17	28
Retailing and trade		
Federated Department Stores	61	72
Kroger	16	47
Marriott	32	51
McDonald's	46	57
Sears, Roebuck	36	55
Media		
ABC (excluding Capital Cities)	36	43
Time	46	54
Times Mirror	27	37
Financial Services		
American Express	37	57
BankAmerica	64	72
Chemical Bank	34	57
Prudential Life Insurance	32	53
Wells Fargo Bank	58	71

 JOBS

In addition to the percentage of women employed in the company, create additional independent variables by using dummy variables to account for the type of industry.

a. Develop the best one-variable model which can be used to predict the percentage of management jobs held by women.

b. Use the stepwise procedure to develop the best model.

c. Use the backward-elimination procedure to develop the best model.

33. Refer to the data in Exercise 31. Consider a model in which only INDUS is used to predict AUDELAY. At the $\alpha = .01$ level of significance, test for any positive autocorrelation in the data.

34. Refer to the data in Exercise 31.

a. Develop an estimated regression equation which can be used to predict AUDELAY using INDUS and ICQUAL.

T A B L E 16.17

	Browser	
Nonbrowser	Light	Heavy
4	5	5
5	6	7
6	5	5
3	4	7
3	7	4
4	4	6
5	6	5
4	5	7

b. Plot the residuals obtained from the model developed in (a) as a function of the order in which the data are presented. Does there appear to be any autocorrelation present in the data? Explain.

c. At the .05 level of significance, test for any positive autocorrelation in the data.

35. Refer to the data in Exercise 32.

a. Develop an estimated regression equation which can be used to predict the percentage of management jobs held by women given the percentage of women employees in the company.

b. Plot the residuals obtained from the model developed in (a) as a function of the order in which the data are presented. Does there appear to be any autocorrelation present in the data? Explain.

c. At the .05 level of significance, test for any positive autocorrelation in the data.

36. A study was conducted to investigate the browsing activity by shoppers (*Journal of the Academy of Marketing Science,* Winter 1989). Shoppers were classified as nonbrowsers, light browsers, and heavy browsers. For each shopper in the study, a measure was obtained to determine how comfortable the shopper was in the store. Higher scores indicated greater comfort. Assume that the data in Table 16.17 is from this study. Use a .05 level of significance to test for differences in comfort levels among the three types of browsers.

Computer Case: *Unemployment Study*

Layoffs and unemployment have affected a substantial number of workers in recent years. A study reported in *Industrial and Labor Relations Review* (April 1988) collected data on variables that may be related to the number of weeks a manufacturing worker has been jobless. The dependent variable in the study (WEEKS) was defined as the number of weeks a worker has been jobless due to a layoff. The independent variables in the study were as follows:

AGE	The age of the worker.
EDUC	The number of years of education.
MARRIED	A dummy variable; 1 if married, 0 otherwise.
HEAD	A dummy variable; 1 if the head-of-household, 0 otherwise.
TENURE	The number of years on the old job.
MGT	A dummy variable; 1 if management occupation, 0 otherwise.
SALES	A dummy variable; 1 if sales occupation, 0 otherwise.

Assume the following data were collected for 50 displaced workers. These data are available on the data disk in the file named LAYOFFS.

 LAYOFFS

WEEKS	AGE	EDUC	MARRIED	HEAD	TENURE	MGT	SALES
37	30	14	1	1	1	0	0
62	27	14	1	0	6	0	0
49	32	10	0	1	11	0	0
73	44	11	1	0	2	0	0
8	21	14	1	1	2	0	0
15	26	13	1	0	7	1	0
52	26	15	1	0	6	0	0
72	33	13	0	1	6	0	0
11	27	12	1	1	8	0	0
13	33	12	0	1	2	0	0
39	20	11	1	0	1	0	0
59	35	7	1	1	6	0	0
39	36	17	0	1	9	1	0
44	26	12	1	1	8	0	0
56	36	15	0	1	8	0	0

—Table continues

WEEKS	AGE	EDUC	MARRIED	HEAD	TENURE	MGT	SALES
—Table continued from previous page							
31	38	16	1	1	11	0	1
62	34	13	0	1	13	0	0
25	27	19	1	0	8	0	0
72	44	13	1	0	22	0	0
65	45	15	1	1	6	0	0
44	28	17	0	1	3	0	1
49	25	10	1	1	1	0	0
80	31	15	1	0	12	0	0
7	23	15	1	0	2	0	0
14	24	13	1	1	7	0	0
94	62	13	0	1	8	0	0
48	31	16	1	0	11	0	0
82	48	18	0	1	30	0	0
50	35	18	1	1	5	0	0
37	33	14	0	1	6	0	1
62	46	15	0	1	6	0	0
37	35	8	0	1	6	0	0
40	32	9	1	1	13	0	0
16	40	17	1	0	8	1	0
34	23	12	1	1	1	0	0
4	36	16	0	1	8	0	1
55	33	12	1	0	10	0	1
39	32	16	0	1	11	0	0
80	62	15	1	0	16	0	1
19	29	14	1	1	12	0	0
98	45	12	1	0	17	0	0
30	38	15	0	1	6	0	1
22	40	8	1	1	16	0	1
57	42	13	1	0	2	1	0
64	45	16	1	1	22	0	0
22	39	11	1	1	4	0	0
27	27	15	1	0	10	0	1
20	42	14	1	1	6	1	0
30	31	10	1	1	8	0	0
23	33	13	1	1	8	0	0

Managerial Report

Use the methods presented in this and previous chapters to analyze this data set. Present a summary of your analysis, including key statistical results, conclusions, and recommendations in a managerial report. Include any technical material that you feel is appropriate (computer output, residual plots, etc.) in an appendix.

17

Index Numbers

Contents

U.S. Department of Labor
Bureau of Labor Statistics

WASHINGTON D.C.

The U.S. Department of Labor, through its Bureau of Labor Statistics, compiles and distributes a number of indexes and statistics which are indicators of business and economic activity in the United States. For instance, the Bureau compiles and publishes the Consumer Price Index, the Producer Price Index, and statistics on average hours and earnings of various groups of workers. Perhaps the most widely quoted index produced by the Bureau of Labor Statistics is the Consumer Price Index.

In March 1992, the Department of Labor reported that the Consumer Price Index increased by .5% over February. The increase was fueled by sharp increases for food, gasoline, and clothing. Grocery store prices rose .7%, fruit and vegetable prices increased substantially due to flooding in Texas and California, and energy prices advanced by .6%. The March increase was the largest in the last 17 months, and economists were divided over what it meant for inflation.

The March increase, if it continued for the entire year, would result in an inflation rate of 6.2%. But, most economists did not believe the trend would continue. Victor Zarnowitz, Professor Emeritus of Economics and Finance at the University of Chicago, was quoted in *The Wall Street Jour-*

nal as saying, "The problem now is slow growth and not really inflation at this point." Many other economists noted that the Producer Price Index had only increased by .2% in March and contended that this showed that future increases in the Consumer Price Index would be modest. Not everyone agreed, however. *The Wall Street Journal* also quoted Marilyn Schaja, an economist with Donaldson, Lufkin, and Jenrette as saying, "These data clearly reinforce that inflation will be 4% or higher this year, as opposed to the consensus estimate of 3% to 3.5%."

The Labor Department also reported, in March 1992, that average weekly earnings had increased to $362.21 a week from $360.13 a week in February. While this appeared to be a plus for workers, the Labor Department indicated that average weekly earnings had failed to keep pace with inflation and that, after adjusting for the price increases reflected in the Consumer Price Index, real wages actually declined by .1% in March.

In this chapter we will see how various indexes, such as the Consumer and Producer Price Indexes, are computed and how they should be interpreted. We shall also learn how actual wages are adjusted by a price index to obtain real wages and a measure of purchasing power.

E ach month the U.S. government publishes a variety of indexes that are designed to help individuals better understand current business and economic conditions. Perhaps the most widely known and cited of these indexes is the Consumer Price Index (CPI). As its name implies, the CPI is an indicator of what is happening to prices consumers are paying for items purchased. Specifically, the CPI measures changes in price over a period of time. With a given starting point or *base period* and its associated index of 100, the CPI can be used to compare current period consumer prices with those in the base period. For example, a CPI of 125 reflects the condition that consumer prices as a whole are running approximately 25% above the base period prices for the same items. Although relatively few individuals know exactly what this number means, they do know enough about the CPI to understand that an increase means higher prices.

The CPI is perhaps the most widely known index. However, many other governmental and private-sector indexes are available to help us measure and understand how economic conditions in one period compare with economic conditions in other periods. The purpose of this chapter is to describe the most widely used types of indexes. We will begin by constructing some index numbers to gain a better understanding of how indexes are computed.

17.1   Price Relatives

The simplest form of a price index shows how the current price per unit for a given item compares to a base period price per unit for the same item. For example, Table 17.1 shows the cost of 1 gallon of unleaded gasoline for the years 1982–1990.

TABLE 17.1
Unleaded Gasoline Costs

Year	Price per Gallon ($)
1982	1.30
1983	1.24
1984	1.21
1985	1.20
1986	.93
1987	.95
1988	.95
1989	1.02
1990	1.16

TABLE 17.2

Price Relatives for 1 Gallon of Unleaded Gasoline (1982–1990)

Year	Price Relative (Base 1982)
1982	(1.30/1.30)100 = 100
1983	(1.24/1.30)100 = 95
1984	(1.21/1.30)100 = 93
1985	(1.20/1.30)100 = 92
1986	(.93/1.30)100 = 72
1987	(.95/1.30)100 = 73
1988	(.95/1.30)100 = 73
1989	(1.02/1.30)100 = 78
1990	(1.16/1.30)100 = 89

To facilitate comparisons with other years, the actual cost-per-gallon figure can be converted to a *price relative,* which expresses the unit price in each period as a percentage of the unit price in a base period.

$$\text{Price relative in period } t = \frac{\text{Price in period } t}{\text{Base period price}}(100) \qquad (17.1)$$

For the gasoline prices in Table 17.1 and with 1982 as the base year, the price relatives for 1 gallon of unleaded gasoline in the years 1982–1990 can be calculated. These price relatives are shown in Table 17.2. Note how easily the price in any one year can be compared with the price in the base year by knowing the price relative. For example, the price relative of 93 in 1984 shows that the gasoline cost in 1984 was 7% below the 1982 base-year cost. Similarly, the 1986 price relative of 72 shows a 28% decrease in gasoline cost in 1986 over the 1982 base-year cost. Price relatives, such as the ones for unleaded gasoline, are extremely helpful in terms of understanding and interpreting changing economic and business conditions over time.

17.2   Aggregate Price Indexes

While price relatives can be used to identify price changes over time for individual items, we are often more interested in the general price change for a group of items taken as a whole. For example, if we want an index that measures the change in the overall cost of living over time, we will want the index to be based on the price changes for a variety of items, including food, housing, clothing, transportation, medical care, and so on. An *aggregate price index* is developed for the specific purpose of measuring the combined change of a group of items.

Consider the development of an aggregate price index for a group of items falling under the heading of normal automotive operating expenses. For purposes of this illustration, we limit the items included in the group to gasoline, oil, tire, and insurance expenses.

TABLE 17.3

Data for Automotive Operating Expense Index (in $)

Item	1982	1990
Gallon of gasoline	1.30	1.16
Quart of oil	1.50	2.10
Tires	80.00	130.00
Insurance policy	300.00	820.00

Table 17.3 provides the data for the four components of our automotive operating-expense index for the years 1982 and 1990. With 1982 as the base period, an aggregate price index for the four components will give us a measure of the change in normal automotive operating expenses over the 1982–1990 period.

An unweighted aggregate index can be developed by simply summing the unit prices in the year of interest (e.g., 1990) and dividing this sum by the sum of the unit prices in the base year (1982). Let

$$P_{it} = \text{unit price for item } i \text{ in period } t$$

$$P_{i0} = \text{unit price for item } i \text{ in the base period}$$

An unweighted aggregate price index in period t, denoted by I_t, is given by

$$I_t = \frac{\Sigma P_{it}}{\Sigma P_{i0}}(100) \tag{17.2}$$

where the sum is over all items in the group.

An unweighted aggregate index for normal automotive operating expenses in 1990 $(t = 1990)$ is given by

$$I_{1990} = \frac{1.16 + 2.10 + 130.00 + 820.00}{1.30 + 1.50 + 80.00 + 300.00}(100)$$

$$= \frac{953.26}{382.80}(100) = 249$$

From the unweighted aggregate price index, we might be tempted to conclude that the price of normal automotive operating expenses increased 149% over the period from 1982 to 1990. But, note that the unweighted aggregate approach to establishing a composite price index for automotive expenses is heavily influenced by the items with large per-unit prices. Consequently, items with relatively low unit prices, such as gasoline and oil, are dominated by the high-unit-price items such as tires and insurance. The unweighted aggregate index for automotive operating expenses tends to be too heavily influenced by price changes in tires and insurance.

The sensitivity of an unweighted index to one or more high-priced items prevents this form of aggregate index from being widely used. A weighted aggregate price index provides a better comparison when usage quantities differ.

The philosophy behind the *weighted aggregate index* is that each item in the group should be weighted according to its importance. In most cases, the *quantity* of usage provides the best measure of importance. Thus, one must obtain a measure of the quantity of usage for the various items in the group. Table 17.4 provides annual usage information for each item of automotive operating expense based on the typical operation of a midsize automobile for approximately 15,000 miles per year. The quantities listed show the expected annual usage for this type of driving situation.

Let Q_i = quantity for item i. The weighted aggregate price index in period t is given by

$$I_t = \frac{\Sigma P_{it}Q_i}{\Sigma P_{i0}Q_i}(100) \tag{17.3}$$

The above sums are over all items in the group. It is based on dividing total operating costs in 1990 by total operating costs in 1982.

Let $t = 1990$, and use the quantity weights in Table 17.4. We obtain a weighted aggregate price index for automotive operating expenses in 1990:

TABLE 17.4

Annual Usage Information for Automotive Operating Expense Index

Item	Quantity Weights*
Gallons of gasoline	1000
Quarts of oil	15
Tires	2
Insurance policy	1

*Based on 15,000 miles per year. Tire usage is based on a 30,000-mile tire life.

$$I_{1990} = \frac{1.16(1000) + 2.10(15) + 130.00(2) + 820.00(1)}{1.30(1000) + 1.50(15) + 80.00(2) + 300.00(1)}(100)$$

$$= \frac{2271.5}{1782.5}(100) = 127$$

From this weighted aggregate price index, we would conclude that the price of automotive operating expenses has increased 27% over the period from 1982 to 1990.

Most individuals will agree that compared with the unweighted aggregate index, the above weighted index provides a more accurate indication of the price change for automotive operating expenses over the 1982–1990 period. When the quantity of usage of gasoline is taken into account, it helps to offset the large increase in insurance costs. The weighted index shows a more moderate increase in automobile operating expenses than the unweighted index. In general, the weighted aggregate index with quantities of usage as weights is the preferred method for establishing a price index for a group of items.

In the weighted aggregate price index formula (17.3), note that the quantity term Q_i does not have a second subscript to indicate the time period. The reason for this is that the quantities Q_i are considered *fixed* and do not vary with time, as the prices do. The fixed weights or quantities are specified by the designer of the index at levels believed to be representative of typical usage. Once established, they are held constant or fixed for all periods of time the index is in use. Indexes for years other than 1990 require the gathering of new price data P_{it}, but the weighting quantities Q_i remain the same.

A special case of the fixed-weight aggregate index is when the quantities are determined from base-year usages. In this case, we write $Q_i = Q_{i0}$, with the 0 subscript indicating base-year quantity weights. In this case, (17.3) would become

$$I_t = \frac{\Sigma P_{it}Q_{i0}}{\Sigma P_{i0}Q_{i0}}(100) \tag{17.4}$$

Whenever the fixed quantity weights are determined from base-year usage, the weighted aggregate index is given the name *Laspeyres index*.

Another option exists for determining quantity weights. This option differs from the Laspeyres index in that the quantities are revised each period. A quantity Q_{it} has to be determined for each year that the index is computed. The weighted aggregate index in period t with these quantity weights is given by

$$I_t = \frac{\Sigma P_{it}Q_{it}}{\Sigma P_{i0}Q_{it}}(100) \tag{17.5}$$

Note that the same quantity weights are used for the base period (period 0) and for period t. However, the weights are based on usage in period t, not the base period. This weighted aggregate index is referred to as the *Paasche index*. It has the advantage of being based on current usage patterns. While use of current quantity weights has some appeal, this method of computing a weighted aggregate index has two disadvantages: The normal usage quantities Q_{it} must be redetermined each year, thus adding to the time and cost of data collection, and each year the index numbers for previous years must be recomputed to reflect the effect of the new quantity weights. Because of these disadvantages, the Laspeyres index is more widely used. The automobile operation-expense index was computed using base-period quantities; thus, it was a Laspeyres index. Had usage figures for 1990 been used, we would have had a Paasche index. Indeed, because of more fuel efficient cars, gasoline usage has decreased, and a Paasche index would differ.

☐ ☐ **Exercises**

Methods

SELF TEST ▶

1. Shown below are prices and usage quantities for two items in 1982 and 1991.

	Quantity		Unit Price ($)	
Item	1982	1991	1982	1991
A	1500	1800	7.50	7.75
B	2	1	630.00	1500.00

a. Compute price relatives for each item in 1991 using 1982 as the base period.
b. Compute an unweighted aggregate price index for the two items in 1991 using 1982 as the base period.
c. Compute a weighted aggregate price index for the two items using the Laspeyres method.
d. Compute a weighted aggregate price index for the two items using the Paasche method.

2. An item with a price relative of 132 cost $10.75 in 1991. Its base year was 1986.
a. What was the percentage increase or decrease in cost of the item over the 5-year period?
b. What did the item cost in 1986?

Applications

SELF TEST ▶

3. A large manufacturer purchases an identical component from three independent suppliers, each of which differs in terms of unit price and quantity supplied. The relevant data for 1989 and 1991 are shown in Table 17.5.

a. Compute the price relatives for each of the component suppliers separately. Compare the price increases by the various suppliers over the 2-year period.
b. Compute an unweighted aggregate price index for the component part in 1991.
c. Compute a 1991 weighted aggregate price index for the component part. What is the interpretation of this index for the manufacturing firm?

T A B L E **17.5**

	Quantity	Unit Price ($)	
Supplier	**(1989)**	1989	1991
A	150	5.45	6.00
B	200	5.60	5.95
C	120	5.50	6.20

4. R&B Beverages, Inc., provides a complete line of beer, wine, and soft-drink products for distribution through retail outlets in central Iowa. Unit-price data and quantities sold in cases are shown below for 1990 and 1991:

	Quantity (1990)	Unit Price ($)	
Item	**(cases)**	1990	1991
Beer	35,000	12.00	13.00
Wine	5,000	47.00	50.00
Soft Drink	60,000	7.85	8.00

Compute a weighted aggregate index for the R&B Beverage sales in 1991, with 1990 as the base period.

5. Under the LIFO inventory valuation method, a price index for inventory must be established for tax purposes. The quantity weights are based on year-ending inventory levels. Use the beginning-of-the-year price per unit as the base-period price and develop a weighted aggregate index for the total inventory value at the end of the year. What type of weighted aggregate price index must be developed for the LIFO inventory valuation?

Product	Ending Inventory	Unit Price ($)	
		Beginning	Ending
A	500	.15	.19
B	50	1.60	1.80
C	100	4.50	4.20
D	40	12.00	13.20

17.3   Computing an Aggregate Index from Price Relatives

In Section 17.1 we defined the concept of a price relative and showed how a price relative could be computed with knowledge of the current-period unit price and the base-period unit price. We now want to show how aggregate price indexes like the ones developed in Section 17.2 can be computed directly from information about the price relative of each item in the group. Because of the limited use of unweighted indexes, we restrict our attention to weighted aggregate price indexes. Let us return to the automotive operating-expense index of the preceding section. The necessary information for the four items is shown in Table 17.6.

Let w_i be the weight applied to the price relative for item i. The general expression for a weighted average of price relatives is given by

$$I_t = \frac{\sum \frac{P_{it}}{P_{i0}} w_i}{\sum w_i}(100) \tag{17.6}$$

The proper choice of weights in (17.6) will allow us to compute a weighted aggregate price index from the price relatives. The proper choice of weights is given by multiplying the base-period price by the quantity of usage:

$$w_i = P_{i0}Q_i \tag{17.7}$$

Substitution of the value for w_i shown in (17.7) into (17.6) provides the following expression for a weighted price relatives index:

$$I_t = \frac{\sum \frac{P_{it}}{P_{i0}}(P_{i0}Q_i)(100)}{\sum P_{i0}Q_i} \tag{17.8}$$

TABLE 17.6
Price Relatives for Automotive Operating-Expense Index

Item	Unit Price ($)		Price Relative $(P_t/P_0)100$	Annual Usage
	1982 (P_0)	1990 (P_t)		
Gallons of gasoline	1.30	1.16	89	1,000
Quarts of oil	1.50	2.10	140	15
Tires	80.00	130.00	163	2
Insurance policy	300.00	820.00	273	1

TABLE 17.7

Automotive Operating-Expense Index (1982–1990) Based on Weighted Price Relatives

Item	Price Relatives	Base Price ($)	Quantity Q_i	Weight $w_i = P_{i0}Q_i$	Weighted Price Relatives $(P_{it}/P_{i0})(100)w_i$
Gasoline	89	1.30	1000	1300.00	115,700
Oil	140	1.50	15	22.50	3,150
Tires	163	80.00	2	160.00	26,080
Insurance	273	300.00	1	300.00	81,900
			Totals	1782.50	226,830

$$I_{1990} = \frac{226,830}{1782.50} = 127$$

With the canceling of the P_{i0} terms in the numerator, the weighted price relatives index becomes

$$I_t = \frac{\Sigma P_{it}Q_i}{\Sigma P_{i0}Q_i}(100)$$

Thus, we see that the weighted price relatives index with $w_i = P_{i0}Q_i$ provides a price index identical to the weighted aggregate index presented in Section 17.2. (See Equation (17.3).) A choice of base-period quantities (i.e., $Q_i = Q_{i0}$) in (17.7) leads to a Laspeyres index. A choice of current-period quantities (i.e., $Q_i = Q_{it}$) in (17.7) leads to a Paasche index.

Let us return to the automotive operating-expense data in Tables 17.3 and 17.4. We can use the price relatives in Table 17.6 and Equation (17.6) to compute a weighted average of price relatives. The results obtained using the weights specified by (17.7) are presented in Table 17.7. The index number of 127 shows a 27% increase in automotive operating expenses. Except for rounding differences, this is the same as the increase identified by the weighted aggregate index computation in Section 17.2.

☐ ☐ Exercises

Methods

SELF TEST ▶

6. Shown in Table 17.8 are price relatives for three items along with base-period prices and usage. Compute a weighted aggregate price index for the current period.

Applications

SELF TEST ▶

7. The Mitchell Chemical Company produces a special industrial chemical that is a blend of three chemical ingredients. The beginning-year cost per pound, the ending-year cost per pound, and the blend proportions are shown:

TABLE 17.8

Item	Price Relative	Base Period Price	Base Period Usage
A	150	22.00	20
B	90	5.00	50
C	120	14.00	40

Ingredient	Cost per Pound ($) Beginning	Cost per Pound ($) Ending	Quantity (pounds) per 100 Pounds of Product
A	2.50	3.95	25
B	8.75	9.90	15
C	.99	.95	60

a. Compute the price relatives for the three ingredients.

b. Compute a weighted average of the price relatives in order to develop a 1-year price index for the product. What is your interpretation of this index value?

8. An investment portfolio consists of four stocks. The purchase price, current price, and number of shares are shown:

Stock	Purchase Price/Share ($)	Current Price/Share ($)	Number of Shares
Holiday Trans	15.50	17.00	500
NY Electric	18.50	20.25	200
KY Gas	26.75	26.00	500
PQ Soaps	42.25	45.50	300

Construct a weighted average of price relatives as an index of the performance of the portfolio to date. Interpret this price index.

9. Compute the price relatives for the R&B Beverages products in Exercise 4. Use a weighted average of price relatives to show that this method provides the same index as the weighted aggregate method.

17.4   Some Important Price Indexes

We have identified the procedures used to compute price indexes for single items or groups of items. Now let us consider some price indexes that are important measures of business and economic conditions. Specifically, we will consider the Consumer Price Index, the Producer Price Index, and the Dow Jones averages.

Consumer Price Index

Perhaps the most widely known and cited measure of change in general economic conditions is the *Consumer Price Index* (CPI). This index, published monthly by the U.S. Bureau of Labor Statistics, is the primary measure of the cost of living in the United States. The group of items used to develop the index consists of a *market basket* of 400 items including food, housing, clothing, transportation, and medical items. The CPI is a weighted aggregate index with fixed weights.* The weight applied to each item in the market basket derives from a usage survey of urban families throughout the United States.

In March 1992, the CPI, computed with a (1982–1984) base index of 100, was 139.3. This means that the cost of purchasing the market basket of goods and services had increased 39.3% since the base period 1982–1984. The 30-year time series of the CPI from 1950 to 1990 is shown in Figure 17.1. Note how the CPI measure reflects the sharp inflationary behavior of the economy in the late 1970s and early 1980s.

---

*There are actually two Consumer Price Indexes. The Bureau of Labor Statistics publishes a Consumer Price Index for all urban consumers (CPI-U) and a revised Consumer Price Index for urban wage earners and clerical workers (CPI-W). The CPI-U is the one most widely quoted, and it is published regularly in *The Wall Street Journal*.

FIGURE 17.1
**Consumer Price Index, 1950–1990
with Base 1982–1984 = 100**

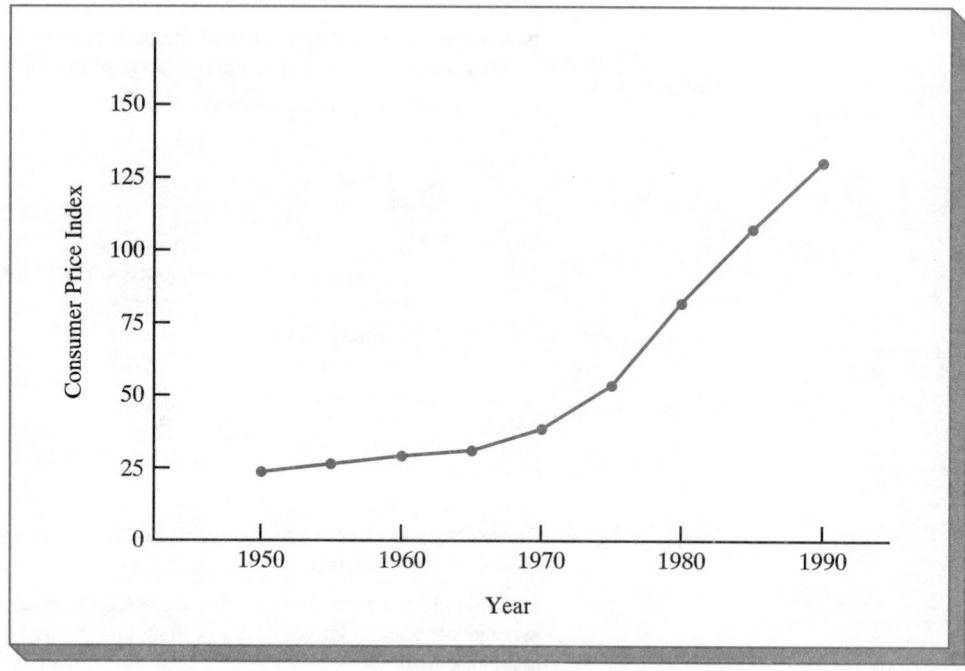

Producer Price Index

The *Producer Price Index* (PPI), also published monthly by the U.S. Bureau of Labor Statistics, measures the monthly changes in prices in primary markets in the United States. This index replaces the old Wholesale Price Index. The PPI is based on prices for the first transaction of each product in nonretail markets. All commodities sold in commercial transactions in these markets are represented, including those imported for sale. The survey includes raw, manufactured, and processed goods at each level of processing and includes the output of industries classified as manufacturing, agriculture, forestry, fishing, mining, gas and electricity, and public utilities. One of the common uses of this index is as a leading indicator of the future trend of consumer prices and the cost of living. An increase in the PPI reflects producer price increases that will eventually be passed on to the consumer through higher retail prices.

Weights for the various items in the PPI are based on the value of shipments, and it is a weighted average of price relatives calculated by the Laspeyres method. In March 1992, the PPI, computed with a 1982 base index of 100, was 122.0.

Dow Jones Averages

The *Dow Jones averages* are indexes which are designed to show price trends and movements on the New York Stock Exchange. The best known of the Dow Jones indexes is the Dow Jones Industrial Average (DJIA), which is based on common stock prices of 30 industrial stocks. It is a weighted average of these stock prices, with the weights revised from time to time to adjust for stock splits and switching of companies in the index. Unlike the other price indexes that we have studied, it is not expressed as a percentage of base-year prices. The specific firms used in April 1992 to compute the DJIA are shown in Table 17.9.

TABLE 17.9

The 30 Industrial Companies Used in the Dow Jones Industrial Average Price Index (April 1992)

Allied Signal	DuPont	Minnesota MnMfg.
Alcoa	Eastman Kodak	J. P. Morgan
American Express	Exxon	Philip Morris
A.T.&T.	General Electric	Procter & Gamble
Bethlehem Steel	General Motors	Sears
Boeing	Goodyear	Texaco
Caterpillar	IBM	Union Carbide
Chevron	International Paper	United Technologies
Coca Cola	McDonalds	Westinghouse
Disney	Merck	Woolworth

Other Dow Jones averages are computed for 20 transportation stocks and for 15 utility stocks. The Dow Jones averages are computed and published daily in *The Wall Street Journal* and other financial publications.

17.5   Deflating a Series by Price Indexes

Many business and economic series reported over time, such as company sales, industry sales, and inventories, are measured in dollar amounts. These time series often show an increasing growth pattern over time, which is generally interpreted as showing an increase in the physical volume associated with these activities. For example, a total dollar amount of inventory up by 10% is generally interpreted to mean that the physical inventory is 10% larger. Such interpretations can be very misleading whenever a time series is measured in terms of dollars, since the total dollar amount is a combination of both price and quantity changes. Thus, in periods where price changes are significant, the changes in the dollar amounts may not be indicative of quantity changes unless we are able to adjust the time series to eliminate the price change effect.

For example, from 1976 to 1980, the total amount of spending in the construction industry increased approximately 75%. At first glance this figure suggests an excellent growth in construction activity. However, during this period of time, construction prices were increasing just as fast as—or sometimes even faster than—this 75% rate. In fact, while total construction spending was increasing, construction activity was staying relatively constant or, as in the case of new housing starts, showing a decrease. Thus, for us to correctly interpret construction spending activity over the 1976–1980 period, it is necessary to adjust the total spending series by a price index to remove the price-increase effect from the time series. Whenever we remove the price-increase effect from the time series, we say we are *deflating the series*.

In the area of personal income and wages, we often hear discussions concerning issues such as "real wages" or the "purchasing power" of wages. These concepts are based on the notion of deflating an hourly wage index. For example, Figure 17.2 shows the pattern of hourly wages of factory workers for the period 1976–1980. At first glance, we see the sharply increasing trend in wages, with excellent growth in wages from $4.90 per hour to $7.00 per hour. Should the factory workers be pleased with this growth in hourly wages? Perhaps yes; but on the other hand, if the cost of living has increased just as fast, maybe the answer is no. If we can compare the purchasing power of the $7.00 hourly wage in 1980 with the purchasing power of the $4.90 hourly wage in 1976, we will have a better idea of the relative improvement in wages.

Table 17.10 shows both the hourly wage rate and the CPI for the period 1976–1980. Note that we are using 1976 here as the base for the CPI. With these data we will show

TABLE 17.10

Hourly Wages of Factory Workers and Consumer Price Index: 1976–1980

Year	Hourly Wage ($)	CPI (1976 Base)
1976	4.90	100
1977	5.40	105
1978	5.85	113
1979	6.40	122
1980	7.00	138

FIGURE 17.2
Hourly Wages of Factory Workers

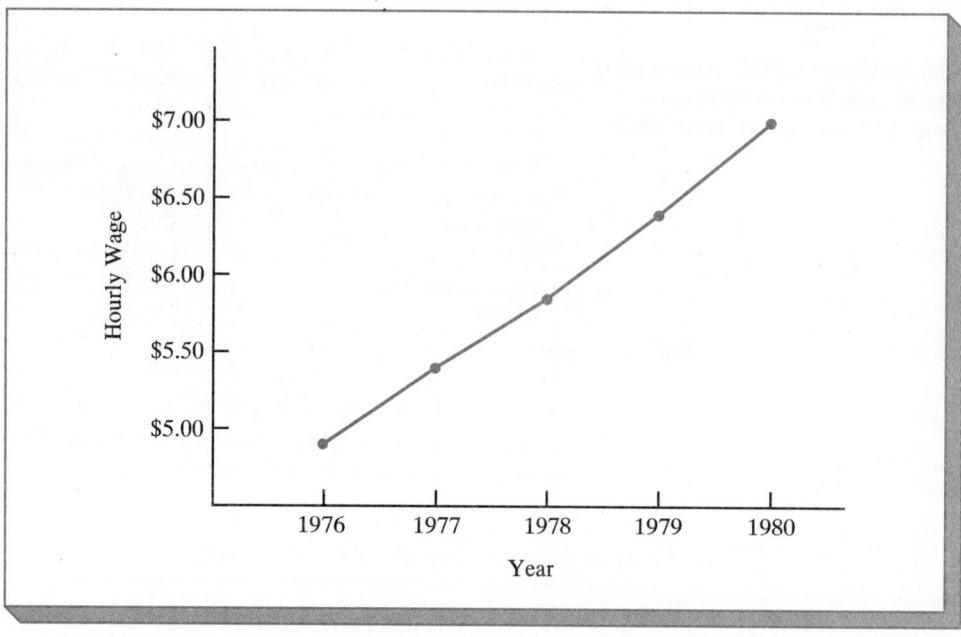

FIGURE 17.2
Hourly Wages of Factory Workers

TABLE 17.11

Deflated Series of Hourly Wages for Factory Workers

Year	Deflated Hourly Wage
1976	($4.90/100)(100) = $4.90
1977	($5.40/105)(100) = $5.14
1978	($5.85/113)(100) = $5.18
1979	($6.40/122)(100) = $5.25
1980	($7.00/138)(100) = $5.07

how the CPI can be used to deflate the index of hourly wages. In effect, we shall be removing the consumer price increases from the hourly wage index in an attempt to measure the change in purchasing power of the wages. Thus, we will be better able to determine what has happened to real wages over the time span.

The calculations used to deflate the hourly wage index are not difficult. The deflated series is found by dividing the hourly wage rate in each year by the corresponding value of the CPI and multiplying by 100. The deflated hourly wage index for factory workers is shown in Table 17.11. A graph showing both the actual wage rates and the deflated or real wages is shown in Figure 17.3.

What does the deflated series of wages tell us about the real wages or purchasing power of workers during the 1976–1980 period? In terms of 1976 dollars, the hourly wage rate has risen from $4.90 to $5.07, or approximately 3.5%. Thus, after we remove the price-increase effect we see that over this period, factory workers have done little more than keep even with inflation. From 1979 to 1980, even with wage increases, the factory workers lost in terms of real wages. Thus, we see that the advantage of using price indexes to deflate a series is that we have a clearer picture of the real dollar changes that are occurring.

This process of deflating a series measured over time has an important application in the computation of the Gross National Product (GNP). The GNP is the total value of all goods and services produced in a given country. Obviously, over time the GNP will show gains which are in part due to price increases if the GNP is not deflated by a price index. Thus, to adjust the total value of goods and services to reflect actual changes in the volume of goods and services produced and sold, the GNP must be computed with a price index deflator. The process is similar to that previously discussed in the real wages computation.

FIGURE 17.3
Actual and 1976 Constant-Dollar Real Wages of Factory Workers

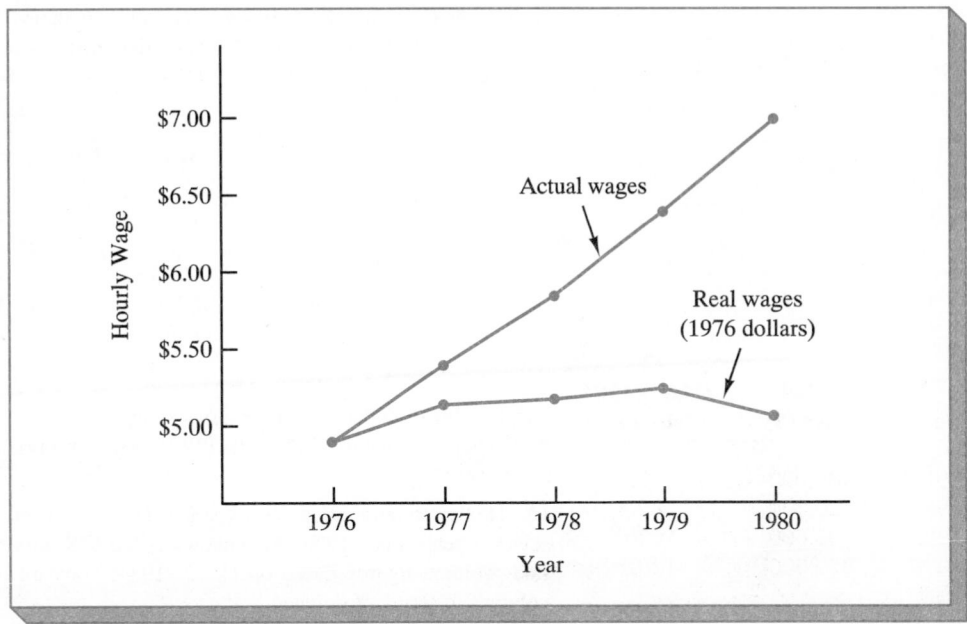

Exercises

Methods

SELF TEST ▶

10. Average hourly wages for factory workers in 1980 were $7.00; in 1990, they were $10.83. Using 1980 as a base, the CPI in 1990 was 159.

a. Use 1980 as a base, and deflate the hourly wage rate in 1990 to find the real wage rate.

b. What is the actual percentage change in hourly wages from 1980 to 1990?

c. What is the percentage change in real wages from 1980 to 1990?

Applications

11. The U.S. Department of Commerce reported total personal income for the 5 years from 1976 to 1980 as shown in Table 17.12. Use the Consumer Price Index information in Table 17.10 to deflate the personal income series. What has been the percent increase in real personal income from 1978 to 1980?

12. The U.S. Department of Commerce reported total inventories of all manufacturers for the 5 years from 1976–1980 as follows:

TABLE 17.12

Year	Total Personal Income (billions of dollars)
1976	1300
1977	1440
1978	1605
1979	1800
1980	2050

Year	Total Inventories (billions of dollars)
1976	155
1977	163
1978	178
1979	198
1980	227

a. Use the Consumer Price Index information in Table 17.10 to deflate this series, and comment on the pattern of manufacturers' inventories in terms of constant dollars.

b. The Producer Price Index for 1976–1980 is given below, with 1976 as a base year:

Year	PPI (1976 base)
1976	100
1977	103
1978	110
1979	120
1980	135

Use the PPI to deflate the inventories series.

c. Do you feel that the CPI or the PPI is most appropriate to use as a deflator for inventory values? Discuss.

13. Dooley Retail Outlets has experienced the total retail sales volumes shown in Table 17.13 for selected years since 1975. Also shown is the CPI with the index base of 1982–1984. Deflate the sales volume figures based on (1982–1984) constant dollars, and comment on the firm's sales volumes in terms of deflated dollars.

TABLE **17.13**

Year	Retail Sales ($)	CPI (1982–1984 base)
1975	380,000	53.8
1980	520,000	82.4
1985	700,000	107.6
1990	870,000	130.7

17.6   Price Indexes: Other Considerations

In the previous sections we described several methods used to compute price indexes, discussed the use of some important indexes, and presented a procedure for using price indexes to deflate a time series. There are several other issues that must be considered to enhance our understanding of how price indexes are constructed and how they are used. Some of these considerations are discussed in this section.

Selection of Items

The primary purpose of a price index is to measure the price change over time for a specified class of items, products, and so on. Whenever the class of items is very large, it is clear that the index cannot be based on all items in the class. Rather a sample of representative items must be used. By collecting price and quantity information for the sampled items, we hope to obtain a good idea of the price behavior of all items that the index is representing. For example, in the Consumer Price Index the total number of items that might be considered in the population of normal purchase items for a consumer could run as high as 2000 or more. However, the index is based on the price-quantity characteristics of just 400 items. The selection of the specific items in the index is not a trivial question. Surveys of user purchase patterns as well as good judgment go into the selection process. A simple random sample of all potential items for the index is not used to select the 400 items.

In addition to the initial selection process, the group of items in the index must be periodically reviewed and revised whenever actual purchase patterns change. Thus, the issue of which items to include in an index is a key question that must be resolved before an index can be developed or revised.

Selection of a Base Period

Many indexes are established with a base-period value of 100 at some specific time. All future values of the index are then related to the base period. But what base period is appropriate for an index? This is not an easy question, and the answer must be based on the judgment of the developer of the index.

Many of the indexes established by the United States government as of 1992 use a 1982 base period. As a general guideline it is believed that the base period should not be too far from the current period. For example, a Consumer Price Index that used a 1945 base period would be difficult for most individuals to relate to because of unfamiliarity with conditions in 1945. Thus, the base period for most indexes is adjusted periodically to a more recent period of time. For the CPI, the base period was moved from 1967 to the 1982–1984 average in 1988.

Quality Changes

The purpose of a price index is to attempt to measure changes in prices over time. Ideally, price data are collected for the same set of items at several times, and then the index is computed. A basic assumption is that the prices are identified for the same items each period. A problem is encountered when a product changes in quality from one period to the next. For example, a manufacturer may alter the quality of a product by using less expensive materials, fewer features, and so on, from year to year. While the price may go up in following years, the price is for a lower quality product. In some instances, this means that the price has actually gone up more than is represented by the list price for the item. However, it is very difficult, if not impossible, to adjust the index for decreases in the quality of an item.

On the other hand, a substantial quality improvement may cause an increase in price for basically the same product. Thus, a portion of the price related to the quality improvement should be excluded from the index computation. Again, however, it is extremely difficult, if not impossible, to adjust the index for the price increase that is related to the higher quality factor.

While quality changes cause some concern in the development of price indexes, common practice is simply to ignore minor quality changes in developing a price index. Major quality changes must be addressed because they can alter the product description from period to period. If a product description is changed, the index must be modified to account for it. For example, the product could be deleted from the index.

17.7   Quantity Indexes

Although the previous sections have emphasized the important area of price indexes, there are other types of indexes. In particular, one other use of index numbers is to measure changes in quantity levels over time. This type of index is called a *quantity index*.

Recall that in the development of the weighted aggregate price index in Section 17.2 to compute an index number for period *t*, we required data on unit prices at a base period (P_0) and period t (P_t). The formula for a weighted aggregate price index is restated below:

$$I_t = \frac{\Sigma P_{it} Q_i}{\Sigma P_{i0} Q_i} (100) \tag{17.3}$$

The numerator, $\Sigma P_{it} Q_i$, represents the total value of fixed quantities of the index items in period t. The denominator, $\Sigma P_{i0} Q_i$, represents the total value of the same fixed quantities of the index items in year 0.

The weighted aggregate quantity index is computed in a fashion quite similar to that for a weighted aggregate price index. Quantities for each item are measured in the base period and period t, with Q_{i0} and Q_{it}, respectively, representing these quantities for item i. The quantities are then weighted by a fixed price, the value added, and so on. Note that the "value added" to a product is the sales value minus the cost of purchased inputs. The formula for computing a weighted aggregate quantity index for period t is

$$I_t = \frac{\Sigma Q_{it} w_i}{\Sigma Q_{i0} w_i}(100) \tag{17.9}$$

In some quantity indexes the weight for item i is taken to be the base-period price (P_{i0}), in which case the weighted aggregate quantity index becomes

$$I_t = \frac{\Sigma Q_{it} P_{i0}}{\Sigma Q_{i0} P_{i0}}(100) \tag{17.10}$$

Quantity indexes can also be computed on the basis of weighted quantity relatives. One formula for this version of a quantity index is as follows:

$$I_t = \frac{\sum \dfrac{Q_{it}}{Q_{i0}}(Q_{i0}P_i)(100)}{\Sigma Q_{i0}P_i} \tag{17.11}$$

This formula is the quantity version of the weighted price relatives as developed in Section 17.3 (see (17.8)).

The *Index of Industrial Production,* developed by the Federal Reserve Board, is probably the best known quantity index. This index is reported monthly and has a base period of 1987. It is designed to measure changes in volume of production levels for a variety of manufacturing classifications in addition to mining and utilities. In November 1991, this index was 107.3.

❑ ❑ Exercises

Methods

SELF TEST ▶ **14.** Data concerning quantities of three items sold in 1987 and 1992 are shown in Table 17.14 along with the sales prices of the items in 1987. Compute a weighted aggregate quantity index for 1992.

TABLE **17.14**

Item	Quantity Sold 1987	Quantity Sold 1992	Price/Unit 1987 ($)
A	350	300	18.00
B	220	400	4.90
C	730	850	15.00

Applications

SELF TEST ▶ **15.** A trucking firm handles four commodities for a particular distributor. Total shipments for the commodities in 1987 and 1991, as well as the 1987 prices, are shown in the table at the top of page 665.

T A B L E **17.15**

	Sales		Mean Price per Sale
Model	1989	1991	**(1989)**
Sedan	200	170	$15,200
Sport	100	80	$17,000
Wagon	75	60	$16,800

	Shipments		Price/Shipment
Commodity	1987	1991	**1987**
A	120	95	$1200
B	86	75	$1800
C	35	50	$2000
D	60	70	$1500

Develop a weighted aggregate quantity index with a 1987 base. Use base-period prices for weights. Comment on the growth or decline in quantities over this period.

16. An automobile dealer reports the 1989 and 1991 sales for three models in Table 17.15. Develop a weighted aggregate quantity index using the 2 years of data.

Summary

Price and quantity indexes are important measures of changes in price and quantity levels within the business and economic environment. Price relatives are simply the ratio of the current unit price of an item to a base-period unit price, with a value of 100 indicating no difference in the current- and base-period prices. Aggregate price indexes are created as a composite measure of the overall change in prices for a given group of items or products. Usually the items in an aggregate price index are weighted by their quantity of usage. A weighted aggregate price index can also be computed by weighting the price relatives for the items in the index.

The Consumer Price Index and the Producer Price Index are both widely quoted indexes using (1982–1984) and 1982, respectively, as base years. The Dow Jones Industrial Average is another widely quoted price index. It is a weighted average of the prices of 30 common stocks listed on the New York Stock Exchange. Unlike many other indexes, it is not stated as a percentage of some base-period value.

Often price indexes are used to deflate some other economic series reported over time. We saw how the CPI could be used to deflate hourly wages to obtain an index of real wages. Issues such as selecting the items to be included in the index, selecting a base period for the index, and adjusting for changes in quality are important additional considerations in the development of an index number. Quantity indexes were briefly discussed, and the Index of Industrial Production was mentioned as an important quantity index.

Glossary

Price relative A price index for a given item which is computed by dividing a current unit price by a base-period unit price and multiplying the result by 100.

Aggregate price index A composite price index based on the prices of a group of items.

Weighted aggregate price index A composite price index where the prices of the items in the composite are weighted by their relative importance.

Laspeyres index A weighted aggregate price index where the weight for each item is its base-period quantity.

Paasche index A weighted aggregate price index where the weight for each item is its current-period quantity.

Consumer Price Index A monthly price index that uses the price changes in a market basket of consumer goods and services to measure the changes in consumer prices over time.

Producer Price Index A monthly price index that is designed to measure changes in prices for goods sold in primary markets (i.e., first purchase of a commodity in nonretail markets).

Dow Jones averages Aggregate price indexes reflecting the prices of stocks listed on the New York Stock Exchange.

Quantity index An index that is designed to measure changes in quantities over time.

Index of Industrial Production A quantity index which is designed to measure changes in the physical volume or production levels of industrial goods over time.

Key Formulas

Price Relative in Period t

$$\frac{\text{Price in period } t}{\text{Base period price}}(100) \tag{17.1}$$

Unweighted Aggregate Price Index in Period t

$$I_t = \frac{\Sigma P_{it}}{\Sigma P_{i0}}(100) \tag{17.2}$$

Weighted Aggregate Price Index in Period t

$$I_t = \frac{\Sigma P_{it}Q_i}{\Sigma P_{i0}Q_i}(100) \tag{17.3}$$

Weighted Average of Price Relatives

$$I_t = \frac{\sum \dfrac{P_{it}}{P_{i0}} w_i}{\Sigma w_i}(100) \tag{17.6}$$

Weighting Factor for (17.6)

$$w_i = P_{i0}Q_i \tag{17.7}$$

Weighted Aggregate Quantity Index

$$I_t = \frac{\Sigma Q_{it}w_i}{\Sigma Q_{i0}w_i}(100) \tag{17.9}$$

❑ ❑ Supplementary Exercises

17. The median purchase price for existing single-family houses in Atlanta, Georgia, for the years 1982–1987 is given in Table 17.16 (*Statistical Abstract of the United States,* 1989).

a. Use 1982 as the base year, and develop a price index for existing single-family homes in Atlanta over this 5-year period.

TABLE 17.16

Year	Price ($1000s)	Year	Price ($1000s)
1982	96.0	1985	106.7
1983	97.5	1986	119.8
1984	98.3	1987	131.1

b. Use 1985 as the base year, and develop a price index for existing single-family homes in Atlanta over this 5-year period.

18. Nickerson Manufacturing Company shows the following data on units shipped and quantities shipped for each of its four products:

Products	Base-Period Quantities (1987)	Mean Shipping Cost per Unit ($)	
		1987	1991
A	2000	10.50	15.90
B	5000	16.25	32.00
C	6500	12.20	17.40
D	2500	20.00	35.50

a. Compute the price relative for each product.
b. Compute a weighted aggregate price index that reflects the shipping cost change over the 4-year period.

19. Use the price data in Exercise 18 to compute a Paasche index for the shipping cost if 1991 quantities are 4000, 3000, 7500, and 3000 for each of the four products.

20. Boran Stockbrokers, Inc., selects four stocks for the purpose of developing its own index of stock market behavior. Costs per share for a 1988 base period, January 1990, and March 1990 are shown below. Base-year quantities have been set based on historical volumes for the four stocks.

Stock	Industry	1988 Quantity	Cost per Share ($)		
			1988 Base	January 1990	March 1990
A	Oil	100	31.50	32.75	32.50
B	Computer	150	65.00	59.00	57.50
C	Steel	75	40.00	42.00	39.50
D	Real Estate	50	18.00	16.50	13.75

Use the 1988 base period to compute the Boran index for January 1990 and March 1990. Comment on what the index tells you about what is happening in the stock market.

21. Compute the price relatives for the four stocks making up the Boran index in Exercise 20. Use the weighted aggregates of price relatives to compute the January 1990 and March 1990 Boran indexes.

22. Consider the price relatives and quantity information for the following grain production in Iowa:

Product	1985 Quantities (millions of bushels)	Base Price per Unit ($)	1985–1987 Price Relatives
Corn	1707	2.02	79
Soybeans	310	4.99	107

What is the weighted aggregates price index for the Iowa grains?

23. Dairy product price and quantity data for the years 1970 and 1980 are shown below. Quantities are based on estimated annual usage for a family of two adults and two children.

Product	1970 Quantities	1970 Price ($)	1980 Price ($)
Milk (gallons)	125	.79	2.09
Eggs (dozens)	50	.49	.95
Butter (pounds)	50	.60	.91
Cheese (pounds)	25	.79	1.49

a. Compute the price relative for each product.

b. Compute a weighted aggregate price index for dairy products.

24. Starting faculty salaries (9-month basis) for assistant professors of business administration at a major Midwestern university are shown in Table 17.17. Use the CPI to deflate the salary data to constant dollars. Comment on the trend in salaries in higher education as indicated by these data.

25. A particular stock shows the following 5-year historical price per share (also shown is the Consumer Price Index with a 1982–1984 base period):

Year	Price per Share ($)	CPI (1982–1984 Base)
1983	51.00	99.6
1984	54.00	103.9
1985	58.00	107.6
1986	59.50	109.6
1987	59.00	113.6

Deflate the stock price series and comment on the investment aspects of this stock. What would the 1987 price per share have been for this stock if it had kept pace with the cost of living as indicated by the Consumer Price Index?

26. A major manufacturing company has reported the quantity and product value information for 1987 and 1991 in Table 17.18. Compute a weighted aggregate quantity index for the data. Comment on what this quantity index means.

TABLE 17.17

Year	Starting Salary ($)	CPI (1982–1984 Base)
1970	14,000	38.8
1975	17,500	53.8
1980	23,000	82.4
1985	37,000	107.6
1990	53,000	130.7

TABLE 17.18

Product	Quantities 1987	Quantities 1991	Value ($)
A	800	1,200	30.00
B	600	500	20.00
C	200	500	25.00

18

Time Series Analysis and Forecasting

Contents

The Cincinnati Gas & Electric Company*

CINCINNATI, OHIO

The Cincinnati Gas & Electric Company (CG&E) is a privately owned public utility serving approximately 370,000 gas customers and 600,000 electric customers. The company's service area covers approximately 3000 square miles in and around the Greater Cincinnati area. Forecasting at CG&E offers some unique perspectives compared to forecasting in other industries. Since there are no finished-goods or in-process inventories of electricity, this product must be generated on demand to meet the needs of customers. Electrical shortages are not merely lost sales; they are brownouts or blackouts, with all the accompanying implications. This situation places an unusual burden on the forecaster. On the positive side, the demand for, and the sale of, energy is more predictable than for many other products. Also, unlike the situation in a multiproduct firm, a great amount of forecasting effort and expertise can be concentrated on the two products, gas and electricity.

Two types of forecasts that CG&E must develop are the long-range forecasts of electric peak load and electric energy. The largest observed electric demand for any period, such as an hour, a day, a month, or a year, is defined as the peak load. The cumulative amount of energy generated and used over the period of an hour is referred to as electric energy. Until the mid 1970s, the seasonal pattern of both electric energy and electric peak load were regular and generally quite predictable. Up to this time, CG&E was able to develop accurate forecasts using a simple trend projection model. In the mid 1970s, however, a variety of government actions, off-and-on energy shortages, and the resulting price signals to the consumer began to affect the consumption of electric energy. As a result, the behavior of the peak load and electric energy time series became more and more unpredictable, and the use of a simple trend projection model was no longer adequate.

A special forecasting model—referred to as an econometric model—was developed by CG&E to better account for the behavior of these time series. Their econometric model forecasts the annual energy consumption by residential, commercial, and industrial classes of service. These initial forecasts are then used to develop predictions of summer and peak winter loads. A number of economic and demographic time series are used in the construction of this model. Simply speaking, the entire forecasting system is a compilation of several, statistically verified, multiple regression equations.

The forecast of the annual electric peak load guides the timing decisions for constructing future generating units. The financial impact of these decisions is great. For example, the last generating unit built by the company cost nearly 600 million dollars. Because of the high cost of borrowed capital, a timing decision that leads to having the unit available no sooner than necessary is crucial.

In this chapter we will introduce several methods that can help predict future aspects of a business operation. The objective of each method is to develop good forecasts or predictions of future values. The long-run success of an organization is closely related to how well it is able to perform this forecasting function.

A Cincinnati Gas & Electric Company line worker repairs a high-voltage transmission tower.

*The authors are indebted to Dr. Richard Evans, The Cincinnati Gas & Electric Company, for providing this Statistics in Practice.

A critical aspect of managing any organization is planning for the future. Indeed, the long-run success of an organization is closely related to how well management is able to foresee the future and develop appropriate strategies. Good judgment, intuition, and an awareness of the state of the economy may give a manager a rough idea or "feeling" of what is likely to happen in the future. However, it is often difficult to convert this feeling into a number that can be used as next quarter's sales volume or next year's raw material cost per unit. The purpose of this chapter is to introduce several methods that can help predict many future aspects of a business operation.

Suppose that we have been asked to provide quarterly estimates of the sales volume for a particular product during the coming 1-year period. Production schedules, raw material purchasing plans, inventory policies, and sales quotas will all be affected by the quarterly estimates we provide. Consequently, poor estimates may result in poor planning and hence result in increased costs for the firm. How should we go about providing the quarterly sales volume estimates?

We will certainly want to review the actual sales data for the product in past periods. Suppose that we have actual sales data for each quarter over the past 3 years. Using these historical data, we can identify the general level of sales and determine whether there is a trend, such as an increase or decrease in sales volume over time. A further review of the data might reveal a seasonal pattern, such as peak sales occurring in the third quarter of each year and sales volume bottoming out during the first quarter. By reviewing historical data over time, we can often develop a better understanding of the pattern of past sales; often this can lead to better predictions of future sales for the product.

The historical sales data form what is called a *time series*. Specifically, a time series is a set of observations measured at successive points in time or over successive periods of time. In this chapter, we will introduce several procedures that can be used to analyze time series data. The objective of this analysis will be to provide good *forecasts* or predictions of future values of the time series.

Forecasting methods can be classified as quantitative or qualitative. Quantitative forecasting methods are based on an analysis of historical data concerning a time series and possibly other related time series. If the historical data used are restricted to past values of the series that we are trying to forecast, the forecasting procedure is called a time series method. In this chapter we discuss three time series methods: smoothing (moving averages and exponential smoothing), trend projection, and trend projection adjusted for seasonal influence. If the historical data used in a quantitative forecasting method involve other time series that are believed to be related to the time series we are trying to forecast, we say that we are using a causal method. We discuss the use of multiple regression analysis as a causal forecasting method.

Qualitative forecasting methods generally utilize the judgment of experts to make forecasts. An advantage of these procedures is that they can be applied in situations where no historical data are available. We discuss some of these approaches in Section 18.6. Figure 18.1 provides an overview of the different types of forecasting methods.

18.1   The Components of a Time Series

To explain the pattern or behavior of the data in a time series, it is often helpful to think of the time series as consisting of several components. The usual assumption is that four separate components—trend, cyclical, seasonal, and irregular—combine to make the time series take on specific values. Let us look more closely at each of these components of a time series.

FIGURE **18.1**
**An Overview of Forecasting
Methods**

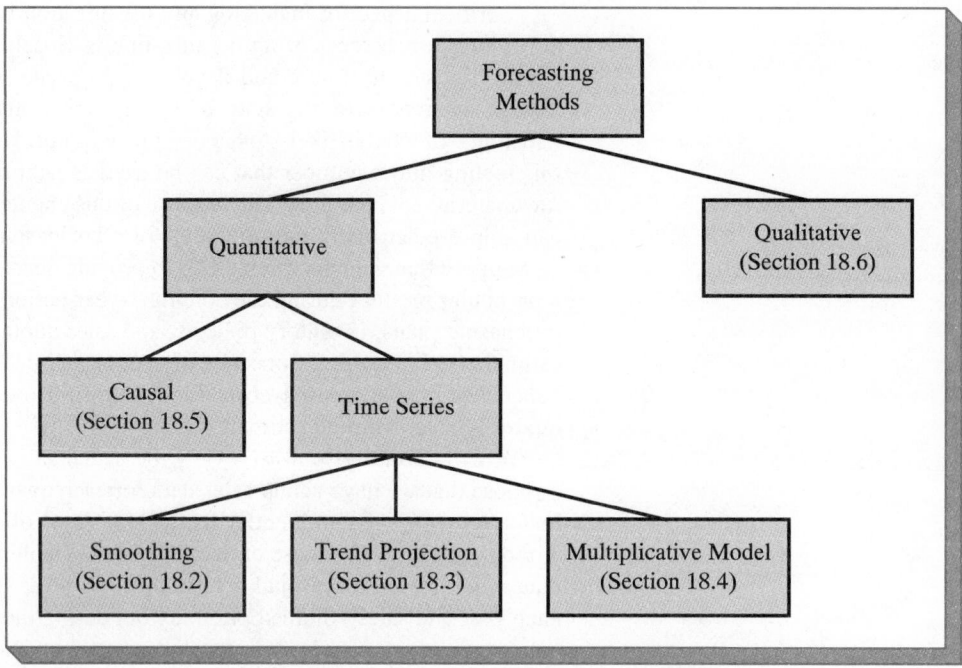

Trend Component

In time series analysis the measurements may be taken every hour, day, week, month, or year or at any other regular interval.* Although time series data generally exhibit random fluctuations, the time series may still show gradual shifts or movements to relatively higher or lower values over a longer period of time. The gradual shifting of the time series, which is usually due to long-term factors such as changes in the population, changes in demographic characteristics of the population, changes in technology, and changes in consumer preferences, is referred to as the *trend* in the time series.

For example, a manufacturer of photographic equipment may see substantial month-to-month variability in the number of cameras sold. However, in reviewing the sales over the past 10–15 years, this manufacturer may find a gradual increase in the annual sales volume. Suppose that the sales volume was approximately 1800 cameras per month in 1982, 2200 cameras per month in 1987, and 2600 cameras per month in 1992. While actual month-to-month sales volumes may vary substantially, this gradual growth in sales over time shows an upward trend for the time series. Figure 18.2 shows a straight line that may be a good approximation of the trend in the sales data. While the trend for camera sales appears to be linear and increasing over time, sometimes the trend in a time series is better described by other patterns.

Figure 18.3 shows some other possible time series trend patterns. In panel A of this figure we see a nonlinear trend. The curve shown describes a time series showing very little growth initially, followed by a period of rapid growth, and then a leveling off. This might be a good approximation to sales for a product from introduction through a growth period and into a period of market saturation. The linear decreasing trend in panel B of

---

*We restrict our attention here to time series where the values of the series are recorded at equal intervals. Treatment of cases where the observations are not made at equal intervals is beyond the scope of this text.

FIGURE 18.2
Linear Trend of Camera Sales

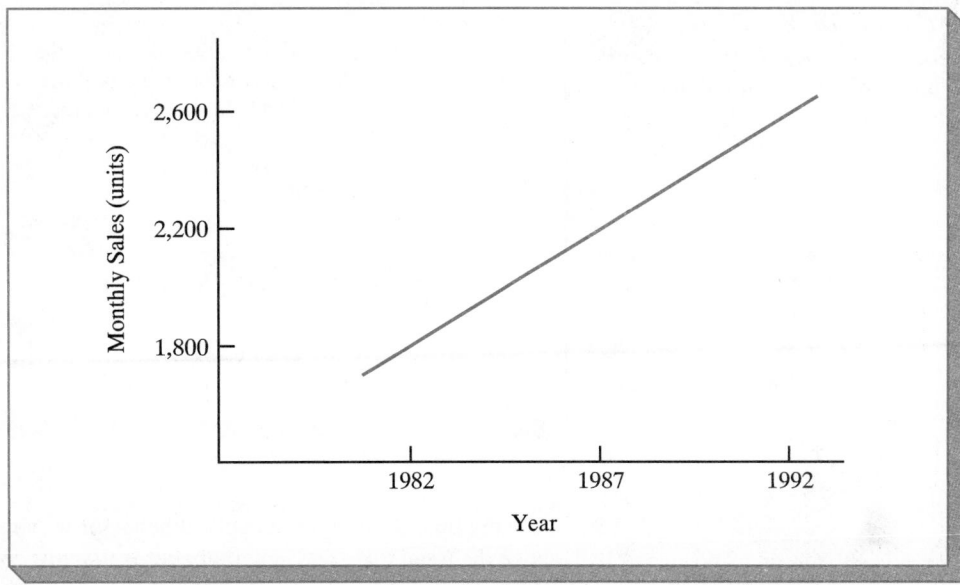

FIGURE 18.2
Linear Trend of Camera Sales

FIGURE 18.3
Examples of Some Possible Time Series Trend Patterns

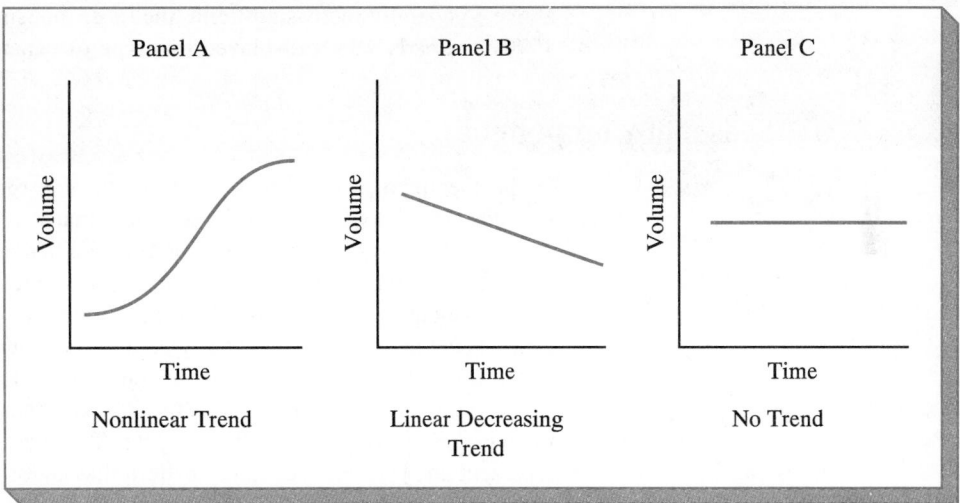

Figure 18.3 is useful for time series displaying a steady decrease over time. The horizontal line in panel C of Figure 18.3 is used for a time series that does not show any consistent increase or decrease over time. It is actually the case of no trend.

Cyclical Component

Although a time series may exhibit a gradual shifting or trend pattern over long periods of time, we cannot expect all future values of the time series to be exactly on the trend line. In fact, time series often show alternating sequences of points below and above the trend line. Any regular pattern of sequences of points above and below the trend line lasting more than 1 year can be attributed to the *cyclical component* of the time series. Figure 18.4 shows the graph of a time series with an obvious cyclical component. The observations are taken at intervals 1 year apart.

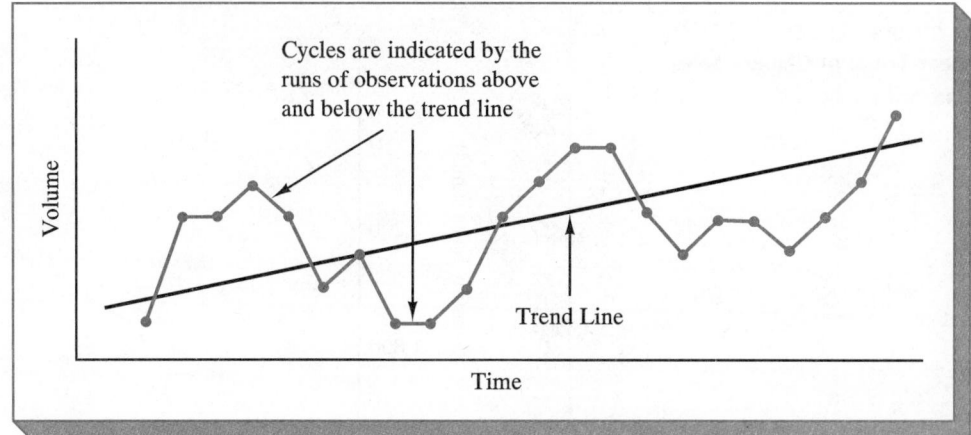

Many time series exhibit cyclical behavior with regular runs of observations below and above the trend line. The general belief is that this component of the time series represents multiyear cyclical movements in the economy. For example, periods of moderate inflation followed by periods of rapid inflation can lead to many time series that alternate below and above a generally increasing trend line (e.g., housing costs). Many time series in the late 1970s and early 1980s displayed this type of behavior.

Seasonal Component

While the trend and cyclical components of a time series are identified by analyzing multiyear movements in historical data, many time series show a regular pattern of variability within 1-year periods. For example, a manufacturer of swimming pools expects low sales activity in the fall and winter months, with peak sales occurring in the spring and summer months. Manufacturers of snow removal equipment and heavy clothing, however, expect just the opposite yearly pattern. It should not be surprising that the component of the time series that represents the variability in the data due to seasonal influences is called the *seasonal component*. Although we generally think of seasonal movement in a time series as occurring within 1 year, the seasonal component can also be used to represent any regularly repeating pattern that is less than 1 year in duration. For example, daily traffic volume data show within-the-day "seasonal" behavior, with peak levels occurring during rush hours, moderate flow during the rest of the day and early evening, and light flow from midnight to early morning.

Irregular Component

The *irregular component* of the time series is the residual, or "catchall," factor that accounts for the deviation of the actual time series value from what we would expect if the trend, cyclical, and seasonal components completely explained the time series. It accounts for the random variability in the time series. The irregular component is caused by the short-term, unanticipated, and nonrecurring factors that affect the time series. Since this component accounts for the random variability in the time series, it is unpredictable. We cannot attempt to predict its impact on the time series in advance.

18.2   Forecasting Using Smoothing Methods

In this section we discuss forecasting techniques that are appropriate for a fairly stable time series—that is, one that exhibits no significant trend, cyclical, or seasonal effects. In such situations, the objective of the forecasting method is to "smooth out" the irregular component of the time series through an averaging process. We begin with a consideration of the method known as moving averages.

Moving Averages

The *moving averages* method uses the average of the *most recent n* data values in the time series as the forecast for the next period. Mathematically, the moving average calculation is made as follows.

Moving Average

$$\text{Moving Average} = \frac{\Sigma(\text{most recent } n \text{ data values})}{n} \qquad (18.1)$$

The term *moving* is used because every time a new observation becomes available for the time series, it replaces the oldest observation in (18.1), and a new average is computed. As a result, the average will change, or move, as new observations become available.

To illustrate the moving averages method, consider the 12 weeks of data presented in Table 18.1 and Figure 18.5. These data show the number of gallons of gasoline sold by a gasoline distributor in Bennington, Vermont, over the past 12 weeks.

To use moving averages to forecast gasoline sales, we must first select the number of data values to be included in the moving average. As an example, let us compute forecasts using a 3-week moving average. The moving average calculation for the first 3 weeks of the gasoline sales time series is as follows:

$$\text{Moving Average (Weeks 1–3)} = \frac{17 + 21 + 19}{3} = 19$$

This moving average value is then used as the forecast for week 4. Since the actual value observed in week 4 is 23, we see that the forecast error in week 4 is 23 − 19 = 4. In general, the error associated with any forecast is the difference between the observed value of the time series and the forecast value.

The calculation for the second 3-week moving average is shown below:

$$\text{Moving Average (Weeks 2–4)} = \frac{21 + 19 + 23}{3} = 21$$

Hence, the forecast for week 5 is 21. The error associated with this forecast is 18 − 21 = −3. Thus, we see that the forecast error can be positive or negative depending upon whether the forecast is too low or too high. A complete summary of the 3-week moving average calculations for the gasoline sales time series is shown in Table 18.2 and Figure 18.6.

Tᴀʙʟᴇ **18.1**
Gasoline Sales Time Series

Week	Sales (1000s of gallons)
1	17
2	21
3	19
4	23
5	18
6	16
7	20
8	18
9	22
10	20
11	15
12	22

FIGURE 18.5
Graph of Gasoline Sales Time
Series

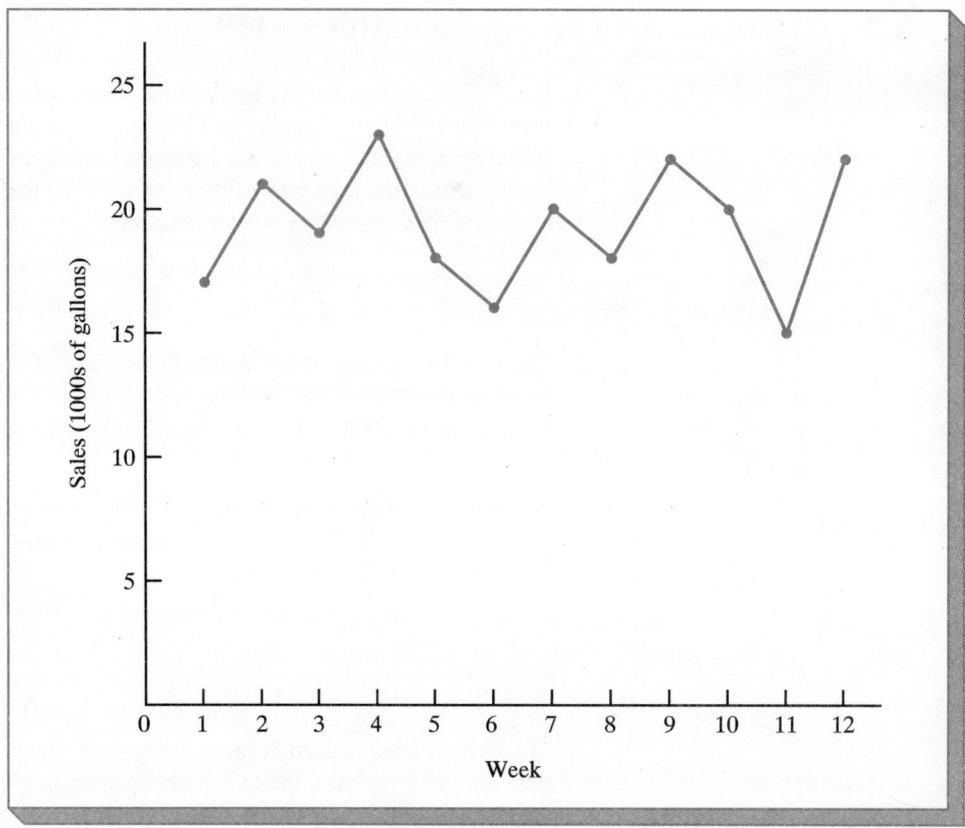

TABLE 18.2
Summary of 3-Week Moving
Average Calculations

Week	Time Series Value	Moving Average Forecast	Forecast Error	Squared Forecast Error
1	17			
2	21			
3	19			
4	23	19	4	16
5	18	21	−3	9
6	16	20	−4	16
7	20	19	1	1
8	18	18	0	0
9	22	18	4	16
10	20	20	0	0
11	15	20	−5	25
12	22	19	3	9
			Totals 0	92

FIGURE 18.6

Graph of Gasoline Sales Time Series and 3-Week Moving Average Forecasts

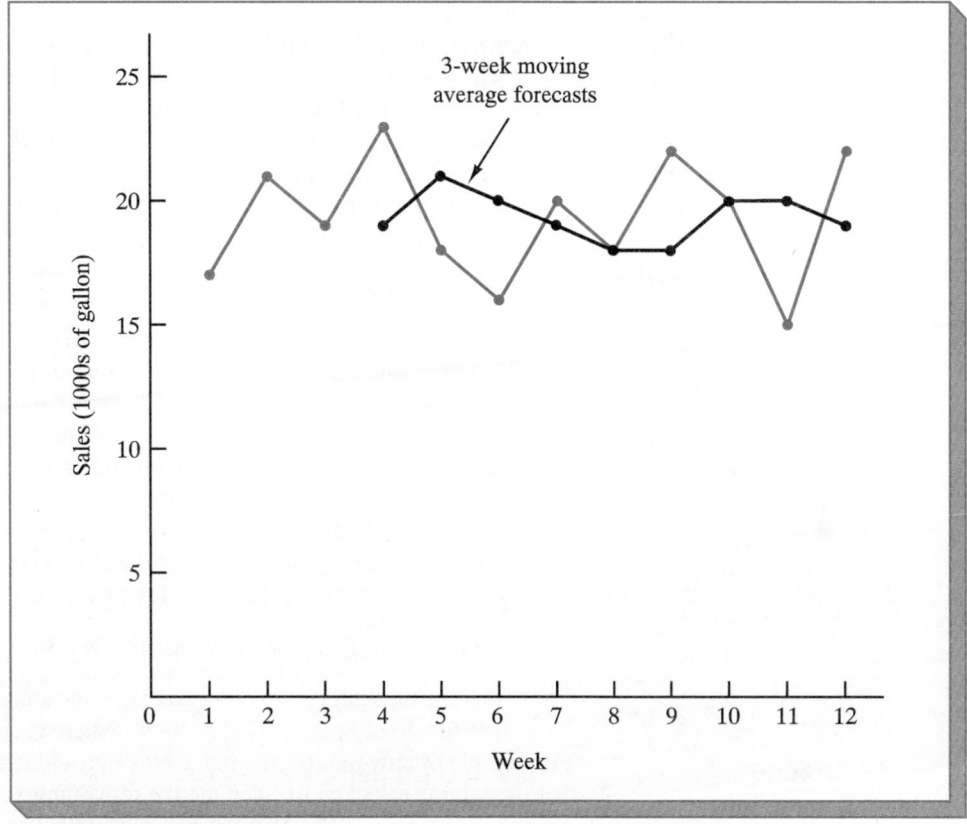

An important consideration in using any forecasting method is the accuracy of the forecast. Clearly, we would like the forecast errors to be small. The last two columns of Table 18.2, which contain the forecast errors and the forecast errors squared, can be used to develop measures of accuracy.

One measure of forecast accuracy you might think of using would be to simply sum the forecast errors over time. The problem with this measure is that if the errors are random (as they should be if the forecasting method selected is appropriate), some errors will be positive and some errors will be negative, resulting in a sum near zero regardless of the size of the individual errors. Indeed, we see from Table 18.2 that the sum of forecast errors for the gasoline sales time series is zero. This difficulty can be avoided by squaring each of the individual forecast errors.

For the gasoline sales time series, we can use the last column of Table 18.2 to compute the average of the sum of the squared errors. Doing so we obtain

$$\text{Average of the Sum of Squared Errors} = \frac{92}{9} = 10.22$$

This average of the sum of squared errors is commonly referred to as the *mean squared error* (MSE). The MSE is an often-used measure of the accuracy of a forecasting method and is the one we use in this chapter.

As we indicated previously, to use the moving averages method, we must first select the number of data values to be included in the moving average. It should not be too surprising that for a particular time series, different length moving averages will differ in their ability to accurately forecast the time series. One possible approach to choosing the number of values to be included is to use trial and error to identify the length that minimizes the MSE. Then, if we are willing to assume that the length that is best for the past will also be best for the future, we would forecast the next value in the time series using the number of data values that minimized the MSE for the historical time series. Exercise 2 at the end of the section will ask you to consider 4-week and 5-week moving averages for the gasoline sales data. A comparison of the MSE for each will indicate the number of weeks of data you may want to include in the moving average calculation.

Weighted Moving Averages

In the moving averages method, each observation in the moving average calculation receives the same weight. One possible variation, known as *weighted moving averages,* involves selecting different weights for each data value and then computing a weighted mean as the forecast. In most cases, the most recent observation receives the most weight, and the weight decreases for older data values. For example, using the gasoline sales time series, let us illustrate the computation of a weighted 3-week moving average, where the most recent observation receives a weight three times as great as that given the oldest observation, and the next oldest observation receives a weight twice as great as the oldest. The weighted moving average forecast for week 4 would be computed as follows:

$$\text{Forecast for Week 4} = \tfrac{3}{6}(19) + \tfrac{2}{6}(21) + \tfrac{1}{6}(17) = 19.33$$

Note that for the weighted moving average, the sum of the weights is equal to 1. This was also true for the simple moving average, where each weight was $\tfrac{1}{3}$. However, recall that the simple or unweighted moving average provided a forecast of 19. Exercise 3 at the end of the section asks you to calculate the remaining values for the 3-week weighted moving average and compare the forecast accuracy with what we have obtained for the unweighted moving average.

Exponential Smoothing

Exponential smoothing is a forecasting technique that uses a weighted average of past time series values to smooth the data; the weighted average is used to forecast the value of the time series in the next period. The basic exponential smoothing model is as follows.

Exponential Smoothing Model

$$F_{t+1} = \alpha Y_t + (1 - \alpha)F_t \qquad (18.2)$$

where

F_{t+1} = the forecast of the time series for period $t + 1$

Y_t = actual value of the time series in period t

F_t = forecast of the time series for period t

α = smoothing constant $(0 \leq \alpha \leq 1)$

To see that the forecast for any period is a weighted average of *all the previous actual values* for the time series, suppose that we have a time series consisting of three periods of data, Y_1, Y_2, and Y_3. To get the exponential smoothing calculations started, we let F_1 equal the actual value of the time series in period 1; that is, $F_1 = Y_1$. Hence, the forecast for period 2 is written as follows:

$$F_2 = \alpha Y_1 + (1 - \alpha)F_1$$

$$= \alpha Y_1 + (1 - \alpha)Y_1$$

$$= Y_1$$

In general, then, the exponential smoothing forecast for period 2 is equal to the actual value of the time series in period 1.

To obtain the forecast for period 3, we substitute $F_2 = Y_1$ in the expression for F_3; the result is

$$F_3 = \alpha Y_2 + (1 - \alpha)Y_1$$

Finally, substituting this expression for F_3 in the expression for F_4, we obtain

$$F_4 = \alpha Y_3 + (1 - \alpha)[\alpha Y_2 + (1 - \alpha)Y_1]$$

$$= \alpha Y_3 + \alpha(1 - \alpha)Y_2 + (1 - \alpha)^2 Y_1$$

Hence, we see that F_4 is a weighted average of the first three time series values. The sum of the coefficients or weights for Y_1, Y_2, and Y_3 equals 1. A similar argument can be made to show that any forecast F_{t+1} is a weighted average of the previous t time series values.

An advantage of exponential smoothing is that it is a simple procedure and requires very little historical data for its use. Once the *smoothing constant* α has been selected, only two pieces of information are required to compute the forecast for the next period. Referring to (18.2), we see that with a given α we can compute the forecast for period $t + 1$ simply by knowing the actual and forecast time series values for period t, that is, Y_t and F_t.

To illustrate the exponential smoothing approach to forecasting, consider the gasoline sales time series presented previously in Table 18.1 and Figure 18.5. As we indicated in the discussion above, the exponential smoothing forecast for period 2 is equal to the actual value of the time series in period 1. Thus, with $Y_1 = 17$, we will set $F_2 = 17$ to get the exponential smoothing computations started. Referring to the time series data in Table 18.1, we find an actual time series value in period 2 of $Y_2 = 21$. Thus, period 2 has a forecast error of $21 - 17 = 4$.

Continuing with the exponential smoothing computations provides the following forecast for period 3:

$$F_3 = .2Y_2 + .8F_2 = .2(21) + .8(17) = 17.8$$

Once the actual time series value in period 3, $Y_3 = 19$, is known, we can generate a forecast for period 4 as follows:

$$F_4 = .2Y_3 + .8F_3 = .2(19) + .8(17.8) = 18.04$$

By continuing the exponential smoothing calculations, we are able to determine the weekly forecast values and the corresponding weekly forecast errors, as shown in Table 18.3. Note that we have not shown an exponential smoothing forecast or the forecast error for period 1, because F_1 was set equal to Y_1 to begin the smoothing computations. For week 12, we have $Y_{12} = 22$ and $F_{12} = 18.48$. Can you use this information to generate a forecast for week 13 before the actual value of week 13 becomes known? Using the exponential smoothing model, we have

$$F_{13} = .2Y_{12} + .8F_{12} = .2(22) + .8(18.48) = 19.18$$

T A B L E 18.3	Week (t)	Time Series Value (Y_t)	Exponential Smoothing Forecast (F_t)	Forecast Error (Y_t − F_t)

T A B L E 18.3

Summary of the Exponential Smoothing Forecasts and Forecast Errors for Gasoline Sales with Smoothing Constant α = .2

Week (t)	Time Series Value (Y_t)	Exponential Smoothing Forecast (F_t)	Forecast Error ($Y_t - F_t$)
1	17		
2	21	17.00	4.00
3	19	17.80	1.20
4	23	18.04	4.96
5	18	19.03	−1.03
6	16	18.83	−2.83
7	20	18.26	1.74
8	18	18.61	−.61
9	22	18.49	3.51
10	20	19.19	.81
11	15	19.35	−4.35
12	22	18.48	3.52

Thus, the exponential smoothing forecast of the amount sold in week 13 is 19.18, or 19,180 gallons of gasoline. With this forecast, the firm can make plans and decisions accordingly. The accuracy of the forecast will not be known until the firm conducts its business through week 13.

Figure 18.7 shows the plot of the actual and the forecast time series values. Note in particular how the forecasts smooth out the irregular fluctuations in the time series.

In the preceding exponential smoothing calculations, we used a smoothing constant of α = .2, although any value of α between 0 and 1 is acceptable. However, some values will yield better forecasts than others. Some insight into choosing a good value for α can be obtained by rewriting the basic exponential smoothing model as follows:

$$F_{t+1} = \alpha Y_t + (1 - \alpha)F_t$$

$$F_{t+1} = \alpha Y_t + F_t - \alpha F_t$$

$$F_{t+1} = F_t + \alpha(Y_t - F_t) \qquad (18.3)$$

Forecast in period t Forecast error in period t

Thus, we see that the new forecast F_{t+1} is equal to the previous forecast F_t plus an adjustment, which is α times the most recent forecast error, $Y_t - F_t$. That is, the forecast in period $t + 1$ is obtained by adjusting the forecast in period t by a fraction of the forecast error. If the time series contains substantial random variability, a small value of the smoothing constant is preferred. The reason for this choice is that since much of the forecast error is due to random variability, we do not want to overreact and adjust the forecasts too quickly. For a time series with relatively little random variability, larger values of the smoothing constant have the advantage of quickly adjusting the forecasts when forecasting errors occur and therefore allowing the forecast to react faster to changing conditions.

The criterion we will use to determine a desirable value for the smoothing constant α is the same as the criterion we proposed earlier for determining the number of periods of data to include in the moving averages calculation. That is, we choose the value of α that minimizes the mean squared error (MSE). A summary of the MSE calculations for the exponential smoothing forecast of gasoline sales with α = .2 is shown in Table 18.4. Note that there is one less squared error term than the number of time periods, because we had

FIGURE 18.7

Graph of Actual and Forecast Gasoline Sales Time Series with Smoothing Constant α = .2

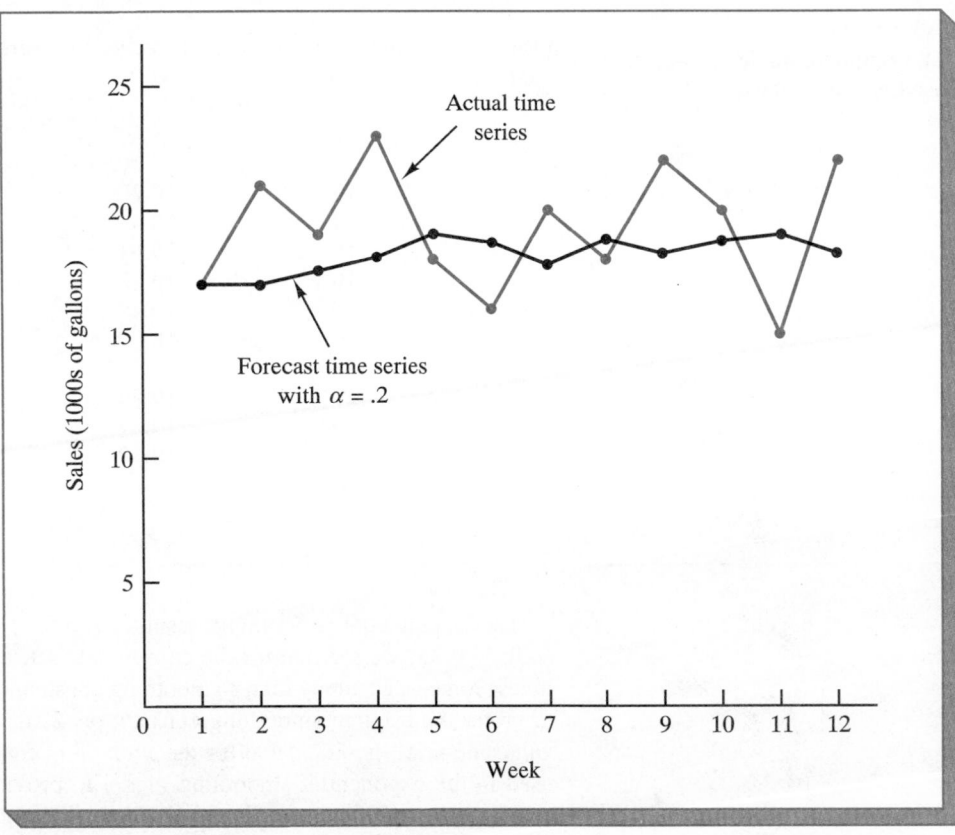

TABLE 18.4

MSE Computations for Forecasting Gasoline Sales with α = .2

Week (t)	Time Series Value (Y_t)	Forecast (F_t)	Forecast Error ($Y_t - F_t$)	Squared Forecast Error ($Y_t - F_t)^2$
1	17			
2	21	17.00	4.00	16.00
3	19	17.80	1.20	1.44
4	23	18.04	4.96	24.60
5	18	19.03	−1.03	1.06
6	16	18.83	−2.83	8.01
7	20	18.26	1.74	3.03
8	18	18.61	−.61	.37
9	22	18.49	3.51	12.32
10	20	19.19	.81	.66
11	15	19.35	−4.35	18.92
12	22	18.48	3.52	12.39
			Total	98.80

$$MSE = \frac{98.80}{11} = 8.98$$

no past values with which to make a forecast for period 1. Would a different value of α have provided better results in terms of a lower MSE value? Perhaps the most straightforward way to answer this question is simply to try another value for α. We will then compare its mean squared error with the MSE value of 8.98 obtained using a smoothing constant of .2.

TABLE 18.5

MSE Computations for Forecasting Gasoline Sales with $\alpha = .3$

Week (t)	Time Series Value (Y_t)	Forecast (F_t)	Forecast Error $(Y_t - F_t)$	Squared Forecast Error $(Y_t - F_t)^2$
1	17			
2	21	17.00	4.00	16.00
3	19	18.20	.80	.64
4	23	18.44	4.56	20.79
5	18	19.81	−1.81	3.28
6	16	19.27	−3.27	10.69
7	20	18.29	1.71	2.92
8	18	18.80	−.80	.64
9	22	18.56	3.44	11.83
10	20	19.59	.41	.17
11	15	19.71	−4.71	22.18
12	22	18.30	3.70	13.69
			Total	102.83

$$\text{MSE} = \frac{102.83}{11} = 9.35$$

The exponential smoothing results with $\alpha = .3$ are shown in Table 18.5. With MSE = 9.35, we see that for the current data set, a smoothing constant of $\alpha = .3$ results in less forecast accuracy than a smoothing constant of $\alpha = .2$. Thus, we would be inclined to prefer the original smoothing constant of .2. In trial-and-error calculations with other values of α, a "good" value for the smoothing constant can be found. This value can be used in the exponential smoothing model to provide forecasts for the future. At a later date, after a number of new time series observations have been obtained, it is good practice to analyze the newly collected time series data to see if the smoothing constant should be revised to provide better forecasting results.

NOTES & COMMENTS

Another commonly used measure of forecast accuracy is the *mean absolute deviation* (MAD). This measure is simply the average of the sum of the absolute values of all the forecast errors. Using the errors given in Table 18.2, we obtain

$$\text{MAD} = \frac{4 + 3 + 4 + 1 + 0 + 4 + 0 + 5 + 3}{9} = 2.67$$

One major difference between MSE and MAD is that the MSE measure is influenced much more by large forecast errors than by small errors (since for the MSE measure, the errors are squared). The selection of the best measure of forecasting accuracy is not a simple matter. Indeed, forecasting experts often disagree as to which measure should be used. In this chapter we will use the MSE measure.

❑ ❑ **Exercises**

Methods

SELF TEST ▶

1. Consider the following time series data:

Week	1	2	3	4	5	6
Value	8	13	15	17	16	9

a. Develop a 3-week moving average for this time series. What is the forecast for week 7?

b. Compute the MSE for the 3-week moving average.

c. Use $\alpha = .2$ to compute the exponential smoothing values for the time series. What is the forecast for week 7?

d. Compare the 3-week moving average forecast with the exponential smoothing forecast using $\alpha = .2$. Which appears to provide the better forecast?

e. Use a smoothing constant of .4 to compute the exponential smoothing values. Does a smoothing constant of .2 or .4 appear to provide the better forecast? Explain.

2. Refer to the gasoline sales time series data in Table 18.1.

a. Compute 4-week and 5-week moving averages for the time series.

b. Compute the MSE for the 4-week and 5-week moving average forecasts.

c. What appears to be the best number of weeks of past data to use in the moving average computation? Remember that the MSE for the 3-week moving average is 10.22.

3. Refer again to the gasoline sales time series data in Table 18.1.

a. Using a weight of ½ for the most recent observation, ⅓ for the second most recent, and ⅙ for third most recent, compute a 3-week weighted moving average for the time series.

b. Compute the MSE for the weighted moving average in (a). Do you prefer this weighted moving average to the unweighted moving average? Remember that the MSE for the unweighted moving average is 10.22.

c. Suppose you are allowed to choose any weight as long as they sum to 1. Could you always find a set of weights that would make the MSE smaller for a weighted moving average than an unweighted moving average? Why or why not?

4. Use the gasoline time series data from Table 18.1 to show the exponential smoothing forecasts using $\alpha = .1$. Using the MSE criterion, would you prefer a smoothing constant of $\alpha = .1$ or $\alpha = .2$ for the gasoline sales time series?

5. Using a smoothing constant of $\alpha = .2$, Equation (18.2) shows that the forecast for the 13th week of the gasoline sales data from Table 18.1 is given by $F_{13} = .2Y_{12} + .8F_{12}$. However, the forecast for week 12 is given by $F_{12} = .2Y_{11} + .8F_{11}$. Thus, we could combine these two results to show that the forecast for the 13th week can be written

$$F_{13} = .2Y_{12} + .8(.2Y_{11} + .8F_{11}) = .2Y_{12} + .16Y_{11} + .64F_{11}$$

a. Making use of the fact that $F_{11} = .2Y_{10} + .8F_{10}$ (and similarly for F_{10} and F_9), continue to expand the expression for F_{13} until it is written in terms of the past data values $Y_{12}, Y_{11}, Y_{10}, Y_9$, and Y_8, and the forecast for period 8.

b. Refer to the coefficients or weights for the past data $Y_{12}, Y_{11}, Y_{10}, Y_9$, and Y_8; what observation can you make about how exponential smoothing weights past data values in arriving at new forecasts? Compare this weighting pattern with the weighting pattern of the moving averages method.

Applications

6. Kings Island Amusement Park, in Kings Island, Ohio, recently reported park attendance figures (in millions) as shown in Table 18.6. (*The Cincinnati Enquirer*, April 5, 1992).

a. Use exponential smoothing with smoothing constants of .2 and .3 to forecast the attendance figures.

b. What is the forecast of park attendance for 1992?

7. Corporate Triple A Bond interest rates for the 12 months of 1988 are shown below (*The Media General Financial Weekly*, April 17, 1989).

9.5, 9.3, 9.4, 9.6, 9.8, 9.7, 9.8, 10.5, 9.9, 9.7, 9.6, 9.6.

a. Develop three-month and four-month moving averages for this time series. Does the three-month or four-month moving average provide the better forecasts? Explain.

b. What is the moving average forecast for January, 1989?

8. Alabama building contracts by month for 1988 are shown below (*Alabama Business*, June 1989). Data are in millions of dollars.

TABLE 18.6

Year	Attendance
1980	2.49
1981	2.82
1982	2.72
1983	2.60
1984	2.81
1985	2.98
1986	2.86
1987	3.13
1988	2.98
1989	3.16
1990	3.20
1991	2.85

SELF TEST ▶

TABLE 18.7

Month	Sales
1	105
2	135
3	120
4	105
5	90
6	120
7	145
8	140
9	100
10	80
11	100
12	110

240, 350, 230, 260, 280, 320, 220, 310, 240, 310, 240, 230.

a. Compare a 3-month moving averages forecast with an exponential smoothing forecast using $\alpha = .2$. Which provides the better forecasts?

b. What is the forecast for January 1989?

9. The time series in Table 18.7 shows the sales of a particular product over the past 12 months.

a. Use $\alpha = .3$ to compute the exponential smoothing values for the time series.

b. Use a smoothing constant of .5 to compute the exponential smoothing values. Does a smoothing constant of .3 or .5 appear to provide the better forecasts?

10. The Dow Jones Industrial Average (DJIA) is based on common stock prices of 30 industrial stocks. This average is used to describe what is happening in the stock market. The weekly closing levels of the DJIA for 12 weeks during the period May–July of 1989 are shown below (*The Wall Street Journal*, July 31, 1989).

Week	DJIA	Week	DJIA
1	2480	7	2520
2	2470	8	2470
3	2475	9	2440
4	2510	10	2480
5	2500	11	2530
6	2480	12	2550

a. Compute the exponential smoothing forecasts using $\alpha = .2$.

b. Compute the exponential smoothing forecasts using $\alpha = .3$.

c. Which exponential smoothing model provides the better forecasts? What is the forecast of the DJIA for week 13?

18.3 Forecasting Time Series Using Trend Projection

TABLE 18.8
Bicycle Sales Data

Year (t)	Sales (1000s) (Y_t)
1	21.6
2	22.9
3	25.5
4	21.9
5	23.9
6	27.5
7	31.5
8	29.7
9	28.6
10	31.4

In this section we will see how to forecast the values of a time series that exhibits a long-term linear trend. Specifically, let us consider the time series data for bicycle sales of a particular manufacturer over the past 10 years, as shown in Table 18.8 and Figure 18.8. Note that 21,600 bicycles were sold in year 1; 22,900 were sold in year 2; and so on. In year 10, the most recent year, 31,400 bicycles were sold. Although the graph in Figure 18.8 shows some up-and-down movement over the past 10 years, the time series seems to have an overall increasing or upward trend in the number of bicycles sold.

We do not want the trend component of a time series to follow each and every "up" and "down" movement. Rather, the trend component should reflect the gradual shifting—in our case, growth—of the time series values. After we review the time series data in Table 18.8 and the graph in Figure 18.8, we might agree that a linear trend as shown in Figure 18.9 has the potential of providing a reasonable description of the long-run movement in the series. Thus, we can now concentrate on finding the linear function that best approximates the trend.

Using the bicycle sales data to illustrate the calculations involved, we will now describe how regression analysis can be used to identify a linear trend for a time series. Recall that in the discussion of simple linear regression in Chapter 14, we described how the least squares method was used to find the best straight line relationship between two variables. This is the methodology we will use to develop the trend line for the bicycle

FIGURE 18.8
Graph of the Bicycle Sales Time Series

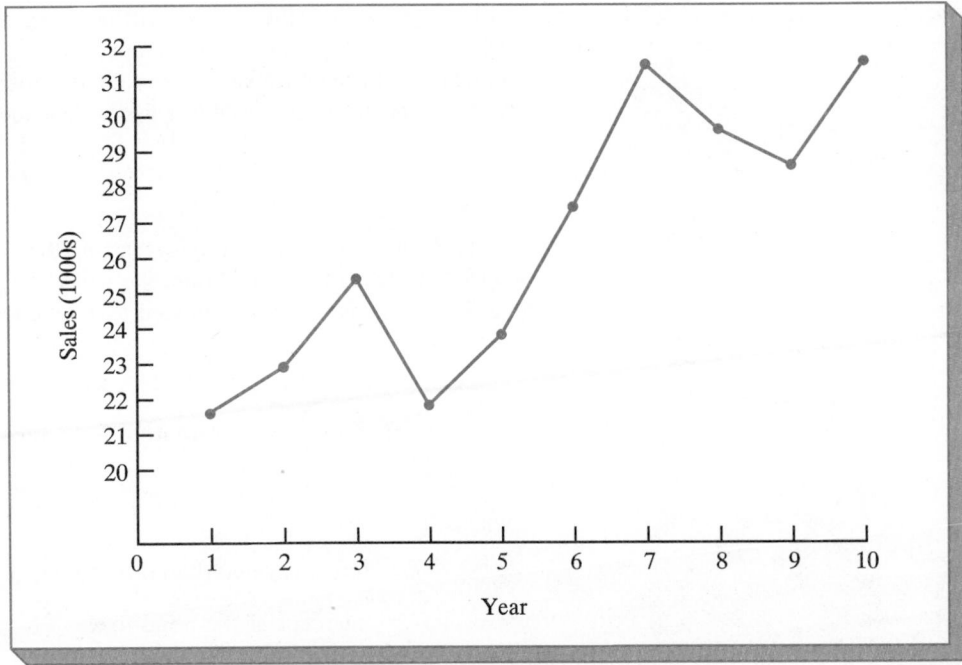

FIGURE 18.9
Trend Represented by a Linear Function for Bicycle Sales

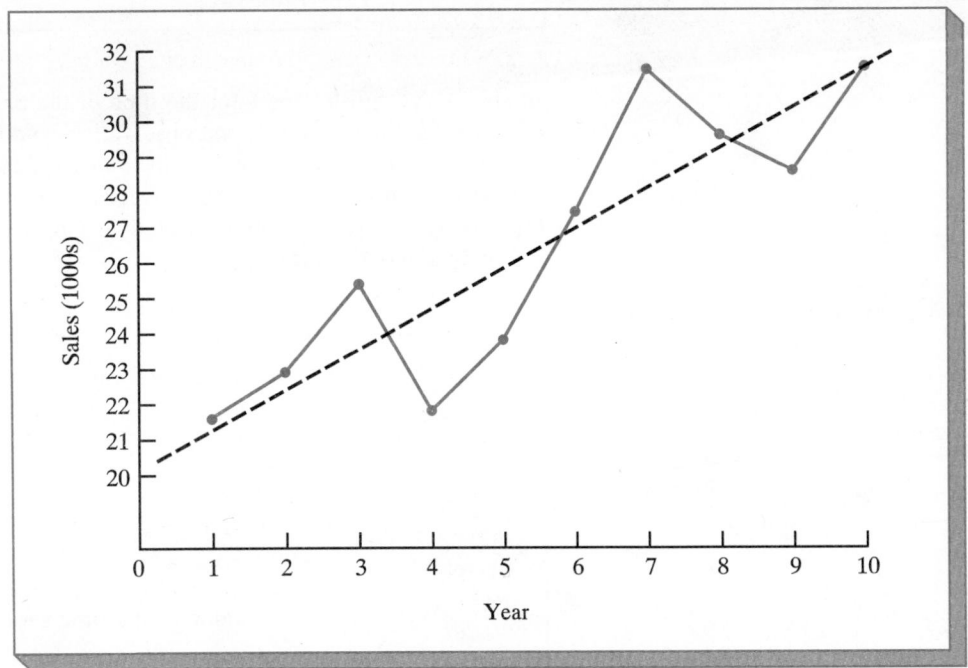

sales time series. Specifically, we will be using regression analysis to estimate the relationship between time and sales volume.

In Chapter 14 the estimated regression equation describing a straight-line relationship between an independent variable x and a dependent variable y was written

$$\hat{y} = b_0 + b_1 x \tag{18.4}$$

To better focus on the fact that in forecasting the independent variable is time, we will use t in (18.4) instead of x; in addition, we will use T_t in place of $\hat{y}$. Thus, for a linear trend, the estimated sales volume expressed as a function of time can be written as follows.

Equation for Linear Trend

$$T_t = b_0 + b_1 t \tag{18.5}$$

where

T_t = forecast value (based on trend) of the time series in period t

b_0 = intercept of the trend line

b_1 = slope of the trend line

t = point in time

In (18.5), we will let $t = 1$ for the time of the first observation on the time series data, $t = 2$ for the time of the second observation, and so on. Note that for the time series on bicycle sales, $t = 1$ corresponds to the oldest time series value, and $t = 10$ corresponds to the most recent year's data. Formulas for computing the estimated regression coefficients (b_1 and b_0) in (18.4) were presented in Chapter 14; they are repeated below, with t replacing x and Y_t replacing y_i.

Computing the Slope (b_1) and Intercept (b_0)

$$b_1 = \frac{\Sigma t Y_t - (\Sigma t \, \Sigma Y_t)/n}{\Sigma t^2 - (\Sigma t)^2/n} \tag{18.6}$$

$$b_0 = \overline{Y} - b_1 \overline{t} \tag{18.7}$$

where

Y_t = actual value of the time series in period t

n = number of periods

$\overline{Y}$ = average value of the time series; that is, $\overline{Y} = \Sigma Y_t / n$

$\overline{t}$ = average value of t; that is, $\overline{t} = \Sigma t / n$

Using both these relationships for b_0 and b_1 and the bicycle sales data of Table 18.8, we have the following calculations:

t	Y_t	tY_t	t^2
1	21.6	21.6	1
2	22.9	45.8	4
3	25.5	76.5	9
4	21.9	87.6	16
5	23.9	119.5	25
6	27.5	165.0	36
7	31.5	220.5	49
8	29.7	237.6	64
9	28.6	257.4	81
10	31.4	314.0	100
Totals 55	264.5	1545.5	385

where

$$\bar{t} = \frac{55}{10} = 5.5 \text{ years}$$

$$\bar{Y} = \frac{264.5}{10} = 26.45 \text{ thousand}$$

$$b_1 = \frac{1545.5 - (55)(264.5)/10}{385 - (55)^2/10} = 1.10$$

$$b_0 = 26.45 - 1.10(5.5) = 20.4$$

Therefore,

$$T_t = 20.4 + 1.1t \tag{18.8}$$

is the expression for the linear trend component for the bicycle sales time series.

Trend Projections

The slope of 1.1 indicates that over the past 10 years the firm has experienced an average growth in sales of around 1100 units per year. If we assume that the past 10-year trend in sales is a good indicator of the future, then (18.8) can be used to project the trend component of the time series. For example, substituting $t = 11$ into (18.8) yields next year's trend projection, T_{11}:

$$T_{11} = 20.4 + 1.1(11) = 32.5$$

Thus, using the trend component only, we would forecast sales of 32,500 bicycles next year.

The use of a linear function to model the trend is common. However, as we discussed earlier, sometimes time series exhibit a curvilinear, or nonlinear, trend similar to those shown in Figure 18.10. In Chapter 16 we discussed how regression analysis can be used

FIGURE 18.10
Some Possible Functional Forms for Nonlinear Trend Patterns

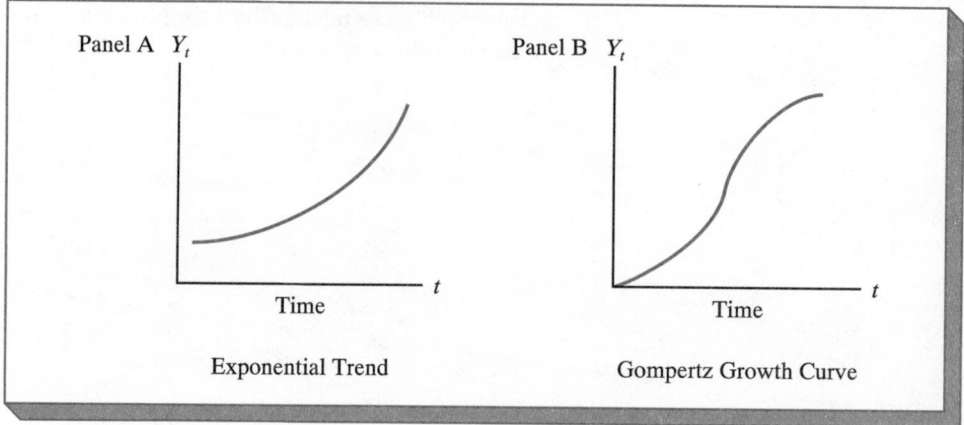

Panel A Y_t Panel B Y_t

Time   Time

Exponential Trend   Gompertz Growth Curve

to model curvilinear relationships of the type shown in panel A of Figure 18.10. More advanced texts discuss in detail how to develop regression models for more complex relationships such as the one shown in panel B of Figure 18.10. For our purposes, it is sufficient to note that the analyst should choose the function that provides the best fit to the data.

Exercises

Methods

SELF TEST ▶ **11.** Consider the following time series:

t	1	2	3	4	5
Y_t	6	11	9	14	15

Develop an equation for the linear trend component for this time series. What is the forecast for $t = 6$?

12. Consider the following time series:

t	1	2	3	4	5	6
Y_t	205	202	195	190	191	188

Develop an equation for the linear trend component for this time series. What is the forecast for $t = 7$?

Applications

13. Data on the number of cellular telephone subscribers in millions over the last 5 years are shown in Table 18.9 (*Business Week*, February 24, 1992).
a. Use a linear trend to forecast the number of cellular telephone subscribers in years 6 and 7.
b. Use the trend equation to estimate the increase in the number of subscribers per year.

SELF TEST ▶ **14.** The trend for the future appears to be more two-income households for married couples (*Business Week*, April 6, 1992). With working wives, a surge in affluent two-income households is anticipated. Data from the Department of Commerce shows the number of two-income married couples earning $50,000 or more annually. The data are in millions of couples with annual income based on constant 1990 dollars.

TABLE 18.9

Year	Subscribers
1	1.2
2	2.5
3	3.3
4	5.3
5	7.4

Year	Millions of Couples
1970	9.5
1975	10.5
1980	14.0
1985	14.9
1990	18.0

Develop a linear trend equation for this time series. If the trend in two-income households continues, what is the forecast of the number of working couples who will make at least $50,000 annually in 1995?

15. Table 18.10 gives average attendance figures at home football games for a major university for the past 7 years. Develop the equation for the linear trend component for this time series.

16. Automobile sales at B. J. Scott Motors, Inc., provided the following 10-year time series:

Year	Sales
1	400
2	390
3	320
4	340
5	270
6	260
7	300
8	320
9	340
10	370

Plot the time series and comment on the appropriateness of a linear trend. What type of functional form do you believe would be most appropriate for the trend pattern of this time series?

17. The president of a small manufacturing firm has been concerned about the continual growth in manufacturing costs over the past several years. Shown in Table 18.11 is a time series of the cost per unit for the firm's leading product over the past 8 years:
a. Show a graph of this time series. Does a linear trend appear to exist?
b. Develop the equation for the linear trend component for the above time series. What is the average cost increase that the firm has been realizing per year?

18. Earnings per share for the Walgreen Company for the most recent 10-year period are as follows (*The Value Line*, April 21, 1989).

.64, .73, .94, 1.14, 1.33, 1.53, 1.67, 1.68, 2.10, 2.50.

a. Use a linear trend projection to forecast this time series for the coming year.
b. What does this time series analysis tell you about the Walgreen Company? Do the historical data indicate the Walgreen Company is a good investment?

19. *The Wall Street Journal* (February 3, 1992) reported the combined number of applications for admission at 10 business schools: Dartmouth, Harvard, Stanford, Pennsylvania, MIT, Columbia, Virginia, Chicago, Northwestern, and UCLA for 1984–1991. The depressed job market of the late 1980s and early 1990s appears to have prompted an unexpected upsurge of interest in graduate programs of business administration. Eager to escape troubled businesses and also enhance their credentials, many recession-weary professionals gave up jobs and headed to business schools. The data with number of applications expressed in thousands are shown in Table 18.12.

TABLE 18.10

Year	Attendance
1	28,000
2	30,000
3	31,500
4	30,400
5	30,500
6	32,200
7	30,800

TABLE 18.11

Year	Cost/Unit ($)
1	20.00
2	24.50
3	28.20
4	27.50
5	26.60
6	30.00
7	31.00
8	36.00

TABLE 18.12

Year	Applications
1984	25.2
1985	28.0
1986	27.5
1987	31.3
1988	33.4
1989	33.1
1990	36.0
1991	36.4

a. Develop a linear trend equation for the time series. What is the average annual increase in the number of applications over the 8-year period?

b. If the trend continues, what is the forecast of the number of applications at these 10 schools for 1992?

20. The gross revenue data for Delta Airlines for a 10-year period are shown below (*Moody's Transportation News Report,* May 26, 1989). Data are in millions of dollars.

Year	Revenue	Year	Revenue
1	2428	6	4264
2	2951	7	4738
3	3533	8	4460
4	3618	9	5318
5	3616	10	6915

a. Develop a linear trend equation for this time series. Comment on what the equation tells about the gross revenue for Delta Airlines for the 10-year period.

b. Provide the forecasts for gross revenue for years 11 and 12.

18.4   Forecasting a Time Series with Trend and Seasonal Components

TABLE 18.13

Quarterly Data for Television Set Sales

Year	Quarter	Sales (1000s)
1	1	4.8
	2	4.1
	3	6.0
	4	6.5
2	1	5.8
	2	5.2
	3	6.8
	4	7.4
3	1	6.0
	2	5.6
	3	7.5
	4	7.8
4	1	6.3
	2	5.9
	3	8.0
	4	8.4

In the previous section we showed how to forecast a time series that had a trend component. In this section we expand the discussion by showing how to forecast a time series that has both trend and seasonal components. The approach we will take is first to remove the seasonal effect or seasonal component from the time series. This step is referred to as *deseasonalizing* the time series. After deseasonalizing, the time series will have only a trend component. As a result, we can use the method described in the previous section to identify the trend component of the time series. Then, using a trend projection calculation, we will be able to forecast the trend component of the time series in future periods. The final step in developing the forecast will be to incorporate the seasonal component by using a seasonal index to adjust the trend projection. In this manner, we will be able to identify the trend and seasonal components and consider both in forecasting the time series.

In addition to a trend component (T) and a seasonal component (S), we will assume that the time series also has an irregular component (I). The irregular component accounts for any random effects in the time series that cannot be explained by the trend and seasonal components. Using T_t, S_t, and I_t to identify the trend, seasonal, and irregular components at time t, we will assume that the actual time series value, denoted by Y_t, can be described by the following *multiplicative time series model:*

$$Y_t = T_t \times S_t \times I_t \tag{18.9}$$

In this model, T_t is the trend measured in units of the item being forecast. However, the S_t and I_t components are measured in relative terms, with values above 1.00 indicating effects above the trend, and values below 1.00 indicating effects below the trend.

In this section we will illustrate the use of the multiplicative model with trend, seasonal, and irregular components by working with the quarterly data presented in Table 18.13 and Figure 18.11. These data show the television set sales (in thousands of units) for a particular manufacturer over the past 4 years. We begin by showing how to identify the seasonal component of the time series.

FIGURE 18.11
Graph of Quarterly Television-Set Sales Time Series

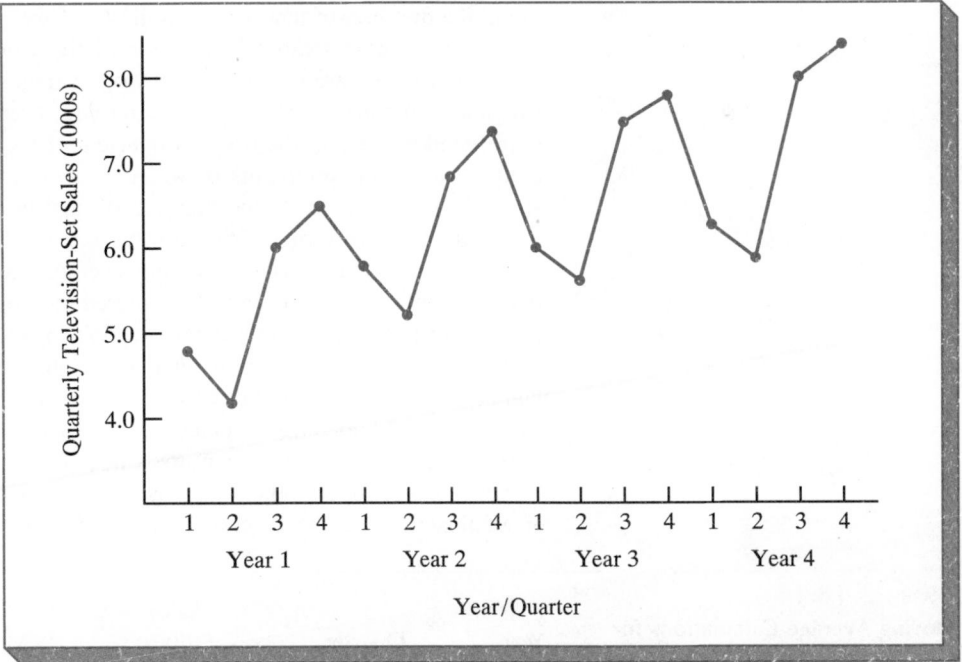

Calculating the Seasonal Indexes

Looking at Figure 18.11, we observe that sales are lowest in the second quarter of each year, followed by higher sales levels in quarters 3 and 4. Thus, we conclude that a seasonal pattern exists for the television-set sales. The computational procedure used to identify each quarter's seasonal influence begins by computing a moving average to isolate the combined seasonal and irregular components, S_t and I_t.

To do this, we use 1 year of data in each calculation. Since we are working with a quarterly series, we will use 4 data values in each moving average. The moving average calculation for the first 4 quarters of the television-set sales data is as follows:

$$\text{First Moving Average} = \frac{4.8 + 4.1 + 6.0 + 6.5}{4} = \frac{21.4}{4} = 5.35$$

Note that the moving average calculation for the first 4 quarters yields the average quarterly sales over the first year of the time series. Continuing the moving average calculation, we next add the 5.8 value for the first quarter of year 2 and drop the 4.8 for the first quarter of year 1. Thus, the second moving average is

$$\text{Second Moving Average} = \frac{4.1 + 6.0 + 6.5 + 5.8}{4} = \frac{22.4}{4} = 5.60$$

Similarly, the third moving average calculation is $(6.0 + 6.5 + 5.8 + 5.2)/4 = 5.875$.

Before we proceed with the moving average calculations for the entire time series, let us return to the first moving average calculation, which resulted in a value of 5.35. The 5.35 value represents an average quarterly sales volume (across all seasons) for year 1. As we look back at the calculation of the 5.35 value, perhaps it makes sense to associate 5.35 with the "middle" quarter of the moving average group. However, note that some difficulty in identifying the middle quarter is encountered; with 4 quarters in the moving average, there is no middle quarter. The 5.35 value corresponds to the last half of quarter

2 and the first half of quarter 3. Similarly, if we go to the next moving average value of 5.60, the middle corresponds to the last half of quarter 3 and the first half of quarter 4.

Recall that we are computing moving averages to isolate the combined seasonal and irregular components. However, the moving average values we have computed do not correspond directly to the original quarters of the time series. We can resolve this difficulty by using the midpoints between successive moving average values. For example, since 5.35 corresponds to the first half of quarter 3 and 5.60 corresponds to the last half of quarter 3, we will use $(5.35 + 5.60)/2 = 5.475$ as the moving average value for quarter 3. Similarly, we associate a moving average value of $(5.60 + 5.875)/2 = 5.738$ with quarter 4. What results is called a centered moving average. A complete summary of the moving average calculations for the television-set sales data is shown in Table 18.14.

Note that if the number of data points in a moving average calculation is an odd number, the middle point will correspond to one of the periods in the time series. In such cases, we would not have to center the moving average values to correspond to a particular time period, as we have done in the calculations in Table 18.14.

Let us pause for a moment to consider what the moving averages in Table 18.14 tell us about this time series. A plot of the actual time series values and the corresponding

TABLE 18.14

Moving Average Calculations for the Television-Set Sales Time Series

Year	Quarter	Sales (1000s)	4-Quarter Moving Average	Centered Moving Average
1	1	4.8		
	2	4.1		
			5.350	
	3	6.0		5.475
			5.600	
	4	6.5		5.738
			5.875	
2	1	5.8		5.975
			6.075	
	2	5.2		6.188
			6.300	
	3	6.8		6.325
			6.350	
	4	7.4		6.400
			6.450	
3	1	6.0		6.538
			6.625	
	2	5.6		6.675
			6.725	
	3	7.5		6.763
			6.800	
	4	7.8		6.838
			6.875	
4	1	6.3		6.938
			7.000	
	2	5.9		7.075
			7.150	
	3	8.0		
	4	8.4		

FIGURE 18.12

Graph of Quarterly Television-Set Sales Time Series and Centered Moving Average

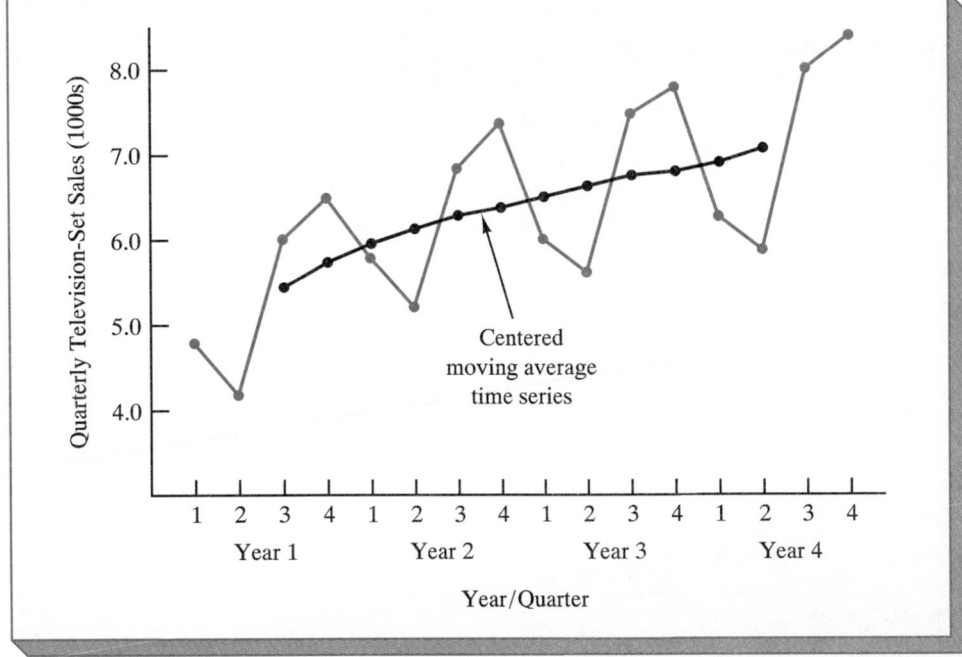

centered moving average is shown in Figure 18.12. Note particularly how the centered moving average values tend to smooth out the fluctuations in the time series. Since the moving average values were computed for 4 quarters of data, they do not include the fluctuations due to seasonal influences. Each point in the centered moving average represents what the value of the time series would be if there were no seasonal or irregular influence.

By dividing each time series observation by the corresponding centered moving average, we can identify the seasonal-irregular effect in the time series. For example, the third quarter of year 1 shows $6.0/5.475 = 1.096$ as the combined seasonal-irregular component. The resulting seasonal-irregular values for the entire time series values are summarized in Table 18.15.

Consider the third quarter. The results from years 1, 2, and 3 show third-quarter values of 1.096, 1.075, and 1.109, respectively. Thus, in all cases, the seasonal-irregular component appears to have an above-average influence in the third quarter. Since the year-to-year fluctuations in the seasonal-irregular component can be attributed primarily to the irregular component, we can average the computed values to eliminate the irregular influence and obtain an estimate of the third-quarter seasonal influence:

$$\text{Seasonal Effect of Third Quarter} = \frac{1.096 + 1.075 + 1.109}{3} = 1.09$$

We refer to 1.09 as the *seasonal index* for the third quarter. In Table 18.16 we summarize the calculations involved in computing the seasonal indexes for the television-set sales time series. Thus, we see that the seasonal indexes for all 4 quarters are as follows: quarter 1, .93; quarter 2, .84; quarter 3, 1.09; and quarter 4, 1.14.

Interpretation of the values in Table 18.16 provides some observations about the seasonal component in television-set sales. The best sales quarter is the fourth quarter, with sales averaging 14% above the average quarterly value. The worst, or slowest, sales quarter is the second quarter, with its seasonal index at .84, showing the sales average 16% below the average quarterly sales. The seasonal component corresponds nicely to the intuitive expectation that television-viewing interest and thus television-purchase patterns

TABLE 18.15
Seasonal-Irregular Factors for the
Television-Set Sales Time Series

Year	Quarter	Sales (1000s)	Centered Moving Average	Seasonal-Irregular Component
1	1	4.8		
	2	4.1		
	3	6.0	5.475	1.096
	4	6.5	5.738	1.133
2	1	5.8	5.975	.971
	2	5.2	6.188	.840
	3	6.8	6.325	1.075
	4	7.4	6.400	1.156
3	1	6.0	6.538	.918
	2	5.6	6.675	.839
	3	7.5	6.763	1.109
	4	7.8	6.838	1.141
4	1	6.3	6.938	.908
	2	5.9	7.075	.834
	3	8.0		
	4	8.4		

TABLE 18.16
Seasonal Index Calculations for the
Television-Set Sales Time Series

Quarter	Seasonal-Irregular Component Values $(S_t I_t)$	Seasonal Index (S_t)
1	.971, .918, .908	.93
2	.840, .839, .834	.84
3	1.096, 1.075, 1.109	1.09
4	1.133, 1.156, 1.141	1.14

tend to peak in the fourth quarter, with its coming winter season and fewer outdoor activities. The low second-quarter sales reflect the reduced interest in television viewing that results from the spring and presummer activities of potential customers.

One final adjustment is sometimes necessary in obtaining the seasonal indexes. The multiplicative model requires that the average seasonal index equal 1.00; that is, the sum of the four seasonal indexes in Table 18.16 must equal 4.00. This is necessary if the seasonal effects are to even out over the year, as they must. The average of the seasonal indexes in our example is equal to 1.00, and hence this type of adjustment is not necessary. In some cases, a slight adjustment may be necessary. The adjustment can be made by simply multiplying each seasonal index by the number of seasons divided by the sum of the unadjusted seasonal indexes. For example, for quarterly data we would multiply each seasonal index by 4/(sum of the unadjusted seasonal indexes). Some of the exercises will require this adjustment to obtain the appropriate seasonal indexes.

Deseasonalizing the Time Series

Often, the purpose of finding seasonal indexes is to remove the seasonal effects from a time series. This process is referred to as *deseasonalizing* the time series. Economic time series adjusted for seasonal variations (deseasonalized time series) are often reported in publications such as the *Survey of Current Business* and *The Wall Street Journal*. Using the notation of the multiplicative model, we have

$$Y_t = T_t \times S_t \times I_t$$

By dividing each time series observation by the corresponding seasonal index, we have removed the effect of season from the time series. The deseasonalized time series for television-set sales is summarized in Table 18.17. A graph of the deseasonalized television-set sales time series is shown in Figure 18.13.

Using the Deseasonalized Time Series to Identify Trend

Looking at Figure 18.13, we see that while the graph shows some up-and-down movement over the past 16 quarters, the time series seems to have an upward linear trend. To identify this trend, we will use the same procedure we introduced for identifying trends when forecasting with annual data; in this case, since we have deseasonalized the data, quarterly sales values can be used. Thus, for a linear trend the estimated sales volume expressed as a function of time can be written

$$T_t = b_0 + b_1 t$$

where

T_t = trend value for television-set sales in period t

b_0 = intercept of the trend line

b_1 = slope of the trend line

As we did before, we will let $t = 1$ for the time of the first observation on the time series data, $t = 2$ for the time of the second observation, and so on. Thus, for the deseasonalized television-set sales time series, $t = 1$ corresponds to the first deseasonalized quarterly sales value, and $t = 16$ corresponds to the most recent deseasonalized quarterly sales value. The formulas for computing the value of b_0 and the value of b_1 are shown again

TABLE 18.17

Deseasonalized Values for the Television-Set Sales Time Series

Year	Quarter	Sales (1000s) (Y_t)	Seasonal Index (S_t)	Deseasonalized Sales $(Y_t/S_t = T_t I_t)$
1	1	4.8	.93	5.16
	2	4.1	.84	4.88
	3	6.0	1.09	5.50
	4	6.5	1.14	5.70
2	1	5.8	.93	6.24
	2	5.2	.84	6.19
	3	6.8	1.09	6.24
	4	7.4	1.14	6.49
3	1	6.0	.93	6.45
	2	5.6	.84	6.67
	3	7.5	1.09	6.88
	4	7.8	1.14	6.84
4	1	6.3	.93	6.77
	2	5.9	.84	7.02
	3	8.0	1.09	7.34
	4	8.4	1.14	7.37

FIGURE 18.13

Deseasonalized Television-Set Sales Time Series

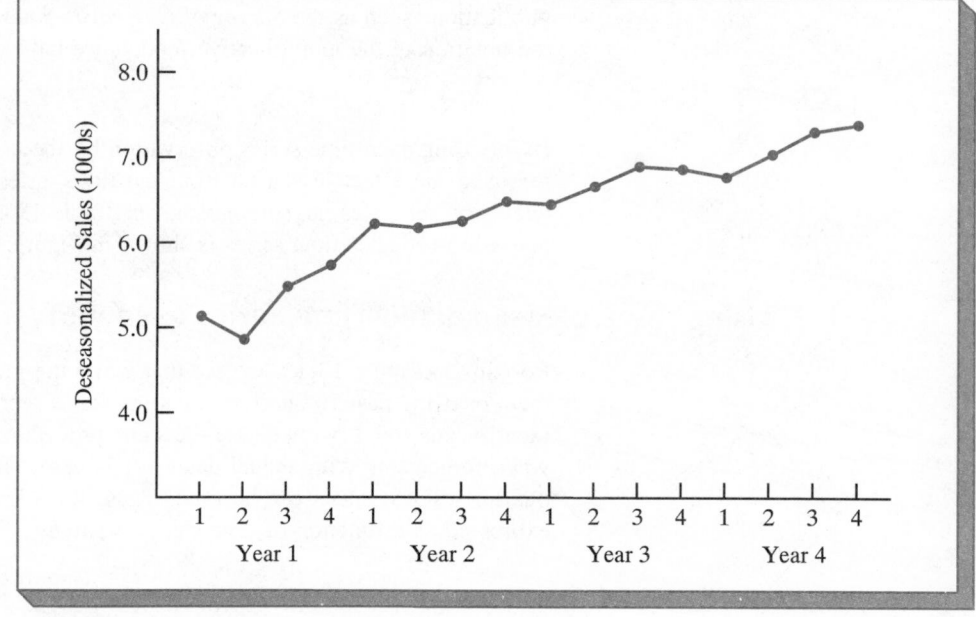

$$b_1 = \frac{\Sigma tY_t - (\Sigma t \Sigma Y_t)/n}{\Sigma t^2 - (\Sigma t)^2/n}$$

$$b_0 = \overline{Y} - b_1 \overline{t}$$

Note, however, that Y_t now refers to the deseasonalized time series value at time t and not to the actual value of the time series. Using the given relationships for b_0 and b_1 and the deseasonalized sales data of Table 18.17, we have the following calculations:

t	Y_t (Deseasonalized)	tY_t	t^2
1	5.16	5.16	1
2	4.88	9.76	4
3	5.50	16.50	9
4	5.70	22.80	16
5	6.24	31.20	25
6	6.19	37.14	36
7	6.24	43.68	49
8	6.49	51.92	64
9	6.45	58.05	81
10	6.67	66.70	100
11	6.88	75.68	121
12	6.84	82.08	144
13	6.77	88.01	169
14	7.02	98.28	196
15	7.34	110.10	225
16	7.37	117.92	256
Totals 136	101.74	914.98	1496

where

$$\bar{t} = \frac{136}{16} = 8.5$$

$$\bar{Y} = \frac{101.74}{16} = 6.359$$

$$b_1 = \frac{914.98 - (136)(101.74)/16}{1496 - (136)^2/16} = .148$$

$$b_0 = 6.359 - .148(8.5) = 5.101$$

Therefore,

$$T_t = 5.101 + .148t$$

is the expression for the linear trend component of the time series.

The slope of .148 indicates that over the past 16 quarters, the firm has experienced an average deseasonalized growth in sales of around 148 sets per quarter. If we assume that the past 16-quarter trend in sales data is a reasonably good indicator of the future, then this equation can be used to project the trend component of the time series for future quarters. For example, substituting $t = 17$ into the equation yields next quarter's trend projection, T_{17}:

$$T_{17} = 5.101 + .148(17) = 7.617$$

Using the trend component only, we would forecast sales of 7617 television sets for the next quarter. In a similar fashion, if we use the trend component only, we would forecast sales of 7765, 7913, and 8061 television sets in quarters 18, 19, and 20, respectively.

Seasonal Adjustments

Now that we have a forecast of sales for each of the next 4 quarters based on trend, we must adjust these forecasts to account for the effect of season. For example, since the seasonal index for the first quarter of year 5 ($t = 17$) is .93, the quarterly forecast can be obtained by multiplying the forecast based on trend ($T_{17} = 7617$) times the seasonal index (.93). Thus, the forecast for the next quarter is 7617(.93) = 7084. Table 18.18 shows the quarterly forecast for quarters 17, 18, 19, and 20. The quarterly forecasts show the high-volume fourth quarter with a 9190 unit forecast, while the low-volume second quarter has a forecast of 6523 units.

TABLE 18.18
Quarterly Forecasts for the
Television-Set Sales Time Series

Year	Quarter	Trend Forecast	Seasonal Index (see Table 18.16)	Quarterly Forecast
5	1	7617	.93	(7617)(.93) = 7084
	2	7765	.84	(7765)(.84) = 6523
	3	7913	1.09	(7913)(1.09) = 8625
	4	8061	1.14	(8061)(1.14) = 9190

Models Based on Monthly Data

The television-set sales example provided in this section used quarterly data to illustrate the computation of seasonal indexes with relatively few computations. Many businesses use monthly rather than quarterly forecasts. In such cases, the procedures introduced in this section can be applied with minor modifications. First, a 12-month moving average replaces the 4-quarter moving average; second, 12 monthly seasonal indexes, rather than 4 quarterly seasonal indexes, will need to be computed. Other than these changes, the computational and forecasting procedures are identical. Exercise 23 at the end of this section asks you to develop monthly seasonal indexes for a situation requiring monthly forecasts.

Cyclical Component

Mathematically, the multiplicative model of (18.9) can be expanded to include a cyclical component as follows:

$$Y_t = T_t \times C_t \times S_t \times I_t \tag{18.10}$$

Just as with the seasonal component, the cyclical component is expressed as a percent of trend. As mentioned in Section 18.1, this component is attributable to multiyear cycles in the time series. It is analogous to the seasonal component, but over a longer period of time. However, because of the length of time involved, it is often difficult to obtain enough relevant data to estimate the cyclical component. Another difficulty is that cycles usually vary in their length. We leave further discussion of the cyclical component to texts on forecasting methods.

❑ ❑ Exercises

Methods

S ELF T EST ▶

21. Consider the time series data shown in Table 18.19.
a. Show the 4-quarter and centered moving average values for this time series.
b. Compute seasonal indexes for the 4 quarters.

Applications

22. The quarterly sales data (number of copies sold) for a college textbook over the past 3 years are as follows:

T ABLE **18.19**

	Year		
Quarter	1	2	3
1	4	6	7
2	2	3	6
3	3	5	6
4	5	7	8

Quarter	Year 1	Year 2	Year 3
1	1690	1800	1850
2	940	900	1100
3	2625	2900	2930
4	2500	2360	2615

a. Show the 4-quarter and centered moving average values for this time series.
b. Compute seasonal indexes for the 4 quarters.
c. When does the textbook publisher experience the largest seasonal index? Does this appear reasonable? Explain.

TABLE 18.20

Month	Year		
	1	2	3
January	170	180	195
February	180	205	210
March	205	215	230
April	230	245	280
May	240	265	290
June	315	330	390
July	360	400	420
August	290	335	330
September	240	260	290
October	240	270	295
November	230	255	280
December	195	220	250

23. Identify the monthly seasonal indexes for the 3 years of expenses for a 6-unit apartment house in southern Florida shown in Table 18.20. Use a 12-month moving average calculation.

24. Air pollution control specialists in southern California monitor the amount of ozone, carbon dioxide, and nitrogen dioxide in the air on an hourly basis (*Los Angeles Times,* July 21, 1989). The hourly time series data exhibit seasonality with the levels of pollutants showing similar patterns over the hours in the day. On July 15, 16, and 17, the observed levels of nitrogen dioxide in the downtown area for the 12 hours from 6:00 A.M. to 6:00 P.M. were as follows:

July 15:	25	28	35	50	60	60	40	35	30	25	25	20
July 16:	28	30	35	48	60	65	50	40	35	25	20	20
July 17:	35	42	45	70	72	75	60	45	40	25	25	25

a. Identify the hourly seasonal factors for the 12-hour daily readings.
b. Using the seasonal factors from (a), the data was deseasonalized; the trend equation developed for the deseasonalized data was $T_t = 32.983 + .3922t$. Using the trend component only, develop forecasts for the 12 hours for July 18.
c. Use the seasonal factors from (a) to adjust the trend forecasts developed in (b).

18.5   Forecasting Time Series Using Regression Models

In our discussion of regression analysis in Chapters 14, 15, and 16, we showed how one or more independent variables could be used to predict the value of a single dependent variable. Looking at regression analysis as a forecasting tool, the time series value that we would like to forecast can be viewed as the dependent variable. Thus, if we can identify a good set of related independent, or predictor, variables we may be able to develop an estimated regression equation for predicting or forecasting the time series.

The approach we used in Section 18.3 to fit a linear trend line to the bicycle sales time series is a special case of regression analysis. In that example, two variables—bicycle sales and time—were shown to be linearly related.* The inherent complexity of most real-world problems necessitates the consideration of more than one variable to predict the

---

*In a purely technical sense, the number of bicycles sold is not thought of as being related to time; instead, time is used as a surrogate for variables that the number of bicycles sold is actually related to but that are either unknown or too difficult or too costly to measure.

variable of interest. The statistical technique known as multiple regression analysis can be used in such situations.

Recall that to develop an estimated regression equation, we need a sample of observations for the dependent variable and all independent variables. In time series analysis the n periods of time series data provide a sample of n observations on each variable that can be used in the analysis. For a function involving k independent variables, we use the following notation:

$$Y_t = \text{actual value of the time series in period } t$$

$$x_{1t} = \text{value of independent variable 1 in period } t$$

$$x_{2t} = \text{value of independent variable 2 in period } t$$

.

.

.

$$x_{kt} = \text{value of independent variable } k \text{ in period } t$$

The n periods of data necessary to develop the estimated regression equation would appear as follows:

Period	Time Series Value (Y_t)	Value of Independent Variables						
		x_{1t}	x_{2t}	x_{3t}	.	.	.	x_{kt}
1	Y_1	x_{11}	x_{21}	x_{31}	.	.	.	x_{k1}
2	Y_2	x_{12}	x_{22}	x_{32}	.	.	.	x_{k2}
.	.	.	.	.	.	.	.	.
.	.	.	.	.	.	.	.	.
.	.	.	.	.	.	.	.	.
n	Y_n	x_{1n}	x_{2n}	x_{3n}	.	.	.	x_{kn}

As you might imagine, there are a number of possible choices for the independent variables in a forecasting model. One possible choice for an independent variable is simply time. This is the choice we made in Section 18.3 when we estimated the trend of the time series using a linear function of the independent variable time. Letting

$$x_{1t} = t$$

we obtain an estimated regression equation of the form

$$\hat{Y}_t = b_0 + b_1 t$$

where $\hat{Y}_t$ is the estimate of the time series value Y_t and where b_0 and b_1 are the estimated regression coefficients. In a more complex model, additional terms could be added corresponding to time raised to other powers. For example, if

$$x_{2t} = t^2$$

and

$$x_{3t} = t^3$$

the estimated regression equation would then become

$$\hat{Y}_t = b_0 + b_1 x_{1t} + b_2 x_{2t} + b_3 x_{3t}$$

$$= b_0 + b_1 t + b_2 t^2 + b_3 t^3$$

Note that this model provides a forecast of a time series with curvilinear characteristics over time.

Other regression-based forecasting models employ a mixture of economic and demographic independent variables. For example, in forecasting the sale of refrigerators, we might select independent variables such as the following:

x_{1t} = price in period t

x_{2t} = total industry sales in period $t - 1$

x_{3t} = number of building permits for new houses in period $t - 1$

x_{4t} = population forecast for period t

x_{5t} = advertising budget for period t

According to the usual multiple regression procedure, an estimated regression equation with five independent variables would be used to develop forecasts.

Whether or not a regression approach provides a good forecast depends largely on how well we are able to identify and obtain data for independent variables that are closely related to the time series. Generally, during the development of an estimated regression equation, we will want to consider many possible sets of independent variables. Thus, part of the regression analysis procedure should focus on the selection of the set of independent variables that provides the best forecasting model.

In the chapter introduction we stated that the *causal forecasting models* utilize time series related to the one being forecast in an effort to better explain the cause of a time series behavior. Regression analysis is the tool most often used in developing these causal models. The related time series become the independent variables, and the time series being forecast is the dependent variable.

Another type of regression-based forecasting model occurs whenever the independent variables are all previous values of the same time series. For example, if the time series values are denoted by $Y_1, Y_2, \ldots, Y_n$, then with a dependent variable Y_t, we might try to find an estimated regression equation relating Y_t to the most recent times series values Y_{t-1}, Y_{t-2}, and so on. With the three most recent periods as independent variables, the estimated regression equation would be

$$\hat{Y}_t = b_0 + b_1 Y_{1-1} + b_2 Y_{t-2} + b_3 Y_{t-3}$$

Regression models where the independent variables are previous values of the time series are referred to as *autoregressive models*.

Finally, another regression-based forecasting approach is one that incorporates a mixture of the independent variables previously discussed. For example, we might select a combination of time variables, some economic/demographic variables, and some previous values of the time series variable itself.

18.6   Qualitative Approaches to Forecasting

In the previous sections we have discussed several types of quantitative forecasting methods. Since each of these techniques requires historical data on the variable of interest, these techniques cannot be applied in situations where no historical data are available.

Furthermore, even when historical data are available, a significant change in environmental conditions affecting the time series may make the use of past data questionable in predicting future values of the time series. For example, a government-imposed gasoline rationing program would cause one to question the validity of a gasoline sales forecast based on past data. Qualitative forecasting techniques offer an alternative in these, and other, cases.

One of the most commonly used qualitative forecasting methods is the *Delphi approach*. This technique, originally developed by a research group at the Rand Corporation, attempts to obtain forecasts through group consensus. In the usual application of this technique, the members of a panel of experts—all of whom are physically separated from and unknown to each other—are asked to respond to a series of questionnaires. The responses from the first questionnaire are tabulated and used to prepare a second questionnaire which contains information and opinions of the whole group. Each respondent is then asked to reconsider and possibly revise his or her previous response in light of the group information that has been provided. This basic process continues until the coordinator feels that some degree of consensus has been reached. Note that the goal of the Delphi approach is not to produce a single answer as output but to produce instead a relatively narrow spread of opinions within which the majority of experts concur.

The qualitative procedure referred to as *scenario writing* consists of developing a conceptual scenario of the future based on a well-defined set of assumptions. Thus, by starting with a different set of assumptions, many different future scenarios can be presented. The job of the decision maker is to decide which scenario is most likely to occur in the future and then to make decisions accordingly.

Subjective or *intuitive qualitative approaches* are based on the ability of the human mind to process a variety of information that is, in most cases, difficult to quantify. These techniques are often used in group work, wherein a committee or panel seeks to develop new ideas or solve complex problems through a series of brainstorming sessions. In such sessions, individuals are freed from the usual group restrictions of peer pressure and criticism, since any idea or opinion can be presented without regard to its relevancy and, even more importantly, without fear of criticism.

Summary

The purpose of this chapter has been to provide an introduction to the basic methods of time series analysis and forecasting. First, we showed that to explain the behavior of a time series, it is often helpful to think of the time series as consisting of four separate components: trend, cyclical, seasonal, and irregular. By isolating these components and measuring their apparent effect, it is possible to forecast future values of the time series.

We discussed how smoothing methods can be used to forecast a time series that exhibits no significant trend, seasonal, or cyclical effect. The moving averages approach consists of computing an average of past data values and then using this average as the forecast for the next period. The exponential smoothing method uses a weighted average of past time series values to compute a forecast.

When the time series exhibits only a long-term trend, we showed how regression analysis could be used to make trend projections. When both trend and seasonal influences are significant, we showed how to isolate the effects of the two factors and prepare better forecasts. Finally, regression analysis was described as a procedure for developing so-called causal forecasting models. A causal forecasting model is one that relates the time series value (dependent variable) to other independent variables that are believed to explain (cause) the time series behavior.

Qualitative forecasting methods were discussed as approaches that could be used when little or no historical data were available. These methods are also considered most appropriate when the past pattern of the time series is not expected to continue into the future.

It is important to realize that time series analysis and forecasting is a major field in its own right. In this chapter we have just scratched the surface of the field of time series and forecasting methodology.

Glossary

Time series A set of observations measured at successive points in time or over successive periods of time.

Forecast A projection or prediction of future values of a time series.

Trend The long-run shift or movement in the time series observable over several periods of time.

Cyclical component The component of the time series model that results in periodic above-trend and below-trend behavior of the time series lasting more than 1 year.

Seasonal component The component of the time series model that shows a periodic pattern over 1 year or less.

Irregular component The component of the time series model that reflects the random variation of the actual time series values beyond what can be explained by the trend, cyclical, and seasonal components.

Moving averages A method of forecasting or smoothing a time series by averaging each successive group of data points. The moving averages method can be used to isolate the seasonal component of the time series.

Mean squared error (MSE) One approach to measuring the accuracy of a forecasting model. This measure is the average of the sum of the squared differences between the forecast values and the actual time series values.

Weighted moving averages A method of forecasting or smoothing a time series by computing a weighted average of past data values. The sum of the weights must equal 1.

Exponential smoothing A forecasting technique that uses a weighted average of past time series values to arrive at smoothed time series values which can be used as forecasts.

Smoothing constant A parameter of the exponential smoothing model which provides the weight given to the most recent time series value in the calculation of the forecast value.

Multiplicative time series model A model that assumes that the separate components of the time series can be multiplied together to identify the actual time series value. When the four components of trend, cyclical, seasonal, and irregular are assumed present, we obtain: $Y_t = T_t \times C_t \times S_t \times I_t$. When the cyclical component is not modeled, we obtain: $Y_t = T_t \times S_t \times I_t$.

Deseasonalized time series A time series that has had the effect of season removed by dividing each original time series observation by the corresponding seasonal index.

Causal forecasting methods Forecasting methods that relate a time series to other variables that are believed to explain or cause its behavior.

Autoregressive model A time series model that uses a regression relationship based on past time series values to predict the future time series values.

Delphi approach A qualitative forecasting method that obtains forecasts through group consensus.

Scenario writing A qualitative forecasting method which consists of developing a conceptual scenario of the future based on a well-defined set of assumptions.

Key Formulas

Moving Average

$$\text{Moving Average} = \frac{\Sigma(\text{most recent } n \text{ data values})}{n} \tag{18.1}$$

Exponential Smoothing Model

$$F_{t+1} = \alpha Y_t + (1 - \alpha)F_t \tag{18.2}$$

or

$$F_{t+1} = F_t + \alpha(Y_t - F_t) \tag{18.3}$$

Equation for Linear Trend

$$T_t = b_0 + b_1 t \tag{18.5}$$

Multiplicative Time Series Model with Seasonal Component

$$Y_t = T_t \times S_t \times I_t \tag{18.9}$$

Multiplicative Time Series Model with Seasonal and Cyclical Components

$$Y_t = T_t \times C_t \times S_t \times I_t \tag{18.10}$$

❏ ❏ Supplementary Exercises

25. Data below show the monthly percentage of all shipments that were received on time during 1988 (*Purchasing*, February 23, 1989).

80, 82, 84, 83, 83, 84, 85, 84, 82, 83, 84, 83.

a. Compare a 3-month moving averages forecast with an exponential smoothing forecast using $\alpha = .2$. Which provides the better forecasts?

b. What is the forecast for January 1989?

26. The number of component parts used in a production process the last 10 weeks are shown in Table 18.21. Using a smoothing constant of $\alpha = .25$, develop the exponential smoothing values for this time series. Indicate your forecast for next week.

27. The percentage of individual investors' portfolios committed to stock varies depending on the state of the economy (*AAII Journal*, April 1992). As of February 1982, a typical portfolio showed stocks (32%) were followed by stock funds (26%), cash (23%), bonds (10%) and bond fund (9%) in terms of where individuals were placing their assets. Shown below are the percentage of funds in stocks for 9 quarters from 1990 to 1992.

TABLE 18.21

Week	Parts
1	200
2	350
3	250
4	360
5	250
6	210
7	280
8	350
9	290
10	320

TABLE 18.22

Week	Demand
1	22
2	18
3	23
4	21
5	17
6	24
7	20
8	19
9	18
10	21

Quarter	Stock %
1st—1990	27.0
2nd—1990	26.0
3rd—1990	26.5
4th—1990	27.5
1st—1991	25.5
2nd—1991	30.0
3rd—1991	28.0
4th—1991	30.5
1st—1992	32.0

a. Use exponential smoothing to forecast this time series. Consider smoothing constants of $\alpha = .2, .3,$ and $.4$. What value of the smoothing constant provides the best forecast?

b. What is the forecast of the percentage of assets committed to stocks for the second quarter of 1992?

28. A chain of grocery stores experienced the weekly demand (in cases) shown in Table 18.22 for a particular brand of automatic-dishwasher detergent. Use exponential smoothing with $\alpha = .2$ to develop a forecast for week 11.

29. United Dairies, Inc., supplies milk to several independent grocers throughout Dade County, Florida. Management at United Dairies would like to develop a forecast of the number of half-gallons of milk sold per week. Sales data for the past 12 weeks are as follows:

Week	Sales (Units)	Week	Sales (Units)
1	2750	7	3300
2	3100	8	3100
3	3250	9	2950
4	2800	10	3000
5	2900	11	3200
6	3050	12	3150

Using exponential smoothing with $\alpha = .4$, develop a forecast of demand for the 13th week.

30. Ten weeks of data on the Commodity Futures Index are shown below (*Security Traders Handbook,* April 21, 1989):

$$7.35, \ 7.40, \ 7.55, \ 7.56, \ 7.60, \ 7.52, \ 7.52, \ 7.70, \ 7.62, \ 7.55.$$

a. Compute the exponential smoothing forecasts using $\alpha = .2$.
b. Compute the exponential smoothing forecasts using $\alpha = .3$.
c. Which exponential smoothing model provides the better forecasts? What is the forecast for the next week?

31. The vacancy rate for office rentals is reported in terms of the percentage of available offices that are not rented. Office vacancy rates for downtown Philadelphia from 1980 to 1987 are shown in Table 18.23 (*Business Review,* May–June 1989).
a. Develop a linear trend for this time series.
b. Provide forecasts of the vacancy rate for 1988, 1989 and 1990.
c. Should city planners be concerned with the forecasts of office vacancy? What conclusion should be reached, and what possible actions should the city planners consider?

T A B L E 18.23

Year	Vacancy Rate
1980	5.9
1981	4.6
1982	6.4
1983	9.5
1984	9.2
1985	9.5
1986	10.8
1987	11.0

32. In 1990 and 1991, the nation's banks were feeling the effects of the recession (*USA Today,* September 11, 1991). In particular, banks were struggling with the problems of bad real estate loans. Bad real estate loans in billions of dollars are shown for the 4 quarters of 1990 and the first 2 quarters of 1991.

Quarter	Bad Loans
1st—1990	24.5
2nd—1990	28.2
3rd—1990	31.0
4th—1990	36.0
1st—1991	40.1
2nd—1991	41.4

T A B L E 18.24

Year	Price Index
1	66.9
2	74.8
3	81.2
4	85.0
5	89.2
6	94.6
7	97.8
8	101.9
9	106.9

Fit a linear trend equation to these data. What is the average increase in bad real estate loans per quarter? What is the projected total amount of bad real estate loans by the third quarter of 1991?

33. The time series for the retail price index of consumer goods and services is shown in Table 18.24 (*Accountancy,* July 1989). Use trend projection to forecast the retail price index for years 10 and 11.

34. Canton Supplies, Inc., is a service firm that employs approximately 100 individuals. Because of the necessity of meeting monthly cash obligations, management of Canton Supplies would like

to develop a forecast of monthly cash requirements. Because of a recent change in operating policy, only the past 7 months of data are considered to be relevant. Use the historical data shown below to develop a forecast of cash requirements for each of the next 2 months using trend projection.

Month	1	2	3	4	5	6	7
Cash Required ($1000s)	205	212	218	224	230	240	246

35. Data below show the time series of the most recent quarterly capital expenditures in billions of dollars for the 1000 largest manufacturing firms (*Manufacturing Investment Outlook,* Spring 1989).

24, 25, 23, 24, 22, 26, 28, 31, 29, 32, 37, 42.

a. Develop a linear trend expression for the above time series.
b. Show a graph of the time series and the linear trend expression.
c. Using the time series, what appears to be happening to the capital expenditures? What is the forecast 1 year or 4 quarters into the future?

36. The Costello Music Company has been in business for 5 years. During this time the sale of electric organs has grown from 12 units in the first year to 76 units in the most recent year. Fred Costello, the firm's owner, would like to develop a forecast of organ sales for the coming year. The historical data are shown below:

Year	1	2	3	4	5
Sales	12	28	34	50	76

a. Show a graph of this time series. Does a linear trend appear to exist?
b. Develop the equation for the linear trend component for the above time series. What is the average increase in sales that the firm has been realizing per year?

37. Hudson Marine has been an authorized dealer for C&D marine radios for the past 7 years. The number of radios sold each year is shown below:

Year	1	2	3	4	5	6	7
Number Sold	35	50	75	90	105	110	130

a. Show a graph of this time series. Does a linear trend appear to exist?
b. Develop the equation for the linear trend component for the above time series.
c. Use the linear trend developed in (b) and prepare a forecast for annual sales in year 8.

38. *Software Magazine* (March, 1992) reported a study that predicted spending for data management software through 1995. The projection of software spending in millions of dollars was made for personal computer (PC) systems and mainframe systems. The data provided are shown in Table 18.25.

a. Fit a linear trend to both the PC systems data and the mainframe systems data.
b. Forecast data management software sales for PC systems and mainframe computer systems in 1996.
c. Using the trend equation, compare what is happening to data management software sales for PC systems and mainframe computer systems.

39. Refer to Exercise 37. Suppose that the quarterly sales values for the 7 years of historical data are as follows:

Tᴀʙʟᴇ **18.25**

Year	PC Systems	Mainframe Systems
1990	21	147
1991	80	182
1992	175	195
1993	215	198
1994	262	210
1995	326	239

Year	Quarter 1	Quarter 2	Quarter 3	Quarter 4	Total Sales
1	6	15	10	4	35
2	10	18	15	7	50
3	14	26	23	12	75
4	19	28	25	18	90
5	22	34	28	21	105
6	24	36	30	20	110
7	28	40	35	27	130

a. Show the 4-quarter moving average values for this time series. Plot both the original time series and the moving average series on the same graph.
b. Compute the seasonal indexes for the 4 quarters.
c. When does Hudson Marine experience the largest seasonal effect? Does this seem reasonable? Explain.

40. Consider the Costello Music Company problem presented in Exercise 36. The quarterly sales data are shown below:

Year	Quarter 1	Quarter 2	Quarter 3	Quarter 4	Total Yearly Sales
1	4	2	1	5	12
2	6	4	4	14	28
3	10	3	5	16	34
4	12	9	7	22	50
5	18	10	13	35	76

a. Compute the seasonal indexes for the 4 quarters.
b. When does Costello Music experience the largest seasonal effect? Does this appear reasonable? Explain.

41. Refer to the Hudson Marine data presented in Exercise 39.
a. Deseasonalize the data, and use the deseasonalized time series to identify the trend.
b. Use the results of (a) to develop a quarterly forecast for next year based on trend.
c. Use the seasonal indexes developed in Exercise 39 to adjust the forecasts developed in (b) to account for the effect of season.

42. Consider the Costello Music Company time series presented in Exercise 40.
a. Deseasonalize the data, and use the deseasonalized time series to identify the trend.
b. Use the results of (a) to develop a quarterly forecast for next year based on trend.
c. Use the seasonal indexes developed in Exercise 40 to adjust the forecasts developed in (b) to account for the effect of season.

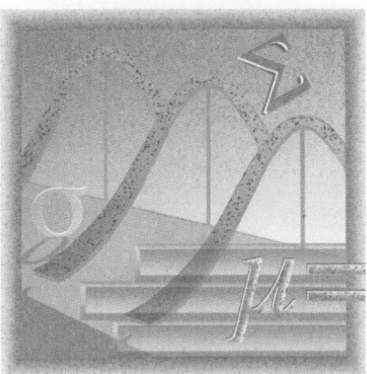

Nonparametric Methods

West Shell Realtors*

CINCINNATI, OHIO

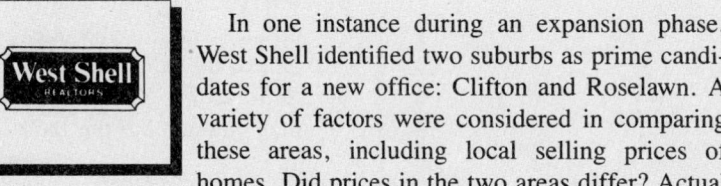

West Shell Realtors was founded in 1958 with one office and a sales staff of three people. The company's first-year sales were $900,000. In 1964, the company began a long-term expansion program, with new offices being added almost yearly. West Shell is now one of the largest realtors in Greater Cincinnati with offices also located in southwest Ohio, southeast Indiana, and northern Kentucky.

Statistical analysis helps real estate firms such as West Shell monitor sales performance in their efforts to remain competitive. Monthly reports are generated for each of West Shell's offices as well as for the total company. Statistical summaries of total sales dollars, number of units sold, and mean selling price per unit are essential in keeping both office managers and the company's top management informed of progress and trouble spots in the organization.

In addition to monthly summaries of ongoing operations, the company uses statistical considerations to guide corporate plans and strategies. West Shell has implemented a strategy of planned expansion in recent years. Each time an expansion plan calls for the establishment of a new sales office, the company must address the question of office location. Selling prices of homes, turnover rates, and forecast sales volumes are the types of data used in evaluating and comparing alternative locations.

In one instance during an expansion phase, West Shell identified two suburbs as prime candidates for a new office: Clifton and Roselawn. A variety of factors were considered in comparing these areas, including local selling prices of homes. Did prices in the two areas differ? Actual sales prices in the two areas were viewed as two populations. West Shell employed nonparametric statistical methods using small samples to help identify any differences.

Samples of 25 sales in the Clifton area and 18 sales in the Roselawn area were taken, and the Mann-Whitney-Wilcoxon rank-sum test was chosen as an appropriate statistical test of the difference in sales prices. At the .05 level of significance, the Mann-Whitney-Wilcoxon test did not allow rejection of the null hypothesis that the two populations of sales prices were identical. Thus, West Shell was able to focus on criteria other than sales price in their selection process.

The real estate business continues to be extremely competitive. At West Shell, statistical considerations play a meaningful role in helping the company maintain its leadership in the industry. In this chapter we will learn how nonparametric statistical tests, such as the Mann-Whitney-Wilcoxon test are applied. We shall also discuss the proper interpretation of such tests.

*The authors are indebted to Rodney Fightmaster of West Shell Realtors for providing this Statistics in Practice.

The statistical methods presented thus far in the text are generally referred to as *parametric methods*. In this chapter we introduce several statistical methods that are referred to as *nonparametric methods*. These nonparametric methods are often applicable in situations where the parametric methods of the preceding chapters are not. Nonparametric methods typically require less restrictive assumptions concerning the level of data measurement and fewer assumptions concerning the form of the probability distributions generating the sample data.

One consideration used to determine whether a parametric or a nonparametric method should be used is the scale of measurement used to generate the data. As discussed in Chapter 1, there are four scales of measurement: nominal, ordinal, interval, and ratio. All data are generated by one of these four scales of measurement; thus, all statistical analyses are conducted with either nominal, ordinal, interval, or ratio data.

Let us review the definition of the four scales of measurement. Examples of each were provided in Chapter 1.

1. *Nominal scale.* The scale of measurement is nominal if the data are simply labels or categories used to define an attribute of the element.
2. *Ordinal scale.* The scale of measurement is ordinal if the data have the properties of nominal data *and* the data can be used to rank, or order, the observations.
3. *Interval scale.* The scale of measurement is interval if the data have the properties of ordinal data *and* the interval between observations is expressed in terms of a fixed unit of measure.
4. *Ratio scale.* The scale of measurement is ratio if the data have the properties of interval data *and* the ratio of observations is meaningful.

Most of the statistical methods referred to as parametric require the use of interval- or ratio-scaled data. With these levels of measurement, means, variances, standard deviations, and so on, can be computed, interpreted, and used in the analysis. With nominal or ordinal data, it is inappropriate to compute means, variances, and standard deviations; hence, parametric methods normally cannot be used. Thus, whenever the data are nominal or ordinal, nonparametric methods are often the only way to analyze the data and draw statistical conclusions.

Another consideration used to determine whether a parametric method or a nonparametric method should be employed is the assumption concerning the population from which the data were obtained. For example, a parametric procedure for testing a hypothesis about the difference between the means of two populations was presented in Chapter 10. In the small-sample case, the *t* distribution can be used for this test provided we are willing to assume that the populations are normally distributed with equal variances. If this assumption about the populations is not appropriate, the parametric method based on the use of the *t* distribution should not be used even if the data are interval or ratio scaled. However, nonparametric methods, which require no assumptions about the population probability distributions are available for testing for differences between two populations. Because of this and other cases in which no population assumptions are required, nonparametric methods are often referred to as *distribution-free* methods.

In general, for a statistical method to be classified as nonparametric, it must satisfy at least one of the following conditions:*

1. The method may be used with nominal data.
2. The method may be used with ordinal data.
3. The method may be used with interval or ratio data when no assumption can be made about the population probability distribution.

If the level of data measurement is interval or ratio and if the necessary probability distribution assumptions for the population are appropriate, parametric methods provide more powerful or more discerning statistical procedures. In many cases where a nonparametric method as well as a parametric method can be applied, the nonparametric method is almost as good or almost as powerful as the parametric method. In cases where the data are nominal or ordinal or in cases where the assumptions required by parametric methods are inappropriate, only nonparametric methods are available. Because of the less restrictive data-measurement requirements and the fewer assumptions needed concerning the population distribution, nonparametric methods are regarded as more generally applicable than parametric methods. The sign test, the Wilcoxon signed-rank test, the Mann-Whitney-Wilcoxon test, the Kruskal-Wallis test, and Spearman rank correlation are the nonparametric methods presented in this chapter.

---

*See W. J. Conover, *Practical Nonparametric Statistics,* 2nd ed. (New York: John Wiley and Sons, 1980).

19.1   Sign Test

A common market research application of the *sign test* involves using a sample of *n* potential customers to identify a preference for one of two brands of a product such as coffee, soft drinks, and detergents. The *n* expressions of preference are nominal data because the consumer simply names, or labels, a preference. Given these data, our objective is to determine whether a difference in preference exists between the two items being compared. As we will see, the sign test is a nonparametric statistical procedure for answering this question.

Small-Sample Case

The small-sample case for the sign test should be used whenever $n \leq 20$. Let us illustrate the use of the sign test for the small-sample case by considering a study conducted for Sun Coast Farms. Sun Coast Farms produces an orange juice product marketed under the name "Citrus Valley." A competitor of Sun Coast Farms has begun producing a new orange juice product known as "Tropical Orange." In a study of consumer preferences for the two brands, 12 individuals were given unmarked samples of the two brands of orange juice. The brand each individual tasted first was randomly selected. After tasting the two products, the individuals were asked to state a preference for one of the two brands. The purpose of the study is to determine whether consumers prefer one product over the other. Letting *p* indicate the proportion of the population of consumers favoring Citrus Valley, we want to test the following hypotheses:

$$H_0: p = .50$$

$$H_a: p \neq .50$$

If H_0 cannot be rejected, there will be no evidence indicating that a difference in preference exists for the two brands of orange juice. However, if H_0 can be rejected, the conclusion can be made that the consumer preferences are different for the two brands. In this case, the brand selected by the greater number of consumers can be considered the most preferred brand.

In the following discussion we will show how the small-sample version of the sign test can be used to test these hypotheses and draw a conclusion about consumer preferences. In recording the preference data for the 12 individuals participating in the study, a + sign will be recorded if the individual expresses a preference for Citrus Valley and a − sign will be recorded if the individual expresses a preference for Tropical Orange. Using this procedure, the data will be recorded in terms of the + or − signs; this nonparametric test is referred to as the sign test.

Under the assumption that H_0 is true ($p = .50$), the number of + values follows a binomial probability distribution with $p = .50$. With a sample size of $n = 12$, Table 5 in Appendix B shows the probabilities for the binomial probability distribution with $p = .50$ as displayed in Table 19.1. A graphical representation of this binomial probability distribution is shown in Figure 19.1. This probability distribution shows the probability of the number of + signs under the assumption that H_0 is true and is therefore the appropriate sampling distribution for the hypothesis test. We use this sampling distribution to determine a rule for rejecting H_0; our approach will be similar to the method we used to develop rejection rules for hypothesis testing in Chapter 9. For example, using $\alpha = .05$, we would place a rejection region or area of approximately .025 in each tail of the distribution shown in Figure 19.1. Starting at the lower end of the distribution, we see that the probability of obtaining 0, 1, or 2 + signs is .0002 + .0029 + .0161 = .0192.

TABLE 19.1

Binomial Probabilities with $n = 12, p = .50$

Number of + Signs	Probability
0	.0002
1	.0029
2	.0161
3	.0537
4	.1208
5	.1934
6	.2256
7	.1934
8	.1208
9	.0537
10	.0161
11	.0029
12	.0002

FIGURE 19.1

Binomial Probabilities for the Number of + Signs when $n = 12$ and $p = .50$

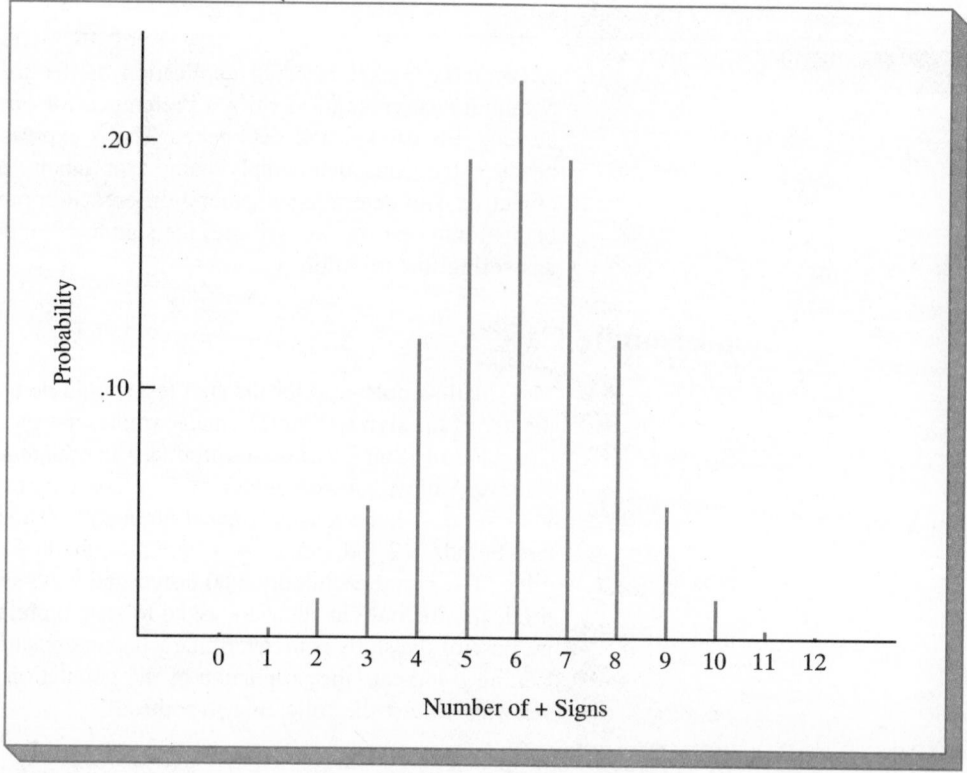

Note that we stop at 2 + signs because adding the probability of 3 + signs would make the area in the lower tail equal to .0192 + .0537 = .0729, which substantially exceeds the desired area of .025. At the upper end of the distribution, we find the same probability of .0192 corresponding to 10, 11, or 12 + signs. Thus, the closest we can come to $\alpha = .05$, without exceeding it, is .0192 + .0192 = .0384. As a result, we adopt the following rejection rule:

Reject H_0 if the number of + signs is less than 3 or greater than 9

The preference data that were obtained for the Sun Coast Farms example are shown in Table 19.2. Since only 2 + signs were observed, the null hypothesis is rejected. There is evidence from this study that consumer preference differs for the two brands of orange juice. We would advise Sun Coast Farms that consumers indicate a preference for the competitor's Tropical Orange brand.

In the Sun Coast Farms example, all 12 individuals in the study were able to state a preference. In many situations, one or more individuals in the sample may not be able to state a definite preference. In such cases, the individual's response of no preference can be removed from the study and the analysis conducted with a smaller sample size.

The binomial probability distribution as shown in Table 5 of Appendix B can be used to provide the decision rule for any sign test up to a sample size of $n = 20$. Using the null hypothesis $p = .50$ and the sample size n, the decision rule can be established for any level of significance. In addition, by considering the probabilities in only the lower or upper tail of the binomial probability distribution, rejection rules can also be developed for one-tailed tests. Appendix B does not provide binomial probability distribution tables for sample sizes greater than 20. In these cases, we can use the large-sample normal approximation of binomial probabilities to determine the appropriate rejection rule for the sign test.

TABLE 19.2
Preference Data for the Sun Coast Farms' Taste Test

Individual	Brand Preference	Recorded Data
1	Tropical Orange	−
2	Tropical Orange	−
3	Citrus Valley	+
4	Tropical Orange	−
5	Tropical Orange	−
6	Tropical Orange	−
7	Tropical Orange	−
8	Tropical Orange	−
9	Citrus Valley	+
10	Tropical Orange	−
11	Tropical Orange	−
12	Tropical Orange	−

Large-Sample Case

Using the null hypothesis H_0: $p = .50$ and a sample size of $n > 20$, the normal approximation of the sampling distribution for the number of + signs is as follows.

Normal Approximation of the Sampling Distribution of the Number of + Signs when No Preference Exists

$$\text{Mean: } \mu = .50n \qquad (19.1)$$

$$\text{Standard Deviation: } \sigma = \sqrt{.25n} \qquad (19.2)$$

Distribution form: approximately normal provided $n > 20$.

Let us consider an application of the sign test to political polling. A poll taken during a recent presidential election campaign asked 200 registered voters to rate the Democratic and Republican candidates in terms of best overall foreign policy. Results of the poll showed 72 rated the Democratic candidate higher, 103 rated the Republican candidate higher, and 25 indicated no difference between the candidates. Does the poll indicate that there is a significant difference between public opinion of the foreign policies of the two candidates?

Using the sign test, we see that $n = 200 - 25 = 175$ individuals were able to indicate the candidate they believed had the best overall foreign policy. Using (19.1) and (19.2), we find the sampling distribution of the number of + signs has the following properties:

$$\mu = .50n = .50(175) = 87.5$$

$$\sigma = .25n = \sqrt{.25(175)} = 6.6$$

In addition, with $n = 175$ we can assume that the sampling distribution is approximately normal. This distribution is shown in Figure 19.2. Since the distribution is approximately normal, we can use the table of areas for the standard normal probability distribution to

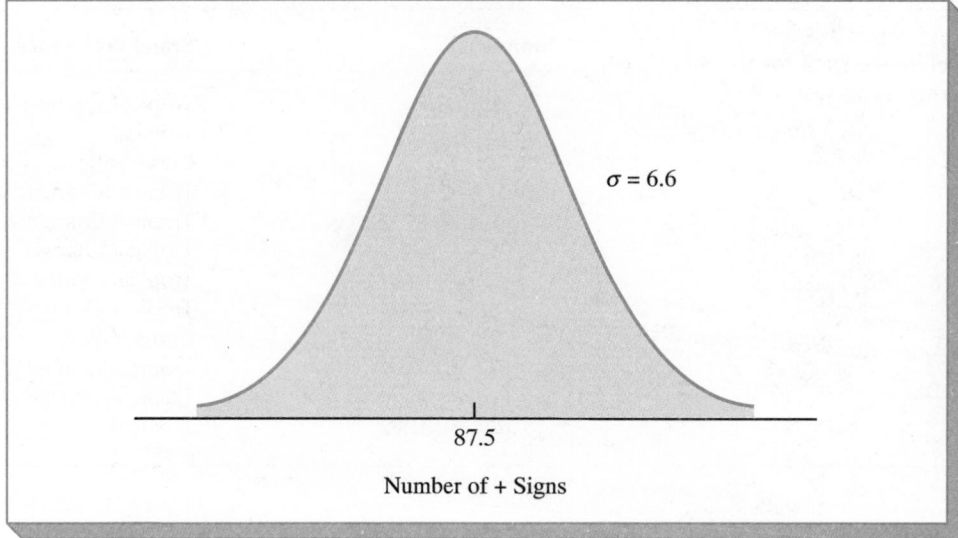

develop the rejection rule for the test. With $\alpha = .05$, the rejection rule for this two-tailed test can be written as follows:

$$\text{Reject } H_0 \text{ if } z < -1.96 \text{ or if } z > +1.96$$

Using the number of times the Democratic candidate received the higher foreign policy rating as the number of + signs ($x = 72$), we have the following value of the test statistic:

$$z = \frac{x - \mu}{\sigma} = \frac{72 - 87.5}{6.6} = -2.35$$

Since $z = -2.35$ is less than -1.96, the hypothesis of no difference in foreign policy for the two candidates should be rejected at the .05 level of significance. Based on this study, the Republican candidate is perceived to have the higher-rated foreign policy.

Hypothesis Tests about a Median

In Chapter 9 we described how hypothesis tests can be used to make an inference about a population mean. We now show how the sign test can be used to conduct hypothesis tests about a population median. Recall that the median splits a population such that 50% of the values fall at the median or above and 50% fall at the median or below. We can apply the sign test to conduct a hypothesis test about the value of a median by using a + sign whenever the data in the sample is above the median and a − sign whenever the data in the sample is below the median. Any data exactly equal to the hypothesized value of the median should be discarded. The computations for the sign test are done in exactly the same manner as before.

For example, the following hypothesis test is being conducted about the median price of new homes in St. Louis, Missouri.

$$H_0: \text{Median} = \$120,000$$

$$H_a: \text{Median} \neq \$120,000$$

From a sample of 62 new homes, 34 had prices above $120,000, 26 had prices below $120,000, and 2 had prices of exactly $120,000.

Using (19.1) and (19.2) for the $n = 60$ homes with prices different than \$120,000, we have

$$\mu = .50n = .50(60) = 30$$

$$\sigma = \sqrt{.25n} = \sqrt{.25(60)} = 3.87$$

Using $x = 34$ as the number of $+$ signs, the test statistic becomes

$$z = \frac{x - \mu}{\sigma} = \frac{34 - 30}{3.87} = 1.03$$

With a two-tailed test and a level of significance of $\alpha = .05$, we reject H_0 if z is less than -1.96 or if z is greater than $+1.96$. Since the test statistic $z = 1.03$, we cannot reject H_0. Based on these data, we are unable to reject the assumption that the median selling price of a new home is \$120,000.

NOTES & COMMENTS

The number of $+$ signs was used in the calculations to determine whether to reject the null hypothesis that $p = .50$. One could just as easily use the number of $-$ signs; the test result would be the same.

Exercises

Methods

SELF TEST ▷ 1. The following data show the preferences indicated by 10 individuals in taste tests involving two brands of a product:

Individual	Brand A Versus Brand B	Individual	Brand A Versus Brand B
1	+	6	+
2	+	7	−
3	+	8	+
4	−	9	−
5	+	10	+

With $\alpha = .05$, test for a significant difference in the preferences for the two brands. A $+$ indicates a preference for brand A over brand B.

SELF TEST ▷ 2. The following hypothesis test is to be conducted:

$$H_0\text{: Median} \leq 150$$

$$H_a\text{: Median} > 150$$

A sample of size 30 yields 22 cases in which a value greater than 150 is obtained, 3 cases in which a value of exactly 150 is obtained, and 5 cases in which a value less than 150 is obtained. Use $\alpha = .01$ and conduct the hypothesis test.

Applications

3. Researchers studied physical contact between parents and children in the same family (*Journal of Marriage and the Family*, August 1984). Assume that a sample of interactions for 20 different families with two children showed 14 cases where the mother was touched more than the father, 4 cases where the father was touched more than the mother, and 2 cases where the touching was judged equal. Suppose a sign test is to be used to see if there is any difference in touching between mothers and fathers.

a. What are the null and alternative hypotheses for the sign test?

b. Using $\alpha = .05$, what is the rejection rule?

c. What is your conclusion?

SELF TEST ▶

4. A Louis Harris poll asked 1253 adults a series of questions about the state of the economy and their children's future (*Business Week*, April 6, 1992). One question was, "Do you expect your children to have a better life than you have had, a worse life, or a life about as good as yours?" The responses were 34% better, 29% worse, 33% about the same, and 4% not sure. Use the sign test and a .05 level of significance to determine if more adults feel their children will have a better future than feel their children will have a worse future. What is your conclusion?

5. In a television preference poll, a sample of 180 individuals was asked to state a preference for one of the two shows aired at the same time on Friday evenings. "Big Town Detective" was favored by 100; 65 favored "The Friday Variety Special"; and 15 were unable to state a preference for one over the other. Is there evidence of a significant difference in the preferences for the two shows? Use $\alpha = .05$ for the test.

6. Menu planning at the Hampshire House Restaurant involves the question of customer preferences for steak and seafood. A sample of 250 customers were asked to state a preference for the two menu items. A preference for steak was stated by 140, and 110 stated a preference for seafood. Use $\alpha = .05$, and test for a difference in the preference for the two menu items.

7. The nationwide median hourly wage for a particular labor group is $14.50 per hour. A sample of 200 individuals in this labor group was taken in one city: 134 individuals had a wage rate less than $14.50 per hour; 54 individuals had a wage rate greater than $14.50 per hour; 12 individuals had a wage rate of $14.50. Test the null hypothesis that the median hourly wage in this city is the same as the nationwide median hourly wage. Use a .02 level of significance.

8. In a sample of 150 college basketball games, it was found that the home team won 98 games. Test to see if this data supports the claim that there is a home-team advantage in college basketball. Use a .05 level of significance. What is your conclusion?

9. The median number of part-time employees at fast-food restaurants in a particular city was known to be 15 last year. City officials think the use of part-time employees may be increasing. A sample of nine fast-food restaurants showed that there were more than 15 part-time employees at seven of the restaurants, one restaurant had exactly 15 part-time employees, and one had fewer than 15 part-time employees. Test at $\alpha = .05$ whether it can be concluded that there has been an increase in the median number of part-time employees.

10. *The Wall Street Journal*, October 22, 1988, reported that the median age at first marriage for men is 25.9 years and for women is 23.6 years. Suppose a sample of 225 first marriages in a certain Ohio county showed 122 cases where men were less than 25.9 years of age and 103 cases where men were more than 25.9 years of age. Test the hypothesis that the median age at first marriage for men in the sampled county is the same as the reported 25.9 years. Use $\alpha = .05$. What is your conclusion?

11. The median hourly wage for the population of blue- and white-collar workers nationwide is $9.00 per hour. A sample of workers was selected in the Los Angeles area (*Newsweek*, February 17, 1992). Use the sample data in Table 19.3 to test the hypotheses H_0: median ≤ 9, H_a: median > 9 for the population of workers in Los Angeles. Using a .05 level of significance, what is your conclusion?

TABLE 19.3

11.50	8.40	11.75
10.05	10.25	8.00
13.65	7.05	9.05
11.90	9.90	6.85
15.35	11.10	14.70
13.15	13.10	6.65
13.10	9.20	9.15
12.05	8.45	5.85
9.80		

19.2   Wilcoxon Signed-Rank Test

The Wilcoxon signed-rank test is the nonparametric alternative to the parametric matched-sample test presented in Chapter 10. In the matched-sample situation, each experimental unit generates two paired or matched observations, one from population 1 and one from population 2. The differences between the matched observations provide insight concerning the differences between the two populations.

The methodology of the parametric matched-sample analysis (the t test on paired differences) requires interval data and the assumption that the population of differences between the pairs of observations is *normally distributed*. With this assumption, the t distribution can be used to test the null hypothesis of no difference between population means. If some question exists concerning the appropriateness of the assumption of normally distributed differences, the nonparametric Wilcoxon signed-rank test can be used. We illustrate this nonparametric test by comparing the effectiveness of two production methods.

A manufacturing firm is attempting to determine if a difference in task-completion times exists for two production methods. A sample of 11 workers was selected, and each worker completed a production task using each of the production methods. The production method that each worker used first was selected randomly. Thus, each worker in the sample provided a pair of observations, as shown in Table 19.4. A positive difference in task-completion times indicates that method 1 required more time, and a negative difference in times indicates that method 2 required more time. Do the data indicate that the methods are significantly different in terms of task-completion times?

The question raised is whether the two methods provide differences in task-completion times. In effect, we have two populations of task-completion times, one population associated with each method. The hypotheses that will be tested are

$$H_0: \text{The populations are identical}$$

$$H_a: \text{The populations are not identical}$$

If H_0 cannot be rejected, there will be insufficient evidence to conclude that the task-completion times differ for the two methods. However, if H_0 can be rejected, we will conclude that the two methods differ in terms of task-completion times.

The first step of the Wilcoxon signed-rank test requires a ranking of the *absolute value* of the differences between the two methods. To do this, we first discard any differences of zero and then rank the remaining absolute differences from lowest to highest. Tied differences are assigned average rank values. The ranking of the absolute values of differences is shown in the fourth column of Table 19.5. Note that the difference of 0 for worker 8 is discarded from the rankings; then the smallest absolute difference of .1 is assigned the rank of 1. This ranking of absolute differences continues with the largest absolute difference of .9 assigned the rank of 10. The absolute differences of .4 for workers 3 and 5 are assigned the average rank of 3.5, while the absolute differences of .5 for workers 4 and 10 are assigned the average rank of 5.5.

Once the ranks of the absolute differences have been determined, the ranks are given the sign of the original difference in the data. For example, the .1 difference for worker 7, which was assigned the rank of 1, is given the value of $+1$ because the observed difference between the two methods was positive. The .2 difference, which was assigned the rank of 2, is given the value of -2 because the observed difference between the two methods was negative for this individual. The complete list of signed ranks, together with their sum, is shown in the last column of Table 19.5.

TABLE 19.4
Production Task-Completion Times (Minutes)

Worker	Method 1	Method 2	Difference
1	10.2	9.5	.7
2	9.6	9.8	−.2
3	9.2	8.8	.4
4	10.6	10.1	.5
5	9.9	10.3	−.4
6	10.2	9.3	.9
7	10.6	10.5	.1
8	10.0	10.0	.0
9	11.2	10.6	.6
10	10.7	10.2	.5
11	10.6	9.8	.8

TABLE 19.5

Ranking of Absolute Differences for
the Production Task-Completion
Time Example

Worker	Difference	Absolute Value of Difference	Rank	Signed Rank
1	.7	.7	8	+ 8
2	− .2	.2	2	− 2
3	.4	.4	3.5	+ 3.5
4	.5	.5	5.5	+ 5.5
5	− .4	.4	3.5	− 3.5
6	.9	.9	10	+10
7	.1	.1	1	+ 1
8	0	0	—	—
9	.6	.6	7	+ 7
10	.5	.5	5.5	+ 5.5
11	.8	.8	9	+ 9
			Sum of Signed Ranks	+44.0

Let us return to the original hypothesis of identical population task-completion times for the two methods. If the populations representing task-completion times for each of the two methods are identical, we would expect the positive ranks and the negative ranks to cancel each other, so that the sum of the signed rank values would be approximately 0. Thus, the test for significance under the Wilcoxon signed-rank test involves determining whether the computed sum of signed ranks ($+44$ in our example) is significantly different than 0.

Let T denote the sum of the signed-rank values in a Wilcoxon signed-rank test. It can be shown that if the two populations are identical and the number of matched pairs of data is 10 or more, the sampling distribution of T can be approximated as follows.

Sampling Distribution of T for Identical Populations

$$\text{Mean: } \mu_T = 0 \tag{19.3}$$

$$\text{Standard Deviation: } \sigma_T = \sqrt{\frac{n(n + 1)(2n + 1)}{6}} \tag{19.4}$$

Distribution form: approximately normal provided $n \geq 10$.

For the example, we have $n = 10$, since we discarded the observation with the difference of 0 (worker 8). Thus, using (19.4), we have

$$\sigma_T = \sqrt{\frac{10(11)(21)}{6}} = 19.62$$

The sampling distribution of T under the assumption of identical populations is shown in Figure 19.3.

FIGURE 19.3
Sampling Distribution of the
Wilcoxon *T* for the Production
Task-Completion Time Example

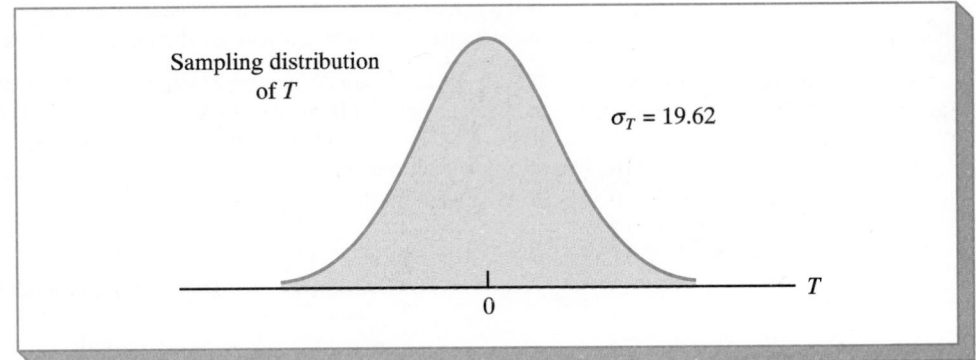

FIGURE 19.3
Sampling Distribution of the Wilcoxon *T* for the Production Task-Completion Time Example

Sampling distribution of *T*

$\sigma_T = 19.62$

The value of the test statistic z is as follows:

$$z = \frac{T - \mu_T}{\sigma_T} = \frac{44 - 0}{19.62} = 2.24$$

Testing the null hypothesis of no difference using a level of significance of $\alpha = .05$, we reject H_0 if $z < -1.96$ or if $z > 1.96$. With the value of $z = 2.24$, we reject H_0 and conclude that the two populations are not identical and that the methods differ in terms of task-completion times. The fact that method 2 showed the shorter completion times for 8 of the 11 workers would lead us to conclude that differences between the two populations indicate method 2 to be the better production method.

□ □ **Exercises**

Applications

12. Two fuel additives are being tested to determine their effect on miles per gallon for passenger cars. Test results using 12 cars are shown below; each car was tested with both fuel additives. Use $\alpha = .05$ and the Wilcoxon signed-rank test to see if there is a significant difference in the additives.

	Additive			Additive	
Car	1	2	**Car**	1	2
1	20.12	18.05	7	16.16	17.20
2	23.56	21.77	8	18.55	14.98
3	22.03	22.57	9	21.87	20.03
4	19.15	17.06	10	24.23	21.15
5	21.23	21.22	11	23.21	22.78
6	24.77	23.80	12	25.02	23.70

SELF TEST ▶ **13.** A sample of 10 individuals was used in a study to test the effects of a relaxant on the time required to fall asleep for male adults. Data for 10 subjects showing the number of minutes required

TABLE 19.6

Subject	Without Relaxant	With Relaxant
1	15	10
2	12	10
3	22	12
4	8	11
5	10	9
6	7	5
7	8	10
8	10	7
9	14	11
10	9	6

to fall asleep with and without the relaxant are given in Table 19.6. Use a .05 level of significance to determine if the relaxant reduces the time required to fall asleep. What is your conclusion?

14. Shown below are the number of baggage-related complaints per 1000 passengers for 10 airlines during the months of December 1988 and January 1989 (*U.S. Department of Transportation*, March 1989). Use $\alpha = .05$ and the Wilcoxon signed-rank test to determine if the data indicate the number of baggage related complaints for the airline industry has *decreased* over the two months studied. What is your conclusion?

Airline	December Complaints	January Complaints
American	8.9	8.0
Delta	8.2	7.9
Continental	7.9	8.2
Eastern	7.5	7.8
Northwest	9.6	6.5
Pan American	5.0	5.1
Piedmont	12.3	11.0
TWA	11.2	10.9
United	7.7	7.4
USAir	8.6	7.9

TABLE 19.7

Delivery	Service 1	Service 2
1	24.5	28.0
2	26.0	25.5
3	28.0	32.0
4	21.0	20.0
5	18.0	19.5
6	36.0	28.0
7	25.0	29.0
8	21.0	22.0
9	24.0	23.5
10	26.0	29.5
11	31.0	30.0

15. A test was conducted of two overnight mail-delivery services. Two samples of identical deliveries were set up such that both delivery services were notified of the need for a delivery at the same time. The number of hours required to make the delivery was recorded in Table 19.7 for each service. Do the data shown suggest a difference in the delivery times for the two services? Use a .05 level of significance for the test.

16. Elsbernd Investors, Inc., provides a 6-week training program for newly hired management trainees. As part of the program-evaluation procedure, the firm gives each trainee a pretest and posttest. Use a one-tailed test with $\alpha = .05$, and analyze the following data as part of the evaluation of the firm's management training program. What is your conclusion?

Trainee	Pretest Score	Posttest Score
1	45	65
2	60	70
3	65	63
4	60	67
5	52	60
6	62	58
7	57	70
8	70	65
9	72	80
10	66	88
11	78	74

17. Ten test-market cities were selected as part of a market research study designed to evaluate the effectiveness of a particular advertising campaign. The sales dollars for each city were recorded for the week prior to the promotional program. Then the campaign was conducted for 2 weeks, with

new sales data collected for the week immediately following the campaign. The resulting data with sales in thousands of dollars are shown.

City	Precampaign Sales	Postcampaign Sales
Kansas City	130	160
Dayton	100	105
Cincinnati	120	140
Columbus	95	90
Cleveland	140	130
Indianapolis	80	82
Louisville	65	55
St. Louis	90	105
Pittsburgh	140	152
Peoria	125	140

Use $\alpha = .05$. What conclusion would you draw concerning the value of the advertising program?

19.3   Mann-Whitney-Wilcoxon Test

In this section we present another nonparametric method that can be used to determine if there is a difference between two populations. This test, unlike the signed-rank test, is not based on a matched sample. Two independent samples, one from each population, are used. This test, developed jointly by Mann, Whitney and Wilcoxon, is sometimes referred to as the *Mann-Whitney test* and is sometimes referred to as the *Wilcoxon rank-sum test*. Both the Mann-Whitney and Wilcoxon versions of this test are equivalent; we refer to it as the *Mann-Whitney-Wilcoxon (MWW) test*.

The MWW test is based on independent random samples from each population. Recall that in Chapter 10 we conducted a parametric test for the difference between the means of two populations. The hypotheses tested were as follows:

$$H_0: \mu_1 - \mu_2 = 0$$

$$H_a: \mu_1 - \mu_2 \neq 0$$

In the small-sample case, the parametric method used was based on two assumptions:

1. Both populations are normally distributed.
2. The variances of the two populations are equal.

The nonparametric MWW test does not require either of the above assumptions. The only requirement of the MWW test is that the measurement scale for the data generated by the two independent random samples is at least ordinal. Instead of testing for the difference between the means of the two populations, the MWW test determines whether the two populations are identical. The hypotheses for the MWW test are as follows:

$$H_0: \text{The two populations are identical}$$

$$H_a: \text{The two populations are not identical}$$

TABLE 19.8
High School Class-Standing Data

| | Garfield Students | | Mulberry Students | |
Student	Class Standing		Student	Class Standing
Fields	8		Hart	70
Clark	52		Phipps	202
Jones	112		Kirkwood	144
Tibbs	21		Abbott	175
			Guest	146

We first demonstrate how the MWW test can be applied by showing an application for the small-sample-size case.

Small-Sample Case

The small-sample-size case for the MWW test should be used whenever the sample sizes for both populations are less than or equal to 10. We illustrate the use of the MWW test for the small-sample case by considering the academic potential of students attending Johnston High School. The majority of students attending Johnston High School previously attended either Garfield Junior High School or Mulberry Junior High School. The question raised by school administrators was whether the population of students that attended Garfield were identical to the population of students that attended Mulberry in terms of academic potential. The hypotheses under consideration were expressed as follows:

H_0: The two populations are identical in terms of academic potential

H_a: The two populations are not identical in terms of academic potential

Using high school records, Johnston High School administrators selected a random sample of four high school students who had attended Garfield Junior High and another random sample of five students who had attended Mulberry Junior High. The current high school class standing was recorded for each of the nine students used in the study. The ordinal class standing for the nine students is shown in Table 19.8.

The first step in the MWW procedure is to rank the *combined* data from the two samples from low to high. The lowest value (class standing 8) receives a rank of 1 and the highest value (class standing 202) receives a rank of 9. The ranking of the nine students is shown in Table 19.9.

The next step is to sum the ranks for each sample separately. This calculation is shown in Table 19.10. The MWW procedure may utilize the sum of the ranks for either sample. In the following discussion, we use the sum of the ranks for the sample of four students from Garfield. We denote this sum by the symbol T. Thus, for our example, $T = 11$.

Let us consider for a moment the properties of the sum of the ranks for the Garfield sample. Since there are four students in the sample, Garfield could have the top four ranking students in the study. If this were the case, $T = 1 + 2 + 3 + 4 = 10$ would be the smallest value possible for the rank sum T. On the other hand, Garfield could have the bottom four ranking students, in which case $T = 6 + 7 + 8 + 9 = 30$ would be the largest value possible for T. Thus, T for the Garfield sample must take on a value between 10 and 30.

TABLE 19.9
Ranking of High School Students

Student	Class Standing	Combined Sample Rank
Fields	8	1
Tibbs	21	2
Clark	52	3
Hart	70	4
Jones	112	5
Kirkwood	144	6
Guest	146	7
Abbott	175	8
Phipps	202	9

TABLE 19.10 **Rank Sums for High School Students from Each Junior High School**		Garfield Students			Mulberry Students	
	Student	Class Standing	Sample Rank	Student	Class Standing	Sample Rank
	Fields	8	1	Hart	70	4
	Clark	52	3	Phipps	202	9
	Jones	112	5	Kirkwood	144	6
	Tibbs	21	2	Abbott	175	8
				Guest	146	7
	Sum of Ranks		11			34

Note that values of T near 10 imply Garfield has the significantly better, or higher ranking, students, whereas values of T near 30 imply Garfield has the significantly weaker, or lower ranking, students. Thus, if the two populations of students were identical in terms of academic potential, we would expect the value of T to be near the average of the above two values, or $(10 + 30)/2 = 20$.

Critical values of the MWW T statistic are provided in Table 10 of Appendix B for cases where both sample sizes are less than or equal to 10.* In these tables, n_1 refers to the sample size corresponding to the sample whose rank sum is being used in the test. The value of T_L is read directly from the tables and the value of T_U is computed from (19.5).

$$T_U = n_1(n_1 + n_2 + 1) - T_L \qquad (19.5)$$

Neither the value of T_L nor the value of T_U is in the rejection region. The null hypothesis of identical populations should be rejected only if T is strictly less than T_L or strictly greater than T_U.

For example, using Table 10 of Appendix B with a .05 level of significance, we see that the lower-tail critical value for the MWW statistic with $n_1 = 4$ (Garfield) and $n_2 = 5$ (Mulberry) is $T_L = 12$. The upper-tail critical value for the MWW statistic is computed using (19.5) as follows:

$$T_U = 4(4 + 5 + 1) - 12 = 28$$

Thus, the MWW decision rule indicates that the null hypothesis of identical populations can be rejected if the sum of the ranks for the first sample (Garfield) is less than 12 or greater than 28. The rejection rule can be written as follows:

$$\text{Reject } H_0 \text{ if } T < 12 \text{ or if } T > 28$$

Referring to Table 19.10, we see that $T = 11$. Thus, the null hypothesis H_0 is rejected, and we can conclude that the population of students at Garfield differs from the population of students at Mulberry in terms of academic potential. The higher class ranking obtained by the sample of Garfield students indicates that Garfield students appear to be better prepared for high school than the Mulberry students.

---

*A more complete table of critical values for the Mann-Whitney-Wilcoxon test can be found in *Practical Nonparametric Statistics,* by W. J. Conover.

Large-Sample Case

In the case where both sample sizes are greater than or equal to 10, a normal approximation of the distribution of T can be used to conduct the analysis for the MWW test. We illustrate the large-sample case by considering a situation at Third National Bank.

Third National Bank has two branch offices. Data collected from two independent simple random samples, one from each branch, are shown in Table 19.11. Do the data indicate whether the populations of checking account balances at the two branch banks are identical?

The first step in the MWW test is to rank the *combined* data from the lowest to the highest values. Using the combined set of 22 observations shown in Table 19.11, we find the lowest data value of $750 (6th item of sample 2) and assign to it a rank of 1. Continuing the ranking, we have the following.

Account Balance ($)	Item	Assigned Rank
750	6th of sample 2	1
800	5th of sample 2	2
805	7th of sample 1	3
850	2nd of sample 2	4
.	.	.
.	.	.
.	.	.
1195	4th of sample 1	21
1200	3rd of sample 1	22

In ranking the combined data, we may find that two or more data values are the same. In this case, these same values are given the *average* ranking of their positions in the

TABLE 19.11
Account Balances for Two Branches of Third National Bank

Branch 1		Branch 2	
Sampled Account	Account Balance ($)	Sampled Account	Account Balance ($)
1	1095	1	885
2	955	2	850
3	1200	3	915
4	1195	4	950
5	925	5	800
6	950	6	750
7	805	7	865
8	945	8	1000
9	875	9	1050
10	1055	10	935
11	1025		
12	975		

combined data set. This situations of *ties* occurs with the ranking of the 22 account balances from the two branch banks. For example, the balance of $945 (8th item of sample 1) will be assigned the rank of 11. However, the next two values in the data set are tied with values of $950 (see the 6th item of sample 1 and the 4th item of sample 2). Since these two values will be considered for assigned ranks of 12 and 13, they are both given the assigned rank of 12.5. At the next highest data value of $955, we continue the ranking process by assigning $955 the rank of 14. Table 19.12 shows the entire data set with the assigned rank of each observation.

The next step in the MWW test is to sum the ranks for each sample. These sums are shown in Table 19.12. The test procedure can be based on the sum of the ranks for either sample. We use the sum of the ranks for the sample from branch 1. Thus, for this example, $T = 169.5$.

Given that the sample sizes are $n_1 = 12$ and $n_2 = 10$, we can use the normal approximation to the sampling distribution of the rank sum T. The appropriate sampling distribution is given by the following.

Sampling Distribution of T for Identical Populations

$$\text{Mean: } \mu_T = \tfrac{1}{2} n_1(n_1 + n_2 + 1) \tag{19.6}$$

$$\text{Standard Deviation: } \sigma_T = \sqrt{\tfrac{1}{12} n_1 n_2(n_1 + n_2 + 1)} \tag{19.7}$$

Distribution form: approximately normal provided $n_1 \geq 10$ and $n_2 \geq 10$.

TABLE 19.12

Combined Ranking of the Data in the Two Samples from Third National Bank

	Branch 1			Branch 2	
Sampled Account	Account Balance ($)	Rank	Sampled Account	Account Balance ($)	Rank
1	1095	20	1	885	7
2	955	14	2	850	4
3	1200	22	3	915	8
4	1195	21	4	950	12.5
5	925	9	5	800	2
6	950	12.5	6	750	1
7	805	3	7	865	5
8	945	11	8	1000	16
9	875	6	9	1050	18
10	1055	19	10	935	10
11	1025	17		Sum of Ranks	83.5
12	975	15			
	Sum of Ranks	169.5			

For branch 1, we have

$$\mu_T = \tfrac{1}{2}\,12(12 + 10 + 1) = 138$$

$$\sigma_T = \sqrt{\tfrac{1}{12}\,12(10)(12 + 10 + 1)} = 15.17$$

The sampling distribution of T is shown in Figure 19.4. Following the usual hypothesis-testing procedure, we compute the test statistic z to determine if the observed value of T appears to be from the sampling distribution of Figure 19.4. If T does not appear to be from this distribution, we will reject the null hypothesis and conclude that the populations are not identical. Computing the test statistic, we have

$$z = \frac{T - \mu_T}{\sigma_T} = \frac{169.5 - 138}{15.17} = 2.08$$

At an $\alpha = .05$ level of significance, we know that, to reject H_0, z must be less than -1.96 or greater than $+1.96$. Since $z = 2.08$, we reject H_0. Thus, we conclude that the two populations are not identical. That is, the populations of account balances at the two branches are not the same.

In summary, the Mann-Whitney-Wilcoxon rank-sum test follows the steps outlined below to determine if two independent random samples are selected from identical populations.

1. Rank the combined sample observations from lowest to highest, with tied values being assigned the average of the tied rankings.
2. Compute T, the sum of the ranks for the first sample.
3. In the large-sample case, make the test for significant differences between the two populations by using the observed value of T and comparing it to the sampling distribution of T for identical populations (see Equations (19.6) and (19.7)). The value of the standardized test statistic z will provide the basis for deciding whether to reject H_0. In the small-sample case, use Table 10 in Appendix B to find the critical values for the test.

FIGURE 19.4

Sampling Distribution of T for the Third National Bank Example

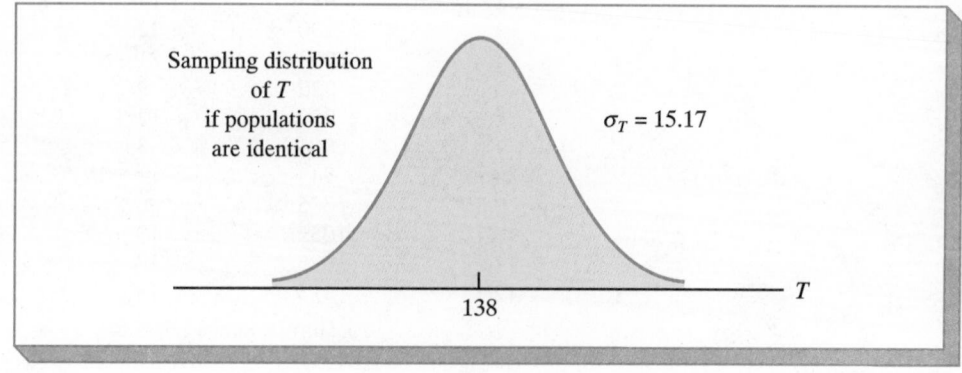

NOTES & COMMENTS

The nonparametric test discussed in this section is used to determine whether two populations are identical. Parametric statistical tests, such as the t test described in Chapter 10, test the equality of two population means. When we reject the hypothesis that the means are equal, we conclude that the populations differ only in their means. When we reject the hypothesis that the populations are identical using the MWW test, we cannot state how they differ. The populations could have different means, different variances, and/or different forms. Nonetheless, if we believe that the populations are the same in every way except for the means, a rejection of H_0 using the nonparametric method implies that the means differ. The major advantages of the MWW test, compared to the parametric t test, are that it does not require any assumptions about the form of the probability distribution from which the measurements come and that the test may be used with ordinal data.

▢ ▢ Exercises

Applications

SELF TEST ▶

18. Two fuel additives are being tested to determine their effect on gas mileage. Seven cars were tested using additive 1; another independent sample of nine cars was tested using additive 2. The data in Table 19.13 show the miles per gallon obtained using the two additives. Use $\alpha = .05$ and the MWW test to see if there is a significant difference in gasoline mileage.

SELF TEST ▶

19. Starting salary data for college graduates is reported by the College Placement Council (*USA Today*, April 6, 1992). Annual salaries in thousands of dollars for a sample of accounting majors and a sample of finance majors are shown below.

TABLE 19.13

Additive 1	Additive 2
17.3	18.7
18.4	17.8
19.1	21.3
16.7	21.0
18.2	22.1
18.6	18.7
17.5	19.8
	20.7
	20.2

Accounting	Finance
28.8	26.3
25.3	23.6
26.2	25.0
27.9	23.0
27.0	27.9
26.2	24.5
28.1	29.0
24.7	27.4
25.2	23.5
29.2	26.9
29.7	26.2
29.3	24.0

a. Use a .05 level of significance to test the null hypothesis that there is no difference between the annual starting salaries for accounting majors and finance majors. What is your conclusion?

b. What are the sample means for accounting majors and finance majors?

20. The Anderson Company has sent two groups of employees to a privately run program providing word-processing training. One group was from the data-processing department; the other was from the typing pool. At the completion of the program, the Anderson Company received a report showing the class rank for each of its employees. Of the 70 persons finishing the program, the class

TABLE 19.14

Data-Processing Group	Typists
1	17
12	26
15	29
23	33
30	45
33	51
	62

ranks of the 13 employees of the Anderson Company are given in Table 19.14. Use $\alpha = .10$ and test to see whether there is a performance difference between the two groups in the word-processing program.

21. Mileage performance tests were conducted for two models of automobiles. Twelve automobiles of each model were randomly selected and a miles-per-gallon rating for each model was developed based on 1000 miles of highway driving. The data are shown below.

Model 1		Model 2	
Automobile	Miles per Gallon	Automobile	Miles per Gallon
1	20.6	1	21.3
2	19.9	2	17.6
3	18.6	3	17.4
4	18.9	4	18.5
5	18.8	5	19.7
6	20.2	6	21.1
7	21.0	7	17.3
8	20.5	8	18.8
9	19.8	9	17.8
10	19.8	10	16.9
11	19.2	11	18.0
12	20.5	12	20.1

Use $\alpha = .10$, and test for a significant difference in the populations of miles-per-gallon ratings for the two models.

22. Insurance costs for men and women were reported in *Newsweek*, May 8, 1989. Assume that the data in Table 19.15 show the annual cost of $100,000, 5-year term insurance policies for nonsmoking men and nonsmoking women. Use the MWW to test for a significant difference between the costs for men and women. Use $\alpha = .05$.

23. The following data from police records show the number of daily crime reports from a sample of days during the winter months and a sample of days during the summer months. Using a .05 level of significance, determine if there is a significant difference between the number of crime reports in the winter and summer months.

TABLE 19.15

Men	Women
167	146
175	162
160	164
165	148
172	166
180	158
185	150
170	150
163	140
184	142

Winter	Summer
18	28
20	18
15	24
16	32
21	18
20	29
12	23
16	38
19	28
20	18

TABLE 19.16

Dallas	San Antonio
445	460
489	451
405	435
485	479
439	475
449	445
436	429
420	434
430	410
405	422
	425
	459
	430

24. A certain brand of microwave oven was priced at 10 stores in Dallas and 13 stores in San Antonio. The data are shown in Table 19.16. Use a .05 level of significance, and test whether prices for the microwave oven are the same in the two cities.

25. Miami University reported the starting salaries by major for graduates of the Richard T. Farmer School of Business Administration (*Miami University Class Profile*, 1990). The starting salaries shown below are reported in thousands of dollars. Use the Mann-Whitney-Wilcoxon test to see if there is evidence to conclude that the starting salaries for majors in management information systems differ from those for general business majors. Use a .05 level of significance.

Management Information Systems	General Business
27.2	25.6
29.0	29.1
28.0	21.7
27.0	24.0
27.5	25.2
30.5	25.0
24.3	25.0
32.5	23.8
26.0	

19.4   Kruskal-Wallis Test

The MWW test in Section 19.3 can be used to test whether two populations are identical. This test has been extended to the case of three or more populations by Kruskal and Wallis. The hypotheses for the *Kruskal-Wallis test* with $k \geq 3$ populations can be written as follows:

$$H_0: \text{All populations are identical}$$

$$H_a: \text{Not all populations are identical}$$

The Kruskal-Wallis test is based on the analysis of independent random samples from each of the k populations.

In Chapter 13 we introduced the analysis of variance (ANOVA) as a procedure that could be used to test for the equality of means among three or more populations. The ANOVA model requires interval or ratio data, the populations to all be normally distributed, and the variances of the populations to be equal.

The nonparametric Kruskal-Wallis test can be used with ordinal data as well as with interval or ratio data. In addition, the Kruskal-Wallis test does not require the assumptions of normality and equal variances that are required by the parametric analysis of variance procedure. Thus, whenever the data from $k \geq 3$ independent random samples is ordinal, or whenever the assumptions of normality and equal variances are questionable, the Kruskal-Wallis test provides an alternate statistical procedure for testing whether the populations are identical. Let us demonstrate the Kruskal-Wallis test by using it in an employee-selection application.

Williams Manufacturing Company hires employees for its management staff from three local colleges. Recently, the company's personnel department has been collecting

Nonparametric Methods

and reviewing annual performance ratings in an attempt to determine if there are differences in performance among the managers hired from these colleges. Performance-rating data are available from independent samples of seven employees from college A, six employees from college B, and seven employees from college C. These data are summarized in Table 19.17; the overall performance rating of each manager is given on a 0–100 scale, with 100 being the highest possible performance evaluation.

Suppose that we are interested in testing whether the three populations are identical with respect to performance evaluations. The Kruskal-Wallis test statistic, which is based on the sum of ranks for each of the samples, can be computed as follows.

TABLE 19.17
Performance Evaluation Ratings for 20 Employees

College A	College B	College C
25	60	50
70	20	70
60	30	60
85	15	80
95	40	90
90	35	70
80		75

Kruskal-Wallis Test Statistic

$$W = \frac{12}{n_T(n_T + 1)} \sum_{i=1}^{k} \frac{R_i^2}{n_i} - 3(n_T + 1) \qquad (19.8)$$

where

k = the number of populations

n_i = the number of items in sample i

$n_T = \Sigma n_i$ = total number of items in all samples

R_i = sum of the ranks for sample i

Kruskal and Wallis were able to show that, under the null hypothesis that the populations are identical, the sampling distribution of W can be approximated by a chi-square distribution with $k - 1$ degrees of freedom. This approximation is generally acceptable if each of the sample sizes is greater than or equal to 5.

To compute the W statistic for our example, we must first rank-order all 20 data items. The lowest data value of 15 from the college B sample receives a rank of 1, whereas the highest data value of 95 from the college A sample receives a rank of 20. The data values, their associated ranks, and the sum of the ranks for the three samples are shown in Table 19.18. Note that we assign the average rank to tied items;* for example, the data values of 60, 70, 80, and 90 had ties.

The sample sizes are:

$$n_1 = 7 \qquad n_2 = 6 \qquad n_3 = 7$$

and

$$n_T = \Sigma n_i = 7 + 6 + 7 = 20$$

Using (19.8), the W statistic is computed as follows:

$$W = \frac{12}{20(21)} \left[\frac{(95)^2}{7} + \frac{(27)^2}{6} + \frac{(88)^2}{7} \right] - 3(20 + 1) = 8.92$$

The chi-square distribution table (Table 3 of Appendix B) shows that with $k - 1 = 2$ degrees of freedom and $\alpha = .05$ in the upper tail of the distribution, the critical

---

*If numerous tied ranks are observed, (19.8) must be modified; the modified formula can be found in *Practical Nonparametric Statistics,* by W. J. Conover.

TABLE 19.18
Rankings for the 20 Employees

College A	Rank	College B	Rank	College C	Rank
25	3	60	9	50	7
70	12	20	2	70	12
60	9	30	4	60	9
85	17	15	1	80	15.5
95	20	40	6	90	18.5
90	18.5	35	5	70	12
80	15.5			75	14
Sum of Ranks	95		27		88

chi-square value is $\chi^2 = 5.99147$. Since the test statistic $W = 8.92$ is greater than 5.99147, we reject the null hypothesis that the three populations are identical. As a result, we conclude that manager performance differs significantly depending on the college attended. Furthermore, since the performance ratings were lowest for college B, it would appear reasonable for the company to either cut back on its recruiting from college B or to at least do a more thorough evaluation of graduates from this college.

NOTES & COMMENTS

The Kruskal-Wallis procedure illustrated in this example began with the collection of interval-scaled data showing employee-performance evaluation ratings. However, the procedure would have also worked had the data been the ordinal rankings of the 20 employees. In this case, the Kruskal-Wallis test could have been applied directly to the original data; the step of constructing the rank orderings from the performance evaluation ratings would have been omitted.

☐ ☐ **Exercises**

Methods

SELF TEST ▶ 26. Three products received the following performance ratings by a panel of 15 consumers.

Product		
A	B	C
50	80	60
62	95	45
75	98	30
48	87	58
65	90	57

Use the Kruskal-Wallis test and $\alpha = .05$ to determine if there is a significant difference in the performance ratings for the products.

27. Three different admission-test-preparation programs are being evaluated. The scores obtained by a sample of 20 people utilizing the test-preparation programs yielded the results shown in Table 19.19. Use the Kruskal-Wallis test to determine if there is a significant difference in the three test preparation programs. Use $\alpha = .01$.

TABLE 19.19

Program		
A	B	C
540	450	600
400	540	630
490	400	580
530	410	490
490	480	590
610	370	620
	550	570

Applications

SELF TEST ▶

28. An American Medical Association survey found that the average annual income for doctors is $155,000 (*St. Petersburg Times*, December 15, 1990). The data below show the annual income in thousands of dollars for samples of physicians specializing in surgery, radiology, and obstetrics. Do the data indicate differences in annual income for the following three specialties? Use a .05 level of significance. What is your conclusion?

Surgery	Radiology	Obstetrics
240	250	200
205	180	175
275	210	185
200	225	220
195	190	188
205	215	202

TABLE 19.20

Automobile

A	B	C
19	19	24
21	20	26
20	22	23
19	21	25
21	23	27

29. In Chapter 13 the ANOVA procedure was used to test for significant differences in gas mileage for three types of automobiles. The miles per gallon data obtained from tests on five automobiles of each type are shown again in Table 19.20.
a. Use the Kruskal-Wallis test with $\alpha = .05$ to determine if there is a significant difference in the gasoline mileage for the three automobiles.
b. What information available in the data is used by the ANOVA procedure and not the Kruskal-Wallis test?

30. A large corporation has been sending many of its first-level managers to an off-site supervisory skills course. Four different management-development centers offer this course, and the corporation is interested in determining if there are differences in the quality of training provided. A sample of 20 employees who have attended these programs has been chosen and the employees ranked with respect to supervisory skills. The results are shown:

Course	Supervisory Skills Rank				
1	3	14	10	12	13
2	2	7	1	5	11
3	19	16	9	18	17
4	20	4	15	6	8

TABLE 19.21

M&Ms	Kit Kat	Milky Way II
230	225	200
210	205	208
240	245	202
250	235	190
230	220	180

Note that the top-ranked supervisor attended course 2 and the lowest-ranked supervisor attended course 4. Use $\alpha = .05$ and test to see if there is a significant difference in the training provided by the four programs.

31. The better selling candies are higher in calories. Hershey's Milk Chocolate bar has an average of 240 calories; Twix has an average of 280 calories; and Reese's Peanut Butter Cups have an average of 250 calories (*USA Today*, April 7, 1992). Assume that the data in Table 19.21 show the calorie content from samples of M&Ms, Kit Kats, and Milky Way II's. Test for significant differences in the calorie content of these three candies. At a .05 level of significance, what is your conclusion?

19.5   Rank Correlation

Correlation was introduced in Chapter 14 as a measure of the linear association between two variables for which interval or ratio data are available. In this section we consider measures of association between two variables when only ordinal data are available. The *Spearman rank-correlation coefficient* r_s has been developed for this purpose.

Spearman Rank-Correlation Coefficient

$$r_s = 1 - \frac{6\Sigma d_i^2}{n(n^2 - 1)} \qquad (19.9)$$

where

n = the number of items or individuals being ranked

x_i = the rank of item i with respect to one variable

y_i = the rank of item i with respect to a second variable

$d_i = x_i - y_i$

Let us illustrate the use of the Spearman rank-correlation coefficient in the following example. A company wants to determine if individuals who were expected at the time of employment to be better salespersons actually turn out to have better sales records. To investigate this question, the vice president in charge of personnel has carefully reviewed the original job interview summaries, academic records, and letters of recommendations for 10 current members of the firm's sales force. Based on the review of this information, the vice president ranked the 10 individuals in terms of their potential for success, basing the assessment solely on the information available at the time of employment. Then a list was obtained of the number of units sold by each salesperson over the first 2 years. Based on actual sales performance, a second ranking of the 10 salespersons was carried out. Table 19.22 shows the relevant data and the two rankings. The statistical question in-

TABLE 19.22

Sales Potential and Actual 2-Year Sales Data for 10 Salespersons

Salesperson	Ranking of Potential	2-Year Sales (units)	Ranking According to 2-Year sales
A	2	400	1
B	4	360	3
C	7	300	5
D	1	295	6
E	6	280	7
F	3	350	4
G	10	200	10
H	9	260	8
I	8	220	9
J	5	385	2

TABLE 19.23

Computation of the Spearman Rank-Correlation Coefficient for Sales Potential and Sales Performance

Salesperson	x_i = Ranking of Potential	y_i = Ranking of Sales Performance	$d_i = x_i - y_i$	d_i^2
A	2	1	1	1
B	4	3	1	1
C	7	5	2	4
D	1	6	-5	25
E	6	7	-1	1
F	3	4	-1	1
G	10	10	0	0
H	9	8	1	1
I	8	9	-1	1
J	5	2	3	9
				$\Sigma d_i^2 = 44$

$$r_s = 1 - \frac{6\Sigma d_i^2}{n(n^2 - 1)} = 1 - \frac{6(44)}{10(100 - 1)} = .73$$

volves determining whether there is agreement between the ranking of potential at the time of employment and the ranking based upon the actual sales performance over the first 2 years.

Let us compute the Spearman rank-correlation coefficient for the data in Table 19.22. The computations for the rank-correlation coefficient are summarized in Table 19.23. Here we see that the rank-correlation coefficient is a positive .73. The Spearman rank-correlation coefficient ranges from -1.0 to $+1.0$, with an interpretation similar to the sample correlation coefficient in that positive values near 1.0 indicate a strong association between the rankings; as one rank increases, the other rank increases. On the other hand, rank correlations near -1.0 indicate a strong negative association in the ranks (as one rank increases, the other rank decreases). The value $r_s = .73$ indicates a positive correlation between potential and actual performance. Individuals ranked high on potential tend to rank high on performance.

A Test for Significant Rank Correlation

At this point, we have seen how sample results can be used to compute the sample rank-correlation coefficient. As with many other statistical procedures, we may wish to use the sample results to make an inference about the population rank correlation ρ_s between two variables. In our example, the population rank-correlation coefficient could be obtained by making the rank-correlation coefficient computations for all members of the sales force. However, we would like to avoid all this data collection and make an inference about the population rank-correlation based on the sample rank-correlation coefficient r_s. To make this inference, we must test the following hypotheses:

$$H_0: \rho_s = 0$$

$$H_a: \rho_s \neq 0$$

Under the null hypothesis of no rank correlation ($\rho_s = 0$), the rankings are independent, and the sampling distribution of r_s is as follows.

$$
\boxed{
\begin{array}{c}
\textbf{Sampling Distribution of } r_s \\[6pt]
\text{Mean: } \mu_{r_s} = 0 \hspace{4cm} (19.10) \\[10pt]
\text{Standard deviation: } \sigma_{r_s} = \sqrt{\dfrac{1}{n-1}} \hspace{2cm} (19.11) \\[10pt]
\text{Distribution form: approximately normal provided } n \geq 10.
\end{array}
}
$$

The sample rank-correlation coefficient for sales potential and sales performance in our example was $r_s = .73$. Using this value, we will test for a significant rank correlation. From (19.10) we have $\mu_{r_s} = 0$, and from (19.11) we have $\sigma_{r_s} = \sqrt{1/(10-1)} = .33$. Using the test statistic, we have

$$
z = \frac{r_s - \mu_{r_s}}{\sigma_{r_s}} = \frac{.73 - 0}{.33} = 2.21
$$

Using a level of significance of $\alpha = .05$, we see that the null hypothesis of no correlation will be rejected if $z < -1.96$ or if $z > 1.96$. Since $z = 2.21 > 1.96$, we reject the hypothesis of no rank correlation. Thus, we can conclude that a significant rank correlation exists between sales potential and sales performance.

☐ ☐ Exercises

Methods

SELF TEST ▶

32. Consider the set of rankings on a sample of 10 elements shown in Table 19.24.
a. Compute the Spearman rank-correlation coefficient for the data.
b. Test for significant rank correlation using $\alpha = .05$, and state your conclusion.

33. Consider the following two sets of rankings for six items.

TABLE **19.24**

Element	x_i	y_i
1	10	8
2	6	4
3	7	10
4	3	2
5	4	5
6	2	7
7	8	6
8	5	3
9	1	1
10	9	9

	Case One			Case Two	
Item	First Ranking	Second Ranking	Item	First Ranking	Second Ranking
A	1	1	A	1	6
B	2	2	B	2	5
C	3	3	C	3	4
D	4	4	D	4	3
E	5	5	E	5	2
F	6	6	F	6	1

Note that in the first case the rankings are identical, whereas in the second case the rankings are exactly opposite. What value should you expect for the Spearman rank-correlation coefficient for each of these cases? Explain. Calculate the rank-correlation coefficient for each case.

Applications

SELF TEST

34. *Financial World* (May, 1992) presented a comparison of a number of educational statistics for the states. Shown below, for a sample of 11 states, are the ranks on pupil-teacher ratio (1 = lowest; 11 = highest) and the ranks on expenditure per pupil (1 = highest; 11 = lowest).

State	Rank Pupil-Teacher Ratio	Rank Expenditure per Pupil	State	Rank Pupil-Teacher Ratio	Rank Expenditure per Pupil
Arizona	10	9	Massachusetts	1	1
Colorado	8	5	Nebraska	2	7
Florida	6	4	North Dakota	7	8
Idaho	11	12	South Dakota	5	10
Iowa	4	6	Washington	9	3
Louisiana	3	11			

At the $\alpha = .05$ level, does there appear to be a relationship between expenditure per pupil and pupil-teacher ratio?

35. Airline passengers file complaints about lost, stolen, damaged, and delayed baggage. How do the airlines compare? According to the U.S. Department of Transportation (March, 1989), 10 airlines can be ranked from fewest to most complaints per 1000 passengers. Table 19.25 shows the rankings of the airlines for December 1988 and January 1989.

a. Compute the rank correlation for the airlines for the two months of data.

b. Test for significant rank correlation using $\alpha = .05$. What is your conclusion?

36. In a poll of men and women television viewers, preferences for the top 10 shows led to the following rankings. Is there a relationship between the rankings by the two groups? Use $\alpha = .10$.

TABLE 19.25

December 1988	January 1989
Pan American	Pan American
Eastern	Northwest
United	United
Continental	USAir
Delta	Eastern
USAir	Delta
American	American
Northwest	Continental
TWA	TWA
Piedmont	Piedmont

Television Show	Ranking by Men	Ranking by Women	Television Show	Ranking by Men	Ranking by Women
1	1	5	6	3	2
2	5	10	7	10	9
3	8	6	8	4	8
4	7	4	9	6	1
5	2	7	10	9	3

TABLE 19.26

Professor	Ranking by Current Students	Ranking by Recent Graduates
1	4	6
2	6	8
3	8	5
4	3	1
5	1	2
6	2	3
7	5	7
8	10	9
9	7	4
10	9	10

37. A student organization surveyed both recent graduates and current students in an attempt to obtain information on the quality of teaching at a particular university. An analysis of the responses provided the rankings shown in Table 19.26 for 10 professors on the basis of teaching ability. Do the rankings given by the current students agree with the rankings given by the recent graduates? Use $\alpha = .10$, and test for a significant rank correlation.

Summary

In this chapter we have presented several statistical procedures that are classified as nonparametric methods. The parametric methods of the earlier chapters generally required interval or ratio data and were often based on assumptions concerning the population; for example, the probability distribution was normal. Since nonparametric methods can be applied to nominal and ordinal data as well as interval and ratio data and since nonparametric methods do not require population-distribution assumptions, nonparametric methods expand the class of problems that can be subjected to statistical analysis.

The sign test provides a nonparametric procedure for identifying differences between two populations when the only data available are nominal data. In the small-sample case, the binominal probability distribution can be used to determine the critical values for the sign test; in the large-sample case, a normal approximation may be used. The Wilcoxon signed-rank test provides a procedure for analyzing matched-sample data whenever interval- or ratio-scaled data are available for each matched pair. No assumptions are made concerning the population distribution. The Wilcoxon procedure tests the hypothesis that the two populations being considered are identical.

The Mann-Whitney-Wilcoxon test provides a nonparametric method for testing for a difference between two populations based on two independent random samples. Tables were presented for the small-sample case, and a normal approximation was provided for the large-sample case. The Kruskal-Wallis test extended the Mann-Whitney-Wilcoxon test to the case of three or more populations. The Kruskal-Wallis test is the nonparametric analog of the parametric ANOVA test for differences among population means.

In the last section of this chapter we introduced the Spearman rank-correlation coefficient as a measure of association for two ordinal or rank-ordered sets of items.

Glossary

Nonparametric methods A collection of statistical methods that generally require very few, if any, assumptions about the population distributions and the level of measurement. These methods can be applied when nominal or ordinal data are available.

Distribution-free methods Another name for nonparametric statistical methods suggested by the lack of assumptions required concerning the population distribution.

Sign test A nonparametric statistical test for identifying differences between two populations based on the analysis of nominal data.

Wilcoxon signed-rank test A nonparametric statistical test for identifying differences between two populations based on the analysis of two matched or paired samples.

Mann-Whitney-Wilcoxon (MWW) test A nonparametric statistical test for identifying differences between two populations based on the analysis of two independent samples.

Kruskal-Wallis test A nonparametric test for identifying differences among three or more populations.

Spearman rank-correlation coefficient A correlation measure based on rank-ordered data for two variables.

Key Formulas

Sign Test (Large-Sample Case)

$$\text{Mean: } \mu = .50n \tag{19.1}$$

$$\text{Standard deviation: } \sigma = \sqrt{.25n} \tag{19.2}$$

Wilcoxon Signed-Rank Test

$$\text{Mean: } \mu_T = 0 \tag{19.3}$$

$$\text{Standard Deviation: } \sigma_T = \sqrt{\frac{n(n+1)(2n+1)}{6}} \tag{19.4}$$

Mann-Whitney-Wilcoxon Test (Large-Sample)

$$\text{Mean: } \mu_T = \tfrac{1}{2} n_1(n_1 + n_2 + 1) \qquad \textbf{(19.6)}$$

$$\text{Standard Deviation: } \sigma_T = \sqrt{\tfrac{1}{12} n_1 n_2 (n_1 + n_2 + 1)} \qquad \textbf{(19.7)}$$

Kruskal-Wallis Test Statistic

$$W = \frac{12}{n_T(n_T + 1)} \sum_{i=1}^{k} \frac{R_i^2}{n_i} - 3(n_T + 1) \qquad \textbf{(19.8)}$$

Spearman Rank-Correlation Coefficient

$$r_s = 1 - \frac{6 \Sigma d_i^2}{n(n^2 - 1)} \qquad \textbf{(19.9)}$$

☐ ☐ **Supplementary Exercises**

38. Mueller Beverage Products of Milwaukee, Wisconsin, has conducted a market research study designed to determine if there is a consumer preference for Mueller's Old Brew Beer over the individual consumer's usual beer. Each individual participating in the test was provided with a glass of his or her usual beer and a glass of Mueller's Old Brew. The two glasses were not labeled, the individuals had no way of knowing beforehand which of the two glasses was Mueller's Old Brew and which was the individual's usual brand. The glass that each individual tasted first was randomly selected. After tasting the beer in each glass, the individuals were asked to indicate their preferred beer. The test results from a sample of 24 individuals are shown:

Individual	Brand Preferred	Value Recorded	Individual	Brand Preferred	Value Recorded
1	Old Brew	+	13	Usual Brand	−
2	Old Brew	+	14	Usual Brand	−
3	Usual Brand	−	15	Old Brew	+
4	Old Brew	+	16	Usual Brand	−
5	Usual Brand	−	17	Old Brew	+
6	Old Brew	+	18	Old Brew	+
7	Usual Brand	−	19	Old Brew	+
8	Old Brew	+	20	Usual Brand	−
9	Old Brew	+	21	Old Brew	+
10	Usual Brand	−	22	Old Brew	+
11	Old Brew	+	23	Usual Brand	−
12	Usual Brand	−	24	Old Brew	+

If an individual selected Mueller's Old Brew as the preferred beer, a + sign was recorded. On the other hand, if the individual stated a preference for his or her usual brand, a − sign was recorded. Do the data for the 24 individuals indicate a significant difference in the preferences for the beers? Use $\alpha = .05$.

39. Two pilots for a prime-time television show (a western and a mystery show) are being tested. Both have been shown to a group of 12 viewers. The viewer preferences are shown.

Viewer	Preference	Viewer	Preference
1	Mystery	7	Mystery
2	Mystery	8	Western
3	Mystery	9	Mystery
4	Western	10	Mystery
5	Mystery	11	Mystery
6	Western	12	Mystery

Using $\alpha = .05$, test to see if there is a significant difference in preferences.

40. In a soft-drink taste test, 48 individuals stated a preference for one of two well-known brands. Results showed 28 favoring brand A, 16 favoring brand B, and 4 undecided. Use the sign test with $\alpha = .10$ and determine whether there is a significant difference in preferences for the two brands of soft-drinks.

41. Use the sign test, and perform the statistical analysis that will help us determine whether the task-completion times for two production methods differ. The data are shown in Table 19.27. Use $\alpha = .05$.

42. The national median price of new homes for 1988 was $123,500 (*U.S. News & World Report*, September 19, 1988). Assume that data on the prices of new homes were obtained from samples of loans recorded in Chicago and Dallas-Fort Worth. Use the data to test the hypothesis that the median price of homes in each of the two cities is the same as the national median price. Use $\alpha = .05$. State your conclusion for each city.

TABLE 19.27

Worker	Method 1 (minutes)	Method 2 (minutes)
1	10.2	9.5
2	9.6	9.8
3	9.2	8.8
4	10.6	10.1
5	9.9	10.3
6	10.2	9.3
7	10.6	10.5
8	10.0	10.0
9	11.2	10.6
10	10.7	10.2
11	10.6	9.8

City	Greater Than $123,500	Equal to $123,500	Less Than $123,500
Chicago	55	6	28
Dallas-Fort Worth	42	3	36

43. Mayfield Products, Inc., has collected data on preferences of 12 individuals concerning cleaning power of two brands of detergent. The individuals and their preferences are shown below. A + indicates a preference for brand A.

Individual	Brand A Versus Brand B	Individual	Brand A Versus Brand B
1	−	7	−
2	+	8	+
3	+	9	+
4	+	10	−
5	−	11	+
6	+	12	+

With $\alpha = .10$, test for a significant difference in the preference for the two brands.

TABLE 19.28

Homemaker	Model 1	Model 2
1	$650	$ 900
2	760	720
3	740	690
4	700	850
5	590	920
6	620	800
7	700	890
8	690	920
9	900	1000
10	500	690
11	610	700
12	720	700

TABLE 19.29

Production Line 1	Production Line 2
13.6	13.7
13.8	14.1
14.0	14.2
13.9	14.0
13.4	14.6
13.2	13.5
13.3	14.4
13.6	14.8
12.9	14.5
14.4	14.3
	15.0
	14.9

TABLE 19.30

No Program	Company Program	Off-Site Program
16	12	7
9	20	1
10	17	4
15	19	2
11	6	3
13	18	8
	14	5

44. Twelve homemakers were asked to estimate the retail selling price of two models of refrigerators. The estimates of selling price provided by the homemakers are shown in Table 19.28. Use these data, and test at the .05 level of significance to determine if there is a difference in the homemaker's perception of selling price for the two models.

45. A study was designed to evaluate the weight-gain potential of a new poultry feed. A sample of 12 chickens was used in a 6-week study. The weight of each chicken was recorded before and after the 6-week test period. The difference between the before and after weights of each chicken are as follows: 1.5, 1.2, $-.2$, .0, .5, .7, .8, 1.0, .0, .6, .2, $-.01$. A negative value indicates a weight loss during the test period, whereas .0 indicates no weight change over the period. Use a .05 level of significance to determine if the new feed appears to provide a weight gain for the chickens.

46. The data in Table 19.29 show product weights for items produced on two production lines. Test for a difference between the product weights for the two lines. Use $\alpha = .10$.

47. A client desires to determine if there is a significant difference in the time required to complete a program evaluation with the three different methods that are in common use. The times (in hours) required for each of 18 evaluators to conduct a program evaluation are given below.

Method 1	Method 2	Method 3
68	62	58
74	73	67
65	75	69
76	68	57
77	72	59
72	70	62

Use $\alpha = .05$ and test to see if there is a significant difference in the time required by the three methods.

48. A sample of 20 engineers, who have been with a company for 3 years, has been taken, and they have been rank-ordered with respect to managerial potential. Some of the engineers have attended the company's management-development course, others have attended an off-site management-development program at a local university, and the remainder have not attended any program. Use the rankings given in Table 19.30 and $\alpha = .025$ to test for a significant difference in the managerial potential of the three groups.

49. Shown below are course evaluation ratings for four instructors. Use $\alpha = .05$ and the Kruskal-Wallis procedure to test for a significant difference in teaching abilities.

Instructor	Course-Evaluation Rating								
Black	88	80	79	68	96	69			
Jennings	87	78	82	85	99	99	85	94	
Swanson	88	76	68	82	85	82	84	83	81
Wilson	80	85	56	71	89	87			

50. Wisman investment analysts ranked 12 companies, first with respect to book value and then with respect to growth potential.

Company	Ranking of Book Value	Ranking of Growth Potential	Company	Ranking of Book Value	Ranking of Growth Potential
1	12	2	7	3	12
2	2	9	8	11	1
3	8	6	9	4	7
4	1	11	10	5	10
5	9	4	11	6	8
6	7	5	12	10	3

For these data, does a relationship exist between the companies' book values and growth potentials? Use $\alpha = .05$.

51. Two individuals provided the preference rankings of seven soft drinks shown in Table 19.31. Compute the rank correlation for the two individuals.

52. A sample of 15 students obtained the following rankings on midterm and final examinations in a statistics course:

T A B L E 19.31

Soft Drink	Ranking by Individual 1	Ranking by Individual 2
A	1	3
B	3	2
C	5	5
D	6	7
E	7	6
F	4	1
G	2	4

Rank		Rank		Rank	
Midterm	Final	Midterm	Final	Midterm	Final
1	4	6	2	11	14
2	7	7	5	12	15
3	1	8	12	13	11
4	3	9	6	14	10
5	8	10	9	15	13

Compute the Spearman rank-correlation coefficient for the data and test for a significant correlation with $\alpha = .10$.

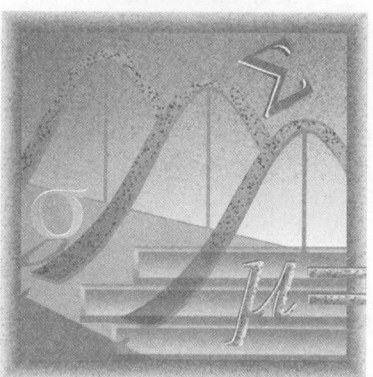

Statistical Methods for Quality Control

Contents

Dow Chemical U.S.A.*

FREEPORT, TEXAS

Dow Chemical U.S.A., Texas Operations, began in 1940 when The Dow Chemical Company purchased 800 acres of Texas land on the Gulf Coast to build a magnesium production facility. That original site has expanded to cover more than 5000 acres and is now one of the largest petrochemical complexes in the world. Among the products from Texas Operations are magnesium, styrene, plastics, adhesives, solvent, glycol, and chlorine. While some products are made solely for use in other processes, many end up as essential ingredients in products such as pharmaceuticals, toothpastes, dog food, water hoses, ice chests, milk cartons, garbage bags, shampoos, and furniture.

Dow's Texas Operations produces over 30% of the world's magnesium, an extremely lightweight metal found in products ranging from tennis racquets to suitcases to "mag" wheels. The Magnesium Department was the first group in Texas Operations to train its technical people and managers in the use of statistical quality control. Some of the earliest successful applications of statistical quality control were in chemical processing.

In one application involving the operation of a drier, samples of the output were taken at periodic intervals; the average value for each sample was computed and recorded on a chart referred to as an $\bar{x}$ chart. Such a chart enabled Dow analysts to monitor trends in the output that might have indicated the process was not operating correctly. In one instance, analysts began to observe values for the sample mean that were not indicative of a process operating within its design limits. On further examination of the control chart and the operation itself, the analysts found that the variation they were experiencing could be traced to problems involving one operator. The $\bar{x}$ chart recorded after this operator was retrained showed a significant improvement in the process quality.

Dow Chemical has experienced quality improvements everywhere statistical quality control has been used. Documented savings of several hundred thousand dollars per year have been realized, and new applications are continually being discovered.

In this chapter we will show how an $\bar{x}$ chart such as the one used by Dow Chemical can be developed. These types of charts are a part of statistical quality control referred to as statistical process control. We will also discuss methods of quality control for situations where a decision has to be made to accept or reject a group of items based on a sample.

*The authors are indebted to Clifford B. Wilson, Magnesium Technical Manager, The Dow Chemical Company, for providing this Statistics in Practice.

The American Society for Quality Control (ASQC) defines *quality* as "the totality of features and characteristics of a product or service that bears on its ability to satisfy given needs." Quality measures how well a product or service meets customer needs. In recent years, the Japanese commitment to producing high-quality products has caused many American companies to reevaluate their views regarding quality and how it can be achieved. Organizations recognize that to be competitive in today's marketplace, they must strive to reach high levels of quality. As a result, there has been an increased emphasis on methods for monitoring and maintaining quality.

Quality assurance refers to the entire system of policies, procedures, and guidelines established by an organization to achieve and maintain quality. Quality assurance consists of two principal functions: quality engineering and quality control. The objective of *quality engineering* is to include quality in the design of products and processes and to identify potential quality problems prior to production. *Quality control* consists of making a series of inspections and measurements to determine if quality standards are being met. If quality standards are not being met, quality control takes corrective and/or preventive action to achieve and maintain conformance. As we will show in this chapter, statistical techniques are extremely useful in quality control.

Traditional manufacturing approaches to quality control have been found to be less than satisfactory and are being replaced by improved managerial tools and techniques. Ironically, it was two U.S. consultants, Dr. W. Edwards Deming and Dr. Joseph Juran, who helped educate the Japanese in quality management. Today, U.S. firms are relearning these lessons from Japan.

Although quality is everybody's job, Deming stresses that quality must be led by management. In this regard, he has developed a list of 14 points which he believes are the key responsibilities of management. For instance, Deming states that management must cease dependence on mass inspection; must end the practice of awarding business solely on the basis of price; must seek continual improvement in all production processes and services; must foster a team-oriented environment; and must eliminate numerical goals, slogans, and work standards that prescribe numerical quotas. Perhaps most importantly, management must create a work environment in which a commitment to quality and productivity is maintained at all times.

In 1987, the U.S. Congress enacted Public Law 107: the Malcolm Baldrige National Quality Improvement Act. The Baldrige Award is given annually to U.S. firms that excel in quality. This award, along with the perspectives of individuals like Dr. Deming and Dr. Juran, has helped top management recognize that improving service quality and product quality are the most critical challenges facing their companies. Previous winners of the Malcolm Baldrige award include Motorola, IBM, Xerox, and Federal Express. In this chapter we present two statistical methods used in quality control. The first method, referred to as *acceptance sampling,* is used in situations where a decision has to be made to accept or reject a group of items based on the quality found in a sample. The second method, referred to as *statistical process control,* uses graphical displays known as *control charts* to monitor a production process; the goal is to determine whether the process can be continued or whether it should be adjusted to achieve a desired quality level.

20.1   Acceptance Sampling

In acceptance sampling the items of interest can be incoming shipments of raw materials or purchased parts as well as finished goods from final assembly. Suppose that we want to decide whether to accept or reject a group of items based on specified quality characteristics. In quality-control terminology, the group of items is referred to as a *lot,* and *acceptance sampling* is a statistical method that enables us to make the accept-reject decision based on the inspection of a sample of items from the lot.

The general steps of acceptance sampling are shown in Figure 20.1. After a lot is received, a sample of items is selected for inspection. The results of the inspection are compared to specified quality characteristics. If the quality characteristics are satisfied, the lot is accepted and sent to production or shipped to customers. If the lot is rejected, management must decide on the disposition of the lot. In some cases, the decision may be to keep the lot and remove the unacceptable or nonconforming items during production. In other cases, the lot may be returned to the supplier at the supplier's expense; the extra work and cost placed on the supplier often provides good motivation for increasing the supplier's commitment to providing high-quality lots. Finally, in those instances where the rejected lot contains finished goods, the goods must be scrapped or reworked to bring them up to acceptable quality standards.

FIGURE **20.1**
Acceptance Sampling Procedure

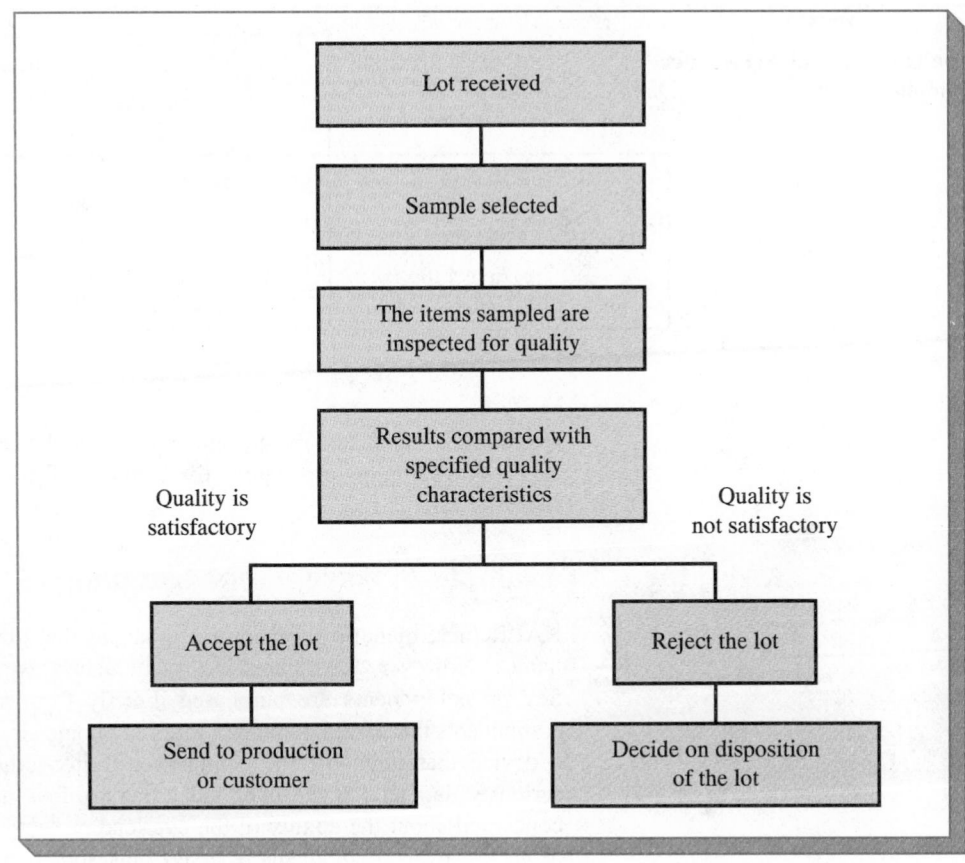

The statistical procedure of acceptance sampling is based on the hypothesis-testing methodology presented in Chapter 9. The null and alternative hypotheses are stated as follows:

$$H_0: \text{Good-quality lot}$$

$$H_a: \text{Poor-quality lot}$$

The states of nature (good-quality lot, poor-quality lot) and the decisions to accept or reject the lot are shown in Table 20.1. This table, just like the one presented in Chapter 9, shows the results of a hypothesis-testing procedure. Note that correct decisions correspond to accepting a good-quality lot and rejecting a poor-quality lot. However, as with other hypothesis-testing procedures, we need to be aware of the possibilities of making a Type I error (rejecting a good-quality lot) or a Type II error (accepting a poor-quality lot).

Since the probability of a Type I error creates a risk for the producer of the lot, it is referred to as the *producer's risk*. For example, a producer's risk of .05 indicates that there is a 5% chance that a good-quality lot will be erroneously rejected and returned to the producer. On the other hand, since the probability of a Type II error creates a risk for the consumer of the lot, it is referred to as the *consumer's risk*. For example, a consumer's risk of .10 means that there is a 10% chance that a poor-quality lot will be erroneously accepted and thus used in production or shipped to the customer. Specific values for the

TABLE 20.1
The Outcomes of Acceptance Sampling

| | | States of Nature | |
		H_0 True Good-Quality Lot	H_0 False Poor-Quality Lot
Decision	Accept the Lot	Correct decision	Type II error (accepting a poor-quality lot)
	Reject the Lot	Type I error (rejecting a good-quality lot)	Correct decision

the producer's risk and the consumer's risk can be controlled by the person designing the acceptance sampling procedure. To illustrate how this is done, let us consider the problem faced by KALI, Inc.

KALI, Inc.: An Example of Acceptance Sampling

KALI, Inc., manufactures home appliances that are marketed under a variety of trade names. However, KALI does not manufacture every component used in its products. Several components are purchased directly from suppliers. For example, one of the components that KALI purchases for use in home air conditioners is an overload protector, a device that turns off the compressor if it overheats. Since the compressor can be seriously damaged if the overload protector does not function properly, KALI is very concerned about the quality of the overload protectors. One way to assure quality would be to test every component received; this approach is referred to as 100% inspection. However, to determine proper functioning of an overload protector, it must be subjected to time-consuming and expensive tests, and KALI cannot justify testing every overload protector it receives.

Instead, KALI uses an acceptance sampling plan to monitor the quality of the overload protectors. The acceptance sampling plan requires that KALI's quality-control inspectors select and test a sample of overload protectors from each shipment. If very few defective units are found in the sample, the lot is probably of good quality and should be accepted. However, if a large number of defective units are found in the sample, the lot is probably of poor quality and should be rejected.

An *acceptance sampling plan* consists of a sample size n and an acceptance criterion c. The *acceptance criterion* is the maximum number of defective items that can be found in the sample and still indicate an acceptable lot. For example, for the KALI problem let us assume for the moment that a sample of 15 items will be selected from each incoming shipment or lot. Furthermore, assume that the manager of quality control states that the lot can be accepted only if no defective items are found. In this case then, the acceptance sampling plan established by the quality-control manager is $n = 15$ and $c = 0$.

This acceptance sampling plan is easy for the quality-control inspector to implement. The inspector simply selects a sample of 15 items, performs the tests, and reaches a conclusion based on the following decision rule:

- *Accept the lot* if 0 defective items are found.
- *Reject the lot* if 1 or more defective items are found.

Before implementing this acceptance sampling plan, the quality-control manager will be interested in evaluating the risks or errors possible under the plan. The plan will be implemented only if both the producer's risk (Type I error) and the consumer's risk (Type II error) are controlled at reasonable levels.

Computing the Probability of Accepting a Lot

The key to analyzing both the producer's risk and the consumer's risk is based on a "What-if?" type of analysis. That is, we will assume that a lot has some known percentage of defective items and compute the probability of accepting the lot for a given sampling plan. By varying the assumed percentage of defective items, we can examine the effect the sampling plan has on both types of risks.

Let us begin by assuming that a large shipment of overload protectors has been received and that 5% of the overload protectors in the shipment are defective. For a shipment or lot with 5% of the items defective, what is the probability that the $n = 15$, $c = 0$ sampling plan will lead us to accept the lot? Since each overload protector tested will be either defective or nondefective and since the lot size is large, the number of defective items in a sample of 15 has a *binomial probability distribution*. The binomial probability function, which was presented in Chapter 5, is as follows.

Binomial Probability Function for Acceptance Sampling

$$f(x) = \frac{n!}{x!(n - x)!} p^x (1 - p)^{(n - x)} \qquad (20.1)$$

where

n = the sample size

p = the proportion of defective items in the lot

x = the number of defective items in the sample

$f(x)$ = the probability of finding x defective items in a sample of n items

---

For the KALI acceptance sampling plan, $n = 15$; thus, for a lot with 5% defective ($p = .05$), we have

$$f(x) = \frac{15!}{x!(15 - x)!} (.05)^x (1 - .05)^{(15 - x)} \qquad (20.2)$$

Using (20.2), $f(0)$ will provide the probability that 0 overload protectors will be defective and the lot will be accepted. In using (20.2), recall that $0! = 1$. Thus, the probability computation for $f(0)$ is as follows:

$$f(0) = \frac{15!}{0!(15 - 0)!} (.05)^0 (1 - .05)^{(15 - 0)}$$

$$= \frac{15!}{0!(15)!} (.05)^0 (.95)^{15} = (.95)^{15} = .4633$$

We now know that the $n = 15$, $c = 0$ sampling plan has a .4633 probability of accepting a lot with 5% defective. Thus, there must be a corresponding $1 - .4633 = .5367$ probability of rejecting a lot with 5% defective.

In Table 20.2 we show the probability that the $n = 15$, $c = 0$ sampling plan will lead to the acceptance of lots with 1%, 2%, 3%, . . . defective. The probabilities in the table were computed by using $p = .01$, $p = .02$, $p = .03$, . . . in the binomial probability function (20.1).

Tables of binomial probabilities (see Table 5, Appendix B) can help reduce the computational effort in determining the probabilities of accepting lots. Selected binomial probabilities for $n = 15$ and $n = 20$ are shown in Table 20.3, and it can be used to verify that if the lot contains 10% defective, there would be a .2059 probability that the $n = 15$, $c = 0$ sampling plan would indicate an acceptable lot.

Using the data in Table 20.2, a graph of the probability of accepting the lot versus the percent defective in the lot can be drawn as shown in Figure 20.2. This graph, or curve, is called the *operating characteristic (OC) curve* for the $n = 15$, $c = 0$ acceptance sampling plan.

Perhaps we should consider other sampling plans, ones with different sample sizes n and/or different acceptance criteria c. First consider the case where the sample size remains $n = 15$ but the acceptance criterion increases from $c = 0$ to $c = 1$. That is, we will now accept the lot if 0 or 1 defective components are found in the sample. For a lot with 5% defective ($p = .05$), the binomial probability function in (20.2) can be used to compute $f(0)$ and $f(1)$. Summing these two probabilities provides the probability that the $n = 15$, $c = 1$ sampling plan will accept the lot. Alternatively, using Table 20.3, we find that with $n = 15$ and $p = .05$, $f(0) = .4633$ and $f(1) = .3658$. Thus, there is a $.4633 + .3658 = .8291$ probability that the $n = 15$, $c = 1$ plan will lead to the acceptance of a lot with 5% defective.

Figure 20.3 shows the operating characteristic curves for four alternative acceptance sampling plans for the KALI problem. Both samples of size 15 and 20 are considered. Note that regardless of the percent defective in the lot, the $n = 15$, $c = 1$ sampling plan provides the highest probabilities of accepting the lot. On the other hand, the $n = 20$, $c = 0$ sampling plan provides the lowest probabilities of accepting the lot; however, this plan also provides the highest probabilities of rejecting the lot.

Selecting an Acceptance Sampling Plan

Now that we know how to use the binomial probability distribution to compute the probability of accepting a lot with a given percent defective, we are ready to select the values of n and c that determine the desired acceptance sampling plan for the application being studied. To do this, management must specify two values for the fraction defective in the lot. One value, denoted p_0, will be used to control for the producer's risk, and the other value, denoted p_1, will be used to control for the consumer's risk.

In showing how this can be done, we will use the following notation:

α = the producer's risk; the probability that a lot with p_0 defective will be rejected

β = the consumer's risk; the probability that a lot with p_1 defective will be accepted

Suppose that for the KALI problem, management specifies that $p_0 = .03$ and $p_1 = .15$. From the OC curve in Figure 20.4, we see that $p_0 = .03$ provides a producer's risk of approximately $1 - .63 = .37$, and $p_1 = .15$ shows a consumer's risk of approximately

---

TABLE 20.2

Probability of Accepting the Lot for the KALI Problem with $n = 15$ and $c = 0$

Percent Defective in the Lot	Probability of Accepting the Lot
1	.8601
2	.7386
3	.6333
4	.5421
5	.4633
10	.2059
15	.0874
20	.0352
25	.0134

TABLE 20.3   Selected Binomial Probabilities for Samples of Sizes 15 and 20

n	x	.05	.10	.15	.20	.25	.30	.35	.40	.45	.50
15	0	.4633	.2059	.0874	.0352	.0134	.0047	.0016	.0005	.0001	.0000
	1	.3658	.3432	.2312	.1319	.0668	.0305	.0126	.0047	.0016	.0005
	2	.1348	.2669	.2856	.2309	.1559	.0916	.0476	.0219	.0090	.0032
	3	.0307	.1285	.2184	.2501	.2252	.1700	.1110	.0634	.0318	.0139
	4	.0049	.0428	.1156	.1876	.2252	.2186	.1792	.1268	.0780	.0417
	5	.0006	.0105	.0449	.1032	.1651	.2061	.2123	.1859	.1404	.0916
	6	.0000	.0019	.0132	.0430	.0917	.1472	.1906	.2066	.1914	.1527
	7	.0000	.0003	.0030	.0138	.0393	.0811	.1319	.1771	.2013	.1964
	8	.0000	.0000	.0005	.0035	.0131	.0348	.0710	.1181	.1647	.1964
	9	.0000	.0000	.0001	.0007	.0034	.0116	.0298	.0612	.1048	.1527
	10	.0000	.0000	.0000	.0001	.0007	.0030	.0096	.0245	.0515	.0916
	11	.0000	.0000	.0000	.0000	.0001	.0006	.0024	.0074	.0191	.0417
	12	.0000	.0000	.0000	.0000	.0000	.0001	.0004	.0016	.0052	.0139
	13	.0000	.0000	.0000	.0000	.0000	.0000	.0001	.0003	.0010	.0032
	14	.0000	.0000	.0000	.0000	.0000	.0000	.0000	.0000	.0001	.0005
	15	.0000	.0000	.0000	.0000	.0000	.0000	.0000	.0000	.0000	.0000
20	0	.3585	.1216	.0388	.0115	.0032	.0008	.0002	.0000	.0000	.0000
	1	.3774	.2702	.1368	.0576	.0211	.0068	.0020	.0005	.0001	.0000
	2	.1887	.2852	.2293	.1369	.0669	.0278	.0100	.0031	.0008	.0002
	3	.0596	.1901	.2428	.2054	.1339	.0716	.0323	.0123	.0040	.0011
	4	.0133	.0898	.1821	.2182	.1897	.1304	.0738	.0350	.0139	.0046
	5	.0022	.0319	.1028	.1746	.2023	.1789	.1272	.0746	.0365	.0148
	6	.0003	.0089	.0454	.1091	.1686	.1916	.1712	.1244	.0746	.0370
	7	.0000	.0020	.0160	.0545	.1124	.1643	.1844	.1659	.1221	.0739
	8	.0000	.0004	.0046	.0222	.0609	.1144	.1614	.1797	.1623	.1201
	9	.0000	.0001	.0011	.0074	.0271	.0654	.1158	.1597	.1771	.1602
	10	.0000	.0000	.0002	.0020	.0099	.0308	.0686	.1171	.1593	.1762
	11	.0000	.0000	.0000	.0005	.0030	.0120	.0336	.0710	.1185	.1602
	12	.0000	.0000	.0000	.0001	.0008	.0039	.0136	.0355	.0727	.1201
	13	.0000	.0000	.0000	.0000	.0002	.0010	.0045	.0146	.0366	.0739
	14	.0000	.0000	.0000	.0000	.0000	.0002	.0012	.0049	.0150	.0370
	15	.0000	.0000	.0000	.0000	.0000	.0000	.0003	.0013	.0049	.0148
	16	.0000	.0000	.0000	.0000	.0000	.0000	.0000	.0003	.0013	.0046
	17	.0000	.0000	.0000	.0000	.0000	.0000	.0000	.0000	.0002	.0011
	18	.0000	.0000	.0000	.0000	.0000	.0000	.0000	.0000	.0000	.0002
	19	.0000	.0000	.0000	.0000	.0000	.0000	.0000	.0000	.0000	.0000
	20	.0000	.0000	.0000	.0000	.0000	.0000	.0000	.0000	.0000	.0000

Operating Characteristic Curve for the *n* = 15, c = 0 Acceptance Sampling Plan

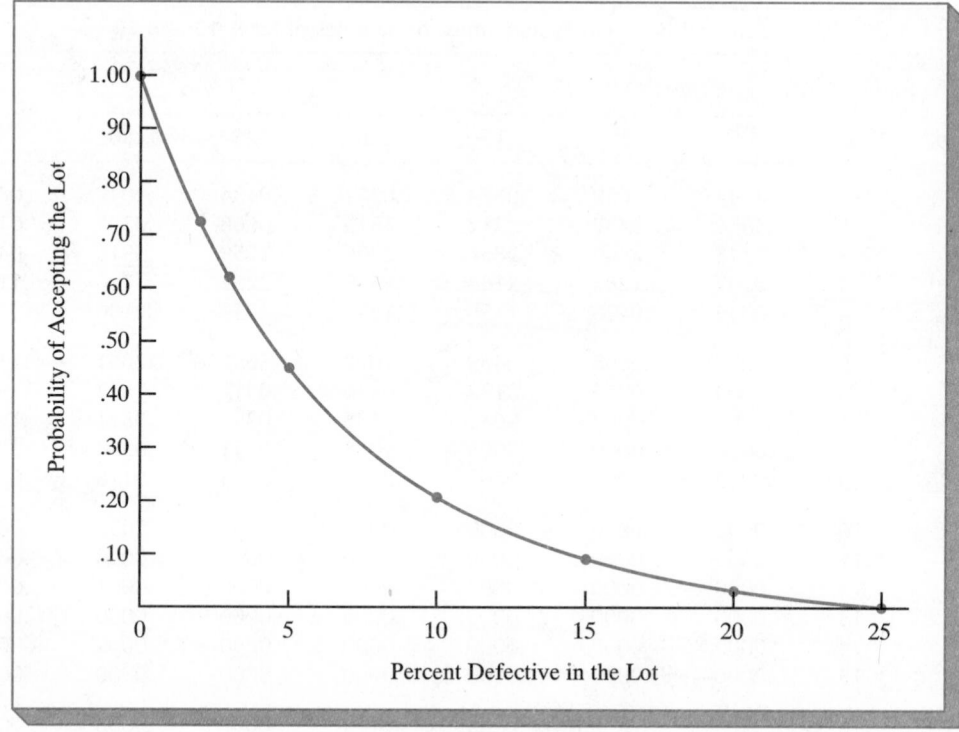

Operating Characteristic Curves for Four Acceptance Sampling Plans

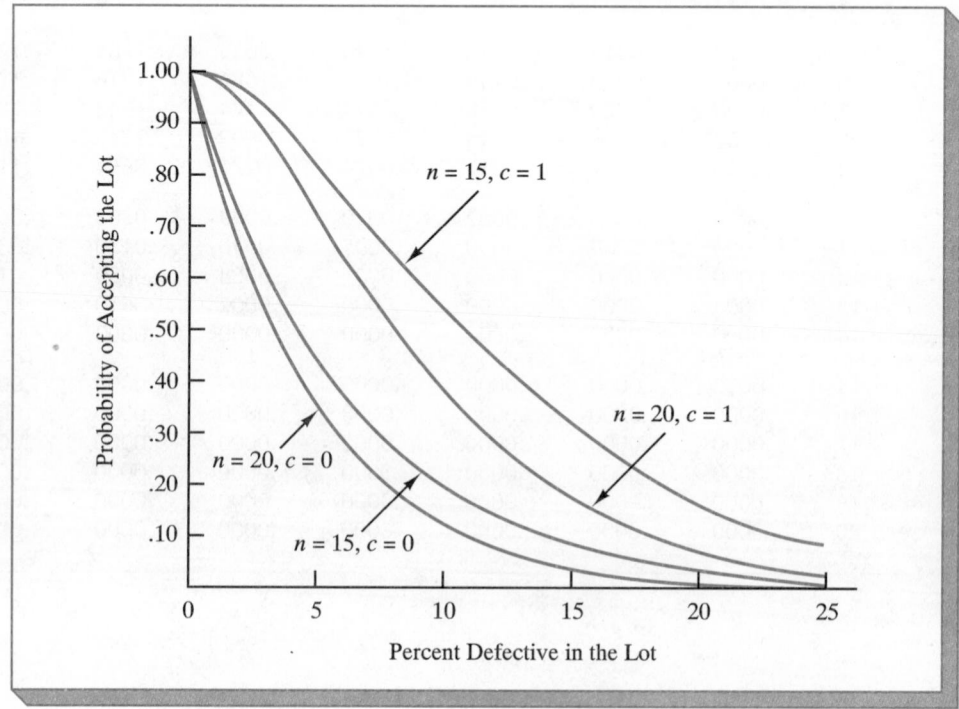

FIGURE **20.4**
Operating Characteristic Curve for
$n = 15$, $c = 0$ with $p_0 = .03$ and
$p_1 = .15$

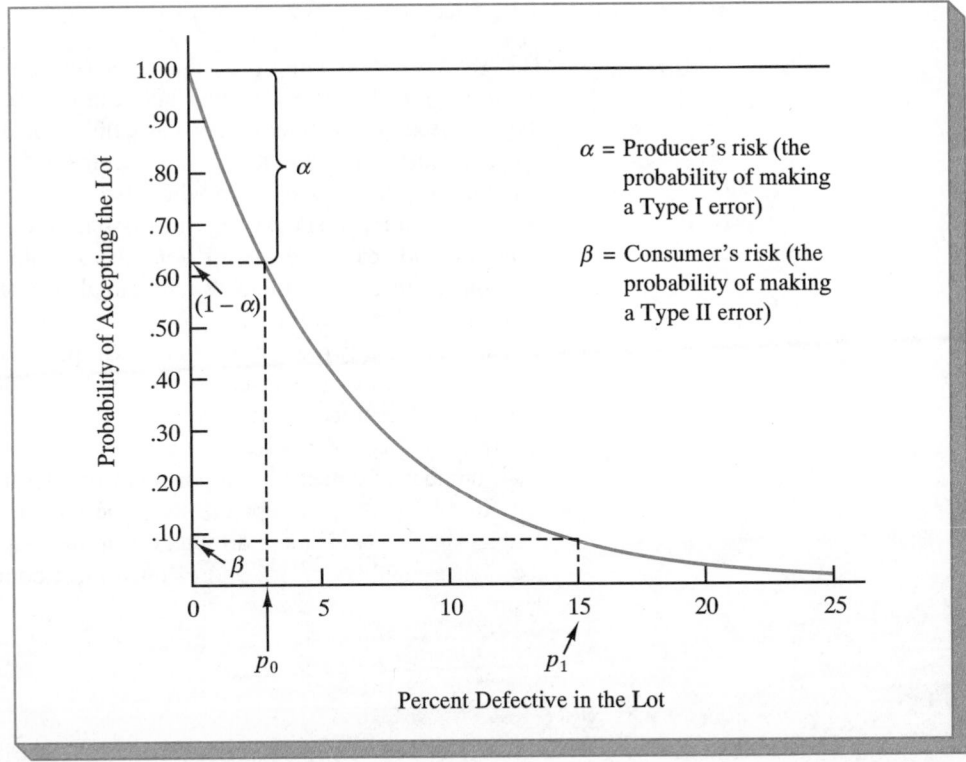

FIGURE **20.4**
Operating Characteristic Curve for
$n = 15$, $c = 0$ with $p_0 = .03$ and
$p_1 = .15$

.09. Thus, if management is willing to tolerate both a .37 probability of rejecting a lot with 3% defective and a .09 probability of accepting a lot with 15% defective, the $n = 15$, $c = 0$ acceptance sampling plan would be acceptable.

Suppose, however, that management requests a producer's risk of $\alpha = .10$ and a consumer's risk of $\beta = .20$. We see that now the $n = 15$, $c = 0$ sampling plan has a better-than-desired consumer's risk but an unacceptably large producer's risk. The fact that $\alpha = .37$ indicates that 37% of the lots will have the error of being rejected when only 3% of the items in the lot are defective. The producer's risk is too high, and a different acceptance sampling plan should be considered.

Using $p_0 = .03$, $\alpha = .10$, $p_1 = .15$, and $\beta = .20$ in Figure 20.3 shows that the acceptance sampling plan with $n = 20$ and $c = 1$ comes the closest to meeting both the producer's and the consumer's risk requirements. Exercise 4 at the end of this section will ask you to compute the producer's risk and the consumer's risk for the $n = 20$, $c = 1$ sampling plan.

As shown in this section, several computations and several operating characteristic curves may need to be considered to determine the sampling plan with the desired producer's and consumer's risk. Fortunately, tables of sampling plans are published. For example, the American Military Standard Table, MIL-STD-105D, provides information helpful in designing acceptance sampling plans. More advanced texts on quality control, such as those described in the bibliography, describe the use of such tables. In addition, these more advanced texts also discuss the role of sampling costs in determining the optimal sampling plan.

Multiple Sampling Plans

The acceptance sampling procedure that we have presented for the KALI problem is a *single-sample* plan. It is called a single-sample plan because only one sample or sampling stage is used. After determining the number of defective components in the sample, a decision must be made to accept or reject the lot. An alternative to the single-sample plan is a multiple sampling plan, in which two or more stages of sampling are used. At each stage a decision is made among three possibilities: stop sampling and accept the lot, stop sampling and reject the lot, or continue sampling. Although more complex, multiple sampling plans often result in a smaller total sample size than single-sample plans with the same α and β probabilities.

The logic of a two-stage, or double-sample, plan is shown in Figure 20.5. Initially, a sample of n_1 items is selected. If the number of defective components x_1 is less than or equal to c_1, accept the lot. If x_1 is greater than or equal to c_2, reject the lot. If x_1 is between c_1 and $c_2 (c_1 < x_1 < c_2)$, select a second sample of n_2 items. Determine the combined, or total, number of defective components from the first sample (x_1) and the second sample (x_2). If $x_1 + x_2 \leq c_3$, accept the lot; otherwise reject the lot. The development of the double-sample plan is more difficult due to the fact that the sample sizes n_1 and n_2 and the acceptance numbers c_1, c_2, and c_3 must meet both the producer's and consumer's risks desired.

**NOTES &
COMMENTS**

1. The use of the binomial probability distribution for acceptance sampling is based on the assumption of large lots. In situations where the lot size is small, the hypergeometric probability distribution is the appropriate distribution. Experts in the field of quality control indicate that the Poisson distribution provides a good approximation for acceptance sampling when the sample size is at least 16, the lot size is at least 10 times the sample size, and p is less than .1.* For larger sample sizes, the normal approximation to the binomial probability distribution can be used.

2. In the MIL-ST-105D sampling tables, p_0 is referred to as the acceptable quality level (AQL). In some sampling tables, p_1 is called the lot tolerance percent defective (LTPD) or the rejectable quality level (RQL). Many of the published sampling plans also use quality indexes such as the indifference quality level (IQL) and the average outgoing quality limit (AOQL). The more advanced texts listed in the bibliography provide a complete discussion of these other indexes.

3. In this section we provided an introduction to *attributes sampling plans*. In these plans each item sampled is classified as nondefective or defective. In *variables sampling plans,* a sample is taken, and a measurement of the quality characteristic is taken. For example, in gold jewelry a measurement of quality may be the amount of gold in the jewelry. A simple statistic such as the average amount of gold in the sample jewelry is computed and compared with an allowable value to determine whether to accept or reject the lot.

*J. M. Juran and Frank M. Gryna, Jr., *Quality Planning and Analysis,* McGraw-Hill, New York, 1980, p. 412.

FIGURE 20.5

A Two-Stage Acceptance Sampling Plan

```
                    ┌─────────────────┐
                    │  Inspect n₁     │
                    │  items          │
                    └────────┬────────┘
                             │
                    ┌────────▼────────┐
                    │  Find x₁        │
                    │  defective items│
                    │  in this sample │
                    └────────┬────────┘
                             │
                        ╱────▼────╲                    ┌──────────────┐
                       ╱    Is     ╲      Yes          │  Accept      │
                      ⟨  x₁ ≤ c₁    ⟩────────────────▶│  the lot     │
                       ╲    ?      ╱                    └──────────────┘
                        ╲────┬────╱
                         No  │
                        ╱────▼────╲
       ┌──────────┐    ╱    Is     ╲
       │ Reject   │◀──⟨  x₁ ≥ c₂    ⟩   Yes
       │ the lot  │Yes ╲    ?      ╱
       └──────────┘     ╲────┬────╱
                         No  │
                    ┌────────▼────────┐
                    │  Inspect n₂     │
                    │  additional items│
                    └────────┬────────┘
                             │
                    ┌────────▼────────┐
                    │  Find x₂        │
                    │  defective items│
                    │  in this sample │
                    └────────┬────────┘
                             │
                        ╱────▼────╲
              No       ╱    Is     ╲    Yes
          ◀───────────⟨ x₁+x₂ ≤ c₃ ⟩──────────▶
                       ╲    ?      ╱
                        ╲─────────╱
```

Caption: Inspect n_1 items → Find x_1 defective items in this sample → Is $x_1 \le c_1$? Yes → Accept the lot; No → Is $x_1 \ge c_2$? Yes → Reject the lot; No → Inspect n_2 additional items → Find x_2 defective items in this sample → Is $x_1 + x_2 \le c_3$? Yes → Accept the lot; No → Reject the lot.

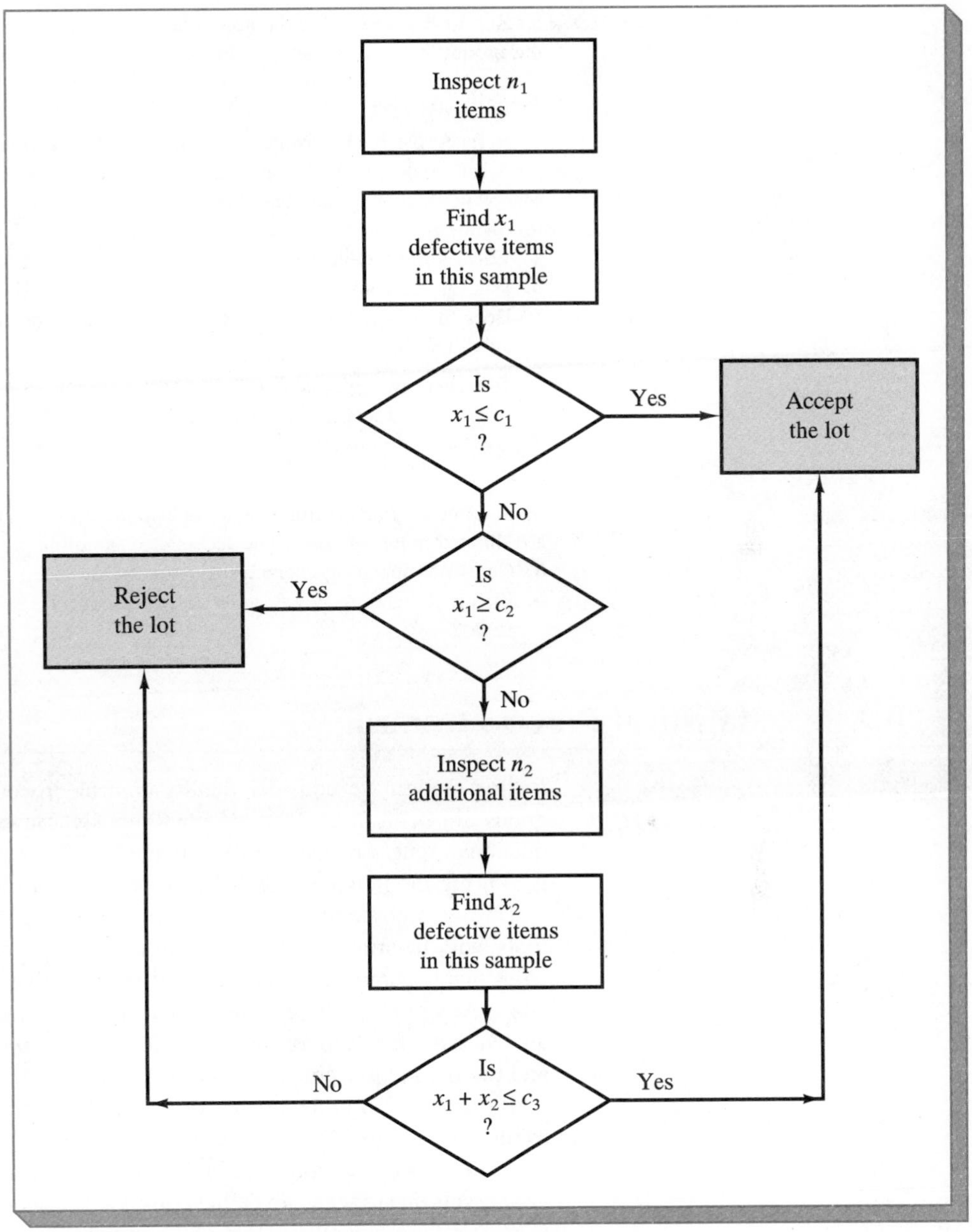

Exercises

Methods

SELF TEST

1. For an acceptance sampling plan with $n = 25$ and $c = 0$, find the probability of accepting a lot that has a defective rate of 2%. What is the probability of accepting the lot if the defective rate is 6%?

2. Consider an acceptance sampling plan with $n = 20$ and $c = 0$. Compute the producer's risk for each of the following cases:

a. The lot has a defective rate of 2%.

b. The lot has a defective rate of 6%.

3. Repeat Exercise 2 for the acceptance sampling plan with $n = 20$ and $c = 1$. What happens to the producer's risk as the acceptance number c is increased? Explain.

Applications

4. Refer to the KALI problem presented in this section. The quality-control manager requested a producer's risk of .10 when p_0 was .03 and a consumer's risk of .20 when p_1 was .15. Consider the acceptance sampling plan based on a sample size of 20 and an acceptance number of 1. Answer the following questions:

a. What is the producer's risk for the $n = 20$, $c = 1$ sampling plan?

b. What is the consumer's risk for the $n = 20$, $c = 1$ sampling plan?

c. Does the $n = 20$, $c = 1$ sampling plan satisfy the risks requested by the quality-control manager? Discuss.

5. To inspect incoming shipments of raw materials, a manufacturer is considering samples of size 10, 15, and 20. Use the binomial probabilities from Table 5 of Appendix B to select a sampling plan that provides a producer's risk of $\alpha = .03$ when p_0 is .05 and a consumer's risk of $\beta = .12$ when p_1 is .30.

6. A domestic manufacturer of watches purchases quartz crystals from a Swiss firm. The crystals are shipped in lots of 1000. The acceptance sampling procedure uses 20 randomly selected crystals.

a. Construct operating characteristic curves for acceptance numbers of 0, 1, and 2.

b. If p_0 is .01 and $p_1 = .08$, what are the producer's and consumer's risks for each sampling plan in (a)?

20.2   Statistical Process Control

In this section we consider quality-control procedures for a production process where goods are being manufactured continuously. Based on sampling and inspection of production output, a decision will be made to either continue the production process or adjust it to bring the items or goods being produced up to acceptable quality standards.

Despite high standards of quality in manufacturing and production operations, machine tools will invariably wear out, vibrations will cause machine settings to fall out of adjustment, purchased materials will be defective, and human operators will make mistakes. Any or all of these factors can result in poor-quality output. Fortunately, procedures are available for monitoring production output so that poor quality can be detected early and the production process adjusted or corrected.

If the variation in the quality of the production output is due to *assignable causes* such as tools wearing out, incorrect machine settings, poor-quality raw materials, or operator error, the process should be adjusted or corrected as soon as possible. Alternatively, if the variation is due to what are called common causes — that is, randomly occurring variations in materials, temperature, humidity, and so on, which the manufacturer cannot possibly account for — the process does not need to be adjusted. The main objective of statistical process control is to determine whether variations in output are due to assignable causes or common causes.

Whenever assignable causes are detected, we will conclude that the process is *out of control*. In this case, corrective action will be taken to bring the process back to an acceptable level of quality. However, if the variation in the output of a production process is due only to common causes, we will conclude that the process is in *statistical control*, or simply *in control;* in such cases, no changes or adjustments are necessary.

The statistical procedures for process control are based on the hypothesis-testing methodology presented in Chapter 9. The null hypothesis H_0 is formulated in terms of the production process being in control. The alternative hypothesis H_a is formulated in terms of the production process being out of control. The states of nature are that the production

TABLE 20.4
The Outcomes of Statistical Process Control

Decision		States of Nature	
		H_0 True Process in Control	H_0 False Process Out of Control
Decision	Continue Process	Correct decision	Type II error (allowing an out-of-control process to continue)
	Adjust Process	Type I error (adjusting an in-control process)	Correct decision

process is in control and that the production process is out of control; the decision alternatives are to continue or to adjust the production process. Table 20.4 shows that correct decisions of continuing an in-control process and adjusting an out-of-control process are possible. However, as with other hypothesis-testing procedures, there also exists the possibility of making a Type I error (adjusting an in-control process) as well as the possibility of making a Type II error (allowing an out-of-control process to continue).

Control Charts

A *control chart* is used to help determine if a process is in control or out of control. As such, a control chart provides a basis for deciding whether the variation in the output is due to common causes (in control) or to assignable causes (out of control). Whenever an out-of-control situation is detected, adjustments and/or other corrective action will be taken to bring the process back into control.

Control charts can be classified by the type of data they contain. For instance, an $\overline{x}$ chart is used for cases where the quality of the output is measured in terms of a variable such as length, weight, temperature, and so on. In this case, the decision to continue or to adjust the production process will be based on the mean value found in a sample of the output. To introduce some of the concepts common to all control charts, let us consider some specific features of an $\overline{x}$ chart.

Figure 20.6 shows the general structure of an $\overline{x}$ chart. The center line of the chart corresponds to the mean of the process when the process is *in control*. The vertical line identifies the scale of measurement for the variable of interest. Each time a sample is taken from the production process, a value of $\overline{x}$ is computed, and a data point showing the value of $\overline{x}$ for the sample is plotted on the control chart.

The two lines labeled UCL and LCL play an important role in determining whether the process is in control or out of control. The lines are referred to as the *upper control limit* and the *lower control limit*, respectively. These limits are chosen so that when the process is in control, there will be a high probability that the value of $\overline{x}$ will fall between the two control limits. Values outside the control limits provide strong statistical evidence that the process is out of control and corrective action should be taken.

Over time, more and more data points will be added to the control chart. The order of the data points will appear from left to right as the process is sampled. In essence, every time a point is plotted on the control chart, we are carrying out a hypothesis test to see if the process is in control.

FIGURE 20.6
$\bar{x}$ Chart Structure

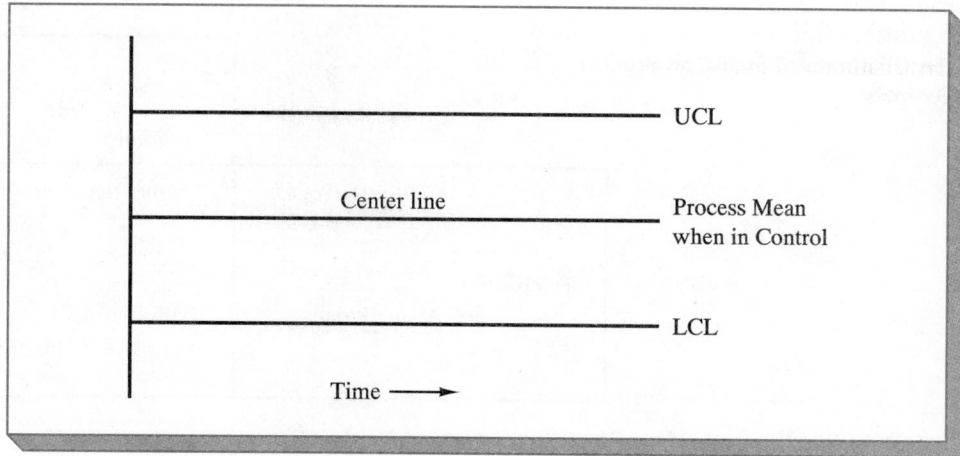

In addition to the $\bar{x}$ chart, there are also control charts that can be used to monitor the range of the measurements in the sample (R chart), the percent defective in the sample (p chart), and the number of defects in the sample (c chart). In each case, the general structure of the control chart follows the format of the $\bar{x}$ chart shown in Figure 20.6. The major difference in each control chart is the measurement scale used; for instance, in a p chart the measurement scale denotes the proportion of defective items in the sample instead of the sample mean. In the following discussion, we will illustrate the construction and use of the $\bar{x}$ chart, R chart, and p chart.

$\bar{x}$ Chart: Process Mean and Standard Deviation Known

To illustrate the construction of an $\bar{x}$ chart, let us consider the situation involving KJW Packaging. This company operates a production line where cartons of cereal are filled. Suppose that KJW knows that when the process is operating correctly—and hence the system is in control—the mean filling weight is $\mu = 16.05$ ounces, and the process standard deviation is $\sigma = .10$ ounces. In addition, suppose that the distribution of filling weights is normally distributed. This distribution is shown in Figure 20.7.

The sampling distribution of $\bar{x}$, as presented in Chapter 7, can be used to determine the variation that can be expected in $\bar{x}$ values for a process that is in control. To show how this is done, let us first briefly review the properties of the sampling distribution of $\bar{x}$. First, recall that the expected value or mean of $\bar{x}$ is equal to μ, the mean filling weight when the production line is in control. For samples of size n, the formula for the standard deviation of $\bar{x}$, referred to as the standard error of the mean, is as follows:

$$\sigma_{\bar{x}} = \frac{\sigma}{\sqrt{n}} \tag{20.3}$$

In addition, since the distribution of filling weights is normally distributed, the sampling distribution of $\bar{x}$ is normal for any sample size. Thus, the sampling distribution of $\bar{x}$ is normally distributed with mean μ and standard deviation $\sigma_{\bar{x}}$. This distribution is shown in Figure 20.8.

The sampling distribution of $\bar{x}$ is used to determine what values of $\bar{x}$ are reasonable to observe if the process is in control. The general practice in quality control is to define

FIGURE 20.7
Distribution of Cereal-Carton Filling Weights

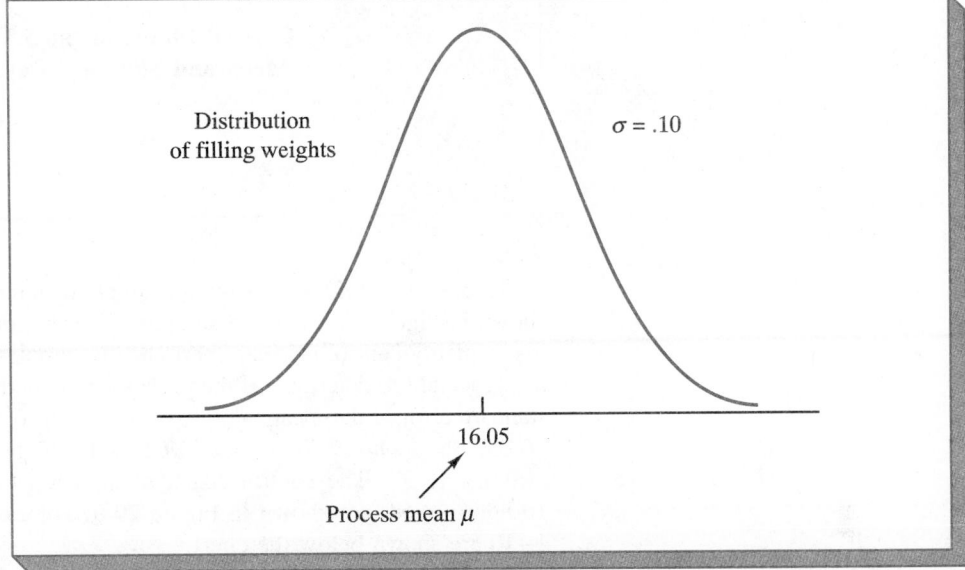

Distribution
of filling weights

$\sigma = .10$

16.05

Process mean μ

FIGURE 20.8
Sampling Distribution of $\bar{x}$

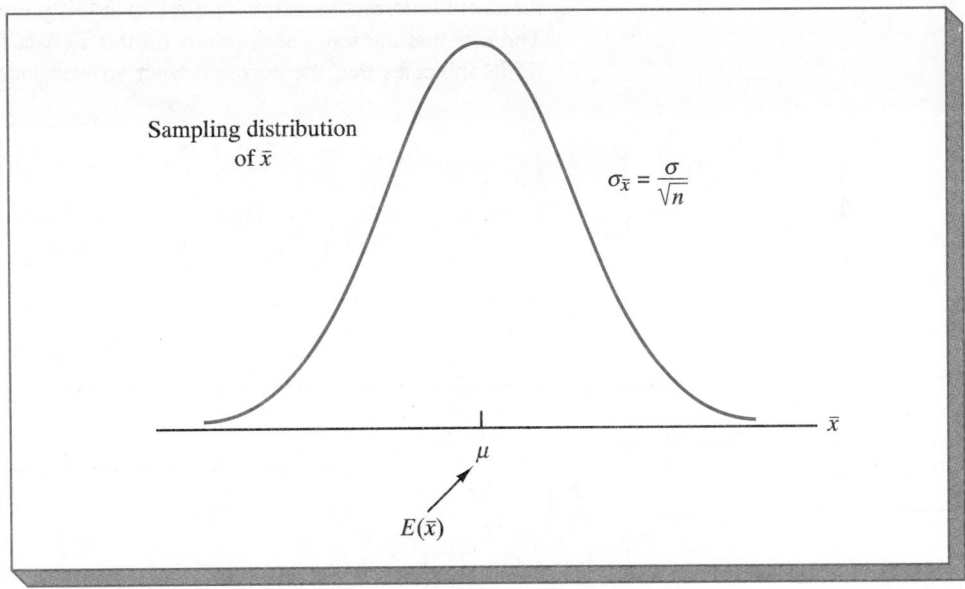

Sampling distribution
of $\bar{x}$

$$\sigma_{\bar{x}} = \frac{\sigma}{\sqrt{n}}$$

$\bar{x}$

μ

$E(\bar{x})$

reasonable as any value of $\bar{x}$ that is within 3 standard deviations above or below the mean value. Recall from the study of the normal probability distribution that approximately 99.7% of the values of a normally distributed random variable lie within ±3 standard deviations of its mean value. Thus, if a value of $\bar{x}$ falls within the interval $\mu - 3\sigma_{\bar{x}}$ to $\mu + 3\sigma_{\bar{x}}$, we will assume that the process is in control. In summary, then, the control limits for an $\bar{x}$ chart are computed as follows.

**Control Limits for an $\bar{x}$ Chart: Process
Mean and Standard Deviation Known**

$$\text{UCL} = \mu + 3\sigma_{\bar{x}} \qquad\qquad (20.4)$$

$$\text{LCL} = \mu - 3\sigma_{\bar{x}} \qquad\qquad (20.5)$$

Reconsider the KJW Packaging example with the process distribution of filling weights shown in Figure 20.7 and the sampling distribution of $\bar{x}$ shown in Figure 20.8. Assume that a quality-control inspector periodically samples six cartons and uses the sample mean filling weight to determine if the process is in control or out of control. Using (20.3), the standard error of the mean is $\sigma_{\bar{x}} = \sigma/\sqrt{n} = .10/\sqrt{6} = .04$. Thus, with the process mean at 16.05, the control limits are UCL = 16.05 + 3(.04) = 16.17 and LCL = 16.05 − 3(.04) = 15.93. The control chart, along with the results of 10 samples taken over a 10-hour period, are shown in Figure 20.9. For ease of reading, the sample numbers of 1–10 are shown below the chart.

Refer to Figure 20.9; note that the mean for the fifth sample shows that the process was out of control. In other words, the sample mean $\bar{x}$ = 15.89 provides an indication that assignable causes in output variation were detected and that underfilling was occurring. As a result, corrective action was taken at this point to bring the process back into control. The fact that the remaining points on the $\bar{x}$ chart fall within the upper and lower control limits indicates that the corrective action was successful.

$\bar{x}$ Chart: Process Mean and Standard Deviation Unknown

In the KJW Packaging example, we showed how an $\bar{x}$ chart can be developed when the mean and standard deviation of the process are known. In most situations, these values

---

FIGURE 20.9

The $\bar{x}$ Chart for the Cereal-Carton Filling Process

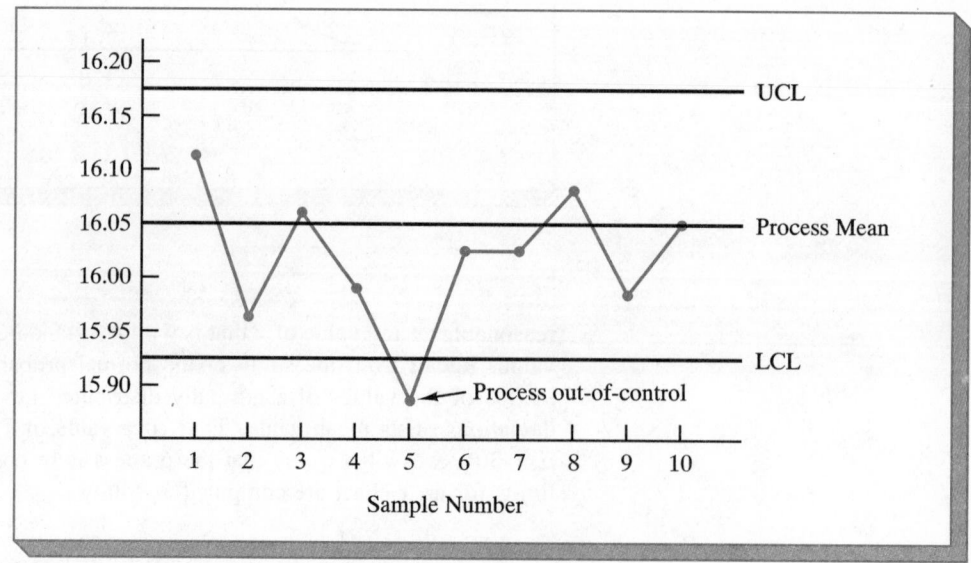

must be estimated using samples that are selected from the process when it is known to be operating in control. For instance, KJW might select a random sample of five boxes each morning and five boxes each afternoon for 10 days of in-control operation. For each subgroup, or sample, the mean and standard deviation of the sample are computed. The overall averages of both the sample means and the sample standard deviations are used to construct control charts for both the process mean and the process standard deviation. Note that this approach recognizes that in most problems it is important to control both the mean of the process as well as the variability of the process.

In practice, it is more common to monitor the variability of the process using the range instead of the standard deviation. In addition to providing good estimates of the process standard deviation when the sample size is small, the range can be used to construct upper and lower control limits for both the $\bar{x}$ chart and the R chart with little computational effort. To illustrate how these control charts are constructed, let us consider the problem facing Jensen Computer Supplies, Inc.

Jensen Computer Supplies (JCS) manufactures 3.5 inch diameter floppy disks that are used in personal computers. Of concern to JCS is the maintenance of strict control over both the process mean and the process variability. Suppose that JCS were to select random samples from the manufacturing process when it is believed to be operating in control. For example, a random sample of five disks could be taken during the first hour of operation, five disks during the second hour of operation, and so on, until 20 samples were selected. Table 20.5 shows the data obtained, including the mean $\bar{x}_j$ and range R_j for each of the samples.

Assume that the diameter of disks produced when the process is in control is a normally distributed random variable with mean μ and standard deviation σ, and that k samples, each of size n, have been selected. The estimate of the process mean μ is given by the overall sample mean.

TABLE 20.5
Data for the Jensen Computer Supplies Problem

Sample Number	Observations					Sample Mean $\bar{x}_j$	Sample Range R_j
1	3.5056	3.5086	3.5144	3.5009	3.5030	3.5065	.0135
2	3.4882	3.5085	3.4884	3.5250	3.5031	3.5026	.0368
3	3.4897	3.4898	3.4995	3.5130	3.4969	3.4978	.0233
4	3.5153	3.5120	3.4989	3.4900	3.4837	3.5000	.0316
5	3.5059	3.5113	3.5011	3.4773	3.4801	3.4951	.0340
6	3.4977	3.4961	3.5050	3.5014	3.5060	3.5012	.0099
7	3.4910	3.4913	3.4976	3.4831	3.5044	3.4935	.0213
8	3.4991	3.4853	3.4830	3.5083	3.5094	3.4970	.0264
9	3.5099	3.5162	3.5228	3.4958	3.5004	3.5090	.0270
10	3.4880	3.5015	3.5094	3.5102	3.5146	3.5047	.0266
11	3.4881	3.4887	3.5141	3.5175	3.4863	3.4989	.0312
12	3.5043	3.4867	3.4946	3.5018	3.4784	3.4932	.0259
13	3.5043	3.4769	3.4944	3.5014	3.4904	3.4935	.0274
14	3.5004	3.5030	3.5082	3.5045	3.5234	3.5079	.0230
15	3.4846	3.4938	3.5065	3.5089	3.5011	3.4990	.0243
16	3.5145	3.4832	3.5188	3.4935	3.4989	3.5018	.0356
17	3.5004	3.5042	3.4954	3.5020	3.4889	3.4982	.0153
18	3.4959	3.4823	3.4964	3.5082	3.4871	3.4940	.0259
19	3.4878	3.4864	3.4960	3.5070	3.4984	3.4951	.0206
20	3.4969	3.5144	3.5053	3.4985	3.4885	3.5007	.0259

Overall Sample Mean

$$\bar{\bar{x}} = \frac{\bar{x}_1 + \bar{x}_2 + \cdots + \bar{x}_k}{k} \qquad (20.6)$$

where

$$\bar{x}_j = \text{mean of the } j\text{th sample}$$

For the JCS data in Table 20.5, the overall sample mean is $\bar{\bar{x}} = 3.4995$. This value will be the center line for the $\bar{x}$ chart. The range of each sample, denoted R_j, is simply the difference between the largest and smallest value in each sample. The average range is shown below.

Average Range

$$\bar{R} = \frac{R_1 + R_2 + \cdots + R_k}{k} \qquad (20.7)$$

For the JCS data in Table 20.5, the average range is $\bar{R} = .0253$.

In the preceding section we showed that the upper and lower control limits for the $\bar{x}$ chart are

$$\bar{\bar{x}} \pm 3 \frac{\sigma}{\sqrt{n}} \qquad (20.8)$$

Thus, to construct the control limits for the $\bar{x}$ chart, we need to estimate the standard deviation of the process. One approach is to calculate the standard deviation of each sample and then use the average of the sample standard deviations as an estimate of σ. Alternatively, an estimate of σ can also be developed using the range of each sample.

It can be shown that an estimator of the process standard deviation σ is the average range divided by d_2, a constant which depends upon the sample size n. That is,

$$\text{Estimator of } \sigma = \frac{\bar{R}}{d_2} \qquad (20.9)$$

Values for d_2 are provided in Table 12, which is found in Appendix B. For instance, when $n = 5$, $d_2 = 2.326$ and the estimate of σ is the average range divided by 2.326. If we substitute $\bar{R}/d_2$ for σ in (20.8), we can write the control limits for the $\bar{x}$ chart as

$$\bar{\bar{x}} \pm 3 \frac{\bar{R}/d_2}{\sqrt{n}} = \bar{\bar{x}} \pm \frac{3}{d_2\sqrt{n}} \bar{R} = \bar{\bar{x}} \pm A_2 \bar{R} \qquad (20.10)$$

Note that $A_2 = 3/d_2\sqrt{n}$ is a constant that depends only on the sample size. Values for A_2 are provided in Table 12 in Appendix B. For $n = 5$, $A_2 = .577$; thus, the control limits for the $\bar{x}$ chart are

$$3.4995 \pm (.577)(.0253) = 3.4995 \pm .0146$$

Thus, LCL = 3.4849 and UCL = 3.5141.

R Chart

To develop the R chart, we need to think of the range of a sample as a random variable with its own mean and standard deviation. The average range $\bar{R}$ provides an estimate of the mean of this random variable. Moreover, it can be shown that an estimate of the standard deviation of the range is

$$\hat{\sigma}_R = d_3 \frac{\bar{R}}{d_2} \tag{20.11}$$

where d_2 and d_3 are constants which depend upon the sample size. Thus, the UCL for the R chart is given by

$$\bar{R} + 3\hat{\sigma}_R = \bar{R} + 3d_3 \frac{\bar{R}}{d_2} \tag{20.12}$$

and the LCL is

$$\bar{R} - 3\hat{\sigma}_R = \bar{R} - 3d_3 \frac{\bar{R}}{d_2} \tag{20.13}$$

If we let

$$D_4 = 1 + 3\frac{d_3}{d_2} \tag{20.14}$$

$$D_3 = 1 - 3\frac{d_3}{d_2} \tag{20.15}$$

we can write the control limits for the R chart as

$$\text{UCL} = \bar{R}D_4 \tag{20.16}$$

$$\text{LCL} = \bar{R}D_3 \tag{20.17}$$

Values for D_3, and D_4 are also provided in Table 12 of Appendix B. Note that for $n = 5$, $D_3 = 0$ and $D_4 = 2.115$. Thus, with $\bar{R} = .0253$, the control limits are

$$\text{UCL} = .0253(2.115) = .0535$$

$$\text{LCL} = .0253(0) = 0$$

The R chart is shown in Figure 20.10. The 20 ranges plotted on this chart do not indicate that the process is out of control. Figure 20.11 shows the $\bar{x}$ chart with the 20 sample means for the JCS data; we note that these points show no indication of an out-of-control condition. Since these data were collected during a period of what was thought to be an in-control state, these data also confirm our assumption that during the time period in which the data were collected, the process was in control.

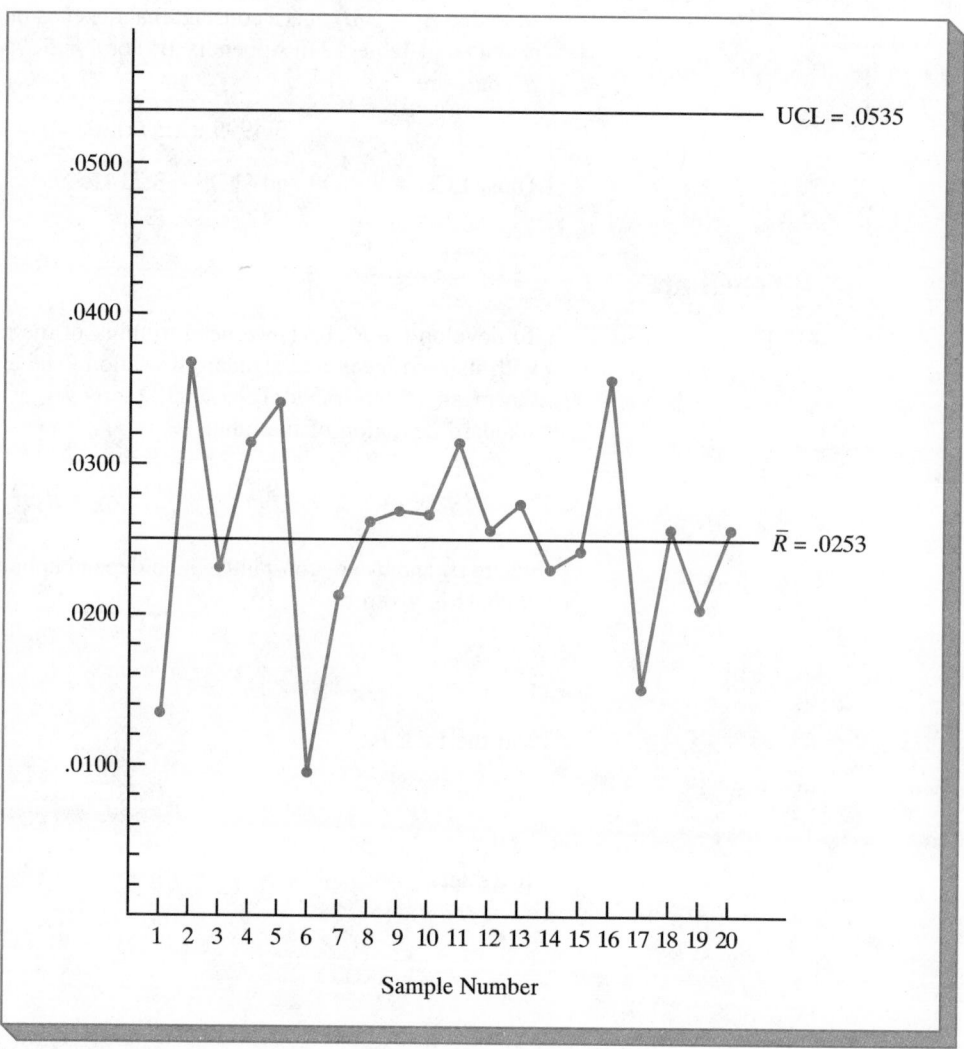

FIGURE **20.10**
R **Chart for the Jensen Computer Supplies Problem**

p Chart

Let us consider the case where the output quality is measured in terms of the items being either nondefective or defective. The decision to continue or to adjust the production process will be based on $\bar{p}$, the proportion of defective items found in a sample of the output. The control chart used for proportion defective data is called a *p chart*.

To illustrate the construction of a *p* chart, consider the situation involving the use of automated mail-sorting machines in a post office. These automated machines scan the zip code on letters and divert each letter to its proper carrier route. Even when the machine is operating properly, some letters are diverted to incorrect routes. Assume that when the machine is operating correctly, or in a state of control, 3% of the letters are incorrectly diverted. Thus *p*, the fraction defective when the process is in control, is .03.

The sampling distribution of $\bar{p}$, as presented in Chapter 7, can be used to determine the variation that can be expected in $\bar{p}$ values for a process that is in control. Recall that the expected value or mean of $\bar{p}$ is *p*, the fraction defective when the process is in control.

F I G U R E **20.11**

$\bar{x}$ **Chart for the Jensen Computer Supplies Problem**

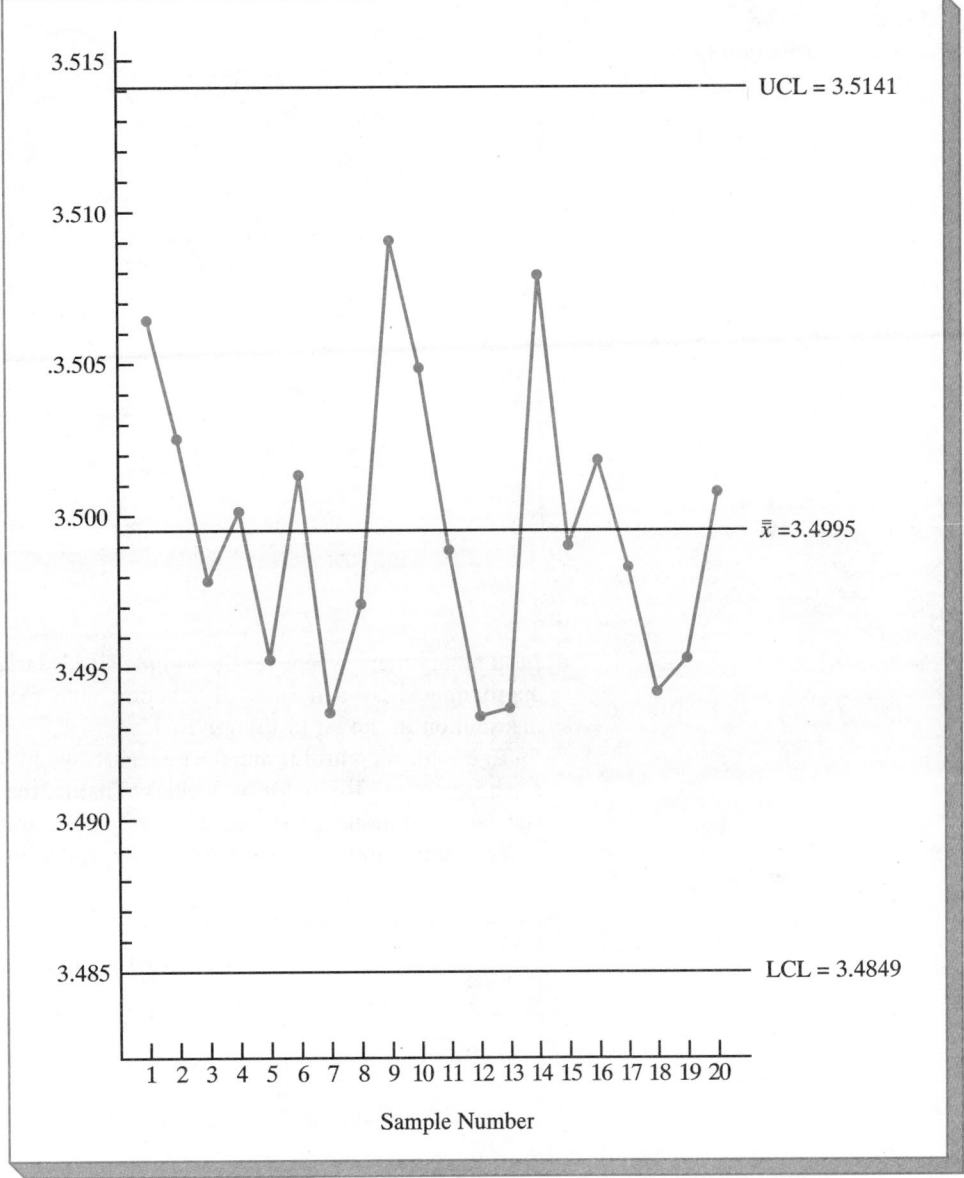

With samples of size n, the formula for the standard deviation of $\bar{p}$, referred to as the standard error of the proportion, is as follows:

$$\sigma_{\bar{p}} = \sqrt{\frac{p(1-p)}{n}} \qquad (20.18)$$

We also learned in Chapter 7 that the sampling distribution of $\bar{p}$ can be approximated by a normal probability distribution whenever the sample size is large. With $\bar{p}$, the sample size can be considered large whenever the following two conditions are satisfied:

$$np \geq 5$$

$$n(1-p) \geq 5$$

Figure 20.12
Sampling Distribution of $\bar{p}$

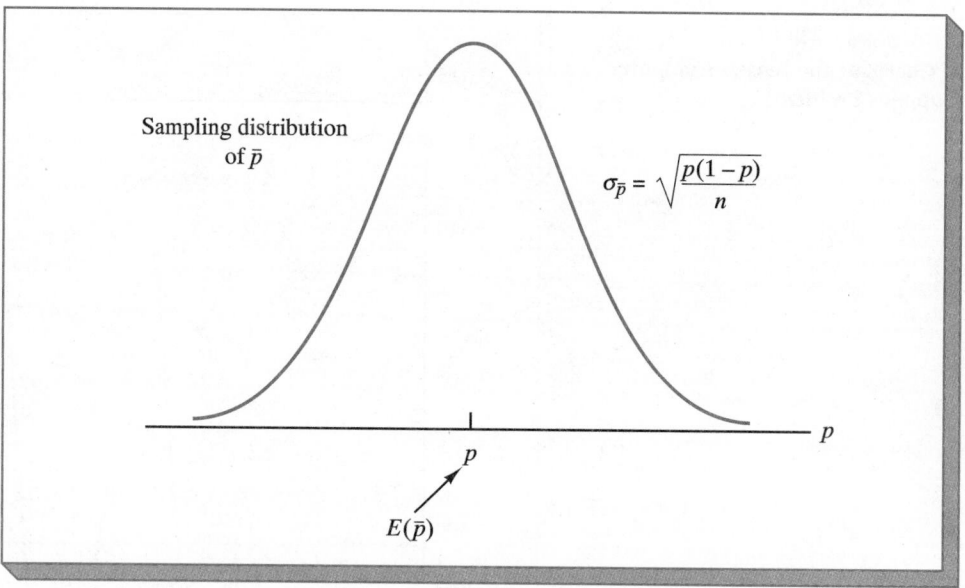

Sampling distribution
of $\bar{p}$

$$\sigma_{\bar{p}} = \sqrt{\frac{p(1-p)}{n}}$$

p

$E(\bar{p})$

p

In summary then, whenever the sample size is large, the sampling distribution of $\bar{p}$ can be approximated by a normal distribution with mean p and standard deviation $\sigma_{\bar{p}}$. This distribution is shown in Figure 20.12.

To establish control limits for a p chart, we follow the same procedure that we used to establish control limits for an $\bar{x}$ chart. That is, the limits for the control chart are set at 3 standard deviations, or standard errors, above and below the percent defective when the process is in control. Thus, we have the following control limits.

Control Limits for a p Chart

$$\text{UCL} = p + 3\sigma_{\bar{p}} \qquad\qquad\qquad (20.19)$$

$$\text{LCL} = p - 3\sigma_{\bar{p}} \qquad\qquad\qquad (20.20)$$

Using $p = .03$ and samples of size $n = 200$, (20.18) shows that the standard error is

$$\sigma_{\bar{p}} = \sqrt{\frac{.03(1 - .03)}{200}} = .0121$$

Thus, the control limits are UCL $= .03 + 3(.0121) = .0662$ and LCL $= .03 - 3(.0121) = -.0063$. Since LCL is negative, LCL is set equal to 0 in the control chart.

The control chart for the mail-sorting process is shown in Figure 20.13. The points plotted show a series of proportion defective found in samples of 200 letters taken from the process. Since all points are within the control limits, there is no evidence to conclude that the sorting process is out of control. In fact, the p chart indicates that the process should continue to operate.

In cases where the percentage of defective items for a process that is in control is not known, this value is first estimated using sample data. Suppose, for example, that M

FIGURE 20.13
p Chart for the Proportion
Defective in a Mail-Sorting Process

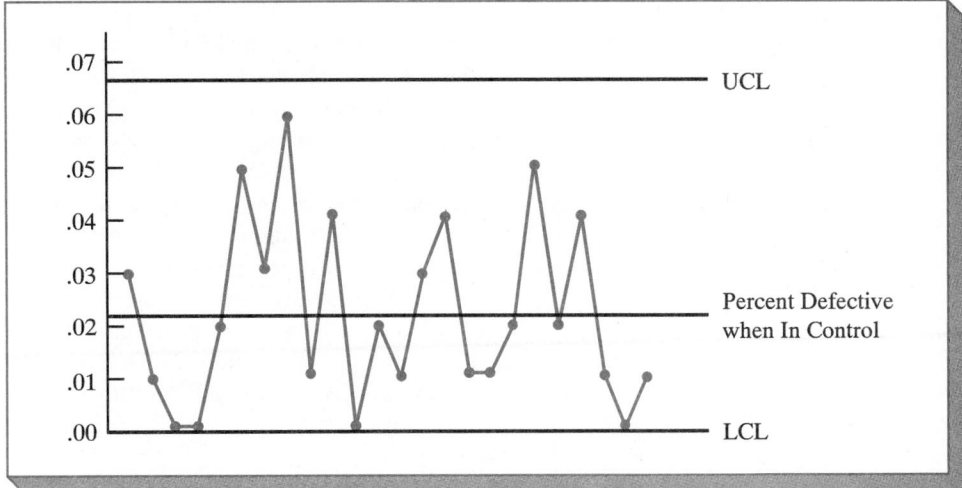

FIGURE 20.13
***p* Chart for the Proportion Defective in a Mail-Sorting Process**

different samples, each of size *n*, are selected from a process that is in control. The fraction or proportion of defective items in each sample is then determined. Treating all the data collected as one large sample, we can determine the average number of defective items for all the data; this value can then be used to provide an estimate of *p*, the percentage of defective items observed when the process is in control. Note that this estimate of *p* also enables us to estimate the standard error of the proportion; upper and lower control limits can then be established.

Interpretation of Control Charts

The location and pattern of points in a control chart enable us to determine, with a small probability of error, whether a process is in statistical control. A primary indication that a process may be out of control occurs whenever a data point is outside the control limits. This situation was observed for point 5 in Figure 20.9. When such a point is found, there is strong statistical evidence that the process is out of control; in such cases, corrective action should be taken as soon as possible.

In addition to points outside the control limits, certain patterns of the points within the control limits can act as warning signals for quality-control problems. For example, assume that all the data points are within the control limits but that a large number of points are on one side of the center line. This may indicate that an equipment problem, a change in materials, or some other assignable cause of a shift in quality has occurred. When this pattern occurs, careful investigation of the production process should be undertaken to determine if indeed a change in quality has occurred.

Another pattern to watch for in control charts is a gradual shift, or trend, over time. For example, as tools wear out, the dimensions of machined parts will gradually deviate more and more from their designed levels. Gradual changes in temperature or humidity, general equipment deterioration, dirt buildup, or operator fatigue may also result in a trend pattern in control charts. About six or seven points in a row that indicate either an increasing or decreasing trend should signal cause for concern, even if the data points are all within the control limits. When such a pattern occurs, the process should be reviewed for possible changes or shifts in quality. Corrective action to bring the process back into control may be necessary.

1. Since the control limits for the $\bar{x}$ chart depend on the value of the average range, these limits will not have much meaning unless the process variability is in control. In practice, the R chart is usually constructed before the $\bar{x}$ chart; if the R chart indicates that the process variability is in control, then the $\bar{x}$ chart is constructed.

2. Standard practice has been to set upper control limits and lower control limits at 3 standard deviations or 3σ above and below the process mean. Motorola suggested that if processes could be designed using control limits more than 3 standard deviations above and below the process mean, the probability of a defect would be reduced substantially. The Motorola Six Sigma (6σ) Quality Level sets a goal of producing no more than 3.4 defects per million operations. (*American Production and Inventory Control Society*, July 1991).

❑ ❑ Exercises

Methods

7. A process that is in control has a mean of $\mu = 12.5$ and a standard deviation of $\sigma = .8$.
    a. Construct an $\bar{x}$ chart if samples of size 4 are to be used.
    b. Repeat (a) for samples of size 8 and 16.
    c. What happens to the limits of the control chart as the sample size is increased? Discuss why this is reasonable.

8. Twenty-five samples, each of size 5, were selected from a process that was in control. The sum of all the data collected was 677.5 pounds.
    a. What is an estimate of the process mean (in terms of pounds per unit) when the process is in control?
    b. Develop the control chart for this process if samples of size 5 will be used. Assume that the process standard deviation is .5 when the process is in control, and that the mean of the process is the estimate developed in (a).

9. Twenty-five samples of 100 items each were inspected when a process was considered to be operating satisfactorily. In the 25 samples, a total of 135 items were found to be defective.
    a. What is an estimate of the proportion defective when the process is in control?
    b. What is the standard error of the proportion if samples of size 100 will be used for statistical process control?
    c. Compute the upper and lower control limits for the control chart.

SELF TEST ▷ 10. A process, sampled 20 times with a sample of size 8 resulted in the following $\bar{\bar{x}}$ and $\bar{R}$ values:

$$\bar{\bar{x}} = 28.5, \bar{R} = 1.6$$

Compute the upper and lower control limits for the R and $\bar{x}$ charts for this process.

Applications

11. Temperature is used to measure the output of a production process. When the process is in control, the mean of the process is $\mu = 128.5$, and the standard deviation is $\sigma = .4$.
    a. Construct an $\bar{x}$ chart if samples of size 6 are to be used.
    b. Is the process in control for a sample providing the following data: 128.8, 128.2, 129.1, 128.7, 128.4, and 129.2?
    c. Is the process in control for a sample providing the following data: 129.3, 128.7, 128.6, 129.2, 129.5, and 129.0?

TABLE 20.6			
Sample	**Tread Wear***		
1	31	42	28
2	26	18	35
3	25	30	34
4	17	25	21
5	38	29	35
6	41	42	36
7	21	17	29
8	32	26	28
9	41	34	33
10	29	17	30
11	26	31	40
12	23	19	25
13	17	24	32
14	43	35	17
15	18	25	29
16	30	42	31
17	28	36	32
18	40	29	31
19	18	29	28
20	22	34	26

*Hundredths of an inch

12. A soap manufacturer uses a weight to measure the output of a laundry detergent powder. The control limits are set at UCL = 20.12 and LCL = 19.90. Samples of size 5 are used for the sampling and inspection process. What are the process mean and process standard deviation for the manufacturing operation?

13. The Goodman Tire and Rubber Company periodically tests its tires for tread wear under simulated road conditions. To study and control its manufacturing process, 20 samples, each containing 3 radial tires, were chosen from different shifts over several days of operation. The results are shown in Table 20.6. Assuming that these data were collected when the manufacturing process was believed to be operating in control, develop the R and $\bar{x}$ charts.

14. Over several weeks of normal, or in-control, operation, 20 samples of 150 packages each of synthetic-gut tennis strings were tested for breaking strength. A total of 141 packages failed to conform to the manufacturer's specifications.
a. What is an estimate of the process proportion defective when the system is in control?
b. Compute the upper and lower control limits for a p chart.
c. What conclusion should be made about the process if tests for a new sample of 150 packages found 12 defective? Do there appear to be assignable causes present in this situation?

15. An automotive industry supplier produces pistons for several models of automobiles. Twenty samples, each consisting of 200 pistons, were selected when the process was known to be operating correctly. The number of defective pistons found in each sample is as follows:

8	10	6	4	5	7	8	12	8	15
14	10	10	7	5	8	6	10	4	8

a. What is an estimate of the process proportion defective for the piston-manufacturing process when it is in control?
b. Construct a p chart for the manufacturing process, assuming each sample will have 200 pistons.
c. What conclusion should be made if a sample of 200 has 20 defective pistons?

Summary

In this chapter we discussed how statistical methods can be used to assist in the control of quality. We first considered the technique referred to as acceptance sampling. With this quality-control procedure, a sample of the lot is selected and inspected. The number of defective items in the sample provides the basis for accepting or rejecting the lot. The sample size and the acceptance criterion can be adjusted to control both the producer's risk (Type I error) and the consumer's risk (Type II error) associated with the acceptance sampling procedure.

The $\bar{x}$, R, and p control charts were presented as graphical aids in monitoring process quality. Control limits are established for each chart, and samples are selected periodically and the data points plotted on the control chart. Data points outside the control limits indicate that the process is out of control and that corrective action should be taken. Patterns of data points within the control limits can also indicate potential quality-control problems and suggest that corrective action may be warranted.

Glossary

Quality control A series of inspections and measurements that determine whether quality standards are being met.

Lot A group of items such as incoming shipments of raw materials or purchased parts as well as finished goods from final assembly.

Acceptance sampling A statistical, quality-control procedure in which the number of defective items found in a sample is used to determine whether a lot should be accepted or rejected.

Producer's risk The risk of rejecting a good-quality lot in acceptance sampling. This is the Type I error.

Consumer's risk The risk of accepting a poor-quality lot in acceptance sampling. This is the Type II error.

Acceptance criterion The maximum number of defective items that can be found in the sample and still enable acceptance of the lot.

Operating characteristic curve A graph showing the probability of accepting the lot as a function of the percent defective in the lot. This curve can be used to help determine if a particular acceptance sampling plan meets both the producer's and the consumer's risk requirements.

Multiple sampling plan A form of acceptance sampling where more than one sample or stage is used. Based on the number of defective items found in a sample, a decision will be made to accept the lot, to reject the lot, or to continue sampling.

Common causes Normal or natural variations in process outputs that are due purely to chance. No corrective action is necessary when output variations are due to common causes.

Assignable causes Variations in process outputs that are due to factors such as machine tools wearing out, incorrect machine settings, poor-quality raw materials, operator error, and so on. Corrective action should be taken when assignable causes of output variation are detected.

Control chart A graphical tool used to help determine if a process is in control or out of control.

$\bar{x}$ Chart A control chart used when the output of a process is measured in terms of the mean value of a variable such as a length, weight, temperature, and so on.

R Chart A control chart used when the output of a process is measured in terms of the range of a variable.

p Chart A control chart used when the output of a process is measured in terms of the proportion defective.

Key Formulas

Binomial Probability Function for Acceptance Sampling

$$f(x) = \frac{n!}{x!(n-x)!} p^x (1-p)^{(n-x)} \tag{20.1}$$

Standard Error of the Mean

$$\sigma_{\bar{x}} = \frac{\sigma}{\sqrt{n}} \tag{20.3}$$

Control Limits for an $\bar{x}$ Chart: Process Mean and Standard Deviation Known

$$UCL = \mu + 3\sigma_{\bar{x}} \tag{20.4}$$

$$LCL = \mu - 3\sigma_{\bar{x}} \tag{20.5}$$

Overall Sample Mean

$$\bar{\bar{x}} = \frac{\bar{x}_1 + \bar{x}_2 + \cdots + \bar{x}_k}{k} \tag{20.6}$$

Average Range

$$\bar{R} = \frac{R_1 + R_2 + \cdots + R_k}{k} \tag{20.7}$$

Control Limits for an $\bar{x}$ Chart: Process Mean and Standard Deviation Unknown

$$\bar{\bar{x}} \pm A_2 \bar{R} \tag{20.10}$$

Control Limits for an R Chart

$$UCL = \bar{R}D_4 \qquad\qquad (20.16)$$

$$LCL = \bar{R}D_3 \qquad\qquad (20.17)$$

Standard Error of the Proportion

$$\sigma_{\bar{p}} = \sqrt{\frac{p(1-p)}{n}} \qquad\qquad (20.18)$$

Control Limits for a p Chart

$$UCL = p + 3\sigma_{\bar{p}} \qquad\qquad (20.19)$$

$$LCL = p - 3\sigma_{\bar{p}} \qquad\qquad (20.20)$$

❏ ❏ **Supplementary Exercises**

16. An $n = 10$, $c = 2$ acceptance sampling plan is being considered; assume that $p_0 = .05$ and $p_1 = .20$.
   a. Compute both the producer's and the consumer's risk for this acceptance sampling plan.
   b. Would either the producer or the consumer, or both, be unhappy with the proposed sampling plan?
   c. What change in the sampling plan, if any, would you recommend?

17. An acceptance sampling plan with $n = 15$ and $c = 1$ has been designed with a producer's risk of .075.
   a. Was the value of p_0 .01, .02, .03, .04, or .05? What does this value mean?
   b. What is the consumer's risk associated with this plan if p_1 is .25?

18. A manufacturer produces lots of a canned food product. Let p denote the proportion of the lot that do not meet the product quality specifications. An $n = 25$, $c = 0$ acceptance sampling plan will be used.
   a. Compute points on the operating characteristic curve when $p = .01, .03, .10,$ and $.20$.
   b. Plot the operating characteristic curve.
   c. What is the probability the acceptance sampling plan will reject a lot that has .01 defective?

19. Sometimes an acceptance sampling plan will be based on a large sample. In this case, the normal approximation to the binomial probability distribution can be used to compute the producer's and the consumer's risk associated with the plan. Referring to Chapter 6, the normal distribution used to approximate binomial probabilities has a mean of np and a standard deviation of $\sqrt{np(1-p)}$. Assume that an acceptance sampling plan is $n = 250$, $c = 10$.
   a. What is the producer's risk if p_0 is .02? As discussed in Chapter 6, a continuity correction factor should be used in this case. Thus, the probability of acceptance is based on the normal probability of the random variable being less than or equal to 10.5.
   b. What is the consumer's risk if p_1 is .08?
   c. What is an advantage of a large sample size for acceptance sampling? What is a disadvantage?

20. Samples of size 5 provided the 20 sample means shown in Table 20.7 for a production process that is believed to be in control.
   a. Based on these data, what is an estimate of the mean when the process is in control?
   b. Assuming that the process standard deviation is $\sigma = .50$, develop a control chart for this production process. Assume that the mean of the process is the estimate developed in (a).
   c. Do any of the 20 sample means given above indicate that the process was out of control?

T A B L E 20.7

95.72	95.24	95.18
95.44	95.46	95.32
95.40	95.44	95.08
95.50	95.80	95.22
95.56	95.22	95.04
95.72	94.82	95.46
95.60	95.78	

21. Product filling weights are normally distributed with a mean of 350 grams and a standard deviation of 15 grams.
   a. Develop the control limits for samples of size 10, 20, and 30.
   b. What happens to the control limits as the sample size is increased?
   c. What happens when a Type I error is made?
   d. What happens when a Type II error is made?
   e. What is the probability of a Type I error for samples of size 10, 20, and 30?
   f. What is the advantage of increasing the sample size for control-chart purposes? What error probability is reduced as the sample size is increased?

22. Twenty-five samples of size 5 resulted in $\bar{\bar{x}} = 5.42$ and $\bar{R} = 2.0$. Compute control limits for the $\bar{x}$ and R charts, and estimate the standard deviation of the process.

23. Construct $\bar{x}$ and R charts for the following sample data. Assume that the sample of size 5 was used.

Sample	$\bar{x}$	R	Sample	$\bar{x}$	R
1	95.72	1.0	11	95.80	.6
2	95.24	.9	12	95.22	.2
3	95.18	.8	13	95.56	1.3
4	95.44	.4	14	95.22	.5
5	95.46	.5	15	95.04	.8
6	95.32	1.1	16	95.72	1.1
7	95.40	.9	17	94.82	.6
8	95.44	.3	18	95.46	.5
9	95.08	.2	19	95.60	.4
10	95.50	.6	20	95.74	.6

24. Develop $\bar{x}$ and R charts for the following data:

Sample	Observations				
	1	2	3	4	5
1	3.05	3.08	3.07	3.11	3.11
2	3.13	3.07	3.05	3.10	3.10
3	3.06	3.04	3.12	3.11	3.10
4	3.09	3.08	3.09	3.09	3.07
5	3.10	3.06	3.06	3.07	3.08
6	3.08	3.10	3.13	3.03	3.06
7	3.06	3.06	3.08	3.10	3.08
8	3.11	3.08	3.07	3.07	3.07
9	3.09	3.09	3.08	3.07	3.09
10	3.06	3.11	3.07	3.09	3.07

25. Consider the following situations. For each, comment on whether there is reason for concern about the quality of the process.
   a. A p chart has LCL = 0 and UCL = .068. When the process is in control, the proportion defective is .033. Plot the following seven sample results: .035, .062, .055, .049, .058, .066, and .055. Discuss.

b. An $\bar{x}$ chart has LCL = 22.2 and UCL = 24.5. The mean is μ = 23.35 when the process is in control. Plot the following seven sample results: 22.4, 22.6, 22.65, 23.2, 23.4, 23.85, and 24.1. Discuss.

26. Managers of 1200 different retail outlets make twice-a-month restocking orders from a central warehouse. Past experience has shown that 4% of the orders contain one or more errors such as wrong item shipped, wrong quantity shipped, and item requested but not shipped. Random samples of 200 orders are selected monthly and checked for accuracy.

a. Construct a control chart for this situation.

b. Six months of data show the following number of orders with one or more errors: 10, 15, 6, 13, 8, and 17. Plot the data on the control chart. What does your plot indicate about the order process?

21

Sample Survey

Contents		

Cincinnati Gas & Electric Company*

CINCINNATI, OHIO

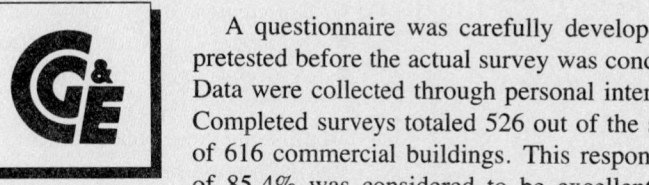

The Cincinnati Gas & Electric Company (CG&E) is a public utility which provides gas and electric power to customers in the Greater Cincinnati area. To improve service to its customers, CG&E continually strives to stay up-to-date with its customers' needs. In 1991, CG&E undertook a sample survey, the Building Characteristics Survey, to learn about the energy requirements of commercial buildings in its service area.

A variety of information concerning commercial buildings was sought, such as the floor space, number of employees, energy end-use, age of the building, type of building materials, and energy conservation measures. During preparations for the survey, CG&E analysts found that there were approximately 27,000 commercial buildings in the CG&E service area. Based on available funds and the precision desired in the results, they recommended that a sample of 616 commercial buildings be surveyed.

The sample design chosen was stratified simple random sampling. Total electrical usage over the past year for each commercial building in the service area was available from company records, and because many of the building characteristics of interest (size, number of employees, etc.) were related to usage, it was the criteria used to divide the population of buildings into six strata.

The first stratum contained the commercial buildings for the 100 largest energy users; each building in this stratum was included in the sample. Although these buildings constituted only .2% of the population, they accounted for 14.4% of the total electrical usage. For the other strata, the number of buildings sampled was determined on the basis of obtaining the greatest precision possible per unit cost.

A questionnaire was carefully developed and pretested before the actual survey was conducted. Data were collected through personal interviews. Completed surveys totaled 526 out of the sample of 616 commercial buildings. This response rate of 85.4% was considered to be excellent. Currently, CG&E is using the survey results to improve the forecasts of energy demand and to improve service to its commercial customers.

In this chapter you will learn about the issues that statisticians consider in the design and execution of a sample survey such as the one conducted by CG&E. Sample surveys are often used to develop profiles of a company's customers; they are also used by the government and other agencies to learn about various segments of the population.

A Cincinnati Gas & Electric Company line worker fixes an electric high-voltage transmission tower.

*The authors are indebted to Mr. Jim Riddle of Cincinnati Gas & Electric for providing this Statistics in Practice.

A *survey* is a process designed to collect data, or facts, about a situation. A survey of a complete population is called a *census;* a survey of a subset of a population is called a *sample survey.* The purpose of a sample survey is to collect data that can be used to develop inferences about population parameters such as the mean, total, and proportion. If properly designed, a sample survey can be quite precise and cost effective, especially when compared to a census. The purpose of this chapter is to provide an overview of several types of sample surveys and to describe the sampling plans that are most commonly employed. We also comment on some of the nonsampling issues associated with sample surveys.

21.1   Terminology Used in Sample Surveys

In Chapter 1 we defined an element, a population, and a sample as follows:

- An *element* is the entity on which data are collected.
- A *population* is the collection of all the elements of interest.
- A *sample* is a subset of the population.

To illustrate these concepts, consider the following situation. Dunning Microsystems, Inc. (DMI), a manufacturer of personal computers and peripherals, would like to collect data about the characteristics of individuals who have purchased a DMI personal computer. To obtain this data, a sample survey of DMI personal computer owners could be conducted. The *elements* in this sample survey would be individuals who have purchased a DMI personal computer. The *population* would be the collection of all people who have purchased a DMI personal computer, and the *sample* would be the subset of DMI personal computer owners that we survey.

In sample surveys it is necessary to distinguish between the *target population* and the *sampled population*. The target population is the population we want to make inferences about, while the sampled population is the population from which the sample is actually selected. It is important to understand that these two populations are not always the same. In the DMI example, the target population consists of all people who have purchased a DMI personal computer. The sampled population, however, might be all owners who had sent warranty registration cards back to DMI. Since every person who buys a DMI personal computer does not send in the warranty card, the sampled population would differ from the target population.

Conclusions drawn from a sample survey apply only to the sampled population. Whether these conclusions can be extended to the target population depends on the judgment of the analyst. The key issue is whether the correspondence between the sampled population and the target population on the elements of interest is close enough to allow this extension.

Before sampling, the population must be divided into *sampling units*. In some cases, the sampling units are simply the elements. In other cases, the sampling units are groups of the elements. For example, suppose we want to survey certified professional engineers who are involved in the design of heating and air conditioning systems for commercial buildings. If a list of all professional engineers involved in this work were available, the sampling units would be the professional engineers we want to survey. If such a list is not available, we must find an alternative approach. A business telephone directory might provide a list of all engineering firms involved in the design of heating and air conditioning systems. Given this list, we could select a sample of the engineering firms to survey; then, for each firm surveyed, we might interview all the professional engineers. In this case, the engineering firms would be the sampling units and the engineers interviewed would be the elements.

A list of the sampling units for a particular study is called a *frame*. In the sample survey of professional engineers, the frame is defined as all engineering firms listed in the business telephone directory; the frame is not a list of all professional engineers because this information is not available. The choice of a particular frame and hence the definition of the sampling units is often determined by the availability and reliability of a list. In practice, the development of the frame can be one of the most difficult and important parts of conducting a sample survey.

21.2   Types of Surveys and Sampling Methods

The three most common types of surveys are mail surveys, telephone surveys, and personal interview surveys; each of these involves the design and administration of a questionnaire. There are also other types of surveys used to collect data that do not involve questionnaires. For example, auditing firms are often hired to sample a company's inventory of goods to estimate the value of inventory on the company's balance sheet. In such surveys, there is no questionnaire; someone simply counts the items and records the results.

In surveys that use questionnaires, the design of the questionnaire is critical. The designer must resist the temptation to add questions that *might* be of interest, since every question adds to the length of the questionnaire. Long questionnaires lead not only to respondent fatigue, but also to interviewer fatigue; this is especially true when mail and telephone surveys are conducted. However, if personal interviews are to be used, a longer and more complex questionnaire can be used. A large body of knowledge exists concerning the phrasing, sequencing, and grouping of questions for a questionnaire. These issues are discussed in more advanced books devoted to survey sampling; several good sources for this type of information are listed in the bibliography.

Sample surveys can also be classified in terms of the sampling method used. With a *probabilistic sampling method,* the probability of obtaining each possible sample can be computed; with a *nonprobabilistic sampling method,* this probability is unknown. Nonprobabilistic sampling methods should not be used if the researcher wants to make statements about the precision of the estimates. In contrast, probabilistic sampling methods can be used to develop confidence intervals which provide bounds on the sampling error. In the following sections, four of the most popular probabilistic sampling methods are discussed: simple random sampling, stratified simple random sampling, cluster sampling, and systematic sampling.

Although statisticians prefer to use a probabilistic sampling method, nonprobabilistic sampling methods often have to be used. The advantages of nonprobabilistic sampling methods are their low expense and ease of implementation. The disadvantage is that statistically valid statements cannot be made about the precision of the estimates. Two of the more common nonprobabilistic methods are convenience sampling and judgment sampling.

With *convenience sampling,* the units included in the sample are chosen because of accessibility. For example, a professor conducting a research study at a university may ask student volunteers to participate in a study simply because they are in the professor's class; in this case, the sample of students is referred to as a convenience sample. In some situations, convenience sampling is the only practical approach. For example, a shipment of oranges might be sampled by an inspector selecting oranges haphazardly from several crates since labeling each orange in the entire shipment to create a frame and employ a probabilistic method of sampling would be impractical. Wildlife captures and volunteer panels for consumer research are other examples of convenience samples.

Although convenience samples provide a relatively easy approach to sample selection and data gathering, it is impossible to evaluate the "goodness" of the sample statistics obtained in terms of their ability to estimate the population parameters of interest. A convenience sample may provide good results or it may not; there is no statistically justified procedure for making any statistical inferences from the sample results. Nevertheless, at times some researchers apply a statistical method designed for a probability sample to the data gathered from a convenience sample. In doing so, the researcher may argue that the convenience sample can be treated as if it were a random sample in the sense that it is representative of the population. However, this argument should be questioned; one should be very cautious in using a convenience sample to make statistical inferences about population parameters.

In using the nonprobability sampling technique referred to as *judgment sampling,* a person knowledgeable on the subject of the study selects sampling units that he or she feels are most representative of the population. Although judgment sampling is often a relatively easy way to select samples, users of the survey results must recognize that the quality of the results is dependent on the judgment of the person selecting the sample. Consequently, caution must be exercised when using judgment samples to make statistical inferences about a population parameter. In general, no statistical statements concerning the precision of the results from a judgment sample should be made.

Both probabilistic and nonprobabilistic sampling methods can be used to select a sample. The advantage of nonprobabilistic methods is that they are usually inexpensive and easy to use. However, if it is necessary to provide statements about the precision of the estimates, a probabilistic sampling method must be used. Almost all large sample surveys employ probabilistic sampling methods.

21.3   Survey Errors

Two types of errors can occur when conducting a survey. One type, *sampling error,* is defined as the magnitude of the difference between the point estimator and the population parameter. In other words, sampling error is the error that occurs because not every element in the population is surveyed. The second type, *nonsampling error,* refers to all other types of errors that can occur when a survey is conducted such as measurement error, interviewer error, and processing error. Although sampling error can only occur in a sample survey, nonsampling errors can occur in both a census and a sample survey.

Nonsampling Error

One of the most common types of nonsampling error occurs whenever we incorrectly measure the characteristic of interest. Measurement error can occur in a census or a sample survey. For either type of survey, the researcher must exercise care to ensure that any measuring instruments (e.g., the questionnaire) are properly calibrated and that the people who take the measurements are properly trained. Attention to detail is the best precaution in most situations.

Errors due to nonresponse are a concern to both the statistician responsible for the design of the survey and the manager using the results. This type of nonsampling error occurs when data cannot be obtained for some of the units surveyed or when only partial data are obtained. This problem is most serious when a bias is created. For example, if interviews were conducted to assess women's attitudes toward working outside the home, housecalls only during the daytime would create an obvious bias since women who work outside the home would be excluded from the sample.

Nonsampling errors due to lack of respondent knowledge occur frequently with technical surveys. For example, suppose building managers were surveyed to obtain detailed information about the types of ventilation systems used in office buildings. Managers of large office buildings may be very knowledgeable about such systems because they may attend training seminars and have support staff to help keep them current. In contrast, managers of smaller office buildings may be less knowledgeable about these systems because of the wide variety of duties they must perform. This difference in knowledge can significantly affect the survey results.

Two other types of nonsampling errors are selection errors and processing errors. Selection errors occur when an inappropriate item is included in the survey. Suppose a sample survey was designed to develop a profile of men with beards; if some interviewers interpreted the statement ''men with beards'' to include men with mustaches while other interviewers did not, the resulting data would be flawed. Processing errors occur whenever data are incorrectly recorded or incorrectly transferred from recording forms such as from questionnaires to computer files.

Although some nonsampling errors will occur in most surveys, they can be minimized by careful planning. Care should be taken to ensure that the sampled population corresponds closely to the target population, good questionnaire design principles should be followed, interviewers should be well trained, and so on. In the final report on a survey, there should be some discussion of the likely impact of nonsampling errors on the results.

Sampling Error

Recall the Dunning Microsystems (DMI) sample survey introduced in Section 21.1. Suppose that DMI wanted to estimate the mean age of people who have purchased a DMI personal computer. If the entire population of DMI personal computer owners could be surveyed (a census), and nonsampling errors were not present, we would know the mean age exactly. On the other hand, suppose that less than 100% of the population of DMI owners can be surveyed. In this case, there will most likely be a difference between the sample mean and the population mean; the absolute value of this difference is the sampling error. In practice, it is not possible to know what the sampling error will be for any one particular sample because the population mean is unknown; however, it is possible to provide probability statements regarding the size of the sampling error.

As stated, sampling error occurs because a sample, and not the entire population, is surveyed. Even though sampling error cannot be avoided, it can be controlled. Selecting an appropriate sampling method or design is one important way to control this type of error. In the following sections we will discuss four probabilistic sampling methods: simple random sampling, stratified simple random sampling, cluster sampling, and systemic sampling.

21.4   Simple Random Sampling

Recall the definition of simple random sampling from Chapter 7:

A simple random sample of size n from a finite population of size N is a sample selected such that every possible sample of size n has the same probability of being selected.

To conduct a sample survey using simple random sampling, we begin by developing a frame or list of all elements in the sampled population. Then a selection procedure, based on the use of random numbers, is used to ensure that each element in the sampled population has the same probability of being selected. In this section we show how estimates of a population mean, total, and proportion are made for sample surveys that use simple random sampling.

Population Mean

In most sample surveys, the form of the probability distribution for the population is unknown. For example, in the DMI sample survey, management wanted to estimate μ, the mean age of people who had purchased a DMI personal computer. Most likely, DMI would not know what the probability distribution of age is for the population of all DMI owners. In most cases, this is not a problem because the properties of the sampling distribution of $\bar{x}$, the point estimator of μ, mostly depend on the choice of the sample design.

In Chapter 7 we indicated that if a *large* $(n \geq 30)$ *simple random sample* is selected, the central limit theorem enables us to conclude that the sampling distribution of $\bar{x}$ can be approximated by a normal probability distribution. In Chapter 8 we showed that for cases where the sampling distribution of $\bar{x}$ can be approximated by a normal probability distribution, an interval estimate of μ is given by

$$\bar{x} \pm z_{\alpha/2}\sigma_{\bar{x}} \tag{21.1}$$

where

$$\sigma_{\bar{x}} = \text{standard error of the mean}$$

Recall that $1 - \alpha$ is the confidence coefficient and $z_{\alpha/2}$ is the z value providing an area of $\alpha/2$ in the upper tail of the standard normal probability distribution. For example, for a 95% confidence interval, $z_{.025} = 1.96$. Note also that the standard error of the mean, $\sigma_{\bar{x}}$, is just the standard deviation of the sampling distribution of $\bar{x}$. In general, whenever we use the term *standard error* in this chapter, we will be referring to the standard deviation of the sampling distribution for the point estimator being considered.

When a simple random sample of size n is selected from a finite population of size N, an estimate of the standard error of the mean is

$$s_{\bar{x}} = \sqrt{\frac{N-n}{N}}\left(\frac{s}{\sqrt{n}}\right) \tag{21.2}$$

In this case, the interval estimate of the population mean becomes

$$\bar{x} \pm z_{\alpha/2}s_{\bar{x}} \tag{21.3}$$

In a sample survey it is common practice to use a value of $z = 2$ when developing interval estimates. Thus, when simple random sampling is used, an approximate 95% confidence interval estimate of the population mean is given by the following expression.

Approximate 95% Confidence Interval Estimate of the Population Mean

$$\bar{x} \pm 2s_{\bar{x}} \tag{21.4}$$

As an example, consider the situation of the publisher of Great Lakes Recreation, a regional magazine specializing in articles on boating and fishing. The magazine currently has $N = 8000$ subscribers. A simple random sample of $n = 484$ subscribers provided a mean annual income of $30,500 with a standard deviation of $7040. An unbiased estimate of the mean annual income of all subscribers is given by $\bar{x} = \$30,500$. Using the sample results and (21.2), an estimate of the standard error of the mean is

$$s_{\bar{x}} = \sqrt{\frac{8000 - 484}{8000}} \left(\frac{7040}{\sqrt{484}} \right) = \$310$$

Therefore, using (21.4), an approximate 95% confidence interval estimate of the mean annual income for the magazine subscribers is

$$30{,}500 \pm 2(310) = 30{,}500 \pm 620$$

or $29,880 to $31,120.

The above procedure can be used to compute an interval estimate for other population parameters such as the population total or the population proportion. In cases where the sampling distribution of the point estimator can be approximated by a normal probability distribution, the approximate 95% confidence interval can be written as

Point Estimator ± 2(Estimate of the Standard Error of the Point Estimator)

For example, in the Great Lakes Recreation sample survey, an estimate of the standard error of the point estimator is $s_{\bar{x}} = \$310$, and the bound on the error of the estimate is 2 ($310) = $620.

Population Total

Consider the problem facing Northeast Electric and Gas (NEG). As part of an energy usage study, NEG needs to estimate the *total* square footage for the 500 public schools in its service area. We will denote the total square footage for the 500 schools as $X;$ in other words, X denotes the population total. Note that if μ, the mean square footage for the 500 public schools, was known, the value of X could be computed by multiplying N times μ. However, since μ is unknown, a point estimate of X is obtained by multiplying N times $\bar{x}$. We will denote the point estimator of X as $\hat{X}$.

Point Estimator of a Population Total

$$\hat{X} = N\bar{x} \tag{21.5}$$

An estimate of the standard error of this point estimator is given by

$$s_{\hat{X}} = N s_{\bar{x}} \tag{21.6}$$

where

$$s_{\bar{x}} = \sqrt{\frac{N - n}{N}} \left(\frac{s}{\sqrt{n}} \right) \tag{21.7}$$

Note that (21.7) is just the formula for the estimated standard error of the mean. Using the above standard error, an approximate 95% confidence interval for the population total is given by the following expression.

Approximate 95% Confidence Interval Estimate of the Population Total

$$N\bar{x} \pm 2 s_{\hat{X}} \tag{21.8}$$

Suppose that in the NEG study a simple random sample of $n = 50$ public schools was selected from the population of $N = 500$ schools; the sample mean was $\bar{x} = 22{,}000$ square feet, and the sample standard deviation was $s = 4000$ square feet. Using (21.5), the point estimator of the population total is

$$\hat{X} = (500)(22{,}000) = 11{,}000{,}000$$

Equation (21.7) can be used to compute an estimate of the standard error of the mean.

$$s_{\bar{x}} = \sqrt{\frac{500 - 50}{500}} \left(\frac{4000}{\sqrt{50}} \right) = 536.66$$

Then, using (21.6), an estimate of the standard error of $\hat{X}$ is

$$s_{\hat{X}} = (500)(536.66) = 268{,}330$$

Therefore, using (21.8), an approximate 95% confidence interval estimate of the total square footage for the 500 public schools in NEG's service area is

$$11{,}000{,}000 \pm 2(268{,}330) = 11{,}000{,}000 \pm 536{,}660$$

or 10,463,340 to 11,536,660 square feet.

Population Proportion

The population proportion p is the fraction of the elements in the population with some characteristic of interest. In a market research study, for example, one might be interested in the proportion of consumers preferring a certain brand of product. The sample proportion $\bar{p}$ is an unbiased point estimator of the population proportion. An estimate of the standard error of the proportion is given by

$$s_{\bar{p}} = \sqrt{\left(\frac{N - n}{N} \right) \left(\frac{\bar{p}(1 - \bar{p})}{n - 1} \right)} \tag{21.9}$$

An approximate 95% confidence interval estimate of the population proportion is given as follows.

**Approximate 95% Confidence Interval Estimate
of the Population Proportion**

$$\bar{p} \pm 2s_{\bar{p}} \tag{21.10}$$

As an illustration, suppose that in the Northeast Electric and Gas sampling problem, NEG would also like to estimate the proportion of the 500 public schools in its service area that use natural gas as fuel for heating. If 35 of the 50 sampled schools indicated that they use natural gas, the point estimate of the proportion of the 500 schools in the population that use natural gas is $\bar{p} = 35/50 = .70$. Using (21.9), we can compute an estimate of the standard error of the proportion.

$$s_{\bar{p}} = \sqrt{\left(\frac{500 - 50}{500} \right) \left(\frac{.7(1 - .7)}{50 - 1} \right)} = .062$$

Therefore, using (21.10), an approximate 95% confidence interval for the population proportion is

$$.7 \pm 2(.062) = .7 \pm .124$$

or .576 to .824.

As this example has shown, the width of the confidence interval can be rather large when estimating a population proportion. In general, large sample sizes are usually needed to obtain precise estimates of population proportions.

Determining the Sample Size

An important consideration in sample design is the choice of sample size. The best choice usually involves a trade-off between cost and precision. Larger samples provide greater precision (tighter bounds on the sampling error), but are more costly. Oftentimes the budget for a study will dictate how large the sample can be. In other cases, the size of the sample must be large enough to provide a specified level of precision.

A common approach to choosing the sample size is to first specify the precision desired and then determine the smallest sample size providing that precision. In this context, the term *precision* refers to the size of the approximate confidence interval; smaller confidence intervals provide more precision. Since the size of the approximate confidence interval depends on the bound B on the sampling error, choosing a level of precision amounts to choosing a value for B. Let us see how this approach works in choosing the sample size necessary to estimate the population mean.

Equation (21.2) showed that the estimate of the standard error of the mean is

$$s_{\bar{x}} = \sqrt{\frac{N - n}{N}} \left(\frac{s}{\sqrt{n}} \right)$$

Recall that the bound on the sampling error is "2 times the estimate of the standard error of the point estimator." Thus,

$$B = 2\sqrt{\frac{N - n}{N}} \left(\frac{s}{\sqrt{n}} \right) \tag{21.11}$$

Solving (21.11) for n will provide a bound on the sampling error equal to B. Doing so yields

$$n = \frac{Ns^2}{N \left(\dfrac{B^2}{4} \right) + s^2} \tag{21.12}$$

Once a desired level of precision has been selected (by choosing a value for B), (21.12) can be used to find the value of n that will provide the desired level of precision. Using (21.12) to choose n for a practical study presents problems, however. In addition to specifying the desired bound on the sampling error B, one must know the sample variance s^2. But, s^2 will not be known until the sample is actually taken.

Cochran* suggests several ways to develop an estimate of s^2 in practice. Three of these are stated below.

1.   Take the sample in two stages. Use the value of s^2 found in stage 1 in (21.12); the resulting value of n is what the size of the total sample must be. Then, select the

---

*William G. Cochran, *Sampling Techniques,* 3rd ed., Wiley, 1977.

number of additional units needed at stage 2 to provide the total sample size determined in stage 1.

2. Use the results of a pilot survey or pretest to estimate s^2.
3. Use information from a previous sample.

Let us now consider an example involving the estimate of the population mean for starting salaries of college graduates at a particular university. Suppose there are $N = 5000$ graduates, and we want to develop an approximate 95% confidence interval with a width of at most $1000. To provide such a confidence interval, $B = 500$. Before using (21.12) to determine the sample size, we need an estimate of s^2. Suppose a study of starting salaries was also conducted last year, and it was found that $s = \$3000$. We can use the data from this previous sample to estimate s^2. Using $B = 500$, $s = 3000$, and $N = 5000$, we can now use (21.12) to determine the sample size.

$$n = \frac{5000(3000)^2}{5000\left(\frac{(500)^2}{4}\right) + (3000)^2}$$

$$= 139.9689$$

Rounding up, we see that a sample size of 140 will provide an approximate 95% confidence interval of width $1000. Keep in mind, however, that this calculation is based on our initial estimate of $s = \$3000$. If s turns out to be larger in this year's sample survey, the resulting approximate confidence interval will have a width greater than $1000. Consequently, if cost considerations permit, the designer of the survey might choose a sample size of, say, 150 to provide added assurance that the final approximate 95% confidence interval will have a width less than $1000.

The formula for determining the sample size necessary for estimating a population total with a bound B on the error of estimate is similar to that for the sample mean.

$$n = \frac{Ns^2}{\frac{B^2}{4N} + s^2} \tag{21.13}$$

In our previous example, we wanted to estimate the mean starting salary with a bound on the sampling error of $B = 500$. Suppose we were also interested in estimating the total salary of all 5000 graduates with a bound of $2 million. Using (21.13) with $B = 2,000,000$, we see that the sample size needed to provide such a bound on the population total is

$$n = \frac{5000(3000)^2}{\frac{(2,000,000)^2}{4(5000)} + (3000)^2}$$

$$= 215.311$$

Rounding up, we see that a sample size of 216 is necessary to provide an approximate 95% confidence interval with a bound of $2 million. We note here that if the same survey is expected to provide a bound of $500 on the population mean and a bound of $2 million on the population total, a sample size of at least 216 must be used. This will provide a tighter bound than necessary on the population mean, while providing the minimum desired precision for the population total.

To choose the sample size for estimating a population proportion, we use a formula very similar to the one for the population mean. We simply substitute $\overline{p}(1 - \overline{p})$ for s^2 in (21.12) to obtain

$$n = \frac{N\,\overline{p}(1 - \overline{p})}{N\left(\dfrac{B^2}{4}\right) + \overline{p}(1 - \overline{p})}$$

(21.14)

To use (21.14), the desired bound B must be specified and an estimate of $\overline{p}$ must be provided. If a good estimate of $\overline{p}$ is not available, we can use $\overline{p} = .5$; this will ensure that the resulting approximate confidence interval will have a bound on the sampling error at least as small as desired.

❑ ❑ Exercises

Methods

SELF TEST ▶

1. Simple random sampling has been used to obtain a sample of $n = 50$ elements from a population of $N = 800$. The sample mean was $\overline{x} = 215$, and the sample standard deviation was found to be $s = 20$.
   a. Estimate the population mean.
   ‚. Estimate the standard error of the mean.
      Develop an approximate 95% confidence interval for the population mean.

2. Simple random sampling has been used to obtain a sample of $n = 80$ elements from a population of $N = 400$. The sample mean was $\overline{x} = 75$, and the sample standard deviation was found to be $s = 8$.
   a. Estimate the population total.
   b. Estimate the standard error of the population total.
   c. Develop an approximate 95% confidence interval for the population total.

3. Simple random sampling has been used to obtain a sample of $n = 100$ elements from a population of $N = 1000$. The sample proportion was $\overline{p} = .30$.
   a. Estimate the population proportion.
   b. Estimate the standard error of the proportion.
   c. Develop an approximate 95% confidence interval for the population proportion.

4. A sample is to be taken to develop an approximate 95% confidence interval estimate of the population mean. The population consists of 450 elements, and a pilot study has resulted in $s = 70$. How large must the sample be if we want to develop an approximate 95% confidence interval with a width of 30?

Applications

SELF TEST ▶

5. In 1987, 153 U.S. banks operated foreign branches (*Statistical Abstract of the United States,* 1989). A sample of 30 of those banks showed average assets of $1.8 billion with a standard deviation of $.4 billion. The mean number of foreign branches for the banks in the sample was 6 with a standard deviation of 1.4.
   a. Develop an approximate 95% confidence interval for the population mean of the assets for the 153 banks.
   b. Develop an approximate 95% confidence interval for the total number of foreign branches of these banks.
   c. Suppose that 25% of the banks in the sample had one or more branches in the United Kingdom. Develop an approximate 95% confidence interval for the population proportion of banks having foreign branches in the United Kingdom.

6. A county in California had 724 corporate tax returns filed. The mean annual income reported was $161.22 thousand with a standard deviation of $31.3 thousand. How large a sample will be necessary next year to develop an approximate 95% confidence interval on mean annual corporate earnings? The precision required is an interval width of no more than $5000.

21.5   Stratified Simple Random Sampling

To use stratified simple random sampling, the population must be divided into H groups, called strata. Then for stratum h, a simple random sample of size n_h is selected; the data from the H simple random samples are then combined to develop an estimate of a population parameter such as the population mean, total, or proportion.

If the variability within each stratum is smaller than the variability across the strata, a stratified simple random sample can lead to greater precision (narrower interval estimates of the population parameters). The basis for forming the various strata depend upon the judgment of the designer of the sample. Depending on the application, a population can be stratified by department, location, age, product type, industry type, sales levels, and so on.

As an example, suppose that the College of Business at Lakeside College wants to conduct a survey of this year's graduating class to learn about their starting salaries. There are five majors in the college: accounting, finance, information systems, marketing, and operations management. Of the $N = 1500$ students who graduated this year, there were $N_1 = 500$ accounting majors, $N_2 = 350$ finance majors, $N_3 = 200$ information systems majors, $N_4 = 300$ marketing majors, and $N_5 = 150$ operations management majors. Based on the analysis of previous salary data, it was believed that there would be more variability in starting salaries across majors than within each major. As a result, a stratified simple random sample of $n = 180$ students was selected; 45 of the 180 students that were sampled majored in accounting ($n_1 = 45$), 40 majored in finance ($n_2 = 40$), 30 majored in information systems ($n_3 = 30$), 35 majored in marketing ($n_4 = 35$), and 30 majored in operations management ($n_5 = 30$).

Population Mean

In stratified sampling an unbiased estimate of the population mean is obtained by computing a weighted average of the sample means for each stratum. The weights used are the fraction of the population in each stratum. The resulting point estimator, denoted $\bar{x}_{st}$, is defined as follows:

Point Estimator of the Population Mean

$$\bar{x}_{st} = \sum_{h=1}^{H} \left(\frac{N_h}{N}\right) \bar{x}_h \qquad (21.15)$$

H = number of strata

$\bar{x}_h$ = sample mean for stratum h

N_h = number of elements in the population in stratum h

N = total number of elements in the population; $N = N_1 + N_2 + \cdots + N_H$

For stratified simple random sampling, the formula for computing an estimate of the standard error of the mean is

$$s_{\bar{x}_{st}} = \sqrt{\frac{1}{N^2} \sum_{h=1}^{H} N_h(N_h - n_h) \frac{s_h^2}{n_h}}$$ (21.16)

Using the above results, an approximate 95% confidence interval estimate of the population mean is as follows:

Approximate 95% Confidence Interval Estimate of the Population Mean

$$\bar{x}_{st} \pm 2 s_{\bar{x}_{st}}$$ (21.17)

Suppose that the survey of 180 graduates of the College of Business at Lakeland College provided the sample results shown in Table 21.1. The sample means for each major, or stratum, are as follows: \$24,000 for accounting, \$22,500 for finance, \$25,500 for information systems, \$21,000 for marketing, and \$25,000 for operations management. Using these results and (21.15), we can compute a point estimate of the population mean.

$$\bar{x}_{st} = \left(\frac{500}{1500}\right)(24,000) + \left(\frac{350}{1500}\right)(22,500) + \left(\frac{200}{1500}\right)(25,500)$$

$$+ \left(\frac{300}{1500}\right)(21,000) + \left(\frac{150}{1500}\right)(25,000) = \$23,350$$

In Table 21.2 we show a portion of the calculations needed to estimate the standard error; note that

$$\sum_{h=1}^{5} N_h(N_h - n_h) \frac{s_h^2}{n_h} = 42,909,037,698$$

Thus,

$$s_{\bar{x}_{st}} = \sqrt{\left(\frac{1}{(1500)^2}\right)(42,909,037,698)} = \sqrt{19,070.68} = 138$$

Hence, using (21.17), an approximate 95% confidence interval estimate of the population mean is $23,350 \pm 2(138) = 23,350 \pm 276$, or \$23,074 to \$23,626.

TABLE 21.1
Lakeland College Sample Survey of Starting Salaries for College Seniors

Major (h)	$\bar{x}_h$	s_h	N_h	n_h
Accounting	\$24,000	2000	500	45
Finance	\$22,500	1700	350	40
Information systems	\$25,500	2300	200	30
Marketing	\$21,000	1600	300	35
Operations management	\$25,000	2250	150	30

TABLE 21.2
Partial Calculations for the Estimate of the Standard Error of the Mean for the Lakeland College Sample Survey

Major	h	$N_h(N_h - n_h)\dfrac{s_h^2}{n_h}$
Accounting	1	$500(500 - 45)\dfrac{(2000)^2}{45} = 20{,}222{,}222{,}222$
Finance	2	$350(350 - 40)\dfrac{(1700)^2}{40} = 7{,}839{,}125{,}000$
Information systems	3	$200(200 - 30)\dfrac{(2300)^2}{30} = 5{,}995{,}333{,}333$
Marketing	4	$300(300 - 35)\dfrac{(1600)^2}{35} = 5{,}814{,}857{,}143$
Operations management	5	$150(150 - 30)\dfrac{(2250)^2}{30} = \underline{3{,}037{,}500{,}000}$
		$42{,}909{,}037{,}698$

$$\sum_{h=1}^{5} N_h(N_h - n_h)\frac{s_h^2}{n_h}$$

Population Total

The point estimator of the population total (X) is obtained by multiplying N times $\overline{x}_{st}$.

Point Estimator of the Population Total

$$\hat{X} = N\overline{x}_{st} \tag{21.18}$$

An estimate of the standard error of this point estimator is

$$s_{\hat{X}} = Ns_{\overline{x}_{st}} \tag{21.19}$$

Thus, an approximate 95% confidence interval for the population total is given by the following:

Approximate 95% Confidence Interval Estimate of the Population Total

$$N\overline{x}_{st} \pm 2s_{\hat{X}} \tag{21.20}$$

Now suppose the College of Business at Lakeland College would also like to estimate the total earnings of the 1500 business graduates in order to estimate their impact on the economy. Using (21.18), an unbiased estimate of the total earnings is

$$\hat{X} = (1500)23{,}350 = \$35{,}025{,}000$$

Using (21.19), an estimate of the standard error of the population total is

$$s_{\hat{X}} = 1500(138) = \$207{,}000$$

Thus, using (21.20) an approximate 95% confidence interval estimate of the total earnings of the 1500 graduates is $35,025,000 \pm 2(207,000) = 35,025,000 \pm 414,000$ or $34,611,000 to $35,439,000.

Population Proportion

An unbiased estimate of the population proportion, p, for stratified simple random sampling is a weighted average of the proportions for each stratum. The weights used are the fraction of the population in each stratum. The resulting point estimator, denoted $\bar{p}_{st}$, is defined as follows.

Point Estimator of the Population Proportion

$$\bar{p}_{st} = \sum_{h=1}^{H} \left(\frac{N_h}{N} \right) \bar{p}_h \qquad (21.21)$$

where

H = the number of strata

$\bar{p}_h$ = the sample proportion for stratum h

N_h = the number of elements in the population in stratum h

N = the total number of elements in the population; $N = N_1 + N_2 + \cdots + N_H$

An estimate of the standard error of $\bar{p}_{st}$ is given by

$$s_{\bar{p}_{st}} = \sqrt{\frac{1}{N^2} \sum_{h=1}^{H} N_h (N_h - n_h) \left[\frac{\bar{p}_h(1 - \bar{p}_h)}{n_h - 1} \right]} \qquad (21.22)$$

Thus, an approximate 95% confidence interval estimate of the population proportion is as follows.

**Approximate 95% Confidence Interval Estimate
of the Population Proportion**

$$\bar{p}_{st} \pm 2 s_{\bar{p}_{st}} \qquad (21.23)$$

In the Lakeland College survey, the college wants to know the proportion of graduates receiving a starting salary of $30,000 or more. The results of the sample survey of 180 graduates show that 20 received starting salaries of $30,000 or more and that 4 of the 20 majored in accounting, 2 majored in finance, 7 majored in information systems, 1 majored in marketing, and 6 majored in operations management.

Using (21.21), the point estimate of the proportion receiving starting salaries of $30,000 or more is computed as follows:

$$\bar{p}_{st} = \left(\frac{500}{1500} \right) \left(\frac{4}{45} \right) + \left(\frac{350}{1500} \right) \left(\frac{2}{40} \right) + \left(\frac{200}{1500} \right) \left(\frac{7}{30} \right) + \left(\frac{300}{1500} \right) \left(\frac{1}{35} \right) + \left(\frac{150}{1500} \right) \left(\frac{6}{30} \right)$$

$$= .0981$$

TABLE 21.3
Partial Calculations for the Estimate of the Standard Error of $\bar{p}_{st}$ for the Lakeland College Sample Survey

Major	h	$N_h(N_h - n_h)\left[\dfrac{\bar{p}_h(1 - \bar{p}_h)}{n_h - 1}\right]$
Accounting	1	$500(500 - 45)\left[\dfrac{(4/45)(41/45)}{45 - 1}\right] = 418.7430$
Finance	2	$350(350 - 40)\left[\dfrac{(2/40)(38/40)}{40 - 1}\right] = 132.1474$
Information systems	3	$200(200 - 30)\left[\dfrac{(7/30)(23/30)}{30 - 1}\right] = 209.7318$
Marketing	4	$300(300 - 35)\left[\dfrac{(1/35)(34/35)}{35 - 1}\right] = 64.8980$
Operations management	5	$150(150 - 30)\left[\dfrac{(6/30)(24/30)}{30 - 1}\right] = \underline{99.3103}$

$$924.8305$$

$$\sum_{h=1}^{5} N_h(N_h - n_h)\left[\frac{\bar{p}_h(1 - \bar{p}_h)}{n_h - 1}\right]$$

In Table 21.3 we show a portion of the calculations needed to estimate the standard error; note that

$$\sum_{h=1}^{5} N_h(N_h - n_h)\left[\frac{\bar{p}_h(1 - \bar{p}_h)}{n_h - 1}\right] = 924.8305$$

Thus,

$$s_{\bar{p}_{st}} = \sqrt{\frac{1}{(1500)^2}(924.8305)}$$

$$= .0203$$

Using (21.23), an approximate 95% confidence interval for the proportion of graduates receiving starting salaries of $30,000 or more is .0981 ± 2(.0203) = .0981 ± .0406, or .0575 to .1387.

Determining the Sample Size

With stratified simple random sampling we can think of choosing a sample size as a two-step process. First, a total sample size n must be chosen. Second, we must decide on how to assign the sampled units to the various strata. Alternately, we could first decide how large a sample to take in each stratum and then sum the strata sample sizes to obtain the total sample size. Since it is often of interest to develop estimates of the mean, total, and proportion for the individual strata, a combination of these two approaches is often employed. An overall sample size n and allocation that will provide the necessary precision for the overall population parameter of interest is found. Then, if the sample sizes in some of the strata are not large enough to provide the precision necessary for the estimates within the strata, the sample sizes for those strata are adjusted upward as necessary. In this section we discuss some of the issues pertinent to allocating the total sample to the various strata and present a method for choosing the total sample size and making the allocation.

The allocation issue has to do with deciding what fraction of the total sample should be assigned to each stratum. This determines how large the simple random sample will be in each stratum. The factors considered most important in making the allocation are:

1. The number of elements in each stratum.
2. The variance of the elements within each stratum.
3. The cost of selecting elements within each stratum.

Generally speaking, larger samples should be assigned to the larger strata and to the strata with larger variances. Conversely, to get the most information for a given cost, smaller samples should be allocated to the strata where the cost per unit of sampling is greatest.

The individual stratum variances often differ greatly. For example, suppose that in a particular study we were interested in determining the mean number of employees per building; since there is greater variability in a stratum with larger buildings than there is in one with smaller buildings, a proportionately larger sample should be taken in such a stratum. The cost of selection can be an important consideration when significant travel is required of the interviewer or survey-taker between sampled units in some of the strata but not in others; this frequently occurs when some of the strata involve rural areas and others involve cities.

In many surveys the cost per unit of sampling is approximately the same for each strata (e.g., mail and telephone surveys); in such cases, the cost of sampling can be ignored in making the allocation. We present here the appropriate formulas for choosing the sample size and making the allocation in such cases. More advanced texts on sampling provide formulas for the case when sampling costs vary significantly across strata. The formulas we present in this section will minimize the total sampling cost for a given level of precision. This method, known as *Neyman allocation,* allocates the total sample n to the various strata as follows:

$$n_h = n \left(\frac{N_h s_h}{\sum\limits_{h=1}^{H} N_h s_h} \right) \qquad (21.24)$$

Equation (21.24) shows that the number of units allocated to a stratum increases with the stratum size and variance. Note that to make this allocation, we need to first determine the total sample size n. Given a specified level of precision B, the formulas for choosing the total sample size when estimating the population mean and the population total are given below:

Sample Size when Estimating the Population Mean

$$n = \frac{\left(\sum\limits_{h=1}^{H} N_h s_h \right)^2}{N^2 \left(\dfrac{B^2}{4} \right) + \sum\limits_{h=1}^{H} N_h s_h^2} \qquad (21.25)$$

Sample Size when Estimating the Population Total

$$n = \frac{\left(\sum\limits_{h=1}^{H} N_h s_h\right)^2}{\dfrac{B^2}{4} + \sum\limits_{h=1}^{H} N_h s_h^2} \qquad (21.26)$$

As an example, suppose a Chevrolet dealer wants to survey the customers who have purchased a Corvette, Geo Prizm, or a Cavalier to obtain information the dealer feels will be helpful in determining future advertising. In particular, suppose the dealer wants to estimate the mean monthly income for these customers with a bound on the sampling error of $100. The dealer's 600 Corvette, Geo Prizm, and Cavalier customers have been divided into three strata: 100 Corvette owners, 200 Geo Prizm owners, and 300 Cavalier owners. A pilot survey was used to estimate the standard deviation in each strata; the results are $s_1 = \$1300$, $s_2 = \$900$, and $s_3 = \$500$ for the Corvette, Geo Prizm, and Cavalier owners, respectively.

The first step in choosing a sample size for this survey is to use (21.25) to determine the total sample size necessary to provide a bound of $B = \$100$ on the estimate of the population mean. First, we compute

$$\sum_{h=1}^{3} N_h s_h = 100(1300) + 200(900) + 300(500) = 460,000$$

Next, we compute

$$\sum_{h=1}^{3} N_h s_h^2 = 100(1300)^2 + 200(900)^2 + 300(500)^2 = 406,000,000$$

Substituting these values into (21.25), we can determine the total sample size needed to provide a bound on the sampling error of $B = \$100$.

$$n = \frac{(460,000)^2}{\dfrac{(600)^2(100)^2}{4} + 406,000,000} = 162$$

Thus, a total sample size of 162 will provide the precision desired. To allocate the total sample to the three strata, we use (21.24).

$$n_1 = 162\left(\frac{100(1300)}{460,000}\right) = 46$$

$$n_2 = 162\left(\frac{200(900)}{460,000}\right) = 63$$

$$n_3 = 162\left(\frac{300(500)}{460,000}\right) = 53$$

We would therefore recommend sampling 46 Corvette owners, 63 Geo Prizm owners, and 53 Cavalier owners for a total sample size of 162 customers.

To determine the sample size when estimating a population proportion, we simply substitute $\sqrt{\bar{p}_h(1 - \bar{p}_h)}$ for s_h in (21.25); the result is

$$n = \frac{\left(\sum_{h=1}^{H} N_h \sqrt{\bar{p}_h(1 - \bar{p}_h)}\right)^2}{N^2\left(\dfrac{B^2}{4}\right) + \sum_{h=1}^{H} N_h \bar{p}_h(1 - \bar{p}_h)} \tag{21.27}$$

Once the total sample size for the population proportion estimate has been determined, allocation to the various strata is again made using (21.24).

NOTES &
COMMENTS

An advantage of stratified simple random sampling is that estimates of population parameters for each stratum are automatically available as a byproduct of the sampling procedure. For example, besides obtaining an estimate of the average starting salary for all graduates in the Lakeland College sampling problem, we also obtained an estimate of the average starting salary for each major. Since each of the starting salary estimates was based on a simple random sample from each stratum, the procedure for developing an approximate confidence interval estimate when a simple random sample is selected (see Equation (21.4)) can be used to compute an approximate 95% confidence interval estimate for the mean in each stratum. In a similar manner, interval estimates for the population total and the population proportion for each stratum can be developed using (21.8) and (21.10), respectively.

Another type of allocation that is sometimes used with stratified simple random sampling is called *proportional allocation*. Using this approach, the sample size allocated to each stratum is given by

$$n_h = n\left(\frac{N_h}{N}\right) \tag{21.28}$$

Proportional allocation is appropriate when the stratum variances are approximately equal and the cost per unit of sampling is about the same across strata. In the case where the strata variances are equal, proportional allocation and the Neyman procedure result in the same allocation.

Exercises

Methods

SELF TEST ▶ **7.** A stratified simple random sample has been taken with the following results:

Stratum (h)	$\bar{x}_h$	s_h	$\bar{p}_h$	N_h	n_h
1	138	30	.50	200	20
2	103	25	.78	250	30
3	210	50	.21	100	25

a. Develop an estimate of the population mean for each stratum.
b. Develop an approximate 95% confidence interval for the population mean in each stratum.
c. Develop an approximate 95% confidence interval for the overall population mean.

8. Reconsider the sample results in Exercise 7.
a. Develop an estimate of the population total for each stratum.
b. Develop a point estimate of the total for all 550 elements in the population.
c. Develop an approximate 95% confidence interval for the population total.

9. Reconsider the sample results in Exercise 7.
a. Develop an approximate 95% confidence interval for the proportion in each stratum.
b. Develop a point estimate of the population proportion for the 550 elements in the population.
c. Estimate the standard error of the population proportion.
d. Develop an approximate 95% confidence interval for the population proportion.

10. A population has been divided into three strata with $N_1 = 300$, $N_2 = 600$, and $N_3 = 500$. From a past survey, the following estimates for the standard deviations in the three strata are available: $s_1 = 150$, $s_2 = 75$, $s_3 = 100$.
a. Suppose an estimate of the population mean with a bound on the error of estimate of $B = 20$ is required. How large must the sample be? How many elements should be allocated to each stratum?
b. Suppose that a bound of $B = 10$ is desired. How large must the sample be? How many elements should be allocated to each stratum?
c. Suppose an estimate of the population total with a bound of $B = 15,000$ is requested. How large must the sample be? How many elements should be allocated to each stratum?

Applications

11. A drug store chain has stores located in four cities; there are 38 stores in Indianapolis, 45 in Louisville, 80 in St. Louis, and 70 in Memphis. Pharmacy sales in the four cities vary considerably due to the competition. The following sales data (in $1000s) are available from a sample survey. Each of the cities was considered a separate stratum, and a stratified simple random sample was taken.

Indianapolis	Louisville	St. Louis	Memphis
50.3	48.7	16.7	14.7
41.2	59.8	38.4	88.3
15.7	28.9	51.6	94.2
22.5	36.5	42.7	76.8
26.7	89.8	45.0	35.1
20.8	96.0	59.7	48.2
	77.2	80.0	57.9
	81.3	27.6	18.8
			22.0
			74.3

a. Estimate the mean sales for each city (stratum).
b. Develop an approximate 95% confidence interval for the mean sales in each city.
c. Estimate the proportion of stores with sales of $50,000 or more.
d. Develop an approximate 95% confidence interval for the proportion of stores with sales of $50,000 or more.

12. Reconsider the sample survey results in Exercise 11.
a. Estimate the population total for St. Louis.
b. Estimate the population total for Indianapolis.

c. Develop an approximate 95% confidence interval for mean pharmacy sales for the drug store chain.

d. Develop an approximate 95% confidence interval for total pharmacy sales for the drug store chain.

13. An accounting firm has a number of clients in the banking, insurance, and brokerage industries. There are $N_1 = 50$ banks, $N_2 = 38$ insurance companies, and $N_3 = 35$ brokerage firms. A marketing research firm has been hired to survey the accounting firm's clients in these three industries. The survey will ask a variety of questions about both the clients' businesses and their satisfaction with services provided by the accounting firm. Suppose an approximate 95% confidence interval is requested for the mean number of employees for the 123 clients with a bound on the error of estimation of $B = 30$.

a. Suppose a pilot study finds $s_1 = 80$, $s_2 = 150$, and $s_3 = 45$. Choose a total sample size, and explain how the sample size should be allocated to the three strata.

b. Suppose the pilot test is called into question and it is decided to assume the stratum standard deviations are all equal to 100 in choosing the sample size. Choose a total sample size, and determine how many elements should be sampled in each stratum.

21.6　　Cluster Sampling

Cluster sampling requires that the population be divided into N groups of elements called clusters such that each element in the population belongs to one and only one cluster. For example, suppose that we wanted to survey registered voters in the state of Ohio. One approach would be to develop a frame consisting of all registered voters in the state of Ohio and then select a simple random sample of voters from this frame. Alternatively, in cluster sampling, we might choose to define the frame as the list of the $N = 88$ counties in the state. In this approach, each county or cluster would consist of a group of registered voters, and each registered voter in the state would belong to one and only one cluster. A map of Ohio showing the 88 counties or clusters is provided in Figure 21.1.

Suppose that we selected a simple random sample of $n = 12$ of the 88 counties. At this point, we could collect data for *all* registered voters in each of the 12 sampled clusters, an approach referred to as *single-stage cluster sampling,* or we could select a simple random sample of registered voters from each of the 12 sampled clusters, an approach referred to as *two-stage cluster sampling.* In either case, formulas are available for using the sample results to develop point and interval estimates of population parameters such as the population mean, total, or proportion. In this chapter, however, we only consider single-stage cluster sampling; more advanced texts on sampling present results for two-stage cluster sampling.

In the sense that both stratified and cluster sampling divide the population into groups of elements, the two sampling procedures are similar. The reasons for choosing cluster sampling, however, differ from the reasons for choosing stratified sampling. Cluster sampling tends to provide better results when the elements within the clusters are heterogeneous (not alike). In the ideal case, each cluster would be a small-scale version of the entire population. In this case, sampling a small number of clusters would provide good information about the characteristics of the entire population.

One of the primary applications of cluster sampling involves area sampling, where the clusters are counties, townships, city blocks, or other well-defined geographic sections of the population. Since data are collected from only a sample of the total geographic areas or clusters available, and since the elements within the clusters are typically close to one another, significant savings in time and cost can be realized when

FIGURE **21.1**

Counties of the State of Ohio Used as Clusters of Registered Voters

a data collector or interviewer is sent to a sampled unit. As a result, even if a larger total sample size is required, cluster sampling may be less costly than either simple random sampling or stratified simple random sampling. In addition, cluster sampling can minimize the time and cost associated with developing the frame or list of elements to be sampled, since cluster sampling does not require that a list of every element in the population be developed. One only needs a list of the elements in the clusters sampled.

To illustrate cluster sampling, let us consider a survey conducted by the CPA (certified public accountant) Society of the 12,000 practicing CPAs in a particular state. As part of the survey, the CPA Society collected information on income, sex, and factors related to the CPA's life style. Because personal interviews were needed to obtain all the desired information, the CPA Society used a cluster sample to minimize the total travel and interviewing cost. The frame consisted of all CPA firms that were registered to practice accounting in the state. Suppose that there are $N = 1000$ clusters or CPA firms registered to practice accounting in the state, and that a simple random sample of $n = 10$ CPA firms is to be selected.

In presenting the formulas for cluster sampling that are needed to develop approximate 95% confidence interval estimates of the population mean, total, and proportion, we will use the following notation:

N = number of clusters in the population

n = number of clusters selected in the sample

M_i = number of elements in cluster i

M = number of elements in the population; $M = M_1 + M_2 + \cdots + M_N$

$\bar{M} = M \div N$ = average number of elements in a cluster

x_i = total of all observations in cluster i

a_i = number of observations in cluster i with a certain characteristic

For the CPA Society sample survey we have

$$N = 1000$$

$$n = 10$$

$$M = 12{,}000$$

$$\bar{M} = 12{,}000/1000 = 12$$

Note that Table 21.4 shows the values of M_i and x_i for each of the sampled clusters as well as the number of female CPAs in the sampled firms (a_i).

Population Mean

The point estimator of the population mean obtained from cluster sampling is given by the following formula.

Point Estimator of the Population Mean

$$\bar{x}_c = \frac{\sum\limits_{i=1}^{n} x_i}{\sum\limits_{i=1}^{n} M_i} \tag{21.29}$$

An estimate of the standard error of this point estimator is

$$s_{\bar{x}_c} = \sqrt{\left(\frac{N-n}{Nn\bar{M}^2}\right) \frac{\sum\limits_{i=1}^{n} (x_i - \bar{x}_c M_i)^2}{n-1}} \tag{21.30}$$

Firm (i)	CPAs (M_i)	Total Salary ($1000s) for Sample ($x_i$)	Females (a_i)
1	8	320	2
2	25	1125	8
3	4	115	0
4	17	714	6
5	7	247	1
6	3	94	2
7	15	634	2
8	4	147	0
9	12	481	5
10	33	1567	9
Totals	128	5444	35

TABLE 21.4
Results of CPA Sample Survey

Thus, an approximate 95% confidence interval estimate of the population mean is

---

Approximate 95% Confidence Interval Estimate of the Population Mean

$$\bar{x}_c \pm 2s_{\bar{x}_c} \qquad\qquad (21.31)$$

Using the data in Table 21.4, an estimate of the mean salary for practicing certified public accountants is

$$\bar{x}_c = \frac{5444}{128} = 42.531$$

Since the salary data in Table 21.4 are in thousands of dollars, an estimate of the mean salary for practicing certified public accountants in the state is $42,531.

In Table 21.5, we show a portion of the calculations needed to estimate the standard error; note that

$$\sum_{i=1}^{n} (x_i - \bar{x}_c M_i)^2 = 39,178.688$$

Thus,

$$s_{\bar{x}_c} = \sqrt{\left[\frac{1000 - 10}{(1000)(10)(12)^2}\right] \frac{39,178.688}{10 - 1}} = 1.730$$

Hence, the standard error is $1730. Using (21.31), an approximate 95% confidence interval estimate for the mean annual salary is $42,531 \pm 2(1730) = 42,531 \pm 3460$ or $39,071 to $45,991.

TABLE 21.5
Partial Calculations for the Estimate of the Standard Error for the Mean for CPA Sample Survey where $\bar{x}_c = 42.531$

Firm (i)	M_i	x_i	$(x_i - 42.531 M_i)^2$
1	8	320	$[320 - 42.531(8)]^2 = \quad 409.982$
2	25	1,125	$[1125 - 42.531(25)]^2 = \quad 3,809.976$
3	4	115	$[115 - 42.531(4)]^2 = \quad 3,038.655$
4	17	714	$[714 - 42.531(17)]^2 = \quad\quad 81.487$
5	7	247	$[247 - 42.531(7)]^2 = \quad 2,572.214$
6	3	94	$[94 - 42.531(3)]^2 = \quad 1,128.490$
7	15	634	$[634 - 42.531(15)]^2 = \quad\quad 15.721$
8	4	147	$[147 - 42.531(4)]^2 = \quad\quad 534.719$
9	12	481	$[481 - 42.531(12)]^2 = \quad\quad 862.714$
10	33	1,567	$[1567 - 42.531(33)]^2 = 26,724.730$
Totals	128	5,444	39,178.688

$$\sum_{i=1}^{n} (x_i - \bar{x}_c M_i)^2$$

Population Total

The point estimator of the population total X is obtained by multiplying M times $\bar{x}_c$.

Point Estimator of the Population Total

$$\hat{X} = M\bar{x}_c \tag{21.32}$$

An estimate of the standard error of this point estimator is

$$s_{\hat{X}} = M s_{\bar{x}_c} \tag{21.33}$$

Thus, an approximate 95% confidence interval estimate for the population total is given by the following expression.

Approximate 95% Confidence Interval Estimate of the Population Total

$$M\bar{x}_c \pm 2s_{\bar{x}_c} \tag{21.34}$$

For the CPA sample survey,

$$\hat{X} = M\bar{x}_c = 12,000(42,531) = \$510,372,000$$

$$s_{\hat{X}} = M s_{\bar{x}_c} = 12,000(1730) = \$20,760,000$$

Thus, using (21.34), an approximate 95% confidence interval is \$510,372,000 ± 2(\$20,760,000) = \$510,372,000 ± \$41,520,000 or \$468,852,000 to \$551,892,000.

Population Proportion

The point estimator of the population proportion obtained from cluster sampling is as follows.

Point Estimator of the Population Proportion

$$\overline{p}_c = \frac{\displaystyle\sum_{i=1}^{n} a_i}{\displaystyle\sum_{i=1}^{n} M_i} \qquad (21.35)$$

where

a_i = number of elements in cluster i with the characteristic of interest

---

An estimate of the standard error of this point estimator is

$$s_{\overline{p}_c} = \sqrt{\left(\frac{N-n}{Nn\overline{M}^2}\right) \frac{\displaystyle\sum_{i=1}^{n} (a_i - \overline{p}_c M_i)^2}{n-1}} \qquad (21.36)$$

Thus, an approximate 95% confidence interval estimate for the population proportion is given by the following expression.

Approximate 95% Confidence Interval Estimate of the Population Proportion

$$\overline{p}_c \pm 2 s_{\overline{p}_c} \qquad (21.37)$$

For the CPA sample survey, we can use (21.35) and the data in Table 21.4 to develop an estimate of the number of practicing certified public accountants that are women.

$$\overline{p}_c = \frac{2 + 8 + \cdots + 9}{8 + 25 + \cdots + 33} = \frac{35}{128} = .2734$$

In Table 21.6 we show a portion of the calculations needed to estimate the standard error; note that

$$\sum_{i=1}^{n} (a_i - \overline{p}_c M_i)^2 - 15.2098$$

TABLE 21.6

Partial Calculations for the Estimate of the Standard Error of $\bar{p}_c$ for the CPA Sample Survey where $\bar{p}_c = .2734$

Firm (i)	M_i	a_i	$(a_i - .2734 M_i)^2$
1	8	2	$[2 - .2734(8)]^2 =$ .0350
2	25	8	$[8 - .2734(25)]^2 =$ 1.3572
3	4	0	$[0 - .2734(4)]^2 =$ 1.1960
4	17	6	$[6 - .2734(17)]^2 =$ 1.8284
5	7	1	$[1 - .2734(7)]^2 =$ .8350
6	3	2	$[2 - .2734(3)]^2 =$ 1.3919
7	15	2	$[2 - .2734(15)]^2 =$ 4.4142
8	4	0	$[0 - .2734(4)]^2 =$ 1.1960
9	12	5	$[5 - .2734(12)]^2 =$ 2.9556
10	33	9	$[9 - .2734(33)]^2 =$ .0005
Totals	128	35	15.2098

$$\sum_{i=1}^{n} (a_i - \bar{p}_c M_i)^2$$

Thus,

$$s_{\bar{p}_c} = \sqrt{\left[\frac{1000 - 10}{(1000)(10)(12)^2}\right]\frac{15.2098}{10 - 1}} = .0341$$

Hence, using (21.37), an approximate 95% confidence interval for the proportion of practicing CPAs that are women is $.2734 \pm 2(.0341) = .2734 \pm .0682$ or .2052 to .3416.

Determining the Sample Size

Once the clusters have been formed, the primary issue in choosing a sample size is selecting the number of clusters n. The procedure for cluster sampling is similar to that for other methods of sampling. An acceptable level of precision is specified by choosing a value for B, the bound on the sampling error. Then a formula is developed for finding the value of n that will provide the desired precision.

The average cluster size and the variance between clusters are key factors in deciding how many clusters to include in the sample. If the clusters are similar, there will be a small variance between them, and the number of clusters sampled can be smaller. Also, if the average number of elements per cluster is larger, the number of clusters sampled can be smaller. The formulas for making an exact determination of sample size are included in more advanced texts on sampling.

❏ ❏ **Exercises**

Methods

SELF TEST ▶ **14.** A sample of four clusters is to be taken from a population with $N = 25$ clusters and $M = 300$ elements. The values of M_i, x_i, and a_i for each cluster in the sample are given in Table 21.7.

TABLE 21.7

Cluster (i)	M_i	x_i	a_i
1	7	95	1
2	18	325	6
3	15	190	6
4	10	140	2
Totals	50	750	15

a. Develop point estimates of the population mean, total, and proportion.
b. Estimate the standard error for the estimates in (a).
c. Develop an approximate 95% confidence interval for the population mean.
d. Develop an approximate 95% confidence interval for the population total.
e. Develop an approximate 95% confidence interval for the population proportion.

15. A sample of six clusters is to be taken from a population with $N = 30$ clusters and $M = 600$ elements in the population. The values of M_i, x_i, and a_i for each cluster in the sample are given below:

Cluster (i)	M_i	x_i	a_i
1	35	3,500	3
2	15	965	0
3	12	960	1
4	23	2,070	4
5	20	1,100	3
6	25	1,805	2
Totals	130	10,400	13

a. Develop point estimates of the population mean, total, and proportion.
b. Develop an approximate 95% confidence interval for the population mean.
c. Develop an approximate 95% confidence interval for the population total.
d. Develop an approximate 95% confidence interval for the population proportion.

Applications

16. A public utility is conducting a survey of mechanical engineers to learn more about the factors influencing the choice of heating, ventilation, and air conditioning (HVAC) equipment for new commercial buildings. There are a total of 120 firms in the utility's service area engaged in designing HVAC systems. The sampling plan is to use cluster sampling with each firm representing a cluster. For each firm in the sample, all of the mechanical engineers will be interviewed. It is believed that there are approximately 500 mechanical engineers employed by the 120 firms.

A sample of 10 firms was taken. Among other things, the age of each respondent was recorded as well as whether or not the respondent had attended the local university.

Cluster (i)	M_i	Total of Respondents' Ages	Number Attending Local University
1	12	520	8
2	1	33	0
3	2	70	1
4	1	29	1
5	6	270	3
6	3	129	2
7	2	102	0
8	1	48	1
9	9	337	7
10	13	462	12
Totals	50	2000	35

a. Estimate the mean age of mechanical engineers engaged in this type of work.

b. Estimate the proportion of mechanical engineers in the utility's service areas that attended the local university.

c. Develop an approximate 95% confidence interval for the mean age of mechanical engineers designing HVAC systems for commercial buildings.

d. Develop an approximate 95% confidence interval for the proportion of mechanical engineers in the utility's service area who attended the local university.

17. A national real estate company has just acquired a smaller firm that has 150 offices and 6000 agents in Los Angeles and other parts of southern California. The national firm has conducted a sample survey to learn about attitudes and other characteristics of its new employees. A sample of eight offices has been taken, and all of the agents at these offices have completed the questionnaire. Results of the survey for the eight offices are given.

Office	Agents	Average Age	College Graduates	Males
1	17	37	3	4
2	35	32	14	12
3	26	36	8	7
4	66	30	38	28
5	43	41	18	12
6	12	52	2	6
7	48	35	20	17
8	57	44	25	26

a. Estimate the mean age of the agents.

b. Estimate the proportion of agents who are college graduates and the proportion who are male.

c. Develop an approximate 95% confidence interval for the mean age of the agents.

d. Develop an approximate 95% confidence interval for the proportion of agents who are college graduates.

e. Develop an approximate 95% confidence interval for the proportion of agents who are male.

21.7   Systematic Sampling

Systematic sampling is often used as an alternative to simple random sampling. In some sampling situations, especially those with large populations, it can be time-consuming to select a simple random sample by first finding a random number and then counting or searching through the frame until the corresponding element is found. Systematic sampling offers an alternative to simple random sampling in such cases. For example, if a sample size of 50 from a population containing 5000 elements is desired, we might sample 1 element for every 5000/50 = 100 elements in the population. A systematic sample for this case would involve randomly selecting 1 of the first 100 elements from the frame. The remaining sample elements are identified by starting with the first sampled element and then selecting every 100th element that follows in the frame. In effect, the sample of 50 is identified by moving systematically through the population and identifying every 100th element after the first randomly selected element. The sample of 50 will often be easier to select in this manner than would be the case if simple random sampling were used. Since the first element selected is a random choice, the assumption that a systematic sample has the properties of a simple random sample is often made. This assumption is usually appropriate when the frame is a random ordering of the elements in the population.

Summary

We have provided a brief introduction to the field of survey sampling in this chapter. The purpose of survey sampling is to collect existing data for the purpose of making estimates of population parameters such as the population mean, total, or proportion. Survey sampling as a method of data collection can be contrasted with conducting experiments to generate data. When survey sampling is used, the design of the sampling plan is of critical importance in determining which existing data is collected. When experiments are employed, experimental design issues are of critical importance in determining which data is generated, or created.

There are two types of errors that can occur with sample surveys: sampling error and nonsampling error. Sampling error is the error that occurs because a sample, and not the entire population, is used to estimate a population parameter. Nonsampling error refers to all the other types of errors that can occur, such as measurement, interviewer, nonresponse, and processing error. Nonsampling errors are controlled by proper questionnaire design, thorough training of interviewers, careful verification of data, and so on. Sampling errors are minimized by a proper choice of sample design and by selecting an appropriate sample size.

We discussed four commonly used sample designs in this chapter: simple random sampling, stratified simple random sampling, cluster sampling, and systematic sampling. The objective of sample design is to get the most precise estimates for the least cost. When the population can be divided into strata so that the elements within each stratum are relatively homogeneous, stratified simple random sampling will provide more precision (smaller approximate confidence intervals) than simple random sampling. When the elements can be grouped in clusters so that all the elements in a cluster are close together geographically, cluster sampling often reduces interviewer cost; in these situations, cluster sampling will often provide the most precision per dollar. Systematic random sampling was presented as an alternative to simple random sampling.

Glossary

Element The entity on which data are collected.

Population The collection of all elements of interest.

Sample A subset of the population.

Sampled population The population from which the sample is taken.

Target population The population about which inferences are made.

Sampling unit The units selected for sampling. A sampling unit may include several elements.

Frame A list of the sampling units for a study. The sample is drawn by selecting units from the frame.

Probabilistic sampling Any method of sampling in which the probability of each possible sample can be computed.

Nonprobabilistic sampling Any method of sampling for which the probability of selecting a sample cannot be computed.

Convenience sampling A nonprobabilistic method of sampling whereby elements are selected on the basis of convenience.

Judgment sampling A nonprobabilistic method of sampling whereby element selection is based on the judgment of the person doing the study.

Sampling error The error that occurs because a sample, and not the entire population, is used to estimate a population parameter.

Nonsampling error All types of errors other than sampling error, such as measurement error, interviewer error, and processing error.

Simple random sample A sample selected in such a manner that each sample of size n has the same probability of being selected.

Bound on sampling error A number added to and subtracted from a point estimate to create an approximate 95% confidence interval. It is given by two times the standard error of the point estimator.

Stratified simple random sampling A method of selecting a sample in which the population is first divided into strata and a simple random sample is then taken from each stratum.

Cluster sampling A probabilistic method of sampling in which the population is first divided into clusters and then one or more clusters is selected for sampling. In single-stage cluster sampling, every element in each selected cluster is sampled; in two-stage cluster sampling, a sample of the elements in each selected cluster is collected.

Area sampling A version of cluster sampling in which the elements are formed into clusters on the basis of their geographic proximity.

Systematic sampling A method of choosing a sample by randomly selecting the first element and then selecting every kth element thereafter.

Key Formulas

Simple Random Sampling

Interval Estimate of the Population Mean

$$\overline{x} \pm z_{\alpha/2}\sigma_{\overline{x}} \qquad (21.1)$$

Estimate of the Standard Error of the Population Mean

$$s_{\overline{x}} = \sqrt{\frac{N-n}{N}}\left(\frac{s}{\sqrt{n}}\right) \qquad (21.2)$$

Interval Estimate of the Population Mean

$$\overline{x} \pm z_{\alpha/2}s_{\overline{x}} \qquad (21.3)$$

Approximate 95% Confidence Interval Estimate of the Population Mean

$$\overline{x} \pm 2s_{\overline{x}} \qquad (21.4)$$

Point Estimator of a Population Total

$$\hat{X} = N\overline{x} \qquad (21.5)$$

Estimate of the Standard Error of $\hat{X}$

$$s_{\hat{X}} = Ns_{\overline{x}} \qquad (21.6)$$

Approximate 95% Confidence Interval Estimate of the Population Total

$$N\overline{x} \pm 2s_{\hat{X}} \qquad (21.8)$$

Estimate of the Standard Error of the Population Proportion

$$s_{\overline{p}} = \sqrt{\left(\frac{N-n}{N}\right)\left(\frac{\overline{p}(1-\overline{p})}{n-1}\right)} \qquad (21.9)$$

Approximate 95% Confidence Interval Estimate of the Population Proportion

$$\overline{p} \pm 2s_{\overline{p}} \qquad (21.10)$$

Sample Size for an Estimate of the Population Mean

$$n = \frac{Ns^2}{N\left(\dfrac{B^2}{4}\right) + s^2} \qquad (21.12)$$

Sample Size for an Estimate of the Population Total

$$n = \frac{Ns^2}{\left(\dfrac{B^2}{4N}\right) + s^2} \tag{21.13}$$

Sample Size for an Estimate of the Population Proportion

$$n = \frac{N\bar{p}(1 - \bar{p})}{N\left(\dfrac{B^2}{4}\right) + \bar{p}(1 - \bar{p})} \tag{21.14}$$

Stratified Simple Random Sampling

Point Estimator of the Population Mean

$$\bar{x}_{st} = \sum_{h=1}^{H} \left(\frac{N_h}{N}\right)\bar{x}_h \tag{21.15}$$

Estimate of the Standard Error of the Population Mean

$$s_{\bar{x}_{st}} = \sqrt{\frac{1}{N^2} \sum_{h=1}^{H} N_h(N_h - n_h)\frac{s_h^2}{n_h}} \tag{21.16}$$

Approximate 95% Confidence Interval Estimate of the Population Mean

$$\bar{x}_{st} \pm 2s_{\bar{x}_{st}} \tag{21.17}$$

Point Estimator of the Population Total

$$\hat{X} = N\bar{x}_{st} \tag{21.18}$$

Estimate of the Standard Error of $\hat{X}$

$$s_{\hat{X}} = Ns_{\bar{x}_{st}} \tag{21.19}$$

Approximate 95% Confidence Interval Estimate of the Population Total

$$N\bar{x}_{st} \pm 2s_{\hat{X}} \tag{21.20}$$

Point Estimator of the Population Proportion

$$\bar{p}_{st} = \sum_{h=1}^{H} \left(\frac{N_h}{N}\right)\bar{p}_h \tag{21.21}$$

Estimate of the Standard Error of $\bar{p}_{st}$

$$s_{\bar{p}_{st}} = \sqrt{\frac{1}{N^2} \sum_{h=1}^{H} N_h(N_h - n_h)\left[\frac{\bar{p}_h(1 - \bar{p}_h)}{n_h - 1}\right]} \tag{21.22}$$

Approximate 95% Confidence Interval Estimate of the Population Proportion

$$\bar{p}_{st} \pm 2s_{\bar{p}_{st}} \tag{21.23}$$

Allocating the Total Sample n to the Strata: Neyman Allocation

$$n_h = n \left(\frac{N_h s_h}{\sum\limits_{h=1}^{H} N_h s_h} \right) \qquad (21.24)$$

Sample Size for the Estimate of the Population Mean

$$n = \frac{\left(\sum\limits_{h=1}^{H} N_h s_h \right)^2}{N^2 \left(\dfrac{B^2}{4} \right) + \sum\limits_{h=1}^{H} N_h s_h^2} \qquad (21.25)$$

Sample Size for the Estimate of the Population Total

$$n = \frac{\left(\sum\limits_{h=1}^{H} N_h s_h \right)^2}{\dfrac{B^2}{4} + \sum\limits_{h=1}^{H} N_h s_h^2} \qquad (21.26)$$

Sample Size for the estimate of the Population Proportion

$$n = \frac{\left(\sum\limits_{h=1}^{H} N_h \sqrt{\overline{p}_h(1 - \overline{p}_h)} \right)^2}{N^2 \left(\dfrac{B^2}{4} \right) + \sum\limits_{h=1}^{H} N_h \overline{p}_h(1 - \overline{p}_h)} \qquad (21.27)$$

Proportional Allocation of Sample n to the Strata

$$n_h = n \left(\frac{N_h}{N} \right) \qquad (21.28)$$

Cluster Sampling

Point Estimator of the Population Mean

$$\overline{x}_c = \frac{\sum\limits_{i=1}^{n} x_i}{\sum\limits_{i=1}^{n} M_i} \qquad (21.29)$$

Estimate of the Standard Error of the Population Mean

$$s_{\overline{x}_c} = \sqrt{ \left(\frac{N - n}{N n \overline{M}^2} \right) \frac{\sum\limits_{i=1}^{n} (x_i - \overline{x}_c M_i)^2}{n - 1} } \qquad (21.30)$$

Approximate 95% Confidence Interval Estimate of the Population Mean

$$\bar{x}_c \pm 2s_{\bar{x}_c} \tag{21.31}$$

Point Estimator of the Population Total

$$\hat{X} = M\bar{x}_c \tag{21.32}$$

Estimate of the Standard Error of $\hat{X}$

$$s_{\hat{X}} = Ms_{\bar{x}_c} \tag{21.33}$$

Approximate 95% Confidence Interval Estimate of the Population Total

$$M\bar{x}_c \pm 2s_{\bar{x}_c} \tag{21.34}$$

Point Estimator of the Population Proportion

$$\bar{p}_c = \frac{\sum_{i=1}^{n} a_i}{\sum_{i=1}^{n} M_i} \tag{21.35}$$

Estimate of the Standard Error of $\bar{p}_c$

$$s_{\bar{p}_c} = \sqrt{\left(\frac{N-n}{Nn\bar{M}^2}\right)\frac{\sum_{i=1}^{n}(a_i - \bar{p}_c M_i)^2}{n-1}} \tag{21.36}$$

Approximate 95% Confidence Interval Estimate of the Population Proportion

$$\bar{p}_c \pm 2s_{\bar{p}_c} \tag{21.37}$$

❑ ❑ Supplementary Exercises

18. A *USA Today*/CNN/Gallup telephone poll of 1421 adults nationwide (*USA TODAY* January 16, 1992) concerned issues that were likely to influence voters. The following questions are based on that survey. (Note: Since the survey sampled a very small fraction of total adults, assume $N - n/N = 1$ in any formulas involving standard error.)

a. Eighty percent indicated they would be more likely to vote for a candidate in favor of "requiring all able-bodied people on welfare to do work for their welfare checks." Develop an approximate 95% confidence interval for the population proportion.

b. Sixty percent indicated they would be more likely to vote for a candidate in favor of "a national health care system paid for by new taxes." Develop an approximate 95% confidence interval for the population proportion.

c. Forty-four percent indicated they would be more likely to vote for a candidate in favor of "giving families a tax credit for each child attending a private or parochial school" while 40% were less likely to vote for such a candidate, and 16% did not express a preference. Develop an approximate 95% confidence interval for the proportion of adults more likely to vote for such a candidate.

d. *USA TODAY* reported that the "margin of sampling error is plus or minus 3 percentage points" for the survey. What does this mean, and how do you think they arrived at this number?

e. How might nonsampling error bias the results of such a survey? The actual polling was done by the Gallup organization.

19. The National Association of Children's Hospitals and Related Institutions and the new Coalition for America's Children (*Cincinnati Enquirer* January 8, 1992) commissioned a nationwide telephone survey of 1083 registered voters concerning children's issues.

a. Seventy-one percent of respondents worried that "the quality of education is declining for children in public schools." Develop an approximate 95% confidence interval for the population proportion.

b. Sixty-two percent worried that "children are not provided with the basics that they need in health care, food and education." Develop an approximate 95% confidence interval for the population proportion.

c. Thirty-four percent worried that "pregnant mothers aren't getting the health care they need to make sure their child is healthy." Develop an approximate 95% confidence interval for this estimate.

d. It was reported that the "percent of error is plus or minus 3.1%." Interpret this statement.

20. A quality of life survey was conducted with employees of a manufacturing firm. Of the firm's 3000 employees, a sample of 300 were sent questionnaires. Two hundred usable questionnaires were obtained for a response rate of 67%.

a. The mean annual salary for the sample was $\bar{x} = \$23,200$ with $s = 3000$. Develop an approximate 95% confidence interval for the mean annual salary of the population.

b. Using the information in (a), develop an approximate 95% confidence interval for the total salary of all 3000 employees.

c. Seventy-three percent of respondents reported that they were "generally satisfied" with their job. Develop an approximate 95% confidence interval for the population proportion.

d. Comment on whether or not you think the results in (c) might be biased. Would your opinion change if you knew the respondents were guaranteed anonymity?

21. A U.S. Senate Judiciary Committee report showed the number of homicides state by state for 1991. In the states of Indiana, Ohio, and Kentucky, the number of homicides were, respectively, 380, 760, and 260. Suppose a stratified random sample with the following results were taken to learn more about the victims and the cause of death.

Stratum	Sample Size	Shootings	Beatings	Black Victims
Indiana	30	10	9	21
Ohio	45	19	12	34
Kentucky	25	7	11	15

a. Develop an approximate 95% confidence interval for the proportion of shooting deaths in Indiana.

b. Develop an estimate for the total number of shooting deaths in Ohio.

c. Develop an approximate 95% confidence interval for the proportion of shooting deaths in Ohio.

d. Develop an approximate 95% confidence interval for the proportion of shooting deaths across all three states.

22. Refer again to the data in Exercise 21.

a. Develop an estimate of the total number of deaths (in the three states) due to beatings.

b. Develop an approximate 95% confidence interval for the proportion of deaths across all three states due to beatings.

c. Develop an approximate 95% confidence interval for the proportion of victims that are black.

d. Develop an estimate of the total number of victims that are black.

23. A stratified simple random sample is to be taken of a bank's customers to learn about a variety of attitudinal and demographic issues. The stratification is to be based on savings account balances as of June 30, 1992. A frequency distribution showing the number of accounts in each stratum, together with the standard deviation of account balances by stratum, is shown below:

Stratum ($)	Accounts	Standard Deviation of Account Balances
0.00– 1,000.00	3,000	80
1,000.01– 2,000.00	600	150
2,000.01– 5,000.00	250	220
5,000.01–10,000.00	100	700
over 10,000.00	50	3000

a. Assuming the cost per unit sampled is approximately equal across strata, determine the total number of persons to include in the sample. Assume we would like a bound on the error of estimate of the population mean for savings account balances of $B = \$20$.

b. Use the Neyman allocation procedure to determine the number to be sampled for each stratum.

24. A public agency is interested in learning more about the persons living in nursing homes in a particular city. There are a total of 100 nursing homes caring for 4800 people in the city and a cluster sample of six homes has been taken. Each person in the six homes has been interviewed. A portion of the sample results are shown below:

Home	Residents	Average Age of Residents	Disabled Residents
1	14	61	12
2	7	74	2
3	96	78	30
4	23	69	8
5	71	73	10
6	29	84	22

a. Develop an estimate of the mean age of nursing home residents in this city.

b. Develop an approximate 95% confidence interval for the proportion of disabled persons in the city's nursing homes.

c. Estimate the total number of disabled persons residing in nursing homes in this city.

22

Decision Analysis

Ohio Edison Company*

AKRON, OHIO

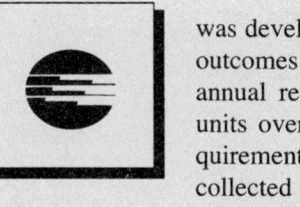

Ohio Edison Company, an investor-owned electric utility headquartered in northeastern Ohio, provides electrical service to over 2 million people. Most of this electricity is generated by coal-fired power plants. To meet evolving pollution-control requirements, Ohio Edison embarked on a program to replace the existing pollution-control equipment at most of its generating plants.

The flue gas emitted by coal-fired power plants contains small ash particles and sulfur dioxide. To meet newer emission limits for sulfur dioxide at one of its largest power plants, Ohio Edison decided to burn low-sulfur coal in four of the smaller units at the plant and to install fabric filters on these units to control particulate emissions. Fabric filters use thousands of fabric bags to filter out particulates and function in much the same way as a household vacuum cleaner.

It was considered likely, although not certain, that the three larger units at this plant would burn medium- to high-sulfur coal. Preliminary studies narrowed the particulate equipment choice for these larger units to fabric filters and electrostatic precipitators (which remove particulates suspended in the flue gas by passing it through a strong electric field). A number of uncertainties would affect the final choice, including uncertainty in the way some air-quality laws and regulations might be interpreted, potential changes in future air-quality laws and regulations, and uncertain construction costs.

Because of the complexity of the problem, the high degree of uncertainty associated with factors affecting the decision, and the cost impact on Ohio Edison, decision analysis was used in the selection process. A graphical description of the problem, referred to as a decision tree

was developed. The measure used to evaluate the outcomes depicted on the decision tree was the annual revenue requirements for the three large units over their remaining lifetime. Revenue requirements are the monies that would have to be collected from the utility customers to recover costs resulting from the installation of the new pollution-control equipment. An analysis of the resulting decision tree led to the following conclusions:

- The expected value of annual revenue requirements for the electrostatic precipitators was approximately $1 million less than that for the fabric filters.
- The fabric filters had a higher probability of high revenue requirements than did the electrostatic precipitators.
- The precipitators had nearly a .8 probability of having lower annual revenue requirements.

These results led Ohio Edison to select the electrostatic precipitators for the generating units in question. Had the decision analysis not been performed, the particulate-control decision might have been based chiefly on capital cost, a decision measure which favored the fabric filter equipment. It was felt that the use of decision analysis resulted in both lower expected revenue requirements and lower risk.

In this chapter we will introduce the methodology of decision analysis that Ohio Edison used in the above application. Our focus will be on showing how decision analysis can be used to identify the best decision alternative given an uncertain or risk-filled pattern of future events.

*The authors are indebted to Thomas J. Madden and M. S. Hyrnick of Ohio Edison Company, Akron, Ohio for providing this Statistics in Practice.

D ecision analysis can be used to determine optimal strategies when a decision maker is faced with several decision alternatives and an uncertain or risk-filled pattern of future events. For example, a manufacturer of a new style or line of seasonal clothing would like to manufacture large quantities of the product if consumer acceptance and consequently demand for the product are going to be high. However, the manufacturer would like to produce much smaller quantities if consumer acceptance and demand for the product are going to be low. Unfortunately, seasonal clothing items require the manufacturer to make a production-volume decision before the demand is known. Actual consumer acceptance of the new product will not be determined until the items have been placed in the stores and buyers have had the opportunity to purchase them. Selection of the best production-volume decision from among several alternatives when the decision maker is faced with the uncertainty of future demand is a problem suited for decision analysis.

We begin the study of decision analysis by considering problems in which there are reasonably few decision alternatives and possible future events. The concepts of a payoff table and a decision tree are introduced to provide a structure for this type of decision situation and to illustrate the fundamentals of decision analysis. The discussion is then extended to show how additional information obtained through experimentation can be combined with the decision maker's preliminary information to develop an optimal decision strategy.

22.1   Structuring the Decision Problem

To illustrate the decision analysis approach, let us consider the case of Political Systems, Inc. (PSI), a newly formed computer service firm specializing in information services such as surveys and data analysis for individuals running for political office. The firm is in the final stages of selecting a computer system for its Midwest branch, located in Chicago. While PSI has decided on a computer manufacturer, it is currently attempting to determine the size of the computer system that would be most economical to lease. We will use decision analysis to help PSI make its computer-leasing decision.

The first step in the decision analysis approach is to identify the alternatives considered by the decision maker. For PSI, the final decision will be to lease one of three computer systems, which differ in size and capacity. The three *decision alternatives,* denoted by d_1, d_2, and d_3, are as follows:

$$d_1 = \text{lease the large computer system}$$

$$d_2 = \text{lease the medium-sized computer system}$$

$$d_3 = \text{lease the small computer system}$$

Obviously, the selection of the *best* decision alternative will depend on what PSI management foresees as the possible market acceptance of the service and consequently the possible demand or load on the PSI computer system. Often, the future events associated with a decision situation are uncertain. That is, while a decision maker may have an idea of the variety of possible future events, the decision maker will be unsure as to which particular event will occur. Thus, the second step in a decision analysis approach is to identify the future events that might occur. These future events, which are not under the control of the decision maker, are referred to as the *states of nature*. It is assumed that the list of possible states of nature includes everything that can happen and that the individual states of nature do not overlap; that is, the states of nature are defined so that one and only one of the listed states of nature will occur.

When asked about the states of nature for the PSI decision problem, management viewed the possible acceptance of the PSI service as an either-or situation. That is, management believed that the firm's overall level of acceptance in the market place would be one of two possibilities: high acceptance or low acceptance. Thus, the PSI states of nature, denoted s_1 and s_2, are as follows:

$$s_1 = \text{high customer acceptance of PSI services}$$

$$s_2 = \text{low customer acceptance of PSI services}$$

Given the three decision alternatives and the two states of nature, which computer system should PSI lease? To answer this question, we will need information on the profit associated with each combination of a decision alternative and a state of nature. For example, what profit would PSI experience if the firm decided to lease the large computer system (d_1) and market acceptance was high (s_1)? What profit would PSI experience if the firm decided to lease the large computer system (d_1) and market acceptance was low (s_2)?

Payoff Tables

In decision analysis terminology, we refer to the outcome resulting from making a certain decision and the occurrence of a particular state of nature as the *payoff*. Using the best information available, management has estimated the payoffs or profits for the PSI leasing problem. These estimates are presented in Table 22.1. A table of this form is referred to as a *payoff table*. In general, entries in a payoff table can be stated in terms of profits, costs, or any other measure of output that may be appropriate for the particular situation being analyzed. The notation we will use for the entries in the payoff table is $V(d_i, s_j)$, which denotes the payoff associated with decision alternative d_i and state of nature s_j. Using this notation we see that $V(d_3, s_1) = \$100,000$.

Decision Trees

A *decision tree* provides a graphical representation of the decision-making process. Figure 22.1 shows a decision tree for the PSI leasing problem. Note that the tree shows the natural or logical progression that will occur over time. First, the firm must make its decision $(d_1, d_2, \text{or } d_3)$; then, once the decision is implemented, the state of nature $(s_1 \text{ or } s_2)$ will occur. The number at each end point of the tree represents the payoff associated with a particular chain of events. For example, the topmost payoff of $\$200,000$ arises whenever management makes the decision to purchase a large system (d_1) and

TABLE 22.1

Payoff Table (Profit in $) for the PSI Leasing Problem

Decision Alternatives		States of Nature	
		High Acceptance s_1	Low Acceptance s_2
Large system	d_1	200,000	−20,000
Medium system	d_2	150,000	20,000
Small system	d_3	100,000	60,000

FIGURE 22.1
Decision Tree for the PSI Leasing Problem (Payoffs in $)

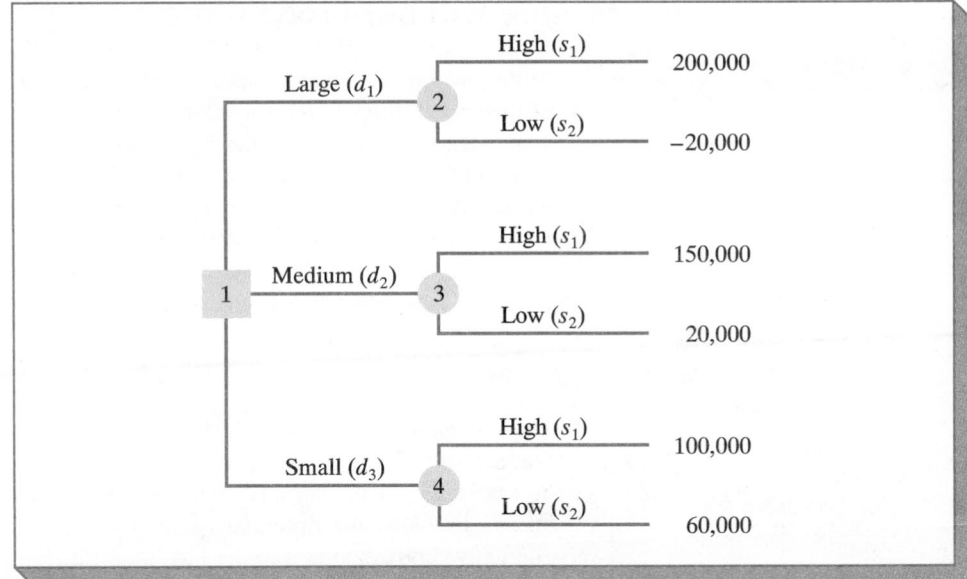

customer acceptance turns out to be high (s_1). The next-lower terminal point of $-\$20,000$ is reached when management has made the decision to lease the large system (d_1) and the state of nature turns out to be a low degree of customer acceptance (s_2). Thus, we see that each possible sequence of events for the PSI leasing problem is represented in the decision tree.

Using the general terminology associated with decision trees, we will refer to the intersection or junction points of the tree as *nodes* and the arcs or connectors between the nodes as *branches*. Figure 22.1 shows the PSI decision tree with the nodes numbered 1 to 4. When the branches *leaving* a given node are decision branches, we refer to the node as a *decision node*. Decision nodes are denoted by squares. Similarly, when the branches leaving a given node are state-of-nature branches, we refer to the node as a *state-of-nature-node*. State-of-nature nodes are denoted by circles. Using this node-labeling procedure, node 1 is a decision node, whereas nodes 2, 3, and 4 are state-of-nature nodes.

The identification of the decision alternatives, the states of nature, and the determination of the payoff associated with each decision alternative and state-of-nature combination are the first three steps in the decision analysis process. The question we now turn to is the following: How can the decision maker best utilize the information presented in the payoff table or the decision tree to arrive at a decision? As we will see, there are several approaches that may be used.

NOTES &
COMMENTS

1. Experts in problem solving approaches agree that the first step in solving a complex problem is to decompose it into a series of smaller subproblems. Decision trees provide a useful means of showing how the problem can be decomposed, as well as showing the sequential nature of the decision process.
2. People often view the same problem from different perspectives. Thus, there is no one correct way to develop a decision tree for a problem. Often the discussion regarding the most appropriate decision tree provides additional insight regarding the problem.

22.2   Decision Making without Probabilities

In this section we consider approaches to decision making that do not require knowledge of the probabilities of the states of nature. These approaches are appropriate in situations where the decision maker has very little confidence in his or her ability to assess the probabilities of the various states of nature, or where it is desirable to consider best- and worst-case analyses that are independent of state-of-nature probabilities. Because different approaches sometimes lead to different decision recommendations, it is important for the decision maker to understand the approaches available and then select the specific approach that, according to the decision maker's judgment, is the most appropriate.

Optimistic Approach

The *optimistic approach* evaluates each decision alternative in terms of the *best* payoff that can occur. The decision alternative that is recommended is the one that provides the best possible payoff. For a problem in which maximum profit is desired, as it is in the PSI leasing problem, the optimistic approach would lead the decision maker to choose the alternative corresponding to the largest profit. For problems involving minimization, this approach leads to choosing the alternative with the smallest payoff.

To illustrate the use of the optimistic approach, we will show how it can be used to develop a recommendation for the PSI leasing problem. First, we determine the maximum payoff possible for each of the decision alternatives; then we select the decision alternative that provides the overall maximum profit. This is just a systematic way of identifying the decision alternative that provides the largest possible profit. Table 22.2 illustrates this process.

Since \$200,000, corresponding to d_1, is the largest profit or payoff, the decision to lease a large system is the recommended decision alternative using the optimistic approach. It is easy to see why this is called an optimistic approach. It simply recommends the decision alternative that provides the best of all payoffs, \$200,000.

Conservative Approach

The *conservative approach* evaluates each decision alternative in terms of the *worst* payoff that can occur. The decision alternative recommended is the one that provides the best of the worst possible payoffs. For a problem in which the output measure is profit, as it is in the PSI leasing problem, the conservative approach would lead the decision maker to choose the alternative that maximizes the minimum possible profit that could be obtained. For problems involving minimization, this approach identifies the alternative that will minimize the maximum payoff.

To illustrate the conservative approach, we will show how it can be used to develop a recommendation for the PSI leasing problem. First, the decision maker would identify the minimum payoff for each of the decision alternatives; then, the decision maker would

Table 22.2
PSI Maximum Payoff (\$) for Each Decision Alternative

Decision Alternatives		Maximum Payoff	
Large system	d_1	200,000 ◄——— Maximum of the maximum payoff values
Medium system	d_2	150,000
Small system	d_3	100,000

TABLE 22.3	**Decision Alternatives**	**Minimum Payoff**

TABLE 22.3
PSI Minimum Payoff ($) for Each Decision Alternative

Decision Alternatives		Minimum Payoff
Large system	d_1	$-20,000$
Medium system	d_2	20,000
Small system	d_3	60,000 ◄——— Maximum of the minimum payoff values

select the alternative that maximizes the minimum payoff. Table 22.3 illustrates this approach for the PSI leasing problem.

Since $60,000, corresponding to d_3, yields the maximum of the minimum payoffs, the decision alternative to lease a small system is recommended. This decision approach is considered conservative because it concentrates on the worst possible payoffs and then recommends the decision alternative that avoids the possibility of extremely ''bad'' payoffs. In using the conservative approach, PSI is guaranteed a profit of at least $60,000. While PSI may still make more, it *cannot* make less than $60,000.

Minimax Regret Approach

Minimax regret is another approach to decision making without probabilities. This approach is neither purely optimistic or purely conservative. Let us illustrate the minimax regret approach by showing how it can be used to select a decision alternative for the PSI leasing problem.

Suppose that we make the decision to lease the small system (d_3) and afterwards learn that customer acceptance of the PSI service is high (s_1). Table 22.1 shows the resulting profit to be $100,000. However, now that we know that state of nature s_1 has occurred, we see that the large-system decision (d_1), yielding a profit of $200,000, would have been the best decision. The difference between the best payoff ($200,000) and the payoff experienced ($100,000) is referred to as the *opportunity loss* or *regret* associated with the d_3 decision when state of nature s_1 occurs ($200,000 $-$ $100,000) = $100,000. If we had made decision d_2 and state of nature s_1 had occurred, the opportunity loss or regret would have been $200,000 $-$ $150,000 = $50,000.

In maximization problems, the general expression for opportunity loss or regret is given by the following formula.

Opportunity Loss or Regret

$$R(d_i, s_j) = V^*(s_j) - V(d_i, s_j) \qquad (22.1)$$

where

$R(d_i, s_j)$ = regret associated with decision alternative d_i and state of nature s_j

$V^*(s_j)$ = best payoff value[†] under state of nature s_j

$V(d_i, s_j)$ = payoff associated with decision alternative d_i and state of nature s_j

---

[†]In cost minimization problems, $V^*(s_j)$ will be the smallest entry in column j. Thus for minimization problems, Equation (22.1) must be changed to $R(d_i, s_j) = V(d_i, s_j) - V^*(s_j)$.

TABLE 22.4

Regret or Opportunity Loss ($) for the PSI Leasing Problem

		States of Nature	
Decision Alternatives		High Acceptance s_1	Low Acceptance s_2
Large system	d_1	0	80,000
Medium system	d_2	50,000	40,000
Small system	d_3	100,000	0

TABLE 22.5

PSI Maximum Regret or Opportunity Loss ($) for Each Decision Alternative

Decision Alternatives		**Maximum Regret or Opportunity Loss**	
Large system	d_1	80,000	
Medium system	d_2	50,000 ◄———— Minimum of the
Small system	d_3	100,000	maximum regret

Using Equation (22.1) and the payoffs in Table 22.1, we can compute the regret associated with all combinations of decision alternatives d_i and states of nature s_j. We simply replace each entry in the payoff table with the value found by subtracting the entry from the largest entry in its column. Table 22.4 shows the regret, or opportunity loss, table for the PSI leasing problem.

The next step in applying the minimax regret approach requires the decision maker to identify the maximum regret for each decision alternative. These data are shown in Table 22.5. The choice of a best decision is made by selecting the alternative corresponding to the *mini*mum of the *max*imum *regret* values; hence the name *minimax regret*. For the PSI leasing problem, the decision to lease a medium-sized computer system, with a corresponding regret of $50,000, is the recommended minimax regret decision.

Note that the three approaches discussed in this section have provided different recommendations. This is not in itself bad. It simply reflects the difference in decision-making philosophies that underlie the various approaches. Ultimately, the decision maker will have to choose the most appropriate approach and then make the final decision accordingly. The major criticism of the approaches discussed in this section is that they do not consider any information about the probabilities of the various states of nature. In the next section we discuss an approach that utilizes probability information in selecting a decision alternative.

❏ ❏ Exercises

Methods

SELF TEST ▶

1. The payoff table showing profit for a decision analysis problem with two decisions and three states of nature is shown at the top of the next page.

		States of Nature		
		s_1	s_2	s_3
Decision Alternatives	d_1	250	100	25
	d_2	100	100	75

a. Construct a decision tree for this problem.

b. If the decision maker knows nothing about the chances or probability of occurrence of the three states of nature, what is the recommended decision using the optimistic, conservative, and minimax regret approaches?

2. Suppose that a decision maker faced with four decision alternatives and four states of nature develops the following profit payoff table:

		States of Nature			
		s_1	s_2	s_3	s_4
Decision	d_1	14	9	10	5
	d_2	11	10	8	7
Alternatives	d_3	9	10	10	11
	d_4	8	10	11	13

a. If the decision maker knows nothing about the chances or probability of occurrence of the four states of nature, what is the recommended decision using the optimistic, conservative, and minimax regret approaches?

b. Which approach do you prefer? Explain. Is it important for the decision maker to establish the most appropriate approach before analyzing the problem? Explain.

c. Assume that the payoff table provides *cost* rather than profit payoffs. What is the recommended decision using the optimistic, conservative, and minimax regret approaches?

Applications

3. Southland Corporation's decision to produce a new line of recreational products has resulted in the need to construct either a small plant or a large plant. The selection of plant size depends on how the marketplace reacts to the new product line. To conduct an analysis, marketing management has decided to view the possible long-run demand as either low, medium, or high. The following payoff table shows the projected profit in millions of dollars:

		Long-Run Demand		
		Low	*Medium*	*High*
Decision	*Small plant*	150	200	200
Alternatives	*Large plant*	50	200	500

a. Construct a decision tree for this problem.

b. Recommend a decision using the optimistic, conservative, and minimax regret approaches.

SELF TEST ▶ 4. McHuffter Condominiums, Inc., of Pensacola, Florida, recently purchased land near the Gulf of Mexico and is attempting to determine the size of the condominium development it should build.

Three sizes of developments are being considered: small, d_1; medium, d_2; and large, d_3. At the same time, an uncertain economy makes it difficult to ascertain the demand for the new condominiums. McHuffter's management realizes that a large development followed by a low demand could be very costly to the company. However, if McHuffter makes a conservative small-development decision and then finds a high demand, the firm's profits will be lower than they might have been. With the three levels of demand—low, medium, and high—McHuffter's management has prepared the following profit ($1000s) payoff table:

		Demand		
		Low	Medium	High
Decision Alternatives	Small	400	400	400
	Medium	100	600	600
	Large	−300	300	900

a. Construct a decision tree for this problem.
b. If nothing is known about the demand probabilities, what are the recommended decisions using the optimistic, conservative, and minimax regret approaches?

22.3   Decision Making with Probabilities

In many decision-making situations, it is possible to obtain probability estimates for each of the possible states of nature. When such probabilities are available, the *expected value approach* can be used to identify the best decision alternative. The expected value approach evaluates each decision alternative in terms of its expected value. The decision alternative that is recommended is the one that provides the best expected value. Let us first define the expected value of a decision alternative and then show how it can be used for the PSI leasing problem.

Let

$$N = \text{the number of possible states of nature}$$

$$P(s_j) = \text{the probability of state of nature } s_j$$

Since one and only one of the N states of nature can occur, the associated probabilities must satisfy the following two conditions:

$$P(s_j) \geq 0 \qquad \text{for all states of nature} \qquad (22.2)$$

$$\sum_{j=1}^{N} P(s_j) = P(s_1) + P(s_2) + \cdots + P(s_N) = 1 \qquad (22.3)$$

The expected value (EV) of decision alternative d_i is defined as follows.

Expected Value of Decision Alternative d_i

$$\text{EV}(d_i) = \sum_{j=1}^{N} P(s_j) V(d_i, s_j) \qquad (22.4)$$

In words, the expected value of a decision alternative is the sum of weighted payoffs for the alternative. The weight for a payoff is the probability of the associated state of nature and therefore the probability that the payoff occurs. Let us now return to the PSI leasing problem to see how the expected value approach can be applied.

Suppose that PSI management believes that s_1, the high-acceptance state of nature, has a .3 probability of occurrence and that s_2, the low-acceptance state of nature, has a .7 probability. Thus, $P(s_1) = .3$ and $P(s_2) = .7$. Using the payoff values $V(d_i, s_j)$ shown in Table 22.1 and Equation (22.4), expected values for the three decision alternatives can be calculated:

$$EV(d_1) = .3(200,000) + .7(-20,000) = \$46,000$$

$$EV(d_2) = .3(150,000) + .7(20,000) \quad = \$59,000$$

$$EV(d_3) = .3(100,000) + .7(60,000) \quad = \$72,000$$

Thus, according to the expected value approach, since d_3 has the highest expected value (\$72,000), d_3 is the recommended decision.

The calculations required to identify the decision alternative with the best expected value can be conveniently carried out on a decision tree. Figure 22.2 shows the decision tree for the PSI leasing problem with state-of-nature branch probabilities. We will now use the branch probabilities and the expected value approach to arrive at the optimal decision for PSI.

Working backward through the decision tree, we first compute the expected value at each state-of-nature node. That is, at each state-of-nature node, we weight each possible payoff by its chance of occurrence. By doing this, we obtain the expected values for nodes 2, 3, and 4 as shown in Figure 22.3.

Since the decision maker controls the branch leaving decision node 1 and since we are trying to maximize expected profits, the best decision branch at node 1 is d_3. Thus, the decision tree analysis leads us to recommend d_3 with an expected value of \$72,000. Note that this is the same recommendation that was obtained using the expected value approach in conjunction with the payoff table.

FIGURE 22.2
PSI Decision Tree with State-of-Nature Branch Probabilities (Payoffs in $)

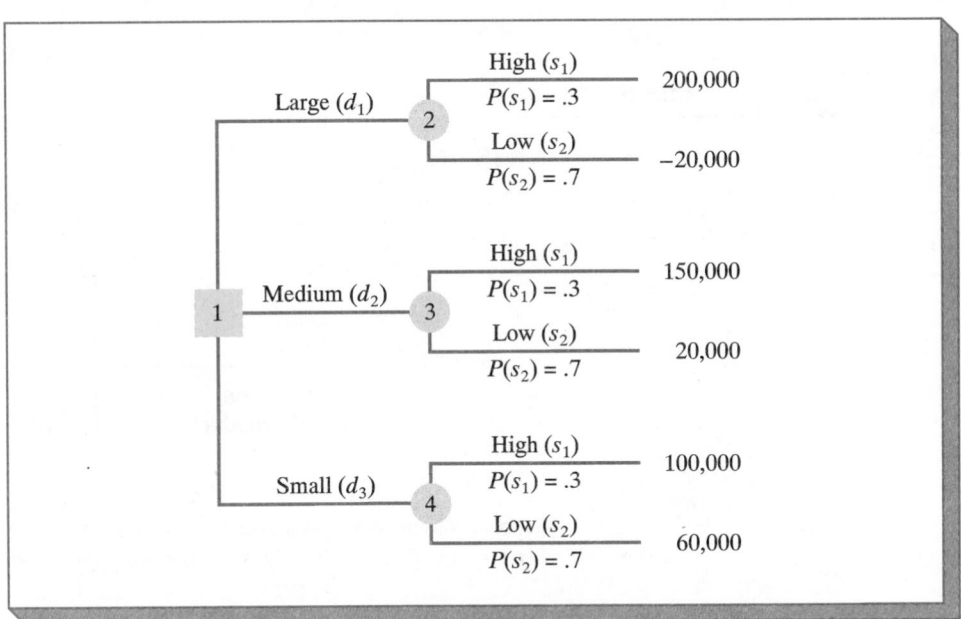

FIGURE 22.3

Applying the Expected Value Approach Using Decision Trees

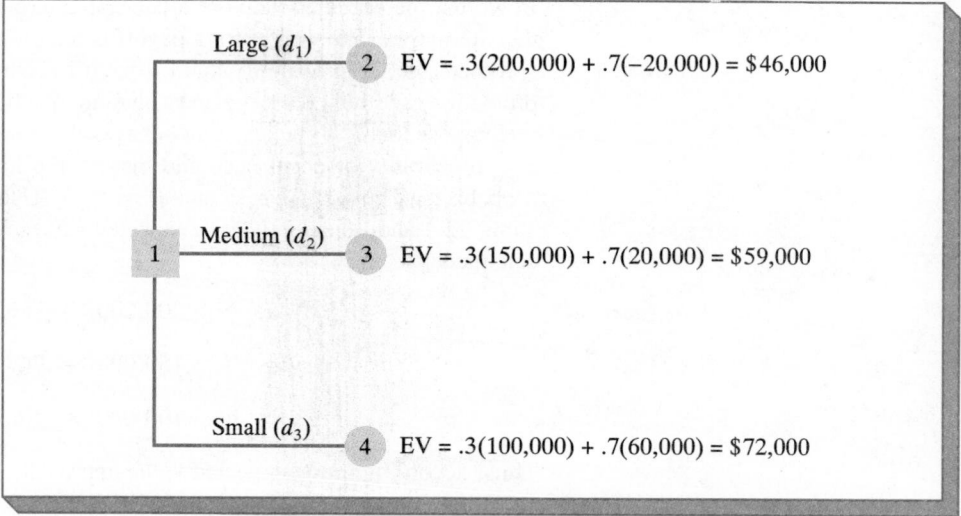

We have seen how decision trees can be used to analyze decisions with state-of-nature probabilities. While other decision problems may be substantially more complex than the PSI leasing problem, if there are a reasonable number of decision alternatives and states of nature, the decision tree approach outlined in this section can be used. First, the analyst must draw a decision tree consisting of decision and state-of-nature nodes and branches that describe the sequential nature of the problem. Assuming that the expected value approach is to be used, the next step is to determine the probabilities for each of the state-of-nature branches and compute the expected value at each state-of-nature node. The decision branch leading to the state-of-nature node with the best expected value is then selected. The decision alternative associated with this branch is the recommended decision.

❏ ❏ **Exercises**

Methods

SELF TEST ▶   5. The payoff table presented in Exercise 1 is repeated here:

		States of Nature		
		s_1	s_2	s_3
Decision	d_1	250	100	25
Alternatives	d_2	100	100	75

Suppose that the decision maker has obtained the following probability estimates: $P(s_1) = .65$, $P(s_2) = .15$, $P(s_3) = .20$. Use the expected value approach to determine the optimal decision.

6. The profit payoff table presented in Exercise 2 is repeated here:

		States of Nature			
		s_1	s_2	s_3	s_4
	d_1	14	9	10	5
Decision	d_2	11	10	8	7
Alternatives	d_3	9	10	10	11
	d_4	8	10	11	13

Suppose that the decision maker obtains information that enables the following probability estimates to be made: $P(s_1) = .5$, $P(s_2) = .2$, $P(s_3) = .2$, $P(s_4) = .1$.

a. Use the expected value approach to determine the optimal decision.

b. Now assume that the entries in the payoff table are costs; use the expected value approach to determine the optimal decision.

Applications

7. Hale's TV Productions is considering producing a pilot for a comedy series for a major television network. While the network may reject the pilot and the series, it may also purchase the program for 1 or 2 years. Hale may decide to produce the pilot or transfer the rights for the series to a competitor for $100,000. Hale's profits are summarized in the following profit ($1000s) payoff table:

		States of Nature		
		Reject	*1 Year*	*2 Years*
Produce pilot	d_1	− 100	50	150
Sell to competitor	d_2	100	100	100

Assuming the probability estimates for the states of nature are $P(\text{reject}) = .2$, $P(1 \text{ year}) = .3$, and $P(2 \text{ years}) = .5$, what should the company do?

 8. Consider the McHuffter Condominiums problem presented in Exercise 4. If $P(\text{low}) = .20$, $P(\text{medium}) = .35$, and $P(\text{high}) = .45$, what is the recommended decision using the expected value approach?

9. Martin's Service Station is considering investing in a heavy-duty snowplow this fall. Martin has analyzed the situation carefully and feels that this would be a very profitable investment if the snowfall is heavy. A small profit could still be made if the snowfall is moderate, but Martin would lose money if snowfall is light. Specifically, Martin forecasts a profit of $7000 if snowfall is heavy and $2000 if it is moderate, and a $9000 loss if it is light. Based on the weather bureau's long-range forecast, Martin estimates that $P(\text{heavy snowfall}) = .4$, $P(\text{moderate snowfall}) = .3$, and $P(\text{light snowfall}) = .3$.

a. Prepare a decision tree for Martin's problem.

b. What is the expected value at each state-of-nature node?

c. Using the expected value approach, would you recommend that Martin invest in the snowplow?

10. Joseph Software, Inc. (JSI), has been investigating the possibility of developing a grammar-and-style checker for use on microcomputers. Based on their experience with other software projects, JSI estimates that the total cost to develop a prototype of the software is $200,000. If the performance of the prototype is somewhat better than existing software, referred to as a *moderate success*, JSI believes they could sell the rights to the software to a larger software developer for

$160,000. If the performance of the prototype is significantly better than existing software, referred to as a *major success*, they believe that the software can be sold for $1.2 million. However, if the performance of the prototype does not exceed the performance of existing software, referred to as a *failure*, JSI will not be able to sell the software and hence will lose all of their development costs.

a. Prepare a decision tree for JSI.

b. If the best estimates of the states of nature are $P(\text{failure}) = .70$, $P(\text{moderate success}) = .20$, and $P(\text{major success}) = .10$, what should JSI do if they use the expected value approach?

11. Six months ago, Doug Reynolds paid $25,000 for an option to purchase a tract of land that he is considering developing. Another investor has offered to purchase Doug's option for $275,000. If Doug does not accept the investor's offer, he will purchase the property, clear the land, and prepare the site for building. He believes that once the site is prepared he can sell the land to a home builder. However, the success of the investment depends upon the real estate market at the time he sells the property. If the real estate market is down, Doug feels that he will lose $1.5 million. If market conditions stay at their current level, he estimates that his profit will be $1 million; if market conditions are up at the time he sells, he estimates a profit of $4 million. Because of other commitments Doug does not consider it feasible to hold the land once he has developed it; thus, the only two alternatives are to sell the option or to develop the land. Suppose that the probabilities of the real estate market being down, at the current level, or up are .6, .3, and .1, respectively. What decision should Doug make following the expected value approach?

22.4   Sensitivity Analysis

For the PSI leasing problem, management provided a .3 probability for s_1, the high-acceptance state of nature, and a .7 probability for s_2, the low-acceptance state of nature. Using these probabilities, we found that decision alternative d_3 had the highest expected value and was the recommended decision. In this section we consider how changes in the probability estimates for the states of nature affect or alter the recommended decision. The study of the effect of such changes is referred to as *sensitivity analysis*.

One approach to sensitivity analysis is to consider different probabilities for the states of nature and then recompute the expected value for each decision alternative. Repeating this several times, we can begin to learn how changes in the probabilities for the states of nature affect the recommended decision. For example, suppose that we consider a change in the probabilities for the states of nature such that $P(s_1) = .6$ and $P(s_2) = .4$. Using these probabilities and repeating the expected value computations, we find the following:

$$EV(d_1) = .6(200,000) + .4(-20,000) = \$112,000$$

$$EV(d_2) = .6(150,000) + .4(20,000) \ = \$ 98,000$$

$$EV(d_3) = .6(100,000) + .4(60,000) \ = \$ 84,000$$

Thus, with these probabilities, the recommended decision alternative is d_1, with an expected value of $112,000.

Obviously, we could continue to modify the probabilities of the states of nature and thus learn more about how such changes affect the recommended decision. The only drawback to this approach is that there will be numerous calculations required to evaluate the effect of several possible changes in the state-of-nature probabilities.

For the special case of two states of nature, the sensitivity analysis computations can be eased substantially through the use of a graphical procedure. Let us demonstrate this procedure by further analyzing the PSI leasing problem. We begin by denoting the probability of state of nature s_1 by p. That is,

$$P(s_1) = p$$

and thus

$$P(s_2) = 1 - P(s_1) = 1 - p$$

The expected value for decision alternative d_1 can then be written as a function of p.

$$\text{EV}(d_1) = P(s_1)(200,000) + P(s_2)(-20,000)$$

$$= p(200,000) + (1 - p)(-20,000)$$

$$= 220,000p - 20,000 \tag{22.5}$$

Repeating the expected value computation for decision alternatives d_2 and d_3, we obtain the following expressions for expected value as a function of p:

$$\text{EV}(d_2) = 130,000p + 20,000 \tag{22.6}$$

$$\text{EV}(d_3) = 40,000p + 60,000 \tag{22.7}$$

Thus, we have developed three linear equations that express the expected value of the three decision alternatives as a function of the probability of state of nature s_1.

Let us continue by developing a graph with values of p on the horizontal axis and the associated expected values on the vertical axis. Since (22.5), (22.6), and (22.7) are all linear equations, we can graph each equation, or line, by finding any two points and drawing a line through the points. Using $\text{EV}(d_1)$ in (22.5) as an example, we first let $p = 0$ and find that $\text{EV}(d_1) = -20,000$. Then, letting $p = 1$, we find that EV $(d_1) = 200,000$. Connecting these two points, $(0, -20,000)$ and $(1, 200,000)$, provides the line labeled $\text{EV}(d_1)$ in Figure 22.4. This figure also shows a line labeled $\text{EV}(d_2)$ and a line labeled $\text{EV}(d_3)$; these are the graphs of (22.6) and (22.7), respectively.

Figure 22.4 can now be used for sensitivity analysis. Recall that PSI is seeking to maximize profit. Note that for small values of p, decision alternative d_3 provides the largest expected value and is thus the recommended decision. Similarly, for large values of p, we see that decision alternative d_1 provides the largest expected value and is thus the recommended decision. Further, note that there is no section of the graph for which decision alternative d_2 provides the largest expected value. Thus, with the exception of the point where the three expected value lines intersect and all three expected values are equal, decision alternative d_2 can never be the decision alternative recommended by the expected value approach.

Referring to Figure 22.4 again, we see that the three lines intersect at a value of p between .4 and .5. In other words, at some point between .4 and .5, all three decision alternatives provide the same expected value.

Whenever two or more lines intersect on a sensitivity analysis graph, we can set the equations for two of the intersecting lines equal to each other and solve for the value of p. Using the equations for decision alternative d_1 and decision alternative d_3, we obtain

$$220,000p - 20,000 = 40,000p + 60,000$$

Thus,

$$180,000p = 80,000$$

$$p = \frac{80,000}{180,000} = .44$$

Hence, whenever $p = .44$, each decision alternative will provide the same expected value. Using this value of p in Figure 22.4, we can now conclude that for $p < .44$,

FIGURE 22.4

Expected Value as a Function of *p*

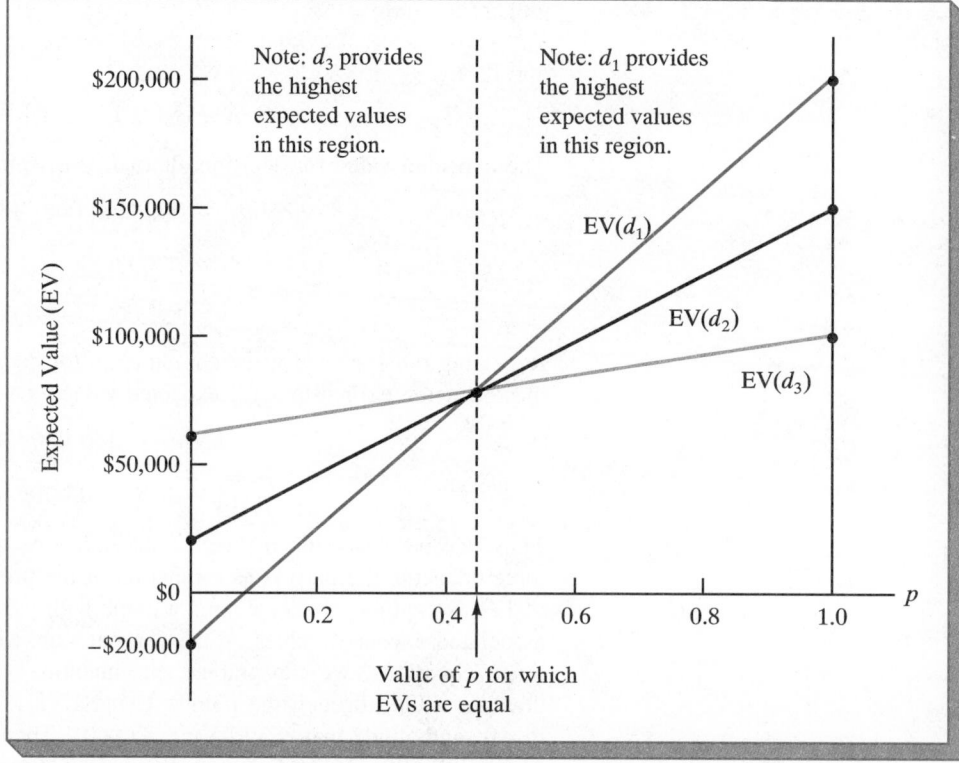

decision alternative d_3 provides the largest expected value and that for $p > .44$, decision alternative d_1 provides the largest expected value. Since p is simply the probability of state of nature s_1 and $(1 - p)$ is the probability of state of nature s_2, we now have the sensitivity analysis information that tells us how changes in the state-of-nature probabilities affect the recommended decision alternative.

The benefit of performing sensitivity analysis is that it can provide a better perspective on management's original judgment regarding the state-of-nature probabilities. Management originally estimated the probability of high customer acceptance as $P(s_1) = .3$. As a result, decision alternative d_3 was recommended. After carrying out the sensitivity analysis, we can now tell management that the original estimate of $P(s_1)$ is not extremely critical for d_3 to be the recommended decision. In fact, as long as $P(s_1) < .44$, the d_3 decision alternative remains optimal.

Note that in the PSI leasing problem, the three lines for the three decision alternatives graphed in Figure 22.4 intersect at the same point ($p = .44$). Similar sensitivity analysis computations for other decision analysis problems with two states of nature and three decision alternatives should not be expected to result in the same type of graph. With a different sensitivity analysis graph, d_1 could be the best decision alternative for certain values of p, d_2 the best decision alternative of other values of p, and d_3 the best decision alternative for the remaining values of p. This situation is demonstrated in the sensitivity analysis computation for Exercise 13 at the end of this section.

The graphical sensitivity analysis procedure we have described for the PSI leasing problem applies only to decision analysis problems with two states of nature. However, sensitivity analysis is important in problems with more than two states of nature. In these cases, a computer software package can be used to assist with the computations. Basically, we return to the approach of testing a variety of likely changes for the state-of-nature probabilities. The software package is helpful in making the necessary expected value computations and providing the decision alternative recommendations with a minimum of time and effort on the part of the analyst.

□ □ **Exercises**

Methods

SELF TEST ▶

12. The payoff table showing profit for a decision problem with two states of nature and two decision alternatives is shown below:

		States of Nature	
		s_1	s_2
Decision	d_1	10	1
Alternatives	d_2	4	3

Use graphical sensitivity analysis to determine the probability of state of nature s_1 for which each of the decision alternatives has the largest expected value.

13. The payoff table showing profit for a decision problem with two states of nature and three decision alternatives is presented below:

		States of Nature	
		s_1	s_2
Decision	d_1	80	50
Alternatives	d_2	65	85
	d_3	30	100

Use graphical sensitivity analysis to determine the values of the probability of state of nature s_1 for which each of the decision alternatives has the largest expected value.

Applications

SELF TEST ▶

14. Milford Trucking, located in Chicago, has requests to haul two shipments, one to St. Louis and one to Detroit. Because of a scheduling problem, Milford will be able to accept only one of these assignments. The St. Louis customer has guaranteed a return shipment, but the Detroit customer has not. Thus, if Milford accepts the Detroit shipment and cannot find a Detroit-to-Chicago return shipment, the truck will return to Chicago empty. The payoff table showing profit is as follows:

			Return Shipment from Detroit	No Return Shipment from Detroit
			s_1	s_2
Shipment	*St. Louis*	d_1	2000	2000
	Detroit	d_2	2500	1000

a. If the probability of a Detroit return shipment is .4, what should Milford do?

b. Use graphical sensitivity analysis to determine the values of the probability of state of nature s_1 for which d_1 has the largest expected value.

15. Refer to Exercise 11. Suppose that the probabilities of the real estate market being down, at the current level, or up are .5, .3, and .2, respectively. What decision should Doug make following the expected value approach? What if the probabilities are .4, .4, and .2? What do the above results suggest regarding the proposed investment?

16. Suppose that after further consideration, Doug Reynolds (Exercises 11 and 15) concludes that .1 is a good estimate of the probability of the real estate market being up. However, Doug is unable to reach any definite conclusions regarding the probabilities for the other two states of nature. What would the value of the probability of the market being down have to be for the expected value approach to recommend that he should sell his option for $275,000? Do you think that this information would help Doug make a decision regarding whether to sell the option or to develop the site? Explain.

22.5   Expected Value of Perfect Information

Suppose that PSI had the opportunity to conduct a market research study that would evaluate consumer need for the PSI service. Such a study could help in the leasing decision by improving the current probability assessments for the states of nature. However, if the cost of obtaining the market research information exceeds its value, PSI should not conduct the market research study.

To determine the maximum possible value that PSI should pay for additional information, let us suppose that PSI could obtain perfect information regarding the states of nature; that is, we will assume that PSI could determine with certainty which state of nature will occur. To make use of perfect information, we need to develop a decision strategy for PSI to follow. As we will show, a decision strategy is simply a policy or decision rule that is to be followed by the decision maker. In computing the *expected value of perfect information* (EVPI), the decision strategy is a rule that specifies which decision alternative should be selected given each state of nature.

To help determine the optimal decision strategy for PSI, we have reproduced PSI's payoff table as Table 22.6. We see that if state of nature s_1 occurs, then the best decision alternative is d_1 with a profit of $200,000. Similarly, if state of nature s_2 occurs, then the best decision alternative is d_3 with a profit of $60,000. Thus, PSI's optimal decision strategy if perfect information were available can be stated as follows:

If s_1 occurs, then select d_1.

If s_2 occurs, then select d_3.

TABLE 22.6
Payoff Table ($) for the PSI Leasing Problem

| | | States of Nature | |
| | | High Acceptance s_1 | Low Acceptance s_2 |
Decision Alternatives			
Large system	d_1	200,000	−20,000
Medium system	d_2	150,000	20,000
Small system	d_3	100,000	60,000

What is the expected value for this decision strategy? Since $P(s_1) = .3$ and $P(s_2) = .7$, we see that there is a .3 probability that PSI will make \$200,000 and a .7 probability PSI will make \$60,000. Thus, the expected value of the decision strategy that uses perfect information is

$$.3(\$200,000) + .7(\$60,000) = \$102,000$$

We refer to \$102,000 as the *expected value with perfect information* (EVwPI).

Recall that when perfect information was not available, the expected value approach resulted in recommending decision alternative d_3, with an expected value of \$72,000. We refer to \$72,000 as the *expected value without perfect information* (EVwoPI). Since \$72,000 is the expected value without perfect information (EVwoPI) and \$102,000 is the expected value with perfect information (EVwPI), \$102,000 − \$72,000 = \$30,000 represents the expected value of perfect information (EVPI); that is,

$$\text{EVPI} = \text{EVwPI} - \text{EVwoPI}$$
$$= \$102,000 - \$72,000$$
$$= \$30,000$$

In other words, \$30,000 represents the additional expected value that can be obtained if perfect information were available about the states of nature. Figure 22.5 provides a summary of the computation of the EVPI for the PSI leasing problem.

Generally speaking, a market research study will not provide ''perfect'' information; however, the information provided might be worth a good portion of the \$30,000. In any case, PSI's management knows it should never pay more than \$30,000 for any information, no matter how good. Provided the market survey cost is reasonably small—say, \$5000 to \$10,000—it appears economically desirable for PSI to consider the market research study.

F IGURE 22.5
The Expected Value of Perfect Information

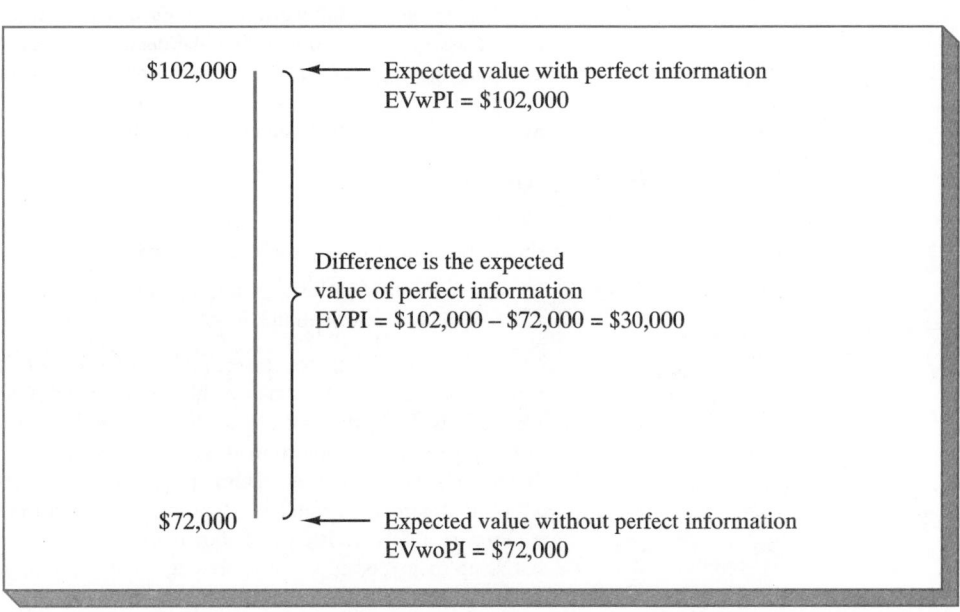

$102,000 ◄—— Expected value with perfect information
EVwPI = $102,000

Difference is the expected
value of perfect information
EVPI = $102,000 – $72,000 = $30,000

$72,000 ◄—— Expected value without perfect information
EVwoPI = $72,000

Exercises

Methods

SELF TEST ▶ 17. The payoff table presented in Exercises 1 and 5 is repeated here:

		States of Nature		
		s_1	s_2	s_3
Decision	d_1	250	100	25
Alternatives	d_2	100	100	75

The probabilities for the states of nature are: $P(s_1) = .65$, $P(s_2) = .15$, $P(s_3) = .20$.
a. What is the optimal decision strategy if perfect information were available?
b. What is the expected value for the decision strategy developed in (a)?
c. Using the expected value approach, what is the recommended decision? What is its expected value?
d. What is the expected value of perfect information?

18. The profit payoff table presented in Exercises 2 and 6 is repeated here:

		States of Nature			
		s_1	s_2	s_3	s_4
	d_1	14	9	10	5
Decision	d_2	11	10	8	7
Alternatives	d_3	9	10	10	11
	d_4	8	10	11	13

The probabilities of the states of nature are: $P(s_1) = .5$, $P(s_2) = .2$, $P(s_3) = .2$, and $P(s_4) = .1$.
a. What is the optimal decision strategy if perfect information were available?
b. What is the expected value for the decision strategy developed in (a)?
c. Using the expected value approach, what is the recommended decision? What is its expected value?
d. What is the expected value of perfect information?

Applications

19. Consider the Hale's TV Productions problem (Exercise 7). What is the maximum that Hale should be willing to pay for inside information on what the network will do?

SELF TEST ▶ 20. Consider the McHuffter Condominiums problem (Exercises 4 and 8). What is the expected value of perfect information?

21. Refer again to the investment problem faced by Martin's Service Station (Exercise 9). Martin can purchase a blade to attach to his service truck that can also be used to plow driveways and parking lots. Since this truck must also be available to start cars, Martin will not be able to generate as much revenue plowing snow if he elects this alternative. But he will keep his loss smaller if there is light snowfall. Under this alternative, Martin forecasts a profit of $3500 if snowfall is heavy and $1000 if it is moderate and a $1500 loss if snowfall is light.
a. Prepare a new decision tree showing all three alternatives.
b. Using the expected value approach, what is the optimal decision?
c. What is the expected value of perfect information?

22. Consider the Milford Trucking problem (Exercise 14). What is the expected value of perfect information that would tell Milford Trucking whether Detroit has a return shipment?

22.6   Decision Analysis with Sample Information

In applying the expected value approach, we have seen how probability information about the states of nature affects the expected value calculations and thus the decision recommendation. Frequently, decision makers have preliminary or prior probability estimates for the states of nature that are initially the best probability values available. However, to make the best possible decision, the decision maker may want to seek additional information about the states of nature. This new information can be used to revise or update the prior probabilities so that the final decision is based on more accurate probability estimates for the states of nature.

The seeking of additional information is most often accomplished through experiments designed to provide sample information or more current data about the states of nature. Raw material sampling, product testing, and test market research are examples of experiments that may enable a revision or updating of the state-of-nature probabilities. In the following discussion, we will reconsider the PSI leasing problem and show how sample information can be used to revise the state-of-nature probabilities. We will then show how the revised probabilities can be used to develop an optimal decision strategy for PSI.

Recall that management had assigned a probability of $P(s_1) = .3$ to state of nature s_1 and a probability of $P(s_2) = .7$ to state of nature s_2. At this point, we will refer to these initial probability estimates, $P(s_1)$ and $P(s_2)$, as the *prior probabilities* for the states of nature. Using these prior probabilities, we found that d_3, the decision to lease the small system was optimal, yielding an expected value of $72,000. Recall also that we showed that the expected value of new information about the states of nature could potentially be worth as much as EVPI = $30,000.

Suppose that PSI decides to consider hiring a market research firm to study the potential acceptance of the PSI service. The market research study will provide new information that can be combined with the prior probabilities through a Bayesian procedure to obtain updated or revised probability estimates for the states of nature. These *revised* probabilities are called *posterior probabilities*. The process of revising probabilities is depicted in Figure 22.6.

We will refer to the new information obtained through research or experimentation as an *indicator*. Since in many cases the experiment conducted to obtain the additional information will consist of taking a statistical sample, the new information is also often referred to as *sample information*.

Using the indicator terminology, we can denote the outcomes of the PSI marketing research study as follows:

I_1 = favorable market research report (i.e., in the market research study, the individuals contacted generally express interest in PSI's services)

I_2 = unfavorable market research report (i.e., in the market research study, the individuals contacted generally express little interest in PSI's services).

Given one of these possible indicators, our objective is to provide improved estimates of the probabilities of the two states of nature. The end result of the *Bayesian revision*

FIGURE 22.6
Probability Revision Based on New Information

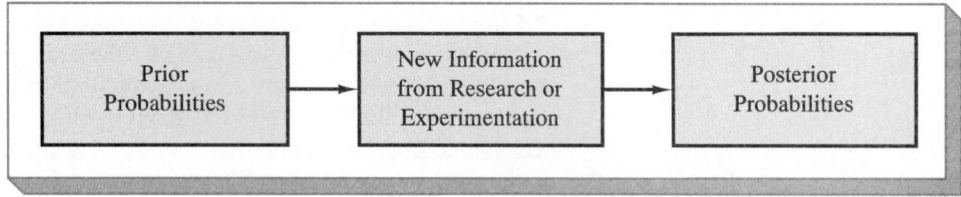

process depicted in Figure 22.6 is a set of posterior probabilities of the form $P(s_j \mid I_k)$, where $P(s_j \mid I_k)$ represents the conditional probability that state of nature s_j will occur given that the outcome of the market research study was indicator I_k.

To make effective use of this indicator information, we must know something about the probability relationships between the indicators and the states of nature. For example, in the PSI leasing problem, given that the state of nature ultimately turns out to be high customer acceptance, what is the probability that the market research study will result in a favorable report? In this case, we are asking about the conditional probability of indicator I_1 given state of nature s_1, written $P(I_1 \mid s_1)$. To carry out the analysis, we will need conditional probabilities for all indicators given all states of nature, that is, $P(I_1 \mid s_1)$, $P(I_1 \mid s_2)$, $P(I_2 \mid s_1)$, and $P(I_2 \mid s_2)$.

In the PSI leasing problem, the marketing research company used the results of similar studies to develop the following estimates of the relevant conditional probabilities:

	Market Research Report	
States of Nature	Favorable I_1	Unfavorable I_2
High acceptance s_1	$P(I_1 \mid s_1) = .8$	$P(I_2 \mid s_1) = .2$
Low acceptance s_2	$P(I_1 \mid s_2) = .1$	$P(I_2 \mid s_2) = .9$

Note that these probability estimates indicate that a good degree of confidence can be placed in the market research report. When the true state of nature is s_1, the market research report will be favorable 80% of the time and unfavorable only 20%. When the true state is s_2, the report will make the correct indication 90% of the time. Now let us see how this additional information can be incorporated into the decision-making process.

22.7   Developing a Decision Strategy

A decision strategy is a policy or decision rule that is to be followed by the decision maker. In the PSI case, with the market research study, a decision strategy is a rule that recommends a particular decision based on whether the market research report is favorable or unfavorable. We will employ a decision tree analysis to find the optimal decision strategy for PSI.

Figure 22.7 shows the decision tree for the PSI leasing problem provided that a market research study is conducted. Note that as you move from left to right, the tree shows the natural or logical order that will occur in the decision-making process. First, the firm will obtain the market research report indicator (I_1 or I_2); then a decision (d_1, d_2, or d_3) will be made; finally, the state of nature (s_1 or s_2) will occur. The decision and the state of nature combine to provide the final profit or payoff.

Using decision tree terminology, we have now introduced an *indicator node*, node 1, and *indicator branches*, I_1 and I_2. Since the branches emanating from indicator nodes are not under the control of the decision maker but are determined by chance, these nodes are depicted by a circle just like the state-of-nature nodes. We see that nodes 2 and 3 are decision nodes, while nodes 4, 5, 6, 7, 8, and 9 are state-of-nature nodes. For decision nodes the decision maker must select the specific branch d_1, d_2, or d_3 that will be taken.

FIGURE 22.7

PSI Decision Tree Incorporating the Results of the Market Research Study (Payoffs in $)

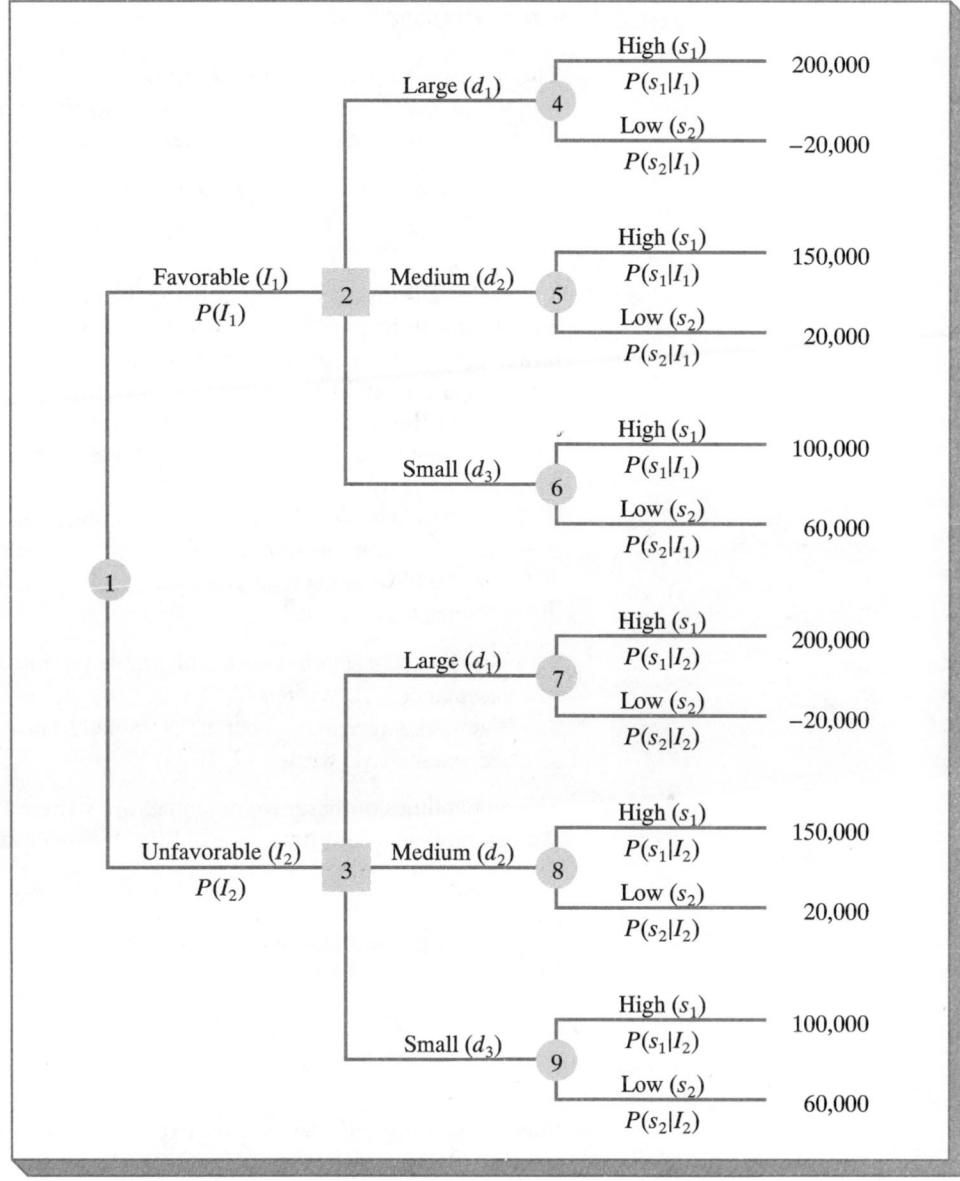

Selecting the best decision branch is equivalent to making the best decision. However, since the indicator and state-of-nature branches are not controlled by the decision maker, the specific branch leaving an indicator or a state-of-nature node will depend on the probability associated with the branch. Thus, before we can carry out an analysis of the decision tree and develop a decision strategy, we must compute the probability of each indicator branch and the probability of each state-of-nature branch. Note from the decision tree that the state-of-nature branches occur *after* the indicator branches. Thus, when we attempt to compute state-of-nature branch probabilities, we will need to consider which indicator was previously observed. That is, we will express the state-of-nature probabilities in terms of the probability of state of nature s_j *given* that indicator I_k was observed. Thus, all state-of-nature probabilities will be expressed in a $P(s_j \mid I_k)$ form.

Computing Branch Probabilities

The prior probabilities for the states of nature in the PSI leasing problem were given as $P(s_1) = .3$ and $P(s_2) = .7$. In Section 22.6 we identified the relationships between the market research indicators and states of nature with the conditional probabilities

$$P(I_1 \mid s_1) = .8 \qquad P(I_2 \mid s_1) = .2$$

$$P(I_1 \mid s_2) = .1 \qquad P(I_2 \mid s_2) = .9$$

To develop a decision strategy utilizing the decision tree in Figure 22.7, we need indicator branch probabilities $P(I_k)$ and state-of-nature branch probabilities $P(s_j \mid I_k)$. The problem now facing us is to determine how to use the given prior probability estimates $P(s_j)$ and conditional probability estimates $P(I_k \mid s_j)$ to calculate the branch probabilities $P(I_k)$ and $P(s_j \mid I_k)$. In this section we will show how the Bayesian revision process discussed in Chapter 4 and referred to in Figure 22.6 can be used to calculate the branch probabilities $P(I_k)$ and $P(s_j \mid I_k)$.

To see how this Bayesian procedure is applied and at the same time understand how the procedure works, let us look closely at the calculation of the indicator branch probability $P(I_1)$ for the PSI market research study. First, note that there are only two ways in which the outcome I_1 can occur:

1. The market research report is favorable (I_1) *and* the state of nature turns out to be high acceptance (s_1), written $(I_1 \cap s_1)$.
2. The market research report is favorable (I_1) *and* the state of nature turns out to be low acceptance (s_2), written $(I_1 \cap s_2)$.

The probabilities of these two outcomes are written $P(I_1 \cap s_1)$ and $P(I_1 \cap s_2)$, respectively. We can now add these two probabilities to obtain the following branch probability:

$$P(I_1) = P(I_1 \cap s_1) + P(I_1 \cap s_2) \tag{22.8}$$

The multiplication law of probability provides the following formulas for $P(I_1 \cap s_1)$ and $P(I_1 \cap s_2)$:

$$P(I_1 \cap s_1) = P(I_1 \mid s_1)P(s_1) \tag{22.9}$$

$$P(I_1 \cap s_2) = P(I_1 \mid s_2)P(s_2) \tag{22.10}$$

Finally, substituting the above expressions for $P(I_1 \cap s_1)$ and $P(I_1 \cap s_2)$ in Equation (22.8), we obtain

$$P(I_1) = P(I_1 \mid s_1)P(s_1) + P(I_1 \mid s_2)P(s_2) \tag{22.11}$$

Generalizing the above expression for any indicator branch probability, $P(I_k)$, and N states of nature, $s_1, s_2, \ldots, s_N$, we have

$$P(I_k) = P(I_k \mid s_1)P(s_1) + P(I_k \mid s_2)P(s_2) + \cdots + P(I_k \mid s_N)P(s_N) \tag{22.12}$$

or

$$P(I_k) = \sum_{j=1}^{N} P(I_k \mid s_j)P(s_j) \tag{22.13}$$

Returning to the PSI leasing problem with the two prior probabilities $P(s_1) = .3$ and $P(s_2) = .7$ and the conditional probabilities $P(I_1 \mid s_1) = .8$, $P(I_1 \mid s_2) = .1$,

$P(I_2 \mid s_1) = .2$, and $P(I_2 \mid s_2) = .9$, we can use Equation (22.13) to compute the two indicator branch probabilities. These calculations are as follows:

$$P(I_1) = P(I_1 \mid s_1)P(s_1) + P(I_1 \mid s_2)P(s_2)$$

$$= (.8)(.3) + (.1)(.7) = .31$$

and

$$P(I_2) = P(I_2 \mid s_1)P(s_1) + P(I_2 \mid s_2)P(s_2)$$

$$= (.2)(.3) + (.9)(.7) = .69$$

The above probabilities indicate that the probability of I_1, a favorable market research report, is .31 and the probability of I_2, an unfavorable market research report, is .69.

Now that we know the indicator branch probabilities, let us show how the Bayesian process enables us to compute the revised, or posterior, state-of-nature branch probabilities $P(s_j \mid I_k)$. We will illustrate this procedure by considering the state-of-nature branch probability $P(s_1 \mid I_1)$, the probability the market acceptance is high (s_1) given that the market research report is favorable (I_1). The fundamental conditional probability relationship as presented in Chapter 4 can be written

$$P(s_1 \mid I_1) = \frac{P(I_1 \cap s_1)}{P(I_1)} \qquad (22.14)$$

Using Equation (22.9) for $P(I_1 \cap s_1)$, we have

$$P(s_1 \mid I_1) = \frac{P(I_1 \mid s_1)P(s_1)}{P(I_1)} \qquad (22.15)$$

With known probabilities $P(I_1 \mid s_1) = .8$, $P(s_1) = .3$, and $P(I_1) = .31$, the revised state-of-nature probability, $P(s_1 \mid I_1)$, becomes

$$P(s_1 \mid I_1) = \frac{(.8)(.3)}{.31} = \frac{.24}{.31} = .7742$$

Recall that the prior probability of a high market acceptance was $P(s_1) = .3$. The preceding probability information now tells us that if the market research indicator is favorable, the probability of a high market acceptance should be revised to $P(s_1 \mid I_1) = .7742$.

Generalizing (22.15) for any state of nature s_j and any indicator I_k provides

$$P(s_j \mid I_k) = \frac{P(I_k \mid s_j)P(s_j)}{P(I_k)} \qquad (22.16)$$

Thus, we can use (22.16) to compute the revised or posterior state-of-nature branch probabilities. For example, the revised probability of low market acceptance s_2 given the market research indicator is favorable I_1 becomes

$$P(s_2 \mid I_1) = \frac{P(I_1 \mid s_2)P(s_2)}{P(I_1)} = \frac{(.1)(.7)}{.31} = \frac{.07}{.31} = .2258$$

Similar calculations for an unfavorable market research indicator I_2 will provide the revised state-of-nature branch probabilities $P(s_1 \mid I_2) = .0870$ and $P(s_2 \mid I_2) = .9130$. Figure 22.8 shows the PSI decision tree after all indicator and all revised state-of-nature branch probabilities have been computed.

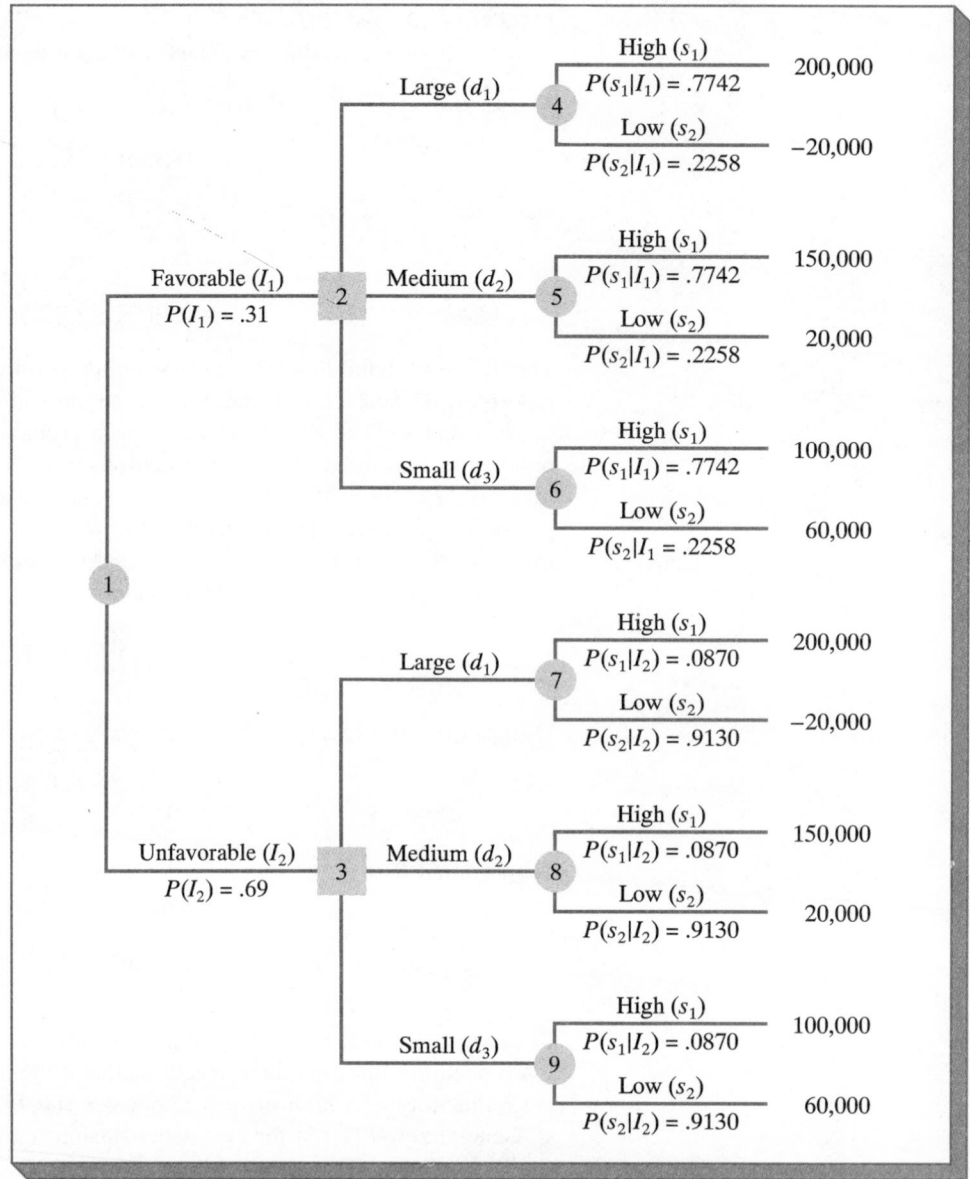

Although the above procedure can be used to compute branch probabilities, the calculations can become quite cumbersome as the problem size grows larger. Thus, to ease computations using Bayes' theorem, especially for large decision analysis problems, we present the following tabular procedure.

Computing Branch Probabilities: A Tabular Procedure

The procedure for computing the probabilities of the indicator and state-of-nature branches can be accomplished via the tabular approach for Bayes' theorem discussed in Chapter 4. First, for each indicator I_k we form a table consisting of the following five column headings:

- Column 1   States of nature s_j
- Column 2   Prior probabilities $P(s_j)$
- Column 3   Conditional probabilities $P(I_k \mid s_j)$
- Column 4   Joint probabilities $P(I_k \cap s_j)$
- Column 5   Posterior probabilities $P(s_j \mid I_k)$

Then, given any indicator I_k, the following procedure can be used to calculate $P(I_k)$ and the $P(s_j \mid I_k)$ values:

Step 1.   In column 1 list the states of nature appropriate to the problem being analyzed.

Step 2.   In column 2 enter the prior probability corresponding to each state of nature listed in column 1.

Step 3.   In column 3 enter the appropriate value of $P(I_k \mid s_j)$ for each state of nature specified in column 1.

Step 4.   To compute each entry in column 4, multiply each entry in column 2 by the corresponding entry in column 3.

Step 5.   Add the entries in column 4. The sum is the value of $P(I_k)$. For convenience, write the sum below column 4.

Step 6.   To compute each entry in column 5, divide the corresponding entry in column 4 by $P(I_k)$.

We will now use this procedure to compute $P(I_1)$ and the revised state-of-nature probabilities $P(s_j \mid I_1)$ for the PSI leasing problem.

Steps 1, 2, and 3.

s_j	$P(s_j)$	$P(I_1 \mid s_j)$	$P(I_1 \cap s_j)$	$P(s_j \mid I_1)$
s_1	.3	.8		
s_2	.7	.1		

Steps 4 and 5.

s_j	$P(s_j)$	$P(I_1 \mid s_j)$	$P(I_1 \cap s_j)$	$P(s_j \mid I_1)$
s_1	.3	.8	.24	
s_2	.7	.1	.07	
			$P(I_1) = .31$	

Step 6.

s_j	$P(s_j)$	$P(I_1 \mid s_j)$	$P(I_1 \cap s_j)$	$P(s_j \mid I_1)$
s_1	.3	.8	.24	.24/.31 = .7742
s_2	.7	.1	.07	.07/.31 = .2258
			$P(I_1) = .31$	

Note that $P(I_1)$, $P(s_1 \mid I_1)$, and $P(s_2 \mid I_1)$ are exactly the same as we calculated by applying Equations (22.13) and (22.16) directly. The preceding tabular computations could be repeated to compute $P(I_2)$ and the revised state-of-nature probabilities $P(s_j \mid I_2)$.

An Optimal Decision Strategy

Regardless of the method used to compute the branch probabilities, we can now use the branch probabilities and the expected value approach to arrive at the optimal decision for PSI. Working *backward* through the decision tree, we first compute the expected value at each state-of-nature node. That is, at each state-of-nature node the possible payoffs are weighted by their chance of occurrence. Thus, the expected values for nodes 4–9 are computed as follows:

$$EV(\text{node } 4) = .7742(200,000) + .2258(-20,000) = 150,324$$

$$EV(\text{node } 5) = .7742(150,000) + .2258(20,000) \quad = 120,646$$

$$EV(\text{node } 6) = .7742(100,000) + .2258(60,000) \quad = 90,968$$

$$EV(\text{node } 7) = .0870(200,000) + .9130(-20,000) = -860$$

$$EV(\text{node } 8) = .0870(150,000) + .9130(20,000) \quad = 31,310$$

$$EV(\text{node } 9) = .0870(100,000) + .9130(60,000) \quad = 63,480$$

Figure 22.9 shows the above calculations directly on the decision tree. Since the decision maker controls the branch leaving a decision node and since we are trying to maximize expected profits, the optimal decision at node 2 is d_1. Thus, since d_1 leads to an expected value of \$150,324, we say that EV(node 2) = \$150,324.

A similar analysis of decision node 3 shows that the optimal decision branch at this node is d_3. Thus, EV(node 3) becomes \$63,480, provided that the optimal decision of d_3 is made.

As a final step, we can continue working backward to the indicator node and establish its expected value. We see that since node 1 has probability branches, we cannot select the best branch. Rather, we must compute the expected value over all possible branches. Thus, we have

$$EV(\text{node } 1) = (.31)EV(\text{node } 2) + (.69)EV(\text{node } 3)$$

$$= .31(150,324) + .69(63,480) = \$90,402$$

The value of \$90,402 is viewed as the expected value of the optimal decision strategy when the market research study is used. In other words, it is the expected value using the sample information provided by the market research report.

Note that the final decision has not yet been determined. We will need to know the results of the market research study before deciding to lease a large system (d_1) or a small system (d_3). The results of the decision analysis at this point, however, have provided us with the following optimal *decision strategy* if the market research study is conducted:

If	Then
Report favorable (I_1)	Lease large system (d_1)
Report unfavorable (I_2)	Lease small system (d_3)

FIGURE 22.9   **Developing a Decision Strategy for the PSI Leasing Problem**

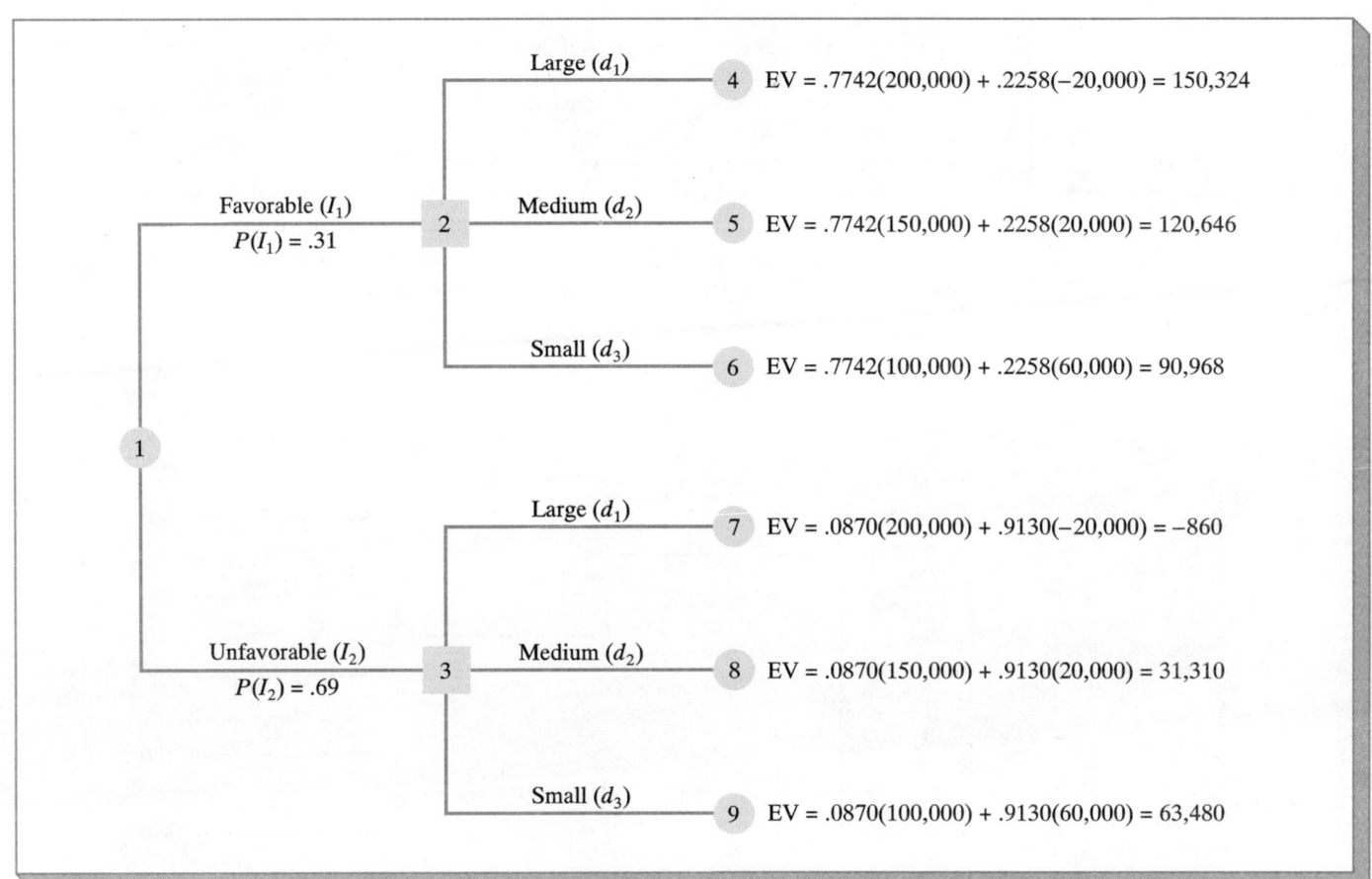

Thus, we have seen how the decision tree approach can be used to develop optimal decision strategies when sample information is available. While other decision analysis problems may not be as simple as the PSI leasing problem, the approach we have outlined is still applicable. First, draw a decision tree consisting of indicator, decision, and state-of-nature nodes and branches such that the tree describes the specific sequence of decisions and chance outcomes. Posterior probability calculations must be made to establish indicator and state-of-nature branch probabilities. Then, by working backward through the tree, computing expected values at state-of-nature and indicator nodes, and selecting the best decision branch at decision nodes, the analyst can determine an optimal decision strategy and its associated expected value.

❏   ❏   **Exercises**

Methods

SELF TEST ▶ **23.** Suppose that you are given a decision situation with three possible states of nature: s_1, s_2, and s_3. The prior probabilities are $P(s_1) = .2$, $P(s_2) = .5$, and $P(s_3) = .3$. Indicator information I is obtained, and it is known that $P(I \mid s_1) = .1$, $P(I \mid s_2) = .05$, and $P(I \mid s_3) = .2$. Compute the revised or posterior probabilities: $P(s_1 \mid I)$, $P(s_2 \mid I)$, and $P(s_3 \mid I)$.

TABLE 22.7

	s_1	s_2
d_1	15	10
d_2	10	12
d_3	8	20

24. The payoff table showing profit for a decision problem with two states of nature and three decision alternatives is presented in Table 22.7. The prior probabilities for s_1 and s_2 are $P(s_1) = .8$ and $P(s_2) = .2$.
   a. Using only the prior probabilities and the expected value approach, find the optimal decision.
   b. Use graphical sensitivity analysis to determine the values of the propability of state of nature s_1 for which each of the decision alternatives has the largest expected value.
   c. Find the EVPI.
   d. Suppose that some indicator information I is obtained with $P(I \mid s_1) = .2$ and $P(I \mid s_2) = .75$. Find the posterior probabilities $P(s_1 \mid I)$ and $P(s_2 \mid I)$. Recommend a decision alternative based on these probabilities.

25. Consider the following decision tree representation of a decision analysis problem with two indicators, two decision alternatives, and two states of nature:

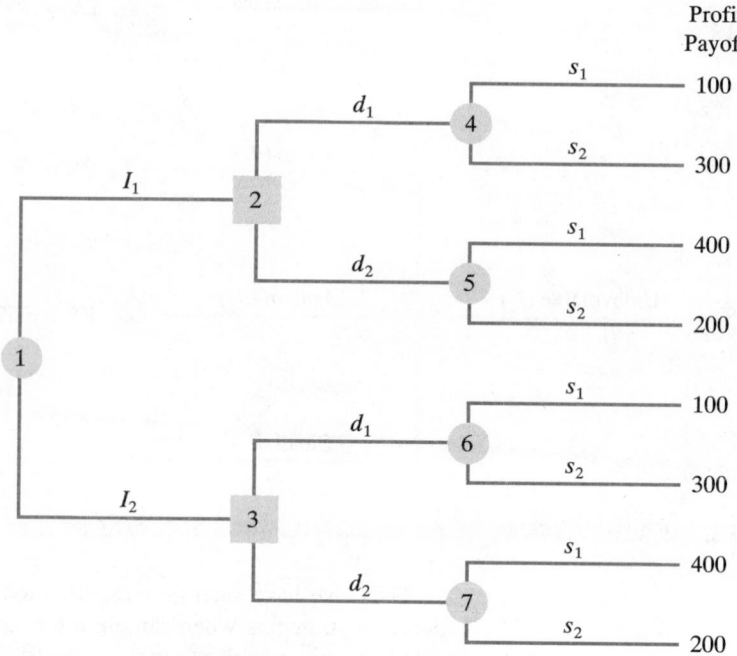

Assume that the following probability information is given:

$$P(s_1) = .4 \qquad P(I_1 \mid s_1) = .8 \qquad P(I_2 \mid s_1) = .2$$

$$P(s_2) = .6 \qquad P(I_1 \mid s_2) = .4 \qquad P(I_2 \mid s_2) = .6$$

   a. What are the values for $P(I_1)$ and $P(I_2)$?
   b. What are the values of $P(s_1 \mid I_1)$, $P(s_2 \mid I_1)$, $P(s_1 \mid I_2)$, and $P(s_2 \mid I_2)$?
   c. Use the decision tree approach, and determine the optimal decision strategy. What is the expected value of your solution?

22.8   Expected Value of Sample Information

In the PSI leasing problem, management now has a decision strategy of leasing the large computer system if the market research report is favorable and leasing the small computer system if the market research report is unfavorable. Since the additional information provided by the market research firm will result in an added cost for PSI in terms of the fee paid to the research firm, PSI management may question the value of this market research information.

The value of sample information is often measured by calculating what is referred to as the *expected value of sample information* (EVSI). For maximization problems,* the following holds.

Expected Value of Sample Information

$$\text{EVSI} = \begin{bmatrix} \text{Expected Value of the} \\ \text{Optimal Decision } \textit{with} \\ \text{Sample Information} \end{bmatrix} - \begin{bmatrix} \text{Expected Value of the} \\ \text{Optimal Decision } \textit{without} \\ \text{Sample Information} \end{bmatrix} \quad (22.17)$$

For PSI the market research information is considered the sample information. The decision tree calculations indicated that the expected value of the optimal decision with the market research information was \$90,402; thus, \$90,402 is the *expected value with sample information* (EVwSI). When sample information is not available, the expected value approach resulted in recommending decision alternative d_3 with an expected value of \$72,000; thus, \$72,000 is the *expected value without sample information* (EVwoSI). Using Equation (22.17), the expected value of the market research report is

$$\text{EVSI} = \$90,402 - \$72,000 = \$18,402$$

Thus, PSI should be willing to pay up to \$18,402 for the market research information.

Efficiency of Sample Information

In Section 22.5 we saw that the expected value of perfect information (EVPI) for the PSI leasing problem was \$30,000. While we never expected the market research report to obtain perfect information, we can use an *efficiency* measure to express the value of the report. With perfect information having an efficiency rating of 100%, the efficiency rating E for sample information is computed as follows.

Efficiency of Sample Information

$$E = \frac{\text{EVSI}}{\text{EVPI}} \times 100 \quad (22.18)$$

---

*In minimization problems, the expected value of the optimal decision with sample information will be less than or equal to the expected value of the optimal decision without sample information. Thus, in minimization problems,

$$\text{EVSI} = \begin{bmatrix} \text{Expected Value of the} \\ \text{Optimal Decision } \textit{without} \\ \text{Sample Information} \end{bmatrix} - \begin{bmatrix} \text{Expected Value of the} \\ \text{Optimal Decision } \textit{with} \\ \text{Sample Information} \end{bmatrix}$$

For the PSI leasing problem,

$$E = \frac{18,402}{30,000} \times 100 = 61.3\%$$

In other words, the information from the market research firm is 61.3% as efficient as perfect information.

Low efficiency ratings for sample information might lead the decision maker to look for other types of information. On the other hand, high efficiency ratings indicate that the sample information is almost as good as perfect information, and additional sources of information should not be worthwhile.

NOTES & COMMENTS

1. The use of microcomputer software packages for decision analysis is becoming commonplace. The ARBORIST, a software package available from Texas Instruments, allows the user to develop a graph of the decision tree on the screen; the package will then perform all the decision analysis calculations. Another product, SUPERTREE, from SDG Decision Systems, requires the user to specify the tree in a manner similar to that used in PERT/CPM problems; that is, the user provides the system with the nodes and their immediate predecessors, and the software package develops the resulting decision tree. Although packages such as the ARBORIST and SUPERTREE are not needed for problems as small as those introduced in this chapter, computer support is necessary for larger decision problems.

2. The expected value without perfect information (EVwoPI) defined in Section 22.5 and the expected value without sample information (EVwoSI) are always equal. This value simply represents the expected value of the best decision using the prior probabilities.

☐ ☐ **Exercises**

Methods

SELF TEST

26. For a particular decision problem, suppose that the expected value of the optimal decision with sample information is $45,000 and that the expected value of the optimal decision without sample information is $37,000. The expected value with perfect information is $60,000.
   a. Compute EVSI.
   b. What is the efficiency of the sample information?
   c. If the study required to obtain the sample information had a cost of $10,000, what would you recommend?

27. Refer to Exercise 25.
   a. What is your decision without the indicator information?
   b. What is the expected value of the indicator or sample information EVSI?
   c. What is the expected value of perfect information EVPI?
   d. What is the efficiency of the indicator information?

Applications

SELF TEST 28. The payoff table (profit in $1000s) for Hale's TV Productions (Exercises 7 and 19) is as follows:

			States of Nature		
			s_1	s_2	s_3
Decision Alternatives	Produce pilot	d_1	-100	50	150
	Sell to competitor	d_2	100	100	100
Probability of states of nature			.2	.3	.5

For a consulting fee of $2500, an agency will review the plans for the comedy series and indicate the overall chances of a favorable network reaction to the series. If the special agency review results in a favorable (I_1) or an unfavorable (I_2) evaluation, what should Hale's decision strategy be? Assume that Hale believes that the following conditional probabilities are realistic appraisals of the agency's evaluation accuracy:

$$P(I_1 \mid s_1) = .3 \qquad P(I_2 \mid s_1) = .7$$

$$P(I_1 \mid s_2) = .6 \qquad P(I_2 \mid s_2) = .4$$

$$P(I_1 \mid s_3) = .9 \qquad P(I_2 \mid s_3) = .1$$

a. Show the decision tree for this problem.
b. What is the recommended decision strategy and the expected value, assuming that the agency information is obtained?
c. What is the EVSI? Is the $2500 consulting fee worth the information? What is the maximum that Hale should be willing to pay for the consulting information?

29. McHuffter Condominiums (Exercises 4, 8, and 20) is conducting a survey that will help evaluate the demand for the new condominium development. McHuffter's payoff table (profit in $1000s) is as follows:

			States of Nature		
			Low	Medium	High
			s_1	s_2	s_3
Decision Alternatives	Small	d_1	400	400	400
	Medium	d_2	100	600	600
	Large	d_3	-300	300	900
Probability of states of nature			.20	.35	.45

TABLE 22.8

	$P(I_k \mid s_k)$		
	I_1	I_2	I_3
s_1	.6	.3	.1
s_2	.4	.4	.2
s_3	.1	.4	.5

The survey will result in three indicators of demand (weak (I_1), average (I_2), or strong (I_3)), where the conditional probabilities are as shown in Table 22.8.
a. What is McHuffter's optimal strategy?
b. What is the value of the survey information?
c. What are the EVPI and the efficiency of the survey information?

30. The payoff table ($) for Martin's Service Station (Exercises 9 and 21) is as follows:

			Snowfall		
			Heavy s_1	Moderate s_2	Light s_3
Decision Alternatives	Purchase snowplow	d_1	7000	2000	−9000
	Do not invest	d_2	0	0	0
	Purchase snowplow with blade	d_3	3500	1000	−1500
Probability of states of nature			.4	.3	.3

Suppose that Martin decides to wait to check the September temperature pattern before making a final decision. Estimates of the probabilities associated with an unseasonably cold September (I_1) are as follows: $P(I_1 \mid s_1) = .30$, $P(I_1 \mid s_2) = .20$, $P(I_1 \mid s_3) = .05$. If Martin observes an unseasonably cold September, what is the recommended decision? If Martin does not observe an unseasonably cold September (I_2), what is the recommended decision?

31. Suppose that Joseph Software, Inc., in Exercise 10, can hire an independent consultant to review their ideas for the new software. For a fee of $5000, the consultant will make a recommendation as to whether or not JSI should develop a prototype. Based on previous experience with this consultant, JSI has assigned the following conditional probabilities:

$$P(I_1 \mid s_1) = .2$$

$$P(I_1 \mid s_2) = .6$$

$$P(I_1 \mid s_3) = .9$$

where

I_1 = recommendation to develop a prototype

s_1 = failure

s_2 = moderate success

s_3 = major success

Should JSI hire the consultant? Explain.

32. Milford Trucking (Exercises 14 and 22) has the following payoff table:

			Return Shipment from Detroit s_1	No Return Shipment from Detroit s_2
Shipment	St. Louis	d_1	2000	2000
	Detroit	d_2	2500	1000
Probabilities			.40	.60

a. Milford can phone a Detroit truck dispatch center and determine if the general Detroit shipping activity is busy (I_1) or slow (I_2). If the report is busy, the chances of obtaining a return shipment will increase. Suppose the following conditional probabilities are given:

$$P(I_1 \mid s_1) = .6 \qquad P(I_2 \mid s_1) = .4$$

$$P(I_1 \mid s_2) = .3 \qquad P(I_2 \mid s_2) = .7$$

What should Milford do?

b. If the Detroit report is busy (I_1), what is the probability that Milford will obtain a return shipment if it makes the trip to Detroit?

c. What is the efficiency of the phone information?

22.9   Other Topics in Decision Analysis

In discussing decision analysis, we have considered only situations where there are a finite number of states of nature. The next step would be to consider situations where the states of nature are so numerous that it would be impractical, if not impossible, to treat the states of nature as a discrete random variable consisting of a finite number of values. For example, let us suppose that we are attempting to price a new product and are concerned with the potential sales volume we might experience at different prices. We might think of the states of nature as being all possible sales volumes from 0 to 200,000 units. Although there are a finite number of states of nature, no units sold, 1 unit sold, and so on, we recognize that attempting to deal with this large number of possible states of nature is extremely impractical. The solution procedure that is used in such circumstances is to treat the state of nature as a continuous random variable. For example, perhaps a reasonable approximation of the state of nature (i.e., sales volume) is that sales are normally distributed with a mean of 100,000 units and a standard deviation of 25,000 units. Although decision analysis techniques have been developed to handle such situations, we shall not attempt to present these procedures in this chapter.

Another area of decision analysis is concerned with alternative measures of the payoffs. In the PSI leasing problem we used profit in dollars as the measure of the payoff. Then the expected value approach was used to select the best decision. While decision analysis applications are often based on payoffs measured in monetary values, perhaps there are other measures of payoff that should be used.

For example, let us consider a situation in which we have two alternative investments. Investment A yields a certain profit of $50,000. Investment B yields a 50% chance of making $100,002 but also a 50% chance of making nothing. Thus the expected value for B is

$$E(B) = .5(100,002) + .5(0)$$

$$= \$50,001$$

Using the expected value approach with profits in dollars as the measure of the payoff, we would select decision alternative B. However, many decision makers, if not most, would select alternative A; that is, some decision makers prefer the no-risk $50,000 profit over the higher expected value but risky 50–50 chance of a $100,002 profit. In decision theory terminology, if alternative A is preferred, we say that alternative A has a higher *utility*, where utility is a measure of the decision maker's preference considering monetary value as well as the risk involved. Ideally, we would like to measure payoffs in terms of the decision maker's utility and select optimal decision strategies based on expected utility rather than expected monetary value. More advanced books on decision analysis present the details associated with this approach to decision making.

Summary

In this chapter we have illustrated how decision analysis can be used to solve problems with a limited number of decision alternatives and a limited number of possible states of nature. The goal of decision analysis is to identify the best decision alternative given an uncertain or risk-filled pattern of future events (i.e., states of nature).

We presented three approaches to decision making without probabilities and discussed the use of the expected value approach for solving problems with probabilities. Then we showed how additional information about the states of nature can be used to revise or update the probability estimates and develop an optimal decision strategy for the problem. The notions of expected value of sample information, expected value of perfect information, and efficiency of information were used to evaluate the contribution of the sample information.

Glossary

States of nature Uncontrollable future events that can affect the payoff associated with a decision.

Payoff The outcome measure, such as profit, cost, time, and so on. Each combination of a decision alternative and a state of nature has an associated payoff.

Payoff table A tabular representation of the payoffs for a decision problem.

Optimistic approach An approach to choosing a decision alternative without using probabilities. For a maximization problem, it leads to choosing the alternative corresponding to the largest payoff; for a minimization problem, it leads to choosing the alternative corresponding to the smallest payoff.

Conservative approach An approach to choosing a decision alternative without using probabilities. For a maximization problem, it leads to choosing the alternative that maximizes the minimum payoff; for a minimization problem, it leads to choosing the alternative that minimizes the maximum payoff.

Minimax regret An approach to choosing a decision alternative without using probabilities. For each alternative, the maximum regret is computed. This approach leads to choosing the alternative that minimizes the maximum regret.

Opportunity loss or regret The amount of loss (lower profit or higher cost) due to not making the best decision for each state of nature.

Decision tree A graphical representation of the decision-making situation from decision to state-of-nature to payoff.

Nodes The intersection or junction points of the decision tree.

Branches Lines or arcs connecting nodes of the decision tree.

Expected value For a decision alternative, it is the weighted average of the payoffs. The weights are the state-of-nature probabilities.

Expected value of perfect information (EVPI) The expected value of information that would tell the decision maker exactly which state of nature was going to occur (i.e., perfect information).

Prior probabilities The probabilities of the states of nature prior to obtaining sample information.

Posterior (revised) probabilities The probabilities of the states of nature after using Bayes' theorem to adjust the prior probabilities based on given indicator information.

Indicators Information about the states of nature. An indicator may be the result of a sample.

Bayesian revision The process of revising prior probabilities to create the posterior probabilities based on sample information.

Expected value of sample information (EVSI) The difference between the expected value of an optimal strategy based on new information and the "best" expected value without any new information. It is a measure of the value of new information.

Efficiency The ratio of EVSI to EVPI; perfect information is 100% efficient.

Key Formulas

Opportunity Loss or Regret

$$R(d_i, s_j) = V^*(s_j) - V(d_i, s_j) \tag{22.1}$$

Expected Value of Decision Alternative d_i

$$EV(d_i) = \sum_{j=1}^{N} P(s_j)V(d_i, s_j) \tag{22.4}$$

Indicator Branch Probability

$$P(I_k) = \sum_{j=1}^{N} P(I_k \mid s_j)P(s_j) \tag{22.13}$$

State-of-Nature Branch Probability

$$P(s_j \mid I_k) = \frac{P(I_k \mid s_j)P(s_j)}{P(I_k)} \tag{22.16}$$

Efficiency of Sample Information

$$E = \frac{EVSI}{EVPI}(100) \tag{22.18}$$

❑ ❑ Supplementary Exercises

33. To save on gasoline expenses, Rona and Jerry agreed to form a carpool for traveling to and from work. After limiting the travel routes to two alternatives, Rona and Jerry could not agree on the best way to travel to work. Jerry preferred the expressway, since it was usually the fastest; however, Rona pointed out that traffic jams on the expressway sometimes led to long delays. Rona preferred the somewhat longer but more consistent Queen City Avenue. While Jerry still preferred the expressway, he agreed with Rona that they should take Queen City Avenue if the expressway had a traffic jam. Unfortunately, they do not know the state of the expressway ahead of time. The following payoff table provides the one-way time estimates in minutes for traveling to or from work:

			States of Nature	
			Expressway Open s_1	*Expressway Jammed* s_2
Route	*Expressway*	d_1	25	45
	Queen City Avenue	d_2	30	30

a. After driving to work on the expressway for 1 month (20 days), they found the expressway jammed three times. Assuming that these days are representative of future days, should they continue to use the expressway for traveling to work? Explain.

b. Use graphical sensitivity analysis to determine the values of the probability of state of nature s_1 for which d_1 has the best expected value.

c. Would it make sense to not use the expected value approach for this particular problem? Explain.

After a period of time Rona and Jerry noted that the weather seemed to affect the traffic conditions on the expressway. They identified three weather conditions (indicators) with the following conditional probabilities:

$$I_1 = \text{clear} \qquad I_2 = \text{overcast} \qquad I_3 = \text{rain}$$

$$P(I_1 \mid s_1) = .8 \qquad P(I_2 \mid s_1) = .2 \qquad P(I_3 \mid s_1) = 0$$

$$P(I_1 \mid s_2) = .1 \qquad P(I_2 \mid s_2) = .3 \qquad P(I_3 \mid s_2) = .6$$

d. Show the decision tree for the problem of traveling to work.
e. What is the optimal decision strategy and the expected travel time?
f. What is the efficiency of the weather information?

34. The Gorman Manufacturing Company must decide whether it should purchase a component from a supplier or manufacture the component at its Milan, Michigan, plant. If demand is high, it would be to Gorman's advantage to manufacture the component. However, if demand is low, Gorman's unit manufacturing cost will be high due to underutilization of equipment. The projected profit in thousands of dollars for Gorman's make-or-buy decision is shown below:

		Demand		
		Low	Medium	High
Decision Alternatives	Manufacture component	−20	40	100
	Purchase component	10	45	70

The states of nature have the following probabilities: $P(\text{low demand}) = .35$, $P(\text{medium demand}) = .35$, and $P(\text{high demand}) = .30$.
a. Use a decision tree to recommend a decision.
b. Use EVPI to determine whether Gorman should attempt to obtain a better estimate of demand.

A test market study of the potential demand for the product is expected to report either a favorable (I_1) or unfavorable (I_2) condition. The relevant conditional probabilities are as follows:

$$P(I_1 \mid s_1) = .10 \qquad P(I_2 \mid s_1) = .90$$

$$P(I_1 \mid s_2) = .40 \qquad P(I_2 \mid s_2) = .60$$

$$P(I_1 \mid s_3) = .60 \qquad P(I_2 \mid s_3) = .40$$

c. What is the probability that the market research report will be favorable?
d. What is Gorman's optimal decision strategy?
e. What is the expected value of the market research information?
f. What is the efficiency of the information?

35. A firm produces a perishable food product at a cost of $10 per case. The product sells for $15 per case. For planning purposes, the company is considering possible demands of 100, 200, or 300 cases. If the demand is less than production, the excess production is lost. If demand is more than production, the firm, in an attempt to maintain a good service image, will satisfy the excess demand with a special production run at a cost of $18 per case. The product, however, always sells at $15 per case.
a. Set up the payoff table for this problem.
b. If $P(100) = .2$, $P(200) = .2$, and $P(300) = .6$, should the company produce 100, 200, or 300 cases?
c. What is the EVPI?

36. Sealcoat, Inc., has a contract with one of its customers to supply a unique liquid chemical product that will be used by the customer in the manufacture of a lubricant for airplane engines.

Because of the chemical process used by Sealcoat, batch sizes for the liquid chemical product must be 1000 pounds. The customer has agreed to adjust manufacturing to the full batch quantities and will order either one, two, or three batches every 3 months. Since an aging process of 1 month exists for the product, Sealcoat will have to make its production (how much to make) decision before the customer places an order. Thus, Sealcoat can list the product demand alternatives of 1000, 2000, or 3000 pounds, but the exact demand is unknown.

Sealcoat's manufacturing costs are $150 per pound, and the product sells at the fixed contract price of $200 per pound. If the customer orders more than Sealcoat has produced, Sealcoat has agreed to absorb the added cost of filling the order by purchasing a higher quality substitute product from another chemical firm. The substitute product, including transportation expenses, will cost Sealcoat $240 per pound. Since the product cannot be stored more than 2 months without spoilage, Sealcoat cannot inventory excess production until the customer's next 3-month order. Therefore, if the customer's current order is less than Sealcoat has produced, the excess production will be reprocessed and is valued at $50 per pound.

The inventory decision in this problem is how much should Sealcoat produce given the above costs and the possible demands of 1000, 2000, or 3000 pounds? Based on historical data and an analysis of the customer's future demands, Sealcoat has assessed the probability distribution for demand shown in Table 22.9.

a. Develop a payoff table for the Sealcoat problem.

b. How many batches should Sealcoat produce every 3 months?

c. How much of a discount should Sealcoat be willing to allow the customer for specifying in advance exactly how many batches will be purchased?

Sealcoat has identified a pattern in the demand for the product based on the customer's previous order quantity. Let

$$I_1 = \text{customer's last order was 1000 pounds}$$

$$I_2 = \text{customer's last order was 2000 pounds}$$

$$I_3 = \text{customer's last order was 3000 pounds}$$

The conditional probabilities are as follows:

$$P(I_1 \mid s_1) = .10 \qquad P(I_2 \mid s_1) = .40 \qquad P(I_3 \mid s_1) = .50$$

$$P(I_1 \mid s_2) = .22 \qquad P(I_2 \mid s_2) = .68 \qquad P(I_3 \mid s_2) = .10$$

$$P(I_1 \mid s_3) = .80 \qquad P(I_2 \mid s_3) = .20 \qquad P(I_3 \mid s_3) = .00$$

d. Develop an optimal decision strategy for Sealcoat.

e. What is the EVSI?

f. What is the efficiency of the information for the most recent order?

37. A quality-control procedure involves 100% inspection of parts received from a supplier. Historical records indicate that the defective rates shown in Table 22.10 have been observed. The cost to inspect 100% of the parts received is $250 for each shipment of 500 parts. If the shipment is not 100% inspected, defective parts will cause rework problems later in the production process. The rework cost is $25 for each defective part.

a. Complete the following payoff table, where the entries represent the total cost of inspection and reworking:

TABLE 22.9

Demand	Probability
1000	.3
2000	.5
3000	.2

TABLE 22.10

Percent Defective	Probability
0	.15
1	.25
2	.40
3	.20

		Percent Defective			
		0	*1*	*2*	*3*
Inspection	*100% inspection*	$250	$250	$250	$250
	No inspection				

b. The plant manager is considering eliminating the inspection process to save the $250 inspection cost per shipment. Do you support this action? Use expected value to justify your answer.

c. Show the decision tree for this problem.

d. Suppose that a sample of 5 parts is selected from the shipment and 1 defect is found. Let $I = 1$ defect in a sample of 5. Use the binomial probability distribution to compute $P(I \mid s_1)$, $P(I \mid s_2)$, $P(I \mid s_3)$, and $P(I \mid s_4)$ where the state of nature identifies the value for p. The binomial probability function is as follows:

$$f(x) = \frac{n!}{x!(n - x)!} \, p^x(1 - p)^{n - x}$$

where

$$n = \text{the sample size}$$

$$x = \text{the number of defects}$$

$$p = \text{the proportion defective}$$

In this problem, $n = 5$, $x = 1$, and $p = 0$, .01, .02, and .03.

e. If I occurs, what are the revised probabilities for the states of nature?

f. Should the entire shipment be 100% inspected whenever 1 defect is found in a sample of size 5?

g. What is the cost savings associated with the sample information?

38. A food processor considers daily production runs of 100, 200, and 300 cases. Possible demands for the product are 100, 200, and 300 cases. The payoff table is as follows:

			Demand		
			100	200	300
			s_1	s_2	s_3
	100	d_1	500	200	−100
Production	200	d_2	−400	800	700
	300	d_3	−1000	−200	1600

a. If $P(s_1) = .2$, $P(s_2) = .2$, and $P(s_3) = .6$, what is your recommended production quantity?

b. On some days the firm receives phone calls for advance orders, and on some days it does not. Suppose that I_1 = advance orders are received and I_2 = no advance orders are received. If $P(I_2 \mid s_1) = .8$, $P(I_2 \mid s_2) = .4$, and $P(I_2 \mid s_3) = .1$, what is your recommended production quantity for days the company does not receive any advance orders?

39. The research and development manager for Beck Company is trying to decide whether to fund the development of a new lubricant. It is assumed that the project will be either a major technical success, a minor success, or a failure. The company has estimated that the value of a major success is $150,000, since the lubricant can be used in a number of products the company is making. If the project is a minor success, its value is $10,000, since Beck feels that the knowledge gained will benefit some other ongoing projects. If the project is a failure, it will cost the company $100,000.

Based on the opinion of the scientists involved and the manager's own subjective assessment, the assigned prior probabilities are as follows:

$$P(\text{major success}) = .15$$

$$P(\text{minor success}) = .45$$

$$P(\text{failure}) = .40$$

a. Using the expected value approach, should the project be funded?

b. Suppose that a group of expert scientists from a research institute could be hired as consultants to study the project and make a recommendation. If this study will cost $30,000, should the Beck Company consider hiring the consultants?

Suppose that an experiment can be conducted to shed some light on the technical feasibility of the project. There are three possible outcomes for the experiment:

I_1 = prototype lubricant works well at all temperatures
I_2 = prototype lubricant works well only at temperatures above 10°F
I_3 = prototype lubricant does not work well at any temperature

Suppose that we can determine the following conditional probabilities:

$P(I_1 \mid \text{major success}) = .70$ $P(I_2 \mid \text{major success}) = .25$ $P(I_3 \mid \text{major success}) = .05$

$P(I_1 \mid \text{minor success}) = .10$ $P(I_2 \mid \text{minor success}) = .70$ $P(I_3 \mid \text{minor success}) = .20$

$P(I_1 \mid \text{failure}) = .10$ $P(I_2 \mid \text{failure}) = .30$ $P(I_3 \mid \text{failure}) = .60$

c. Assuming that the experiment is conducted and the prototype lubricant works well at all temperatures, should the development project be funded?

d. Assuming that the experiment is conducted and the prototype lubricant works well only at temperatures above 10°F, should the project be funded?

e. Develop a decision strategy that Beck's R&D manager can use to recommend a funding decision based on the outcome of the experiment.

f. Find the EVSI for the experiment. How efficient is the information in the experiment?

Appendixes

A
References and Bibliography

B
Tables

C
Summation Notation

D
Data Disk and Its Use

E
Answers to Even-Numbered Exercises

F
Solutions to Self-Test Exercises

APPENDIX

References and Bibliography

General

Freedman, D., R. Pisani, and R. Purves, *Statistics,* New York, W. W. Norton, 1978.

Freund, J. E., and R. E. Walpole, *Mathematical Statistics,* 4th ed., Englewood Cliffs, N.J., Prentice-Hall, 1987.

Hoaglin, D. C., F. Mosteller, and J. W. Tukey, *Understanding Robust and Exploratory Data Analysis,* New York, Wiley, 1983.

Hogg, R. V., and A. T. Craig, *Introduction to Mathematical Statistics,* 4th ed., New York, Macmillian, 1978.

Mood, A. M., F. A. Graybill, and D. C. Boes, *Introduction to the Theory of Statistics,* 3rd ed., New York, McGraw-Hill, 1974.

Neter, J., W. Wasserman, and G. A. Whitmore, *Applied Statistics,* 3rd ed., Boston, Allyn & Bacon, 1987.

Roberts, H., *Data Analysis for Managers,* Redwood City, Calif., Scientific Press, 1988.

Ryan, T. A., B. L. Joiner, and B. F. Ryan, *Minitab Handbook,* 2nd ed., Boston, PWS-Kent, 1985.

Tanur, J. M., et al., *Statistics: A Guide to the Unknown,* 2nd ed., San Francisco, Holden-Day, 1978.

Tukey, J. W., *Exploratory Data Analysis,* Reading, Mass., Addison-Wesley, 1977.

Winkler, R. L., and W. L. Hays, *Statistics: Probability, Inference, and Decision,* 2nd ed., New York, Holt, Rinehart & Winston, 1975.

Probability

Barr, D. R., and P. W. Zehna, *Probability: Modeling Uncertainty,* Reading, Mass., Addison-Wesley, 1983.

Feller, W., *An Introduction to Probability Theory and Its Applications,* Vol. I, 3rd ed., New York, Wiley, 1968.

Feller, W., *An Introduction to Probability Theory and Its Applications,* Vol. II, 2nd ed., New York, Wiley, 1971.

Hoel, P. G., S. C. Port, and C. J. Stone, *Introduction to Probability Theory,* Boston, Houghton Mifflin, 1971.

Mendenhall, W., R. L. Scheaffer, and D. Wackerly, *Mathematical Statistics with Applications,* 4th ed., Boston, PWS-Kent, 1990.

Parzen, E., *Modern Probability Theory and Its Applications,* New York, Wiley, 1960.

Wadsworth, G. P., and J. G. Bryan, *Applications of Probability and Random Variables,* 2nd ed., New York, McGraw-Hill, 1974.

Zehna, P. W., *Probability Distributions and Statistics,* Boston, Allyn & Bacon, 1970.

Experimental Design

Anderson, V. L., and R. A. McLean, *Design of Experiments: A Realistic Approach,* New York, Marcel Dekker, 1974.

Box, G. E. P., W. G. Hunter, and J. S. Hunter, *Statistics for Experimenters,* New York, Wiley, 1978.

Cochran, W. G., and G. M. Cox, *Experimental Designs,* 2nd ed., New York, Wiley, 1957.

Hicks, C. R., *Fundamental Concepts in the Design of Experiments,* 3rd ed., New York, Holt, Rinehart & Winston 1982.

Maxwell, S. E. and H. D. Delaney, *Designing Experiments and Analyzing Data,* Belmont, CA, Wadsworth, 1990.

Mendenhall, W., *Introduction to Linear Models and the Design and Analysis of Experiments,* Belmont, Calif., Wadsworth, 1968.

Montgomery, D. C., *Design and Analysis of Experiments,* New York, Wiley, 1976.

Winer, B. J., *Statistical Principles in Experimental Design,* 2nd ed., New York, McGraw-Hill, 1971.

Regression Analysis

Belsley, D. A., E. Kuh, and R. Welsch, *Regression Diagnostics: Identifying Influential Data and Sources of Collinearity,* New York, Wiley, 1980.

Chatterjee, S., and B. Price, *Regression Analysis by Example,* New York, Wiley, 1978.

Cook, R. D., and S. Weisberg, *Residuals and Influence in Regression,* New York, Chapman and Hall, 1982.

Daniel, C., and F. Wood, *Fitting Equations to Data,* 2nd ed., New York, Wiley, 1980.

Draper, N. R., and H. Smith, *Applied Regression Analysis,* 2nd ed., New York, Wiley, 1981.

Gunst, R. F., and R. L. Mason, *Regression Analysis and Its Application: A Data-Oriented Approach,* New York, Marcel Dekker, 1980.

Kleinbaum, D. G., and L. L. Kupper, *Applied Regression Analysis and Other Multivariable Methods,* North Scituate, Mass., Duxbury Press, 1978.

Mendenhall, W., *Introduction to Linear Models and the Design and Analysis of Experiments,* Belmont, Calif., Wadsworth, 1968.

Mosteller, F., and J. W. Tukey, *Data Analysis and Regression: A Second Course in Statistics,* Reading, Mass., Addison-Wesley, 1977.

Neter, J., W. Wasserman, and M. H. Kutner, *Applied Linear Statistical Models,* 2nd ed., Homewood, Ill., Richard D. Irwin, 1985.

Weisberg, S., *Applied Linear Regression,* 2nd ed., New York, Wiley, 1985.

Wesolowsky, G. O., *Multiple Regression and Analysis of Variance,* New York, Wiley, 1976.

Wonnacott, T. H., and R. J. Wonnacott, *Regression: A Second Course in Statistics,* New York, Wiley, 1981.

Index Numbers

U.S. Dept. of Commerce *Survey of Current Business,* December, 1991.

U.S. Department of Labor, Bureau of Labor Statistics, *CPI Detailed Report,* October, 1991.

U.S. Department of Labor, *Producer Price Indexes,* Data for December, 1991.

Forecasting

Bowerman, B. L., and R. T. O'Connell, *Time Series Forecasting,* 2nd ed., Boston, PWS-Kent, 1987.

Box, G. E. P., and G. M. Jenkins, *Time Series Analysis, Forecasting and Control,* rev. ed., San Francisco, Holden-Day, 1976.

Brown, R. G., *Smoothing, Forecasting, and Prediction,* Englewood Cliffs, N.J., Prentice-Hall, 1963.

Gilchrist, W. G., *Statistical Forecasting,* New York, Wiley, 1976.

Makridakis, S., and S. C. Wheelwright, *Forecasting: Methods and Applications,* New York, Wiley, 1978.

Nelson, C. R., *Applied Time Series Analysis for Managerial Forecasting,* San Francisco, Holden-Day, 1973.

Thomopoulos, N. T., *Applied Forecasting Methods,* Englewood Cliffs, N.J., Prentice-Hall, 1980.

Wheelwright, S. C., and S. Makridakis, *Forecasting Methods for Management,* 3rd ed., New York, Wiley, 1980.

Nonparametric Methods

Conover, W. J., *Practical Nonparametric Statistics,* 2nd ed., New York, Wiley, 1980.

Daniel, W. W., *Applied Nonparametric Statistics,* Boston, Houghton Mifflin, 1978.

Gibbons, J. D., *Nonparametric Statistical Inference,* New York, McGraw-Hill, 1971.

Gibbons, J. D., I. Olkin, and M. Sobel, *Selecting and Ordering Populations: A New Statistical Methodology,* New York, Wiley, 1977.

Hollander, M., and D. A. Wolfe, *Nonparametric Statistical Methods,* New York, Wiley, 1973.

Lehmann, E. L., *Nonparametrics: Statistical Methods Based on Ranks,* San Francisco, Holden-Day, 1975.

Mosteller, F., and R. E. K. Rourke, *Sturdy Statistics,* Reading, Mass., Addison-Wesley, 1973.

Siegel, S., *Nonparametric Statistics for the Behavioral Sciences,* New York, McGraw-Hill, 1956.

Quality Control

Deming, W. E., *Quality, Productivity, and Competitive Position,* Cambridge, Mass., MIT Center for Advanced Engineering Study, 1982.

Duncan, A. J., *Quality Control and Industrial Statistics,* 5th ed. Homewood, Ill., Irwin, 1986.

Evans, J. R., and W. M. Lindsay, *The Management and Control of Quality,* St. Paul, Minn., West, 1989.

Gitlow, H., S. Gitlow, A. Oppenheim, and R. Oppenheim, *Tools and Methods for the Improvement of Quality,* Homewood, Ill., Irwin, 1989.

Ishikawa, Kaoru, *Guide to Quality Control,* 2nd rev. ed., New York, Quality Resources, 1986.

Juran, J. M., and F. M. Gryna, Jr., *Quality Planning and Analysis,* 2nd ed., New York, McGraw-Hill, 1980.

Montgomery, D. C., Introduction to Statistical Quality Control, 2nd ed., New York, Wiley, 1991.

Sampling Methods

Cochran, W. G., *Sampling Techniques,* 3rd ed., New York, Wiley, 1977.

Deming, W. E., *Sample Design in Business Research,* New York, Wiley, 1960.

Kish, L., *Survey Sampling,* New York, Wiley, 1965.

Levy, P. S., and S. Lemeshow, *Sampling of Populations: Methods and Applications,* New York, Wiley, 1991.

Scheaffer, R. L., W. Mendenhall, and L. Ott, *Elementary Survey Sampling,* 4th ed., Boston, PWS-Kent, 1990.

Williams, B., *A Sampler on Sampling,* New York, Wiley, 1978.

Warwick, D. P., and C. Lininger, *The Sample Survey: Theorey and Practice,* New York, McGraw-Hill, 1975.

Decision Analysis

Behn, R. D., and J. W. Vaupel, *Quick Analysis for Busy Decision Makers,* New York, Basic Books, 1982.

Brown, R. V., A. S. Kahn, and C. Peterson, *Decision Analysis for the Manager,* New York, Holt, Rinehart & Winston, 1974.

Hadley, G., *Introduction to Probability and Statistical Decision Theory,* San Francisco, Holden-Day, 1967.

Luce, R. D., and H. Raiffa, *Games and Decisions: Introduction and Critical Survey,* New York, Wiley, 1957.

Raiffa, H., *Decision Analysis: Introductory Lectures on Choices under Uncertainty,* Reading, Mass., Addison-Wesley, 1968.

Winkler, R. L., *An Introduction to Bayesian Inference and Decision,* New York, Holt, Rinehart & Winston, 1972.

APPENDIX

B

Tables

Table 1
Standard Normal Distribution

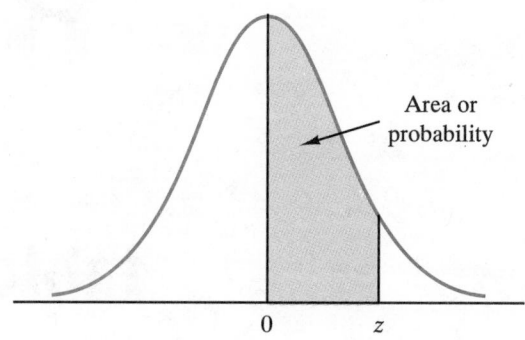

Entries in the table give the area under the curve between the mean and z standard deviations above the mean. For example, for $z = 1.25$ the area under the curve between the mean and z is .3944.

z	.00	.01	.02	.03	.04	.05	.06	.07	.08	.09
.0	.0000	.0040	.0080	.0120	.0160	.0199	.0239	.0279	.0319	.0359
.1	.0398	.0438	.0478	.0517	.0557	.0596	.0636	.0675	.0714	.0753
.2	.0793	.0832	.0871	.0910	.0948	.0987	.1026	.1064	.1103	.1141
.3	.1179	.1217	.1255	.1293	.1331	.1368	.1406	.1443	.1480	.1517
.4	.1554	.1591	.1628	.1664	.1700	.1736	.1772	.1808	.1844	.1879
.5	.1915	.1950	.1985	.2019	.2054	.2088	.2123	.2157	.2190	.2224
.6	.2257	.2291	.2324	.2357	.2389	.2422	.2454	.2486	.2518	.2549
.7	.2580	.2612	.2642	.2673	.2704	.2734	.2764	.2794	.2823	.2852
.8	.2881	.2910	.2939	.2967	.2995	.3023	.3051	.3078	.3106	.3133
.9	.3159	.3186	.3212	.3238	.3264	.3289	.3315	.3340	.3365	.3389
1.0	.3413	.3438	.3461	.3485	.3508	.3531	.3554	.3577	.3599	.3621
1.1	.3643	.3665	.3686	.3708	.3729	.3749	.3770	.3790	.3810	.3830
1.2	.3849	.3869	.3888	.3907	.3925	.3944	.3962	.3980	.3997	.4015
1.3	.4032	.4049	.4066	.4082	.4099	.4115	.4131	.4147	.4162	.4177
1.4	.4192	.4207	.4222	.4236	.4251	.4265	.4279	.4292	.4306	.4319
1.5	.4332	.4345	.4357	.4370	.4382	.4394	.4406	.4418	.4429	.4441
1.6	.4452	.4463	.4474	.4484	.4495	.4505	.4515	.4525	.4535	.4545
1.7	.4554	.4564	.4573	.4582	.4591	.4599	.4608	.4616	.4625	.4633
1.8	.4641	.4649	.4656	.4664	.4671	.4678	.4686	.4693	.4699	.4706
1.9	.4713	.4719	.4726	.4732	.4738	.4744	.4750	.4756	.4761	.4767
2.0	.4772	.4778	.4783	.4788	.4793	.4798	.4803	.4808	.4812	.4817
2.1	.4821	.4826	.4830	.4834	.4838	.4842	.4846	.4850	.4854	.4857
2.2	.4861	.4864	.4868	.4871	.4875	.4878	.4881	.4884	.4887	.4890
2.3	.4893	.4896	.4898	.4901	.4904	.4906	.4909	.4911	.4913	.4916
2.4	.4918	.4920	.4922	.4925	.4927	.4929	.4931	.4932	.4934	.4936
2.5	.4938	.4940	.4941	.4943	.4945	.4946	.4948	.4949	.4951	.4952
2.6	.4953	.4955	.4956	.4957	.4959	.4960	.4961	.4962	.4963	.4964
2.7	.4965	.4966	.4967	.4968	.4969	.4970	.4971	.4972	.4973	.4974
2.8	.4974	.4975	.4976	.4977	.4977	.4978	.4979	.4979	.4980	.4981
2.9	.4981	.4982	.4982	.4983	.4984	.4984	.4985	.4985	.4986	.4986
3.0	.4986	.4987	.4987	.4988	.4988	.4989	.4989	.4989	.4990	.4990

T A B L E 2
t Distribution

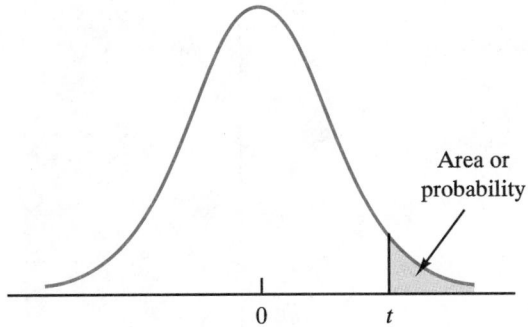

Area or
probability

0   *t*

Entries in the table give *t* values for an area or probability in the upper tail of the *t* distribution. For example, with 10 degrees of freedom and a .05 area in the upper tail, $t_{.05} = 1.812$.

Degrees of Freedom	Area in Upper Tail				
	.10	.05	.025	.01	.005
1	3.078	6.314	12.706	31.821	63.657
2	1.886	2.920	4.303	6.965	9.925
3	1.638	2.353	3.182	4.541	5.841
4	1.533	2.132	2.776	3.747	4.604
5	1.476	2.015	2.571	3.365	4.032
6	1.440	1.943	2.447	3.143	3.707
7	1.415	1.895	2.365	2.998	3.499
8	1.397	1.860	2.306	2.896	3.355
9	1.383	1.833	2.262	2.821	3.250
10	1.372	1.812	2.228	2.764	3.169
11	1.363	1.796	2.201	2.718	3.106
12	1.356	1.782	2.179	2.681	3.055
13	1.350	1.771	2.160	2.650	3.012
14	1.345	1.761	2.145	2.624	2.977
15	1.341	1.753	2.131	2.602	2.947
16	1.337	1.746	2.120	2.583	2.921
17	1.333	1.740	2.110	2.567	2.898
18	1.330	1.734	2.101	2.552	2.878
19	1.328	1.729	2.093	2.539	2.861
20	1.325	1.725	2.086	2.528	2.845
21	1.323	1.721	2.080	2.518	2.831
22	1.321	1.717	2.074	2.508	2.819
23	1.319	1.714	2.069	2.500	2.807
24	1.318	1.711	2.064	2.492	2.797
25	1.316	1.708	2.060	2.485	2.787
26	1.315	1.706	2.056	2.479	2.779
27	1.314	1.703	2.052	2.473	2.771
28	1.313	1.701	2.048	2.467	2.763
29	1.311	1.699	2.045	2.462	2.756
30	1.310	1.697	2.042	2.457	2.750
40	1.303	1.684	2.021	2.423	2.704
60	1.296	1.671	2.000	2.390	2.660
120	1.289	1.658	1.980	2.358	2.617
∞	1.282	1.645	1.960	2.326	2.576

[handwritten annotations: 80% under 1.282; 90% under 1.645; 95% under 1.960; 98% under 2.326; 99% to the right of 2.576]

TABLE 3 **Chi-Square Distribution**

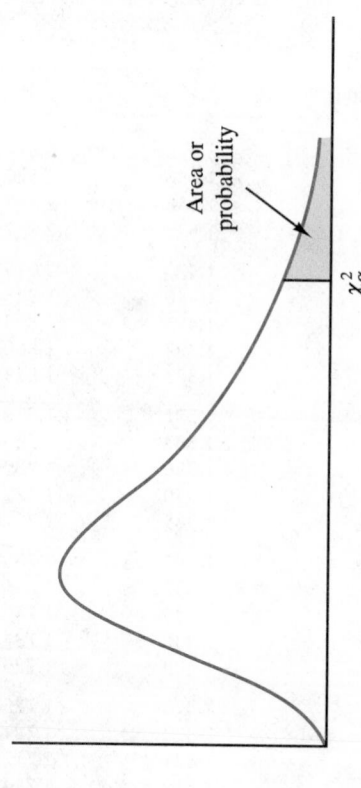

Area or probability

χ_α^2

Entries in the table give χ_α^2 values, where α is the area or probability in the upper tail of the chi-square distribution. For example, with 10 degrees of freedom and a .01 area in the upper tail, $\chi_{.01}^2 = 23.2093$.

Area in Upper Tail

Degrees of Freedom	.995	.99	.975	.95	.90	.10	.05	.025	.01	.005
1	392.704×10^{-10}	157.088×10^{-9}	982.069×10^{-9}	393.214×10^{-8}	.0157908	2.70554	3.84146	5.02389	6.63490	7.87944
2	.0100251	.0201007	.0506356	.102587	.210720	4.60517	5.99147	7.37776	9.21034	10.5966
3	.0717212	.114832	.215795	.351846	.584375	6.25139	7.81473	9.34840	11.3449	12.8381
4	.206990	.297110	.484419	.710721	1.063623	7.77944	9.48773	11.1433	13.2767	14.8602
5	.411740	.554300	.831211	1.145476	1.61031	9.23635	11.0705	12.8325	15.0863	16.7496
6	.675727	.872085	1.237347	1.63539	2.20413	10.6446	12.5916	14.4494	16.8119	18.5476
7	.989265	1.239043	1.68987	2.16735	2.83311	12.0170	14.0671	16.0128	18.4753	20.2777
8	1.344419	1.646482	2.17973	2.73264	3.48954	13.3616	15.5073	17.5346	20.0902	21.9550
9	1.734926	2.087912	2.70039	3.32511	4.16816	14.6837	16.9190	19.0228	21.6660	23.5893

(TABLE CONTINUES)

10	2.15585	2.55821	3.24697	3.94030	4.86518	15.9871	18.3070	20.4831	23.2093	25.1882
11	2.60321	3.05347	3.81575	4.57481	5.57779	17.2750	19.6751	21.9200	24.7250	26.7569
12	3.07382	3.57056	4.40379	5.22603	6.30380	18.5494	21.0261	23.3367	26.2170	28.2995
13	3.56503	4.10691	5.00874	5.89186	7.04150	19.8119	22.3621	24.7356	27.6883	29.8194
14	4.07468	4.66043	5.62872	6.57063	7.78953	21.0642	23.6848	26.1190	29.1413	31.3193
15	4.60094	5.22935	6.26214	7.26094	8.54675	22.3072	24.9958	27.4884	30.5779	32.8013
16	5.14224	5.81221	6.90766	7.96164	9.31223	23.5418	26.2962	28.8454	31.9999	34.2672
17	5.69724	6.40776	7.56418	8.67176	10.0852	24.7690	27.5871	30.1910	33.4087	35.7185
18	6.26481	7.01491	8.23075	9.39046	10.8649	25.9894	28.8693	31.5264	34.8053	37.1564
19	6.84398	7.63273	8.90655	10.1170	11.6509	27.2036	30.1435	32.8523	36.1908	38.5822
20	7.43386	8.26040	9.59083	10.8508	12.4426	28.4120	31.4104	34.1696	37.5662	39.9968
21	8.03366	8.89720	10.28293	11.5913	13.2396	29.6151	32.6705	35.4789	38.9321	41.4010
22	8.64272	9.54249	10.9823	12.3380	14.0415	30.8133	33.9244	36.7807	40.2894	42.7958
23	9.26042	10.19567	11.6885	13.0905	14.8479	32.0069	35.1725	38.0757	41.6384	44.1813
24	9.88623	10.8564	12.4011	13.8484	15.6587	33.1963	36.4151	39.3641	42.9798	45.5585
25	10.5197	11.5240	13.1197	14.6114	16.4734	34.3816	37.6525	40.6465	44.3141	46.9278
26	11.1603	12.1981	13.8439	15.3791	17.2919	35.5631	38.8852	41.9232	45.6417	48.2899
27	11.8076	12.8786	14.5733	16.1513	18.1138	36.7412	40.1133	43.1944	46.9630	49.6449
28	12.4613	13.5648	15.3079	16.9279	18.9392	37.9159	41.3372	44.4607	48.2782	50.9933
29	13.1211	14.2565	16.0471	17.7083	19.7677	39.0875	42.5569	45.7222	49.5879	52.3356
30	13.7867	14.9535	16.7908	18.4926	20.5992	40.2560	43.7729	46.9792	50.8922	53.6720
40	20.7065	22.1643	24.4331	26.5093	29.0505	51.8050	55.7585	59.3417	63.6907	66.7659
50	27.9907	29.7067	32.3574	34.7642	37.6886	63.1671	67.5048	71.4202	76.1539	79.4900
60	35.5346	37.4848	40.4817	43.1879	46.4589	74.3970	79.0819	83.2976	88.3794	91.9517
70	43.2752	45.4418	48.7576	51.7393	55.3290	85.5271	90.5312	95.0231	100.425	104.215
80	51.1720	53.5400	57.1532	60.3915	64.2778	96.5782	101.879	106.629	112.329	116.321
90	59.1963	61.7541	65.6466	69.1260	73.2912	107.565	113.145	118.136	124.116	128.299
100	67.3276	70.0648	74.2219	77.9295	82.3581	118.498	124.342	129.561	135.807	140.169

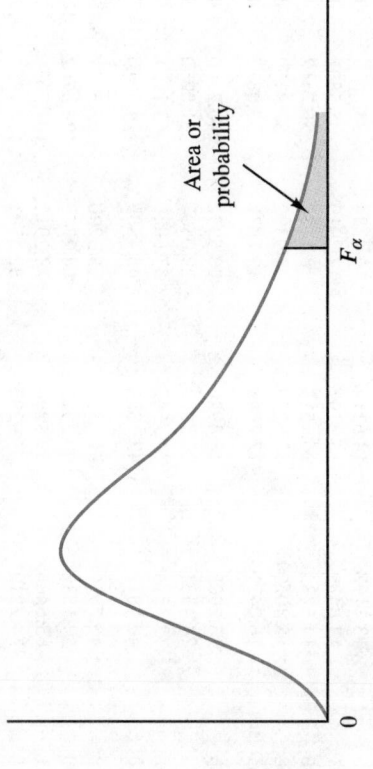

Area or
probability

F_α

Entries in the table give F_α values, where α is the area or probability in the upper tail of the F distribution. For example, with 12 numerator degrees of freedom, 15 denominator degrees of freedom, and a .05 area in the upper tail, $F_{.05} = 2.48$.

Table of $F_{.05}$ Values

Denominator Degrees of Freedom	Numerator Degrees of Freedom																		
	1	2	3	4	5	6	7	8	9	10	12	15	20	24	30	40	60	120	∞
1	161.4	199.5	215.7	224.6	230.2	234.0	236.8	238.9	240.5	241.9	243.9	245.9	248.0	249.1	250.1	251.1	252.2	253.3	254.3
2	18.51	19.00	19.16	19.25	19.30	19.33	19.35	19.37	19.38	19.40	19.41	19.43	19.45	19.45	19.46	19.47	19.48	19.49	19.50
3	10.13	9.55	9.28	9.12	9.01	8.94	8.89	8.85	8.81	8.79	8.74	8.70	8.66	8.64	8.62	8.59	8.57	8.55	8.53
4	7.71	6.94	6.59	6.39	6.26	6.16	6.09	6.04	6.00	5.96	5.91	5.86	5.80	5.77	5.75	5.72	5.69	5.66	5.63
5	6.61	5.79	5.41	5.19	5.05	4.95	4.88	4.82	4.77	4.74	4.68	4.62	4.56	4.53	4.50	4.46	4.43	4.40	4.36

(TABLE CONTINUES)

6	5.99	5.14	4.76	4.53	4.39	4.28	4.21	4.15	4.10	4.06	4.00	3.94	3.87	3.84	3.81	3.77	3.74	3.70	3.67
7	5.59	4.74	4.35	4.12	3.97	3.87	3.79	3.73	3.68	3.64	3.57	3.51	3.44	3.41	3.38	3.34	3.30	3.27	3.23
8	5.32	4.46	4.07	3.84	3.69	3.58	3.50	3.44	3.39	3.35	3.28	3.22	3.15	3.12	3.08	3.04	3.01	2.97	2.93
9	5.12	4.26	3.86	3.63	3.48	3.37	3.29	3.23	3.18	3.14	3.07	3.01	2.94	2.90	2.86	2.83	2.79	2.75	2.71
10	4.96	4.10	3.71	3.48	3.33	3.22	3.14	3.07	3.02	2.98	2.91	2.85	2.77	2.74	2.70	2.66	2.62	2.58	2.54
11	4.84	3.98	3.59	3.36	3.20	3.09	3.01	2.95	2.90	2.85	2.79	2.72	2.65	2.61	2.57	2.53	2.49	2.45	2.40
12	4.75	3.89	3.49	3.26	3.11	3.00	2.91	2.85	2.80	2.75	2.69	2.62	2.54	2.51	2.47	2.43	2.38	2.34	2.30
13	4.67	3.81	3.41	3.18	3.03	2.92	2.83	2.77	2.71	2.67	2.60	2.53	2.46	2.42	2.38	2.34	2.30	2.25	2.21
14	4.60	3.74	3.34	3.11	2.96	2.85	2.76	2.70	2.65	2.60	2.53	2.46	2.39	2.35	2.31	2.27	2.22	2.18	2.13
15	4.54	3.68	3.29	3.06	2.90	2.79	2.71	2.64	2.59	2.54	2.48	2.40	2.33	2.29	2.25	2.20	2.16	2.11	2.07
16	4.49	3.63	3.24	3.01	2.85	2.74	2.66	2.59	2.54	2.49	2.42	2.35	2.28	2.24	2.19	2.15	2.11	2.06	2.01
17	4.45	3.59	3.20	2.96	2.81	2.70	2.61	2.55	2.49	2.45	2.38	2.31	2.23	2.19	2.15	2.10	2.06	2.01	1.96
18	4.41	3.55	3.16	2.93	2.77	2.66	2.58	2.51	2.46	2.41	2.34	2.27	2.19	2.15	2.11	2.06	2.02	1.97	1.92
19	4.38	3.52	3.13	2.90	2.74	2.63	2.54	2.48	2.42	2.38	2.31	2.23	2.16	2.11	2.07	2.03	1.98	1.93	1.88
20	4.35	3.49	3.10	2.87	2.71	2.60	2.51	2.45	2.39	2.35	2.28	2.20	2.12	2.08	2.04	1.99	1.95	1.90	1.84
21	4.32	3.47	3.07	2.84	2.68	2.57	2.49	2.42	2.37	2.32	2.25	2.18	2.10	2.05	2.01	1.96	1.92	1.87	1.81
22	4.30	3.44	3.05	2.82	2.66	2.55	2.46	2.40	2.34	2.30	2.23	2.15	2.07	2.03	1.98	1.94	1.89	1.84	1.78
23	4.28	3.42	3.03	2.80	2.64	2.53	2.44	2.37	2.32	2.27	2.20	2.13	2.05	2.01	1.96	1.91	1.86	1.81	1.76
24	4.26	3.40	3.01	2.78	2.62	2.51	2.42	2.36	2.30	2.25	2.18	2.11	2.03	1.98	1.94	1.89	1.84	1.79	1.73
25	4.24	3.39	2.99	2.76	2.60	2.49	2.40	2.34	2.28	2.24	2.16	2.09	2.01	1.96	1.92	1.87	1.82	1.77	1.71
26	4.23	3.37	2.98	2.74	2.59	2.47	2.39	2.32	2.27	2.22	2.15	2.07	1.99	1.95	1.90	1.85	1.80	1.75	1.69
27	4.21	3.35	2.96	2.73	2.57	2.46	2.37	2.31	2.25	2.20	2.13	2.06	1.97	1.93	1.88	1.84	1.79	1.73	1.67
28	4.20	3.34	2.95	2.71	2.56	2.45	2.36	2.29	2.24	2.19	2.12	2.04	1.96	1.91	1.87	1.82	1.77	1.71	1.65
29	4.18	3.33	2.93	2.70	2.55	2.43	2.35	2.28	2.22	2.18	2.10	2.03	1.94	1.90	1.85	1.81	1.75	1.70	1.64
30	4.17	3.32	2.92	2.69	2.53	2.42	2.33	2.27	2.21	2.16	2.09	2.01	1.93	1.89	1.84	1.79	1.74	1.68	1.62
40	4.08	3.23	2.84	2.61	2.45	2.34	2.25	2.18	2.12	2.08	2.00	1.92	1.84	1.79	1.74	1.69	1.64	1.58	1.51
60	4.00	3.15	2.76	2.53	2.37	2.25	2.17	2.10	2.04	1.99	1.92	1.84	1.75	1.70	1.65	1.59	1.53	1.47	1.39
120	3.92	3.07	2.68	2.45	2.29	2.17	2.09	2.02	1.96	1.91	1.83	1.75	1.66	1.61	1.55	1.50	1.43	1.35	1.25
∞	3.84	3.00	2.60	2.37	2.21	2.10	2.01	1.94	1.88	1.83	1.75	1.67	1.57	1.52	1.46	1.39	1.32	1.22	1.00

(TABLE CONTINUES ON NEXT PAGE)

This table is reprinted by permission of the Biometrika Trustees from Table 18, Percentage Points of the *F* Distribution, by E. S. Pearson and H. O. Hartley, *Biometrika Tables for Statisticians*, Vol. I, 3rd Edition, 1966.

TABLE 4 (Continued)

Table of $F_{.01}$ Values

Denominator Degrees of Freedom	Numerator Degrees of Freedom																		
	1	2	3	4	5	6	7	8	9	10	12	15	20	24	30	40	60	120	∞
1	4,052	4,999.5	5,403	5,625	5,764	5,859	5,928	5,982	6,022	6,056	6,106	6,157	6,209	6,235	6,261	6,287	6,313	6,339	6,366
2	98.50	99.00	99.17	99.25	99.30	99.33	99.36	99.37	99.39	99.40	99.42	99.43	99.45	99.46	99.47	99.47	99.48	99.49	99.50
3	34.12	30.82	29.46	28.71	28.24	27.91	27.67	27.49	27.35	27.23	27.05	26.87	26.69	26.60	26.50	26.41	26.32	26.22	26.13
4	21.20	18.00	16.69	15.98	15.52	15.21	14.98	14.80	14.66	14.55	14.37	14.20	14.02	13.93	13.84	13.75	13.65	13.56	13.46
5	16.26	13.27	12.06	11.39	10.97	10.67	10.46	10.29	10.16	10.05	9.89	9.72	9.55	9.47	9.38	9.29	9.20	9.11	9.06
6	13.75	10.92	9.78	9.15	8.75	8.47	8.26	8.10	7.98	7.87	7.72	7.56	7.40	7.31	7.23	7.14	7.06	6.97	6.88
7	12.25	9.55	8.45	7.85	7.46	7.19	6.99	6.84	6.72	6.62	6.47	6.31	6.16	6.07	5.99	5.91	5.82	5.74	5.65
8	11.26	8.65	7.59	7.01	6.63	6.37	6.18	6.03	5.91	5.81	5.67	5.52	5.36	5.28	5.20	5.12	5.03	4.95	4.86
9	10.56	8.02	6.99	6.42	6.06	5.80	5.61	5.47	5.35	5.26	5.11	4.96	4.81	4.73	4.65	4.57	4.48	4.40	4.31
10	10.04	7.56	6.55	5.99	5.64	5.39	5.20	5.06	4.94	4.85	4.71	4.56	4.41	4.33	4.25	4.17	4.08	4.00	3.91
11	9.65	7.21	6.22	5.67	5.32	5.07	4.89	4.74	4.63	4.54	4.40	4.25	4.10	4.02	3.94	3.86	3.78	3.69	3.60
12	9.33	6.93	5.95	5.41	5.06	4.82	4.64	4.50	4.39	4.30	4.16	4.01	3.86	3.78	3.70	3.62	3.54	3.45	3.36
13	9.07	6.70	5.74	5.21	4.86	4.62	4.44	4.30	4.19	4.10	3.96	3.82	3.66	3.59	3.51	3.43	3.34	3.25	3.17
14	8.86	6.51	5.56	5.04	4.69	4.46	4.28	4.14	4.03	3.94	3.80	3.66	3.51	3.43	3.35	3.27	3.18	3.09	3.00
15	8.68	6.36	5.42	4.89	4.56	4.32	4.14	4.00	3.89	3.80	3.67	3.52	3.37	3.29	3.21	3.13	3.05	2.96	2.87
16	8.53	6.23	5.29	4.77	4.44	4.20	4.03	3.89	3.78	3.69	3.55	3.41	3.26	3.18	3.10	3.02	2.93	2.84	2.75
17	8.40	6.11	5.18	4.67	4.34	4.10	3.93	3.79	3.68	3.59	3.46	3.31	3.16	3.08	3.00	2.92	2.83	2.75	2.65
18	8.29	6.01	5.09	4.58	4.25	4.01	3.84	3.71	3.60	3.51	3.37	3.23	3.08	3.00	2.92	2.84	2.75	2.66	2.57
19	8.18	5.93	5.01	4.50	4.17	3.94	3.77	3.63	3.52	3.43	3.30	3.15	3.00	2.92	2.84	2.76	2.67	2.58	2.49
20	8.10	5.85	4.94	4.43	4.10	3.87	3.70	3.56	3.46	3.37	3.23	3.09	2.94	2.86	2.78	2.69	2.61	2.52	2.42
21	8.02	5.78	4.87	4.37	4.04	3.81	3.64	3.51	3.40	3.31	3.17	3.03	2.88	2.80	2.72	2.64	2.55	2.46	2.36
22	7.95	5.72	4.82	4.31	3.99	3.76	3.59	3.45	3.35	3.26	3.12	2.98	2.83	2.75	2.67	2.58	2.50	2.40	2.31
23	7.88	5.66	4.76	4.26	3.94	3.71	3.54	3.41	3.30	3.21	3.07	2.93	2.78	2.70	2.62	2.54	2.45	2.35	2.26
24	7.82	5.61	4.72	4.22	3.90	3.67	3.50	3.36	3.26	3.17	3.03	2.89	2.74	2.66	2.58	2.49	2.40	2.31	2.21
25	7.77	5.57	4.68	4.18	3.85	3.63	3.46	3.32	3.22	3.13	2.99	2.85	2.70	2.62	2.54	2.45	2.36	2.27	2.17
26	7.72	5.53	4.64	4.14	3.82	3.59	3.42	3.29	3.18	3.09	2.96	2.81	2.66	2.58	2.50	2.42	2.33	2.23	2.13
27	7.68	5.49	4.60	4.11	3.78	3.56	3.39	3.26	3.15	3.06	2.93	2.78	2.63	2.55	2.47	2.38	2.29	2.20	2.10
28	7.64	5.45	4.57	4.07	3.75	3.53	3.36	3.23	3.12	3.03	2.90	2.75	2.60	2.52	2.44	2.35	2.26	2.17	2.06
29	7.60	5.42	4.54	4.04	3.73	3.50	3.33	3.20	3.09	3.00	2.87	2.73	2.57	2.49	2.41	2.33	2.23	2.14	2.03
30	7.56	5.39	4.51	4.02	3.70	3.47	3.30	3.17	3.07	2.98	2.84	2.70	2.55	2.47	2.39	2.30	2.21	2.11	2.01
40	7.31	5.18	4.31	3.83	3.51	3.29	3.12	2.99	2.89	2.80	2.66	2.52	2.37	2.29	2.20	2.11	2.02	1.92	1.80
60	7.08	4.98	4.13	3.65	3.34	3.12	2.95	2.82	2.72	2.63	2.50	2.35	2.20	2.12	2.03	1.94	1.84	1.73	1.60
120	6.85	4.79	3.95	3.48	3.17	2.96	2.79	2.66	2.56	2.47	2.34	2.19	2.03	1.95	1.86	1.76	1.66	1.53	1.38
∞	6.63	4.61	3.78	3.32	3.02	2.80	2.64	2.51	2.41	2.32	2.18	2.04	1.88	1.79	1.70	1.59	1.47	1.32	1.00

(TABLE CONTINUES)

TABLE 4 (Continued)

Table of $F_{.025}$ Values

Numerator Degrees of Freedom

Denominator Degrees of Freedom	1	2	3	4	5	6	7	8	9	10	12	15	20	24	30	40	60	120	∞
1	647.8	799.5	864.2	899.6	921.8	937.1	948.2	956.7	963.3	968.6	976.7	984.9	993.1	997.2	1,001	1,006	1,010	1,014	1,018
2	38.51	39.00	39.17	39.25	39.30	39.33	39.36	39.37	39.39	39.40	39.41	39.43	39.45	39.46	39.46	39.47	39.48	39.49	39.50
3	17.44	16.04	15.44	15.10	14.88	14.73	14.62	14.54	14.47	14.42	14.34	14.25	14.17	14.12	14.08	14.04	13.99	13.95	13.90
4	12.22	10.65	9.98	9.60	9.36	9.20	9.07	8.98	8.90	8.84	8.75	8.66	8.56	8.51	8.46	8.41	8.36	8.31	8.26
5	10.01	8.43	7.76	7.39	7.15	6.98	6.85	6.76	6.68	6.62	6.52	6.43	6.33	6.28	6.23	6.18	6.12	6.07	6.02
6	8.81	7.26	6.60	6.23	5.99	5.82	5.70	5.60	5.52	5.46	5.37	5.27	5.17	5.12	5.07	5.01	4.96	4.90	4.85
7	8.07	6.54	5.89	5.52	5.29	5.12	4.99	4.90	4.82	4.76	4.67	4.57	4.47	4.42	4.36	4.31	4.25	4.20	4.14
8	7.57	6.06	5.42	5.05	4.82	4.65	4.53	4.43	4.36	4.30	4.20	4.10	4.00	3.95	3.89	3.84	3.78	3.73	3.67
9	7.21	5.71	5.08	4.72	4.48	4.32	4.20	4.10	4.03	3.96	3.87	3.77	3.67	3.61	3.56	3.51	3.45	3.39	3.33
10	6.94	5.46	4.83	4.47	4.24	4.07	3.95	3.85	3.78	3.72	3.62	3.52	3.42	3.37	3.31	3.26	3.20	3.14	3.08
11	6.72	5.26	4.63	4.28	4.04	3.88	3.76	3.66	3.59	3.53	3.43	3.33	3.23	3.17	3.12	3.06	3.00	2.94	2.88
12	6.55	5.10	4.47	4.12	3.89	3.73	3.61	3.51	3.44	3.37	3.28	3.18	3.07	3.02	2.96	2.91	2.85	2.79	2.72
13	6.41	4.97	4.35	4.00	3.77	3.60	3.48	3.39	3.31	3.25	3.15	3.05	2.95	2.89	2.84	2.78	2.72	2.66	2.60
14	6.30	4.86	4.24	3.89	3.66	3.50	3.38	3.29	3.21	3.15	3.05	2.95	2.84	2.79	2.73	2.67	2.61	2.55	2.49
15	6.20	4.77	4.15	3.80	3.58	3.41	3.29	3.20	3.12	3.06	2.96	2.86	2.76	2.70	2.64	2.59	2.52	2.46	2.40
16	6.12	4.69	4.08	3.73	3.50	3.34	3.22	3.12	3.05	2.99	2.89	2.79	2.68	2.63	2.57	2.51	2.45	2.38	2.32
17	6.04	4.62	4.01	3.66	3.44	3.28	3.16	3.06	2.98	2.92	2.82	2.72	2.62	2.56	2.50	2.44	2.38	2.32	2.25
18	5.98	4.56	3.95	3.61	3.38	3.22	3.10	3.01	2.93	2.87	2.77	2.67	2.56	2.50	2.44	2.38	2.32	2.26	2.19
19	5.92	4.51	3.90	3.56	3.33	3.17	3.05	2.96	2.88	2.82	2.72	2.62	2.51	2.45	2.39	2.33	2.27	2.20	2.13
20	5.87	4.46	3.86	3.51	3.29	3.13	3.01	2.91	2.84	2.77	2.68	2.57	2.46	2.41	2.35	2.29	2.22	2.16	2.09
21	5.83	4.42	3.82	3.48	3.25	3.09	2.97	2.87	2.80	2.73	2.64	2.53	2.42	2.37	2.31	2.25	2.18	2.11	2.04
22	5.79	4.38	3.78	3.44	3.22	3.05	2.93	2.84	2.76	2.70	2.60	2.50	2.39	2.33	2.27	2.21	2.14	2.08	2.00
23	5.75	4.35	3.75	3.41	3.18	3.02	2.90	2.81	2.73	2.67	2.57	2.47	2.36	2.30	2.24	2.18	2.11	2.04	1.97
24	5.72	4.32	3.72	3.38	3.15	2.99	2.87	2.78	2.70	2.64	2.54	2.44	2.33	2.27	2.21	2.15	2.08	2.01	1.94
25	5.69	4.29	3.69	3.35	3.13	2.97	2.85	2.75	2.68	2.61	2.51	2.41	2.30	2.24	2.18	2.12	2.05	1.98	1.91
26	5.66	4.27	3.67	3.33	3.10	2.94	2.82	2.73	2.65	2.59	2.49	2.39	2.28	2.22	2.16	2.09	2.03	1.95	1.88
27	5.63	4.24	3.65	3.31	3.08	2.92	2.80	2.71	2.63	2.57	2.47	2.36	2.25	2.19	2.13	2.07	2.00	1.93	1.85
28	5.61	4.22	3.63	3.29	3.06	2.90	2.78	2.69	2.61	2.55	2.45	2.34	2.23	2.17	2.11	2.05	1.98	1.91	1.83
29	5.59	4.20	3.61	3.27	3.04	2.88	2.76	2.67	2.59	2.53	2.43	2.32	2.21	2.15	2.09	2.03	1.96	1.89	1.81
30	5.57	4.18	3.59	3.25	3.03	2.87	2.75	2.65	2.57	2.51	2.41	2.31	2.20	2.14	2.07	2.01	1.94	1.87	1.79
40	5.42	4.05	3.46	3.13	2.90	2.74	2.62	2.53	2.45	2.39	2.29	2.18	2.07	2.01	1.94	1.88	1.80	1.72	1.64
60	5.29	3.93	3.34	3.01	2.79	2.63	2.51	2.41	2.33	2.27	2.17	2.06	1.94	1.88	1.82	1.74	1.67	1.58	1.48
120	5.15	3.80	3.23	2.89	2.67	2.52	2.39	2.30	2.22	2.16	2.05	1.94	1.82	1.76	1.69	1.61	1.53	1.43	1.31
∞	5.02	3.69	3.12	2.79	2.57	2.41	2.29	2.19	2.11	2.05	1.94	1.83	1.71	1.64	1.57	1.48	1.39	1.27	1.00

TABLE 5
Binomial Probabilities

Entries in the table give the probability of x successes in n trials of a binomial experiment, where p is the probability of a success on one trial. For example, with six trials and p = .05, the probability of two successes is .0305.

n	x	.01	.02	.03	.04	.05	.06	.07	.08	.09
2	0	.9801	.9604	.9409	.9216	.9025	.8836	.8649	.8464	.8281
	1	.0198	.0392	.0582	.0768	.0950	.1128	.1302	.1472	.1638
	2	.0001	.0004	.0009	.0016	.0025	.0036	.0049	.0064	.0081
3	0	.9703	.9412	.9127	.8847	.8574	.8306	.8044	.7787	.7536
	1	.0294	.0576	.0847	.1106	.1354	.1590	.1816	.2031	.2236
	2	.0003	.0012	.0026	.0046	.0071	.0102	.0137	.0177	.0221
	3	.0000	.0000	.0000	.0001	.0001	.0002	.0003	.0005	.0007
4	0	.9606	.9224	.8853	.8493	.8145	.7807	.7481	.7164	.6857
	1	.0388	.0753	.1095	.1416	.1715	.1993	.2252	.2492	.2713
	2	.0006	.0023	.0051	.0088	.0135	.0191	.0254	.0325	.0402
	3	.0000	.0000	.0001	.0002	.0005	.0008	.0013	.0019	.0027
	4	.0000	.0000	.0000	.0000	.0000	.0000	.0000	.0000	.0001
5	0	.9510	.9039	.8587	.8154	.7738	.7339	.6957	.6591	.6240
	1	.0480	.0922	.1328	.1699	.2036	.2342	.2618	.2866	.3086
	2	.0010	.0038	.0082	.0142	.0214	.0299	.0394	.0498	.0610
	3	.0000	.0001	.0003	.0006	.0011	.0019	.0030	.0043	.0060
	4	.0000	.0000	.0000	.0000	.0000	.0001	.0001	.0002	.0003
	5	.0000	.0000	.0000	.0000	.0000	.0000	.0000	.0000	.0000
6	0	.9415	.8858	.8330	.7828	.7351	.6899	.6470	.6064	.5679
	1	.0571	.1085	.1546	.1957	.2321	.2642	.2922	.3164	.3370
	2	.0014	.0055	.0120	.0204	.0305	.0422	.0550	.0688	.0833
	3	.0000	.0002	.0005	.0011	.0021	.0036	.0055	.0080	.0110
	4	.0000	.0000	.0000	.0000	.0001	.0002	.0003	.0005	.0008
	5	.0000	.0000	.0000	.0000	.0000	.0000	.0000	.0000	.0000
	6	.0000	.0000	.0000	.0000	.0000	.0000	.0000	.0000	.0000
7	0	.9321	.8681	.8080	.7514	.6983	.6485	.6017	.5578	.5168
	1	.0659	.1240	.1749	.2192	.2573	.2897	.3170	.3396	.3578
	2	.0020	.0076	.0162	.0274	.0406	.0555	.0716	.0886	.1061
	3	.0000	.0003	.0008	.0019	.0036	.0059	.0090	.0128	.0175
	4	.0000	.0000	.0000	.0001	.0002	.0004	.0007	.0011	.0017
	5	.0000	.0000	.0000	.0000	.0000	.0000	.0000	.0001	.0001
	6	.0000	.0000	.0000	.0000	.0000	.0000	.0000	.0000	.0000
	7	.0000	.0000	.0000	.0000	.0000	.0000	.0000	.0000	.0000
8	0	.9227	.8508	.7837	.7214	.6634	.6096	.5596	.5132	.4703
	1	.0746	.1389	.1939	.2405	.2793	.3113	.3370	.3570	.3721
	2	.0026	.0099	.0210	.0351	.0515	.0695	.0888	.1087	.1288
	3	.0001	.0004	.0013	.0029	.0054	.0089	.0134	.0189	.0255
	4	.0000	.0000	.0001	.0002	.0004	.0007	.0013	.0021	.0031
	5	.0000	.0000	.0000	.0000	.0000	.0000	.0001	.0001	.0002
	6	.0000	.0000	.0000	.0000	.0000	.0000	.0000	.0000	.0000
	7	.0000	.0000	.0000	.0000	.0000	.0000	.0000	.0000	.0000
	8	.0000	.0000	.0000	.0000	.0000	.0000	.0000	.0000	.0000

(TABLE CONTINUES)

TABLE 5
(Continued)

						p				
n	x	.01	.02	.03	.04	.05	.06	.07	.08	.09
9	0	.9135	.8337	.7602	.6925	.6302	.5730	.5204	.4722	.4279
	1	.0830	.1531	.2116	.2597	.2985	.3292	.3525	.3695	.3809
	2	.0034	.0125	.0262	.0433	.0629	.0840	.1061	.1285	.1507
	3	.0001	.0006	.0019	.0042	.0077	.0125	.0186	.0261	.0348
	4	.0000	.0000	.0001	.0003	.0006	.0012	.0021	.0034	.0052
	5	.0000	.0000	.0000	.0000	.0000	.0001	.0002	.0003	.0005
	6	.0000	.0000	.0000	.0000	.0000	.0000	.0000	.0000	.0000
	7	.0000	.0000	.0000	.0000	.0000	.0000	.0000	.0000	.0000
	8	.0000	.0000	.0000	.0000	.0000	.0000	.0000	.0000	.0000
	9	.0000	.0000	.0000	.0000	.0000	.0000	.0000	.0000	.0000
10	0	.9044	.8171	.7374	.6648	.5987	.5386	.4840	.4344	.3894
	1	.0914	.1667	.2281	.2770	.3151	.3438	.3643	.3777	.3851
	2	.0042	.0153	.0317	.0519	.0746	.0988	.1234	.1478	.1714
	3	.0001	.0008	.0026	.0058	.0105	.0168	.0248	.0343	.0452
	4	.0000	.0000	.0001	.0004	.0010	.0019	.0033	.0052	.0078
	5	.0000	.0000	.0000	.0000	.0001	.0001	.0003	.0005	.0009
	6	.0000	.0000	.0000	.0000	.0000	.0000	.0000	.0000	.0001
	7	.0000	.0000	.0000	.0000	.0000	.0000	.0000	.0000	.0000
	8	.0000	.0000	.0000	.0000	.0000	.0000	.0000	.0000	.0001
	9	.0000	.0000	.0000	.0000	.0000	.0000	.0000	.0000	.0001
	10	.0000	.0000	.0000	.0000	.0000	.0000	.0000	.0000	.0001
12	0	.8864	.7847	.6938	.6127	.5404	.4759	.4186	.3677	.3225
	1	.1074	.1922	.2575	.3064	.3413	.3645	.3781	.3837	.3827
	2	.0060	.0216	.0438	.0702	.0988	.1280	.1565	.1835	.2082
	3	.0002	.0015	.0045	.0098	.0173	.0272	.0393	.0532	.0686
	4	.0000	.0001	.0003	.0009	.0021	.0039	.0067	.0104	.0153
	5	.0000	.0000	.0000	.0001	.0002	.0004	.0008	.0014	.0024
	6	.0000	.0000	.0000	.0000	.0000	.0000	.0001	.0001	.0003
	7	.0000	.0000	.0000	.0000	.0000	.0000	.0000	.0000	.0000
	8	.0000	.0000	.0000	.0000	.0000	.0000	.0000	.0000	.0000
	9	.0000	.0000	.0000	.0000	.0000	.0000	.0000	.0000	.0000
	10	.0000	.0000	.0000	.0000	.0000	.0000	.0000	.0000	.0000
	11	.0000	.0000	.0000	.0000	.0000	.0000	.0000	.0000	.0000
	12	.0000	.0000	.0000	.0000	.0000	.0000	.0000	.0000	.0000
15	0	.8601	.7386	.6333	.5421	.4633	.3953	.3367	.2863	.2430
	1	.1303	.2261	.2938	.3388	.3658	.3785	.3801	.3734	.3605
	2	.0092	.0323	.0636	.0988	.1348	.1691	.2003	.2273	.2496
	3	.0004	.0029	.0085	.0178	.0307	.0468	.0653	.0857	.1070
	4	.0000	.0002	.0008	.0022	.0049	.0090	.0148	.0223	.0317
	5	.0000	.0000	.0001	.0002	.0006	.0013	.0024	.0043	.0069
	6	.0000	.0000	.0000	.0000	.0000	.0001	.0003	.0006	.0011
	7	.0000	.0000	.0000	.0000	.0000	.0000	.0000	.0001	.0001
	8	.0000	.0000	.0000	.0000	.0000	.0000	.0000	.0000	.0000
	9	.0000	.0000	.0000	.0000	.0000	.0000	.0000	.0000	.0000
	10	.0000	.0000	.0000	.0000	.0000	.0000	.0000	.0000	.0000
	11	.0000	.0000	.0000	.0000	.0000	.0000	.0000	.0000	.0000
	12	.0000	.0000	.0000	.0000	.0000	.0000	.0000	.0000	.0000
	13	.0000	.0000	.0000	.0000	.0000	.0000	.0000	.0000	.0000
	14	.0000	.0000	.0000	.0000	.0000	.0000	.0000	.0000	.0000
	15	.0000	.0000	.0000	.0000	.0000	.0000	.0000	.0000	.0000

(TABLE CONTINUES ON NEXT PAGE)

TABLE 5
(Continued)

n	x	.01	.02	.03	.04	.05	.06	.07	.08	.09
18	0	.8345	.6951	.5780	.4796	.3972	.3283	.2708	.2229	.1831
	1	.1517	.2554	.3217	.3597	.3763	.3772	.3669	.3489	.3260
	2	.0130	.0443	.0846	.1274	.1683	.2047	.2348	.2579	.2741
	3	.0007	.0048	.0140	.0283	.0473	.0697	.0942	.1196	.1446
	4	.0000	.0004	.0016	.0044	.0093	.0167	.0266	.0390	.0536
	5	.0000	.0000	.0001	.0005	.0014	.0030	.0056	.0095	.0148
	6	.0000	.0000	.0000	.0000	.0002	.0004	.0009	.0018	.0032
	7	.0000	.0000	.0000	.0000	.0000	.0000	.0001	.0003	.0005
	8	.0000	.0000	.0000	.0000	.0000	.0000	.0000	.0000	.0001
	9	.0000	.0000	.0000	.0000	.0000	.0000	.0000	.0000	.0000
	10	.0000	.0000	.0000	.0000	.0000	.0000	.0000	.0000	.0000
	11	.0000	.0000	.0000	.0000	.0000	.0000	.0000	.0000	.0000
	12	.0000	.0000	.0000	.0000	.0000	.0000	.0000	.0000	.0000
	13	.0000	.0000	.0000	.0000	.0000	.0000	.0000	.0000	.0000
	14	.0000	.0000	.0000	.0000	.0000	.0000	.0000	.0000	.0000
	15	.0000	.0000	.0000	.0000	.0000	.0000	.0000	.0000	.0000
	16	.0000	.0000	.0000	.0000	.0000	.0000	.0000	.0000	.0000
	17	.0000	.0000	.0000	.0000	.0000	.0000	.0000	.0000	.0000
	18	.0000	.0000	.0000	.0000	.0000	.0000	.0000	.0000	.0000
20	0	.8179	.6676	.5438	.4420	.3585	.2901	.2342	.1887	.1516
	1	.1652	.2725	.3364	.3683	.3774	.3703	.3526	.3282	.3000
	2	.0159	.0528	.0988	.1458	.1887	.2246	.2521	.2711	.2818
	3	.0010	.0065	.0183	.0364	.0596	.0860	.1139	.1414	.1672
	4	.0000	.0006	.0024	.0065	.0133	.0233	.0364	.0523	.0703
	5	.0000	.0000	.0002	.0009	.0022	.0048	.0088	.0145	.0222
	6	.0000	.0000	.0000	.0001	.0003	.0008	.0017	.0032	.0055
	7	.0000	.0000	.0000	.0000	.0000	.0001	.0002	.0005	.0011
	8	.0000	.0000	.0000	.0000	.0000	.0000	.0000	.0001	.0002
	9	.0000	.0000	.0000	.0000	.0000	.0000	.0000	.0000	.0000
	10	.0000	.0000	.0000	.0000	.0000	.0000	.0000	.0000	.0000
	11	.0000	.0000	.0000	.0000	.0000	.0000	.0000	.0000	.0000
	12	.0000	.0000	.0000	.0000	.0000	.0000	.0000	.0000	.0000
	13	.0000	.0000	.0000	.0000	.0000	.0000	.0000	.0000	.0000
	14	.0000	.0000	.0000	.0000	.0000	.0000	.0000	.0000	.0000
	15	.0000	.0000	.0000	.0000	.0000	.0000	.0000	.0000	.0000
	16	.0000	.0000	.0000	.0000	.0000	.0000	.0000	.0000	.0000
	17	.0000	.0000	.0000	.0000	.0000	.0000	.0000	.0000	.0000
	18	.0000	.0000	.0000	.0000	.0000	.0000	.0000	.0000	.0000
	19	.0000	.0000	.0000	.0000	.0000	.0000	.0000	.0000	.0000
	20	.0000	.0000	.0000	.0000	.0000	.0000	.0000	.0000	.0000

(TABLE CONTINUES)

TABLE 5
(Continued)

						p				
n	x	.10	.15	.20	.25	.30	.35	.40	.45	.50
2	0	.8100	.7225	.6400	.5625	.4900	.4225	.3600	.3025	.2500
	1	.1800	.2550	.3200	.3750	.4200	.4550	.4800	.4950	.5000
	2	.0100	.0225	.0400	.0625	.0900	.1225	.1600	.2025	.2500
3	0	.7290	.6141	.5120	.4219	.3430	.2746	.2160	.1664	.1250
	1	.2430	.3251	.3840	.4219	.4410	.4436	.4320	.4084	.3750
	2	.0270	.0574	.0960	.1406	.1890	.2389	.2880	.3341	.3750
	3	.0010	.0034	.0080	.0156	.0270	.0429	.0640	.0911	.1250
4	0	.6561	.5220	.4096	.3164	.2401	.1785	.1296	.0915	.0625
	1	.2916	.3685	.4096	.4219	.4116	.3845	.3456	.2995	.2500
	2	.0486	.0975	.1536	.2109	.2646	.3105	.3456	.3675	.3750
	3	.0036	.0115	.0256	.0469	.0756	.1115	.1536	.2005	.2500
	4	.0001	.0005	.0016	.0039	.0081	.0150	.0256	.0410	.0625
5	0	.5905	.4437	.3277	.2373	.1681	.1160	.0778	.0503	.0312
	1	.3280	.3915	.4096	.3955	.3602	.3124	.2592	.2059	.1562
	2	.0729	.1382	.2048	.2637	.3087	.3364	.3456	.3369	.3125
	3	.0081	.0244	.0512	.0879	.1323	.1811	.2304	.2757	.3125
	4	.0004	.0022	.0064	.0146	.0284	.0488	.0768	.1128	.1562
	5	.0000	.0001	.0003	.0010	.0024	.0053	.0102	.0185	.0312
6	0	.5314	.3771	.2621	.1780	.1176	.0754	.0467	.0277	.0156
	1	.3543	.3993	.3932	.3560	.3025	.2437	.1866	.1359	.0938
	2	.0984	.1762	.2458	.2966	.3241	.3280	.3110	.2780	.2344
	3	.0146	.0415	.0819	.1318	.1852	.2355	.2765	.3032	.3125
	4	.0012	.0055	.0154	.0330	.0595	.0951	.1382	.1861	.2344
	5	.0001	.0004	.0015	.0044	.0102	.0205	.0369	.0609	.0938
	6	.0000	.0000	.0001	.0002	.0007	.0018	.0041	.0083	.0156
7	0	.4783	.3206	.2097	.1335	.0824	.0490	.0280	.0152	.0078
	1	.3720	.3960	.3670	.3115	.2471	.1848	.1306	.0872	.0547
	2	.1240	.2097	.2753	.3115	.3177	.2985	.2613	.2140	.1641
	3	.0230	.0617	.1147	.1730	.2269	.2679	.2903	.2918	.2734
	4	.0026	.0109	.0287	.0577	.0972	.1442	.1935	.2388	.2734
	5	.0002	.0012	.0043	.0115	.0250	.0466	.0774	.1172	.1641
	6	.0000	.0001	.0004	.0013	.0036	.0084	.0172	.0320	.0547
	7	.0000	.0000	.0000	.0001	.0002	.0006	.0016	.0037	.0078
8	0	.4305	.2725	.1678	.1001	.0576	.0319	.0168	.0084	.0039
	1	.3826	.3847	.3355	.2670	.1977	.1373	.0896	.0548	.0312
	2	.1488	.2376	.2936	.3115	.2965	.2587	.2090	.1569	.1094
	3	.0331	.0839	.1468	.2076	.2541	.2786	.2787	.2568	.2188
	4	.0046	.0185	.0459	.0865	.1361	.1875	.2322	.2627	.2734
	5	.0004	.0026	.0092	.0231	.0467	.0808	.1239	.1719	.2188
	6	.0000	.0002	.0011	.0038	.0100	.0217	.0413	.0703	.1094
	7	.0000	.0000	.0001	.0004	.0012	.0033	.0079	.0164	.0312
	8	.0000	.0000	.0000	.0000	.0001	.0002	.0007	.0017	.0039

(**TABLE CONTINUES ON NEXT PAGE**)

TABLE 5
(Continued)

n	x	.10	.15	.20	.25	.30	.35	.40	.45	.50
9	0	.3874	.2316	.1342	.0751	.0404	.0207	.0101	.0046	.0020
	1	.3874	.3679	.3020	.2253	.1556	.1004	.0605	.0339	.0176
	2	.1722	.2597	.3020	.3003	.2668	.2162	.1612	.1110	.0703
	3	.0446	.1069	.1762	.2336	.2668	.2716	.2508	.2119	.1641
	4	.0074	.0283	.0661	.1168	.1715	.2194	.2508	.2600	.2461
	5	.0008	.0050	.0165	.0389	.0735	.1181	.1672	.2128	.2461
	6	.0001	.0006	.0028	.0087	.0210	.0424	.0743	.1160	.1641
	7	.0000	.0000	.0003	.0012	.0039	.0098	.0212	.0407	.0703
	8	.0000	.0000	.0000	.0001	.0004	.0013	.0035	.0083	.0176
	9	.0000	.0000	.0000	.0000	.0000	.0001	.0003	.0008	.0020
10	0	.3487	.1969	.1074	.0563	.0282	.0135	.0060	.0025	.0010
	1	.3874	.3474	.2684	.1877	.1211	.0725	.0403	.0207	.0098
	2	.1937	.2759	.3020	.2816	.2335	.1757	.1209	.0763	.0439
	3	.0574	.1298	.2013	.2503	.2668	.2522	.2150	.1665	.1172
	4	.0112	.0401	.0881	.1460	.2001	.2377	.2508	.2384	.2051
	5	.0015	.0085	.0264	.0584	.1029	.1536	.2007	.2340	.2461
	6	.0001	.0012	.0055	.0162	.0368	.0689	.1115	.1596	.2051
	7	.0000	.0001	.0008	.0031	.0090	.0212	.0425	.0746	.1172
	8	.0000	.0000	.0001	.0004	.0014	.0043	.0106	.0229	.0439
	9	.0000	.0000	.0000	.0000	.0001	.0005	.0016	.0042	.0098
	10	.0000	.0000	.0000	.0000	.0000	.0000	.0001	.0003	.0010
12	0	.2824	.1422	.0687	.0317	.0138	.0057	.0022	.0008	.0002
	1	.3766	.3012	.2062	.1267	.0712	.0368	.0174	.0075	.0029
	2	.2301	.2924	.2835	.2323	.1678	.1088	.0639	.0339	.0161
	3	.0853	.1720	.2362	.2581	.2397	.1954	.1419	.0923	.0537
	4	.0213	.0683	.1329	.1936	.2311	.2367	.2128	.1700	.1208
	5	.0038	.0193	.0532	.1032	.1585	.2039	.2270	.2225	.1934
	6	.0005	.0040	.0155	.0401	.0792	.1281	.1766	.2124	.2256
	7	.0000	.0006	.0033	.0115	.0291	.0591	.1009	.1489	.1934
	8	.0000	.0001	.0005	.0024	.0078	.0199	.0420	.0762	.1208
	9	.0000	.0000	.0001	.0004	.0015	.0048	.0125	.0277	.0537
	10	.0000	.0000	.0000	.0000	.0002	.0008	.0025	.0068	.0161
	11	.0000	.0000	.0000	.0000	.0000	.0001	.0003	.0010	.0029
	12	.0000	.0000	.0000	.0000	.0000	.0000	.0000	.0001	.0002
15	0	.2059	.0874	.0352	.0134	.0047	.0016	.0005	.0001	.0000
	1	.3432	.2312	.1319	.0668	.0305	.0126	.0047	.0016	.0005
	2	.2669	.2856	.2309	.1559	.0916	.0476	.0219	.0090	.0032
	3	.1285	.2184	.2501	.2252	.1700	.1110	.0634	.0318	.0139
	4	.0428	.1156	.1876	.2252	.2186	.1792	.1268	.0780	.0417
	5	.0105	.0449	.1032	.1651	.2061	.2123	.1859	.1404	.0916
	6	.0019	.0132	.0430	.0917	.1472	.1906	.2066	.1914	.1527
	7	.0003	.0030	.0138	.0393	.0811	.1319	.1771	.2013	.1964
	8	.0000	.0005	.0035	.0131	.0348	.0710	.1181	.1647	.1964
	9	.0000	.0001	.0007	.0034	.0116	.0298	.0612	.1048	.1527
	10	.0000	.0000	.0001	.0007	.0030	.0096	.0245	.0515	.0916
	11	.0000	.0000	.0000	.0001	.0006	.0024	.0074	.0191	.0417
	12	.0000	.0000	.0000	.0000	.0001	.0004	.0016	.0052	.0139
	13	.0000	.0000	.0000	.0000	.0000	.0001	.0003	.0010	.0032
	14	.0000	.0000	.0000	.0000	.0000	.0000	.0000	.0001	.0005
	15	.0000	.0000	.0000	.0000	.0000	.0000	.0000	.0000	.0000

(TABLE CONTINUES)

TABLE 5
(Continued)

n	x	.10	.15	.20	.25	.30	.35	.40	.45	.50
18	0	.1501	.0536	.0180	.0056	.0016	.0004	.0001	.0000	.0000
	1	.3002	.1704	.0811	.0338	.0126	.0042	.0012	.0003	.0001
	2	.2835	.2556	.1723	.0958	.0458	.0190	.0069	.0022	.0006
	3	.1680	.2406	.2297	.1704	.1046	.0547	.0246	.0095	.0031
	4	.0700	.1592	.2153	.2130	.1681	.1104	.0614	.0291	.0117
	5	.0218	.0787	.1507	.1988	.2017	.1664	.1146	.0666	.0327
	6	.0052	.0301	.0816	.1436	.1873	.1941	.1655	.1181	.0708
	7	.0010	.0091	.0350	.0820	.1376	.1792	.1892	.1657	.1214
	8	.0002	.0022	.0120	.0376	.0811	.1327	.1734	.1864	.1669
	9	.0000	.0004	.0033	.0139	.0386	.0794	.1284	.1694	.1855
	10	.0000	.0001	.0008	.0042	.0149	.0385	.0771	.1248	.1669
	11	.0000	.0000	.0001	.0010	.0046	.0151	.0374	.0742	.1214
	12	.0000	.0000	.0000	.0002	.0012	.0047	.0145	.0354	.0708
	13	.0000	.0000	.0000	.0000	.0002	.0012	.0045	.0134	.0327
	14	.0000	.0000	.0000	.0000	.0000	.0002	.0011	.0039	.0117
	15	.0000	.0000	.0000	.0000	.0000	.0000	.0002	.0009	.0031
	16	.0000	.0000	.0000	.0000	.0000	.0000	.0000	.0001	.0006
	17	.0000	.0000	.0000	.0000	.0000	.0000	.0000	.0000	.0001
	18	.0000	.0000	.0000	.0000	.0000	.0000	.0000	.0000	.0000
20	0	.1216	.0388	.0115	.0032	.0008	.0002	.0000	.0000	.0000
	1	.2702	.1368	.0576	.0211	.0068	.0020	.0005	.0001	.0000
	2	.2852	.2293	.1369	.0669	.0278	.0100	.0031	.0008	.0002
	3	.1901	.2428	.2054	.1339	.0716	.0323	.0123	.0040	.0011
	4	.0898	.1821	.2182	.1897	.1304	.0738	.0350	.0139	.0046
	5	.0319	.1028	.1746	.2023	.1789	.1272	.0746	.0365	.0148
	6	.0089	.0454	.1091	.1686	.1916	.1712	.1244	.0746	.0370
	7	.0020	.0160	.0545	.1124	.1643	.1844	.1659	.1221	.0739
	8	.0004	.0046	.0222	.0609	.1144	.1614	.1797	.1623	.1201
	9	.0001	.0011	.0074	.0271	.0654	.1158	.1597	.1771	.1602
	10	.0000	.0002	.0020	.0099	.0308	.0686	.1171	.1593	.1762
	11	.0000	.0000	.0005	.0030	.0120	.0336	.0710	.1185	.1602
	12	.0000	.0000	.0001	.0008	.0039	.0136	.0355	.0727	.1201
	13	.0000	.0000	.0000	.0002	.0010	.0045	.0146	.0366	.0739
	14	.0000	.0000	.0000	.0000	.0002	.0012	.0049	.0150	.0370
	15	.0000	.0000	.0000	.0000	.0000	.0003	.0013	.0049	.0148
	16	.0000	.0000	.0000	.0000	.0000	.0000	.0003	.0013	.0046
	17	.0000	.0000	.0000	.0000	.0000	.0000	.0000	.0002	.0011
	18	.0000	.0000	.0000	.0000	.0000	.0000	.0000	.0000	.0002
	19	.0000	.0000	.0000	.0000	.0000	.0000	.0000	.0000	.0000
	20	.0000	.0000	.0000	.0000	.0000	.0000	.0000	.0000	.0000

(TABLE CONTINUES ON NEXT PAGE)

TABLE 5
(Continued)

						p				
n	x	.55	.60	.65	.70	.75	.80	.85	.90	.95
2	0	.2025	.1600	.1225	.0900	.0625	.0400	.0225	.0100	.0025
	1	.4950	.4800	.4550	.4200	.3750	.3200	.2550	.1800	.0950
	2	.3025	.3600	.4225	.4900	.5625	.6400	.7225	.8100	.9025
3	0	.0911	.0640	.0429	.0270	.0156	.0080	.0034	.0010	.0001
	1	.3341	.2880	.2389	.1890	.1406	.0960	.0574	.0270	.0071
	2	.4084	.4320	.4436	.4410	.4219	.3840	.3251	.2430	.1354
	3	.1664	.2160	.2746	.3430	.4219	.5120	.6141	.7290	.8574
4	0	.0410	.0256	.0150	.0081	.0039	.0016	.0005	.0001	.0000
	1	.2005	.1536	.1115	.0756	.0469	.0256	.0115	.0036	.0005
	2	.3675	.3456	.3105	.2646	.2109	.1536	.0975	.0486	.0135
	3	.2995	.3456	.3845	.4116	.4219	.4096	.3685	.2916	.1715
	4	.0915	.1296	.1785	.2401	.3164	.4096	.5220	.6561	.8145
5	0	.0185	.0102	.0053	.0024	.0010	.0003	.0001	.0000	.0000
	1	.1128	.0768	.0488	.0284	.0146	.0064	.0022	.0005	.0000
	2	.2757	.2304	.1811	.1323	.0879	.0512	.0244	.0081	.0011
	3	.3369	.3456	.3364	.3087	.2637	.2048	.1382	.0729	.0214
	4	.2059	.2592	.3124	.3601	.3955	.4096	.3915	.3281	.2036
	5	.0503	.0778	.1160	.1681	.2373	.3277	.4437	.5905	.7738
6	0	.0083	.0041	.0018	.0007	.0002	.0001	.0000	.0000	.0000
	1	.0609	.0369	.0205	.0102	.0044	.0015	.0004	.0001	.0000
	2	.1861	.1382	.0951	.0595	.0330	.0154	.0055	.0012	.0001
	3	.3032	.2765	.2355	.1852	.1318	.0819	.0415	.0146	.0021
	4	.2780	.3110	.3280	.3241	.2966	.2458	.1762	.0984	.0305
	5	.1359	.1866	.2437	.3025	.3560	.3932	.3993	.3543	.2321
	6	.0277	.0467	.0754	.1176	.1780	.2621	.3771	.5314	.7351
7	0	.0037	.0016	.0006	.0002	.0001	.0000	.0000	.0000	.0000
	1	.0320	.0172	.0084	.0036	.0013	.0004	.0001	.0000	.0000
	2	.1172	.0774	.0466	.0250	.0115	.0043	.0012	.0002	.0000
	3	.2388	.1935	.1442	.0972	.0577	.0287	.0109	.0026	.0002
	4	.2918	.2903	.2679	.2269	.1730	.1147	.0617	.0230	.0036
	5	.2140	.2613	.2985	.3177	.3115	.2753	.2097	.1240	.0406
	6	.0872	.1306	.1848	.2471	.3115	.3670	.3960	.3720	.2573
	7	.0152	.0280	.0490	.0824	.1335	.2097	.3206	.4783	.6983
8	0	.0017	.0007	.0002	.0001	.0000	.0000	.0000	.0000	.0000
	1	.0164	.0079	.0033	.0012	.0004	.0001	.0000	.0000	.0000
	2	.0703	.0413	.0217	.0100	.0038	.0011	.0002	.0000	.0000
	3	.1719	.1239	.0808	.0467	.0231	.0092	.0026	.0004	.0000
	4	.2627	.2322	.1875	.1361	.0865	.0459	.0185	.0046	.0004
	5	.2568	.2787	.2786	.2541	.2076	.1468	.0839	.0331	.0054
	6	.1569	.2090	.2587	.2965	.3115	.2936	.2376	.1488	.0515
	7	.0548	.0896	.1373	.1977	.2670	.3355	.3847	.3826	.2793
	8	.0084	.0168	.0319	.0576	.1001	.1678	.2725	.4305	.6634

(TABLE CONTINUES)

$x \geq 6$ add together

TABLE 5
(Continued)

		p								
n	x	.55	.60	.65	.70	.75	.80	.85	.90	.95
9	0	.0008	.0003	.0001	.0000	.0000	.0000	.0000	.0000	.0000
	1	.0083	.0035	.0013	.0004	.0001	.0000	.0000	.0000	.0000
	2	.0407	.0212	.0098	.0039	.0012	.0003	.0000	.0000	.0000
	3	.1160	.0743	.0424	.0210	.0087	.0028	.0006	.0001	.0000
	4	.2128	.1672	.1181	.0735	.0389	.0165	.0050	.0008	.0000
	5	.2600	.2508	.2194	.1715	.1168	.0661	.0283	.0074	.0006
	6	.2119	.2508	.2716	.2668	.2336	.1762	.1069	.0446	.0077
	7	.1110	.1612	.2162	.2668	.3003	.3020	.2597	.1722	.0629
	8	.0339	.0605	.1004	.1556	.2253	.3020	.3679	.3874	.2985
	9	.0046	.0101	.0207	.0404	.0751	.1342	.2316	.3874	.6302
10	0	.0003	.0001	.0000	.0000	.0000	.0000	.0000	.0000	.0000
	1	.0042	.0016	.0005	.0001	.0000	.0000	.0000	.0000	.0000
	2	.0229	.0106	.0043	.0014	.0004	.0001	.0000	.0000	.0000
	3	.0746	.0425	.0212	.0090	.0031	.0008	.0001	.0000	.0000
	4	.1596	.1115	.0689	.0368	.0162	.0055	.0012	.0001	.0000
	5	.2340	.2007	.1536	.1029	.0584	.0264	.0085	.0015	.0001
	6	.2384	.2508	.2377	.2001	.1460	.0881	.0401	.0112	.0010
	7	.1665	.2150	.2522	.2668	.2503	.2013	.1298	.0574	.0105
	8	.0763	.1209	.1757	.2335	.2816	.3020	.2759	.1937	.0746
	9	.0207	.0403	.0725	.1211	.1877	.2684	.3474	.3874	.3151
	10	.0025	.0060	.0135	.0282	.0563	.1074	.1969	.3487	.5987
12	0	.0001	.0000	.0000	.0000	.0000	.0000	.0000	.0000	.0000
	1	.0010	.0003	.0001	.0000	.0000	.0000	.0000	.0000	.0000
	2	.0068	.0025	.0008	.0002	.0000	.0000	.0000	.0000	.0000
	3	.0277	.0125	.0048	.0015	.0004	.0001	.0000	.0000	.0000
	4	.0762	.0420	.0199	.0078	.0024	.0005	.0001	.0000	.0000
	5	.1489	.1009	.0591	.0291	.0115	.0033	.0006	.0000	.0000
	6	.2124	.1766	.1281	.0792	.0401	.0155	.0040	.0005	.0000
	7	.2225	.2270	.2039	.1585	.1032	.0532	.0193	.0038	.0002
	8	.1700	.2128	.2367	.2311	.1936	.1329	.0683	.0213	.0021
	9	.0923	.1419	.1954	.2397	.2581	.2362	.1720	.0852	.0173
	10	.0339	.0639	.1088	.1678	.2323	.2835	.2924	.2301	.0988
	11	.0075	.0174	.0368	.0712	.1267	.2062	.3012	.3766	.3413
	12	.0008	.0022	.0057	.0138	.0317	.0687	.1422	.2824	.5404
15	0	.0000	.0000	.0000	.0000	.0000	.0000	.0000	.0000	.0000
	1	.0001	.0000	.0000	.0000	.0000	.0000	.0000	.0000	.0000
	2	.0010	.0003	.0001	.0000	.0000	.0000	.0000	.0000	.0000
	3	.0052	.0016	.0004	.0001	.0000	.0000	.0000	.0000	.0000
	4	.0191	.0074	.0024	.0006	.0001	.0000	.0000	.0000	.0000
	5	.0515	.0245	.0096	.0030	.0007	.0001	.0000	.0000	.0000
	6	.1048	.0612	.0298	.0116	.0034	.0007	.0001	.0000	.0000
	7	.1647	.1181	.0710	.0348	.0131	.0035	.0005	.0000	.0000
	8	.2013	.1771	.1319	.0811	.0393	.0138	.0030	.0003	.0000
	9	.1914	.2066	.1906	.1472	.0917	.0430	.0132	.0019	.0000
	10	.1404	.1859	.2123	.2061	.1651	.1032	.0449	.0105	.0006
	11	.0780	.1268	.1792	.2186	.2252	.1876	.1156	.0428	.0049
	12	.0318	.0634	.1110	.1700	.2252	.2501	.2184	.1285	.0307
	13	.0090	.0219	.0476	.0916	.1559	.2309	.2856	.2669	.1348
	14	.0016	.0047	.0126	.0305	.0668	.1319	.2312	.3432	.3658
	15	.0001	.0005	.0016	.0047	.0134	.0352	.0874	.2059	.4633

(TABLE CONTINUES ON NEXT PAGE)

TABLE 5
(Continued)

n	x	.55	.60	.65	.70	.75	.80	.85	.90	.95
18	0	.0000	.0000	.0000	.0000	.0000	.0000	.0000	.0000	.0000
	1	.0000	.0000	.0000	.0000	.0000	.0000	.0000	.0000	.0000
	2	.0001	.0000	.0000	.0000	.0000	.0000	.0000	.0000	.0000
	3	.0009	.0002	.0000	.0000	.0000	.0000	.0000	.0000	.0000
	4	.0039	.0011	.0002	.0000	.0000	.0000	.0000	.0000	.0000
	5	.0134	.0045	.0012	.0002	.0000	.0000	.0000	.0000	.0000
	6	.0354	.0145	.0047	.0012	.0002	.0000	.0000	.0000	.0000
	7	.0742	.0374	.0151	.0046	.0010	.0001	.0000	.0000	.0000
	8	.1248	.0771	.0385	.0149	.0042	.0008	.0001	.0000	.0000
	9	.1694	.1284	.0794	.0386	.0139	.0033	.0004	.0000	.0000
	10	.1864	.1734	.1327	.0811	.0376	.0120	.0022	.0002	.0000
	11	.1657	.1892	.1792	.1376	.0820	.0350	.0091	.0010	.0000
	12	.1181	.1655	.1941	.1873	.1436	.0816	.0301	.0052	.0002
	13	.0666	.1146	.1664	.2017	.1988	.1507	.0787	.0218	.0014
	14	.0291	.0614	.1104	.1681	.2130	.2153	.1592	.0700	.0093
	15	.0095	.0246	.0547	.1046	.1704	.2297	.2406	.1680	.0473
	16	.0022	.0069	.0190	.0458	.0958	.1723	.2556	.2835	.1683
	17	.0003	.0012	.0042	.0126	.0338	.0811	.1704	.3002	.3763
	18	.0000	.0001	.0004	.0016	.0056	.0180	.0536	.1501	.3972
20	0	.0000	.0000	.0000	.0000	.0000	.0000	.0000	.0000	.0000
	1	.0000	.0000	.0000	.0000	.0000	.0000	.0000	.0000	.0000
	2	.0000	.0000	.0000	.0000	.0000	.0000	.0000	.0000	.0000
	3	.0002	.0000	.0000	.0000	.0000	.0000	.0000	.0000	.0000
	4	.0013	.0003	.0000	.0000	.0000	.0000	.0000	.0000	.0000
	5	.0049	.0013	.0003	.0000	.0000	.0000	.0000	.0000	.0000
	6	.0150	.0049	.0012	.0002	.0000	.0000	.0000	.0000	.0000
	7	.0366	.0146	.0045	.0010	.0002	.0000	.0000	.0000	.0000
	8	.0727	.0355	.0136	.0039	.0008	.0001	.0000	.0000	.0000
	9	.1185	.0710	.0336	.0120	.0030	.0005	.0000	.0000	.0000
	10	.1593	.1171	.0686	.0308	.0099	.0020	.0002	.0000	.0000
	11	.1771	.1597	.1158	.0654	.0271	.0074	.0011	.0001	.0000
	12	.1623	.1797	.1614	.1144	.0609	.0222	.0046	.0004	.0000
	13	.1221	.1659	.1844	.1643	.1124	.0545	.0160	.0020	.0000
	14	.0746	.1244	.1712	.1916	.1686	.1091	.0454	.0089	.0003
	15	.0365	.0746	.1272	.1789	.2023	.1746	.1028	.0319	.0022
	16	.0139	.0350	.0738	.1304	.1897	.2182	.1821	.0898	.0133
	17	.0040	.0123	.0323	.0716	.1339	.2054	.2428	.1901	.0596
	18	.0008	.0031	.0100	.0278	.0669	.1369	.2293	.2852	.1887
	19	.0001	.0005	.0020	.0068	.0211	.0576	.1368	.2702	.3774
	20	.0000	.0000	.0002	.0008	.0032	.0115	.0388	.1216	.3585

TABLE 6
Values of $e^{-\mu}$

μ	$e^{-\mu}$	μ	$e^{-\mu}$	μ	$e^{-\mu}$
.00	1.0000				
.05	.9512	2.05	.1287	4.05	.0174
.10	.9048	2.10	.1225	4.10	.0166
.15	.8607	2.15	.1165	4.15	.0158
.20	.8187	2.20	.1108	4.20	.0150
.25	.7788	2.25	.1054	4.25	.0143
.30	.7408	2.30	.1003	4.30	.0136
.35	.7047	2.35	.0954	4.35	.0129
.40	.6703	2.40	.0907	4.40	.0123
.45	.6376	2.45	.0863	4.45	.0117
.50	.6065	2.50	.0821	4.50	.0111
.55	.5769	2.55	.0781	4.55	.0106
.60	.5488	2.60	.0743	4.60	.0101
.65	.5220	2.65	.0707	4.65	.0096
.70	.4966	2.70	.0672	4.70	.0091
.75	.4724	2.75	.0639	4.75	.0087
.80	.4493	2.80	.0608	4.80	.0082
.85	.4274	2.85	.0578	4.85	.0078
.90	.4066	2.90	.0550	4.90	.0074
.95	.3867	2.95	.0523	4.95	.0071
1.00	.3679	3.00	.0498	5.00	.0067
1.05	.3499	3.05	.0474	5.05	.0064
1.10	.3329	3.10	.0450	5.10	.0061
1.15	.3166	3.15	.0429	5.15	.0058
1.20	.3012	3.20	.0408	5.20	.0055
1.25	.2865	3.25	.0388	5.25	.0052
1.30	.2725	3.30	.0369	5.30	.0050
1.35	.2592	3.35	.0351	5.35	.0047
1.40	.2466	3.40	.0334	5.40	.0045
1.45	.2346	3.45	.0317	5.45	.0043
1.50	.2231	3.50	.0302	5.50	.0041
1.55	.2122	3.55	.0287	5.55	.0039
1.60	.2019	3.60	.0273	5.60	.0037
1.65	.1920	3.65	.0260	5.65	.0035
1.70	.1827	3.70	.0247	5.70	.0033
1.75	.1738	3.75	.0235	5.75	.0032
1.80	.1653	3.80	.0224	5.80	.0030
1.85	.1572	3.85	.0213	5.85	.0029
1.90	.1496	3.90	.0202	5.90	.0027
1.95	.1423	3.95	.0193	5.95	.0026
2.00	.1353	4.00	.0183	6.00	.0025
				7.00	.0009
				8.00	.000335
				9.00	.000123
				10.00	.000045

TABLE 7
Poisson Probabilities

Entries in the table give the probability of x occurrences for a Poisson process with a mean μ. For example, when $\mu = 2.5$, the probability of four occurrences is .1336.

					μ					
x	0.1	0.2	0.3	0.4	0.5	0.6	0.7	0.8	0.9	1.0
0	.9048	.8187	.7408	.6703	.6065	.5488	.4966	.4493	.4066	.3679
1	.0905	.1637	.2222	.2681	.3033	.3293	.3476	.3595	.3659	.3679
2	.0045	.0164	.0333	.0536	.0758	.0988	.1217	.1438	.1647	.1839
3	.0002	.0011	.0033	.0072	.0126	.0198	.0284	.0383	.0494	.0613
4	.0000	.0001	.0002	.0007	.0016	.0030	.0050	.0077	.0111	.0153
5	.0000	.0000	.0000	.0001	.0002	.0004	.0007	.0012	.0020	.0031
6	.0000	.0000	.0000	.0000	.0000	.0000	.0001	.0002	.0003	.0005
7	.0000	.0000	.0000	.0000	.0000	.0000	.0000	.0000	.0000	.0001

					μ					
x	1.1	1.2	1.3	1.4	1.5	1.6	1.7	1.8	1.9	2.0
0	.3329	.3012	.2725	.2466	.2231	.2019	.1827	.1653	.1496	.1353
1	.3662	.3614	.3543	.3452	.3347	.3230	.3106	.2975	.2842	.2707
2	.2014	.2169	.2303	.2417	.2510	.2584	.2640	.2678	.2700	.2707
3	.0738	.0867	.0998	.1128	.1255	.1378	.1496	.1607	.1710	.1804
4	.0203	.0260	.0324	.0395	.0471	.0551	.0636	.0723	.0812	.0902
5	.0045	.0062	.0084	.0111	.0141	.0176	.0216	.0260	.0309	.0361
6	.0008	.0012	.0018	.0026	.0035	.0047	.0061	.0078	.0098	.0120
7	.0001	.0002	.0003	.0005	.0008	.0011	.0015	.0020	.0027	.0034
8	.0000	.0000	.0001	.0001	.0001	.0002	.0003	.0005	.0006	.0009
9	.0000	.0000	.0000	.0000	.0000	.0000	.0001	.0001	.0001	.0002

					μ					
x	2.1	2.2	2.3	2.4	2.5	2.6	2.7	2.8	2.9	3.0
0	.1225	.1108	.1003	.0907	.0821	.0743	.0672	.0608	.0550	.0498
1	.2572	.2438	.2306	.2177	.2052	.1931	.1815	.1703	.1596	.1494
2	.2700	.2681	.2652	.2613	.2565	.2510	.2450	.2384	.2314	.2240
3	.1890	.1966	.2033	.2090	.2138	.2176	.2205	.2225	.2237	.2240
4	.0992	.1082	.1169	.1254	.1336	.1414	.1488	.1557	.1622	.1680

(TABLE CONTINUES)

T A B L E 7
(Continued)

x	2.1	2.2	2.3	2.4	2.5	2.6	2.7	2.8	2.9	3.0
5	.0417	.0476	.0538	.0602	.0668	.0735	.0804	.0872	.0940	.1008
6	.0146	.0174	.0206	.0241	.0278	.0319	.0362	.0407	.0455	.0504
7	.0044	.0055	.0068	.0083	.0099	.0118	.0139	.0163	.0188	.0216
8	.0011	.0015	.0019	.0025	.0031	.0038	.0047	.0057	.0068	.0081
9	.0003	.0004	.0005	.0007	.0009	.0011	.0014	.0018	.0022	.0027
10	.0001	.0001	.0001	.0002	.0002	.0003	.0004	.0005	.0006	.0008
11	.0000	.0000	.0000	.0000	.0000	.0001	.0001	.0001	.0002	.0002
12	.0000	.0000	.0000	.0000	.0000	.0000	.0000	.0000	.0000	.0001

μ

x	3.1	3.2	3.3	3.4	3.5	3.6	3.7	3.8	3.9	4.0
0	.0450	.0408	.0369	.0344	.0302	.0273	.0247	.0224	.0202	.0183
1	.1397	.1304	.1217	.1135	.1057	.0984	.0915	.0850	.0789	.0733
2	.2165	.2087	.2008	.1929	.1850	.1771	.1692	.1615	.1539	.1465
3	.2237	.2226	.2209	.2186	.2158	.2125	.2087	.2046	.2001	.1954
4	.1734	.1781	.1823	.1858	.1888	.1912	.1931	.1944	.1951	.1954
5	.1075	.1140	.1203	.1264	.1322	.1377	.1429	.1477	.1522	.1563
6	.0555	.0608	.0662	.0716	.0771	.0826	.0881	.0936	.0989	.1042
7	.0246	.0278	.0312	.0348	.0385	.0425	.0466	.0508	.0551	.0595
8	.0095	.0111	.0129	.0148	.0169	.0191	.0215	.0241	.0269	.0298
9	.0033	.0040	.0047	.0056	.0066	.0076	.0089	.0102	.0116	.0132
10	.0010	.0013	.0016	.0019	.0023	.0028	.0033	.0039	.0045	.0053
11	.0003	.0004	.0005	.0006	.0007	.0009	.0011	.0013	.0016	.0019
12	.0001	.0001	.0001	.0002	.0002	.0003	.0003	.0004	.0005	.0006
13	.0000	.0000	.0000	.0000	.0001	.0001	.0001	.0001	.0002	.0002
14	.0000	.0000	.0000	.0000	.0000	.0000	.0000	.0000	.0000	.0001

μ

x	4.1	4.2	4.3	4.4	4.5	4.6	4.7	4.8	4.9	5.0
0	.0166	.0150	.0136	.0123	.0111	.0101	.0091	.0082	.0074	.0067
1	.0679	.0630	.0583	.0540	.0500	.0462	.0427	.0395	.0365	.0337
2	.1393	.1323	.1254	.1188	.1125	.1063	.1005	.0948	.0894	.0842
3	.1904	.1852	.1798	.1743	.1687	.1631	.1574	.1517	.1460	.1404
4	.1951	.1944	.1933	.1917	.1898	.1875	.1849	.1820	.1789	.1755
5	.1600	.1633	.1662	.1687	.1708	.1725	.1738	.1747	.1753	.1755
6	.1093	.1143	.1191	.1237	.1281	.1323	.1362	.1398	.1432	.1462
7	.0640	.0686	.0732	.0778	.0824	.0869	.0914	.0959	.1002	.1044
8	.0328	.0360	.0393	.0428	.0463	.0500	.0537	.0575	.0614	.0653
9	.0150	.0168	.0188	.0209	.0232	.0255	.0280	.0307	.0334	.0363

(TABLE CONTINUES ON NEXT PAGE)

TABLE 7
(Continued)

					μ					
x	4.1	4.2	4.3	4.4	4.5	4.6	4.7	4.8	4.9	5.0
10	.0061	.0071	.0081	.0092	.0104	.0118	.0132	.0147	.0164	.0181
11	.0023	.0027	.0032	.0037	.0043	.0049	.0056	.0064	.0073	.0082
12	.0008	.0009	.0011	.0014	.0016	.0019	.0022	.0026	.0030	.0034
13	.0002	.0003	.0004	.0005	.0006	.0007	.0008	.0009	.0011	.0013
14	.0001	.0001	.0001	.0001	.0002	.0002	.0003	.0003	.0004	.0005
15	.0000	.0000	.0000	.0000	.0001	.0001	.0001	.0001	.0001	.0002

					μ					
x	5.1	5.2	5.3	5.4	5.5	5.6	5.7	5.8	5.9	6.0
0	.0061	.0055	.0050	.0045	.0041	.0037	.0033	.0030	.0027	.0025
1	.0311	.0287	.0265	.0244	.0225	.0207	.0191	.0176	.0162	.0149
2	.0793	.0746	.0701	.0659	.0618	.0580	.0544	.0509	.0477	.0446
3	.1348	.1293	.1239	.1185	.1133	.1082	.1033	.0985	.0938	.0892
4	.1719	.1681	.1641	.1600	.1558	.1515	.1472	.1428	.1383	.1339
5	.1753	.1748	.1740	.1728	.1714	.1697	.1678	.1656	.1632	.1606
6	.1490	.1515	.1537	.1555	.1571	.1584	.1594	.1601	.1605	.1606
7	.1086	.1125	.1163	.1200	.1234	.1267	.1298	.1326	.1353	.1377
8	.0692	.0731	.0771	.0810	.0849	.0887	.0925	.0962	.0998	.1033
9	.0392	.0423	.0454	.0486	.0519	.0552	.0586	.0620	.0654	.0688
10	.0200	.0220	.0241	.0262	.0285	.0309	.0334	.0359	.0386	.0413
11	.0093	.0104	.0116	.0129	.0143	.0157	.0173	.0190	.0207	.0225
12	.0039	.0045	.0051	.0058	.0065	.0073	.0082	.0092	.0102	.0113
13	.0015	.0018	.0021	.0024	.0028	.0032	.0036	.0041	.0046	.0052
14	.0006	.0007	.0008	.0009	.0011	.0013	.0015	.0017	.0019	.0022
15	.0002	.0002	.0003	.0003	.0004	.0005	.0006	.0007	.0008	.0009
16	.0001	.0001	.0001	.0001	.0001	.0002	.0002	.0002	.0003	.0003
17	.0000	.0000	.0000	.0000	.0000	.0001	.0001	.0001	.0001	.0001

					μ					
x	6.1	6.2	6.3	6.4	6.5	6.6	6.7	6.8	6.9	7.0
0	.0022	.0020	.0018	.0017	.0015	.0014	.0012	.0011	.0010	.0009
1	.0137	.0126	.0116	.0106	.0098	.0090	.0082	.0076	.0070	.0064
2	.0417	.0390	.0364	.0340	.0318	.0296	.0276	.0258	.0240	.0223
3	.0848	.0806	.0765	.0726	.0688	.0652	.0617	.0584	.0552	.0521
4	.1294	.1249	.1205	.1162	.1118	.1076	.1034	.0992	.0952	.0912
5	.1579	.1549	.1519	.1487	.1454	.1420	.1385	.1349	.1314	.1277
6	.1605	.1601	.1595	.1586	.1575	.1562	.1546	.1529	.1511	.1490
7	.1399	.1418	.1435	.1450	.1462	.1472	.1480	.1486	.1489	.1490
8	.1066	.1099	.1130	.1160	.1188	.1215	.1240	.1263	.1284	.1304
9	.0723	.0757	.0791	.0825	.0858	.0891	.0923	.0954	.0985	.1014

(TABLE CONTINUES)

TABLE 7
(Continued)

					μ					
x	6.1	6.2	6.3	6.4	6.5	6.6	6.7	6.8	6.9	7.0
10	.0441	.0469	.0498	.0528	.0558	.0588	.0618	.0649	.0679	.0710
11	.0245	.0265	.0285	.0307	.0330	.0353	.0377	.0401	.0426	.0452
12	.0124	.0137	.0150	.0164	.0179	.0194	.0210	.0227	.0245	.0264
13	.0058	.0065	.0073	.0081	.0089	.0098	.0108	.0119	.0130	.0142
14	.0025	.0029	.0033	.0037	.0041	.0046	.0052	.0058	.0064	.0071
15	.0010	.0012	.0014	.0016	.0018	.0020	.0023	.0026	.0029	.0033
16	.0004	.0005	.0005	.0006	.0007	.0008	.0010	.0011	.0013	.0014
17	.0001	.0002	.0002	.0002	.0003	.0003	.0004	.0004	.0005	.0006
18	.0000	.0001	.0001	.0001	.0001	.0001	.0001	.0002	.0002	.0002
19	.0000	.0000	.0000	.0000	.0000	.0000	.0000	.0001	.0001	.0001

					μ					
x	7.1	7.2	7.3	7.4	7.5	7.6	7.7	7.8	7.9	8.0
0	.0008	.0007	.0007	.0006	.0006	.0005	.0005	.0004	.0004	.0003
1	.0059	.0054	.0049	.0045	.0041	.0038	.0035	.0032	.0029	.0027
2	.0208	.0194	.0180	.0167	.0156	.0145	.0134	.0125	.0116	.0107
3	.0492	.0464	.0438	.0413	.0389	.0366	.0345	.0324	.0305	.0286
4	.0874	.0836	.0799	.0764	.0729	.0696	.0663	.0632	.0602	.0573
5	.1241	.1204	.1167	.1130	.1094	.1057	.1021	.0986	.0951	.0916
6	.1468	.1445	.1420	.1394	.1367	.1339	.1311	.1282	.1252	.1221
7	.1489	.1486	.1481	.1474	.1465	.1454	.1442	.1428	.1413	.1396
8	.1321	.1337	.1351	.1363	.1373	.1382	.1388	.1392	.1395	.1396
9	.1042	.1070	.1096	.1121	.1144	.1167	.1187	.1207	.1224	.1241
10	.0740	.0770	.0800	.0829	.0858	.0887	.0914	.0941	.0967	.0993
11	.0478	.0504	.0531	.0558	.0585	.0613	.0640	.0667	.0695	.0722
12	.0283	.0303	.0323	.0344	.0366	.0388	.0411	.0434	.0457	.0481
13	.0154	.0168	.0181	.0196	.0211	.0227	.0243	.0260	.0278	.0296
14	.0078	.0086	.0095	.0104	.0113	.0123	.0134	.0145	.0157	.0169
15	.0037	.0041	.0046	.0051	.0057	.0062	.0069	.0075	.0083	.0090
16	.0016	.0019	.0021	.0024	.0026	.0030	.0033	.0037	.0041	.0045
17	.0007	.0008	.0009	.0010	.0012	.0013	.0015	.0017	.0019	.0021
18	.0003	.0003	.0004	.0004	.0005	.0006	.0006	.0007	.0008	.0009
19	.0001	.0001	.0001	.0002	.0002	.0002	.0003	.0003	.0003	.0004
20	.0000	.0000	.0001	.0001	.0001	.0001	.0001	.0001	.0001	.0002
21	.0000	.0000	.0000	.0000	.0000	.0000	.0000	.0000	.0001	.0001

(TABLE CONTINUES ON NEXT PAGE)

TABLE 7
(Continued)

x	8.1	8.2	8.3	8.4	8.5	8.6	8.7	8.8	8.9	9.0
0	.0003	.0003	.0002	.0002	.0002	.0002	.0002	.0002	.0001	.0001
1	.0025	.0023	.0021	.0019	.0017	.0016	.0014	.0013	.0012	.0011
2	.0100	.0092	.0086	.0079	.0074	.0068	.0063	.0058	.0054	.0050
3	.0269	.0252	.0237	.0222	.0208	.0195	.0183	.0171	.0160	.0150
4	.0544	.0517	.0491	.0466	.0443	.0420	.0398	.0377	.0357	.0337
5	.0882	.0849	.0816	.0784	.0752	.0722	.0692	.0663	.0635	.0607
6	.1191	.1160	.1128	.1097	.1066	.1034	.1003	.0972	.0941	.0911
7	.1378	.1358	.1338	.1317	.1294	.1271	.1247	.1222	.1197	.1171
8	.1395	.1392	.1388	.1382	.1375	.1366	.1356	.1344	.1332	.1318
9	.1256	.1269	.1280	.1290	.1299	.1306	.1311	.1315	.1317	.1318
10	.1017	.1040	.1063	.1084	.1104	.1123	.1140	.1157	.1172	.1186
11	.0749	.0776	.0802	.0828	.0853	.0878	.0902	.0925	.0948	.0970
12	.0505	.0530	.0555	.0579	.0604	.0629	.0654	.0679	.0703	.0728
13	.0315	.0334	.0354	.0374	.0395	.0416	.0438	.0459	.0481	.0504
14	.0182	.0196	.0210	.0225	.0240	.0256	.0272	.0289	.0306	.0324
15	.0098	.0107	.0116	.0126	.0136	.0147	.0158	.0169	.0182	.0194
16	.0050	.0055	.0060	.0066	.0072	.0079	.0086	.0093	.0101	.0109
17	.0024	.0026	.0029	.0033	.0036	.0040	.0044	.0048	.0053	.0058
18	.0011	.0012	.0014	.0015	.0017	.0019	.0021	.0024	.0026	.0029
19	.0005	.0005	.0006	.0007	.0008	.0009	.0010	.0011	.0012	.0014
20	.0002	.0002	.0002	.0003	.0003	.0004	.0004	.0005	.0005	.0006
21	.0001	.0001	.0001	.0001	.0001	.0002	.0002	.0002	.0002	.0003
22	.0000	.0000	.0000	.0000	.0001	.0001	.0001	.0001	.0001	.0001

μ

x	9.1	9.2	9.3	9.4	9.5	9.6	9.7	9.8	9.9	10
0	.0001	.0001	.0001	.0001	.0001	.0001	.0001	.0001	.0001	.0000
1	.0010	.0009	.0009	.0008	.0007	.0007	.0006	.0005	.0005	.0005
2	.0046	.0043	.0040	.0037	.0034	.0031	.0029	.0027	.0025	.0023
3	.0140	.0131	.0123	.0115	.0107	.0100	.0093	.0087	.0081	.0076
4	.0319	.0302	.0285	.0269	.0254	.0240	.0226	.0213	.0201	.0189
5	.0581	.0555	.0530	.0506	.0483	.0460	.0439	.0418	.0398	.0378
6	.0881	.0851	.0822	.0793	.0764	.0736	.0709	.0682	.0656	.0631
7	.1145	.1118	.1091	.1064	.1037	.1010	.0982	.0955	.0928	.0901
8	.1302	.1286	.1269	.1251	.1232	.1212	.1191	.1170	.1148	.1126
9	.1317	.1315	.1311	.1306	.1300	.1293	.1284	.1274	.1263	.1251

(TABLE CONTINUES)

TABLE 7

(Continued)

	μ									
x	9.1	9.2	9.3	9.4	9.5	9.6	9.7	9.8	9.9	10.0
10	.1198	.1210	.1219	.1228	.1235	.1241	.1245	.1249	.1250	.1251
11	.0991	.1012	.1031	.1049	.1067	.1083	.1098	.1112	.1125	.1137
12	.0752	.0776	.0799	.0822	.0844	.0866	.0888	.0908	.0928	.0948
13	.0526	.0549	.0572	.0594	.0617	.0640	.0662	.0685	.0707	.0729
14	.0342	.0361	.0380	.0399	.0419	.0439	.0459	.0479	.0500	.0521
15	.0208	.0221	.0235	.0250	.0265	.0281	.0297	.0313	.0330	.0347
16	.0118	.0127	.0137	.0147	.0157	.0168	.0180	.0192	.0204	.0217
17	.0063	.0069	.0075	.0081	.0088	.0095	.0103	.0111	.0119	.0128
18	.0032	.0035	.0039	.0042	.0046	.0051	.0055	.0060	.0065	.0071
19	.0015	.0017	.0019	.0021	.0023	.0026	.0028	.0031	.0034	.0037
20	.0007	.0008	.0009	.0010	.0011	.0012	.0014	.0015	.0017	.0019
21	.0003	.0003	.0004	.0004	.0005	.0006	.0006	.0007	.0008	.0009
22	.0001	.0001	.0002	.0002	.0002	.0002	.0003	.0003	.0004	.0004
23	.0000	.0001	.0001	.0001	.0001	.0001	.0001	.0001	.0002	.0002
24	.0000	.0000	.0000	.0000	.0000	.0000	.0000	.0001	.0001	.0001

	μ									
x	11	12	13	14	15	16	17	18	19	20
0	.0000	.0000	.0000	.0000	.0000	.0000	.0000	.0000	.0000	.0000
1	.0002	.0001	.0000	.0000	.0000	.0000	.0000	.0000	.0000	.0000
2	.0010	.0004	.0002	.0001	.0000	.0000	.0000	.0000	.0000	.0000
3	.0037	.0018	.0008	.0004	.0002	.0001	.0000	.0000	.0000	.0000
4	.0102	.0053	.0027	.0013	.0006	.0003	.0001	.0001	.0000	.0000
5	.0224	.0127	.0070	.0037	.0019	.0010	.0005	.0002	.0001	.0001
6	.0411	.0255	.0152	.0087	.0048	.0026	.0014	.0007	.0004	.0002
7	.0646	.0437	.0281	.0174	.0104	.0060	.0034	.0018	.0010	.0005
8	.0888	.0655	.0457	.0304	.0194	.0120	.0072	.0042	.0024	.0013
9	.1085	.0874	.0661	.0473	.0324	.0213	.0135	.0083	.0050	.0029
10	.1194	.1048	.0859	.0663	.0486	.0341	.0230	.0150	.0095	.0058
11	.1194	.1144	.1015	.0844	.0663	.0496	.0355	.0245	.0164	.0106
12	.1094	.1144	.1099	.0984	.0829	.0661	.0504	.0368	.0259	.0176
13	.0926	.1056	.1099	.1060	.0956	.0814	.0658	.0509	.0378	.0271
14	.0728	.0905	.1021	.1060	.1024	.0930	.0800	.0655	.0514	.0387
15	.0534	.0724	.0885	.0989	.1024	.0992	.0906	.0786	.0650	.0516
16	.0367	.0543	.0719	.0866	.0960	.0992	.0963	.0884	.0772	.0646
17	.0237	.0383	.0550	.0713	.0847	.0934	.0963	.0936	.0863	.0760
18	.0145	.0256	.0397	.0554	.0706	.0830	.0909	.0936	.0911	.0844
19	.0084	.0161	.0272	.0409	.0557	.0699	.0814	.0887	.0911	.0888

(TABLE CONTINUES ON NEXT PAGE)

TABLE 7
(Continued)

x	μ									
	11	12	13	14	15	16	17	18	19	20
20	.0046	.0097	.0177	.0286	.0418	.0559	.0692	.0798	.0866	.0888
21	.0024	.0055	.0109	.0191	.0299	.0426	.0560	.0684	.0783	.0846
22	.0012	.0030	.0065	.0121	.0204	.0310	.0433	.0560	.0676	.0769
23	.0006	.0016	.0037	.0074	.0133	.0216	.0320	.0438	.0559	.0669
24	.0003	.0008	.0020	.0043	.0083	.0144	.0226	.0328	.0442	.0557
25	.0001	.0004	.0010	.0024	.0050	.0092	.0154	.0237	.0336	.0446
26	.0000	.0002	.0005	.0013	.0029	.0057	.0101	.0164	.0246	.0343
27	.0000	.0001	.0002	.0007	.0016	.0034	.0063	.0109	.0173	.0254
28	.0000	.0000	.0001	.0003	.0009	.0019	.0038	.0070	.0117	.0181
29	.0000	.0000	.0001	.0002	.0004	.0011	.0023	.0044	.0077	.0125
30	.0000	.0000	.0000	.0001	.0002	.0006	.0013	.0026	.0049	.0083
31	.0000	.0000	.0000	.0000	.0001	.0003	.0007	.0015	.0030	.0054
32	.0000	.0000	.0000	.0000	.0001	.0001	.0004	.0009	.0018	.0034
33	.0000	.0000	.0000	.0000	.0000	.0001	.0002	.0005	.0010	.0020
34	.0000	.0000	.0000	.0000	.0000	.0000	.0001	.0002	.0006	.0012
35	.0000	.0000	.0000	.0000	.0000	.0000	.0000	.0001	.0003	.0007
36	.0000	.0000	.0000	.0000	.0000	.0000	.0000	.0001	.0002	.0004
37	.0000	.0000	.0000	.0000	.0000	.0000	.0000	.0000	.0001	.0002
38	.0000	.0000	.0000	.0000	.0000	.0000	.0000	.0000	.0000	.0001
39	.0000	.0000	.0000	.0000	.0000	.0000	.0000	.0000	.0000	.0001

TABLE 8
Random Numbers

63271	59986	71744	51102	15141	80714	58683	93108	13554	79945
88547	09896	95436	79115	08303	01041	20030	63754	08459	28364
55957	57243	83865	09911	19761	66535	40102	26646	60147	15702
46276	87453	44790	67122	45573	84358	21625	16999	13385	22782
55363	07449	34835	15290	76616	67191	12777	21861	68689	03263
69393	92785	49902	58447	42048	30378	87618	26933	40640	16281
13186	29431	88190	04588	38733	81290	89541	70290	40113	08243
17726	28652	56836	78351	47327	18518	92222	55201	27340	10493
36520	64465	05550	30157	82242	29520	69753	72602	23756	54935
81628	36100	39254	56835	37636	02421	98063	89641	64953	99337
84649	48968	75215	75498	49539	74240	03466	49292	36401	45525
63291	11618	12613	75055	43915	26488	41116	64531	56827	30825
70502	53225	03655	05915	37140	57051	48393	91322	25653	06543
06426	24771	59935	49801	11082	66762	94477	02494	88215	27191
20711	55609	29430	70165	45406	78484	31639	52009	18873	96927
41990	70538	77191	25860	55204	73417	83920	69468	74972	38712
72452	36618	76298	26678	89334	33938	95567	29380	75906	91807
37042	40318	57099	10528	09925	89773	41335	96244	29002	46453
53766	52875	15987	46962	67342	77592	57651	95508	80033	69828
90585	58955	53122	16025	84299	53310	67380	84249	25348	04332
32001	96293	37203	64516	51530	37069	40261	61374	05815	06714
62606	64324	46354	72157	67248	20135	49804	09226	64419	29457
10078	28073	85389	50324	14500	15562	64165	06125	71353	77669
91561	46145	24177	15294	10061	98124	75732	00815	83452	97355
13091	98112	53959	79607	52244	63303	10413	63839	74762	50289
73864	83014	72457	22682	03033	61714	88173	90835	00634	85169
66668	25467	48894	51043	02365	91726	09365	63167	95264	45643
84745	41042	29493	01836	09044	51926	43630	63470	76508	14194
48068	26805	94595	47907	13357	38412	33318	26098	82782	42851
54310	96175	97594	88616	42035	38093	36745	56702	40644	83514
14877	33095	10924	58013	61439	21882	42059	24177	58739	60170
78295	23179	02771	43464	59061	71411	05697	67194	30495	21157
67524	02865	39593	54278	04237	92441	26602	63835	38032	94770
58268	57219	68124	73455	83236	08710	04284	55005	84171	42596
97158	28672	50685	01181	24262	19427	52106	34308	73685	74246
04230	16831	69085	30802	65559	09205	71829	06489	85650	38707
94879	56606	30401	02602	57658	70091	54986	41394	60437	03195
71446	15232	66715	26385	91518	70566	02888	79941	39684	54315
32886	05644	79316	09819	00813	88407	17461	73925	53037	91904
62048	33711	25290	21526	02223	75947	66466	06232	10913	75336
84534	42351	21628	53669	81352	95152	08107	98814	72743	12849
84707	15885	84710	35866	06446	86311	32648	88141	73902	69981
19409	40868	64220	80861	13860	68493	52908	26374	63297	45052
57978	48015	25973	66777	45924	56144	24742	96702	88200	66162
57295	98298	11199	96510	75228	41600	47192	43267	35973	23152
94044	83785	93388	07833	38216	31413	70555	03023	54147	06647
30014	25879	71763	96679	90603	99396	74557	74224	18211	91637
07265	69563	64268	88802	72264	66540	01782	08396	19251	83613
84404	88642	30263	80310	11522	57810	27627	78376	36240	48952
21778	02085	27762	46097	43324	34354	09369	14966	10158	76089

TABLE 9
Critical Values for the Durbin-Watson Test for Autocorrelation

Entries in the table give the critical values for a one-tailed Durbin-Watson test for autocorrelation. For a two-tailed test, the level of significance is doubled.

Significance Points of d_L and d_U: $\alpha = .05$
Number of Independent Variables

k	1		2		3		4		5	
n	d_L	d_U	d_L	d_U	d_L	d_U	d_L	d_U	d_L	d_U
15	1.08	1.36	0.95	1.54	0.82	1.75	0.69	1.97	0.56	2.21
16	1.10	1.37	0.98	1.54	0.86	1.73	0.74	1.93	0.62	2.15
17	1.13	1.38	1.02	1.54	0.90	1.71	0.78	1.90	0.67	2.10
18	1.16	1.39	1.05	1.53	0.93	1.69	0.82	1.87	0.71	2.06
19	1.18	1.40	1.08	1.53	0.97	1.68	0.86	1.85	0.75	2.02
20	1.20	1.41	1.10	1.54	1.00	1.68	0.90	1.83	0.79	1.99
21	1.22	1.42	1.13	1.54	1.03	1.67	0.93	1.81	0.83	1.96
22	1.24	1.43	1.15	1.54	1.05	1.66	0.96	1.80	0.86	1.94
23	1.26	1.44	1.17	1.54	1.08	1.66	0.99	1.79	0.90	1.92
24	1.27	1.45	1.19	1.55	1.10	1.66	1.01	1.78	0.93	1.90
25	1.29	1.45	1.21	1.55	1.12	1.66	1.04	1.77	0.95	1.89
26	1.30	1.46	1.22	1.55	1.14	1.65	1.06	1.76	0.98	1.88
27	1.32	1.47	1.24	1.56	1.16	1.65	1.08	1.76	1.01	1.86
28	1.33	1.48	1.26	1.56	1.18	1.65	1.10	1.75	1.03	1.85
29	1.34	1.48	1.27	1.56	1.20	1.65	1.12	1.74	1.05	1.84
30	1.35	1.49	1.28	1.57	1.21	1.65	1.14	1.74	1.07	1.83
31	1.36	1.50	1.30	1.57	1.23	1.65	1.16	1.74	1.09	1.83
32	1.37	1.50	1.31	1.57	1.24	1.65	1.18	1.73	1.11	1.82
33	1.38	1.51	1.32	1.58	1.26	1.65	1.19	1.73	1.13	1.81
34	1.39	1.51	1.33	1.58	1.27	1.65	1.21	1.73	1.15	1.81
35	1.40	1.52	1.34	1.58	1.28	1.65	1.22	1.73	1.16	1.80
36	1.41	1.52	1.35	1.59	1.29	1.65	1.24	1.73	1.18	1.80
37	1.42	1.53	1.36	1.59	1.31	1.66	1.25	1.72	1.19	1.80
38	1.43	1.54	1.37	1.59	1.32	1.66	1.26	1.72	1.21	1.79
39	1.43	1.54	1.38	1.60	1.33	1.66	1.27	1.72	1.22	1.79
40	1.44	1.54	1.39	1.60	1.34	1.66	1.29	1.72	1.23	1.79
45	1.48	1.57	1.43	1.62	1.38	1.67	1.34	1.72	1.29	1.78
50	1.50	1.59	1.46	1.63	1.42	1.67	1.38	1.72	1.34	1.77
55	1.53	1.60	1.49	1.64	1.45	1.68	1.41	1.72	1.38	1.77
60	1.55	1.62	1.51	1.65	1.48	1.69	1.44	1.73	1.41	1.77
65	1.57	1.63	1.54	1.66	1.50	1.70	1.47	1.73	1.44	1.77
70	1.58	1.64	1.55	1.67	1.52	1.70	1.49	1.74	1.46	1.77
75	1.60	1.65	1.57	1.68	1.54	1.71	1.51	1.74	1.49	1.77
80	1.61	1.66	1.59	1.69	1.56	1.72	1.53	1.74	1.51	1.77
85	1.62	1.67	1.60	1.70	1.57	1.72	1.55	1.75	1.52	1.77
90	1.63	1.68	1.61	1.70	1.59	1.73	1.57	1.75	1.54	1.78
95	1.64	1.69	1.62	1.71	1.60	1.73	1.58	1.75	1.56	1.78
100	1.65	1.69	1.63	1.72	1.61	1.74	1.59	1.76	1.57	1.78

(TABLE CONTINUES)

This table comes from J. Durbin and G. S. Watson, "Testing for serial correlation in least square regression II," *Biometrika*, 38, 1951, 159–178.

TABLE 9
(Continued)

Significance Points of d_L and d_U: $\alpha = .025$
Number of Independent Variables

k	1		2		3		4		5	
n	d_L	d_U	d_L	d_U	d_L	d_U	d_L	d_U	d_L	d_U
15	0.95	1.23	0.83	1.40	0.71	1.61	0.59	1.84	0.48	2.09
16	0.98	1.24	0.86	1.40	0.75	1.59	0.64	1.80	0.53	2.03
17	1.01	1.25	0.90	1.40	0.79	1.58	0.68	1.77	0.57	1.98
18	1.03	1.26	0.93	1.40	0.82	1.56	0.72	1.74	0.62	1.93
19	1.06	1.28	0.96	1.41	0.86	1.55	0.76	1.72	0.66	1.90
20	1.08	1.28	0.99	1.41	0.89	1.55	0.79	1.70	0.70	1.87
21	1.10	1.30	1.01	1.41	0.92	1.54	0.83	1.69	0.73	1.84
22	1.12	1.31	1.04	1.42	0.95	1.54	0.86	1.68	0.77	1.82
23	1.14	1.32	1.06	1.42	0.97	1.54	0.89	1.67	0.80	1.80
24	1.16	1.33	1.08	1.43	1.00	1.54	0.91	1.66	0.83	1.79
25	1.18	1.34	1.10	1.43	1.02	1.54	0.94	1.65	0.86	1.77
26	1.19	1.35	1.12	1.44	1.04	1.54	0.96	1.65	0.88	1.76
27	1.21	1.36	1.13	1.44	1.06	1.54	0.99	1.64	0.91	1.75
28	1.22	1.37	1.15	1.45	1.08	1.54	1.01	1.64	0.93	1.74
29	1.24	1.38	1.17	1.45	1.10	1.54	1.03	1.63	0.96	1.73
30	1.25	1.38	1.18	1.46	1.12	1.54	1.05	1.63	0.98	1.73
31	1.26	1.39	1.20	1.47	1.13	1.55	1.07	1.63	1.00	1.72
32	1.27	1.40	1.21	1.47	1.15	1.55	1.08	1.63	1.02	1.71
33	1.28	1.41	1.22	1.48	1.16	1.55	1.10	1.63	1.04	1.71
34	1.29	1.41	1.24	1.48	1.17	1.55	1.12	1.63	1.06	1.70
35	1.30	1.42	1.25	1.48	1.19	1.55	1.13	1.63	1.07	1.70
36	1.31	1.43	1.26	1.49	1.20	1.56	1.15	1.63	1.09	1.70
37	1.32	1.43	1.27	1.49	1.21	1.56	1.16	1.62	1.10	1.70
38	1.33	1.44	1.28	1.50	1.23	1.56	1.17	1.62	1.12	1.70
39	1.34	1.44	1.29	1.50	1.24	1.56	1.19	1.63	1.13	1.69
40	1.35	1.45	1.30	1.51	1.25	1.57	1.20	1.63	1.15	1.69
45	1.39	1.48	1.34	1.53	1.30	1.58	1.25	1.63	1.21	1.69
50	1.42	1.50	1.38	1.54	1.34	1.59	1.30	1.64	1.26	1.69
55	1.45	1.52	1.41	1.56	1.37	1.60	1.33	1.64	1.30	1.69
60	1.47	1.54	1.44	1.57	1.40	1.61	1.37	1.65	1.33	1.69
65	1.49	1.55	1.46	1.59	1.43	1.62	1.40	1.66	1.36	1.69
70	1.51	1.57	1.48	1.60	1.45	1.63	1.42	1.66	1.39	1.70
75	1.53	1.58	1.50	1.61	1.47	1.64	1.45	1.67	1.42	1.70
80	1.54	1.59	1.52	1.62	1.49	1.65	1.47	1.67	1.44	1.70
85	1.56	1.60	1.53	1.63	1.51	1.65	1.49	1.68	1.46	1.71
90	1.57	1.61	1.55	1.64	1.53	1.66	1.50	1.69	1.48	1.71
95	1.58	1.62	1.56	1.65	1.54	1.67	1.52	1.69	1.50	1.71
100	1.59	1.63	1.57	1.65	1.55	1.67	1.53	1.70	1.51	1.72

(TABLE CONTINUES ON NEXT PAGE)

TABLE 9
(Continued)

Significance Points of d_L and d_U: $\alpha = .01$
Number of Independent Variables

n	k = 1 d_L	d_U	2 d_L	d_U	3 d_L	d_U	4 d_L	d_U	5 d_L	d_U
15	0.81	1.07	0.70	1.25	0.59	1.46	0.49	1.70	0.39	1.96
16	0.84	1.09	0.74	1.25	0.63	1.44	0.53	1.66	0.44	1.90
17	0.87	1.10	0.77	1.25	0.67	1.43	0.57	1.63	0.48	1.85
18	0.90	1.12	0.80	1.26	0.71	1.42	0.61	1.60	0.52	1.80
19	0.93	1.13	0.83	1.26	0.74	1.41	0.65	1.58	0.56	1.77
20	0.95	1.15	0.86	1.27	0.77	1.41	0.68	1.57	0.60	1.74
21	0.97	1.16	0.89	1.27	0.80	1.41	0.72	1.55	0.63	1.71
22	1.00	1.17	0.91	1.28	0.83	1.40	0.75	1.54	0.66	1.69
23	1.02	1.19	0.94	1.29	0.86	1.40	0.77	1.53	0.70	1.67
24	1.04	1.20	0.96	1.30	0.88	1.41	0.80	1.53	0.72	1.66
25	1.05	1.21	0.98	1.30	0.90	1.41	0.83	1.52	0.75	1.65
26	1.07	1.22	1.00	1.31	0.93	1.41	0.85	1.52	0.78	1.64
27	1.09	1.23	1.02	1.32	0.95	1.41	0.88	1.51	0.81	1.63
28	1.10	1.24	1.04	1.32	0.97	1.41	0.90	1.51	0.83	1.62
29	1.12	1.25	1.05	1.33	0.99	1.42	0.92	1.51	0.85	1.61
30	1.13	1.26	1.07	1.34	1.01	1.42	0.94	1.51	0.88	1.61
31	1.15	1.27	1.08	1.34	1.02	1.42	0.96	1.51	0.90	1.60
32	1.16	1.28	1.10	1.35	1.04	1.43	0.98	1.51	0.92	1.60
33	1.17	1.29	1.11	1.36	1.05	1.43	1.00	1.51	0.94	1.59
34	1.18	1.30	1.13	1.36	1.07	1.43	1.01	1.51	0.95	1.59
35	1.19	1.31	1.14	1.37	1.08	1.44	1.03	1.51	0.97	1.59
36	1.21	1.32	1.15	1.38	1.10	1.44	1.04	1.51	0.99	1.59
37	1.22	1.32	1.16	1.38	1.11	1.45	1.06	1.51	1.00	1.59
38	1.23	1.33	1.18	1.39	1.12	1.45	1.07	1.52	1.02	1.58
39	1.24	1.34	1.19	1.39	1.14	1.45	1.09	1.52	1.03	1.58
40	1.25	1.34	1.20	1.40	1.15	1.46	1.10	1.52	1.05	1.58
45	1.29	1.38	1.24	1.42	1.20	1.48	1.16	1.53	1.11	1.58
50	1.32	1.40	1.28	1.45	1.24	1.49	1.20	1.54	1.16	1.59
55	1.36	1.43	1.32	1.47	1.28	1.51	1.25	1.55	1.21	1.59
60	1.38	1.45	1.35	1.48	1.32	1.52	1.28	1.56	1.25	1.60
65	1.41	1.47	1.38	1.50	1.35	1.53	1.31	1.57	1.28	1.61
70	1.43	1.49	1.40	1.52	1.37	1.55	1.34	1.58	1.31	1.61
75	1.45	1.50	1.42	1.53	1.39	1.56	1.37	1.59	1.34	1.62
80	1.47	1.52	1.44	1.54	1.42	1.57	1.39	1.60	1.36	1.62
85	1.48	1.53	1.46	1.55	1.43	1.58	1.41	1.60	1.39	1.63
90	1.50	1.54	1.47	1.56	1.45	1.59	1.43	1.61	1.41	1.64
95	1.51	1.55	1.49	1.57	1.47	1.60	1.45	1.62	1.42	1.64
100	1.52	1.56	1.50	1.58	1.48	1.60	1.46	1.63	1.44	1.65

TABLE 10

T_L Values for the Mann-Whitney-Wilcoxon Test

Reject the hypothesis of identical populations if the sum of the ranks for the n_1 items is *less than* the value T_L shown in the following table or if the sum of the ranks for the n_1 items is *greater than* the value T_U where

$$T_U = n_1(n_1 + n_2 + 1) - T_L$$

$\alpha = .05$		n_2								
		2	3	4	5	6	7	8	9	10
	2	3	3	3	3	3	3	4	4	4
	3	6	6	6	7	8	8	9	9	10
	4	10	10	11	12	13	14	15	15	16
	5	15	16	17	18	19	21	22	23	24
n_1	6	21	23	24	25	27	28	30	32	33
	7	28	30	32	34	35	37	39	41	43
	8	37	39	41	43	45	47	50	52	54
	9	46	48	50	53	56	58	61	63	66
	10	56	59	61	64	67	70	73	76	79

$\alpha = .10$		n_2								
		2	3	4	5	6	7	8	9	10
	2	3	3	3	4	4	4	5	5	5
	3	6	7	7	8	9	9	10	11	11
	4	10	11	12	13	14	15	16	17	18
	5	16	17	18	20	21	22	24	25	27
n_1	6	22	24	25	27	29	30	32	34	36
	7	29	31	33	35	37	40	42	44	46
	8	38	40	42	45	47	50	52	55	57
	9	47	50	52	55	58	61	64	67	70
	10	57	60	63	67	70	73	76	80	83

TABLE 11 Critical Values of the Studentized Range Distribution

$\alpha = .05$

Degrees of Freedom	Number of Populations																		
	2	3	4	5	6	7	8	9	10	11	12	13	14	15	16	17	18	19	20
1	18.0	27.0	32.8	37.1	40.4	43.1	45.4	47.4	49.1	50.6	52.0	53.2	54.3	55.4	56.3	57.2	58.0	58.8	59.6
2	6.08	8.33	9.80	10.9	11.7	12.4	13.0	13.5	14.0	14.4	14.7	15.1	15.4	15.7	15.9	16.1	16.4	16.6	16.8
3	4.50	5.91	6.82	7.50	8.04	8.48	8.85	9.18	9.46	9.72	9.95	10.2	10.3	10.5	10.7	10.8	11.0	11.1	11.2
4	3.93	5.04	5.76	6.29	6.71	7.05	7.35	7.60	7.83	8.03	8.21	8.37	8.52	8.66	8.79	8.91	9.03	9.13	9.23
5	3.64	4.60	5.22	5.67	6.03	6.33	6.58	6.80	6.99	7.17	7.32	7.47	7.60	7.72	7.83	7.93	8.03	8.12	8.21
6	3.46	4.34	4.90	5.30	5.63	5.90	6.12	6.32	6.49	6.65	6.79	6.92	7.03	7.14	7.24	7.34	7.43	7.51	7.59
7	3.34	4.16	4.68	5.06	5.36	5.61	5.82	6.00	6.16	6.30	6.43	6.55	6.66	6.76	6.85	6.94	7.02	7.10	7.17
8	3.26	4.04	4.53	4.89	5.17	5.40	5.60	5.77	5.92	6.05	6.18	6.29	6.39	6.48	6.57	6.65	6.73	6.80	6.87
9	3.20	3.95	4.41	4.76	5.02	5.24	5.43	5.59	5.74	5.87	5.98	6.09	6.19	6.28	6.36	6.44	6.51	6.58	6.64
10	3.15	3.88	4.33	4.65	4.91	5.12	5.30	5.46	5.60	5.72	5.83	5.93	6.03	6.11	6.19	6.27	6.34	6.40	6.47
11	3.11	3.82	4.26	4.57	4.82	5.03	5.20	5.35	5.49	5.61	5.71	5.81	5.90	5.98	6.06	6.13	6.20	6.27	6.33
12	3.08	3.77	4.20	4.51	4.75	4.95	5.12	5.27	5.39	5.51	5.61	5.71	5.80	5.88	5.95	6.02	6.09	6.15	6.21
13	3.06	3.73	4.15	4.45	4.69	4.88	5.05	5.19	5.32	5.43	5.53	5.63	5.71	5.79	5.86	5.93	5.99	6.05	6.11
14	3.03	3.70	4.11	4.41	4.64	4.83	4.99	5.13	5.25	5.36	5.46	5.55	5.64	5.71	5.79	5.85	5.91	5.97	6.03
15	3.01	3.67	4.08	4.37	4.59	4.78	4.94	5.08	5.20	5.31	5.40	5.49	5.57	5.65	5.72	5.78	5.85	5.90	5.96
16	3.00	3.65	4.05	4.33	4.56	4.74	4.90	5.03	5.15	5.26	5.35	5.44	5.52	5.59	5.66	5.73	5.79	5.84	5.90
17	2.98	3.63	4.02	4.30	4.52	4.70	4.86	4.99	5.11	5.21	5.31	5.39	5.47	5.54	5.61	5.67	5.73	5.79	5.84
18	2.97	3.61	4.00	4.28	4.49	4.67	4.82	4.96	5.07	5.17	5.27	5.35	5.43	5.50	5.57	5.63	5.69	5.74	5.79
19	2.96	3.59	3.98	4.25	4.47	4.65	4.79	4.92	5.04	5.14	5.23	5.31	5.39	5.46	5.53	5.59	5.65	5.70	5.75
20	2.95	3.58	3.96	4.23	4.45	4.62	4.77	4.90	5.01	5.11	5.20	5.28	5.36	5.43	5.49	5.55	5.61	5.66	5.71
24	2.92	3.53	3.90	4.17	4.37	4.54	4.68	4.81	4.92	5.01	5.10	5.18	5.25	5.32	5.38	5.44	5.49	5.55	5.59
30	2.89	3.49	3.85	4.10	4.30	4.46	4.60	4.72	4.82	4.92	5.00	5.08	5.15	5.21	5.27	5.33	5.38	5.43	5.47
40	2.86	3.44	3.79	4.04	4.23	4.39	4.52	4.63	4.73	4.82	4.90	4.98	5.04	5.11	5.16	5.22	5.27	5.31	5.36
60	2.83	3.40	3.74	3.98	4.16	4.31	4.44	4.55	4.65	4.73	4.81	4.88	4.94	5.00	5.06	5.11	5.15	5.20	5.24
120	2.80	3.36	3.68	3.92	4.10	4.24	4.36	4.47	4.56	4.64	4.71	4.78	4.84	4.90	4.95	5.00	5.04	5.09	5.13
∞	2.77	3.31	3.63	3.86	4.03	4.17	4.29	4.39	4.47	4.55	4.62	4.68	4.74	4.80	4.85	4.89	4.93	4.97	5.01

(TABLE CONTINUES)

Reprinted by permission of the Biometrika Trustees from *Biometrika Tables for Statisticians,* E. S. Pearson and H. O. Hartley, Vol. 1, 176–77.

TABLE 11　(Continued)

									$\alpha = .01$										

Number of Populations

Degrees of Freedom	2	3	4	5	6	7	8	9	10	11	12	13	14	15	16	17	18	19	20
1	90.0	135.	164.	186.	202.	216.	227.	237.	246.	253.	260.	266.	272.	277.	282.	286.	290.	294.	298.
2	14.0	19.0	22.3	24.7	26.6	28.2	29.5	30.7	31.7	32.6	33.4	34.1	34.8	35.4	36.0	36.5	37.0	37.5	37.9
3	8.26	10.6	12.2	13.3	14.2	15.0	15.6	16.2	16.7	17.1	17.5	17.9	18.2	18.5	18.8	19.1	19.3	19.5	19.8
4	6.51	8.12	9.17	9.96	10.6	11.1	11.5	11.9	12.3	12.6	12.8	13.1	13.3	13.5	13.7	13.9	14.1	14.2	14.4
5	5.70	6.97	7.80	8.42	8.91	9.32	9.67	9.97	10.2	10.5	10.7	10.9	11.1	11.2	11.4	11.6	11.7	11.8	11.9
6	5.24	6.33	7.03	7.56	7.97	8.32	8.61	8.87	9.10	9.30	9.49	9.65	9.81	9.95	10.1	10.2	10.3	10.4	10.5
7	4.95	5.92	6.54	7.01	7.37	7.68	7.94	8.17	8.37	8.55	8.71	8.86	9.00	9.12	9.24	9.35	9.46	9.55	9.65
8	4.74	5.63	6.20	6.63	6.96	7.24	7.47	7.68	7.87	8.03	8.18	8.31	8.44	8.55	8.66	8.76	8.85	8.94	9.03
9	4.60	5.43	5.96	6.35	6.66	6.91	7.13	7.32	7.49	7.65	7.78	7.91	8.03	8.13	8.23	8.32	8.41	8.49	8.57
10	4.48	5.27	5.77	6.14	6.43	6.67	6.87	7.05	7.21	7.36	7.48	7.60	7.71	7.81	7.91	7.99	8.07	8.15	8.22
11	4.39	5.14	5.62	5.97	6.25	6.48	6.67	6.84	6.99	7.13	7.25	7.36	7.46	7.56	7.65	7.73	7.81	7.88	7.95
12	4.32	5.04	5.50	5.84	6.10	6.32	6.51	6.67	6.81	6.94	7.06	7.17	7.26	7.36	7.44	7.52	7.59	7.66	7.73
13	4.26	4.96	5.40	5.73	5.98	6.19	6.37	6.53	6.67	6.79	6.90	7.01	7.10	7.19	7.27	7.34	7.42	7.48	7.55
14	4.21	4.89	5.32	5.63	5.88	6.08	6.26	6.41	6.54	6.66	6.77	6.87	6.96	7.05	7.12	7.20	7.27	7.33	7.39
15	4.17	4.83	5.25	5.56	5.80	5.99	6.16	6.31	6.44	6.55	6.66	6.76	6.84	6.93	7.00	7.07	7.14	7.20	7.26
16	4.13	4.78	5.19	5.49	5.72	5.92	6.08	6.22	6.35	6.46	6.56	6.66	6.74	6.82	6.90	6.97	7.03	7.09	7.15
17	4.10	4.74	5.14	5.43	5.66	5.85	6.01	6.15	6.27	6.38	6.48	6.57	6.66	6.73	6.80	6.87	6.94	7.00	7.05
18	4.07	4.70	5.09	5.38	5.60	5.79	5.94	6.08	6.20	6.31	6.41	6.50	6.58	6.65	6.72	6.79	6.85	6.91	6.96
19	4.05	4.67	5.05	5.33	5.55	5.73	5.89	6.02	6.14	6.25	6.34	6.43	6.51	6.58	6.65	6.72	6.78	6.84	6.89
20	4.02	4.64	5.02	5.29	5.51	5.69	5.84	5.97	6.09	6.19	6.29	6.37	6.45	6.52	6.59	6.65	6.71	6.76	6.82
24	3.96	4.54	4.91	5.17	5.37	5.54	5.69	5.81	5.92	6.02	6.11	6.19	6.26	6.33	6.39	6.45	6.51	6.56	6.61
30	3.89	4.45	4.80	5.05	5.24	5.40	5.54	5.65	5.76	5.85	5.93	6.01	6.08	6.14	6.20	6.26	6.31	6.36	6.41
40	3.82	4.37	4.70	4.93	5.11	5.27	5.39	5.50	5.60	5.69	5.77	5.84	5.90	5.96	6.02	6.07	6.12	6.17	6.21
60	3.76	4.28	4.60	4.82	4.99	5.13	5.25	5.36	5.45	5.53	5.60	5.67	5.73	5.79	5.84	5.89	5.93	5.98	6.02
120	3.70	4.20	4.50	4.71	4.87	5.01	5.12	5.21	5.30	5.38	5.44	5.51	5.56	5.61	5.66	5.71	5.75	5.79	5.83
∞	3.64	4.12	4.40	4.60	4.76	4.88	4.99	5.08	5.16	5.23	5.29	5.35	5.40	5.45	5.49	5.54	5.57	5.61	5.65

TABLE 12
Factors for $\bar{x}$ and R Control Charts

Observations in Sample, n	d_2	A_2	d_3	D_3	D_4
2	1.128	1.880	0.853	0	3.267
3	1.693	1.023	0.888	0	2.574
4	2.059	0.729	0.880	0	2.282
5	2.326	0.577	0.864	0	2.114
6	2.534	0.483	0.848	0	2.004
7	2.704	0.419	0.833	0.076	1.924
8	2.847	0.373	0.820	0.136	1.864
9	2.970	0.337	0.808	0.184	1.816
10	3.078	0.308	0.797	0.223	1.777
11	3.173	0.285	0.787	0.256	1.744
12	3.258	0.266	0.778	0.283	1.717
13	3.336	0.249	0.770	0.307	1.693
14	3.407	0.235	0.763	0.328	1.672
15	3.472	0.223	0.756	0.347	1.653
16	3.532	0.212	0.750	0.363	1.637
17	3.588	0.203	0.744	0.378	1.622
18	3.640	0.194	0.739	0.391	1.608
19	3.689	0.187	0.734	0.403	1.597
20	3.735	0.180	0.729	0.415	1.585
21	3.778	0.173	0.724	0.425	1.575
22	3.819	0.167	0.720	0.434	1.566
23	3.858	0.162	0.716	0.443	1.557
24	3.895	0.157	0.712	0.451	1.548
25	3.931	0.153	0.708	0.459	1.541

Adapted from Table 27 of ASTM STP 15D *ASTM Manual on Presentation of Data and Control Chart Analysis.*
Copyright 1976 American Society for Testing and Materials, Philadelphia, Pa.

Summation Notation

Summations

Definition

$$\sum_{i=1}^{n} x_i = x_1 + x_2 + \cdots + x_n \qquad \text{(C.1)}$$

Example for $x_1 = 5$, $x_2 = 8$, $x_3 = 14$:

$$\sum_{i=1}^{3} x_i = x_1 + x_2 + x_3$$
$$= 5 + 8 + 14$$
$$= 27$$

Result 1

For a constant c:

$$\sum_{i=1}^{n} c = \underbrace{(c + c + \ldots + c)}_{n \text{ times}} = nc \qquad \text{(C.2)}$$

Example for $c = 5$, $n = 10$:

$$\sum_{i=1}^{10} 5 = 10(5) = 50$$

Example for $c = \bar{x}$:

$$\sum_{i=1}^{n} \bar{x} = n\bar{x}$$

Result 2

$$\sum_{i=1}^{n} cx_i = cx_1 + cx_2 + \cdots + cx_n$$
$$= c(x_1 + x_2 + \cdots + x_n) = c \sum_{i=1}^{n} x_i \qquad \text{(C.3)}$$

Example for $x_1 = 5$, $x_2 = 8$, $x_3 = 14$, $c = 2$:

$$\sum_{i=1}^{3} 2x_i = 2 \sum_{i=1}^{3} x_i = 2(27) = 54$$

Result 3

$$\sum_{i=1}^{n} (ax_i + by_i) = a \sum_{i=1}^{n} x_i + b \sum_{i=1}^{n} y_i \qquad\qquad (C.4)$$

Example for $x_1 = 5$, $x_2 = 8$, $x_3 = 14$, $a = 2$, $y_1 = 7$, $y_2 = 3$, $y_3 = 8$, $b = 4$:

$$\sum_{i=1}^{3} (2x_i + 4y_i) = 2 \sum_{i=1}^{3} x_i + 4 \sum_{i=1}^{3} y_i$$

$$= 2(27) + 4(18)$$

$$= 54 + 72$$

$$= 126$$

Double Summations

Consider the following data involving the variable x_{ij}, where i is the subscript denoting the row position and j is the subscript denoting the column position:

		Column		
		1	2	3
Row	1	$x_{11} = 10$	$x_{12} = 8$	$x_{13} = 6$
	2	$x_{21} = 7$	$x_{22} = 4$	$x_{23} = 12$

Definition

$$\sum_{i=1}^{n} \sum_{j=1}^{m} x_{ij} = (x_{11} + x_{12} + \cdots + x_{1m}) + (x_{21} + x_{22} + \cdots + x_{2m})$$

$$+ (x_{31} + x_{32} + \cdots + x_{3m}) + + \cdots + (x_{n1} + x_{n2} + \cdots + x_{nm}) \qquad (C.5)$$

Example:

$$\sum_{i=1}^{2} \sum_{j=1}^{3} x_{ij} = x_{11} + x_{12} + x_{13} + x_{21} + x_{22} + x_{23}$$

$$= 10 + 8 + 6 + 7 + 4 + 12$$

$$= 47$$

Definition

$$\sum_{i=1}^{n} x_{ij} = x_{1j} + x_{2j} + \cdots + x_{nj} \qquad\qquad (C.6)$$

Example:

$$\sum_{i=1}^{2} x_{i2} = x_{12} + x_{22}$$
$$= 8 + 4$$
$$= 12$$

Shorthand Notation

Sometimes when a summation is for all values of the subscript, we use the following shorthand notations:

$$\sum_{i=1}^{n} x_i = \Sigma x_i \tag{C.7}$$

$$\sum_{i=1}^{n} \sum_{j=1}^{m} x_{ij} = \Sigma \Sigma x_{ij} \tag{C.8}$$

$$\sum_{i=1}^{n} x_{ij} = \sum_{i} x_{ij} \tag{C.9}$$

A P P E N D I X

D

The Data Disk and Its Use

The Data Disk contains most of the larger data sets presented in the text. A list of the data set filenames is shown below. To retrieve a data set using Minitab or The Data Analyst software package, you only need to know the filename of the data set. For example, the data set corresponding to Exercise 16 in Chapter 2 is named AUTOCOST; thus, if you are using Minitab, you would access these data by entering the Minitab command Retrieve 'AUTOCOST'. If you are using The Data Analyst, a list of each data set filename is displayed whenever you select Choice 2, Retrieve a Previously Saved Data Set, from the Data Set Selection Menu; in this case, you simply select the filename corresponding to the data set that you want to retrieve. Note that each data set on The Data Disk is also identified in the text with a logo that appears in the margin.

Chapter 2

AUTOCOST	Exercise 16
EXSALARY	Exercise 18
MPGDATA	Exercise 20
COMPUTER	Exercise 21
APTEST	Table 2.13
JOBSAT	Exercise 27
LAWAGES	Exercise 34
COMSTOCK	Exercise 36
GRADEAVE	Exercise 37
ROOMCOST	Exercise 38
STATES	Exercise 40
CITIES	Exercise 42
CONSOLID	Computer Case

Chapter 3

STARTSAL	Exercise 5
LAWAGES	Exercise 26
GROWTH	Exercise 44
INJURY	Exercise 47
WORLD	Exercise 48
PRICES	Exercise 56
MORTGAGE	Exercise 59
EXAM	Exercise 65
DUKE	Exercise 70
CONSOLID	Computer Case 1
HEALTH-1	Computer Case 2
HEALTH-2	Computer Case 2

Chapter 8

AUTO	Computer Case

Chapter 9

QUALITY	Computer Case

Chapter 10

GOLF	Computer Case

Chapter 11

TRAINING	Computer Case

Chapter 13

INFO	Exercise 8
MACHINES	Exercise 10
JUDGMENT	Exercise 29
PAINT	Exercise 30
SERVICE	Exercise 54
ASSEMBLY	Exercise 55
DOWJONES	Exercise 58
BROWSE	Exercise 61
MEDICAL1	Computer Case
MEDICAL2	Computer Case

Chapter 14

PRESCRIP	Exercise 45
HOME1	Exercise 46
HOME2	Exercise 55

TEMPSC	Exercise 56
HIGHLOW	Exercise 65
SAFETY	Computer Case

Chapter 15

BUTLER	Table 15.2
MOVIE	Exercise 6
MOWER	Exercise 7
PHARMACY	Exercise 8
SCHOOLS1	Exercise 9
HOUSING	Exercise 10
FORBES1	Exercise 23
REPAIR	Exercise 33
FORBES2	Exercise 35
STROKE	Exercise 36
MOVIE	Exercise 39
JOBS	Exercise 50

| FOOTBALL | Exercise 52 |
| CONSUMER | Computer Case |

Chapter 16

REYNOLDS	Table 16.1
TYLER	Table 16.2
MPG	Table 16.4
POLAROID	Exercise 8
HOUSING	Exercise 9
FOOTBALL	Exercise 12
STROKE	Exercise 14
CRAVENS	Table 16.9
SCHOOLS2	Exercise 15
FORBES2	Exercise 19
AUDIT	Exercise 31
JOBS	Exercise 32
LAYOFFS	Computer Case

Note Regarding the Files on The Data Disk

If you look at the directory of the files on The Data Disk you will find that there are two files corresponding to each data set. Files that can be accessed by Minitab have the filename extension MTW (for example, AUTOCOST.MTW), and files that can be accessed by The Data Analyst have the filename extension ASW (e.g., AUTOCOST.ASW). You *do not* have to use these filename extensions when using Minitab or The Data Analyst.

Answers to Even-Numbered Exercises

CHAPTER 1

2. **a.** 7
   b. 4
   c. 7
   d. Pay versus corporation profit rating is qualitative; others quantitative
   e. Pay versus corporation profit rating is ordinal; others are ratio
4. **a.** 10
   b. *Fortune* 500 companies
   c. $2673.40
   d. $2673.40
6. **a**, **c**, and **d** are quantitative; **b** and **e** are qualitative
8. **a.** 1000
   b. Nominal
   c. Percentages
   d. 17%
10. **a.** Quantitative, ratio
    b. Qualitative, nominal
    c. Qualitative, ordinal
    d. Qualitative, nominal
    e. Quantitative, ratio
12. **a.** Visitors to Hawaii
    b. Yes; vast majority travel to Hawaii by air
    c. 1 and 4 are quantitative; 2 and 3 are qualitative
14. **a.** Product taste and preference data
    b. Actual test data from young adults
16. **a.** 40% in the sample died of heart disease
    b. Qualitative; nominal scale
20. **a.** All adult viewers reached by the television station
    b. The viewers contacted in the telephone survey
    c. A sample requires less time and a lower cost compared to contacting all viewers in the population
22. **a.** Correct
    b. Challenge
    c. Correct
    d. Challenge
    e. Challenge

CHAPTER 2

2. **a.** .20
   b. 40
   c.

Class	Frequency
A	44
B	36
C	80
D	40

4. **a.** Scale is nominal
   b.

Beverage	Frequency	Relative Frequency
Milk	3	.10
Fruit juice	4	.13
Soft drink	13	.43
Beer	8	.27
Bottled water	2	.07
Totals	30	1.00

   d. Soft drink

6. **a.**

Video	Frequency	Relative Frequency
B	9	.150
E	13	.217
F	14	.233
H	8	.133
L	6	.100
T	10	.167
Totals	60	1.000

   b. Fantasia

8. **a.**

Position	Frequency	Relative Frequency
P	17	.309
H	4	.073
1	5	.091
2	4	.073
3	2	.036
S	5	.091
L	6	.109
C	5	.091
R	7	.127
Totals	55	1.000

   b. Pitcher
   c. 3rd base
   d. Rightfield
   e. Infielders 16 to outfielders 18

10. a. Ordinal

b.

Rating	Frequency	Relative Frequency
Poor	2	.03
Fair	4	.07
Good	12	.20
Very good	24	.40
Excellent	18	.30
Totals	60	1.00

12.

Class	Cumulative Frequency	Cumulative Relative Frequency
≤ 19	8	.200
≤ 29	20	.500
≤ 39	35	.875
≤ 49	40	1.000

14. b, c.

Class	Frequency	Relative Frequency
6.0–7.9	4	.20
8.0–9.9	2	.10
10.0–11.9	8	.40
12.0–13.9	3	.15
14.0–15.9	3	.15
Totals	20	1.00

16.

Prices	New Freq.	New Rel. Freq.	Used Freq.	Used Rel. Freq.
0.0–4.9	0	.000	3	.25
5.0–9.9	1	.071	6	.50
10.0–14.9	5	.357	3	.25
15.0–19.9	5	.357	0	.00
20.0–24.9	3	.214	0	.00
Totals	14		12	

18.

Salary ($1000s)	Freq.	Rel. Freq.	Cum. Freq.
0–499	1	.04	1
500–999	9	.36	10
1000–1499	11	.44	21
1500–1999	2	.08	23
2000–2499	2	.08	25
Totals	25	1.00	

20. a, b.

Miles per Gallon	Frequency	Relative Frequency
24.0–25.9	2	.07
26.0–27.9	5	.17
28.0–29.9	10	.33
30.0–31.9	9	.30
32.0–33.9	3	.10
34.0–35.9	1	.03
Totals	30	1.00

d. Civic

22.
```
5 | 7  8
6 | 4  5  8
7 | 0  2  2  5  5  6  8
8 | 0  2  3  5
```

24.
```
11 | 6
12 | 0  2
13 | 0  6  7
14 | 2  2  7
15 | 5
16 | 0  2  8
17 | 0  2  3
```

26.
```
-1 | 3  7
 0 | 1  8
 1 | 3  4  5  6
 2 | 0  1  4  5  5  6
 3 | 0  3  3
 4 | 5  9
 . | .
 . | .
 . | .
12 | 0
```

28. a.
```
1 | 5  6  8  9
2 | 1  2
3 | 4
4 | 1  6  8
5 |
6 | 5  6  8
7 | 1
8 | 9  9
9 | 2  8
```

30. a, b

Industry	Frequency	Relative Frequency
Beverage	2	.10
Chemicals	3	.15
Electronics	6	.30
Food	7	.35
Aerospace	2	.10
Totals	20	1.00

32. a.

Party Affiliation	Frequency	Relative Frequency
Democrat	17	.425
Republican	17	.425
Independent	6	.150
Totals	40	1.000

34.

Hourly Wage	Freq.	Rel. Freq.	Cum. Freq.	Cum. Rel. Freq.
4.00–5.99	1	.04	1	.04
6.00–7.99	3	.12	4	.16
8.00–9.99	8	.32	12	.48
10.00–11.99	6	.24	18	.72
12.00–13.99	5	.20	23	.92
14.00–15.99	2	.08	25	1.00
Totals	25	1.00		

36.

Closing Price	Freq.	Rel. Freq.	Cum. Freq.	Cum. Rel. Freq.
0–9⅞	9	.225	9	.225
10–19⅞	10	.250	19	.475
20–29⅞	5	.125	24	.600
30–39⅞	11	.275	35	.875
40–49⅞	2	.050	37	.925
50–59⅞	2	.050	39	.975
60–69⅞	0	.000	39	.975
70–79⅞	1	.025	40	1.000
Totals	40	1.000		

38. a, b.

Cost	Frequency	Relative Frequency
2000–2499	2	.04
2500–2999	7	.14
3000–3499	16	.32
3500–3999	11	.22
4000–4499	5	.10
4500–4999	4	.08
5000–5499	2	.04
5500–5999	3	.06
Totals	50	1.00

40.

Income	Frequency	Relative Frequency
12,000–13,999	1	.020
14,000–15,999	10	.196
16,000–17,999	17	.333
18,000–19,999	11	.216
20,000–21,999	6	.118
22,000–23,999	3	.059
24,000–25,999	3	.059
Totals	51	1.000

42. a.

```
4 | 4
4 | 7
5 | 0  0  0  2  2
5 | 5  6  6  8
6 | 1  1  2  3
6 | 6  7  8  9
7 | 1  3
7 | 9
8 | 0  0
```

b.
```
2 | 5  8  8  9  9  9
3 | 2
3 | 5  5  5  5  5  7  9
4 | 0  1  1
4 | 9
5 | 0  0  3
5 | 6  7  8
```

d. 7

e.

	Frequency	
Temperature	High Temp.	Low Temp.
20–29	0	6
30–39	0	8
40–49	2	4
50–59	9	6
60–69	8	0
70–79	3	0
80–89	2	0

CHAPTER 3

2. 16, 16.5
4. 2 from each end
6. a. 171.25, 175, 145
   b. 145, 202.5
8. a. 38.75, 29
   b. 38.61, 38.38
   c. 38.5
   d. 29.5, 47.5
   e. 31
10. a. 178
    b. 178
    c. Do not report a mode
    d. 184
12. a. 48.33, 49; do not report a mode
    b. 45, 55
    c. 45, 55
14. City: mean = 15.58, median = 15.9, mode = 15.3
     Country: mean = 18.92, median = 18.7, mode = 18.6 and 19.4
16. a. 13.13, 14.5, 15
    b. 5, 17
    c. Q_1 = 8.5, Q_3 = 16.5
       L. Hinge = 8.5, U. Hinge = 16.5
18. 16, 4
20. a. New: $16.05, $16.40
       Used: $7.53, $7.35
    b. New: 14.9, 4.0, 4.21
       Used: 7.5, 5.15, 2.80

c. New cars are $8000–9000 more expensive
22. a. Range = 32, IQR = 10
    b. 92.75, 9.63
24. Dawson: range = 2, s = .67
    Clark: range = 8, s = 2.58
26. a. 10.40
    b. 10.05
    c. 9.50
    d. 3.60
    e. 6.75
    f. 2.60
28.
Quarter milers: s = .056, Coef. of Var. = 5.8
    Milers: s = .130, Coef. of Var. = 2.9
30. .20, 1.50, 0, − .50, − 2.20
32. a. 95%
    b. Almost all
    c. 68%
34. a. $\bar{x}$ = 77.5, s = 9.86
    b. z = 3.30, an outlier
    c. 16%, 2.5%
36. a. At least 75%
    b. At least 89%
    c. 95%; almost all
38. a. 75%, 84%, 89%
    b. 95%; almost all
40. 15, 22.5, 26, 29, 34
42. 5, 8, 10, 15, 18
44. a. 1.9, 10.0, 14.5, 20.4, 55.9
    b. Inner fences: − 5.6, 36.0
       Outer fences: − 21.2, 51.6
    c. 46.1 and 49.9 are mild outliers; 55.9 is an extreme outlier
46. a. 79.31, 78.5
    b. 76.5, 80.5
    c. 72, 76.5, 78.5, 80.5, 90
    d. Camcorders have higher variation
    e. Yes; Hitachi and Mitsubishi are mild outliers
48. a. 9.32, 9.8
    b. 6.5, 14.6
    c. Inner fences: − 5.65, 26.75
       Outer fences: − 17.8, 38.9
       Six mild outliers plus
       Hong Kong is an extreme outlier
50. 25, 5
52. 74.02, 158.78, 12.6
54. 10.74, 25.63, 5.06
56. a. 3.19, 3.30, 3.70
    b. 2.90, 3.50
    c. 2.80, .60
    d. .30, .54

e. *Inner fences:* 2.00, 4.40
   Two mild outliers
58. a. $\bar{x} = 1028.18$, median = 1000, no mode
  b. Range = 510, IQR = 220
  c. $s^2 = 24{,}256.36$, $s = 155.74$
  d. No outliers
60. a. 12.33
  b. 12
  c. 18
  d. 10
  e. 17
  f. 30.75
  g. 5.55
62. a. 68%
  b. 95%
  c. $z = 4.56$, yes
64. a. Public: 32; auto: 32
  b. Public: 4.64; auto: 1.83
  c. Auto has less variability
66. a. 400, 624, 836, 999, 1278
  c. *Inner fences:* 61.5, 1561.5
    No outliers
68. 51.50, 227.37, 15.08
70. a.   *Duke:* 87.97, 88.50
    Opponent: 72.64, 71.50
  b.   *Duke:* 50, 21.5
    Opponent: 52, 15.5

CHAPTER 4

2. 20
4. a. 9
6. a. 9
  c. 5
  d. 1
8. 2,598,960
10. 35
12. a. 1,000,000
  b. 5,760,000
14. .40, .26, .34; relative frequency method
16. a. 52
  b. Classical
  c. $\frac{1}{52}$
18. a. 6
  b. Relative frequency
  c. .12, .24, .30, .20, .10, .04; requirements are satisfied.

20. No, probabilities do not sum to 1
22. No
24. a. $\frac{1}{4}$
  b. $\frac{1}{2}$
  c. $\frac{3}{4}$
26. a. 36
  c. $\frac{1}{6}$
  d. $\frac{5}{18}$
  e. No; $P(\text{odd}) = P(\text{even}) = \frac{1}{2}$
  f. Classical
28. a. $P(0) = .05$
  b. $P(4 \text{ or } 5) = .20$
  c. $P(0, 1, \text{ or } 2) = .55$
30. a. .31
  b. .69
32. a. .40, .40, .60
  b. .80, yes
  c. $A^c = \{E_3, E_4, E_5\}$; $C^c = \{E_1, E_4\}$; $P(A^c) = .60$; $P(C^c) = .40$
  d. $\{E_1, E_2, E_5\}$; .60
  e. .80
34. .26
36. .50
38. a. .33
  b. .28
  c. .21
40. a. $A =$ first car starts, $B =$ second car starts
    $P(A) = .80$, $P(B) = .40$, $P(A \cap B) = .30$.
  b. .90
  c. .10
42. a. .67
  b. .80
  c. No
44. a.

	Single	Married	Total
Under 30	.55	.10	.65
30 or over	.20	.15	.35
Total	.75	.25	1.00

  b. Higher probability of under 30
  c. Higher probability of single
  d. .55
  e. .8462
  f. No

46. a.

		Own U.S. Car		
		Yes	No	**Total**
Own	Yes	.15	.05	.20
Foreign Car	No	.75	.05	.80
Total		.90	.10	1.00

  b. 20% own a foreign car; 90% own a U.S. car
  c. .15
  d. .95
  e. .17
  f. .75
  g. No
48. c. .72
  d. .40
  e. No
50. a. .5541
  c. .67
52. a. .197
  b. .121
  c. No
54. a. .10, .20, .09
  b. .51
  c. .26, .51, .23
56. a. .21
  b. Yes
58. .6007
60. a. Three
  b. .10, .50, .40
62. a. $\{E_1, E_2, E_3, E_4\}$
  b. $\{E_1, E_5, E_6, E_7, E_8\}$
  c. $\{E_2\}$
  d. $\{E_7, E_8\}$
  e. Empty
  f. $\{E_4, E_5, E_6, E_7, E_8\}$
  g. $\{E_1, E_2, E_3, E_4\}$
  h. $\{E_1, E_2, E_3, E_4\}$
  i. $\{E_1, E_2, E_3\}$
  j. No
  k. Yes
64. a. .02
  b. .64
  c. .93
  d. Yes
66. a. .76
  b. .24
68. b. .2022
  c. .4618
  d. .4005

70. a. .35, .80, .67
  b. No
72. a. .90, 0
  b. No
74. a.

	Smoker	Non-smoker	Total
Record of Heart Disease	.10	.08	.18
No Record of Heart Disease	.20	.62	.82
Total	.30	.70	1.00

  b. .10
  c. 30% smokers; 70% nonsmokers; 18% with record of heart disease; 82% with no record of heart disease
  d. .333
  e. .114
  f. No
  g. Smokers have a higher probability of having a record of heart disease
76. a. .25
  b. Yes
  c. No
78. a. .25, .40, .10
  b. .25
  c. B and S are independent; program appears to have no effect
80. a. .20
  b. .35
  c. 65%
82. Call back since $P(\text{sale} \mid \text{call back}) = .21$
84. 3.44%
86. a. .12
  b. .625
  c. .305
88. a. .0625
  b. .0132
  c. Three

CHAPTER 5

2. a. x = time in minutes to assemble product
  b. Any positive value: $x > 0$
  c. Continuous
4. 0, 1, 2, 3, 4, 5
6. a. 0, 1, 2, . . . , 20; discrete
  b. 0, 1, 2, . . . ; discrete
  c. 0, 1, 2, . . . , 50; discrete

  d. $0 \le x \le 8$; continuous
  e. $x > 0$; continuous
8. a.

x	1	2	3	4
$f(x)$	.15	.25	.40	.20

  c. $f(x) \ge 0$, $\Sigma f(x) = 1$
10. a. It is a proper probability distribution
  b. .60
12. a. Yes
  b. .65
14. a. .05
  b. .70
  c. .40
16. a. 5.20
  b. 4.56, 2.14
  c. 11.20
  d. 9.06, 3.01
18. a. 13.474%
  b. $\sigma^2 = 71.79$, $\sigma = 8.47$
  c. 6.964, 1.7401
20. a. 166
  b. -94; concern is to protect against the expense of a big accident
22. a. 445
  b. 1250 loss
24. a. Medium: 145; large: 140
  b. Medium: 2725; large: 12,400
26. a. 17
  b. 3.33, 1.82
28. a. $f(0) = .3487$
  b. $f(2) = .1937$
  c. .9298
  d. .6513
  e. 1
  f. $\sigma^2 = .9000$, $\sigma = .9487$
30. a. .2301
  b. .3410
  c. .8784
32. a. Probability of a defective part must be .03 for each trial; trials must be independent
  c. 2
  d.

Number of defects	0	1	2
Probability	.9409	.0582	.0009

34. a. $P(x \ge 8) = .2402$
  b. $f(1) = .2488$
36. a. .90
  b. .99
  c. .999
  d. Yes

38. a. .5905
  b. .1937
  c. .6082
42. 212.5, 31.88
44. a. $f(x) = \dfrac{3^x e^{-3}}{x!}$
  b. .2241
  c. .1494
  d. .8008
46. a. .1952
  b. .1048
  c. .0183
  d. .0907
48. a. .0821
  b. .0653
  c. .3840
50. a. .000045
  b. .010245
  c. .0821
  d. .9179
52. a. .50
  b. .067
  c. .4667
  d. .30
54. .00582
56. a. .01
  b. .07
  c. .92
  d. .07
58. a. 1.6
  b. $120
60. a. 2.2
  b. 1.16
62. a. $f(x) \ge 0$ and $\Sigma f(x) = 1$
  b. $17.25
  c. $1.25, 7.8%
  d. 1.3875
64. a. 3 hours
  b. 1
66. a. 22.80, 5.625, 2.37
  b. 228, 56.25, 7.5
68. a. .1712
  b. .2529
70. a. .2785
  b. .3417
72. a. .9510
  b. .0480
  c. .0490
74. .1912
76. a. .2240
  b. .5767

CHAPTER 6

2. **b.** .50
  c. .60
  d. 15
  e. 8.33
4. **b.** .50
  c. .30
  d. .40
6. **a.** .40
  b. .64
  c. .68
10. **a.** .3413
  b. .4332
  c. .4772
  d. .4938
12. **a.** .2967
  b. .4418
  c. .3300
  d. .5910
  e. .8849
  f. .2388
14. **a.** $z = 1.96$
  b. $z = .61$
  c. $z = 1.12$
  d. $z = .44$
16. **a.** $z = 2.33$
  b. $z = 1.96$
  c. $z = 1.645$
  d. $z = 1.28$
18. **a.** .1814
  b. .9656
  c. $12,816 or more
20. **a.** 50.77%
  b. 15.87%
  c. 23.88%
22. **a.** .7745
  b. 36.32
  c. 19%
24. **a.** 2.865 years
  b. .6247
26. **a.** 11
  b. .5780
  c. .1170
  d. .6950
28. **a.** .1151
  b. .2852
  c. 49 or less
30. **a.** .8336
  b. .0049
32. **a.** $1 - e^{-x_0/3}$
  b. .4866
  c. .3679

  d. .8111
  e. .3245
34. **b.** .6321
  c. .3935
  d. .0821
36. **a.** 4 hours
  b. $1/4\, e^{-x/4}$
  c. .7788
  d. .1353
38. **b.** .30
  c. .15
  d. .40
  e. 2.50 minutes
40. **b.** .25
42. **a.** .8106
  b. .0833
  c. .6955
  d. .9500
44. **a.** .3174, 317.4 defects
  b. .0028, 2.8 defects
46. .0062
48. **a.** .5899
  b. 30 or more
50. **a.** 47.06%
  b. .0475
  c. 42,480
52. **a.** .0228
  b. Do not use the die
  c. 3.36733
54. **a.** 5.16%
  b. 57.87%
  c. 99.55%
  d. Approximately 0
56. **a.** 2 minutes
  b. .2212
  c. .3935
  d. .0821

CHAPTER 7

2. 22, 147, 229, 289
4. **a.** L.A. Gear, New Balance, Adidas, Avia, Reebok
  b. 0
  c. 252
6. 2782, 493, 825, 1807, 289
8. 447, 348, 499, 568, 055, 392
  126, 036, 599, 294, 570, 159
10. **a.** Randomly select a page (1–853), and then randomly select a line (1–400) on the sampled page

  b. Skip or ignore inappropriate lines, and repeat the sampling procedure of part (a)
12. Finite, infinite, infinite, infinite, finite
14. **a.** .50
  b. .3667
16. **a.** 8.875
  b. 1.31
18. .0288
20. **a.** 200
  b. 5
  c. Normal with $E(\bar{x}) = 200$ and $\sigma_{\bar{x}} = 5$
  d. The probability distribution of $\bar{x}$
22. **a.** .6826
  b. .9544
24. 3.54, 2.50, 2.04, 1.77
  $\sigma_{\bar{x}}$ decreases as n increases
26. **a.** Only for $n = 30$ and $n = 40$
  b. $n = 30$; normal with $E(\bar{x}) = 400$ and $\sigma_{\bar{x}} = 9.13$
     $n = 40$; normal with $E(\bar{x}) = 400$ and $\sigma_{\bar{x}} = 7.91$
28. **a.** Normal with $E(\bar{x}) = 51,800$ and $\sigma_{\bar{x}} = 516.40$
  b. $\sigma_{\bar{x}}$ decreases to 365.15
  c. $\sigma_{\bar{x}}$ decreases as n increases
30. **a.** Normal with $E(\bar{x}) = 12.55$ and $\sigma_{\bar{x}} = .6325$
  b. .8858
  c. .5704
32. 170, 4.43
34. **a.** .6528
  b. .3616
36. **a.** Normal with $E(\bar{x}) = 16,012$ and $\sigma_{\bar{x}} = 420$
  b. .9826
  c. .7660, .4514, .1896
  d. Increase the sample size
38. **a.** Normal with $E(\bar{x}) = 14.25$ and $\sigma_{\bar{x}} = .2828$
  b. .9441
  c. .6212
  d. .9878, .7888
40. **a.** Normal with $E(\bar{x}) = 320$ and $\sigma_{\bar{x}} = 13.69$
  b. 13.69
  c. .8558
  d. .3557
42. **a.** .6156
  b. .8530
44. **a.** .6156
  b. .7814

c. .9488
d. .9942
e. High probability with larger n
46. **a.** .9756
b. .7372
48. **a.** .7062
b. .1469
c. .0025
50. **a.** Normal with $E(\bar{p}) = .80$ and $\sigma_{\bar{p}} = .02$
b. .8664
c. .9596
52. **a.** Normal with $E(\bar{p}) = .15$ and $\sigma_{\bar{p}} = .0505$
b. .4448
c. .8389
54. 4324, 2875, 318, 538, 4771
56. **a.** .4714
b. .7198
c. 693
58. **a.** Normal with $E(\bar{x}) = 49,000$ and $\sigma_{\bar{x}} = 1200$
b. .5934
c. .7620
d. .9050
e. 553
60. **a.** 67
b. 1.5
c. Normal with $E(\bar{x}) = 67$ and $\sigma_{\bar{x}} = 1.5$
d. .9082
e. .4972
62. **a.** No, since $n/N = .01$
b. Use $\sigma_{\bar{x}} = .0566$
c. .9232
64. 246
66. **a.** 625
b. .7888
68. **a.** Assume population has a normal distribution
b. .9266
c. Increase n to at least 30
70. **a.** .5990
b. .7660
c. 385
72. .9525
74. .4714

CHAPTER 8

2. **a.** 30.60 to 33.40
b. 30.34 to 33.66
c. 29.82 to 34.18

4. 62
6. 397.60 to 322.40
8. **a.** 23,480 to 24,520
b. 23,380 to 24,620
c. 23,186 to 24,814
10. **a.** 46.15 to 47.85
b. 42.53 to 45.47
c. Men due to larger n
12. **a.** 6
b. 96
14. **a.** 24.31 to 26.15
b. 24.14 to 26.32
16. **a.** 1.734
b. −1.321
c. 3.365
d. −1.761 and 1.761
e. −2.048 and 2.048
18. **a.** 15.97 to 18.53
b. 15.71 to 18.79
c. 15.14 to 19.36
20. **a.** 13.2
b. 7.8
c. 7.62 to 18.78
d. Wide interval; larger sample desirable
22. 6.28 to 6.78
24. **a.** 21.15 to 23.65
b. 21.12 to 23.68
c. Intervals are essentially the same
26. 4.51 to 6.59
28. **a.** 9
b. 35
c. 78
30. **a.** 50, 89, 200
b. Only if $E = 1$ is essential
32. 384
34. 59
36. **a.** 49
b. 30 to 34
38. **a.** .6733 to .7267
b. .6682 to .7318
40. 1067
42. .3233 to .3767
44. .6007 to .7393
46. **a.** .2271 to .3529
b. .2537 to .3263
c. .2619 to .3181
d. The interval width decreases indicating a better precision
48. **a.** .4081 to .5319
b. 383
50. **a.** 72
b. 129

c. 289
d. 801
e. n becomes larger
52. 2.11 to 2.39
54. 55.15 to 56.05
56. **a.** 1483.1
b. 374.9
c. 1136.4 to 1829.8
58. 9.20 to 14.80
60. 710
62. 37
64. 54
66. .0438 to .2362
68. .5165 to .6035
.2216 to .2984
A greater proportion use recognition
70. **a.** 1267
b. 1508
72. .5134 to .5866
74. **a.** 406
b. .7356 to .8044
c. .6008 to .6792
d. Yes

CHAPTER 9

2. **a.** $H_0: \mu \le 14$
$H_a: \mu > 14$
4. **a.** $H_0: \mu \ge 220$
$H_a: \mu < 220$
6. **a.** $H_0: \mu \le 1$
$H_a: \mu > 1$
b. Claiming $\mu > 1$ when it is not true
c. Claiming $\mu \le 1$ when it is not true
8. **a.** $H_0: \mu \ge 220$
$H_a: \mu < 220$
b. Claiming $\mu < 220$ when it is not true
c. Claiming $\mu \ge 220$ when it is not true
10. **a.** Reject H_0 if $z > 2.05$
b. 1.36
c. .0869
d. Do not reject H_0
12. **a.** .0344; reject H_0
b. .3264; do not reject H_0
c. .0668; do not reject H_0
d. Approximately 0; reject H_0
e. .8413; do not reject H_0
14. **a.** $H_0: \mu \le 6.5$; $H_a: \mu > 6.5$
b. $z = 5.91$; reject H_0
c. Current cars being driven longer
16. **a.** $z = 1.77$; do not reject H_0

18. $z = -2.74$; reject H_0
    p-value $= .0031$
20. **a.** Reject H_0 if $z < -1.645$
    b. $z = -1.98$; reject H_0
    c. .0239
22. **a.** Reject H_0 if $z < -2.33$ or $z > 2.33$
    b. 1.13
    c. .2584
    d. Do not reject H_0
24. **a.** .0718; do not reject H_0
    b. .6528; do not reject H_0
    c. .0404; reject H_0
    d. Approximately 0; reject H_0
    e. .3174; do not reject H_0
26. **a.** $z = -1.06$; do not reject H_0
    b. .2892
28. $z = 6.37$; reject H_0
30. **a.** \$50,367 to \$53,633
    b. Reject H_0
32. **a.** \$8.66 or less
    b. Reject H_0
34. **a.** 18
    b. 1.41
    c. Reject H_0 if $t < -2.571$ or $t > 2.571$
    d. -3.47
    e. Reject H_0
36. **a.** .01; reject H_0
    b. .10; do not reject H_0
    c. Between .025 and .05; reject H_0
    d. Greater than .10; do not reject H_0
    e. Approximately 0; reject H_0
38. $t = 1.20$; do not reject H_0
40. **a.** $t = -3.33$; reject H_0
    b. p-value is less than .005
42. $\bar{x} = 2.4$, $s = .52$, $t = 2.43$
    Reject H_0
44. **a.** Reject H_0 if $z < -1.96$ or $z > 1.96$
    b. -1.25
    c. .2112
    d. Do not reject H_0
46. .1118; do not reject H_0
48. $z = -1.25$; do not reject H_0
    p-value $= .1056$
50. **a.** $z = 1.38$; reject H_0
    b. .0838
52. $z = -3.29$; reject H_0; yes
54. $z = 1.63$; do not reject H_0
    p-value $= .0516$
56. **a.** .2912
    b. Type II error
    c. .0031

58. **a.** Concluding $\mu \leq 15$ when it is not true
    b. .2676
    c. .0179
60. **a.** Concluding $\mu = 28$ when it is not true
    b. .0853, .6179, .6179, .0853
    c. .9147
62. .1151, .0015
    Increasing n reduces β
64. 214
66. 109
68. 324
70. **a.** $z = 2.68$; reject H_0
    b. .0037
    c. 27.16 to 28.04
72. $z = 2.26$; reject H_0
74. **a.** $z = 1.64$; do not reject H_0
    b. .0505
    c. 349 to 375
76. **a.** .0143
    b. Reject H_0
78. $z = 2.52$; reject H_0
    p-value $= .0059$
80. **a.** Show $p < .50$
    b. $z = -6.62$; reject H_0
82. $z = 1.07$; do not reject H_0
    p-value $= .1423$
84. $z = -1.66$; do not reject H_0
    p-value $= .0485$
86. 219

CHAPTER 10

2. **a.** 2.4
   b. 5.27
   c. .91 to 4.71
4. **a.** \$4800
   b. \$3651 to \$5949
6. 2548.73 to 2951.27
8. **a.** 1200
   b. 438 to 1962
   c. Populations normal with equal variances
10. **a.** 1.63
    b. 1.48 to 3.52
12. **a.** $z = 2.03$; reject H_0
    b. .0212
14. .1446; do not reject H_0
16. $z = 4.99$; reject H_0
18. $z = 2.66$; reject H_0
20. **a.** $t = 2.22$; reject H_0
    b. .11 to 3.27 (thousands)

22. **a.** 3, -1, 3, 5, 3, 0, 1
    b. 2
    c. 2.08
    d. 2
    e. .08 to 3.92
24. **a.** $t = 2.38$; reject H_0
    b. .03 to 1.39
26. $\bar{d} = 3$; $t = 2.23$; reject H_0
28. **a.** $t = 5.88$; reject H_0
    b. 1.4 to 3.0
30. **a.** .12
    b. .0586 to .1814
    c. .0469 to .1931
32. .2122 to .2878
34. **a.** $z = 2.42$; reject H_0
    p-value $= .0078$
    b. Be aware of potential for less spending
    c. .0156 to .1444
36. $z = 5.67$; reject H_0
38. **a.** $z = 2.33$; reject H_0
    b. .02 to .22
40. 1354 to 2646
42. $t = -1.69$; do not reject H_0
44. $t = 2.30$; reject H_0
46. **a.** $H_0: p_1 - p_2 \leq 0$
       $H_a: p_1 - p_2 > 0$
    b. $z = 1.80$; reject H_0
       p-value $= .0359$
48. .0174; reject H_0

CHAPTER 11

2. **a.** 15.76 to 46.95
   b. 14.46 to 53.33
   c. 3.8 to 7.3
4. **a.** .22 to .71
   b. .47 to .84
6. **a.** 7.56 to 12.77
   b. 7.95 to 11.81
   c. 8.35 to 11.03
   d. The estimate is more precise for larger n
8. **a.** $\chi^2 = 32.39$; do not reject H_0
   b. 40.83 to 133.02
   c. 6.39 to 11.53
10. $\chi^2 = 31.5$; reject H_0
12. $\chi^2 = 10.16$; do not reject H_0
14. $F = 2.42$; reject H_0
16. $F = 1.19$; do not reject H_0
18. $F = 1.13$; do not reject H_0
20. $F = 2.20$; reject H_0

22. **a.** $F = 4$; reject H_0
    b. Drive carefully on wet pavement
24. 10.72 to 24.68
26. **a.** $\chi^2 = 27.44$; reject H_0
    b. .00012 to .00042
28. **a.** 15
    b. 6.25 to 11.13
30. $F = 1.39$; do not reject H_0
32. $F = 2.08$; reject H_0

CHAPTER 12

2. $\chi^2 = 15.33$, $\chi^2_{.05} = 7.81473$, reject H_0
4. $\chi^2 = 12.51$, $\chi^2_{.05} = 7.81473$; reject H_0
6. $\chi^2 = 6.24$, $\chi^2_{.10} = 4.60517$; reject H_0
8. $\chi^2 = 8.89$, $\chi^2_{.05} = 9.48773$; do not reject H_0
10. $\chi^2 = 19.78$, $\chi^2_{.05} = 9.48773$; reject H_0
12. $\chi^2 = 6.31$, $\chi^2_{.05} = 9.48773$; do not reject H_0
14. $\chi^2 = 14.72$, $\chi^2_{.05} = 5.99147$; reject H_0
16. $\chi^2 = 37.17$, $\chi^2_{.01} = 9.21034$; reject H_0
18. $\chi^2 = 7.96$ $\chi^2_{.05} = 9.48773$; do not reject H_0
20. **a.** Do not reject $\chi^2 = .39$
    b. Do not reject $\chi^2 = 1.57$
    c. 9 males and 3 females, $\chi^2 = 3.53$
22. $\chi^2 = 3.20$, $\chi^2_{.025} = 9.34840$; do not reject H_0
24. $\chi^2 = 4.98$, $\chi^2_{.10} = 7.77944$; do not reject H_0
26. $\chi^2 = 11.2$, $\chi^2_{.01} = 9.21034$; reject H_0
28. $\chi^2 = 3.55$, $\chi^2_{.05} = 7.81473$; do not reject H_0
30. $\chi^2 = 7.44$, $\chi^2_{.05} = 9.48773$; do not reject H_0
32. $\chi^2 = 5.26$, $\chi^2_{.05} = 5.99147$; do not reject H_0
34. $\chi^2 = 22.87$, $\chi^2_{.05} = 9.48773$; reject H_0
36. $\chi^2 = 6.20$, $\chi^2_{.05} = 12.5916$; do not reject H_0
38. $\chi^2 = 7.78$, $\chi^2_{.05} = 7.81473$; do not reject H_0

CHAPTER 13

2. **a.** MSB = 268
    b. MSW = 92.04
    c. Cannot reject H_0 since $F = 2.91 < F_{.05} = 4.26$

d.

Source of Variation	Sum of Squares	Degrees of Freedom	Mean Square	F
Between	536	2	268	2.91
Within	828.39	9	92.04	
Total	1364.39	11		

4. **b.** Reject H_0 since $F = 80 > F_{.05} = 2.76$
6. Reject H_0 since $F = 10.63 > F_{.05} = 4.26$
8. Source of information does not significantly affect the dissemination of the information; $F = .27 < F_{.05} = 3.47$
10. Significant difference; $F = 19.86 > F_{.05} = 3.10$
12. **a.** Significant difference; $F = 7.87 > F_{.05} = 4.26$
    b. 1 and 2; 2 and 3
    c. 1 and 2
    d. 1 and 2; 2 and 3
14. −9.37 to −.63
16. Reject the hypothesis that the means are equal
18. 2 and 3
20. **a.**

Source of Variation	Sum of Squares	Degrees of Freedom	Mean Square	F
Treatment	1488	2	744	5.50
Error	2029.55	15	135.3	
Total	3517.55	17		

    b. Significant difference between A and C
22. **a.** H_0: $\mu_1 = \mu_2 = \mu_3 = \mu_4 = \mu_5$
    H_a: Not all the population means are equal
    b. Reject H_0 since $F = 14.07 > 2.69$
24. Significant difference; $F = 43.99$ exceeds the critical value, which is between 3.15 and 3.23
26. **b.** Significant difference; $F = 9.87 > F_{.05} = 3.35$
28. Not significant; $F = 1.78 < F_{.05} = 3.89$
30. Not significant; $F = 2.54 < F_{.05} = 3.24$
32. Means are all different
34. Significant; $F = 6.60 > F_{.05} = 4.46$
36. Significant; $F = 12.60 > F_{.05} = 3.07$
38. Significant; $F = 7.12 > F_{.05} = 3.26$
40. Factor A is not significant since $F = 2.05 < F_{.05} = 5.99$

Factor B is not significant since $F = 4.06 < F_{.05} = 5.14$
Interaction is significant since $F = 7.66 > F_{.05} = 5.14$
42. Factor A is significant since $F = 10.75 > F_{.05} = 5.14$
Factor B is not significant since $F = 3 < F_{.05} = 5.99$
Interaction is not significant since $F = 1.75 < F_{.05} = 5.14$
44. Size is not significant since $F = 1.17 < F_{.05} = 5.99$
Price is significant since $F = 6.15 > F_{.05} = 5.14$
Interaction is not significant since $F = 4.71 < F_{.05} = 5.14$
46. The means of the k populations have to be equal
48. MSTR is based on the variation between sample means whereas MSE is computed based on the variation within each sample
50. **a.** Not significant since $t = .42 < t_{.025} = 2.477$
    b. Not significant since $F = .18 < F_{.05} = 5.99$
52. Significant; $F = 10.59$ exceeds the critical value, which is between 2.84 and 2.92
54. Significant; $F = 11.65 > F_{.05} = 3.13$
56. Significant; $F = 7.23 > F_{.05} = 4.26$
58. Not significant; $F = 1.66 < F_{.05} = 3.01$
60. Significant; $F = 5.19 > F_{.05} = 4.26$
62. Significant; $F = 6.99 > F_{.05} = 4.46$
64. Not significant; $F = 1.67 < F_{.01} = 10.92$
66. Type of machine is significant; type of loading system and interaction are not significant

CHAPTER 14

2. **b.** There appears to be a linear relationship between x and y
    d. $\hat{y} = 30.33 - 1.88x$
    e. 19.05
4. **b.** There appears to be a linear relationship between x and y
    d. $\hat{y} = -240.5 + 5.5x$
    e. 106 pounds
6. **c.** $\hat{y} = .38 + .35x$
    d. 9.13
8. **a.** $\hat{y} = 49.5 + 75.5x$
    b. 276

10. d. $\hat{y} = 6.38 + 45.68x$, where x and y are measured in 1000s

 e. $129,716

12. c. $\hat{y} = -55.84 + 1.67x$

 d. 44.36%

 e. $\hat{y} = 36.01$; predicted value is almost the same as the observed value

14. b. $\hat{y} = 192.77 + 59.07x$

 c. 783.47; estimated value is much greater than the observed value

16. a. SSE = 6.3325, SST = 114.80, SSR = 108.47

 b. $r^2 = .945$

18. a. SSE = 85,135.14, SST = 335,000, SSR = 249,864.86

 b. $r^2 = .746$

20. a. $\hat{y} = -54.85 + 98.80x$

 b. $r^2 = .992$

 c. 38.13 mgs

22. a. 4.133

 b. 2.033

 c. .643

 d. Reject H_0 since $t = 4.04 > t_{.05} = 3.182$

 e. Reject H_0 since $F = 16.36 > F_{.05} = 10.13$

24. Not significant since $t = 1.82 < t_{.05} = 3.182$

26. They are related since $F = 48.17 > F_{.01} = 34.12$

28. They are related since $F = 106.92 > F_{.05} = 5.32$

30. a. 1.11

 b. 7.07 to 14.13

 c. 2.32

 d. 3.22 to 17.98

32. *Confidence interval:* $-.4$ to 4.98
 Prediction interval: -2.27 to 7.31

34. $1443.51 to $1824.05

36. $1186.32 to $2081.24

38. a. $\hat{y} = 39.0162 - .2861x$

 b. They are related since $F = 38.72 > F_{.05} = 5.32$

 c. $r^2 = .829$; a good fit

 d. 23.10 to 26.32

 e. 20.10 to 29.32

40.

Source of Variation	Sum of Squares	Degrees of Freedom	Mean Square	F
Regression	108.47	1	108.47	51.41
Error	6.33	3	2.11	
Total	114.80	5		

42. a. 9

 b. $\hat{y} = 20.0 + 7.21x$

 c. 1.3626

 d. Significant relationship since $F = 28 > F_{.05} = 5.59$

 e. $380,500

44. a. $\hat{y} = 80.0 + 50.0x$

 b. 30

 c. Significant relationship since $F = 83.17 > F_{.05} = 4.20$

 d. $680,000

46. b. Yes

 c. $\hat{y} = 17.3 + 1.32x$

 d. Significant relationship; p-value = 0.000

 e. $r^2 = .537$; not a very good fit

 f. $60,940 to $65,903

 g. $53,493 to $65,903

48. a. $\hat{y} = 2.32 + .64x$

 b. No; the variance does not appear to be the same for all values of x

50. b. Yes

52. a. Yes; $x = 135$, $y = 145$ may be an outlier

 b. Yes

 c. Yes

54. a. $\hat{y} = -.23 + .049x$

 b. Minitab identifies observation 10 as an influential observation; the standardized residual plot shows an unusual trend in the residuals

56. a. $\hat{y} = -10.9 + 1.04x$

 b. Observation 10 is a possible outlier

58. b. There appears to be a linear relationship between x and y

 c. $s_{xy} = 26.5$

 d. $r_{xy} = .69$

60. b. Yes; linear or even a possible curvilinear relationship

 c. $s_{xy} = 36.8$

 d. $r_{xy} = .92$

62. b. Yes

 c. $r_{xy} = .87$

64. a. Yes

 b. $r_{xy} = .85$

 c. Repeat H_0 since $4.56 > t_{.005} = 3.355$

70. a. $\hat{y} = 22.173 - .1478x$

 b. Significant relationship since $F = 11.33 > F_{.05} = 7.71$

 c. $r^2 = .739$; a reasonably good fit

 d. 12.296 to 17.270

72. a. $\hat{y} = 10.5 + .953x$

 b. Significant relationship; p-value = 0.000

 c. $2874 to $4952

 d. Yes

74. a. $\hat{y} = 3767 - 322x$

 b. Significant relationship; p-value = 0.000

 c. $r^2 = .944$; an excellent fit

 d. $2115.90 to $2839.70

76. a. $\hat{y} = 20.7 - .234x$

 b. Not significant; p-value = .096

 c. Linear relationship is not appropriate

78. a. $\hat{y} = 220 + 132x$

 b. Significant relationship; p-value = 0.000

 c. $r^2 = .873$; a very good fit

 d. $559.50 to $933.90

80. a. $\hat{y} = 5.85 + .830x$

 b. Significant relationship; p-value = 0.000

 c. 84.65

 d. 65.35 to 103.96

CHAPTER 15

2. a. $\hat{y} = 45.06 + 1.94x_1$; $\hat{y} = 132.36$

 b. $\hat{y} = 85.22 + 4.32x_2$; $\hat{y} = 150.02$

 c. $\hat{y} = -18.37 + 2.01x_1 + 4.74x_2$; $\hat{y} = 143.18$

4. a. $255,000

6. a. $\hat{y} = 83.2 + 2.29$ TVADV $+ 1.30$ NEWSADV

 b. No

8. a. $\hat{y} = 2.1 + .0138x_1 + .00584x_2$

 b. Advertising expenditure, per capita income, store size

10. a. $\hat{y} = -5.7 + 1.54$ STARTS $+ 1.81$ INCOME

 b. 96.96

12. a. .926

 b. .905

 c. Yes

14. a. .75

 b. .68

16. a. $R^2 = .377$, $R_a^2 = .282$

 b. The fit is not very good

18. a. Significant; p-value = .000

 b. Significant; p-value = .000

 c. Significant; p-value = .000

20. a. SSE = 4000, $s^2 = 571.43$,

  MSR = 6000

 b. Significant; $F = 10.50 > F_{.05} = 4.74$

22. Significant; $F = 6.58 > F_{.05} = 4.74$

24. a. Significant; p-value $= .046 < \alpha = .05$

 b. Not significant; p-value = .219

 c. Not significant; p-value = .099

26. a. 132.16 to 154.15

 b. 111.15 to 175.17

28. a. 71,130 to 112,350

 b. 42,870 to 140,610

30. a. $x_2 = 0$ if level 1; $x_2 = 1$ if level 2

  $E(y) = \beta_0 + \beta_1 x_1 + \beta_2 x_2$

 b. $E(y) = \beta_0 + \beta_1 x_1$

 c. $E(y) = \beta_0 + \beta_1 x_1 + \beta_2$

 d. $\beta_2 = E(y \mid \text{level 2}) - E(y \mid \text{level 1})$

32. a. $15,300

 b. $56,100

 c. $41,600

34. a. $\beta_2 = E(y \mid \text{electrical problem}) -$

  $E(y \mid \text{mechanical problem})$

 b. $\hat{y} = .930 + .388 x_1 + 1.26 x_2$

 c. Significant; p-value $= .001 < \alpha = .05$

 d. Yes, $R_a^2 = .819$

 e. $b_2 = 1.26$ hours

36. a. $\hat{y} = -91.8 + 1.08$ AGE

  $+ .252$ PRESSURE

  $+ 8.74$ SMOKER

 b. Significant; p-value $= .01 < \alpha = .05$

 c. 95% prediction interval is 21.35 to 47.19 or a probability of .2135 to .4719; quit smoking and begin some type of treatment to reduce his blood pressure

38. a. $\hat{y} = -53.3 + 3.11 x$

 b. $-1.40, -.15, 1.36, .47, -1.39$; no

 c. .38, .28, .22, .20, .98; no

 d. .60, .00, .26, .03, 11.09; yes, the fifth observation

40. a. Shape of the plot supports the assumptions

 b. Observation 13 is an outlier

 c. No

42. b. 3.19

44. a. $R^2 = .95$, $R_a^2 = .93$

 b. Significant; $F = 57.84 > F_{.01} = 5.21$

46. a. 3.04, 3.61, 5.08

 b. Both are significant

 d. .91

48. b. Significant; $F = 22.79 > F_{.05} = 5.79$

 c. $R_a^2 = .861$; good fit

 d. Both are significant

50. a. $\hat{y} = -16.7 + 1.02$ %WOMEN

 b. $R^2 = .868$; very good fit

 c.

D_1	D_2	D_3	D_4	D_5	Industry
0	0	0	0	0	Industrial
1	0	0	0	0	Technology
0	1	0	0	0	Consumer
0	0	1	0	0	Retailing
0	0	0	1	0	Media
0	0	0	0	1	Financial

  $\hat{y} = 11.5 + 7.75 D_1 + 7.75 D_2 +$

  $26.7 D_3 + 24.8 D_4 + 33.5 D_5$

 d. %WOMEN is a better predictor than the type of industry

 e. $\hat{y} = -20.0 + 1.18$ %WOMEN $-$

  $1.96 D_1 - 5.49 D_2 - 8.20 D_3 +$

  $3.74 D_4 - 7.99 D_5$

 f. Adding type of industry to the model involving %WOMEN is not significant

52. a. POINTS $= 170 + 6.61$ TEAMINT

 b. Pattern is acceptable; observation 13 looks unusual

 c. No outliers

 d. Yes; observation 13

CHAPTER 16

2. a. $\hat{y} = 9.32 + .424 x$; p-value = .117 indicates a weak relationship between x and y

 b. $\hat{y} = -8.10 + 2.41 x -$ $.0480 x^2$; $R_a^2 = .932$, a very good fit

 c. 20.965

4. a. $\hat{y} = 943 + 8.71 x$

 b. Significant; p-value $= .005 < \alpha = .01$

6. b. No, the relationship appears to be curvilinear

 c. Several possible models; e.g., $\hat{y} = 2.90 - .185 x + .00351 x^2$

8. b. Simple linear model may be appropriate; however, there may be some curvilinear effect in the data

 c. $\hat{y} = -6.60 - 14.2 x + .508 x^2$

10. a. Significant; $t = 4.34 > 2.069$

 b. Not significant; $t = 1.48 < 2.069$

 c. Significant; $t = -4.46 < -2.069$

 d. x_2 can be dropped

12. a. $\hat{y} = 170 + 6.61$ TEAMINT

 b. $\hat{y} = 280 + 5.18$ TEAMINT $-$ .0037 RUSHING $- 3.92$ OPPONINT

 c. Addition of the two independent variables is not significant

14. a. $\hat{y} = -111 + 1.32$ AGE $+$ .296 PRESSURE

 b. $\hat{y} = -123 + 1.51$ AGE $+$ .448 PRESSURE $+ 8.87$ SMOKER $-$ .00276 AGEXEXPRES

 c. Significant

16. a. $\hat{y} = 780 + 3.06$ %COLLEGE

 b. $\hat{y} = 780 + 3.06$ %COLLEGE

18. a. $\hat{y} = 383 - 12.2$ WINS

 b. Use WINS, PASSING, and OPPONINT

20. $\hat{y} = -91.8 + 1.08$ AGE $+ .252$ PRESSURE $+ 8.74$ SMOKER

22. $d = 1.60$; test is inconclusive

24.

x_1	x_2	Treatment
0	0	1
1	0	2
0	1	3

 $x_3 = 0$ if block 1; $x_3 = 1$ if block 2

 $E(y) = \beta_0 + \beta_1 x_1 + \beta_2 x_2 + \beta_3 x_3$

26. a.

D_1	D_2	Manufacturer
0	0	1
1	0	2
0	1	3

 $E(y) = \beta_0 + \beta_1 D_1 + \beta_2 D_2$

 b. $\hat{y} = 23.0 + 5.00 D_1 - 2.00 D_2$

 c. $H_0: \beta_1 = \beta_2 = 0$

 d. Mean time is not the same for each manufacturer; p-value = .004

28. Significant difference between the two analyzers

30. b. There may be some curvilinear effect

 c. $\hat{y} = 2081 + 21.9$ TIME $- .0417$ TIMESQ

32. a. $\hat{y} = -16.7 + 1.02$ %WOMEN
  b. Use %WOMEN and D4, where
    $D_4 = 1$ if media, $D_4 = 0$ otherwise
34. a. $\hat{y} = 70.6 + 12.7$ INDUS
    $- 2.92$ ICQUAL
  b. No obvious pattern
  c. Test is inconclusive
36. Significant differences between comfort
    levels for the three types of browsers

CHAPTER 17

2. a. 32%
  b. $8.14
4. $I_{1991} = 105$
6. $I = 125$
8. $I = 105$; portfolio is up 5%
10. a. $6.81
  b. 54.7%
  c. -2.7%
12. a. 155, 155, 158, 162, 164
  b. 155, 158, 162, 165, 168
14. $I_{1992} = 110$
16. $I = 83$
18. a. 151, 197, 143, 178
  b. $I = 170$
20. $I_{Jan} = 96$ $I_{Mar} = 92$
22. $I = 88$
24. 14,000, 12,621, 10,830, 13,342, 15,734
26. $I = 143$; quantity is up 43%

CHAPTER 18

2. a.

Week	4-Week	5-Week
10	19.00	18.80
11	20.00	19.20
12	18.75	19.00

  b. 9.65, 7.41
  c. 5-week
4. Weeks 10, 11, and 12: 18.48, 18.63,
  18.27
  MSE = 9.25; $\alpha = .2$ is better
6. a.

Week	$\alpha = .2$	$\alpha = .3$
10	2.85	2.92
11	2.91	2.99
12	2.97	3.05

  b. Using $\alpha = .3$, we obtain 2.99
8. a.

Month	3-Month	$\alpha = .2$
10	256.67	265.51
11	286.67	274.41
12	263.33	267.53

Using months 4 to 12 for both, $\alpha = .2$ is
better
MSE (3-month) = 1998.72
MSE ($\alpha = .2$) = 2528.95
  b. 260
10. a. Months 10, 11, and 12: 2478.44,
    2478.75, 2489.00
  b. Months 10, 11, and 12: 2474.31,
    2476.02, 2492.21
  c. MSE ($\alpha = .2$) = 1075;
    MSE ($\alpha = .3$) = 1092; use $\alpha = .2$;
    $F_{13} = 2501$
12. $T_t = 207.467 - 3.514t$; 182.87
14. $T_t = 6.96 + 2.14t$; 19.8
16. Consider a nonlinear trend
18. a. $T_t = .365 + .193t$; $2.49
  b. EPS increasing by an average of $.193
    per year
20. a. $T_t = 1997.6 + 397.545t$
  b. $T_{11} = 6371$, $T_{12} = 6768$
22. a. 1938.75, 1966.25, 1956.25, 2025.00,
    1990.00, 2002.50, 2052.50, 2060.00,
    2123.75
  b. .900, .486, 1.396, 1.217
  c. 3rd quarter; yes, fall back-to-school
    demand
24. a. .771, .864, .954, 1.392, 1.571, 1.667,
    1.207, .994, .850, .647, .579, .504
  b. 47.49, 47.89, 48.28, 48.67, 49.06,
    49.46, 49.85, 50.24, 50.63, 51.02,
    51.42, 51.81
  c. 37, 41, 46, 68, 77, 82, 60, 50, 43, 33,
    30, 26
26. *Forecast for weeks 8, 9, 10, and 11:*
    258.64, 281.48, 283.61, and 292.71
28. 20.26
30. a. *Weeks 8, 9, and 10:* 7.48, 7.53, 7.55
  b. *Weeks 8, 9, and 10:* 7.51, 7.57, 7.58
  c. $\alpha = .3$, $F_{11} = 7.57$
32. $T_t = 21.0135 + 3.5771t$
    $T_7 = 46.05$
34. $T_8 = 252.28$, $T_9 = 259.10$
36. a. Yes
  b. $T_t = -5 + 15t$

38. a.   PC:
      $T_t = -31.267 + 60.314t$
      Mainframe:
      $T_t = 140.467 + 15.629t$
  b.   PC: 390.93
    Mainframe: 249.87
40. a. 1.271, .613, .498, 1.619
  b. Quarter 4
42. a. $T_t = -.345 + .995t$
  b. 20.55, 21.55, 22.54, 23.54
  c. 26.10, 13.15, 11.27, 38.13

CHAPTER 19

2. $z = 3.27$; reject H_0
4. $z = 3.15$; reject H_0
6. $z = 1.90$; do not reject H_0
8. $z = 3.76$; reject H_0
10. $z = 1.27$; do not reject H_0
12. $z = 2.43$; reject H_0
14. $z = 1.89$; reject H_0
16. $z = -2.05$; reject H_0
18. $T = 34$; reject H_0
20. $T = 28.5$; reject H_0
22. $z = 3.33$; reject H_0
24. $z = -.25$; do not reject H_0
26. $W = 10.22$; reject H_0
28. $W = 3.06$; do not reject H_0
30. $W = 8.03$; reject H_0
32. a. .68
  b. $z = 2.06$; reject H_0
34. $z = .72$; do not reject H_0
36. $r_s = .04$; $z = .12$; do not reject H_0
38. $z = .82$; do not reject H_0
40. $z = 1.81$; reject H_0
42. $z = 2.96$; reject H_0
    $z = .68$; do not reject H_0
44. $z = -2.59$; reject H_0
46. $z = -2.97$; reject H_0
48. $W = 12.61$; reject H_0
50. $r_s = -.92$; $z = -3.05$; reject H_0
52. $r_s = .76$; $z = 2.84$; reject H_0

CHAPTER 20

2. a. .3324
  b. .7099
4. a. .1198
  b. .1756

6. b.

c	Producer's Risk	Consumer's Risk
0	.1821	.1887
1	.0169	.5169
2	.0010	.7880

8. a. 5.42
   b. UCL = 6.09, LCL = 4.75

10.

	R Chart	$\bar{x}$ Chart
UCL	2.98	29.10
LCL	.22	27.90

12. 20.01, .082
14. a. .0470
    b. UCL = .0989, LCL = −.0049 (use LCL = 0)
    c. $\bar{p}$ = .08; in control
16. a. Producer's risk = .0116
       Consumer's risk = .6778
18. c. .2222
20. a. 95.4
    b. UCL = 96.07, LCL = 94.73
    c. No

22.

	R Chart	$\bar{x}$ Chart
UCL	4.23	6.57
LCL	0	4.27

Estimate of standard deviation = .86

24.

	R Chart	$\bar{x}$ Chart
UCL	.11	3.11
LCL	0	3.05

26. a. UCL = .0817, LCL = −.0017 (use LCL = 0)

CHAPTER 21

2. a. 30,000
   b. 320
   c. 29,360 to 30,640
4. 73
6. 337
8. a. *stratum 1:* 27,600
      stratum 2: 25,750
      stratum 3: 21,000

b. 74,350
c. 70,599.88 to 78,100.12
10. a. $n = 93$, $n_1 = 30$, $n_2 = 30$, $n_3 = 33$
    b. $n = 306$, $n_1 = 98$, $n_2 = 98$, $n_3 = 109$
    c. $n = 275$, $n_1 = 88$, $n_2 = 88$, $n_3 = 98$
12. a. $3,617,000
    b. $1,122,265
    c. $41,066 to $56,499
    d. $9,568,261 to $13,164,197
14. a. 15, 4500, .30
    b. 1.4708, 441.24, .0484
    c. 12.0584 to 17.9416
    d. 3617.52 to 5382.48
    e. .2032 to .3968
16. a. 40
    b. .70
    c. 35.8634 to 44.1366
    d. .5234 to .8766
18. a. .7788 to .8212
    b. .5740 to .6260
    c. .4136 to .4664
20. a. $22,790 to $23,610
    b. $68,370,366 to $70,829,634
    c. .6692 to .7908
22. a. 431
    b. .2175 to .3983
    c. .6230 to .8002
    d. 996
24. a. 75.275
    b. .198 to .502
    c. 1680

CHAPTER 22

2. a.   *Optimistic:* d_1
         Conservative: d_3
         Minimax regret: d_3
   c.   *Optimistic:* d_1
         Conservative: d_2 or d_3
         Minimax regret: d_2
4. b.   *Optimistic:* large
         Conservative: small
         Minimax regret: medium
6. a. d_1
   b. d_4
8. d_2 (medium)
10. b. Develop the software
12. p = .25
14. a. d_1 (St. Louis)

b.

p	Best Decision(s)
$0 \leq p < .67$	d_1
$p = .67$	d_1 or d_2
$.67 < p \leq 1.0$	d_2

16. .42
18. a. s_1: d_1; s_2: d_2, d_3, or d_4; s_3: d_4; s_4: d_4
    b. 12.5
    c. 11.3
    d. 1.2
20. 195
22. $200
24. a. d_1
    b.

p	Best Decision(s)
$0 \leq p < .59$	d_3
$p = .59$	d_1 or d_3
$.59 < p \leq 1.0$	d_1

   c. 2
   d. d_3
26. a. $8000
    b. 13.13%
    c. Not worth doing the study
28. b. If I_1, then d_1; if I_2, then d_2
       EV(node 1) = $101.50
    c. EVSI = $1500
30. If unseasonably cold, then purchase snowplow
    If not unseasonably cold, then purchase blade
32. a. If I_1, then d_1
       If I_2, then d_1
       EV = 2000
    b. .57
    c. 0%
34. a. Purchase component
    b. EVPI = $9000
    c. .355
    d. If I_1, then manufacture
       If I_2, then purchase
       EV = 43.9
    e. EVSI = $3650
    f. 40.6%

36. a.

		Demand		
		1000	2000	3000
Amount Produced	1000	50,000	10,000	−30,000
	2000	−50,000	100,000	60,000
	3000	−150,000	0	150,000

  b. $d_2 = 2000$ pounds

  c. $48,000

  d. If I_1, then d_3

     If I_2, then d_2

     If I_3, then d_1

  e. $10,900

  f. 22.7%

38. a. d_3

  b. d_1

Solutions to Self-Test Exercises

CHAPTER 1

5. **a.** Five variables: sales, rank, profit, assets, and industry
   b. Rank and industry are qualitative; sales, profit, and assets are quantitative
   c. Sales—ratio; rank—ordinal; profit—ratio; assets—ratio; industry—nominal

15. **a.** All adults in the United States
   b. Favorite way to spend time in the evening
   c. Qualitative since the variable is nominal
   d. 1500
   e. 70% of the adults in the sample selected "staying at home with the family"
   f. Statistical inference would be the inference that 70% of the population of adults prefer "staying at home with the family" as the favorite way to spend time in the evening

CHAPTER 2

3. **a.** $360° \times 58/120 = 174°$
   b. $360° \times 42/120 = 126°$
   c.

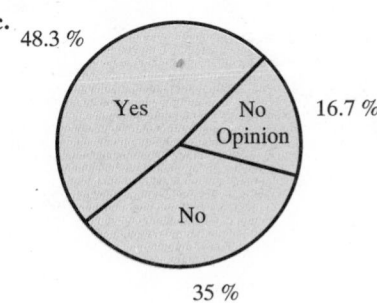

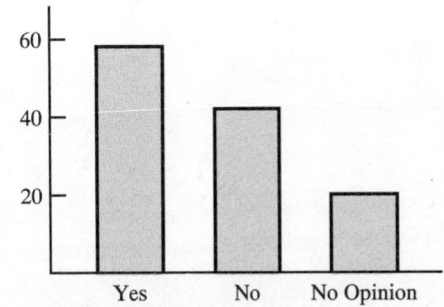

7.

Rating	Frequency	Relative Frequency
Outstanding	19	.38
Very good	13	.26
Good	10	.20
Average	6	.12
Poor	2	.04

Management should be pleased with these results: 64% of the ratings are very good to outstanding, and 84% of the ratings are good or better; comparing these ratings to previous results will show whether or not the restaurant is making improvements in its customers' ratings of food quality

12.

Class	Cumulative Frequency	Cumulative Relative Frequency
≤ 19	8	.200
≤ 29	20	.500
≤ 39	35	.925
≤ 49	40	1.000

15. a, b.

Waiting Time	Frequency	Relative Frequency
0–4	4	.20
5–9	8	.40
10–14	5	.25
15–19	2	.10
20–24	1	.05
Totals	20	1.00

c, d.

Waiting Time	Cumulative Frequency	Cumulative Relative Frequency
≤ 4	4	.20
≤ 9	12	.60
≤ 14	17	.85
≤ 19	19	.95
≤ 24	20	1.00

e. $^{12}/_{20} = .60$

23.

6	3
7	5 5 7
8	1 3 4 8
9	3 6
10	0 4 5
11	3

25.

9	8 9
10	2 4 6 6
11	4 5 7 8 8 9
12	2 4 5 7
13	1 2
14	4
15	1

CHAPTER 3

3. Arrange data in order: 15, 20, 25, 25, 27, 28, 30, 34.

$i = \dfrac{20}{100}(8) = 1.6$; round up to position 2

20th percentile = 20

$i = \dfrac{25}{100}(8) = 2$; use positions 2 and 3

25th percentile $= \dfrac{20 + 25}{2} = 22.5$

$i = \dfrac{65}{100}(8) = 5.2$; round up to position 6

65th percentile = 28

$i = \dfrac{75}{100}(8) = 6$; use positions 6 and 7

75th percentile $= \dfrac{28 + 30}{2} = 29$

8. a. $\bar{x} = \dfrac{\Sigma x_i}{n} = \dfrac{775}{20} = 38.75$

Mode = 29 (appears three times)

b. .05(20) = 1; trim lowest (22) and highest (58) values

5% trimmed mean $= \dfrac{695}{18} = 38.61$

.10(20) = 2; trim two lowest (22, 24) and two highest (57, 58) values

10% trimmed mean $= \dfrac{614}{16} = 38.38$

c. Data in order: 22, 24, 29, 29, 29, 30, 31, 31, 32, 37, 40, 41, 44, 44, 46, 49, 50, 52, 57, 58

Median (10th and 11th positions)

$\dfrac{37 + 40}{2} = 38.5$

At home workers are slightly younger

d. $i = \dfrac{25}{100}(20) = 5$; use positions 5 and 6

$Q_1 = \dfrac{29 + 30}{2} = 29.5$

$i = \dfrac{75}{100}(20) = 15$; use positions 15 and 16

$Q_3 = \dfrac{46 + 49}{2} = 47.5$

e. $i = \dfrac{32}{100}(20) = 6.4$; round up to position 7

32nd percentile = 31

At least 32% of the people are 31 or younger

19. Range = 34 − 15 = 19

Arrange data in order: 15, 20, 25, 25, 27, 28, 30, 34

$i = \dfrac{25}{100}(8) = 2$; $Q_1 = \dfrac{20 + 25}{2} = 22.5$

$i = \dfrac{75}{100}(8) = 6$; $Q_3 = \dfrac{28 + 30}{2} = 29$

IQR $= Q_3 - Q_1 = 29 - 22.5 = 6.5$

$\bar{x} = \dfrac{\Sigma x_i}{n} = \dfrac{204}{8} = 25.5$

x_i	$(x_i - \bar{x})$	$(x_i - \bar{x})^2$
27	1.5	2.25
25	−.5	.25
20	−5.5	30.25
15	−10.5	110.25
30	4.5	20.25
34	8.5	72.25
28	2.5	6.25
25	−.5	.25
		242.00

$$s^2 = \frac{\Sigma(x_i - \bar{x})^2}{n - 1} = \frac{242}{8 - 1} = 34.57$$

$$s = \sqrt{34.57} = 5.88$$

25. **a.** Range $= 190 - 168 = 22$

   b. $\bar{x} = \dfrac{\Sigma x_i}{n} = \dfrac{1068}{6} = 178$

   $$s^2 = \frac{\Sigma(x_i - \bar{x})^2}{n - 1}$$

   $$= \frac{4^2 + (-10)^2 + 6^2 + 12^2 + (-8)^2 + (-4)^2}{6 - 1}$$

   $$= \frac{376}{5} = 75.2$$

   c. $s = \sqrt{75.2} = 8.67$

   d. $\dfrac{s}{\bar{x}}(100) = \dfrac{8.67}{178}(100) = 4.87$

31. Chebyshev's theorem: *at least* $(1 - 1/k^2)$

   a. $k = \dfrac{40 - 30}{5} = 2;\ (1 - \tfrac{1}{2}^2) = .75$

   b. $k = \dfrac{45 - 30}{5} = 3;\ (1 - \tfrac{1}{3}^2) = .89$

   c. $k = \dfrac{38 - 30}{5} = 1.6;\ (1 - \tfrac{1}{1.6}^2) = .61$

   d. $k = \dfrac{42 - 30}{5} = 2.4;\ (1 - \tfrac{1}{2.4}^2) = .83$

   e. $k = \dfrac{48 - 30}{5} = 3.6;\ (1 - \tfrac{1}{3.6}^2) = .92$

34. **a.** $\bar{x} = \dfrac{\Sigma x_i}{n} = 77.5$

   $$s = \sqrt{\frac{\Sigma(x_i - \bar{x})^2}{n - 1}} = \sqrt{\frac{874.5}{10 - 1}} = 9.86$$

   b. $z = \dfrac{x_i - \bar{x}}{s} = \dfrac{110 - 77.5}{9.86} = 3.30;$ yes

c. $z = \dfrac{87 - 77.5}{9.86} = .96$ or approximately 1

68% are within ± 1;
32% are outside ± 1 (half above $+1$ and half below -1);
therefore, 32%/2 = 16% should be 87 or above

$$z = \frac{58 - 77.5}{9.86} = -1.98 \text{ or approximately } -2$$

95% are within ± 2; 5% are outside ± 2 (half above $+2$ and half below -2); therefore, 5%/2 = 2.5% should be 58 or below

42. Arrange data in order: 5, 6, 8, 10, 10, 12, 15, 16, 18

   $i = \dfrac{25}{100}(9) = 2.25$; round up to position 3

   $Q_1 = 8$

   Median (5th position) $= 10$

   $i = \dfrac{75}{100}(9) = 6.75$; round up to position 7

   $Q_3 = 15$

   5-number summary: 5, 8, 10, 15, 18

   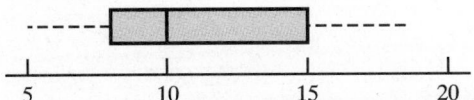

45. **a.** Arrange data in order: 484, 598, 636, 1049, 1061, 1220, 2188, 2366, 2472, 2731, 3122, 3261, 4514, 8030, 8258, 15,980, 32,249

   $i = \dfrac{25}{100}(17) = 4.25$; round up to 5th position

   $Q_1 = 1061$

   Median (9th position) $= 2472$

   $i = \dfrac{75}{100}(17) = 12.75$; round up to 13th position

   $Q_3 = 4514$

   5-number summary: 484, 1061, 2472, 4514, 32,249

   b. IQR $= Q_3 - Q_1 = 4514 - 1061 = 3453$
   Inner Fences: $Q_1 - 1.5$IQR $= 1061 - 1.5(3453) = -4118$
   $\qquad\qquad\quad Q_3 + 1.5$IQR $= 4514 + 1.5(3453) = 9694$
   Outer Fences: $Q_1 - 3$IQR $= 1061 - 3(3453) = -9298$
   $\qquad\qquad\quad Q_3 + 3$IQR $= 4514 + 3(3453) = 14873$

   c. Dow Chemical and DuPont are extreme outliers with very large sales compared to the rest of the industry.

d.

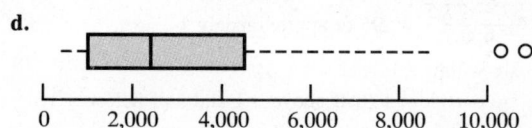

0	2,000	4,000	6,000	8,000	10,000

49.

f_i	M_i	f_iM_i
4	5	20
7	10	70
9	15	135
5	20	100
25		325

$$\bar{x} = \frac{\Sigma f_iM_i}{n} = \frac{325}{25} = 13$$

50.

f_i	M_i	$(M_i - \bar{x})$	$(M_i - \bar{x})^2$	$f_i(M_i - \bar{x})^2$
4	5	− 8	64	256
7	10	− 3	9	63
9	15	2	4	36
5	20	7	49	245
25				600

$$s^2 = \frac{\Sigma f_i(M_i - \bar{x})^2}{n - 1} = \frac{600}{25 - 1} = 25$$

$$s = \sqrt{25} = 5$$

CHAPTER 4

2. $\binom{6}{3} = \frac{6!}{3!3!} = \frac{6 \cdot 5 \cdot 4 \cdot 3 \cdot 2 \cdot 1}{(3 \cdot 2 \cdot 1)(3 \cdot 2 \cdot 1)} = 20$

ABC	ACE	BCD	BEF
ABD	ACF	BCE	CDE
ABE	ADE	BCF	CDF
ABF	ADF	BDE	CEF
ACD	AEF	BDF	DEF

9. $\binom{50}{4} = \frac{50!}{4!46!} = \frac{50 \cdot 49 \cdot 48 \cdot 47}{4 \cdot 3 \cdot 2 \cdot 1} = 230,300$

14. $P(E_1) = .40$, $P(E_2) = .26$, $P(E_3) = .34$
The relative frequency method was used

17. $P(\text{never married}) = \frac{1106}{2038}$

$P(\text{married}) = \frac{826}{2038}$

$P(\text{other}) = \frac{106}{2038}$

Note that the sum of the probabilities equals 1

25. a. S = {ace of clubs, ace of diamonds, ace of hearts, ace of spades}

b. S = {2 of clubs, 3 of clubs, . . . , 10 of clubs, J of clubs, Q of clubs, K of clubs, A of clubs}

c. There are 12; jack, queen, or king in each of the four suits

d. *For(a)*: 4/52 = 1/13 = .08
For(b): 13/52 = 1/4 = .25
For(c): 12/52 = .23

27. a. (4, 6), (4, 7), (4, 8)

b. .05 + .10 + .15 = .30

c. (2, 8), (3, 8), (4, 8)

d. .05 + .05 + .15 = .25

e. .15

33. a. $P(A) = P(E_1) + P(E_4) + P(E_6) = .05 + .25 + .10 = .40$

$P(B) = P(E_2) + P(E_4) + P(E_7) = .20 + .25 + .05 = .50$

$P(C) = P(E_2) + P(E_3) + P(E_5) + P(E_7)$

$= .20 + .20 + .15 + .05 = .60$

b. $A \cup B = \{E_1, E_2, E_4, E_6, E_7\}$

$P(A \cup B) = P(E_1) + P(E_2) + P(E_4) + P(E_6) + P(E_7)$

$= .05 + .20 + .25 + .10 + .05$

$= .65$

c. $A \cap B = \{E_4\}$, $P(A \cap B) = P(E_4) = .25$

d. Yes, they are mutually exclusive

e. $B^c = \{E_1, E_3, E_5, E_6\}$

$P(B^c) = P(E_1) + P(E_3) + P(E_5) + P(E_6)$

$= .05 + .20 + .15 + .10$

$= .50$

38. Let: B = rented a car for business reasons
P = rented a car for personal reasons

a. $P(B \cap P) = P(B) + P(P) - P(B \cup P)$
$= .54 + .51 - .72$
$= .33$

b. $P(\text{did not rent}) = 1 - P(B \cup P)$
$= 1 - .72$
$= .28$

c. $P(\text{business only}) = P(B \cap P^c)$
$= P(B) - P(B \cap P)$
$= .54 - .33$
$= .21$

42. a. $P(A \mid B) = \dfrac{P(A \cap B)}{P(B)} = \dfrac{.40}{.60} = .6667$

b. $P(B \mid A) = \dfrac{P(A \cap B)}{P(A)} = \dfrac{.40}{.50} = .80$

c. No, because $P(A \mid B) \neq P(A)$

45. a.

| | Reason for Applying | | | |
	Quality	Cost/Convenience	Other	Total
Full-time	.218	.204	.039	.461
Part-time	.208	.307	.024	.539
Total	.426	.511	.063	1.00

b. It is most likely a student will cite cost or convenience as the first reason (probability = .511); school quality is the first reason cited by the second largest number of students (probability = .426)

c. $P(\text{quality} \mid \text{full-time}) = .218/.461 = .473$

d. $P(\text{quality} \mid \text{part-time}) = .208/.539 = .386$

e. For independence, we must have $P(A)P(B) = P(A \cap B)$; from the table,

$$P(A \cap B) = .218, \ P(A) = .461, \ P(B) = .426$$
$$P(A)P(B) = (.461)(.426) = .196$$

Since $P(A)P(B) \neq P(A \cap B)$, the events are not independent

53. a. Yes, since $P(A_1 \cap A_2) = 0$

b. $P(A_1 \cap B) = P(A_1)P(B \mid A_1) = .40(.20) = .08$
$P(A_2 \cap B) = P(A_2)P(B \mid A_2) = .60(.05) = .03$

c. $P(B) = P(A_1 \cap B) + P(A_2 \cap B) = .08 + .03 = .11$

d. $P(A_1 \mid B) = \dfrac{.08}{.11} = .7273$

$P(A_2 \mid B) = \dfrac{.03}{.11} = .2727$

56. M = missed payment
D_1 = customer defaults
D_2 = customer does not default
$P(D_1) = .05, \ P(D_2) = .95, \ P(M \mid D_2) = .2, \ P(M \mid D_1) = 1$

a. $P(D_1 \mid M) = \dfrac{P(D_1)P(M \mid D_1)}{P(D_1)P(M \mid D_1) + P(D_2)P(M \mid D_2)}$

$= \dfrac{(.05)(1)}{(.05)(1) + (.95)(.2)}$

$= \dfrac{.05}{.24} = .21$

b. Yes, the probability of default is greater than .20

CHAPTER 5

1. a. Head, Head (H, H)
Head, Tail (H, T)
Tail, Head (T, H)
Tail, Tail (T, T)

b. x = number of heads on two coin tosses

c.

Outcome	Values of x
(H, H)	2
(H, T)	1
(T, H)	1
(T, T)	0

3. Let: Y = position is offered
N = position is not offered

a. $S = \{(Y, Y, Y), (Y, Y, N), (Y, N, Y), (Y, N, N), (N, Y, Y), (N, Y, N), (N, N, Y), (N, N, N)\}$

b. Let N = number of offers made; N is a discrete random variable

c.

Experimental Outcome	(Y, Y, Y)	(Y, Y, N)	(Y, N, Y)	(Y, N, N)	(N, Y, Y)	(N, Y, N)	(N, N, Y)	(N, N, N)
Value of N	3	2	2	1	2	1	1	0

7. a. $f(x) \geq 0$ for all values of x
$\Sigma f(x) = 1$; therefore, it is a proper probability distribution

b. Probability $x = 30$ is $f(30) = .25$

c. Probability $x \leq 25$ is $f(20) + f(25) = .20 + .15 = .35$

d. Probability $x > 30$ is $f(35) = .40$

8. a.

x	$f(x)$
1	3/20 = .15
2	5/20 = .25
3	8/20 = .40
4	4/20 = .20
Total	1.00

b.

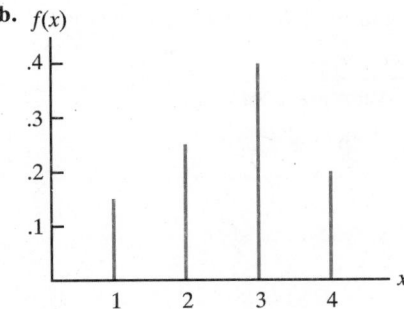

c. $f(x) \geq 0$ for $x = 1, 2, 3, 4$
$\Sigma f(x) = 1$

16. a.

y	f(y)	yf(y)
2	.20	.40
4	.30	1.20
7	.40	2.80
8	.10	.80
Totals	1.00	5.20

$$E(y) = \mu = 5.20$$

b.

y	$y - \mu$	$(y - \mu)^2$	f(y)	$(y - \mu)^2 f(y)$
2	−3.20	10.24	.20	2.048
4	−1.20	1.44	.30	.432
7	1.80	3.24	.40	1.296
8	2.80	7.84	.10	.784
			Total	4.560

$$Var(y) = 4.56$$
$$\sigma = \sqrt{4.56} = 2.14$$

c. $E(x + y) = E(x) + E(y) = 6.00 + 5.20 = 11.20$

d. $Var(x + y) = Var(x) + Var(y) = 4.50 + 4.56 = 9.06$
  $\sigma_{x+y} = \sqrt{9.06} = 3.01$

Note: The standard deviation of the sum is not the sum of the standard deviations

18. a. $E(x) = 26.0(.31) + 8.6(.23) + 7.7(.45) - 2.9(.01)$
  $= 13.474$
  The expected return on $1.00 for an individual investor is 13.474%

b.

x − 13.474	(x − 13.474)²	f(x)	(x − 13.474)²f(x)
12.526	156.900680	.31	48.639211
−4.874	23.755876	.23	5.463851
−5.774	33.339076	.45	15.002584
16.374	268.107880	.01	2.681079
		Total	71.786725

$$Var(x) = 71.7867$$
$$\sigma = 8.4727$$

c.

x	f(x)	xf(x)	$(x - \mu)$	$(x - \mu)^2$	$(x - \mu)^2 f(x)$
5.0	.31	1.550	−1.964	3.857296	1.195762
8.6	.23	1.978	1.636	2.676496	.615594
7.7	.45	3.465	.736	.541696	.243763
−2.9	.01	−.029	−9.864	97.298496	.972985
	Totals	6.964			3.028104

$$E(x) = 6.964$$
$$Var(x) = 3.028104$$
$$\sigma = 1.7401$$

27. a.

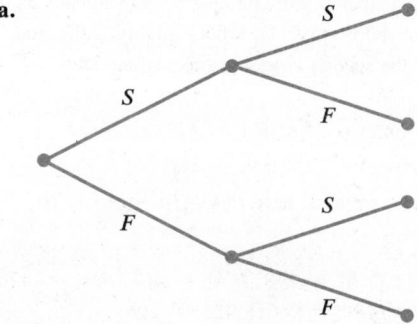

b. $f(1) = \binom{2}{1}(.4)^1(.6)^1 = \dfrac{2!}{1!1!}(.4)(.6) = .48$

c. $f(0) = \binom{2}{0}(.4)^0(.6)^2 = \dfrac{2!}{0!2!}(1)(.36) = .36$

d. $f(2) = \binom{2}{2}(.4)^2(.6)^0 = \dfrac{2!}{2!0!}(.16)(1) = .16$

e. $P(x \geq 1) = f(1) + f(2) = .48 + .16 = .64$

f. $E(x) = np = 2(.4) = .8$
  $Var(x) = np(1 - p) = 2(.4)(.6) = .48$
  $\sigma = \sqrt{.48} = .6928$

32. a. Probability of a defective part being produced must be .03 for each trial; trials must be independent

b. Let: D = defective

G = not defective

1st part	2nd part	Experimental Outcome	Number Defective

```
                              D ●  (D, D)        2
               D  ●
                              G ●  (D, G)        1
    ●
                              D ●  (G, D)        1
               G  ●
                              G ●  (G, G)        0
```

c. Two outcomes result in exactly one defect

d. $P(\text{no defects}) = (.97)(.97) = .9409$

$P(1 \text{ defect}) = 2(.03)(.97) = .0582$

$P(2 \text{ defects}) = (.03)(.03) = .0009$

45. a. $f(x) = \dfrac{2^x e^{-2}}{x!}$

b. $\mu = 6$ for 3 time periods

c. $f(x) = \dfrac{6^x e^{-6}}{x!}$

d. $f(2) = \dfrac{2^2 e^{-2}}{2!} = \dfrac{4(.1353)}{2} = .2706$

e. $f(6) = \dfrac{6^6 e^{-6}}{6!} = .1606$

f. $f(5) = \dfrac{4^5 e^{-4}}{5!} = .1563$

46. a. $\mu = 48(5/60) = 4$

$f(3) = \dfrac{4^3 e^{-4}}{3!} = \dfrac{(64)(.0183)}{6} = .1952$

b. $\mu = 48(15/60) = 12$

$f(10) = \dfrac{12^{10} e^{-12}}{10!} = .1048$

c. $\mu = 48(5/60) = 4$; one can expect 4 callers to be waiting after 5 minutes

$f(0) = \dfrac{4^0 e^{-4}}{0!} = .0183$; the probability none will be waiting after 5 minutes is .0183

d. $\mu = 48(3/60) = 2.4$

$f(0) = \dfrac{2.4^0 e^{-2.4}}{0!} = .0907$; the probability of no interruptions in 3 minutes is .0907

52. a. $f(1) = \dfrac{\dbinom{3}{1}\dbinom{10-3}{4-1}}{\dbinom{10}{4}} = \dfrac{\left(\dfrac{3!}{1!2!}\right)\left(\dfrac{7!}{3!4!}\right)}{\dfrac{10!}{4!6!}}$

$= \dfrac{(3)(35)}{210} = .50$

b. $f(2) = \dfrac{\dbinom{3}{2}\dbinom{10-3}{2-2}}{\dbinom{10}{2}} = \dfrac{(3)(1)}{45} = .067$

56. $N = 60$, $n = 10$

a. $r = 20$, $x = 0$

$f(0) = \dfrac{\dbinom{20}{0}\dbinom{40}{10}}{\dbinom{60}{10}} = \dfrac{(1)\left(\dfrac{40!}{10!30!}\right)}{\dfrac{60!}{10!50!}} = \left(\dfrac{40!}{10!30!}\right)\left(\dfrac{10!50!}{60!}\right)$

$= \dfrac{40\cdot39\cdot38\cdot37\cdot36\cdot35\cdot34\cdot33\cdot32\cdot31}{60\cdot59\cdot58\cdot57\cdot56\cdot55\cdot54\cdot53\cdot52\cdot51}$

$\approx .01$

b. $r = 20$, $x = 1$

$f(1) = \dfrac{\dbinom{20}{1}\dbinom{40}{9}}{\dbinom{60}{10}} = 20\left(\dfrac{40!}{9!31!}\right)\left(\dfrac{10!50!}{60!}\right)$

$\approx .07$

c. $1 - f(0) - f(1) = 1 - .08 = .92$

d. Same as the probability one will be from Hawaii; in part (b) that was found to equal approximately .07

CHAPTER 6

1. a.

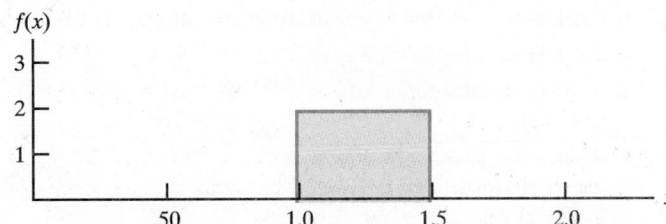

b. $P(x = 1.25) = 0$; the probability of any single point is zero since the area under the curve above any single point is zero

c. $P(1.0 \le x \le 1.25) = 2(.25) = .50$

d. $P(1.20 < x < 1.5) = 2(.30) = .60$

2. a.

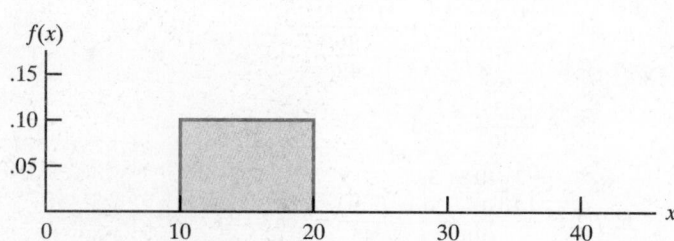

b. $P(x < 15) = .10(5) = .50$

c. $P(12 \le x \le 18) = .10(6) = .60$

d. $E(x) = \dfrac{10 + 20}{2} = 15$

e. $\text{Var}(x) = \dfrac{(20 - 10)^2}{12} = 8.33$

4. a.

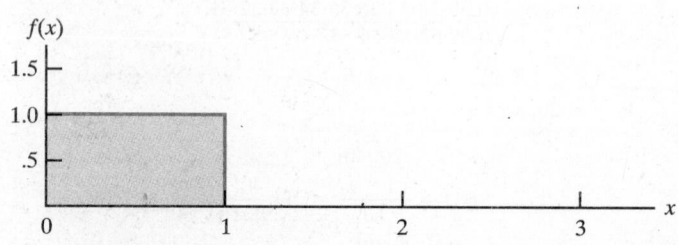

b. $P(.25 < x < .75) = 1(.50) = .50$

c. $P(x \le .30) = 1(.30) = .30$

d. $P(x > .60) = 1(.40) = .40$

13. a. $.4761 + .1879 = .6640$

b. $.3888 - .1985 = .1903$

c. $.4599 - .3508 = .1091$

15. a. Look in the table for an area of $.5000 - .2119 = .2881$; since the value we are seeking is below the mean, the z value must be negative; thus, for an area of .2881, $z = -.80$

b. Look in the table for an area of $.9030/2 = .4515$; $z = 1.66$

c. Look in the table for an area of $.2052/2 = .1026$; $z = .26$

d. Look in the table for an area of $.9948 - .5000 = .4948$; $z = 2.56$

e. Look in the table for an area of $.6915 - .5000 = .1915$; since the value we are seeking is below the mean, the z value must be negative; thus, $z = -.50$

18. a. With $z = \dfrac{(12,000 - 10,000)}{2200} = .91$,

$P(x \le 12,000) = P(z \le .91) = .5000 + .3186 = .8186$

$P(x \ge 12,000) = 1 - P(x \le 12,000) = 1 - .8186 = .1814$

So, the probability that a rehabilitation program will cost at least $12,000 is .1814

b. With $z = \dfrac{(6000 - 10,000)}{2200} = -1.82$,

$P(x \ge 6000) = P(z \ge -1.82) = .4656 + .5000 = .9656$

The probability that a rehabilitation program will cost at least $6000 is .9656

c. First, find the value of z that cuts off an area of .10 in the upper tail of the standard normal distribution; a value of $z = 1.28$ does this

Now find the value of x corresponding to $z = 1.28$

$\dfrac{x - 10,000}{2200} = 1.28$

$x = 10,000 + 1.28(2200) = 12,816$

The cost range for the 10% most expensive programs is $12,816 or more

32. a. $P(x \le x_0) = 1 - e^{-x_0/3}$

b. $P(x \le 2) = 1 - e^{-2/3} = 1 - .5134 = .4866$

c. $P(x \ge 3) = 1 - P(x \le 3) = 1 - (1 - e^{-3/3}) = e^{-1} = .3679$

d. $P(x \le 5) = 1 - e^{-5/3} = 1 - .1889 = .8111$

e. $P(2 \le x \le 5) = P(x \le 5) - P(x \le 2) = .8111 - .4866$
$= .3245$

34. a.

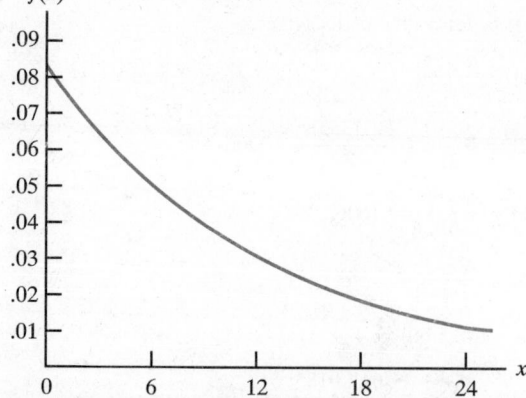

b. $P(x \le 12) = 1 - e^{-12/12} = 1 - .3679 = .6321$

c. $P(x \le 6) = 1 - e^{-6/12} = 1 - .6065 = .3935$

d. $P(x \ge 30) = 1 - P(x < 30)$
$= 1 - (1 - e^{-30/12})$
$= .0821$

CHAPTER 7

1. a. AB, AC, AD, AE, BC, BD, BE, CD, CE, DE

   b. With 10 samples, each has a $\frac{1}{10}$ probability

   c. E and C because 8 and 0 do not apply; 5 identifies E; 7 does not apply; 5 is skipped since E is already in the sample; 3 identifies C; 2 is not needed since the sample of size 2 is complete

3. 554, 459, 147, 385, 689, 640, 113, 340, 756, 953, 401, 827

13. a. $\bar{x} = \dfrac{\Sigma x_i}{n} = \dfrac{54}{6} = 9$

   b. $s^2 = \sqrt{\dfrac{\Sigma(x_i - \bar{x})^2}{n - 1}}$

   $\Sigma(x_i - \bar{x})^2 = (-4)^2 + (-1)^2 + 1^2 + (-2)^2 + 1^2 + 5^2 = 48$

   $s = \sqrt{\dfrac{48}{6 - 1}} = 3.1$

15. a. $\bar{x} = \dfrac{\Sigma x_i}{n} = \dfrac{465}{5} = 93$

x_i	$(x_i - \bar{x})$	$(x_i - \bar{x})^2$
94	+1	1
100	+7	49
85	−8	64
94	+1	1
92	−1	1
Totals 465	0	116

   $s = \sqrt{\dfrac{\Sigma(x_i - \bar{x})^2}{n - 1}} = \sqrt{\dfrac{116}{4}} = 5.39$

22. a. The sampling distribution is normal with:

   $E(\bar{x}) = \mu = 200$

   $\sigma_{\bar{x}} = \dfrac{\sigma}{\sqrt{n}} = \dfrac{50}{\sqrt{100}} = 5$

   For ± 5, $(\bar{x} - \mu) = 5$,

   $z = \dfrac{\bar{x} - \mu}{\sigma_{\bar{x}}} = \dfrac{5}{5} = 1$

   Area = .3413 × 2 = .6826

   b. For ± 10, $(\bar{x} - \mu) = 10$,

   $z = \dfrac{\bar{x} - \mu}{\sigma_{\bar{x}}} = \dfrac{10}{5} = 2$

   Area = .4772 × 2 = .9544

29. a.

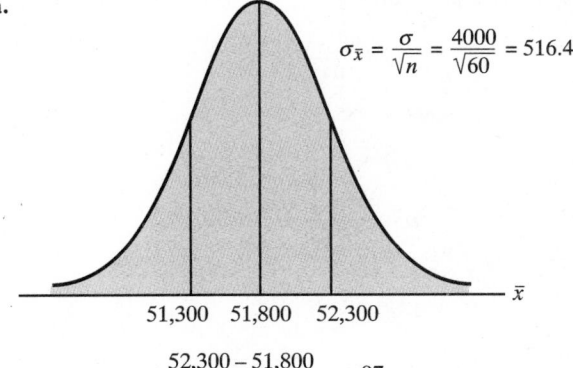

$\sigma_{\bar{x}} = \dfrac{\sigma}{\sqrt{n}} = \dfrac{4000}{\sqrt{60}} = 516.40$

$z = \dfrac{52,300 - 51,800}{516.40} = +.97$

Area = .3340 × 2 = .6680

   b. $\sigma_{\bar{x}} = \dfrac{\sigma}{\sqrt{n}} = \dfrac{4000}{\sqrt{120}} = 365.15$

   $z = \dfrac{52,300 - 51,800}{365.15} = +1.37$

   Area = .4147 × 2 = .8294

42. a. $E(\bar{p}) = .40$

   $\sigma_{\bar{p}} = \sqrt{\dfrac{p(1 - p)}{n}} = \sqrt{\dfrac{(.40)(.60)}{200}} = .0346$

   $z = \dfrac{\bar{p} - p}{\sigma_{\bar{p}}} = \dfrac{.03}{.0346} = .87$

   Area = .3078 × 2 = .6156

   b. $z = \dfrac{\bar{p} - p}{\sigma_{\bar{p}}} = \dfrac{.05}{.0346} = 1.45$

   Area = .4265 × 2 = .8530

45. a.

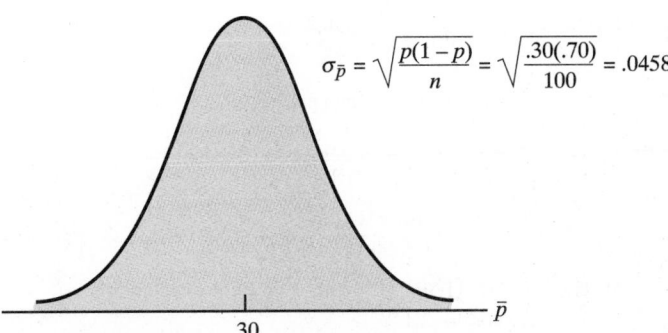

$\sigma_{\bar{p}} = \sqrt{\dfrac{p(1 - p)}{n}} = \sqrt{\dfrac{.30(.70)}{100}} = .0458$

The normal distribution is appropriate because $np = 100(.30) = 30$ and $n(1 - p) = 100(.70) = 70$ are both greater than 5

b. $P(.20 \le \bar{p} \le .40) = ?$

$$z = \frac{.40 - .30}{.0458} = 2.18$$

Area $= .4854 \times 2 = .9708$

c. $P(.25 \le \bar{p} \le .35) = ?$

$$z = \frac{.35 - .30}{.0458} = 1.09$$

Area $= .3621 \times 2 = .7242$

Chapter 8

2. Use $\bar{x} \pm z_{\alpha/2}(\sigma/\sqrt{n})$ with the sample standard deviation s used to estimate σ

a. $32 \pm 1.645(6/\sqrt{50})$
   32 ± 1.4; (30.6 to 33.4)

b. $32 \pm 1.96(6/\sqrt{50})$
   32 ± 1.66; (30.34 to 33.66)

c. $32 \pm 2.575(6/\sqrt{50})$
   32 ± 2.18; (29.82 to 34.18)

5. a. $\sigma_{\bar{x}} = \sigma/\sqrt{n} = 2.50/\sqrt{49} = .3571$

b. Sampling error less than or equal to $1.96\sigma_{\bar{x}} = .70$

c. $12.60 \pm .70$ or (11.90 to 13.30)

17. a. $\bar{x} = \dfrac{\Sigma x_i}{n} = \dfrac{80}{8} = 10$

b. $s = \sqrt{\dfrac{\Sigma(x_i - \bar{x})^2}{n - 1}} = \sqrt{\dfrac{84}{8 - 1}} = 3.46$

c. With 7 degrees of freedom, $t_{.025} = 2.365$

$$\bar{x} \pm t_{.025}\frac{s}{\sqrt{n}}$$

$$10 \pm 2.365\frac{3.464}{\sqrt{8}}$$

$$10 \pm 2.89; (7.11 \text{ to } 12.89)$$

19. At 90%, $80 \pm t_{.05}(s/\sqrt{n})$ with degrees of freedom =
   17, $t_{.05} = 1.740$
   $80 \pm 1.740(10/\sqrt{18})$
   80 ± 4.10; (75.90 to 84.10)

At 95%, $80 \pm 2.11(10/\sqrt{18})$ with degrees of freedom =
   17, $t_{.025} = 2.110$
   80 ± 4.97; (75.03 to 84.97)

28. a. Planning value of $\sigma = \dfrac{\text{Range}}{4} = \dfrac{36}{4} = 9$

b. $n = \dfrac{z_{.025}^2 \sigma^2}{E^2} = \dfrac{(1.96)^2(9)^2}{(3)^2} = 34.6 \approx 35$

c. $n = \dfrac{(1.96)^2(9)^2}{(2)^2} = 77.8 \approx 78$

29. Use $n = \dfrac{z_{\alpha/2}^2 \sigma^2}{E^2}$,

$$n = \frac{(1.96)^2(6.82)^2}{(1.5)^2} = 79.4 \text{ or } 80$$

$$n = \frac{(1.645)^2(6.82)^2}{(2)^2} = 31.5 \text{ or } 32$$

37. a. $\bar{p} = \dfrac{100}{400} = .25$

b. $\sqrt{\dfrac{\bar{p}(1 - \bar{p})}{n}} = \sqrt{\dfrac{.25(.75)}{400}} = .0217$

c. $\bar{p} \pm z_{.025}\sqrt{\dfrac{\bar{p}(1 - \bar{p})}{n}}$

$.25 \pm 1.96(.0275)$

$.25 \pm .0539$; (.1961 to .3039)

41. a. $\bar{p} = 248/400 = .62$

b. $\bar{p} \pm 1.645\sqrt{\dfrac{.62(.38)}{400}}$

$.62 \pm .04$; (.58 to .66)

Chapter 9

2. a. $H_0: \mu \le 14$
   $H_a: \mu > 14$; research hypothesis

b. There is no statistical evidence that the new bonus plan increases sales volume

c. The research hypothesis that $\mu > 14$ is supported; we can conclude that the new bonus plan increases the mean sales volume

5. a. Claiming $\mu > 6.08$ when it is not; the researcher would claim an annual rate higher than the national average when this is not the case

b. Concluding $\mu \le 6.08$ when it is not; the researcher would not detect the fact that adults in Des Moines have a higher annual rate

10. a. $z = 2.05$
   Reject H_0 if $z > 2.05$

b. $z = \dfrac{\bar{x} - \mu}{\sigma/\sqrt{n}} = \dfrac{16.5 - 15}{7/\sqrt{40}} = 1.36$

c. Area at $z = 1.36 = .4131$
   p-value $= .5000 - .4131 = .0869$

d. Do not reject H_0

13. a. H_0: $\mu \geq 26{,}100$

H_a: $\mu < 26{,}100$; research hypothesis

b. Reject H_0 if $z < -1.645$

$$z = \frac{\bar{x} - \mu_0}{\sigma/\sqrt{n}} = \frac{25{,}000 - 26{,}100}{2400/\sqrt{36}} = -2.75$$

Reject H_0 and conclude that the mean cost is less than $26,100

c. p-value $= .5000 - .4970 = .0030$

22. a. Reject H_0 if $z < -2.33$ or $z > 2.33$

b. $z = \dfrac{\bar{x} - \mu}{\sigma/\sqrt{n}} = \dfrac{14.2 - 15}{5/\sqrt{50}} = 1.13$

c. p-value $= 2(.5000 - .3708) = .2584$

d. Do not reject H_0

25. a. H_0: $\mu = 9.70$

H_a: $\mu \neq 9.70$

Reject H_0 if $z < -1.96$ or if $z > 1.96$

$$z = \frac{\bar{x} - \mu_0}{\sigma/\sqrt{n}} = \frac{9.30 - 9.70}{1.05/\sqrt{49}} = -2.67$$

Reject H_0; conclude that wage rates in the city are not $\mu = 9.70$

b. p-value $= 2(.5000 - .4962) = .0076$

34. a. $\bar{x} = \dfrac{\Sigma x_i}{n} = \dfrac{108}{6} = 18$

b. $s = \sqrt{\dfrac{\Sigma(x_i - \bar{x})}{n-1}} = \sqrt{\dfrac{10}{6-1}} = 1.41$

c. Reject H_0 if $t < -2.571$ or $t > 2.571$

d. $t = \dfrac{\bar{x} - \mu}{s/\sqrt{n}} = \dfrac{18 - 20}{1.41/\sqrt{6}} = -3.47$

e. Reject H_0; conclude H_a is true

37. a. H_0: $\mu \leq 200$

H_a: $\mu > 200$

With 9 degrees of freedom, reject H_0 if $t > 1.833$

$$\bar{x} = \frac{\Sigma x_i}{n} = 218$$

$$s = \sqrt{\frac{\Sigma(x_i - \bar{x})^2}{n-1}} = \sqrt{\frac{6210}{9}} = 26.27$$

$$t = \frac{\bar{x} - \mu_0}{s/\sqrt{n}} = \frac{218 - 200}{26.27/\sqrt{10}} = 2.17$$

Reject H_0; conclude the population mean rental rate exceeds the $200 per month rate in Baltimore

b. Using the t distribution table with 9 degrees of freedom, we find

$$t_{.05} = 1.833 \quad \text{and} \quad t_{.025} = 2.262$$

Interpolation will show the approximate p-value associated with $t = 2.17$ is:

$$p\text{-value} = .05 - .025\frac{(2.17 - 1.833)}{(2.262 - 1.833)} = .03$$

44. a. Reject H_0 if $z < -1.96$ or $z > 1.96$

b. $\sigma_{\bar{p}} = \sqrt{\dfrac{.20(.80)}{400}} = .02$

$$z = \frac{\bar{p} - p}{\sigma_{\bar{p}}} = \frac{.175 - .20}{.02} = -1.25$$

c. p-value $= 2(.5000 - .3944) = .2112$

d. Do not reject H_0

47. H_0: $p \leq .113$

H_a: $p > .113$

Reject H_0 if $z > 1.645$

$$\sigma_{\bar{p}} = \sqrt{\frac{p(1-p)}{n}} = \sqrt{\frac{.113(.887)}{200}} = .0224$$

$$\bar{p} = \frac{29}{200} = .145$$

$$z = \frac{\bar{p} - p_0}{\sigma_{\bar{p}}} = \frac{.145 - .113}{.0224} = 1.43$$

Do not reject H_0; there is no statistical evidence to justify concluding $p > .113$

56. $\sigma_{\bar{x}} = \dfrac{\sigma}{\sqrt{n}} = \dfrac{5}{\sqrt{120}} = .46$

$$c = 10 - 1.645(5/\sqrt{120}) = 9.25$$

Reject H_0 if $\bar{x} < 9.25$

a. When $\mu = 9$,

$$z = \frac{9.25 - 9}{5/\sqrt{120}} = .55$$

$$P(H_0) = (.5000 - .2088) = .2912$$

b. Type II error

c. When $\mu = 8$,

$$z = \frac{9.25 - 8}{5/\sqrt{120}} = 2.74$$
$$\beta = (.5000 - .4969) = .0031$$

59. a. $H_0: \mu \geq 25$
$H_a: \mu < 25$
Reject H_0 if $z < -2.05$

$$z = \frac{\bar{x} - \mu_0}{\sigma/\sqrt{n}} = \frac{\bar{x} - 25}{3/\sqrt{30}} = -2.05$$

Solve for $\bar{x} = 23.88$
Decision Rule: Accept H_0 if $\bar{x} \geq 23.88$
Reject H_0 if $\bar{x} < 23.88$

b. For $\mu = 23$,

$$z = \frac{23.88 - 23}{3/\sqrt{30}} = 1.61$$
$$\beta = .5000 - .4463 = .0537$$

c. For $\mu = 24$,

$$z = \frac{23.88 - 24}{3/\sqrt{30}} = -.22$$
$$\beta = .5000 + .0871 = .5871$$

d. The Type II error cannot be made in this case; note that when $\mu = 25.5$, H_0 is true: the Type II error can only be made when H_0 is false

64. $n = \frac{(z_\alpha + z_\beta)\sigma^2}{(\mu_0 - \mu_a)^2} = \frac{(1.645 + 1.28)^2(5)^2}{(10 - 9)^2} = 214$

67. At $\mu_0 = 400$, $\alpha = .02$; $z_{.02} = 2.05$
At $\mu_a = 385$, $\beta = .10$; $z_{.10} = 1.28$
With $\sigma = 30$,

$$n = \frac{(z_\alpha + z_\beta)^2\sigma^2}{(\mu_0 - \mu_a)^2} = \frac{(2.05 + 1.28)^2(30)^2}{(400 - 385)^2} = 44.4 \text{ or } 45$$

CHAPTER 10

1. a. $\bar{x}_1 - \bar{x}_2 = 13.6 - 11.6 = 2$

b. $s_{\bar{x}_1 - \bar{x}_2} = \sqrt{\frac{s_1^2}{n_1} + \frac{s_2^2}{n_2}} = \sqrt{\frac{(2.2)^2}{50} + \frac{3^2}{35}} = .59$

$2 \pm 1.645(.59)$

$2 \pm .98$; (1.02 to 2.98)

c. $2 \pm 1.96(.59)$

2 ± 1.17; (.83 to 3.17)

8. a. $\bar{x}_1 - \bar{x}_2 = 15,700 - 14,500 = 1200$

b. Pooled variance

$$s^2 = \frac{7(700)^2 + 11(850)^2}{18} = 632,083$$

$$s_{\bar{x}_1 - \bar{x}_2} = \sqrt{632,083\left(\frac{1}{8} + \frac{1}{12}\right)} = 362.88$$

With 18 degrees of freedom $t_{.025} = 2.101$,

$$1200 \pm 2.101(362.88)$$
$$1200 \pm 762, \text{ or } 438 \text{ to } 1962$$

c. Populations are normally distributed with equal variances

11. a. $s_{\bar{x}_1 - \bar{x}_2} = \sqrt{\frac{s_1^2}{n_1} + \frac{s_2^2}{n_2}} = \sqrt{\frac{(5.2)^2}{40} + \frac{6^2}{50}} = 1.18$

$$z = \frac{(\bar{x}_1 - \bar{x}_2) - (\mu_1 - \mu_2)}{s_{\bar{x}_1 - \bar{x}_2}} = \frac{(25.2 - 22.8)}{1.18} = 2.03$$

Reject H_0 if $z > 1.645$; therefore reject H_0; conclude H_a is true and $\mu_1 > \mu_2$

b. p-value $= .5000 - .4788 = .0212$

15. $H_0: \mu_1 - \mu_2 = 0$
$H_a: \mu_1 - \mu_2 \neq 0$

Reject H_0 if $z < -1.96$ or if $z > 1.96$

$$z = \frac{(\bar{x}_1 - \bar{x}_2) - 0}{\sqrt{\sigma_1^2/n_1 + \sigma_2^2/n_2}} = \frac{40 - 35}{\sqrt{(9)^2/36 + (10)^2/49}}$$

$$= \frac{5}{2.07} = 2.42$$

Reject H_0; customers at the two stores differ in terms of mean ages

21. a. 1, 2, 0, 0, 2

b. $\bar{d} = \frac{\Sigma d_i}{n} = \frac{5}{5} = 1$

c. $s_d = \sqrt{\frac{\Sigma(d_i - \bar{d})^2}{n - 1}} = \sqrt{\frac{4}{5 - 1}} = 1$

d. With 4 degrees of freedom, $t_{.05} = 2.132$; reject H_0 if $t > 2.132$

$$t = \frac{\bar{d} - \mu_d}{s_d/\sqrt{n}} = \frac{1 - 0}{1/\sqrt{5}} = 2.24$$

Reject H_0; conclude $\mu_d > 0$

23. d (difference) = rating after − rating before
$H_0: \mu_d \leq 0$
$H_a: \mu_d > 0$

With 7 degrees of freedom, reject H_0 if $t > 1.895$; when $\bar{d} = .63$ and $s_d = 1.3025$,

$$t = \frac{\bar{d} - \mu_d}{s_d/\sqrt{n}} = \frac{.63 - 0}{1.3025/\sqrt{8}} = 1.37$$

Do not reject H_0; we cannot conclude that seeing the commercial improves the potential to purchase

31. a. $\bar{p} = \dfrac{n_1\bar{p}_1 + n_2\bar{p}_2}{n_1 + n_2} = \dfrac{200(.22) + 300(.16)}{200 + 300} = .184$

$s_{\bar{p}_1 - \bar{p}_2} = \sqrt{(.184)(.816)\left(\dfrac{1}{200} + \dfrac{1}{300}\right)} = .0354$

Reject H_0 if $z > 1.645$

$z = \dfrac{(.22 - .16) - 0}{.0354} = 1.69$

Reject H_0

b. p-value $= (.5000 - .4545) = .0455$

33. $\bar{p}_1 = 270/500 = .54$; $\bar{p}_2 = 162/360 = .45$

$\bar{p}_1 - \bar{p}_2 = .54 - .45 = .09$

$.09 \pm 1.96\sqrt{\dfrac{.54(.46)}{500} + \dfrac{.45(.55)}{360}}$

$.09 \pm .0675$, or $.0225$ to $.1575$

CHAPTER 11

2. $s^2 = 25$

a. With 19 degrees of freedom, $\chi^2_{.05} = 30.1435$ and $\chi^2_{.95} = 10.1170$

$\dfrac{19(25)}{30.1435} \le \sigma^2 \le \dfrac{19(25)}{10.1170}$

$15.76 \le \sigma^2 \le 46.95$

b. With 19 degrees of freedom, $\chi^2_{.025} = 32.8523$ and $\chi^2_{.975} = 8.90655$

$\dfrac{19(25)}{32.8523} \le \sigma^2 \le \dfrac{19(25)}{8.90655}$

$14.46 \le \sigma^2 \le 53.33$

c. $3.8 \le \sigma \le 7.3$

9. $H_0\colon \sigma^2 \le .0004$

$H_a\colon \sigma^2 > .0004$

$n = 30$

$\chi^2_{.05} = 42.5569$ (29 degrees of freedom)

$\chi^2 = \dfrac{(29)(.0005)}{.0004} = 36.25$

Do not reject H_0; the product specification does not appear to be violated

15. We recommend placing the larger sample variance in the numerator; with $\alpha = .05$, $F_{.025,20,24} = 2.33$: reject H_0 if $F > 2.33$

$F = \dfrac{8.2}{4.0} = 2.05$; do not reject H_0

Or, had we used the lower tail F value,

$F_{.025,20,24} = \dfrac{1}{F_{.025,24,20}} = \dfrac{1}{2.46} = .41$

$F = \dfrac{4.0}{8.2} = .49$

$F > .41$; do not reject H_0

17. a. Let: $\sigma_1^2 =$ variance in repair costs (4-year-old automobiles)

$\sigma_2^2 =$ variance in repair costs (2-year-old automobiles)

$H_0\colon \sigma_1^2 \le \sigma_2^2$

$H_a\colon \sigma_1^2 > \sigma_2^2$

b. $s_1^2 = (170)^2 = 28,900$

$s_2^2 = (100)^2 = 10,000$

$F = \dfrac{s_1^2}{s_2^2} = \dfrac{28,900}{10,000} = 2.89$

$F_{.01,24,24} = 2.66$

Reject H_0; conclude that automobiles 4 years old have a larger variance in annual repair costs compared to automobiles 2 years old; this is expected due to the fact that older automobiles are more likely to have some very expensive repairs that lead to greater variance in the annual repair costs

CHAPTER 12

1. *Expected frequencies:* $e_1 = 200(.40) = 80$, $e_2 = 200(.40) = 80$, $e_3 = 200(.20) = 40$

Actual frequencies: $f_1 = 60$, $f_2 = 120$, $f_3 = 20$

$\chi^2 = \dfrac{(60 - 80)^2}{80} + \dfrac{(120 - 80)^2}{80} + \dfrac{(20 - 40)^2}{20}$

$= \dfrac{400}{80} + \dfrac{1600}{80} + \dfrac{400}{20}$

$= 5 + 20 + 20 = 45$

$\chi^2_{.01} = 9.21034$ with $k - 1 = 3 - 1 = 2$ degrees of freedom

Since $\chi^2 = 45 > 9.21034$, reject the null hypothesis; that is, the population proportions are not as stated in the null hypothesis

3. *Expected frequencies:* $300(.29) = 87$, $300(.28) = 84$, $300(.25) = 75$, $300(.18) = 54$

$e_1 = 87$, $e_2 = 84$, $e_3 = 75$, $e_4 = 54$

Actual frequencies: $f_1 = 95$, $f_2 = 70$, $f_3 = 89$, $f_4 = 46$

$$\chi^2_{.05} = 7.81 \text{ (3 degrees of freedom)}$$

$$\chi^2 = \frac{(95 - 87)^2}{87} + \frac{(70 - 84)^2}{84} + \frac{(89 - 75)^2}{75}$$

$$+ \frac{(46 - 54)^2}{54} = 6.87$$

Do not reject H_0; there is no significant change in the viewing audience proportions

9. H_0: The column factor is independent of the row factor
H_a: The column factor is not independent of the row factor

Expected frequencies:

	A	B	C
P	27.5	38.5	44
Q	22.5	31.5	36

$$\chi^2 = \frac{(20 - 27.5)^2}{27.5} + \frac{(44 - 38.5)^2}{38.5} + \frac{(50 - 44)^2}{44} + \frac{(30 - 22.5)^2}{22.5}$$

$$+ + \frac{(26 - 31.5)^2}{31.5} + \frac{(30 - 36)^2}{36}$$

$$= 8.1097$$

$\chi^2_{.025} = 7.37776$ with $(2 - 1)(3 - 1) = 2$ degrees of freedom

Since $\chi^2 = 8.1097 > 7.37776$, reject H_0; that is, conclude that the column factor is not independent of the row factor

11. H_0: There is no difference in shooting percentage among the teams
H_a: There is a difference in shooting percentage among the teams

Row 1 total = 629; row 2 total = 988
Column 1 total = 374; column 2 total = 341
Column 3 total = 369; column 4 total = 533

Overall total = 1617

Using these totals, we compute the expected frequencies

Expected frequencies:

	Duke	Mich.	Ind.	Cin.
Made	145.4830	132.6463	143.5380	207.3327
Missed	228.5170	208.3537	225.4620	325.6673

$$\chi^2 = \frac{(160 - 145.4830)^2}{145.4830} + \frac{(113 - 132.6463)^2}{132.6463} + \cdots$$

$$+ \frac{(331 - 325.6673)^2}{325.6673}$$

$$= 1.4486 + 2.9098 + .7625$$

$$+ .1372 + .9222 + 1.8525 + .4855 + .0873$$

$$= 8.6056$$

$$\chi^2_{.05} = 7.81473 \text{ with 3 degrees of freedom}$$

Since $\chi^2 = 8.6056 > 7.81473$, reject H_0; that is, conclude that there is a difference in 3-point shooting ability for the teams

21. First estimate μ from the sample data (sample size = 120)

$$\mu = \frac{0(39) + 1(30) + 2(30) + 3(18) + 4(3)}{120}$$

$$= \frac{156}{120} = 1.3$$

Therefore, we use Poisson probabilities with $\mu = 1.3$ to compute expected frequencies

x	Observed Frequency	Poisson Probability	Expected Frequency	Difference $(f_i - e_i)$
0	39	.2725	32.700	6.300
1	30	.3543	42.516	−12.516
2	30	.2303	27.636	2.364
3	18	.0998	11.976	6.024
4 or more	3	.0430	5.160	−2.160

$$\chi^2 = \frac{(6.300)^2}{32.700} + \frac{(-12.516)^2}{42.516} + \frac{(2.364)^2}{27.636} + \frac{(6.024)^2}{11.976}$$

$$+ \frac{(-2.160)^2}{5.160} = 9.0348$$

$$\chi^2_{.05} = 7.81473 \text{ with } 5 - 1 - 1 = 3 \text{ degrees of freedom}$$

Since $\chi^2 = 9.0348 > 7.81473$, reject H_0; that is, conclude that the data do not follow a Poisson probability distribution

Chapter 13

1. **a.** See Figure F13.1a

b. $\bar{\bar{x}} = (30 + 45 + 36)/3 = 37$

$$SSB = \sum_{j=1}^{k} n_j(\bar{x}_j - \bar{\bar{x}})^2$$

$$= 5(30 - 37)^2 + 5(45 - 37)^2$$

$$+ 5(36 - 37)^2 = 570$$

$$MSB = \frac{SSB}{k - 1} = \frac{570}{2} = 285$$

FIGURE F13.1a

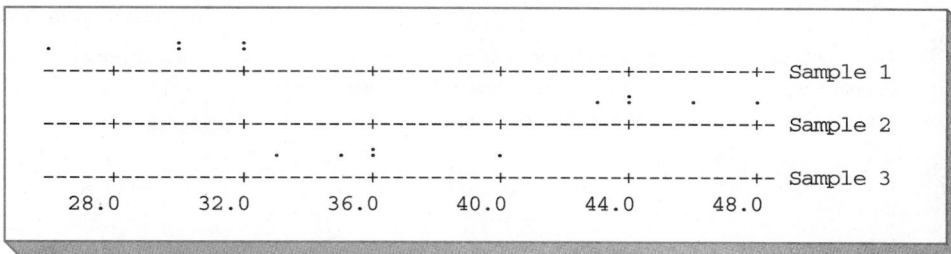

c. $\text{SSW} = \sum_{j=1}^{k} (n_j - 1)s_j^2$

$= 4(6) + 4(4) + 4(6.5) = 66$

$\text{MSW} = \dfrac{\text{SSW}}{n_T - k} = \dfrac{66}{15 - 3} = 5.5$

d. $F = \dfrac{\text{MSB}}{\text{MSW}} = \dfrac{285}{5.5} = 51.82$

$F_{.05} = 3.89$ (2 degrees of freedom numerator and 12 denominator)

Since $F = 51.82 > F_{.05} = 3.89$, we reject the null hypothesis that the means of the three populations are equal

e.

Source of Variation	Sum of Squares	Degrees of Freedom	Mean Square	F
Between	570	2	285	51.82
Within	66	12	5.5	
Total	636	14		

6.

	Manufacturer 1	Manufacturer 2	Manufacturer 3
Sample mean	23	28	21
Sample variance	6.67	4.67	3.33

$\bar{\bar{x}} = (23 + 28 + 21)/3 = 24$

$\text{SSB} = \sum_{j=1}^{k} n_j(\bar{x}_j - \bar{\bar{x}})^2$

$= 4(23 - 24)^2 + 4(28 - 24)^2$

$+ 4(21 - 24)^2 = 104$

$\text{MSB} = \dfrac{\text{SSB}}{k - 1} = \dfrac{104}{2} = 52$

$\text{SSW} = \sum_{j=1}^{k} (n_j - 1)s_j^2$

$= 3(6.67) + 3(4.67) + 3(3.33) = 44.01$

$\text{MSW} = \dfrac{\text{SSW}}{n_T - k} = \dfrac{44.01}{12 - 3} = 4.89$

$F = \dfrac{\text{MSB}}{\text{MSW}} = \dfrac{52}{4.89} = 10.63$

$F_{.05} = 4.26$ (2 degrees of freedom numerator and 9 denominator)

Since $F = 10.63 > F_{.05} = 4.26$, we reject the null hypothesis that the mean time needed to mix a batch of material is the same for each manufacturer

11. a. $\text{LSD} = t_{\alpha/2}\sqrt{\text{MSW}\left(\dfrac{1}{n_i} + \dfrac{1}{n_j}\right)}$

$= t_{.025}\sqrt{5.5\left(\dfrac{1}{5} + \dfrac{1}{5}\right)}$

$= 2.776\sqrt{2.2} = 4.12$

$|\bar{x}_1 - \bar{x}_2| = |30 - 45| = 15 > \text{LSD}$; significant difference

$|\bar{x}_1 - \bar{x}_3| = |30 - 36| = 6 > \text{LSD}$; significant difference

$|\bar{x}_2 - \bar{x}_3| = |45 - 36| = 9 > \text{LSD}$; significant difference

b. $\bar{x}_1 - \bar{x}_2 \pm t_{\alpha/2}\sqrt{\text{MSW}\left(\dfrac{1}{n_1} + \dfrac{1}{n_2}\right)}$

$(30 - 45) \pm 2.776\sqrt{5.5\left(\dfrac{1}{n_1} + \dfrac{1}{n_2}\right)}$

$-15 \pm 4.12 = -19.12$ to -10.88

c. $\alpha = .05/3 = .017$

$t_{.017/2} = t_{.0085}$, which is approximately $t_{.01} = 3.747$

$\text{BSD} = 3.747\sqrt{5.5\left(\dfrac{1}{5} + \dfrac{1}{5}\right)} = 5.56$

Thus, if the absolute value of the difference between any two sample means exceeds 5.56, there is sufficient evidence to reject the hypothesis that the corresponding population means are equal; since the differences in absolute value are each greater than 5.56, all three means appear to be different

d. $\text{TSD} = q\sqrt{\dfrac{\text{MSW}}{n}} = 5.04\sqrt{\dfrac{5.5}{5}} = 5.29$

Since the absolute value of the differences between any two sample means exceeds 5.29, there is sufficient evidence to conclude that the three means are different

e. $\bar{x}_i - \bar{x}_j \pm q\sqrt{\dfrac{\text{MSW}}{n}}$

$\bar{x}_i - \bar{x}_j \pm 5.04\sqrt{\dfrac{5.5}{5}}$

$\bar{x}_i - \bar{x}_j \pm 5.29$

Populations 1 and 2: $30 - 45 \pm 5.29 = -20.29 \text{ to } -9.71$
Populations 1 and 3: $30 - 36 \pm 5.29 = -11.29 \text{ to } -.71$
Populations 2 and 3: $45 - 36 \pm 5.29 = 3.71 \text{ to } 14.29$

13. $\text{LSD} = t_{\alpha/2}\sqrt{\text{MSW}\left(\dfrac{1}{n_1} + \dfrac{1}{n_3}\right)}$

$= t_{.025}\sqrt{4.89\left(\dfrac{1}{4} + \dfrac{1}{4}\right)}$

$= 2.262\sqrt{2.45} = 3.54$

Since $|\bar{x}_1 - \bar{x}_3| = |23 - 21| = 2 < 3.54$, there does not appear to be any significant difference between the means of populations 1 and 2

14. $\bar{x}_1 - \bar{x}_2 \pm q\sqrt{\dfrac{\text{MSW}}{n}}$

$23 - 28 \pm 3.95\sqrt{\dfrac{4.89}{4}}$

$-5 \pm 4.37 = -9.37 \text{ to } -.63$

19. a. $\bar{\bar{x}} = (156 + 142 + 134)/3 = 144$

$\text{SSTR} = \displaystyle\sum_{j=1}^{k} n_j(\bar{x}_j - \bar{\bar{x}})^2$

$= 6(156 - 144)^2 + 6(142 - 144)^2 + 6(134 - 144)^2 = 1488$

b. $\text{MSTR} = \dfrac{\text{SSTR}}{k-1} = \dfrac{1488}{2} = 744$

c. $s_1^2 = 164.35, \quad s_2^2 = 131.10, \quad s_3^2 = 110.46$

$\text{SSE} = \displaystyle\sum_{j=1}^{k} (n_j - 1)s_j^2$

$= 5(164.35) + 5(131.10) + 5(110.46)$
$= 2029.55$

d. $\text{MSE} = \dfrac{\text{SSE}}{n_T - k} = \dfrac{2029.55}{18 - 3} = 135.3$

e. $F = \dfrac{\text{MSTR}}{\text{MSE}} = \dfrac{744}{135.3} = 5.50$

$F_{.05} = 3.68$ (2 degrees of freedom numerator and 15 denominator)

Since $F = 5.50 > F_{.05} = 3.68$, we reject the hypothesis that the means for the three treatments are equal

34. ***Treatment Means***

$\bar{x}_{.1} = 13.6 \qquad \bar{x}_{.2} = 11.0 \qquad \bar{x}_{.3} = 10.6$

Block Means

$\bar{x}_{1.} = 9 \qquad \bar{x}_{2.} = 7.67 \qquad \bar{x}_{3.} = 15.67 \qquad \bar{x}_{4.} = 18.67$

$\bar{x}_{5.} = 7.67$

Overall Mean

$\bar{\bar{x}} = 176/15 = 11.73$

Step 1

$\text{SST} = \displaystyle\sum_i \sum_j (x_{ij} - \bar{\bar{x}})^2$

$= (10 - 11.73)^2 + (9 - 11.73)^2 + \cdots + (8 - 11.73)^2$
$= 354.93$

Step 2

$\text{SSTR} = b\displaystyle\sum_j (\bar{x}_{.j} - \bar{\bar{x}})^2$

$= 5[(13.6 - 11.73)^2 + (11.0 - 11.73)^2$
$+ (10.6 - 11.73)^2] = 26.53$

Step 3

$\text{SSBL} = k\displaystyle\sum_i (\bar{x}_{i.} - \bar{\bar{x}})^2$

$= 3[(9 - 11.73)^2 + (7.67 - 11.73)^2 + (15.67 - 11.73)^2$
$+ (18.67 - 11.73)^2 + (7.67 - 11.73)^2] = 312.32$

Step 4

$\text{SSE} = \text{SST} - \text{SSTR} - \text{SSBL}$

$= 354.93 - 26.53 - 312.32 = 16.08$

Source of Variation	Sum of Squares	Degrees of Freedom	Mean Square	F
Treatments	26.53	2	13.27	6.60
Blocks	312.32	4	78.08	
Error	16.08	8	2.01	
Total	354.93	14		

$F_{.05} = 4.46$ (2 numerator degrees of freedom and 8 denominator)

Since $F = 6.60 > F_{.05} = 4.46$, we reject the null hypothesis that the means of the three treatments are equal

40. See Table F13.40

Step 1

$$\text{SST} = \sum_i \sum_j \sum_k (x_{ijk} - \bar{\bar{x}})^2$$

$$= (135 - 111)^2 + (165 - 111)^2 + \cdots$$
$$+ (136 - 111)^2 = 9028$$

TABLE F13.40

		Factor B			Factor A
		Level 1	*Level 2*	*Level 3*	**Means**
Factor A	*Level 1*	$\bar{x}_{11} = 150$	$\bar{x}_{12} = 78$	$\bar{x}_{13} = 84$	$\bar{x}_{1.} = 104$
	Level 2	$\bar{x}_{21} = 110$	$\bar{x}_{22} = 116$	$\bar{x}_{23} = 128$	$\bar{x}_{2.} = 118$
Factor B Means		$\bar{x}_{.1} = 130$	$\bar{x}_{.2} = 97$	$\bar{x}_{.3} = 106$	$\bar{\bar{x}} = 111$

Step 2

$$SSA = br \sum_i (\bar{x}_{i.} - \bar{\bar{x}})^2$$

$$= 3(2)[(104 - 111)^2 + (118 - 111)^2] = 588$$

Step 3

$$SSB = ar \sum_j (\bar{x}_{.j} - \bar{\bar{x}})^2$$

$$= 2(2)[(130 - 111)^2 + (97 - 111)^2$$
$$+ (106 - 111)^2] = 2328$$

Step 4

$$SSAB = r \sum_i \sum_j (\bar{x}_{ij} - \bar{x}_{i.} - \bar{x}_{.j} + \bar{\bar{x}})^2$$

$$= 2[(150 - 104 - 130 + 111)^2$$
$$+ (78 - 104 - 97 + 111)^2 + \cdots$$
$$+ (128 - 118 - 106 + 111)^2] = 4392$$

Step 5

$$SSE = SST - SSA - SSB - SSAB$$

$$= 9028 - 588 - 2328 - 4392 = 1720$$

Source of Variation	Sum of Squares	Degrees of Freedom	Mean Square	F
Factor A	588	1	588	2.05
Factor B	2328	2	1164	4.06
Interaction	4392	2	2196	7.66
Error	1720	6	286.67	
Total	9028	11		

$F_{.05} = 5.99$ (1 degree of freedom numerator and 6 denominator)

$F_{.05} = 5.14$ (2 degrees of freedom numerator and 6 denominator)

Since $F = 2.05 < F_{.05} = 5.99$, factor A is not significant
Since $F = 4.06 < F_{.05} = 5.14$, factor B is not significant
Since $F = 7.66 > F_{.05} = 5.14$, interaction is significant

CHAPTER 14

1. **a.**

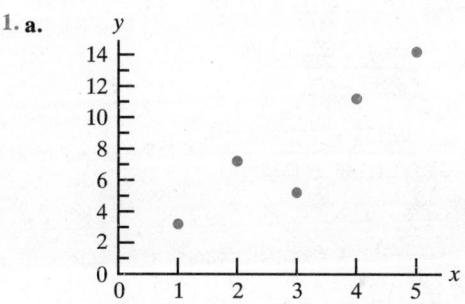

b. There appears to be a linear relationship between x and y

c. Many different straight lines can be drawn to provide a linear approximation of the relationship between x and y; in part (d) we will determine the equation of a straight line that "best" represents the relationship according to the least squares criterion

d. $\Sigma x_i = 15$, $\Sigma y_i = 40$, $\Sigma x_i y_i = 146$, $\Sigma x_i^2 = 55$

$$b_1 = \frac{\Sigma x_i y_i - (\Sigma x_i \Sigma y_i)/n}{\Sigma x_i^2 - (\Sigma x_i)^2/n}$$

$$= \frac{146 - (15)(40)/5}{55 - (15)^2/5} = 2.6$$

$$b_0 = \bar{y} - b_1 \bar{x}$$

$$= 8 - 2.6(3) = .2$$

$$\hat{y} = .2 + 2.6x$$

e. $\hat{y} = .2 + 2.6x = .2 + 2.6(4) = 10.6$

4. a.

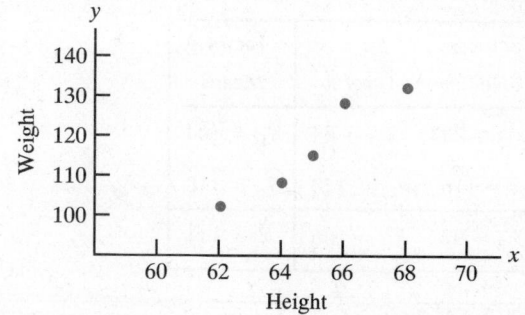

b. It indicates there may be a linear relationship between the variables

c. Many different straight lines can be drawn to provide a linear approximation of the relationship between x and y; in part (d) we will determine the equation of a straight line that "best" represents the relationship according to the least squares criterion

d. $\Sigma x_i = 325$, $\Sigma y_i = 585$; $\Sigma x_i y_i = 38{,}135$, $\Sigma x_i^2 = 21{,}145$

$$b_1 = \frac{\Sigma x_i y_i - (\Sigma x_i \, \Sigma y_i)/n}{\Sigma x_i^2 - (\Sigma x_i)^2/n}$$

$$= \frac{38{,}135 - (325)(585)/5}{21{,}145 - (325)^2/5} = 5.5$$

$$b_0 = \bar{y} - b_1 \bar{x}$$

$$= 117 - 5.5(65) = -240.5$$

$$\hat{y} = -240.5 + 5.5x$$

e. $\hat{y} = -240.5 + 5.5(63) = 106$

The estimate of weight is 106 pounds

15. a. $\hat{y}_i = .2 + 2.6x$ and $\bar{y} = 8$

x_i	y_i	$\hat{y}_i$	$y_i - \hat{y}_i$	$(y_i - \hat{y}_i)^2$	$y_i - \bar{y}$	$(y_i - \bar{y})^2$
1	3	2.8	.2	.04	−5	25
2	7	5.4	1.6	2.56	−1	1
3	5	8.0	−3.0	9.00	−3	9
4	11	10.6	.4	.16	3	9
5	14	13.2	.8	.64	6	36
				SSE = 12.40		SST = 80

$$\text{SSR} = \text{SST} - \text{SSE} = 80 - 12.4 = 67.6$$

b. $r^2 = \dfrac{\text{SSR}}{\text{SST}} = \dfrac{67.6}{80} = .845$

The least squares line provided a very good fit; 84.5% of the variability in y has been explained by the least squares line

c. $\Sigma x_i = 15$, $\Sigma y_i = 40$, $\Sigma x_i y_i = 146$,

$$\Sigma x_i^2 = 55, \quad \Sigma y_i^2 = 400$$

$$\text{SSR} = \frac{[\Sigma x_i y_i - (\Sigma x_i \Sigma y_i)/n]^2}{\Sigma x_i^2 - (\Sigma x_i)^2/n}$$

$$= \frac{[146 - (15)(40)/5]^2}{55 - (15)^2/5} = 67.6$$

$$\text{SST} = \Sigma y_i^2 - \frac{(\Sigma y_i)^2}{n}$$

$$= 400 - \frac{(40)^2}{5} = 80$$

Note that these are the same values shown in part (a)

18. a. $\Sigma x_i = 19.2$, $\Sigma y_i = 10{,}500$, $\Sigma x_i y_i = 34{,}030$
$\Sigma x_i^2 = 62.18$, $\Sigma y_i^2 = 18{,}710{,}000$

$$\text{SSR} = \frac{[\Sigma x_i y_i - (\Sigma x_i \Sigma y_i)/n]^2}{\Sigma x_i^2 - (\Sigma x_i)^2/n}$$

$$= \frac{[34{,}030 - (19.2)(10{,}500)/6]^2}{62.18 - (19.2)^2/6} = 249{,}864.86$$

$$\text{SST} = \Sigma y_i^2 - \frac{(\Sigma y_i)^2}{n}$$

$$= 18{,}710{,}000 - \frac{(10{,}500)^2}{6} = 335{,}000$$

$$\text{SSE} = \text{SST} - \text{SSR} = 335{,}000 - 249{,}864.86$$

$$= 85{,}135.14$$

b. $r^2 = \dfrac{\text{SSR}}{\text{SST}} = \dfrac{249{,}864.86}{335{,}000} = .746$

The least squares line accounted for 74.6% of the total sum of squares

22. a. $s^2 = \text{MSE} = \dfrac{\text{SSE}}{n-2} = \dfrac{12.4}{3} = 4.133$

b. $s = \sqrt{\text{MSE}} = \sqrt{4.133} = 2.033$

c. $\Sigma x_i = 15$, $\Sigma x_i^2 = 55$

$$s_{b_1} = \frac{s}{\sqrt{\Sigma x_i^2 - (\Sigma x_i)^2/n}}$$

$$= \frac{2.033}{\sqrt{55 - (15)^2/5}} = .643$$

d. $t = \dfrac{b_1 - \beta_1}{s_{b_1}} = \dfrac{2.6 - 0}{.643} = 4.04$

$t_{.025} = 3.182$ (3 degrees of freedom)
Since $t = 4.04 > t_{.05} = 3.182$, we reject H_0: $\beta_1 = 0$

e. $\text{MSR} = \dfrac{\text{SSR}}{1} = 67.6$

$$F = \frac{\text{MSR}}{\text{MSE}} = \frac{67.6}{4.133} = 16.36$$

$F_{.05} = 10.13$ (1 degree of freedom numerator
and 3 denominator)

Since $F = 16.36 > F_{.05} = 10.13$, we reject $H_0: \beta_1 = 0$

25. a. $s^2 = \text{MSE} = \dfrac{\text{SSE}}{n-2} = \dfrac{85,135.14}{4} = 21,283.79$

$s = \sqrt{\text{MSE}} = \sqrt{21,283.79} = 145.89$

$\Sigma x_i = 19.2, \ \Sigma x_i^2 = 62.18$

$$s_{b_1} = \frac{s}{\sqrt{\Sigma x_i^2 - (\Sigma x_i)^2/n}}$$

$$= \frac{145.89}{\sqrt{62.18 - (19.2)^2/6}} = 169.59$$

$$t = \frac{b_1 - \beta_1}{s_{b_1}} = \frac{581.08 - 0}{169.59} = 3.43$$

$t_{.025} = 2.776$ (4 degrees of freedom)

Since $t = 3.43 > t_{.025} = 2.776$, we reject $H_0: \beta_1 = 0$

b. $\text{MSR} = \dfrac{\text{SSR}}{1} = \dfrac{249,864.86}{1} = 249,864.86$

$$F = \frac{\text{MSR}}{\text{MSE}} = \frac{249,864.86}{21,283.79} = 11.74$$

$F_{.05} = 7.71$ (1 degree of freedom numerator and 4
denominator)

Since $F = 11.74 > F_{.05} = 7.71$, we reject $H_0: \beta_1 = 0$

30. a. $s = 2.033$

$\Sigma x_i = 15, \quad \Sigma x_i^2 = 55$

$$s_{\hat{y}_p} = s\sqrt{\frac{1}{n} + \frac{(x_p - \bar{x})^2}{\Sigma x_i^2 - (\Sigma x_i)^2/n}}$$

$$= 2.033\sqrt{\frac{1}{5} + \frac{(4-3)^2}{55 - (15)^2/5}} = 1.11$$

b. $\hat{y} = .2 + 2.6x = .2 + 2.6(4) = 10.6$

$\hat{y}_p \pm t_{\alpha/2}s_{\hat{y}_p}$

$10.6 \pm 3.182(1.11)$

10.6 ± 3.53 or 7.07 to 14.13

c. $s_{\text{ind}} = s\sqrt{1 + \dfrac{1}{n} + \dfrac{(x_p - \bar{x})^2}{\Sigma x_i^2 - (\Sigma x_i)^2/n}}$

$$= 2.033\sqrt{1 + \frac{1}{5} + \frac{(4-3)^2}{55 - (15)^2/5}} = 2.32$$

d. $\hat{y}_p \pm t_{\alpha/2}s_{\hat{y}_p}$

$10.6 \pm 3.182(2.32)$

10.6 ± 7.38 or 3.22 to 17.98

34. $s = 145.89, \quad \Sigma x_i = 19.2, \quad \Sigma x_i^2 = 62.18$

$\hat{y} = -109.46 + 581.08x = -109.46 + 581.08(3)$

$\quad = 1633.78$

$$s_{\hat{y}_p} = s\sqrt{\frac{1}{n} + \frac{(x_p - \bar{x})^2}{\Sigma x_i^2 - (\Sigma x_i)^2/n}}$$

$$= 145.89\sqrt{\frac{1}{6} + \frac{(3 - 3.2)^2}{62.18 - (19.2)^2/6}} = 68.54$$

$\hat{y}_p \pm t_{\alpha/2}s_{\hat{y}_p}$

$1633.78 \pm 2.776(68.54)$

1633.78 ± 190.27 or \$1443.51 to \$1824.05

36. $s = 145.89, \quad \Sigma x_i = 19.2, \quad \Sigma x_i^2 = 62.18$

$\hat{y} = -109.46 + 581.08x = -109.46 + 581.08(3)$

$\quad = 1633.78$

$$s_{\text{ind}} = s\sqrt{1 + \frac{1}{n} + \frac{(x_p - \bar{x})^2}{\Sigma x_i^2 - (\Sigma x_i)^2/n}}$$

$$= 145.89\sqrt{1 + \frac{1}{6} + \frac{(3 - 3.2)^2}{62.18 - (19.2)^2/6}} = 161.19$$

$\hat{y}_p \pm t_{\alpha/2}s_{\text{ind}}$

$1633.78 \pm 2.776(161.19)$

1633.78 ± 447.46 or \$1186.32 to \$2081.24

39.

Source of Variation	Sum of Squares	Degrees of Freedom	Mean Square	F
Regression	67.6	1	67.6	16.36
Error	12.4	3	4.133	
Total	80.0	4		

42. a. 9

b. $\hat{y} = 20.0 + 7.21x$

c. 1.3626

d. $\text{SSE} = \text{SST} - \text{SSR} = 51,984.1 - 41,587.3 = 10,396.8$

$\text{MSE} = 10,396.8/7 = 1485.3$

$$F = \frac{\text{MSR}}{\text{MSE}} = \frac{41,587.3}{1485.3} = 28.0$$

$F_{.05} = 5.59$ (1 degree of freedom numerator and 7 denominator)

Since $F = 28 > F_{.05} = 5.59$, we reject $H_0: \beta_1 = 0$

e. $\hat{y} = 20.0 + 7.21(50) = 380.5$ or \$380,500

47. a. $\Sigma x_i = 70$, $\Sigma y_i = 76$, $\Sigma x_i y_i = 1264$, $\Sigma x_i^2 = 1106$

$$b_1 = \frac{\Sigma x_i y_i - (\Sigma x_i \, \Sigma y_i)/n}{\Sigma x_i^2 - (\Sigma x_i)^2/n}$$

$$= \frac{1264 - (70)(65)/5}{1106 - (70)^2/5} = 1.5873$$

$$b_0 = \bar{y} - b_1 \bar{x}$$

$$= 15.2 - 1.5873(14) = -7.0222$$

$$\hat{y} = -7.02 + 1.59x$$

b.

x_i	y_i	$\hat{y}_i$	$y_i - \hat{y}_i$
6	6	2.52	3.48
11	8	10.47	−2.47
15	12	16.83	−4.83
18	20	21.60	−1.60
20	30	24.78	5.22

c.

With only five observations, it is difficult to determine whether the assumptions are satisfied; however, the plot does suggest curvature in the residuals, which would indicate that the error term assumptions are not satisfied; the scatter diagram for these data also indicates that the underlying relationship between x and y may be curvilinear

d. $s^2 = 23.78$

$$h_i = \frac{1}{n} + \frac{(x_i - \bar{x})^2}{\Sigma x_i^2 - (\Sigma x_i)^2/n}$$

$$= \frac{1}{5} + \frac{(x_i - 14)^2}{1106 - (70)^2/5} = \frac{1}{5} + \frac{(x_i - 14)^2}{126}$$

x_i	h_i	$s_{y_i - \hat{y}_i}$	$y_i - \hat{y}_i$	Standardized Residuals
6	.7079	2.64	3.48	1.32
11	.2714	4.16	−2.47	−.59
15	.2079	4.34	−4.83	−1.11
18	.3270	4.00	−1.60	−.40
20	.4857	3.50	5.22	1.49

e.

The plot of the standardized residuals against $\hat{y}$ has the same shape as the original residual plot; as stated in part (c), the curvature observed indicates that the assumptions regarding the error term may not be satisfied

49. $\Sigma x_i = 57$, $\Sigma y_i = 294$, $\Sigma x_i y_i = 2841$,

$\Sigma x_i^2 = 753$, $\Sigma y_i^2 = 13{,}350$

$$b_1 = \frac{\Sigma x_i y_i - (\Sigma x_i \Sigma y_i)/n}{\Sigma x_i^2 - (\Sigma x_i)^2/n}$$

$$= \frac{2841 - (57)(294)/7}{753 - (57)^2/7} = 1.5475$$

$$b_0 = \bar{y} - b_1 \bar{x}$$

$$= 42 - (1.5475)(8.1492) = 29.3989$$

$$\hat{y} = 29.40 + 1.55x$$

b. From Exercise 19 we have SSR = 691.72 and SST = 1002; therefore SSE = 1002 − 691.72 = 310.28

$$F = \frac{MSR}{MSE} = \frac{691.72}{310.28/5} = 11.15$$

$$F_{.05} = 6.61 \ (1 \ \text{degree of freedom}$$

$$\text{numerator and 5 denominator})$$

Since $F = 11.47 > F_{.05} = 6.61$, we reject H_0: $\beta_1 = 0$; the relationship is significant at the .05 level

c.

x_i	y_i	$\hat{y}_i = 29.40 + 1.55x_i$	$y_i - \hat{y}_i$
1	19	30.95	−11.95
2	32	32.50	−.50
4	44	35.60	8.40
6	40	38.70	1.30
10	52	44.90	7.10
14	53	51.10	1.90
20	54	60.40	−6.40

d. The residuals plot shown below leads us to question the assumption of a linear relationship between x and y; even though the relationship is significant at the $\alpha = .05$ level, it would be extremely dangerous to extrapolate beyond the range of the data (e.g., $x > 20$)

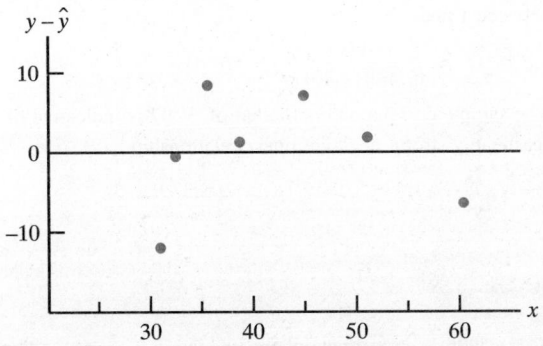

52. a. Using Minitab, we obtained the estimated regression equation $\hat{y} = 66.1 + .4023x$; a portion of the Minitab output is shown in Figure F14.52a

The fitted values and standardized residuals are shown below:

x_i	y_i	$\hat{y}_i$	Standardized Residuals
135	145	120.41	2.11
110	100	110.35	−1.08
130	120	118.40	.14
145	120	124.43	−.38
175	130	136.50	−.78
160	130	130.47	−.04
120	110	114.38	−.41

b. Standardized Residuals

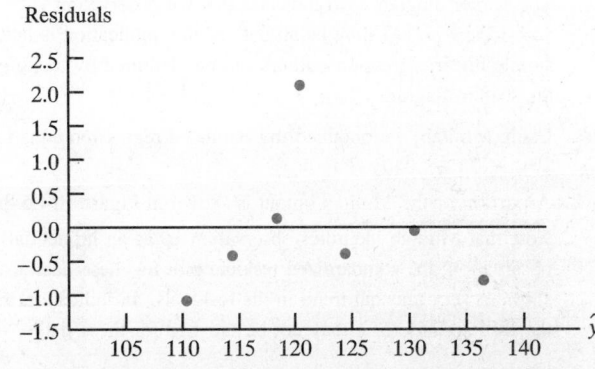

The standardized residual plot indicates that the observation $x = 135$, $y = 145$ may be an outlier; note that this observation has a standardized residual of 2.11

FIGURE F14.52a

```
Predictor      Coef       Stdev     t-ratio       p
Constant       66.10      32.06        2.06    0.094
X              0.4023     0.2276       1.77    0.137

s = 12.62      R-sq = 38.5%     R-sq(adj) = 26.1%

Analysis of Variance

SOURCE        DF          SS          MS        F       p
Regression     1       497.2       497.2     3.12   0.137
Error          5       795.7       159.1
Total          6      1292.9

Unusual Observations
Obs.      X          Y       Fit  Stdev.Fit  Residual  St.Resid
   1    135     145.00    120.42      4.87     24.58     2.11R
```

c. The scatter diagram is shown below:

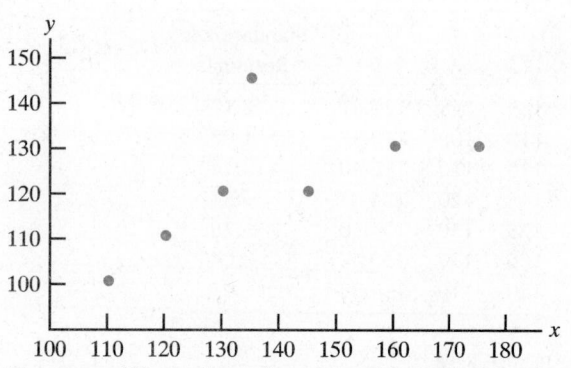

The scatter diagram also indicates that the observation $x = 135$, $y = 145$ may be an outlier; the implication is that for simple linear regression outliers can be identified by looking at the scatter diagram

54. a. Using Minitab, we obtained the estimated regression equation $\hat{y} = -.23 + .049x$

b. A portion of the Minitab output is shown in Figure F14.54b.

Note that Minitab identifies observation 10 as an influential observation; the standardized residual plot for these data also shows a very unusual trend in the residuals, an indication that the assumptions for ϵ may not be satisfied for these data

57.

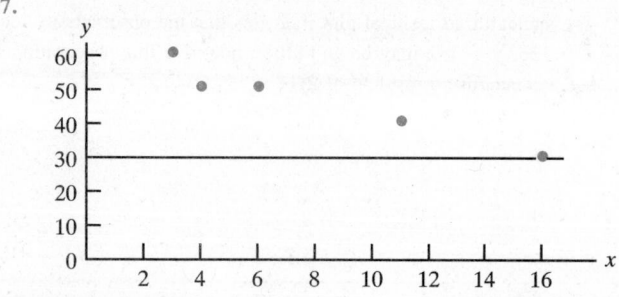

b. There appears to be a negative linear relationship between x and y

c.

x_i	y_i	$x_i - \bar{x}$	$y_i - \bar{y}$	$(x_i - \bar{x})(y_i - \bar{y})$
4	50	−4	4	−16
6	50	−2	4	−8
11	40	3	−6	−18
3	60	−5	14	−70
16	30	8	−16	−128
40	230	0	0	−240

$\bar{x} = 8$; $\bar{y} = 46$

$$s_{xy} = \frac{\Sigma(x_i - \bar{x})(y_i - \bar{y})}{n - 1} = \frac{-240}{4} = -60$$

The sample covariance indicates a negative linear association between x and y

d. $r_{xy} = \dfrac{s_{xy}}{s_x s_y} = \dfrac{-60}{(5.43)(11.40)} = -.97$

The sample correlation coefficient of $-.97$ is indicative of a reasonably strong negative linear relationship

CHAPTER 15

1. a. $b_1 = .5906$ is an estimate of the change in y corresponding to a 1-unit change in x_1 when x_2 is held constant

$b_2 = .4980$ is an estimate of the change in y corresponding to a 1-unit change in x_2 when x_1 is held constant

6. a. The Minitab output is shown in Figure F15.6a

b. No, it is 1.60 in 5a and 2.29 above; in this exercise it represents the marginal change in revenue due to an increase in television advertising with newspaper advertising held constant

FIGURE F14.54b

```
Predictor        Coef        Stdev      t-ratio        p
Constant      -0.2303       0.1138       -2.02      0.078
X            0.048987      0.002769      17.69      0.000

s = 0.1086      R-sq = 97.5%       R-sq(adj) = 97.2%

Analysis of Variance

SOURCE          DF         SS            MS          F         p
Regression       1       3.6946        3.6946     313.04    0.000
Error            8       0.0944        0.0118
Total            9       3.7890

Unusual Observations
Obs.        X           Y        Fit  Stdev.Fit  Residual    St.Resid
  10      67.0      3.0000     3.0518     0.0843   -0.0518      -0.76 X
```

FIGURE F15.6a

```
The regression equation is
REVENUE = 83.2 + 2.29 TVADV + 1.30 NEWSADV

Predictor       Coef        Stdev      t-ratio          p
Constant      83.230       1.574        52.88      0.000
TVADV          2.2902      0.3041         7.53      0.001
NEWSADV        1.3010      0.3207         4.06      0.010

s = 0.6426      R-sq = 91.9%      R-sq(adj) = 88.7%

Analysis of Variance

SOURCE        DF          SS           MS          F         p
Regression     2      23.435       11.718      28.38     0.002
Error          5       2.065        0.413
Total          7      25.500
```

12. a. $R^2 = \dfrac{SSR}{SST} = \dfrac{14{,}052.2}{15{,}182.9} = .926$

b. $R_a^2 = 1 - (1 - R^2)\dfrac{n-1}{n-p-1}$

$\qquad = 1 - (1 - .926)\dfrac{10-1}{10-2-1} = .905$

c. Yes; after adjusting for the number of independent variables in the model, we see that 90.5% of the variability in y has been accounted for

14. a. $R^2 = \dfrac{SSR}{SST} = \dfrac{12{,}000}{16{,}000} = .75$

b. $R_a^2 = 1 - (1 - R^2)\left(\dfrac{n-1}{n-p-1}\right)$

$\qquad = 1 - (.25)\left(\dfrac{9}{7}\right)$

$\qquad = .68$

c. The adjusted coefficient of determination shows that 68% of the variability has been explained by the two independent variables; thus, we conclude that the model does explain a large amount of the variability

17. a. $\text{MSR} = \dfrac{SSR}{p} = \dfrac{6216.375}{2} = 3108.188$

$\quad\;\; \text{MSE} = \dfrac{SSE}{n-p-1} = \dfrac{507.75}{10-2-1} = 72.536$

b. $F = \dfrac{\text{MSR}}{\text{MSE}} = \dfrac{3108.188}{72.536} = 42.85$

$F_{.05} = 4.74$ (2 degrees of freedom numerator and 7 denominator)

Since $F = 42.85 > F_{.05} = 4.74$, the overall model is significant

c. $t = \dfrac{b_1}{s_{b_1}} = \dfrac{.5906}{.0813} = 7.26$

$t_{.025} = 2.365$ (7 degrees of freedom)

Since $t = 7.26 > t_{.025} = 2.365$, β_1 is significant

d. $t = \dfrac{b_2}{s_{b_2}} = \dfrac{.4980}{.0567} = 8.78$

Since $t = 8.78 > t_{.025} = 2.365$, β_2 is significant

20. a. $\text{SSE} = \text{SST} - \text{SSR} = 16000 - 12000 = 4000$

$\qquad s^2 = \dfrac{\text{SSE}}{n-p-1} = \dfrac{4000}{7} = 571.43$

$\qquad \text{MSR} = \dfrac{\text{SSR}}{p} = \dfrac{12{,}000}{2} = 6000$

b. $F = \dfrac{\text{MSR}}{\text{MSE}} = \dfrac{6000}{571.43} = 10.50$

$F_{.05} = 4.74$ (2 degrees of freedom numerator and 7 denominator)

Since $F = 10.50 > F_{.05} = 4.74$, reject H_0; there is a significant relationship among the variables

26. a. Using Minitab, the 95% confidence interval is 132.16 to 154.15

b. Using Minitab, the 95% prediction interval is 111.15 to 175.17

28. a. Using Minitab, the 95% confidence interval is 71.13 to 112.35; in terms of units: 71,130 to 112,350

b. Using Minitab, the 95% prediction interval is 42.87 to 140.61; in terms of units: 42,870 to 140,610

This prediction interval estimate would probably not be of interest to Heller since they are a manufacturer of mowers and not a retailer

30. a. $\qquad E(y) = \beta_0 + \beta_1 x_1 + \beta_2 x_2$

$\qquad$ where $x_2 = \begin{cases} 0 \text{ if level 1} \\ 1 \text{ if level 2} \end{cases}$

b. $E(y) = \beta_0 + \beta_1 x_1 + \beta_2(0) = \beta_0 + \beta_1 x_1$

c. $E(y) = \beta_0 + \beta_1 x_1 + \beta_2(1) = \beta_0 + \beta_1 x_1 + \beta_2$

d. $\beta_2 = E(y \mid \text{level 2}) - E(y \mid \text{level 1})$

β_1 is the change in $E(y)$ for a 1-unit change in x_1 holding x_2 constant

32. a. $15,300, since $b_3 = 15.3$

b. $\hat{y} = 10.1 - 4.2(2) + 6.8(8) + 15.3(0)$

$= 10.1 - 8.4 + 54.4$

$= 56.1$

Sales prediction: $56,100

c. $\hat{y} = 10.1 - 4.2(1) + 6.8(3) + 15.3(1)$

$= 10.1 - 4.2 + 20.4 + 15.3$

$= 41.6$

Sales prediction: $41,600

37. a. The Minitab output is shown in Figure F15.37a

b. Using Minitab, we obtained the following values:

x_i	y_i	$\hat{y}_i$	Standardized Residual
1	3	2.8	.16
2	7	5.4	.94
3	5	8.0	−1.65
4	11	10.6	.24
5	14	13.2	.62

Standardized Residuals

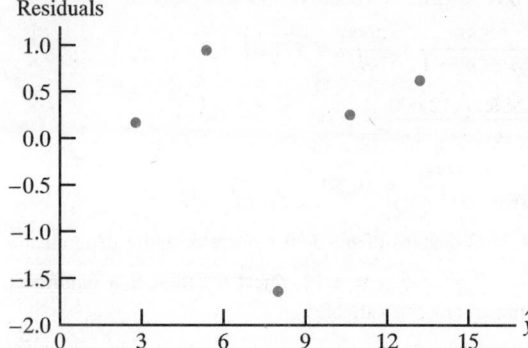

The point (3,5) does not appear to follow the trend of the remaining data; however, the value of the standardized residual for this point, −1.65, is not large enough for us to conclude that (3,5) is an outlier

c. Using Minitab, we obtained the following values:

x_i	y_i	Studentized Deleted Residual
1	3	.13
2	7	.92
3	5	−4.42
4	11	.19
5	14	.54

$t_{.025} = 4.303$ ($n - p - 2 = 5 - 1 - 2 = 2$ degrees of freedom)

Since the studentized deleted residual for (3,5) is $-4.42 < -4.303$, we conclude that the 3rd observation is an outlier

39. a. The Minitab output appears in Figure F15.6a; the estimated regression equation is

$$\text{REVENUE} = 83.2 + 2.29 \text{ TVADV} + 1.30 \text{ NEWSADV}$$

b. Using Minitab, we obtained the following values:

$\hat{y}_i$	Standardized Residual	$\hat{y}_i$	Standardized Residual
96.63	−1.62	94.39	1.10
90.41	−1.08	94.24	−.40
94.34	1.22	94.42	−1.12
92.21	−.37	93.35	1.08

FIGURE **F15.37a**

```
The regression equation is
Y = 0.20 + 2.60 X

Predictor       Coef        Stdev       t-ratio        p
Constant        0.200       2.132        0.09       0.931
X               2.6000      0.6429       4.04       0.027

s = 2.033       R-sq = 84.5%     R-sq(adj) = 79.3%

Analysis of Variance

SOURCE          DF          SS           MS          F          p
Regression      1           67.600       67.600      16.35      0.027
Error           3           12.400       4.133
Total           4           80.000
```

Standardized Residuals

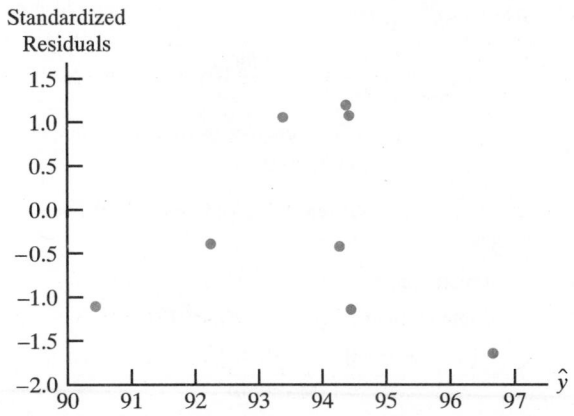

With relatively few observations, it is difficult to determine if any of the assumptions regarding ϵ have been violated; for instance, an argument could be made that there does not appear to be any pattern in the plot; alternatively, an argument could be made that there is a curvilinear pattern in the plot

c. The values of the standardized residuals are greater than -2 and less than $+2$; thus, using this test, there are no outliers

As a further check for outliers, we used Minitab to compute the following studentized deleted residuals:

Observation	Studentized Deleted Residual	Observation	Studentized Deleted Residual
1	−2.11	5	1.13
2	−1.10	6	−.36
3	1.31	7	−1.16
4	−.33	8	1.10

$t_{.025} = 2.776$ $(n - p - 2 = = 8 - 2 - 2 = 4$ degrees of freedom)

Since none of the studentized deleted residuals is less than -2.776 or greater than 2.776, we conclude that there are no outliers in the data

d. Using Minitab, we obtained the following values:

Observation	h_i	D_i
1	.63	1.52
2	.65	.70
3	.30	.22
4	.23	.01
5	.26	.14
6	.14	.01
7	.66	.81
8	.13	.06

The critical leverage value is

$$\frac{3(p + 1)}{n} = \frac{3(2 + 1)}{8} = 1.125$$

Since none of the values exceed 1.125, we conclude that there are no influential observations

However, using Cook's distance measure, we see that $D_1 > 1$ (rule of thumb critical value); thus, we conclude the first observation is influential

Final conclusion: observation 1 is an influential observation

CHAPTER 16

1. a. The Minitab output is shown in Figure F16.1a

   b. Since the p-value corresponding to $F = 6.85$ is $.059 > \alpha = .05$, the relationship is not significant

FIGURE F16.1a

```
The regression equation is
Y = - 6.8 + 1.23 X

Predictor        Coef       Stdev     t-ratio         p
Constant        -6.77       14.17       -0.48     0.658
X              1.2296      0.4697        2.62     0.059

s = 7.269      R-sq = 63.1%      R-sq(adj) = 53.9%

Analysis of Variance

SOURCE        DF          SS          MS        F        p
Regression     1      362.13      362.13     6.85    0.059
Error          4      211.37       52.84
Total          5      573.50
```

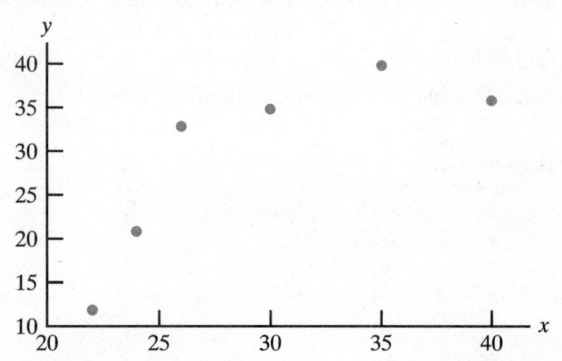

The scatter diagram suggests that a curvilinear relationship may be appropriate

d. The Minitab output is shown in Figure F16.1d

e. Since the p-value corresponding to $F = 25.68$ is $.013 < \alpha = .05$, the relationship is significant

f. $\hat{y} = -168.88 + 12.187(25) - .17704(25)^2 = 25.145$

5. **a.** The Minitab output is shown in Figure F16.5a

b. Since the relationship between y and x was significant in Exercise 4, the relationship which includes x^2 must also be significant; looking at the p-value, we see that $.003 < .01$, and thus we would reject $H_0: \beta_1 = \beta_2 = 0$

c. See Figure F16.5c

11. **a.** $R^2 = \dfrac{SSR}{SST} = \dfrac{1760}{1805} = .975$

b. $R_a^2 = 1 - (1 - R^2)\left(\dfrac{n-1}{n-p-1}\right)$

$= 1 - (.025)\left(\dfrac{29}{25}\right)$

$= .971$

c. $SSE = 1805 - 1760 = 45$

$$F = \frac{MSR}{MSE} = \left(\frac{1760/4}{45/25}\right) = 244.44$$

$F_{.05} = 2.76$ (4 degrees of freedom numerator and 25 denominator)

Since $244.44 > 2.76$, reject H_0, and conclude the relationship is significant

12. **a.** The Minitab output is shown in Figure F16.12a

b. The Minitab output is shown in Figure F16.12b

c. $F = \dfrac{[SSE(\text{reduced}) - SSE(\text{full})]/(\# \text{ extra terms})}{MSE(\text{full})}$

$= \dfrac{(23,157 - 14,317)/2}{1432} = 3.09$

$F_{.05} = 4.10$ (2 degrees of freedom numerator and 10 denominator)

Since $F = 3.09 < F_{.05} = 4.10$, the addition of the two independent variables is not significant

Note: Suppose that we consider adding only the number of interceptions made by the opponents; the corresponding Minitab output is shown in Figure F16.12c; in this case,

$$F = \frac{(23,157 - 14,335)/1}{1303} = 6.77$$

$F_{.05} = 4.84$ (1 degree of freedom numerator and 11 denominator)

Since $F = 6.77 > F_{.05} = 4.84$, the addition of the number of interceptions made by the opponents is significant

---

FIGURE F16.1d

```
The regression equation is
Y = - 169 + 12.2 X - 0.177 XSQ

Predictor        Coef        Stdev      t-ratio         p
Constant      -168.88        39.79        -4.24       0.024
X              12.187         2.663         4.58       0.020
XSQ          -0.17704       0.04290        -4.13       0.026

s = 3.248       R-sq = 94.5%       R-sq(adj) = 90.8%

Analysis of Variance

SOURCE        DF           SS           MS           F          p
Regression     2        541.85       270.92       25.68      0.013
Error          3         31.65        10.55
Total          5        573.50
```

FIGURE **F16.5a**

```
The regression equation is
Y = 433 + 37.4 X -0.383 XSQ

Predictor        Coef        Stdev       t-ratio          p
Constant        432.6        141.2          3.06       0.055
X              37.429        7.807          4.79       0.017
XSQ           -0.3829       0.1036         -3.70       0.034

s = 15.83        R-sq = 98.0%      R-sq(adj) = 96.7%

Analysis of Variance

SOURCE          DF           SS           MS          F          p
Regression       2         36643        18322       73.15     0.003
Error            3           751          250
Total            5         37395
```

FIGURE **F16.5c**

```
      Fit     Stdev.Fit              95% C.I.                  95% P.I.
   1302.01         9.93     (1270.41, 1333.61)     (1242.55, 1361.47)
```

FIGURE **F16.12a**

```
The regression equation is
POINTS = 170 + 6.61 TEAMINT

Predictor        Coef        Stdev       t-ratio          p
Constant        170.13       44.02          3.86       0.002
TEAMINT          6.613        2.258          2.93       0.013

s = 43.93        R-sq = 41.7%      R-sq(adj) = 36.8%

Analysis of Variance

SOURCE          DF           SS           MS          F          p
Regression       1         16546        16546        8.57     0.013
Error           12         23157         1930
Total           13         39703

Unusual Observations
Obs.  TEAMINT     POINTS       Fit Stdev.Fit  Residual    St.Resid
 13      33.0      340.0      388.4      34.2     -48.4       -1.75 X

X denotes an obs. whose X value gives it large influence.
```

FIGURE F16.12b

```
The regression equation is
POINTS = 280 + 5.18 TEAMINT - 0.0037 RUSHING - 3.92 OPPONINT

Predictor        Coef        Stdev    t-ratio        p
Constant       280.34        81.42       3.44    0.006
TEAMINT         5.176        2.073       2.50    0.032
RUSHING      -0.00373      0.03336      -0.11    0.913
OPPONINT       -3.918        1.651      -2.37    0.039

s = 37.84       R-sq = 63.9%     R-sq(adj) = 53.1%

Analysis of Variance

SOURCE        DF          SS          MS         F        p
Regression     3       25386        8462      5.91    0.014
Error         10       14317        1432
Total         13       39703

SOURCE        DF      SEQ SS
Regression     1       16546
Error          1         776
Total          1        8064
```

FIGURE F16.12c

```
The regression equation is
POINTS = 274 + 5.23 TEAMINT - 3.96 OPPONINT

Predictor        Coef        Stdev    t-ratio        p
Constant       273.77        53.81       5.09    0.000
TEAMINT         5.227        1.931       2.71    0.020
OPPONINT       -3.965        1.524      -2.60    0.025

s = 36.10       R-sq = 63.9%     R-sq(adj) = 57.3%

Analysis of Variance

SOURCE        DF          SS          MS         F        p
Regression     2       25386       12684      9.73    0.004
Error         11       14335        1303
Total         13       39703

SOURCE        DF      SEQ SS
TEAMINT        1       16546
OPPONINT       1        8822
```

15. **a.** The Minitab output is shown in Figure F16.15a

  b. Stepwise procedure (see Figure F16.15b)

  c. Backward elimination procedure (see Figure F16.15c)

  d. Best subsets regression of %COLLEGE (see Figure F16.15d)

21. **a.** The Minitab output is shown in Figure F16.21a

  b. Residual plot as a function of the order in which the data are presented is shown below; there does not appear to be any pattern indicative of positive autocorrelation

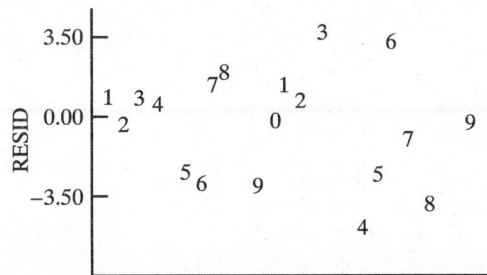

  c. The Durban-Watson statistic (obtained from Minitab) is $d = 2.34$; at $\alpha = .05$, $d_L = 1.18$ and $d_U = 1.39$; since $d > d_U$, there is no significant positive autocorrelation

23.

x_1	x_2	x_3	Treatment
0	0	0	A
1	0	0	B
0	1	0	C
0	0	1	D

$$E(y) = \beta_0 + \beta_1 x_1 + \beta_2 x_2 + \beta_3 x_3$$

26. **a.**

D_1	D_2	Manufacturer
0	0	1
1	0	2
0	1	3

$$E(y) = \beta_0 + \beta_1 D_1 + \beta_2 D_2$$

  b. See Figure F16.26b

  c. $H_0: \beta_1 = \beta_2 = 0$

  d. Since the p-value is $.004 < \alpha = .05$, we conclude that the mean time to mix a batch of material is not the same for each manufacturer

FIGURE **F16.15a**

```
The regression equation is
%COLLEGE = -26.6 + 0.0970 SATSCORE

Predictor        Coef        Stdev      t-ratio          p
Constant       -26.61        37.22        -0.72      0.485
SATSCORE      0.09703       0.03734         2.60      0.019

s = 12.83      R-sq = 29.7%      R-sq(adj) = 25.3%

Analysis of Variance

SOURCE           DF          SS           MS          F          p
Regression        1      1110.8       1110.8       6.75      0.019
Error            16      2632.3        164.5
Total            17      3743.1
```

FIGURE **F16.15b**

```
STEP                    1           2
CONSTANT           -26.61      -26.93

SATSCORE            0.097       0.084
t-RATIO              2.60        2.46

%TAKESAT                        0.204
t-RATIO                          2.21

s                    12.8        11.5
R-sq                29.68       46.93
```

FIGURE **F16.15c**

STEP	1	2	3	4
CONSTANT	33.71	17.46	-32.47	-26.93
SIZE	-1.56	-1.39		
t-RATIO	-1.43	-1.42		
STUDENT$	-0.0024	-0.0026	-0.0019	
t-RATIO	-1.47	-1.75	-1.31	
SALARY	-0.00026			
t-RATIO	-0.40			
SATSCORE	0.077	0.081	0.095	0.084
t-RATIO	2.06	2.36	2.77	2.46
%TAKESAT	0.285	0.274	0.291	0.204
t-RATIO	2.47	2.53	2.60	2.21
s	11.2	10.9	11.2	11.5
R-sq	59.65	59.10	52.71	46.93

FIGURE **F16.15d**

```
                                    S   S %
                                    T   A T
                                    U S T A
                                    D A S K
                                  S E L C E
                                  I N A O S
                   Adj.            Z T R R A
      Vars  R-sq   R-sq     S      E $ Y E T

         1  29.7   25.3   12.826       X
         1  25.5   20.8   13.203         X
         2  46.9   39.9   11.508       X X
         2  38.2   30.0   12.417   X   X
         3  52.7   42.6   11.244     X X X
         3  49.5   38.7   11.618     X   X X
         4  59.1   46.5   10.852   X X   X X
         4  52.8   38.3   11.660     X X X X
         5  59.6   42.8   11.219   X X X X X
```

FIGURE F16.21a

```
The regression equation is
P/E = 6.51 + 0.569 %PROFIT

Predictor        Coef        Stdev      t-ratio         p
Constant        6.507        1.509        4.31       0.000
%PROFIT        0.5691       0.1281        4.44       0.000

s = 2.580        R-sq = 53.7%      R-sq(adj) = 51.0%

Analysis of Variance

SOURCE          DF           SS           MS          F          p
Regression       1        131.40       131.40      19.74      0.000
Error           17        113.14         6.66
Total           18        244.54
```

FIGURE F16.26b

```
The regression equation is
Time = 23.0 + 5.00 D1 - 2.00 D2

Predictor        Coef        Stdev      t-ratio         p
Constant       23.000        1.106       20.80       0.000
D1              5.000        1.563        3.20       0.011
D2             -2.000        1.563       -1.28       0.233

s = 2.211        R-sq = 70.3%      R-sq(adj) = 63.7%

Analysis of Variance

SOURCE          DF           SS           MS          F          p
Regression       2       104.000       52.000      10.64      0.004
Error            9        44.000        4.889
Total           11       148.000
```

CHAPTER 17

1. a.

Item	Price Relative
A	103
B	238

b. $I_{1991} = \dfrac{7.75 + 1500.00}{7.50 + 630.00} = \dfrac{1507.75}{637.50} = 237$

c. $I_{1991} = \dfrac{7.75(1500) + 1500.00(2)}{7.50(1500) + 630.00(2)} = \dfrac{14{,}625.00}{12{,}510.00} = 117$

d. $I_{1991} = \dfrac{7.75(1800) + 1500.00(1)}{7.50(1800) + 630.00(1)} = \dfrac{15{,}450.00}{14{,}130.00} = 109$

6.

Item	Price Relative	Base Period Price	Base Period Usage	Weight	Weighted Price Relative
A	150	22.00	20	440	66,000
B	90	5.00	50	250	22,500
C	120	14.00	40	560	67,200
			Totals	1250	155,700

$$I = \frac{155{,}700}{1250} = 125$$

7. a. Price relatives for A $= (3.95/2.50)100 = 158$
$\qquad\qquad\qquad\quad$ B $= (9.90/8.75)100 = 113$
$\qquad\qquad\qquad\quad$ C $= (.95/.99)100 = 96$

b.

Item	Price Relative	Base Price	Quantity	Weight $P_{i0}Q_i$	Weighted Price Relative
A	158	2.50	25	62.5	9,875
B	113	8.75	15	131.3	14,837
C	96	.99	60	59.4	5,702
			Totals	253.2	30,414

$$I = \frac{30414}{253.2} = 120$$

Cost is up 20% for the chemical

10. a. $\dfrac{\$10.83}{159}(100) = \$6.81$

b. $\dfrac{10.83}{7.00}(100) = 154.7$

The actual percentage increase is 54.7%

c. $\dfrac{6.81}{7.00}(100) = 97.3$

The change in real wages is a 2.7% decrease

15. $I = \dfrac{95(1200) + 75(1800) + 50(2000) + 70(1500)}{120(1200) + 86(1800) + 35(2000) + 60(1500)}(100) = 99$

Quantities are down slightly

CHAPTER 18

1. a.

Week	Time Series Value	Forecast	Forecast Error	Squared Forecast Error
1	8			
2	13			
3	15			
4	17	12	5	25
5	16	15	1	1
6	9	16	−7	49
			Total	75

Forecast for week 7 is $(17 + 16 \div 9)/3 = 14$

b. MSE $= 75/3 = 25$

c.

Week (t)	Time Series Value (Y_t)	Forecast F_t	Forecast Error $Y_t - F_t$	Squared Error $(Y_t - F_t)^2$
1	8			
2	13	8.00	5.00	25.00
3	15	9.00	6.00	36.00
4	17	10.20	6.80	46.24
5	16	11.56	4.44	19.71
6	9	12.45	−3.45	11.90
			Total	138.85

Forecast for week 7 is $.2(9) + .8(12.45) = 11.76$

d. For the $\alpha = .2$ exponential smoothing forecast

$$MSE = \frac{138.85}{5} = 27.77$$

Since the 3-week moving average has a smaller MSE, it appears to provide the better forecasts

e.

Week (t)	Time Series Value (Y_t)	Forecast F_t	Forecast Error $Y_t - F_t$	Squared Error $(Y_t - F_t)^2$
1	8			
2	13	8.0	5.0	25.00
3	15	10.0	5.0	25.00
4	17	12.0	5.0	25.00
5	16	14.0	2.0	4.00
6	9	14.8	−5.8	33.64
			Total	112.64

$$MSE = \frac{112.64}{5} = 22.53$$

A smoothing constant of .4 appears to provide the better forecasts; for week 7 the forecast using $\alpha = .4$ is $.4(9) + .6(14.8) = 12.48$

8. a.

Month	Time Series Value	3-Month Moving Average Forecast	(Error)^2	$\alpha = .2$ Forecast	(Error)^2
1	240				
2	350			240.00	12,100.00
3	230			262.00	1,024.00
4	260	273.33	177.69	255.60	19.36
5	280	280.00	0.00	256.48	553.19
6	320	256.67	4,010.69	261.18	3,459.79
7	220	286.67	4,444.89	272.95	2,803.70
8	310	273.33	1,344.69	262.36	2,269.57
9	240	283.33	1,877.49	271.89	1,016.97
10	310	256.67	2,844.09	265.51	1,979.36
11	240	286.67	2,178.09	274.41	1,184.05
12	230	263.33	1,110.89	267.53	1,408.50
		Totals	17,988.52		27,818.49

MSE (3-month) = 17,988.52/9 = 1998.72

MSE (α = .2) = 27,818.49/11 = 2528.95

Based on the above MSE values, the 3-month moving average appears better; however, exponential smoothing was penalized by including month 2, which was difficult for any method to forecast. Using only the errors for months 4–12, the MSE for exponential smoothing is revised to

$$\text{MSE}(\alpha = .2) = 14,694.49/9 = 1632.72$$

Thus, exponential smoothing was better considering months 4–12

b. Using exponential smoothing,

$$F_{13} = \alpha Y_{12} + (1 - \alpha)F_{12}$$

$$= .20(230) + .80(267.53) = 260$$

11. $\Sigma t = 15$, $\Sigma t^2 = 55$, $\Sigma Y_t = 55$, $\Sigma tY_t = 186$

$$b_1 = \frac{\Sigma tY_t - (\Sigma t \, \Sigma Y_t)/n}{\Sigma t^2 - (\Sigma t)^2/n}$$

$$= \frac{186 - (15)(55)/5}{55 - (15)^2/5} = 2.1$$

$$b_0 = \bar{Y} - b_1\bar{t} = 11 - 2.1(3) = 4.7$$

$$T_t = 4.7 + 2.1t$$

$$T_6 = 4.7 + 2.1(6) = 17.3$$

14. $\Sigma t = 15$, $\Sigma t^2 = 55$, $\Sigma Y_t = 66.9$, $\Sigma tY_t = 222.1$

$$b_1 = \frac{\Sigma tY_t - (\Sigma t \, \Sigma Y_t)/n}{\Sigma t^2 - (\Sigma t)^2/n}$$

$$= \frac{222.1 - (15)(66.9)/5}{55 - (15)^2/5} = 2.14$$

$$b_0 = \bar{Y} - b_1\bar{t} = 13.38 - 2.14(3) = 6.96$$

$$T_t = 6.96 + 2.14t$$

$T_6 = 6.96 + 2.14(6)$ = 19.8 or 19.8 million working couples

21. a.

Year	Quarter	Y_t	Four-Quarter Moving Average	Centered Moving Average
1	1	4		
	2	2		
			3.50	
	3	3		3.750
			4.00	
	4	5		4.125
			4.25	
2	1	6		4.500
			4.75	
	2	3		5.000
			5.25	
	3	5		5.375
			5.50	
	4	7		5.875
			6.25	
3	1	7		6.375
			6.50	
	2	6		6.625
			6.75	
	3	6		
	4	8		

b.

Year	Quarter	Y_t	Centered Moving Average	Seasonal—Irregular Component
1	1	4		
	2	2		
	3	3	3.750	.8000
	4	5	4.125	1.2121
2	1	6	4.500	1.3333
	2	3	5.000	.6000
	3	5	5.375	.9302
	4	7	5.875	1.1915
3	1	7	6.375	1.0980
	2	6	6.625	.9057
	3	6		
	4	8		

Quarter	Seasonal—Irregular Component Values	Seasonal Index
1	1.3333, 1.0980	1.2157
2	.6000, .9057	.7529
3	.8000, .9302	.8651
4	1.2121, 1.1915	1.2018
	Total	4.0355

Adjustment for seasonal index $= \dfrac{4}{4.0355} = .9912$

Quarter	Adjusted Seasonal Index
1	1.2050
2	.7463
3	.8575
4	1.1912

CHAPTER 19

1. Binomial probabilities for $n = 10$, $p = .50$

x	Probability	x	Probability
0	.0010	6	.2051
1	.0098	7	.1172
2	.0439	8	.0439
3	.1172	9	.0098
4	.2051	10	.0010
5	.2461		

$P(0) + P(1) = .0108$; Adding $P(2)$, exceeds .025 required in the tail; therefore, reject H_0 if the number of plus signs is less than 2 or greater than 8;

Number of plus signs is 7

Do not reject H_0; conclude that there is no indication that a difference exists

2. $n = 27$ cases in which a value different from 150 is obtained

Use normal approximation with $\mu = np = .5(27) = 13.5$ and $\sigma = \sqrt{.25n} = \sqrt{.25(27)} = 2.6$

Use $x = 22$ as the number of plus signs and obtain the following test statistic:

$$z = \frac{x - \mu}{\sigma} = \frac{22 - 13.5}{2.6} = 3.27$$

With $\alpha = .01$, we reject if $z > 2.33$;

Since $z = 3.27 > 2.33$, reject H_0 and, conclude the median is greater than 150

4. We need to determine the number of "better" responses and the number of "worse" responses; the sum of the two is the sample size used for the study

$$n = .34(1253) + .29(1253) = 789.4$$

Use the large-sample test using the normal distribution; this means the value of $n(n = 789.4$ above) need not be integer. Use

$$\mu = .5n = .5(789.4) = 394.7$$
$$\sigma = \sqrt{.25n} = \sqrt{.25(394.7)} = 9.93$$

Let: $p =$ proportion of adults who feel children will have a better future

H_0: $p \le .50$

H_a: $p > .50$

$$x = .34(1253) = 426.0$$
$$z = \frac{x - \mu}{\sigma} = \frac{426.0 - 394.7}{9.93} = 3.15$$

With $\alpha = .05$, we reject if $z > 1.645$;

Since $z = 3.15 > 1.645$, reject H_0 and, conclude that more than half of the adults feel their children will have a better future

12. H_0: The populations are identical

H_a: The populations are not identical

Additive			Absolute		Signed
1	2	Difference	Value	Rank	Rank
20.12	18.05	2.07	2.07	9	+9
23.56	21.77	1.79	1.79	7	+7
22.03	22.57	−.54	.54	3	−3
19.15	17.06	2.09	2.09	10	+10
21.23	21.22	.01	.01	1	+1
24.77	23.80	.97	.97	4	+4
16.16	17.20	−1.04	1.04	5	−5
18.55	14.98	3.57	3.57	12	+12
21.87	20.03	1.84	1.84	8	+8
24.23	21.15	3.08	3.08	11	+11
23.21	22.78	.43	.43	2	+2
25.02	23.70	1.32	1.32	6	+6
					$T = 62$

$$\mu_T = 0$$

$$\sigma_T = \sqrt{\frac{n(n+1)(2n+1)}{6}} = \sqrt{\frac{12(13)(25)}{6}} = 25.5$$

$$z = \frac{T - \mu_T}{\sigma_T} = \frac{62 - 0}{25.5} = 2.43$$

Two-tailed test; reject H_0 if $z < -1.96$ or $z > 1.96$

Since $z = 2.43 > 1.96$, reject H_0 and, conclude that there is a significant difference in the additives

13.

Without Relaxant	With Relaxant	Difference	Rank of Absolute Difference	Signed Rank
15	10	5	9	9
12	10	2	3	3
22	12	10	10	10
8	11	−3	6.5	−6.5
10	9	1	1	1
7	5	2	3	3
8	10	−2	3	−3
10	7	3	6.5	6.5
14	11	3	6.5	6.5
9	6	3	6.5	6.5
				$T = 36$

$$\mu_T = 0$$

$$\sigma_T = \sqrt{\frac{n(n+1)(2n+1)}{6}} = \sqrt{\frac{10(11)(21)}{6}} = 19.62$$

$$z = \frac{T - \mu_T}{\sigma_T} = \frac{36}{19.62} = 1.83$$

One-tailed test; reject H_0 if $z > 1.645$

Reject H_0; there is a significant difference in favor of the relaxant

18. Rank the combined samples and find rank sum for each sample; this is a small-sample test since $n_1 = 7$ and $n_2 = 9$

Additive 1		Additive 2	
MPG	Rank	MPG	Rank
17.3	2	18.7	8.5
18.4	6	17.8	4
19.1	10	21.3	15
16.7	1	21.0	14
18.2	5	22.1	16
18.6	7	18.7	8.5
17.5	3	19.8	11
	34	20.7	13
		20.2	12
			102

$T = 34$

With $\alpha = .05$, $n_1 = 7$, and $n_2 = 9$

$$T_L = 41 \text{ and } T_U = 7(7 + 9 + 1) - 41 = 78$$

Since $T = 34 < 41$, reject H_0; and, conclude that there is a significant difference in gasoline mileage

19. a. Rank the combined samples, and find rank sum for each sample; with $n_1 = 12$ and $n_2 = 12$, this is a large-sample case

H_0: There are no differences in the distribution of starting salaries

H_a: There is a difference between the distributions of starting salaries

We reject H_0 if $z < -1.96$ or $z > 1.96$

Accounting		Finance	
Salary	Rank	Salary	Rank
28.8	20	26.3	13
25.3	9	23.6	3
26.2	11	25.0	7
27.9	17.5	23.0	1
27.0	15	27.9	17.5
26.2	11	24.5	5
28.1	19	29.0	21
24.7	6	27.4	16
25.2	8	23.5	2
29.2	22	26.9	14
29.7	24	26.2	11
29.3	23	24.0	4
Totals 327.6	185.5	307.3	114.5

$$\mu_T = \tfrac{1}{2}n_1(n_1 + n_2 + 1) = \tfrac{1}{2}12(12 + 12 + 1) = 150$$

$$\sigma_T = \sqrt{\tfrac{1}{12}n_1 n_2(n_1 + n_2 + 1)} = \sqrt{\tfrac{1}{12}(12)(12)(25)} = 17.32$$

$$T = 185.5$$

$$z = \frac{T - \mu_T}{\sigma_T} = \frac{185.5 - 150}{17.32} = 2.05$$

Since $z = 2.05 > 1.96$, reject H_0; and, conclude that there is a difference in starting salaries

26. Rankings:

	Product A	Product B	Product C
	4	11	7
	8	14	2
	10	15	1
	3	12	6
	9	13	5
Sums	34	65	21

$$W = \frac{12}{(15)(16)}\left[\frac{(34)^2}{5} + \frac{(65)^2}{5} + \frac{(21)^2}{5}\right] - 3(16)$$

$$= 58.22 - 48 = 10.22$$

$\chi^2_{.05} = 5.99147$ (2 degrees of freedom)

Reject H_0; and conclude the ratings for the products differ

28. Specialty rankings:

	Surgery	Radiology	Obstetrics
	16	17	7.5
	10.5	2	1
	18	12	3
	7.5	15	14
	6	5	4
	10.5	13	9
Totals	68.5	64	38.5

$$W = \frac{12}{(18)(19)}\left[\frac{(68.5)^2}{6} + \frac{(64)^2}{6} + \frac{(38.5)^2}{6}\right] - 3(19)$$

$$= 60.06 - 57 = 3.06$$

$\chi^2_{.05} = 5.99147$ (2 degrees of freedom)

Since $3.06 \leq 5.99147$, do not reject H_0; for these three specialties, we cannot conclude that there are significant differences in salary

32. a. $\Sigma d_i^2 = 52$

$$r_s = 1 - \frac{6\Sigma d_i^2}{n(n^2 - 1)} = 1 - \frac{6(52)}{10(99)} = .68$$

b. $\sigma_{r_s} = \sqrt{\dfrac{1}{n - 1}} = \sqrt{\dfrac{1}{9}} = .33$

$$z = \frac{r_s - 0}{\sigma_{r_s}} = \frac{.68}{.33} = 2.06$$

Reject H_0 if $z < -1.96$ or $z > 1.96$;
Since $z = 2.06 > 1.96$, reject H_0 and conclude that significant rank correlation exists

34. $\Sigma d_i^2 = 170$

$$r_s = 1 - \frac{6\Sigma d_i^2}{n(n^2 - 1)} = 1 - \frac{6(170)}{11(120)} = .23$$

$$\sigma_{r_s} = \sqrt{\frac{1}{n - 1}} = \sqrt{\frac{1}{10}} = .32$$

$$z = \frac{r_s - 0}{\sigma_{r_s}} = \frac{.23}{.32} = .72$$

Reject H_0 if $z < -1.96$ or $z > 1.96$; since $z = .72$, do not reject H_0; we cannot conclude that there is a significant relationship between the rankings

CHAPTER 20

1. $f(x) = \dfrac{n!}{x!(n - x)!}p^x(1 - p)^{n - x}$

When $p = .02$, the probability of accepting the lot is

$$f(0) = \frac{25!}{0!(25 - 0)!}(.02)^0(1 - .02)^{25} = .6035$$

When $p = .06$, the probability of accepting the lot is

$$f(0) = \frac{25!}{0!(25 - 0)!}(.06)^0(1 - .06)^{25} = .2129$$

10. *R chart:*
UCL $= \bar{R}D_4 = 1.6(1.864) = 2.98$
LCL $= \bar{R}D_3 = 1.6(.136) = .22$
$\bar{x}$ *chart:*
UCL $= \bar{\bar{x}} + A_2\bar{R} = 28.5 + .373(1.6) = 29.10$
LCL $= \bar{\bar{x}} - A_2\bar{R} = 28.5 - .373(1.6) = 27.90$

14. a. $p = \dfrac{141}{20(150)} = .0470$

b. $\sigma_{\bar{p}} = \sqrt{\dfrac{p(1 - p)}{n}} = \sqrt{\dfrac{(.0470)(.9530)}{150}} = .0173$

UCL $= p + 3\sigma_{\bar{p}} = .0470 + 3(.0173) = .0989$
LCL $= p - 3\sigma_{\bar{p}} = .0470 - 3(.0173) = -.0049$
Use LCL $= 0$

c. $\bar{p} = \dfrac{12}{150} = .08$

Process should be considered in control

CHAPTER 21

1. **a.** $\bar{x} = 215$ is an estimate of the population mean

   b. $s_{\bar{x}} = \dfrac{20}{\sqrt{50}} \sqrt{\dfrac{800 - 50}{800}} = 2.7386$

   c. $215 \pm 2(2.7386)$ or 209.5228 to 220.4772

5. **a.** $\bar{x} = 1.8$ and $s = .4$

$$s_{\bar{x}} = \left(\frac{.4}{\sqrt{30}}\right) \sqrt{\frac{153 - 30}{153}} = .0655$$

   Approximate 95% confidence interval:

$$1.8 \pm 2(.0655)$$

   or \$1.669 to \$1.931 billion

   b. $\bar{x} = 6$ and $s = 1.4$

$$N\bar{x} = 153(6) = 918$$

$$Ns_{\bar{x}} = 153\left(\frac{1.4}{\sqrt{30}}\right) \sqrt{\frac{153 - 30}{153}} = 35.0643$$

   Approximate 95% confidence interval:

$$918 \pm 2(35.0643)$$

   or 847.8714 to 988.1286

   c. $\bar{p} = .25$

$$s_{\bar{p}} = \sqrt{\left(\frac{153 - 30}{153}\right)\frac{(.25)(.75)}{29}} = .0721$$

   Approximate 95% confidence interval:

$$.25 \pm 2(.0721)$$

   or $.1058$ to $.3942$

This is a rather large interval; sample sizes must be rather large to obtain tight confidence intervals on a population proportion

7. **a.** *Stratum 1:* $\bar{x}_1 = 138$
   Stratum 2: $\bar{x}_2 = 103$
   Stratum 3: $\bar{x}_3 = 210$

   b. *Stratum 1*

$$\bar{x}_1 = 138; \; s_{\bar{x}_1} = \left(\frac{30}{\sqrt{20}}\right) \sqrt{\frac{200 - 20}{200}} = 6.3640$$

   Approximate 95% confidence interval:

$$138 \pm 2(6.3640)$$

   or 125.272 to 150.728

   Stratum 2

$$\bar{x}_2 = 103; \; s_{\bar{x}_2} = \left(\frac{25}{\sqrt{30}}\right) \sqrt{\frac{250 - 30}{250}} = 4.2817$$

Approximate 95% confidence interval:

$$103 \pm 2(4.2817)$$

or 94.4366 to 111.5634

Stratum 3

$$\bar{x}_3 = 210; \; s_{\bar{x}_3} = \left(\frac{50}{\sqrt{25}}\right) \sqrt{\frac{100 - 25}{100}} = 8.6603$$

Approximate 95% confidence interval:

$$210 \pm 2(8.6603)$$

or 192.6794 to 227.3206

c. $\bar{x}_{st} = \left(\dfrac{200}{550}\right)138 + \left(\dfrac{250}{550}\right)103 + \left(\dfrac{100}{550}\right)210$

$$= 50.1818 + 46.8182 + 38.1818 = 135.1818$$

$$s_{\bar{x}_{st}} = \sqrt{\left(\frac{1}{(550)^2}\right)\left(200(180)\frac{(30)^2}{20} + 250(220)\frac{(25)^2}{30} + 100(75)\frac{(50)^2}{25}\right)}$$

$$= \sqrt{\left(\frac{1}{(550)^2}\right)3{,}515{,}833.3} = 3.4092$$

Approximate 95% confidence interval:

$$135.1818 \pm 2(3.4092)$$

or 128.3634 to 142.0002

14. **a.** $\bar{x}_c = \dfrac{\Sigma x_i}{\Sigma M_i} = \dfrac{750}{50} = 15$

$$\hat{X} = M\bar{x}_c = 300(15) = 4500$$

$$\bar{p}_c = \frac{\sum a_i}{\sum M_i} = \frac{15}{50} = .30$$

   b. $\sum(x_i - \bar{x}_c M_i)^2 = [95 - 15(7)]^2 + [325 - 15(18)]^2$

$$+ [190 - 15(15)]^2 + [140 - 15(10)]^2$$

$$= (-10)^2 + (55)^2 + (-35)^2 + (-10)^2 = 4450$$

$$s_{\bar{x}_c} = \sqrt{\left(\frac{25 - 4}{(25)(4)(12)^2}\right)\left(\frac{4450}{3}\right)} = 1.4708$$

$$s_{\hat{X}} = Ms_{\bar{x}_c} = 300(1.4708) = 441.24$$

$$\sum (a_i - \bar{p}_c M_i)^2 = [1 - .3(7)]^2 + [6 - .3(18)]^2$$

$$+ [6 - .3(15)]^2 + [2 - .3(10)]^2$$

$$= (-1.1)^2 + (.6)^2 + (1.5)^2 + (-1)^2 = 4.82$$

$$s_{\bar{p}_c} = \sqrt{\left(\frac{25 - 4}{(25)(4)(12)^2}\right)\left(\frac{4.82}{3}\right)} = .0484$$

c. Approximate 95% confidence interval
for population mean:

$$15 \pm 2(1.4708)$$

or 12.0584 to 17.9416

d. Approximate 95% confidence interval
for population total:

$$4500 \pm 2(441.24)$$

or 3617.52 to 5382.48

e. Approximate 95% confidence interval for
population proportion:

$$.30 \pm 2(.0484)$$

or .2032 to .3968

CHAPTER 22

1.

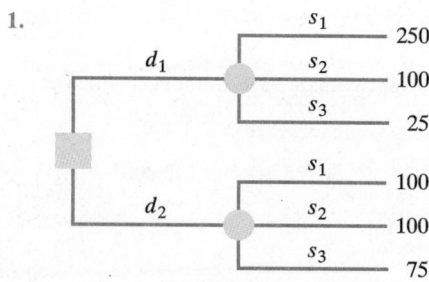

b.

	Profit	
Decision	*Maximum*	*Minimum*
d_1	250	25
d_2	100	75

Optimistic approach: select d_1
Conservative approach: select d_2

Regret or Opportunity Loss Table

	s_1	s_2	s_3	**Maximum Regret**
d_1	0	0	50	50
d_2	150	0	0	150

Minimax regret approach: select d_1

4. a.

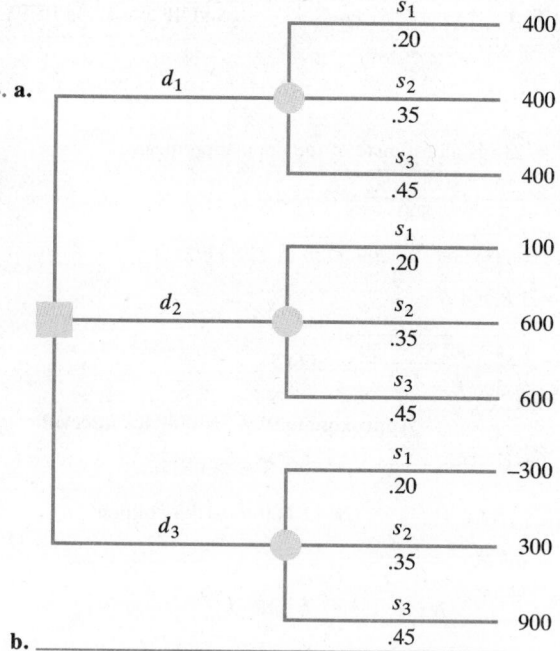

b.

	Profit		Maximum
Decision	*Maximum*	*Minimum*	**Regret**
d_1	400	400	500
d_2	600	100	300
d_3	900	−300	700

Optimistic approach: select d_3
Conservative approach: select d_1
Minimax regret approach: select d_2

Note: The maximum regret was obtained using the following
regret or opportunity loss table

	s_1	s_2	s_3
d_1	0	200	500
d_2	300	0	300
d_3	700	300	0

5. EV(d_1) = .65(250) + .15(100) + .20(25) = 182.5
EV(d_2) = .65(100) + .15(100) + .20(75) = 95
The optimal decision is d_1

8. a. EV(d_1) = .20(400) + .35(400) + .45(400) = 400
EV(d_2) = .20(100) + .35(600) + .45(600) = 500
EV(d_3) = .20(−300) + .35(300) + .45(900) = 450
Recommended decison: d_2 (medium)

12. $EV(d_1) = p(10) + (1 - p)(1) = 9p + 1$

     $EV(d_2) = p(4) + (1 - p)(3) = 1p + 3$

Expected
Values

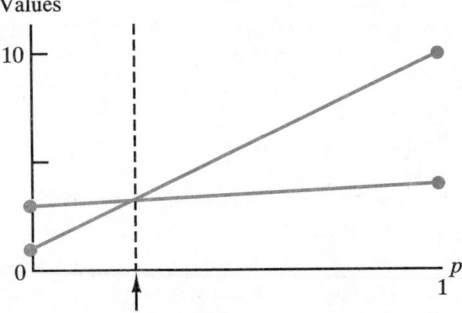

Value of p for which
EVs are equal

     $9p + 1 = 1p + 3$

         $8p = 2$

         $p = .25$

14. a. $EV(d_1) = .4(2000) + .6(2000) = 2000$

       $EV(d_2) = .4(2500) + .6(1000) = 1600$

       Recommended decision: d_1(St. Louis)

   b.   Let: p = probability of s_1

           $EV(d_1) = 2000$

           $EV(d_2) = p(2500) + (1 - p)1000 = 1500p + 1000$

     Solving for the point where $EV(d_1)$ and $EV(d_2)$ are equal

           $EV(d_1) = EV(d_2)$

           $2000 = 1500p + 1000$

           $\therefore p = 0.67$

Value(s) of p	Best Decision(s)
$0 \le p < .67$	d_1
$p = .67$	d_1 or d_2
$.67 < p \le 1.0$	d_2

17. a.   If s_1, then d_1; if s_2, then d_1 or d_2; if s_3, then d_2

   b. $EVwPI = .65(250) + .15(100) + .20(75) = 192.5$

   c.   From the solution to Exercise 5, we know that $EV(d_1) = 182.5$ and $EV(d_2) = 95$; thus, the recommended decision is d_1; hence, $EVwoPI = 182.5$

   d. $EVPI = EVwPI - EVwoPI$

           $= 192.5 - 182.5 = 10$

20. Optimal decision with perfect information:

     If s_1, then d_1

     If s_2, then d_2

     If s_3, then d_3

     Expected value of this strategy is

         $.2(400) + .35(600) + .45(900) = 695$

     $EVPI = 695 - 500 = 195$

23.

State of Nature	$P(s_j)$	$P(I \mid s_j)$	$P(I \cap s_j)$	$P(s_j \mid I)$
s_1	.2	.10	.020	.1905
s_2	.5	.05	.025	.2381
s_3	.3	.20	.060	.5714
	1.0		$P(I) = .105$	1.0000

26. a. $EVSI = EVwSI - EVwoSI$

          $= 45{,}000 - 37{,}000 = \$8000$

   b. $E = \dfrac{EVSI}{EVPI} \times 100$

         $= \dfrac{8000}{60{,}000} \times 100 = 13.13\%$

   c.   Since the cost required to obtain the sample information exceeds the expected value of the sample information, the study should not be conducted

28. a.

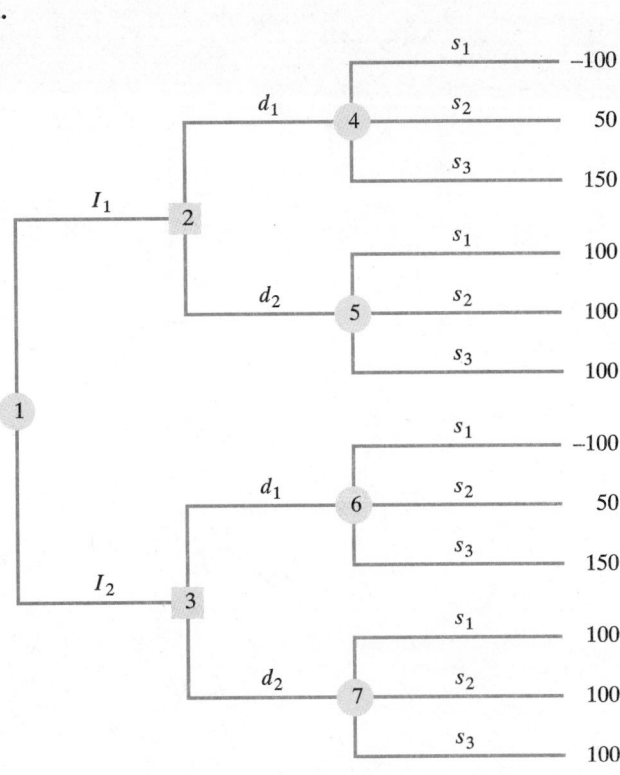

b. For I_1—favorable,

States of Nature	$P(s_j)$	$P(I_1 \mid s_j)$	$P(I_1 \cap s_j)$	$P(s_j \mid I_1)$
s_1	.2	.3	.06	.08696
s_2	.3	.6	.18	.26087
s_3	.5	.9	.45	.65217
			$P(I_1) = .69$	

For I_2—unfavorable,

States of Nature	$P(s_j)$	$P(I_2 \mid s_j)$	$P(I_2 \cap s_j)$	$P(s_j \mid I_2)$
s_1	.2	.7	.14	.45161
s_2	.3	.4	.12	.38710
s_3	.5	.1	.05	.16129
			$P(I_2) = .31$	

EV(node 4) $= .087(-100) + .261(50) + .652(150)$

$= 102.17$

EV(node 5) $= 100$

EV(node 6) $= .452(-100) + .387(50) + .161(150)$

$= -1.7$

EV(node 7) $= 100$

Decision strategy:

If I_1, then select d_1 since EV(node 4) > EV(node 5)
If I_2, then select d_2 since EV(node 7) > EV(node 6)

c. From exercise 7, we know that the expected value of the best decision (d_2) is $100; thus, EVSI $= 101.5 - 100 = 1.5$, or $1500; the consulting information is not worthwhile since the cost of $2500 is worth more than the expected gain

Index

t Distribution

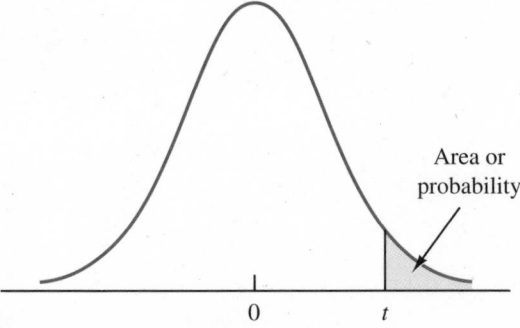

Area or probability

0   *t*

Entries in the table give *t* values for an area or probability in the upper tail of the *t* distribution. For example, with 10 degrees of freedom and a .05 area in the upper tail, $t_{.05} = 1.812$.

Degrees of Freedom	Area in Upper Tail				
	.10	.05	.025	.01	.005
1	3.078	6.314	12.706	31.821	63.657
2	1.886	2.920	4.303	6.965	9.925
3	1.638	2.353	3.182	4.541	5.841
4	1.533	2.132	2.776	3.747	4.604
5	1.476	2.015	2.571	3.365	4.032
6	1.440	1.943	2.447	3.143	3.707
7	1.415	1.895	2.365	2.998	3.499
8	1.397	1.860	2.306	2.896	3.355
9	1.383	1.833	2.262	2.821	3.250
10	1.372	1.812	2.228	2.764	3.169
11	1.363	1.796	2.201	2.718	3.106
12	1.356	1.782	2.179	2.681	3.055
13	1.350	1.771	2.160	2.650	3.012
14	1.345	1.761	2.145	2.624	2.977
15	1.341	1.753	2.131	2.602	2.947
16	1.337	1.746	2.120	2.583	2.921
17	1.333	1.740	2.110	2.567	2.898
18	1.330	1.734	2.101	2.552	2.878
19	1.328	1.729	2.093	2.539	2.861
20	1.325	1.725	2.086	2.528	2.845
21	1.323	1.721	2.080	2.518	2.831
22	1.321	1.717	2.074	2.508	2.819
23	1.319	1.714	2.069	2.500	2.807
24	1.318	1.711	2.064	2.492	2.797
25	1.316	1.708	2.060	2.485	2.787
26	1.315	1.706	2.056	2.479	2.779
27	1.314	1.703	2.052	2.473	2.771
28	1.313	1.701	2.048	2.467	2.763
29	1.311	1.699	2.045	2.462	2.756
30	1.310	1.697	2.042	2.457	2.750
40	1.303	1.684	2.021	2.423	2.704
60	1.296	1.671	2.000	2.390	2.660
120	1.289	1.658	1.980	2.358	2.617
∞	1.282	1.645	1.960	2.326	2.576